TEXAS/ESO–CERN SYMPOSIUM ON RELATIVISTIC ASTROPHYSICS, COSMOLOGY, AND FUNDAMENTAL PHYSICS

ANNALS OF THE NEW YORK ACADEMY OF SCIENCES
Volume 647

TEXAS/ESO–CERN SYMPOSIUM ON RELATIVISTIC ASTROPHYSICS, COSMOLOGY, AND FUNDAMENTAL PHYSICS

Edited by John D. Barrow, Leon Mestel, and Peter A. Thomas

The New York Academy of Sciences
New York, New York
1991

♾ *The paper used in this publication meets the minimum requirements of American National Standard for Information Sciences–Permanence of Paper for Printed Library Materials, ANSI Z39.48-1984.*

Library of Congress Cataloging-in-Publication Data

Texas/ESO-CERN Symposium on Relativistic Astrophysics, Cosmology, and Fundamental Physics (1990 : Brighton, England)
Texas/ESO-CERN Symposium on Relativistic Astrophysics, Cosmology, and Fundamental Physics / edited by John D. Barrow, Leon Mestel, and Peter A. Thomas.
p. cm.—(Annals of the New York Academy of Sciences, ISSN 0077-8923; v. 647)
"Papers . . . presented at the combined Fifteenth [Texas] Symposium and the Fourth ESO-CERN Symposium, which was held on December 16–21, 1990, in Brighton, U.K."—P.
Includes bibliographical references and index.
ISBN 0-89766-707-7 (cloth: alk. paper).—ISBN 0-89766-708-5 (pbk.: alk. paper)
1. Astrophysics—Congresses. 2. General relativity (Physics)—Congresses. 3. Cosmology—Congresses. 4. Physics—Congresses. 5. Hove, L. van (Léon) I. Barrow, John D., 1952– . II. Mestel, Leon. III. Thomas, Peter A. (Peter Andrew), 1961– . IV. European Southern Observatory. V. European Organization for Nuclear Research. VI. Title. VII. Series.
Q11.N5 vol. 647
[QB462.65]
500 s—dc20
[523.01]

SP
Printed in the United States of America
ISBN 0-89766-707-7 (cloth)
ISBN 0-89766-708-5 (paper)
ISSN 0077-8923

ANNALS OF THE NEW YORK ACADEMY OF SCIENCES

Volume 647
December 28, 1991

TEXAS/ESO–CERN SYMPOSIUM ON RELATIVISTIC ASTROPHYSICS, COSMOLOGY, AND FUNDAMENTAL PHYSICS[a]

Editors and Local Organizing Committee
JOHN D. BARROW, LEON MESTEL, and PETER A. THOMAS

CONTENTS

[a]The papers in this volume were presented at the combined Fifteenth Texas Symposium and the Fourth ESO–CERN Symposium, which was held on December 16–21, 1990, in Brighton, U.K. The ESO–CERN Symposia are organized biennially under the aegis of the Theory Division of CERN and the Science Division of ESO.

Evening Lecture

Part II. Particle Astronomy Minisymposium

Part III. Neutron Stars and Black Hole Astrophysics Minisymposium

Part IV. Lunar and Space-based Astrophysics Minisymposium

Part V. Large-scale Structure and Galaxy Formation Minisymposium

Financial assistance was received from:

- CERN
- ESO
- ICSC WORLD LABORATORY, SWITZERLAND
- INTERNATIONAL CENTER FOR THEORETICAL PHYSICS, TRIESTE
- LONDON MATHEMATICAL SOCIETY
- NASA
- NATIONAL SCIENCE FOUNDATION
- NEW YORK ACADEMY OF SCIENCES
- ROYAL ASTRONOMICAL SOCIETY
- ROYAL SOCIETY OF LONDON
- UNIVERSITY OF TEXAS AT DALLAS

LÉON VAN HOVE
1924–1990

Dedication

One day in July 1982 I was in the office of Professor Léon van Hove at CERN to discuss the outlines of the first ESO/CERN Symposium. The basic idea was that in view of the exciting developments in particle physics, astronomy, and cosmology, it was important to hold a meeting at which physicists, astrophysicists, and cosmologists could effectively interact and exchange new scientific information, at the same time trying to establish a common language that could overcome the barriers imposed by technical terms, or jargon, that are so deeply rooted in the various specialized fields. Léon's wide knowledge of astrophysics and cosmology, and the fact that he was always so well informed of the important advances in these fields, largely compensated for my ignorance of particle physics and made our interaction and common work very fruitful from the start.

Following the success of the first ESO/CERN Symposium, held at CERN in the autumn of 1983, we came to the conclusion that the initiative should be continued and we embarked on the organization of the second and third ESO/CERN Symposia. I thus had the opportunity and the good fortune to be associated with Léon van Hove for many years and to become acquainted with his great personality. His exceptionally clear mind and vast culture made it always a pleasure and an intellectual reward to talk with him about any subject, not to mention scientific matters where he has been a master and a teacher. These traits of his personality were so characteristic that they would stand out even at one's first encounter with him. What came as a really pleasant surprise was to discover his warm, rich, and very human personality, well matched by a fine sense of humor.

But, of course, he was first of all a great scientist who has made fundamental contributions to the field of quantum mechanics, from work on statistical mechanics applied to phase transitions, to condensed matter physics, and more recently to multiparticle production and quark–gluon plasma dynamics. After 1961 he was associated with CERN, of which he became Research Director General in 1976. In this capacity he was able to show yet another aspect of his exceptional qualities as a leader of science who contributed so much to bring European high-energy research to the forefront.

His qualities as a leader of science also reflected positively on the development of astronomical research in Europe where his interest was not only limited to the interaction with ESO. First as a member of the team that shaped the long-term science program, known as "Horizon 2000," of the European Space Agency, and then as Chairman of the Science Programme Committee of this Agency, he had a direct impact on the forging of European astronomical research from space.

The third ESO/CERN Symposium, held in Bologna in 1988, was the last one to see Léon van Hove involved as an organizer, since he was to retire shortly. On that occasion he delivered a memorable "Concluding Lecture" from which I would like to quote one paragraph giving his evaluation of the difficult task facing physics and astronomy:

> In very different ways, astronomy and particle physics are confronted with the fact that the most fundamental processes are not observable and must be reconstructed from

> "relics". This reconstruction is practically impossible in a model-independent way and it requires much guesswork in the form of theoretical assumptions. These are apt to be influenced by a priori choices or convictions which are not necessarily correct. The "theory of everything" is not yet here, and if it were we would be naive to believe that it is unique. But meanwhile our two disciplines progress steadily through experimentation, observation and theoretical interpretation, each one being helped by the advances of the other. This is why meetings such as the present one are of great interest and value.

However, enough time remained before his official retirement to enable him to be directly involved in establishing the Joint Texas-ESO/CERN Symposium.

Léon van Hove died on the evening of September 2, 1990. His death has been a great loss for science and of a respected friend whose intellectual honesty and human integrity will remain as a bright example. It will be impossible to replace his outstanding figure in European science.

We proposed, and our friends from the Texas Conference kindly agreed, that this Symposium should be dedicated to his memory.

GIANCARLO SETTI
European Southern Observatory

Preface

JOHN D. BARROW, LEON MESTEL, AND
PETER A. THOMAS

Astronomy Centre
Division of Physics and Astronomy
University of Sussex
Falmer, Brighton, BN1 9QH, England

The 1978 Texas Symposium at Munich and the 1984 Symposium at Jerusalem established the precedent that every third such meeting should be held outside the United States. The choice of the United Kingdom was made at the 1986 Chicago meeting; the final decision for Brighton followed the generous offer from the Brighton Borough Council of free use of lecture rooms in the Conference Centre and of the hotel booking services of their Accommodation Bureau. The suggestion that the meeting be held jointly with the fourth ESO-CERN Symposium was accepted by Dr. Giancarlo Setti and Dr. Harry van der Laan at ESO and by Dr. Léon van Hove and Dr. Carlo Rubbia at CERN.

The five mornings were devoted to plenary lectures, and the four afternoons to three parallel minisymposia on neutron star and black hole astrophysics, particle astronomy—status and prospects, and large-scale structure and galaxy formation, organized respectively by Roger Blandford, Peter Smith and Bernard Sadoulet, and George Efstathiou and Simon White. On the evenings of Monday 17 and Thursday 20 two sessions of the minisymposium on space-based and lunar-based astrophysics were held, organized by the late Harlan Smith. The Symposium dinner was held in the Refectory Building, University of Sussex, on Wednesday 19, with Roger Tayler in the Chair and Sir Fred Hoyle as postprandial speaker. On the evening of Tuesday 18, Stephen Hawking gave an open lecture on "The Beginning of the Universe." A large number of poster papers were displayed during the meeting. It was a particular pleasure to have the chair taken at one of the sessions by Dr. Fang Li Zhi, who was forcibly prevented from participating in the 1988 Texas Symposium in Dallas.

In the opening ceremony, the Symposium participants were welcomed by Sir Leslie Fielding, Vice-Chancellor, University of Sussex, Sir Michael Atiyah, Master of Trinity College, Cambridge, and President of the Royal Society, and Dr. Giancarlo Setti, who paid tribute to Dr. Léon van Hove, to whose memory the present volume is dedicated.

Nearly all the plenary lectures and the contributions to the four minisymposia are collected in this volume. We wish to thank the authors for the careful preparation of the manuscripts. Special thanks are due to Lynne Stuart both for her efficient organizing of the meeting and her assistance in the editorial preparation of the material. From that point, the staff of the Editorial Department of the New York Academy of Sciences saw the volume through the press with its usual professionalism and skill; in that regard special thanks are due to Trumbull Rogers. The Symposium poster (reproduced on the soft cover edition of this issue) was designed by Harry Kroto of the University of Sussex.

The support in kind by the Brighton Borough Council has already been mentioned. The valuable help from Ian Taylor, Celia Adams and colleagues of the Brighton Centre, and of Barbara Bridge and colleagues of the Accommodation Bureau is gratefully acknowledged. On the evening of Sunday 16, the Council held a Civic Reception, with addresses by the Mayor of Brighton, Councillor Mrs. Christine Simpson, and by Sir Leslie Fielding.

Altogether, some 540 scientists participated. Generous financial support came from the University of Texas at Dallas; ESO; CERN; New York Academy of Sciences; Royal Society of London; Royal Astronomical Society; London Mathematical Society; ICSC World Laboratory, Lausanne, Switzerland; International Center for Theoretical Physics (ICTP), Trieste; NASA; and the National Science Foundation. The support from the World Lab and ICTP was particularly appreciated as it enabled many colleagues from the Third World to attend.

New Implications of SN1987A

DAVID ARNETT

Steward Observatory and Department of Physics
University of Arizona
Tucson, Arizona 85721

INTRODUCTION

With the first euphoria following the brightest supernova since Kepler, it became clear that many features of the phenomena followed easily from existing theory. For the history of astrophysics, SN1987A will be a spectacular confirmation of the general notion of core collapse and weak interactions in supernovae. This is comforting to those of us who were advised early in our careers to change to "some topic more attuned to reality, like stellar atmospheres." Much effort has been, and will continue to be expended in quantifying how well the pre–SN1987A theory works. Such confirmation of existing ideas is crucial to science; we must validate our theories. However, we must also refine and extend them. This review will emphasize those things we have learned from SN1987A that are *new,* that require us to modify pre-1987 ideas.

THE BLUE PROGENITOR

After recovering from the fact that theories—based so much on arguments of microphysics and so little upon observational crutches—could work at all, the surprise most commented on was the fact that the progenitor, Sanduleak $-69°$ 202, was a blue supergiant, not a red one. This was not a question of supernova theory, but of stellar evolution. Just over ten years prior to SN1987A, at the Eighth Texas Symposium on Relativistic Astrophysics, I speculated on what an explosion in a presupernova of small radius might be like (Arnett[1]). The early blueness, the lower luminosity, and the later dominance of the heating due to radioactive decay from $0.1M_\odot$ of ^{56}Ni (^{56}Co) were predicted. The difficulty was that the stellar evolutionary calculations did not seem to give core collapse except for red supergiants (however, see Brunish and Truran[2]). Struggling to give astronomical credence to what seemed an otherwise plausible physical possibility, I suggested that the evolutionary origin might be mass lossing stars that were stripped to their bare He cores, and encouraged observers to search for these dim supernovae. Observers knew that finding bright supernovae was hard enough, and probably thought it was the theorist who was dim.

It might be argued that SN1987A was a special case, and unusual circumstances might be involved in its origin. This cannot be rejected out of hand, but one feels more comfortable with the idea that Sanduleak $-69°$ 202 is as typical of its sibling supergiants of the Large Magellanic Cloud (LMC) as its preexplosion data suggest. If this were true, we would have a splendid clue about the evolution of massive stars, which is a thorny problem (e.g., Brunish and Truran,[2] Chiosi and Maeder,[3] Barkat and Wheeler,[4] Fitzpatrick and Garmany,[5] Tuchman and Wheeler,[6,7] Arnett[8]).

With the initial news of SN1987A, presupernova models were in demand as initial states for modeling the early outburst. By simply using existing procedures, Arnett[9] easily found a blue to "yellow" evolution, with core collapse happening in a star with a position in the HR diagram similar to Sanduleak −69° 202. This proved to be too easy; see below. Using the Schwarzschild criterion for convection, Hillebrandt *et al.*[10] and Truran and Weiss[11,12] found such a blue to "yellow" evolution; these models had difficulty in explaining the red giant phase prior to explosion as a B star, and in reproducing the observed HR diagram for the LMC (Renzini[13]).

It was known that extensive mass loss could move a red supergiant back from the red to the blue (see Chiosi and Maeder[3]). Maeder[14] and Wood and Faulkner[15] suggested such a model for SN1987A; it now seems that far too much hydrogen must be lost to be consistent with the observational data.

Woosley *et al.*,[16] Woosley,[17] and Weiss[18] used the Ledoux criterion for convection (semiconvective mixing was inhibited) and generated progenitor models that were previously red supergiants. The missing physical motivation for such a procedure had been provided earlier by Langer *et al.*;[19] they had used solar abundances and saw no B-type progenitors. Using LMC abundances, Langer *et al.*[20] have found blue progenitors that were red supergiants.

Alternatively, Saio *et al.*[21] have used an *ad hoc* mixing of helium into the hydrogen-rich layers to obtain a red to blue evolution in a sequence constructed with the Schwarzschild criterion. This is consistent with the high He/H ratio observed in other LMC supergiants (Kudritzki *et al.*[22]), but lacks a convincing mechanism.

A new set of stellar evolutionary sequences, for mass $M/M_\odot = 15$, 20, and 25 and metallicity $z = 0.002, 0.005, 0.007, 0.010$, and 0.020 is available (Arnett[8]); semiconvection is restricted to operating slower than the local thermal timescale (as previously done). The lack of a red supergiant phase in Arnett[9] was due to errors in the envelope integration. The presupernova models for SN1987A are virtually identical; the trip to the red supergiant region had little effect on the interior. FIGURE 1 shows the evolutionary paths in the HR diagram for these stars in the $z = 0.020$ case (solar abundances). The stars evolve to the red, and die on the Hayashi track as red supergiants. However, as FIGURE 2 shows, for $z = 0.007$, which represents abundances of the LMC, the behavior is more complicated. All three stars ($M/M_\odot = 15$, 20, and 25) evolve back toward the blue after being red supergiants. The $M/M_\odot = 25$ has core collapse after returning to the Hayashi track (red supergiant), while $M/M_\odot = 15$ and 20 die as blue supergiants much like Sanduleak −69° 202. Despite the variety of envelope behavior, the structure of the core at collapse is rather similar for the stars of a given mass. Variations due to different rates of mass loss are likely to be larger than those due to composition. Simple models of the massive star content of the LMC agree moderately well with the observational data of Fitzpatrick and Garmany.[5] They are at least as successful as interpretations based upon use of the Schwarschild criterion to describe semiconvection; they imply that we have found few of the main sequence supergiants in the LMC. There is an increasing tendency for dredge up of helium core material with increasing heavy element abundance, but these simulations give a smaller effect to that assumed by Saio *et al.*[21] Untangling the effects of various uncertainties will require more rigor in both observation and theory.

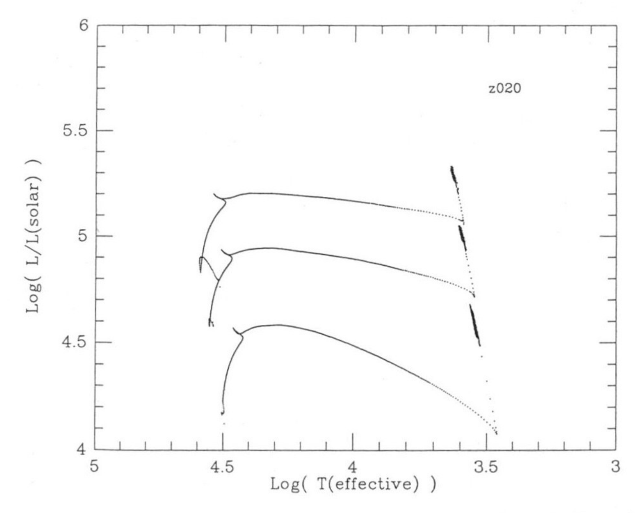

FIGURE 1. The evolution in the HR diagram for stars of $m/msol$ = 15, 20, and 25, for solar abundances (Arnett[8]). The stars all die as red supergiants.

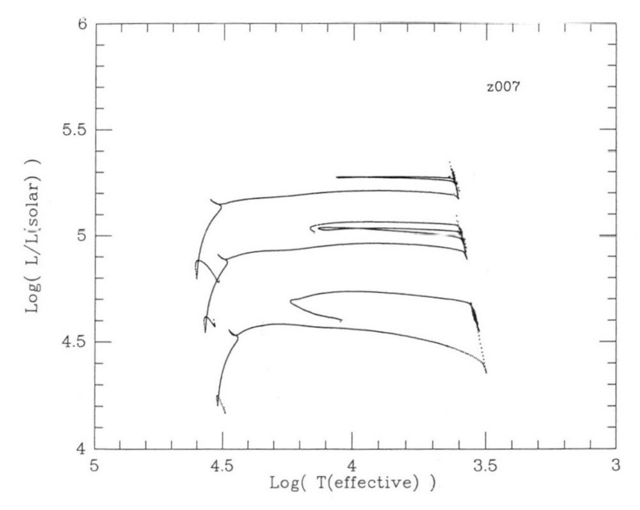

FIGURE 2. The evolution in the HR diagram for stars of $m/msol$ = 15, 20, and 25, for abundances of the Large Magellanic Cloud (Arnett[8]). Only the $m/msol$ = 25 dies as a red supergiant. Both the $m/msol$ = 15 and 20 stars die as progenitors similar to Sanduleak −69° 202 after spending time as red supergiants.

NONSPHERICITY

A startling feature of SN1987A that is obscured in the literature is that spherically symmetric models *cannot* reproduce the observations. It is necessary to introduce extensive *ad hoc* mixing (that is, nonspherical motions) to fit the data. In particular, the early onset of γ and x-ray luminosity (Dotani *et al.*,[23] Sunyaev *et al.*,[24]

Wilson *et al.*,[25] Matz *et al.*[26]), the smoothness of the bolometric light curve from two weeks to 120 days, the Bochum spectral event (Hanuschik and Dachs[27]), and the detailed line shapes in various wavelength regions (from γ to infrared) are all indications of mixing. Any such mixing is, by definition, not spherically symmetric. This point is usually lost in discussion of mixing in the context of one-dimensional models.

There is an elementary type of motion that cannot be properly represented by a spherically symmetric hydrodynamic computation: vortex motion with a center not at the origin. When compacted onto a radial coordinate, such motion involves interpenetration of material, and it becomes difficult to understand what is meant by such a procedure. Such vortex motion is a consequence of the conservation of angular momentum, which in turn is a consequence of the isotropy of space (Landau and Lifshitz,[28] p. 18). To presume that a hydrodynamic system does *not* exhibit such fundamental motions is a drastic assumption. Further, even if a system starts from a state without such circulatory motion, Kelvin's theorem (Landau and Lifshitz,[29] p. 15) shows that the nonuniform generation of entropy can cause such motion to develop.

In a supernova, such generation of entropy can occur by a variety of processes. To name a few: neutrino heating, thermonuclear burning, radioactive decay, and shock heating. Therefore, from the theoretical point of view, it is quite likely that vortex motions will occur and give rise to nonradial velocities and macroscopic mixing of material.

Observations of SN1987A give strong confirmation of this view. The analytic model of Type I and II light curves of Arnett,[30] when modified to include the effects of recombination (Arnett,[31] Arnett and Fu[32]), provides a striking representation of the bolometric light curve from about day 14 through day 800 (the theoretical curve fits the observed to within 20 percent over that range; see FIG. 3). This differs from the curve of Arnett[31] only in that leakage of x-rays from the ejecta has been included. The UVOIR data shown in FIGURE 3 is from SAAO (Whitelock,[33] Menzies *et al.*,[34] Catchpole *et al.*[35]). After 500 days there is a large far IR flux that must be included for comparison with the theoretical "thermal" curve. With this correction, there is good agreement from day 14 to day 800. As Danziger has mentioned (this issue), there are significant deviations after day 900. This may be evidence for the long-sought pulsar. The theoretical curves for x-rays (small dashes) and γ-rays (large dashes) are in reasonable agreement with the observational data, at least for the first 500 days (the theoretical curves shown here of these fluxes break down at later times due to neglect of clumping and asymmetry). The simple analytic model works very well; notice the good fit to the peak around day 14 to day 200 or so. Only later did one-dimensional numerical models represent this phase adequately, and then only with the inclusion of *ad hoc* mixing.

The constraint necessary to obtain a separation of variables in the analytic solution at first seemed peculiar: it was necessary to assume that the radioactive source was distributed like the thermal energy (Arnett[30]). A second necessary assumption was that the opacity, and therefore the composition, was slowly varying except at and above the recombination region; this was contrary to the layering of composition characteristic of presupernova models. We now know that both these conditions are naturally satisfied by hydrodynamic instability during the explosion (Arnett, Fryxell, and Müller[36]); calculations indicate that the helium shell is shredded in less than 3 hours after core collapse, and ^{56}Ni may be spread through this region. Without such mixing the light curves show "running waves," which give rise to bumps and dips in luminosity that are not observed; these may be seen in early numerical calculations of the light curves.

The ejecta are unlikely to be microscopically mixed. If the ionic collision cross section is $\sigma = 10^{-16}$ cm^2, then $6M_\odot$, expanding homologously at velocities up to

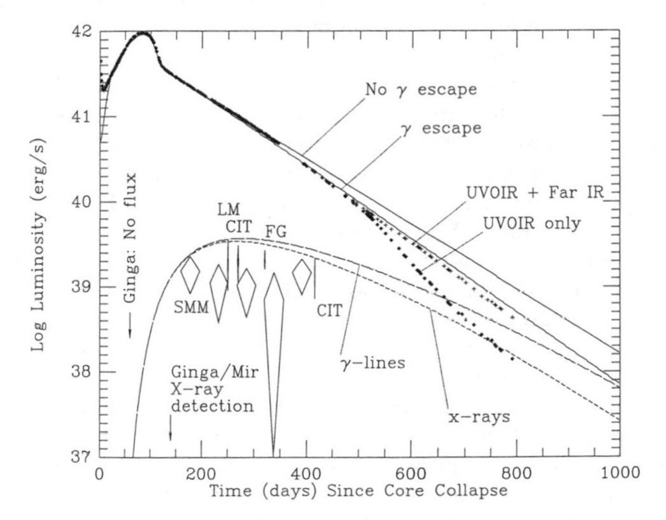

FIGURE 3. A summary of theoretical and observational luminosities from SN1987A. The analytic model of Type I and II light curves of Arnett,[30] when modified to include the effects of recombination (Arnett,[31] Arnett and Fu[32]), is shown as a *solid line* labeled γ-escape. This differs from the curve of Arnett[31] only in that leakage of x-rays from the ejecta has been included. The UVOIR data shown are from SAAO. After 500 days there is a large far IR flux that must be included for comparison with the theoretical "thermal" curve. With this correction, there is good agreement from day 14 to day 800; there are significant deviations after day 900 (Danziger, this conference). This may be evidence for the long-sought pulsar. The theoretical curves for x-rays (*small dashes*) and γ-rays (*large dashes*) are in reasonable agreement with the observational data, at least for the first 500 days (the theoretical curves shown here of these fluxes break down at later times due to neglect of clumping and asymmetry). Notice the good fit to the peak around day 14 to day 200 or so; this requires the layering of the presupernova to be destroyed.

$v = 3 \times 10^8$ cm s^{-1}, will have a diffusion time equal to the expansion time at 600 years. Long before that the thermal velocities that drive the diffusion will be less than the expansion velocities, so that mixing is primarily a phenomenon of the later, "remnant" evolution. *This means that a measurement of the composition identifies matter from each burning shell, and allows us to probe the explosion by comparing observational and numerical maps of composition-velocity-position.*

The following section deals with theoretical problems that arise when attempting to answer this challenge.

EXPLOSION MODELING

The challenge of SN1987A has brought about numerical simulations of a level of realism unprecedented in this area. This technical development will be of direct importance to other areas of astrophysics; the techniques can be applied there as well.

The explosive behavior is complex, and several sorts of instabilities have been found. First is the well-known radial instability that causes core collapse. The first demonstration of nonradial instabilities in realistic progenitor models was by Arnett, Fryxell, and Müller;[36] several possible instabilities are mentioned there. As the shock moves out past the burning shells, a compressible version of the classical Rayleigh–Taylor (RT) instability develops; this has been the focus of most numerical work.

Low resolution gives rise to numerical errors that act like a dissipation or viscosity, and such errors tend to inhibit growth of an instability. These RT instabilities depend in a relatively sensitive way upon the presupernova structure. Consequently, it becomes important to distinguish between these possible causes for differing computational results.

Insofar as we are aware, all calculations beginning from even a vaguely plausible initial model show an RT instability at the H/He interface. Fryxell, Müller, and Arnett[37] and Herant and Benz[38] find another RT instability at the CO/He interface for the initial model used by Arnett, Fryxell, and Müller.[36] This model was a $15M_\odot$ star after carbon burning; with no core removed (to be a neutron star) this was thought to be a reasonable representation of the mantle and envelope of a slightly more massive star. Hachisu *et al.*,[39] using a different initial model (a $20M_\odot$ evolved to core instability), find only the instability at the H/He interface. Herant and Benz[38] attribute this to the difference in stellar structure, which is consistent with the results of linear stability analysis (Shigeyama *et al.*,[40] Müller, Arnett, and Fryxell[41]).

To see if the instability were due to the neglect of the final evolution to core collapse, we (Fryxell, Müller, and Arnett[42]) have used a $20M_\odot$ initial model (Arnett[8]) after core bounce. This is later than the Hachisu *et al.*[39] initial model, and should be more realistic. Then the inner $1.28M_\odot$ were removed, and an explosion of 2×10^{51} ergs added to the inner $0.1M_\odot$ of the mantle. This produced about $0.09M_\odot$ of ^{56}Ni by explosive burning in the ^{16}O shell, in reasonable agreement with that observed in SN1987A; modifications of the precise nature of the explosion would alter this yield. The energy was that found appropriate to the early light curve for such a progenitor model (see Arnett[9,31]).

We find two regions of strong instability: (1) the H/He interface, and (2) the He/heavy interface. The He/heavy interface region includes the newly synthesized ^{56}Ni from explosive oxygen burning. We are not yet able to specify the source of this instability more precisely (is it the He/CO region, the Ni region, or . . .). Once started, the instability spreads over all these regions. The two instabilities arise as the reverse shock moves through those regions, and thus are at least partially related.

THE NI INSTABILITY

Because the mean life of ^{56}Ni is short (8.8 days) compared to the diffusion time for γ-rays from ^{56}Ni decay (months at that epoch), most of the decay energy is

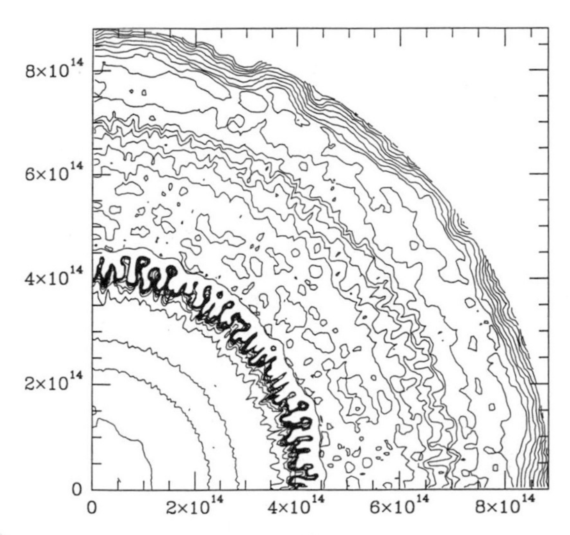

FIGURE 4. Contours of density for the simplest ^{56}Ni model (see text) after about two months. The instability is weak, and this model fails to provide a sufficiently extensive mixing of ^{56}Ni.

thermalized by the local matter. For 2 MeV per decay, this corresponds to $Q \sim 4 \times 10^{16}$ erg g^{-1} of ^{56}Ni, or 7×10^{49} erg/$M_\odot$ of ^{56}Ni. While small compared to supernova energies, this is enough to affect the velocities of slowly moving matter near the mass cut.

To understand the magnitudes involved, let us consider some idealized models. First, assume that an opaque sphere of ^{56}Ni is accelerated by the increased pressure resulting from the heat released by decay. The root mean square velocity is $\sqrt{2Q} = 2.8 \times 10^3$ km s^{-1}.

This is a bit high because there should be some interaction with matter that is not ^{56}Ni. As a second case, suppose the same opaque sphere, with an initial outer velocity of 1000 km s^{-1}, snowplows into homologously expanding matter ($v \propto r$). Then

$$m = m_{\rm Ni}\left(\frac{v}{v_0}\right)^3,$$

and

$$\tfrac{1}{2}m_{\rm Ni}v_0^2 + m_{\rm Ni}Q = \tfrac{1}{2}mv^2,$$

so that $v_{\rm rms} = 1.5 \times 10^3$ km s^{-1}.

Observations suggest velocities closer to the first value. To approach case (1) we need:

1. ^{56}Ni energy to be deposited in the ^{56}Ni itself. This is true, since γ attenuation lengths are short for times $t \leq \tau_{\rm Ni}$.
2. Holes in the mass distribution of overlying matter for $t \leq \tau_{\rm Ni}$. Something similar may be needed for "early" γ- and x-ray luminosity.

3. A new physical effect that accelerates the ^{56}Ni. An obvious candidate for the source of this effect is the new neutron star, whose rotational energy at least might be available.

Because of the high degree of interest in the last possibility, it is vital to see if the observational data demand it, by attempting simpler explanations.

The simplest is a pure "nickel bubble" model, in which the RT instabilities are deemed irrelevant, and ignored along with possibility (3). FIGURE 4 shows density contours for such a model at a time about two months after the core collapse. It was constructed from a one-dimensional shock calculation to 10^5 s, followed onward in time by a two-dimensional calculation with the PROMETHEUS code (see Fryxell, Müller, and Arnett[37]). A low-density, hot ^{56}Ni bubble does form. While there is clear evidence for an instability, it is far too weak to break through the snowplow layer, much less to give a significantly greater dispersion in ^{56}Ni velocities. Lower velocities result, like the snowplow case given previously. This model fails.

COMBINING RT AND NI INSTABILITIES

Can we reproduce the observations with the combined action of RT and ^{56}Ni instabilities? The RT instability happens on a timescale of hours, not days, after core collapse. It may mix some ^{56}Ni outward to regions of larger velocity. This would give a different "initial" state for the ^{56}Ni decay to work on. Perhaps what we see is the result of a two-stage process, with the RT instability acting as a sort of "booster."

This is a more subtle question than that of the nature of the pure ^{56}Ni bubble; in order to begin we must first have the RT instability well in hand. As was mentioned before, that is not yet the case, so that any uncertainties there are propagated into this problem.

For example, low-resolution calculations on an Eulerian grid will have numerical diffusion errors that *overestimate* the mixing of ^{56}Ni. However, low-resolution calculations of any sort fail to resolve gradients that drive the mixing and breakup of blobs of ^{56}Ni, which *underestimates* the mixing. This is by no means a complete list of the difficulties, but is merely illustrative.

In order to maintain the highest possible resolution, but have sufficient computational efficiency to explore the problem, we have modified PROMETHEUS to use a moving grid. The natural choice is a spherical coordinate system with a radial expansion of the grid so that it moves with the matter. Thus it is, on average, somewhat like a Lagrange grid, but allows for complex vortex motion that a Lagrange grid cannot handle. This gives some of the "motion with the matter" advantages of smooth particle hydrodynamics (SPH) methods without their loss of resolution.

Preliminary results were shown at the conference; at that time it was not possible to determine whether or not the combined instabilities would give adequate ^{56}Ni velocities (although the numerical results tended to be too low). More complete results are being prepared for publication.

CONCLUSION

A final point: The observations of SN1987A are providing us with an unprecedented view of the mass loss phenomena in a massive star. Already from the light echo, and subsequently from the shock, this ejected matter will be investigated. This will lead into the *very* early evolution of a supernova remnant. The list of new things from SN1987A is continuing to increase.

The differences between various groups of observers concerning the light curve after day 900 are exacerbated by the difficulties of the observations (Danziger, this issue). But because these observations may relate to the newly formed condensed object that was signaled by the neutrino burst, or at least to the detailed nucleosynthesis yield, it is vital that we continue to sort out exactly what is implied.

Numerical calculations of aspherical explosions are beginning to constrain simple theoretical models of SN1987A, and to provide a fertile set of quantitative data for comparison with observation. We hope to be able to relate, for example, the overall redshift of the ^{56}Co lines, the birth velocity of the neutron star, and the observed mixing and clumping of the ejecta with the still perplexing puzzle of the explosion mechanism.

Besides the obviously deepened and improved perspective on supernova explosions, a lasting legacy of SN1987A may be the qualitative improvement in observational and theoretical procedures—of wider applicability in astrophysics—that resulted from its challenge and that were tested on its reality.

REFERENCES

1. ARNETT, W. D. 1982. Ann. N.Y. Acad. Sci. **302:** 90 (Eighth Texas Symposium on Relativistic Astrophysics).
2. BRUNISH, W. & J. W. TRURAN. 1982. Astrophys. J. **256:** 247.
3. CHIOSI, C. & A. MAEDER. 1986. Ann. Rev. Astron. Astrophys. **24:** 329.
4. BARKAT, Z. & J. C. WHEELER. 1988. Astrophys. J. **322:** 247.
5. FITZPATRICK, E. L. & C. D. GARMANY. 1990. Astrophys. J. **363:** 119.
6. TUCHMAN, Y. & J. C. WHEELER. 1989. Astrophys. J. **344:** 835.
7. TUCHMAN, Y. & J. C. WHEELER. 1989. Astrophys. J. **346:** 417.
8. ARNETT, W. D. 1991. submitted for publication in Astrophys. J.
9. ARNETT, W. D. 1987. Astrophys. J. **319:** 136.
10. HILLEBRANDT, W., P. HÖFLICH, J. W. TRURAN & A. WEISS. 1987. Nature **327:** 597.
11. TRURAN, J. W. & A. WEISS. 1987. *In* Proc. ESO Workshop of the SN1987A, J. Danziger, Ed.: 271.
12. TRURAN, J. W. & A. WEISS. 1988. *In* Supernova 1987A in the Large Magellanic Cloud, M. Kafatos and A. Michalitsianos, Eds.: 322. Cambridge University Press. Cambridge, England.
13. RENZINI, A. 1987. *In* Proc. ESO Workshop of the SN1987A, J. Danziger, Ed.: 295.
14. MAEDER, A. 1987. *In* Proc. ESO Workshop of the SN1987A, J. Danziger, Ed.: 251.
15. WOOD, P. R. & D. H. FAULKNER. 1987. Proc. Astron. Soc. Aust. **7:** 75.
16. WOOSLEY, S. E., P. A. PINTO & T. WEAVER. 1988. Proc. Astron. Soc. Aust. **7:** 355.
17. WOOSLEY, S. E. 1988. *In* Supernova 1987A in the Large Magellanic Cloud, M. Kafatos and A. Michalitsianos, Eds.: 289. Cambridge University Press. Cambridge, England.
18. WEISS, A. 1988. Astrophys. J. **339:** 365.
19. LANGER, N., M. F. EL EID & K. J. FRICKE. 1985. Astron. Astrophys. **145:** 179.
20. LANGER, N., M. F. EL EID & I. BARAFFE. 1990. In preparation.

21. SAIO, H., M. KATO & K. NOMOTO. 1988. Astrophys. J. **331:** 388.
22. KUDRITZKI, R. P., H. G. GROTH, K. BUTLER, D. HUSFELD, S. BECKER, F. EBER & E. FITZPARTICK. 1987. *In* Proc. ESO Workshop of the SN1987A, J. Danziger, Ed.: 39; Saio, H., M. Kato, & K. Nomoto. 1988. Astrophys. J. **331:** 388.
23. DOTANI, T., *et al.* 1987. Nature **330:** 230.
24. SUNYAEV, R. A., A. KANIOVSKY, V. EFREMOV, M. GILFANOV, E. CHURAZOV, *et al.* 1987. Nature **330:** 227.
25. WILSON, R. B., *et al.* 1988. *In* Nuclear Spectroscopy of Astrophysical Sources, N. Gehrels and G. Share, Eds.: 66. American Institute of Physics. New York.
26. MATZ, S. M., G. H. SHARE, M. D. LEISING, E. L. CHUPP, W. T. VESTRAND, W. R. PURCELL, M. S. STRICKMAN & C. REPPIN. 1988. Nature **331:** 416.
27. HANUSCHIK, R. W. & J. DACHS. 1987. Astron. Astrophys., Lett. **182:** L29.
28. LANDAU, L. D. & E. M. LIFSHITZ. 1960. Mechanics. Pergamon. Oxford.
29. LANDAU, L. D. & E. M. LIFSHITZ. 1959. Fluid Mechanics. Pergamon. Oxford.
30. ARNETT, W. D. 1982. Astrophys. J. **253:** 785.
31. ARNETT, W. D. 1988. *In* Supernova 1987A in the Large Magellanic Cloud, M. Kafatos and A. Michalitsianos, Eds: 301. Cambridge University Press. Cambridge, England.
32. ARNETT, W. D. & A. FU. 1989. Astrophys. J. **340:** 396.
33. WHITELOCK, P. A., *et al.* 1989. Mon. Not. R. Astron. Soc. **240:** 7P.
34. MENZIES, J. W., *et al.* 1987. Mon. Not. R. Astron. Soc. **227:** 39P.
35. CATCHPOLE, R. M., *et al.* 1988. Mon. Not. R. Astron. Soc. **229:** 15P.
36. ARNETT, W. D., B. A. FRYXELL & E. MÜLLER. 1989. Astrophys. J. **341:** L63.
37. FRYXELL, B. A., E. MÜLLER & W. D. ARNETT. 1991. Astrophys. J. In press.
38. HERANT, M. & W. BENZ. 1991. Astrophys. J., Lett. In press.
39. HACHISU, I., T. MATSUDA, K. NOMOTO & T. SHIGEYAMA. 1990. Astrophys. J. **358:** L57.
40. SHIGEYAMA, T., K. NOMOTO, T. TSUJIMOTO & M. HASHIMOTO. 1990. Astrophys. J. **361:** L23.
41. MÜLLER, E., W. D. ARNETT & B. A. FRYXELL. 1991. *In* Proc. of the Elba Workshop on SN1987A and Other Supernovae, J. Danziger, Ed. European Southern Observatory. Garching, Germany.
42. FRYXELL, B. A., E. MÜLLER & W. D. ARNETT. 1991. Submitted for publication.

Solar Neutrinos: New Physics?[a]

JOHN N. BAHCALL

Institute for Advanced Study
Princeton, New Jersey 08540

INTRODUCTION

At the conference, I discussed several topics: (1) a summary of the existing and proposed solar neutrino experiments; (2) the predictions for solar neutrino experiments of the standard solar and electroweak model; (3) a statistical analysis of the suggested variations in the chlorine experiment; (4) two theoretical arguments that suggest—independent of particular solar or particle physics models—that solar neutrino experiments reveal new physics beyond the standard electroweak model; and (5) an attractive MSW solution of the solar neutrino problem.

Progress in the different experiments was described clearly by the speakers in an afternoon session at the conference.

I will concentrate in this written report on the fourth point, since it has not been as widely recognized as, in my opinion, it deserves. Much of what I said about the existing and proposed experiments, about the standard model, about variability in the chlorine experiment, and about the MSW effect is contained in the book *Neutrino Astrophysics*.[1] The newer material, which included a discussion of the effect of the diffusion of heavier elements, of more recent opacity calculations, and of new subroutines for nuclear energy generation and for neutrino production, is being written up now in a more complete form.

NEW PHYSICS FROM THE CHLORINE AND KAMIOKANDE II EXPERIMENTS

The standard solar and electroweak model predicts an event rate in solar neutrino units (SNU) for the chlorine experiment of[1]

$$\langle \phi\sigma \rangle_{\text{Cl, theory}} = (7.9 \pm 2.6)\ \text{SNU}, \tag{1}$$

where the indicated error refers to the total theoretical uncertainty. The rate at which Ray Davis and his collaborators detect neutrinos in the chlorine experiment is[2,3]

$$\langle \phi\sigma \rangle_{\text{Cl, exp}} = (2.1 \pm 0.3)\ \text{SNU}, \qquad (1\sigma\ \text{error}) \tag{2}$$

for neutrinos above the 0.81-MeV threshold energy. The difference between the experimental and calculated rates given in **(1)** and **(2)** constituted for two decades the "solar neutrino problem."

[a] This work was supported in part by National Science Foundation Grant PHY-86-20266 at the Institute for Advanced Study.

More recently, the nature of the solar neutrino problem has been deepened and extended by the result from the Kamiokande II neutrino–electron scattering experiment,[4] which is,

$$\langle \phi\sigma \rangle_{e-\nu} + [0.46 \pm 0.05(\text{stat}) \pm 0.06(\text{syst})] \langle \phi\sigma \rangle_{\text{STND}} \quad \textbf{(3)}$$

for recoil electrons with energies greater than 7.5 MeV.

I now want to give a general argument that Hans Bethe and I developed last summer[5] that makes it seem very unlikely that the standard model (standard solar model plus standard electroweak theory) can account for the results of both these experiments. The *shape* of the neutrino energy spectrum from an individual source like ^{8}B is essentially independent of any solar physics. I described at the conference the small corrections to the shape that depend upon temperature and density, but they are negligible in any practical sense. Therefore, if one knows the rate at which ^{8}B neutrinos are observed in the Kamiokande II experiment, one can predict the corresponding event rate in the chlorine experiment. Using the calculated cross sections for each detector,[1] the Kamiokande rate,[4] and an unmodified neutrino spectrum,[1] we found a rate of 2.8 SNU in the chlorine detector from ^{8}B neutrinos alone. In addition, there are significant contributions expected[1] from ^{7}Be (1.1 SNU from the standard model) and from *pep* (0.2 SNU from the standard model). The fluxes from ^{7}Be and *pep* can be calculated much more reliably than the ^{8}B fluxes. For example, the flux of ^{8}B neutrinos depends upon the central temperature of the sun, T_c, approximately like Flux $\approx T_c^{18}$, whereas the flux of ^{7}Be neutrinos varies only like Flux $\approx T_c^{8}$. The total uncertainty in the *pep* neutrinos is only about 2 percent.

We concluded that any solar models that are consistent with **(3)** will predict event rates for the chlorine experiment that are larger than the measured value, **(2)**. The logical inference is: Solar neutrino experiments reveal new physics.

A RIGOROUS LOWER LIMIT FOR THE GALLIUM EVENT RATE

The gallium experiment provides an independent argument that suggests that new physics is required to explain solar neutrino observations.

The rigorous argument is given in [1, chap. 11]. The only astronomical ingredient in this argument is the general statement that the sun is currently producing energy by nuclear reactions at a rate equal to its luminosity. Given a nuclear origin for the solar energy, the event rate in a gallium detector is minimized by having all of the nuclear fusion reactions terminate through the ^{3}He–^{3}He reaction. In this way, the only neutrinos that are produced are the low-energy *pp* neutrinos (and a few associated *pep* neutrinos) whose absorption cross section is small.

The ^{3}He–^{4}He reaction is the gateway to the higher energy, more easily detected neutrinos produced by ^{8}B decay and to the moderate-energy neutrinos from electron capture on ^{7}Be. If, in a solar model calculation, one artificially sets equal to zero the cross section of the ^{3}He–^{4}He reaction, then no ^{8}B or ^{7}Be neutrinos are produced. We know from laboratory measurements[6] that the rate of this reaction is not zero, and, in fact, is competitive with the rate of the ^{3}He–^{3}He reaction at the calculated temperature given by solar models. But, in order to calculate an absolute lower limit to the neutrino event rate, we can consider what would happen if the ^{3}He–^{4}He rate were zero.

If nuclear fusion provides the solar luminosity and nothing happens to the solar neutrinos after they are created in the sun, then the lowest rate that could be found in the gallium experiment is obtained if only low-energy *pp* and *pep* neutrinos are produced. This rate is 80 SNU. The preliminary results from the GALLEX experiment[7] are less than this value, at least at the level of one standard deviation, providing an independent indication that new physics is required.

COMMENTS AND DISCUSSION

It seems likely that the astronomy, chemistry, and physics communities have been lucky and have stumbled onto something important concerning microscopic physics while trying to test accurately stellar evolution theory. If so, we will need many more experiments in order to establish uniquely the range of weak interaction parameters (mixing angles, masses, magnetic moments, and the like) that are viable.

REFERENCES

1. BAHCALL, J. N. 1989. Neutrino Astrophysics. Cambridge University Press. Cambridge, England.
2. DAVIS, R., JR. 1987. *In* Proc. of the 7th Workshop on Grand Unification, ICOBAHN '86, J. Arafune, Ed.: 237. Singapore. World Scientific; DAVIS, R., JR., K. LANDE, C. K. LEE, P. WILDENHAIN, A. WEINBERGER, T. DAILY, B. CLEVELAND & J. ULLMAN. 1990. *In* Proc. of the 21st International Cosmic Ray Conference. Adelaide. Australia.
3. ROWLEY, J. K., B. T. CLEVELAND & R. DAVIS, JR. 1985. *In* Solar Neutrinos and Neutrino Astronomy, M. L. Cherry, W. A. Fowler, and K. Lande, Eds.: 1. American Institute of Physics. New York.
4. HIRATA, K. S., *et al.* 1990. Phys. Rev. Lett. **65:** 1297.
5. BAHCALL, J. N. & H. A. BETHE. 1990. Phys. Rev. Lett. **65:** 2233.
6. PARKER, P. D. 1986. *In* Physics of the Sun, Vol. I., P. A. Sturrock, T. E. Holzer, D. M. Mihalas, and R. K. Ulrich, Eds.: 15. Reidel. New York.
7. GAVRIN, V. N., *et al.* 1991. Nucl. Phys. B, Proc. Suppl. Sec. In press.

The Quasar Population at High Redshifts[a]

B. J. BOYLE

Institute of Astronomy
University of Cambridge
Madingley Road
Cambridge, CB3 0HA, England

1. INTRODUCTION

Quasars are the extremely bright centers of certain galaxies. Their luminosities typically range from 10^{45} erg/s to 10^{48} erg/s; a few times to many hundred times the total luminosity of their host galaxy. As a result of their high luminosities, they are the only class of astronomical object that can be found in significant numbers at high redshifts ($z > 3$) when the Universe was less than $\sim$ 20 percent of its present age. The evolution in the properties of the quasar population (e.g., space density, clustering) at these epochs can provide valuable constraints on models of galaxy formation and the growth of structure in the Universe.

Information on the evolution of the quasar population is drawn primarily from the statistical analysis of large quasar catalogues. Unfortunately, the construction of such catalogues is not straightforward. Quasars must first be selected from the vast majority of galactic stars, and then observed spectroscopically to obtain their redshifts. The initial selection process has to be extremely efficient; at faint apparent magnitudes ($B \sim 20$ mag) quasars are typically outnumbered 50:1 by galactic stars, and for $z > 3$ quasars this ratio increases to 2000:1. Traditionally, selection techniques have been based on the identification of quasars by their anomalous optical broadband colors. The simplest method uses U–B colors to discriminate between quasars and galactic stars, but is only successful for quasars with $z < 2.2$. At higher redshifts, more sophisticated methods using additional broadband colors are required. A comprehensive review of these techniques is given by Warren and Hewett.[1]

Difficulties in identifying high-redshift quasars and the prohibitive amounts of telescope time required to obtain spectra for large numbers of faint quasar candidates had, until recently, limited most quasar surveys to $B < 20$ mag and $z < 2.2$ [see Fig. 1(a)]. At these magnitudes no discrimination between different models for quasar evolution is possible (see Section 2.1), and the space density of quasars is too low to conduct a detailed analysis of their clustering properties at scales less than $\lesssim$ 100 Mpc (Osmer[2]). At $z > 3$ possible incompleteness in quasar catalogues and small number statistics plagued most attempts to determine the evolution of high-redshift quasars. In particular, the reality of any decline in the space density of

[a]The author received a SERC/Isaac Newton fellowship during the course of this work.

quasars at high redshift was difficult to establish from the few samples of quasars with $z > 3$ (Osmer[3]).

Over the last few years, however, advances in instrumentation and search techniques have resulted in a vast increase in the numbers of spectroscopically identified quasars both at faint magnitudes and high redshifts [FIG. 1(b)]. This review discusses the most recent results on quasar clustering and evolution derived from such samples.

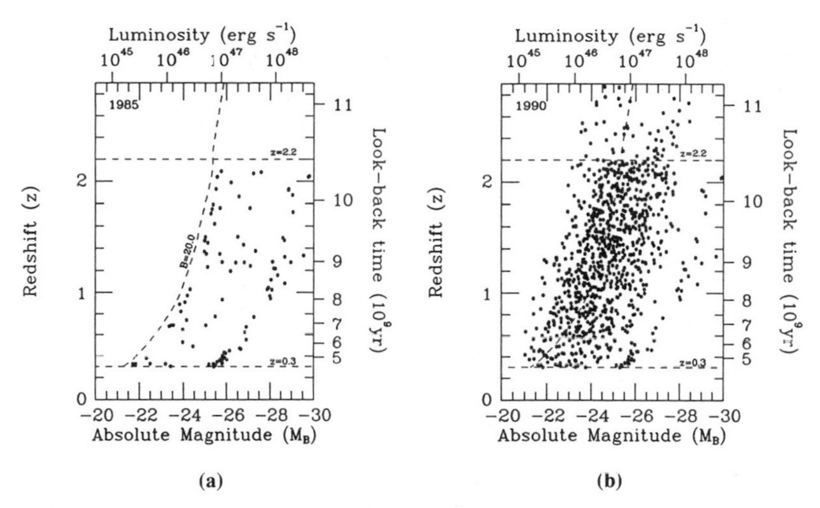

FIGURE 1. (**a**) Absolute magnitude vs. redshift diagrams for all optically selected quasars identified in complete spectroscopic surveys published prior to 1986. Quasars with $z < 0.3$ are not indicated because of possible incompleteness problems at low redshift. The total luminosites were calculated using the approximate relation $L \simeq 4 \times 10^{36}\ \mathrm{dex}(-0.4M_B)$ erg/s. Look-back times have been derived for $H_0 = 50$ km/s/Mpc and $q_0 = 0.5$. (**b**) As in part (**a**), but now including all such surveys published prior to December 1990.

2. QUASAR EVOLUTION

2.1 The Luminosity Function

It is been known since the pioneering work of Schmidt[4] that the properties of the quasar population change with redshift. Schmidt[4] demonstrated that the steep number–magnitude relation for quasars could be explained by a rapid increase in their comoving space density[b] with redshift. Later, Mathez[5] showed that the quasar number–magnitude relation could equally be interpreted as an increase in the luminosity (rather than the space density) of quasars with redshift. The similarity

[b]The comoving space density for any class of astronomical object is simply the space density corrected for the expansion of the Universe. Thus the comoving space density of a nonevolving population remains constant with redshift.

between these two apparently contradictory evolutionary models can be understood as follows. The evolution in the number density or luminosity of quasars with cosmic epoch is most directly established from the redshift dependence of the quasar luminosity function (LF). As illustrated by the schematic diagram in FIGURE 2, for any given redshift interval the quasar LF simply represents the comoving space density of quasars as a function of luminosity. It is clear from the example in FIGURE 2(a) that, if the LF exhibits no conspicuous features in the magnitude range surveyed, the indicated redshift dependence of the quasar LF can be interpreted either as a uniform increase in space density (a vertical shift in the LF) or as a uniform increase in luminosity (a horizontal shift in the LF) toward higher redshift.

Unfortunately, while quasar surveys were limited to $B < 20$ mag, the derived LF exhibited just such a featureless power law form (see, e.g., Marshall *et al.*[6]), and consequently little discrimination between the density and luminosity evolution

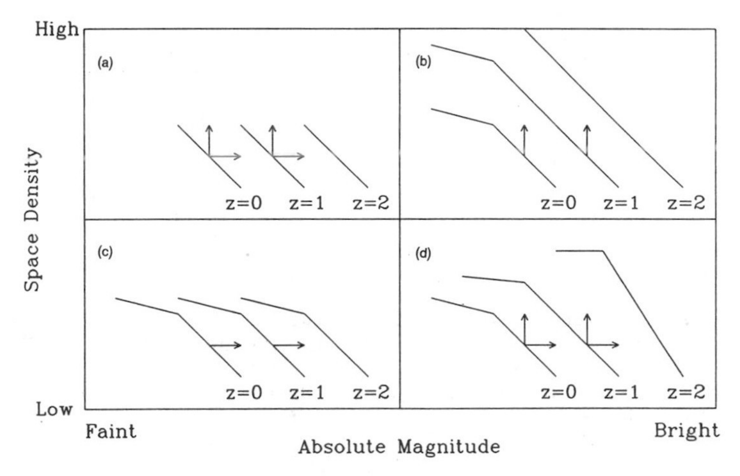

FIGURE 2. A schematic representation of the quasar luminosity function. (**a**) Pure power law LF (no discrimination possible). (**b**) Density evolution. (**c**) Luminosity evolution. (**d**) Density and luminosity evolution. See the text for more details.

models was possible. The first indications that a break from a pure power law LF existed at low luminosities came from Koo,[7] who was able to carry out a limited spectroscopic survey of quasar candidates to $B < 22.5$ mag. The importance of a break in the LF is demonstrated from FIGURE 2(b)–(d), where such a feature in the LF affords a straightforward determination of whether the LF evolves with redshift by increasing in comoving space density [FIG. 2(b)], luminosity [FIG. 2(c)], or a more complicated combination of both density and luminosity [FIG. 2(d)]. More details on the techniques employed in the derivation of the quasar LF and its evolution can be found in two recent reviews of this subject by Hartwick and Schade[8] and Warren and Hewett.[1]

2.2 *Low Redshifts* ($z < 2.2$)

In the last few years, the dramatic increase in the number of quasars identified at faint magnitudes (see, e.g., FIG. 1) has led to a greatly improved understanding of quasar evolution at low redshifts ($z < 2.2$). Using a survey of 420 quasars with $B < 21$ mag, Boyle *et al.*[9] were able to demonstrate that the quasar LF at $z < 2.2$ does indeed exhibit a break at low luminosities whose redshift dependence could be expressed as a uniform power law increase in *luminosity* with redshift

$$L(z) \propto (1 + z)^{k_L},$$

where $k_L = 3.34 \pm 0.1$ and $k_L = 3.15 \pm 0.1$ for $q_0 = 0.0$ and $q_0 = 0.5$ universes, respectively. In such a model, the space density of the quasar population remains constant while their luminosity declines by a factor of more than 30 from $z = 2$ to the present epoch. Boyle *et al.*[9] also showed that the extrapolation of quasar LF to $z = 0$ gave good agreement with estimates of the Seyfert LF by Cheng *et al.*[10] and Marshall.[11] This general form of evolution has also been confirmed by a number of other authors (Marano *et al.*,[12] Koo and Kron[13]), including a detailed reanalysis by Hartwick and Schade[8] of all surveys published prior to 1990.

The simplest physical interpretation of this evolution is that it results from a single generation of long-lived (10^{10} yr) quasars. However, under the conventional supermassive black hole model for quasar energy generation (Rees[14]), this interpretation predicts larger remnant black hole masses (10^9–10^{10} $M_\odot$) and smaller Eddington ratios (~ 0.001) in low-redshift quasars and Seyferts than are inferred from emission line–continuum studies (Wandel and Mushotsky,[15] Padovani[16]). Smaller black hole masses (10^7–10^8 $M_\odot$) are predicted in models that invoke successive generations of short-lived (10^8 yr) quasars to explain the observed evolution (Koo[17]), but such models require a great deal to "fine tuning" to account for the constant comoving number density of quasars at $z < 2.2$ (Cavaliere and Padovani[18]). A third explanation is the "recurrent" model proposed by Cavaliere and Padovani[18] in which quasar activity occurs recurrently in a significant fraction of the galaxy population, possibly fueled by declining episodes of interactions with companion galaxies (Smith *et al.*[19]). In this case, the resulting black hole masses would be 10^8–10^9 $M_\odot$. Since all models make different predictions for the remnant black hole mass, it is clear that detailed measurements of the masses of the central regions of nearby galaxies and active galactic nuclei will play a fundamental role in discriminating between them. At present the evidence for supermassive black holes in nearby galaxies is inconclusive (Dressler[20]), although a fully operational Hubble space telescope may help to resolve this issue.

2.3 *Intermediate Redshifts* ($2.2 < z < 2.9$)

At $z > 2.2$ the evolution of quasars is much less well determined. This is principally due to the lower number of quasars identified at these redshifts (see FIG. 1). Although partly caused by the more frequent use in the past of the simpler U–B selection to identify $z < 2.2$ quasars, the recent introduction of successful multicolor (e.g., Koo and Kron,[13] Warren *et al.*[21]) and grism (e.g., Schmidt *et al.*[22]) techniques to search for high-redshift quasars has made it apparent that their rarity is not simply due to the lack of suitable methods of searching for such quasars. Rather, the low numbers of $z > 2.2$ quasars indicate that the strong power-law evolution in luminosity witnessed at low redshifts cannot be extrapolated to redshifts much greater than $z \sim 2$. However, the nature and amount of a change required in the evolution of quasars at this redshift is still controversial. Warren *et al.*[21] claim that the comoving space density of quasars declines by a factor of 3 between $z = 2$ and $z = 3$, in broad agreement with the result derived by Crampton *et al.*,[23] but in disagreement with Koo and Kron[13] and Osmer,[24] who found no evidence for any decline in the comoving number density of quasars in this redshift range.

Much of the apparent discrepancy between these results arises from the small number of low-luminosity quasars identified in the redshift range $2 < z < 3$, particularly those fainter than the break in the LF at these redshifts ($M_B \gtrsim -26$). This problem has been partially alleviated by two recent surveys for faint ($B < 22$ mag) quasars at $z < 2.9$ (Boyle *et al.*[25] and Zitelli *et al.*[26]), which contain 25 quasars in the redshift range $2 < z < 2.9$ with $M_B > -26$. The LFs obtained from the combination of the Boyle *et al.*[25] survey with other published surveys for $z \lesssim 3$ quasars (see TABLE 1) are presented in FIGURE 3. Separate estimates for the LF in a $q_0 = 0.1$

TABLE 1. Surveys Used to Derive the $z < 2.9$ LF

Reference	Magnitude Limit	Adopted z Range	Number of Quasars
Schmidt and Green[54]	$B \lesssim 16.2$	$0.3 < z < 2.2$	45
Mitchell *et al*[55]	$B \lesssim 17.6$	$0.3 < z < 2.2$	26
Foltz *et al.*[56]	$B \lesssim 18.6$	$0.3 < z < 2.9$	139
Cristiani[42]	$B \lesssim 19.6$	$0.3 < z < 2.9$	61
Marshall *et al.*[6]	$B \lesssim 19.8$	$0.3 < z < 2.2$	50
Crampton *et al.*[23]	$B \lesssim 20.5$	$0.3 < z < 2.9$	95
Boyle *et al.*[41]	$B \lesssim 20.9$	$0.3 < z < 2.2$	394
Marano *et al.*[12]	$B \lesssim 21.0$	$0.3 < z < 2.9$	24
Boyle *et al*[25]	$B \lesssim 22.1$	$0.3 < z < 2.9$	60
Koo and Kron[13]	$B \lesssim 22.6$	$0.3 < z < 2.9$	28

and $q_0 = 0.5$ universe are shown. Note that continuum absolute magnitudes (M_B) have been calculated using the Cristiani and Vio[27] spectral energy distribution for quasars. The redshift bins in the diagrams have been chosen to give equal intervals in $\log(1 + z)$ between $z = 0.3$ and $z = 2.9$, so that the $(1 + z)$ power-law luminosity evolution witnessed at low redshifts is represented by a constant shift in absolute magnitude between successive redshift bins. It is evident from the low redshift LFs plotted in FIGURE 3 that the power-law luminosity evolution model is a good fit to the data at $z < 2$. It is also apparent from the smaller shift between the $1.25 < z < 2.0$ LF and the $2.0 < z < 2.9$ LF that this strong evolution does not continue on beyond $z \sim 2$. However, there is no evidence from the LFs for any decrease in the space density of quasars between $z \sim 2$ and $z \sim 3$ in either a $q_0 = 0.1$ or $q_0 = 0.5$ universe.

A more quantitative model for the evolution of quasars at $z < 3$ can be obtained from a maximum likelihood fit to the data used to construct the LFs in FIGURE 3. Following Boyle *et al.*,[9] the quasar LF $\Phi(M_B, z)$ can be represented by a smooth two

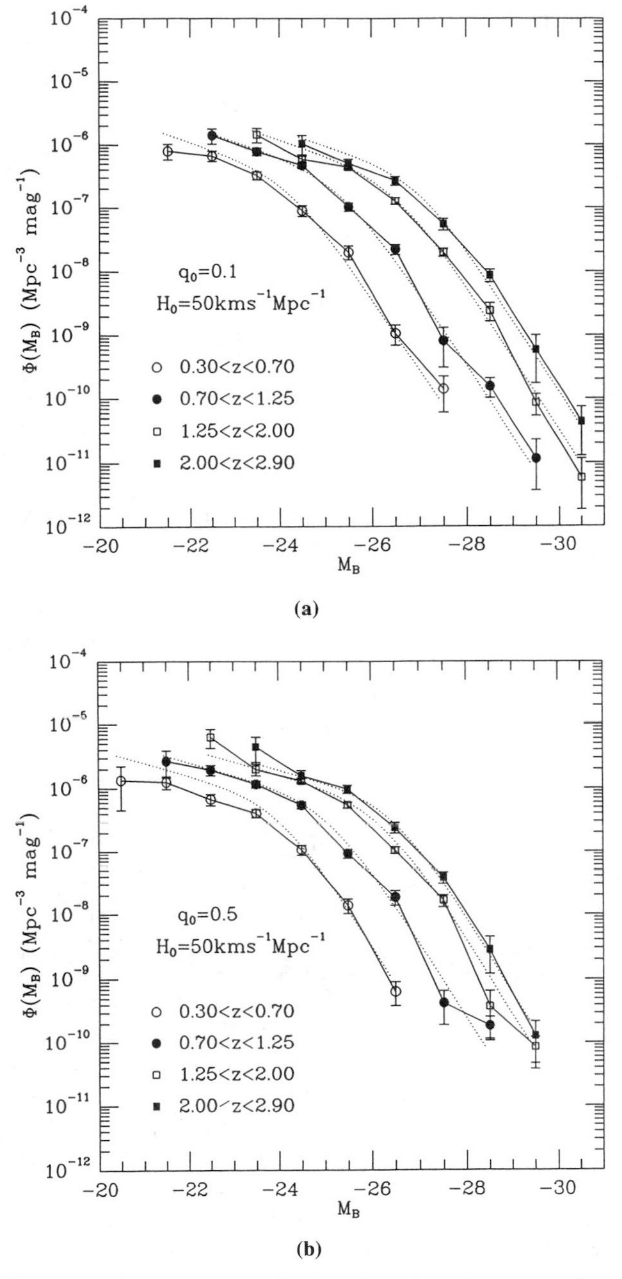

FIGURE 3. The $z < 3$ quasar luminosity function based on the samples listed in TABLE 1. Redshift bins represent equal intervals in $\log(1 + z)$. *Error bars* are based on Poisson statistics. The *dotted lines* indicate the derived model fit to the data. **(a)** $q_0 = 0.1$. **(b)** $q_0 = 0.5$.

power-law function:

$$\Phi(M_B, z) = \frac{\Phi^*}{10^{0.4[M_B - M_B(z)](\alpha+1)} + 10^{0.4[M_B - M_B(z)](\beta+1)}},$$

where the evolution of the LF is given by

$$M_B(z) = M_B^* - 2.5k_L \log(1 + z) \qquad z < z_{max}$$

$$M_B(z) = M_B(z_{max}) \qquad z > z_{max}$$

the additional parameter (z_{max}) introduced here representing the redshift at which the luminosity evolution "switches off"; at redshifts greater than z_{max} no further evolution in either space density or luminosity takes place. The "best-fit" values for the model parameters α, β, M_B^*, k_L, z_{max}, and Φ^* obtained from the maximum likelihood analysis are given in TABLE 2 for $q_0 = 0.1$ and $q_0 = 0.5$. In both cases, the model gives a satisfactory fit to the data as measured by the two-dimensional KS statistic (see TABLE 2). The values for the LF parameters (α, β, and M_B^*) obtained here are similar to those derived by Boyle *et al.*[9] The values of k_L derived in this model are slightly higher, reflecting the introduction of luminosity evolution cutoff parameter (z_{max}), which has a value $z_{max} \sim 2$ for both $q_0 = 0.1$ and $q_0 = 0.5$ cosmological models. These results may therefore be interpreted as evidence for a simple cutoff in the strong luminosity evolution of quasars at $z \gtrsim 2$, with no additional decrease in the comoving space density of quasars in the range $2 < z < 2.9$. The results presented here represent a preliminary analysis of data that will be reported in full elsewhere.

2.4. High Redshifts (z > 3)

Prior to 1987, no quasars with $z > 4$ were known. As of December 1990, 20 quasars with $z > 4$ have been identified (see Irwin and McMahon[28]), the highest quasar redshift currently being $z = 4.74$ (Schneider *et al.*[29]). [NOTE ADDED IN PROOF: Schneider *et al.*[57] have since identified a quasar with $z = 4.89$.] A similar number of quasars in the redshift range $3.5 < z < 4.0$ have also been discovered since 1987, a significant improvement on the eight known prior to 1987 (see Hazard *et al.*[30]). Thus, while there are still relatively few $z \gtrsim 3$ quasars known, it has only been in the past 3 to 4 years that any realistic determination of the high-redshift quasar LF and its evolution has been possible. Unfortunately, many of the majors surveys for high-redshift quasars are still in progress (Schmidt *et al.*,[22] Warren *et al.*,[21] Irwin and McMahon[28]), and so it is difficult to obtain precise details (e.g., area surveyed, completeness) of the individual surveys with which to estimate accurate comoving space densities. A recent review of the current state of these surveys is given by Warren and Hewett.[1]

The general picture that emerges from the present survey work is that the space density of the lowest luminosity quasars ($M \sim -26$) currently observed at high redshift appears to decline dramatically at $z > 3$. From a sample of 15 quasars with $z > 3$, Schmidt *et al.*[22] derived a factor of 7 decrease in the comoving space density of

TABLE 2. "Best-fit" Parameters for the Quasar Evolution Models

Universe Model	Redshift Range	Magnitude Range	Number of Quasars	α	β	M^*	k_L	z_{max}	Φ ($mag^{-1}Mpc^{-3}$)	P_z
$q_0 = 0.1$	$0.3 < z < 2.9$	$M_B < -21$	923	−3.8	−1.6	−22.6	3.55	2.1	3.5×10^{-7}	0.14
$q_0 = 0.5$	$0.3 < z < 2.9$	$M_B < -21$	922	−3.9	−1.5	−22.4	3.45	1.9	6.5×10^{-7}	0.10

TABLE 3. Observed and Predicted Surface Densities for $z > 3$ Quasars

Reference	Redshift Range	Magnitude Range	Area (deg^{-2})	Number of Quasars Observed	Predicted $q_0 = 0.1$	Predicted $q_0 = 0.5$
Schmidt *et al.*[22]	$3.0 < z < 4.5$	$R < 20.5$	14	15	92	86
Hazard *et al.*[30]†	$3.3 < z < 3.8$	$R < 18.0$	150	4	1.4	2.5
Miller *et al.*[31]	$3.4 < z < 4.1$	$17.0 < R < 18.5$	145*	3	4.1	5.2
Warren *et al.*[21]	$3.0 < z < 4.0$	$R < 20.0$	15*	32	37	38
	$4.0 < z < 4.5$	$R < 20.0$	15*	3	4.8	6.7
Irwin and McMahon[28]‡	$4.0 < z < 4.5$	$I < 19.0$	~250	5	6.6	9.4

†Revised survey area and magnitude limit from McMahon (private communication).
‡Reported by Warren and Hewett.[1]
*Survey area corrected for incompleteness.

$M_B < -26$ quasars between $z = 2.2$ and $z = 3.3$.[c] Warren *et al.*[21] obtained a similar factor for the decrease in the space density of $M_B < -26.5$ quasars between $z = 2$ and $z = 4$, based on a sample of 24 quasars with $3 < z < 4.5$. In contrast, no evidence is found by Miller *et al.*[31] for any decrease between $z = 2$ and $z \sim 3.5$ in the comoving space density of quasars at higher luminosities ($M_B < -28$). A previous claim for a continuing increase at $z > 3$ in the space density of $M_B < -28$ quasars (Hazard *et al.*[30]) is now probably also consistent with a constant comoving space density at $z \lesssim 3.5$ (Warren and Hewett[1]).

Warren and Hewett[1] have shown that a model incorporating continuing luminosity evolution but declining density evolution at $z > 3$ is consistent with the preceding results. In another review, Hartwick and Schade[8] also concluded that a decline in the comoving space density of quasars at $z > 3$ is required to fit the observed number of high-redshift quasars, although they note that this conclusion is dependent on the choice of quasar continuum slope and q_0.

In order to investigate this caveat in more detail, the observed numbers of $z > 3$ quasars are compared in TABLE 3 with those predicted from an extrapolation of the model for quasar evolution at $z < 3$ derived in the previous section. Recall that this model does not invoke any decrease in the comoving space density of quasars at high redshift. Where possible the survey areas have been corrected for known incompleteness in the original survey, but it should be noted that many of these surveys here are only partially completed, and surface densities quoted here may change when more accurate corrections for incompleteness, magnitude limit, etc., are known. The Warren *et al.*[21] numbers are obtained from the report of 34 quasars identified in a field at South Galactic Pole field with an estimated completeness of 60 percent (Warren and Hewett[1]). It is apparent from TABLE 3 that the only significant ($> 3\sigma$) discrepancy between the model and the observations occurs with the Schmidt *et al.*[22] survey where the model predicts ~ 6 times more quasars than are observed. Note also that the predicted numbers of high-redshift quasars in the $q_0 = 0.1$ model are, in most cases, slightly lower than those for $q_0 = 0.5$, and consequently are closer to the

[c]A private communication from Schmidt (1989) reported by Hartwick and Schade[8] suggests that the decrease in comoving space density may be less severe than this.

observed number of high-redshift quasars. This is in general agreement with Hartwick and Schade,[8] who found that the decline in the comoving space density at high redshifts was less marked for low values of q_0.

Approximate comoving space densities for the high-redshift quasars are plotted with the modeled $z = 2$ LF in FIGURE 4. In the absence of detailed information on the completeness levels of the various surveys, it is not yet possible to plot meaningful error bars on this diagram. Nevertheless, this figure clearly demonstrates that, with the exception of the Schmidt *et al.*[22] survey, there is no evidence for any significant decrease in the comoving density of $M_B < -26$ quasars at $z > 3$. Indeed, the factor of 7 discrepancy between the Schmidt *et al.*[22] data and the model would also disappear if the estimate for M_B in this survey were revised 0.5 mag brighter and a correction for incompleteness of $\sim$ 60 percent applied. At these redshifts such a shift in M_B could be achieved by a decrease of only 0.3 in the value adopted for α, the mean quasar spectral index ($f_\nu \propto \nu^\alpha$) used in the calculation of M_B.

This example is only intended to serve as a cautionary note. It would probably require a conspiracy between a number of factors to bring the Schmidt *et al.*[22] results into agreement with a model that required no decrease in the space density of quasars at $z > 3$. Nevertheless, it does highlight the difficulty of comparing the observations with model predictions on the steep slope of the LF. At these luminosities uncertainties in the adopted value of M_B (dominated by the choice of spectral index) can lead to significant differences in the predicted number of quasars. Indeed, it is likely that the Cristiani and Vio[27] quasar spectral energy distribution used in the

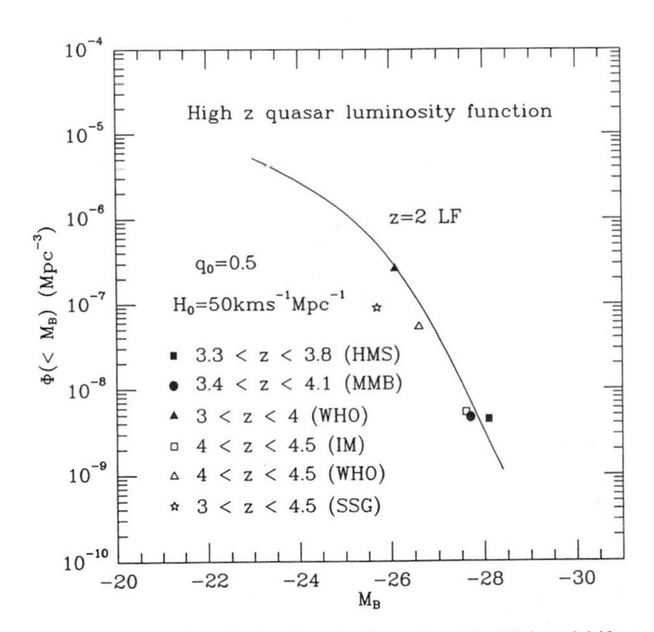

FIGURE 4. The $z > 3$ quasar luminosity function based on the high-redshift quasar surveys listed in TABLE 3. References are as follows: HMS—Hazard *et al.*;[30] MMB—Miller *et al.*;[31] WHO—Warren *et al.*;[21] IM—Irwin and McMahon;[28] SSG—Schmidt *et al.*[22] The model fit for the $z = 2$ LF is also shown.

calculation of M_B in this paper is largely responsible for the different conclusions drawn here to those obtained by Warren *et al.*[21] By comparison, predictions of density evolution models on the flat portion of the high-redshift LF will be considerably less dependent on the adopted spectral index and surveys for such quasars at these luminosities ($M_B \gtrsim -26$) and may provide the only unequivocal detection of a decline in the comoving number density of quasars at high redshift. Surveys of this type will have to cover large areas ($\gtrsim$ 5 deg^2) of sky at faint magnitudes ($R > 21$ mag) to identify a significant number of high-redshift quasars. With the recent development of large-format CCDs such surveys are already underway (e.g., Schade[32]) and should be completed within the next few years.

In summary, a decline in the comoving space density of quasars at $z > 3$ is only clearly seen at the low luminosities ($M_B \gtrsim -26$) probed by the Schmidt *et al.*[22] survey.

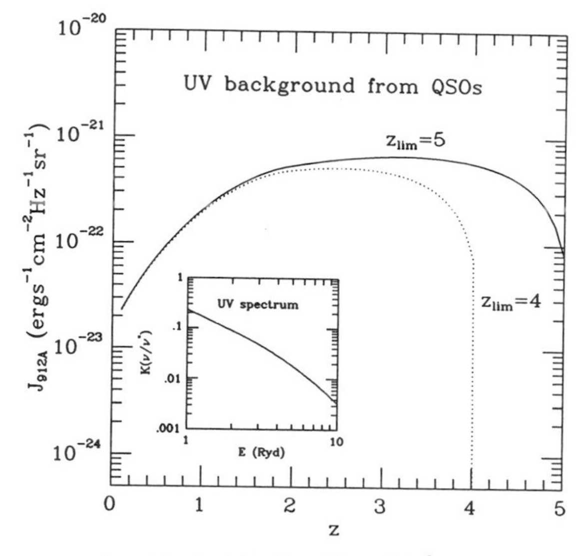

FIGURE 5. The quasar ultraviolet ionizing flux (J_ν) at 912 Å predicted from the model for quasar evolution derived in the text. The adopted UV spectrum is shown in the *inset.*

Even in this case, correction for incompleteness and the adopted spectral index could significantly reduce the amount of a decline required. Models with no decrease in the comoving number density of quasars at high redshift are certainly consistent with the number density of bright ($M_B < -26$) quasars at high redshift. Such models also have the perhaps appealing feature that quasars could be responsible for ionizing the intergalactic medium at $z > 4$ (see FIG. 5). To derive the curve plotted in this figure, the "no absorption model" and "medium UV spectrum" of Bechtold *et al.*[33] have been employed with the model for quasar evolution derived in this paper. For a sudden cutoff in the number density of quasars at $z_{\rm lim} = 5$, a constant background flux at 912 Å of $J_\nu \sim 7 \times 10^{-22}$ erg/s/cm^2 /Hz/sr is obtained over the redshift range $2 < z < 4.5$. This is within 30 percent of the value ($J_\nu \sim 10^{-21}$ erg/s/cm^2/Hz/sr) required to ionize the Universe at these redshifts as implied by the absence of the Gunn–Peterson effect in high-redshift quasars.

The lack of any significant decline in the comoving number density of quasars at high redshifts would pose significant problems for models of galaxy formation that predict a "late" epoch of galaxy formation, in particular the cold dark matter (CDM) model. Efstathiou and Rees[34] have demonstrated that a constant comoving number density of quasars over the range $2 < z < 4$ is only consistent with CDM if quasars are short-lived and radiate at about the Eddington limit, contrary to some physical models of quasar evolution at $z < 2$ (see Section 2.2). Efstathiou and Rees[34] also conclude that the discovery of a significant number of quasars with $z > 5$ would begin to pose serious problems for the CDM hypothesis. As yet none have been found, but the strenuous efforts of researchers in this field should identify such objects, if they exist, within the next year or two.

3. QUASAR CLUSTERING

In recent years there has been rapid progress in the measurement of quasar clustering. This has primarily arisen as a result of the availability of the large, faint ($B < 20$ mag) quasar catalogues whose surface densities and sizes are sufficiently high to permit a detailed study of quasar clustering at scales $< 50\ h^{-1}$ Mpc ($H_0 = 100$ h km/s/Mpc). Previously, studies of quasar clustering were restricted to larger scales with no significant detection of clustering (Osmer,[2] Chu and Zhu[35]).

A useful and frequently employed tool in the measurement of quasar clustering is the 2-point correlation function $\xi(r)$ (Peebles[36]). The correlation function simply measures the strength of clustering by counting pairs of quasars. At any given comoving separation r, the correlation function is given by

$$\xi(r) = \frac{\text{Number of pairs observed}}{\text{Number of pairs expected if random}} - 1.$$

Thus $\xi(r)$ is positive if quasars are clustered, zero if they are randomly distributed, or negative if quasars are anticlustered with respect to each other. The first tentative detection of quasar clustering was made by Shaver[37] from an analysis of the heterogeneous Veron-Cetty and Veron[38] catalogue. This clustering was confirmed at the 4.2σ level by Shanks *et al.*[39,40] using the homogeneous Boyle *et al.*[41] sample of 392 quasars with $z < 2.2$. Shanks *et al.*[40] identified 25 quasar pairs with comoving separations $< 10\ h^{-1}$ Mpc,[d] whereas only 11.1 pairs would have been expected from a random distribution. At small scales, Shanks *et al.*[40] also found that the 2-point quasar correlation function was consistent with the -1.8 power law form derived for galaxies:

$$\xi(r) = \left(\frac{r}{r_0}\right)^{-1.8}$$

with an observed scale length $r_0 = 7 \pm 1.5\ h^{-1}$ Mpc at $z = 1.4$, a level of quasar clustering consistent with that obtained by Shaver.[37] Similar strengths of quasar

[d] Unless otherwise stated, $q_0 = 0.5$ will be assumed throughout this section.

clustering at these scales have also been identified in the samples of Cristiani[42] and Crampton *et al.*[23] by Iovino and Shaver,[43] although at a lower significance level.

Having established that clustering among quasars does indeed exist, it is clear that the redshift dependence of the quasar correlation function provides a unique measure of the growth of structure in the Universe with cosmic epoch. Following Shanks *et al.*,[40] this can be demonstrated by considering two very simple models for the evolution of quasar correlation function. First, a model in which the clustering is stable and the scale length is large. In such a "stable" model the strength of the clustering decreases at higher redshifts where the proper quasar background density is higher by $(1 + z)^3$. This will be reflected by a corresponding decrease in the scale length or amplitude of the quasar correlation function with increasing redshift. However, for models of biased galaxy formation in which galaxies and quasars are more strongly clustered than the mass (Davis *et al.*[44]), the quasar clusters may still be expanding almost as fast as the Hubble flow. In such a "comoving" model the amplitude of the quasar correlation function will remain constant with redshift.

Although Shanks *et al.*[40] found no evidence for any significant evolution with redshift in the amplitude of the quasar correlation function, the errors on the estimates of the amplitude of the correlation function in separate redshift bins were sufficiently large to prevent any discrimination between the "stable" and "comoving" models for the evolution of the quasar correlation function. In contrast, analyses of several smaller homogeneous samples by Kruszewski[45] and Shaver[46] have yielded apparent support for a strong decrease in the strength of quasar clustering with redshift. However, both these claims rest predominantly on the high values of the amplitude derived at $z < 0.5$, where the samples used may be contaminated by radio galaxies that are known to exhibit strong clustering (Peacock *et al.*[47]). By combining various surveys, Iovino and Shaver[43] found some evidence for a decrease in the strength of quasar clustering with redshift, but their data were insufficient to rule out either model of quasar clustering. Evidence for mild evolution in the strength of quasar clustering with redshift was also found from a reanalysis of the Veron-Cetty and Veron[38] catalogue by Anderson *et al.*[48]

The latest information on the quasar correlation function can be obtained from an analysis of the ~ 500 quasars identified in the Boyle *et al.*[25,41] and Zitelli *et al.*[26] surveys. The correlation function derived from these survey is plotted in FIGURE 6. Over the redshift range $0.3 < z < 2.9$, the strong clustering signal at $< 10\ h^{-1}$ Mpc is clearly seen; 32 quasar pairs are observed at these separations, whereas only 14.5 pairs are predicted on random hypothesis. On splitting the number of quasars equally into high ($1.5 < z < 2.9$) and low ($0.3 < z < 1.5$) redshift intervals, no significant difference in the amplitude of the quasar correlation function is observed. An equal number of quasar pairs with comoving separations $< 10\ h^{-1}$ Mpc are observed in each redshift interval; for a random distribution the predicted number of pairs are 7.8 and 8.2 for the low ($0.3 < z < 1.5$) and high ($1.5 < z < 2.9$) redshift intervals, respectively. From this sample, at least, quasar clustering at $z \sim 2$ is confirmed at a moderately high level of significance; in direct contrast to previous estimates of level of quasar clustering at high redshift (Iovino and Shaver[43]). Moreover, there is no evidence for any evolution in the strength of quasar clustering with redshift. Such an observation supports the "comoving" model for the evolution of structure in the Universe, but cannot yet rule out the "stable" model. Full details of this analysis will be published elsewhere.

Despite the availability of much larger samples of faint quasars, it is evident from FIGURE 6 that the quasar correlation function still exhibits no significant features at scales $\gtrsim$ 10 h^{-1} Mpc (see also Shanks *et al.*,[40] Osmer and Hewett[49]). Some isolated

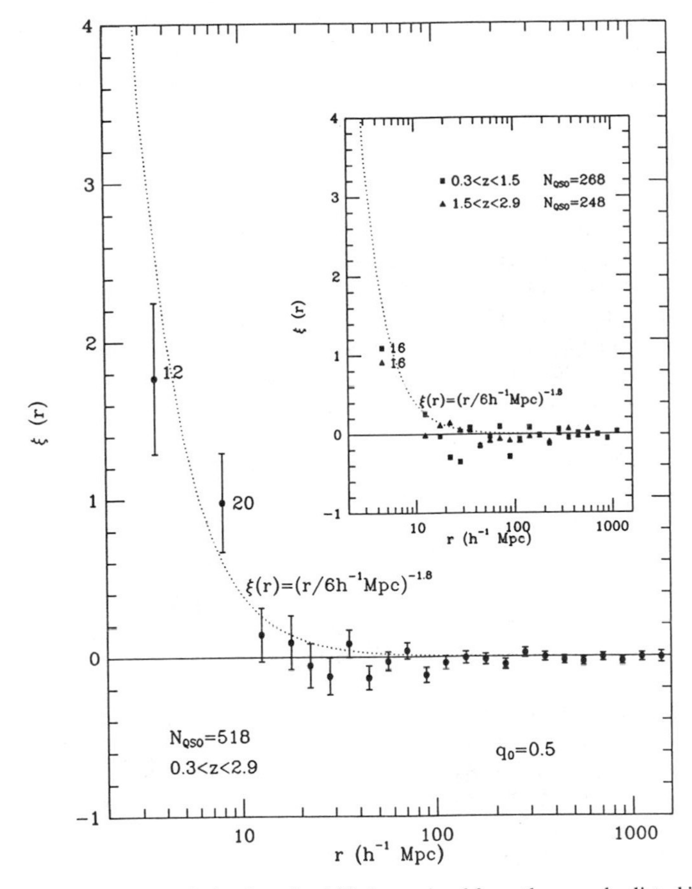

FIGURE 6. The quasar correlation function $\xi(r)$ determined from the samples listed in the text. The numbers presented by the data points indicate the number of quasar pairs found at these separations. The high ($1.5 < z < 2.9$) and low ($0.3 < z < 1.5$) redshift correlation functions are shown in the *inset.* A -1.8 power-law correlation function with $r_0 = 6$ h^{-1} Mpc is indicated by the *dotted line.*

groups of quasars with comoving sizes $r \sim 100$ h^{-1} Mpc have been identified (Webster,[50] Crampton *et al.*,[51] Clowes & Campusano[52]), but their statistical significance is hard to assess since they do not appear to contribute significantly toward a signal in the quasar correlation function at these scales. However, the appearance of any large-scale structure in the quasar correlation function may have significant cosmological importance. In gravitational instability models, any small features in $\xi(r)$ at large scales will grow as $(1 + z)^{-2}$ until $z = 0$ in an $\Omega_0 = 1$ Universe, whereas little growth is expected in a low Ω_0 Universe (Peebles[36]). Any observed increase in the large-scale amplitude of $\xi(r)$ toward low redshift would therefore imply $\Omega_0 = 1$. In contrast, the position of any large-scale feature is not expected to evolve significantly with redshift. The redshift dependence of such a feature in models with different values of q_0 would therefore provide a sensitive q_0 test, since differences in calculated comoving separations amount to 40 percent between $q_0 = 0.1$ and $q_0 = 0.5$ models, even over the limited redshift range $0 < z < 1.4$. Although Shanks *et al.*[40] found no significant evidence for the quasar correlation function to develop systematically positive correlations toward lower redshift, an increased scatter about $\xi(r) = 0$ was observed at lower redshifts.

It is clear that quasar clustering provides a direct means by which to study the evolution of structure in the Universe. Definitive tests, in particular the redshift evolution of the correlation function, will, however, require at least an order of magnitude more data at faint magnitudes than the $\sim$ 500 quasars currently available. Although a catalogue of 5000–10,000 quasars with $B \lesssim < 21$ may not seem feasible at present, continuing developments in instrumentation may allow such a catalogue to be realized in the not too distant future. Wide-field multiobject spectrographs catering for several hundred objects over a few square degrees (see, e.g., Taylor and Gray[53]) would enable such catalogues to be constructed in only a few nights on a large telescope. Coupled with the advance in large-format CCDs, which now allow the high redshift LF to be probed to significantly lower luminosities, the field of quasar clustering and evolution looks set for another significant advance, providing further fundamental insights into the structure and evolution of the Universe at high redshift.

ACKNOWLEDGMENTS

Many of the results reported here would not have been possible without the skill and dedication of the staff at the U.K. Schmidt Telescope, the Anglo-Australian Observatory, and the COSMOS and APM measuring machines in the United Kingdom. I am indebted to Max Pettini for suggesting the UV background calculation, and would also like to thank my colleagues at the Institute of Astronomy, Mike Irwin, Richard McMahon, Paul Hewett, and David Schade for many useful discussions.

REFERENCES

1. WARREN, S. J. & P. S. HEWETT. 1990. Rev. Mod. Phys. **53:** 1093.
2. OSMER, P. S. 1981. Astrophys. J. **247:** 762.
3. OSMER, P. S. 1982. Astrophys. J. **253:** 28.
4. SCHMIDT, M. 1968. Astrophys. J. **151:** 393.
5. MATHEZ, G. 1976. Astron. Astrophys. **53:** 15.
6. MARSHALL, H. L., Y. AVNI, A. BRACCESI, J. P. HUCHRA, H. TANENBAUM, G. ZAMORANI & V. ZITELLI. 1984. Astrophys. J. **283:** 50.

7. Koo, D. C. 1983. *In* Quasars and Gravitational Lenses, 24th Liege Astrophysical Colloquium: 240. University of Liege, Liege, Belgium.
8. Hartwick, F. D. A. & D. Schade. 1990. Ann. Rev. Astron. Astrophys. **28:** 437.
9. Boyle, B. J., T. Shanks & B. A. Peterson. 1988. Mon. Not. R. Astron. Soc. **235:** 935.
10. Cheng, F. Z., J. Danese, G. De Zotti & A. Franceschini. 1985. Mon. Not. R. Astron. Soc. **212:** 857.
11. Marshall, H. L. 1987. Astron. J. **94:** 628.
12. Marano, B., G. Zamorani & V. Zitelli. 1988. Mon. Not. R. Astron. Soc. **232:** 111.
13. Koo, D. C. & R. G. Kron. 1988. Astrophys. J. **325:** 92.
14. Rees, M. J. 1984. Ann. Rev. Astron. Astrophys. **22:** 471.
15. Wandel, A. & R. F. Mushotsky. 1986. Astrophys. J. **306:** L61.
16. Padovani, P. 1989. Astron. Astrophys. **209:** 27.
17. Koo, D. C. 1986. *In* Structure and Evolution of Active Galactic Nuclei, G. Giuricin *et al.,* Eds.: 317. Reidel. Dordrecht, the Netherlands.
18. Cavalieri, A. & P. Padovani. 1988. Astrophys. J. **333:** L33.
19. Smith, E. P., T. M. Heckman, G. D. Bothun, W. Romanshin & B. Balik. 1986. Astrophys. J. **306:** 64.
20. Dressler, A. 1989. *In* Active Galactic Nuclei, IAU Symposium No. 134, D. E. Osterbrock and J. S. Miller, Eds.: 217. Reidel. Dordrecht, the Netherlands.
21. Warren, S. J., P. S. Hewett & P. S. Osmer. 1988. *In* Proceedings of a Workshop on Optical Surveys for Quasars, ASP Conference Series No. 2, P. S. Osmer *et al.,* Eds.: 98. Brigham Young. Provo, Utah.
22. Schmidt, M., D. P. Schneider & R. P. Green. 1988. *In* Proceedings of a Workshop on Optical Surveys for Quasars, ASP Conference Series No. 2, P. S. Osmer *et al.,* Eds.: 87. Brigham Young. Provo, Utah.
23. Crampton, D., A. P. Cowley & F. D. A. Hartwick. 1987. Astrophys. J. **314:** 129.
24. Osmer, P. S. 1980. Astrophys. J. Suppl. **42:** 523.
25. Boyle, B. J., L. R. Jones & T. Shanks. 1991. Submitted for publication in Mon. Not. R. Astron. Soc.
26. Zitelli, V., M. Mignoli, B. Marano, G. Zamorani & B. J. Boyle. 1991. Submitted for publication in Mon. Not. R. Astron. Soc.
27. Cristiani, S. & Vio. 1990. Astron. Astrophys. **227:** 385.
28. Irwin, M. & R. G. McMahon. 1990. Gemini **30:** 6.
29. Schneider, D. P., M. Schmidt & J. E. Gunn. 1989. Astron. J. **98:** 1951.
30. Hazard, C., R. G. McMahon & W. Sargent. 1986. Nature **322:** 38.
31. Miller, L., P. S. Mitchell & B. J. Boyle. 1990. Mon. Not. R. Astron. Soc. **244:** 1.
32. Schade, D. 1991. Submitted for publication in Astron. J.
33. Bechtold, J., R. J. Weymann, Z. Lin & M. A. Malkan. 1987. Astrophys. J. **315:** 180.
34. Efstathiou, G. & M. J. Rees. 1988. Mon. Not. R. Astron. Soc. **235:** 5P.
35. Chu, Y.-Q. & X.-F. Zhu. 1983. Astrophys. J. **267:** 4.
36. Peebles, P. J. E. 1980. The Large Scale Structure of the Universe. Wiley. Princeton, N.J.
37. Shaver, P. A. 1984. Astron. Astrophys. **136:** L9.
38. Veron-Cetty, M. P. & P. Veron. 1984. A Catalogue of Quasars and Active Nuclei. ESO Scientific Report No. 1.
39. Shanks, T., R. Fong, B. J. Boyle & B. A. Peterson. 1987. Mon. Not. R. Astron. Soc. **227:** 739.
40. Shanks, T., B. J. Boyle & B. A. Peterson. 1988. *In* Proceedings of a Workshop on Optical Surveys for Quasars, ASP Conference Series No. 2, P. S. Osmer *et al.,* Eds.: 244. Brigham Young. Provo, Utah.
41. Boyle, B. J., R. Fong, T. Shanks & B. A. Peterson. 1990. Mon. Not. R. Astron. Soc. **243:** 1.
42. Cristiani, S. 1989. Astron. Astrophys. Suppl. **77:** 161.
43. Iovino, A. & P. Shaver. 1988. Astrophys. J. **330:** L13.
44. Davis, M., G. Efstathiou, C. S. Frenk & S. D. M. White. 1985. Astrophys. J. **221:** 371.
45. Kruszewski, A. 1988. Acta Astron. **38:** 155.
46. Shaver, P. A. 1988. *In* Proceedings of a Workshop on Optical Surveys for Quasars, P. S. Osmer *et al.,* Eds.: 265. Brigham Young. Provo, Utah.

47. PEACOCK, J. A., L. MILLER, C. A. COLLINS, D. NICHOLSON & S. J. LILLY. 1988. *In* Large Scale Structures in the Universe, IAU Symposium No. 130, Audouze *et al.,* Eds.: 579. Reidel. Dordrecht, the Netherlands.
48. ANDERSON, N., D. KUNTH & W. L. W. SARGENT. 1988. Astron. J. **95:** 644.
49. OSMER, P. S. & P. C. HEWETT. 1991, Astrophys. J. Supp. **75:** 273.
50. WEBSTER, A. 1982. Mon. Not. R. Astron. Soc. **199:** 683.
51. CRAMPTON, D., A. P. COWLEY & F. D. A. HARTWICK. 1989. Astrophys. J. **345:** 59.
52. CLOWES, R. G. & L. T. CAMPUSANO. 1991. Mon. Not. R. Astron. Soc. In press.
53. TAYLOR, K. T. & P. M. GRAY. 1989. AAO Design Study Report, A Wide-Field Multi-Fibre Prime Focus for the AAT.
54. SCHMIDT, M. & R. F. GREEN. 1983. Astrophys. J. **269:** 352.
55. MITCHELL, K. J., A. WARNOCK & P. D. USHER. 1984. Astrophys. J. **287:** L3.
56. FOLTZ, C. B., F. H. CHAFFEE, P. C. HEWETT, R. J. WEYMANN, S. F. ANDERSON & G. M. MACALPINE. 1989. Astron. J. **98:** 1959.
57. SCHNEIDER, D. P., M. SCHMIDT & J. E. GUNN. 1991. Preprint.

Young Galaxies[a]

LENNOX L. COWIE

Institute for Astronomy
University of Hawaii at Manoa
Honolulu, Hawaii 96822

INTRODUCTION

Traditionally, it has been thought worthwhile to study high-redshift galaxies first of all to understand the formation[b] and the history of galaxies as a whole and to get clues about the fluctuation spectrum from which they grew, while a second goal has been to try to find out the geometry and age of the Universe. Observationally, we have made great strides: we now know of galaxies at redshifts of four to five, potentially far enough in the past to study volume histories and put constraints on available timescales. But unfortunately this type of work is never straightforward because we always have to understand the questions of galaxy formation and evolution in order to address the questions of cosmological geometry and timescales and vice versa.

The largest redshift object we know of today is the quasar PC1158 + 4635 at $z = 4.73$ (Schneider *et al.*[1]). It is natural (though perhaps not essential) to suppose that this quasar resides in an underlying galaxy, since we know lower redshift quasars do and, because the high-redshift quasars are extremely luminous, that underlying galaxy is likely to be very massive, perhaps at least comparable to massive ($10^{11}\ M_{\odot}$) local galaxies (Efstathiou and Rees[2]). And, significantly, because the time required to grow a massive central black hole is long, it is likely that this galaxy is already quite old.

The highest redshift galaxy directly detected is the massive galaxy hosting the radio source 4C 41.17 at $z = 3.8$ (Chambers *et al.*[3]). Initially it was thought that the colors of these high-redshift radio galaxies (we now know of some dozen of them at $z > 2$) implied that they were already very old (≥ 1 Gyr; Lilly[4]), which would have really started to push the time available even in an open universe. However, these objects (and others) have recently inspired a burst of improved observations and galaxy evolution modeling that has shown that the galaxies can be quite young (as little as 0.3 Gyr old) and still have their observed spectral energy distributions (Chambers and Charlot[5]). This has not removed the problem entirely; it could come back to trouble us again if individual portions of these morphologically complex objects are found to be redder and hence older than the average. The only high-redshift radio galaxy that looks "young" is the galaxy Herc202 at $z = 2.390$

[a]This work was supported in part by NASA Grant NAGW 959.

[b]From an observational viewpoint it is simplest to consider galaxy formation as the time of first star formation in the galaxies, which is the time at which they become potentially observable by most techniques. This is not necessarily the time of turnaround of the density fluctuation, which would perhaps be a more natural definition of formation.

(Windhorst *et al.*[6]). This blue galaxy was found among a much fainter radio sample than those where the redder high-z radio galaxies were detected, and this may imply strong selection effects in the radio samples as regards galaxy type.

The radio galaxies and quasars provide evidence that massive galaxies were in existence at z = 4–5, but tell us nothing about how much of the population was already present by that time. Indeed, they could easily constitute just a small fraction of the galaxies, with the bulk forming considerably later. It is very hard to be definite about this because quasars and highly luminous radio galaxies are extremely rare objects.[c]

The very distant quasars can also be used to find out about the intergalactic gas at high redshift, where one of the most interesting recent discoveries has been the extension of the Gunn–Peterson effect (the absence of a lot of neutral diffuse intergalactic gas), even to the very highest redshift quasars (z = 4–5). (The Gunn–Peterson effect [Gunn and Peterson[8]] is looked for in the spectra of quasars, where the Lyman α opacity from neutral gas would kill the flux at wavelengths shorter than the quasar's redshifted Lyman α.) This is actually quite a surprising result because the quasars are not sufficiently populous at these redshifts to themselves ionize any reasonable intergalactic medium (that is, one containing a good fraction of present-day baryons) (e.g., Bechtold *et al.*[9]). The best alternative explanation might be that galaxy formation at these redshifts provides the necessary ionization, which, if true, would mean that several percent of it must already have taken place at $z > 5$ (e.g., Miralda-Escudé and Ostriker;[10] Songaila *et al.*[11]).

Another very direct and interesting result is that there *is* a very large amount of baryonic mass present in the form of neutral hydrogen clumps at z = 2–3 (Wolfe *et al.*[12]), detected as radiatively damped Lyman α absorption lines in the spectra of some quasars. These systems cover a large fraction of the sky and the damping wings imply a considerable column density of neutral hydrogen ($N_H \geq 10^{20}$ cm^{-2}). The neutral gas density in these clouds alone corresponds to a local density

$$\rho_H = \frac{(1 + 2q_0 z)^{1/2}}{(1 + z)} \left(\frac{H_0 m_H}{c}\right) \left(\frac{dN_H}{dz}\right), \qquad (1)$$

or approximately 7×10^{-33} (H_0/50 km s^{-1} Mpc^{-1}) g cm^{-3} for $q_0 = 0.5$, where m_H is the mass of the hydrogen atom. This means that these systems alone are the sites of a large fraction of the current baryonic mass in normal galaxies, which is roughly 2×10^{-32} (H_0/50 km s^{-1} Mpc^{-1})^2 g cm^{-3}, though with quite large uncertainty (e.g., Cowie[13]). This extremely interesting result of Wolfe's clearly shows that there was still a lot of cold baryonic material around at $z \sim 2$ that had not yet turned into galaxies.

[c] One curious result is that there is a very tight relationship between infrared flux and redshift for radio flux-selected galaxies (less than 40 percent variation; Lilly[7]). This result seems to suggest that these galaxies have a very small range of masses at any redshift and must indeed be a very tightly selected population. It is remarkable that the active galactic nuclei responsible for the radio activity and quasar properties are so sensitive to the overall properties of the galaxy.

OPTICAL AND INFRARED GALAXY COUNTS

Turning now to direct observation of field galaxies, I show a comparison of optical and near IR images of a typical region of sky in FIGURE 1. The current status of the differential galaxy counts in B (4500 Å) and K is given in FIGURE 2. Consider first the B band counts, which I am going to compare with number-count models, both with and without galaxy evolution, that I have adapted from those of Yoshii and Takahara.[14] There are numerous assumptions in these models about the present population mix and galaxy luminosity function and the nature of galaxy evolution (that is, how its luminosity and spectral type changed with time, younger galaxies generally being more luminous and bluer); these are well described by Ellis.[15] However, there are some very basic features that are more or less model independent. First, it is possible to roughly match the observed counts to $B = 21$ even with a

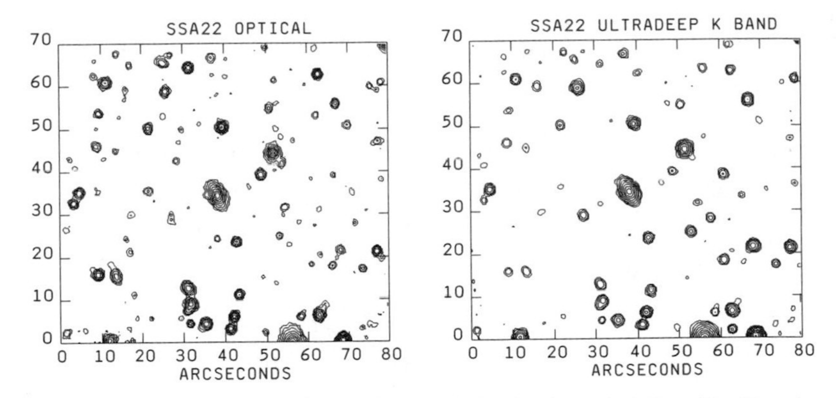

FIGURE 1. Very deep images of a small area of the sky in optical ($B + V + I$) and near infrared (K) light are shown as contour plots. The figure shows the extremely high surface density of galaxies on the sky, and also demonstrates that there are only a very small number of objects that are very red and detected in K but not the optical.

model with no galaxy luminosity evolution (though as Maddox *et al.*[28] stress, and as can be seen clearly in FIG. 2, these models do not actually fit the shape of the counts, even at the bright end). Beyond this point actual counts tend to rise more rapidly than any of the nonevolving predictions, even including the most extreme effects of geometry: an open universe has more volume at high redshift than a closed one, and a zero-curvature universe with $\Omega_0 = 0.1$ and positive cosmological constant has even more; these models therefore predict more faint-end counts, but this effect is small compared with the count discrepancy.

Luminosity evolution has a much larger effect, as can be seen from FIGURE 2 where the evolving models are shown as solid lines. With the luminosity evolution included, one sees galaxies to much larger distances, and consequently the models predict many more counts, but of course they predict that the excess counts should be produced by extended tails or secondary peaks *at high redshift.* Prior to the advent of

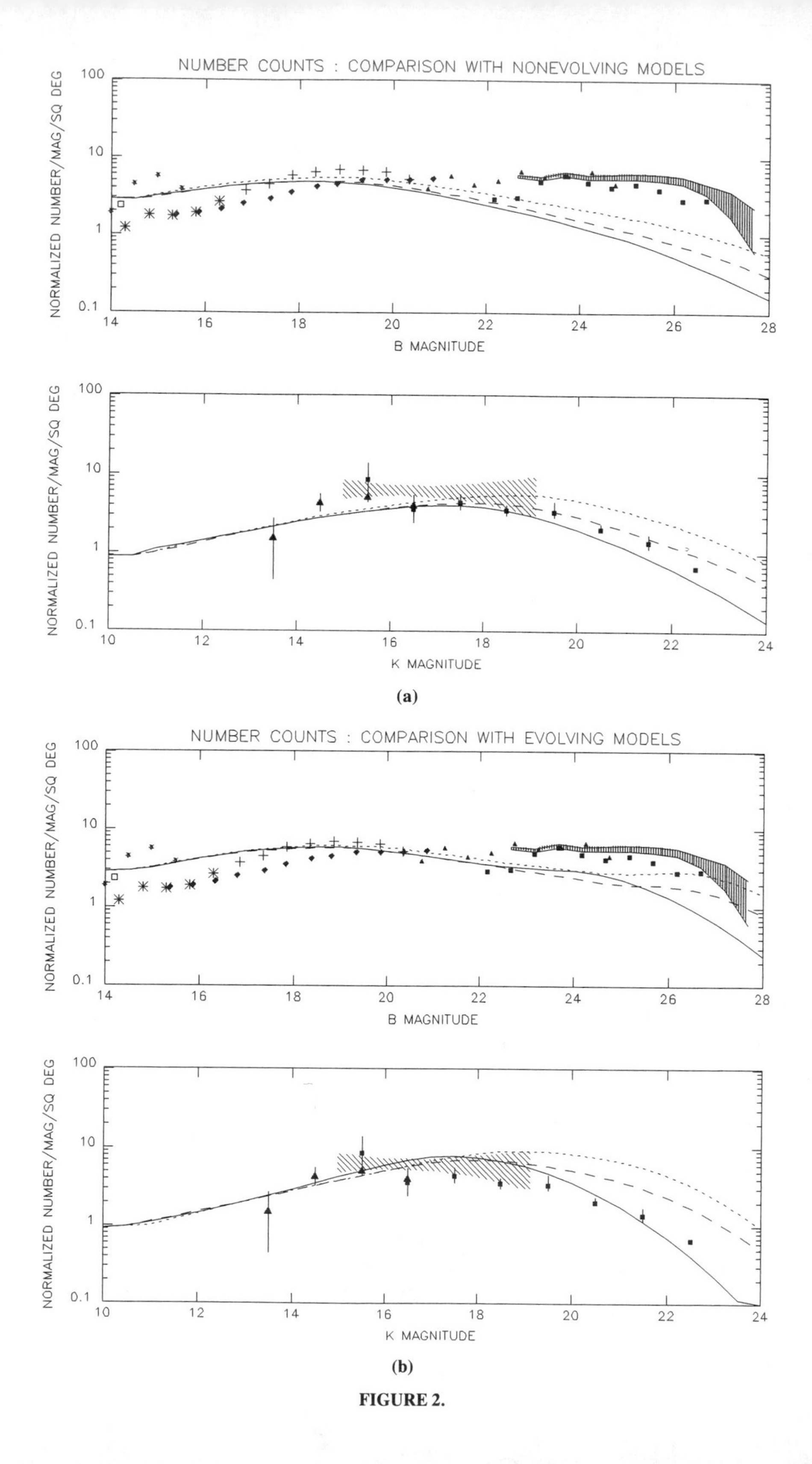

NUMBER COUNTS : COMPARISON WITH NONEVOLVING MODELS
NORMALIZED NUMBER/MAG/SQ DEG
100
10
1
0.1
14
16
18
20
22
24
26
28
B MAGNITUDE
NORMALIZED NUMBER/MAG/SQ DEG
100
10
1
0.1
10
12
14
16
18
20
22
24
K MAGNITUDE
(a)
NUMBER COUNTS : COMPARISON WITH EVOLVING MODELS
NORMALIZED NUMBER/MAG/SQ DEG
B MAGNITUDE
NORMALIZED NUMBER/MAG/SQ DEG
K MAGNITUDE
(b)

FIGURE 2.

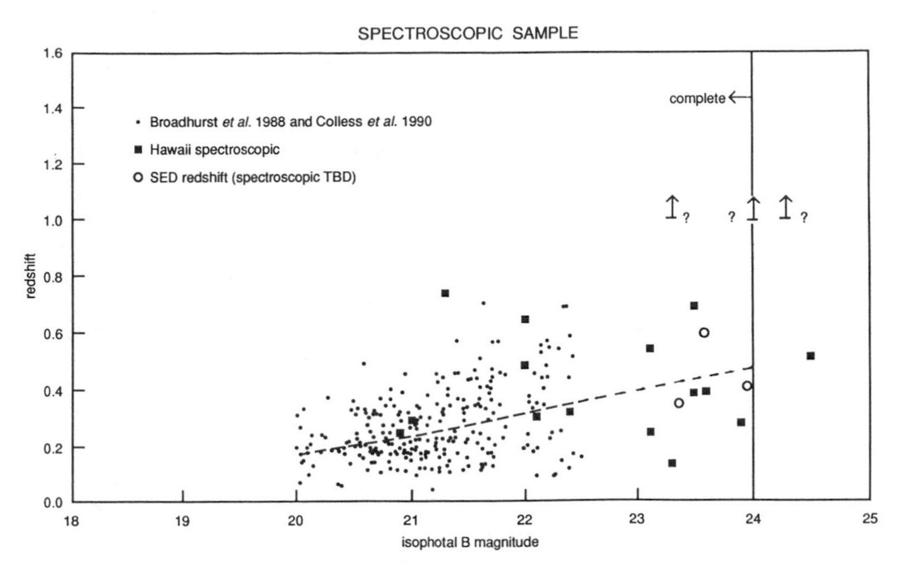

FIGURE 3. Spectroscopic samples for B magnitude limited galaxies. The *dots* show the $B < 21$ and $B < 22.5$ samples of Broadhurst *et al.*[19] and Colless *et al.*[27] The latter is about 80 percent complete. The sold squares show the Hawaii spectroscopic data (e.g., Lilly *et al.*[16]), which are nearly complete to $B = 24$. (*Open circles* are three objects that remain to be observed spectroscopically shown at redshifts estimated from their colors.) Only one object at $B < 24$ defied spectroscopic identification and could be at high z (*upward-pointing arrow*); nearly all objects are at $z < 0.6$. The *dashed line* shows a predicted mean redshift for a model with no galaxy evolution; it provides a remarkably good fit.

spectroscopic data, this seemed a good explanation for the excess counts, but as the faint-object spectra have become available in the last couple of years, it has become clear that it must be ruled out (cf. Lilly *et al.*[16]). To demonstrate this I show data from a number of complete spectroscopic samples in FIGURE 3. To $B = 24$ there is only one possible object in these samples that could even conceivably be at $z > 1$; the majority have $z < 0.6$.

←

FIGURE 2. A comparison of blue and infrared counts with model predictions. The blue number counts are taken from the compilation of Metcalfe *et al.*[21] in addition to those of Maddox *et al.* (*diamonds*), Tyson[22] (*shaded region*), and Lilly *et al.*[16] (*solid squares*). Following Maddox *et al.* we normalize the counts with a power law of slope 0.45 and unit value at $B = 16$. The infrared counts are from Glazebrook *et al.*[23] (*triangles*), Jenkins and Reid[24] (*shaded region*), and Cowie *et al.*[25,26] (*squares and diamonds*). For the IR data, I have shown error limits. The IR counts are normalized with a power law of slope 0.45 and a unit value at $K = 11$. Two classes of model are shown. Part (**a**) shows models in which galaxies do not change at all in luminosity or type. The *solid curve* has $q_0 = 0.5$, the *dashed curve,* $q_0 = 0.02$, and the *dotted curve,* a zero-curvature model with $\Omega_0 = 0.1$ and a cosmological constant. Part (**b**) shows models with the same three geometries as above, but in which the galaxies evolve with z roughly following the prescription of Yoshii and Takahara.[14] The galaxies are assumed to form at $z_f = 5$ in all cases. The K-band counts fit best to a $q_0 = 0.5$ cosmological geometry with evolution or a $q_0 = 0.02$ model with no evolution, but the latter model is quite physically implausible. All the models that fit the K-band counts grossly underpredict the faint-end blue counts.

Until the faint K-band counts became available there remained two possible explanations for the combined faint blue counts and blue-selected spectroscopy, one cosmological and one astrophysical, with no clear-cut way to decide between them. That is, either there is a major population of blue dwarf galaxies that was present at $z \sim 0.3$, but that was not included in the assumed present-day luminosity functions (this is discussed in much more detail below), or alternatively there must be much more volume at low redshift than even a $q_0 = 0.02$ Friedman model would predict. The latter could be the case, for example, in a zero-curvature $\Omega = 0.1$ universe with a cosmological constant, a possibility that has many other attractions, such as solving the cosmological timescale problem and easing the difficulties of large-scale structure. As is shown in FIGURE 2, such a model (with some galaxy evolution included) can roughly fit the number counts (e.g., Lilly *et al.*,[16] Fukugita *et al.*[17]) and yet not violate the spectroscopic constraints.

However, we now have the K-band number counts that, as can be seen from FIGURE 2, are much more sensitive to cosmological geometry than the blue counts are. This effort has produced the result that can be seen in FIGURE 2. Basically, at around $K = 19$ the counts stop rising rapidly, and they move to a much shallower slope. What is suggested here is that the fainter magnitudes no longer sample to larger distances; we must have reached at last the magnitude at which we have broken through to the end of the galaxy-occupied cosmological volume. At this point the flat or slowly rising counts represent the faint-end shape of the galaxy luminosity function averaged through this volume.

If this is the case, we are in a much stronger position to decide what the cosmological geometry is than when we only had the ambiguous still-rising B-band counts to work with. Comparing the model predictions with the data, the absolute K number density at the faint end favors a $q_0 = 0.5$ universe (FIG. 2), and this is a relatively robust result. To see why, suppose the galaxy population completely occupies a volume to a redshift z_f where galaxy formation occurred. Then the number density of galaxies in a given observed magnitude interval per comoving volume is $f(M) = N(M)/V(z_f)$, where $N(M)$ is the observed number of counts (just under 10^5 deg^{-2} mag^{-1} from $K = 20$ to $= 23$) and $V(z_f)$ is the comoving volume to redshift z_f corresponding to the observed solid angle. When the counts are flat, not rising with magnitude, we must be sampling all populations at the faint end of the luminosity function, and so we can approximate $f(M)$ with the local density of galaxies at several magnitudes below standard luminosity, a number that is quite well determined at about 2×10^{-3} $(H_0/50$ km s^{-1} Mpc$^{-1})^3$ Mpc^{-3} (e.g., Efstathiou *et al.*[18]) and depends only weakly on luminosity at these magnitudes. The local number density of counts that would be inferred from the observed counts at $K = 21.5$ and 22.5 is shown in FIGURE 4 as a function of cosmological geometry and z_f. What can be seen here is that in a $q_0 = 0.5$ geometry we get reasonably good agreement for a wide range of z_f, while for an open model we require $z_f \sim 2$, and for zero curvature cosmological constant models, $z_f \sim 1.5$. Given the amount of evidence we have that galaxy formation began beyond these epochs, the data strongly favors a $q_0 = 0.5$ model.

One might reasonably wonder whether luminosity evolution or some other astrophysical effect could change the number density of galaxies to explain this result, but the problem is that we have here a *deficiency* in the number of objects observed compared to local density, not an excess as we have in the optical, and this is much harder to find excuses for. If we add populations, include luminosity evolution, or have galaxies in fragments at higher z, we predict *more* counts, not fewer, and are forced even further toward a low-volume geometry. Even worse, the stars that currently give the K band light were there for the whole cosmological lifetime of the galaxy, so there is no way to "dim" the galaxy relative to its current luminosity.

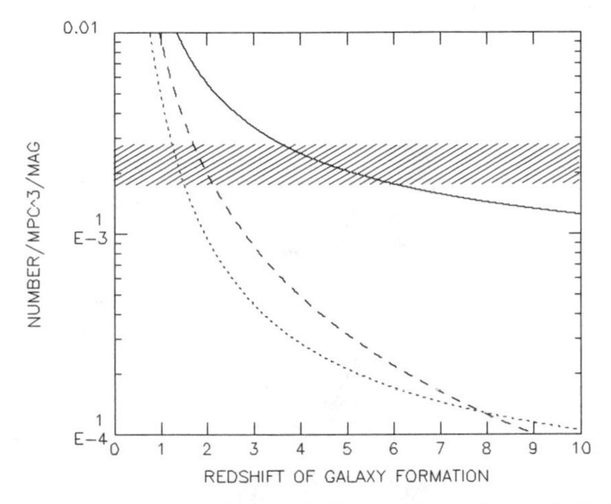

FIGURE 4. The average number density of galaxies per observed magnitude is shown for the K = 21.5 counts (*lower line* in each case) and K = 22.5 counts (*upper line* in each case) as a function of the epoch of formation and the cosmological geometry: q_0 = 0.5 (*solid lines*); q_0 = 0.02 (*dashed lines*); zero-curvature, Ω_0 = 0.1, cosmological constant (*dotted lines*). The *shaded region* shows the local density of galaxies at 2.5 to 3.5 magnitudes below the Schechter luminosity, based on the summary of Efstathiou *et al.*[18] The closed case can be roughly consistent over a wide range of formation epoch, but models with larger volume elements fail radically unless the bulk of galaxies formed very recently.

DWARF GALAXIES AND THE FAINT BLUE NUMBER COUNTS

For any model that roughly fits the K-band counts, we can see from FIGURE 2 that we now have the problem of explaining a huge excess of faint blue galaxies. Another way of expressing the relative overabundance of faint blue galaxies and underabundance of faint IR galaxies is to note that the K-band galaxy sample shows a rapid trend to bluer colors at fainter magnitudes, and that there are relatively few faint objects that are red in (B–K). This is shown in FIGURE 5, which is a histogram of the (B–K) color in various magnitude ranges. Objects that are not detected at the 1 σ level in B are shown as open regions on the histogram and could lie anywhere to the red. Compared to the observations the models we discussed previously predict redder distributions. The most extreme are the nonevolving models (we show the q_0 = 0.02 case), but even a q_0 = 0.5 evolving model predicts too many red galaxies.

The blueness of the faint galaxy population is of course the cause of the discrepancy between the blue and IR number counts. What must be explained in any model is why there are so many very blue faint galaxies. However, there is a more

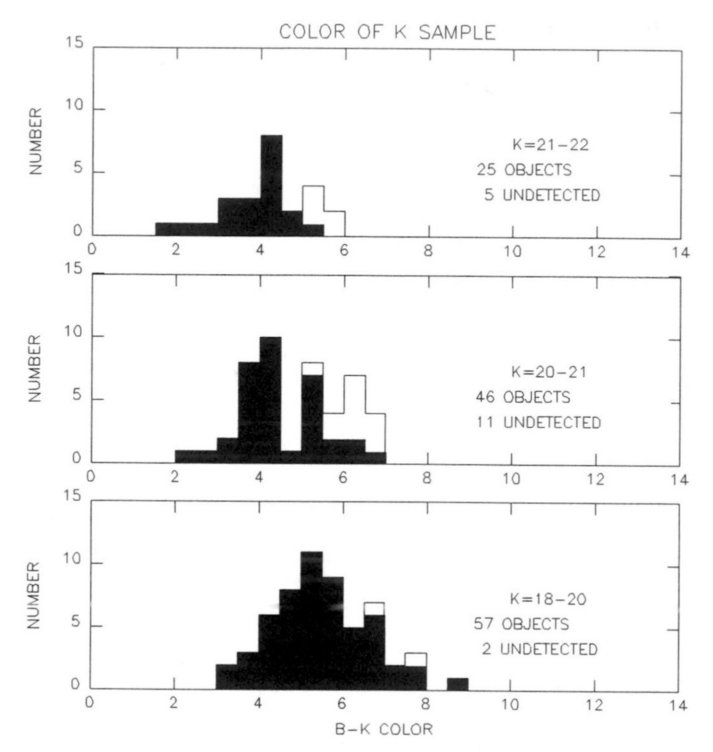

FIGURE 5. Histograms of the (B–K) distribution of the K sample as a function of magnitude. Galaxies that are not detected at the one-sigma level in B are shown as the open area. They could lie anywhere to the red of their position in the graph.

profound conclusion that can be drawn from FIGURE 5, namely that only those galaxies with the redder colors are being correctly understood in the conventional smoothly evoloving models; the blue galaxies must correspond to some other population or to episodic star formation in the fainter end of the normal galaxy population (e.g., Broadhurst *et al.*[19]).

We can combine the redshift data with the K-band magnitudes to determine what the population producing the faint B counts actually is. That is, the absolute K magnitudes should give a good estimate of the galaxy mass except for the most extreme star-bursting cases where it constitutes an upper limit. (They can be roughly used without a color correction because of the flatness of galaxy spectral energy distributions (SEDs) in the near infrared.) The absolute K magnitudes for the galaxies giving the $B < 24$ counts are shown in FIGURE 6. At brighter B magnitudes, most of the galaxies are at or near the K_* of -25.3 typical of the most luminous elliptical galaxies ($H_0 = 50$ km s^{-1} Mpc^{-1}). At the faint end, which is also the position at which the counts begin to rise rapidly above predictions, we begin to see many much smaller galaxies that are typically about 4 magnitudes fainter in K. This is an unexpected effect in a magnitude-limited sample where we expect the counts to be dominated by the near-L_* galaxies that can be seen to the limits of largest volumes.

Roughly two-thirds of the faintest galaxies appear to be dwarf galaxies of this type. If we try to turn this into a luminosity function, we find that the total K luminosity density is 8.1×10^{8} $(H_0/50)L_{\odot}/\mathrm{Mpc}^3$, with roughly half coming from the dwarfs and half from normal galaxies.

Since the amount of K light in the dwarfs is essentially equal to that in the normal galaxies, the dwarfs must contain a comparable amount of baryonic matter in low-mass stars. Given that they are likely to have considerably more gas than stars, they are almost certain to dominate the baryonic mass. In particular if they contain approximately four times as much mass in gas as in stars, typical of local blue dwarf irregular galaxies, then the baryonic density would be roughly 1.4×10^{-31} g cm^{-3}.

An alternative way to estimate the baryonic mass in the dwarfs is to consider the amount they contribute on average to the sky surface brightness in the blue or ultraviolet. This quantity is a direct measure of the average density of the metals produced by the population *regardless of cosmology or the details of the star formation process* (Songaila *et al.*[11]). The reason for this is fairly clear: the ultraviolet light density is an integrated history of massive star formation in a given local volume, and these massive stars also generate the released metals. The detailed relationship is

$$S_{\nu} = 3.6 \times 10^{-25} \left(\frac{\rho Z}{10^{-34}\ \mathrm{g\ cm^{-3}}} \right) \mathrm{ergs\ cm^{-2}\ s^{-1}\ Hz^{-1}\ deg^{-2}}, \qquad (2)$$

where S_{ν} is the average sky surface brightness at frequency ν, and ρZ is the local density of metals.

From the spectroscopic data previously discussed we know that the bulk of the sky light (probably at least two-thirds) at $B = 22$–24 comes from the dwarf galaxies at

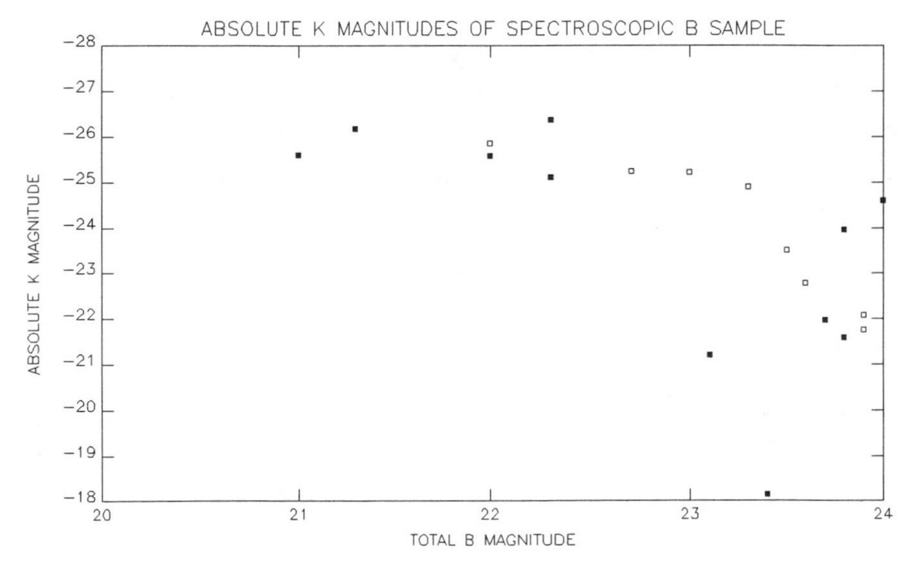

FIGURE 6. Absolute K magnitudes for the $B < 24$ sample. The *solid symbols* show objects with spectroscopic redshifts and the *open symbols* those where the redshifts have been estimated from colors. The *dashed line* shows the approximate magnitude of a typical giant galaxy.

$z \sim 0.3$–0.4. This "blue sky" contribution dominates the ultraviolet sky light with $S_\nu \approx 10^{-24}$ ergs cm^{-2} s^{-1} Hz^{-1} and corresponds to $\rho Z \approx 3 \times 10^{-34}$ g cm^{-3}. Even if the fractional metal content Z were as high as 0.02, typical of disks or the centers of massive elliptical galaxies, the density of material in this new population would be $\sim 10^{-32}$ g cm^{-3}, nearly equal to that in all known galaxies. On the other hand, it is much more likely that Z is considerably lower in these systems, which may be closer in properties to local giant extragalactic HII regions and blue irregular galaxies; in this case, their average baryon density would be around 10^{-31} g cm^{-3}, consistent with the more direct estimate.

These estimates are quite approximate, but can be compared directly with homogeneous Big Bang nucleosynthesis, which predicts a baryon density of $2 - 9 \times 10^{-31}$ g cm^{-3} (e.g., Yang *et al.*[20]), and it is apparent that the dwarfs can provide the required baryon density without any problem. The interesting thing is that it is entirely plausible, physically, for the blue galaxy population to contain the bulk of the baryons. The dwarfs could also be more uniformly distributed than, or even distributed preferentially away from, normal galaxies at the same epoch, which would result in quite significant biasing of baryonic matter from present-day luminous material.

SUMMARY

Our understanding of galaxy formation and evolution is improving, but as it does so it is clear that the complexity is more than we expected. The population of blue dwarfs at low redshift is a major episode of star formation, and this population may be more important in mass terms than the normal galaxies. However, we still have no handle on when the normal galaxies formed, except that we know some were present before $z = 5$. The complexity of multiple populations confuses the interpretation of galaxy number counts. Despite this the lowness of the faint infrared number counts strongly suggests a low-volume geometry even when only normal galaxies are used in predicting the counts. It remains to be seen how this constraint will hold up to close examination; nevertheless we may be hopeful that we are moving toward an understanding of galaxy evolution and with it some measure of the cosmological geometry.

ACKNOWLEDGMENTS

I am very grateful to Toni Songaila for many interesting discussions that greatly clarified my thinking on these problems. I would also like to thank John Kormendy and Ken Freeman for extremely illuminating discussions about dwarf galaxies, and my student Jon Gardner and colleagues Klaus Hodapp, Esther Hu, Simon Lilly, and Richard Wainscoat on whose work this talk is partly based.

REFERENCES

1. SCHNEIDER, D. P., M. SCHMIDT & J. E. GUNN. 1989. Astron. J. **98:** 1951.
2. EFSTATHIOU, G. & M. J. REES. 1988. Mon. Not. R. Astron. Soc. **230:** 5p.

3. Chambers, K. C., G. K. Miley & W. J. M. van Breugel. 1990. Astrophys. J. **363:** 21.
4. Lilly, S. J. 1988. Astrophys. J. **161:** 139.
5. Chambers, K. & S. Charlot. 1990. Astrophys. J., Lett. **348:** L1.
6. Windhorst, R., D. Mathis & L. Neuschaefer. 1989. *In* The Evolution of the Universe of Galaxies, R. G. Kron, Ed.: 389. Astronomical Society of the Pacific. San Francisco.
7. Lilly, S. J. 1989. *In* The Evolution of the Universe of Galaxies, R. G. Kron, Ed.: 344. Astronomical Society of the Pacific. San Francisco.
8. Gunn, J. E. & B. A. Peterson. 1965. Astrophys. J. **142:** 1633.
9. Bechtold, J., R. J. Weymann, Z. Lin & M. A. Malkan. 1987. Astrophys. J. **315:** 180.
10. Miralda-Escude, J. & J. P. Ostriker. 1990. Astrophys. J. **350:** 1.
11. Songaila, A., L. L. Cowie & S. J. Lilly. 1990. Astrophys. J. **348:** 371.
12. Wolfe, A. M., D. A. Turnshek, H. E. Smith & R. D. Cohen. 1986. Astrophys. J. Suppl. **61:** 249.
13. Cowie, L. L. 1988. *In* The Post-Recombination Universe (NATO Advanced Science Institute Series), N. Kaiser and A. Lasenby, Eds.: 1.
14. Yoshii, Y. & F. Takahara. 1988. Astrophys. J. **326:** 1.
15. Ellis, R. S. 1983. *In* The Origin and Evolution of Galaxies, B. T. Jones and J. Jones, Eds.: 255. Reidel. Dordrecht, the Netherlands.
16. Lilly, S. J., L. L. Cowie & J. P. Gardner. 1990. Astrophys. J. Suppl. In press.
17. Fukugita, M., F. Takahara, K. Yamashita & Y. Yoshii. 1990. Preprint.
18. Efstathiou, G., R. S. Ellis & B. A. Peterson. 1988. Mon. Not. R. Astron. Soc. **232:** 431.
19. Broadhurst, T. J., R. S. Ellis & T. Shanks. 1988. Mon. Nat. R. Astron. Soc. **235:** 827.
20. Yang, J., M. S. Turner, G. Steigman, D. N. Schramm & K. A. Olive. 1984. Astrophys. J. **281:** 493.
21. Metcalfe, N., T. Shanks & R. Fong. 1990. Preprint.
22. Tyson, A. J. 1988. Astron. J. **96:** 1.
23. Glazebrook, K., J. A. Peacock, C. A. Collins & L. Miller. 1990. Preprint.
24. Jenkins, C. & J. Reid. 1990. Astron. J. In press.
25. Cowie, L. L., J. P. Gardner, S. J. Lilly & I. McLean. 1990. Astrophys. J., Lett. **360:** L1.
26. Cowie, L. L., J. P. Gardner, R. J. Wainscoat & K. Hodapp. 1990. In preparation.
27. Colless, M. M., R. S. Ellis, K. Taylor & R. N. Hook. 1990. Mon. Nat. R. Astron. Soc. **244:** 408.
28. Maddox, S. J., W. J. Sutherland & G. Efstathiou. 1990. Mon. Not. R. Astron. Soc. **247:** 1p.

SN 1987A: Observations of the Later Phases

I. J. DANZIGER, P. BOUCHET, C. GOUIFFES, AND L. B. LUCY

European Southern Observatory
Karl-Schwarzschild-Strasse 2
D-8046 Garching bei München, Germany

INTRODUCTION

The formation of dust in the expanding envelope of SN 1987A has affected the subsequent evolution of this object. This is because the dust has absorbed much of the harder radiation and thermalized it. It has affected our ability to measure the line radiation from the various ionic species, and has undercut our opportunity to follow directly the radioactive decay of ^{56}Co and ^{57}Co. Because the dust radiates predominantly at IR wavelengths, the determination of the bolometric light curve involves accurate measurement of the IR flux beyond 3.5 microns.

All of these aspects, as well as some recent results from observations of the circumstellar material, are discussed in the following review.

DUST FORMATION

The presence of dust in the expanding ejecta of SN 1987A has important consequences for the determination of the bolometric light curve, and so we review briefly what is known at the present time about the dust. We believe that we know precisely when it formed, and also something about the dust composition, the particle sizes, the large-scale distribution in the envelope, its clumpiness, and its temperature evolution.

Unequivocal evidence for dust formation in the expanding material of the supernova starting at day 530 was provided by an apparent blueward shift of the broad line profiles.[1] This apparent shift results when dust distributed in the envelope absorbs more of the radiation from the far receding (redshifted) side than the near approaching side. In subsequent analyses of the profiles of selected emission lines of [OI], [MgI], and [CI] by Lucy *et al.*[2] it was shown that, to first order, the dust was probably distributed uniformly through a major part of the envelope. The dust formation process appears to have continued over a period of 240 days, by which time the mean optical depth at the visual wavelength approached 1.25. Because a wavelength-dependent absorption was apparent, one could conclude that the dust was not dominated by large grains. In fact, if we ascribe the large *unexpected* (beyond what one would anticipate from radioactive decay alone) decrease of the [CoII] 10.52-μ line observed at day 600 and beyond, to selective extinction at this wavelength it, combined with extinction at shorter wavelengths, is much more consistent with absorption by small grains of silicate dust (≤ 0.5 μ) than with that by graphite or iron.

Until now we have been unable to determine the mass of dust because it has never been observed in an optically thin state. We have one clue about the mass which, however, is not unambiguous. The rate of decrease of the strength of the [SiI] 1.64-μ line after day 530 is much greater than that of other lines and continuum fluxes at nearby or shorter wavelengths. If silicon were a major constituent of the dust, but only a minor constituent of the overall nuclear abundances, then if sufficient dust were formed, this extra observed decrease of the silicon line strength might result from depletion of silicon into grains. A similar effect in oxygen would not be observed because its mass is a dominant one. This being the case, then the mass of dust would be some significant fraction of the expected mass of silicon in the envelope, viz., 0.1 $M_{\odot}$. Thus the implied condensation efficiency is very high. One should note that the temperature sensitivity of this [SiI] line might suggest other reasons for its more rapid decline. Nevertheless, there is a further piece of observational evidence that the low condensation efficiencies computed from the spectroscopically determined optical depths may be underestimates because large-scale clumping of the dust has occurred. An analysis of extinction by Lucy *et al.*,[3] who used both emission line fluxes and continuum points from both broad- and narrow-band observations, revealed two components of extinction—the wavelength-dependent part previously discussed, and a wavelength-independent part. This latter could be provided by very optically thick clumps that are responsible for wavelength-independent geometrical obscuration. FIGURE 1 is a representation of the sort of situation that may obtain. The presence of optically thick clumps providing a wavelength-independent extinction provides a natural explanation for the blue shifts reported for IR emission lines[4,5] that would not be otherwise very much affected by the small-grain silicate dust. The presence of the large dense clumps may also be responsible for obscuration of whatever source of energy exists at the center. This is conceivably relevant to the behavior of the bolometric light curve after day 900.

Initially, the presence of dust has had an effect on the photometry, particularly, but not only, noticeable at IR wavelengths. This is because the dust is absorbing shorter wavelength radiation and thermally reemitting it at the temperature imposed on the dust grains by the dilute radiation field. An excess at infrared wavelengths began to be apparent after day 530, and has developed by day 1300 to the point of dominating the total radiative energy output of the SN. The black body temperature obtained by fitting to data points between 2 and 20 μ has evolved during the same time from approximately 700 K to 150 K.

The interpretation of optical and infrared photometry and the various contributions to the bolometric light curve have been discussed by, for example, Whitelock *et al.*,[6] Suntzeff and Bouchet,[7] Bouchet *et al.*,[8,9] and Wooden.[5]

Lucy *et al.*[2] showed that, after day 530 when a significant fraction of the radiative luminosity was emitted at 20 μ and even beyond, in order that reasonable agreement be achieved between the observed bolometric light and the theoretical ones derived for either mixed or unmixed models, all of the observed radiation had to be included in the bolometric luminosity. This would not have been the case had the IR excess after day 530 been ascribed to an IR echo from dust lying behind the supernova, as proposed by Roche *et al.*[10] Subsequent analysis of this point by Suntzeff and Bouchet[7] using a longer temporal base line, confirms this conclusion beyond doubt.

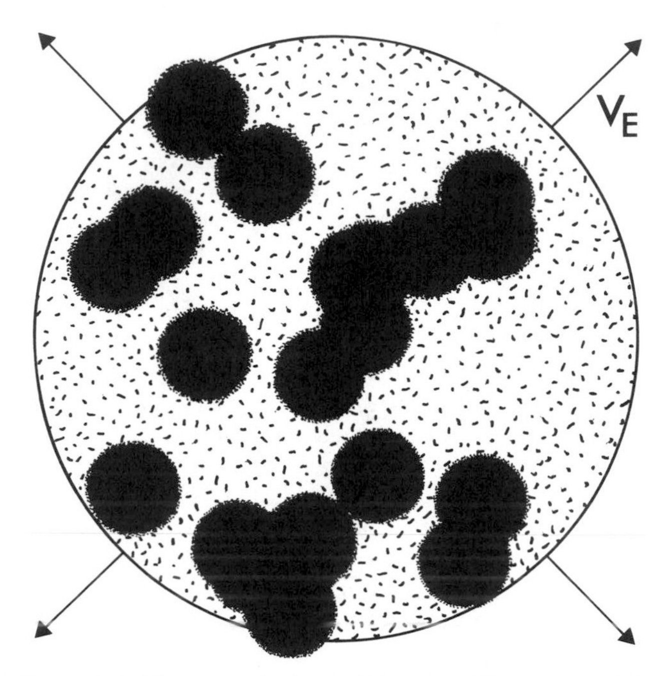

FIGURE 1. The model with opaque clouds and diffuse dust. The number of clouds (20) and their radii are chosen to reproduce the observed occultation optical depth.

That dust is still present throughout the envelope of SN 1987A beyond day 1400 can be seen by inspection of the spectrum of SN 1987A taken on Jan. 5, 1991. It can be seen in FIGURE 2 that the peak (indicated by the vertical dashed line) of the broad [OI] 6300 line emitted by the expanding envelope is blue shifted relative to the

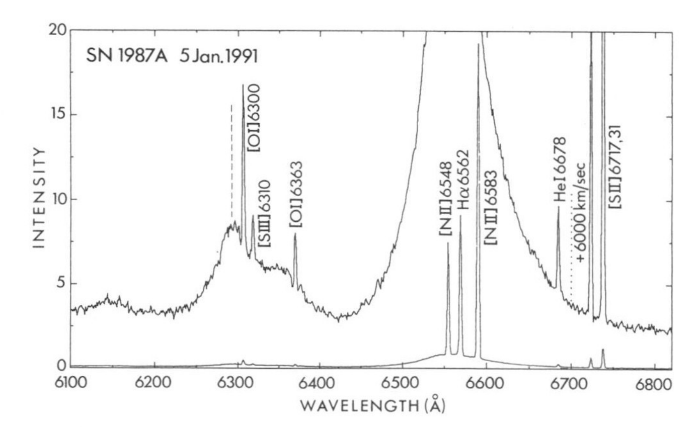

FIGURE 2. A low-resolution spectrum of SN 1987A taken at ESO on Jan. 5, 1991. Both *broad lines* from the envelope, and *narrow lines* from the circumstellar matter are evident. A blue shift of the broad component of [OI] 6300 is evident.

narrow line, which falls at the systemic velocity of the Large Magellanic Cloud (LMC). This shift is as large as or larger than at any previous epoch.

NUCLEOSYNTHESIS AND ABUNDANCES

That nucleosynthesis has occurred in the supernova and in the precursor star that had an original mass of 16–20 $M_\odot$ could be tested in some detail after the expanding envelope and ejecta became optically thin for the emission line spectrum. We present here some first-order results taken from the paper of Danziger *et al.*[11] Strong caveats should be observed in literal use of these results. They result from analysis of emission line spectra taken at or near day 410 when our Monte Carlo code assuming the Sobolev approximation takes account of any optical depth. Sobolev optical depths were, in fact, negligible for all of the utilized lines at this epoch. The envelope was assumed to be isothermal with a temperature of 2700 K [derived from ratios of lines of [FeII] assuming local thermodynamic equlibrium (LTE)].

Clearly, if the material in the envelope is clumped—and there are several independent pieces of evidence that it is—and if the clumping is different for different elements—which might be expected if only because of the onion-skin structure of the precursor star—the assumption of one excitation temperature for all ions will surely be inaccurate.

Deviations from LTE will almost certainly exist for the population of many higher levels in many ions for a number of reasons. For example, there is strong evidence that radioactive cobalt, which provides the source of heating, is mixed through the envelope. Ultimately, the γ-rays emitted in the decay produce thermal electrons as the radiation is degraded. However, the chain of degradation passes through a stage of production of nonthermal electrons that can cause population of higher levels not described by LTE. Also, line coincidences can give rise to fluorescent-type pumping mechanisms in the manner identified by Lucy *et al.*[12] for certain lines of NaI.

Nevertheless, the results presented in TABLE 1 do encourage the belief that one is demonstrating that nucleosynthesis has occurred in SN 1987A and in the precursor star. This is in part because, for the elements that one has been able to measure, there is encouraging agreement within uncertainties with nucleosynthetic models of such objects described by Thielemann *et al.*[13] and others, including Arnett and Woosley and coworkers. Furthermore, although it is not explicitly shown in the table one should note that all the derived masses are significantly higher than one would expect in ten solar masses of material with a uniform composition appropriate to the LMC.

There are currently some very large uncertainties, particularly so, unfortunately, for oxygen, one of the dominant elements. This uncertainty arises mostly because of the temperature sensitivity for the transition giving rise to the [OI] 6300,63 line. A future, more accurate determination of its mass will allow an important discrimination amongst some classes of models. There is some suggestion in the table that nickel may not be in agreement with the theoretical results. Here there appears to be disagreement among the observers that remains to be clarified before a stronger conclusion can be drawn.

Abundance results for cobalt might have been presented in this table, but a separate discussion seems justified because the cobalt we observe is radioactive and therefore the mass is decreasing with time. In fact, direct observations of the temporal decay of the CoII 10.52-μ line have been made since day 195. Observations as early as day 100 may be seriously compromised by the blending of the CoII line with other lines of comparable strength. Mass determinations are more reliable for this element because the line in question arises from transitions between fine structure levels in the ground state of CoII, which is also the predominant ion of cobalt. More details are given by Danziger *et al.*[11]

Our observations in the time interval day 190–300 are consistent with the production of an original mass of ^{56}Co of 0.075 $M_\odot$. ^{56}Co has a half-life of 77 days. This is in accord with the less direct observation of the bolometric light curve.

TABLE 1. SN 1987A

	Abundances Day 410 April 1988			
	Observed $M/M_\odot$†		Recent Model* $M/M_\odot$	
C I (9823,49)	0.036	(×2)	0.114	
O I (6300,64)	3.0 (0.2)	(×?)	1.48	
Si I (1.644 μ)	0.034	(×3)	0.1	
Ar II (6.98 μ)	0.0005	(× ≥ 1.5)	0.004	(1)
Ca II (7291,7324)	<0.007	(×1.5)	0.0033	
Fe I (1.423-1.462 μ)	0.043	(×3)		
Fe II (1.248-1.27 μ) (17.93 μ)	0.098 0.055	(×1.5)	0.075	(2)
Co II (10.52 μ)	0.00193	(×1.5)	0.00188	^{56}Co only
Ni I (3.12 μ) Ni II (6.63 μ)	0.00074 0.0014	(×3) (×1.5)	0.0177	(1)

Note: Isothermal LTE = 2700 K; expansion velocity = 2000 km/s.
*Thielemann, Hashimoto, and Nomoto.[13]
†Correction factor for unseen stages of ionization.
(1) Line strengths from Witteborn *et al.,* IAU Circ. No. 4592.
(2) Line strengths from Moseley *et al.,* IAU Circ. No. 4576.

In the interval day 370–520 the observations indicate a slightly higher mass. It is just at this epoch that ^{57}Co (with a half-life of 271 days) would begin to contribute a significant fraction to the line strength if the ^{57}Co/^{56}Co production were the same as or higher than the ratio of stable isotopes of the same mass found in the solar system, viz., 1/45. It should be noted that the opportunity of observing ^{57}Co by direct measurement of an emission line occurs earlier than the capability of detecting its effect on the bolometric light curve. This is because compared to ^{56}Co, the energy released from the β-decay of ^{57}Co is much lower and it occurs 3.5 times more slowly. Indeed our observations suggest that the ^{57}Co/^{56}Co ratio is 1.5 times the solar system ratio. It might even be less than 1.5 if our estimate of the correction for the proportion of CoI is too high. Other published estimates of the CoI/CoII ratio indicate that to date our estimate of this ratio is the highest. The value of 1.5 is in good agreement with an analysis of the CoII 1.54-μ line by Varani *et al.,*[14] and the upper limit from the x-ray emission following the Comptonized γ-ray lines from the

^{57}Co decay reported by Sunyaev *et al.*[15] We believe that the uncertainties in our determination are not large enough to encompass the possibility that ^{57}Co/^{56}Co is as high as 4, a value quoted (IAU Circ. No. 4996) to explain the flattening of the late-time bolometric luminosity.

THE BOLOMETRIC LIGHT CURVE

Since the early (<400 days) bolometric light curve provides evidence for an energy input resulting from the radioactive decay[16] of 0.07 $M_\odot$ of ^{56}Co, one is in a position to examine in greater detail whether there are significant deviations from this result at a later date when longer lived isotopes may make a significant contribution. In fact, starting at about day 900 a significant flattening of the bolometric light curve became apparent. Some feeling for what is involved in computing a bolometric light curve can be obtained from a study of FIGURE 3, which provides results for day 1316. The observed points corresponding to the N and Q_0 photometric bands (10 and 20 μ, respectively) are fitted by a black body temperature of 155 K, which presumably corresponds to the temperature of the dust. Near that date a detection of continuum emission at 1.3 mm by Chini *et al.*[17] with a flux of 7.59 ± 2.45 mJy seemed to fit a black body curve corresponding to a temperature a little lower than 155 K, but probably consistent with this value within the uncertainties. However, it must be realized that the sort of astronomical dust we have described previously is not a good absorber/emitter at 1.3 mm, and therefore it is difficult to imagine that it could be optically thick at such a wavelength and therefore be incorporated in a black body fit. An escape from this dilemma (shown in FIG. 3) is provided by ascribing the radiation at 1.3 mm to free-free (ff) radiation from the ionized gas in the envelope. A suitable choice of temperature (2389 K) provides a natural explanation for what otherwise would be excess radiation at 1.3 mm and at the L photometric band (3.5 μ). This ff emission contributes very little to the total radiation flux from the supernova.

Note that at this late date a large fraction of the radiation from the supernova is being emitted by the dust at wavelengths upward of 20 μ, which is the longest IR wavelength observable from the ground. This is, of course, a reason to be concerned about accuracy in the determination of the bolometric light curve where fitting of black body curves is involved.

The observed bolometric light curve from day 1 to day 1316 is shown by the individual points in FIGURE 4. It should be noted that as a result of differences in the N and Q_0 magnitudes measured by the CTIO team,[18] a less pronounced flattening of the bolometric light curve is obtained by them. The reason for the differences is not yet understood.

There are a number of different possible sources of the energy required for the flattening of the light curve. Bouchet *et al.*[19] have already pointed out that the amount of ^{57}Co required would be too great (by a factor of 10–20) to be compatible with the direct observations of the Co II 10.52-μ line in the period day 400–520. If we leave ^{57}Co at the value determined at that time, but increase the amount of radioactive ^{44}Ti (half-life 54.2 years) and ^{22}Na (half-life 2.6 years) by factors of 20 and 30, respectively, relative to the expected amounts,[20] we can obtain an acceptable fit to the observed data. However, these arbitrarily large factors seem difficult to justify on the basis of nucleosynthetic calculations.

Other plausible energy sources involve a compact object at the center, a neutron star, radiating as a pulsar, or surrounded by an accretion disk depositing matter either continuously or at varying intervals onto the collapsed object. Note that these possibilities are explored in the theoretical fits shown in FIGURE 4. The excess energy

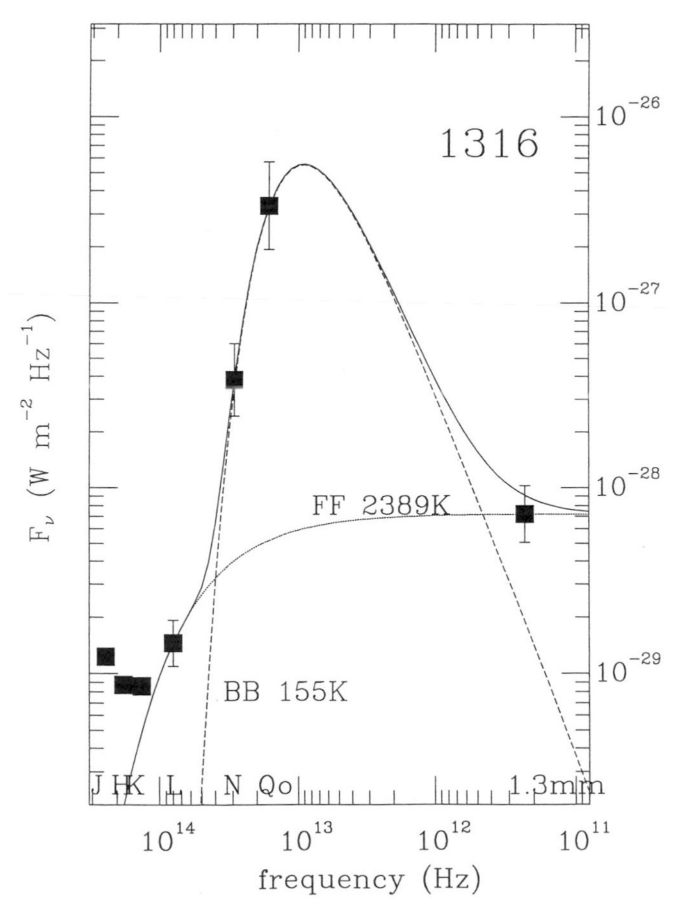

FIGURE 3. The energy spectrum of SN 1987A at day 1316 represented by the points. The fits described in the text are shown by the various lines.

input is approximately 10^{38} ergs/sec, which is only slightly less than the Eddington luminosity for a neutron star. While this does not prove the accretion disk scenario, the coincidence is, of course, suggestive, as it is in the case of x-ray pulsators. Whether the apparent fluctuations amongst the individual points are real or due to the observational uncertainties is impossible to say at the moment.

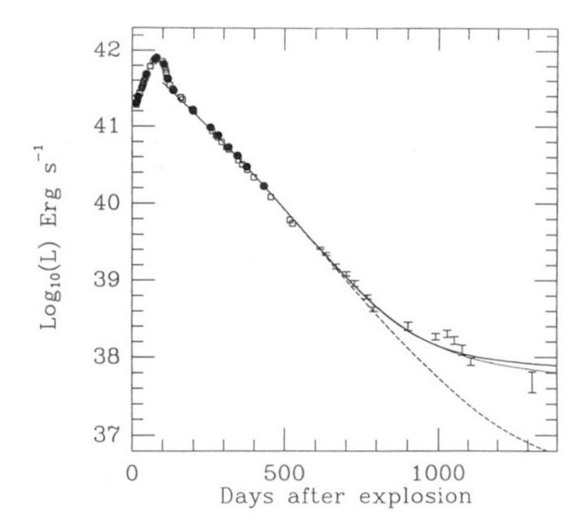

FIGURE 4. The bolometric light curve of SN 1987A. Points represent observations.

These IR observations are near the limit of what is possible at the present time, so it is desirable that confirming observations be attempted as often as possible.

CIRCUMSTELLAR MATTER

The presence of a circumstellar ring of material emitting strongly in the line [OIII] 5007 was revealed by direct images taken with the European Southern Observatory (ESO) New Technology Telescope (NTT) in Dec. 1989.[21] In FIGURE 5 we show the first Hubble Space Telescope (HST) image of SN 1987A taken in OIII 5007 light.[22,23] Restoration of the original data to remove the effects of the mirror aberration has been made by H.-M. Adorf, using the algorithm of Lucy.[24] The central object is the supernova surrounded by a ring structure. The authors of the HST

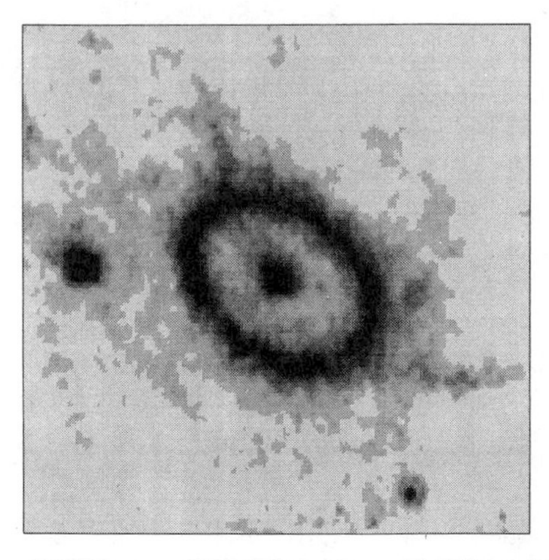

FIGURE 5. A restored HST image of SN 1987A taken with HST and described in the text.

results have resolved the supernova image at 0.17 ± 0.03 arcsec. At the distance of the LMC (50 kpc) this size corresponds to an average expansion velocity of 6000 km/sec. This is consistent with the continuing visibility of emitting hydrogen at 6000 km/sec shown by the dotted line on the long wavelength side of Hα in FIGURE 2. This figure also shows some of the narrow emission lines coming from the ring. The fact that the narrow [NII] lines are very strong relative to narrow Hα already qualitatively shows that the nitrogen abundance in this object is much higher than in the ambient ionized gas of the LMC. An observation of this kind is a signature of the origin of this material in some sort of presupernova mass-loss phenomenon.

As a result of the new restoration it has been shown that the brightness contrast between the ring and the inner regions is ~100/1. Therefore, since virtually no emission is found inside the ring, it may be concluded that a true ring structure exists and not a limb-brightened shell. The structure in the ring and the faint extensions

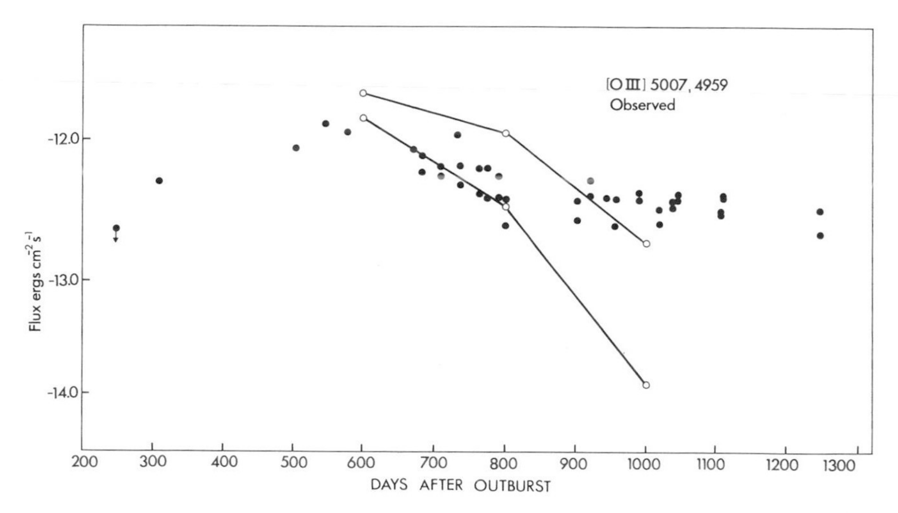

FIGURE 6. The observed temporal variation of the narrow emission lines [OIII] 4959, 5007. The *open circles connected by lines* represent theoretical predictions with an arbitrary displacement in the vertical direction.

inwards and outwards appear to be real. There is still the possibility that the ring structure formed after the mass-loss event had occurred, rather than being ejected as such.

There is a further puzzle concerning this ring if its excitation is due solely to the initial UV burst. According to the models of Fransson and Lundquist,[25] and also intuitively, one might expect the emissivity of the ring in most lines to pass through a maximum and then undergo a steady decrease. In FIGURE 6 we show a plot of the temporal behavior of the narrow [OIII] 4959, 5007 line strength. It does appear to pass through a maximum near day 550 with a decrease to day 850. Since that time any further decrease has been much slower than expected. One might conjecture that there is another source of energy keeping the gas in the ring excited—a pulsar or radioactive decay?

ACKNOWLEDGMENT

We are grateful to H.-M. Adorf for providing the reconstructed HST image of SN 1987A.

REFERENCES

1. DANZIGER, I. J., P. BOUCHET, C. GOUIFFES & L. B. LUCY. 1989. IAU Circ. No. 4746.
2. LUCY, L. B., I. J. DANZIGER, C. GOUIFFES & P. BOUCHET. 1989. *In* Structure and Dynamics of Interstellar Medium. Lecture Notes in Physics, G. Tenorio-Tagle, M. Moles, and J. Melnick, Eds.: 164. Springer-Verlag. New York/Berlin.
3. LUCY, L. B., I. J. DANZIGER, C. GOUIFFES & P. BOUCHET. 1991. *In* Supernovae, Proceedings of the Tenth Santa-Cruz Summer Workshop in Astronomy and Astrophysics (University of California at Santa Cruz, July 10–21, 1989), S. Woosley, Ed.: 52. Springer-Verlag. New York/Berlin.
4. SPYROMILIO, J., W. P. S. MEIKLE & D. A. ALLEN. 1990. Mon. Not. R. Astron. Soc. **242:** 669.
5. WOODEN, D. H. 1989. Ph.D. Thesis, University of California at Santa Cruz, Dec. 1989.
6. WHITELOCK, P. A., R. CATCHPOLE & M. FEAST. 1991. *In* Supernovae, Proceedings of the Tenth Santa-Cruz Summer Workshop in Astronomy and Astrophysics (University of California at Santa Cruz, July 10–21, 1989), S. Woosley, Ed.: 15. Springer-Verlag. New York/Berlin.
7. SUNTZEFF, N. B. & P. BOUCHET. 1990. Astron. J. **99:** 650. (Paper I)
8. BOUCHET, P., M. M. PHILLIPS, N. B. SUNTZEFF & C. GOUIFFES. 1991. Astron. Astrophys. In press. (Paper II)
9. BOUCHET, P., I. J. DANZIGER & L. B. LUCY. 1991. *In* Supernovae, Proceedings of the Tenth Santa-Cruz Summer Workshop in Astronomy and Astrophysics (University of California at Santa Cruz, July 10–21, 1989), S. Woosley, Ed.: 49. Springer-Verlag. New York/Berlin.
10. ROCHE, P. F., D. K. AITKEN, C. H. SMITH & S. D. JAMES. 1989. Nature **337:** 533.
11. DANZIGER, I. J., L. B. LUCY, P. BOUCHET & C. GOUIFFES. 1991. *In* Supernovae, Proceedings of the Tenth Santa-Cruz Summer Workshop in Astronomy and Astrophysics (University of California at Santa Cruz, July 10–21, 1989), S. Woosley, Ed.: 69. Springer-Verlag. New York/Berlin.
12. LUCY, L. B., I. J. DANZIGER & C. GOUIFFES. 1991. Astron. Astrophys. **243:** 223.
13. THIELEMANN, F.-K., M. HASHIMOTO & K. NOMOTO. 1990. Astrophys. J. **349:** 222.
14. VARANI, G.-F., W. P. S. MEIKLE, J. SPYROMILIO & D. A. ALLEN. 1990. Mon. Not. R. Astron. Soc. **245**(3): 570.
15. SUNYAEV, A. E., *et al.* 1990. Sov. Astron. Lett. **16:** 403.
16. BOUCHET, P., I. J. DANZIGER & L. B. LUCY. 1991. Astron. J. **102:** 1135.
17. CHINI, R., C. G. T. HASLAM, E. KREYSA, R. LEMKE & A. SIEVERS. 1991. In preparation.
18. SUNTZEFF, N. B., M. M. PHILLIPS, D. L. DEPOY, J. H. ELIAS & A. R. WALKER. 1991. Astron. J. **102:** 1118.
19. BOUCHET, P., I. J. DANZIGER & L. B. LUCY. 1989. IAU Circ. No. 4933.
20. WOOSLEY, S. E., P. A. PINTO & D. HARTMANN. 1989. Astrophys. J. **346:** 395.
21. WAMPLER, E. J., L. WANG, D. BAADE, K. BANSE, S. D'ODORICO, C. GOUIFFES & M. TARENGHI. 1990. Astrophys. J., Lett. **362:** L13.
22. PANAGIA, N., R. GILMOZZI, F. D. MACCHETTO, H.-M. ADORF & R. P. KIRSHNER. 1991. Bull. Am. Astron. Soc., 177th Am. Astron. Soc. Meeting, Philadelphia, Pa.
23. JAKOBSEN, P., *et al.* 1991. Astrophys. J., Lett. **369:** L63.
24. LUCY, L. B. 1991. *In* The Restoration of HST Images and Spectra (Space Telescope Science Institute, Baltimore, Md., Aug. 21–22, 1990), R. L. White and R. J. Allen, Eds.: 80. Space Telescope Science Institute. Baltimore, Md.
25. FRANSSON, C. & P. LUNDQUIST. 1989. Astrophys. J., Lett. **341:** L59.

Streaming Velocities and the Formation of Large-scale Structure[a]

AVISHAI DEKEL

Racah Institute of Physics
The Hebrew University of Jerusalem
Jerusalem, 91904 Israel
and
DAEC
Observatoire de Paris-Meudon
92195 Meudon, France

1. INTRODUCTION

1.1. Prolog

The observed peculiar velocities of galaxies provide the only direct evidence for the dynamics of matter on large scales, which can constrain theory independently of the relationship between galaxies and mass. I first describe reconstruction methods for bringing the observed radial velocities to common grounds with theoretical predictions, focusing on the *POTENT* procedure that recovers the three-dimensional velocity field assuming a potential flow and then determines the underlying mass-density field. I discuss error analysis, alternative methods, and possible flaws in the general procedure.

I then report on several results concerning the confrontation of theory and observation: (1) A correlation between the mass distribution and the *IRAS* galaxy density, both featuring the Great Attractor (GA) phenomenon. The comparison yields $\Omega^{0.6}/b_I = 1.35 \pm 0.25$, where b_I is the biasing factor for *IRAS* galaxies. Nonlinear analysis favors $\Omega \simeq b_I \simeq 1$ (but low values of Ω are also possible, provided that the galaxies are severely antibiased). (2) The shape of the mass fluctuation power spectrum in the wavelength range 10–100 h^{-1} Mpc is consistent with an initially scale-invariant spectrum modified by cold dark matter (CDM). (3) The one-point probability distribution of the velocities reflects the initial distribution. The velocity distributions by *POTENT* and by *IRAS* are consistent with Gaussian initial fluctuations. (4) Our local GA region, if put at the last-scattering surface, is equivalent to a gradient of $\Delta T/T \simeq 10^{-5}$ over 1° in the cosmic microwave background (CMB).

I conclude from these results that there is no *dynamical* evidence against the "standard" model of the formation of large-scale structure, that is, a homogeneous universe with $\Omega = 1$ and Gaussian, scale-invariant initial fluctuations that grew via gravitational instability under CDM. The indicated large-scale excess of power in the *galaxy* distribution could be attributed to nontrivial biasing of galaxies relative to the

[a]This research was supported by U.S.-Israel Binational Science Foundation Grants 86-00190 and 89-00194, and by Israeli Academy Basic Research Grant 316/87.

mass distribution. The alternative, an $\Omega \sim 0.1$, baryonic universe with isocurvature fluctuations, is not convincingly ruled out either, but it requires nontrivial antibiasing on $\sim h^{-1}$ Mpc.

1.2. Discussion

The formation of large-scale structure has become a phenomenological science in which the main effort is directed at confronting theory with observation. FIGURE 1 schematically illustrates this confrontation.

A theoretical model, within the "standard" framework, consists of the following five basic ingredients:

- The *cosmological model.* For example, a homogeneous Robertson–Walker universe with an inflation phase. This model is characterized by the parameters Ω and Λ, which affect the rate of cosmic evolution and the characteristic scale of structure—the horizon when the universe-turned-matter dominated.
- The *initial fluctuations.* Their probability distribution $P(\rho)$ (e.g., Gaussian or not), their power spectrum P_k (e.g., scale-invariant or not), and the way they are distributed among the different components of energy density (e.g., adiabatic versus isocurvature).
- The nature of the dominant *dark matter,* which filters the fluctuation power spectrum in a characteristic way during the radiation/plasma era. For example, the dark matter could be baryonic or nonbaryonic, "hot" or "cold" (CDM), depending on when it became nonrelativistic.

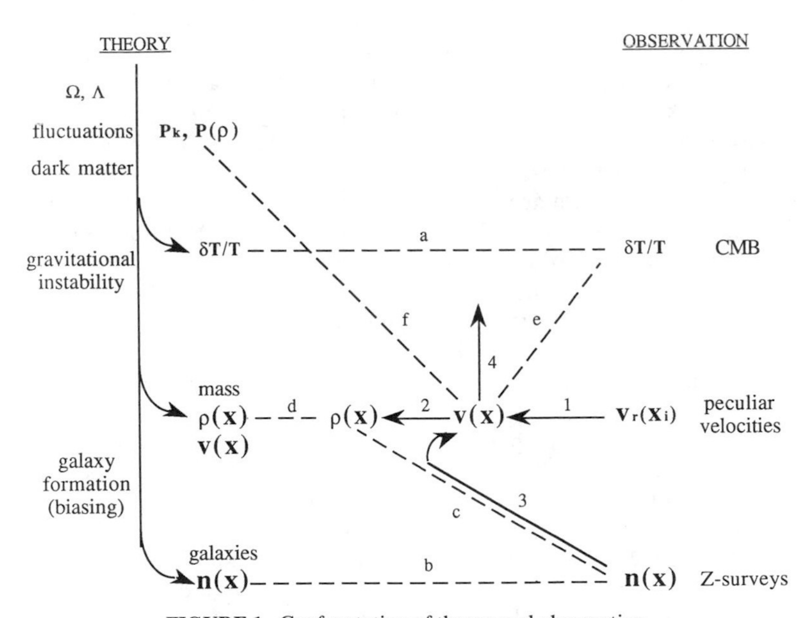

FIGURE 1. Confrontation of theory and observation.

- The way the fluctuations evolve under *gravitational instability*. The linear evolution yields a prediction for $\Delta T/T$ fluctuations in the cosmic microwave background (CMB). Further nonlinear evolution, using simulations or approximate methods, predicts the statistical properties of the *dynamical* fields today: the fluctuations in the *mass* density $\rho(\vec{x})$ and the associated peculiar velocity $\vec{v}(\vec{x})$.
- The process of *galaxy formation,* or a "biasing" scheme, which determines where galaxies form given the dynamical fields, and yields the properties of the galaxy number density field $n(\vec{x})$. The relation between the galaxy and mass distributions is commonly described by a linear biasing factor, $\delta n/n = b\delta\rho/\rho$. This is the most uncertain component among the model ingredients.

Until a few years ago, the relevant observations primarily consisted of two kinds: upper limits on $\Delta T/T$ on different angular scales,[1] and maps of galaxies on the sky and in redshift space,[2] from which a uniform galaxy number density field $n(\vec{x})$ can be obtained either by weighting each galaxy inversely with the selection function or by cutting the sample into a volume-limited sample. While $\Delta T/T$ is directly comparable to the observed upper limits[1] (arrow "a" in the figure), the galaxy distribution could be compared to the corresponding theoretical prediction ("b") only after specific assumptions are made concerning biasing.

The new, very promising kind of data consist of *peculiar velocities* of galaxies.[3] In principle, they are related to the predicted dynamical fields so their comparison with theory can bypass the complex issue of biasing. It is safer to assume that the galaxies move as test bodies in a large-scale gravitational field, independently of how they were selected, than to assume that they are honest tracers of the mass distribution. However, the raw data consist of only the radial components of the velocities along the lines of sight, $v_r(\vec{r}_i)$, and it is sparse and noisy, so *reconstruction* is needed before a comparison with theory is possible. Methods have been developed to reconstruct a smoothed, three-dimensional *velocity* field out of this data (arrow "1"). The original method of this kind is *POTENT,*[4–6] which uses the potential nature of the flow predicted by gravity. One can then reconstruct the underlying *mass-density* field,[7] assuming a value for Ω ("2"). I describe below (Section 2) the basic ideas of *POTENT,* focusing on new developments concerning calculating the density field in the quasi-linear regime (Section 2.2), the error analysis (Section 2.3), alternative methods (Section 2.4), and tests of the basic assumptions made in the analysis (Section 2.5).

The dynamical fields can also be reconstructed from a galaxy redshift survey ("3"), provided that it is fairly complete and uniform over the sky, such as the *IRAS* samples. One can first translate $\delta n/n$ into a mass-density field by assuming a certain biasing scheme, and then reconstruct the peculiar velocity field assuming Ω and a dynamical scheme (e.g., linear theory).[8,9]

One can also attempt to operate a "time machine" to trace the dynamical fields back into the linear regime ("4"), but this is beyond the scope of this paper.[9]

The ingredients of the theoretical models can be tested and constrained by comparing the dynamical fields against each other and against the theoretical predictions. I will discuss several such applications including a comparison ("c" in FIG. 1) of the *IRAS* and *POTENT* reconstructed fields, aiming at determining Ω and b, and a similar comparison with optical surveys (Section 3.1), a comparison of the reconstructed dynamical fields with the theoretical ones ("d") via the fluctuation power spectrum (Section 3.2), an attempt to address directly the statistical nature of the primordial fluctuations ("f") using the one-point probability distribution of the velocities (Section 3.3), and an attempt to relate the velocity field to the CMB anisotropies ("e") by producing a map of $\Delta T/T$, pretending that our local neighborhood is a patch on the last-scattering surface (Section 3.4). My conclusion from the current results are summarized in Section 4.

2. RECONSTRUCTION OF THE DYNAMICAL FIELDS

2.1. Potent Velocity Reconstruction

Potential analysis has been used to reconstruct the three-dimensional velocity field from its observed radial components. The basic *POTENT* method developed by Bertschinger and myself[4–7] has been tested using idealized N-body simulations,[4–7] and under nonuniform, sparse sampling and measurement errors.[5,6]

The raw data are distances r_i and redshifts z_i for a set of objects (galaxies or clusters) in direction $\hat{r}_i$. The basic sample[3,6] contains a total of 973 galaxies in 493 objects. The distances have been measured based on the Tully–Fisher relation for spirals and the D_n–σ relation for ellipticals and stellar objects (SOs). These empirical scaling laws allow an estimation of the intrinsic luminosity of a galaxy from the observed internal velocities in it, and the distance is obtained from the ratio of this intrinsic luminosity to the observed apparent luminosity. The redshifts are corrected to the CMB frame and the corresponding radial peculiar velocities are $v_{r_i} = cz_i - H_0 r_i$. (H_0 is set to unity so distances can be measured in km s^{-1}.)

Given the radial velocities of a sparse sample, *POTENT* first smooths and interpolates the data into a spherical grid using a tensor window function to produce a smoothed radial velocity field, $v_r(\vec{r})$. This is the most likely value of the velocity averaged over a local spherical Gaussian window of radius R_s centered on $\vec{x}$. In the current analysis we use $R_s = 1200$ km s^{-1}. The smoothing is done with a weighting scheme that mimics volume averaging in order to minimize the errors due to sampling gradients, and the effect of measurement errors is reduced by weighting inversely with the individual expected errors σ_i^2.

In order to obtain the tangential components of the velocity field $\vec{v}(\vec{r})$, we make the crucial assumption that the velocity field is derived from a scalar *potential*: $\vec{v}(\vec{r}) = -\vec{\nabla}\Phi(\vec{r})$, that is, it has zero vorticity. According to gravitational instability theory in the linear regime, this velocity potential is proportional to the gravitational potential, and the no-vorticity mode is the only growing mode. Based on Kelvin's circulation theorem, the flow remains vorticity-free even in the quasi-linear regime, as long as orbit mixing does not occur. Finally, we have demonstrated that the flow remains a potential flow, to a good approximation, when regions of collapse and orbit mixing are properly smoothed over.[4]

The velocity potential at each point on a spherical grid can therefore be calculated by integrating the radial velocity along radial rays,

$$\Phi(\vec{r}) = -\int_0^r v_r(r', \theta, \phi)\, dr'. \tag{1}$$

Differentiating this potential in the transverse directions then recovers the two missing components of the velocity.

2.2. *Potent Mass-Density Reconstruction*

Given the smoothed velocity field $\vec{v}(\vec{r})$, we now wish to reconstruct the underlying mass-density fluctuation field, $\delta(\vec{r})$. Our method of doing that has been improved drastically since the original version of *POTENT.* In the linear regime the relation is simple: $\delta_d = -\vec{\nabla}\cdot\vec{v}/Hf(\Omega)$, where $f(\Omega) = d(\ln\delta)/d(\ln a)$ and a is the universal expansion factor. But our system is not linear, so we need a practical quasi-linear approximation. By testing several methods against N-body simulations we[7] found the best approximation to be the exact solution of the continuity equation under the Zel'dovich assumption that particle displacements grow in a universal rate.[10] Despite the Langrangian nature of the Zel'dovich approximation, this density can be expressed in terms of the partial derivatives of the *Eulerian* velocity field, which makes it very useful.

The approximation works as follows. Let $\vec{q}$ be the initial (Langrangian) comoving position of a particle. The Zel'dovich approximation assumes that the comoving position of that particle at time t (the Eulerian position) is

$$\vec{r}(\vec{q},t) = \vec{q} + D(t)\vec{\psi}(\vec{q}). \quad (2)$$

The approximation is in writing the displacement as a product of a spatial perturbation function, $\vec{\psi}(\vec{q})$, and a universal time-dependent function, $D(t)$. The comoving peculiar velocity is then

$$\vec{v}(\vec{q},t) = \dot{D}(t)\,\vec{\psi}(\vec{q}) = DHf(\Omega)\,\vec{\psi}(\vec{q}). \quad (3)$$

In the second equation we have used the fact that $D(t)$, in the Zel'dovich approximation, is the linear growth rate of density fluctuations, $\delta \propto D(t)$, so $\dot{D}/D = Hf(\Omega)$.

If the velocity field has been smoothed over the scale where collapse and orbit mixing have occurred, one can assume that there is a one-to-one correspondence between $\vec{r}$ and $\vec{q}$. This allows us to write the Zel'dovich displacement in *Eulerian* space,

$$\vec{q}(\vec{r}) = \vec{r} - D\vec{\psi}[\vec{q}(\vec{r})]. \quad (4)$$

The continuity equation, in the Zel'dovich language, can be written as $\rho_r(\vec{r})\,d^3r = \rho_q\,d^3q$, where ρ_r is the Eulerian density and ρ_q is the Langrangian (mean) density. The density contrast is therefore

$$\delta_c(\vec{r}) = \left\|\frac{\partial\vec{q}}{\partial\vec{r}}\right\| - 1 = \left\|I - (Hf)^{-1}\frac{\partial\vec{v}}{\partial\vec{r}}\right\| - 1, \quad (5)$$

where the double vertical bars denote the Jacobian determinant and I is the unit matrix. Note that this is a nonlinear expression, but it still involves only the first partial derivatives.

This approximation was tested[7] against N-body simulations of CDM and neutrino cosmologies, with $\Omega = 1$ and 0.2, smoothed with a Gaussian of radius 500 km s^{-1} or 1000 km s^{-1}. FIGURE 2 shows the "linear" and the quasi-linear approximations derived from the velocity field versus the true density contrast in the $\Omega = 1$, CDM simulation with 500 km s^{-1} smoothing. The rms error of δ_c is less than 0.1 over the range $-0.8 \leq \delta \leq 4.5$. It thus provides a very useful tool for reconstructing the mass-density field from the *POTENT* velocity field.

The inverse problem of extracting the velocity field from a given density field (e.g., from a redshift survey like *IRAS*) also needs nonlinear corrections. This problem

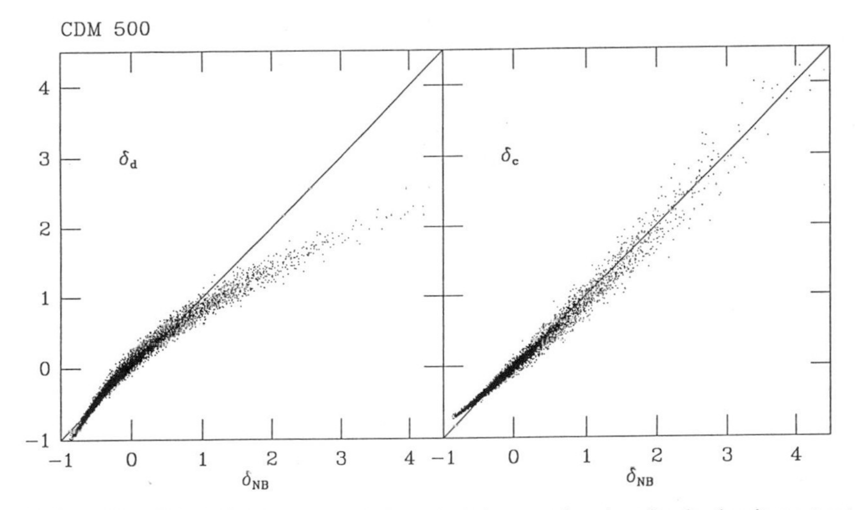

FIGURE 2. The "linear" (**left**) and quasi-linear (**right**) approximations for the density contrast, derived from the velocity field, versus the true density in an $\Omega = 1$ CDM N-body simulation, Gaussian smoothed with radius 500 km s^{-1}.

is harder because, already in the linear regime, it requires nonlocal integration of the relation $\delta_d \propto -\vec{\nabla} \cdot \vec{v}$. The tight correlation seen in the N-body simulations between the true and the "linear" densities that are derived from the true velocities suggests the following scheme for introducing nonlinear corrections. Use an empirical fit based on the N-body simulations to translate the given density to an equivalent "linear" density, δ_d, and then integrate the linear relation for the velocity as usual, but using δ_d instead. A useful empirical fit of this sort, good to an accuracy of about 0.1 in the range $-0.8 < \delta < 4.5$ and applicable even outside this range, is given by

$$\delta_d = \frac{\delta}{1 + 0.18\delta}. \tag{6}$$

This method has been tested and it is now being used in the POTENT–IRAS comparison.[11]

Back to POTENT, the local "cosmography" as deduced from the peculiar velocities is displayed in FIGURE 3, which shows the POTENT reconstructed velocity field and the corresponding mass-density field in the supergalactic plane.[6]

2.3. Potent Error Analysis

What makes the POTENT output suitable for quantitative studies is the error analysis.[5,6] The largest source of error is the random *distance errors* of the individual galaxies, whose effect on the reconstructed fields is assessed using Monte Carlo simulations. We construct 100 artificial redshift-distance samples in which the distance (and hence peculiar velocity) of each object is scattered using an independent Gaussian random number of standard deviation σ_i (plus a $\sigma_f = 150$ km s^{-1} perturbation to z_i to mimic small-scale sampling fluctuations). For each artificial sample we compute the potential, velocity and density fields and then construct maps of the means and standard deviations (σ_Φ, σ_v, and σ_δ) of these noisy fields at each grid point. The standard deviation provides our error estimate while the mean is useful for diagnosing Malmquist bias. In the well-sampled regions [out to 3000 km s^{-1} in most directions and nearly 6000 km s^{-1} toward the Great Attractor (GA)] the rms errors are $\sigma_v < 250$ km s^{-1} and $\sigma_\delta < 0.2$, but the errors exceed 1000 km s^{-1} and 1.0, respectively, in some poorly sampled, noisy regions. To exclude noisy regions, we limit any quantitative analysis to points where σ_v or σ_δ are smaller than a certain critical value.

Our reconstructed fields are also subject to *biases* such as the Malmquist bias and a sampling gradient bias. We estimated the Malmquist bias in the velocity field, using *N*-body artificial data, to be less than 200 km s^{-1} nearly everywhere for the 1200 km s^{-1} smoothing. This error is thus small compared to the random errors. The sampling gradient bias arises from the coupling of gradients in the true velocity field and the selection function. It is difficult to assess from the data, but we showed using Monte Carlo simulations of the *N*-body "data" that with our volume-weighting scheme (weight of the contribution from each object $\propto R_4^3$, where R_4 is the distance from the object to the fourth nearest object) this bias is generally much smaller than the scatter due to distance errors. There are, however, a few very empty regions where sampling bias might be severe enough to generate coherent flows of no physical reality. These include regions of low galactic latitude (small supergalactic $|Y|$) and voids in the galaxy distribution away from the supergalactic plane. Being unable to quantify the sampling gradient biases in these regions at this stage, we simply exclude very empty regions from any quantitative analysis, by rejecting all grid points where the distance to the fourth nearest object, R_4, is greater than a certain critical value.

The inclusion of more data in any given region can improve the signal-to-noise ratio significantly, as is demonstrated in FIGURE 4. Shown is the "old" POTENT density map based on the original data[6] in comparison with a preliminary POTENT reconstruction of an extended data set. The new data include 501 new spirals, partly grouped in 183 objects, both in the GA region (by A. Dressler and S. M. Faber) and in the Pisces region (by J. Willick), which have not been carefully tested, matched, and merged with the old data set yet. The point of showing this is to demonstrate how the error

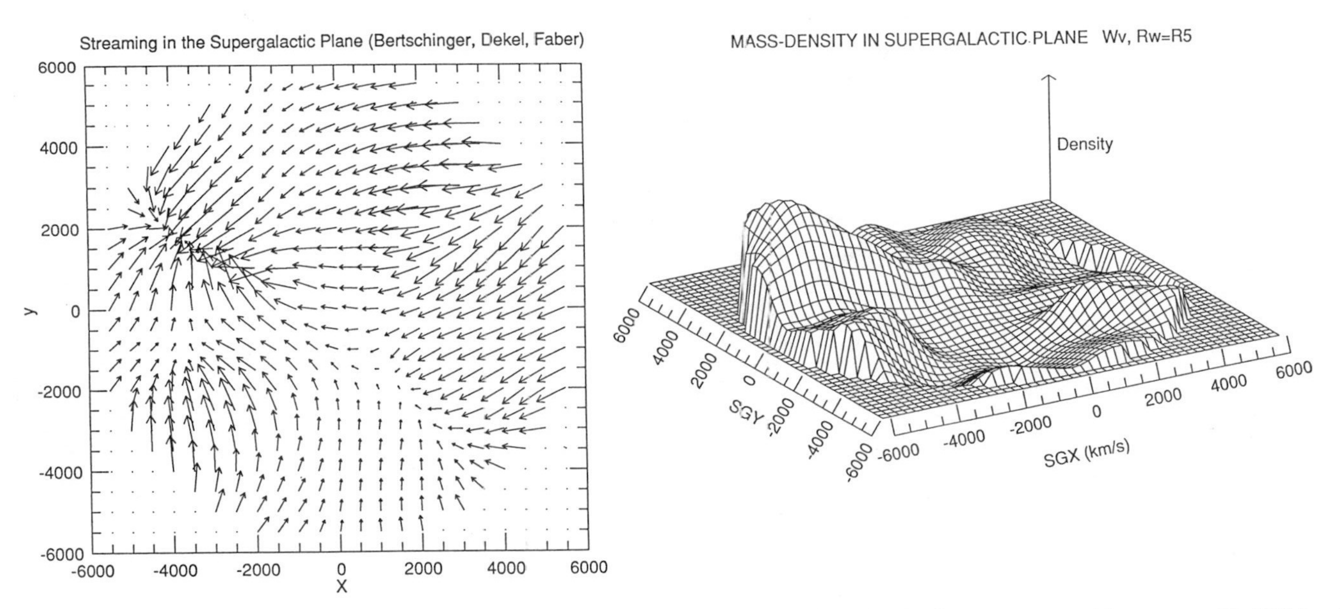

FIGURE 3. Cosmography: The velocity field **(left)** and mass-density field **(right)** in the supergalactic plane as recovered by *POTENT* with 1200 km s^{-1} smoothing. The local group is at the origin and the Great Attractor is centered at about $X = -4000$, $Y = 1000$. The vectors shown are projections of the 3-dimensional velocities. The height of the surface plotted is proportional to the density fluctuation.

contours move outwards with the inclusion of more data. The density peak of the GA now shows up at a level of $\delta = 1.4 \pm 0.3$, and it has moved slightly toward the galactic plane ($X = -4500, Y = 0$). The Pisces peak reaches $\delta = 1.0 \pm 0.4$, but the still poorly sampled Perseus region (near $X = 5000, Y = 0$) is dominated by noise; it carries large errors both in terms of σ_δ and R_4.

2.4. *Omnipotent—An Alternative to Potent?*

The constraint of no-vorticity can be applied to the data in several different ways. For example, one alternative to *POTENT* is now being developed by G. Blumenthal and myself. It promises to have three basic advantages: the data are smoothed simultaneously with the derivation of the full velocity field and not as a preceding

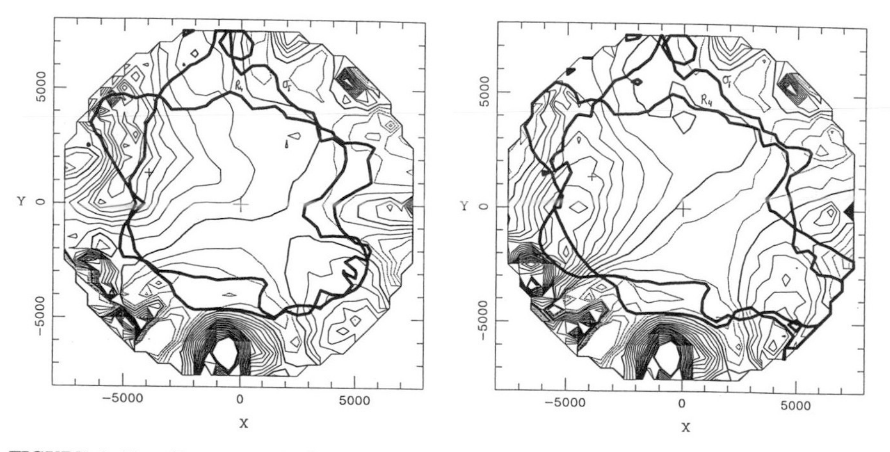

FIGURE 4. Density contrast in the supergalactic plane. (**Left**) From the original data input to *POTENT*. (**Right**) Based on a preliminary extended sample with 501 new galaxies in the Great Attractor and in the Pisces regions. The smoothing radius is 1200 km s^{-1}. Positive and negative contours are *solid* and *dotted*, respectively, with the *zero contour slightly heavier* and contour spacing 0.1. The *heavy contours* mark errors: $\sigma_\delta = 0.4$ and $R_4 = 1500$.

step, the sampling-gradient bias is reduced, and the derivation of the density does not require numerical differentiation.

In order to evaluate the peculiar velocity and its spatial derivatives at a point $\vec{r}$, expand the ith component of the velocity at any nearby point $\vec{r} + \vec{x}$ in terms of powers of x_j,

$$v_i(\vec{x}) = B_i + L_{ij}x_j + Q_{ijk}x_jx_k + C_{ijkl}x_jx_kx_l + \cdots . \tag{7}$$

The requirement $\vec{\nabla} \times \vec{v} = 0$ simply means that the matrices L_{ij}, Q_{ijk}, and C_{ijkl} are all *symmetric*. The expansion can be truncated at any order, making a model for the local flow with a finite number, p, of free parameters (the various matrix elements) to be determined by the data. The zeroth order model with only the first term, the bulk flow, has 3 parameters, the linear model has 9, the quadratic model has 24, the cubic model has 66, etc.

Given a set of N objects with measured radial peculiar velocities v_{r_n} at positions $\vec{r}_n$ $(n = 1, N)$, we determine these parameters by minimizing the sum

$$\chi^2 = \sum_{n=1}^{N} [v_{r_n} - \vec{v}(\vec{r}_n) \cdot \hat{r}_n]^2 W(\vec{r}_n, \vec{r}), \tag{8}$$

provided that $p \leq N$. W is a window function, which could be a spherical Gaussian window as in *POTENT*. Then the desired velocity at $\vec{r}$ is $\vec{v}(\vec{r}) = \vec{B}$, and the density contrast, using the continuity density under the Zel'dovich approximation (**5**), is simply $\delta_c(\vec{r}) = \|I - L_{ij}\| - 1$.

The zeroth order scheme of this sort is somewhat analogous to the *POTENT* procedure, which first fits a local bulk flow in the neighborhood of each point. The linear term provides a straightforward density determination. A model of higher order allows more degrees of freedom, and therefore more flexibility in the local fit. It eliminates several biases that *POTENT* suffers from, including the sampling-gradient bias that arises from the averaging of the velocities sampled inhomogeneously within the window. But there are also disadvantages that become severe when the model has too many parameters: a bias is introduced because the data have to be interpolated from a larger volume about each point, and the small-scale noise in the data is picked up and may eventually dominate the fit. We find that the optimal model for the current data is either the quadratic or the cubic model.

Another method for reconstructing the density field from the peculiar velocities has been suggested by Stebbins and Kaiser.[12] It is based on "maximum probability" estimation of the density Fourier components. The method works nicely and it produces informative maps over an extended volume, but it is somewhat model-dependent. Because the resultant density in each region is modulated by the quality of the data there, it is not straightforward to use in a quantitative comparison with other data or theory.

2.5. Impotent—Where Could We Go Wrong?

• *Systematic Distance Errors.* Several authors have raised the worry that the large-scale flow might be an artifact of a systematic zero-point variation in the galaxy scaling laws that are used in the distance measurements.[13] Evidence is accumulating against this possibility.[14] In particular, it seems that any such systematic effect would require unnatural large-scale coherence across the sky that needs an explanation as much as the GA phenomena does.

A promising test for the reality of the velocity measurements can be provided by comparing the results from the Tully–Fisher and the $D_n - \sigma$ relations. Bertschinger showed[15] to what extent the *POTENT* smoothed radial velocities, as deduced by the two different methods at the same points in space, agree with each other. We find that the velocity and density fields extracted by *POTENT* from the two orthogonal sets of data in the regions where the uncertainties in both are acceptable reveal the same main features, including the GA, the big underdense region between us and Pisces, and the bulk flow. This similarity provides an encouraging confirmation for the reality of the velocity field.

• *Potential Flow.* The ansatz of potential flow is based on theoretical considerations. It is impossible to confirm its validity by observation independently of further assumptions because vorticity could always be hidden in the transverse velocities that we cannot observe. Given the radial components of any velocity field, with or without vorticity, *POTENT* would recover the one potential flow that corresponds to the given radial velocities. The resulting velocity and density maps could be very different from the true ones.

This could be illustrated by amusing examples. Consider, for example, a "true" velocity field that mimics a huge "galaxy" in the *XY* plane (centered, say, at $X = 3000, Y = 0$): a circular motion about an axis (normal to the *XY* plane) with a flat rotation curve (500 km s^{-1}, say) extending all the way to infinity. Such a velocity field

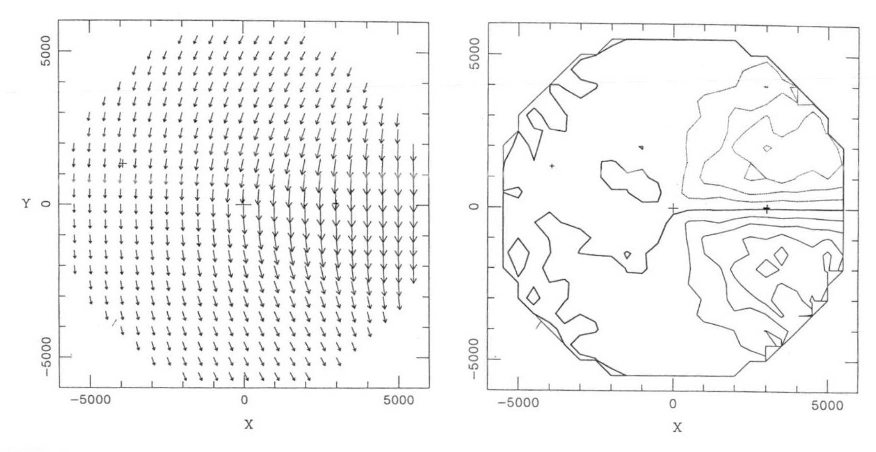

FIGURE 5. The velocity (**left**) and density (**right**) fields reconstructed by *POTENT* from the line-of-sight velocities of a rotating "galaxy" with a flat rotation curve centered at $X = 3000$, $Y = 0$.

corresponds to a spherical distribution of mass about the center of rotation with an r^{-2} density falloff. Given the radial components of this field, *POTENT* recovers a monotonic potential gradient along the *Y*-axis, perpendicular to the line of sight near the center of the "galaxy." The derived velocity and density maps are shown in FIGURE 5, with the latter revealing fictitious density maximum and minimum on the two sides of the true density peak (at about $X = 3000, Y = \pm 2000$).

We are in the process of applying *POTENT* to more general velocity fields with large-scale vorticity, for example, as generated by explosions or by cosmic turbulence. But the preceding simple case already reveals a characteristic detectable imprint of vorticity: that the mass distribution as recovered by *POTENT* does not follow the light distribution, under the *assumption* that the light does trace the true mass distribution (e.g., the true "galaxy") via some monotonic biasing scheme. The fact that the *IRAS* and *POTENT* densities are, in general, monotonically correlated with each other (see Section 3.1 below) can serve as a reassuring consistency test for the potential flow ansatz.

Another test of the potential flow ansatz is possible under the *assumption* of large-scale isotropy. If there is vorticity such that the velocity recovered by POTENT is not real, there is no reason for it to be isotropic. In fact, the POTENT procedure imposes a preferred direction at each point—the radial direction to the origin—so the radial and tangential velocities could violate the isotropy condition $\langle v^2_{\rm tan}\rangle = 2\langle v^2_{\rm rad}\rangle$. We have found that the POTENT velocity field inside 4000 km s^{-1} obeys this condition well within the 1σ error bars derived from the Monte Carlo simulations.

Also reassuring is the fact that the main features in the velocity field recovered by POTENT are also recovered by the "maximum probability" method,[12] which is not explicitly dependent on the ansatz of potential flow.

Finally, since vorticity does exist on a small scale where collapse has occurred, it is important to formally confirm our hypothesis, so far based on N-body simulations,[4] that the velocity field smoothed on sufficiently large scales is indeed vorticity-free to a satisfactory accuracy.

• *Sparse and Noisy Data.* In order to distinguish between signal and noise, it is crucial to estimate the uncertainties in the analysis. As discussed in Section 2.3, we have confidence in our analysis of the uncertainties due to the distance measurement errors and discrete sampling (including Malmquist bias and shot noise), but we do not yet have a satisfactory way of estimating the remaining biases due to sampling gradients, except of eliminating too-empty regions altogether. Nevertheless, we already have a fairly good indication for where the results should be trusted and where they are dominated by errors, and every use of these results should take the local uncertainties into account. The errors are expected to be reduced significantly when more data fill poorly sampled regions and penetrate deeper with improved distance measurement tools.

3. CONFRONTATION OF OBSERVATION AND THEORY

3.1. Comparison to IRAS Galaxy Density—Estimating Ω *and* b_I

The POTENT density, δ_P, is Ω dependent, proportional to $f(\Omega)^{-1}$ in the linear regime. The galaxy density extracted from the IRAS catalog,[8,9] δ_I, is only slightly affected by Ω and the specific relation between galaxies and mass, only through the predicted peculiar velocities used to correct the redshifts to real positions. If we parameterize the relation between galaxy and mass density fluctuations by a universal "biasing" factor,[16] $\delta_I = b_I\delta$, then we expect in the linear regime a linear relation of the sort $\delta_P = [f(\Omega)/b_I]\delta_I$. Given the uncertainties in the two datasets, we can ask whether the IRAS data are consistent with being a noisy version of the POTENT data, or vice versa, and obtain, via linear regression, the best-fit value for $f(\Omega)/b_I$ with associated confidence limits.[11]

The degeneracy of Ω and b_I is, in principle, broken in nonlinear regions, where $\delta(\vec{v})$ is no longer simply proportional to $f(\Omega)^{-1}$. If Ω and b_I are of order unity or less, we find that the local volume where the comparison is possible does contain regions where $|\delta| \sim 1$, even after 1200 km s^{-1} smoothing. The compatible quasi-linear treatments of *POTENT* and *IRAS* previously described (Section 2.2) allow a first attempt at determining Ω and b_I separately, but still assuming the simplest biasing model b_I = const.

A similar comparison could be made at the level of velocities, but it is complicated by an unknown quadrupole moment.[17] I focus here on the comparison of densities, which is local, independent of reference frame, and can be more easily corrected for nonlinear effects.

FIGURE 6 compares density maps in the supergalactic plane. Gaussian smoothing of radius 1200 km s^{-1} has been applied and $\Omega = b_I = 1$ has been assumed for this plot. Despite our efforts to minimize the effects of nonuniform sampling, it does introduce a nonnegligible bias into the results of the *POTENT* smoothing procedure. Instead of comparing the *POTENT* density (P) to the raw *IRAS* density (I), we should compare it to an *IRAS* density field biased in a similar way: the *IRAS* velocities (with 500 km s^{-1} smoothing) were sampled nonuniformly as in the velocity sample and then passed through the *POTENT* machinery with 1200 km s^{-1} smoothing [$P(I)$]. To be on the conservative side, we limit the comparison to the volume where the *POTENT* noise and emptiness are below $\sigma_\delta = 0.2$ and $R_4 = 1500$ km s^{-1}. The effective volume for this comparison is $\sim(5300 \text{ km s}^{-1})^3$.

Both maps feature the general GA phenomenon as a ramp that peaks beyond the Hydra–Centaurus clusters at about $X \simeq -4000$, $Y \simeq 1000$ and falls off gradually toward Virgo ($X \simeq -300$, $Y \simeq 1300$) and toward Pavo across the galactic plane. The GA peak in the *IRAS* map in fact coincides with the original prediction of the Seven Samurai[3] for the center of the velocity flow (the small cross). So the GA certainly exists, both as a dynamical entity and as a supercluster of galaxies. (Any statements to the contrary simply reflect differences in terminology: Is the dynamical attractor associated with known concentrations of galaxies, or is part of it lying away from known clusters, perhaps hidden in the galactic zone of obscuration?) The two fields also show an adjacent "great void" in the region $X < 2000$, $Y < 0$. There is an apparent discrepancy in the Perseus region ($X > 3000$, $-2000 < Y < 3000$), where the *IRAS* map shows a density enhancement, while the *POTENT* map indicates an underdensity, but this region is both too noisy and undersampled in the current *POTENT* data. In fact, *POTENT* does recover a significant overdensity near Pisces ($X \simeq 3000$, $Y \simeq -3000$); the preliminary *POTENT* analysis of new data by J. Willick (FIG. 3) recovers a significant density peak on the order of $\delta = 1$ there.

The *POTENT* and *IRAS* densities are plotted against each other at grid points of spacing 500 km s^{-1} in FIGURE 7. Our model is a linear relation: $\delta_P = \lambda\delta_I + s$. (A shift s is possible because δ_I is normalized by the mean *IRAS* density within its survey volume, while the input velocities to *POTENT* are calculated relative to a mean Hubble expansion in a different volume.) The parameters λ and s are obtained by linear regression of δ_P on δ_I, with the weights σ_δ^{-2}. The lines drawn are the best fits: the naive comparison of P and I yields $\lambda = 0.81$, and the bias-corrected comparison of P and $P(I)$ yields $\lambda = 1.15$. In general, there is a strong correlation between the *POTENT* and *IRAS* fields, with $f(\Omega)/b_I$ of order unity, and with a small systematic variation of mass-to-light ratio in space. The bias does not change the growth features, only the details. But the exact value of $f(\Omega)/b_I$ determined this way should not be regarded as a fully self-consistent estimate, because these density fields were calculated assuming $\Omega = b_I = 1$.

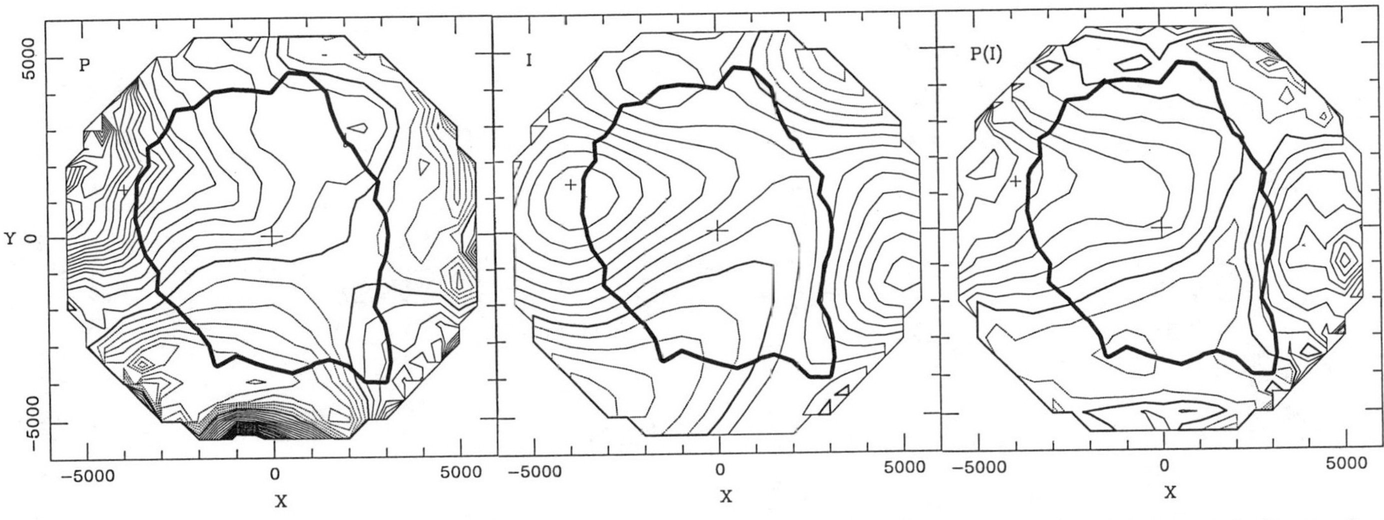

FIGURE 6. Density fluctuation fields in the supergalactic plane (the local group at the origin): P refers to mass density by POTENT, I refers to galaxy density by IRAS, and $P(I)$ is the result of POTENT when fed with the IRAS velocities sampled nonuniformly. Contours are spaced by 0.1 in δ, with *positive values solid, negative ones dashed,* and the *zero contour slightly thicker.* The *heavy line* marks the $\sigma_\delta = 0.2$ error contour.

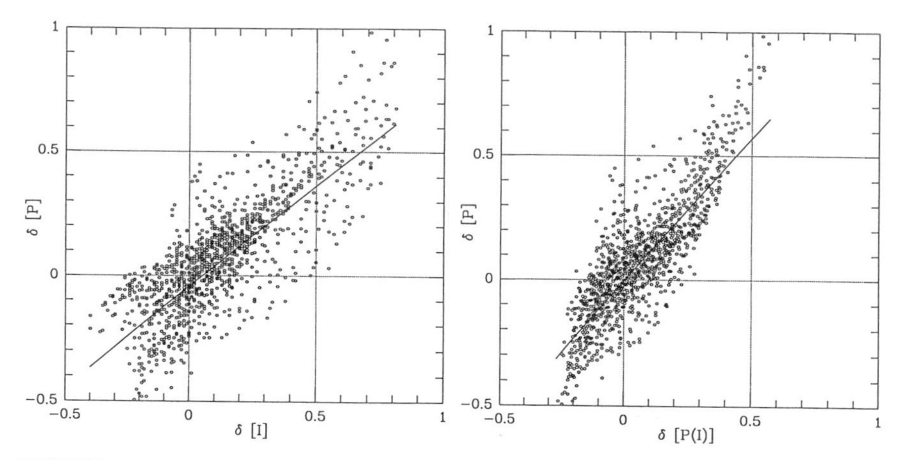

FIGURE 7. *POTENT* vs. *IRAS* density fluctuations at the points of a cubic grid of spacing 500 km s^{-1} inside the volume where $\sigma_8 < 0.2$ and $R_4 < 1500$ km s^{-1}. (**Left**) P vs. I. (**Right**) P vs. $P(I)$.

To properly correlate the nonlinear *POTENT* and *IRAS* fields and determine Ω and b_I in a bias-independent way, given the measurement errors, we[11] have pursued a likelihood analysis using Monte Carlo simulations of data plus noise, spanning the parameter plane $\Omega - b_I$. I just wish to show here, in FIGURE 8, a preliminary result from this ongoing project, of likelihood contours in the $\Omega - b_I$ plane. Our robust result, which is quite insensitive to nonlinear effects, is that $\Omega^{0.6}/b_I$ is of order unity. As long as b_I is not much smaller then unity, I can quote $\Omega^{0.6}/b_I = 1.35 \pm 0.25$. The nonlinear effects seem to introduce apparent lower limits on Ω and b_I, but the results

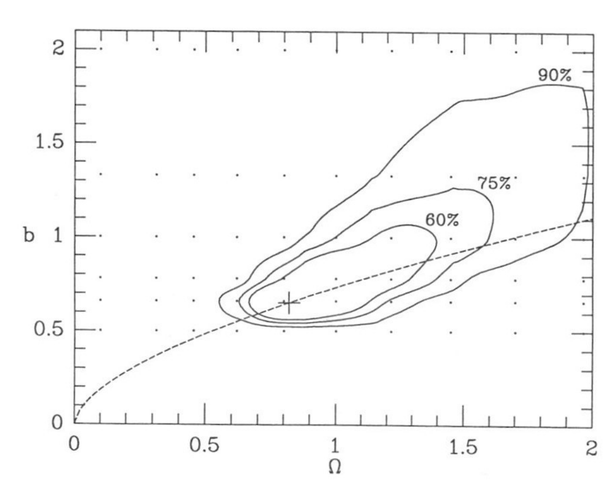

FIGURE 8. Likelihood contours in the $b - \Omega$ parameter plane. The *cross* marks the most likely point: $\Omega = 0.82$, $b_I = 0.65$. The *dashed line* is $f(\Omega)/b_I = 1.36$. The *dots* mark the cases sampled. The results for $b_I < 0.5$ are still highly uncertain.

for low values of b_I should be considered uncertain at this stage because they involve extreme nonlinear corrections that are still under investigation. Another reason for care in the interpretation of the current nonlinear analysis is that it assumes that b_I is independent of position.

Preliminary results[18] from a similar comparison with the galaxy distribution in optical catalogs (e.g., the CfA and SSRS surveys and the *Atlas of Galaxies* by Tully and Fisher), but in a smaller volume and with less smoothing, indicate a similar general correlation between light and mass, with $f(\Omega)/b_o \sim 0.5$. Since Ω is a universal factor, this indicates that the biasing factor for optical galaxies, b_o, is somewhat larger than the biasing factor for *IRAS* galaxies, which is in agreement with common theoretical ideas.[19]

3.2. The Fluctuation Power Spectrum

What does the observed velocity field tell us about the power spectrum of the mass-density fluctuations? Is is consistent with a scale-invariant tail of slope $n = 1$ on scales larger than 100 h^{-1} Mpc and gradual flattening corresponding to "standard" CDM on smaller scales?

The left side of FIGURE 9 shows the bulk flow of the *POTENT* velocity field smoothed by a Gaussian of 1200 km s^{-1}, inside top-hat spheres of radius R about the local group. For example, the bulk flow in a sphere of radius 4000 km s^{-1} is $V = 388 \pm 67$ km s^{-1}. This is fairly consistent with the rms prediction for CDM, $\langle V^2 \rangle^{1/2} = 287\, b^{-1}$ km s^{-1} (based on the standard normalization of $\delta M/M = b^{-1}$ in spheres of 800 km s^{-1}), even if $b \simeq 2$.

An analysis of the "Mach number" of this velocity field, defined by the ratio of bulk velocity to the standard deviation from it, gives $M = 1.2 \pm 0.2$ in a sphere of radius 4000 km s^{-1}. Based on Monte Carlo simulations,[20] this is consistent at the 1σ level with the rms value predicted for CDM, again indicating a general agreement of the shape of the power spectrum with the "standard" model.

Finally, the power spectrum, P_k, has been determined[21] by the fast Fourier transform (FFT) of the *POTENT* potential field, which has been artificially and gradually damped to zero at radii 6000–8000 km s^{-1} to make periodic boundary conditions at the edges of a cube of side 16,000 km s^{-1}. The resulting power spectrum for the density is shown in the right side of FIGURE 9. Its shape agrees nicely with the shape predicted for CDM out to wavelength ~ 100 h^{-1} Mpc, where the logarithmic slope of P_k bends to $n \geq 0$. This is consistent with a Zel'dovich scale-invariant spectrum ($n = 1$) on larger scales, but we cannot explore such large scales yet in terms of the dynamical fields.

3.3. Evidence for Gaussian Initial Fluctuations

In the "standard" model, the fluctuations are assumed to be a random field, $\delta(\vec{x})$, that is, a set of random variables, one for each point in space, which is fully specified by the m-point joint probability distribution functions,[22] $P_m(\delta)$. The $m \geq 2$ functions involve the spatial correlation functions, or the power spectrum, whose

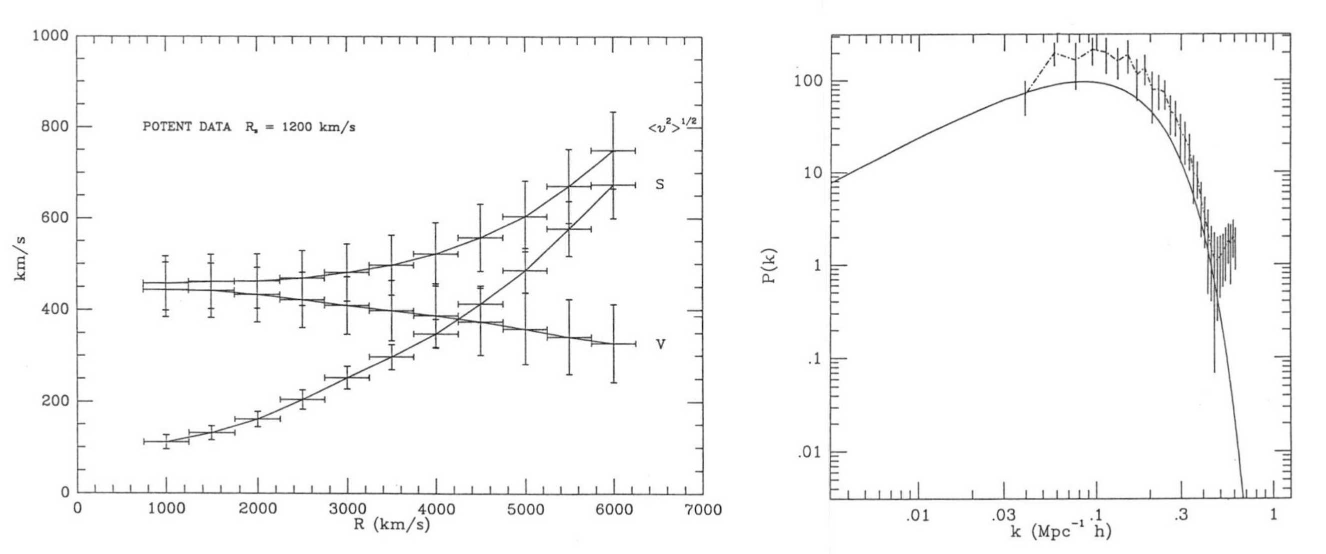

FIGURE 9. Statistics of the POTENT velocity field, smoothed 1200 km s^{-1}. *Error bars* mark the standard deviation in Monte Carlo noise simulations. (**Left**) Bulk velocity V, and the dispersion about it S, in spheres of radius R. (**Right**) The power spectrum of the mass-density fluctuations derived inside a sphere of radius 6000 km s^{-1}. The *solid line* is the rms CDM spectrum ($\Omega = 1, h = 0.5, b = 1$), Gaussian smoothed on small scales as in POTENT.

time evolution is affected by the nature of the dark matter. But $P_1(\delta)$ is not explicitly sensitive to it, so present observations can directly reflect the initial one-point distribution without speculating about the dark matter.

The natural choice is a Gaussian random field, where $P_1(\delta) \propto \exp(-\delta^2/2\sigma^2)$ and the joint probabilities are multidimensional Gaussians. But non-Gaussian initial fluctuations have also been considered in recent years as a possible explanation for the indicated structure on very large scales in excess of the predictions of the "standard" CDM model.[23] In fact, the general inflation picture permits non-Gaussian fluctuations.[24] These fluctuations also arise in scenarios where they originate from topological defects such as cosmic strings[25] and textures,[26] or from explosions.[27] The statistical nature of the fluctuations is therefore an important distinguishing feature between the competing theories.

The distribution functions do not change during linear evolution, but the recent nonlinear evolution of structure could introduce strong non-Gaussian features that

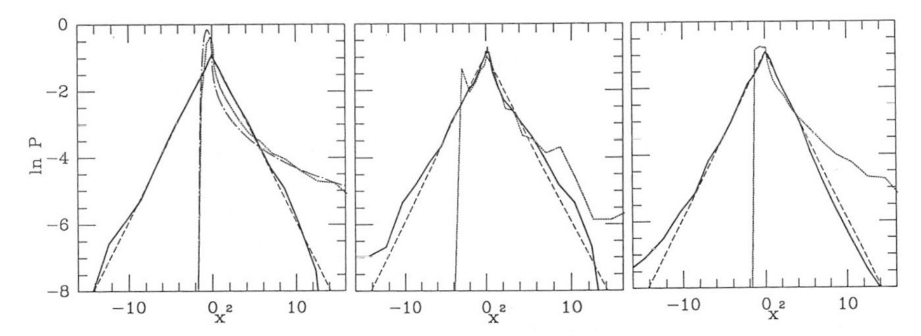

FIGURE 10. The point probability distribution functions compared with a Gaussian (*dashed*) and with the Zel'dovich approximation (*dashed-dotted*). The random variable x is $\tilde{v}$ (*solid*) or $\tilde{\delta}$ (*dotted*). To stress the behavior at the tails of the distribution, ln P_1 is plotted against x^2 (the minus signs refer to x, not x^2). The distributions are based on N-body simulation (**left**), peculiar velocities via POTENT (**middle**), and IRAS galaxy distribution (**right**).

might blur any initial non-Gaussianities. The nonlinear evolution of an initially Gaussian *density* distribution can be followed using the Zel'dovich approximation or N-body simulations,[28] but its present deviation from Gaussianity depends on the amplitude of the fluctuations that are ill determined, due to the unknown "biasing" of galaxies.

We find,[28] however, that for Gaussian initial fluctuations the probability distribution of *velocities* (scaled properly in time) is rather insensitive to nonlinear evolution. This can be shown using the Zel'dovich formalism, which properly approximates a cosmological system as long as there is no orbit mixing, or when smoothed over a large enough scale. This means that the observed, smoothed velocity field could almost transparently reflect the statistical properties of the initial fluctuations.

To test the nonlinear effects, I show on the left side of FIGURE 10 the distributions $P_1(v)$ and $P_1(\delta)$ from a cosmological N-body simulation of "standard" CDM (with 128^3 grid cells and 128^3 particles in a cubic periodic box of comoving size 160 h^{-1}

Mpc, normalized such that $\langle \delta^2 \rangle = 1$ in spheres of radius 800 km s^{-1} today). The fields are smoothed with a Gaussian of radius 500 km s^{-1}. The dimensionless variable x is either $\tilde{v}$ or $\tilde{\delta}$, with a zero mean and a standard deviation of unity. The velocity components along the three axes are regarded as one random variable with $3N$ entries. The figure demonstrates that while $P_1(\tilde{\delta})$ develops a characteristic non-Gaussianity, $P_1(\tilde{v})$ is hardly affected by nonlinear effects.

The distributions of the observed smoothed fields from *POTENT* and *IRAS* are shown in the middle and right side of FIGURE 10, respectively. In *POTENT* we use only points with Monte Carlo errors in the velocities and densities smaller than 400 km s^{-1} and 0.4, respectively. The resultant volume corresponds to a sphere of radius 2300 km s^{-1}. In *IRAS* we use only points with bootstrap errors smaller than 200 km s^{-1} and 0.2, respectively. The resultant volume corresponds to a radius 6000 km s^{-1}. The finite volume affects the distribution mostly via the random $\sqrt{N}$ effect; some of the small deviations from a pure Gaussian distribution can be due to this noise, especially in the *POTENT* output that is limited to a rather small volume.

Given the errors, the observed $P_1(\tilde{v})$ are indistinguishable from the corresponding N-body distributions and from a Gaussian distribution. The observed density distributions are also consistent with the N-body simulations of an initially Gaussian field. Based on these results we can probably rule out non-Gaussian velocity distributions such as are predicted by the version of the cosmic strings scenario, where the velocity field in each $\sim 40\ h^{-1}$ Mpc region is typically dominated by a single "wake." But it remains to be seen whether other non-Gaussian models, such as textures and explosions, do not produce nearly Gaussian velocity distributions despite the non-Gaussian nature of their density distribution, simply as a result of the central limit theorem when summing over several weakly correlated sources of gravity. Note also that our data, especially in *POTENT*, are noisy, they are heavily smoothed, and they are limited to a finite volume. Therefore, the distribution functions carry nonnegligible random errors until the samples are extended and improved.

3.4 Cosmic Microwave Background Map of Our Neighborhood

Fluctuations in the gravitational potential and in the velocity field cause fluctuations in the temperature of the cosmic microwave background (CMB). We can therefore reconstruct maps of the CMB anisotropy $\Delta T/T$ as seen by a distant observer viewing our region of the universe as a patch on his or her last scattering surface. These maps,[29] for the first time, are based on the measured gravitational potential and on the basics of gravitational instability rather than on a specific theoretical model.

In the linear approximation, applicable on large scales, the peculiar gravitational potential $\phi(\vec{x})$ is proportional to the velocity potential recovered by *POTENT*: $\phi(\vec{x}) \simeq (3/2)\Omega^{0.4}\Phi(\vec{x})$. The current data suffice to recover the potential out to $\lesssim$ 6000 km s^{-1} with a signal-to-noise ratio of nearly[6] 10. The local potential field is dominated by one huge potential well, the GA. The flow toward the GA is coherent over a scale of nearly 10,000 km s^{-1}, with a magnitude of ~ 500 km s^{-1}, and the corresponding maximum variation in ϕ is $\sim 8 \times 10^{-5} c^2$.

Gravitational potential fluctuations on the last-scattering surface cause CMB anisotropy via the Sachs–Wolfe and Doppler effects.[30] For a matter-dominated universe, the CMB brightness temperature fluctuation in direction $\vec{n}$ is

$$\frac{\Delta T}{T} = \frac{1}{3c^2}[\phi(\vec{x}_e) - \phi(\vec{x}_o)] - \frac{1}{c}\vec{n} \cdot (\vec{v}_e - \vec{v}_o), \tag{9}$$

where $\vec{x}_o$ and $\vec{v}_o$ are the observer's comoving position and peculiar velocity, and $\vec{x}_e$ and $\vec{v}_e$ are the comoving position on the last-scattering surface in the direction $\vec{n}$ and the effective peculiar velocity of the photon–baryon fluid there at decoupling. (The finite thickness of the last-scattering surface has been accounted for approximately by the *POTENT* smoothing of the potential with a Gaussian of radius 1200 km s^{-1}, or 7′.) The first term, the Sachs–Wolfe gravitational redshift, is dominant on scales larger than 2° (for $\Omega = 1$), while the second, the Doppler contribution, is important on smaller scales. A comoving separation $|\Delta\vec{x}_e|$ on the last scattering surface corresponds to an angular separation $2\Omega \arcsin(|\Delta\vec{x}_e|/4c)$. The Sachs–Wolfe anisotropy is generic, while the Doppler anisotropy, as well as the adiabatic contribution that we have neglected, depends on the amounts of baryonic and nonbaryonic matter and on the mode of perturbation (e.g., adiabatic or isocurvature).

For $\Omega = 1$, in the linear regime, the gravitational potential remains constant in time. We can therefore compute the Sachs–Wolfe contribution to the anisotropy in the radiation that decoupled at $z_{\rm dec} \simeq 1300$ directly from the potential field derived by *POTENT* today. The Doppler contribution may be estimated crudely by assuming that the photon–baryon fluid at decoupling has the same peculiar velocity field as the dark matter, that is, smaller than the present velocities by a factor $(1 + z_{\rm dec})^{-1/2}$. Radiation last scattered locally would now be reaching a distant observer nearly at the edge of our horizon. FIGURE 11 shows maps of the relative brightness temperature fluctuations caused by the Sachs–Wolfe and Doppler effects for such an observer looking upon the local universe from the direction of the supergalactic north pole. The GA corresponds to the potential well just beyond the left edge of the map. (If $\Omega < 1$, the angular scale would decrease and the amplitude of $\Delta T/T$ would increase, with the model-dependent Doppler and adiabatic contributions dominating the Sachs–Wolfe anisotropy.)

Although FIGURE 11 shows the CMB anisotropy viewed by a distant observer in one small region of his or her last-scattering surface, the Copernican principle argues that we should see similar regions of anisotropy on our last-scattering surface. The maximum Sachs–Wolfe anisotropy, caused by the GA, is about 1.7×10^{-5} on a scale of 1° (for $\Omega = 1$ and a 30-arcmin beam). The maximum Doppler contribution, 4.9×10^{-5}, is overestimated because the photon–baryon fluid typically oscillates with a smaller velocity than the dark matter. We are currently trying to estimate the Doppler contribution more accurately.[31] Anisotropy as large as that shown in FIGURE 11 would be seen only by a carefully placed observer with properly oriented beams directed right across the GA. The typical Sachs–Wolfe anisotropy on a 1° scale is $\sim 5 \times 10^{-6}$. Until accurate velocity measurements are made for galaxies beyond 6000 km s^{-1}, we cannot estimate how frequently anisotropies as large as 2×10^{-5} should be seen, but unless we live in a special place, anisotropies exceeding 10^{-5} should be fairly common.

If $\Omega = 1$, the angular scale of the GA potential well is an excellent match to the two-beam subtraction experiment of Lubin *et al.*,[32] with beams of 30′ FWHM and a beam throw of 1°. Present anisotropy limits on this scale ($\Delta T/T \lesssim 3.5 \times 10^{-5}$ at 90 percent confidence[1]) are a factor of 2 or 3 above the Sachs–Wolfe anisotropies predicted here. These predictions, leaving aside the Doppler and adiabatic anisotropies, are minimal, unless the fluctuations were smeared out by late reionization. If gravitational instability is responsible for the generation of large-scale peculiar motions, and if our neighborhood is not unique, the precursors of large-scale structure should soon be detected in CMB anisotropy experiments.

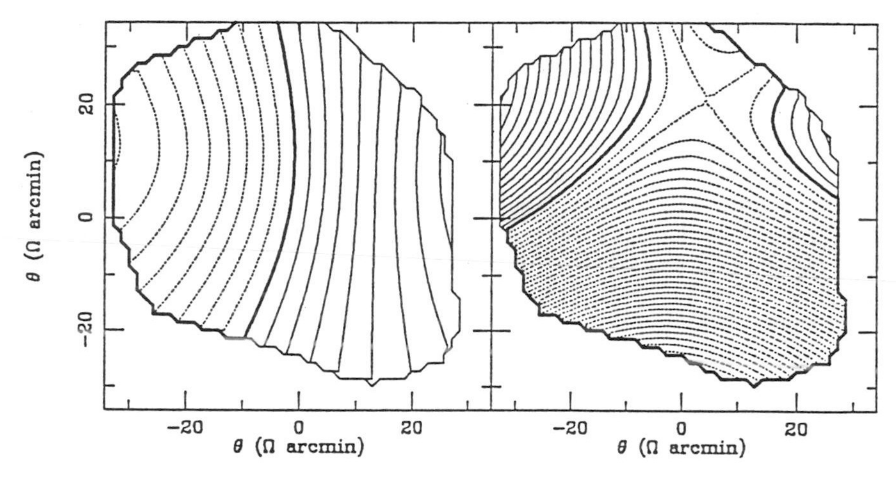

FIGURE 11. Contour maps of $\Delta T/T$ for a distant observer who sees the last scattering of radiation occurring in the supergalactic plane. (**Left**) Sachs–Wolfe contribution. (**Right**) An overestimate of the Doppler contribution. The contour spacing is $\Delta T/T = 10^{-6}$, with negative contours *dotted* and the zero contour (arbitrarily chosen) *heavy*. Regions with Sachs–Wolfe standard errors (arising from distance errors) greater than 2×10^{-6}, determined from Monte Carlo simulations, have been excluded. Each map has been convolved with a Gaussian beam of 30 arcmin FWHM (for $\Omega = 1$) to mimic the experiment of Lubin *et al.*,[17] but no subtraction has been performed.

4. CONCLUSIONS

I have highlighted several aspects of the methods for reconstructing the full dynamical fields from the observed peculiar velocities, focusing on recent developments in the *POTENT* analysis, and reported on several results concerning the dynamical fields and the comparison with theory, which could be summarized as follows.

- *On the Method.* The consistency of the results obtained using different distance estimators, for spirals and for ellipticals and SOs, indicates that the measured velocities are real. The agreement with the fields reconstructed by different methods,[12] the monotonic correlation between fluctuations in the light and mass, and the apparent isotropy of the reconstructed velocity field all support the basic ansatz of potential flow, which originated from theoretical argumentation based on gravitational instability.

• *Cosmography.*[6] Smoothed with a 1200 km s^{-1} Gaussian, the Great Attractor is an extended overdensity, centered near $X = -4500$, $Y = 0$ (± 500 km s^{-1}), with $\delta = 1.4 \pm 0.3$ at the peak. A back flow is marginally detected, with a velocity of 600 ± 400 km s^{-1}. The Pisces peak is $\delta = 1.0 \pm 0.4$, but a large fraction of the Perseus–Pisces (PP) supercluster is still very poorly sampled. An extended, deep underdense region lies between the local group (LG) and PP.

• *Cosmological Parameters and Galaxy Formation.* The *IRAS* galaxy distribution is correlated with the mass fluctuations. A robust linear comparison yields $b_I/\Omega^{0.6} \simeq 1$. A preliminary comparison with optical data yields $b_o/\Omega^{0.6} \simeq 2$. These values are consistent with $\Omega = 1$ and typical b values of order unity. An open model, with $\Omega \simeq 0.1$, say, is also acceptable, but only if the *IRAS* galaxies are severely "antibiased": $b_I \leq 0.25$. Our preliminary attempt to determine Ω and b_I independently, using nonlinear corrections, seems to favor the standard option: $\Omega \simeq b_I \simeq 1$, but this result is still highly uncertain. These results are insensitive to the value of Λ.

• *Dynamical Power Spectrum.* The bulk flow in a top-hat sphere of radius 4000 km s^{-1} is $V = 388 \pm 67$ km s^{-1}. Contrary to previous belief, this bulk flow is fairly consistent with the rms prediction for CDM ($287b^{-1}$ km s^{-1}), even if the normalization is $b \simeq 2$. An analysis of the Mach number of the smoothed *POTENT* velocity field gives values of order unity, which, again, are consistent at the 1σ level with the rms value predicted for CDM. A direct recovery of the fluctuation power spectrum, P_k, from the *POTENT* output reveals a detailed agreement of shape with the power spectrum of CDM out to wavelength ~ 100 h^{-1} Mpc, where the logarithmic slope has gradually bent to $n \simeq 0$. This is consistent with a Harrison–Zel'dovich scale-invariant spectrum ($n = 1$) on larger scales, but these scales cannot be explored dynamically yet.

• *Probability Distribution.* We found that the one-point probability distribution function of the smoothed peculiar velocity field, $P_1(v)$, remains quite constant even through nonlinear evolution. The $P_1(v)$ of the observed (one-dimensional) velocities, as processed through *POTENT* or as predicted by the *IRAS* analysis, is found to be indistinguishable from a normal distribution, suggesting that the initial fluctuations were Gaussian and that extreme non-Gaussian models, such as cosmic strings where 40 h^{-1} Mpc wakes dominate the formation of large-scale structure, are ruled out.

• *Microwave Background and Gravitational Instability.* If our basic ideas concerning gravitational instability are valid, the GA indicates that $\Delta T/T$ fluctuations on the order 10^{-5} should be detected (unless reionization has smeared out the fluctuations). This is still consistent with the observed upper limits, but the constraints are getting close.

These results lead me to a rather conservative conclusion. The observed velocities and the corresponding mass-density field are *consistent* with the "standard" model: a homogeneous universe with an inflation phase, $\Omega = 1$ and $\Lambda = 0$, Gaussian, adiabatic initial fluctuations with a scale-invariant spectrum, cold dark matter with ~ 10 percent baryons, and gravitational growth of fluctuations to the present structure. In this case, the indications[22] for an excess of power in the *galaxy* distribution on scales ≥ 40 h^{-1} Mpc must be accounted for by a nontrivial relation between galaxies and mass, where the biasing factor is both a function of scale and density.

The alternative of an $\Omega \sim 0.1$ ($\Lambda \sim 0.9$) baryon-dominated universe,[33] with isocurvature fluctuations and late reionization, is not convincingly ruled out either. But it also requires a nontrivial relation between galaxies and mass: severe antibiasing of *IRAS* galaxies on scales of order 10 h^{-1} Mpc.

ACKNOWLEDGMENT

This paper describes work done in collaboration with Ed Bertschinger. The collaborators on various parts of the method are G. Blumenthal, S. Faber, and A. Nusser. The main velocity data are due to D. Burstein, A. Dressler, S. Faber, and J. Willick. The comparison with *IRAS* is a collaboration with M. Davis, M. Strauss, and A. Yahil. The spectrum has been analyzed with T. Kolatt, the Mach number with T. Kolatt and Y. Hoffman, the probability distribution with G. Gelb and L. Kofman, and the CMB map with K. Gorski.

REFERENCES

1. Melchiori, F., B. Partridge, G. F. Smoot & N. Vittorio. 1991. *All in* Physical Cosmology, M. Lachieze-Rey, Ed. Editor Frontier.
2. Maddox, S. J., G. Efstathiou, W. J. Sutherland & J. Loveday. 1990. Mon. Not. R. Astron. Soc. **242:** 43; de Lapparent, V., M. J. Geller & J. P. Huchra. 1986. Astrophys. J., Lett. **302:** L1.
3. Originally Lynden Bell, D., S. M. Faber, D. Burstein, R. L. Davies, A. Dressler, R. J. Terlevich & G. Wegner. 1988. Astrophys. J. **326:** 19.
4. Bertschinger, E. & A. Dekel. 1989. Astrophys. J., Lett. **336:** L5.
5. Dekel, A., E. Bertschinger & S. M. Faber. 1990. Astrophys. J. **364:** 349.
6. Bertschinger, E., A. Dekel, S. M. Faber, A. Dressler & D. Burstein. 1990. Astrophys. J. **364:** 370.
7. Nusser, A., A. Dekel, E. Bertschinger & G. R. Blumenthal. 1991. Astrophys. J. In press.
8. Strauss, M. A., M. Davis, A. Yahil & J. P. Huchra. 1990. Astrophys. J. In press; Yahil, A., M. A. Strauss, M. Davis & J. Huchra. 1991. Astrophys. J. In press.
9. Nusser, A. & A. Dekel. 1991. Astrophys. J. In press.
10. Zel'dovich, Ya. B. 1970. Astron. Astrophys. **5:** 20.
11. Dekel, A., E. Bertschinger, A. Yahil, M. Strauss & M. Davis. 1991. In preparation.
12. Stebbins, A. & N. Kaiser. 1990. Preprint.
13. Djorgovski, S., R. de Carvalho & M.-S. Han. 1989. *In* The Extragalactic Distance Scale, S. van den Bergh and C. J. Pritchet, Eds.: 329. Astronomical Society of the Pacific. Provo, Utah; Silk, J. 1989. Astrophys. J., Lett. **345:** L11.
14. Burstein, D. 1990. Rep. Prog. Phys. In press; Lucey, J. R. 1991. *In* Observational Tests of Inflation, T. Shanks, Ed. Kluwer. Dordrecht, the Netherlands. In press.
15. Bertschinger, E. 1991. *In* Physical Cosmology, M. Lachieze-Rey, Ed. Editor Frontier.
16. Dekel, A. & M. J. Rees. 1987. Nature **326:** 455.
17. Yahil, A., E. Bertschinger, A. Dekel, M. Strauss & M. Davis. 1991. In preparation.
18. Bistoles, V., A. Dekel, O. Lahav, B. Tully & E. Bertschinger. 1991. In preparation.
19. For example, Dekel, A. & J. Silk. 1986. Astrophys. J. **303:** 39.
20. Kolatt, T., A. Dekel, Y. Hoffman & E. Bertschinger. 1991. In preparation.
21. Kolatt, T., A. Dekel & E. Bertschinger. 1991. In preparation.
22. Bardeen, J., J. R. Bond, N. Kaiser & A. Szalay. 1986. Astrophys. J.
23. Maddox, S. J., G. Efstathiou, W. J. Sutherland & J. Loveday. 1990. Mon. Not. R. Astron. Soc. **242:** 43; Kaiser, N. 1991. The density and clustering of mass in the universe. This issue.; Efstathiou, G., N. Kaiser, W. Saunders, A. Lawrence, M.

ROWAN-ROBINSON, R. S. ELLIS & C. S. FRENK. 1990. Mon. Not. R. Astron. Soc. In press; Bahcall, N. 1988. Ann. Rev. Astron. Astrophys. **26:** 631; BROADHURST, T. J., R. S. ELLIS, D. C. KOO & A. S. SZALAY. 1990. Nature **343:** 726.
24. KOFMAN, L., G. R. BLUMENTHAL, H. HODGES & J. R. PRIMACK. 1990. *In* Large Scale Structure and Peculiar Motions in the Universe, D. W. Latham and L. N. da Costa, Eds. Astronomical Society of the Pacific. San Francisco.
25. BERTSCHINGER, E. 1989. Ann. N.Y. Acad. Sci. **571:** 151.
26. TUROK, N. 1990. Paper presented at the Texas/ESO-CERN Symposium on Relativistic Astrophysics, Cosmology and Fundamental Physics, Brighton, England, December 16–21, 1990.
27. OSTRIKER, J. P. 1988. *In* IAU Symp. No. 130, J. Audouze and A. Szalay, Eds. Reidel. Dordrecht, the Netherlands.
28. KOFMAN, L., E. BERTSCHINGER, J. GELB, A. NUSSER & A. DEKEL. 1991. In preparation.
29. BERTSCHINGER, E., K. GORSKI & A. DEKEL. 1990. Nature **345:** 507.
30. SACHS, R. K. & A. M. WOLFE. 1967. Astrophys. J. **147:** 73.
31. VITTORIO, N., A. DEKEL & E. BERTSCHINGER. 1991. In preparation.
32. LUBIN, P., *et al.* 1990. *In* The Cosmic Microwave Background 25 Years Later, N. Mandolesi and N. Vittorio, Eds. In press.
33. PEEBLES, P. J. E. 1987. Nature **327:** 210; BLUMENTHAL, G. R., A. DEKEL & J. R. PRIMACK. 1988. Astrophys. J. **326:** 539.

Cosmic Nucleosynthesis with Fluctuations[a]

CRAIG J. HOGAN

Astronomy and Physics Departments
FM-20
University of Washington
Seattle, Washington 98195

INTRODUCTION

Standard big bang nucleosynthesis (SBBN) works extremely well for a theory that describes events as early as the first few seconds of cosmic time.[1] It is an overdetermined theory with fewer parameters than predictions; it provides evidence for a simple history of the universe back to the very early epoch of weak decoupling ($T > 1$ MeV, $t < 1$ sec); it confirms rough statistical homogeneity at that time; it confirms the expansion law $t(T)$ predicted by general relativity; it informs us of rough entropy conservation since early times; and it confirms standard model physics at early times. If its assumptions are taken literally, it also yields very tight constraints on the cosmic baryon abundance and some unique constraints on higher energy physics—new species, generations, and decays. It seems indulgent to ask for more than this, but the continuing success and longevity of the standard model suggests the more ambitious goal of seeking additional information about the early universe. Higher energy events at $T \gg 1$ MeV could lead to departures from the standard model that produce observable effects. Indeed, departures from SBBN may turn out to be mandated by standard model physics if instabilities are present at high energy. It is important to know what these effects might be so that we can look for them. It is also important to test how robust are the standard constraints on baryon density, since these are used in many areas of extragalactic astronomy.

In the past few years these goals have been vigorously pursued in both observational and theoretical work extending the scope of the standard model. A recent review of a multitude of nonstandard nucleosynthesis effects can be found in Malaney and Mathews.[2] The present brief review emphasizes in particular the effects of small-scale fluctuations in baryon number, which may be caused by a first-order quantum chromodynamics (QCD) phase transition at ≈ 100 MeV or inhomogeneous baryon generation at ≈ 100 GeV. In the last year there have been a number of important current developments that were not included in previous reviews.[3,4] Most importantly, there is better quantitative information on how far SBBN constraints $\Omega_b h^2 \simeq 0.02 \pm 0.1$ or $N_\nu \simeq 3 \pm 0.2$ can be relaxed, and on the conditions under which one expects observable new effects in cosmic abundances.

The standard model, if taken literally, with current data leads to almost a unique value of the cosmic baryon density: $\Omega_b h_{100}^2 = 0.02$, where h_{100}, the current value of

[a]This work was supported by National Science Foundation Grant AST-9012690, and by an Alfred P. Sloan Foundation fellowship.

Hubble's constant in units of 100 km/sec/Mpc, probably lies between 0.5 and 1, and Ω_b is the fraction of the cosmic critical density in baryons. These are less than fairly reliable dynamical determinations of the mean density ($\Omega \gtrsim 0.2$), leading to the standard conclusion[5] that there must be a substantial component of the total mass density that is not made of ordinary matter: at least 50 percent, and possibly as much as 98 percent if $\Omega_{tot} = 1$, as many cosmologists believe, and $h \simeq 1$, as current observations indicate. On the other hand, since it is much more than all the accounted forms of baryons (from absorption line studies or stellar population accounting, which both yield $\Omega_b \simeq 10^{-3} h_{100}^{-1}$), it also implies that the bulk of ordinary matter (as much as 95 percent of it) is also dark and unobserved. SBBN also places restrictive limits on the number of allowed light neutrino species and other properties of the underlying physical theory. Even if these conclusions turn out to be valid, the arguments used to reach them may not be, so one of the themes here is to examine whether these conclusions are still warranted in a model with plausible amounts of inhomogeneity. In particular, an attractive hypothesis to investigate is whether Ω_b can be 1, so that no exotic forms of dark matter are required.[5,6]

Modifications of SBBN are only well motivated if physics makes such modifications mandatory. It is possible that standard model physics will predict that even a universe that starts out uniform will undergo instabilities that spontaneously and unavoidably produce fluctuations. For some instabilities, our most reliable probe is precisely their imprint on the distribution of matter in the early universe and the abundances it produces. The early universe is very large and long-lived (by about a factor of $M_{\rm Planck}/T \simeq 10^{19}$) compared to a laboratory heavy-ion collision of the same characteristic particle energy, and therefore is much better suited to study collective statistical effects than a laboratory experiment. The amount of spatial extrapolation from the event is actually less in the cosmological case, where for example the universe has expanded by a factor of 10^{12} since the QCD epoch, than in the laboratory, where there is a ratio of 10^{15} between the size of the collision region and the size of the detectors.

PRODUCTION OF FLUCTUATIONS

Inflation provides a natural mechanism for starting the expansion of the universe and giving it an enormous size.[7] The nonuniformity of the universe produced by this generic process is, however, theoretically almost unconstrained. The working hypothesis of most cosmologists is that it is just sufficient to introduce binding-energy fluctuations on a galactic scale that later develop into galaxies, but have a small enough amplitude to leave the homogeneous nucleosynthesis approximation intact.

Inflation leaves a neutral medium behind—one with equal numbers of particles and antiparticles. At some later time, the cosmic baryon number was produced. It is not known how this occurred, although some mechanisms lead inevitably to spatial fluctuations in entropy per baryon. Even if the baryon number is created homogeneously, it is possible that entropy fluctuations appear afterward due to instabilities. These fluctuations could significantly modify element production if they had large amplitude on any scale large enough to survive until nucleosynthesis. Nucleosynthesis thereby probes events at very early times, much earlier than the nominal epoch of microscopic nuclear statistical equilibrium. Two phase transitions may appear in standard model physics, one associated with electroweak symmetry breaking at $\simeq 100$ GeV, the other associated with the quark-to-hadron gas transition at $\simeq 100$ MeV, which could create such fluctuations. There may also be other, earlier transitions, but they are likely to leave fluctuations on too small a scale to affect nucleosynthesis.

The standard model electroweak transition could create the cosmic baryon asymmetry, since it includes the three necessary and sufficient conditions for generating matter–antimatter asymmetry: CP violation, departure from thermal equilibrium, and processes that violate baryon number conservation. For quantitatively plausible scenarios,[8–11] a "nonminimal" Higgs sector is probably required. Unwinding of δn_b-violating knots in $SU(2) \times U(1)$ gauge fields can be guided by a Higgs field that in turn has large-amplitude spatial fluctuations from spinodal decomposition. Since the gradient energy of the Higgs tends to align the field into domains with horizon-scale coherence, one generically expects baryon production to be inhomogeneous with significant amplitude and correlations up to the horizon scale.

The rest mass of baryons contained in the Hubble radius at temperature T is[7]

$$M_{\mathrm{Hb}} \simeq (\hbar c/Gm_p^2)^{3/2} S^{-1} (kT/m_p c^2)^{-3} m_p \simeq 3 \times 10^{-10} M_{\odot} \Omega_b h^2 g_{\mathrm{eff}}^{-1/2} T_{\mathrm{GeV}}^{-3}.$$

Here $S = \eta^{-1} \simeq 10^9$ denotes the mean cosmic ratio of photons to baryons, approximately the "specific entropy," and g_{eff} is the effective number of relativistic degrees of freedom of matter at temperature T. Electroweak baryon production can therefore generate fluctuations with $(\delta\rho/\bar{\rho}) > 1$ on scales up to about $10^{-16}\, \Omega_b h^2 M_{\odot}$; there may also be smaller structures with larger amplitude, depending on the model.

Structures could be produced during a QCD phase transition by a different effect, the concentration of preexisting baryon number.[12] Because their size is determined by nucleation rather than spinodal decomposition, these structures are typically at least a factor of a million smaller than M_{Hb}, so although the horizon is larger, the structures turn out to be similar or even smaller in scale than the earlier electroweak fluctuations.[12–15] They can, however, have very large amplitude, because of a unique "sieve effect."[12] Very high concentrations of baryon number form during the quasi-static part of the transition, when the universe is quietly expanding with both phases coexisting at the critical temperature, and material is gradually being converted from high-density quark matter to low-density hadron matter. Because stable hadrons are heavier than kT at the QCD epoch, baryon number is more "soluble" in the quark phase than in the hadron phase. As in fast quenches in metallurgy (or in the freezing of sea ice), the solute, in this case baryon number, tries to adopt a concentration that will equilibrate its thermodynamic potential in the two phases, so that as material moves from one phase to the other the baryons increase their concentration just behind the interface until their rate of "evaporation" across the interface just keeps up with the fluid flow. This sharp concentration spike (a factor of one or two orders of magnitude) can diffuse (and in some circumstances even convect) back into the quark phase, which can therefore build up a large concentration of baryons. In effect the phase boundary can transmit entropy more effectively than baryon number. The result is large-amplitude fluctuations in entropy.[6,12,15–20]

None of this would happen if there is no first-order phase transition.[21,22] But even if we knew the transition physics well, it would be difficult to estimate exactly how concentrated the baryons get, and how many of them get concentrated, because of several complicating factors, such as the critical role played by (possibly turbulent) fluid transport of baryon number behind the interface. The geometry is not spherical, but a system of percolating networks of both phases with no particular simplifying symmetry. The sieve effect occurs simultaneously with a rapid fluid flow through the phase interface and heat conduction by diffusing species between fluid elements.[6] If conduction were to dominate the transport, dendritic instabilities would also come into play, further complicating the geometry.[5,23]

The rough scale of these perturbations is easier to compute than their detailed structure, as long as it is controlled by classical nucleation theory. Simple estimates for a variety of nucleation models have been made,[15,16] generally yielding a scale of between 30 cm and 300 m for the bubbles. This range of lengthscales corresponds to about $M_b \simeq 10^{-23}$ to $10^{-14}\ M_\odot$. It could be much smaller if nucleation occurs around contaminants, as it almost always does in the laboratory, or if the surface tension is much smaller than expected, in which case the length scale could be much smaller. It is unlikely on general grounds to be larger, however.[14]

DESTRUCTION OF FLUCTUATIONS

Once fluctuations are generated, a variety of processes compete to destroy them.[3,6,24] On the small scales considered here, the basic tendency is for baryons to diffuse toward uniformity; on larger scales, baryonic gravity tends to cause the collapse of baryons into isolated high-density clumps. Destruction occurs at a rate determined by the diffusion of various species: the baryons themselves, photons, and neutrinos. To describe what happens to nonlinear entropy perturbations, one is forced to adopt a simplified nonlinear model; for example, one can follow the behavior of spherical clouds of baryons of a certain baryon mass M_b and baryon density enhancement δ above the mean value. The behavior can be understood through a variety of effects, some of which have been included in detailed computer modeling and others not.

Neutrino Inflation

Low-mass, high-baryon-density regions in pressure equilibrium with their surroundings have a lower photon temperature, so that heat conduction by neutrinos will conduct entropy into such a region and inflate it away. This affects all species equally, since the frame in which the total pressure appears isotropic is slowly expanding in comoving coordinates. It operates only until ν decouples at $T \sim 10^6$ eV, but sets a mass-dependent upper limit on the overdensity of lumps after that time.

Baryon Diffusion

Protons are electrically charged, and couple closely to the electron–photon plasma, whereas neutrons interact with the electron plasma mainly through their magnetic moment and by scattering off of protons.[6,25] As a result, after $T_{weak} \simeq 1$ MeV when inverse beta decays turn off and protons stop spending half of their time as neutrons, the two species move through the plasma at different rates and will start to be distributed differently, moving to fill in areas of low concentration. In general, the neutrons fill more of space, so they can even come to dominate in some areas. Significant neutron–proton segregation can occur if the scale of the lumps is bigger than the baryon diffusion length at $T \simeq 1$ MeV, which is about $10^{-23}\, M_\odot$, but smaller than the neutron diffusion length at $T \simeq 0.1$ MeV, which is about $10^{-17}\, M_\odot$. Interestingly, this range of scales includes those expected from the candidate standard model instabilities just discussed. Since the segregation effect has striking consequences for nucleosynthesis, there is some hope of finding a relic of events as early as $T \simeq 100$ GeV.

Photon Inflation

As in neutrino inflation, photons tend to leak into the low-mass baryon concentrations, inflating the denser ones to a time- and mass-dependent maximum density contrast.[24] Like neutrino inflation this affects all species equally. However, unlike neutrino inflation this process is occurring during nucleosynthesis itself, and operates on scales comparable to baryon diffusion. The possible nucleosynthetic effects of this "photon bomb" have been studied especially by the Livermore group,[26] but is only now being incorporated into numerical studies. This process is efficient at erasing lumps up to $\sim 1\, M_\odot$ before recombination, guaranteeing that the final abundances are spatially uniform.

Bulk Fluid Effects

Significantly overdense and underdense lumps, which occur both in very dense baryon concentrations after the QCD transition and in the segregated phases of different density during the phase transition, accelerate each other due to peculiar gravitational fields. This effect depends on the distribution and correlations between lumps. The drift can cause ablation, and possibly turbulent mixing. As in stars and accretion disks, this can lead to much more efficient particle transport than microscopic diffusion. Needless to say, no exact calculation of such effects is available.

NUCLEOSYNTHESIS WITH FLUCTUATIONS

All of these transport processes are occurring before, during, and after the epoch of nucleosynthesis, which occurs at $T \lesssim 0.1$ MeV, when photodissociation rates become low enough for deuterium to survive and form heavier elements. Lumps larger than about $1S^{1/2}\, M_\odot$ collapse before or at recombination, so that whatever nuclei are made in them do not necessarily get out into the material we see in visible objects; they might even end up in black holes. So, one can arrange to have observed abundances produced in the portion of the matter not in such lumps, which are just like the abundances produced in a uniform, higher entropy universe.[27] Such fluctuations are perhaps implausible as they are too large in size to be produced by noninflationary, standard model physics.

Lumps smaller than this mix with the generally distributed gas we observe today, so their abundance patterns affect what we see. If they are larger than about $10^{-17}M_\odot$ there is not much difference between the neutron and the proton distributions at nucleosynthesis, so the abundances that emerge from the lumps will be like those from a universe of some lower entropy.[28] (Although, if the lump is dense enough to be matter-dominated at nucleosynthesis, this is not strictly true because its expansion rate, fixed by pressure equilibrium to match that of the background universe, is not the same as a real low-entropy universe.)

The most interesting regime is for lumps in the range of sizes $10^{-23} \lesssim M \lesssim 10^{-17} M_\odot$. Below this range, baryons are uniformly distributed before T_{weak}, and no evidence of the lumps gets left behind to affect nucleosynthesis. But in this range of scales there is a significant segregation effect. Neutrons can diffuse across and out of a lump, or across the space between them, after T_{weak} and before or during nucleosynthesis, but protons cannot. In this regime of size one can therefore expect some nucleosynthesis to occur in neutron-dominated regions, a qualitative difference from the usual scenario that leads to some striking differences in the products.[29–44]

SBBN has the virtue of the simplicity of a homogeneous medium: it is a problem of pure statistical physics to compute the abundances. In the nonstandard models, one is suddenly confronted with a qualitative change—a host of fluid dynamical and particle transport processes need to be computed at the same time, and the nuclear reaction network needs to be computed separately for each position in space. The presence of neutron-rich regions also places more rigorous demands on the reaction networks, requiring more neutron-rich species and also heavier ones. In some cases, this has even required the measurement of new laboratory cross sections.[45,46]

For calculational convenience and some analytical clarity, the segregation effects just described were first analyzed in the context of simple two-zone toy models, in which the neutron- and proton-rich regions were assumed to be separately homogeneous and "disconnected" during nucleosynthesis. In reality, of course, neutrons are diffusing during nucleosynthesis itself. This means that a realistic computation must include diffusion for all the reactants, and other species as well, such as photons that can transport heat (the inflation effect previously described), at the same time that the entire network of nuclear reactions is separately computed for every point in the system. Multizone calculations are required so that within each zone conditions are homogeneous enough for a conventional reaction network approximation. State of the art codes now incorporate both good spatial zoning and modern expanded reaction networks.[36,37,47–50] It is found that indeed such sophistication is necessary in order to follow the spatial composition gradients that form as the system evolves. Even these codes do not yet include all of the diffusion effects, but for some abundances interesting results emerge from these studies that are nevertheless likely to stay.

It appears at present that these models can substantially enlarge the acceptable range of entropy per baryon beyond the standard model while at the same time producing acceptable abundances. The main reason for the higher allowed Ω_b is that the presence of neutron-rich regions tends to reduce the ^{4}He abundance below that of a homogeneous model of the same mean entropy, because some neutrons decay before being incorporated into nuclei. This effect is partially canceled by the fact that nucleosynthesis begins earlier in very dense regions, when fewer neutrons have decayed, so that the "back-diffusion" of neutrons into these regions during nucleosynthesis tends to raise the ^{4}He abundance. The most serious problem for the hypothesis that $\Omega_b = 1$ in these models still appears at present to be ^{4}He overproduction. Recent multizone computations[47] show that the reduction of ^{4}He is not sufficient for any choice of overdensity or scale. Although it is possible that this result will change quantitatively with improved diffusion and reaction networks, the basic conclusion seems fairly robust.

The second conspicuous effect of neutron–proton segregation is a much larger production of deuterium at a given entropy than in standard models, again allowing higher Ω_b. This effect arises because nucleosynthesis in a neutron-rich region takes longer than usual, since reactions are paced by the 10-minute neutron-decay timescale. That is, all the protons are quickly consumed, and new nuclei can form only as fast as new protons are introduced by β-decay. Thus some deuterium is produced at $t \gtrsim 1000$ sec, when the temperature and density are low enough to let it survive.

A third effect, a much larger production of elements heavier than Be, also occurs because of the relative paucity of protons. Neutron-mediated reaction chains allow elements to build up to the neutron-rich side of the valley of beta stability, which cannot occur with many protons around; in particular, a proton hitting a ^{11}B gives an unstable ^{12}C, which flies apart into α's, but a neutron can give ^{12}B, which can lead to heavier things. Thus one has for example

$$^4\mathrm{He}(^3\mathrm{H}, \gamma)^7\mathrm{Li}(n, \gamma)^8\mathrm{Li}(\alpha, n)^{11}\mathrm{B}(n, \gamma)^{12}\mathrm{B}(\beta)^{12}\mathrm{C}(n, \gamma)^{13}\mathrm{C}(n, \gamma)^{14}\mathrm{C},$$

instead of the competing process

$$^{11}\mathrm{B}(p, 2\alpha)2\alpha.$$

No triple-alpha process is needed to synthesize heavy elements this way. The famous "^{8}Be gap" that prevents standard big bangs from synthesizing heavy elements via two-body reactions is no longer relevant; the reactions can go around it without having to jump it.

Although it was thought that ^{9}Be might be an interesting relic,[32,34,46] it has now been shown to be easily destroyed by back-diffusion.[49,50] Recent results moreover indicate that although abundances of primordial heavy elements and even unusual isotope ratios are in the observable range for $\Omega_b = 1$ models,[33,35] they appear at present to be unobservable for models with a low enough baryon density to have acceptably low helium production.

The most dramatic effect of segregation is that extremely heavy elements, such as gold or platinum, can be produced in observable quantities.[51,30] The big bang produces a larger neutron flux for a longer time than almost any other astrophysical environment, and might provide a site for "fission cycling" to occur. Analogous to a fission chain reaction, this occurs when a heavy nucleus absorbs so many neutrons so quickly that it grows to the nuclear mass limit and undergoes fission, which leads to two nuclei absorbing the neutrons, and so on until the neutron supply runs out. A tiny initial abundance of seed nuclei can grow to prominence this way by multiple doubling, absorbing more and more of the remaining neutrons, with a final abundance proportional to 2^N, where N is the number of fission cycles. The most striking feature of the resulting abundance pattern is that only elements heavier than a typical fission product (e.g., barium) experience the doubling effect;[52] one finds, for example, substantially more elements heavier than barium, such as gold, platinum, and rare earth elements, than lighter r-process elements, such as strontium. Non-r-process heavy elements, such as iron, are practically negligible. Unfortunately the exponential behavior here makes calculation of the abundances extremely sensitive to uncertainties in nuclear physics, particularly the decay rates of heavy neutron-rich nuclei that determine the time it takes for one seed to grow to the fission limit, and hence determine N.

A recent improved calculation[53,54] includes a large reaction network with modern updated rates and is considerably more reliable than earlier approximate treatments. According to the current state of the art, cosmic r-process is not quite efficient enough to produce easily observable quantities at neutron densities low enough to get the light elements right. In addition, the abundance pattern is not quite that of the classical r-process, since the neutron flux is not high enough to stay far to the right of the valley of β stability.

Since the efficiency of these effects is an exponential function of density, there are certainly some geometries where r-process products would be conspicuous without conflicting with light element abundances; for example, very high-amplitude fluctuations of appropriate scale in a very small fraction of the matter. Such fluctuations are, however, not "natural" for the particular production mechanisms we discussed before.

A controversial issue surrounding these models involves ^{7}Li, which is minimally produced ($\simeq 10^{10}$ by number) in the standard model, agreeing with the abundance in metal poor stars.[55] The controversy previously involved both observational and theoretical uncertainty. Observationally, the situation is much clearer than it was; the evidence has steadily mounted that the small population II abundance is indeed the primordial one.[56–58] This has also been confirmed by stellar evolution models.[59,60] On the cosmological side, however, it is now known that the Li abundance is very sensitive to photon diffusion effects during nucleosynthesis, which have not been included in the calculations, so at present the test is inconclusive.[26]

CONCLUSIONS

Most results so far indicate that nonstandard models have the flexibility to allow both higher and lower Ω_b than SBBN, up to the observed dynamical density and down to the observed baryonic density. The SBBN conclusions that there must be abundant baryonic and nonbaryonic dark matter can thus be significantly changed without adding nonstandard physics or unnatural initial conditions. On the other hand, given the wide range of abundance patterns that has emerged from these calculations, the one that fits everything most convincingly is still something close to SBBN.

The range of acceptable model parameters will be better quantified by new codes that include photon inflation, which mitigates the Li problem, as well as neutrino inflation, which eliminates very large density contrasts and therefore aggravates the

He problem. The conclusion that ^{4}He is not consistent with $\Omega_b = 1$ for any parameters of current models is likely to remain true when photon and neutrino inflation effects are included. Moreover, detectable primordial CNO is produced if and only if $\Omega_b \approx 1$, and even then *r*-process doubling, although it occurs, is also not detectable unless the character of fluctuations is unexpectedly non-Gaussian. This is bad news if we want to use the abundances not just as a probe of the mean entropy but as a probe of the cosmic transitions; if Ω_b is small, the distinctive isotopic abundances of the primordial heavy elements may be too scarce to be found. The program of obtaining information from the first microseconds may be limited to the small departures, if any, in the light element abundances from SBBN. The future course of these studies is thus likely to be dictated by increasingly precise abundance data. Nonstandard effects may be indicated if small tuning is needed in SBBN—for example, to reduce the He abundance by 1 or 2 percent, as tentatively indicated by current data. Small but significant departures from SBBN predictions should not be regarded as cause for despair, but for rejoicing; such a small failure in SBBN, while preserving our essential understanding of the first few minutes, would expand the scope and reach of cosmological nucleosynthesis to even earlier times.

REFERENCES

1. PAGEL, B. E. J. 1991. Primordial nucleosynthesis and observed light element abundances. This issue.
2. MALANEY, R. A. & G. J. MATHEWS. 1991. Phys. Rep. In press.
3. HOGAN, C. 1990. *In* Workshop on Primordial Nucleosynthesis, W. J. Thompson, B. W. Carney, and H. J. Karwowski, Eds.: 15. World Scientific. Singapore.
4. ———. 1987. *In* Nearly Normal Galaxies: From the Planck Time to the Present, S. Faber, Ed.: 378. Springer-Verlag. New York/Berlin.
5. LYNDEN-BELL, D. & G. GILMORE, Eds. 1990. Baryonic Dark Matter. Kluwer. Dordrecht, the Netherlands.
6. APPLEGATE, J. H. & C. J. HOGAN. 1985. Phys. Rev. D. **31:** 3037.
7. KOLB, E. & M. S. TURNER. 1990. The Early Universe. Addison-Wesley. Reading, Mass.
8. BOCHKAREV, A. I., S. YU. KHLEBNIKOV & M. E. SHAPOSHNIKOV. 1990. Nucl. Phys. **B329:** 493.
9. MCLERRAN, L. 1989. Phys. Rev. Lett. **62:** 1075.
10. MCLERRAN, L., M. SHAPOSHNIKOV, N. TUROK & M. VOLOSHIN. 1991. Princeton preprint, PUPT 1224-90, Princeton, N.J.
11. TUROK, N. & J. ZADROZNY. 1990. Phys. Rev. Lett. **65:** 2331.
12. WITTEN, E. 1984. Phys. Rev. D **30:** 272.
13. HOGAN, C. 1982. Astrophys. J. **252:** 418.
14. ———. 1983. Phys. Lett. **133B:** 172.
15. KAJANTIE, K. & H. KURKI-SUONIO. 1986. Phys. Rev. **D34:** 1719.
16. FULLER, G. M., G. J. MATHEWS & C. R. ALCOCK. 1988. Phys. Rev. D. **37:** 1380.
17. BONOMETTO, S. A., P. A. MARCHETTI & S. MATARRESE. 1985. Phys. Lett. **157B:** 216.
18. KURKI-SUONIO, H. 1985. Nucl. Phys. **B255:** 231.
19. MILLER, J. C. & O. PANTANO. 1989. Phys. Rev. D **40:** 1789; ———. 1990. Phys. Rev. D **42:** 3334.
20. SUMIYOSHI, K., K. KUSAKA, T. KAMIO & T. KAJINO. 1989. Phys. Lett. B **225:** 10.
21. SATZ, H. 1991. The quark–hadron phase transition. This issue.
22. BROWN, F. R., *et al.* 1990. Phys. Rev. Lett. **65:** 2491.
23. FREESE, K. & F. C. ADAMS. 1990. Phys. Rev. D **41:** 2449.
24. HOGAN, C. 1978. Mon. Not. R. Astron. Soc. **185:** 889.
25. BANERJEE, B. & S. M. CHITRE. 1991. Phys. Lett. B. In press.

26. ALCOCK, C. R., D. S. DEARBORN, G. M. FULLER, G. J. MATHEWS & B. S. MEYER. 1990. Phys. Rev. Lett. **64:** 2607.
27. SALE, K. E. & G. J. MATHEWS. 1986. Astrophys. J., Lett. **309:** L1.
28. EPSTEIN, R. & V. PETROSIAN. 1975. Astrophys. J. **197:** 281.
29. ALCOCK, C. R., G. M. FULLER & G. J. MATHEWS. 1987. Astrophys. J. **320:** 439.
30. APPLEGATE, J. H., C. J. HOGAN & R. J. SCHERRER. 1988. Astrophys. J. **329:** 572.
31. ———. 1987. Phys. Rev. D. **35:** 1151.
32. BOYD, R. N. & T. KAJINO. 1989. Astrophys. J., Lett. **336:** L55.
33. KAJINO, T., G. J. MATHEWS & G. M. FULLER. 1990. Astrophys. J. **364:** 7.
34. KAJINO, T. & R. N. BOYD. 1991. Astrophys. J. In press.
35. KAWANO, L. H., W. A. FOWLER, R. W. KAVANAGH & R. A. MALANEY. 1990. Astrophys. J. In press.
36. KURKI-SUONIO, H., R. A. MATZNER, J. M. CENTRELLA, T. ROTHMAN & J. R. WILSON. 1988. Phys. Rev. D. **38:** 1091.
37. KURKI-SUONIO, H. & R. A. MATZNER. 1989. Phys. Rev. D. **39:** 1046.
38. KURKI-SUONIO, H., R. A. MATZNER, K. A. OLIVE & D. N. SCHRAMM. 1990. Astrophys. J. **353:** 406.
39. KURKI-SUONIO, H., R. A. MATZNER, J. M. CENTRELLA, T. ROTHMAN & J. R. WILSON. 1988. Phys. Rev. D. **38:** 1091.
40. MALANEY, R. A. & W. A. FOWLER. 1987. *In* The Origin and Distribution of the Elements, G. J. Mathews, Ed.: 76. World Scientific. Singapore.
41. MALANEY, R. A. & W. A. FOWLER. 1988. Astrophys. J. **333:** 14.
42. ———. 1989. Astrophys. J., Lett. **345:** L5.
43. TERASAWA, N. & K. SATO. 1989. Phys. Rev. D **39:** 2893.
44. ———. 1989. Prog. Theor. Phys. **81:** 1085.
45. BUNE, C. R., R. W. KAVANAGH, S. E. KELLOG & T. R. WANG. 1991. Phys. Rev. C. In press.
46. FOWLER, W. A. 1990. *In* Baryonic Dark Matter, D. Lynden-Bell and G. Gilmore, Eds.: 257. Kluwer. Dordrecht, the Netherlands.
47. KURKI-SUONIO, H. & R. A. MATZNER. 1990. Phys. Rev. D. **42:** 1047.
48. MATHEWS, G. J., B. MEYER, C. R. ALCOCK & G. M. FULLER. 1990. Astrophys. J. **358:** 36.
49. TERASAWA, N. & K. SATO. 1990. Astrophys. J., Lett. **362:** L47.
50. ———. 1990. Astrophys. J. **355:** 18.
51. APPLEGATE, J. H. 1988. Phys. Rep. **163:** 41.
52. SEEGER, P. A., W. A. FOWLER & D. D. CLAYTON. 1965. Astrophys. J., Suppl. Ser. **11:** 121.
53. THIELEMANN, F.-K., J. H. APPLEGATE, J. J. COWAN & M. WIESCHER. 1991. Astrophys. J. In press.
54. COWAN, J. J., F.-K. THIELEMANN & J. W. TRURAN. 1991. Phys. Rep. In press.
55. SPITE, F. & M. SPITE. 1982. Astron. Astrophys. **115:** 357.
56. REEVES, H., J. RICHER, K. SATO & N. TERASAWA. 1990. Astrophys. J. **355:** 18.
57. REBOLO, R., P. MOLARO, C. ABIA & J. E. BECKMAN. 1988. Astron. Astrophys. **193:** 193.
58. REBOLO, R., P. MOLARO & J. E. BECKMAN. 1988. Astron. Astrophys. **192:** 192.
59. DELIYANNIS, C. P., P. DEMARQUE, S. D. KAWALER, L. M. KRAUSS & P. ROMANELLI. 1989. Phys. Rev. Lett. **62:** 1583.
60. DELIYANNIS, C. P., P. DEMARQUE & S. D. KAWALER. 1990. Astrophys. J., Supp. Ser. **73:** 21.

X-rays from Galactic Black Hole Candidates and Active Galactic Nuclei

H. INOUE

Institute of Space and Astronautical Science
3-1-1, Yoshinodai, Sagamihara
Kanagawa 229, Japan

INTRODUCTION

There are three binary X-ray sources thus far for which the lower mass limit of the compact object exceed three solar masses (a theoretical upper limit of neutron star mass[1]). Hence, these sources are suspected to be black holes and are Cyg X − 1 ($>9.5M_{\odot}$[2]), LMC X − 3 ($>7M_{\odot}$[3]), and A0620 − 00 ($>3.2M_{\odot}$[4]). LMC X − 1 is also included in the list of black hole candidates, according to the considerably large mass lower limit of $2.6M_{\odot}$.[5] For the details of the mass determinations, see the review by McClintock.[6]

Two distinct characteristics of Cyg X − 1 have been suspected as possible signatures of black hole sources. One is the existence of rapid and chaotic intensity fluctuations. The other consists of two distinct states: a high-intensity state (ultrasoft spectrum + hard tail) and a low-intensity state (hard, power-law spectrum). On this basis, GX339 − 4 and Cir X − 1 have been considered to be black hole candidates, although their lower mass limits have never been obtained. However, recent observations with Tenma and EXOSAT have shown that neutron star sources sometimes exhibit those characteristics that have been believed as black hole signatures. In particular, Cir X − 1 was found to produce type I bursts,[7] which is considered solid evidence for a neutron star.

The high throughput and the low background of the large-area counter on board Ginga[8] have enabled us to obtain, highly accurately, temporal and spectral properties of not only galactic black hole candidates (GBHCs) but also active galactic nuclei (AGNs), which are believed to contain black holes at the center (for the GBHC and AGN observations with Ginga, see reviews by Tanaka[9] and by Inoue,[10] respectively). Based on these results, we shall reexamine the characteristics of black hole candidates by comparing X-ray properties of GBHCs with those of neutron star sources, and also with those of AGNs.

GALACTIC BLACK HOLE CANDIDATES

Ultrasoft Component in the High State

Spectra of black hole candidates in the high state generally comprise two components: a soft component represented by a disk blackbody spectrum[11] and a hard component approximated by a power law.

Ginga repeatedly observed the black hole candidate LMC X-3 for about 3 years. FIGURE 1 shows the light curve of LMC X − 3 and the ratio of 9–19-keV flux to 5–9-keV flux as a function of the 1–19-keV flux.[12] As seen in the lower panel of FIGURE 1, the hardness ratio is almost independent of the total flux. This means that the soft and hard components change independently of each other, since the fluxes below and above 9 keV roughly represent the fluxes of the soft and hard components, respectively.

Ebisawa *et al.*[12] fitted a number of spectra at different epochs from LMC X − 3 to a model comprising a disk blackbody spectrum for the soft component and a power

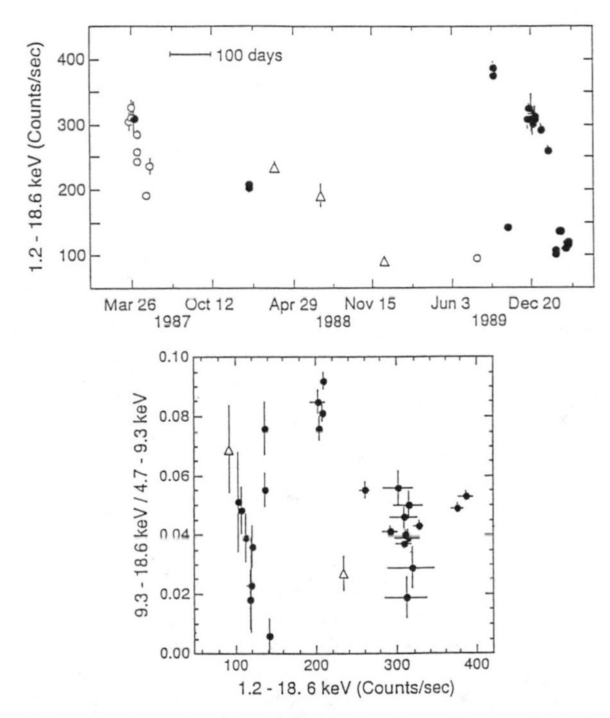

FIGURE 1. A light curve (**upper**) and an intensity-hardness diagram (**lower**) of LMC X − 3 observed with Ginga.[12] *Filled circles, open circles,* and *triangles* denote pointing, scanning, and maneuvering observations, respectively. Errors are 90 percent confidence limits.

law, including a broad absorption feature (which is discussed later) for the hard component, and obtained the histories of the disk parameters during the observation (see FIG. 2). A remarkable finding is that the size of the region that emits the soft component remains virtually constant against the large flux change. Furthermore, the typical radius of the soft-component-emitting region is close to the Schwartzschild radius of a black hole with several solar masses, with the assumption that the distance to the source is 50 kpc, although there remain ambiguities in the deviation of the observed spectrum from the true blackbody spectrum and in the inclination angle of the disk to the observer. Ebisawa, Mitsuda, and Hanawa[13] and Ebisawa[14] showed

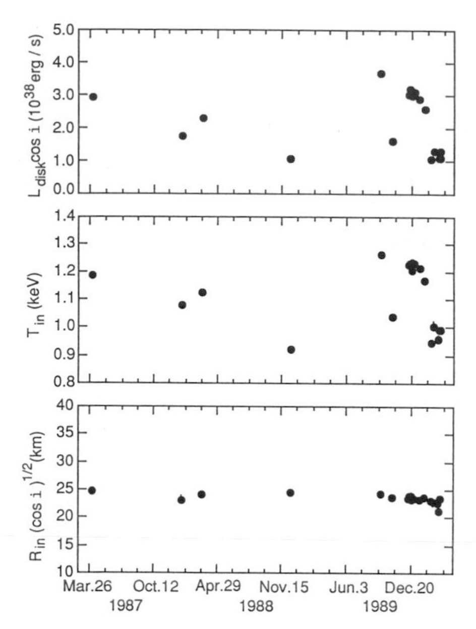

FIGURE 2. Time histories of best fit parameters to the spectra of LMC X − 3; (**top**) bolometric flux of the disk blackbody component, (**middle**) apparent temperature at the innermost radius of the multicolor disk, and (**bottom**) apparent innermost radius of the disk for the distance of 50 kpc. Errors (90 percent confidence limit) are, in most cases, included in *filled circles; i* is the inclination angle of the disk toward us.

that the spectrum of the soft component is consistent with the spectrum expected from an optically thick accretion disk around a black hole and the innermost radius of the disk derived from the spectral fitting is consistent with the innermost Keplerian circular orbit around the black hole.

The constancy of the innermost radius of the optically thick disk emitting the soft component has been observed also from a soft X-ray transient GS2000 + 25 discovered by Ginga.[15]

The X-ray light curve[15] of GS2000 + 25 is very similar to that of the black hole candidate A0620 − 00.[16,17] The spectral change of GS2000 + 25 is shown in FIGURE 3, where the spectra apparently comprise an ultrasoft and a hard, power-law component. These properties suggest that GS2000 + 25 is a possible black hole candidate.

Each of the observed spectra was fitted to a model composed of a disk blackbody component and a power-law component, and the obtained time histories of the disk parameters are shown in FIGURE 4.[18] As seen in this figure, the disk blackbody flux decays fairly smoothly, while the hard component changes almost independently of the disk blackbody flux. Furthermore, we can also see that the innermost radius of the optically thick accretion disk remains virtually constant against the flux change over about two orders of magnitude. The radius is again consistent with the innermost Keplerian circular orbit around a black hole with several solar masses[14] on the assumption that the distance to GS2000 + 25 is 2–3 kpc based on a comparison of the optical counterpart with that of A0620 − 00 ($\sim$1 kpc).

For that distance, the luminosity at the peak of the outburst exceeds 10^{38} erg s^{-1} and is close to the Eddington limit of a several-solar-masses star. The smooth decay of the disk blackbody component of GS2000 + 25 suggests that the flux of the soft component is most likely proportional to the accretion rate from far outside. In the canonical model of accretion disks, the innermost radius of the optically thick accretion disk recedes as the accretion rate increases close to the critical rate (reference 19 and references therein). However, these observations are inconsistent with this model.

Hard Component in the High State

The preceding discussion suggests that an optically thick accretion disk always extends to the innermost Keplerian circular orbit, independently of the accretion rate, when the source is in the high state. A question arises as to where the hard, power-law component comes from. If the hard component came from the outer side of the optically thick accretion disk, the flux of the hard component should have been proportional to the accretion rate, that is, the flux of the soft component.

On a time scale as short as or shorter than a few hundred seconds, the hard component is highly variable, whereas the soft component is relatively stable (see FIG. 5). The larger variability of the higher energy X-rays is also observed from weakly magnetized neutron stars [low-mass X-ray binaries (LMXB)] in the normal

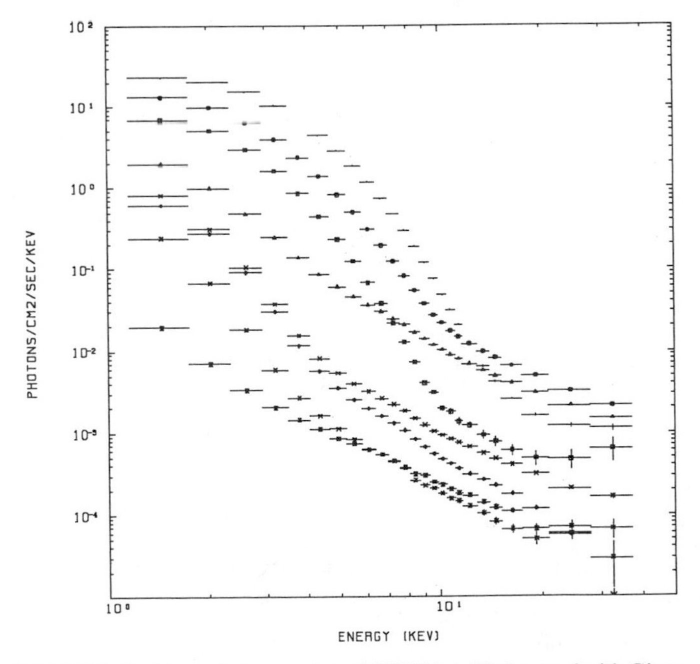

FIGURE 3. Incident photon spectra of GS2000 + 25 observed with Ginga.

branch.[11] Including this, there are several similarities between GBHC in the high states and LMXB in the normal branch.

1. The spectrum consists of two components; a soft component and a hard component.
2. The soft component can be reproduced by a disk blackbody spectrum, and the innermost radius of the emission region is consistent with the innermost Keplerian circulrar orbit.[13,14]
3. The hard component is highly variable, whereas the soft component is

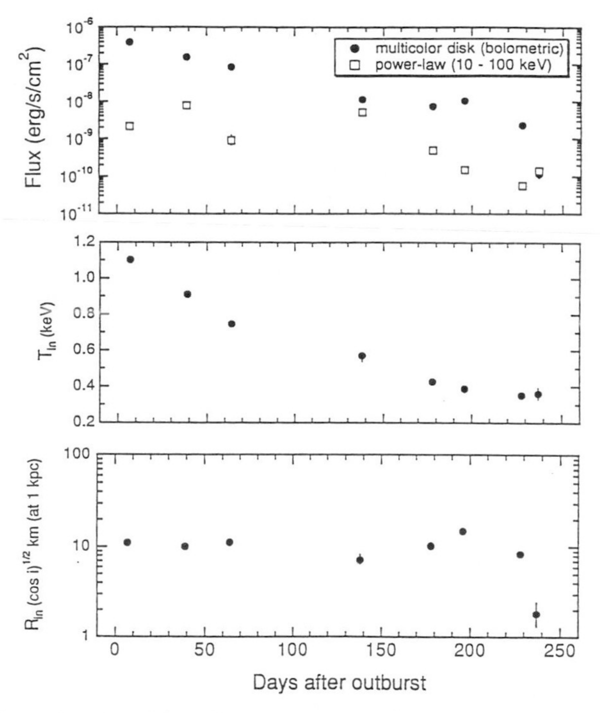

FIGURE 4. Time histories of best fit parameters to the spectra of GS2000 + 25; (**top**) bolometric flux of the disk blackbody component and the integrated (10–100-keV) flux of the power-law component, (**middle**) apparent temperature at the innermost radius of the disk, and (**bottom**) innermost radius of the disk for an assumed distance of 1 kpc.

relatively stable on a time scale as short as or shorter than a few hundred seconds.

4. Quasi-periodoc oscillations are often observed (for neutron star sources, see the review by Lewin, Van Paradijs, and van der Klis;[21] for LMC X − 1, Ebisawa, Mitsuda, and Inoue;[22] for GX339-4, Motch *et al.*[23] and Kitamoto[24]).

The main difference between them seems to exist only in the spectrum of the hard component: a power law with an energy index of about 1 in the case of GBHC, and a blackbody with temperature of about 2 keV in the case of LMXB.

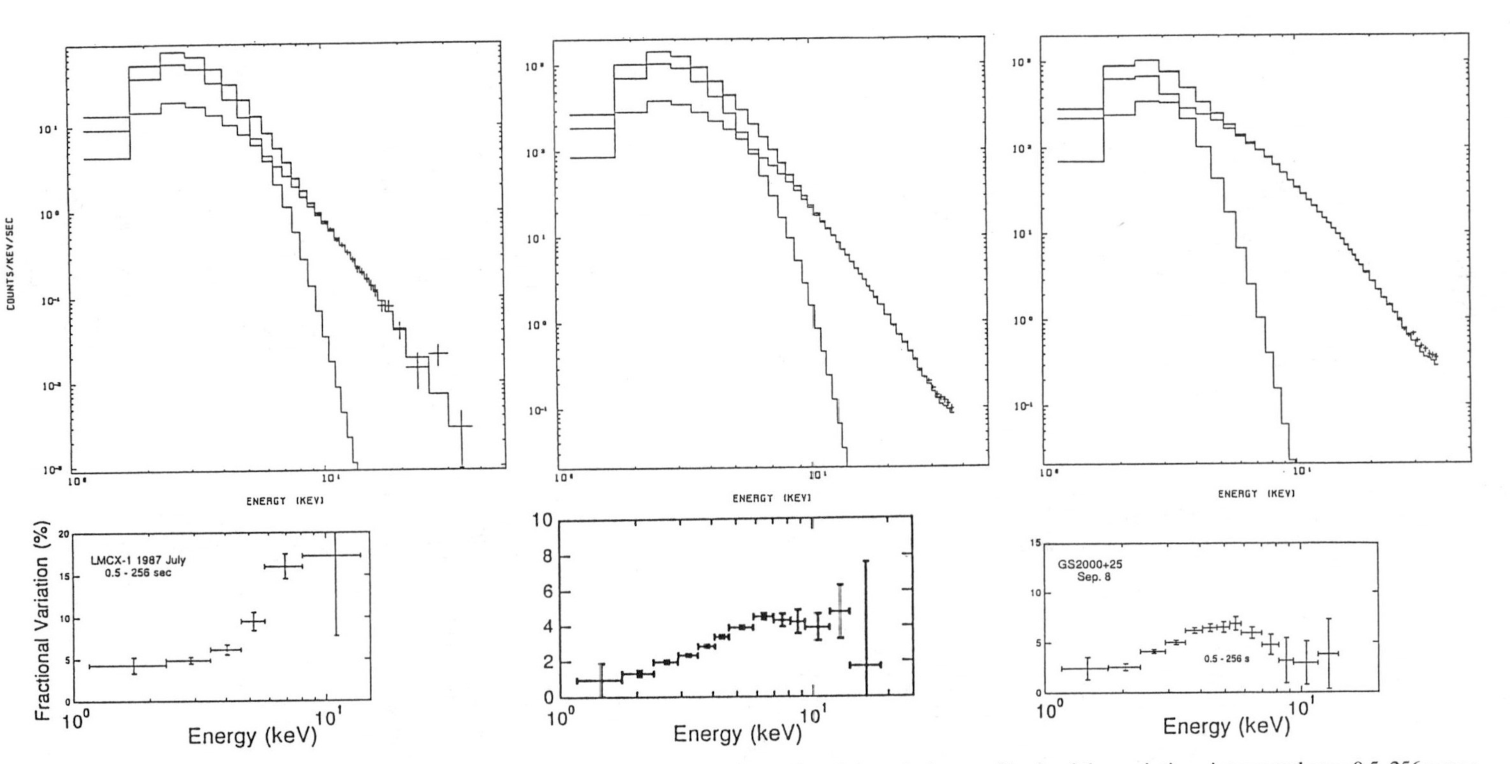

FIGURE 5. The two component fits to the observed pulse-height spectra (**upper panel**) and the relative amplitude of the variations integrated over 0.5–256 sec as a function of energy (**lower panel**) for LMC X − 1,[14] GX339-4,[14,23] and GS2000 + 25.[18]

In the case of LMXB, the 2-keV blackbody emission is understood to come from the surface of a neutron star.[11] Matter circulating along Keplerian orbits in the accretion disk gradually falls inward, and the released gravitational energy is converted in equal amounts to thermal and rotational energy according to the virial theorem. If the accretion disk is optically thick, the thermal energy will be radiated away from the disk surface. However, most of the rotational energy of accreted matter can be released only when the matter lands on the neutron star surface. The former and the latter probably correspond to the soft, disk blackbody component and the hard, 2-keV blackbody component, respectively.

A straightforward analogy from the neutron star case to the black hole case is that the power-law component from GBHC comes from inside of the innermost Keplerian circular orbit. The essential difference in the accretion flow around a black hole compared with a neutron star is the absence of a solid surface. This could be the reason why the 2-keV blackbody component is replaced by the power-law component. However, two serious questions remain.

1. What converts the gravitational energy of the accreted matter to the radiation inside the innermost Keplerian circular orbit? Since centrifugal force cannot prevent matter from free-fall any longer in this region, some other force may be necessary to do it.
2. What causes the variability of the power-law component, independent of the mass accretion rate from far outside? Since the time scale of the variability is much longer than the free-fall time scale near the Schwartzschild radius, the variability should be generated in an optically thick disk.[25] However, the independence of the soft and hard components is clearly seen even on a time scale longer than a few days (see FIGS. 1 and 4). It will be very unlikely for an optically thick disk near the innermost Keplerian circular orbit to modulate the accretion rate on such a long time scale.

Low State

In the low state, the variable power-law component still exists, whereas the stable ultrasoft component seems to disappear. The absence of the ultrasoft component can be understood as due to the receding of the inner boundary of the optically thick accretion disk. In fact, some features superposed on the power-law spectra, as discussed later, are consistent with the presence of an optically thick accretion disk far outside the central X-ray source.

The accretion rate has been believed to determine whether a source is in the high or low state. This was the reason why the states were named as the high and low (-intensity) states. However, recent Ginga observations of two bright transient sources, GS2000 + 25 and GS2023 + 33,[26] have undermined this belief.

As already shown, the peak luminosity of GS2000 + 25 is considered to be close to the Eddington limit. Hence, GS2000 + 25 is regarded to have always been in the high state while the luminosity has decayed by about two orders of magnitude from near the Eddington limit (see FIG. 4).

On the other hand, GS2023 + 33 seems to have always stayed in the low state while the luminosity changed by over three orders of magnitude. FIGURE 6 shows the

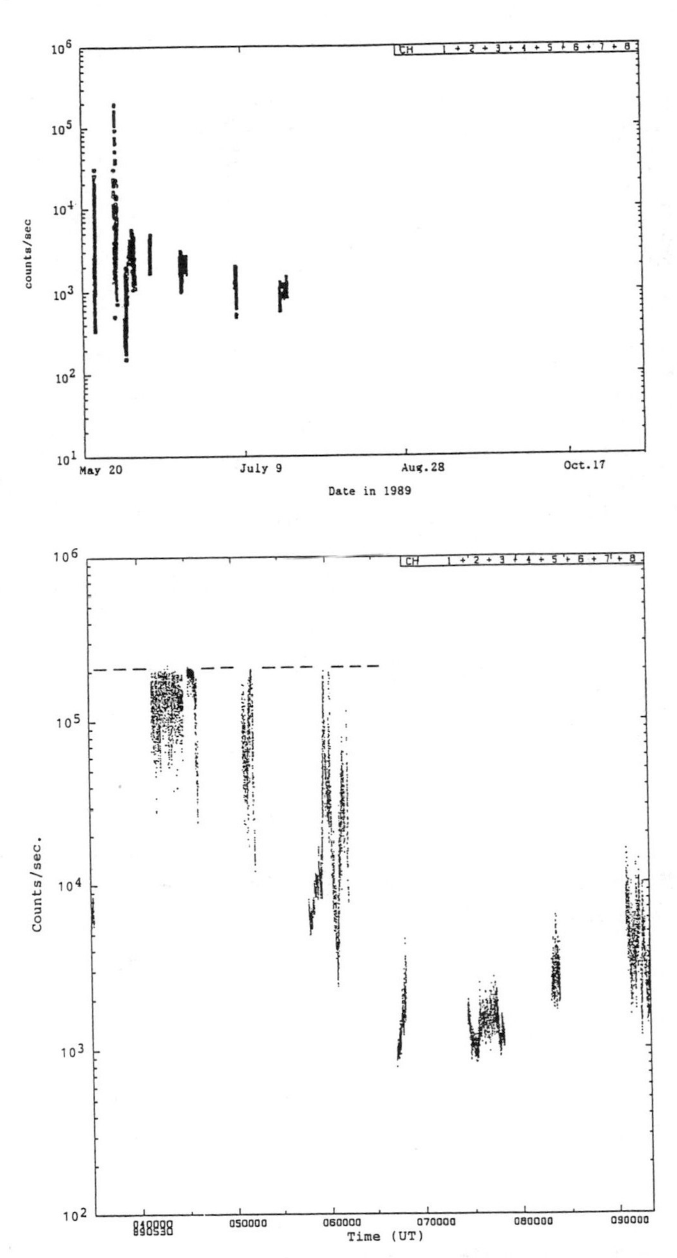

FIGURE 6. X-ray light curves of GS2023 + 33 over six months (**upper panel**) and on May 30 (**lower panel**) in 1989 observed with Ginga.[27] A *dashed line* indicates a possible luminosity saturation.

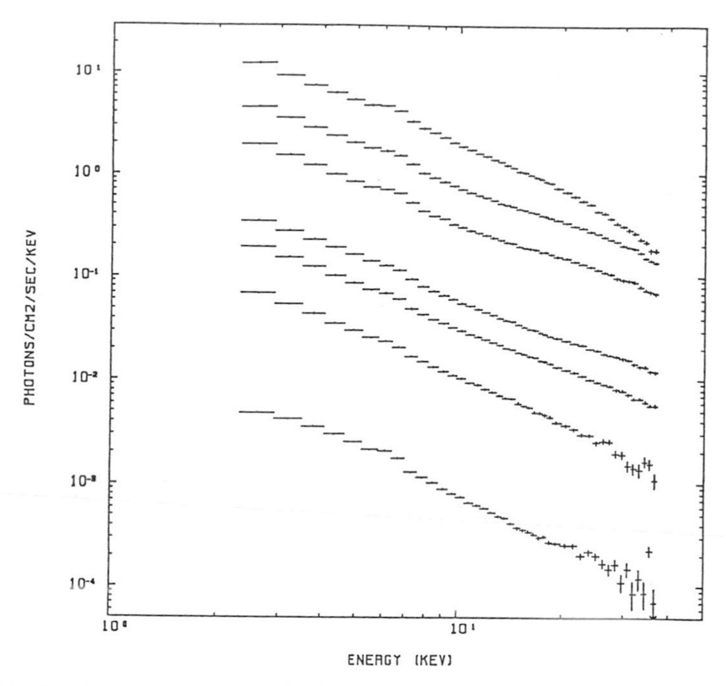

FIGURE 7. Incident photon spectra of GS2023 + 33 observed with Ginga when the effect of absorption is relatively small.

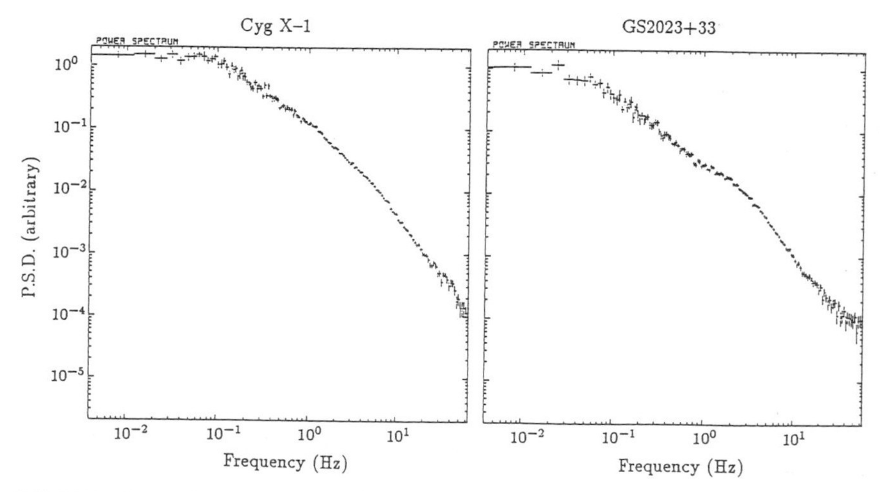

FIGURE 8. Power density spectra of X-rays from Cyg X − 1 in the low state and GS2023 + 33 observed with Ginga.

light curve of GS2023 + 33 on two different time scales.[27] As seen in the lower panel of FIGURE 6, the flux apparently saturated at the peak of the outburst. For the distance of 1–3 kpc evaluated from optical observation,[28] the saturation level is consistent with the Eddington limit of a several-solar-masses star. However, GS2023 + 33 never showed a high-state spectrum throughout the decay by over three orders of magnitude (see FIG. 7), although GS2023 + 33 sometimes suffered from heavy absorption and revealed a complicated spectrum (for an example, see FIG. 12) during the first ten days after the discovery. Both energy and power-density spectra are very similar to those of Cyg X − 1 when heavy absorption is absent (see FIGS. 8 and 10), and thus this source is also regarded as a possible black hole candidate.

SPECTRAL COMPONENTS COMMON TO GBHC AND AGN

Power-law Component

As shown before, a power-law component always exists in the spectra of GBHCs irrespective of whether the source is in the high state or in the low state and also in those of AGNs.

The spectral energy-indices of the power-law component in the 2–30-keV range are listed for some GBHCs in TABLE 1, and are plotted for AGNs in FIGURE 9 as a function of the redshift factor. The indices are all distributed in a fairly narrow range around 0.5–1.0 except for some BL Lac objects and are suggested to be independent of the mass of the central black holes. The spectral slope seems to be independent also of the mass accretion rate. As seen in FIGURE 7, this rate remained virtually constant in GS2023 + 33 during the flux change over about three orders of magnitude.

The power-law spectrum is known to extend down to the optical range in the case of GX339 − 4 in the low state,[29] as well as in the case of AGNs.[30] At the same time, it extends up to at least a few hundred kiloelectronvolts [for GBHCs, Suyaev (presented at this conference, but not included in this issue); for AGNs, e.g., Rothschild *et al.*[31]]. These suggest that the energy range of photons in the power-law component is also independent of the mass of the central black holes.

The extension of the spectrum up to a few hundred kiloelectron volts implies the presence of an optically thin hot region, where most of the electrons have energies higher than several hundred kiloelectronvolts around the central black hole. If the gravitational potential near the Schwartzschild radius directly governs the emission

TABLE 1. Spectral Energy-Indices of Black Hole Candidates

Source Name	Energy Index
Cyg X − 1 (low state)	0.6–0.8
GX339 − 4 (low state)	0.6
GS2000 + 25	0.8–1.2
GS2023 + 33	0.3–0.5
AGNs	0.3–1.3 (Average ~ 0.7)

mechanism, the independence of the power-law component from the central mass and from the mass accretion rate can be understood.

Comptonization of soft photons by the high-energy electrons is one possibility. However, Miyamoto *et al.*[32] discovered a large time lag in the variations between soft and hard X-rays from Cyg X − 1, and found that it is long enough to reject the Comptonization model for the power-law component. Furthermore, it will be difficult not only for the Comptonization model but also for any models to explain the independence of the spectral slope from the mass of the central black hole and from the accretion rate. The problem for the emission mechanism of the power-law spectrum is still open to future study.

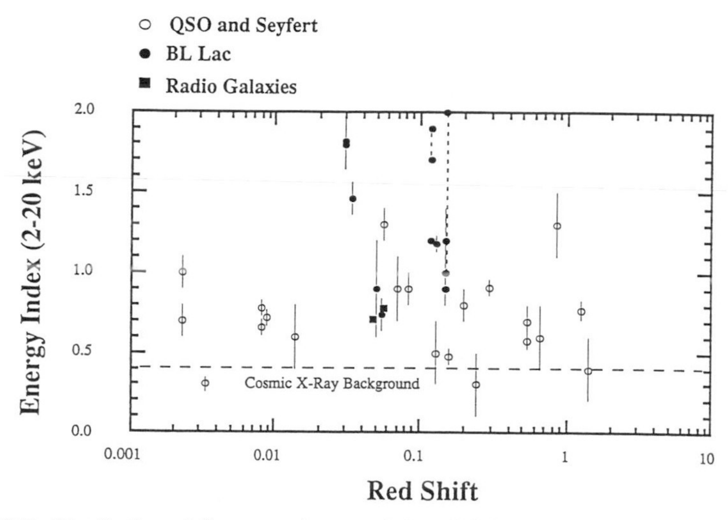

FIGURE 9. Distribution of the spectral energy-index of AGNs observed with Ginga vs. the cosmological redshift, *z*.

Reflected Component

Ginga observations of GBHCs and Seyfert galaxies have revealed that the power-law component cannot be described by a single power law alone, but that in general an emission line feature at 6–7 keV is required as well as a shallow and broad absorptionlike feature at energies from 7 to 15 keV and that the power law flattens at higher energies. FIGURE 10 shows residual plots of the observed spectrum minus a single power-law model for Cyg X − 1, GS2023 + 33, and the average of twelve AGNs;[33] the presence of these features is evident.

These spectral features can be reproduced very well by superposition of a component reflected by cool Thomson-thick matter from a single power law. The importance of X-ray reflection by cool matter around the central engine has been pointed out for AGN spectra by Lightman and White[34] and Guilbert and Rees.[35] FIGURE 11 shows results of a Monte Carlo simulation for the X-ray reflection by neutral matter with cosmic abundances of the elements. When X-rays illuminate

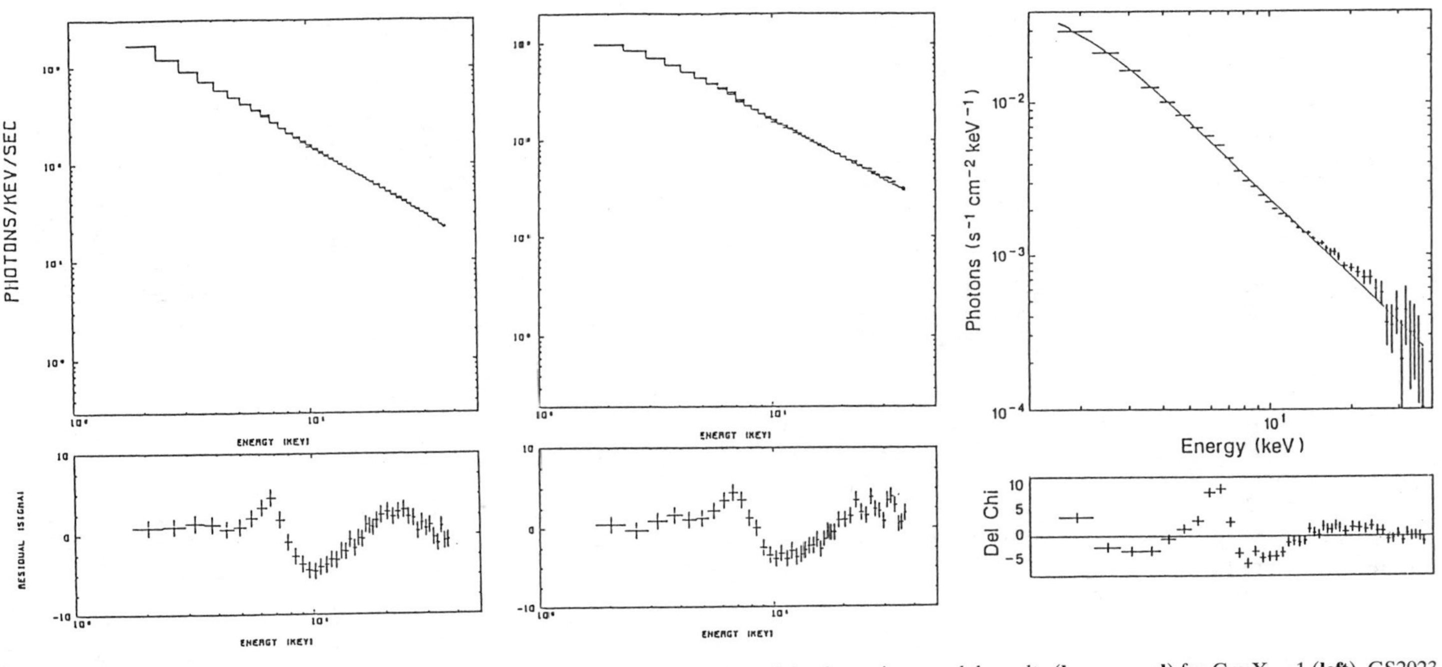

FIGURE 10. Simple power-law fits to the spectra (**upper panel**) and residual plots of the data minus model results (**lower panel**) for Cyg X − 1 (**left**), GS2023 + 33 (**center**), and an average of twelve AGNs (**right,** Pounds *et al.*[33]) observed with Ginga.

cool matter, a certain fraction of them are reflected through Thomson scatterings. Since the fraction is, roughly speaking, in proportion to the ratio of the Thomson scattering cross section to the total cross section, including photoionization, the albedo becomes higher for higher X-ray energies, and an absorption edge appears at the *K*-edge energy of iron. When the absorption-edge structure in the reflected component is superposed on the direct, power-law component, the apparent depth of the absorption edge should saturate at the flux level of the direct component, and

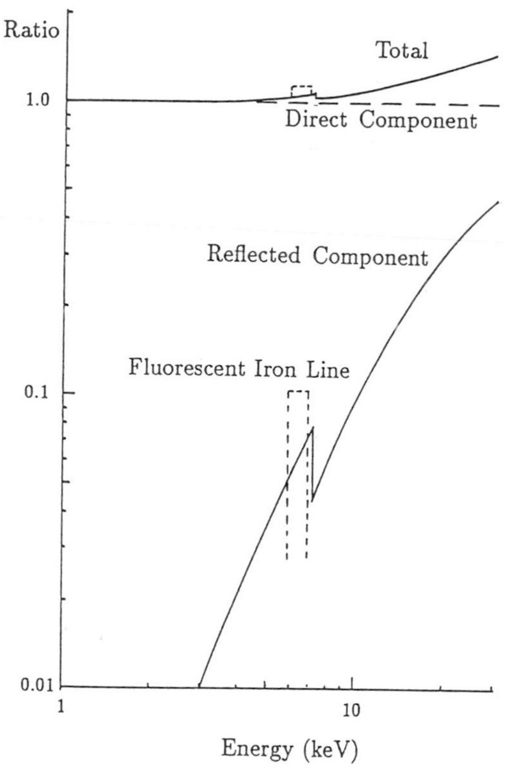

FIGURE 11. Ratio of X-rays reflected from a slab to incident X-rays as a function of energy, for an incident angle of 80 degrees and a solid angle of 2π.[10] The contribution of fluorescent iron line photons are also indicated, assuming the linewidth to be 1 keV.

the absorption feature will be shallow and broad in the energy range above the *K*-edge energy. At the same time, a fraction of illuminated X-rays with energy above the iron *K*-edge energy is absorbed by iron, and is reemitted as fluorescent iron line photons. The emission line feature at 6–7 keV, the shallow and broad absorption feature in the 7–15-keV range, and the excess above 15 keV, all of which are commonly observed in GBHCs and AGNs, are all consistent with the presence of a reflected component in their spectra.

The dominant appearance of the reflection spectrum can be realized when the central region that emits the power-law component is highly obscured by thick matter. The ratio between the reflection and the direct fluxes from Seyfert I galaxies indicates that cool, thick matter around the central engine has a covering factor as large as or larger than 50 percent. This suggests that there is a fairly large probability that the thick matter responsible for the reflection hides the central engine from us. Lack of optical broad lines of Seyfert II galaxies is understood by introducing obscuration of the central part of the AGN, and, in fact, X-ray spectra of these galaxies observed with Ginga show clear evidence for high obscuration.[36,37]

As seen in FIGURE 7, the reflected component seems to always exist in the spectrum of GS2023 + 33, while the flux changes by about three orders of magnitude. The permanent presence of the cool reflector against the large accretion rate change indicates that the accretion disk is a most likely candidate for the reflector. In fact, rapid variability of the iron fluorescent line from the Seyfert galaxy NGC6814[38] suggests that the reflector is fairly close to the Schwartzshild radius. The iron line flux has been found to follow very well the continuum flux, and to be delayed by less than 256 sec. The upper limit on the time delay indicates that the iron line emitting region is no farther than 10^{13} cm from the central X-ray source, while the Schwartzschild radius is no less than about 3×10^{10} cm, unless the luminosity exceeds the Eddington limit. Hence, the reflector should be located at a distance of less than about a few hundred times the Schwartzschild radius. However, the fairly large covering factor requires a geometrically thick disk as the reflector, while the disk should also be optically thick for the Thomson scattering. It is very difficult for the accretion disk at that distance to satisfy these requirements at the same time.

Soft Excess Component

The presence of a soft excess component in AGN spectra was first pointed out for NGC4151 by Holt *et al.*[39] It was also observed from Cen A.[40,41] These sources suffer from heavy absorption by cool matter with a column density of about 10^{23} H atoms cm^{-2}, and the soft excess is observed in the energy range below 3 keV where such thick matter should have absorbed X-rays almost completely. Hence, a partial coverage of the absorber has been proposed to explain the soft excess of these sources.[39] However, EXOSAT has revealed that the soft excess is common to most Seyfert I galaxies even though they are free from heavy absorption.[42] The soft excess component is more likely to come from another X-ray emitter than from the central X-ray source.

A soft excess component is also observed from GBHCs. Cyg X − 1 frequently exhibits irregular intensity dips near the time of superior conjunction (reference 43 and references therein). The spectra during the dips reveal the K-absorption edge of neutral iron, which provides clear evidence for the dips to be due to absorption by intervening clouds, but they show an excess flux below 4 keV that is significantly greater than that expected for absorption by a neutral gas.[43] Since the soft X-ray excess has also been found at nondip phases,[44,45] it is again considered to be an extended emission free from absorption.

GS2023 + 33 has shown further clear evidence for the presence of a soft X-ray

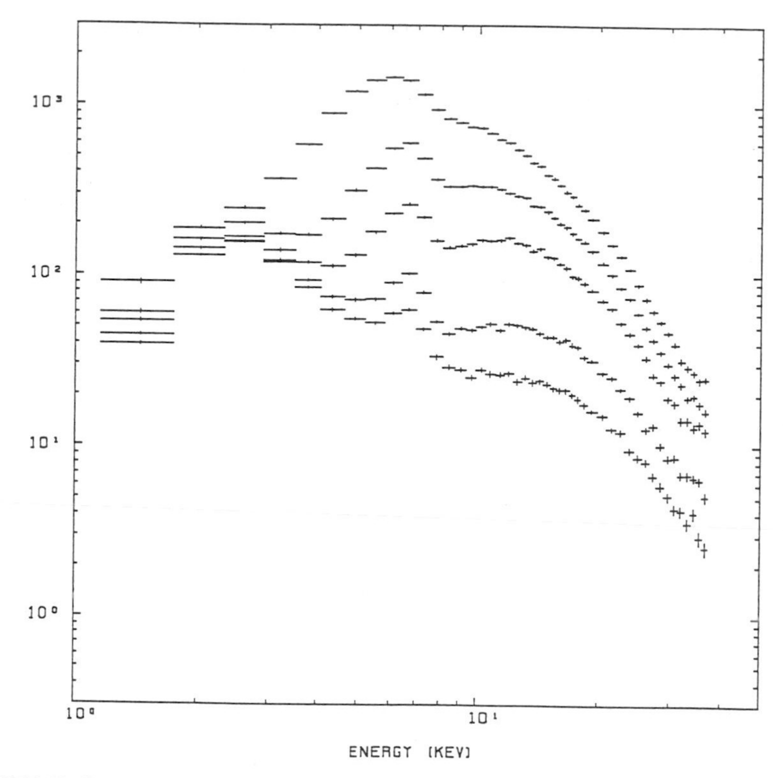

FIGURE 12. Successive pulse-height spectra of GS2023 + 33 from 12:06 to 12:21 (UT) on May 30, 1989.

emitter far outside of the central X-ray emitter. FIGURE 12 shows spectra successively obtained every few minutes from GS2023 + 33 when the central source suffered from heavy obscuration.[27] Evidently, a relatively stable, soft component is present independently of a variable, heavily absorbed component observed at higher energies.

As shown previously, soft excess emission is often observed from both GBHCs and AGNs, and probably comes from far outside of the central engine. However, the observations also imply the physical association of the soft X-ray emitter with the central engine. In the cases of NGC4151[46] and Cen A,[47] the soft X-ray flux roughly correlates with the power-law flux probably from the central engine (see FIG. 13). In the case of GS2023 + 33, the soft component showed time variation on time scales longer than several hours, and decayed in association with the power-law component.

If these soft excess emissions from various sources have a common origin, a thin thermal emission may be suggested for it. The temperature of the soft excess emission seems to be in a fairly narrow range irrespective of whether the source is a GBHC or an AGN. However, if the emitter was optically thick, the temperature would have a dependence on the mass of the central black hole. On the other hand, if the temperature is in proportion to the depth of the potential well at a certain distance in a unit of the Schwartzschild radius, it will be independent of the mass of the gravity center.

Here, it is very interesting to note that jets in SS433 reveal thin thermal emission in X-rays. Doppler-shifted iron *K*-lines have been detected from SS433,[48,49] and are consistent with thin thermal emission from the precessing jets. In fact, the continuum

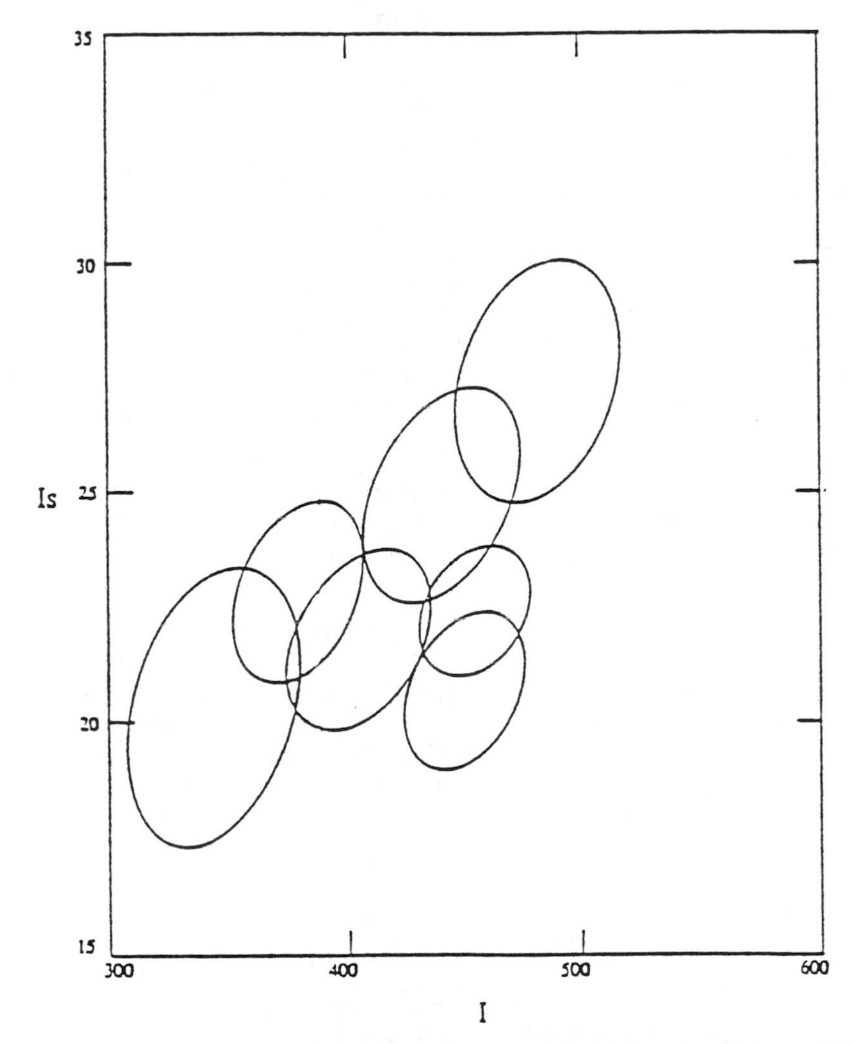

FIGURE 13. Correlation between the soft excess component and the power-law component that suffers from heavy absorption, of Cen A observed with Ginga.[47] The ordinate and abcissa indicate the normalization factors (arbitrary units) of bremsstrahlung emission with a temperature of 2 keV for the soft excess component and of a power-law component with a photon index of 1.8, respectively. *Solid lines* represent 90 percent confidence limits obtained from spectra averaged over several hours.

spectrum can always be fitted well with a thermal bremsstrahlung model with neutral absorption.[50] A remarkable finding by Ginga is spectral softening during eclipses,[50–52] and this can be interpreted by the presence of a temperature gradient along the jet.[52] It is tempting to speculate from this that the soft excess emission may come from parts of jets far from the central engine.

SUMMARY

Recent progress in the observational study of X-rays from galactic black hole candidates (GBHC) and active galactic nuclei (AGN) has been reviewed, based mainly on results from Ginga. Although GBHCs show two spectral states—a high and a low state—these sources as well as AGNs commonly reveal a power-law component with an energy index of around 0.5–1.0 in their spectra, irrespective of their states. A comparison of the temporal and spectral behaviors of GBHCs in the high state with those of weakly magnetized neutron stars in the normal branch suggests that this power-law component, at least in the high state, comes from a region inside the innermost Keplerian circular orbit. The power-law component in the spectra of GBHCs and AGNs is often accompanied by some spectral features imprinted on it. These features can be reproduced by superposition of a component reflected by cool Thomson-thick matter from a single power law. The ratio between the reflection and direct fluxes from the central X-ray source indicates that cool, thick matter around the central engine has a covering factor that is as large as or larger than 50 percent. A soft excess component has also been found to exist in addition to the power-law spectrum of AGNs.[42] Some observational evidence suggests the presence of the soft excess component in the GBHC spectra, too.

ACKNOWLEDGMENTS

The author would like to thank many of the Ginga team members for providing him with a number of results before publication. He also wishes to acknowledge Dr. K. Ebisawa and Messrs. M. Takizawa and K. Yoshida for assistance in preparing the manuscript. He is grateful to Dr. T. Yaqoob for his careful reading of the manuscript.

REFERENCES

1. BAYM, G. & C. PETHICK. 1979. Annu. Rev. Astron. Astrophys. **17:** 415.
2. PACZYNSKI, B. 1974. Astron. Astrophys. **34:** 161.
3. COWLEY, A. P., *et al.* 1983. Astrophys. J. **272:** 118.
4. MCCLINTOCK, J. E. & R. A. REMILLARD. 1986. Astrophys. J. **308:** 110.
5. HUTCHINGS, J. B., D. CRAMPTON & A. P. COWLEY. 1983. Astrophys. J., Lett. **275:** L43.
6. MCCLINTOCK, J. 1991. Black holes in the galaxy. This issue.
7. TENNANT, A. F., A. C. FABIAN & R. A. SHAFER. 1986. Mon. Not. R. Astron. Soc. **219:** 871.
8. TURNER, M. J. L., *et al.* 1989. Publ. Astron. Soc. Japan **41:** 345.
9. TANAKA, Y. 1989. *In* Proc. 23rd ESLAB Symposium on Two Topics in X-ray Astronomy (ESTEC, Noordwijk, the Netherlands), p. 3.
10. INOUE, H. 1989. *In* Proc. 23rd ESLAB Symposium on Two Topics in X-ray Astronomy (ESTEC, Noordwijk, the Netherlands), p. 783.

11. MITSUDA, K., *et al.* 1984. Publ. Astron. Soc. Japan **36:** 741.
12. EBISAWA, K., *et al.* 1991. Private communication.
13. EBISAWA, K., K. MITSUDA & T. HANAWA. 1991. Astrophys. J. **367:** 213.
14. EBISAWA, K. 1991. Ph.D. Thesis, University of Tokyo, Tokyo, Japan.
15. TSUNEMI, H., *et al.* 1989. Astrophys. J., Lett. **337:** L81.
16. ELVIS, M., *et al.* 1975. Nature **257:** 656.
17. KALUZIENSKI, L. J., *et al.* 1977. Astrophys. J. **212:** 203.
18. TAKIZAWA, M., *et al.* 1991. In preparation.
19. HOSHI, R. 1984. Publ. Astron. Soc. Japan **36:** 785.
20. MIYAMOTO, S., *et al.* 1991. Submitted for publication in Astrophys. J.
21. LEWIN, W. H. G., J. VAN PARADIJS & M. VAN DER KLIS. 1988. Space Sci. Rev. **46:** 273.
22. EBISAWA, K., K. MITSUDA & H. INOUE. 1989. Publ. Astron. Soc. Japan **41:** 59.
23. MOTCH, C., *et al.* 1983. Astron. Astrophys. **119:** 171.
24. KITAMOTO, S. 1989. *In* Proc. 23rd ESLAB Symposium on Two Topics in X-ray Astronomy (ESTEC, Noordwijk, the Netherlands), p. 231.
25. INOUE, H. 1990. Adv. Space Res. **10**(2): 153.
26. KITAMOTO, S., *et al.* 1989. Nature **342:** 518.
27. YOSHIDA, K., *et al.* 1991. In preparation.
28. CHARLES, P. A., *et al.* 1989. *In* Proc. 23rd ESLAB Symposium on Two Topics in X-ray Astronomy (ESTEC, Noordwijk, the Netherlands), p. 103.
29. MOTCH, C., S. A. ILOVAISKY & C. CHEVALIER. 1985. Space Sci. Rev. **43:** 219.
30. MAKINO, F. 1989. *In* Proc. 23rd ESLAB Symposium on Two Topics in X-ray Astronomy (ESTEC, Noordwijk, the Netherlands), p. 803.
31. ROTHSCHILD, R. E., *et al.* 1983. Astrophys. J. **269:** 423.
32. MIYAMOTO, S., *et al.* 1988. Nature **336:** 450.
33. POUNDS, K. A., *et al.* 1990. Nature **344:** 132.
34. LIGHTMAN, A. P. & T. R. WHITE. 1988. Astrophys. J. **335:** 57.
35. GUILBERT, P. W. & M. J. REES. 1988. Mon. Not. R. Astron. Soc. **233:** 475.
36. KOYAMA, K., *et al.* 1989. Publ. Astron. Soc. Japan **41:** 731.
37. AWAKI, H., *et al.* 1990. Nature **346:** 544.
38. KUNIEDA, H., *et al.* 1990. Nature **345:** 786.
39. HOLT, S. S., *et al.* 1980. Astrophys. J., Lett. **241:** L13.
40. FEIGELSON, E. D., *et al.* 1981. Astrophys. J. **251:** 31.
41. WANG, B., *et al.* 1986. Publ. Astron. Soc. Japan **38:** 685.
42. TURNER, T. J. & K. A. POUNDS. 1989. Mon. Not. R. Astron. Soc. **240:** 833.
43. KITAMOTO, S., *et al.* 1984. Publ. Astron. Soc. Japan **36:** 731.
44. BALUCINSKA, M. & G. HASINGER. 1989. *In* Proc. 23rd ESLAB Symposium on Two Topics in X-ray Astronomy (ESTEC, Noordwijk, the Netherlands), p. 269.
45. BARR, P. & H. VAN DER WOERD. 1989. *In* Proc. 23rd ESLAB Symposium on Two Topics in X-ray Astronomy (ESTEC, Noordwijk, the Netherlands), p. 275.
46. YAQOOB, T. & R. S. WARWICK. 1991. Mon. Not. R. Astron. Soc. **248:** 773.
47. KANO, T. & H. INOUE. 1991. In preparation.
48. WATSON, M. G., *et al.* 1986. Mon. Not. R. Astron. Soc. **222:** 261.
49. MATSUOKA, M., S. TAKANO & K. MAKISHIMA. 1986. Mon. Not. R. Astron. Soc. **222:** 605.
50. KAWAI, K. 1989. *In* Proc. 23rd ESLAB Symposium on Two Topics in X-ray Astronomy (ESTEC, Noordwijk, the Netherlands), p. 453.
51. KAWAI, K., *et al.* 1989. Publ. Astron. Soc. Japan **41:** 491.
52. STEWART, G. C., H. C. PAN & N. KAWAI. 1989. *In* Proc. 23rd ESLAB Symposium on Two Topics in X-ray Astronomy (ESTEC, Noordwijk, the Netherlands), p. 163.

Black Hole Evaporation: An Open Question[a]

THEODORE JACOBSON

Department of Physics
University of Maryland
College Park, Maryland 20742

INTRODUCTION

Black hole evaporation due to Hawking radiation is a part of quantum gravity that is already well established by quantum gravity standards, and that even has potential observational consequences: explosions of mini–black holes. These would presumably be primordial black holes, formed in the early universe with a mass in the range 10^{14}–10^{17} grams, and that have today nearly completely evaporated.[1] The question I wish to pose is this: Can we really trust our derivations that predict the existence of black hole evaporation, in spite of our ignorance about the physics of ultrashort distances? Before addressing this question, let us consider what is at stake.

Observing a black hole explosion would be a great boon for quantum gravity. Not only would it confirm at least the general ideas of curved-space quantum field theory, it would also reveal something of the nature of true quantum gravity, as regards the final stages of the evaporation process. But even without such observations, black hole evaporation is very interesting from a theoretical point of view. It extends the analogy between thermodynamics and classical black hole mechanics[2] to a remarkable interplay of gravity, quantum field theory, and thermodynamics that culminates in the generalized second law of thermodynamics,[3,4]

$$\delta(S_{\text{outside}} + \tfrac{1}{4}A_{\text{horizon}}) \geq 0, \tag{1}$$

where the first term is the entropy of matter outside the horizon and the second term is one-fourth the area of the horizon, measured in square Planck lengths. This generalized second law (GSL) holds only if one takes seriously the state of the quantum fields around the black hole that give rise to the Hawking radiation.[4] Without the Hawking effect the GSL would *fail,* because one could slowly lower a box containing entropy to the horizon, hold it there, and then drop it in the hole. The entropy of the outside would go down, but the horizon area would remain unchanged, because all of the mass–energy of the box would have been converted to work on whatever was lowering the rope from far away. The Hawking radiation (in combination with the acceleration radiation seen by the static box along its accelerated world line) alters the situation, because it produces a buoyancy force requiring the box to be actually pushed toward the horizon.

From the viewpoint of the "membrane paradigm"[5] for black holes, Zurek and

[a]This work was supported by the National Science Foundation under Grant PHY-8910226 and by the University of Maryland.

Thorne[5,6] have somewhat demystified the GSL by showing a sense in which it is just another instance of the ordinary second law. According to this viewpoint, the entropy of the hole is associated not with a geometrical quantity, horizon area, but with the state of the fields within a thin layer above the horizon but below the "stretched horizon." To an outside observer, this layer is frozen in time, and an outside observer will never see any entropy actually cross the horizon. It should be emphasized that the thermal nature of the horizon and the precise value of the temperature is an *input* to this membrane viewpoint, not an output, so while the viewpoint is consistent with ordinary thermodynamics, it cannot really be said to give a full "explanation" for the validity of the GSL.

One reason for the feeling that there remains some mystery lurking in black hole thermodynamics is the fact, not often stressed anymore, that *classical* black hole mechanics already admits a thermodynamical analogy with horizon area playing the role of entropy.[2] This seems to hint that there is an underlying thermodynamic origin of classical gravitation itself, that is, that gravity is fundamentally somehow an *incoherent* rather than a coherent quantum effect. Such an origin would fit nicely with the idea that gravity may play a crucial role in the "collapse of the wavefunction," that is, in some nonunitary process that supersedes ordinary coherent quantum evolution and either decoheres quantum alternatives or actually selects one macroscopic alternative from among many.[7,8,9]

Mostly because of the tie-in with thermodynamics, Hawking's prediction of black hole radiation is considered by many to have a certain ring of truth. In fact, it seems there is very little doubt in the community that black holes do in principle evaporate. Yet there is some cause for doubt, due to the role apparently played by ultrahigh frequencies or ultrashort distances in the derivation. In the remainder of this paper, I will elaborate on this cause for doubt, discuss the cluster of issues that it raises, and finally present some evidence that, in spite of the reasonable doubt about Hawking's *derivation*, a "safe" derivation that avoids our ignorance of ultrashort-distance physics can likely be formulated.

HAWKING RADIATION AND SHORT DISTANCES

According to Hawking's analysis,[1] if an object collapses to form a nonrotating black hole of mass M, it will radiate to infinity as if it were a hot body at a temperature

$$T_H = \hbar c^3/8\pi kGM \simeq 10^{-7}\,(1.5\ \mathrm{km}/r_s)\ \mathrm{K} \simeq 1\,(10^{16}\ \mathrm{g}/M)\ \mathrm{MeV}/k, \qquad (2)$$

where r_s is the Schwarzschild radius $2GM/c^2$. This is a result of linear quantum field theory in the time-dependent curved spacetime of the hole, and it is widely expected to occur for interacting fields as well. The initial vacuum state of the quantum fields evolves to a state that is not the vacuum state far from the hole in the future. After Hawking's original derivation, other derivations of this effect were given, often in the analytically extended vacuum Scharwzschild spacetime of an "eternal black hole." Since my goal is to understand the physical justification of the derivation, I prefer to stick to the realistic case of the collapsing body.

The essential feature, for our purposes, of the spacetime of a collapsing object is the infinite time-dilation effect: an observer at rest with respect to the hole and far away will measure a finite time interval between the passage of any ingoing null geodesic and the passage of a later ingoing null geodesic that forms a generator of the horizon, whereas after propagating through the object the corresponding outgoing geodesics will be separated by an infinite time interval, as measured by the same distant observer (since the geodesic on the horizon is *never* seen). One finds that in order for a wave packet of a linear massless field to emerge from the hole with a fixed frequency ω_{out} at large radius at time t, it must begin its journey into the collapsing matter with a blueshifted frequency ω_{in} that grows exponentially with time as $\exp(t/4M)$, where M is the mass of the hole. (Here and hereafter we use units with $c = \hbar = k = G = 1$, unless otherwise indicated, so in particular $2M$ is the time it takes for light to travel one Schwarzschild radius.)

Thus for example at a time t since the hole formed, an outgoing mode with frequency equal to the Hawking temperature $1/8\pi M$ originated as an ingoing mode with frequency above the Planck frequency if t is greater than $4M \ln(M/M_{Planck})$. If there were a Planck frequency cutoff on the ingoing modes, the Hawking radiation would seemingly be extinguished on this relatively short timescale.[10]

Another way of looking at the role played by ultrashort distances in black hole evaporation is provided by a recent paper of Fredenhagen and Haag,[11] in which it is shown that the existence of Hawking radiation for free fields can be derived using only the form of the short-distance singularity of the two-point function $\langle\phi(x)\phi(y)\rangle$ at the sphere where the horizon exits the collapsing matter. That is, the *only* assumption needed concerning the quantum state of the fields is the assumption that, in an infinitesimally thin shell surrounding the horizon at one particular time, the strongest singularity in the two-point function has the same form as it has in the Minkowski vacuum, namely σ^{-1}, with σ the square of the geodesic interval between x and y. This condition would rule out, for example, the Boulware vacuum, which has infinite stress at the horizon. The condition says essentially that the very short distance behavior of the vacuum fluctuations "appears to freely falling observers to be the same as in the Minkowski space vacuum." This approach brings out very clearly the sense in which the Hawking effect seems to hinge on ultrashort distance behavior of the fields.

It is important to determine whether the Hawking effect *requires* arbitrarily high-frequency modes and short distances, or whether their role can be eliminated in a more circumspect analysis. In particle physics, when there is ignorance about what is going on at short distances, one aims to extract predictions that are insensitive to the short-distance physics. In the same spirit, one should not be satisfied with a derivation of the Hawking effect unless it is independent of our ignorance about short distances.

There is an essential flaw in the analogy, however. In the particle physics context, to speak of "short-distance physics" or a "short-distance cutoff" on the validity of an interacting theory in no way implies a lack of Lorentz invariance. The fact that the center-of-mass energy or the momentum transfer of an interaction is large can be characterized by Lorentz-invariant scalar quantities. In contrast to this, the Hawking effect has nothing to do with interactions of the quantum fields. The high-frequency modes whose role is being questioned here can always be locally transformed to low frequency by an appropriate Lorentz transformation. In particular, no matter how high the frequency of an ingoing mode may be in the rest frame of the hole, the frequency in a frame that chases after it with sufficient speed is arbitrarily low.

If one is willing to *assume* exact local Lorentz invariance, then this cause for doubt about the Hawking effect is removed. However, such an assumption is unwise for at least two reasons. First, just as we have no experience with interactions at ultrahigh energy, we have no experience with physics in reference frames moving ultrafast relative to us. The fact that Lorentz-invariant *theories* agree with present observations serves only to place limits on possible deviations from Lorentz invariance.[12,13] Although today it may seem almost paradoxical to imagine a preferred state of "rest," one should keep in mind the fact that the universe as a whole does define such a rest frame: that of the microwave background radiation. It is not inconceivable that this cosmic rest frame also plays some role in local physics at short distances, even though nothing of the sort is true in currently accepted theory.

The second reason it is unwise to assume exact Lorentz invariance is that the assumption commits us to assigning an infinite number of degrees of freedom to the fields in any finite spatial volume. This idealization leads inexorably to the divergences in quantum field theory and the nonrenormalizability of quantum gravity. While it is possible that the solution to these problems will come in the form of a *deus ex machina* such as string theory, it is also possible that the solution lies in the removal of the offending assumption, exact Lorentz invariance.

Not knowing what form physics might take at ultrashort distances, and in order to explore the consequences of postulating only a finite number of degrees of freedom in a finite volume, let us frame our question about the Hawking effect this way: Would a black hole radiate if there were a mode cutoff at the Planck frequency in the rest frame of the hole? Although the rest frame is well defined only far from the hole, we will later extend it inward using radial timelike geodesics.

The first thing to notice is the fact that the presence of a Planck frequency cutoff on initial modes of a free field theory not only eliminates the Hawking radiation—according to the usual derivation—but it eliminates the corresponding outgoing modes themselves from the field degrees of freedom! That is, it is not just that the modes are not excited to the Hawking temperature, but that they are not present at all. A classical antenna attempting to radiate away from the hole would emit no waves at those frequencies. This observation poses a puzzle, however: it is implausible that just by a little thought one can arrive at the profound conclusion that either there is no cutoff on field modes at any scale, or there are bizarre effects due to missing modes outside black holes. It seems clear that there must be yet another possibility, namely, some mechanism, due to field interactions or perhaps quantum gravity effects, that could "regenerate" the mode frequencies that have been lost by redshifting. This idea of mode regeneration may at first seem hopelessly vague; however, there is in fact a model from condensed matter physics that is analogous to the black hole situation, and that exhibits just this behavior. We turn now to a discussion of this model, with the aim of clarifying the notion of mode regeneration and making it more plausible.

QUANTUM FLUID-FLOW MODEL OF A BLACK HOLE

In a paper published in 1981, Unruh[14] invented a fluid-flow model of a black hole. His motivation was in part to eventually consider the issue of short-distance physics being discussed here, as well as the quantum back-reaction problem. Although in the paper the fluid is treated exclusively as a continuum, a very interesting result is obtained, namely, that "the same arguments which lead to black-hole evaporation also predict that a thermal spectrum of sound waves should be given out from the sonic horizon in transsonic fluid flow. . . ."

It turns out that the linearized perturbations of an irrotational flow can be described by a massless scalar field propagating in a curved background spacetime whose metric is determined by the background fluid velocity field. For a spherically symmetric, static convergent flow that exceeds the speed of sound at some radius, the metric has approximately the form of a black hole metric. Quantizing the sound field, and *assuming* the field is in the comoving ground state near the horizon for all times, Unruh patched onto the Hawking argument outside the horizon to conclude that there is a thermal flux of phonons at the temperature

$$T = \left(\frac{\hbar}{2\pi k}\right)\left(\frac{\partial v}{\partial r}\right) \simeq 10^{-7}\left(\frac{\partial v}{\partial r}\bigg/\frac{100\ \mathrm{m/s}}{1\ \mathrm{mm}}\right)\mathrm{K}, \qquad \mathbf{(3)}$$

where $\partial v/\partial r$ is the gradient of the background velocity field at the horizon. Comparison with (**2**) indicates that $\partial v/\partial r$ plays the role of the black hole surface gravity $\kappa/c = c/2r_s$.

Taking into account the atomic nature of the fluid, one has a model in which there is both a preferred comoving local rest frame, and a high-frequency/short-wavelength cutoff on the phonon field modes. Let us see how, in this model, the mode regeneration effect referred to in the previous section can occur. To make the model concrete the fluid will be taken to be helium-4 at zero temperature and pressure. The cutoff arises because the phonons are really collective excitations of the helium atoms, and when the wavelength becomes much smaller than the average interatomic spacing, the field-theoretic description of the excitations is no longer valid.

An elementary excitation is a stable state of the fluid with definite energy and momentum. The energy ϵ and momentum $\mathbf{p}$ of the excitations for helium-4 satisfy a particular dispersion relation $\epsilon = \epsilon(p)$, shown in FIGURE 1 for $T = 1.1$ K and zero pressure.[b] States lying on this curve cannot decay, as can be seen from energy and momentum conservation. For wavelengths longer than about 10 Å$(p/\hbar = 2\pi/\lambda \sim 0.6\ \text{Å}^{-1})$, the dispersion relation for these excitations is approximately linear, with slope equal to the speed of sound, $s = 238$ m/s. These are the true "phonons." For shorter wavelengths the slope of the dispersion curve varies (and even becomes negative for a while), so the group velocity $v_g = d\epsilon/dp$ deviates from the sound velocity s. Excitations near the minimum are called *rotons,* and the spectrum cuts off at a wavelength of about 2.3 Å. The cutoff occurs when the single excitation description fails, presumeably because the corresponding state can decay into a pair of lower energy rotons or has some other decay modes.[c]

In Unruh's black hole model, the fluid has a spherically symmetric convergent flow that is faster toward the center and exceeds the speed of sound at some radius,

[b] This graph is adapted from [15, fig. 9].

[c] The experimental information on the form of the dispersion relation at the highest momenta is somewhat cloudy. What is actually measured is neutron scattering cross sections, and to find the dispersion relation for elementary excitations one has to infer the energy of an excitation from the location of the peak it contributes to the cross section at fixed momentum transfer as a function of energy transfer. Near the cutoff, this peak broadens due to the presence of decay modes that are not elementary excitations.

forming a sonic horizon.[d] Now excitations that are propagating upstream loose energy as they climb away from the sonic horizon, just as in the black hole case. It is the energy ϵ in the comoving rest frame that is decreasing, while the energy ϵ_0 with respect to the asymptotic rest frame is constant.[e] The relation between the two is given by

$$\epsilon_0 = \epsilon + \mathbf{p} \cdot \mathbf{v} = \epsilon - pv, \tag{4}$$

where $\mathbf{v}$ is the velocity of the fluid, and in the second equality we have assumed $\mathbf{p}$ is antiparallel to $\mathbf{v}$.

Consider an excitation propagating away from the horizon upstream in the fluid. Having swum against the flow gradient, the excitation must have had a shorter wavelength the closer it was to the horizon. If the cutoff—which is at fixed wavelength in the comoving frame—is reached before we can extrapolate all the way back to the horizon, then we can infer that some sort of mode reconstruction must have taken place to produce this excitation. It may seem that this will certainly happen; however, one must take into account the fact that the location of what we might call

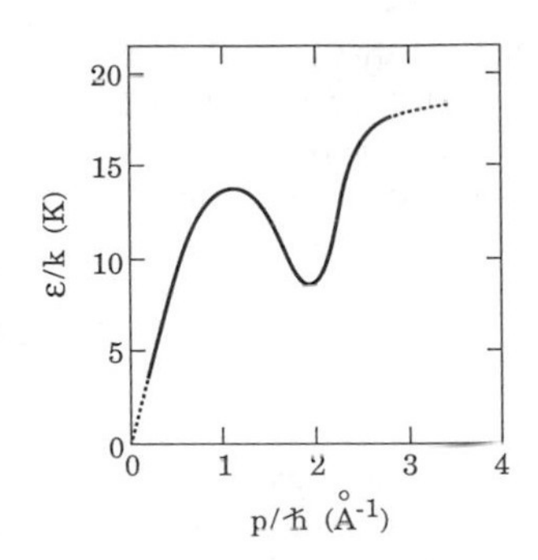

FIGURE 1. Dispersion curve for elementary excitations of liquid helium-4 at $T = 1$ K and approximately zero pressure.

the "effective horizon" depends on the group velocity v_g and therefore on the momentum. For phonons with momenta below the first maximum in the dispersion relation, the effective horizon is reached *before* the cutoff, but for excitations with momenta greater than that of the roton minimum when far from the horizon, the

[d]One might instead consider a flow through a cylindrical capillary tube. To model the collapse to a black hole one could begin with a uniform flow in the tube, and then contract the walls over some length, producing a thinner stretch of tube in which the speed of the fluid exceeds the speed of sound. Although perhaps simpler from an experimental point of view, the lower symmetry of this model would probably make a detailed theoretical analysis more complicated, except to the extent that it could be treated as a one-dimensional problem.

[e]The constancy of ϵ_0 is due to the fact that the excitation cannot exchange energy with the superfluid condensate. In a realistic situation this can be expected to be approximately true. An example of propagation of excitations in an inhomogeneous background superfluid flow is provided by the scattering of rotons by vortices.[16] In this example conservation of ϵ_0 would fail to the extent that the motion of the vortex core can absorb energy. In the black hole model, there is no such localized inhomogeneity, and conservation of ϵ_0 can be expected to hold more accurately.[17]

cutoff is reached first.[f,18] Such a high-momentum excitation propagating freely upstream could not have originated as an elementary excitation arbitrarily near to the horizon. This leads to a puzzle, the analog of which previously led us to doubt the consistency of imposing a cutoff on initial modes in the collapsing black hole spacetime. The puzzle is, how did the outgoing excitation or mode come to exist?

The answer for the superfluid must be that the mode arises from an interaction of two (or more) lower energy modes, for example, in a process that is the time reverse of the decay to a roton pair. Even in the ground state at zero temperature, the effective field-theory description of helium is an interacting one. A detailed analysis of the issue of horizon radiation would require finding the time evolution of the quantum state of the fluid as the inhomogeneous background flow is established and thereafter. This cannot be expected to be easy, on account of the interactions and also due to an instability in the flow mentioned in the Discussion section.

Fortunately, it is not really necessary to understand the details of the corresponding process in superfluid helium to draw from the model an important lesson for the black hole situation. We learn that the existence of a high-frequency cutoff on ingoing modes *can* be consistent with the existence of a full spectrum of outgoing modes (below the cutoff), provided there is some nonlinear process involving the short-distance physics that can "regenerate" the outgoing modes just outside the horizon. Moreover, we learn that one should not rely on any derivation of black hole radiation that involves free propagation through this "zone of ignorance" outside the horizon, where the unknown mode regeneration process is presumably taking place.

DO BLACK HOLES EVAPORATE?

In view of our ignorance of physics at ultrashort distances, we seek a derivation of black hole evaporation that does not assume "business as usual" outside the horizon in the "zone of ignorance" just mentioned. To get a feeling for the physical issues involved, I will discuss here two arguments that assume the existence of a cutoff at the Planck scale. Neither argument is conclusive; however, they should provide guidance in the search for a *bona fide* derivation.

The Cutoff

We are entertaining the hypothesis of a cutoff that, for reasons discussed earlier, must violate Lorentz invariance. It is therefore necessary to specify the reference frame in which the cutoff is applied. The least arbitrary assumption would be that in quasi-flat regions of spacetime, the cutoff is uniform in the rest frame of the cosmos, that is, that of the microwave background radiation. Now we must face the question of how the cutoff is to be extended into the spacetime surrounding a black hole that is approximately at rest with respect to the cosmic rest frame.

As already discussed, it is probably *not* correct simply to impose the cutoff on the field degrees of freedom before the hole forms, and then propagate it using the free field equations, for this leads to an implausibly redshifted cutoff on outgoing modes.

[f]This is true provided the group velocity is still positive at the cutoff.[c]

Instead, we must make an assumption for the nature of the cutoff in each region of spacetime that somehow takes into account whatever mode regeneration processes may have occurred. Lacking the fundamental cutoff theory, the only assumption that suggests itself is that the preferred rest frame is carried down into the hole by observers who fall freely from far away at rest. The resulting frame will be called the "falling frame." Note that in the fluid-flow analog, the falling frame corresponds to the comoving rest frame of the fluid.

Of course this definition of the local rest frame is unambiguous only in the spherically symmetric case, and then only if one restricts to radial free-fall trajectories that are always infalling, ignoring those that pass through the center of the collapsing matter and later fall back. In more general situations, the fundamental cutoff theory would be needed to even formulate the nature of the cutoff. But in order to explore the qualitative implications of a cutoff, it seems reasonable to stick to the spherically symmetric case and to impose the cutoff in the falling frame.

There is another frame that has a preferred status, namely, the frame of the static observers at constant radius. To interpret the consequences of the cutoff it is useful to know how the cutoff is viewed in this "static frame." This is determined from the relative velocity of the frames, by the Doppler shift factor. One finds[18] for massless modes (the frequencies that are important in our discussion are much larger than any mass we might wish to consider) that if the cutoff on both outgoing and ingoing modes is at the Planck frequency in the falling frame, $\omega_c^{\text{fall}} = \omega_p$, then in the static frame near the horizon it will be given by

$$\omega_{c,\text{out}}^{\text{stat}} \simeq \tfrac{1}{2} N\, \omega_p$$

$$\omega_{c,\text{in}}^{\text{stat}} \simeq 2\, N^{-1} \omega_p. \tag{5}$$

Here $N = (1 - 2M/r)^{1/2}$ is the "lapse" function that gives the relation $ds = N\, dt$ between proper time and Schwarzschild time t at fixed Schwarzschild radial coordinate r. Note that $N \rightarrow 0$ at the horizon, and $N \rightarrow 1$ at spatial infinity, and we have used $N \ll 1$ in (5). The cutoff on outgoing modes therefore approaches *zero* near the horizon in the static frame.

Recall that the Hawking radiation is in some sense related to the thermal excitation of static detectors at a temperature that diverges at the horizon. However, these detectors cannot be thermally excited at a temperature above the cutoff, so the fact that the cutoff $\omega_{c,\text{out}}^{\text{stat}}$ approaches zero near the horizon precludes their being excited and calls into question the existence of the Hawking effect itself. In order to get a feeling for the implications of this line of reasoning, we now review the relation between accelerated detector respose and Hawking radiation, and then consider the effect of imposing a cutoff as just described.

Accelerated Detectors

A detector with uniform acceleration a in the usual vacuum state of flat Minkowski space will be thermally excited[19–21] to a temperature $T_U = (\hbar/2\pi k)(a/c)$. This result is easily seen by considering the two-point function $\langle 0|\phi(x(s))\phi(0)|0\rangle$, where $x(s)$ is the accelerated world line and s is the proper time along it. In the Minkowski vacuum state the two-point function depends only on the invariant interval $x^2(s)$. For uniformly accelerated motion, $x(s)$ is the result of exponentiating a boost, so it is periodic in the translation of s by the imaginary quantity $i2\pi/a$, where a is the proper acceleration. Now a two-point function periodic in imaginary time is a thermal two-point function, and one concludes that the detector feels a temperature $T_U = a/2\pi$.

Following Unruh[19] and DeWitt,[20] this result can be applied to a static detector outside the horizon of a black hole. The acceleration of a world line at constant radius is given by $a = N^{-1}(1 - N^2)^2\,\kappa$, where $\kappa = 1/4M$ is the "surface gravity" of the hole. At the horizon $N \to 0$, so the acceleration diverges, and at infinity $N \to 1$, and the acceleration vanishes.

If the two-point function along the world line of the static detector were the same function of proper time as it is along an accelerated world line in the flat-space Minkowski vacuum, one could conclude that the detector is thermally excited at the Unruh temperature

$$T_U = a/2\pi = N^{-1}(1 - N^2)^2\,T_H, \tag{6}$$

where $T_H = \kappa/2\pi$ is the Hawking temperature. As viewed from infinity, this temperature suffers a redshift given by the lapse function N, so at infinity it is given by

$$T_{U,\infty} = (1 - N^2)^2 T_H, \tag{7}$$

which agrees with the Hawking temperature of the hole, provided the detector is placed just outside the horizon where $N \to 0$.

Now let us consider how the argument is affected by the presence of a Planck frequency cutoff in the falling frame. The essential point is that the static cutoff frequency for outgoing modes (**5**) goes to zero at the horizon, whereas the acceleration temperature (**6**) diverges. To apply the argument one must therefore stay far enough from the horizon that the acceleration temperature is below the cutoff. The question that arises is whether this is *so* far that the argument breaks down completely. In fact, we see from (**5**) and (**6**) that the outgoing static cutoff is equal to the acceleration temperature when the lapse function is given by

$$N_{\min} \simeq (T_H/\omega_p)^{1/2} \simeq (l_p/r_s)^{1/2}, \tag{8}$$

provided $l_p \ll r_s$, so that the higher order terms in N can be neglected. At this lapse, the temperature "at infinity" indicated by (**7**) differs by a term of order $O(l_p/r_s)$ from the Hawking temperature.

The preceding reasoning suggests that as long as the Schwarzschild radius r_s is much greater than the Planck length l_p, the cutoff will not make a significant difference for observations far from the hole. More specifically, it suggests that the spectrum of emitted radiation would differ from Hawking's thermal spectrum by terms of order $O(l_p/r_s)$. Another indication of the analysis is that for a given mass hole, there is a maximum local temperature to which a static detector would be excited, given by the acceleration temperature at the position where it is equal to the cutoff. From (**5**) and (**8**), this temperature is seen to be

$$T^{\text{local}}_{\max} \simeq N_{\min}\,\omega_p \simeq (T_H\omega_p)^{1/2}. \tag{9}$$

Since the existence of this limit violates the usual redshift relation for the variation of temperature in a static gravitational field, it may have some implications for the existence or properties of a thermal equilibrium state surrounding a black hole.

Stress-Energy Tensor

In making the argument that a static detector just outside the horizon of a black hole is thermally excited, it was necessary to assume that the two-point function along the detector world line behaves like that along a uniformly accelerated world line in the Minkowski vacuum. A plausibility argument can be advanced to support this assumption for the *outgoing* mode contribution,[18] however it is an admittedly weak one, especially because to actually *calculate* the two-point function from initial conditions before the collapse would involve whatever physics takes place in the zone of ignorance where the mode reconstruction process presumably occurs. We seek instead a derivation that uses assumptions which can be justified on general physical grounds alone. A promising strategy in this regard is based on conservation of the stress tensor.

Consider a conformally invariant field in the two-dimensional black hole spacetime that results when the angular coordinates are dropped. Employing a slight modification of the analysis of Christensen and Fulling,[22] it can be shown[18] that the outgoing flux of energy from the hole at late times is completely determined just by conservation and finiteness of the renormalized stress tensor, together with the value of the anomalous trace. The flux so determined agrees with the Hawking result for the net flux, and it is notable that the result is obtained without evaluating the Bogoliubov coefficients connecting the ingoing and outgoing modes, without specific assumptions regarding the initial state, and without assuming time independence of the final state.

Now let us consider the preceding argument in the context of a cutoff theory. We assume there exists a quantity corresponding to the expectation value of the renormalized stress tensor, even if it is not calculated by the usual rules of quantum field theory but by some other theory that remains unknown. We assume moreover that it is conserved, and that it is regular at the horizon; however, we do not assume the usual trace anomaly. Nevertheless, to proceed we need to know something about the trace. Since it cannot be computed in an unknown theory, I have simply computed it[18] using standard quantum field theory with an *ad hoc* cutoff, employing Fujikawa's method,[23,24] because that method seems best suited to accommodate the effects of a cutoff. In other calculational schemes, the dimensionful scale comes in "through the back door," and I do not know how to relate it to a physical scale.

This method of estimating the trace leads to the conclusion that the effect of a Planck-length cutoff is to modify the result by a term of relative order $(l_p/r_s)^2$ near the hole, which leads to a change in the flux at infinity $\Delta T^{(\infty)}_{uu}$ of the same relative order,

$$\Delta T^{(\infty)}_{uu} = O(l_p^2/r_s^2)\, T^{(\infty)}_{uu}. \tag{10}$$

The analysis does not apply directly to the four-dimensional case, for which supplementary analysis regarding the modification of the tangential stress would be required. Supposing qualitatively similar results, we can conclude that the modification (**10**) is negligible for any but the tiniest black holes. In particular, it would not affect the order of magnitude estimate[1] of the energy released in a mini-black-hole explosion, because most of that energy would be released before the hole gets anywhere near Planck size.

DISCUSSION

So, do black holes evaporate? The analysis presented here supports the viewpoint that black hole evaporation is not really an ultrashort-distance effect, despite the role played by ultrashort distances in its usual derivations. Thus it seems that the existence of a Planck-scale cutoff or other unknown Planck-scale physics would probably not affect the existence of black hole radiation. Nevertheless, many open questions remain.

The notion of mode regeneration was introduced in this paper to account for the compatibility of a short-distance cutoff with the infinite redshift effect of black holes. The implication is that nonlinear field propagation is crucial to a sound physical understanding of black hole radiation. This is somewhat disturbing, since it leads one to question what happens if the regeneration process is not complete. Why wouldn't the final state depend upon some coupling constants, and how could this agree with the Hawking result? A possible answer goes like this. Since the regeneration process is reestablishing the full spectrum of vacuum fluctuations locally, and the Minkowski vacuum is statistically like a zero-temperature thermal state,[21,25] perhaps the regeneration should be thought of as an equilibration-type process, which leads to a state independent of coupling constants, provided enough time passes. This would suffice, since the "pure" Hawking radiation is, strictly speaking, what emerges from the hole at very late times after the collapse.

Another question is whether the mode regeneration process can take place via self-interactions of the quantum fields alone, or whether the quantum structure of spacetime geometry must also play a role. One approach to addressing this question is to study interacting cutoff quantum field theory in a time-dependent background geometry, to see if there is a well-defined sense in which self-interactions can keep the cutoff at a fixed proper value. This question, which is also important in the cosmological context where one would need to understand why the expansion of the universe does not redshift the cutoff, is currently under investigation.

Lacking the fundamental short-distance theory, our present goal should perhaps be to identify the minimal and most physically sound assumptions needed to infer the existence of black hole radiation. This was the motivation for the stress tensor argument of the last section. To improve on that, one needs to generalize from conformal invariant fields in two dimensions to any fields in four dimensions, but it is difficult to see how this can be done since the trace and tangential components of the stress are no longer determined by general considerations.

An alternative is to modify the approach of Fredenhagen and Haag.[11] If a cutoff is imposed on the field modes in their free field calculation, the Hawking effect will disappear. In their calculation, everything comes from infinite magnification of the short-distance structure of the two-point function at the horizon at one time, and linear propagation precludes the necessary mode regeneration. To avoid the need to understand the details of the mode regeneration one might do the following. Instead of imposing a boundary condition on the two-point function at the horizon at one time, one could impose a condition at a spherical surface just *outside* the horizon at *every* time. Points in such a region are not causally independent, so one cannot impose an *initial* condition there; however, it may be possible to identify a condition that, if met, would yield the Hawking effect. Further work could then focus on whether or not this condition is met in any particular theory.

Finally, returning to the fluid-flow model of a black hole, it is interesting to ask whether Unruh's conclusion that a sonic horizon will radiate thermally remains true when the fluid is treated not as a continuum but as the realistic superfluid, helium-4. If the temperature were high enough for something other than the long wavelength phonons to emerge far from the horizon, then due to the form of the excitation spectrum, FIGURE 1, the effective horizon might occur at a speed substantially less than the speed of sound. Unfortunately, this is impossible, for if the temperature were to be 1 K, then according to (**3**) the gradient of the velocity at the horizon would need to be 100 m/s per angström!

For long wavelength phonons, the effective horizon occurs where the flow speed is close to the speed of sound. In particular, when long wavelength outgoing phonons are extrapolated backwards, the effective horizon is reached before the phonon reaches the first maximum of the dispersion curve, FIGURE 1, and therefore well *before* the cutoff is reached. Thus it seems that, due to the nonlinearity of the dispersion relation, the issue of the requisite modes being cut off does not arise for a prospective low-temperature sonic horizon. On the other hand, since the flow speed must be so high, the flow will be subject to various instabilities. If the instability to vortex creation at very low speeds could somehow be suppressed, there is still a critical speed,[26] about 60 m/s, at which the superfluid is unstable to the appearance of a periodic roton condensate[27] that would complicate the flow pattern. Ignoring such complications, Unruh's argument suggests that a low-temperature sonic horizon emits thermal phonons. However, we expect that each mode would be populated at a slightly different temperature, determined by the gradient of the velocity field evaluated at the effective horizon for that mode.

ACKNOWLEDGMENTS

For helpful discussions on various aspects of this research I would like to thank R. Ferrell, L. Ford, D. Garfinkle, B. Kay, L. Pitaevskii, J. Polchinski, D. Samuels, L. Smolin, and R. Wald.

REFERENCES

1. HAWKING, S. W. 1974. Nature **248:** 30; ———. 1975. Commun. Math. Phys. **43:** 199.
2. BARDEEN, J. M., B. CARTER & S. W. HAWKING. 1973. Commun. Math. Phys. **31:** 181.
3. BECKENSTEIN, J. D. 1973. Phys. Rev. D **7:** 2333; ———. 1974. Phys. Rev. D **9:** 3292.
4. UNRUH, W. G. & R. M. WALD. 1982. Phys. Rev. D **25:** 942; ———. 1983. Phys. Rev. D **27:** 2271; RADZIKOWSKI, M. J. & W. G. UNRUH. 1988. Correction. Phys. Rev. D **37:** 3059.
5. THORNE, K. S., R. H. PRICE & D. A. MACDONALD, Eds. 1986. Black Holes: The Membrane Paradigm. Yale University Press. New Haven, Conn.

6. ZUREK, W. H. & K. S. THORNE. 1985. Phys. Rev. Lett. **54:** 2171.
7. FEYNMAN, R. P. 1963. Lectures on Gravitation, 1962–63. Unpublished notes. California Institute of Technology. Pasadena.
8. KÁROLYHÁZY, F. 1966. *Nuovo Cim.* **A42:** 390; KÁROLYHÁZY, F., A. FRENKEL & B. LUKÁCS. 1982. *In* Physics as Natural Philosophy: Essays in Honor of Laszlo Tisza, A. Shimony & H. Feshbach, Eds. MIT Press. Cambridge, Mass.; ———. 1986. *In* Quantum Concepts in Space and Time, C. J. Isham & R. Penrose, Eds. Oxford University Press. Oxford.
9. PENROSE, R. 1986. *In* Quantum Concepts in Space and Time, C. J. Isham & R. Penrose, Eds. Oxford University Press. Oxford; ———. 1987. *In* 300 Years of Gravitation, S. W. Hawking & W. Israel, Eds. Cambridge University Press. Cambridge, England.
10. GIBBONS, G. W. 1977. *In* Proceedings of the First Marcel Grossmann Meeting on General Relativity, R. Ruffini, Ed. North-Holland. Amsterdam, the Netherlands; WALD, R. M. 1976. Phys. Rev. D **13:** 3176.
11. FREDENHAGEN, K. & R. HAAG. 1990. Commun. Math. Phys. **127:** 273.
12. BLOKHINTSEV, D. I. 1964. Phys. Lett. **12:** 272.
13. NIELSEN, H. B. & I. PICEK. 1983. Nucl. Phys. **B211:** 269.
14. UNRUH, W. G. 1981. Phys. Rev. Lett. **46:** 1351.
15. WOODS, A. D. B. & R. A. COWLEY. 1973. Rep. Prog. Phys. **36:** 1135.
16. SAMUELS, D. C. & R. J. DONNELLY. 1990. Phys. Rev. Lett. **65:** 187.
17. SAMUELS, D. C. 1991. Personal communication.
18. JACOBSON, T. 1991. Phys. Rev. D **44:** 1731.
19. UNRUH, W. G. 1976. Phys. Rev. D **14:** 870.
20. DEWITT, B. S. 1982. *In* General Relativity, An Einstein Centenary Survey. S. W. Hawking and W. Israel, Eds. Cambridge University Press. Cambridge, England.
21. SCIAMA, D. W., P. CANDELAS & D. DEUTSCH. 1981. Adv. Phys. **30:** 327.
22. CHRISTENSEN, S. M. & S. A. FULLING. 1977. Phys. Rev. D **15:** 2088.
23. FUJIKAWA, K. 1981. Phys. Rev. D **23:** 2262.
24. ———. 1982. Phys. Rev. D **25:** 2584.
25. SCIAMA, D. W. 1981. *In* Quantum Gravity 2, C. J. Isham, R. Penrose, and D. W. Sciama, Eds. Clarendon Press. Oxford.
26. LANDAU, L. D. 1941. Zh. Eksp. Teor. Fiz. **5:** 71.
27. PITAEVSKII, L. P. 1984. JETP Lett. **39:** 511.

Gravitational Lensing[a]

RAMESH NARAYAN[b,c] AND ROGER BLANDFORD[d]

[c]*Steward Observatory*
University of Arizona
Tucson, Arizona 85721

[d]*Theoretical Astrophysics, 130-33*
California Institute of Technology
Pasadena, California 91125

INTRODUCTION

The basic ideas of gravitational lensing go all the way back to the beginnings of general relativity, when Einstein predicted that light rays are deflected by gravitating masses.[1] This prediction was spectacularly confirmed in the eclipse expedition of 1919.[2] In the same year, Lodge[3] argued that the sun could multiply-image background sources by introducing opposite deflections to rays passing on either side of it. This idea was applied to nearby stars by Chwolson[4] and Einstein.[5] The angular separation between the two images is, however, too small to be resolved by any available instrument.

Zwicky[6] made the important point that not only stars, but also galaxies, could cause multiple images of cosmologically distant sources through the gravitational lens effect. In this case, the image separations will be large enough to be resolvable by conventional telescopes. There was sporadic, but important, theoretical activity in the decades that followed,[7–10] but the subject of gravitational lensing really took off only after the first multiply-imaged quasar was discovered by Walsh, Carswell, and Weymann.[11] Six more noncontroversial examples of gravitationally lensed quasars have since been discovered, as well as seven other good candidates. In addition, there are eleven cases of lensed optical galaxies and three examples of lensed extended radio sources. The variety of lensing phenomenology observed today is very rich, and a large body of theoretical ideas has developed.

In the next two sections we briefly summarize the observational data and the relevant elements of ray optics needed for understanding the phenomena. The three following sections then discuss various astrophysical applications of gravitational lensing in the study of (1) cosmography, (2) dark matter in the universe, and (3) cosmologically distant sources. Because of the length restriction, this article is of necessity somewhat brief and incomplete. More information (in particular, a more extensive bibliography) can be obtained from the forthcoming monograph by Schneider, Ehlers, and Falco[12] or in other recent reviews.[13,14]

[a]This work was supported in part by National Science Foundation Grants AST-8957107 [Presidential Young Investigator Award (R.N.)] and AST-8917765 (R.B.).

[b]Present address: Harvard-Smithsonian Center for Astrophysics, 60 Garden Street, Cambridge, Massachusetts 02138.

OBSERVED LENS CANDIDATES

Multiply-Imaged Quasars

The first example of a multiply-imaged quasar, viz. Q0957 + 561,[11] consists of two quasar images (A and B) at redshift $z = 1.41$ separated by 6". A galaxy at $z = 0.36$, presumably the primary lens, is found about 1" from B and approximately on the line joining the two images. This galaxy is a member of a rich cluster whose mass is also believed to contribute to the lensing.

A number of other examples of multiply-imaged quasars are now known.[15] In FIGURE 1, we exhibit what are in our view the seven best cases. There is a wide variety of image and lens configurations seen in these objects, with geometries ranging from simple (e.g., Q0142 − 100) to rather complex (e.g., Q2016 + 112, which in addition to the three images and two galaxies shown in FIGURE 1 has also two resolved Ly-α clouds). Most cases generally seem to fall into two classes, those with two images (e.g., Q0142 − 100 and Q0957 + 561) and those with four (e.g., Q0414 + 453, Q1115 + 080, Q1413 + 117, and Q2237 + 031). In the latter class, the four images always lie roughly on a circle. In those cases where the lensing galaxy is seen, it is found roughly on the line joining a double image or at the center of four images.

Among the examples shown in FIGURE 1, Q0957 + 561 and Q2237 + 031 have turned out to be particularly important from the point of view of astrophysical applications, while Q2016 + 112 has been the hardest to model. There are legitimate doubts as to whether the simplest lens candidates, such as Q1635 + 267 and Q1120 + 019, which have two quasar images and no luminous lens, are really examples of lensing or merely twin quasars that happen to be physically associated.[16]

Luminous Arcs

In 1986, Lynds and Petrosian[17] and Soucail *et al.*[18] independently discovered blue luminous arcs, each about 20" long, near the centers of the rich galaxy clusters A370 and C12244 − 02. Paczyński[19] proposed that these arcs were the images of distant background galaxies that had been gravitationally distorted by the foreground clusters. This idea was shown to be reasonable in terms of the required optics[20–22] and received firm observational confirmation when the redshift of the arc in A370 was measured to be 0.73,[23] nearly twice the redshift of the cluster (0.37).

A number of large arcs have since been discovered in other clusters.[24] Also, miniarcs/arclets (ranging from ~4" to 10" long) were found, first in A370, and later in several other clusters. Indeed, Tyson[25] reports 50–70 arclets in a number of clusters, notably A1689 and 3C295.

Radio Rings

Hewitt *et al.*[26] discovered the radio source, MG1131 + 0456, which has a very unusual ring-like morphology (see FIG. 2). They proposed that this is an example of a distant "normal" radio source that is being gravitationally lensed by a foreground galaxy. The particular geometry is often referred to as an "Einstein ring".[5] The

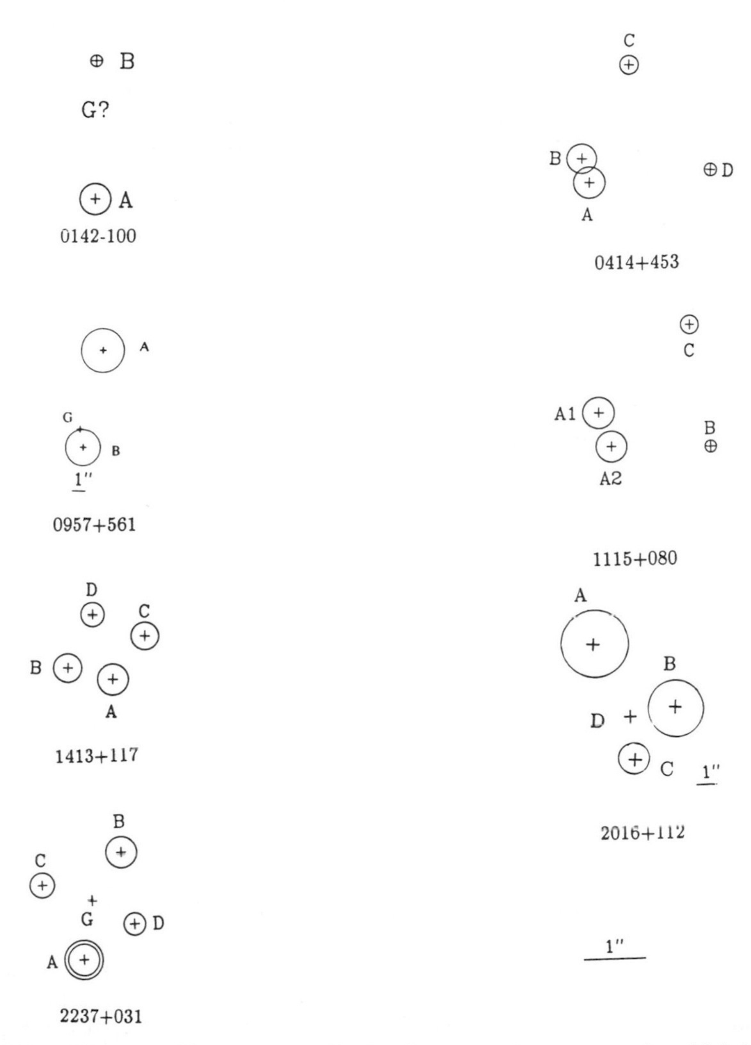

FIGURE 1. Montage of image geometries for the seven strong cases of multiply-imaged quasars.[15,48] The images of Q0957 + 561 and Q2016 + 112 are to quarter and half scale, respectively. The areas of the circles are approximately proportional to the image fluxes. The A image of Q2237 + 031 is highlighted to indicate the detection of microlensing.[42]

gravitational lens hypothesis for MG1131 + 0456 received strong confirmation when a detailed model was developed[27] that could explain most of the radio observations, including polarization, at 5 GHz as well as 15 GHz.

Two other radio rings have since been discovered: MG1654 + 1346[28] is a weaker source than MG1131 + 0456. However, in contrast to MG1131 + 0456, both the background source (z = 1.75) and the lensing galaxy (z = 0.25) have been optically identified, which helps in the modeling. The third ring, 1830 − 211,[29] has a flux of

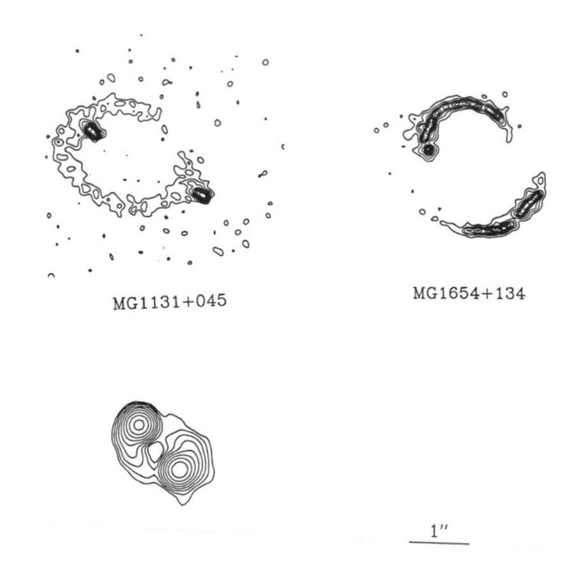

FIGURE 2. The three radio rings.[25,27,28] (From Kochanek.[48] Adapted and reproduced by permission.)

10 Jy and is the brightest ring source by far. There are excellent prospects of obtaining radio very long baseline interferometry (VLBI) observations, and hence tight constraints on the lens model.

LENSING OPTICS

Ray Deflections and the Lens Equation

The basic ray geometry is shown in FIGURE 3. A light ray from a source S is incident on a lens L with impact parameter $\mathbf{r}$ relative to some fiducial lens "center." Assuming the lens is *thin* compared to the total path length, the influence of the lens can be described by a deflection angle $\boldsymbol{\alpha}(\mathbf{r})$ (a two-vector) suffered by the ray on crossing the "lens plane." The deflected ray reaches the observer, who sees the image of the source apparently at position $\boldsymbol{\theta}_I$ on the sky. The true direction of the source, that is, its position on the sky in the absence of the lens, is indicated by $\boldsymbol{\theta}_S$. Using the angular diameter distances, D_{OL}, D_{LS}, and D_{OS} (FIG. 3), the following *lens equation* is easily derived,

$$\boldsymbol{\theta}_S = \boldsymbol{\theta}_I - \boldsymbol{\alpha}'(\boldsymbol{\theta}_I), \tag{1}$$

where the *reduced* deflection angle $\boldsymbol{\alpha}'$ is defined by

$$\boldsymbol{\alpha}'(\boldsymbol{\theta}_I) = \frac{D_{LS}}{D_{OS}}\,\boldsymbol{\alpha}(D_{OL}\boldsymbol{\theta}_I). \tag{2}$$

Deflection angles of interest in astrophysical applications are always small. We are therefore justified in using the weak field approximation of general relativity. For a thin lens, the deflection angle depends only on the projected surface mass density, $\Sigma(\mathbf{r})$, of the lens. It can be shown that

$$\boldsymbol{\alpha}(\mathbf{r}) = \frac{1}{c^2} \nabla\psi(\mathbf{r}), \tag{3}$$

where $\psi(\mathbf{r})$ is *twice* the two-dimensional Newtonian potential due to $\Sigma(\mathbf{r})$, obtained by solving the two-dimensional Poisson equation,

$$\nabla^2\psi(\mathbf{r}) = 8\pi G\Sigma(\mathbf{r}). \tag{4}$$

Note that $\psi(\mathbf{r})$, as well as the deflection angle, are independent of the wavelength of the light.

From the structure of (**1**) it is clear that there is a unique source position $\boldsymbol{\theta}_S$ for each image position $\boldsymbol{\theta}_I$ for a given lens. However, the converse is not true. For a nontrivial deflection law $\boldsymbol{\alpha}'(\boldsymbol{\theta}_I)$, it is possible to have more than one value of $\boldsymbol{\theta}_I$ satisfying (**1**) for a given $\boldsymbol{\theta}_S$. This is the reason for multiple imaging in gravitational lensing. Note also that, in general, $\boldsymbol{\alpha}'(\boldsymbol{\theta}_I)$ is not a linear function of $\boldsymbol{\theta}_I$, and so gravitational lenses behave differently from simple laboratory optical lenses. Whereas man-made lenses are carefully designed to be linear and stigmatic so as to focus a bundle of rays to a point, natural gravitational lenses behave more generically and focus on caustic *sheets*. Caustics play an important role in the gravitational lensing phenomenon.[30]

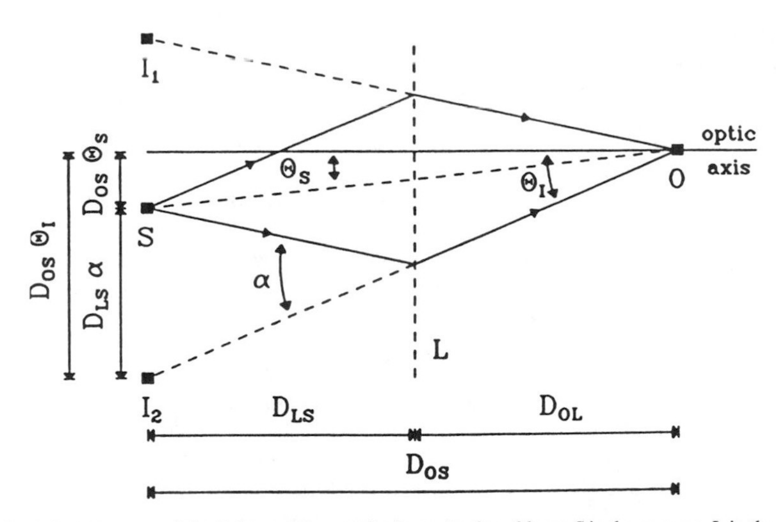

FIGURE 3. Diagram of the light rays in a typical gravitational lens. S is the source, L is the lens, and O is the observer. I_1 and I_2 are two images of the source. D_{OS}, D_{OL}, D_{LS} are angular diameter distances, and $\boldsymbol{\alpha}$ is the angle of deflection.

Simple Lens Models

The deflection angle for impact parameter r relative to a point mass M is given by

$$\alpha(r) = \frac{4GM}{c^2 r}. \quad (5)$$

This is an appropriate model for computing deflections by individual stars.

A galaxy or galaxy cluster, on the other hand, has an extended mass distribution, typically extending beyond the projected positions of the images. A reasonable model for the purposes of rough estimates is to use a singular isothermal sphere, parametrized by the line-of-sight velocity dispersion σ. The deflection angle is then given by

$$\alpha(r) = 4\pi \frac{\sigma^2}{c^2}. \quad (6)$$

Two further improvements are usually needed. First, one can avoid the unphysical singularity at the lens center by introducing a finite core radius. A second necessary elaboration is to break the circular symmetry and allow the lens to be astigmatic,[31] either because the mass distribution is elliptical or because of the presence of an external tidal potential.

These two modifications suffice to reproduce all the *qualitative* features observed in the multiply-imaged quasars, the luminous arcs, and the radio ring.[14,32] In summary, we find that most of the secure examples of gravitational lensing in the universe are attributable to galaxies and galaxy clusters, or at least to mass distributions that resemble these objects. An interesting point is that there is a theorem that the number of images has to be odd,[33] whereas most observed examples of quasar lensing have two or four images. The "missing" image is believed to be located within the core of the lens and should be highly demagnified, or indeed can vanish altogether, whenever the core radius is small enough.[34,35] The absence of this image thus provides an upper bound on the core radius.

Fitting a Lens Model to Observations

When it comes to fitting the observations *quantitatively,* one needs to pick a sufficiently general "elliptical" lens mass distribution with a number of adjustable parameters. The parameters are then fixed by using the constraints imposed by the observations.

First, the lens equation **(1)** provides one vector equation (two components) for each image. Since the true position of the source is not known, this leads to $2(n - 1)$ constraints when there are n images. When the image is resolved, then one replaces these relations by the condition that the surface brightness $I(\boldsymbol{\theta}_I)$ of the image at the position $\boldsymbol{\theta}_I$ is the same as the true surface brightness of the source $I_S(\boldsymbol{\theta}_S)$ for each resolution element. This provides a large number of constraints whenever there is multiple imaging, a fact that was exploited in modeling the radio ring MG1131 + 0456.[27]

The magnification of the ith image relative to the source may be formally written as

$$M_i = \frac{\partial \boldsymbol{\theta}_I}{\partial \boldsymbol{\theta}_S}, \tag{7}$$

which is a symmetric 2×2 matrix. Although this is not directly measurable, the *relative* magnification, M_{ij}, between two images,

$$M_{ij} = M_j M_i^{-1}, \tag{8}$$

is potentially observable whenever the image is resolved. In general, this is a nonsymmetric matrix, described by four parameters for each independent image pair. The relative magnification matrix between the A and B images in Q0957 + 561 has been measured using VLBI observations on the radio cores of the two images.[36] More commonly, the matrices M_{ij} are not available, but only their determinants $|M_{ij}|$, which are given by the ratios of the observed fluxes of the images. These introduce additional constraints on the lens model.

Another potential observable is the time delay between images. The excess time associated with an image at $\boldsymbol{\theta}_I$ relative to the direct ray in the absence of the lens is given by

$$t(\boldsymbol{\theta}_I) = (1 + z_L)\left[\frac{D_{OL}D_{OS}}{2cD_{LS}}(\boldsymbol{\theta}_I - \boldsymbol{\theta}_S)^2 - \frac{\psi}{c^3}\right], \tag{9}$$

where z_L is the redshift of the lens and ψ is twice the two-dimensional Newtonian potential [cf. (4)] evaluated at the ray position. The term proportional to $(\boldsymbol{\theta}_I - \boldsymbol{\theta}_S)^2$ is due to the geometrical excess path length and the term proportional to ψ is the relativistic time delay, known as the "Shapiro effect" in the solar system. If the source happens to be variable, the differential time delay between two images can be measured by cross-correlating the flux variations. A time delay of 1.4 yr has been reported between the two images in Q0957 + 561.[37]

Finally, when the lens is visible, a measurement of the velocity dispersion σ, or the light distribution (coupled with an assumed M/L ratio), can significantly limit the range of possibilities for the lens model.

Microlensing

Soon after the discovery of Q0957 + 561, Chang and Refsdal[38] showed that the presence of individual stars in the lensing galaxy can modify the properties of the observed images relative to that expected with a smoothly distributed lens. Further work[39–42] showed that when a perturbing star moves across the line-of-sight of an image, the magnification and hence the observed flux would vary systematically with an observationally interesting timescale. This phenomenon is termed *microlensing.*

A star of mass M has a significant influence on the magnification of a background source only if the impact parameter r satisfies

$$r < b_o \equiv \left(\frac{4GMD_{\rm eff}}{c^2}\right)^{1/2} = 4 \times 10^{16}\left(\frac{M}{M_\odot}\right)^{1/2}\left(\frac{D_{\rm eff}}{1\ \rm Gpc}\right)^{1/2}\ \rm cm, \tag{10}$$

where $D_{\rm eff} = D_{OL}D_{LS}/D_{OS}$. If the perpendicular velocity of the star relative to the source–earth line (probably dominated by motion of the galaxy with respect to the Hubble flow) is V, then the variability timescale is

$$t_{\rm var} \sim \frac{b_o}{V} = 10 \left(\frac{M}{M_\odot}\right)^{1/2} \left(\frac{D_{\rm eff}}{1\ {\rm Gpc}}\right)^{1/2} \left(\frac{V}{10^3\ {\rm km\ s^{-1}}}\right)^{-1} {\rm yr}. \tag{11}$$

Note that actual flux variations will be observed only if the source is compact enough, that is, its angular size satisfies $\theta_S \lesssim b_o/D_{OL}$.

Irwin *et al.*[43] have presented a strong case for the detection of microlensing in the A image of Q2237 + 0305. This image brightened by about 0.6 magnitude in less than a year. Since the relative time delays in this case are expected to be of the order of days, and since no variations were seen in the other images, the case for a microlensing explanation is persuasive.[44]

APPLICATIONS IN COSMOGRAPHY

Measuring the Hubble Constant

Long before the discovery of the first gravitational lens candidate, Refsdal[7] recognized that gravitational lenses could be used to measure the absolute distance to the lens and the source, and hence to measure the Hubble constant H_o.

The basic idea is as follows. Suppose, for simplicity, that we have a doubly imaged quasar with images at positions $+\theta_1$ and $-\theta_2$ with respect to the center of a circularly symmetric lens. Suppose further that the lens is known to be an isothermal sphere so that the magnitude of the deflection α is independent of the impact parameter. From **(1)** and **(2)** we then find that

$$\theta_1 + \theta_2 = 2\alpha' = \frac{2D_{LS}}{D_{OS}}\alpha, \tag{12}$$

which leads to a determination of α. Furthermore, it is clear that $\theta_S = (\theta_1 - \theta_2)/2$, and so the differential time delay between the two images is determined solely by the potential term in **(9)**,

$$\Delta t = t(\theta_1) - t(\theta_2) = -(1 + z_L)\alpha D_{OL}(\theta_1 - \theta_2)/c. \tag{13}$$

A measurement of Δt thus gives D_{OL} and an estimate of H_o (for an assumed value of q_o in a Friedmann–Robertson–Walker universe). The argument is more general than indicated here and works for other lens models as well, for example, a point mass (which was the case originally considered by Refsdal), where Δt may have a contribution from the geometrical as well as the potential time delay.

One complication is that if the lens has a component of dark matter of unknown strength (as is the case with the surrounding cluster in Q0957 + 561), then there is no unique solution for H_o.[45] Specifically, a *family* of models may be constructed in which α of the primary lens is varied continuously, and at the same time the surface density of a dark matter sheet is adjusted such that the image positions and relative magnifications are left invariant. These models give a range of predictions for $H_o\Delta t$; consequently, a measurement of Δt does not lead to a unique solution for H_o. It is, however, possible to obtain an upper bound on H_o because the dark matter sheet cannot have negative density. Using the measured time delay of $\Delta t = 1.4$ yr in Q0957 + 561,[37] this gives $H_o \lesssim 80$ km s^{-1} Mpc^{-1}. Also, it turns out that Δt determines the lens mass much more tightly than it does H_o.[46]

The model degeneracy discussed previously is broken whenever there is any direct information about the lens. In particular, if the velocity dispersion σ of the lensing galaxy in Q0957 + 561 is measured, then there is a unique relation[47] between Δt and H_o, viz.,

$$H_o = 90 \pm 10 \left(\frac{\sigma}{390 \text{ km s}^{-1}}\right)^2 \left(\frac{\Delta t}{1 \text{ yr}}\right)^{-1} \text{ km s}^{-1} \text{ Mpc}^{-1}. \quad (14)$$

Recently, Rhee[48] has measured σ to be 303 ± 50 km s^{-1}, which gives

$$H_o \sim 40 \text{ km s}^{-1} \text{ Mpc}^{-1}. \quad (15)$$

Thus, we seem to have finally achieved Refsdal's dream of determining H_o through gravitational lensing. However, we must caution that, quite apart from uncertainties in the measured values of σ and Δt, there are other imponderables. First, the formula **(14)** is based on a particular model of the lensing galaxy, viz., a nonsingular isothermal sphere with a compact core and an added dark matter sheet that is allowed to contribute both convergence and shear. Although this is quite a reasonable model one needs to investigate how much the coefficient in **(14)** will change with other "reasonable" models.[49] Second, if the universe is inhomogeneous, then the relation between H_o and Δt can be significantly modified[50] by the action of random mass clumps between the observer and lens or the lens and source. Fortunately, it appears that if the inhomogeneities are sufficiently smooth on the scale of the image separations and contribute only a variable convergence, then the power of the method is hardly diminished.[51] Specifically, under these conditions, there is still a unique relation between Δt and the angular diameter distance to the lens D_{OL} along the particular line-of-sight. In particular, large inhomogeneities between the lens and the source, the part of the line-of-sight that is most hidden from the observer, have no effect at all.

Other Cosmographic Parameters

Models of the universe with different geometries (different q_o) may be discriminated using gravitational lensing.[52] The basic idea is that gravitational lensing depends on the angular diameter distances, D_{OL}, D_{LS}, and D_{OL}. Therefore, since the ratios of these distances for given lens and source redshift depend on q_o, one should be able to use lensing to determine q_o. The best cases are when one has two or more sources at different redshifts behind a single well-studied lens. The arcs and arclets produced by clusters of galaxies may be particularly well-suited for this application.[53]

Some suggestions have also been made for using gravitational lensing to determine the (normalized) cosmological constant λ. Inflation mandates that the universe be flat so that $\Omega_0 + \lambda = 1$. Models of the universe with nonzero λ, produce less focusing of the rays than Einstein–De Sitter models, and so are more likely to produce high redshift gravitational lenses. This will be reflected in the distribution of source redshifts.[54] Also, in λ universes, one can have antipodes at a finite redshift, leading to curious effects.[55]

NATURE AND DISTRIBUTION OF DARK MATTER

Halo of the Milky Way

Paczyński[56] pointed out that if the dark matter in the halo of the Milky Way consists of solar- or planetary-sized compact objects, then they should occasionally make their presence known by microlensing distant stars. He suggested observing $\sim 10^6$ stars in the Large Magellanic Cloud (LMC) on a nightly basis in order to look for variability in any of the stars with the appropriate signature characteristic of microlensing. If such observations are carried out for a year or more, it should be possible to detect a significant population of compact objects in the mass range 10^{-6} $M_\odot$–$10^2\, M_\odot$ provided that microlensing can be distinguished from intrinsic variability. Alternatively, the lack of such detections would rule out such a population as a major constituent of the halo. An observational program based on this idea is now being pursued seriously,[57] and results should be reported within the next few years.

Another technique for finding dark compact objects in the halo is through their distortions of resolved background sources. Hubble space telescope (HST) images of the galactic center or the Andromeda galaxy (preferably at the resolution of the Hubble space telescope) can be searched for characteristic lensing patterns.[58] Alternatively, "phantom" images near to bright radio galaxies may be seen.[59] These techniques will be sensitive to compact halo objects in the mass range $\gtrsim 10^6\, M_\odot$.

Cores of Galaxies

Gravitational lensing can provide straightforward bounds on the core radii of the lensing galaxies. As we have seen earlier, the absence of a detectable odd image at the position of the lens center is most directly interpreted as due to a small core radius. Typically, one requires $r_c \lesssim 0.5$–1 kpc.[34,35] The alternatives are to invoke a central black hole or other compact object of significant mass ($\gtrsim 10^9\, M_\odot$), or stochastic image disappearance because of microlensing by a significant population of stars.[60] Interestingly, the absence of multiple imaging in several BL Lacs that apparently lie behind the cores of foreground galaxies leads to rather large *lower* bounds on the corresponding galaxy core radii, $r_c \gtrsim 1$ kpc.[61] This may be in conflict with the previously mentioned upper bounds.

The detection of microlensing in Q2237 + 031 leads to interesting information on the stellar content of the core of the corresponding lensing galaxy ($z_L = 0.04$). The short variability timescale ($t_{\rm var} < 1$ yr) requires a low-mass star moving with a fairly large transverse velocity. The observations appear to be consistent with a population of hydrogen-burning stars ($M \gtrsim 0.1\, M_\odot$).[62]

Halos of Galaxies

A great deal of information about the dark matter distribution in the halos of galaxies can be obtained through gravitational lensing. As described earlier, the observations provide a number of constraints in each of the known multiply-imaged quasars or radio rings. These constraints can be used to fit parametrized models of the lens surface density. At the very least one can estimate the mass enclosed within the images, though the mass outside the images is poorly constrained.[49] Often one also estimates the ellipticity. A comparison of the mass model with the light distribution of the lens can reveal how well the mass traces the light. Particularly interesting are the examples where no lensing galaxy is apparently seen down to very faint magnitudes (e.g., Q2345 + 007, Q1635 + 267). If these really are gravitational lenses, and on this matter there is still some doubt, the existence of dark halos in the universe with no associated luminous matter is indicated.

An alternative approach to studying halos is statistical.[63–65] One uses the observed frequency of multiple imaging in quasars as a function of magnitude and redshift, as well as the distribution of image separations, and seeks a distribution of halo masses that can reproduce the observations. The data are still not extensive enough for this approach to be very fruitful, but we anticipate that it will gain in importance as the database grows in future.

Perhaps the most intriguing evidence on galaxy halos has come from the many claims of preferential alignment of high z quasars/galaxies with foreground galaxies.[66] If these are shown to be statistically significant, then the most natural explanation is in terms of gravitational lensing.

Dark Matter on Large Scales

As described in greater detail in the article by Tyson,[25] luminous arcs and arclets are uniquely capable of tracing the dark matter distribution in galaxy clusters. Parameters such as core radii and ellipticities have been estimated for a few clusters,[67] but in principle one may be able to map out the complete surface density distribution of clusters. On the larger scale, an inhomogeneous distribution of mass on length scales $\sim$100 Mpc, comparable to the length scales of the "voids" and "walls" reported in galaxy velocity surveys [e.g., Jones (paper presented at this conference, but not included in this issue)], can be detected by mapping the correlated ellipticity in the images of high z galaxies over large areas of the sky (say degrees).[68] These are some of the most exciting future applications of gravitational lensing.

GALAXIES/GALACTIC NUCLEI AT HIGH REDSHIFT

Sizes of Quasar Emission Regions

Microlensing is uniquely able to probe the angular sizes of quasars because the phenomenon requires the source to be smaller in projection than the critical impact parameter b_o of **(10)**. For microlensing by solar-mass stars, the critical size is $\sim 10^{16}$ cm, a very interesting scale for the continuum emission of quasars in current accretion disk models. Indeed, the detection of microlensing in Q2237 + 031 indicates that this quasar has an optical continuum size $\lesssim$ few $\times$ 10^{15} cm.[62] The nature of the continuum emission mechanism can also be constrained, and it can be shown that microlensing is incompatible with a thermal accretion disk model.[69]

If one had a detailed record of the microlensing intensity variation, particularly if one had multistation records (say Earth and a satellite near Saturn), then one could even do a sort of tomography and extract the spatial brightness profile of the source.[70] This works best if the source crosses a simple fold caustic, in which case the "transfer function" through the lens is well understood.

Magnified Images of Galaxies

Zwicky[6] suggested that gravitational lensing would provide a means of studying very distant galaxies through the magnification effect. In essence, the lens acts as a giant cosmic telescope to provide a magnified view of the source. This possibility is beginning to be realized, particularly with the luminous arcs, which are highly stretched out images of presumably normal background galaxies. Already, redshifts have been measured for several high redshift arcs (including $z = 2.23$ for the arc in C12244), which would be impossible if the galaxies had not been magnified. An even more interesting application is in the straight arc in A2390 ($z = 0.92$), where the velocity profile has been mapped along the length of the arc.[71] This is an example of measuring the rotation curve of a $z = 0.92$ galaxy, a very impressive achievement. The data indicate that the arc may consist of two images of the galaxy juxtaposed next to each other with opposite parities.[72]

If an arc galaxy has a supernova in a multiply-imaged region, then one will see the supernova at two or more locations in the arc *at different times* (because of the variable time delay).[73] Indeed, if one had a sufficiently good model of the lens, one could predict exactly where and when each successive flash would occur, and thus obtain a complete light curve of a cosmologically distant supernova!

Distortion of Luminosity Functions

If gravitational lensing is common enough, then it will lead to distortions in the luminosity functions of quasars and galaxies, particularly at high redshift. In general, if $\phi_{true}(L)d \log L$ is the true luminosity function of a population of sources and $P(M)d \log M$ is the probability distribution of gravitational magnification, then the observed luminosity function, $\phi_{obs}(L)d \log L$, is the convolution of $\phi_{true}(L)$ and $P(M)$. Under quite general circumstance, $P(M)$ has a power-law tail varying as M^{-2} at large M. This is a generic feature due to caustics and will be present so long as the source is sufficiently compact. If $\phi_{true}(L)$ is steeper than L^{-2}, then because of the convolution with $P(M)$, the observed $\phi_{obs}(L)$ will develop an L^{-2} tail at the bright end. Most of the sources in this tail will be gravitationally magnified.

The quasar optical luminosity function appears to vary as $L^{-2.6}$ (Boyle[74]) at bright magnitudes, and the data are not yet good enough to tell whether this levels off to L^{-2} at the brightest end. Nevertheless, several groups have used the preceding theoretical expectation to search for multiply-imaged quasars within the sample of high-redshift high-luminosity optical quasars.[15] The success of this approach in finding several new lens candidates appears to confirm that the effect is indeed present.

CONCLUDING REMARKS

A decade of hard observational and theoretical work has produced about twenty examples of gravitational lensing and a far deeper understanding of the optics of the associated lenses than we had before. We also have now a deeper appreciation of the selection effects and systematic errors that are present in the observations. Nevertheless, despite their obvious fascination, gravitational lenses have not yet lived up to their promise as tools for extragalactic astronomy. We can point to no secure, fundamental discovery or measurement by gravitational lens studies, only elegant verifications of what we already knew. However, as we hope this review makes clear, we believe that the study of gravitational lenses is now entering a new phase when distance measurement, mapping of dark matter, and direct and indirect magnification of distant galaxies and quasars will become routine. Our one prediction is that a review given at the end of the next decade will be as rich in applications as Taylor's report on pulsars (which were discovered a decade earlier than gravitational lenses) has been at this meeting.

ACKNOWLEDGMENT

We thank Kevin Rauch for comments on the manuscript.

REFERENCES

1. EINSTEIN, A. 1915. Preuss. Akad. Wiss. Berlin, Sitzber. **47:** 831.
2. DYSON, F. W. 1919. Observatory **42:** 389.
3. LODGE, O. 1919. Nature **104:** 354.
4. CHWOLSON, O. 1919. Astron. Nachr. **221:** 329.
5. EINSTEIN, A. 1936. Science **84:** 506.
6. ZWICKY, F. 1937. Phys. Rev. Lett. **51:** 290.
7. REFSDAL, S. 1964. Mon. Not. R. Astron. Soc. **128:** 295.
8. BARNOTHY, J. 1965. Astron. J. **70:** 666.
9. PRESS, W. H. & J. E. GUNN. 1973. Astrophys. J. **185:** 397.
10. BOURASSA, R. R. & R. KANTOWSKI. 1975. Astrophys. J. **195:** 13.
11. WALSH, D., R. F. CARSWELL & R. WEYMANN. 1979. Nature **279:** 381.
12. SCHNEIDER, P., J. EHLERS & E. FALCO. 1992. Gravitational Lenses. Springer-Verlag. Berlin/New York. In press.
13. BLANDFORD, R. D. & C. S. KOCHANEK. 1987. Dark Matter in the Universe, J. Bahcall, T. Piran, and S. Weinberg, Eds.: 133. World Scientific. Singapore.
14. BLANDFORD, R. D., C. S. KOCHANEK, I. KOVNER & R. NARAYAN. 1989. Science **245:** 824.
15. SURDEJ, J. 1990. Gravitational Lensing. Y. Mellier, B. Fort, and G. Soucail, Eds.: 57. Springer-Verlag. Berlin/New York.
16. PHINNEY, E. S. & R. D. BLANDFORD. 1986. Nature **321:** 569.
17. LYNDS, R. & V. PETROSIAN. 1986. Bull. Am. Astron. Soc. **18:** 1014.
18. SOUCAIL, G., B. FORT, Y. MELLIER & J.-P. PICAT. 1987. Astron. Astrophys. **172:** 414.
19. PACZYŃSKI, B. 1987. Nature **325:** 572.
20. GROSSMAN, S. & R. NARAYAN. 1988. Astrophys. J., Lett. **324:** L37.
21. KOVNER, I. 1988. The Post-Recombination Universe, N. Kaiser and A. N. Lasenby, Eds.: 315. Kluwer. Dordrecht, the Netherlands.
22. BLANDFORD, R. D. & I. KOVNER. 1988. Phys. Rev. A **38:** 4028.
23. SOUCAIL, G., Y. MELLIER, B. FORT, G. MATHEZ & M. CAILLOUX. 1988. Astron. Astrophys. **191:** L19.
24. FORT, B. 1990. Gravitational Lensing, Y. Mellier, B. Fort, and G. Soucail, Eds.: 221. Springer-Verlag. Berlin/New York.
25. TYSON, J. A. 1991. Faint galaxies and dark matter. This issue.
26. HEWITT, J. N., E. L. TURNER, D. P. SCHNEIDER, B. F. BURKE, G. I. LANGSTON & C. R. LAWRENCE. 1988. Nature **333:** 537.
27. KOCHANEK, C. S., R. D. BLANDFORD, C. R. LAWRENCE & R. NARAYAN. 1989. Mon. Not. R. Astron. Soc. **238:** 43.

28. LANGSTON, G. I., *ET AL.* 1989. Astron. J. **97:** 1283.
29. RAO, A. P. & R. SUBRAHMANYAN. 1988. Mon. Not. R. Astron. Soc. **231:** 229.
30. BLANDFORD, R. D. & R. NARAYAN. 1986. Astrophys. J. **310:** 568.
31. BLANDFORD, R. D. & C. S. KOCHANEK. 1987. Astrophys. J. **321:** 658.
32. NARAYAN, R. & S. GROSSMAN. 1989. Gravitational Lenses, J. M. Moran, J. N. Hewitt, and K.-Y. Lo, Eds.: 31. Springer-Verlag. Berlin/New York.
33. MCKENZIE, R. H. 1985. J. Math. Phys. **26:** 1592.
34. NARAYAN, R., R. D. BLANDFORD & R. NITYANANDA. 1984. Nature **310:** 112.
35. NARASIMHA, D., K. SUBRAMANIAN & S. M. CHITRE. 1986. Nature **321:** 45.
36. GORENSTEIN, M. V., N. L. COHEN, I. I. SHAPIRO, A. E. E. ROGERS, R. J. BONOMETTI, E. E. FALCO, N. BARTEL & J. M. MARCAIDE. 1988. Astrophys. J. **334:** 42.
37. ROBERTS, D. H., J. LEHAR, J. N. HEWITT & B. F. BURKE. 1991. Nature. **352:** 43.
38. CHANG, K. & S. REFSDAL. 1979. Nature **282:** 561.
39. YOUNG, P. J. 1981. Astrophys. J. **244:** 756.
40. GOTT, J. R. 1981. Astrophys. J. **243:** 140.
41. KAYSER, R., S. REFSDAL & R. STABELL. 1986. Astron. Astrophys. **166:** 36.
42. SCHNEIDER, P. & A. WEISS. 1987. Astron. Astrophys. **171:** 49.
43. IRWIN, M. J., R. L. WEBSTER, P. C. HEWETT, R. T. CORRIGAN & R. I. JEDRZEJEWSKI. 1989. Astron. J. **89:** 1989.
44. WAMBSGANSS, J., B. PACZYŃSKI & P. SCHNEIDER. 1990. Astrophys. J., Lett. **358:** L33.
45. GORENSTEIN, M. V., E. E. FALCO & I. I. SHAPIRO. 1988. Astrophys. J. **327:** 693.
46. BORGEEST, U. 1986. Astrophys. J. **309:** 467.
47. FALCO, E. E., M. V. GORENSTEIN & I. I. SHAPIRO. 1991. Astrophys. J. **372:** 364.
48. RHEE, G. 1991. Nature. **350:** 211.
49. KOCHANEK, C. S. 1991. Astrophys. J. In press.
50. ALCOCK, C. & N. ANDERSON. 1985. Astrophys. J., Lett. **291:** L29.
51. NARAYAN, R. 1991. Astrophys. J., Lett. **378:** L5.
52. REFSDAL, S. 1966. Mon. Not. R. Astron. Soc. **132:** 101.
53. SOUCAIL, G. & B. FORT. 1991. Preprint.
54. TURNER, E. L. 1991. Astrophys. J., Lett. **365:** L43.
55. GOTT, J. R., M. G. PARK & H. M. LEE. 1989. Astrophys. J. **338:** 1.
56. PACZYŃSKI, B. 1986. Astrophys. J. **304:** 1.
57. GRIEST, K. M., *ET AL.* 1991. Astrophys. J., Lett. **372:** L79.
58. TURNER, E. L., M. WARDLE & D. SCHNEIDER. 1991. Astron. J. **100:** 146.
59. BLANDFORD, R. D. & K. RAUCH. 1992. In preparation.
60. PACZYŃSKI, B. 1986. Astrophys. J. **301:** 503.
61. NARAYAN, R. & P. SCHNEIDER. 1990. Mon. Not. R. Astron. Soc. **243:** 192.
62. WAMBSGANSS, J., B. PACZYŃSKI & P. SCHNEIDER. 1990. Astrophys. J., Lett. **358:** L33.
63. TURNER, E. L., J. P. OSTRIKER & J. R. GOTT III. 1984. Astrophys. J. **284:** 1.
64. NARAYAN, R. & S. D. M. WHITE. 1988. Mon. Not. R. Astron. Soc. **231:** 97P.
65. KOCHANEK, C. S. 1991. Astrophys. J. **379:** 517.
66. WEBSTER, R. L., P. C. HEWETT, M. E. HARDING & G. A. WEGNER. 1988. Nature **336:** 358.
67. TYSON, J. A., F. VALDES & R. WENK. 1990. Astrophys. J., Lett. **349:** L1.
68. BLANDFORD, R. D., A. B. SAUST, T. G. BRAINERD & J. V. VILLUMSEN. 1991. Mon. Not. R. Astron. Soc. **251:** 600.
69. RAUCH, K. & R. D. BLANDFORD. 1991. Astrophys. J. In press.
70. REFSDAL, S. 1990. Gravitational Lensing, Y. Mellier, B. Fort, and G. Soucail, Eds.: 13. Springer-Verlag. Berlin/New York.
71. PELLO, R., J.-F. LE BORGNE, G. SOUCAIL, Y. MELLIER & B. SANAHUJA. 1991. Astrophys. J. **366:** 405.
72. KASSIOLA, A., I. KOVNER & R. D. BLANDFORD. 1991. Submitted for publication in Astrophys. J.
73. KOVNER, I. & B. PACZYŃSKI. 1988. Astrophys. J., Lett. **335:** L9.
74. BOYLE, B. J. 1991. The quasar population at high redshifts. This issue.

Primordial Nucleosynthesis and Observed Light Element Abundances

B. E. J. PAGEL
NORDITA
Blegdamsvej 17
Dk-2100 Copenhagen Ø, Denmark

INTRODUCTION

Standard big bang nucleosynthesis theory (SBBN),[1–6] assuming a homogeneous Friedman universe and small lepton numbers,[7] leads to quite accurate predictions of the outcome of nuclear reactions that set in 100 sec or so after the big bang (ABB), when the temperature has fallen to about 0.1 MeV, as a function of a single cosmic parameter that can be expressed as the ratio η of baryons to photons by number. The ratio η has remained unchanged from the epoch of $e^{\pm}$ annihilation a few seconds ABB to the present day, and is related through the known temperature of the microwave background (MWB) to the present-day mean mass density of baryons, or equivalently to the quantity $\Omega_{b0}h_{100}^2$, where Ω_{b0} is the present-day baryon density in units of the closure density corresponding to the Einstein–de Sitter cosmological model and h_{100} is the Hubble constant in units of 100 kms^{-1}Mpc^{-1} or $(10^{10}$ yr$)^{-1}$, generally agreed to lie somewhere in the range 0.7 ± 0.3. Primordial abundances so predicted[4,6,8] are shown by curves in FIGURE 1, together with horizontal lines indicating limits derived from observation as discussed below. The trace elements D, ^{3}He, and ^{7}Li survive in small quantities because expansion and cooling stop nuclear reactions before their destruction to form ^{4}He is complete, but ^{7}Li has bimodal behavior because at the higher densities it results from later K-capture by the relatively robust nucleus ^{7}Be, leading to a minimum just in the interesting range. On the other hand, ^{4}He increases only very slowly with baryon density because virtually all neutrons surviving until the onset of synthesis are soaked up in its production. Equally important factors influencing the primordial ^{4}He mass fraction Y_P are the number N_ν of families of light (compared to 1 Mev) left-handed neutrinos and antineutrinos (plus any hypothetical exotic light particles coupling to photons), which fixes the equation of state and hence the expansion rate of the universe at a given temperature, and the half-life $\tau_{1/2}$ of the neutron, which measures the weak-interaction constant. Larger values of either one of these two parameters lead to higher neutrino decoupling temperatures, and hence to larger neutron–proton ratios, and accordingly more primordial helium at a given baryon density, so that the postulate of consistency of Y_P (as derived from observation) with SBBN theory and other light element abundances has led to interesting predictions concerning these purely physical constants. The dependence on N_ν was first pointed out by Hoyle and Tayler[9] in 1964 when the MWB had not yet been discovered and only two kinds of neutrino were known. In 1977, when the τ-lepton had been discovered and the big bang theory much better established, Steigman, Schramm, and Gunn[10] used improved knowledge then available about Y_P to make the definite prediction from

SBBN theory that N_ν (and hence the number of families of quarks and leptons to which at that time laboratory physics suggested no particular limit) could not exceed 7, and subsequent improvements in helium determinations enabled YTSSO[4] to bring this limit down to 4. In the meantime Tayler,[11] followed by others, stressed the significance of the neutron half-life (cf. reference 12), then believed to be 10.8 min, which was itself a reduction from earlier estimates; the best experimental values now are close to 10.25 min.[13] Pagel in a semipopular article[14] used still lower observational

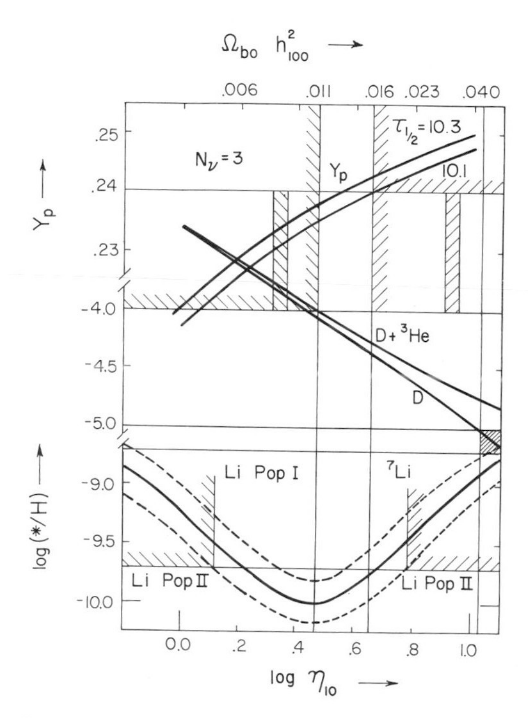

FIGURE 1. Primordial abundances predicted from SBBN theory after YTSSO,[4] Olive *et al.*,[6] and Deliyannis *et al.*,[8] ^{7}Li from the latter reference being shown with $\pm 2\ \sigma$ error limits, as functions of η and $\Omega_{b0}\, h^2_{100}$. Helium abundances are given for two possible values of the neutron half-life assuming $N_\nu = 3$. *Horizontal lines* show limits based on observations together with reasonable arguments from galactic chemical evolution. *Tall vertical lines* show corresponding limits on the baryon density parameter from SBBN, while the *shorter double vertical lines* give approximate limits in the framework of mildly inhomogeneous models adapted from Kurki-Suonio *et al.*[24]

estimates of Y_P (as reported below), together with arguments[4] that place an upper limit on primordial D + ^{3}He, to claim both that N_ν cannot exceed 3 and that $\tau_{1/2}$ cannot exceed 10.4 min. Both claims are supported by laboratory experiments on the Z^0 (e.g., reference 15) and on the neutron half-life.[13] The cosmological argument, however, eliminates certain loopholes left open by the accelerator experiments (and vice versa) in that it also excludes any hypothetical light stable exotic particles coupling to photons but not to the Z^0, whereas the accelerator experiments exclude heavier particles up to 45 GeV, existence of which could still have been consistent with SBBN.

In recent years there has been active discussion of alternative, nonstandard BBN models that postulate isothermal density fluctuations resulting from the quark–hadron phase transition, which could be first-order, and leading through differential diffusion to variations in the neutron–proton ratio.[16–19] In particular, it has been suggested that such models could fit observed light-element abundances with $\Omega_b = 1$. The analysis involves a number of more or less free parameters (density contrast, volume filling factors, and length scales), and the proper treatment of all diffusion effects is difficult and controversial, but at the present time models with $\Omega_b = 1$ do not seem to be viable because they predict too much helium and probably too much lithium for any combination of the parameters.[20–24] But mildly inhomogeneous models could fit the data for somewhat wider upper and lower limits to η than those given by SBBN.[23,24] These wider limits (corrected for $Y_P \leq 0.24$) are shown roughly by the shorter double vertical lines in FIGURE 1. The limits on N_ν and $\tau_{1/2}$ are only slightly affected in these models, but a remote possibility exists that there could be detectable primordial abundances of beryllium, boron, and CNO (for differing views on this, see references 19, 21, 25), *r*-process,[18] and conceivably other heavy elements. So far, however, no evidence exists for detectable primordial abundances of any element except hydrogen, helium, and ^{7}Li.[26]

DEUTERIUM AND HELIUM-3

These elements are well reviewed in the existing literature.[4,5,27] Because deuterium is generally destroyed and not created when diffuse material is cycled through stars, the abundance in the interstellar medium ($D/H \simeq 10^{-5}$ from *Copernicus* observations of Lyman lines) or in the early solar system ($D/H \simeq 2 \times 10^{-5}$ from ancient and modern samples of the solar wind) provides a lower limit to the primordial value, and hence a conservative upper limit to Ω_{b0} (see FIG. 1). The precise factor by which the deuterium abundance has been cut down by "astration" is very model-dependent,[28] but a less model-dependent upper limit to primordial D + ^{3}He was derived by YTSSO[4] and by Olive *et al.*[6] on the grounds that when deuterium is destroyed it becomes ^{3}He, of which at least one-quarter survives further stellar processing. In this way one obtains an upper limit of 10^{-4} to the primordial ratio (D + ^{3}He)/H, shown by a horizontal line in FIGURE 1. Numerical models of galactic chemical evolution[29] confirm this limit, which gives the most stringent lower limits to the baryon density (shown in FIG. 1 by the left tall vertical line for SBBN and by the left shorter double vertical line for reasonable inhomogeneous models[24]), implying the existence of substantial baryonic dark matter.[4,30]

LITHIUM-7

Spite and Spite[31] made the remarkable discovery that, over quite a wide range in effective temperature, extreme subdwarf stars, which are deficient in carbon and heavier elements by factors of 100 to 1000 relative to the sun and most nearby stars, have a significant and nearly constant abundance of lithium-7 in their atmospheres, only one order of magnitude below that in the solar system and young stars like those of the Pleiades cluster, although in the sun itself photospheric lithium is depleted by a factor of 100 as a result of mixing with deeper layers where ^{7}Li is destroyed. FIGURE 2 shows an ensemble of the data for ^{7}Li abundance in stellar atmospheres plotted against their iron abundance, which suggests the existence of a primordial component similar to that seen in the subdwarfs (presumably composed of virtually pristine material very little affected by stellar nucleosynthesis) together with enhancements at higher metal abundances due to stellar synthesis (e.g., in giant stars where large overabundances of lithium are sometimes seen[33]) combined with various amounts of depletion in the more metal-rich stars depending on their age, effective temperature,

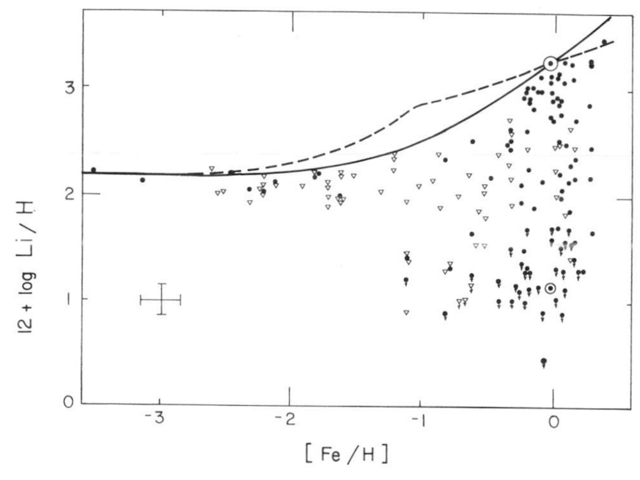

FIGURE 2. Plot of stellar ^{7}Li abundances versus iron abundance, adapted from Rebolo, Molaro, and Beckman.[32] The *solid curve* shows the prediction of a simple model in which ^{7}Li is the sum of a primordial component and an additional component proportional to iron. The *broken curve* is similar, but with the additional component proportional to oxygen. [Fe/H] is the logarithm of stellar iron abundance in units of the solar iron abundance, and the *two "sun" symbols* refer to meteoritic (identified as protosolar) and photospheric (depleted) values, respectively.

and probably other, less readily quantifiable effects. A possibility remains that the photospheric lithium in subdwarfs is substantially less than the primordial amount owing to mixing with hotter layers below,[34-36] but it is also quite likely that such effects may be insignificant and, if so, then the lithium in subdwarfs provides an impressive confirmation of SBBN predictions.

HELIUM-4

Helium is the second most abundant element in the visible universe and there is accordingly a vast amount of information about its distribution from optical and radio emission lines in ionized nebulae, absorption lines in optical spectra of hot stars, scale heights of the atmospheres of the major planets, and the influence of initial helium content on stellar structure and evolution.[4,5,27,28,37,38] However, helium has been enhanced by stellar nucleosynthesis to varying degrees in different objects, so that the observed mass fractions Y ranging from about 0.23 to 0.30 (and thus showing a very distinct "floor") are in general only estimates of upper limits to Y_P, and it is a nontrivial problem to deduce Y_P to the 5 percent accuracy (i.e., ± 0.01) that is needed in order to constrain BBN models significantly.[26,28]

The most precise values of Y come from ionized nebulae (planetary nebulae and HII regions) where hydrogen and helium emission lines come predominantly from a well-understood recombination process,[39] and the extrapolation to Y_P is best done by observing extragalactic HII regions in which the abundance of readily measurable heavy elements like oxygen and nitrogen, synthesized by nuclear processes in stars, is much less than in Galactic objects like the sun or the Orion Nebula. These HII regions are found in dwarf galaxies such as the Magellanic Clouds, in HII galaxies discovered in objective-prism surveys, in some of the blue compact galaxies discovered by Zwicky and Haro on direct photographs or in the outer regions of late-type spirals like M101. Peimbert and Torres-Peimbert[40,41] noted a small but significant trend for helium abundance to increase with that of oxygen in the Magellanic Clouds studied by them and in the blue compact galaxies IZw 18 and IIZw 40 previously observed by Searle and Sargent,[42] and accordingly suggested that Y_P could be derived from observations of a number of such objects by plotting a linear regression of the form

$$Y = Y_P + Z(dY/dZ) = Y_P + (\mathrm{O/H})dY/d(\mathrm{O/H}),$$

where Z ($\simeq 25(\mathrm{O/H})$) is the mass fraction of heavy elements, and extrapolating to $Z = \mathrm{O/H} = 0$. This program was carried out by Lequeux *et al.*,[43] leading to $Y_P = 0.23$, $dY/dZ \simeq 3$, results which I believe to have been essentially correct, although there has been much controversy about them in the meantime,[5,44] regarding both the data themselves and their interpretation. The latter involves uncertainties about various corrections including allowance for unobservable neutral helium in the H^+ zone, the possibility of a collisional contribution to the emission lines, and a few other effects.[45,46]

Since then, understanding of these problems has greatly improved and new, high-quality data have become available;[47–53] cf. FIGURE 3 showing high-resolution data for the red line $\lambda 6678$ that in combination with spectra in the blue region previously obtained by Terlevich and his colleagues[49,53] gives an excellent check on both the data and the underlying assumptions.[52] Combining these results with selected and consistently reprocessed data from the literature,[26,27] we obtain regressions against both oxygen and nitrogen (FIG. 4) with maximum-likelihood and equivalent $\pm 1\,\sigma$ fits[54]

$$\begin{aligned} Y = &\ 0.225 + 169\ \ (\mathrm{O/H}) \\ &\pm 5 \quad \pm 45 \end{aligned}$$

$$\begin{aligned} Y = &\ 0.229 + 3310\ \ (\mathrm{N/H}). \\ &\pm 4 \quad \pm 940 \end{aligned}$$

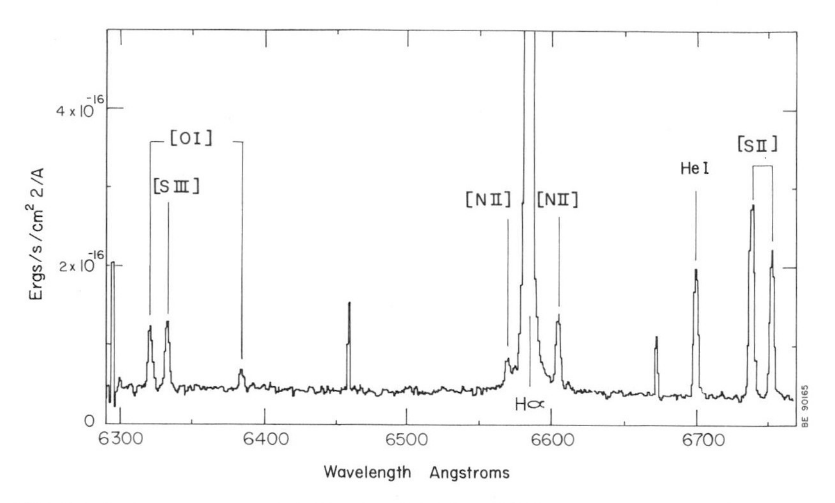

FIGURE 3. Red spectrum of the HII galaxy Michigan 461[53] taken with the Anglo-Australian telescope in April 1988, with a spectral resolving power of about 2000.[52] Emission lines are identified; narrower spikes are due to cosmic-ray events in the CCD detector.

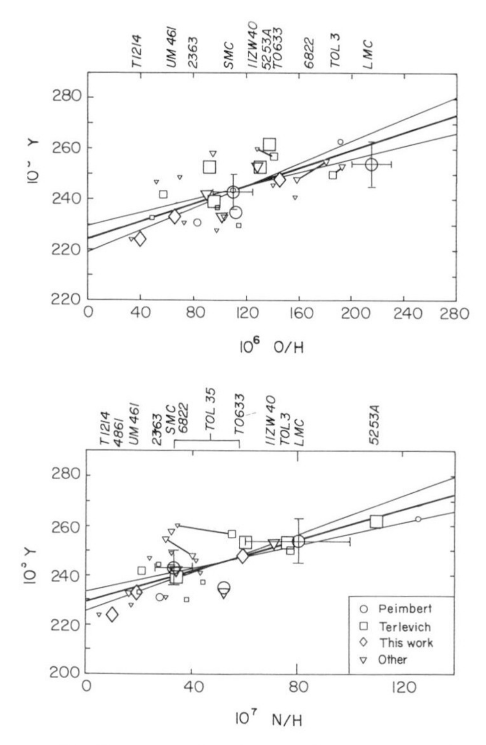

FIGURE 4. Regressions of helium mass fraction against oxygen and nitrogen abundance in irregular and blue compact or HII galaxies with oxygen up to 1/4 solar. Maximum likelihood regression lines are shown with alternatives equivalent to $\pm 1\ \sigma$ errors. Sizes of plotted symbols indicate their weights, typical error bars being shown for two high-weighted objects.[26,52,54]

The regression against oxygen shows deviations among some high-weighted points that are marginally significant, and its slope is very steep; alternatively, one could fit a curve that starts off steep and then flattens. The regression against nitrogen, which leads to our adopted value of Y_P, is smoother (within errors), although it cannot continue linearly beyond the frame of the diagram (e.g., Orion has 10^7 N/H = 800, $Y = 0.27$). The nitrogen connection could be due to local environmental pollution by winds from Wolf–Rayet stars (or their red-giant progenitors) that inject both nitrogen and fresh helium,[50,55,56] although this is not certain.[54] In any case, both regressions lead to consistent values of Y_P with a standard error of but a few parts in the third decimal place, and in excellent agreement both with the earlier work of Peimbert and colleagues[43] and with the values derived by totally different means from studies of globular clusters.[37] TABLE 1 gives a comparison with previous work that shows that the estimates of Y_P have remained quite stable over a decade despite many arguments.

Systematic errors, which could go either way, are inevitably a matter of judgement; errors of 1 or 2 percent can hardly be avoided from atomic theory, flux calibration, residual neutral helium, underlying stellar absorption lines, etc., but the agreement between the three different emission lines and different objects encourages me to believe that they do not amount to more than 0.005 in Y_P. A statistical one-tailed test then gives 95 percent confidence that the true value of Y_P does not exceed 0.240.

TABLE 1. Determinations of Primordial Helium with $\pm 1\ \sigma$ Errors

Lequeux *et al.*[43]	0.230 ± 0.004
Kunth and Sargent[44]	0.243 ± 0.010
Kunth and Sargent without IIZw 40[a]	0.234 ± 0.008
Peimbert[58]	0.232 ± 0.004
Pagel, Terlevich, and Melnick[50]	0.237 ± 0.005
Pagel[27]	0.232 ± 0.004
Torres-Peimbert, Peimbert, and Fierro[51]	0.230 ± 0.006
Pagel and Simonson[52]	0.229 ± 0.004

[a] For which HeI λ5876 suffers absorption by galactic NaI.[57]

IMPLICATIONS OF THE PRIMORDIAL HELIUM ABUNDANCE

Since an additional unit in N_ν increases Y_P by 0.014,[6] it is clear from FIGURE 1 that SBBN theory (or the mildly inhomogeneous version) excludes any contribution to relativistic degrees of freedom at neutrino decoupling from any light particles over and above the electron, muon, and tauon neutrino–antineutrino pairs. Furthermore, SBBN theory places very tight limits on the baryon density, although these could be relaxed somewhat in the mildly inhomogeneous case, and the latter may indeed become more plausible should it be possible to prove that Y_P is significantly less than 0.24, although this possibility remains wholly speculative for the time being. Specific limits on baryon density from SBBN are

$$0.011 \leq \Omega_{b0} h_{100}^2 \leq 0.017,$$

the lower limit coming from (D + ^{3}He)/H and the upper limit from Y_P. While it is universally agreed that h_{100} is between 0.4 and 1.0, it now seems likely to exceed 0.7,[59] which would rule out an Einstein–de Sitter universe with $\Omega = 1$, because of the ages of globular clusters. If $0.7 \leq h_{100} \leq 1.0$, then in SBBN

$$0.011 \leq \Omega_{b0} \leq 0.035.$$

The lower limit (valid also for smaller h_{100}) calls for baryonic dark matter, since visible matter in spiral and irregular galaxies, deduced from its mass–luminosity ratio of about 3 in solar units, corresponds to $\Omega_{vis} = 0.002h_{100}^{-1}$, and the larger mass–luminosity ratio found in ellipticals is itself probably due in large part to white dwarfs and neutron stars.[60] This amount of dark matter could be present in dark halos of spiral and dwarf galaxies deduced mainly from flat 21-cm rotation curves and the dynamics of the local group; alternatively, it might be there in the form of low surface-brightness galaxies not counted in conventional optical surveys.[30] The upper limit is less than most estimates of Ω_0(total) = 0.1 to 0.2 deduced from galaxy clustering dynamics,[61] leaving room for nonbaryonic dark matter.

In the (mildly) inhomogeneous BBN case, the corresponding limits are (roughly)

$$0.009 \leq \Omega_{b0}h_{100}^2 \leq 0.04,$$

or with $0.7 \leq h_{100} \leq 1.0$

$$0.009 \leq \Omega_{b0} \leq 0.08,$$

which leaves the case for dark baryonic matter virtually unchanged, but seriously weakens the case for any nonbaryonic matter. Consequently, it would be of great interest if one could pin down the primordial helium abundance still more accurately; the need for better estimates of Ω_0 and h_{100} is obvious.

ACKNOWLEDGMENTS

I thank the UK PATT for assigning time on the AAT for work described here and the director and staff of the Anglo-Australian Observatory for willing and expert assistance. I also thank Roberto Terlevich, Mike Edmunds, and Ed Simonson, who all played an essential role in our quest for more certainty about primordial helium.

REFERENCES

1. PEEBLES, P. J. E. 1966. Astrophys. J. **146:** 542.
2. WAGONER, R. V., W. A. FOWLER & F. HOYLE. 1967. Astrophys. J. **148:** 3.
3. SCHRAMM, D. N. & R. V. WAGONER. 1977. Ann. Rev. Nucl. Sci. **27:** 37.
4. YANG, J., M. S. TURNER, G. STEIGMAN, D. N. SCHRAMM & K. A. OLIVE. 1984. Astrophys. J. **281:** 493.
5. BOESGAARD, A. M. & G. STEIGMAN. 1985. Ann. Rev. Astron. Astrophys. **23:** 319.
6. OLIVE, K. A., D. N. SCHRAMM, G. STEIGMAN & T. P. WALKER. 1990. Phys. Lett. B **236:** 454.
7. TAYLER, R. J. 1982. *In* The Big Bang and Element Creation, D. Lynden-Bell, Ed.: 19. Philos. Trans. R. Soc. London A, Vol. 307.
8. DELIYANNIS, C. P., P. DEMARQUE & S. D. KAWALER. 1990. Astrophys. J., Suppl. Ser. **73:** 21.

9. HOYLE, F. & R. J. TAYLER. 1964. Nature **203:** 1108.
10. STEIGMAN, G., D. N. SCHRAMM & J. E. GUNN. 1977. Phys. Lett. B **66:** 202.
11. TAYLER, R. J. 1968. Nature **217:** 433.
12. ———. 1990. Quart. J. R. Astron. Soc. **31:** 371.
13. MAMPE, W., P. AGERON, C. BATES, J. M. PENDLEBURY & A. STEYERL. 1989. Phys. Rev. Lett. **63:** 593.
14. PAGEL, B. E. J. 1988. Gemini (Royal Greenwich Obs.), Jan.
15. ELLIS, J., P. SALATI & P. SHAVER, EDS. 1990. Proc. ESO-CERN Topical Workshop: LEP and the Universe, Geneva, CERN:TH 5709/90.
16. APPLEGATE, J. H. & C. J. HOGAN. 1985. Phys. Rev. D **31:** 3037.
17. ALCOCK, C., G. M. FULLER & G. J. MATHEWS. 1987. Astrophys. J. **320:** 439.
18. APPLEGATE, J. H., C. J. HOGAN & R. J. SCHERRER. 1988. Astrophys. J. **329:** 572.
19. KAWANO, L. H., W. A. FOWLER & R. A. MALANEY. 1991. Astrophys. J. **372:** 1.
20. TERASAWA, N. & K. SATO. 1989. Phys. Rev. D **39:** 2893.
21. ———. 1990. Astrophys. J., Lett. **362:** L47.
22. REEVES, H. 1988. *In* Dark Matter, J. Audouze and J. T. T. Van, Eds.: 287. Ed. Frontieres. Paris.
23. ———. 1991. Phys. Rep. **201:** 337.
24. KURKI-SUONIO, H., R. A. MATZNER, K. A. OLIVE & D. N. SCHRAMM. 1990. Astrophys. J. **353:** 406.
25. KAJINO, T. & R. N. BOYD. 1990. Astrophys. J. **359:** 267.
26. PAGEL, B. E. J. 1991. Phys. Scr. **T36:** 7.
27. ———. 1987. *In* A Unified View of the Macro- and Micro Cosmos (First International School on Astro-particle Physics, Erice, Sicily), A. de Rujula, D. V. Nanopoulos, and P. A. Shaver, Eds.: 399. World Scientific. Singapore.
28. PAGEL, B. E. J. 1982. *In* The Big Bang and Element Creation, D. Lynden-Bell, Ed.: 19. Philos. Trans. R. Soc. London A, Vol. 307.
29. VANGIONI-FLAM, E. & J. AUDOUZE. 1988. Astron. Astrophys. **193:** 81.
30. PAGEL, B. E. J. 1990. *In* Baryonic Dark Matter, D. Lynden-Bell and G. Gilmore, Eds.: 237. Kluwer. Dordrecht, the Netherlands.
31. SPITE, F. & M. SPITE. 1982. Astron. Astrophys. **115:** 357.
32. REBOLO, R., P. MOLARO & J. E. BECKMAN. 1988. Astron. Astrophys. **192:** 192.
33. SMITH, V. V. & D. L. LAMBERT. 1990. Astrophys. J., Lett. **361:** L69.
34. VAUCLAIR, S. 1988. Astrophys. J. **335:** 971.
35. MATHEWS, G. J., C. ALCOCK & G. M. FULLER. 1990. Astrophys. J. **349:** 449.
36. KRAUSS, L. M. & P. ROMANELLI. 1990. Astrophys. J. **358:** 47.
37. SHAVER, P. A., D. KUNTH & K. KJÄR, Eds. 1983. Primordial Helium. ESO. Garching.
38. PAGEL, B. E. J. 1989. *In* Evolutionary Phenomena in Galaxies, J. E. Beckman and B. E. J. Pagel, Eds.: 368. Cambridge Univ. Press. Cambridge, England.
39. BROCKLEHURST, M. 1972. Mon. Not. R. Astron. Soc. **157:** 211.
40. PEIMBERT, M. & S. TORRES-PEIMBERT. 1974. Astrophys. J. **193:** 327.
41. ———. 1976. Astrophys. J. **203:** 581.
42. SEARLE, L. & W. L. W. SARGENT. 1972. Astrophys. J. **173:** 25.
43. LEQUEUX, J., M. PEIMBERT, J. F. RAYO, A. SERRANO & S. TORRES-PEIMBERT. 1979. Astron. Astrophys. **80:** 155.
44. KUNTH, D. & W. L. W. SARGENT. 1983. Astrophys. J. **273:** 81.
45. DAVIDSON, K. & T. D. KINMAN. 1985. Astrophys. J., Suppl. Ser. **58:** 321.
46. CLEGG, R. E. S. 1987. Mon. Not. R. Astron. Soc. **229:** 31P.
47. DINERSTEIN, H. & G. SHIELDS. 1986. Astrophys. J. **311:** 45.
48. PEIMBERT, M., S. TORRES-PEIMBERT & M. PEÑA. 1986. Astron. Astrophys. **158:** 266.
49. CAMPBELL, A., R. J. TERLEVICH & J. MELNICK. 1986. Mon. Not. R. Astron. Soc. **223:** 811.
50. PAGEL, B. E. J., R. J. TERLEVICH & J. MELNICK. 1986. Publ. Astron. Soc. Pac. **98:** 1005.
51. TORRES-PEIMBERT, S., M. PEIMBERT & J. FIERRO. 1989. Astrophys. J. **345:** 186.
52. PAGEL, B. E. J. & E. A. SIMONSON. 1989. Rev. Mex. Astron. Astrofis. **18:** 153.
53. TERLEVICH, R. J., J. MELNICK, J. MASEGOSA & M. MOLES. 1991. Astron. Astrophys. Suppl. In press.
54. SIMONSON, E. A. 1990. Ph.D. thesis, Sussex University, Sussex, England.

55. PAGEL, B. E. J. 1987. *In* Starbursts and Galaxy Evolution, T. Montmerle and J. T. T. Van, Eds.: 227. Ed. Frontieres. Paris.
56. WALSH, J. R. & J.-R. ROY. 1989. Mon. Not. R. Astron. Soc. **239:** 297.
57. FRENCH, H. B. 1980. Astrophys. J. **240:** 41.
58. PEIMBERT, M. 1985. *In* Star Forming Dwarf Galaxies, D. Kunth, T. X. Thuan, and J. T. T. Van, Eds.: 403. Ed. Frontieres. Paris.
59. TULLY, R. B. 1990. *In* Astrophysical Ages and Dating Methods, E. Vangioni-Flam, M. Cassé, J. Audouze, and J. T. T. Van, Eds.: 3. Ed. Frontieres. Paris.
60. YOSHII, Y. & N. ARIMOTO. 1987. Astron. Astrophys. **188:** 13.
61. PEEBLES, P. J. E. 1986. Nature **321:** 27.

ROSAT: Early Results[a]

JOACHIM E. TRÜMPER

Max-Planck-Institut für Extraterrestrische Physik
8046 Garching bei München
Germany

INTRODUCTION

The general scientific objectives of ROSAT[1] are to perform (a) the first all-sky surveys using imaging X-ray and extreme ultraviolet (EUV) telescopes and (b) detailed investigations of interesting sources in a guest investigator program.

The survey operations commenced in August 1990 and at the time of the 15th Texas/ESO–CERN Conference about 75 percent of the sky had been scanned. In February 1991 we started the first half year (AO-1) guest observer program for which in total 738 proposals were received. A glimpse of what can be achieved by pointed observations was obtained during the calibration and verification measurements in the early phase of the mission (June/July 1990).

Before I give a brief summary of the results obtained with the X-ray telescope so far, a few remarks on the scientific instruments may be appropriate.

ROSAT

The ROSAT spacecraft is the largest scientific satellite so far built in Western Europe (2.4 tons). It carries two main instruments. The large one is a fourfold nested X-ray Wolter telescope (XRT) with two position-sensitive proportional counters (PSPC) and one high-resolution imager (HRI) that can be brought alternatively into the focus. The smaller one is the threefold nested Wolter–Schwarzschild–mirror system (WFC) with two-channel plate detectors in the focal plane. Both instruments are orientated parallel and observe simultaneously in adjacent energy bands (XRT: 0.1–2.4 keV; WFC: 0.03–0.1 keV).

All instruments are working well. The main new aspects compared with previous X-ray telescope missions (EINSTEIN, EXOSAT) are as follows:

- ROSAT performs the first all-sky survey in X-rays with an imaging telescope having a sensitivity of $\sim 100 \times$ HEAO-1.
- it performs the first all-sky survey in the EUV.
- it provides an increased sensitivity (a factor of 3–10) because of the large collecting area of the X-ray telescope.
- the spectral resolution of the PSPC ($E/_{\Delta E} \sim 2.5$) allows us to perform broad-band spectroscopy: four color bands.

[a]The ROSAT project is supported by the Bundesministerium für Forschung und Technologie (BMFT), by NASA, and by SERC.

- the low intrinsic background of the PSPC ($\sim 1.5 \times 10^{-5}$ cts/cm^{2}s), which together with the extreme uniformity of the detector response, allows us to image directly the diffuse galactic and extragalactic X-ray backgrounds.

ROSAT RESULTS

The present status of the sky survey is shown in FIGURE 1, in which the total PSPC count rate is plotted as a function of galactic coordinates. The map shows some of the large-scale features of the X-ray sky, such as the Cygnus superbubble, with the adjacent Cygnus Loop, the North Polar Spur, the Vela Supernova Remnant, and the Crab Nebula.

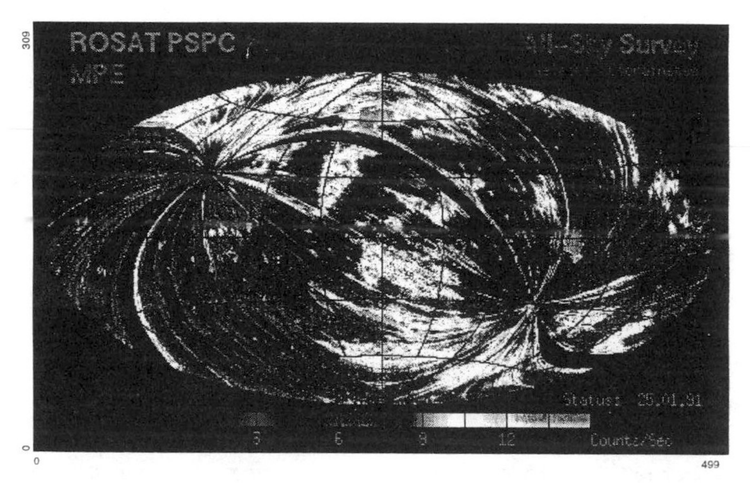

FIGURE 1. Status of the ROSAT all-sky survey in X-rays. Plotted is the PSPC count rate in galactic coordinates with the galactic center in the middle.

Preliminary estimates of the total number of sources visible in the sky survey is close to 60,000.[2] Most of them are expected to be AGNs (twenty to thirty thousand) followed by normal stars. The accuracy of location will be much better from 1 arcmin. FIGURE 2 shows the ratio of X-ray to optical flux as a function of visual magnitude for a compilation of a large number of sources that have been identified with SIMBAD catalog sources.

GALACTIC SOURCES

The ROSAT survey provides us, for the first time, with a view of the X-ray sky that is both wide and deep. FIGURE 3 shows a rather large region (250 square degrees) near the galactic center observed in the XRT survey. Besides the well-known bright X-ray binaries it contains Kepler's supernova remnant (SNR), and the

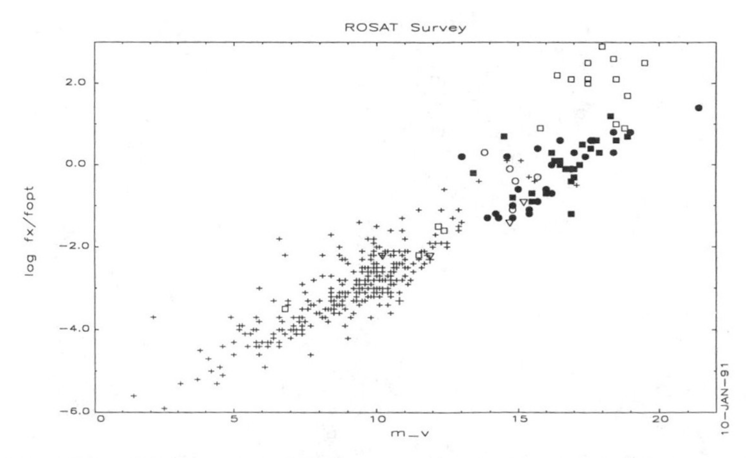

FIGURE 2. Ratio of X-ray to optical flux for a sample of a large number of sources derived from the ROSAT all-sky survey and identified with SIMBAD catalog sources: + stars, ○ white dwarfs, △ cataclysmic variables, □ neutron star binaries, ■ AGN, ● cluster of galaxies.

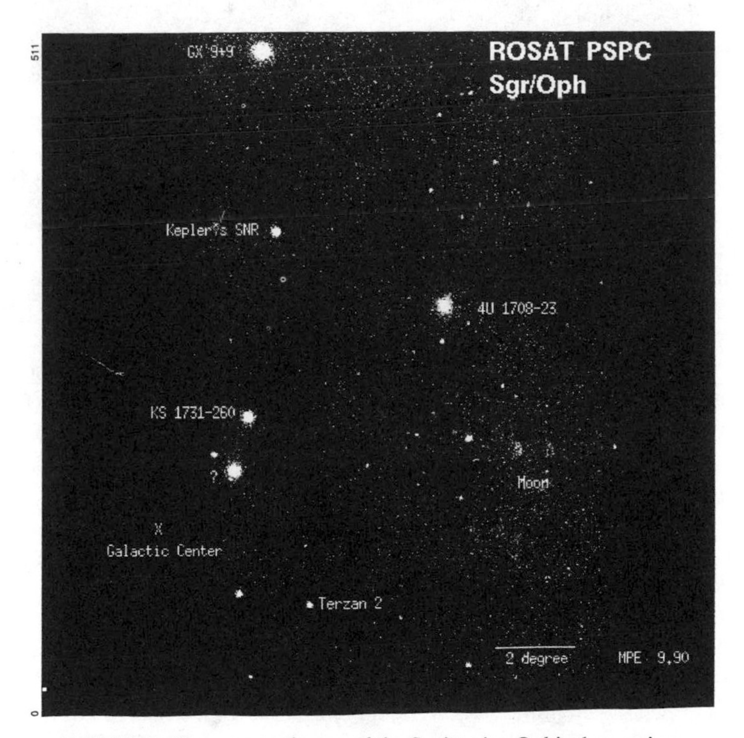

FIGURE 3. X-ray survey image of the Sagittarius Ophiuchus region.

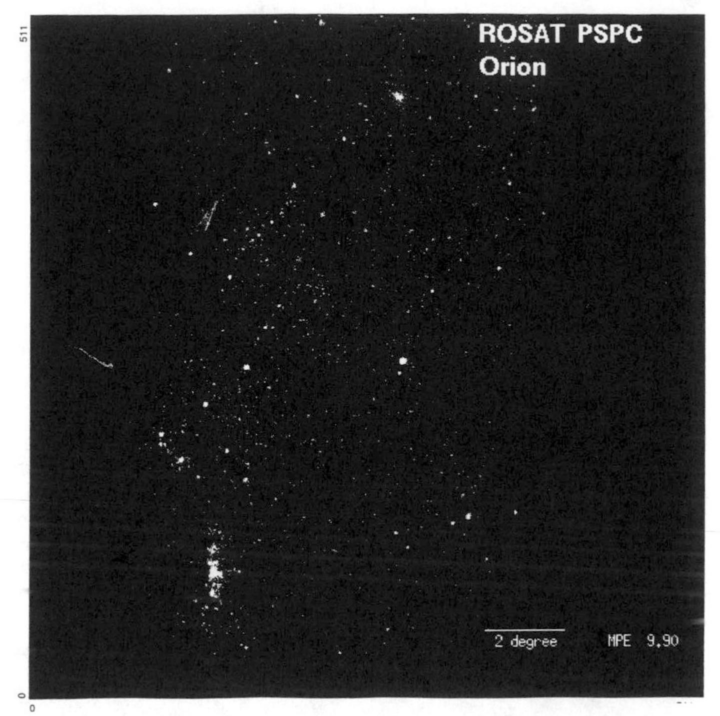

FIGURE 4. X-ray survey image of the Orion region.

moon, which is scanned every fortnight in two consecutive ROSAT orbits. Some diffuse emission regions and many new faint X-ray sources are seen as well. FIGURE 4 shows the Orion region with the bright O stars of the belt and diffuse emission from the well-known star-formation region. Preliminary estimates indicate that fifteen to twenty thousand normal stars of all spectral types will be visible in the survey.

The most photogenic sources are, of course, the old and extended supernova remnants that are imaged by ROSAT for the first time in a homogeneous way. FIGURE 5 shows the Cygnus Loop as an example. The first of our newly discovered supernova remnants is shown in FIGURE 6. It has an almost perfect spherical shape with a diameter of 108 arcmin and delivers an X-ray flux of $\sim 2 \times 10^{-10}$ erg/cm^{2}s (0.1–2.4 keV) ranking it among the 10 X-ray brightest galactic SNRs. A preliminary Sedov analysis[3] indicates that it exploded $\sim 3 \times 10^4$ years ago in a low-density region (0.01 cm^{-3}). It is neither visible in the Condon radio survey or on the optical sky plates. Deep observations with the 100-m Effelsberg telescope have revealed a shell radio source of low surface brightness. The X-ray/radio brightness ratio of this object is a factor of 4 lower than for any other known SNR. It is clear that ROSAT is very powerful in picking up new SNRs. We expect to discover a few dozen of them in the all-sky survey.

LARGE MAGELLANIC CLOUD

The Large Magellanic Cloud (LMC) was our "first light" target because we were very eager to have a look at the SN 1987A. Our rocket flight in August 1979 had given an upper limit of the soft X-ray flux.[4] Despite the much larger collecting power and longer observation time with ROSAT, again only an upper limit could be obtained ($L_X < 3 \times 10^{34}$ erg/s), which was confirmed by later scanning observations. The null result is consistent with the fact that the X-ray emission from the SN shock wave is still weak because of the low density of the progenitor's stellar wind. At the same time it means that no neutron star activity[5] can be seen so far.

On the other hand, of a total number of 45 sources detected in the 8 degree^2 field,[6] 34 had been known from EINSTEIN and EXOSAT observations. Among the 11 new sources there is an extremely soft bright source that is a transient because it was not seen by the Einstein observatory. The spectra of this source (RX I0527.8-6954) and that of CAL 83, which is probably a low-mass X-ray binary, are very similar. The spectra of both sources indicate temperatures in the range 20–40 eV, and hydrogen column densities consistent with a LMC membership (FIG. 7). The luminosities of both sources are close to the Eddington limit. They may represent a new class of low-mass X-ray binaries that remained largely undetected so far. If true, this could help to solve the millisecond birthrate problem.[7]

FIGURE 5. X-ray survey image of the Cygnus Loop (0.1–0.28 keV).

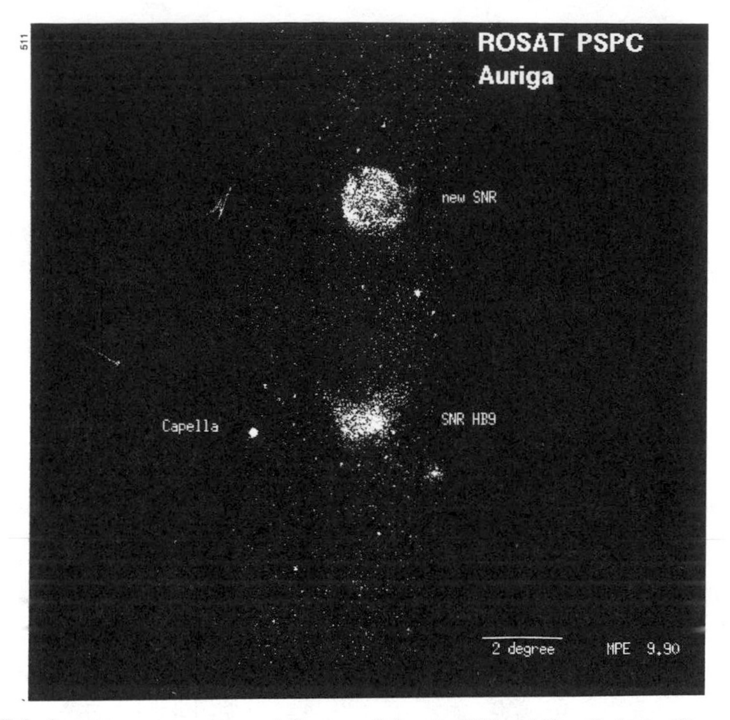

FIGURE 6. A new supernova remnant discovered close to HB9 and Capella in the X-ray survey.

CLUSTERS OF GALAXIES

In the ROSAT all-sky survey we expect to detect five to eight thousand clusters of galaxies.[8] Of course, their identification and redshift determination require optical observations, and therefore this will be a long-range program that we just started. The general goal of this program is to determine the distribution of clusters of galaxies as tracers of the density peaks in our universe.

Another important aspect is the morphological investigation of individual clusters. The first one studied in detail is the PSPC pointing observation of Abell 2256.[9] On the basis of EINSTEIN observations (FIG. 8), this object had been considered as a relaxed Coma-type cluster that is dynamically well-evolved. In the ROSAT image we see a clear double peak structure (FIG. 8).

The eastern peak is coincident with a central cD galaxy, while the morphology of the western part shows indications that it is merging with the main cluster body. This observation provides the first direct observational evidence for a merging event. There are other examples in the ROSAT sample that support this evidence.

EXTRAGALACTIC X-RAY BACKGROUND

The diffuse X-ray background was discovered (along with SCO X-1) accidentally by an experiment that aimed at the detection of lunar soft X-rays.[10] Twenty-eight years later this goal has now been achieved by ROSAT.

The first X-ray image of the moon clearly shows the expected occultation of the X-ray background by the moon.[11] The nature of the X-ray background is still unresolved. In particular it is unclear to what extend it is truely diffuse, that is, produced by thermal emission from a hot intergalactic medium, or whether it is the summed emission of many faint-point sources like quasars. EINSTEIN observations showed that at least ~20 percent of the background above 2 keV was due to a discrete source.[12]

The new aspect offered by ROSAT is that the same instrument can be used to image both the diffuse X-ray background and the sources. The hitherto deepest ROSAT X-ray image has been taken at the north ecliptic pole (NEP). In this field and seven other less deep fields a total of 184 sources were detected in the flux interval 10^{-14}–2×10^{-13} erg/cm² s.[13]

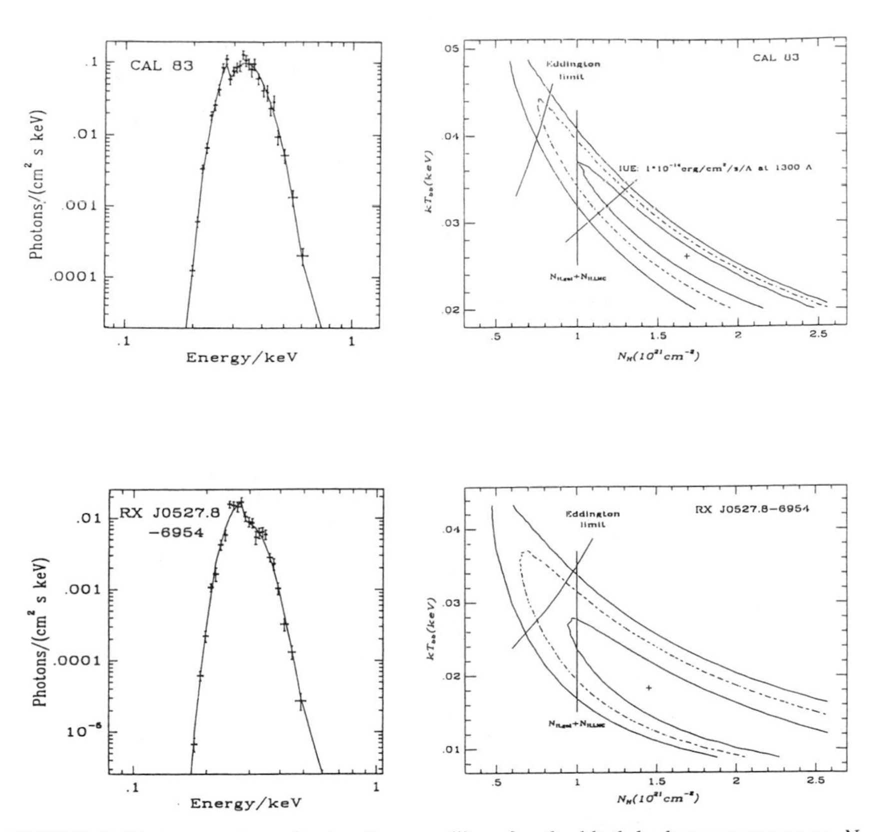

FIGURE 7. Photon spectrum (**top**) and error ellipse for the black body temperature vs. N_H (**right**) of CAL 83. The Eddington and the IUE limit assume an LMC distance. The *contours* in (**b**) correspond to the 68, 95, and 99 percent confidence levels. Photon spectrum (**bottom**) and error ellipse for the black body temperature vs. N_H (**right**) of the new supersoft source RX J0527.8-6954. The Eddington limit assumes an LMC distance. The contours in (**b**) correspond to the 68, 95, and 99 percent confidence levels.

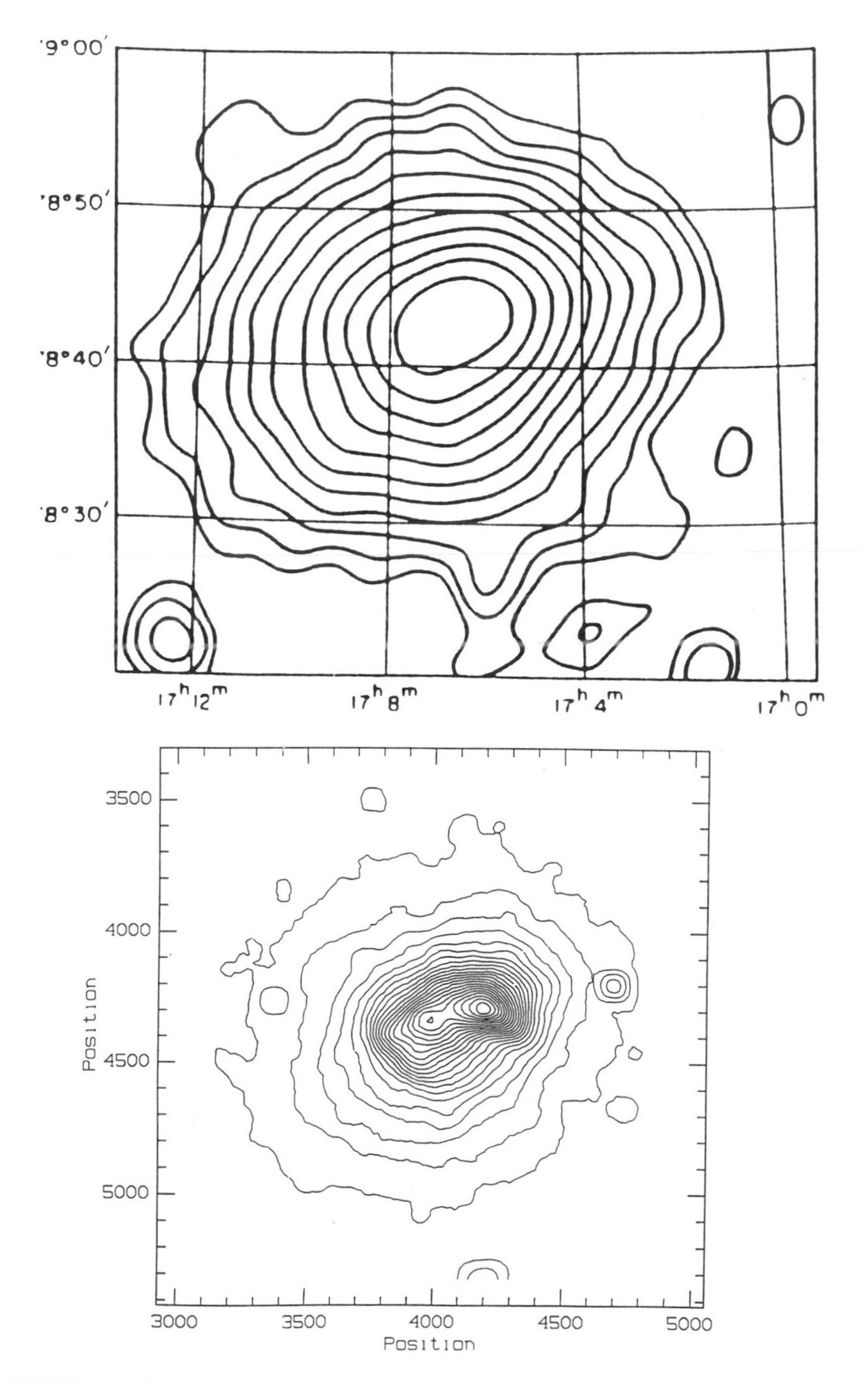

FIGURE 8. Diffuse X-ray emission from Abell 2256. **Top:** Einstein IPC image. **Bottom:** ROSAT PSPC image (same scale).

Here I would like to summarize briefly a few results:

1. ROSAT resolves ~30 percent of the X-ray background into point sources at its present sensitivity limit (10^{-14} erg/cm^{2}s).
2. The log-*N*–log-*S* curve, which has a slope of ~1.5 down to the EINSTEIN deep survey limit, flattens for faint source fluxes (FIG. 9).
3. The spectrum of the X-ray background in the 0.8–2.4-keV band is given by $E^{-\alpha}$, with $\alpha = 1.2 \pm 0.1$, which means that it is significantly steeper than at ≈3 keV. The integral spectrum of the sources has the same slope as the X-ray background spectrum within the error limits ($\alpha = 1.3 \pm 0.1$).
4. We find evidence for significant structures in the residual background in the NEP field.

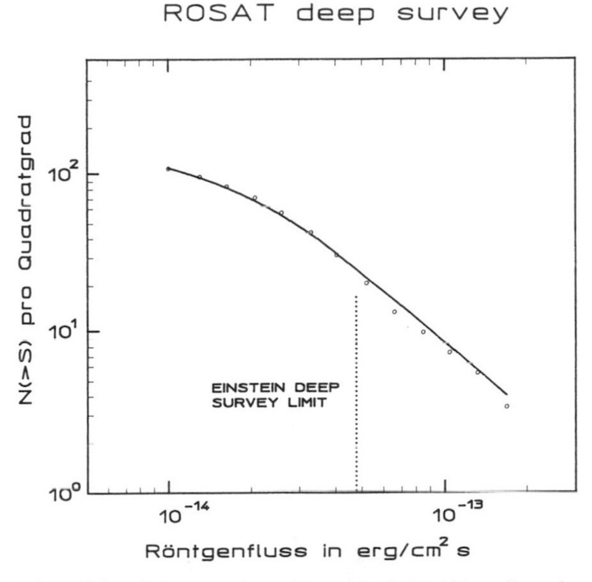

FIGURE 9. Log-*N*–log-*S* diagram derived from the ROSAT medium-deep surveys.

During the forthcoming months we intend to perform deep survey observations that penetrate to substantially deeper (a factor of ~5) limits.

ACKNOWLEDGMENT

I acknowledge the contributions of many people in DARA, in DLR, in industry, and in the participating scientific institutes who helped to make ROSAT possible.

I gratefully acknowledge the efficient mission support provided by the ROSAT team of the German Space Operations Center (GSOC). In particular, I would like to thank many colleagues at the Max-Planck-Institut für Extraterrestrische Physik who worked with dedication on the ROSAT hardware and software.

REFERENCES

1. TRÜMPER, J. 1983. Adv. Space Res. **2:** 241.
2. VOGES, W. 1991. Private communication.
3. PFEFFERMANN, E. *et al.* 1991. Astron. Astrophys. **246:** 228.
4. ASCHENBACH, B. *et al.* 1987. Nature **330:** 232.
5. PACINI, F. 1987. ESO Conf. on SN 1987A.
6. TRÜMPER, J. *et al.* 1991. Nature **349:** 579.
7. GREINER, J. *et al.* 1991. Astron. Astrophys. **246:** 17.
8. BÖHRINGER, H. 1991. Private communication.
9. BRIEL, U. *et al.* 1991. Astron. Astrophys. **246:** 10.
10. GIACCONI, R. *et al.* 1962. Phys. Rev. Lett. **9:** 439.
11. SCHMITT, J. *et al.* 1991. Nature. **349:** 583.
12. GIACCONI, R. *et al.* 1979. Astrophys. J., Lett. **234:** L1.
13. HASINGER, G. *et al.* 1991. Astron. Astrophys. **246:** 2.

Deep ROSAT Observations in a Faint Quasi-stellar Object Survey Field

T. SHANKS,[a] I. GEORGANTOPOULOS,[a] G. C. STEWART,[b]
K. A. POUNDS,[b] B. J. BOYLE,[c] AND R. E. GRIFFITHS[d]

[a]*Physics Department*
University of Durham
South Road
Durham DH1 3LE, England

[b]*X-ray Astronomy Group*
Physics Department
Leicester University
Leicester LE1 7RH, England

[c]*Institute of Astronomy*
Madingley Road
Cambridge CB3 OHA, England

[d]*Space Telescope Science Institute*
3700 San Martin Drive
Baltimore, Maryland 21218

INTRODUCTION

The source of the diffuse X-ray background is still controversial. In particular, estimates of the contribution of discrete AGN sources to the background has varied greatly, with some authors suggesting that they accounted for up to 100 percent of the soft extragalactic background,[1,2] whereas others have suggested upper limits of 20–30 percent.[3] Other suggested contributors to the diffuse X-ray background have included hot gas and starburst galaxies.

The most direct way to identify the source of the X-ray background is by obtaining deep X-ray exposures with good resolution, identifying the optical counterparts, and then following them up with optical spectroscopy. Previously, the best attempt at this was made by Griffiths *et al.*[4] using the Einstein X-ray satellite. They made a 100,000-s HRI exposure on each of four adjoining fields in Pavo. They identified 17 sources that they followed up optically to identify 14 QSOs in a 0.44 deg^2 area for a background X-ray quasi-stellar object (QSO) density of $N_{xqso} = 32 \pm 9$ deg^{-2}. This density was regarded as controversially high; others had not discovered such a high QSO density, albeit in less high-resolution X-ray data (Giacconi *et al.*[5]). Also the X-ray QSO statistics obtained were based on data from only one field.

We describe below a very recent attempt to check and improve on the Griffiths result in a different field.

ROSAT OBSERVATIONS

We have been allocated Röntgenstrahlen Satellit (ROSAT) time to make deep exposures in five of the QSO survey fields of Boyle *et al.*[6] The fields are part of a larger sample that has been surveyed for QSOs to $B = 21^m$ using the optical, ultraviolet excess technique that selects complete samples of QSOs with $z < 2.2$. The aim is to resolve the X-ray background with better sensitivity than previously possible in a large area of sky where good-quality optical data already exist.

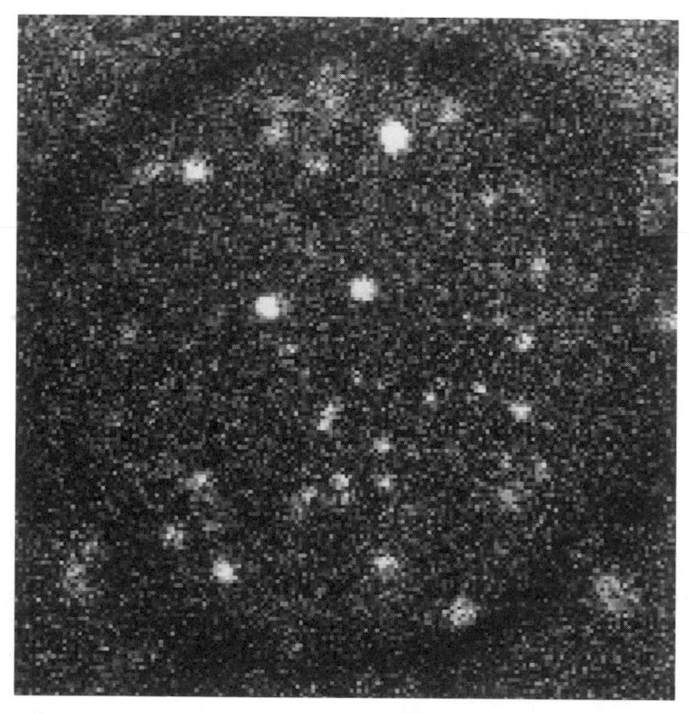

FIGURE 1. The central, 40′ diameter of the ROSAT PSPC field. The energy band is 0.1–2.4 keV and the flux limit is approximately 2×10^{-14} ergs cm^{-2} s^{-1}. Thirty-nine sources have been detected in this area, leading to a sky density of sources of 110 deg^{-2}. The brightest source is a $B < 11^m$F star, and the next two brightest are UVX QSOs with $z < 1$.

Currently, we have obtained one ROSAT PSPC exposure of 30,000 s in the QSF3 (03h40,-44) field of the UVX survey. The ROSAT PSPC detector[7] is sensitive in the 0.1–2-keV waveband, a slightly softer range than the Einstein IPC/HRI energy range. The spatial resolution of the PSPC detector is 25″ full width at half maximum (FWHM) on axis. The quality of the data is very high, with virtually no detector-induced background. The ROSAT data are displayed in FIGURE 1. Although the field of the PSPC is 2° in diameter, the PSPC has best sensitivity and resolution within a 40′ diameter field, and that is what is shown here. This is almost exactly the same area as previously searched for $B < 21^m$ QSOs in the spectroscopic survey.

When the ROSAT data were obtained they were first analyzed for point sources. Thirty-nine such sources were found in the central 40′-diameter field. This corresponds to a very high sky density of 110 ± 18 deg^{-2}. We next discuss the question of the nature of these sources.

OPTICAL OBSERVATIONS

In the previous UVX survey 19 $z < 2.2$ QSOs had been detected in the QSF3 field to $B < 21.^m2$. Using the ROSAT coordinates, it was obvious that 12 of these UVX sources had been picked up, once systematic offsets in the ROSAT right ascention (RA) and declination (dec) had been removed. The optical QSOs have positions accurate to ±1″ and, using these, we derived the error on the relative ROSAT positions to be ±15″. We then cross-correlated the remaining source positions with the COSMOS machine object list from the measurement of the original UK Schmidt Telescope (UKST) plate and found optical counterparts for 34 out of 39 sources to $B = 22.^m3$, including the previously surveyed QSOs. In most cases there was only one counterpart within ±15″.

With the positions of the optical counterparts good to 1″ accuracy, we were then able to use the Anglo-Australian telescope's (AAT's) automated fiber-optic coupler system, AUTOFIB[8] to observe the remaining candidates. The AAT observations were carried out on November 14–16, 1990, and in a service observation on December 11, 1990. The AUTOFIB field of view was 40′ in diameter, and therefore well matched to the PSPC field, which allowed all 34 sources to be observed simultaneously.

From these spectroscopic observations, 11 further QSOs were identified along with two possible QSOs. The spectra of 10 newly confirmed QSOs are shown in FIGURE 2. Taking these along with the 12 QSOs that had been previously identified, means that 25 X-ray sources have been identified as QSOs on this field for an overall sky density of X-ray QSOs of $N_{xqso} = 75 \pm 15$ deg^{-2}. *Thus, not only have we confirmed the high-density of X-ray QSOs found by Griffiths et al. but, in this new field, the X-ray QSO density seems more than a factor of 2 higher.*

Some particular QSOs deserve specific mention. A double QSO with a 5″ separation has been found that was at first thought to be a candidate gravitational lens. However, subsequent observations have proved that, although there are two QSOs, they have the quite different redshifts of $z = 1.59$ and $z = 2.08$.

We have also found a QSO at $z = 2.55$. We believe this to be the highest redshift QSO so far detected in an X-ray-selected sample.

We have also identified other candidate sources from the optical spectroscopy that are not QSOs. Two bright ($B < 11^m$) F stars were identified; one was the brightest X-ray source in the area. Another ($B \approx 18^m$) binary composite star was positively identified, as well as another probable identification of a fainter star. The spectra of one of the F stars is shown in FIGURE 2. Two further sources were identified by COSMOS as galaxies. One has a sharp feature in its spectrum, which if identified as the CaII $H + K$ break gives a redshift of 0.18 (see FIG. 2). In the other

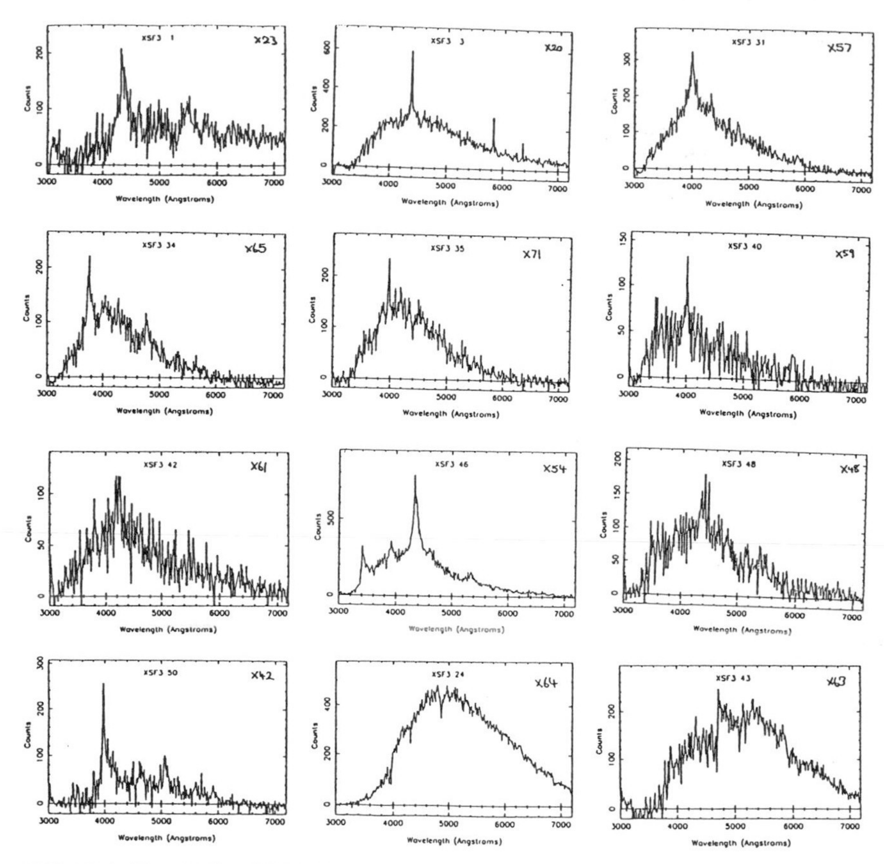

FIGURE 2. The AAT AUTOFIB spectra of 10 newly confirmed QSO X-ray sources. The optical brightness of these QSOs lies past the UVX survey limit of $B = 21.^{m}2$. One of the three confirmed galactic stars is also shown in the *middle of the bottom panel,* and the confirmed galaxy with a strong CaII $H + K$ break giving $z = 0.18$, is shown as the *rightmost object of the bottom panel.*

case the spectrum does not give a reliable redshift. Apart from these there are nine objects for which no spectroscopic identification is possible, including the five sources for which no optical counterpart is seen to the UKST plate limit. The following list summarizes the optical identifications on this field.

UVX QSO	12
Newly identified QSO	11
Possible QSO	2
Galaxy	1
Galaxy?	1
Star	3
Star?	1
Unidentified	9

RESULTS FROM THE SPECTROSCOPIC SURVEY

We first look at the QSO $n(z)$ relation (see FIG. 3). We note that the $n(z)$ extends up to $z = 2.5$. This is unlike the behavior seen for the bright X-ray-selected samples that have very few $z > 1$ QSOs. Our $n(z)$ appears very much more similar to the Einstein $n(z)$ of Griffiths *et al.*[4] The X-ray QSO $n(z)$ is also very similar to the UVX survey $n(z)$ of Boyle *et al.*,[6] which is unsurprising given the large degree of overlap between the two samples.

The X-ray data are confirming the high degree of completeness of the UVX survey; only two QSOs with $z < 2.2$ and $B < 21.^m2$ were picked up in the QSF3 field in the X-ray survey. One was the double QSO that was omitted because it was misclassified by COSMOS as a galaxy. The other had $B = 19.73$ and $z = 0.564$. This QSO therefore lies in a redshift range where QSOs are known to be slightly less UVX; its color is $u - b = -0.12$, which placed it just outside the UVX limit for this field.

Twelve out of nineteen UVX QSOs were detected by ROSAT. Of the seven that were missed, in at least three cases this was understandable, with one UVX QSO lying close to a brighter X-ray source, another being obscured by the telescope optics, and a third, which was previously classed as Q?, would now seem to be stellar on the basis of a longer exposure spectrum. The other three QSOs all have $B > 20.^m5$ and it may not be surprising that some such QSOs may be undetected in the X-ray. Thus at least two-thirds of the optical QSOs have proved to be X-ray loud, and an almost very optically bright QSO has been detected in the X-ray.

No major contributor to the extragalactic X-ray background other than AGN has been found. Only two galaxies were identified; neither of these galaxies showed emission lines, indicating that they are not starburst galaxies. No obvious rich galaxy clusters are associated with any of the X-ray sources.

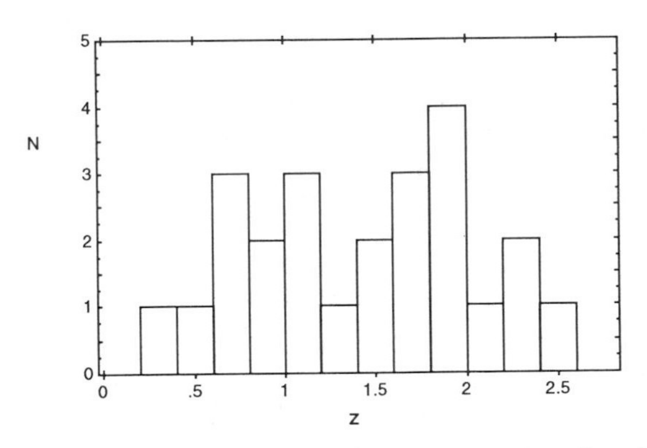

FIGURE 3. The number–redshift relation for the 24 X-ray-selected, confirmed QSOs. The $n(z)$ relation extends to larger redshifts than that for brighter X-ray-selected samples, with the highest redshift QSO having $z = 2.5$. The ROSAT QSO $n(z)$ is more similar to the optically selected QSO $n(z)$ of the UVX samples.

CONTRIBUTION OF DISCRETE SOURCES TO THE X-RAY BACKGROUND

The total contribution of sources to the extragalactic X-ray background can be directly estimated by dividing the summed flux from the sources by the total flux detected in the central 40′-diameter area. This gives a lower limit to the point source contribution of ≈ 20 percent in the 0.1–2.4-keV band, with the 25 detected QSOs contributing around half of it.

The PSPC energy resolution allows us to make a similar estimate at the monochromatic energy of 1 keV. A lower limit to the QSO contribution at 1 keV is $\approx 30 \pm 5$ percent. An increasing fraction of the observed PSPC background is expected to be galactic as we move to lower energies. Extrapolating the 3–40-keV spectrum observed by Marshall *et al.*[9] down to 1 keV suggests that only 60 percent of the ROSAT flux at 1 keV may be extragalactic. This would result in a QSO contribution of ≈ 50 percent to the extragalactic component of the X-ray background.

CONCLUSIONS

A deep ROSAT PSPC observation has been made in a high galactic latitude field previously surveyed for faint QSOs. ROSAT detected a large number of X-ray sources corresponding to a sky density of 110 deg^{-2}. The X-ray observations have been followed up with optical spectroscopy at the Anglo-Australian telescope. Thirty out of thirty-nine sources have been spectroscopically identified; twenty-four have been identified as QSOs. This corresponds to an X-ray QSO sky density of ≈ 70 deg^{-2}. Four galactic stars and two galaxies have also been identified. Thus the faint X-ray sources seem to be dominated by QSOs. The X-ray QSO $n(z)$ extends to $z = 2.5$. Finally, the identified QSOs contribute between 30 and 50 percent of the extragalactic X-ray background at 1 keV.

ACKNOWLEDGMENTS

We thank the ROSAT PSPC team at the Max-Planck Institute für Extraterrestriche Physik for obtaining the ROSAT data. We thank the Director and staff of the Anglo-Australian telescope for their flexibility in allowing us to use the AUTOFIB fiber-optic coupler at short notice.

REFERENCES

1. Piccinotti, G., R. F. Mushotzky, E. A. Boldt, S. S. Holt, F. E. Marshall, P. J. Serlemitsos & R. A. Shafer. 1982. Astrophys. J. **253:** 485–503.
2. Schmidt, M. & R. F. Green. 1986. Astrophys. J. **305:** 68.
3. Barcons, X. & A. C. Fabian. 1989. Mon. Not. R. Astron. Soc. **237:** 119.
4. Griffiths, R. E., I. R. Tuohy, R. J. V. Brissenden, M. Ward, S. S. Murray & R. Burg. 1988. *In* Proceedings of a Workshop on Optical Surveys for Quasars, P. S. Osmer, A. C. Porter, R. F. Green, and C. B. Foltz, Eds.: 351–360. Astronomic Society of the Pacific. San Francisco.

5. GIACCONI, R., *et al.* 1979. Astrophys. J., Lett. **234:** L1.
6. BOYLE, B. J., R. FONG, T. SHANKS & B. A. PETERSON. 1990. Mon. Not. R. Astron. Soc., **243:** 1–56.
7. BRIEL, *et al.* 1990. Proceedings of the Society of Photo-Optical Instrumentation Engineers, Vol. 1344. In press.
8. PARRY, I. R. & R. M. SHARPLES. 1988. *In* Fibre Optics in Astronomy, S. C. Barden, Ed.: 93–98. Astronomic Society of the Pacific. San Francisco.
9. MARSHALL, F. E., E. A. BOLDT, S. S. HOLT, R. B. MILLER & R. F. MUSHOTZKY. 1980. Astrophys. J. **235:** 935.

Numerical Calculations of Black Holes and Naked Singularities[a]

STUART L. SHAPIRO AND SAUL A. TEUKOLSKY

Center for Radiophysics and Space Research
Departments of Astronomy and Physics
Cornell University
Ithaca, New York 14853

INTRODUCTION

Many of the papers presented at this symposium have begun by noting that we are celebrating the anniversary of some important theoretical or observational discovery in astrophysics. It is thus appropriate to point out that it is exactly 25 years since the pioneering paper of May and White[1] started the field of numerical relativity. May and White developed a computer code to solve problems involving spherical gravitational collapse. They had to confront theoretical difficulties that are still with us today. Their code is fine for configurations that do not collapse all the way to black holes. Indeed, modern supernova codes handle general relativity in essentially the same way as May and White. However, black hole formation causes the code to crash. This difficulty results from a poor choice of coordinates to describe spacetime when a singularity forms inside the black hole.

In the past 25 years, we have become more ambitious. We would like to solve nonspherical problems, in two and even three spatial dimensions. A new feature of nonspherical gravitational collapse is the production of gravitational waves. With the advent of large-scale gravitational wave detectors now likely, one of our goals is to be able to calculate reliably the spectrum of gravitational radiation from various astrophysical sources. How close are we to achieving this goal? What is the status of numerical calculations of black hole formation?

While much progress has been made, we are still a long way from achieving our goal. The "holy grail" of numerical relativity is to find a code that will run forever, enabling us to follow the evolution of a given distribution of matter arbitrarily far into the future. The code should not crash when black holes and their associated singularities form. Rather, it should use the coordinate freedom of general relativity to hold back the lapse of proper time in regions of very strong gravitational fields, and thereby avoid integrating into a singularity. This goal has essentially been reached for spherical gravitational collapse.[2] However, the situation in higher dimensions is pretty bleak.

Thus far there has been only one successful integration of Einstein's equations with reasonable accuracy for the collapse of a rotating configuration to a black hole.[3] However, even this case required a fine tuning of the polytropic index, initial radius, and other parameters, to see the black hole form. The code crashed very soon

[a] This work was supported in part by National Foundation Grants AST 87-14475 and PHY 90-07834 at Cornell University.

thereafter. Clearly there is a lot of work to be done before we are able to tackle problems like the fully three-dimensional spiraling together of binary black hole systems!

Much of the work in nonspherical numerical relativity has focused on nonrotating axisymmetric configurations.[4] What interesting results have emerged recently from these calculations? We will describe a tantalizing result we have found[5] in our ongoing studies involving relativistic star clusters.

Besides their potential astrophysical applications, relativistic star clusters have a very desirable property for theoretical studies in numerical relativity. The equations of motion for a star cluster are simply the geodesic equations of motion for its constituent stars. Thus the right-hand side of Einstein's equation is very easy to compute to high accuracy. Contrast this with the case of a fluid matter source, where one must deal with all the complexities of multidimensional relativistic fluid dynamics with shocks, etc.

THE CALCULATIONS

One motivation of our work has been to study the question of cosmic censorship in general relativity. It is well-known that general relativity admits solutions with singularities, and that such solutions can be produced by the gravitational collapse of nonsingular, asymptotically flat initial data. The *cosmic censorship hypothesis*[6] states that such singularities will always be clothed by event horizons, and hence can never be visible from the outside (no naked singularities). If cosmic censorship holds, then there is no problem with predicting the future evolution outside the event horizon. If it does not hold, then the formation of a naked singularity during collapse would be a disaster for general relativity theory. In this situation, one cannot say anything precise about the future evolution of any region of space containing the singularity since new information could emerge from it in a completely arbitrary way.

Are there guarantees that an event horizon will always hide a naked singularity? No definitive theorems exist. Counterexamples[7] are all restricted to spherical symmetry and typically involve shell crossing, shell focusing, or self-similarity. Are these singularities accidents of spherical symmetry?

For nonspherical collapse Thorne[8] has proposed the *hoop conjecture:* Black holes with horizons form when and only when a mass M gets compacted into a region whose circumference in *every* direction is $\mathscr{C} \lesssim 4\pi M$. If the hoop conjecture is correct, aspherical collapse with one or two dimensions appreciably larger than the others might then lead to naked singularities.

For example, consider the Lin–Mestel–Shu instability[9] for the collapse of a nonrotating, homogeneous spheroid of collisionless matter in Newtonian gravity. Such a configuration remains homogeneous and spheroidal during collapse. If the spheroid is slightly oblate, the configuration collapses to a pancake, while if the spheroid is slightly prolate, it collapses to a spindle. While in both cases the density becomes infinite, the formation of a spindle during prolate collapse is particularly worrisome. The gravitational potential, gravitational force, tidal force, kinetic, and potential energies all blow up. This behavior is far more serious than mere shell-crossing, where the density alone becomes momentarily infinite. For collisionless matter, prolate evolution is forced to terminate at the singular spindle state. For oblate evolution the matter simply passes through the pancake state, but then becomes prolate and also evolves to a spindle singularity.

Does this Newtonian example have any relevance to general relativity? We already know that *infinite* cylinders do collapse to singularities in general relativity, and, in accord with the hoop conjecture, are not hidden by event horizons.[8,10] But what about *finite* configurations in asymptotically flat spacetimes?

Previously, we constructed[11] an analytic sequence of momentarily static, prolate, and oblate collisionless spheroids in full general relativity. We found that in the limit of large eccentricity the solutions all become singular. In agreement with the hoop conjecture, extended spheroids have no apparent horizons. Can these singularities arise from the collapse of nonsingular initial data? To answer this, we have performed[5] fully relativistic dynamical calculations of the collapse of these spheroids, starting from nonsingular initial configurations.

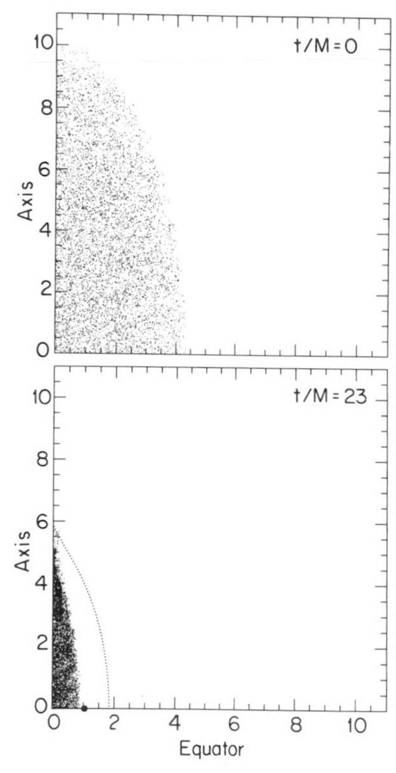

FIGURE 1. Snapshots of the particle positions at the initial and final times for prolate collapse. The minimum exterior polar circumference is shown by a *dotted line.* The minimum equatorial circumference, which is a circle, is indicated by a *solid dot.* The minimum polar circumference is $\mathscr{C}^{\min}_{\text{pole}}/4\pi M = 2.8$. There is no apparent horizon, in agreement with the hoop conjecture. This is a good candidate for a naked singularity, which would violate the cosmic censorship hypothesis.

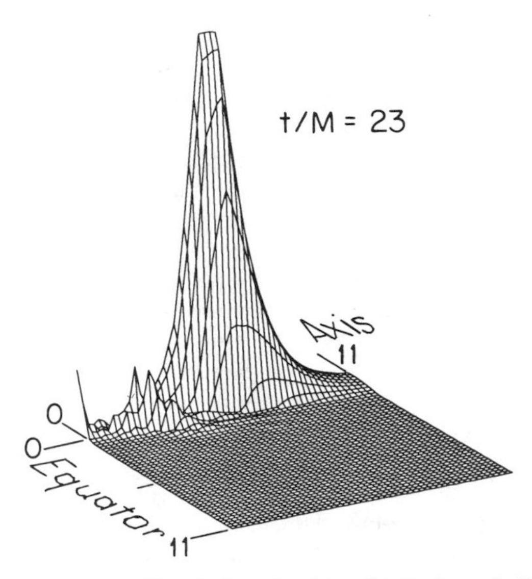

FIGURE 2. Profile of I in a meridional plane for the collapse shown in FIGURE 1. The peak value of I is $24/M^4$ and occurs on the axis just outside the matter.

We find that the collapse of a prolate spheroid with sufficiently large semimajor axis leads to a spindle singularity without an apparent horizon. *Our numerical computations suggest that the hoop conjecture is valid, but that cosmic censorship does not hold because a naked singularity may form in nonspherical relativistic collapse.*

FIGURE 1 shows our candidate for the formation of a naked singularity. It describes the collapse of a prolate spheroid whose initial semimajor axis is $10M$ and initial eccentricity is 0.9. The configuration collapses to a spindle singularity at the pole without the appearance of an apparent horizon.

To measure the growth of any singularity that might arise, we compute the *Riemann invariant*

$$I \equiv R_{\alpha\beta\gamma\delta}R^{\alpha\beta\gamma\delta}$$

at every spatial grid point. Here $R_{\alpha\beta\gamma\delta}$ is the Riemann curvature tensor, which is a relativistic generalization of the Newtonian tidal force. *If* I *blows up at any point, the spacetime has a singularity.*

FIGURE 2 shows the profile of the Riemann invariant I in a meridional plane for the collapse shown in FIGURE 1. The key point is that its peak value is $\gg 1$ and that the peak occurs *outside* the matter on the pole. This is a much more serious type of singularity than those arising from simple shell crossing in spherical symmetry. *The absence of an apparent horizon suggests that the spindle is a naked singularity.*

FUTURE WORK

We have based our evaluation of cosmic censorship and the possibility of naked singularities on the presence or absence of an apparent horizon when a singularity forms. Now the presence of an apparent horizon guarantees the presence of an event horizon, that is, the presence of a black hole. However, the converse is not true: the absence of an apparent horizon does not necessarily imply the absence of an event horizon. Whenever they form, singularities cause our numerical integrations to terminate. Consequently, once we have encountered a singularity, we cannot map out the spacetime arbitrarily far into the future. Such a mapping would be necessary to completely rule out the formation of an event horizon.

In spite of this logical possibility, we do not think that an event horizon forms in those cases where we have found singularities but no apparent horizons. The main reason is that for collapse leading to an apparent horizon, outward light rays always turn around near the singularity. By contrast, for collapse where no apparent horizon forms, outward light rays are still propagating freely away from the vicinity of the singularity up to the time our integrations terminate. Continued outward propagation would imply no event horizon and no black hole.

It is an interesting question for future research whether a different choice of coordinates can be found that will map out the region of spacetime away from the singularity further into the future. Such coordinates would enable one to confirm that all outward light rays, even those originating close to the singularity, propagate to large distances.

Further evidence for the nakedness of the singularity is the similarity of the spindle singularity to the naked singularity that occurs during collapse of an *infinite* cylinder.[8] In both cases, the proper length of a given segment of matter along the axis grows slowly, while its proper circumference and surface area shrink to zero much more rapidly. Also, the singularity is an extended region along the axis and not just a point. An infinite cylinder is unphysical, and hence does not by itself constitute a violation of cosmic censorship. However, if the gravitational field of a finite spindle indeed asymptotes locally to the field of an infinite cylinder, we can be quite confident that no event horizon forms.

While the matter treated here has kinetic pressure, it is collisionless, not fluid. We do not regard the collisional properties of the matter as crucial for settling the question of cosmic censorship. First, the formation of naked singularities should not depend on the particular details of the fundamental interactions affecting matter at high densities. The Einstein field equations alone should be sufficient to rule out naked singularities, at least outside matter, for true cosmic censorship. All the other great theorems of black hole physics have a global geometric character and make minimal assumptions about the properties of matter. For example, Hawking's Area Theorem, the No-Hair Theorem, and the Singularity Theorems have this character.[12]

The second reason why the assumption of collisionless matter may not be crucial is that collisional effects may even accelerate the formation of singularities. Since all forms of energy gravitate in relativity, the added attraction due to interaction energy could overcome the repulsive effect of any additional pressure due to interactions. This relativistic effect is known as "pressure regeneration" (reference 10, p. 605), and numerous examples exist. Future studies of highly asymmetric collapse with fluid matter will help to clarify this issue.

The candidate counterexamples to cosmic censorship implied by this work have no angular momentum. In Newtonian theory the presence of angular momentum prevents an infinitesimally thin spindle singularity from forming on the axis. It is not at all clear that angular momentum will have the same effect in general relativity. For example, we know that angular momentum does not prevent the formation of a singularity when a rotating black hole forms. This shows that strong gravitational fields can overcome centrifugal repulsion. We plan to investigate whether rotation can prevent naked singularity formation by generalizing our work to include angular momentum.

CONCLUSIONS

We have presented numerical evidence that the hoop conjecture is a valid criterion for the formation of black holes during gravitational collapse. More significantly, we have found the first numerical candidates for the formation of naked singularities from nonspherical collapse of well-behaved initial configurations. The singularities we find arise during prolate collapse to a thin spindle. The singular region occurs near the pole and extends into the vacuum outside the matter. These examples are in contrast to any cases of singularities that may arise during spherical collapse. They strongly suggest that the unqualified cosmic censorship hypothesis cannot be valid.

ACKNOWLEDGMENTS

We thank A. Abrahams, M. Choptuik, C. Evans, and L. S. Finn for helpful discussions. Computations were performed on the Cornell National Supercomputer Facility.

REFERENCES

1. MAY, M. & R. H. WHITE. 1966. Phys. Rev. **141:** 1232.
2. See, e.g., SHAPIRO, S. L. & S. A. TEUKOLSKY. 1986. Astrophys. J. **307:** 575, and references therein.
3. STARK, R. F. & T. PIRAN. 1979. Phys. Rev. Lett. **55:** 891.
4. See, e.g., SMARR, L., Ed. 1979. Sources of Gravitational Radiation. Cambridge University Press. Cambridge, England; CENTRELLA, J., Ed. 1986. Dynamical Spacetimes and Numerical Relativity. Cambridge University Press. Cambridge, England; EVANS, C. R., L. S. FINN & D. W. HOBILL, Eds. 1989. Frontiers in Numerical Relativity. Cambridge University Press. Cambridge, England.
5. SHAPIRO, S. L. & S. A. TEUKOLSKY. Phys Rev. Lett. In press.
6. PENROSE, R. 1969. Riv. Nuovo Cim. (Numero Special). **1:** 252.
7. SEE, e.g., GOLDWRITH, D. S., A. ORI & T. PIRAN. 1989. *In* Frontiers in Numerical Relativity, C. R. Evans, L. S. Finn, and D. W. Hobill, Eds.: 414 for discussion and references. Cambridge University Press. Cambridge, England.
8. THORNE, K. S. *In* Magic Without Magic: John Archibald Wheeler, J. Klauder, Ed.: 1. Freeman. San Francisco.
9. LIN, C. C., L. MESTEL & F. H. SHU. 1965. Astrophys. J. **142:** 1431.
10. MISNER, C. W., K. S. THORNE & J. A. WHEELER. 1973. Gravitation: 867. Freeman. San Francisco.
11. NAKAMURA, T., S. L. SHAPIRO & S. A. TEUKOLSKY. 1988. Phys. Rev. **D38:** 2972.
12. HAWKING, S. W. & G. F. R. ELLIS. 1973. The Large Scale Structure of Space-Time. Cambridge University Press. Cambridge, England.

Faint Galaxies and Dark Matter

J. A. TYSON
AT&T Bell Labs
Murray Hill, New Jersey 07974

INTRODUCTION

During the past several years ultradeep optical imaging using charge-coupled devices (CCDs) over the wavelength range 0.3–1 micron have revealed a dense population of faint blue galaxies. At a flux level corresponding to 1 photon/pixel/minute collected in a 4-meter telescope there are over 300,000 galaxies per square degree on the sky. Most of these galaxies have apparent magnitudes between 24 and 27 B mag. Although too faint for spectroscopic redshift determination, several tests indicate that the redshift of this population of galaxies extends between 0.7 and 3. Observations of these galaxies suggest that we are seeing these galaxies at the epoch of formation of much of their stellar content. The resulting UV-bright spectrum, when redshifted to redshifts of 1–3, would produce the observed blue spectral shape. Although interesting in their own right, this backdrop of distant galaxies may be used as a tool for studying foreground mass concentrations.

It is likely that over 90 percent of the matter in the universe by its very nature cannot be seen directly. However, the gravitational influence of sufficiently large mass concentrations on passing photons can be detected. Gravitational lens distortions of background objects [high redshift galaxies or quasi-stellar objects (QSOs)] can serve as a useful tool for the study of the total mass distribution of the foreground lens. These studies have the potential of going beyond being a complementary technique to dynamical studies: surveys for statistical lens distortions of background galaxies can reveal concentrations of dark matter where there may be no luminous matter. It is possible that dark matter clumping exists in the space between stellar systems, and that the resulting shear could be studied using this lens-inversion technique.

BACKGROUND GALAXIES

Optical images of the extragalactic sky show a variety of galaxies with various luminosities and distances, all seen in projection. Although galaxies in rich clusters are understood to have undergone a unique evolution, whether the more numerous isolated "field" galaxies form a single population has been a longstanding open question. Photographic surveys to a surface brightness of 26 magnitude arcsec^{-2} showed as many as 17,000 faint galaxies per square degree at 24th magnitude.[1,2] Deep CCD imaging surveys using new observing and data processing techniques[3] have revealed a population of faint blue resolved galaxies that uniformly cover the sky.[4,5] CCDs at the prime focus of the NOAO 4-m telescopes and the CFHT have been used. There are over 200,000 per square degree in a one magnitude bin at 27th *B* magnitude. FIGURE 1 shows a monochrome image of part of one of several survey fields, 3.8 arcminutes on a side. Between 20 and 26 B magnitude the average number of galaxies per magnitude per square degree is given approximately by Log $N = 0.45B_J - 6.55$, where the blue B_J passband (close to the photographic IIIaJ + GG385 passband) is effectively 3700–5100 Å.[5] However, for B_J magnitudes fainter than 23 the $N(B_J)$ data and models (discussed below) show significant departures from this simple relation. This may be seen in FIGURE 2 where faint galaxy count data are

FIGURE 1. This is a portion of one of the deep fields produced in the continuing CCD survey. A 3.8-by-3.8-arcmin section of the sky is shown. The limiting broadband magnitude in this image is about 27th magnitude. Over 90 percent of these galaxies appear to be at redshifts less than 3. This kind of ultradeep imaging has been carried out in 38 fields over the sky, totaling 32,000 blue galaxies in an area of 630 arcmin^2 to 27th magnitude. Most of these galaxies are 10 magnitudes (10^4 times) fainter than the bright cluster galaxies in a cluster at redshift 0.2–0.4.

shown over the range 20–27 B_J magnitude. A more accurate representation of the blue counts over the range 21–27 B_j magnitude, averaged over the survey fields, is given by the following polynomial least-squares fit:

$$\text{Log } N = a + bB_J + cB_J^2 + dB_J^3,$$

$$21 < B_J < 27 \text{ mag}, \tag{1}$$

where N has units of $deg^{-2}\ mag^{-1}$, and with maximum (upper curve in FIG. 2) and minimum (lower) coefficient sets given by: [max: $a = 76.02, b = -10.389, c = 0.4715, d = -6.791E - 3$] and [min: $a = 89.95, b = -12.397, c = 0.5677, d = -8.320E - 3$]. These experimental error bounds for the *average* counts arise primarily from statistical fluctuations in the crowding correction Monte Carlo simulations that were performed on the real data images. Fluctuations in galaxy number count from field to field in CCD surveys is inversely correlated with limiting magnitude because brighter galaxies are more clustered on the sky. These field-to-field fluctuations range from 50 percent at 20th mag to 10 percent at 27th mag. At a surface brightness of 29 B magnitude $arcsec^{-2}$, the sky is over 15 percent covered with faint blue galaxies. This population has an unusually blue color near $B–R = 0.3$, which makes it particularly

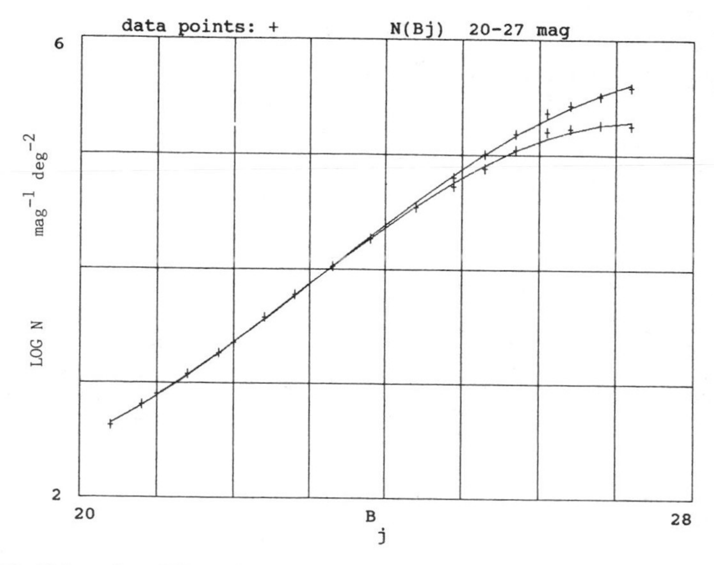

FIGURE 2. Faint galaxy differential number counts $dN(m)$ as a function of blue B_j magnitude over the range 20–27 magnitude. The two branches correspond to the experimental range permitted by the crowding corrected data in 12 CCD survey fields.[4] Units for N are Log_{10} number $deg^{-2}\ mag^{-1}$.

easy to distinguish from red cluster galaxies at lower redshift. The number counts rise less steeply with magnitude at longer wavelengths.[4]

Several lines of evidence suggest that most of these blue galaxies, between 24 and 27 B magnitude, are distributed smoothly over a range of redshifts between 0.7 and 3.[6] A lower redshift limit for this faint population can be derived from its response to a gravitational lens placed in front of it. A good test is provided by the dark matter lens in the clusters 3C295 ($z = 0.46$) and 0024+16 ($z = 0.39$). Roughly the same number of faint blue galaxies per square arcminute are seen behind these red clusters, but most of their images are stretched by the gravitational lens and are aligned orthogonal to the vector to the cluster center. We take this as evidence that most of the background galaxies are at redshifts larger than 0.7. Indeed, simulations fail to reproduce the observed distortions for most of the faint background galaxies in the field of these two high-redshift clusters for background mean redshift of less than 0.9.

Most of these faint galaxies appear to have a redshift of less than 3. An upper redshift limit for this population of faint blue galaxies is immediately apparent from its spectral energy distribution over the wavelength range 0.3–1 micron, and the discontinuity in the spectrum of stars and galaxies at the Lyman break of hydrogen at 912 Å.[6] If a significant fraction of these galaxies were at redshift greater than 2.8, the Lyman break would be shifted through the 0.32–0.36 micron "*U*" passband, causing these galaxies to either drop out in this *U* band or to have a drop in flux between B_J (0.37–0.5 micron) and *U*. Lyman breaks from most stars in these primeval galaxies are probably at least a factor of 2. In addition, these galaxies would have even more hydrogen than present galaxies, which would absorb all the Lyman continuum photons, causing these galaxies to be black in the *U*-band for redshifts larger than 3. This Lyman break has recently been seen in the spectrum of a bright galaxy associated with a radio source at redshift 3.8.

The angular resolution is degraded by atmospheric turbulence ("seeing"): stars and galaxies are blurred on a 0.5–1 arcsecond scale. It is possible to get some statistical information on the average radial profile of the faint blue galaxies, at a given magnitude, by making use of the observed stellar point spread function obtained from brighter stars in the same image. The observed ellipticity distribution of this population (after seeing deconvolution) is similar to that of nearby field galaxies. Some knowledge of the preseeing ellipticity distribution[7] is required for accurate lens Monte Carlo simulations, discussed below. In addition, the redshifts of individual background faint blue galaxies are generally not known, although rough statistical limits have been set in the case of several foreground massive clusters with independently determined masses. The dark matter investigations for galaxy clusters described here will use only clusters of redshift less than 0.3, so that uncertainties in the background source redshifts become relatively unimportant compared with noise and the ellipticity distribution of the background sources.

COSMOLOGY FROM $N(m)$?

Will it be possible to constrain cosmological models with number–magnitude $N(m)$ counts of these faint galaxies? We have run models, using different evolution scenarios including various stellar formation histories and merging, galaxy luminosity functions, stellar initial mass functions, and cosmological parameters. We find that without number–redshift data little if any cosmological information can be uniquely determined from the number–magnitude counts alone. Various evolutionary scenarios for the stellar populations in galaxies have been proposed, all of which have many adjustable parameters: the stellar initial mass function, the star formation rate for each population, dust content as a function of redshift, the galaxy luminosity function (shape and characteristic luminosity for the various galaxy types as a function of look-back time), bulge-to-disk ratio, dissipation and collapse histories in the primeval galaxies, merging, etc. Under very restricted model scenarios in which the shape of the galaxy luminosity function is assumed invariant, stellar formation begins at

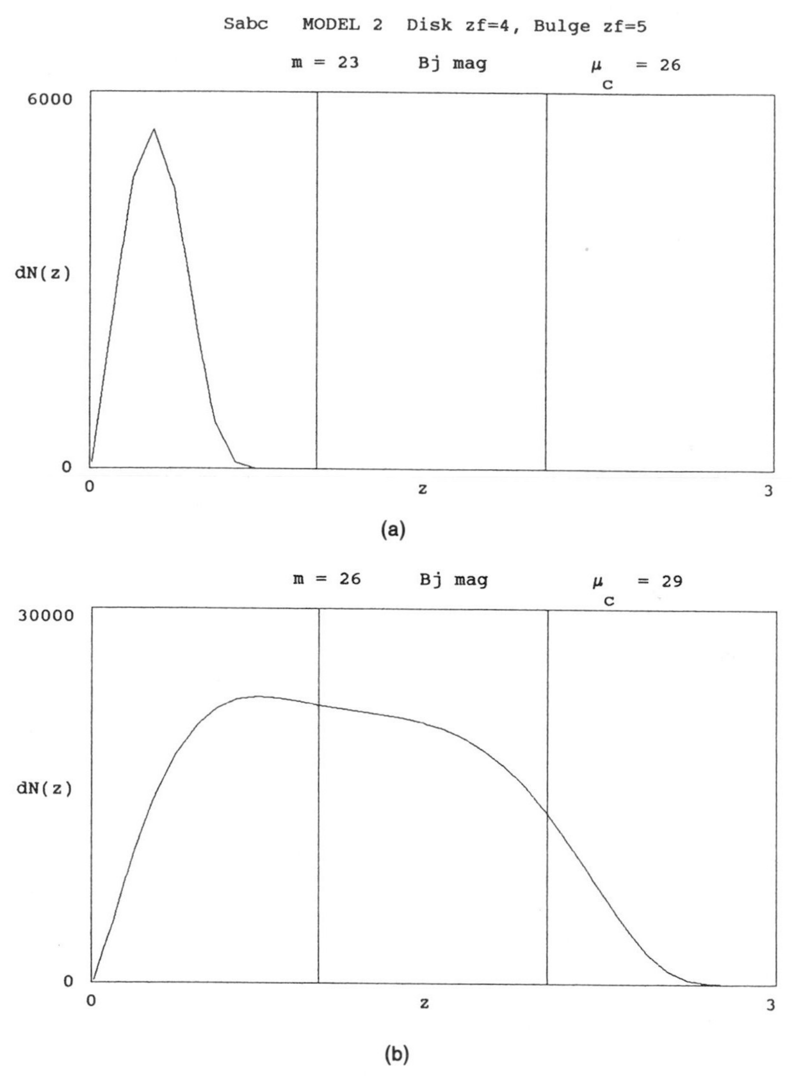

FIGURE 3. Predicted differential number–redshift relation $dN(z, m)$ at two apparent magnitudes, $dN(z, 23)$ and $dN(z, 26)$, for Sab and Sbc spiral galaxies in a modified Bruzual model, which is consistent with the observed $N(m, \lambda)$ and metalicity. This is one of many possible models. Part (**a**) shows that expected redshifts of 23rd magnitude Sabc field galaxies peak around $z = 0.4$, consistent with existing faint galaxy spectroscopic data. Part (**b**) shows that the bulk of the faint blue population is over 16 times fainter ($m > 26$ mag) and samples redshifts up to 3. Units are number per square degree per magnitude per unit redshift. Model $dN(z)$ for other galaxy types generally shows this trend with magnitude. Elliptical galaxy models exhibit a somewhat stronger high redshift component at 26 magnitude.

some universal epoch (different for spirals and ellipticals), and the arbitrary assumptions of no merging and no episodic star formation,[8,9] various authors have remarked that it is difficult to fit the observed galaxy counts with a high value of the cosmological parameter Ω, the number–magnitude counts exceeding evolution model counts for $\Omega = 1$ at 27th magnitude by a factor of 10.[4,10]

This cosmological result is evolution model dependent. Changes in star formation scenario affect the apparent brightness [hence $N(m)$] of galaxies more than changes in cosmology. Changes in cosmology affect the numbers of galaxies seen at some redshift $N(z)$. In general, galaxy number–magnitude counts $N(m)$ are relatively more sensitive to evolution, and number–redshift counts are more sensitive to cosmology. The result is that we will never be able to conclude much of interest cosmologically from galaxy number–magnitude counts without also having reasonably good number–redshift information. Even then, we will have to know something about the shape of the galaxy luminosity function versus redshift.[11] We have experimented with Bruzual-type evolution models, relaxing the assumption of invariant luminosity function, and introducing episodic SFR, and/or merging. While it is possible to fit the observed $N(m, \lambda)$ with low Ω and the traditional models, it is also possible to get a good fit by varying the faint tail of the luminosity function versus look-back time and/or introducing galaxy number nonconservation.[12] Galaxy evolution models have more adjustable parameters than necessary for a good fit to the $N(m, \lambda)$ data. This situation changes when we have redshift constraints on the $N(z)$ distribution, and it will perhaps be possible in coming years to obtain corresponding rough constraints on Ω using these $N(z)$ constraints together with some limits on the luminosity evolution.

What should we expect for the redshift of a randomly chosen galaxy at 23rd or 26th magnitude? Does a survey to 23rd magnitude sample a significant fraction of the population of faint galaxies to 26th magnitude? It is instructive to examine model predictions for $N(z, m)$ for $23 < m < 26$ magnitude. FIGURE 3 shows differential number–redshift counts at 23rd B magnitude and at 26th B magnitude, $dN(z, 23)$ and $dN(z, 26)$, for Sab and Sbc galaxies in a modified Bruzual model that is consistent with the observed $N(m, \lambda)$. Using the largest telescopes, spectroscopy is generally limited to galaxies brighter than 22–23 magnitude. As seen in FIGURE 3(a), expected redshifts of 23 mag Sabc field galaxies peak around $z = 0.4$, consistent with existing galaxy spectroscopic data. On the other hand, the bulk of the faint blue population is over 16 times fainter ($m > 26$ mag) and samples redshifts up to 3, as shown in Figure 3(b). Galaxies of 26th magnitude are too faint for spectroscopic determination of redshift using the current generation of large telescopes and efficient CCD spectrographs.

THE FIRST LENSES

The idea that a foreground massive object would significantly distort the image of a background object is an old one. In the 1930s Zwicky argued that galaxies would act as gravitational lenses, distorting and amplifying background objects.[13] The light deflection is proportional to the square of the velocity dispersion in an isothermal lens, and is about 2 arcsec for a lens with 300 km sec^{-1} velocity dispersion.[14] In 1973

Press and Gunn[15] showed how a cosmologically significant density of compact objects would create small arcs and split images of background objects. In 1979 the first gravitational lens was discovered: a distant QSO split into at least two images by the gravity of a foreground galaxy.[16] After a decade of searching 4681 QSOs for multiple images, nine unambiguous gravitationally lensed QSOs have been found. Several of the lensed QSOs found to date are caused by some isolated foreground galaxy dominated by its dark matter, and the data are consistent with galaxy masses of about 10^{12} solar masses, the number density of foreground galaxies, and the small numbers of background QSOs. The three best cases (2237, 1654, and 0142) yield values of M/r for the lensing galaxy around $3 \times 10^{10}\, M_{\odot}\, \mathrm{kpc}^{-1}$, typical of a heavy galaxy dominated by dark matter.[17]

In the case of a lensed QSO, there are at most a few images, so that there is little information on the mass distribution in the lens, due to the undersampling of the lens plane. The small number of strongly lensed background QSOs (typical redshifts of 2) are due mostly to the relative scarcity of QSOs compared with galaxies at similar redshifts. Although there are several QSOs known that are multiply-imaged by isolated intervening galaxies, distant galaxies are 4 million times more numerous than QSOs, and thus can sample a lens at multiple positions.

GALAXIES LENSED BY GALAXIES

The probability of galaxy–galaxy lensing is higher than for galaxy–QSO lensing. If all galaxies have massive dark halos sufficient to collectively close the Universe, then the systematic distortion of background galaxies within 10–20 arcseconds of these foreground galaxies would be detectable. Automated detection of galaxy–galaxy lensing has been used to set limits to the dark halo of individual foreground galaxies.[18,19] As in the lensed QSO case, there are at most a few background lens-distorted galaxies near the foreground galaxy, so its lens is sparsely sampled.

Twelve hundred foreground galaxies with background galaxies within 30 arcseconds were studied, out of a photographic sample of 50,000 galaxies. No systematic background galaxy image gravitational lens distortion was found. That experiment was designed to test the hypothesis that all galaxies have massive dark halos, independent of absolute luminosity. Although suffering photographic nonlinearities and unknown background galaxy redshifts, those data were sufficient to rule out that hypothesis, for massive isothermal halos with greater than 200 km sec^{-1} circular velocity extending beyond 50 h^{-1} kpc radius. Image distortions at radii larger than a few arcseconds are small, and the measurement of isolated galaxy mass distributions using this technique must wait for large-area deep surveys using CCD mosaics.

GALAXY CLUSTERS AS LENSES

Evidence for gravitational lensing in galaxy clusters came with the discovery in 1988 of several bright blue "arcs" in compact, rich galaxy clusters.[20,21] A large blue arc occurs if a distant evolving galaxy falls near a caustic. Some theoretical work on the possibility of occasional bright arcs near caustics in lenses had been done by 1988.[22-25] Some of these highly magnified background galaxies are bright enough for spectroscopic study, and it is interesting (see FIG. 3) that the five redshifts of bright arcs, corresponding to unmagnified apparent magnitudes of about 25 mag, obtained to date are in the range 0.72–2.2. However, these caustic projections of background objects are rare, occurring only when a background object of the correct redshift coincides with a point of infinite magnification in the lens. The large mass associated with rich clusters of galaxies distorts background galaxy images over a large area of the sky. All the faint background galaxies within 1–2 arcminutes of the cluster will be measurably distorted by the shear. Foreground galaxy clusters at redshifts 0.2–0.5 with radial velocity dispersions above 700 km s^{-1} have sufficient mass density to significantly distort background galaxies of redshift greater than 0.4–1. To accurately measure this effect, deep multiwavelength imaging to 29th magnitude must be done on fields containing bright 17th magnitude cluster galaxies.

We initially searched for this effect in several rich clusters, with deep CCD multiband imaging in 1988–1989 on the CTIO 4-m and CFHT 3.6-m telescopes. Systematic gravitational lens alignment of 10–60 faint background galaxy images, centered on several foreground galaxy clusters of high-velocity dispersion was found.[26] The background galaxy population is selected by its extreme blue color relative to the red cluster galaxies. In deep imaging total exposure times of 2 hr per wavelength band with CCDs on large telescopes there are 30–100 background galaxies arcminute^{-2} anywhere in the sky, which is sufficient to statistically map the dark matter distribution in a foreground cluster. The 20–30 high redshift background galaxies within the central square arcminute of a foreground compact cluster of galaxies will be strongly lensed by the cluster mass distribution, distorting background galaxy images into faint arcs 1–30 arcseconds long. These are not the relatively rare caustic alignments: every background galaxy above some critical redshift will be seen as stretched along a circle centered on the lens. By going faint, the improved statistics of these distortions permits the construction of a map of the dark matter distribution.[26] This new method is immune to the problems associated with unbound galaxy contamination and dynamical assumptions in cluster virial mass estimates.

We have measured multiple background galaxy gravitational lens distortions for 8 clusters of known high-velocity dispersion and X-ray luminosity, in a sample of 24 clusters deep-imaged to date. In order to study the distribution of dark matter under a variety of states of cluster evolution the deep cluster imaging sample is distributed over a wide range of redshift and optical/X-ray properties. Each cluster is imaged to 29 B and 28 R magnitude arcsec^{-2}. This imaging survey of rich clusters is a collaboration between groups at Bell Labs, the Institute for Advanced Study, Princeton University, Toulouse Observatory, and NOAO. Of the 24 clusters imaged to 29 B mag arcsec^{-2} and 28 R mag arcsec^{-2} to date, 19 have Abell richness greater than 3, 5 are poor (for control), 8 have X-ray luminosities greater than 6×10^{44} erg sec^{-1}, 9 are optically compact (relaxed), 5 have redshift less than 0.2, and 7 have redshift greater than 0.4. Larger area CCDs covering over 60 arcmin^2 are now being used for this cluster imaging program. This will ultimately permit the construction of dark matter maps going out beyond 1 Mpc radius in the cluster.

CLUSTER LENS DISTORTIONS

The effect of a cluster gravitational lens is to stretch the apparent size of the image tangent to a circle centered on the lens. The distorted images of background galaxies appear as miniarcs, small parts of circles centered on the foreground cluster. A significant excess of faint blue background galaxies are aligned tangent to circles centered on the foreground cluster lens by the distortion of the gravitational lens. The cumulative systematic alignment of the numerous faint blue (background) galaxy ellipticities gives a strong gravitational signal. This stretch is measured using two scale lengths: the intensity weighted second moment of the galaxy image orthogonal and along the radius relative to the lens center. To detect the coherent stretching from the lens and cancel effects of noise and the intrinsic ellipticities of the galaxies, the scale lengths are averaged over the background population. From the two mean scale lengths we form a dimensionless quantity similar to an ellipticity, the difference of the lengths divided by the sum of the lengths, which we call the net tangential alignment or lens distortion D. This lens distortion is related to the projected mass density and is defined[18] at each point (x, y) in the lens plane by

$$D(x, y) = (a^2 - b^2)/(a^2 + b^2), \tag{2}$$

where $a(x, y)$ and $b(x, y)$ are the (r, θ) principal-axis-transformed second moments of the background galaxy images averaged over the field out to some maximum radius from the point (x, y). Referred to position (x, y) the radial moment is $b(x, y)$ and the tangential moment is $a(x, y)$:

$$a_i = M_{20} \sin^2 \phi + M_{02} \cos^2 \phi - 2M_{11} \sin \phi \cos \phi,$$

$$b_i = M_{20} \cos^2 \phi + M_{02} \sin^2 \phi + 2M_{11} \sin \phi \cos \phi, \tag{3}$$

where ϕ is the position angle of the vector from the point (x, y) to the background galaxy, relative to the x-axis, and the intensity-weighted second moment $M_{\mathrm{lm},i}$ of background galaxy i is defined by

$$M_{\mathrm{lm},i} = M_{0,i}^{-1} \int (x - \langle x \rangle_i)^l (y - \langle y \rangle_i)^m [I_i(x, y) - I_0]\, dxdy,$$

$$M_{0,i} = \int [I_i(x, y) - I_0]\, dxdy, \tag{4}$$

where the sky intensity near this background galaxy is given by I_0. A random population of galaxies randomly oriented will give a net alignment $D(x, y)$ of zero at every point in the lens plane, while a population of lensed galaxies will give a positive value.

Preprocessing and FOCAS automated detection, splitting, and photometry reduced the multicolor images to a position-matched catalog.[4] This matched catalog may be filtered for objects with reproducible position, which are classed as galaxy, and their colors and second moments obtained. Those galaxies fainter than 22 B_J magnitude and with B–R color bluer than the blue tail of the cluster color distribution (the background galaxies) have their positions, total magnitudes, and second moments sent to the lens orientation–correlation software *findlens,* which implements **(2)–(4).** A simpler method that has a wider dynamic range operates on the blue–red difference image directly:[26] Since most cluster galaxies are found to have much redder B–R colors than the faint background population, the difference between a registered deep B image and a suitably scaled deep R image will nearly cancel most of the red cluster galaxies. Such a scaled blue–red difference image is shown in FIGURE 4 for the rich cluster A1689. FOCAS is then run on this $B–\alpha R$ image, and the resulting second moments of the faint blue background galaxies are passed to *findlens.*

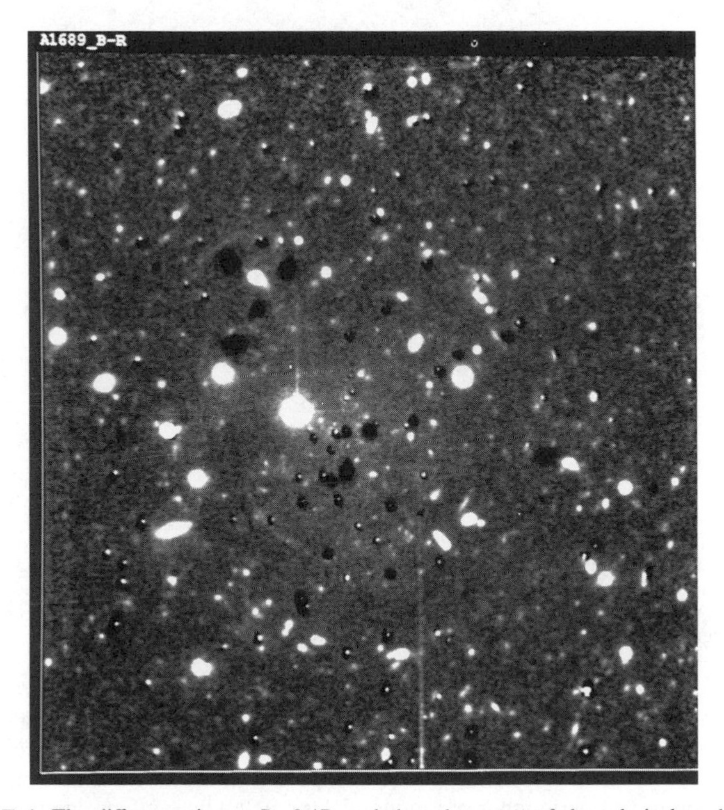

FIGURE 4. The difference image $B_J–0.4R$, scaled so that most of the relatively red cluster galaxies are nulled out, for the cluster A1689. Aside from some bright blue objects like foreground stars, this reveals many faint stretched blue arcs that are images of blue background galaxies that have been gravitationally distorted by the mass of this cluster.

To detect if a gravitational lens is present, the alignment statistic $D(x, y)$ is computed over a grid of positions as candidate lens centers. The alignment statistic is computed using galaxies at various radii from the cluster center [the average in (**2**)]. All faint blue background galaxies with ellipticities > 0.2 within 300 kpc of that point (the radius, which must be less than half the field size, is arbitrary and not critical) are included in the tangential and radial moment average. A real lens produces an alignment value that is a maximum at the true center. Such a dark matter map for the

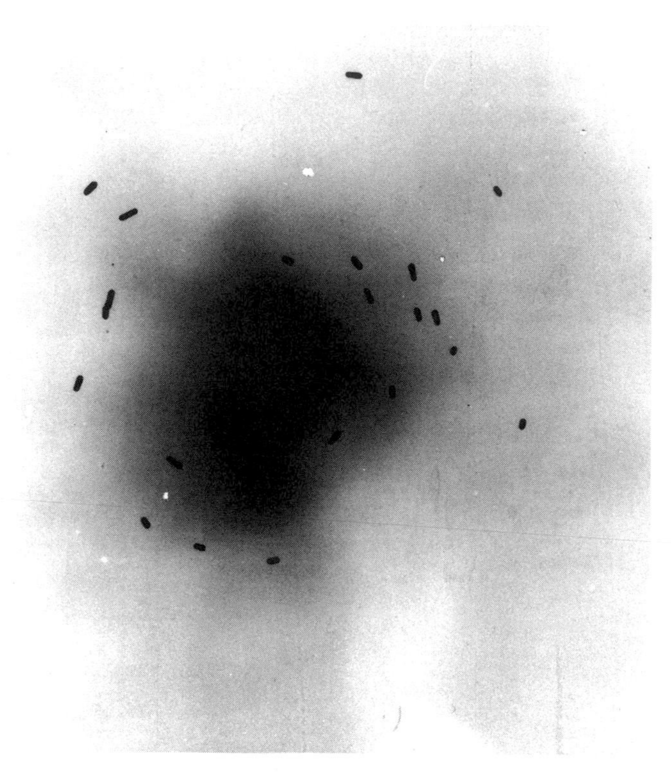

FIGURE 5. Clumps of dark matter can form imperfect but powerful gravitational lenses, giving us a magnified and distorted view of the distant universe. This figure shows the alignment of some faint blue galaxies behind a foreground massive cluster of galaxies, and the dark matter map $D(x, y)$ for this cluster, obtained from hundreds of these systematic lens distortions via (**2**). The foreground cluster, A1689, has a redshift of 0.18.

cluster Abell 1689 is shown in FIGURE 5, along with the orientation pattern of some of the brighter shear-distorted background blue galaxies. The alignment strengths at the lens centers are highly significant (>99 percent likelihood) as determined by comparison to a large number of random fields created by randomizing the sample galaxy positions without changing their measured shapes. We study the alignment strength $D(x, y)$ as a function of radius from the centroid of the mass distribution found by *findlens,* and compare it to the light distribution. FIGURE 6 shows the red luminosity profile compared with the alignment strength of faint blue galaxies, and the dark matter surface density for the cluster A1689. The best fit projected surface density of the following form is also plotted: this best fit surface density profile is of the form

$$\Sigma = \Sigma_0[3/2 - 1/2(r/r_c)^2], \qquad r < r_c,$$
$$\Sigma = \Sigma_0[r_c/r], \qquad r > r_c. \tag{5}$$

For a given background source redshift distribution, the coefficient Σ_0 is proportional to the square of the one-dimensional velocity dispersion:[14] $\Sigma_0 = \sigma_v^2/2Gr_c$. Here the core radius r_c is a 67 percent peak density radius, and for A1689 has the value $r_c = 40$ h_{75}^{-1} kpc, shown as a dotted curve in FIGURE 6. The corresponding 50 percent peak density radius is 50 h_{75}^{-1} kpc = 40 h^{-1} kpc. It is interesting that this is significantly smaller than some observed X-ray core radii in nearby clusters, suggesting that the X-ray gas may be less relaxed dynamically than the dark matter.

The gravitational image distortions appear to extend radially at least as far as the cluster red light. The lens distortions are roughly correlated with cluster red light both in central position and radial extent. This implies that the dark matter distribution follows the smoothed cluster luminosity, with most of the mass interior to 100 h_{75}^{-1} kpc. A null result is found in 12 control fields devoid of any foreground cluster of galaxies, each with a surveyed area similar to that of our cluster data fields. Statistical significance of the lens alignments (7–10σ per cluster) and the integral distortion statistic is particularly high for the optically compact clusters we have examined.

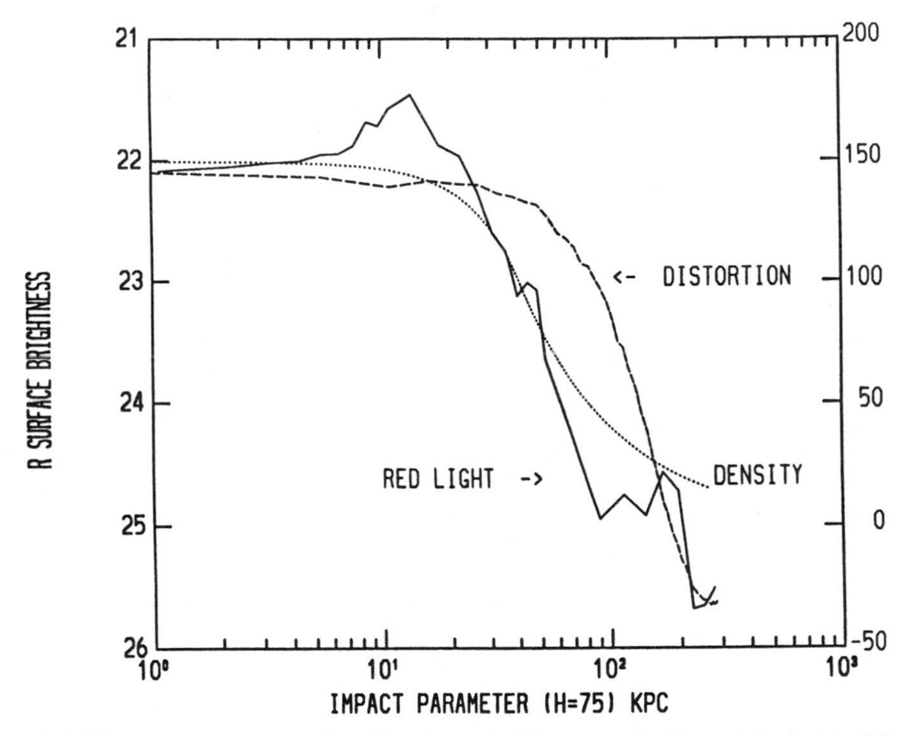

FIGURE 6. Red luminosity profile $I_R(r)$ in cluster A1689, compared with strength of tangential alignment $D(r)$ of faint blue background galaxies. Observed log red flux (subtracting bright stars in foreground) vs. log radius is plotted, together with the radial dependence of the net tangential alignment $a^2 - b^2/a^2 + b^2$ [see **(2)**]. The best fit dark matter surface density profile, **(5)**, with soft core radius $r_c = 50$ h_{75}^{-1} kpc is also plotted as a dotted curve. The right scale is arbitrary units, normalizing all plots at zero radius.

CLUSTER LENS SIMULATIONS

We have simulated cluster lens distortions of background galaxies, by ray-tracing real images, as a function of cluster and dark matter (DM) parameters, and compared them with the observations. The background galaxy fields are synthesized to have the same properties (number–magnitude counts, ellipticity distribution, galaxy intensity profiles) as found in the seeing-deconvolved deep surveys. These background galaxies are assigned redshifts in three model schemes: all at $z = 1$, all at $z = 2$, and equally at $z = 1$ and 2. The cluster galaxies (found by their color, brightness, or redshifts) are assigned truncated isothermal mass distributions obtained from individual galaxy surface photometry to 28 B_J mag, K-corrections derived from the observed colors, and the Faber–Jackson relation $\sigma_v = 220$ dex $[(20.2–B_J)/10]$. These cluster galaxies make a small contribution to the background galaxy image distortions beyond several arcseconds from the cluster galaxies. A soft core isothermal mass distribution is superposed on the cluster galaxies. The velocity dispersion (500–2000 km sec^{-1}) and core 67-percent-surface-density radius (25–100 h_{75}^{-1} kpc) are varied in different Monte Carlo simulations. After the distortion from all components of the lens, the image is convolved with the seeing profile, binned down to the CCD resolution, and noise is added. FIGURE 7 shows a simulation of a galaxy cluster lens with velocity dispersion 1500 km sec^{-1}. A synthetic image of background galaxies placed at redshifts 0.7–2, with all the statistical properties of seeing-deconvolved deep imaging observations, is distorted with a 1500 km sec^{-1} isothermal mass distribution with a core radius of 45 h_{75}^{-1} kpc. Masses are added for all the bright cluster galaxies, breaking the lens symmetry.

To relate the observations to the simulations the net tangential alignment of background galaxies over the extent of the cluster is computed for both the observations and the simulations in exactly the same way (including FOCAS processing). This is necessary because this distortion statistic is weakly dependent on the parent background galaxy ellipticity distribution. The alignment strength is found to be proportional to the radial velocity dispersion and weakly dependent on DM core size ($R_{core} < 100\ h_{75}^{-1}$) and the background redshift ($z_{bg} > 1$). The cluster velocity dispersion, being an integral over the distortion function, is the best determined parameter. The mass core radius (if any) is more difficult to determine, and we can often only obtain rough upper limits. The best agreement with the observed lens statistics for CL3C295, CL0024 + 16, and A1689 is obtained for $30 < R_{core} < 50\ h_{75}^{-1}$ kpc, $\sigma_{v,DM} = 1100–1500$ km sec^{-1}, for $z_{bg} = 2$. Note these "DM" parameters refer to all mass, dark + luminous, *other* than that associated with the modeled 20–50 brighter cluster galaxies. Thus, these velocity dispersions refer primarily to the dark matter component. These are three of the most optically compact clusters in our sample. For many of our clusters, the best fit DM parameters correspond to as large a mass, but distributed over a larger volume (larger core sizes).

Based on these initial clusters, we find that the DM distribution is similar to the smoothed red cluster light. This is consistent either with baryonic DM or baryonic–nonbaryonic dissipative coupling. The scale of the subclumping of the DM within clusters of galaxies is also of interest. It appears that the DM cannot all be clumped on galaxy mass scales; otherwise, the long arcs seen in several relaxed cluster lenses would be broken up on 1–2 arcsecond scales. We have also done an experiment to see how much of the DM can possibly be ascribed to cluster galaxies. We assigned a mass to each of several hundred cluster galaxies, from the Faber–Jackson relation, and then scaled all these galaxy masses up to very large value (σ_{gal} < 500 km sec^{-1}). The resulting number of blue rings around individual cluster galaxies was found to be much larger than observed, and none of these simulations showed arc patterns resembling those found in the real data. Little more can be said of the DM in clusters, based on the small number we have imaged. There appears to be a large

FIGURE 7. A simulation of a galaxy cluster lens with velocity dispersion 1500 km sec^{-1}. A synthetic image of background galaxies placed at redshifts 0.7–2, with all the statistical properties of seeing deconvolved deep imaging observations, is distorted with a 1500-km sec^{-1} isothermal mass distribution with a core radius of 45 h_{75}^{-1} kpc. Reasonable masses were used for the observed cluster galaxies. The mean properties of many of these Monte Carlo simulations are used to define the dark matter parameter space consistent with the cluster lens observations.

scatter in the DM properties within our sample, generally correlated with optical compactness. A study of a large sample of galaxy cluster fields will provide a clearer picture of the forms of DM in clusters, and may further constrain the nature of the dark matter.

The best fit lens data for the rich cluster A1689 imply a total velocity dispersion (dark + luminous matter) of about 1400 km sec^{-1}, somewhat smaller than the measured velocity dispersion of 68 galaxies assumed to be associated with the cluster

(1800 km sec^{-1}). For the optically compact clusters we find the DM peak densities are about 1–2 × 10^5 times ρ_c, with core radii up to 50 h_{75}^{-1} kpc. We have no data as yet on the dark matter beyond 300 h^{-1} kpc. Larger CCDs are required in order to address the question of the possible outer cutoff of the mass. CCDs with 1024 or 2048 pixels per side are capable of addressing the outer cutoff problem, and data on this interesting question are now becoming available.

SUMMARY AND DISCUSSION

The nature and dynamical history of dark matter will be constrained by its clumping distance scales and epochs, and the spatial relation between dark matter and luminous baryonic matter. Prospects are good for a direct measure of the dark matter distribution on scales of stars (from "microlensing"), galaxies, clusters of galaxies, and eventually, intercluster scales. The first large-scale application of this technique has been to rich clusters of galaxies. Using the observed characteristic pattern of background galaxy shapes, automated pattern recognition software generates a two-dimensional lens-distortion map.

To date, we have studied 24 of these clusters to gather statistics on how the dark matter is distributed. The clusters in this sample cover a wide range of optical morphologies and X-ray luminosity. The distribution of dark matter in rich galaxy clusters is found to be similar to the distribution of total luminosity (although on a smoother scale of about 100 kpc). This is consistent with a baryonic origin for this dark matter, conceivably brown dwarf stars or primeval black holes. Currently, our data for clusters are also consistent with dark matter composed of remnant neutrinos of about 10-eV rest mass. A few of the poor unrelaxed clusters studied show no distortion of the background and, thus, have velocity dispersions smaller than 700 km sec^{-1} and/or dark matter core radii larger than 200 h^{-1} kpc.

These shear distortion studies measure mass internal to a radius, M/r. With the exception of the cluster 3C295, we find good agreement with dynamical measures of M/r for those few clusters for which both types of data exist. This suggests that, baring cases of accidental superposition of unbound galaxies, the assumptions behind the application of the virial theorem to clusters may be generally valid for the compact clusters studied so far. This large mass density in dark matter in clusters is not a prediction of current biased cold dark matter theory.[27,28] In addition to the many faint blue arcs, blue rings are occasionally seen around individual bright cluster galaxies in deep CCD images and in simulated clusters, offering the possibility of measurement of the relative mass in cluster galaxies and the diffuse DM mass component. Preliminary results suggest that individual cluster galaxies are no more massive than field galaxies, and that the dark matter in clusters is primarily in the diffuse component. The upper limit to the core radius for this diffuse dark matter in some optically compact and relaxed rich clusters appears to be smaller than typical X-ray core radii for rich clusters, suggesting that at redshifts of a few tenths the X-ray gas is not dynamically relaxed in some compact rich clusters.

If $\Omega \sim 1$, dark matter will fill the Universe and will likely exist in places where there is no current star formation activity. Larger scale applications of this dark matter mapping may eventually discover clumped dark matter unrelated to galaxies or clusters of galaxies. Previous limits to large-scale shear from rotation of the

Universe or Gpc scale mass clumping were set using photographic plate data.[29] Calculations of the faint galaxy orientation correlations induced by clumped dark matter in various theoretical scenarios[30] indicate that interesting limits may be set by ultradeep imaging over fields of up to several degrees. Larger CCDs make such a large-scale search for coherent alignment in the distant faint galaxies particularly attractive, and dark matter on angular scales up to degrees can in principle be studied in this way.

ACKNOWLEDGMENT

It is a pleasure to acknowledge my collaborators in the cluster deep-imaging and image-processing software development: G. Bernstein, B. Fort, R. Guhathakurta, J. Jarvis, Y. Mellier, J. Miralda-Escude, R. Pello, G. Soucail, D. Schneider, E. Turner, F. Valdes, and R. Wenk. I would also like to acknowledge enjoyable discussions with J. Gunn, D. Spergel, G. Evrard, K. Kochanek, R. Narayan, A. Dressler, S. White, C. Hogan, R. Blandford, R. Ellis, C. Norman, and A. Szalay.

REFERENCES

1. KRON, R. G. 1982. Vistas Astron. **26:** 37.
2. JARVIS, J. F. & J. A. TYSON. 1981. Astron. J. **86:** 476.
3. TYSON, J. A. 1990. *In* CCDs in Astronomy, G. Jackoby, Ed.: 1. Astronomical Society of the Pacific. San Francisco.
4. ———. 1988. Astron. J. **96:** 1.
5. TYSON, J. A. & P. SEITZER. 1988. Astrophys. J. **335:** 552.
6. GUHATHAKURTA, P., J. A. TYSON & S. MAJEWSKI. 1990. Astrophys. J., Lett. **357:** L19.
7. MIRALDA-ESCUDE, J. 1991. Astrophys. J. **370:** 1.
8. BRUZUAL, G. 1983. Astrophys. J., Supp. Ser. **53:** 497; ———. 1987. Private communication.
9. ARIMOTO, N. & Y. YOSHII. 1987. Astron. Astrophys. **173:** 23; GUIDERONI, B. & B. ROCCA-VOLMERANGE. 1989. Institute d'Astrophysique de Paris, No. 282. Preprint.
10. YOSHII, Y. & F. TAKAHARA. 1988. Astrophys. J. **326:** 1.
11. CADITZ, D. & V. PETROSIAN. 1989. Astrophys. J., Lett. **337:** L65.
12. GUHATHAKURTA, P. 1991. Astrophys. J.
13. ZWICKY, F. 1937. Phys. Rev. Lett. **51:** 290.
14. TURNER, E. L., J. P. OSTRIKER & J. R. GOTT. 1984. Astrophys. J. **284:** 1.
15. PRESS, W. & J. E. GUNN. 1973. Astrophys. J. **185:** 397.
16. WALSH, D., R. F. CARSWELL & R. J. WEYMANN. 1979. Nature **279:** 381–384.
17. KOCHANEK, C. 1990. Preprint.
18. TYSON, J. A., F. VALDES, J. F. JARVIS & A. P. MILLS. 1984. Astrophys. J., Lett. **281:** L59.
19. TYSON, J. A. 1987. *In* Theory and Observational Limits in Cosmology, W. R. Stoeger, Ed.: 441. Specola Vaticana.
20. SOUCAIL, G., Y. MELLIER, B. FORT, G. MATHEZ & M. CAILLOUX. 1988. Astron. Astrophys. **191:** L19.
21. LYNDS, R. & V. PETROSIAN. 1989. Astrophys. J. **336:** 1.
22. BLANDFORD, R. D. & R. NARAYAN. 1986. Astrophys. J. **310:** 568.
23. PACZYNSKI, B. P. 1987. Nature **325:** 572.
24. BLANDFORD, R. D., E. S. PHINNEY & R. NARAYAN. 1987. Astrophys. J. **313:** 28.
25. GROSSMAN, S. A. & R. NARAYAN. 1988. Astrophys. J., Lett. **324:** L37.
26. TYSON, J. A., F. VALDES & R. A. WENK. 1990. Astrophys. J., Lett. **349:** L1.
27. EVRARD, A. E. 1989. Astrophys. J., Lett. **341:** L71.
28. PEEBLES, P. J. E., R. A. DALY & R. JUSZKIEWICZ. 1990. Astrophys. J. **347:** 563.
29. VALDES, F., J. A. TYSON & J. F. JARVIS. 1983. Astrophys. J. **271:** 431.
30. MIRALDA-ESCUDE, J. 1991. Astrophys. J. **380.**

Strings and Gravity

G. VENEZIANO
Theory Division, CERN
CH-1211 Geneva 23, Switzerland

INTRODUCTION

The subject of this paper is the fascinating, highly nontrivial relationship between strings and gravity. Such a relation is not just interesting, it is the very *raison d'être* of string theory as a (candidate) consistent theory of classical and quantum gravity.

There are (at least) two sides to this subject:

1. A *classical* side that we can subtitle "Strings *in* Gravity"
2. A *quantum* side with the subtitle "Gravity *from* Strings"

As we shall argue in a moment: For 1., strings are *not* that special; for 2., strings are *very* special.

CLASSICAL STRINGS

As elementary, classical systems, strings come next to points as the least complicated objects. Denoting by d the intrinsic dimensions of the system, we have $d = 0$ for points, $d = 1$ for strings (which can be open or closed), $d = 2$ for membranes (open or closed, again), etc. Each one of these systems can be embedded in the D-dimensional spacetime (or target space) in which it moves. If we endow spacetime with a metric $g_{\mu\nu}$ ($\mu, \nu = 0, 1, \ldots, D-1$), that is, with a gravitational field, the classical motion of each elementary object will be given by a generalized geodesic: a geodetic line for the point, a geodetic surface for a string, a geodetic volume for a membrane, etc., corresponding to a minimal length, area, volume, respectively.[a]

What is then so special about $d = 1$, that is, about strings? Apparently nothing! Points, if any, look special. Indeed, there are questions that we may only ask to extended objects ($d \geq 1$), not to point–particles.

Consider, for instance, the metric $g_{\mu\nu}$ to be of a cosmological type, for example, describing an "expanding" universe

$$ds^2 = dt^2 - R^2(t)\, d\mathbf{x}^2; \qquad dR/dt > 0.$$

We may ask of a string (or of a membrane):

1. How does the *size* of the string change with t?
2. Is the expansion *real* in the sense that the ratio (distance between strings)/(size of each) grows? If yes, does it grow like $R(t)$?
3. Can we have new phenomena occur because of the extended nature of strings?

[a] In these geometrical quantities time intervals are multiplied by c (relativity), and are then assimilated to space intervals.

A basic (and trivial) point to remember before answering these questions is the following.

Classical strings have *no* characteristic *size* and, in particular, *no minimal* size. A corollary is that the size L of a classical string, being the only scale in the problem, is irrelevant in flat spacetime.

This is not so, however, in a cosmological background, since this has a scale, $H = (1/R)(dR/dt)$. The dimensionless combination LH does matter and, indeed, one finds that:

If $LH \ll 1$, that is, if the string is "small," it behaves like a physical point, its size keeps constant during the motion and the ratio distance/size grows like R. A solution can be constructed in terms of a systematic expansion around the point–particle geodesic.[1]

If $LH \gg 1$, that is, if the string is "large," its behavior is markedly different from that of a point. In particular, if $\ddot{R} > 0$ (inflation), the string develops "Jeans" instabilities,[1–3] and its size starts to grow with R. In the limiting case, $LH \gg 1$, a new expansion is valid[4,5] around an asymptotic (large R) solution in which the comoving size of the string gets frozen, the oscillatory behavior stops, and proper sizes grow like the scale factor (so that distance/size → constant).

An interesting question concerning this latter regime of high instability arises: If fast inflation can give rise to string instabilities, is it possible that, vice versa, highly unstable strings may provide a suitable source for inflation? Scenarios in which the answer to this question is in the affirmative have been given the name of "self-sustained inflation" (in that no cosmological constant is needed to drive the inflationary process), and examples have been provided.[6,7]

In the asymptotic expansion of references 4 and 5 it is easy to see that, in the perfect fluid approximation, the equation of state of unstable strings is of the type[4,5]

$$p = -\rho/(D-1) + \text{(positive nonleading terms)}.$$

Thus the pressure, though opposite to that of point particles, is not sufficiently negative to drive inflation for $D \geq 4$. The statement follows from the D-dimensional Einstein equations, which imply

$$(D-1)(D-2)\frac{d^2R/dt^2}{R} = -8\pi G_N[\rho(D-3) + p(D-1)]$$

$$= -8\pi G_N \rho(D-4) + \text{neg. terms}.$$

For $\rho > 0, D \geq 4$ this implies $\ddot{R} < 0$, that is, no inflation.

The appealing idea of a self-sustained inflation is not necessarily dead, however. There are (at least) two possible ways out that are worth exploring.

1. Add a "viscosity" term to the equation of state in order to reflect the quantum creation of strings.[7,8] This can be made to work, albeit at the cost of choosing an *ad hoc* viscosity.
2. Start from $D > 4$ (after all, that is what one has to do, at the quantum level, with string theory: see below) and look for solutions describing anisotropic self-sustained inflation of three spatial dimensions accompanied by the con-

traction of $(D - 4)$ internal dimensions. Solutions of this kind have been recently found.[9] Certain initial conditions have to be fulfilled in order to obtain sufficient inflation (for solving the usual difficulties of standard cosmology).

Before turning our attention from classical to quantum strings, I wish to mention an important topic that I call "Playing with the Dilaton". All known string theories allow a string to interact not only with an external "metric" $g_{\mu\nu}(x)$ but also with the so-called "dilaton" $\phi(x)$. Some arguments suggest that the rôle of Einstein's $g_{\mu\nu}$ is played in string theory by the combination:

$$g_{\mu\nu}^{(E)} = g_{\mu\nu} \exp(2\phi) \qquad (D = 4).$$

If this is the case, a way of getting an isotropically expanding universe is to keep $g_{\mu\nu} = \eta_{\mu\nu}$ and give t-dependence to ϕ (see, e.g., Antoniadis *et al.*[10]).

Unfortunately, things turn out not to be so easy. A careful analysis shows[3] that it is actually the $R(t)$ appearing in the original $g_{\mu\nu}$ that determines how distances change relative to *string* sizes. Time dependence in ϕ implies rather time dependence in the Newton and fine-structure constants.

Basically the same observation apparently excludes[11] so-called extended-inflation models based on the dilaton.

The conclusion that T. R. Taylor and I reached a while ago,[12] is that the best thing to wish for the dilaton is that it gets a mass and stays constant (at least in late epochs).

QUANTUM STRINGS

Before turning to quantum strings, let us recall what has happened with quantum point–particles, say in the last 60 to 70 years. That has proved to be a treacherous road. From the Schrödinger equation it is not hard to proceed to relativistic wave equations, but their interpretation is difficult. Historically, one has preferred to start anew, taking the relativistic wave equation (Klein–Gordon, Dirac) as representing a *classical* field theory that is then quantized by the methods of quantum field theory (QFT). One talks of a *second* quantization.

This program has led to infinities associated with loop-corrections, but also, fortunately, to recipes for handling these (in *most* cases), and to end up with finite, testable predictions. The culminating point of this program is the standard model of electroweak and strong interactions, a smashing success indeed.

However—"all that glitters is not gold," and the presence of infinities should not be played down. They take their revenge in the case of quantum gravity.

Indeed, if the logarithmic divergences of $D = 4$ gauge theories can be reabsorbed in the redefinition of the (unobservable) bare masses and couplings, the quadratic (plus quartic, logarithmic, etc.) divergences of $D = 4$ quantum gravity cannot be all reabsorbed into redefinition of G_N, m, etc. The situation is summarized in FIGURE 1. Instead of proceeding "upward" from classical to quantum relativistic point particles, a logical "jump" was made first to classical gauge theory and to classical general relativity. This was followed by QFT with methods that have proved only partly successful (i.e., on the left but not on the right half of FIG. 1).

Let us contrast now this state of affairs with the string theory way. The latter proceeds "upward" without logical jumps, and it is hardly necessary to distinguish in it a first and a second quantization stage.

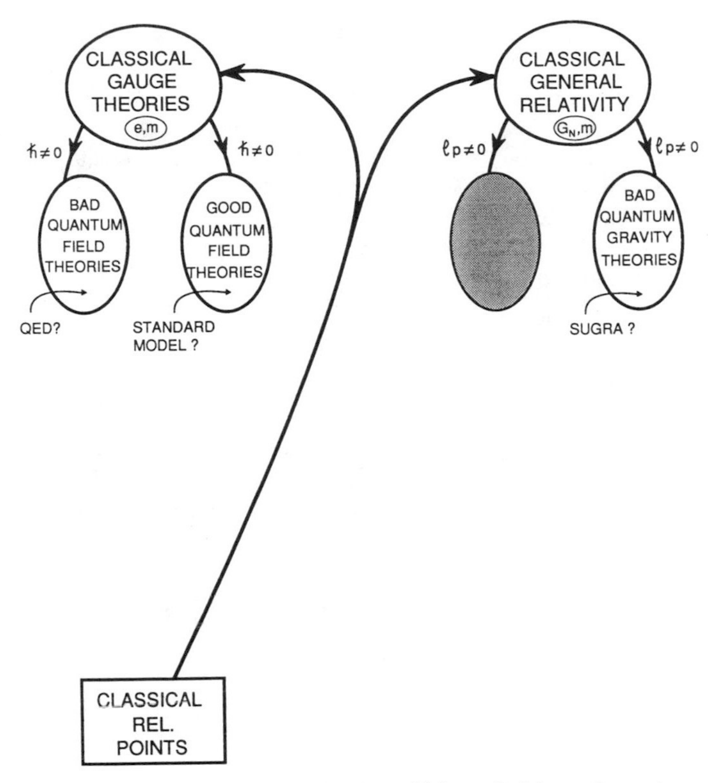

FIGURE 1. The quantum field theory way of combining relativity and quantum mechanics contains a logical jump from classical relativistic points to classical fields, which are then quantized. The procedure is quite successful for gauge interactions, but fails for gravity.

I shall discuss successively three topics

1. (First) quantization
2. Spectrum (e.g., the graviton)
3. Interactions and loops (e.g., quantum gravity)

1. As I said, there is just *one* quantization in string theory. Compared to the point–particle case, one would call it first quantization, since it is the string position and momentum X^μ, P_μ that are quantized [and *not* some $\phi(x)$].

Although not necessary, it is most instructive to use Feynman's path integral quantization, which is achieved through the replacement

$$\text{Classical (stationary} - \text{action) path} \rightarrow \sum_{\text{Paths}} \exp\left(i\,\frac{S(\text{Path})}{\hbar}\right).$$

The crucial object for quantization is thus the (dimensionless) quantity $\hbar^{-1} S$, a *functional* associating a number to each path. Recalling that, for strings, the classical action is proportional to the area of the surface swept by the string (the constant of proportionality being irrelevant at the classical level) we conclude that

$$\hbar^{-1}S = (\text{Area})/\lambda_s^2,$$

where λ_s is a fundamental length. Since only $\hbar^{-1} S$ is relevant, we see that, as appropriate for a geometric theory of . . . everything (including gravity), string theory only needs *two* dimensionful, geometric constants,[13] c (for relativity) and λ_s (for quantization). The appearance of a quantum length is of paramount importance to string theory and to its possible relevance as a unified theory of all interactions. Let us see first how λ_s influences the *spectrum* of string theory

2. The classical spectrum of string theory is continuous. However, angular momentum J and rest mass M satisfy an inequality:

$$\alpha' M^2 \geq J,$$

where α' is a constant of dimensions length/energy, which converts cgs units into the natural units of string theory for which energy is length (cf. Boltzmann constant $K_B T$ = energy). At the quantum level angular momentum gets quantized:

$$J = n\hbar, \qquad 2\alpha' J = n\lambda_s^2 \quad (\lambda_s^2 \equiv 2\alpha'\hbar)$$

in cgs and in string units, respectively. Not surprisingly also M^2 is quantized

$$(\alpha' M)^2 = m\lambda_s^2,$$

but while classically we would have expected $m \geq n$, a careful calculation yields

$$m \geq n - a_0,$$

where $a_0 = 1, 2, \frac{1}{2}, \frac{3}{2}$ for open bosonic, closed bosonic, open fermionic, and closed fermionic strings, respectively. The origin of a_0 is similar to that of the constant $\frac{1}{2}$ in the harmonic oscillator spectrum, $E_n = (n + \frac{1}{2})\hbar\omega$, and comes from normal ordering. A string is a collection of infinitely many harmonic oscillators with their zero-point energies summing up to . . . $-a_0$, a *negative* constant! The shift $(-a_0)$ has miraculous consequences. It allows for the existence of classically forbidden string states that are massless and yet have nonzero angular momentum. Furthermore the massless states have spin between 0 and 2, and thus include gauge bosons, gravitons (as well as their supersymmetric partners), that is, the mediators of all known (and not yet known) fundamental interactions. That is really a good start for a candidate TOE (Theory of Everything)!

Another effect of quantization ($\lambda_s \neq 0$) that is at least as important is the appearance of a *minimal* string size. Again the analogy with the harmonic oscillator is quite appropriate since

$$\Delta x, \Delta p \propto \hbar^{1/2}$$

for the harmonic oscillator and

$$\Delta x \sim \alpha' \Delta p \gtrsim \lambda_s \equiv (2\hbar\alpha')^{1/2}$$

for the string. The implications of this minimal size for string theory are again most welcome. The minimal spacial extent provides a form factor that cuts off high virtual momenta rendering quantum string theory (QST) ultraviolet finite (as we shall discuss further below). Incidentally, λ_s also provides a clear-cut meaning to concepts like "large" distances (meaning $\gg \lambda_s$) or "short" distances ($\lesssim \lambda_s$), and to the limit of large strings (size $\gg \lambda_s$) for which the classical considerations presented at the beginning of this paper retain an almost exact validity.

That is all for free strings. Let us now turn to their interactions.

3. *Interactions.* Unlike the case for point particles, string interactions are unique and geometrical. A typical interaction is the fission of one string into two (or the reversed process). If we make a (closed) string split and then the two strings fuse back again we get, in string theory, the analogue of a self-energy one-loop diagram. Its magnitude (relative to the free propagator) is the square of the coupling constant, that is, the fine-structure constant of string theory α_{SL} (*SL* for string loop). In order to estimate α_{SL} we have to make a digression.

It is easy to see that this self-energy diagram has the topology of a sphere with one hole (handle), that is, of a torus. As we mentioned earlier, there is always, in string theory, a dilaton field $\phi(x)$. It enters the string action in a peculiar form, first pointed out by Fradkin and Tseytlin,[14]

$$\delta S_{\text{dilaton}} = \int d^2z \chi(z)\phi(X(z)),$$

where $\chi(t)$ is the Euler number density of the surface swept by the string. In other words, if we take $\phi(x) = \phi$ (independent of x),

$$\delta S_{\text{dilaton}} = \phi\chi \equiv \phi(2 - 2g),$$

where g is the number of holes (handles). Thus, for the sphere, torus, etc., $g = 0, 1,$ etc.

Since the action appears in the exponent, we deduce that each extra handle (loop) gives a factor $\exp(-2\phi) \equiv \alpha_{SL}$. Thus in string theory the coupling constant is not a free parameter: it is the vacuum expectation value (vev) of a field.[15] The tree level ($g = 0$) is no exception, and one finds

$$\Gamma_{\text{eff}}(g_{\mu\nu}, \phi)\big|_{\text{tree}} = \lambda_s^{-2} \int d^4x\sqrt{-g}\, e^{2\phi}[R + 4\partial_\mu\phi\partial^\mu\phi + \lambda_s^2 F_{\mu\nu}^2 + \cdots].$$

This equation represents the tree-level ($g = 0$), low-energy (the dots represent higher derivatives) effective field theory coming from QST. We see that such effective theory resembles a conventional *classical* field theory (CFT), more specifically a Jordan–Brans–Dicke gravity theory coupled to a Yang–Mills theory. Furthermore, it reveals a very interesting property of tree-level QST, something we may call G^2U (grand square unification). By looking at the coefficients of F^2 and of R, we deduce

$$\alpha_{\text{GUT}} \simeq \alpha_{SL};$$
$$l_p^2 = (G_N\hbar)^{1/2} \simeq \alpha_{SL} \cdot \lambda_s^2.$$

Both gauge and gravitational interaction strengths are controlled by α_{SL}, and indeed

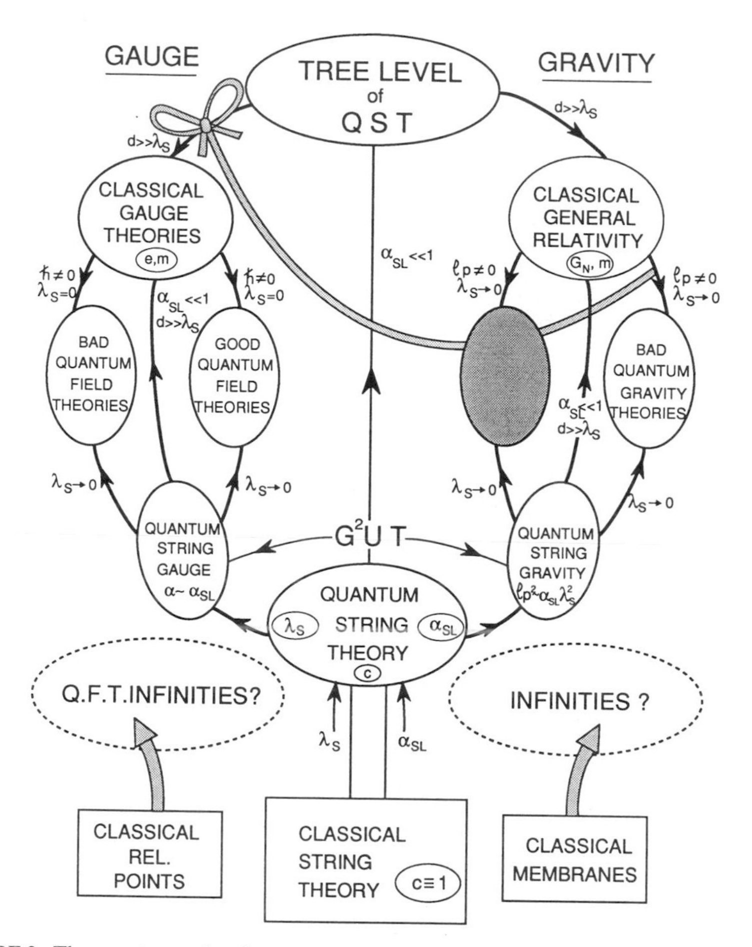

FIGURE 2. The quantum string theory way proceeds instead directly from classical to quantum relativistic strings without any (second) quantization of the spacetime background fields. Classical field theories, including general relativity, emerge as effective theories, representing the low energy, small coupling limit of an ultraviolet–finite, quantum–relativistic, unified description of gauge and gravitational forces.

in such a way that these interactions are *identical* (up to Clebshes) for massive, charged strings at tree level.

What about loop effects? Or, more generally, having reproduced at tree-level CFT, how do we go on to QFT?

If we proceed blindly and apply to Γ_{eff} the usual rules of second quantization, we get back to the UV problems of QFT. In order to be safe we have to keep $\lambda_s \neq 0$ till the very end, and should *not* neglect the dots in Γ_{eff}. This looks like a formidable task, but fortunately there is an elegant, simple shortcut.

No second quantization (i.e., no quantization of $g_{\mu\nu}$, ϕ) is necessary. One only needs to add $g > 0$ terms to the $g = 0$ approximation in order to recover the usual effects of second quantization.

Looking at FIGURE 2, we may say that the real, full-fledged thing lies at the bottom of the circle and includes a finite, unified, quantum theory of gauge and gravitational interactions. The tree level of QST is the $\alpha_{SL} \gg 1$ limit, while conventional classical field theories (e.g., Maxwell or Einstein) require the further limit of large distances ($d \gg \lambda_s$) to be taken. The usual problems with some quantum field theories (including the infinities of quantum gravity or the Landau poles of nonasymptotically free gauge theories) only occur if illicit exchanges of limits are performed (e.g., letting $\lambda_s \to 0$, while $l_p = (G_N \hbar^{1/2}$ is kept finite).

If we proceed in the correct, string way, loop effects are instead finite and calculable. What happens, for instance, to the tree-level G^2U relation? One finds rather easily[16]

$$\text{Gauge loop} = 0\left(\alpha \int_{\lambda_s}^{l} \frac{dx}{x}\right) = 0\left(\alpha \ln \frac{l}{\lambda_s}\right)$$

$$\text{Gravity loop} = 0\left(G_N\hbar \int_{\lambda_s}^{l} \frac{dx}{x^3}\right) = 0\left(\frac{G_N\hbar}{\lambda_s^2} = \alpha\right),$$

where λ_s is the UV cutoff and l, the typical scale of the quantity under study, plays the rôle of an IR cutoff. We see that radiative corrections to α_{GUT} and G_N are similar, except for the logarithmic behavior of the former. This is due to the different IR properties of gravitational and gauge interactions *in* $D = 4$. Thus gauge couplings run (as usual), while G_N does not. As a consequence, G^2U holds strictly speaking at the scale λ_s, and is broken by radiative corrections at large distances [for instance, quantum chronodynamics (QCD) presumably confines, while gravity does not].

PHYSICS BELOW λ_s

In analogy with the fact that the GSW theory of electroweak interactions differs from QED + Fermi's theory at distances $0(G_F^{1/2} \approx 10^{-15}$ cm), we expect string gravity to depart from Einstein's gravity at scales $0(\lambda_s \approx 10^{-32}\text{–}10^{-33}$ cm).

Of course, no direct measurement of these departures is accessible. Rather, there has been a lot of activity in trying to analyze the result of *Gedanken experiments*, such as Planckian-energy string collisions[17] or Planckian-temperature string thermodynamics[18] (actually, these conditions may have been realized in nature at the big bang).

Gedanken experiments have given evidence for large deviations from classical general relativity (and QFT) notably in the form of:

1. A minimal explorable scale $0(\lambda_s)$, suggesting that equal-time commutators of local spacetime fields are actually incorrect at short distances.
2. An extended uncertainty principle, reading

$$\Delta X > \hbar/\Delta p + 2\alpha'\Delta p \quad (> \lambda_s),$$

where the second, stringy term comes physically from the size of the "exchanged" string (which carries a momentum Δp).

3. An extended equivalence principle stating that structures in $g_{\mu\nu}$, ϕ that live on scales shorter than λ_s are irrelevant (more or less like gauge degrees are irrelevant in gauge theories). Thus the physical number of degrees of freedom is sharply reduced at short distances.
4. The minimal length of string theory could also be related to a very amusing quantum symmetry of certain string theories, known as target-space duality.[19]

In its simplest form, duality says that a closed string moving on a circle of radius R cannot be distinguished from one moving on a circle of radius λ_s^2/R. Thus, effectively, saying that R goes from zero to infinity is an illusion. The physically inequivalent values of R only go from λ_s to infinity (or from zero to λ_s) and the fixed point $R = \lambda_s$ plays the rôle of a minimal scale. An application of this idea to string cosmology was proposed by Brandenberger and Vafa.[20] Note, incidentally, an amusing analogy with ΔX of the extended uncertainty principle that is invariant under

$$(2\alpha'\Delta p) \rightarrow \lambda_s^2/(2\alpha'\Delta p)$$

(recall $\lambda_s^2 \equiv 2\alpha'\hbar$). These kinds of symmetries appear also elsewhere and are ultimately related to the two-dimensional nature of string theory and to a symmetry[13] under the interchange of σ with τ (X' with $\dot{X}$). This leads one naturally to address the last two "burning" questions.

1. Why $d = 1$?

There is something special to strings that is neither shared by points nor by higher dimensional systems:

- $d = 0$ (points) give arbitrary and too singular interactions (local vertices) that eventually lead to the incurable ultraviolet divergences of QFT.
- $d \geq 2$ (membranes) meet with quantization problems as $(d + 1)$ quantum field theories (they are UV infinite).
- $d = 1$ (strings) appears indeed as the best (only possible?) compromise. Also, iff $d = 1$, one can combine space and time into a complex number $z = \sigma + i\tau$. Many miracles of two-dimensional field theories, and in particular of string theory, come from the magic properties of analytic functions of z!

2. Why $D = 4$?

String theories make sense in $D > 4$ and actually often "start" in $D > 4$. Naturally, they are Kaluza–Klein-like theories, and the question for string theory is: Why should only 3 + 1 dimensions be macroscopic?

A tentative answer (hint) was given some time ago[21] using the fact that $D = 4$ is the critical dimension for gauge interactions. As we have seen, α's run in $D = 4$ (not in $D > 4$) and, for asymptotically free gauge theories, grow at low energy. This makes it possible for nonperturbative, strong coupling, phenomena (dynamical breaking of supersymmetry, of global and local symmetries) to occur whenever they lower the vacuum energy. In toy models this can be shown to drive an instability on the compactification radii of the extra dimensions that contract to smaller and smaller circles until the self-dual radius $R = \lambda_s$ is attained. But a lot of work remains to be done in order to make the toy model more realistic (for recent progress, see reference 22).

CONCLUSIONS

A few words since the game is not . . . concluded:

1. Gravity needs string theory . . . in order to cure its classical and quantum problems.
2. String theory needs gravity . . . as its most sensitive and maybe unique testing ground, for example, Is the dilaton a string killer? What about the cosmological constant? What about inflation and the problems of standard cosmology?

Obviously, what we need is more cross-talk-fertilization between string and gravity people, of which the symposium was a perfectly good example.

REFERENCES

1. DE VEGA, H. & N. SÀNCHEZ. 1987. Phys. Lett. **B197:** 320.
2. TUROK, N. & P. BHATTACHARJEE. 1984. Phys. Rev. **D29:** 1557.
3. SÀNCHEZ, N. & G. VENEZIANO. 1990. Nucl. Phys. **B333:** 253.
4. GASPERINI, M., N. SÀNCHEZ & G. VENEZIANO. 1991. Int. J. Med. Phys. **6A:** 3853.
5. GASPERINI, M. 1991. Kinematic interpretation of string instability in a background gravitational field. Torino University Preprint DFTT-38/90. Phys. Lett. **B258:** 70; HAN, N. S. & G. VENEZIANO. 1991. Mod. Phys. Lett. **6:** 1993.
6. AHARONOV, Y., F. ENGLERT & J. ORLOFF. 1987. Phys. Lett. **B199:** 366.
7. TUROK, N. 1988. Phys. Rev. Lett. **60:** 549.
8. BARROW, J. D. 1988. Nucl. Phys. **B310:** 743.
9. GASPERINI, M., N. SÀNCHEZ & G. VENEZIANO. 1991. Nucl. Phys. **B364:** 365.
10. ANTONIADIS, I., C. BACHAS, J. ELLIS & D. V. NANOPOULOS. 1988. Nucl. Phys. **B238:** 117.
11. CAMPBELL, B., A. LINDE & K. A. OLIVE. 1991. Nucl. Phys. **B355:** 146.
12. TAYLOR, T. R. & G. VENEZIANO. 1988. Phys. Lett. **B213:** 450.
13. VENEZIANO, G. 1986. Europhys. Lett. **2:** 133; ———. 1990. *In* The Challenging Questions (The Subnuclear Series), Vol. 27, A. Zichichi, Ed.: 199. Plenum. New York.
14. FRADKIN, E. & A. TSEYTLIN. 1985. Phys. Lett. **158B:** 316.
15. WITTEN, E. 1984. Phys. Lett. **149B:** 351.
16. See, e.g., VENEZIANO, G. 1988. DST Workshop on Superstring Theory, H. S. Mani and R. Ramachandran, Eds.: 1. World Scientific. Singapore.
17. For reviews, see, VENEZIANO, G. 1990. Proceedings of Superstring '89 Workshop, Texas A&M University, College Station, R. Arnowill *et al.,* Eds.: 86. World Scientific. Singapore; ———. 1991. Proceedings of the 1st International Conference on Particles, Strings, and Cosmology (Northeastern University, Boston, March 1990), P. Nath and S. Reucroft, Eds.: 486. World Scientific. Singapore.
18. ATICK, J. J. & E. WITTEN. 1988. Nucl. Phys. **B310:** 291.
19. KIKKAWA, K. & M. YAMASAKI. 1984. Phys. Lett. **149B:** 357; SAKAI, N. & I. SENDA. 1986. Prog. Theor. Phys. **75:** 692.
20. BRANDENBERGER, R. & C. VAFA. 1989. Nucl. Phys. **B316:** 391.
21. TAYLOR, T. R. & G. VENEZIANO. 1988. Phys. Lett. **B212:** 147.
22. FONT, A., L. E. IBÁÑEZ, D. LÜST & F. QUEVEDO. 1990. Phys. Lett. **B245:** 401; FERRARA, S., N. MAGNOLI, T. R. TAYLOR & G. VENEZIANO. 1990. Phys. Lett. **B245:** 409; NILLES, H. P. & M. OLECHOWSKI. 1990. Phys. Lett. **B248:** 268; BINÉTRUY, P. & M. K. GAILLARD. 1991. Phys. Lett. **B253:** 119.

COBE[a]

EDWARD L. WRIGHT

Department of Astronomy
University of California at Los Angeles
Los Angeles, California 90024-1562

INTRODUCTION

The Cosmic Background Explorer (*COBE*) was launched on November 18, 1989. NASA's first satellite designed primarily for observational cosmology, *COBE* is performing detailed studies of the cosmic microwave background radiation (CMBR) from the big bang, and is searching for the cosmic infrared background (CIB) radiation from the first objects to form after the big bang. Mather[1] and Gulkis *et al.*[2] provide prelaunch descriptions of the instruments and the mission goals, and Mather *et al.*[3,4] have given a description of the early results, including the first results of the spectrum experiment. Cheng *et al.*[5] presented the spectrum of the dipole anisotropy from 1 to 20 cm^{-1} and found that both the anisotropy experiment, DMR, and absolute far infrared spectrophotometer (FIRAS) measurements are consistent with the Doppler shift of a blackbody interpretation of the dipole. Since these early results have been reported many times in several conference proceedings (Hauser *et al.*,[6] Mather *et al.*,[7] Smoot *et al.*[8]), I will concentrate here on effects that will be important in the final analysis of *COBE* data.

ULTIMATE SPECTRAL LIMITS

FIRAS is a polarizing Michelson interferometer, which measures the difference between the spectrum of the sky, I_ν, and the spectrum of an internal reference source with a controllable temperature T_{cal} and an emissivity close to 1. The data transmitted to the ground consist of co-added interferograms, which are Fourier transformed to give the spectrum. A simple idealization of the FIRAS sensitivity is to define

$$\delta(\nu) = \frac{I_\nu - B_\nu(T_{cal})}{B_{\nu,max}(T_0)}$$

and state that the uncertainty in $\delta(\nu)$ will be less than 10^{-3}. For historical reasons I have used $T_0 = 2.8$ to normalize the plots shown here. The noise in the spectrum would be a constant $\delta(I_\nu)$ if the efficiency were constant. This form of $\delta(\nu)$ was used by Mather *et al.*[3,4] with conservative error bars of 10^{-2}. The actual FIRAS specification is stated in terms of νI_ν: after averaging over 5 percent wide frequency bins, the uncertainty in νI_ν should be less than 10^{-3} of the peak of $\nu B_\nu(T_0)$. The *COBE* Science Working Group hopes that the ultimate FIRAS spectra will have accuracies better than both of these specifications. However, even with perfect measurements of the

[a] This work was supported by NASA.

spectrum, "local" astrophysical sources will limit one's ability to separate out a given cosmological effect. For example, Daly[9] and Barrow and Coles[10] have written papers using the preliminary FIRAS limits on spectral distortions of $|y| < 0.001$ and $|\mu| < 0.01$ to place limits on the slope of the spectrum of primordial density perturbations at small scales that are damped out before recombination. These authors speculate about the meaning of improved limits such as $|y| < 10^{-4}$ that might be obtained using the ultimate sensitivity of FIRAS. I will now consider three effects that might interfere with such precise determinations of y.

GALACTIC DUST

The spectrum of dust in our galaxy is still somewhat uncertain. FIRAS has observed this spectrum, but the calibration for high-frequency data taken far from a nulled condition is still quite poor. So for the current discussion I will use the

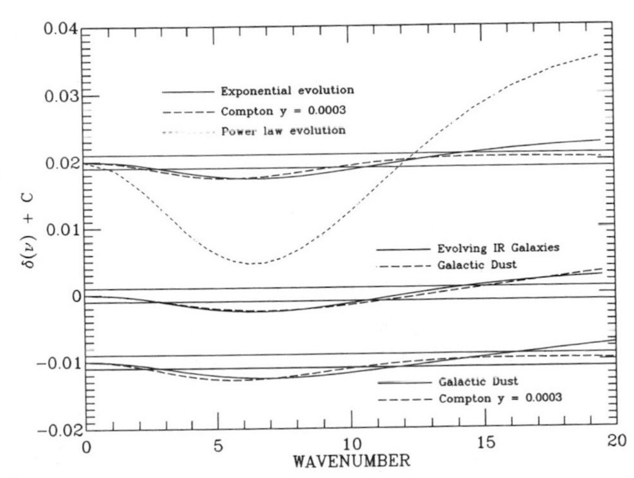

FIGURE 1. Bottom: Comparison of galactic-dust model (*solid curve*) to a y-distorted model with $y = 0.0003$ (*dashed curve*). **Middle:** Comparison of galactic-dust (*dashed curve*) model to exponentially evolving IR galaxies with $\gamma = 4$ (*solid curve*). **Top:** Comparison of exponentially evolving IR galaxies with $\gamma = 4$ (*solid curve*) to a y-model with $y = 0.0003$ (*dashed curve*), and power-law evolving IR galaxies with $n = 4$ and $z_{max} = 5$ (*dotted curve*), which can be ruled out by the preliminary *COBE* results.

spectrum derived by Lubin[11] who connected the IRAS 100-μm flux and a measured 3-mm flux: $I_\nu \propto \nu^{1.36} B_\nu(20K)$. I will scale this to the "galactic gradient," or slope of the IRAS flux versus csc b, of 3 MJy/sr given by de Bernardis *et al.*[12] Note that while the infrared experiment, DIRBE, gives a much smaller total 100-μm flux at the ecliptic poles, most of this change is due to an offset that does not affect the galactic gradient. The actual gain correction is a function of the angular scale, but it is a much smaller effect. Since the low frequencies are tied to the 3-mm data, the change is small for the frequencies less than 20 cm^{-1} that I consider here.

The bottom panel in FIGURE 1 shows a spectrum consisting of a 2.735 K blackbody observed through galactic dust radiating 3 MJy/sr at 100 μm as the solid curve. The double solid lines on this figure show the limits $\delta(\nu) = \pm 0.001$. The dashed curve shows the result of fitting a Compton distortion to this input. The best fit value of y is 0.0003. The difference between the galactic-dust model and the best fit y-model is small, but $\delta(\nu) > 0.001$, so it is marginally detectable. Clearly, the difference between a spectrum with $y = 0.0001$ and a galactic-dust spectrum with three times less dust would not exceed the FIRAS sensitivity specification. In order to detect spectral distortion as small as $y = 0.0001$ one will need not only FIRAS spectra reaching the specified sensitivity but also an independent galactic-dust detector. The FIRAS high-frequency channel and the DIRBE 160- and 240-μm channels are designed to provide this independent dust monitor.

CLUSTERS OF GALAXIES

The y-weighted cross section of a cluster of galaxies is

$$yA = \int y dA = \int\int \frac{\sigma_T n_e kT_e}{m_e c^2} ds\, dA$$

$$yA = \frac{\sigma_T}{m_e c^2} \int n_e kT_e\, dV = \sigma_T \left(\frac{n_e}{n_{\mathrm{TOT}}}\right)\left(\frac{E_{\mathrm{th}}}{m_e c^2}\right).$$

Now I need to know $\phi(E_{\mathrm{th}})dE_{\mathrm{th}}$, the "thermal energy function" for clusters of galaxies. There are much better data on the X-ray luminosity function of galaxies, so I use $E_{\mathrm{th}} = L_x \tau_{\mathrm{cool}}$. For $n_e = 0.001$ and $kT_e = 8$ keV, $\tau_{\mathrm{cool}} \simeq 10^{18}$ sec. The X-ray luminosity function of clusters of galaxies is

$$f(L_x) = A_e (H_0/50)^5 \exp(-L_x/L_{x0})$$

(Ulmer *et al.*[13]) with $A_e = 10^{-7}$ clusters/(Mpc^3 10^{44} erg/sec), and $I_{x0} = 6(50/H_0)^2 \times 10^{44}$ erg/sec. The total energy density in hot gas in clusters of galaxies is thus

$$U = \tau_{\mathrm{cool}} \int f(L_x) L_x\, dL_x \simeq 36 \times 10^{55}\ \mathrm{ergs/Mpc^3}.$$

Now $\sigma_T = 7 \times 10^{-74}$ Mpc^2, so

$$\langle y \rangle = \sigma_T \left(\frac{n_e}{n_{\mathrm{TOT}}}\right)\left(\frac{c}{H_0}\right)\left(\frac{U}{m_e c^2}\right) \simeq 10^{-7}.$$

Of course, the formal thermal energy of the isothermal sphere model used for clusters of galaxies is infinite because the low-density regions outside the core have very large cooling times. But there is clearly no reason to expect clusters of galaxies to contribute more than 10^{-5} to $\langle y \rangle$, so they will not affect *COBE*. However, dust in the intracluster gas would be heated by electrons and photons and would also radiate in the far infrared. Survival of the dust in the hot gas depends on shock heating and sputtering processes. If it survives, then it would be a significant source of far infrared background light, similar to dusty galaxies.

DUSTY EXTERNAL GALAXIES

The deep 60-μm counts done by IRAS show some evidence for evolution in the luminosity function of galaxies. Since the mean redshift observed by IRAS is small, this evolution is not well determined. Beichman and Helou[14] have considered the effect of the integrated light from IR galaxies on the spectrum measured by FIRAS. The top panel in FIGURE 1 shows $\delta(\nu)$ for two models of evolving IR galaxies. The curve with the large deviations corresponds to a density evolution $\propto (1 + z)^4$ cut off at $z_{max} = 5$. This model is clearly ruled out by the current 1 percent limits on spectral deviations.

I have computed more modestly evolving models using density evolution $\propto \exp(-\gamma t/t_0)$. The solid curve in the top panel of FIGURE 1 is the resultant spectrum for $\gamma = 4$. The dashed curve shows the result of fitting a y model to this spectrum. Once again I get $y = 0.0003$. The deviation between the models in this figure is barely detectable. The middle panel in FIGURE 1 shows an even closer match: the evolving galaxy model with $\gamma = 4$ is fit by a galactic dust model. These two curves cannot be distinguished by FIRAS. The only way to decide whether such a distortion is due to dust in our galaxy or the superposition of many distant IR galaxies is to extend the galaxy count data started by IRAS to much lower flux levels. The difference in the number counts reaches a factor of 4 at 60 μm fluxes of 1 mJy. The Space Infrared Telescope Facility (SIRTF) will be able to measure galaxies down to a diffraction-limited, confusion-limited sensitivity of a few tenths of a mJy (Wright and Taylor[15]).

ULTIMATE ANISOTROPY LIMITS

Smoot *et al.*[16] have published preliminary limits on anisotropies of the CMBR, based on six months of data with the DMR instrument (Smoot *et al.*[17]). The quadrupole limit is $\delta T(Q) < 70$ μK RMS, with 95 percent confidence. These limits are somewhat better than all previous all-sky anisotropy studies, but are still limited by estimates of many systematic error upper limits that will be removed in later data processing. This limit does not eliminate the most popular cosmological models such as inflation with cold dark matter (CDM). When normalized to the galaxy correlation function, CDM predicts $\delta T(Q) \approx (30/b)$ μK, where b is the bias factor. When normalized to large-scale streaming motions, Abbott and Wise[18] showed that inflation predicts $\delta T(Q) = 10$ μK. Gorski[19] finds that the minimum possible quadrupole, given the streaming motions, is 5 μK RMS if inflation is correct. If all systematic errors can be modeled and removed, the integration time necessary to achieve a 95 percent confidence limit of <5 μK on the RMS quadrupole is 2 to 3 years.

One important systematic error term that may well ultimately limit the detectable anisotropy is the galactic emission. FIGURE 2 shows the RMS quadrupole amplitude in Rayleigh–Jeans temperature versus frequency for all-sky maps: the galactic plane is not cut out. The Relikt 1 point departs from the trend of the DMR channels, but

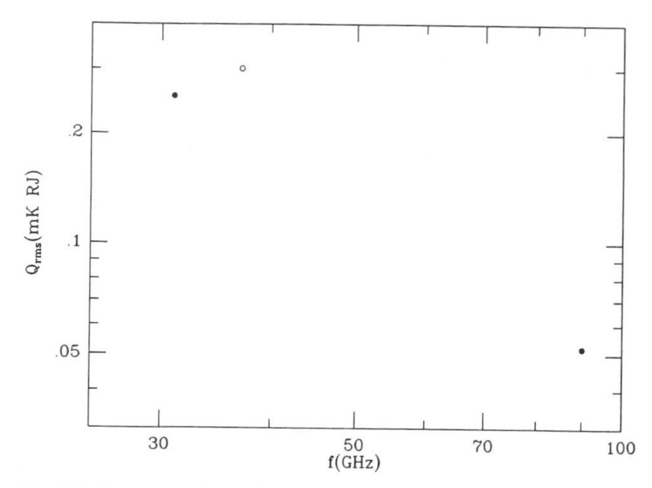

FIGURE 2. The RMS quadrupole in Rayleigh–Jeans mK versus frequency for *COBE* (*solid circles*) and Relikt 1 (*open circle*), with no galactic latitude cut. Systematic errors as large as 0.035 mK could exist in these preliminary *COBE* results.

the deviation is consistent with the noise in the Relikt 1 experiment. The 53- and 90-GHz DMR frequencies bracket the best frequency for observing CMBR anisotropy in the sense of having the lowest ratio of (RMS galaxy quadrupole):dipole. Obviously, the galaxy is always much larger than the large-scale anisotropy predicted by CDM. A factor of at least 10, and preferably 20, reduction in the galactic quadrupole will be needed to test inflation to the ultimate Gorksi limit. FIGURE 3 shows the RMS quadrupole in arbitrary units versus galactic cutoff for two different all-sky maps: the 408-MHz map and a dust map produced from the IRAS 100-μm

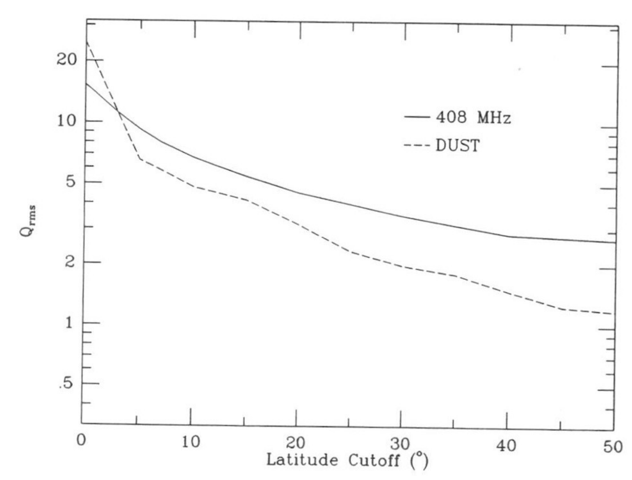

FIGURE 3. The RMS quadrupole in arbitrary units versus galactic latitude cutoff for 408 MHz and dust (IRAS 100 μm) maps.

channel. The RMS quadrupole actually increases for cutoffs larger than those shown because the quadrupole components become very highly correlated, and hence very noisy. But it seems that a simple latitude cutoff will probably not be sufficient at the 5–10-μK level. Some form of spectral fitting will be necessary; however, it is quite clear that the morphology of the galaxy is different at the two frequencies shown. Several galactic components, including synchrotron, free-free, and dust emission, will have to be fitted to the maps and subtracted. The DMR was planned with several frequencies in order to facilitate this fitting.

If I make the optimistic assumption that a cosmic anisotropy will be detected, then the analysis will shift from setting limits to testing models. In the case of inflation the spectrum is predicted, so the quadrupole amplitude will have to be compared to measurements sampling the perturbation spectrum at smaller scales. However, the quadrupole amplitude squared is the sum of the squares of five independent Gaussian components with equal variances in the inflationary model, and is thus proportional to χ^2 with five degrees of freedom. The normalization of the perturbation spectrum to the large-scale streaming motions is less model dependent than the normalization to the galaxy correlation function, but the square of the streaming motion is proportional to χ^2 with only three degrees of freedom. The ratio of Q^2/V^2 thus follows the F distribution with five and three degrees of freedom, which unfortunately is very broad.

$$\frac{Q_{\text{RMS}}}{V_{\text{pec}}} = K\sqrt{F(5,3)},$$

where K is a constant that depends on the cosmological model. Even if Q and V are measured perfectly, the 90 percent confidence interval on K is ± a factor of 2.6, which makes it hard to test theories. To reduce this spread one needs to increase the number of degrees of freedom in both the anisotropy measurement and the peculiar velocity measurement.

To increase the number of degrees of freedom in the peculiar velocity one needs to measure the peculiar velocity of distant groups of galaxies with respect to the CMBR frame of reference. One can use submillimeter measurements of the Sunyaev–Zel'dovich[20] effect to measure the peculiar velocities of clusters. With six clusters one could add six degrees of freedom giving a total of nine degrees of freedom. This measurement will be difficult, since the optical depths in the few clusters with measured Sunyaev–Zel'dovich effects are all about 10^{-2}. Hence to measure the radial peculiar velocity with an accuracy of 100 km s^{-1} requires measuring a ΔT of 9 μK. Separating the velocity effect from the Sunyaev–Zel'dovich effect requires observations at or above the crossover of the Sunyaev–Zel'dovich effect at 1.37-mm wavelength. While the atmosphere is reasonably transparent at 1.37 mm, measuring such small temperature differences may require the stability of a cold telescope in space, such as SIRTF. SIRTF plans to carry bolometers able to measure the short wavelength part of the Sunyaev–Zel'dovich effect. The proposed ESA mission FIRST and the proposed NASA submillimeter mission (SMME or SMILS) will also be able to measure the Sunyaev–Zel'dovich effect. Note that this experiment requires that the optical depth, and hence electron temperature of the intracluster gas be known, so with the X-ray flux one knows the Hubble constant.

One can increase the degrees of freedom in the anisotropy measurement by adding the octupole moment. According to the inflationary model, the weighted sum

$$\hat{Q}^2 = \left(\frac{5}{12}\right)\left(\sum |a_{2m}^2| + 2\sum |a_{3m}^2|\right)$$

should have the same mean value as

$$Q^2 = 4\pi Q_{\rm RMS}^2 = \sum |a_{2m}^2|,$$

but follow the χ^2 distribution with 12 degrees of freedom, since var $(a_{2m}) = 2$ var (a_{3m}).

But for noise, var $(a_{2m}) =$ var (a_{3m}), so the distribution of $\hat{Q}^2$ is not a standard one. Monte Carlo calculations show that the 95 percent confidence level for $\hat{Q}^2$ is 14 var (a_{2m}), while the 95 percent confidence level for Q^2 is 11 var (a_{2m}), when random noise is the only source of variation. Thus, adding more moments to the determination of the anisotropy increases the effect of random noise, and will not be useful unless the cosmic anisotropy is larger than Gorski's minimum.

INFRARED BACKGROUND

The light from many superimposed distant galaxies will provide an isotropic component to the infrared background observed by the DIRBE experiment. But even in the "holes" in the local background at 3.5 and 240 μm this cosmic component is expected to be only a small fraction of the total flux. For example, Matsumoto *et al.*[21] measured 184 kJy/sr at 3.5 μm at the north galactic pole, while the DIRBE measured 122 ± 25 kJy/sr there with the sun in the same relative position. The very deep near infrared galaxy counts by Cowie *et al.*[22] only account for 4 kJy/sr at 2.2 μm. Most of the flux seen by DIRBE at 3.5 μm is clearly due to the zodiacal light, and must be removed before cosmic conclusions are made. One very interesting outcome of the DIRBE experiment will be a comparison of the flux derived by counting up the contributions of galaxies to the flux derived by subtracting out local diffuse sources from the DIRBE flux. An accuracy of 1 percent in modeling the zodiacal light will allow an interesting test at 3.5 μm, but this will require deep galaxy counts at 3.5 μm that are possible from SIRTF but impossible from the ground. The zodiacal light is brighter at 2.2 μm, so an accuracy of 0.5 percent in modeling is required at this wavelength.

The DIRBE measurements of the South Ecliptic Pole can easily be fit by models of the zodiacal dust cloud, but the possible existence of a population of small hot interplanetary dust grains makes it impossible to predict the thermal component of the zodiacal light at 3.5 μm from the well-measured emission at 5–25 μm. Thus geometric modeling of the zodiacal dust cloud will be needed. The DIRBE experiment surveys one-half of the sky each day, covering the range of 64° to 124° from the sun. The hundreds of observations of each pixel made over the course of a year need to be collected and fitted to the zodiacal light model. This work is only now beginning.

DISCUSSION

The results I have shown in this paper are quite preliminary. The immense task of sorting the millions of FIRAS interferograms and the billions (U.S. billion!) of DIRBE data samples is underway. While the spectrum experiment is now turned off, the short wavelength channels of DIRBE and the DMR anisotropy experiment are still collecting excellent data. I expect *COBE* will continue to produce interesting results for several years to come, both from the new data still being collected and from further analysis of the existing data. The examples I have shown here show that the *COBE* experiments will probably be limited by our lack of knowledge of the local astrophysical foregrounds, such as the zodiacal dust cloud, interstellar dust in the Milky Way, and even IR galaxies at redshifts of 1–5. To obtain the full benefit of the *COBE* mission, the results of all three *COBE* instruments must be studied together, and followup observations will be needed from the powerful new infrared and submillimeter facilities that will become available in the coming decade.

ACKNOWLEDGMENT

I would like to thank the other 18 members of the *COBE* Science Working Group and the hundreds of workers at Goddard Space Flight Center whose efforts over many years have made *COBE* a big success.

REFERENCES

1. Mather, J. C. 1982. Opt. Eng. **21**(4): 769–774.
2. Gulkis, S., P. M. Lubin, S. S. Meyer & R. F. Silverberg. 1990. Sci. Am. **262**(1): 132–139.
3. Mather, J. C., *et al.* 1990. Astrophys. J., Lett. **354:** L37–L41.
4. Mather, J. C., *et al.* 1990. *In* Proceedings of the IAU Symposium 123 on Space Astronomy, Y. Kondo, Ed. To be published.
5. Cheng, E., J. Mather, R. Shafer, S. Meyer, R. Weiss, E. Wright, R. Eplee, R. Isaacman & G. Smoot. 1990. Bull. Am. Phys. Soc. **35:** 971.
6. Hauser, M., T. Kelsall, S. H. Moseley, Jr., R. F. Silverberg, T. Murdock, G. Toller, W. Spiesman & J. Weiland. 1991. The Diffuse Infrared Background: COBE and other observations, After the first three minutes. *In* Proceedings of the Cosmology Workshop (University of Maryland, College Park, Oct. 15–17, 1990), S. S. Holt, C. L. Bennett, and V. Trimble, Eds. American Institute of Physics. New York.
7. Mather, J. C., *et al.* 1991. Early results from the Absolute Far Infrared Spectrophotometer (FIRAS), After the first three minutes. *In* Proceedings of the Cosmology Workshop (University of Maryland, College Park, Oct. 15–17, 1990), S. S. Holt, C. L. Bennett, and V. Trimble, Eds. American Institute of Physics. New York.
8. Smoot, G. F., *et al.* 1991. Preliminary DMR measurements of the CMB isotropy, After the first three minutes. *In* Proceedings of the Cosmology Workshop (University of Maryland, College Park, Oct. 15–17, 1990), S. S. Holt, C. L. Bennett, and V. Trimble, Eds. American Institute of Physics. New York.
9. Daly, R. A. 1991. Astrophys. J. **371:** 14.
10. Barrow, J. D. & P. Coles. 1991. Mon. Not. R. Astron. Soc. **248:** 52.
11. Lubin, P. M. 1990. Private communication.
12. de Bernardis, P., S. Masi, F. Melchiorri, G. Moreno, R. Vannoni & S. Aiello. 1988. Astrophys. J. **326:** 941–946.

13. ULMER, M. P., *et al.* 1981. Astrophys. J. **243:** 681.
14. BEICHMAN, C. A. & G. HELOU. 1990. Astrophys. J., Lett. **370:** L1.
15. WRIGHT, E. L. & G. B. TAYLOR. 1988. Bull. Am. Astron. Soc. **20:** 649.
16. SMOOT, G. F., *et al.* 1991. Astrophys. J., Lett. **371:** L1.
17. SMOOT, G., *et al.* 1990. Astrophys. J. **360:** 685–695.
18. ABBOTT, L. F. & M. B. WISE. 1984. Astrophys. J., Lett. **282:** L47.
19. GORSKI, K. M. 1990. Astrophys. J., Lett. **370:** L5.
20. SUNYAEV, R. A. & YA. B. ZEL'DOVICH. 1980. Ann. Rev. Astron. Astrophys. **18:** 537.
21. MATSUMOTO, T., M. AKIBA & H. MURAKAMI. 1988. Astrophys. J. **332:** 575.
22. COWIE, L. L., J. P. GARDNER, S. J. LILLY & I. MCLEAN. 1990. Astrophys. J., Lett. **360:** L1–L4.

Seismic Constraints on the Solar Neutrino Problem

D. O. GOUGH

Institute of Astronomy and Department of Applied Mathematics and Theoretical Physics
University of Cambridge
Cambridge, CB3 0HA England
and
School of Mathematical Sciences
Queen Mary and Westfield College
University of London
London, England

INTRODUCTION

That a so-called standard theoretical model of the sun with a plausible chemical composition cannot produce a flux of neutrinos as low as that which the observations imply raises one of the most important problems in modern astrophysics. A great deal of ingenuity has been brought to bear on that problem, and of the many suggestions for resolving the issue that have been made, most have already been vitiated. It is, however, incumbent upon us to scrutinize the survivors, in the hope of eliminating more of the considerable doubt that still clouds the subject, for the issue is not only one of astrophysical concern but is becoming increasingly pertinent to current discussion in particle physics.

Proposed resolutions of the solar neutrino problem fall naturally into one of two categories: astrophysical remedies, in which the faithfulness of standard solar models is denied, and microphysical remedies, in which it is postulated that standard nuclear or particle physics requires modification. Proposals of the latter kind are currently the more fashionable, and are perhaps of greater concern to the other principal topics discussed at this the 15th Texas/ESO-CERN Symposium. In particular, matter-induced neutrino transitions—the MSW effect—are of considerable interest, and in view of the almost complete lack of *a priori* knowledge of the parameters of the theory that control predictions of the final neutrino states, it is not implausible that the phenomenon might eventually be shown to provide the complete explanation of the problem. Nevertheless, the astrophysicists' task is not merely to produce a theoretical model of the sun that is not obviously at variance with observation, but to learn what the internal structure of the sun actually is, and to understand why it is so. To this end one must examine as much relevant data as it is possible to bring to bear on the subject. Of course, amongst such data are the fluxes of neutrinos, and in the long run it might be possible to use them to learn about conditions inside the sun. But in the meanwhile it is much more likely that inference will be in the reverse direction, the predicted neutrino fluxes being compared with observation in order to constrain microphysical hypotheses. Indeed, this has already begun, with attempts to rationalize the data from the Homestake ^{37}Cl detector, from Kamiokande, and from SAGE

(Bahcall and Bethe;[1] Krastev, Mikheyev, and Smirnov[2]). However, the validity of such analyses rests on the assumption that our knowledge of the conditions under which the neutrinos are produced, currently provided by standard models, is secure; that is surely not the case. It behoves us astrophysicists to determine more reliably the state of the solar neutrino source, and thereby to reduce the uncertainty in the microphysical inferences.

A new and powerful procedure that has recently become available to us for probing the solar interior is seismological analysis. My charge here is briefly to review what has been learned that is pertinent to the neutrino problem by that procedure, and to discuss the implications. Helioseismology is in its infancy, and although several very important properties of the sun's envelope have been revealed, it has not yet been possible to gauge the state of the energy-generating core sufficiently reliably to satisfy even the current requirements of particle physics. However, seismological constraints have been able to rule out some theoretical models, and have thus already substantially reduced the domain of uncertainty.

STANDARD SOLAR MODELS, AND THE REACTIONS OF THE pp CHAIN

Seismological analysis has confirmed that, broadly speaking, the gross properties of standard solar models are indeed representative of conditions in the sun. This is hardly surprising, because the theory of stellar evolution has successfully explained a wide variety of astronomical observations. It is therefore expedient to discuss seismological inferences in terms of the small deviations from a standard model. The model I use for reference is model 2 of Christensen-Dalsgaard *et al.*,[3] which was computed essentially by the procedure described by Christensen-Dalsgaard.[4] It uses the so-called Mihalas–Hummer–Däppen (MHD) equation of state (Hummer and Mihalas;[5] Mihalas, Däppen, and Hummer[6]), the opacity tables of Cox and Tabor,[7] and nuclear-reaction rates as presented by Fowler, Caughlan, and Zimmerman;[8] it is evolved to an age of 4.75×10^9 y, from an initial helium abundance $Y = 0.237$ with a heavy-element abundance $Z = 0.02$, to reproduce the observed present luminosity $L_\odot$ and radius $R_\odot$ of the sun. Of course, there are many other standard models I could have used, but I chose this one partly because it is numerically accurate, in the sense that a consistently accurate difference scheme was used to represent the governing differential equations, and partly because the location of the base of its convection zone, at a radius $r = 0.7149R$, is in accord with the solar value ($r = 0.713 \pm 0.003R$), which has been determined seismologically (Christensen-Dalsgaard *et al.*[3]). The hope, therefore, would be that amongst standard models this model deviates from the sun by less than most.

It is perhaps worth remarking that the nuclear reaction rates and opacities used in the reference model are not the most recently available. However, standard models with similar values of Z computed with more modern published reaction rates and opacities have shallower convection zones, and appear to represent the structure of the solar envelope less well. This is the case, for example, of the standard model of Bahcall and Ulrich,[9] which was computed with opacities from the Los Alamos Opacity Library (LAOL),[10] and whose convection zone extends downward only to $r = 0.74R$; it is true also of model 4 of Christensen-Dalsgaard *et al.*,[3] which was computed with the same opacities and reaction rates as that of Bahcall and Ulrich[9] and essentially the same heavy-element abundance, but with the MHD equation of state, and whose convection zone extends to $r = 0.7332R$. As I discuss later, the discrepancy might be removed by revising the LAOL opacities upward at temperatures of a few million degrees.

I shall dwell on the structural properties of standard models no more, since Bahcall provides further information in his review.[11] However, I do wish to mention the nuclear reactions briefly, to establish a framework for subsequent discussion. The dominant thermonuclear reactions in the solar core are those of the pp chain, the most important of which are

$$p(p, \beta^+\nu)D(p, \gamma)^3He(^3He, 2p)^4He \qquad (85\%)$$

$$p(e^- p, \nu)D \qquad ^3He(^4He, \gamma)^7Be(e^-, \nu)^7Li(p, ^4He)^4He \qquad (15\%)$$

$$^7Be(p, \gamma)^8B(\beta^+\nu)^8Be^*(^4He)^4He \qquad (0.02\%).$$

The figures in parenthesis on the right-hand side denote roughly the percentages of ^4He produced by the three branches of the chain (called, respectively, ppI, ppII, and ppIII) at the center of the sun. Although the ppIII chain produces relatively little

TABLE 1.

Reaction	Q(MeV)	Q_ν(MeV)	η	Timescale (y)
$p(p, \beta^+\nu)D(p, \gamma)^3He$	6.67	0.27	4	10^{10}
$p(e^-p, \nu)D(p, \gamma)^3He$	5.49	1.44	4	10^{12}
$^3He(^3He, 2p)\alpha$	12.86		16	10^5
$^3He(^4He, \gamma)^7Be$	1.59		17	10^6
$^7Be(e^-, \nu)^7Li(p, \alpha)\alpha$	17.39	0.81	$-\frac{1}{2}$	10^{-1}
$^7(Be(p, \gamma)^8B$	0.14		13	10^2
$^8B(\beta^+\nu)^8Be^*(\alpha)\alpha$	11.36	6.62	0	10^{-8}

Note: Mean thermal energy Q and neutrino energy Q_ν per reaction; the temperature exponent η of the reaction rate and the characteristic time scale are for conditions in the center of the sun (obtained from Bahcall and Ulrich[9]).

energy, its importance is in its production of high-energy neutrinos from the decay of ^8B, to which the ^{37}Cl detector and the Kamiokande detector are by far the most sensitive. The pep reaction is rare compared with $p(p, \beta^+ \nu)D$, contributing about 0.25 percent, but is included because it too contributes to neutrinos detected by the ^{37}Cl experiment. Some relevant properties of the reactions are listed in TABLE 1: Q is the mean thermal energy produced per reaction and Q_ν is the mean energy of the neutrino emitted; the rates of all reactions per unit volume are proportional to $\rho^\beta T^\eta$, where ρ is density and T is temperature, with $\beta = n$ for those reactions involving n-particle collisions and $\beta = 1$ for spontaneous decay; the exponents η and the characteristic reaction times are given for conditions at the center of the sun—in the cases of multiple reactions the slowest reaction, which controls the rate, is in all cases the first.

With this information one can estimate how the neutrino fluxes depend on central conditions in solar models. Since all the reactions in the chain except the very first proceed on a timescale much less than the characteristic evolution time of the sun, the reactions are essentially in balance, at least in the central regions of the sun where most of the reactions take place. Therefore, ignoring for simplicity the relatively small contributions to ^{3}He destruction from the ppII and ppIII branches, the ^{3}He abundance can be determined by balancing the ^{3}He-producing reaction with the ^{3}He-destroying ppI branch:

$$X^2\rho^2T^4 \propto X_3^2\rho^2T^{16}, \tag{1}$$

where X and X_3 are the abundances of ^{1}H and ^{3}He, respectively, from which

$$X_3 \propto XT^{-6}. \tag{2}$$

The abundance X_7 of ^{7}Be can now be determined from balancing the ppII branch, ignoring the small contribution from ppIII:

$$X_3Y\rho^2T^{17} \propto (1 + X)X_7\rho^2T^{-1/2}, \tag{3}$$

where Y is the ^{4}He abundance, $(1 + X)\rho$ being roughly proportional to the electron density in the essentially fully ionized plasma. It follows from the proportionalities **(2)** and **(3)** that

$$X_7 \propto \frac{1 - X}{1 + X} X\, T^{11.5}, \tag{4}$$

since $Z = 1 - X - Y \ll 1$. Finally, by balancing the reactions of the ppIII branch, one obtains the abundance X_8 of ^{8}B:

$$X_8 \propto \frac{1 - X}{1 + X} X^2 \rho\, T^{24.5}. \tag{5}$$

The rates $\epsilon_{\nu 1}$, $\epsilon_{\nu 7}$, and $\epsilon_{\nu 8}$ of pep, ^{7}Be, and ^{8}B neutrinos per unit mass therefore satisfy the following relations:

$$\epsilon_{\nu 1} \propto (1 + X)X^2\rho^2T^4, \tag{6}$$

$$\epsilon_{\nu 7} \propto (1 - X)X\rho T^{11}, \tag{7}$$

$$\epsilon_{\nu 8} \propto \frac{1 - X}{1 + X} X^2 \rho\, T^{24.5}. \tag{8}$$

We note, in addition, that the thermal energy generation rate ϵ is given essentially by the rate of the p(p, β^+ ν)D reaction:

$$\epsilon \propto X^2\rho T^4. \tag{9}$$

The proportionalities **(6)–(9)** indicate how the contributions to the fluxes are distributed through the star. In a spherically symmetrical solar model the total neutrino luminosity $L_{\nu i}$ from source i is given by $\int f_{\nu i}\,dr$, where $f_{\nu i} = 4\pi r^2 \rho\epsilon_{\nu i}$. Similarly, the photon (thermal) luminosity is $L = \int f\,dr$, where $f = 4\pi r^2 \rho\epsilon$. Evidently, since ϵ, $\epsilon_{\nu 7}$, and $\epsilon_{\nu 8}$ are successively more rapidly increasing functions of temperature, the contribution functions f, $f_{\nu 7}$ and $f_{\nu 8}$ are successively more concentrated toward the center of the sun; they are illustrated in FIGURE 1. Notice that, according to the proportionality **(2)**, the ^{3}He abundance increases outward through the core; further out, in the envelope where the reactions are too slow to have attained balance, X_3 declines to its primordial value.

Finally, in this discussion of nuclear reactions, permit me to remark that for a homologous sequence of solar models, the photon luminosity L and the neutrino luminosities $L_{\nu i}$ are proportional to ϵ and $\epsilon_{\nu i}$, respectively. Consequently, since L is fixed at the observed luminosity $L_\odot$, it follows that for such a sequence,

$$L_{\nu 1} \propto 1 + X, \tag{10}$$

$$L_{\nu 7} \propto (1 - X)X^{-1}T^7, \tag{11}$$

$$L_{\nu 8} \propto \frac{1 - X}{1 + X}T^{20.5}. \tag{12}$$

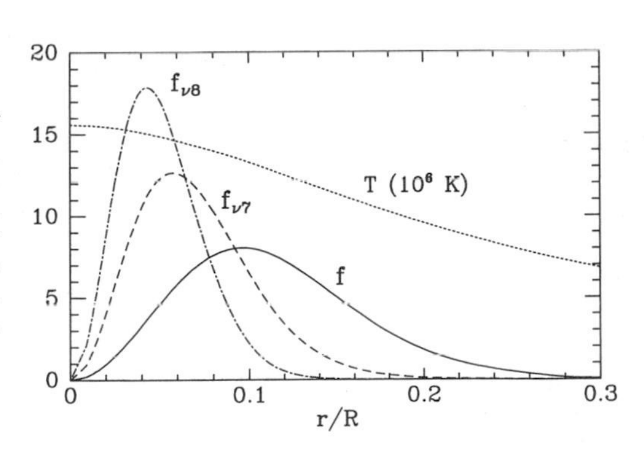

FIGURE 1. Contribution functions f (*continuous curve*), $f_{\nu 7}$ (*dashed curve*), and $f_{\nu 8}$ (*dot dashed curve*), representing the generation of thermal energy, ^{7}Be neutrinos and ^{8}B neutrinos per unit radius. They are normalized to have unit integral with respect to r/R. The *dotted curve* is the temperature, measured in units of 10^6 K.

Before terminating this discussion of standard solar models and their nuclear reactions, I must mention that almost all modern stability calculations find the models to be unstable to g modes in the core (Christensen-Dalsgaard *et al.*;[12] Boury *et al.*;[13] Shibahashi *et al.*;[14] Saio;[15] Kosovichev and Severny[16]). The instability is driven by the heat generated by the relatively temperature-sensitive ^{3}He-destroying reactions that are thrown out of equilibrium by the motion, and occurs as a result of the positive X_3 gradient in the core whose existence I have just explained. Therefore, strictly speaking, it appears that the sun cannot possibly be precisely in static equilibrium in the manner described by the standard models. How much it deviates is a matter of considerable doubt. What is not in doubt, however, is that until the implications of the instability are understood, theoretical predictions of solar neutrino fluxes cannot be trusted.

THE SOLAR NEUTRINO PROBLEM

Standard solar models need to be calibrated to reproduce the observed luminosity $L_\odot$ and the observed photospheric radius $R_\odot$ at the present solar age to $t_\odot$; for simplicity I shall assume $t_\odot$ to be known. Roughly speaking, the radius is calibrated by adjusting an undetermined constant α in a formalism for relating the heat flux to the temperature gradient in the convection zone, and since that formalism is quite uncertain, the resultant value of α does not have a precise physical significance. The luminosity is determined by the chemical composition. If the heavy-element abundance Z is lowered, opacity would tend to decrease; this must be compensated for by augmenting the electron density, which is accomplished by having a higher value of X, at the expense of Y. As is evident from the proportionality (**9**), in order to keep ϵ more-or-less unchanged, to maintain $L = L_\odot$, T must therefore be reduced, and hence the neutrino fluxes $L_{\nu 7}$ and $L_{\nu 8}$ decline. Of course, there is a redistribution of matter to achieve a new hydrostatic balance, the influence on the pressure of the reduction of the mean molecular mass μ resulting from the increase in X being offset by the decline in T.

It seems likely that in view of the steep temperature dependence of $L_{\nu 7}$ and $L_{\nu 8}$, particularly the latter, that dominates the theoretical contribution to the ^{37}Cl neutrino capture rate, it ought to be possible to find values of X, Y, and Z that reproduce the observed neutrino flux. Indeed, that is actually the case. The trouble is that the value of Y so obtained is small, less than 0.2, which is contrary to astronomical experience. In particular, it is less than the helium abundance measured in the photospheres of other stars; moreover, it is contrary to cosmological models of the big bang in which the observed cosmic abundances of ^2H and ^3He are synthesized. So confident are astronomers of this constraint that, rather than dubbing this flaw in the theory the *solar helium problem,* it has more aptly been called the *solar neutrino problem.*

How might the problem be resolved? As I have already mentioned, much ingenious effort has been expended on trying to solve the problem, but few suggestions remain unscathed. What is evident is that, if neutrinos do not undergo some kind of transition between the sun and the earth, the integral of $\epsilon_{\nu 8}$ and, to a lesser extent, $\epsilon_{\nu 7}$ must be reduced without changing the integral of ϵ. An idea that has received some attention in recent years is that the sun has captured weakly interacting massive particles (wimps) that enhance the thermal energy transport through the core, thereby reducing the temperature gradient. In this way the temperature in the central regions of the core where most of the neutrinos are emitted is reduced relative to the mean value, and the neutrino luminosity of the sun is thereby diminished. Another possibility is that perhaps as a result of a ^3He-driven motion such as the g-mode instability mentioned in the previous section, the core is varying with time on a timescale less than the equilibration time of the ppI chain, which invalidates the static relations (**1**)–(**9**) and modifies the nuclear balance. I shall have more to say about that later. In any case, it is one of the goals of helioseismology to determine whether any suggestions of that kind could be correct, or alternatively whether we are forced to conclude that the resolution of the problem must lie in the realm of particle physics.

HELIOSEISMIC INVERSION

The internal structure of the sun is inferred from observed frequencies of oscillation. The amplitudes of oscillation are small enough for linearized theory to be an adequate approximation. Moreover, throughout all but the surface layers of the star, the oscillatory motion can be considered to be adiabatic. Setting aside consideration of nonadiabatic processes for the moment, the cyclic frequency ν_i of a mode $\boldsymbol{i}$ satisfies a variational principle from which the small relative difference $\delta\nu_i$ between the solar oscillation frequency ν_i and the corresponding frequency ν_{0i} of the reference theoretical model can be computed, which itself can be linearized in what is hoped is the small difference $\delta \boldsymbol{S} = \boldsymbol{S} - \boldsymbol{S}_0$ between the structure $\boldsymbol{S}(\boldsymbol{r})$ of the sun and the structure $\boldsymbol{S}_0(\boldsymbol{r})$ of the reference, yielding

$$\frac{\delta\nu_i}{\nu_i} = V_0^{-1} \int_{V_0} \boldsymbol{K}(\boldsymbol{\xi}_{0i}; \boldsymbol{S}_0) \cdot \delta\boldsymbol{S}\, dV, \tag{13}$$

the integral being over the volume V_0 of the reference model. In this equation $\boldsymbol{\xi}_{0i}$ is the displacement oscillation eigenfunction of the reference model, and $\boldsymbol{i} = (n, l, m)$, where n is (principal) order, l is degree, and m is azimuthal order, labels the mode. The kernels $\boldsymbol{K}$ depend only on $\boldsymbol{S}_0$ and $\boldsymbol{\xi}_{0i}$, and fortunately not explicitly on the eigenfunctions of $\boldsymbol{S}$.

Adiabatic oscillations are produced by pressure–gradient fluctuations accelerating material with inertia density ρ; the dynamics therefore depends only on pressure p, density ρ, and the factor γ in the constitutive law $d \ln p = \gamma d \ln \rho$ relating infinitesimal adiabatic perturbations of p and ρ of any given element of material. Consequently, the variation of only p, ρ, and γ through the star is all that can be ascertained directly from the oscillation frequencies; the equations of motion contain no other quantities. Other thermodynamical variables, such as temperature, are secondary, and can be estimated only by invoking some other, nonseismic relation.

An auxiliary relation that can be employed, and that indeed has already been assumed of both $\boldsymbol{S}_0$ and $\boldsymbol{S}$ in the derivation of **(13)**, is the equation of hydrostatic support. That, together with the constraint that both the masses M and the radii R of the sun and the reference model are the same, permits one to relate the pressure difference δp between the sun and the reference to the density difference $\delta\rho$, reducing the dimension of the vector $\delta\boldsymbol{S}$ to two. Any two of the structure variables p, ρ, and γ can then be chosen for $\boldsymbol{S}$, or any two independent functions of them. Convenient choices are the helium abundance Y together with either ρ or $u = p/\rho$, the latter being proportional to the square of the sound speed, which is the dominant quantity determining the propagation of p waves. Moreover, in view of the perfect gas law, which provides at least a good guide to thinking about the solar core because electron degeneracy, though not negligible, is rather small, the quantity u is approximately proportional to temperature. However, u is also inversely proportional to μ; it is necessary to recognize that fact, especially in the core where there is a composition gradient resulting from nuclear transmutation.

Equation **(13)** expresses the relative frequency difference, which is an observed quantity, as an average of the structure difference $\delta\boldsymbol{S}$ weighted by the kernel $\boldsymbol{K}$. With a set of modes associated with which is a sufficient variety of kernels $\boldsymbol{K}$, the averages can be combined to provide relatively localized information, which is easier to digest. Specifically, one can seek a set of parameters $\alpha_i(\boldsymbol{r}_0)$ such that the averaging kernel $\boldsymbol{A}$ defined by the combination

$$\sum_i \alpha_i \frac{\delta\nu_i}{\nu_i} = V_0^{-1}\int_{V_0}\left(\sum_i \alpha_i \boldsymbol{K}_i\right)\cdot \delta\boldsymbol{S}\, dV =: V_0^{-1}\int_{V_0} \boldsymbol{A}(\boldsymbol{r}, \boldsymbol{r}_0)\cdot \delta\boldsymbol{S}\, dV, \tag{14}$$

in which I have used the abbreviation $\boldsymbol{K}_i$ for $\boldsymbol{K}(\xi_{0i}; \boldsymbol{S}_0)$, has essentially only one component and that that component is strongly localized in space near $\boldsymbol{r} = \boldsymbol{r}_0$. There are various techniques for obtaining suitable coefficients α_i, many of which have been pioneered by geoseismologists and which require some tradeoff between the degree of localization of $\boldsymbol{A}$ and unwanted magnification of data errors relative to the exact value of the combination on the left-hand side of **(14)**; the magnification results from large values of α_i causing near cancellation amongst the actual relative frequency differences, but of course not amongst the errors. Application of these so-called inverse techniques to helioseismic problems is discussed, for example, by Gough,[17] Christensen-Dalsgaard *et al.*,[18] and by Gough and Thompson;[19] I shall not discuss them here.

I should remark now that to quite a high degree of approximation the sun is spherically symmetrical. Therefore the frequencies are nearly degenerate with respect to azimuthal order m. The slight departure from spherical symmetry splits the degeneracy, measurement of which provides information about the small aspherical part of the structure. The uniformly weighted averages $\bar{\nu}_i$ of ν_i over m at fixed n and l provide data for the dominant spherically symmetrical part $\bar{\boldsymbol{S}}(r)$ of $\boldsymbol{S}$, defined to be the average of $\boldsymbol{S}$ over spherical surfaces. It is with these averages that I shall here be concerned. They are related by equations of the type **(13)** and **(14)**, except that the integral is now simply over radius r, the factor V_0^{-1} being appropriately replaced by R^{-1}, and the kernels $\boldsymbol{K}_i$ being replaced by appropriate averages $\bar{\boldsymbol{K}}_i$.

The solar modes of oscillation that have been observed and identified are all acoustic modes (p modes). The frequency spectrum of those p modes that are directly useful for the purpose of diagnosis of the deep interior of the sun ranges from 0.26 mHz, the frequency of the fundamental mode of degree 0, up to the acoustic cutoff frequency of the atmosphere, about 5.5 mHz. The graver modes have very low amplitude, and have not yet been observed; we have reliable data only for modes with cyclic frequencies ν above about 1.5 mHz. Modes for which frequencies have been measured accurately have degrees l ranging from zero to $O(10^3)$.

Components of a kernel $\bar{\boldsymbol{K}}$ for such a p mode are illustrated by the continuous curves in FIGURE 2, the basis being such that $\delta\bar{\boldsymbol{S}} = (\delta \ln \bar{u}, \delta\bar{Y})$, though from now on I shall omit the overbar on the structure variables. As is the case for all the observed modes, the first component has a substantial amplitude only between radii r_1 and r_2, the so-called turning points of the governing wave equation, beyond which the eigenfunctions and the associated kernel decay exponentially. The lower turning point of a mode of frequency $\omega = 2\pi\nu_i$ is given approximately by the condition $\omega/(l + \frac{1}{2}) = c(r_1)/r_1$, where $c = (\gamma p/\rho)^{1/2}$ is the adiabatic sound speed; its value is about $0.5R$ for the mode illustrated. The upper turning point depends essentially on frequency alone: $\omega_c(r_2) = \omega$, where ω_c is the acoustic cutoff frequency that, broadly speaking, can be considered for the relevant modes to be an increasing function of r.

Its value is $r_2 = 0.9954R$ for the p mode illustrated, too close to the surface to be readily discernible in the figure. The amplitude of the second component of $\overline{\boldsymbol{K}}$ is substantial only in the ionization zones of H and He, provided that they too lie between r_1 and r_2.

Perhaps the most striking feature of FIGURE 2 (upper panel) is that even above the lower turning point r_1, the amplitude of the ln u component $\overline{K}_{u,Y}$ of $\overline{\boldsymbol{K}}$ is quite small, except in the very outer region of the domain. Physically speaking, the reason is that $\overline{K}_{u,Y}$ contains a factor that is proportional to the time spent per unit volume by a propagating acoustic wave whose resonant interference with itself constitutes the mode; the factor is the product of the wave slowness c^{-1} and a geometrical term that

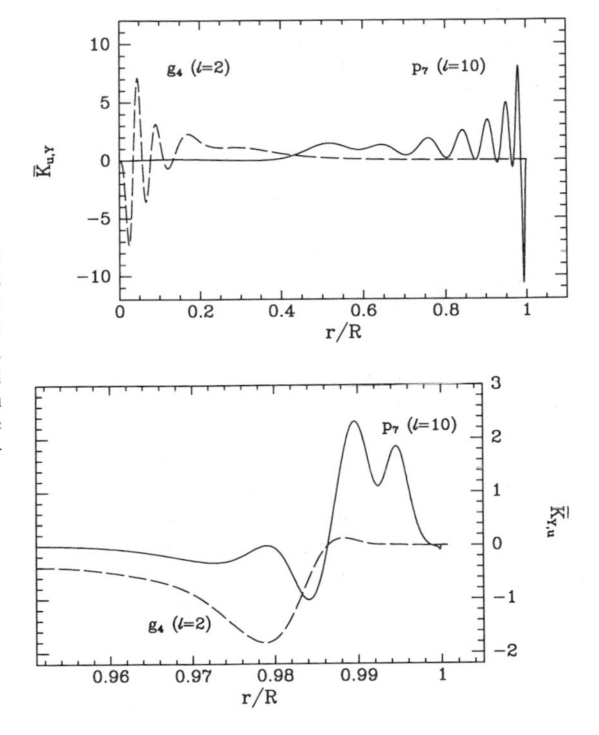

FIGURE 2. Kernels $\overline{K}_{u,Y}$ and $\overline{K}_{Y,u}$, defined such that **(13)** takes the form $\nu_i^{-1}\,\delta\nu_i = R^{-1}\int(\overline{K}_{u,Y}\,\delta\ln u + \overline{K}_{Y,u}\,\delta Y)\,dr$, plotted against r/R for a p mode and a g mode. The kernels $\overline{K}_{Y,u}$ are substantial only in the hydrogen and helium ionization zones, where the abundances influence the adiabatic exponent γ.

depends on l. Because sound speed is relatively high deep inside the sun, the wave spends relatively little time there, and consequently the structure $\boldsymbol{S}$ deep in the sun contributes relatively little to the oscillation frequency. It follows that it is not easy to measure $\boldsymbol{S}$, via $\delta\boldsymbol{S}$, in the core from only p-mode frequencies. Evidently it would be much simpler if frequencies of g modes were available, for their ln u kernels, an example of which is illustrated by the dashed curve in FIGURE 2 (upper panel), are greatest near the center of the sun. However, as can also be seen in FIGURE 2, g-mode kernels are very small in the outer envelope. This is because the eigenfunctions too are small, which renders the oscillations extremely difficult to detect

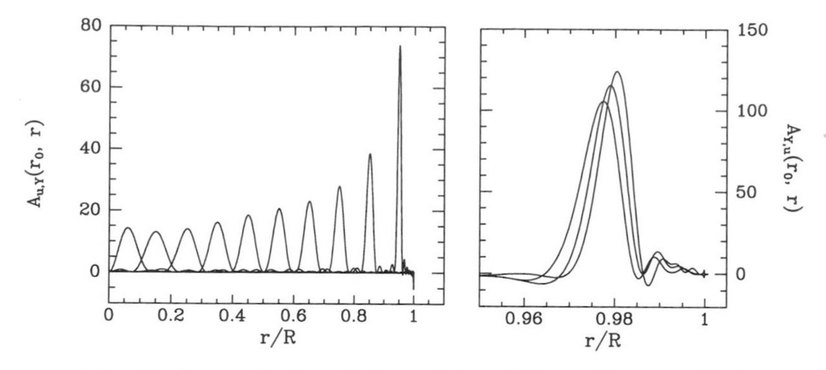

FIGURE 3. Localized averaging kernels $A_{u,Y}$ and $A_{Y,u}$, which are linear combinations of the corresponding kernels $\overline{K}_{u,Y}$ and $\overline{K}_{Y,u}$. For the coefficients α_i defining the combinations illustrated in the **left-hand panel,** the corresponding averaging kernels $A_{Y,u}$ are everywhere small. Similarly, for the combinations producing the kernels $A_{Y,u}$ in the **right-hand panel,** the corresponding kernels $A_{u,Y}$ are everywhere small.

observationally. It is one of the most important challenges of observational helioseismology to measure g modes.

At first sight it might seem to be hopeless to try to infer core structure from p modes alone. Indeed, that attitude was prevalent in the early days of the subject, and has still not been entirely eradicated, particularly amongst heliophysicists not familiar with inversion techniques. However, observations are now so accurate that one can afford a very substantial degree of cancellation in the data and still be left with combinations **(14)** that are not dominated by errors. Some examples of averaging kernels $A_{u,Y}$ and $A_{Y,u}$ constructed from kernels $\overline{\boldsymbol{K}}_i$ of modes that have been observed are illustrated in FIGURE 3. It is evident from the figure that useful information can be obtained about the energy-generating core, though, as can be seen by comparison with FIGURE 1, the widths of the averaging kernels are not yet small enough to resolve the detailed structure within the region where ^{8}B neutrinos are produced. Essentially a single average of that region is all that we have at present, but that is very much better than nothing.

RESULTS OF THE INVERSION

Several inversions by the methods outlined in the previous section have been carried out to determine the deviations $\delta \ln u$ and $\delta \ln \rho$ from a reference model (Gough and Kosovichev;[20,21] Dziembowski *et al.*;[22] Däppen *et al.*[23]). The latest results are reproduced in FIGURE 4, and are broadly representative of the others. They are the inversion of a homogeneous set of frequencies obtained at Big Bear Solar Observatory in 1986 (Libbrecht *et al.*[24]), and kindly made available prior to publication. An inversion of similar data obtained by Libbrecht *et al.* in 1988 yields very similar results. Perhaps the most prominent feature in FIGURE 4 (left panel) is that throughout the entire radiative envelope beneath the convection zone and outside the core, $\delta \ln u$ is positive. Christensen-Dalsgaard *et al.*[25] noticed this in one of the first sound-speed determinations, which was accomplished by inverting an asymptotic approximation to the raw p-mode eigenfrequencies rather than the differential constraints (**13**). They pointed out, after comparison with solar-model calculations, that the deviation could be explained by there being a deficiency of about 20 percent in the published opacities at temperatures below about 4×10^6 K in the radiative interior. Korzennik and Ulrich[26] subsequently found a similar result as one of two possible outcomes of an inversion for opacity, and Cox *et al.*[27] confirmed the result by forward calculations. At about the same time, this physical prediction was confirmed at least qualitatively by new, careful opacity calculations by Iglesias and Rogers,[28] who determined opacities higher than those of the LAOL by some 12 percent at $T = 2.2 \times 10^6$ K, the temperature at the base of the convection zone, declining to 7 percent at $T = 4 \times 10^6$ K.

The second feature to which I wish to draw attention is the dip in $\delta \ln u$ at the edge of the core, and the suggestion that $\delta \ln u$ is positive at the very center of the sun. This too is a robust feature of all the recent inversions, including, in addition to the inversions of the constraints (**13**) previously mentioned, direct asymptotic inversions by Shibahashi and Sekii,[29] Vorontsov,[30] Sekii and Shibahashi,[31] and Vorontsov and Shibahashi,[32] and an inversion using an asymptotic differential relation of the type (**13**) by Christensen-Dalsgaard *et al.*[33] The precise forms of the inferred $\delta \ln u$ vary somewhat. This is due partly to the different methods of inversion that have been used, which, in particular, account differently for the contributions to the eigenfrequencies from the outer layers of the sun in and above the superadiabatic convective boundary layer where the representation of the oscillations as a simple adiabatic perturbation about a horizontally uniform static equilibrium configuration is invalid. In the hope of not being too susceptible to errors resulting from misrepresenting this region, the inversions depicted in FIGURE 4 are based on only modes whose frequencies do not exceed 3 mHz, and whose amplitudes in those outer layers are not so great as those of modes with higher frequency.

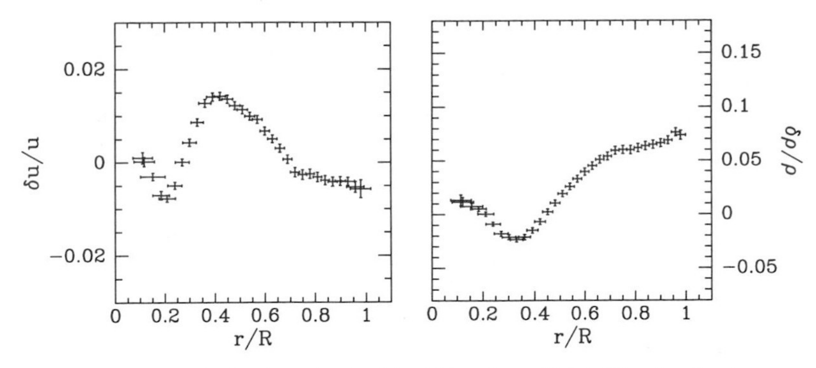

FIGURE 4. Inversion of solar data obtained by Libbrecht *et al.*,[24] yielding localized averages of the relative differences $\delta \ln u$ and $\delta \ln \rho$ between the sun and the reference model. The *horizontal bars* denote characteristic widths of the averaging kernels, some of which are illustrated in FIGURE 3; the *vertical bars* denote standard errors of those averages, computed from the standard errors in the data under the assumption that they are independent. (From Däppen *et al.*[23] Reproduced by permission.)

The negative derivative of $\delta \ln \rho$ interior to $r \simeq 0.3R$, followed by a consequent rise necessary to preserve total mass, is a feature also of earlier inversions (Gough and Kosovichev;[20] Korzennik and Ulrich;[26] Kosovichev[34]), though there is considerable quantitative variation amongst the results. I suspect that part of that variation arises from the use of different data sets, and particularly from combining inconsistent data sets obtained by different observers at different epochs. Indeed, we know that there are frequency changes, of order 10^{-2} percent, associated with the solar cycle (e.g., Woodard and Noyes;[35] Gelly *et al.*;[36] Pallé *et al.*;[37] Elsworth *et al.*[38]), some, if not all, of which result from structural changes in the relatively cool outer layers of the sun (Libbrecht and Woodard[39,40]) where the influence on p-mode eigenfrequencies is greatest. It was with this in mind that Däppen *et al.*[23] used a single data set for the inversions depicted in FIGURE 4.

MODIFIED STANDARD MODELS, AND WIMP ACCRETION

As I have already pointed out, the dominant deviation of the solar value of u from that of the reference model appears to have been explained by the existence of an error in the opacities used for the model. Christensen-Dalsgaard[41] has shown what differences are made to the model when the opacity is modified in the manner suggested by the seismological analysis. The increase in opacity (at fixed ρ, T, and composition) in the outer layers of the radiative envelope has a tendency to increase the magnitude of the temperature gradient, reducing the stability to convection and thereby increasing the depth of the convection zone. In order to compensate globally for the local opacity augmentation, the solar calibration at fixed Z requires a lower electron density, which implies a reduction in X and a corresponding increase in Y. Thus, in view of the proportionality (**9**), ρT^4 must be greater in the core in order to maintain the integrated nuclear-energy generation. The numerical computations show that this is accomplished by having augmentations of both ρ and T at $r = 0$, which decline to zero at the edge of the core. Outside the core, $\delta \ln T$ rises to meet the positive value resulting from the steeper temperature gradient near the base of the convection zone; $\delta \ln \rho$ changes sign, and then becomes positive again in the region $r \gtrsim 0.5R$. The deviation $\delta \ln u$ is superficially like the deviation in $\ln T$, except that there is an additional, negative contribution from $\delta \ln \mu$ resulting from the reduction in X. Thus, superficially, $\delta \ln u$ and $\delta \ln \rho$ resemble the inversions shown in FIGURE 4. However, quantitatively, they are rather different.

What one is left wondering now is whether the opacity error at a few million degrees is all that is wrong with the standard model. In an attempt to answer that question I compare in FIGURE 5 model 13 of Christensen-Dalsgaard *et al.*[3] with the reference model 2. Model 13 was computed using the more modern nuclear reaction rates of Parker[42] and Bahcall and Ulrich[9] and opacities from the LAOL, except that the latter were supplemented by additional opacity sources at the solar surface and also beneath the convection zone, predominantly at temperatures below 4×10^6 K. The model is not fully consistently evolved, but has simply had its helium-abundance profile scaled from a consistent model computed with no opacity augmentation $\Delta\kappa$ beneath the convection zone. Model 13, which has a maximum opacity augmentation of 10 percent, was chosen because the base of its convection zone, like that of the reference model, is located at approximately the same radius as that of the sun.

The similarity between the model differences and the differences illustrated in FIGURE 4 between the sun and the same reference model is startling; with our current precision it seems possible that, with a more appropriate opacity modification, almost complete seismic agreement between a standard theoretical solar model and the sun might well be achieved.

Model 13, like the modified model discussed by Christensen-Dalsgaard, has a higher helium abundance than the reference model: $Y = 0.285$. Consequently μ is greater, and therefore $\delta \ln T$ exceeds $\delta \ln u$ by about 0.04. Thus it is clear from FIGURE 5 that throughout the region of thermonuclear reactions model 13 is both hotter and denser than the reference. The integrated thermal-energy generation rate is unchanged, of course, the tendency to increase ϵ from the increase in ρ and T

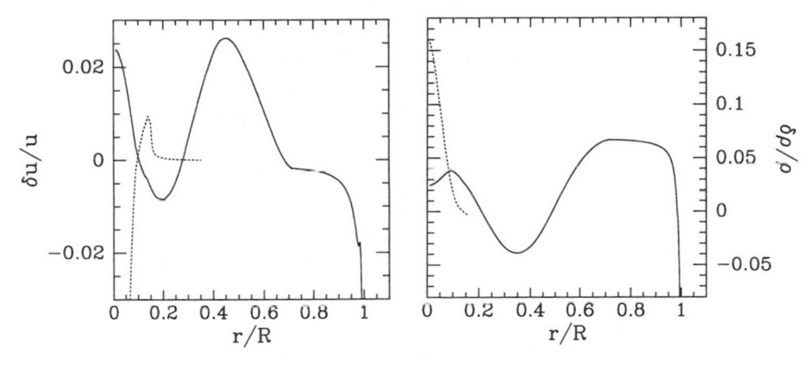

FIGURE 5. The *continuous curves* represent the relative differences $\delta \ln u$ and $\delta \ln \rho$ between the modified standard model (model 13 of Christensen-Dalsgaard *et al.*[3]) discussed in the text and the same reference model that was used for FIGURE 4. The *dashed curves* illustrate the relative deviations brought about by an accretion of wimps in the manner of Gilliland *et al.*;[43] the central value of $\delta \ln u$, not shown in the diagram, is −0.04.

having been canceled by the decrease in X. But the more temperature-sensitive neutrino generation rates, $\epsilon_{\nu 7}$ and $\epsilon_{\nu 8}$, are evidently greater.

Included in FIGURE 5 are the differences between the favored wimp-accreting model of Gilliland *et al.*,[43] which has a ^{37}Cl neutrino flux of 2.0 snu, and their standard model. These are not in agreement with the inversions of FIGURE 4. The disparity has already been pointed out by Gough,[44] who compared the sound–speed differences with some inversions carried out previously by Gough and Kosovichev[20] using only low-degree modes of Jiménez *et al.*[45] and Henning and Scherrer.[46] A subsequent direct comparison of low-degree frequencies (actually the frequency differences $d_{nl} := \nu_{n,l} - \nu_{n-1,l+2}$, which are particularly sensitive to conditions in the core) of a more recently compiled set of whole-disk measurements by Elsworth *et al.*[47] has also suggested that the wimp model is incorrect, whereas the values of d_{nl} from some standard solar models agree with observation.

MACROSCOPIC MOTION IN THE CORE

It is extremely important to realize that it does not follow from the fact that a standard solar model with an appropriately modified opacity is seismically similar to the sun, that that model necessarily represents the sun in all respects. Indeed, as I have already pointed out, standard solar models are unstable and cannot remain precisely in hydrostatic equilibrium. Motion in the core must therefore ensue.

There have been several discussions of the implications of such motion. In particular, the nonlinear development of the g modes is a matter of considerable concern. The most recent discussions suggest that the final state is either one of finite-amplitude nonradial oscillations (Roxburgh;[48] Merryfield *et al.*[49,50]) or slow steady convection concentrated in a shell within which the ^3He gradient is greatest (Ghosal and Spiegel[51]). In either case, there is a significant influence on the neutrino fluxes. Whether the motion is oscillatory or steady, the associated fluctuations in density and temperature increase outward from zero at $r = 0$, and decline yet further out, possibly beyond the region where a significant amount of thermal energy is generated. Because the nuclear reaction rates are nonlinear functions of ρ and T, the mean rates (averaged over spheres and, if necessary, in time over a period of oscillation) are quite different from the reaction rates computed simply with the mean values of ρ and T, and therefore the neutrino fluxes in particular are modified. The thermal fluctuations tend to enhance the mean energy generation rate, which must be compensated by a modification of the mean structure, including a decrement in the initial hydrogen abundance. Roxburgh[48,52] has pointed out that because the peaks on the ^8B and ^7Be neutrino production rates occur closer to the center of the sun than the peak in the thermal-energy generation rate (see FIG. 1), the temperature and density fluctuations to which they are subjected are relatively small, and the diminution caused by the modification of the mean state could exceed the enhancement caused directly by the fluctuations. Indeed, Roxburgh,[48,52] and subsequently Ghosal and Spiegel,[51] have estimated that with maximum temperature fluctuations of order 20–25 percent, the neutrino flux could be reduced to the observed level.

To see more explicitly what the outcome would be, consider a simplified representation of the temperature of the form

$$T = \overline{T}[1 + Af\psi(\theta, \phi, t)] \qquad \textbf{(15)}$$

with respect to spherical polar coordinates (r, θ, ϕ), where $\overline{T}(r)$ is the mean temperature and $f(r; r_c)$ is an amplitude function that vanishes at $r = 0$, achieves a maximum at $r = r_c$, and then declines as r increases beyond the radius r_c. The function ψ, which is independent of time t for the case of slow convection, has zero mean and rms unity. The absolute amplitude of the fluctuation is determined by the constant A. Merryfield *et al.*[49] found that in the nonlinear regime of a highly simplified model of the situation, the fluctuation of temperature has the dominant influence on the fluxes, and for simplicity I shall ignore the associated fluctuations of the form **(15)** in ρ and X, as did Ghosal and Spiegel.[51] To include the terms that have been ignored in this discussion would be straightforward, once it had been determined how they are related to the temperature fluctuation. To estimate the influence on the fluxes it is now adequate, to a first approximation, simply to compute the effect of the fluctuations on the mean values of $\epsilon_{\nu i}$ and ϵ, from the proportionalities $\rho^\beta T^\eta$ discussed in the second section. Thus, for example, the mean rate of generation of energy per unit mass is given by

$$\bar{\epsilon}(r) = \epsilon(X, \rho, \overline{T})\overline{(1 + Af\psi)^{\eta}}, \qquad (16)$$

where $\epsilon(X, \rho, T)$ is the exact formula for the generation rate in material of density ρ, temperature T, and hydrogen abundance X, and the overbar denotes a spherical and temporal average.

In order to calculate the influence on the mean state of the sun, I assume A to be small enough for linearized perturbation theory to be valid for all aspects of the structure except $\bar{\epsilon}$. Thus I set

$$\bar{\epsilon}(r) = \epsilon(X, \rho, \overline{T})[1 + \epsilon_f(r)], \qquad (17)$$

where

$$\epsilon_f = \overline{(1 + Af\psi)^{\eta}} - 1, \qquad (18)$$

and solve the linearized perturbation equations for a star in thermal balance, namely,

$$\frac{dp_1}{dr} = -\frac{Gm_0\rho_0}{r^2 p_0}(m_1 + \rho_1 - p_1), \qquad (19)$$

$$\frac{dm_1}{dr} = \frac{4\pi r^2\rho_0}{m_0}(\rho_1 - m_1), \qquad (20)$$

$$\frac{dL_1}{dr} = \frac{4\pi r^2\rho_0\epsilon_0}{L_0}(\rho_1 - L_1 + \epsilon_1 + \epsilon_f), \qquad (21)$$

together with

$$\frac{dT_1}{dr} = -\frac{3\kappa_0\rho_0 L_0}{16\pi a\tilde{c}r^2 T_0^4}(\kappa_1 + \rho_1 + L_1 - 4T_1), \qquad (22)$$

in the radiative zone, where a is the radiation-density constant and $\tilde{c}$ is the speed of light, subject to the boundary conditions

$$\rho_1 - M_1 = 0, \qquad \rho_1 - L_1 + \epsilon_1 + \epsilon_f = 0 \qquad \text{at } r = 0, \qquad (23)$$

$$m_1 = 0, \qquad L_1 = 4T_1 = 0 \qquad \text{at } r = R. \qquad (24)$$

In these equations, m is the usual mass variable, L is luminosity at radius r, and κ is opacity; subscripts 0 denote values in the unperturbed, standard model, and subscripts 1 denote relative perturbations of the mean state: thus, $p = p_0(1 + p_1)$, etc. The quantities κ_1 and ϵ_1 are approximated by $\kappa_1 = \rho_1 - 3.5T_1 + [X_0/(1 + X_0)]X_1$, appropriate to Kramers' law: $\kappa \propto (1 + X)\rho T^{-3.5}$, and $\epsilon_1 = \rho_1 + 4T_1 + 2X_1$, in accordance with the scaling **(9)**. I assume, also for simplicity, the perfect gas law, so that $\rho_1 = p_1 - T_1 + \mu_1$, where μ is the mean molecular mass, which takes the value $4/(5X + 3)$ for a fully ionized gas. The absolute hydrogen-abundance perturbation, $\delta X \equiv X_0X_1$, required to maintain the surface luminosity at $L_{\odot}$, represents the recalibration of the model, and was taken to be independent of r. The boundary conditions **(23)** are regularity conditions, and the conditions **(24)** require mass and photospheric luminosity to be invariant and the photosphere to remain at $r = R$. Since it is unlikely that conditions in the very outer layers of the star have a severe impact on the structure of the core, I assumed for simplicity **(21)** to hold even in the convection zone. Then **(19)–(24)** constitute a linear inhomogeneous eigenvalue problem, with eigenvalue δX, which is easily solved by any of the standard methods. Once the eigenvalue and the eigenfunctions are determined, the neutrino fluxes can be computed by evaluating directly the integrals of quantities having the form **(16)**, with appropriate values of η.

Before presenting the result, a remark about appropriate values of η is perhaps not out of place. Assuming the characteristic timescale associated with the motion to be less than all the timescales of the reaction chains, listed in TABLE I, which is certainly the case for g modes, then the abundances X_3 and X_7 do not fluctuate with the motion, and $\eta = -\frac{1}{2}$ for the fluctuation of $\epsilon_{\nu 7}$ and $\eta = 13$ for $\epsilon_{\nu 8}$, as can be read from the table. Thus, the augmentation of both $\epsilon_{\nu 7}$ and $\epsilon_{\nu 8}$ by the fluctuations is less than would have been the case were the motion to have been so slow that all the

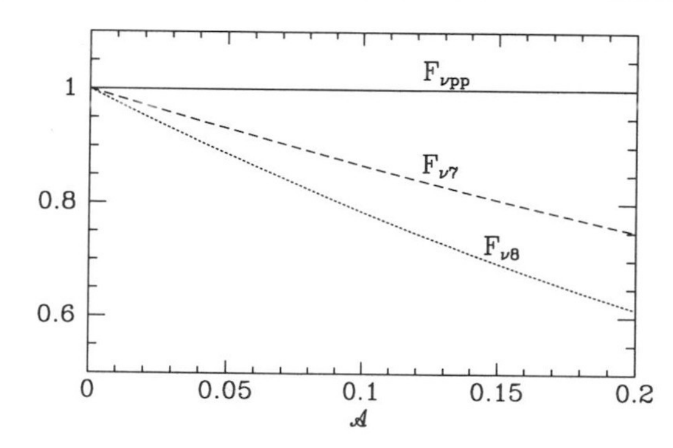

FIGURE 6. Temporally averaged neutrino fluxes of a solar model with a nonradially oscillating core relative to the fluxes of the corresponding hydrostatic standard model, plotted against the oscillation amplitude A. The *continuous curve* represents the pp flux, the *dashed curve* the ^{7}Be flux, and the *dot–dashed* curve represents the ^{8}B flux.

reactions had remained in equilibrium. Moreover, since the ^{3}He-destroying reactions are thrown out of equilibrium with the ^{3}He-producing reactions, and because about the same amount of thermal energy is liberated in each, the effective exponent η in **(18)** is approximately the average of the exponents of the reactions that create and destroy ^{3}He; hence, for fluctuations in the thermal-energy generation rate, $\eta \simeq 10$.

FIGURE 6 shows the expected mean neutrino fluxes from such a calculation, where the function $f(x)$ in **(18)** was taken to be xe^{1-x}, appropriate, perhaps, to a mode of degree $l = 1$, and r_c was taken to be $0.15R$. The fluxes are plotted against the amplitude A of the oscillation. As was suggested by the arguments given previously, both the ^{7}Be and the ^{8}B fluxes are reduced in the presence of the oscillation. The results are not very sensitive to the precise values of r_c and l, provided r_c does not become too small and l is not large. Superposed on these mean fluxes is a fluctuation that, particularly if periodic, one might expect to be detectable by the future super-Kamiokande and Sudbury neutrino experiments. One might also expect to be able to detect the core oscillation, at least if it were periodic, as a frequency modulation of the low-degree p-mode signal, though a preliminary (unpublished) search in currently available whole-disk data (of Pallé *et al.*[53]) has not revealed it.

The details of the modification to the structure of the core are somewhat different if the motion is in the form of slow convection. It is clear from the discussion that the precise result must depend on the magnitude of the convective velocity, upon which the degree of imbalance of the nuclear reactions depend. However, as one can learn from the discussion by Ghosal and Spiegel,[51] the possible convective motion is likely to be much more tightly confined to a shell in the outer reaches of the core than is the motion due to g modes, so the reduction in the neutrino fluxes could be substantially greater.

CONCLUSION

The helioseismic evidence shows that the state of the thermonuclear-energy generating core of the sun is not exactly as predicted by current standard solar models. Part, perhaps most of the discrepancy arises as a response to an error in the opacity in the outer layers of the radiative envelope, which itself was first detected seismologically. Whether there is a residual discrepancy is yet to be ascertained. What seems likely, however, is that if the core of the sun is in hydrostatic and thermal balance, and if no redistribution of the products of the nuclear reactions in the core has taken place, then the density and sound speed, and also probably the temperature, are higher in the core of the sun than they are in a standard theoretical model in which the opacity error has not been taken into account. Therefore, the neutrino fluxes predicted are even higher than those of the standard model, which exacerbates the neutrino problem. However, standard solar models are unstable, and therefore precise hydrostatic and thermal balance cannot be maintained. A possible consequence is substantial macroscopic fluid flow in the core, which not only modifies somewhat the mean stratification of temperature and density, but also changes substantially the neutrino fluxes. Until it can be determined whether or not such a flow is present, or has taken place in the past, one must be wary of using the sun as a physics laboratory to calibrate theories of neutrinos.

ACKNOWLEDGMENT

I am grateful to Dr. A. G. Kosovichev for his assistance in producing the diagrams.

REFERENCES

1. BAHCALL, J. N. & H. A. BETHE. 1990. Phys. Rev. Lett. **65:** 2233–2235.
2. KRASTEV, P. I., S. P. MIKHEYEV & A. YU SMIRNOV. 1990. *In* Proceedings of the School on Low-energy Weak Interactions (Dubno, USSR, Sept. 3–14). In press.
3. CHRISTENSEN-DALSGAARD, J., D. O. GOUGH & M. J. THOMPSON. 1991. Astrophys. J. **378:** 413–437.
4. CHRISTENSEN-DALSGAARD, J. 1982. Mon. Not. R. Astron. Soc. **199:** 735–761.
5. HUMMER, D. G. & D. MIHALAS. 1988. Astrophys. J. **331:** 794–814.

6. MIHALAS, D., W. DÄPPEN & D. G. HUMMER. 1988. Astrophys. J. **331:** 815–825.
7. COX, A. N. & J. E. TABOR. 1976. Astrophys. J., Suppl. Ser. **31:** 271–312.
8. FOWLER, W. A., G. R. CAUGHLIN & B. A. ZIMMERMAN. 1975. Ann. Rev. Astron. Astrophys. **13:** 69–112.
9. BAHCALL, J. N. & R. K. ULRICH. 1988. Rev. Mod. Phys. **60:** 297–372.
10. HEUBNER, W. F., A. L. MERTS, N. H. MAGEE & M. F. ARGO. 1977. Astrophysical Opacity Library. Los Alamos Scientific Lab Rep. LA-6760-M.
11. BAHCALL, J. N. 1991. Solar neutrinos: New physics? This issue.
12. CHRISTENSEN-DALSGAARD, J., F. W. W. DILKE & D. O. GOUGH. 1974. Mon. Not. R. Astron. Soc. **169:** 429–445.
13. BOURY, A., M. GABRIEL, A. NOELS, R. SCUFLAIRE & P. LEDOUX. 1975. Astron. Astrophys. **41:** 279–285.
14. SHIBAHASHI, H., Y. OSAKI & W. UNNO. 1975. Publ. Astron. Soc. Jap. 401–410.
15. SAIO, H. 1980. Astrophys. J. **240:** 685–692.
16. KOSOVICHEV, A. G. & A. B. SEVERNY. 1985. Izv. Krym. Astrofiz. Obs. **72:** 188–198.
17. GOUGH, D. O. 1985. Sol. Phys. **100:** 65–99.
18. CHRISTENSEN-DALSGAARD, J., J. SCHOU & M. J. THOMPSON. 1990. Mon. Not. R. Astron. Soc. **242:** 353–369.
19. GOUGH, D. O. & M. J. THOMPSON. 1991. Solar Interior and Atmosphere, A. N. Cox, W. C. Livingston, and M. Matthews, Eds.: In press. University of Arizona Press. Tucson.
20. GOUGH, D. O. & A. G. KOSOVICHEV. 1988. Seismology of the Sun and Sun-like Stars, V. Domingo and E. J. Rolfe, Eds.: 195–201. ESA SP-286. Noordwijk, the Netherlands.
21. GOUGH, D. O. & A. G. KOSOVICHEV. 1990. Inside the Sun, IAU Colloquium **121,** G. Berthomieu and M. Cribier, Eds.: 327–340. Kluwer. Dordrecht, the Netherlands.
22. DZIEMBOWSKI, W. A., A. A. PAMYATNYKH & R. SIENKIEWICZ. 1990. Mon. Not. R. Astron. Soc. **24:** 542–550.
23. DÄPPEN, W., D. O. GOUGH, A. G. KOSOVICHEV & M. J. THOMPSON. 1991. Challenges to Theories of the Structure of Moderate-mass Stars, Lecture Notes in Physics, D. O. Gough and J. Toomre, Eds.: In press. Springer-Verlag. Heidelberg/New York.
24. LIBBRECHT, K. G., M. F. WOODARD & J. M. KAUFMAN. 1990. Astrophys. J., Suppl. Ser. **74:** 1129–1149.
25. CHRISTENSEN-DALSGAARD, J., T. L. DUVALL, JR., D. O. GOUGH, J. W. HARVEY & E. J. RHODES, JR. 1985. Nature **315:** 378–382.
26. KORZENNIK, S. G. & R. K. ULRICH. 1989. Astrophys. J. **339:** 1144–1155.
27. COX, A. N., J. A. GUZIK & R. B. KIDMAN. 1989. Astrophys. J. **342:** 1187–1206.
28. IGLESIAS, C. A. & F. J. ROGERS. 1990. Inside the Sun, IAU Colloquium **121,** G. Berthomieu and M. Cribier, Eds.: 81–90. Kluwer. Dordrecht, the Netherlands; ———. 1991. Astrophys. J., Lett. **371:** L73–L75.
29. SHIBAHASHI, H. & T. SEKII. 1988. Seismology of the Sun and Sun-like Stars, V. Domingo and E. J. Rolfe, Eds.: 471–474. ESA SP-286. Noordwijk, the Netherlands.
30. VORONTSOV, S. V. 1988. Seismology of the Sun and Sun-like Stars, V. Domingo and E. J. Rolfe, Eds.: 475–480. ESA SP-286. Noordwijk, the Netherlands.
31. SEKII, T. & H. SHIBAHASHI. 1989. Publ. Astron. Soc. Jap. **41:** 311–331.
32. VORONTSOV, S. & H. SHIBAHASHI. 1990. Progress of Seismology of the Sun and Stars, Lecture Notes in Physics **367,** Y. Osaki and H. Shibahashi, Eds.: 325–328. Springer-Verlag. Heidelberg/New York.
33. CHRISTENSEN-DALSGAARD, J., D. O. GOUGH & M. J. THOMPSON. 1988. Seismology of the Sun and Sun-like Stars, V. Domingo and E. J. Rolfe, Eds.: 493–497. ESA SP-286. Noordwijk, the Netherlands.
34. KOSOVICHEV, A. G. 1990. Progress of Seismology of the Sun and Stars, Lecture Notes in Physics **367,** Y. Osaki and H. Shibahashi, Eds.: 319–323. Springer-Verlag. Heidelberg/New York.
35. WOODARD, M. F. & R. W. NOYES. 1985. Nature **318:** 449–450.
36. GELLY, B., E. FOSSAT, G. GREC & M. POMERANTZ. 1988. Advances in Helio- and Asteroseismology, IAU Symposium **123,** J. Christensen-Dalsgaard and S. Frandsen, Eds.: 21–23. Reidel. Dordrecht, the Netherlands.
37. PALLÉ, P. L., C. RÉGULO & T. ROCA CORTÉS. 1990. Progress of Seismology of the Sun and

Stars, Lecture Notes in Physics **367,** Y. Osaki and H. Shibahashi, Eds.: 129–134. Springer-Verlag. Heidelberg/New York.

38. ELSWORTH, Y., R. HOWE, G. R. ISAAK, C. P. MCLEOD & R. NEW. 1990. Nature **345:** 322–324.
39. LIBBRECHT, K. G. & M. F. WOODARD. 1990. Nature **345:** 779–782.
40. LIBBRECHT, K. G. & M. F. WOODARD. 1990. Progress of Seismology of the Sun and Stars, Lecture Notes in Physics **367,** Y. Osaki and H. Shibahashi, Eds.: 145–156. Springer-Verlag. Heidelberg/New York.
41. CHRISTENSEN-DALSGAARD, J. 1988. Seismology of the Sun and Sun-like Stars, V. Domingo and E. J. Rolfe, Eds.: 431–450. ESA SP-286. Noordwijk, the Netherlands.
42. PARKER, P. D. 1986. Physics of the Sun, Vol. 1, P. A. Sturrock, T. E. Holzer, D. Mihalas, and R. K. Ulrich, Eds.: 15–32. Reidel. Dordrecht, the Netherlands.
43. GILLILAND, R., J. FAULKNER, W. H. PRESS & D. N. SPERGEL. 1986. Astrophys. J. **306:** 703–709.
44. GOUGH, D. O. 1990. Inside the Sun, IAU Colloquium. **121,** G. Berthomieu and M. Cribier, Eds.: 451–475. Kluwer. Dordrecht, the Netherlands.
45. JIMÉNEZ, A., P. L. PALLÉ, T. ROCA CORTÉS & V. DOMINGO. 1988. Astron. Astrophys. **193:** 298–302.
46. HENNING, H. M. & P. H. SCHERRER. 1986. Seismology of the Sun and the Distant Stars, D. O. Gough, Ed.: 55–62. Reidel. Dordrecht, the Netherlands.
47. ELSWORTH, Y., R. HOWE, G. R. ISAAK, C. P. MCLEOD & R. NEW. 1990. Nature **345:** 536–539.
48. ROXBURGH, I. W. 1985. Sol. Phys. **100:** 21–51.
49. MERRYFIELD, W. J., J. TOOMRE & D. O. GOUGH. 1990. Astrophys. J. **353:** 678–697.
50. MERRYFIELD, W. J., J. TOOMRE & D. O. GOUGH. 1991. Astrophys. J. **367:** 658–665.
51. GHOSAL, S. & E. A. SPIEGEL. 1991. Geophys. Astrophys. Fluid Dyn. In press.
52. ROXBURGH, I. W. 1987. The Internal Solar Angular Velocity, B. R. Durney and S. Sofia, Eds.: 1–5. Reidel. Dordrecht, the Netherlands.
53. PALLÉ, P. L., F. PÉREZ HERNÁNDEZ, T. ROCA CORTÉS & G. R. ISAAK. 1989. Astron. Astrophys. **216:** 253–258.

Astroparticle Physics and Superstrings[a,b]

D. V. NANOPOULOS

Center for Theoretical Physics
Department of Physics
Texas A & M University
College Station, Texas 77843-4242
and
Astroparticle Physics Group
Houston Advanced Research Center (HARC)
The Woodlands, Texas 77381

INTRODUCTION

It has been recognized for some time now, that particle physics and cosmology have a lot to learn from each other if they can be looked at from an angle emphasizing their common properties. Astroparticle physics tries to encompass these two fascinating disciplines. It has become apparent that further progress in particle physics has to involve cosmology. The unification idea,[1] beyond the standard $SU(3) \times SU(2) \times U(1)$ model, leads to rather high energies (of the order of the Planck scale: $M_{\rm Pl} \simeq 10^{19}$ GeV $\simeq 10^{32}$ K), which go far beyond the accelerator energies at present (e.g., LEP: $E \simeq 100$ GeV) or future (e.g., SSC: $E \simeq 40$ TeV). Using the standard big bang cosmology[2] (BBC), we find that at the "beginning" it was very "hot," ($T \simeq M_{\rm Pl}$) and thus "studying" the embryonic Universe, we may learn about the unification idea. The word *studying* needs some explanation. Clearly, we were not around $t_P \sim 10^{-43}$ sec to study the Universe, but if we believe in a dynamical evolution of the Universe, then by studying the "large scale" structure available to us today, we can extrapolate back and learn about the embryonic Universe, and thus among other things "test" our unification ideas.

On the other hand, cosmology badly needs particle physics. It should be obvious that the fundamental particle interactions play a major role in the evolution of the Universe. After all, gravity is one of our fundamental particle interactions and furthermore, energy densities, interaction rates, and the like, all emerge from our particle physics models. So the synergetics of particle physics and cosmology is compulsory. Luckily, their symbiosis during the last 10 years has been proved to be a very happy one!

In this paper I will review some of the highlights of this fascinating field, with more emphasis on some very recent developments involving superstrings.[3] The method I have chosen to proceed is a bit unconventional. I will analyze the developments in particle physics and cosmology in a unified way and stress at every stage the impact that they have on each other. In the following section I give a quick review of the two standard models, the particle physics one (SM) and the cosmologi-

[a] This work was supported by Department of Energy Grant DE-AS05-81ER40039.
[b] Dedicated to the memory of Léon Van Hove, cofounder of the ESO-CERN Symposia.

cal one (BBC). The third section discusses grand unified theories (GUTs) and the corresponding cosmology, the grand bang, while their supersymmetric versions are discussed in the fourth section. The fifth section is devoted to superstrings as applied to particle physics and cosmology, while the final section is devoted to conclusions.

THE TWO STANDARD MODELS

The modern way of describing a bunch of particles and their interactions is to try to exploit all possible symmetries that such a system may exhibit. The invariance of the "action" under these symmetries restricts considerably the possible terms that one can write down and thus makes life a bit easier. An archetypal example is gauge invariance, a local invariance under some group G, that can act as an "internal" symmetry (e.g., $SU(3)_C \times SU(2)_L \times U(1)_Y$, describing strong and electroweak interactions) or a "spacetime" symmetry (e.g., $SO(3, 1)$ describing *local* Poincaré invariance or general coordinate transformation invariance, i.e., general relativity). Of course, eventually we will like to stop making this artificial separation between "internal" and "spacetime" symmetries and see it all as a unified picture, á la Kaluza–Klein or superstrings.

The standard model (SM) of particle physics,[1] describes the electroweak and strong interactions of all fundamental particles observed in nature, including 5 + (1) quarks $[u, d; c, s; b, (t)]$, 6 leptons $(\nu_e e; \nu_\mu \mu; \nu_\tau \tau)$, 12-gauge bosons (8 gluons g, 3 weak bosons W^+ W^-, Z^0, and the photon γ), and the elusive [still to be discovered (?)] Higgs boson. The success of the SM is stunning. Until now, no cracks have been found, it simply "fits" everything! The standard model of cosmology, the hot big bang cosmology[2] (BBC), based upon the Einstein–Hilbert Lagrangian ($\mathcal{L}_{E-H} = (M_{\rm Pl}^2/16\pi) R + \cdots$, with R the scalar-curvature) and assuming homogeneity and isotropy (i.e., the Robertson–Walker–Friedmann metric) also has met with spectacular successes. The "derivative," simple, "magic" equation,

$$H^2 \left(\equiv \left(\frac{\dot{\alpha}}{\alpha} \right)^2 \right) = \frac{8\pi G_N}{3} \rho - \frac{k_s}{\alpha^2} + \Lambda_c, \tag{1}$$

where H is the Hubble parameter or the expansion rate of the Universe, α is the scale factor, ρ the energy density, k_s the space-curvature parameter (± 1 or 0), and Λ_c the notorious cosmological constant ($G_N \equiv (1/M_{\rm Pl}^2)$, the Newton constant), describes "pretty-well" most of the evolution of the Universe. Indeed, using some "normal" energy density ρ and no cosmological constant, it predicts an $H \neq 0$ (thus an expanding Universe), which when combined with a "normal" adiabatic evolution predicts some relic background radiation with a blackbody spectrum [as has been spectacularly verified recently by (COBE)[4]], and also leads to the "observed" light element abundances,[5] if one uses only three neutrino species (as has been successfully verified by LEP[6]). It should be stressed here that the impact of the SM on BBC is rather important. Because both the strong ($SU(3)_c$) and weak interactions ($SU(2)_L$) are non-Abelian, the coupling constants fall off logarithmically with energy, so that at high energies (and thus high temperatures), quarks, leptons, and other species may be considered (asymptotically) free, thus $\rho \sim T^4$ (or mT^3), that is, "normal" densities, which have been used in BBC. Also, the electroweak interactions play a rather drastic role in nucleosynthesis, since they provide the interaction rates for different processes, which are essential in estimating the "light" element abundancies.

Despite the phenomenal successes of the two standard models, people are not very happy. Both models "describe" or "accommodate" all the data, but still they fall short of "explaining" naturally what is going on.

In the case of the particle physics SM, there are too many free parameters, too many fundamental particles, no "real" unification (we still have three different coupling constants g_3, g_2, g_1), etc., that makes us believe that the SM is not the whole story. In the case of the BBC there are even more fundamental problems: no explanation of the baryon asymmetry observed in the Universe, no explanation of the observed homogeneity and isotropy, no explanation of the kind of "big" energy density (not very different from the critical density that signals the transition from an open to a closed Universe), no explanation of the large-scale structure, etc. I have left separately the biggest problem plaguing both models: Why is the cosmological constant so small (at least ~120 orders of magnitude smaller than its "expected" value $\sim M_{\rm Pl}^4$), which defies any other fine-tuning problems we have ever seen in physics!

It sounds like both our standard models need some revision and/or extension.

GRAND UNIFICATION—GRAND BANG

The spectacular success of unification of weak and electromagnetic interactions into electroweak interactions shows clearly the way to the next step. Grand unified theories[1] (GUTs) are theories that try to unify strong and electroweak interactions. One assumes the existence of a big simple group G that contains the standard model group $SU(3)_C \times SU(2)_L \times U(1)$, so that one starts with *only one* gauge coupling constant g and some representations (R_i) of the group G (that contain all the particles of the SM and more), all happening at very high energies (not far below the Planck scale). At lower energies the group G suffers "spontaneous breakdown" to the SM, where g splits to the three coupling constants (g_3, g_2, g_1) and the representations R_i break to the representations of the SM. Schematically, one has something like

$$\begin{array}{cccc} G \overset{M_x}{\rightarrow} & SU(3)_C \times & SU(2)_L \times & U(1)_Y \\ (g, R_i) & (g_3, r_i^3) & (g_2, r_i^2) & (g_1, r_i^1), \end{array} \tag{2}$$

where M_X is the GUT scale, and the $r_i^{1,2,3}$ may share common particles (e.g., a *left-handed* up-quark is a **3** of $SU(3)_C$ and a **2** of $SU(2)_L$, so it will appear in some r_3 and some r_2, etc.). For example, one can identify G with $SU(5)$,[7] in which case the fifteen states of each generation ($u_L^\alpha, d_L^\alpha, u_R^\alpha, d_R^\alpha, \nu_{e_L}, e_L, e_R$; where $\alpha = 1, 2, 3$ is the color degree of freedom) fit nicely into a **5** and **10** representations of $SU(5)$:

$$\bar{\mathbf{5}} \equiv \begin{pmatrix} d_1^c \\ d_2^c \\ d_3^c \\ e \\ \nu_e \end{pmatrix}_L ; \qquad \mathbf{10} \equiv \left(\begin{pmatrix} u^\alpha \\ d^\alpha \end{pmatrix}_L, u_\alpha^c, e^c \right)_L, \tag{3}$$

where we have replaced right-handed fields (ψ_R) by their equivalent left-handed-charge-conjugated fields (ψ_L^c); and ditto for the other two generations.

This innocent-looking extension of the standard model has far-reaching consequences.[1]

A. *Group-Theoretical:* Because we are forced to put quarks and leptons in the same representations (big groups have big representations that cannot be filled in exclusively with quarks or leptons), we are bound to get relations between quark and lepton quantum numbers, including masses, as well new types of interactions transforming directly quarks into leptons, and vice versa. Indeed:

1. Since the electric charge operator Q is inside a simple group G, it better be traceless ($\text{Tr}\ Q = 0$); thus when it acts on the representations of **(3),** it gives

$$Q_d = \tfrac{1}{3} Q_e; \qquad Q_u = 1 + Q_d = -\tfrac{2}{3} Q_e. \tag{4}$$

   In other words, we explain, for the first time, electric *charge quantization.* The big mystery of why $|Q_e + Q_p| \leq 10^{-20}$ is resolved. Furthermore, we understand why $(u, d)_L$ and $(\nu_e, e)_L$ transform equivalently under $SU(2)_L$: otherwise, they would not fit into **(3)!**
2. The fermion masses will be created by Yukawa couplings of the $\mathbf{\overline{5}}$ and **10** of **(3)** to a $\mathbf{\overline{5}}_H(\mathbf{5}_H)$ of Higgs, that is, $\mathbf{\overline{5}} \cdot \mathbf{10} \cdot \mathbf{\overline{5}}_H$, $\mathbf{10} \cdot \mathbf{10} \cdot \mathbf{5}_H$, which will provide *equal Yukawa couplings* for down-quarks (d, s, b) and charged leptons (e, μ, τ), at the scale M_X, for example,

$$h_b(M_X) = h_\tau(M_X). \tag{5}$$

3. New interactions mediated by superheavy gauge (and/or), Higgs bosons will turn quarks into leptons and vice versa; thus, two sacred, conserved global quantum numbers of the standard model: baryon number (B) and lepton number (L). There is nothing anymore to keep the proton stable and indeed it is predicted to decay with a lifetime[1]

$$\tau_{\text{proton}} \simeq \left(\frac{1}{\alpha_G^2}\right) \cdot \frac{M_X^4}{m_{\text{proton}}^5}, \tag{6}$$

   where $\alpha_G \equiv g^2/4\pi$ is the GUT "fine-structure" constant. The decay modes vary from theory to theory, but $p \to e^+\pi^0$, is a "classic" mode for protons to decay. Combining **(5)** with the currently available lower bound[8] of $\tau_p \geq O(10^{33}\ \text{years})$ implies

$$M_X \geq O(10^{15}\ \text{GeV}), \tag{7}$$

   a rather high GUT unification scale.

B. *Dynamics: Renormalization Group Equations (RGE):* While all the preceding results look promising, we do not "live" at energy scales of 10^{15} GeV! We have to "renormalize" down to lower scales and compare then with experiment. Having an explicit dynamical theory makes it possible, by the use of RGE, to connect with low energies. The change of the coupling strengths with momentum transfers (q) are encoded in the simple RGE:[9]

$$q\,\frac{\partial g_i(q)}{\partial q} \simeq \beta_i g_i^3(q) + O(g_i^5) \qquad (i = 1, 2, 3), \tag{8}$$

which imply (assuming unification at M_X)

$$\frac{1}{g_i^2(q)} = \frac{1}{g_{\text{GUT}}^2} + 2\beta_i \ln\left(\frac{M_X}{q}\right), \tag{9}$$

where

$$\beta_1 = \frac{1}{16\pi^2}\left[\frac{2}{3}f\right]; \qquad \beta_2 = -\frac{1}{16\pi^2}\left[\frac{22}{3} - \frac{2}{3}f\right]; \qquad \beta_3 = -\frac{1}{16\pi^2}\left[11 - \frac{2}{3}f\right] \tag{10}$$

and $f \equiv$ number of quarks = number of flavors = ½ number of generations ($\equiv Ng$).

Using (7) with ($q \simeq M_Z$) gives:

$$\ln\left(\frac{M_X}{M_Z}\right) = \frac{\pi}{11\alpha(M_Z)}\left(1 - \frac{\alpha(M_Z)}{\alpha_3(M_Z)}\cdot\frac{8}{3}\right)$$
$$\sin^2\theta_w = \frac{1}{6} + \frac{5}{9}\frac{\alpha(M_Z)}{\alpha_3(M_Z)}, \tag{11}$$

where θ_w is the electroweak mixing angle: $e = g_2 \sin\theta_w$. Using $\alpha(M_Z) \simeq 1/127$ and $\alpha_3(M_Z) \simeq 0.11$, we get from this very "naive," "kindergarten" use of RGEs

$$M_X \simeq O(10^{14}\ \text{GeV}); \qquad \sin^2\theta_w|_{M_Z} \simeq 0.21, \tag{12}$$

which, at this stage of analysis, are not far off the available experimental results **(7)** and $\sin^2\theta_w|_{M_Z} \simeq 0.23$, as extracted[10] from LEP data.[6] For the moment, because of the crudeness of our analysis, we take these results as very encouraging. After all, we are extrapolating at least *12 orders of magnitude* and we are almost there.

We can play the same game for the Yukawa couplings:[11,12]

$$\frac{\partial h_b(q)}{\partial \ln q} = -\left(\frac{1}{4\pi}\right)h_b\cdot\alpha_3^2 + O(\alpha_2, \alpha_1)$$
$$\frac{\partial h_\tau(q)}{\partial \ln q} = 0 + O(\alpha_2, \alpha_1), \tag{13}$$

where the "zero" in the τ-lepton Yukawa coupling emphasizes the fact that leptons don't feel the strong interactions. Again, elementary algebra gives

$$\frac{m_b}{m_\tau} = \frac{h_b(q = 2m_b)}{h_\tau(q = 2m_\tau)} = \frac{h_b(M_X)}{h_\tau(M_X)}\cdot\left[\frac{\alpha_3(q)}{\alpha_3(M_X)}\right]^{12/(33-2f)} \tag{14}$$

which implies, using **(5)**, which is a natural output of any reasonable GUT, and $(m_b/m_\tau) \simeq 2.95$ as measured experimentally,

$$f = 6 \quad \text{or} \quad N_g = 3. \qquad \textbf{(15)}$$

This result **(15)**, derived first in 1977 by A. Buras *et al.*,[12] has been shown to hold true, even strengthened by more sophisticated calculations done by D. A. Ross and myself,[13] including higher loops and supersymmetry, and also holds true in superstring theories.[14] Its remarkable stability through the years does make it one of the most reliable prediction of unification beyond the standard model, and its stunning verification by LEP experiments[6] in 1989, indicates clearly that some basic ideas involved in its derivation must have something to do with reality. Incidentally, the number of neutrino species coming from primordial nucleosynthesis in the late 1970s was less than 7.[15] Since I do believe that the successful prediction **(15)** carries a lot of weight for unified theories, I would like to digress a bit on its *stability*. Notice that there is a big difference between **(11)** and **(13)**: the proton lifetime **(6)** grows as the fourth power of M_X, a rather strong dependence on M_X, while **(14)** has a very weak logarithmic dependence, as also does $\sin^2\theta_w$. So, while it is possible for a GUT theory to fail badly [as does minimal "classic"[7] $SU(5)$] in predicting a very short proton lifetime, and an increase in M_X is needed (through, e.g., supersymmetry), clearly, m_b/m_τ, as given by **(14)**, would not change much! Furthermore, the usual criticism "what about the other mass relations m_d/m_e and m_s/m_μ?" which seem not to work does not "hold water," for the following simple reason. The first two generations' masses (d, e, s, μ) are at most in the $O(100\text{ MeV})$ range and it is highly probable, and we have explicit examples for this,[16] that the Yukawa coupling already at M_X get large corrections $O(100\text{ MeV})$, and thus **(5)** does not apply to them. On the contrary, because (b, τ) are in the (2–5-GeV) range, the $O(100\text{ MeV})$ corrections are unimportant, and thus **(5)** still holds. In addition, every respectable unified theory, for example, superstrings, provides no Yukawa couplings for the first two generations at the classical (tree)-level and create Yukawa couplings only by quantum effects, thus explaining the observed hierarchical fermion mass-spectrum, and avoiding relations of the type **(5)** for the d, e, and s, μ systems.[17]

I emphasize once more that the successful predictions of $\sin^2\theta_w$ and (m_b/m_τ) (or $N_g = 3$) make the unification program much more reliable, and thus the whole astroparticle connection, on which we now turn, much firmer.

Every and each of the preceding novelties of GUTs have an immediate impact on cosmology, or in this case on the *grand bang*.

1. *Baryon-Number Violation:* It does not take much to think, that we have in our hands a wonderful opportunity to explain the baryon asymmetry in the Universe, as is quantitatively expressed by $n_B/n_\gamma \simeq 10^{-10}$, where, as usual, n_B and n_γ refer to the number densities of bosons and photons in the Universe. Since GUTs contain C and CP-violation, baryon-number violation, and the Universe is expanding, we meet all the Sakharov conditions[18] for creating a substantial baryon-asymmetry. Indeed, as the original calculations[19] showed more than 10 years ago, we can "easily" explain the observed baryon asymmetry in the Universe. Since then, several new ways have been proposed[2] that confirm the initial excitement, and it is fair to say that in the grand bang this problem is basically resolved.[2]

2. *Lepton-Number Violation:* It is no secret that with the violation of lepton number, the road to make neutrinos massive is wide open. Many ways to get neutrino masses have been proposed,[1] and generally lead to masses of few electron volts or less. This is remarkable because, on one hand, they do not overclose the Universe, and on the other hand, we can use them as seeds for galaxy formation[2] (hot dark matter). Furthermore, allowing neutrinos to have masses implies the possibility of neutrino mixing, and thus the interesting effect of neutrino oscillations. An immediate resolution of the solar neutrino problem is provided by the celebrated MSW effect,[20] where ν_e starting in the center of the sun transform to ν_μ by the time they reach the surface, and thus they cannot excite Cl or Ga, in accordance with experiment.[21] Improved experiments[21] will soon tell us about this fascinating explanation. It also should be stressed that because of the apparent quark–lepton symmetry in GUTs, one gets[22] the same kind of number for the ratio of the neutrino-number density to the photon-number density, that is, $n_\nu/n_\gamma \simeq 10^{-10}$, which explains one of the tacit assumptions made in primordial nucleosynthesis calculations, that is, *no neutrino degeneracy.*

Of course, the good news about primordial nucleosynthesis is that by using the GUT prediction[12,13] **(15):** $N_g = 3$, it predicts the correct "light element" abundances:[5] H: ^{4}He : D + ^{3}He : ^{7}Li $\simeq 3/4 : 1/4 : 10^{-5} : 10^{-9}$, which is one of the greatest triumphs of grand bang cosmology. I believe that the GUT prediction[12,13] **(15)**, the primordial nucleosynthesis results,[5] and the LEP determination of the number of light neutrino species,[6]

$$N_\nu = 2.93 \pm 0.09, \tag{16}$$

constitute a milestone for astroparticle physics. Actually, as shown in FIGURE 1 (taken from Ellis *et al.*[23]), the LEP number for neutrino species **(16)** gives new interesting lower bounds on the masses, and couplings to the Z^0-boson, of Majorana and Dirac "neutrinos."[23] In the case of Majorana "neutrinos" $m_M \geq 14$ GeV and $\sin^2 \phi_z$ [Z^0-(coupling strength)^2 relative to a conventional neutrino] ≤ 0.3, while for Dirac "neutrinos" $m_D \geq 20$ GeV and $\sin^2 \phi_z \leq 0.03$, when combined with the constraints from the underground ^{76}Ge experiments.[24] If one insists on "normal" massive neutrinos ($\sin^2 \phi_z = 1$), then the bad news is that:[23] $m_{\nu_M} \geq 38$ GeV and $\Omega_{\nu_M} \cdot h_0^2 \leq 2 \cdot 10^{-3}$ and $m_{\nu_D} \geq 44$ GeV and $\Omega_{\nu_D} h_0^2 \leq 2 \cdot 10^{-5}$, where, as usual, $\Omega_\nu \equiv \rho_\nu/\rho_{\text{crit}}$ and $\rho_{\text{crit}} = 1.88 \cdot 10^{-29}\, h_0^2 (gr/cm^3)$, $h_0 \in [1/2, 0.7]$. In other words, this kind of massive neutrinos *cannot close* the Universe.

3. *Phase Transitions/Inflation:* A fundamental property of gauge theories is the possibility of obtaining *massive* gauge bosons without at the same time destroying renormalizability. The magic trick that does it is called "spontaneous breakdown" (SB). In SB gauge theories, while the Lagrangian or the equations of motion respect all gauge symmetries, it is possible that some "solutions" break some gauge symmetries and thus lead to massive gauge bosons.[1] This phenomenon is widely used in particle physics, for example, as indicated in **(2),** where we start with a big gauge group G, at *very high energies,* and then around the scale M_X it suffers SB down to the standard model, which in turn around M_Z suffers another SB down to $SU(3)_C \times U(1)_{E.M}$, which seem to be exact gauge symmetries. It must be clear by now that SB is a key element in constructing realistic unified theories of particle physics. Further-

more, it also seems to be a key element in cosmology:[25] as the Universe expands starting from a supersmall size and *very high temperatures,* it cools down and thus gives the opportunity for several SBs or *phase-transitions.*[25] The potential existence of phase transitions in the early Universe has far-reaching consequences. The crown jewel of these implications is *inflation.*[25,26] There are three tough problems in BBC that need explanation, the *horizon problem,* the *flatness problem,* and the *primordial energy density perturbations problem.* The Universe is too big (or too old) in order to show such an incredible homogeneity and isotropy at large scales, as for example measured by COBE[4] $|\Delta T^0/T^0| \leq 10^{-5}$ (for "all sky"), where $T_0 = 2.73 \pm 0.06$ K, the cosmic background temperature. Assuming adiabatic (isoentropic) expansion, we can extrapolate back to find out that, for example, at the time of primordial nucleosynthesis, there were about 10^{30} *disconnected causal regions* that were con-

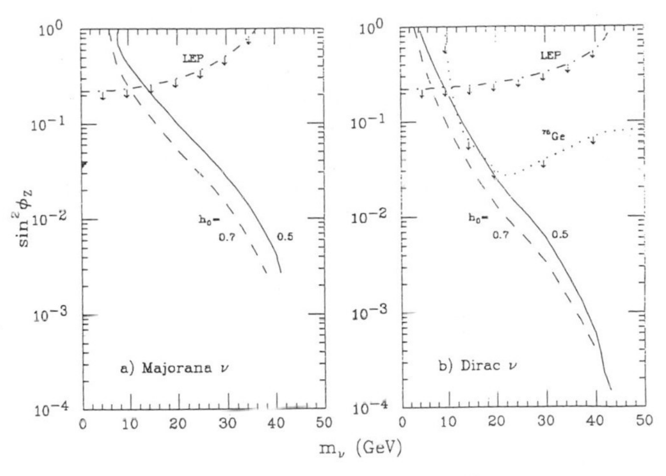

FIGURE 1. Values of mass m and Z^0 coupling strength $\sin^2 \phi_Z$ relative to a conventional neutrino consistent with LEP and giving $\Omega = 1$ for $h_0 = 0.5$ or 0.7 (**a**) for Majorana fermions, and (**b**) for Dirac fermions. In the latter case the constraint from background ^{76}Ge experiments is also shown. (From Ellis *et al.*[23] Reproduced by permission.)

tained in our Universe then! How they all merge to give us such a smooth Universe today causes the *horizon problem.* The observed (mean) energy density of the Universe today, even if not very accurately measured, is very close (within a factor of 10) to the critical density ρ_c. Again, extrapolating adiabatically back at the time of primordial nucleosynthesis, we find $[(\rho - \rho_{crit})/\rho] \sim 10^{-20}$, that is, an extraneous fine-tuning has to be performed, the *flatness problem.* Finally, while the Universe may be homogeneous and isotropic at large scales, it certainly has nontrivial structure at smaller scales, superclusters, clusters of galaxies, galaxies, stars, ... How did all this structure originate, and how were the needed *primordial energy density perturbations* created? Inflation[26] may provide a nice solution to all three aforementioned problems. The common origin of these problems seems to be the assumption of adiabatic expansion during the *whole* evolution of the Universe. This assumption may hold true for *most* of the evolution of the Universe, except for periods, in general, when the Universe suffers a phase transition. In such periods the Universe may enter a "supercooling" period during which the energy density remains *constant* instead of falling off in the "normal" way (T^4 or T^3) with temperature. Then as **(1)** shows, the Universe enters an *exponentially expanding* period (inflation), which if it is sufficiently long may automatically resolve the horizon and flatness problems.[27] In simple terms, the size of the horizon at early times is exponentially larger than what naive adiabatic expansion predicts, and thus we could "fit" all of our present Universe into a causaly connected region at very early times. At the same time, as **(1)** shows, while the energy density ρ remains constant during inflation, the space-curvature term (k_s/R^2) is blasted away due to the exponential increase of R. Thus, whatever the value of k_s was initially (± 1), after inflation we are effectively dealing with a $k_s = 0$ Universe, and thus $\rho = \rho_{crit}$, without any fine-tuning. Actually, one of the observable consequences of inflation is that we are living in a $k_s = 0$ (or $\rho_r = \rho_{crit}$) Universe, which can be taken as independent evidence for the existence of dark matter[28] in order to help closing the Universe. Furthermore, during the phase transition, the inflaton, the field that drives inflation, suffers quantum fluctuations that are encoded as primordial energy density perturbations, which are adiabatic and scale-invariant[25,26] (when entering their corresponding horizon). In other words we are getting the Harisson–Zeldowich[29] energy-density perturbation spectrum that is purportedly what is needed for galaxy formation. After the phase transition is completed, the inflationary era comes to an end and the inflaton oscillates around its "ground state" transforming the "stored" energy to radiation, thus reheating the Universe to a temperature not far below the one that existed just before inflation started. Thus, the Universe reenters again its "normal" adiabatic expansion era, but with a much broader horizon!

While the basic principles of inflation[27] are reasonable and simple, the construction of a realistic (non–ad hoc) model has proved to be cumbersome. We went[26] from "old" to "new" to "primordial" to "chaotic" to "extended" to . . ., with "stochastic," "multiple" . . . in between inflation. It looks to me that inflation should come out as a "derivative" of the big picture and not as an *ad hoc* barroque addition to it. Anyway, unified theories through phase transitions have given, in principle, a very satisfactory explanation of three of the toughest problems of BBC.

Phase transitions may lead to other "good" and "bad" phenomena.[25] During phase transitions, as is well known from condensed-matter physics topological defects may be created, including pointlike objects (monopoles), linelike objects (cosmic strings), or surfacelike (domain walls). With the exception of cosmic strings, all other defects mean trouble and usually we use inflation to blow them away and make sure they are not coming back in subsequent phase transitions. Thus, inflation may not only help BBC, but also unified theories by diluting a lot of problematic and highly undesirable objects, monopoles, and the like.[25,26]

It should be clear by now that GUTs and the grand bang have very nice properties satisfying some of our criteria for a real unified theory and a unified picture of the micro- and macrocosmos. Nevertheless, under closer, detailed scrutiny they face some grave problems when we compare their predictions with experiment, as well as some stability problems when we take into account quantum corrections. So our next move is to resolve these problems, while keeping the nice features of GUTs, which is achieved in the framework of supersymmetric (SUSY) theories.[30,1]

SUPERUNIFICATION—SUPERCOSMOLOGY

As I have previously stressed, while GUTs explain nicely quite a few puzzles of the standard model, their quantitative predictions, while not grossly in error, fall short of getting a satisfactory agreement with experiment. Indeed, one may use **(11)** to predict $\sin^2 \theta_w$, and M_X, both as functions of the strong fine structure construct α_3 (M_Z). As FIGURE 2 (taken from Ellis *et al.*[31]) shows clearly, the minimal $SU(5)$ model fails to reproduce recent LEP data in the $(\sin^2 \theta_w, \alpha_3(M_Z))$ plane, while at the same time the predicted value of M_X, and thus [through **(6)**] the predicted value of the

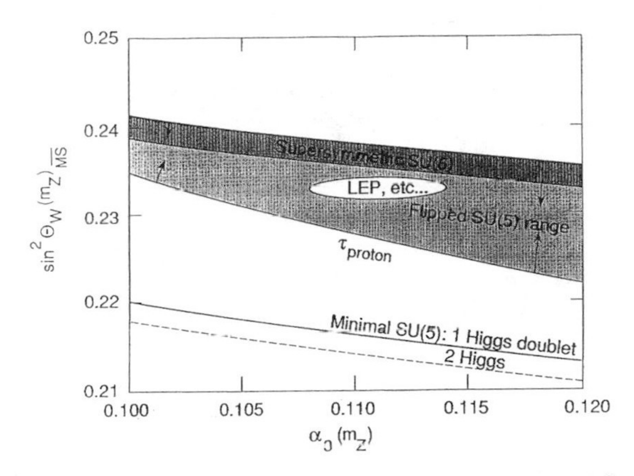

FIGURE 2. Comparison of the experimentally allowed region in $(\alpha_3(M_Z), \sin^2 \theta_w)$ space and the region predicted by minimal supersymmetric $SU(5)$ without extra representations, that allowed by supersymmetric flipped $SU(5) \times U(1)$ and proton decay experiments, and that predicted by nonsupersymmetric $SU(5)$ models. (From Ellis *et al.*[31] Reproduced by permission.)

proton lifetime, falls short when compared to the experimental lower bound[8]

$$\tau(p \to e^+\pi^0) \geq 5.5 \cdot 10^{32} y. \tag{17}$$

The origin of such a disagreement may be traced back to the specific form of the β-function used in these calculations, as given by **(10)**. If there is a richer particle spectrum than the simplest one used above **(10)**, predictions may tend to agree more with experiment. This is the first glimpse of the necessity of extending the minimal particle spectrum to contain new degrees of freedom. There is another very fundamental reason in the same direction. In GUTs, there are two scales $M_X(\approx 10^{15}$ GeV) and $M_Z(\simeq 90$ GeV) related, respectively, to the GUT and electroweak unification. The vast difference that separates the two scales is not only aesthetically unappealing but technically problematic. Quantum corrections mix the two scales and tend to bring them together, toward the bigger scale, and thus render the theory unacceptable. This is the so-called *gauge-hierarchy*[1] problem. One very satisfactory way out of this problem is the softening of quantum corrections, through the inclusion of new degrees of freedom interacting in a very restricted way, or in other words, invoking a new symmetry, beyond the gauge symmetry concept, fermion-boson symmetry, or *Supersymmetry* (SUSY).

SUSY theories treat fermions and bosons on equal footing and in a very symmetric way, and so by exploiting the basic difference between fermions and bosons (statistics), one is able to tame quantum corrections. Anytime that there is a boson loop, a fermion loop also exists, which all other things being equal, has an overall sign difference with the boson loop; thus they cancel each other out. So, the dangerous mixing between M_X and M_Z does not occur, and the gauge hierarchy is stabilized. Nevertheless, the price is high. The particle spectrum has to be at least doubled; for every fermion there should exist a boson, and vice versa, that is, its superpartner. So there should exist sleptons, squarks, photinos, gluinos, and the likes.[30] Their masses should not be above, say, 1 TeV, otherwise the solution to the gauge hierarchy problem is jeopardized. So we end up with a very rich low-energy particle spectrum that may be responsible for the resolution of the gauge hierarchy problem, and may also resolve the previously mentioned problems of ordinary GUTs concerning $\sin^2\theta_w$ and τ_p. Indeed, a similar analysis,[10,31,32] as in ordinary GUTs, shows that, with the β-functions **(10)** modified á la SUSY, one may get satisfactory results,[10,31,32] as shown in FIGURE 2 in the $(\sin^2\theta_w, \alpha_3(M_Z))$ plane and at the same time satisfy the lower bound on proton lifetime **(17).** It is remarkable that SUSY GUTs not only may resolve the toughest problem of ordinary GUTs, the gauge hierarchy problem, but at the same time they satisfy all phenomenological constraints. It should be stressed that until now we have no direct experimental evidence for the existence of SUSY particles. Nevertheless, a lot of us believe that we have not yet used enough energy to produce them, and that hopefully LHC and/or SSC will unveil their existence, which clearly is indirectly supported from the above analysis.[10,31,32] On a more theoretical level, it is interesting that the existence of global SUSY, invoked here to resolve the gauge hierarchy problem, and of gravity, automatically implies the existence of local SUSY or supergravity (SUGRA).[30] Thus, SUGRA GUTs are the first attempt to unify particle physics with gravity in a nontrivial and substantial way.[30] Then we should not be surprised that SUSY theories make a big impact on cosmology, too. Indeed, in the very early Universe supersymmetric phase transitions are more controllable than ordinary phase transitions, because of the taming of quantum fluctuations.[33] So, instead of $\delta\rho/\rho$ of order one, which we get in ordinary GUTs, it is relatively easy[33] to get $\delta\rho/\rho \sim 10^{-4} - 10^{-5}$ in SUSY GUTs, which is *observationally needed* for galaxy formation. Again the magic fermion–boson loop cancellation, which protects the gauge hierarchy, is at work and protects $\delta\rho/\rho$ from being too large. There are several other interesting effects related to supersymmetric phase transitions,[26] but these go beyond the scope of this paper. The existence of a richer particle spectrum for SUSY theories, as compared to ordinary theories, implies the existence of a richer variety of candidates for dark matter.[28] Beyond the ordinary [masses less than $O(100\text{ eV})$] neutrinos and axions (?), SUSY theories contain stable, neutral, weakly interacting massive particles (wimps) that may be our best candidates today for dark matter (DM).[28] While the superpartners of neutrinos, called sneutrinos, seem to have been experimentally excluded as candidates for DM, a neutralino, the lightest linear combination of zino, photino, and the two higgsinos (the spin 1/2 superpartners of Z-boson, photon, and Higgs bosons, respectively), may well fit the bill for DM candidates.[34] It should be stressed that in sharp contrast with other DM candidates (like axions), the closure of the Universe due to neutralinos is a natural and dynamically calculable[34] possibility, and not an *ad hoc* fitting. It is remarkable that recent calculations[23,35] taking into account all recent LEP and other experimental constraints on SUSY particles, show that there are large regions, in the acceptable parameter space, where $\Omega_n \equiv \rho_{\text{neutralino}}/\rho_\gamma$ is close to one. In FIGURE 3 (taken from Ellis *et al.*[23]), the lightest supersymmetric particle (LSP) mass density Ωh_0^2 as a function of m_{LSP} is shown along the ray $\mu/M_2 = 5$, where μ is the higgsino mass mixing parameter and M_2 the gaugino mass. The different lines correspond to

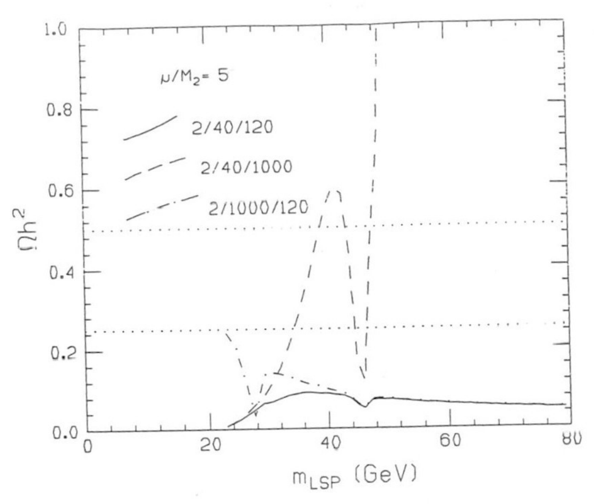

FIGURE 3. The LSP mass density Ωh_0^2 as a function of m_{LSP} along the ray $\mu/M_2 = 5$. The different lines correspond to different choices of tan $\beta/m_A/m_{\tilde{q}}$, discussed in the text. (From Ellis *et al.*[23] Reproduced by permission.)

different choices of tan β $|m_A|m_{\tilde{q}}$, the ratio of the two Higgs expectation values (v_t/v_b), the pseudoscalar Higgs mass, and squark mass, respectively. As FIGURE 3 shows clearly, $\frac{1}{4} \leq \Omega h_0^2 \leq \frac{1}{2}$ (for $h_0 \in [1/2, 0.7]$) is a perfectly allowable region within the "acceptable" (with respect to particle physics constraints) parameter space.[23,35] *Supercosmology* seems to be a viable possibility, and resolves many puzzles of the grand bang type of cosmology.

Once more, despite all the successes of SUGRA GUTs and supercosmology, we do not have the whole story. Supergravity theories are not finite (not even renormalizable) as one would expect for a consistent quantum theory of gravity, so they may be used, at least, as effective theories at low energies. The big question, of course, remains, which type of theory can be used as the fundamental theory encompassing all energies?

SUPERSTRINGS

The identity of a pointlike fundamental particle consists of a set of quantum numbers that can be either "internal" or "external." Electric charge, color, are some internal quantum numbers, while mass and spin are some external (spacetime related) quantum numbers. By now we have used all types of possible "operations" that transform one type of particle to another by changing any of its internal or external quantum numbers. Gauge symmetries and supersymmetries [including, possibly, some global or discrete symmetries (?)] exhaust all possible ways that pointlike particles interact. We may even increase the space dimensions and then make the extra dimensions more compact, á la Kaluza–Klein. Still, in the end, we get the same old story. *It seems impossible* to construct a unified theory of all interactions (gravity included), which is "infinity" free and based upon pointlike particles. Pointlike particles carry the anathema of creating quantum field theories with infinities, due to their very own nature of being pointlike. It is about time to accept this as a fact of life and move to extended objects as more suitable building blocks of a theory of everything (TOE).

Superstrings,[3] one-dimensional extended objects, seem to fit the bill for providing a TOE. As one usually expects, by giving an extension to our fundamental particles, we make the quantum field theory much smoother and actually even *finite,* no "infinities" whatsoever. A very interesting property of superstrings is the fact that, from the point of view of someone "living" on the string worldsheet, there is no separation between "internal" and spacetime symmetries as observed in four dimensions, and thus all interactions are treated on the same footing and share a lot of common properties: a real unification. It is remarkable that the massless superstring spectrum *always* contains the graviton multiplet, in addition to any other "gauge" and "matter" multiplets, that is, unification with gravity is natural and automatic.[3] Actually, the string tension T_{string} or the Regge slope $\alpha'(T_{\text{string}} = 1/\alpha')$ are simply related to the Newton constant $(G_N = 1/M_{\text{Pl}}^2 \simeq \alpha')$, thus fixing the length of the fundamental strings to be of the order of the Planck scale $l_{\text{Pl}} \simeq 1/M_{\text{Pl}} \simeq 10^{-33}$ cm. Clearly, at energies much below the Planck scale, $M_{\text{Pl}} \sim 10^{19}$ GeV, it is safe to treat fundamental particles as pointlike, but we should always remember that this is an "approximate" or "effective" description.

Particle Physics

The question that immediately arises is whether it is possible to construct a realistic string model for particle physics that encompasses all the goodies of SUGRA GUTs, and by being "stringy," avoids the difficulties that SUGRA GUTs face. The answer is yes.

Let us assume that we "live" in flat spacetime, which is a very reasonable assumption for particle physics. One can readily prove[3] that a consistent string theory emerges, if and only if we have 2-dimensional (worldsheet) superconformal invariance.[3] One should be careful to take care of conformal anomalies. Every 2-dimensional free boson carries one unit of conformal anomaly while every 2-dimensional free fermion carries one-half unit. Then we should make suitable combinations of 2-dimensional bosons and/or fermions in order to "kill" the -26 units of anomaly, endemic in "ghosts" of the bosonic string, or −15 units of anomaly in the case of the fermionic string. In other words, in the case of the bosonic string the *central charge* should be[36]

$$c_B = 4 + c_I^B = 26, \tag{18}$$

while in the case of the superstring

$$c_S = 4 + 4/2 + c_I^S = 15. \tag{19}$$

In **(18)** we have exhibited the contribution of the four spacetime dimensions (corresponding to four free 2-dimensional bosons), while **(19)** shows the additional contribution of the superparticles of the four 2-dimensional free bosons (four 2-dimensional free fermions). Both equations show the need of an "internal" contribution[36] to the conformal anomaly cancellation ($c_I^B = 22, c_I^S = 9$). There are two obvious, "extreme" cases among the many conformal field theories available to satisfy **(18)** and **(19)**. In the one case, one may use 22 extra 2-dimensional free bosons for the bosonic string or 9 extra 2-dimensional free bosons for the superstring. Since 2-dimensional free bosons may be identified with spacetime coordinates, we have rediscovered the famous statements:[3] bosonic strings "live" in $D_{\text{crit}} = 26$ dimensions and superstrings "live" in $D_{\text{crit}} = 10$ dimensions. In the other case, one may use 44 extra 2-dimensional free fermions for the bosonic string and 18 extra 2-dimensional free fermions for the superstring. In such a case, both types of strings "live" in four dimensions, and there is really *no reason to invoke extra space-dimensions.* Of course, any D between 4 and 26 or 4 and 10 for the bosonic or superstring is also acceptable and achievable through a suitable mixture of 2-dimensional free bosons and fermions. In other words, there is nothing "magic" or "sacred" about D = 26 or 10, except that they may be identified with the maximal dimensions that a bosonic or superstring may "live," and this restriction is *valid* only in the case of *flat spacetime.* Of course, we are not bound to use only 2-dimensional free fields, since any conformal field theory with suitable "central charge" would do it. Furthermore, one may use a mixture of a bosonic string and a superstring, that is, a closed string on which the left-moving modes correspond to a superstring, while the right-moving modes correspond to a bosonic string. Such a hybrid string is called the *heterotic string.*[37] It turns out that heterotic string theories are the only ones that may lead to realistic, phenomenologically acceptable particle physics models.[3] Let us concentrate now on 4-dimensional heterotic string theories, using the 2-dimensional free fermion way of canceling the anomaly. In other words, we need 18 2-dimensional fermions for the left sector and 44 for the right sector. In this free fermionic formulation[38] (FFF) of the heterotic string, different models are characterized by the different boundary conditions (phases) that the 2-dimensions fermions get when parallel transported along noncontractible loops of the world sheet. During the last few years, the ground rules for constructing such models have been spelled out,[38] as well the rules[39] for calculating "observables." The FFF turns out to be a very powerful formulation,[38,39] when one tries to extract phenomenological information from the heterotic string.

Among the plethora of available models, there is a specific model that *really sticks out:* **Flipped** $\boldsymbol{SU(5)}$[40] ($\boldsymbol{FSU(5)}$). There is a powerful theorem[41] that shows that in the massless spectra of a large number of heterotic string vacua there are no particles that can be identified with Higgs particles belonging to the adjoint or higher representations of GUT groups,[1] like $SU(5)$, $SU(10)$, E_6, ... That is too bad, because the step exposed in **(2)** cannot be completed and we will stick either with G or some other "wrong type" of low-energy group. One possible way to interprete this result is to assume that there is no grand unification and try to construct the standard model directly.[42,43] It turns out that this is not a very easy enterprise,[42,43] and we still have some way to go before we get a completely satisfactory model.[43] In any case, grand unification has several good points and it seems hard to abandon it. Luckily enough, there is an exception. There is *only one GUT:* FSU(5)[40] that does not need adjoints or higher representations to accomplish the step shown in **(2).** Clearly, $FSU(5)$ looks like the *chosen* theory. The reason that $FSU(5)$ does not need adjoints is simple. $FSU(5)$ is a nontrivial variant of the $SU(5)$ theory[7] where in **(3),** $u^c \leftrightarrow d^c$; $e^c \leftrightarrow \nu^c$ (thus the name "flipped"). We have to introduce a right-hand neutrinolike field in the **10** representation, which can get a vacuum expectation value (vev) and complete the step shown in **(2),** thus avoiding any need for adjoints that do not exist in many string vacua, and using 10's that do exist.

We have constructed a realistic $FSU(5)$ string model[44] with three generations, a hierarchical fermion spectrum,[45,46,17] two Higgs doublets, automatic doublet–triplet Higgs splitting, as well as an automatic seesaw mechanism for generating acceptable neutrino masses, sufficiently stable proton decay[47] usually to $e + \pi^0$ with $\tau_p \sim 10^{33} - 10^{35}$ y, as well as acceptable values[10,31] for $\sin^2 \theta_w$ and $\alpha_3(M_Z)$, as shown in FIGURE 2. Furthermore, we have shown by explicit calculations[48] that our model is not ruined by nonrenormalizable effects[39] created when the massive string modes are taken into account, that is, our model is *stable.*[48] In addition, $FSU(5)$ has some very interesting and novel cosmological characteristics. There are no "normal" magnetic monopoles, and *charge quantization* has a dynamical origin.[49] Actually, there is a whole class of new particles with electric charges $\pm 1/2$, but there is also a confining group $SU(4)$ á la quarks with charge ⅔ or $-$⅓ and $SU(3)_C$ as the confining group, so that always the "observable" particles have integer charges.[49] Like the case of quarks, where we have mesons, baryons, etc., as the "observables," in our case we have a whole new class of "observables," (hidden) mesons, baryons, etc., that we call *cryptons.*[49] These particles constitute a new form of matter, and it is even possible that the lightest crypton, which is neutral and (meta) stable, may be a good candidate for dark matter.[49] The analogy with the protons is striking: protons, the lightest, (meta) stable (thanks to baryon-number conservation at the level of the SM), confining states of $SU(3)_C$ constitute the bulk of "observable" matter, while cryptons, the lightest, (meta) stable, confining states of $SU(4)$ may constitute the bulk of the dark matter. While the confining scale of $SU(3)_C$ is $\Lambda_{QCD} \sim$ (1 GeV), which also gives the proton mass, the confining scale of $SU(4)$ is[45,50] $\Lambda_4 \sim (10^{10}$ GeV), which should be taken as the approximate mass scale for cryptons. Cosmological and astrophysical constraints on cryptons, and possible ways to detect them, are currently being worked out.[51]

It should be stressed that the stringy $FSU(5)$[44] or "flipped string"[17] is an archetypal example of a realistic string model that may easily serve as a TOE. It is the first time in particle physics that we may have a model that provides a consistent quantum theory of gravity, a generic property of any superstring derived model, and at the same time, a highly successful SUGRA GUT in agreement with all currently available phenomenological constraints.[17]

Cosmology

One of the *raisons d'être* of superstrings, if not the most fundamental one, is that they seem to provide a consistent quantum theory of gravity. As such, we should be able to answer questions and resolve puzzles concerning gravitational effects at very small distances (close to the Planck scale), or, more generally, strongly gravitational environments (black holes and the like). One expects to get satisfactory answers to a vast range of questions, extending from the origin and the embryonic stages of the Universe, to possible quantum gravitational spacetime topology changes[52] (e.g., wormholes, etc.), to the impact of strong gravitational effects on fundamental quantum mechanics,[53] to explaining the laws of black holes.

Clearly, in order to study these phenomena we have to go away from "flat-land," that is, trivial, flat spacetime, static, string solutions and look for string solutions in an arbitrary "gravitational" background. In the following we will always work, usually out of simplicity, with the bosonic string, injecting, when necessary, some vital information for superstrings.

The motion of a string "living" in an arbitrary gravitational background in D-spacetime dimensions is described by the 2-dimensional action[3]

$$S_{2\text{-}d} = \int d^2\sigma\left(\sqrt{h}G_{\mu\nu}h^{\alpha\beta}\partial_\alpha X^\mu\partial_\beta X^\nu + B_{\mu\nu}\epsilon^{\alpha\beta}\partial_\alpha X^\mu\partial_\beta X^\nu + \frac{\sqrt{h}}{2}\Phi R^{(2)} + \cdots\right), \qquad \textbf{(20)}$$

where, as usual, $h_{\alpha\beta}$ is the 2-dimensional metric and $R^{(2)}$ the 2-dimensional scalar curvature, $X^\mu(\sigma)$ the spacetime coordinates of the string ($\mu = 0, 1, \ldots, D-1$), and $(G_{\mu\nu}(X), B_{\mu\nu}(X), \Phi(X))$ the graviton multiplet background containing the spacetime metric tensor, the antisymmetric tensor, and dilaton fields, respectively, while the dots refer to other field contributions that are irrelevant to our present discussion. It should be stressed right away that the generic form of action **(20)** is an *unavoidable* part of any string theory action; thus, our results will hold true in *any* specific string model.

From the point of view of 2-dimensional physics, $G_{\mu\nu}$, $B_{\mu\nu}$, and Φ are just "coupling functions." As such, in order to ensure conformal invariance (discussed previously), their corresponding β-functions should vanish.[3] Actually, (global) scale invariance implies $\beta_{G_{\mu\nu}} = 0$ and $\beta_{B_{\mu\nu}} = 0$, in turn enforcing β_Φ = constant, which vanishes if and only if conformal invariance is imposed.[3] In our case, though, because we allow for arbitrary gravitational background, the condition **(18)**, which corresponds to $\beta_\Phi(=4 + c_I^B - 26) = 0$, is modified to

$$c'_B = (4 - \delta c) + c'^B_I = 26, \qquad \textbf{(21)}$$

where $\delta c \geq 0$ is the "central charge deficit."[54] We have assumed in **(21)**, as usual, that we are living in four large (uncompacted) spacetime dimensions, and allow some "internal" conformal system of appropriate central charge c'^B_I to take care of the rest of the conformal anomaly. We are entitled, if we so wish, to interpret

$$D_{\max} \equiv 4 + c'^B_I = 26 + \delta c \geq 26 \qquad \textbf{(22)}$$

as the maximal spacetime dimension that a string is allowed to "live" in when it moves in an arbitrary gravitational background. Clearly, there is nothing "critical" or "magic" about $D = 26$ (or $D = 10$ in the case of superstring) that we have been *brainwashed* to believe for about 20 years! Any dimension goes, if we do not confine ourselves to flatland! In other words, "noncritical" strings are equivalent to "critical" strings moving in curved spacetime.[54]

Another way to see this is to work directly with the D-dimensional effective action, involving the graviton multiplet fields, derivable from the $\beta_i = 0$ conditions, which are nothing else but the equations of motion (EOM) of $G_{\mu\nu}$, $B_{\mu\nu}$, and Φ.[3] It is given by[3]

$$S^D_{\text{effective}} \sim \int d^DX \sqrt{G} e^{-\Phi} \left\{ R^{(D)} + (D_\mu \Phi)^2 - \frac{1}{12} H^2 - \frac{1}{3} \delta c + \cdots \right\}, \tag{23}$$

where $R^{(D)}$ is the scalar curvature in D-dimensions, D_μ is the covariant D-dimensional derivative, and $H_{ijl} \equiv 3D_{[i}B_{jl]}$, the $B_{\mu\nu}$ tensor field-strength. It is apparent that whatever way you look at it, either as a dilaton potential or as an "effective" cosmological constant, the existence of a nonzero δc imposes nontrivial dynamical changes in our string theory.[54] Actually, the vanishing of δc looks like an extraneous fine-tuning, analogous to the standard cosmological constant problem. In other words, strings moving in D_{crit} flat spacetime dimensions ($\delta c = 0$) are one out of many nontrivial possibilities, and we should prefer to find a dynamical reason why, if we start in general with $\delta c \neq 0$, the physically acceptable "ground state" has $\delta c = 0$.

The simplest, nontrivial time-dependent, homogeneous, and isotropic string solution has been found only recently:[54,55]

$$G_{\mu\nu} = \eta_{\mu\nu}; \qquad B_{\mu\nu} = 0; \qquad \Phi = -2QX^0 + \Phi^0, \tag{24}$$

where $\eta_{\mu\nu}$ is the standard Minkowski metric, Φ^0 an integrating constant, with

$$\delta c = 12Q^2, \tag{25}$$

and we have identified, as usual, the worldsheet time variable τ with the target spacetime variable X^0. The existence of such a solution has far-reaching consequences.[54] Let us discuss first some of its basic properties:

1. In the absence of dynamical matter, the solution **(24)** is the *unique* solution of the 1-loop, 2-dimensional β-functions with $\delta c \neq 0$, in the following sense:
   (i) Every possible solution approaches asymptotically (large τ or X^0) **(24)**.
   (ii) Even if we include moduli fields of the conformal system(s) coupled to the dilaton Φ, again all possible solutions for large X^0 approach asymptotically **(24)** and constant values for the moduli fields.[56]
2. The solution **(24)** is an *exact* solution of the 2-dimensional σ-model, that is, the 2-dimensional β-functions are vanishing for this solution to all orders in perturbation theory in α'. Actually, varying the action **(20)** in the background described by **(24)**, one gets[54] for the 2-dimensional energy-momentum tensor

$$T = -\frac{1}{2}(\partial_\alpha X)^2 + \frac{1}{2}(\partial_\alpha X^0)^2 - Q\partial_\alpha^2 X^0, \tag{26}$$

   which is nothing else but the energy-momentum tensor of the "Coulomb gas" representation of universal conformal models with "background charge" 2iQ at infinity for the time coordinate X^0.[57] In other words, not only do we know that our solution **(24)** is exact, but we have identified[54] the 2-dimensional conformal system representing it! Let us next try to get some physics out of this nice and simple solution **(24).**

To discuss physical applications of the solution **(24),** we first need its "spacetime" interpretation.[54] Indeed, by inserting **(24)** into **(23),** we get the "physical" metric $g_{\mu\nu}$ (i.e., with properly normalized scalar curvature $R^{(D)}$)

$$g_{\mu\nu} = e^{4QX^0/(D-2)}\eta_{\mu\nu} \quad \textbf{(27)}$$

with "physical" time t (as opposed to "conformal" time X^0)

$$t = \frac{D-2}{2Q} e^{2QX^0/(D-2)}. \quad \textbf{(28)}$$

The "physical" line-element is then[54]

$$ds^2 = -(dt)^2 + (t^2)dX^i\, dX^i; \quad \textbf{(29)}$$

in other words, *a linearly expanding (contracting), spatially flat* ($k_S = 0$), *D*-dimensional, homogeneous, and isotropic Universe for $Q > 0(Q < 0)$. Actually, *if and only if* $D = 4$, there is another solution[54]

$$ds^2 = -dt^2 + t^2\left[\frac{dr^2}{1-k_s r^2} + r^2[d\theta^2 + \sin^2\theta d\phi^2]\right], \quad \textbf{(30)}$$

with

$$k_s = \frac{1}{2Q^2k} \geq 0, \qquad (k = 1, 2, \ldots); \qquad R^{(4)} = \frac{6(1+k_s)}{t^2}, \quad \textbf{(31)}$$

and with $B_{\mu\nu} \neq 0$ in **(24),** that is,

$$b = 2Q^2\sqrt{k_s}\,t, \quad \textbf{(32)}$$

where b is defined through $H_{\lambda\mu\nu} = e^{2\Phi}\epsilon_{\lambda\mu\nu\rho}D^\rho b$. This solution corresponds to a specific conformal theory in two dimensions, namely, the Wess–Zumino–Witten (WZW) model,[58] at level $k(k = 1, 2, \ldots)$ of an $SO(3)$ [or $SU(2)$] group manifold. The uniqueness of the solution **(31), (32)** for $D = 4$, is due to the fact that only for $D = 4$ or only for 3-space dimensions, the corresponding "maximal symmetric space" (S_3) is a group manifold [$SO(3)$]. In such a case, the conformal anomaly cancellation condition **(21)** takes the very specific form[54]

$$\left(3 - \frac{6}{k+2}\right) + (1 - 12Q^2) + c_I^{\prime B} = 26. \quad \textbf{(33)}$$

One of the many fascinating implications[54] of **(33)** is that it leads to a "quantized" spatial curvature k_s **(31).** Indeed, all known conformal field theories have "central changes" of the form $c_n = c_\infty - 0(1/n), c_\infty \epsilon Z/2$, implying that $c_I^{\prime B}$ can only take discrete values, and thus deducting from **(33),** using the fact that the $k \epsilon Z^+$, that Q^2k takes only discrete values, that is, k_s in **(31)** *is quantized!* In the case $Q = 0$, one recovers a *static, Einstein Universe* á la **(30),** but with no t-scale factor, while **(31)** becomes [$\Phi = \Phi_0$ from **(24)**]

$$k_s^{\text{Einstein}} = \frac{e^{\Phi_0}}{2k} \geq 0; \qquad R^{(4)} = 6k_s, \tag{34}$$

again with a *"quantized" spatial curvature,*[54] while **(32)** turns to

$$b = 2e^{-\Phi_0}\sqrt{k_s}X^0. \tag{35}$$

We have discovered something really remarkable here. Consistency of string theories demands conformal invariance, which needs conformal anomaly cancellation, which implies, using the discreteness of "central charges" of 2-dimensional conformal theories, "quantization" of the *spatial curvature* k_s. The outcome of all this is that for large spacetimes k_s is always zero! Indeed, if we use the "Coulomb gas" representation of conformal minimal models to represent our solution **(24),** then as **(29)** shows explicitly, $k_s = 0$. If we insist on a WZW type of solution ($D = 4$ is available!) because k_s is "quantized" in string (or Planck) units, after a few steps (after a few Planck lengths) it is going to hit zero! Thus we have proved that the simplest, nontrival, time-dependent, *unique,* string solution **(24)** always leads, whatever way interpreted, to a *spatially flat* universe after the Universe scale factor exceeds a few Planck lengths (in the worst case!)

$$(k_s)_{\alpha \gg l_{\text{Planck}}} = 0. \tag{36}$$

It sounds to me that we have a dynamical, stringy explanation[54] of the *flatness problem,* discussed in the third section.

It should be emphasized that when the string is moving in the background **(24),** there are some drastic changes in the basic formulas for the masses, plane waves, and asymptotic density of string states. Indeed, the Einstein–Virasoro formula now reads[54]

$$E^2 - \mathbf{p}^2 = \begin{cases} m^2 - Q^2 & \text{for (spacetime) bosons} \\ m^2 & \text{for (spacetime) fermions,} \end{cases} \tag{37}$$

where, as usual, in the case of the bosonic string

$$m^2 = -2 + 2N, \tag{38}$$

with N the integer level of the state.

The "standard" plane waves are multiplied[54] by a factor of e^{-QX^0}, indicating the "expansion" of the Universe, and the asymptotic density of string states becomes[54] $\rho(m) \underset{m\to\infty}{\longrightarrow} e^{2\pi\sqrt{2+q^2}m}$. Actually, the "diluting" factor e^{-QX^0} takes care of the growing modes in the propagator of "tachyonic" rescaled fields **(37)** in the forward light cone. Thus the "tachyonic" mass shifts *need not,* and *a priori,* signal classical instabilities. On the other hand, we may use[59] the "tachyonic" mass shifts in a creative way! Indeed, as is clearly indicated in **(37),** we have $-Q^2$ mass shifts, which imply that the "phase velocities" *increase* during the linear-dilaton epoch **(24).** Taking, for instance, the photon ($m^2 = 0$), we then have[59]

$$\frac{E^2}{|\mathbf{p}^2|} = \frac{|\mathbf{p}^2|}{|\mathbf{p}^2|} = 1 \rightarrow \frac{|\mathbf{p}^2| - Q^2}{|\mathbf{p}^2|} \leq 1, \quad \textbf{(39)}$$

which implies that light can travel *arbitrarily fast* for low frequencies. In such a case, the light cone "opens" [$\theta \geq (\pi/4)$], and can be even at right angle [$\theta \simeq (\pi/2)$]. Thus we are able to *transmit information* all over the place, and hence to solve the *horizon problem* discussed in the third section. It should be stressed that the existence of such stringy *superhorizons* provides a novel,[59] and different from inflation, resolution of the horizon problem. While in inflation one concentrates on sufficiently blowing up of a "normally" small, causally connected region, here we let light transmit information over a big, "normally" causally disconnected region. This *effectively* acausal behavior is a *purely stringy* phenomenon. Indeed, it is due to the nontrivial dilaton–photon field strength coupling ($e^{-\Phi}F_{\mu\nu}F_{\mu\nu}$) contained in the D-dimensional effective action, which is completely analogous to the $e^{-\Phi}H^2$ term contained in **(23)**. Noticeably, this is not a property shared either by the Einstein–Maxwell or the Brans–Dicke type of (point-like) field theories. It sounds to me that we have[59] a dynamical, stringy explanation of the *horizon problem,* as discussed in the third section.

While the classical instabilities due to the $-Q^2$ "tachyonic" mass shifts of **(37)** can, as previously mentioned, be presumably taken care of, the standard "tachyon" instability seems to be real. Indeed, in both cases of the bosonic string and of the superstring, as it is clearly indicated by combining **(37)** and **(38)**, we got a tachyon, $T(\sigma)$. Even in the case of the superstring, this tachyon cannot be GSO-ized[60] away, and we have to live with the consequences.[59] The potential for this tachyon $T(\sigma)$ looks like $V(T) = 4Q^2 - 2T^2 + \cdots$, which clearly implies that the duration of the linear-dilaton epoch **(24)** is finite, indicated by the amount of time that the tachyon spends around $T \simeq 0$, that is, around its local maximum $V_0(\equiv 4Q^2)$. As T moves toward large negative values, the string partition function diverges $\exp[-T(\text{Area})_{2\text{-}d}]$, signaling a phase transition, and accompanied by a lot of *entropy generation* and *energy density fluctuations,* by dissipating the tachyon energy. This tachyonic "happening" may occur one or more times until we reach the level of $V_0 = 0$ or, equivalently, $Q^2 = 0$! In other words, the *effective* tachyonic modes that are created because $Q^2 \neq 0$, destabilize our "vacuum" **(24)**, create a lot of entropy and energy density perturbations, because they trigger a phase transition, and thus dynamically "drive" Q^2 to zero. Thus, not only do we get a nice way of creating a lot of entropy and $\delta\rho/\rho$, but at the same time, we understand *dynamically* why, finally, $Q^2 = 0$ or $\delta c = 0$ in **(23)**. It sounds to me like we have[59] a potential solution of string origin to the *primordial entropy and* $\delta\rho/\rho$ *problem,* discussed in the third section. Before moving any further, let us stress the fact that **(36)**, **(39)**, and the remarks just made, constitute a *credible, dynamical, stringy alternative*[59] to inflation. Incidentally, the preceding resolution of how to reach the $Q^2 = 0$ (or $\delta c = 0$) phase is in accord with Zamolodchikov's c-theorem,[61] which shows that, thanks to the existence of "relevant" operators, conformal theories tend to reduce their central charge, that is, $\delta c \rightarrow 0$.

There is another very interesting reason that indicates that the "linear-dilaton epoch" solution **(24)** is really special: *the quantum origin of time.* Usually, when one uses the classical (genus zero) 2-dimensional action **(20)**, one assumes that the 2-dimensional metric $h_{\alpha\beta}$ can always take the form $h_{\alpha\beta} = e^{\phi}\hat{h}_{\alpha\beta}$, with ϕ the conformal factor and $\hat{h}_{\alpha\beta}$ some fixed metric. Thus, the conformal factor decouples from the classical 2-dimensional action, although this decoupling is not automatic when quantum corrections are taken into account. Only *if and only if* $\delta c = 0$, does the decoupling occur *also* at the quantum level; otherwise, ϕ becomes a dynamical degree of freedom.[62] Indeed, there is a quantum correction to the classic action **(20)** given by[63]

$$S_{2\text{-}d}^{\text{quantum}} = \int d^2\sigma\sqrt{\hat{h}}\{-\hat{h}^{\alpha\beta}\partial_\alpha\phi\partial_\beta\phi + \sqrt{c_m - 25}\,\phi R^{(2)} + \cdots\}, \quad \textbf{(40)}$$

where $c_m = 3 + c_I^{\prime B}$ stands for "matter" central charge. It is impressive that quantum corrections create a kinetic term for the conformal factor field ϕ, thus turning it into a new dynamical degree of freedom. It is even more impressive that this kinetic term has the wrong sign: *minus!* We just have to notice[64,54,59] that **(40)** can be readily absorbed in **(20)** *if and only if*

$$\phi = X^0; \qquad \phi \sim \sqrt{c_m - 25}\phi = \sqrt{c_m - 25}X^0, \quad \textbf{(41)}$$

that is, *if and only if* the string started in the "linear-dilaton epoch" vacuum of **(24)!** Furthermore, **(40)** implies a conformal anomaly contribution of the ϕ-field: $c_\phi = 1 - (c_m - 25)$, or

$$c_m + c_\phi = 26, \quad \textbf{(42)}$$

which is nothing else but the conformal anomaly cancellation condition **(21)**, reexpressed in terms of $c_m (\equiv 3 + c_I^{\prime B})$, and thus $c_m - 25 = \delta c = 12Q^2$, the last equality due to **(25)**. Remarkably, turning ϕ to a dynamical degree of freedom by moving to $c_m > 25$, not only do we "create" time by quantum corrections **(40)** but *also* conformal invariance is imposed *dynamically* **(42)**, not put in by hand. This extra bonus was the missing link for identifying string theory with 2-dimensional quantum gravity, and makes the whole picture much more logical and aesthetically appealing. Using this 2-dimensional quantum gravity language,[63] one can readily show[64,54,59] that the "matter" part of the vertex operator ($e^{ip\cdot X}$) has to be "gravitationally dressed" by $e^{i(p^0+iQ)\phi}$ in order to render a consistent reparametrization invariant conformal field theory in two dimensions. Thus, we explicitly see[64,54,59] how time ($\sim\phi$) is created and "runs," while **(37)** guarantees that the total conformal weight is 1!

The *cosmic evolution scenario* that emerges is not hard to spell. Take as the natural, initial state of the Universe the "linear-dilaton epoch" vacuum **(24)**, with $Q^2 \neq 0 (c_m > 25)$. All space dimensions are Euclidean, while *quantum corrections create dynamically time* [i.e., a 2-dimensional boson field with negative metric *and* background charge ($\sim$2iQ) at infinity]. During this "era" light can travel arbitrarily fast, and thus can transmit information into otherwise "classically" forbidden large regions, while the "tachyon" destabilizes this vacuum and brings it "down" to a $Q^2 = 0 (c_m = 25)$, large, spatially flat, homogeneous, and isotropic, 4-dimensional Universe, seeded with primordial energy density perturbations ($\delta\rho/\rho$)! Eventually, even the uniqueness of $D = 4$ may be explained, if arguments related to the necessity of a WZW-type conformal theory are beefed up along the lines described just before **(33)**.

Black Holes

Black holes are the ideal testing ground for studying the quantum mechanical effects of strong gravitational fields.[65] We expect a consistent quantum theory of gravity to shed "light" on some of the dark sides of black hole (BH) physics. We would like to understand if the Hawking effect,[66] that is, the thermal radiation emitted by (nonrotating, uncharged) black holes, leads to *complete* BH evaporation, or it stops at some point and leads to stable, primordial, mini-BHs. The answer to this problem has very profound consequences. If BHs evaporate completely, that is, a pure state evolves into a mixed state, then we are in for some drastic revisions of quantum mechanics,[53] including the basic assumption of the existence of a unitary time evolution operator, the Hamiltonian! In addition, the large entropy, S_{BH}, content of a BH (proportional to its event horizon area) is kind of difficult to reconcile with the fact that BHs have "no hair."[67] In other words, if mass, angular momentum, and electric charge are the only "quantum numbers" of a BH, how is it possible to give a standard statistical mechanical interpretation of such a large S_{BH} when there are not that many internal states available? If superstring theory is the correct theory of quantum gravity, it should provide a satisfactory answer to the preceding puzzles.

Indeed, very recently we have[68] put forward a rather "bold" proposal of a *very stringy nature* that may lead to a resolution of the BH paradoxes. We have based our arguments on several very interesting recent developments in very different, and at first sight, uncorrelated areas inside and outside string theories, namely, the following.

1. It has been observed[69] that "discrete gauge symmetries," that is, discrete symmetries that are relics of spontaneously broken gauge theories, may lead to *quantum hair*[70] of BH, that is, "hair" not observable by classical means, but observable through quantum effects. Indeed, while "classical hair" needs a long-range classical field (nonvanishing field-strength outside the BH event horizon), "quantum hair" needs long-range "vector-potential" by making use of the celebrated Bohm–Aharanov effect[71] (B–A). Just as an electron outside a solenoid feels the existence of the vector potential, even if the magnetic field *vanishes* outside the solenoid, in the same way a test string carrying some "discrete gauge" charge, and being outside the BH event horizon, may feel, and thus *measure*, the BH "discrete gauge" charge. In other words, *discrete gauge charge* can be used as *quantum hair.*[70]
2. It has been recently shown that the effective 10-D superstring action has all kinds of solitonic solutions (macroscopic strings,[72] five-branes,[73] black p-branes,[74] . . .). Exact solutions, once more based on **(24)** and **(31)** are also available.[75] Furthermore, it also has been shown that,[76] starting from a *similar* 10-D effective action for five-branes, superstrings emerge as a solitonic solution. It has been strongly argued[77,74] that a *new duality symmetry* between superstrings and five-branes has been discovered, analogous to the one conjectured a long time ago by Montonen and Olive[78] between massive gauge bosons and magnetic monopoles (solitons) of a spontaneously broken $SO(3)$ gauge symmetry, or its supersymmetrized versions.[79]

3. Superstring theories contain, among the plethora of their symmetries, a special kind called "target-space duality".[80] In its simplest form, that is, compactification of one space dimension on a circle of radius R, it implies that the string interactions and string spectrum are invariant under $R \leftrightarrow \alpha'/R$, that is, a discrete Z_2 symmetry! This effect has been readily generalized[80] to more interesting and nontrivial cases, implying the existence of rather large, nontrivial discrete symmetries. Furthermore, one is able to show[80] that these discrete symmetries are indeed *discrete gauge symmetries!*

Our proposal[68] makes direct use of the preceding three phenomena. We view BH as solitonic solutions of superstrings and thus through the string-membrane duality,[77] they carry a stringy "discrete gauge" charge, that is, they possess "quantum hair"! Actually, thanks to the large number of available stringy "discrete gauge" charges, it is not inconceivable that we can match the large BH entropy, S_{BH}, with a large number of internal states. After all, maybe the exponential explosion[81] for large m of the degeneracy of string modes ($e^{\beta m}$) is not accidental, implying an $S \propto \ln e^{\beta m} \propto m$. Furthermore, as in the case of electric charge, these "discrete gauge charges" (available to all black holes) may stop the complete evaporation of BHs, thus providing primodial mini-BHs and saving quantum mechanics.

Despite the fact that all these "bold ideas" are, at best, in infancy, they clearly show a possible way to understand quantum gravitational effects in superstring theories.

CONCLUSIONS

After years of speculation, astroparticle physics has turned into a "hard" science. The synergy of particle physics with cosmology has had, and continues to have, a profound impact on both fields. While both the standard models (SM and BBC) are doing very well, we already have *indications* signaling the need for their extension. The exciting new LEP results concerning the number of neutrino species ($N_\nu = 3$) and the very precise values of $\sin^2 \theta_w$ and $\alpha_3(M_Z)$, as shown in FIGURE 2, *clearly* indicate unification beyond the SM. Actually, if we take them at face value, as FIGURE 2 shows, ordinary GUTs are excluded, while SUSY GUTs are the *big winners.* A lot of us feel, even more strongly now, that the discovery of SUSY particles is just a matter of available energy, that is, LHC or SSC will find them! Furthermore, the very small, allowable deviation from the number of three neutrino species puts severe limits on, and excludes a lot of DM candidates. Clearly, the applications to cosmology get firmer, and primodial nucleosynthesis in the BBC is another *big winner.* We have predictable "bumps" in our way, for example, *large-scale structures,* and data from COBE, the Hubble telescope, and other "observations" (great: "walls," "attractors," . . .) are eagerly expected. On a more theoretical front, superstrings start showing their "muscle" both on the particle physics and cosmology fronts, where new ideas have been put forward, including flipped $SU(5)$, alternative to inflation, quantum origin of time, and black hole dynamics.

All in all, the 1990s promise to be very exciting for astroparticle physics, both theoretically and experimentally.

REFERENCES

1. QUIGG, C. 1983. Gauge Theories of the Strong, Weak and Electromagnetic Interactions. Benjamin-Cummings, Mass.; LANGACKER, P. 1981. Phys. Rep. **72:** 185; ROSS, G. G. 1983. Grand Unified Theories. Benjamin-Cummings, Mass.; KOUNNAS, C., A. MASIERO, D. V. NANOPOULOS & K. A. OLIVE. 1984. Grand Unification with and without Supersymmetry and Cosmological Implications. World Scientific. Singapore.
2. KOLB, E. W. & M. S. TURNER. 1990. The Early Universe. Addison-Wesley. Reading, Mass.
3. GREEN, M. B., J. H. SCHWARZ & E. WITTEN. 1987. Superstring Theory. Cambridge University Press. Cambridge, England.
4. MATHEZ, J., *et al.* 1990. Astrophys. J., Lett. **354:** L37; CHENG, T., *et al.* 1990. Bull. Am. Phys. Soc. **35:** 937; SMOOT, G. 1991. Lectures Given at the International School of Astroparticle Physics, Houston Advanced Research Center (HARC), The Woodlands, Tex., January 1991.
5. YANG, J., M. S. TURNER, G. STEIGMAN, D. N. SCHRAMM & K. A. OLIVE. 1984. Astrophys. J. **281:** 493; OLIVE, K. A., D. N. SCHRAMM, G. STEIGMAN & T. WALKER. 1990. Phys. Lett. **B236:** 454.
6. AMALDI, U. 1991. LEP, the Laboratory for Electrostrong Physics, one year later. This issue.
7. GEORGI, H. & S. L. GLASHOW. 1974. Phys. Rev. Lett. **32:** 438.
8. IMB-3 COLLABORATION, BECKER-SZENDY, R., *et al.* 1990. Phys. Rev. **D42:** 2974.
9. GEORGI, H., H. QUINN & S. WEINBERG. 1974. Phys. Rev. Lett. **33:** 451.
10. ELLIS, J., S. KELLEY & D. V. NANOPOULOS. 1990. Phys. Lett. **B249:** 441.
11. CHANOWITZ, M. S., J. ELLIS & M. K. GAILLARD. 1977. Nucl. Phys. **B128:** 506.
12. BURAS, A. J., J. ELLIS, M. K. GAILLARD & D. V. NANOPOULOS. 1978. Nucl. Phys. **B135:** 66.
13. NANOPOULOS, D. V. & D. A. ROSS. 1979. Nucl. Phys. **B157:** 273; NANOPOULOS, D. V. & D. A. ROSS. 1982. Phys. Lett. **108B:** 351; NANOPOULOS, D. V. & D. A. ROSS. 1982. Phys. Lett. **118B:** 99.
14. ELLIS, J., J. S. HAGELIN, S. KELLEY & D. V. NANOPOULOS. 1988. Nucl. Phys. **B311:** 1.
15. STEIGMAN, G., D. N. SCHRAMM & J. GUNN. 1977. Phys. Lett. **66B:** 202.
16. ELLIS, J. & M. K. GAILLARD. 1979. Phys. Lett. **B88:** 315; NANOPOULOS, D. V. & M. SREDNICKI. 1983. Phys. Lett. **B124:** 37.
17. For a recent review, see LOPEZ, J. L. & D. V. NANOPOULOS. 1990. *In* Proceedings of Strings '90: 377. World Scientific. Singapore.
18. SAKHAZOV, A. D. 1967. JETP Lett. **5:** 24.
19. For a review, see KOLB, E. W. & M. S. TURNER. 1983. Annu. Rev. Nucl. Part. Sci. **33:** 645.
20. WOLFENSTEIN, L. 1979. Phys. Rev. **D17:** 2369; MIKHEYEV, S. P. & A. YU. SMIRNOV. 1986. Sov. JETP **64:** 4.
21. For recent reviews, see KENNEDY, D. C. 1990. University of Pennsylvania Preprint UPR0442T; LANDE, K. 1991. Lectures Given at the International School of Astroparticle Physics, Houston Advanced Research Center (HARC), The Woodlands, Tex., January 1991.
22. NANOPOULOS, D. V., D. SUTHERLAND & A. YILDIZ. 1980. Nuovo Cimento Lett. **28:** 205.
23. ELLIS, J., D. V. NANOPOULOS, L. ROSZKOWSKI & D. N. SCHRAMM. 1990. Phys. Lett. **B245:** 251.
24. CADWELL, D., *et al.* 1988. Phys. Rev. Lett. **61:** 510.
25. LINDE, A. 1990. Particle Physics and Inflationary Cosmology, Contemporary Concepts in Physics, Vol. 5. Harwood Academic Press.
26. For a review, see OLIVE, K. A. 1990. Phys. Rep. 190, p. 307.
27. GUTH, A. H. 1981. Phys. Rev. **D23:** 347.
28. For reviews, see PRIMACK, J., D. SECKEL & B. SADULET. 1988. Annu. Rev. Nucl. Part. Sci. **38:** 751; SREDNICKI, M., Ed. 1989. Dark Matter. North-Holland. Amsterdam, the Netherlands.
29. HARRISON, E. R. 1970. Phys. Rev. **D1:** 2726; ZELDOVICH, YA. B. 1972. Mon. Not. R. Astron. Soc. **160:** 1P.
30. For reviews, see ARNOWITT, R. & P. NATH. 1983. Applied $N = 1$ Supergravity. World Scientific. Singapore; HABER, H. E. & G. L. KANE. 1985. Phys. Rep. **117:** 75; NILLES,

H. P. 1984. Phys. Rep. **110:** 1; LAHANAS, A. B. & D. V. NANOPOULOS. 1987. Phys. Rep. **145:** 1.
31. ELLIS, J., S. KELLEY & D. V. NANOPOULOS. 1990. Phys. Lett. **B260:** 131.
32. AMALDI, U., W. DE BOER & H. FÜRSTENAU. 1991. Phys. Lett. **B260:** 447.
33. ELLIS, J., D. V. NANOPOULOS, K. A. OLIVE & K. TAMVAKIS. 1982. Phys. Lett. **B118:** 335; ELLIS, J., D. V. NANOPOULOS, K. A. OLIVE & K. TAMVAKIS. 1983. Nucl. Phys. **B221:** 524.
34. ELLIS, J., J. S. HAGELIN, D. V. NANOPOULOS & K. A. OLIVE. 1984. Nucl. Phys. **B238:** 453.
35. LOPEZ, J. L., D. V. NANOPOULOS & K. YUAN. 1991. Texas A&M University Preprint CTP-TAMU-13/91, and references therein.
36. For a review, see SCHWARZ, J. H. 1989. Int. J. Mod. Phys. **A4:** 2653.
37. GROSS, D. J., J. HARVEY, E. MARTINEC & R. ROHM. 1985. Phys. Rev. Lett. **54:** 502; GROSS, D. J., J. HARVEY, E. MARTINEC & R. ROHM. 1985. Nucl. Phys. **B256:** 253; GROSS, D. J., J. HARVEY, E. MARTINEC & R. ROHM. Nucl. Phys. 1986. **B267:** 75.
38. ANTONIADIS, I., C. BACHAS & C. KOUNNAS. 1987. Nucl. Phys. **B289:** 87; ANTONIADIS, I. & C. BACHAS. 1988. Nucl. Phys. **B298:** 586; KAWAI, H., D. C. LEWELLEN & S. H.-H. TYE. 1986. Phys. Rev. Lett. **57:** 1832; KAWAI, H., D. C. LEWELLEN & S. H.-H. TYE. 1986. Phys. Rev. **D34:** 3794; KAWAI, H., D. C. LEWELLEN & S. H.-H. TYE. 1987. Nucl. Phys. **B288:** 1; BLUHM, R., L. DOLAN & P. GODDARD. 1988. Nucl. Phys. **B309:** 330; DREINER, H., J. L. LOPEZ, D. V. NANOPOULOS & D. REISS. 1989. Nucl. Phys. **B320:** 401.
39. KALARA, S., J. LOPEZ & D. V. NANOPOULOS. 1990. Phys. Lett. **B245:** 421; KALARA, S., J. LOPEZ & D. V. NANOPOULOS. 1991. Nucl. Phys. **B353:** 650.
40. BARR, S. 1982. Phys. Lett. **B112:** 219; DEREDINGER, J., J. KIM & D. V. NANOPOULOS. 1984. Phys. Lett. **B139:** 170; ANTONIADIS, I., J. ELLIS, J. S. HAGELIN & D. V. NANOPOULOS. 1987. Phys. Lett. **B194:** 231.
41. DREINER, H., J. LOPEZ, D. V. NANOPOULOS & D. REISS. 1989. Phys. Lett. **B216:** 283.
42. FARAGGI, A., D. V. NANOPOULOS & K. YUAN. 1990. Nucl. Phys. **B335:** 347.
43. FARAGGI, A. & K. YUAN. 1991. Texas A&M University, Ph.D. Theses.
44. ANTONIADIS, I., J. ELLIS, J. S. HAGELIN & D. V. NANOPOULOS. 1989. Phys. Lett. **B231:** 65.
45. LOPEZ, J. & D. V. NANOPOULOS. 1990. Nucl. Phys. **B338:** 73; LOPEZ, J. & D. V. NANOPOULOS. 1990. Phys. Lett. **B251:** 73.
46. RIZOS, J. & K. TAMVAKIS. 1990. Phys. Lett. **B251:** 369.
47. ELLIS, J., J. LOPEZ & D. V. NANOPOULOS. 1990. Phys. Lett. **B252:** 53.
48. LOPEZ, J. & D. V. NANOPOULOS. 1991. Phys. Lett. **B256:** 150; RIZOS, J. & K. TAMVAKIS. 1990. Phys. Lett. **B251:** 369.
49. ELLIS, J., J. LOPEZ & D. V. NANOPOULOS. 1990. Phys. Lett. **B245:** 375; ELLIS, J., J. LOPEZ & D. V. NANOPOULOS. 1990. Phys. Lett. **B247:** 257.
50. LEONTARIS, G., J. RIZOS & K. TAMVAKIS. 1990. Phys. Lett. **B243:** 220.
51. ELLIS, J., G. GELMINI, J. LOPEZ, D. V. NANOPOULOS & S. SARKAR. 1991. CERN Preprint CERN-TH-5853/90.
52. For a review, see STROMINGER, A. Baby Universes. 1988. Lectures Presented at the TASI Summer School, Brown University, Providence, R.I., June 1988.
53. HAWKING, S. W. 1976. Phys. Rev. **D14:** 2460; HAWKING, S. W. 1982. Commun. Math. Phys. **87:** 395; ELLIS, J., J. S. HAGELIN, D. V. NANOPOULOS & M. SREDNICKI. 1983. Nucl. Phys. **B241:** 381; ELLIS, J., S. MOHANTY & D. V. NANOPOULOS. 1989. Phys. Lett. **221B:** 113; ELLIS, J., S. MOHANTY & D. V. NANOPOULOS. 1990. **235B:** 305.
54. ANTONIADIS, I., C. BACHAS, J. ELLIS & D. V. NANOPOULOS. 1988. Phys. Lett. **B211:** 393; ANTONIADIS, I., C. BACHAS, J. ELLIS & D. V. NANOPOULOS. 1989. Nucl. Phys. **B328:** 117.
55. MYERS, R. C. 1987. Phys. Lett. **B199:** 37.
56. MUELLER, M. 1990. Nucl. Phys. **B337:** 37.
57. FEIGIN, B. L. & D. B. FUCHS. 1983. Moscow Preprint; DOTSENKO, V. & V. FATEEV. 1984. Nucl. Phys. **B240:** 319; DOTSENKO, V. & V. FATEEV. 1985. **B251:** 691.
58. WITTEN, E. 1984. Commun. Math. Phys. **92:** 451; KNIZHNIK, V. & A. B. ZAMOLODCHIKOV. 1984. Nucl. Phys. **B278:** 83; GEPNER, D. & E. WITTEN. 1986. Nucl. Phys. **B278:** 493.
59. ANTONIADIS, I., C. BACHAS, J. ELLIS & D. V. NANOPOULOS. 1991. Phys. Lett. **B257:** 278.
60. GLIOZZI, F., J. SCHERK & D. OLIVE. 1976. Phys. Lett. **B65:** 282; GLIOZZI, F., J. SCHERK & D. OLIVE. 1977. Nucl. Phys. **B122:** 253.
61. ZAMOLODCHIKOV, A. B. 1986. JETP Lett. **43:** 731.

62. GERVAIS, J. L. & A. NEVEU. 1985. Nucl. Phys. **B257:** 59; GERVAIS, J. L. & A. NEVEU. 1986. **B264:** 557.
63. POLYAKOV, A. M. 1981. Phys. Lett. **B103:** 207; POLYAKOV, A. M. 1987. Mod. Phys. Lett. **A2:** 899; KNIZHNIK, V. F., A. M. POLYAKOV & A. A. ZAMOLODCHIKOV. 1988. Mod. Phys. Lett. **A3:** 812; DAVID, F. 1988. Mod. Phys. Lett. **A3:** 1651; DISTLER, J. & H. KAWAI. 1989. Nucl. Phys. **B321:** 509.
64. POLCHINSKI, J. 1989. Nucl. Phys. **B324:** 123; DAS, S. R., S. NAIK & S. R. WADIA. 1989. Mod. Phys. Lett. **A4:** 1033.
65. CALLAN, C. G., R. C. MYERS & M. J. PERRY. 1988. Nucl. Phys. **B311:** 673; HOOFT, G.'T. 1990. Nucl. Phys. **B335:** 138, and references therein.
66. HAWKING, S. W. 1975. Commun. Math. Phys. **43:** 199; BEKENSTEIN, J. D. 1973. Phys. Rev. **D7:** 2333.
67. WALD, R. M. 1984. General Relativity. University of Chicago Press. Chicago.
68. KALARA, S. & D. V. NANOPOULOS. 1991. Phys. Lett. **B267:** 343.
69. KRAUSS, L. M. & F. WILCZEK. 1989. Phys. Rev. Lett. **62:** 1221; PRESKILL, J. & L. M. KRAUSS. 1990. Nucl. Phys. **B341:** 50.
70. For a review, see PRESKILL, J. 1991. Ann. Phys. **210:** 323.
71. AHARONOV, Y. & D. BOHM. 1959. Phys. Rev. **115:** 485.
72. DABHOLKAR, A. & J. HARVEY. 1989. Phys. Rev. Lett. **63:** 719; DABHOLKAR, A., G. GIBBONS, J. HARVEY & F. R. RUIZ. 1990. Nucl. Phys. **B340:** 33.
73. STROMINGER, A. 1990. Nucl. Phys. **B343:** 167.
74. HOROWITZ, G. T. & A. STROMINGER. 1991. Nucl. Phys. **B360:** 197.
75. CALLAN, C., J. HARVEY & A. STROMINGER. 1991. Princeton University Preprint PUPT-1233.
76. DUFF, M. J. & J. X. LU. 1991. Nucl. Phys. **B354:** 141; DUFF, M. J. & J. X. LU. Nucl. Phys. **B357:** 534.
77. DUFF, M. J. 1988. Cl. Quantum Gravity **5:** 189; DUFF, M. J. & J. X. LU. 1991. Nucl. Phys. **B354:** 129.
78. MONTONEN, C. & D. OLIVE. 1977. Phys. Lett. **B72:** 117.
79. WITTEN, E. & D. OLIVE. 1978. Phys. Lett. **B78:** 37; OSBORN, H. 1979. Phys. Lett. **B83:** 321.
80. For reviews, see SCHWARZ, J. H. 1989 and 1990. Caltech Preprints CALT-68-1581 and CALT-68-1609; FERRARA, S. 1989. CERN Preprint CERN-TH-5293/89.
81. HAGEDORN, R. 1965. Nuovo Cimento Supp. **3:** 147; HUANG, K. & S. WEINBERG. 1970. Phys. Rev. Lett. **25:** 895; ANTONIADIS, I., J. ELLIS & D. V. NANOPOULOS. 1987. Phys. Lett. **B199:** 402; AXENIDES, M., S. D. ELLIS & C. KOUNNAS. 1988. Phys. Rev. **D37:** 2964.

LEP, the Laboratory for Electrostrong Physics, One Year Later

UGO AMALDI
CERN
Geneva, Switzerland

INTRODUCTION

In August 1989, LEP, the CERN electron–positron collider, produced the first events in which Z-bosons were created at rest and then decayed. In September the energy of the machine was varied in steps of 1 GeV around the value of 45.5 GeV, and data were collected by the four detectors: ALEPH, DELPHI, L3, and OPAL. In October 1989, the determination of the number of neutrino species, as obtained from a two-parameter fit to the Z-resonance, was announced in a public meeting

$$N_\nu = 3.15 \pm 0.20. \qquad (1)$$

The results of the four Collaborations, which have been averaged in **(1)**, were published immediately afterwards.[1]

In 1990 the luminosity of LEP increased up to a maximum value of

$$L \simeq 6\, 10^{30}\,\mathrm{cm}^{-2}\,\mathrm{s}^{-1}, \qquad (2)$$

and beams have been circulating in the 27-km-long storage ring for up to a day. At such a luminosity, a Z-boson is produced about every five seconds. By the end of 1990, the four Collaborations had collected a total of 700,000 events, somewhat unevenly distributed because, due to a machine asymmetry discovered too late in the year, in the interaction regions occupied by L3 and DELPHI the luminosity was ~ 25 percent lower than in the regions where ALEPH and OPAL are installed. The many results obtained in these first runs have been published in about one hundred reports and were reviewed recently for the Singapore Conference.[2]

Seen from far away, *two* results stand out: the accurate measurement of the number of neutrino species and the fact that no standard Higgs particle (to be called "higgson" in the following) was found in the mass range explored until now ($0 < M \lesssim 50$ GeV). In the following two sections these two results, and only a few others, are described. In the fourth and fifth sections I underline a *third* result that, in my mind, is not less important, but was neglected until now: by using as input the LEP precision results on the electroweak and strong coupling constants (discussed in the second and fourth sections), it is now possible to perform meaningful consistency checks of the grand unified theories (GUT). In this way some interesting conclusions can be drawn on the physics to be expected at energies ten times larger than the one available at present at LEP.

AROUND THE Z-PEAK

The four LEP detectors have been described many times, but the interested reader can find all the necessary details in reference 3.

FIGURE 1 shows, as an example, DELPHI, the experiment in which I am personally involved. ALEPH and OPAL have similar external dimensions ($10 \times 10 \times 10$ m^3), while L3 is much larger because the penetrating muons are measured in an air solenoidal magnet having an inner diameter of about 20 m.

The four detectors can easily distinguish the final states that are of importance for the measurement of the parameters of the shape of the Z-resonance. The final states are either *leptonic* (e^+e^-, $\mu^+\mu^-$, $\tau^+\tau^-$) or *hadronic* (which are due to the Z decaying in a quark–antiquark pair). Leptonic decays are characterized by the very small number of charged tracks: typically two for the e^+e^- and $\mu^+\mu^-$ channels, and two or four in the $\tau^+\tau^-$ channel. The hadronic events contain on average twenty charged hadrons grouped in two or more jets.

The main parameters of the Z-boson are its mass M_Z and its partial decay widths in the various channels, which sum up to the total width

$$\Gamma_Z = \Gamma_e + \Gamma_\mu + \Gamma_\tau + \Gamma_h + \Gamma_\nu + \Gamma_{\text{unknown}}. \quad (3)$$

The notations are self-evident. Γ_ν is, in particular, the partial width for the Z-boson to decay in the different types of "light" neutrino–antineutrino pairs, that is, neutrinos having much smaller mass than $M_Z/2$, so that energy conservation does not prohibit the decay. Before LEP three types of neutrinos were known (ν_e, ν_μ, and ν_τ) and the important question was: Are other (light) neutrinos produced in Z-decays so as to increase proportionally Γ_ν? Moreover, any other decay process not present in the generally accepted standard model (SM) of the weak, electromagnetic, and strong interactions, would contribute to Γ_{unknown}.

Once M_Z is measured, the standard model predicts values for all the other quantities, some of them with very high accuracy, other less precisely, mainly because other not well-known parameters intervene. Among them the *two* most important ones are the mass M_t of the yet undiscovered top quark and the mass M_h of the neutral higgson h^0. In fact, the dependence is such that, by using precision measurements, one can reverse the argument and get from the data useful information on the top mass and eventually, when many million events have been collected, on M_h.

By "scanning," that is, by varying the energy of LEP around the value at which the event rate is maximum because of the resonance production of Z-bosons, the four Collaborations have measured the energy dependence of the cross sections for the production of hadrons (due to quark–antiquark creation) and of leptonic pairs. Some of the results of such a scanning are shown in FIGURE 2. The measured cross sections are plotted on the vertical axis, and the energy, which is computed from the accurately known strength of the magnetic field bending the particles in LEP, on the horizontal one. To transform an observed event rate into a cross section, a detector has to register *at the same time* the rate of a process that has a well-known cross section, so that the integrated luminosity can be measured. Electron–positron scattering events at angles around 1° are used for this purpose. In fact, at such small angles the cross section for these so-called "Bhabha scattering" events can be

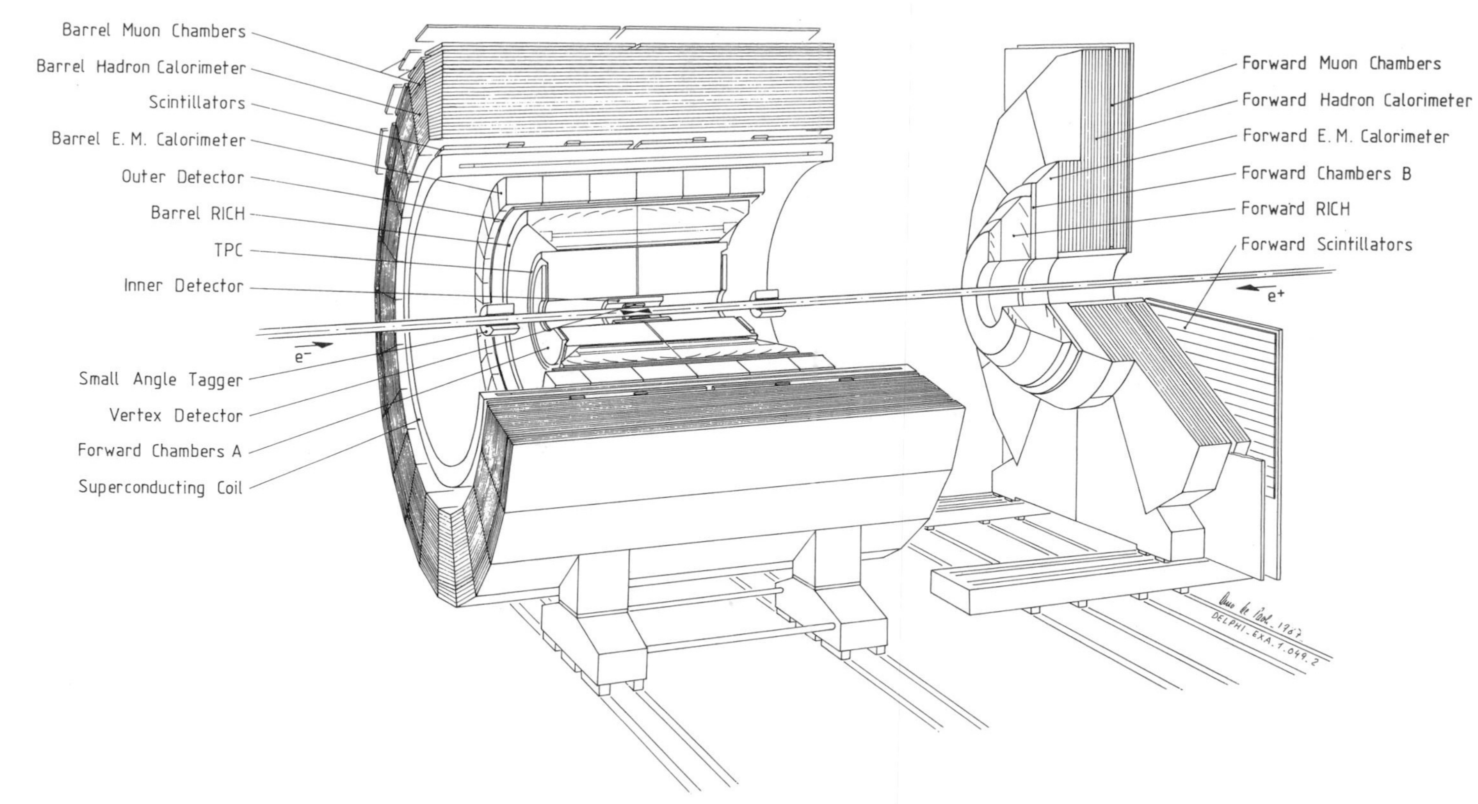

FIGURE 1. Four detectors are taking data at LEP: ALEPH, DELPHI, L3, and OPAL. Here one of them is shown. Its overall height is about 12 m. The diameter of the tube in which the electrons and positrons circulate is 10 cm. Around it many subdetectors of cylindrical shape detect and tag the charged and neutral particles produced in the decay of the *Z*-bosons, which are created at the center of the detector by the annihilations of electron–positron pairs.

computed with quantum electrodynamics (QED) to accuracies that are better than 1 percent. The first great (positive) surprise is that the monitoring systems of the LEP experiments, designed for ~2 percent systematic accuracy, have been working much better than foreseen. Precisions in the range of 0.5–0.8 percent were achieved after a few months of running. This technical prowess is at the origin of the incredible accuracy with which the number of neutrinos have been determined in a short time.

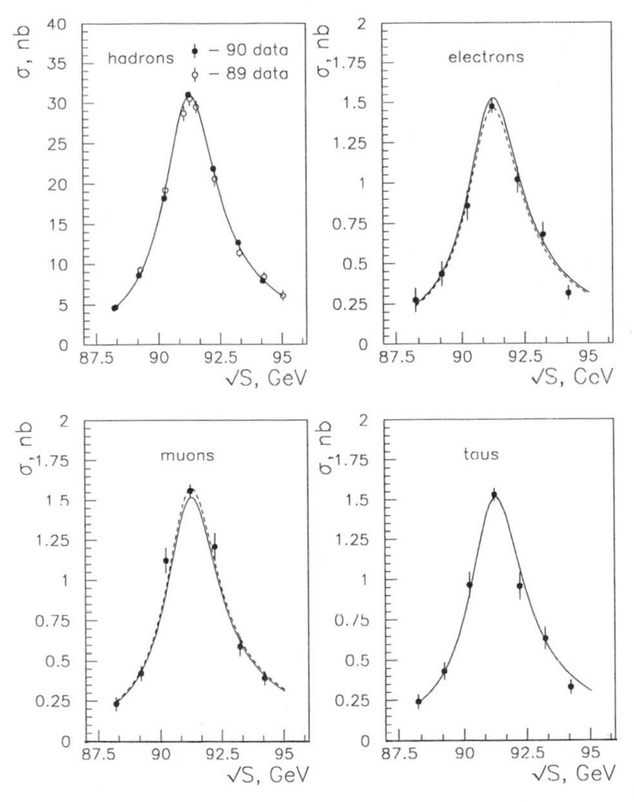

FIGURE 2. Typical results of the energy "scanning." The different figures correspond to different decay channels of the Z-boson. Each one has the characteristic bell shape of a resonant phenomenon of width Γ_Z. The peak cross sections are proportional to the quantity $\Gamma_e\Gamma_i/\Gamma_Z$, where Γ_i is the partial width for the decay in the ith channel. With a fit to all the cross sections one obtains M_Z, Γ_Z, Γ_h, Γ_e, Γ_μ and Γ_τ.

FIGURES 3 to 6 summarize some of the results obtained in the 1989 and 1990 runs by the four Collaborations.[4] For each quantity the horizontal bar represents the final result presented by the Collaboration, after unfolding the systematic error that is common to the four measurements. The weighted average and the χ^2/(degree of freedom) of the four experimental results are computed by using these errors. The average value and its error are represented on the line below, (AVG.W.COM.SYST.ERROR = average with common systematic error); here the

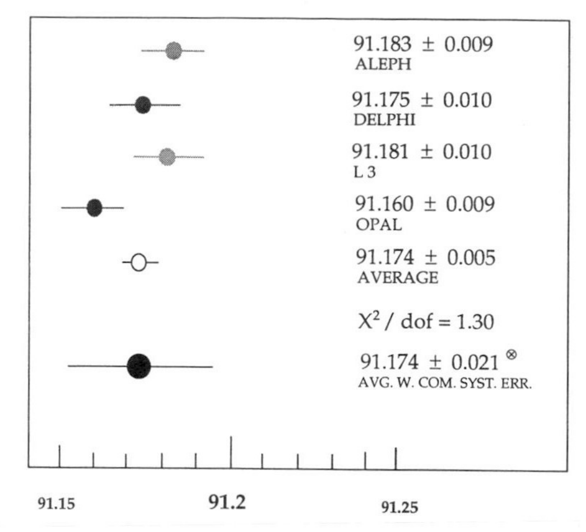

FIGURE 3. Summary of 1989–1990 results obtained by the four experiments on the mass M_Z of the Z-particle. See the text of the explanation of the error treatment. *Note:* ⊗ ± 20 MeV from LEP energy.

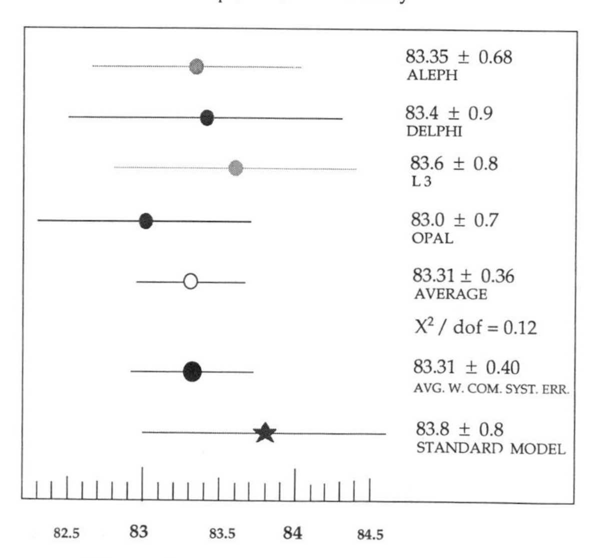

FIGURE 4. Summary of the results on the partial leptonic width Γ_l. Universality of the leptonic couplings is assumed.

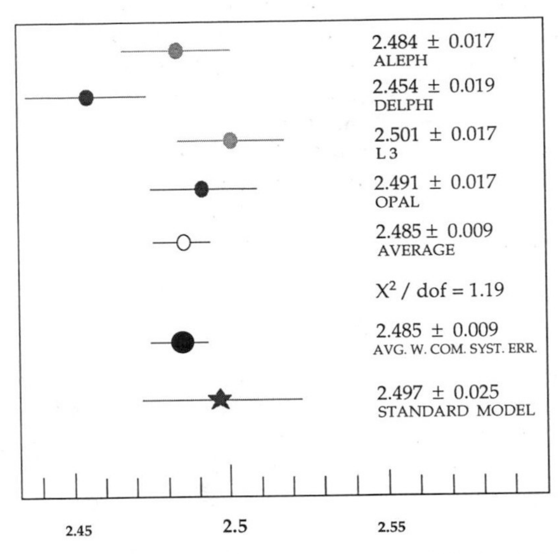

FIGURE 5. Summary of the four measurements of the total width Γ_Z.

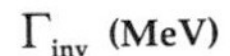

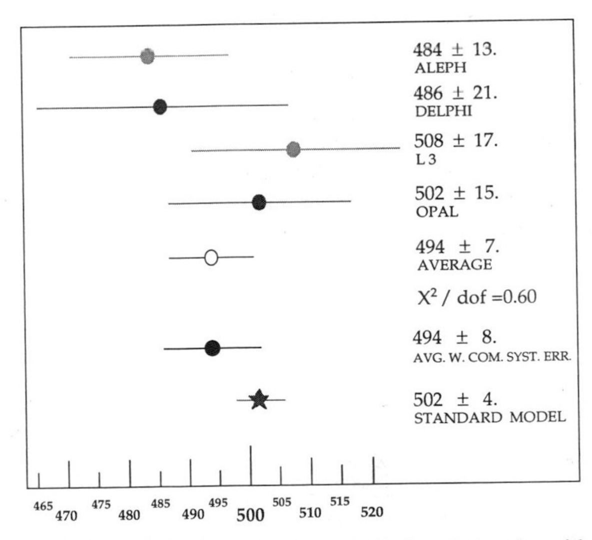

FIGURE 6. The invisible width Γ_{inv} is obtained essentially by substraction with the assumption $\Gamma_{unknown} = 0$ [see (**3**)].

common systematic error is reintroduced. (A typical error of this kind is the ~0.5 percent error in the luminosity due to the theoretical calculation of the Bhabha cross section previously mentioned.) The lowest line in each figure is the expectation of the standard model. For the mass of the Z–particle, this entry does not exist because it is an input to the standard model and, at present, only experiment can tell what it is.

The most striking feature of these results, and of the many others that are not shown, is the agreement with the standard model, once the value of M_Z is known. The partial width Γ_1 represented in FIGURE 4 is obtained by assuming universality of the leptonic couplings, that is, by assuming $\Gamma_1 = \Gamma_e = \Gamma_\mu = \Gamma_\tau$. From the average value of Γ_1, and other quantities, one can obtain a very accurate number for a most important parameter of the model, the *weak mixing angle.* One gets at present[4]

$$\sin^2 \Theta(M_Z) = 0.2320 \pm 0.0010, \tag{4}$$

where the exact definition of $\sin^2 \Theta\ (M_Z)$ is somewhat complicated. With a certain degree of approximation one can say that Θ is an angle such that $\cos \Theta = M_W/M_Z$, where M_W is the mass of the charged weak boson W. Note that before LEP the error on $\sin^2 \Theta$ was about *five* times larger. Combining this result with the outputs of lower energy experiments (in particular on neutrino scattering and direct measurements of M_W) one can obtain, in the framework of the SM, an estimate of the top mass: $M_t = (130 \pm 35)$ GeV.

The total width Γ_Z of the Z-boson is about 2.5 GeV, which means that, according to the indetermination principle, a still-standing Z-particle, once produced in LEP, lives about 10^{-24} sec. FIGURE 6 shows the value of the invisible width Γ_{inv}, which, according to **(3)**, equals the sum $\Gamma_\nu + \Gamma_{\text{unknown}}$. By making the SM hypothesis that $\Gamma_{\text{unknown}} = 0$, one can divide the measured invisible width by the SM expectation for the width of Z into *one* (light) neutrino–antineutrino pair and obtain the number of (light) neutrino species in a three-parameter fit:

$$N_\nu = 2.98 \pm 0.05. \tag{5}$$

By comparing **(1)** and **(5)**, it appears that in one year of running the error was reduced by a factor of 4. By now the error is dominated by systematics effects, and in particular by the accuracy in the determination of the Bhabha scattering rate. Further improvements are possible with more statistics and better control of the systematics. By hard work in the coming years the error of **(5)** will be divided by a factor of 2 to 3.

SEARCHES FOR NEW PARTICLES

Searching for new particles is a must at the opening of a new energy frontier. LEP is no exception, with the additional advantage that the annihilation of an electron–positron pair is a very clear process. The approximately 90 GeV liberated in the annihilation are transferred in a resonant way uniquely to the Z-field. Moreover, the produced Z-boson couples democratically to all known and, presumably, also unknown particles, so that the search for new particles is limited only by the energy conservation law. For this reason, in about one year the four experiments have been able to exclude the existence of most of the particles—imagined by theorists—that, coupled to the Z-boson, have a mass that goes from zero to the kinematical limit of about $M_Z/2 \simeq 45$ GeV. The list of exclusions is very long: new neutrinos, new quarks, heavy leptons, supersymmetric partners of the known matter particles (as sleptons and squarks), leptoquarks, and so on. Here I discuss only the unsuccessful searches for neutral higgson(s), the spin zero boson(s) supposed to be responsible for the fact that both the matter-particles (leptons and quarks) and some of the force-particles (the intermediate bosons W and Z) have a mass different from zero.

Only the neutral higgson h^0 should exist in the minimal SM of the fundamental interactions, which is based on the gauge group $SU(3) \times SU(2) \times U(1)$ and contains *one* Higgs doublet that gives mass to all the particles. Before LEP, one had doubtful limits for the existence of the standard higgson with mass smaller than about 5 GeV. Nowadays we know for sure that a neutral h-boson with mass smaller than about 50 GeV does *not* exist. This is enormous progress, even if many would have liked to see the discovery of such a fundamental particle. I want to remind those who think that because of this lack of discoveries, "nothing new came from LEP," of the popperian view of science: advances come from showing the established theories to be *false*. From this point of view, the negative results of about one year of running at LEP, by showing a number of possible theoretical models to be false, have greatly restricted the number of still open possibilities.

The lower limits (at 95 percent confidence level) on the mass M_h of the standard model higgson obtained by the four experiments are:[5] 48 GeV (ALEPH), 40 GeV (DELPHI), 41.8 GeV (L3), 45 GeV (OPAL). Combining limits is not as easy as combining measured numbers. Still, the preceding results, taken together, certainly exclude the possibility, at a 95 percent confidence level, that the h^0-boson of the SM has a mass smaller than 50 GeV.

The better limits achieved by ALEPH and OPAL are due to a combination of higher statistics (see the first section) with the hermeticity of the detectors. This is an essential aspect in these kinds of experiments, which are based on the search for the decay of the Z-boson into a higgson h^0 and a pair of leptons. The higgson, once produced, decays immediately in a quark–antiquark pair, which appears as two jets of hadrons, so that one is looking for events that contain *two* leptons and *two* jets.

Before LEP start-up, it was thought that the only reactions that could be used to put limits on the production of Higgs particles would be those in which either a pair of electrons or a pair of muons are produced together with the two jets coming from the higgson. The hermeticity of the LEP detectors has instead allowed the use of the much more frequent reaction in which the higgson is produced together with an invisible neutrino–antineutrino pair. In this case the process is characterized by a large missing energy, carried away by neutrinos, and it is very important to use detectors that—being "hermetic"—are able to accurately detect all the energy of the produced particles (apart from neutrinos). This has been the second (positive) surprise: the results have surpassed the most optimistic expectations. Beforehand it was thought that each experiment had to collect more than 10^6 Z decays to put a limit as large as $M_h \geq 45$ GeV. By using the neutrino–antineutrino channel, this limit has been achieved with less than 200,000 events. In the next couple of years with more statistics the limit should be pushed to at least 60 GeV.

In the minimal supersymmetric (SUSY) model the spectrum of the higgsons is richer[6] than in the minimal SM: there are three neutral higgsons (usually indicated with the symbols h^0, H^0, and A^0) and a pair of charged Higgs particles. This comes from the fact that the interaction with the Higgs fields has to give masses both to the matter-particles (leptons and quarks) and to their supersymmetric particles, the "sparticles," so that *two* Higgs doublets have to be introduced. The symmetry between particles and sparticles of all SUSY models has the much needed consequence of curing the divergences that push the mass of the only higgson of the SM to unreasonably large values. Moreover SUSY has many other appealing features like, for example, its compatibility with theories of quantum gravity. For these reasons, minimal SUSY is *the model* that subatomic physicists would really like to falsify. For the moment, LEP data have been able to exclude only a limited, though nonnegligible, fraction of its parameter space.

There are *two* independent parameters in the neutral Higgs sector of the minimal SUSY model: often the masses M_h and M_A of the lightest scalar and pseudoscalar higgsons are used. To lowest order the following very important inequalities hold:[7]

$$M_h \leq M_Z, \tag{6}$$

$$M_H \geq M_Z, \tag{7}$$

$$M_h \leq M_A \leq M_H, \tag{8}$$

which, however, are modified by recently calculated higher order corrections.[8] Before these calculations, **(6)** appeared to be *the equation* that would allow the minimal SUSY model to be proved false. Indeed, in the years 1994–1995 LEP will be upgraded from a total center-of-mass energy of ~100 GeV (LEP 100) to ~180–190 GeV, within the so-called "LEP 200" program. At energies not smaller than 190 GeV a higgson h^0 of mass ≤ 90 GeV would be seen either in the already discussed neutrino channel ($h^0\nu\bar{\nu}$) *or* in the complementary process (h^0A^0), which has four jets in the final state. The nonobservation of the h particle in the mass range up to M_Z would prove the minimal SUSY to be *wrong* if and only if **(6)** is valid. Higher order corrections smear this sharp prediction[8] and will force us to continue in the painful work of gradually eroding the parameter space of the model without coming to a firm conclusion with LEP 200. Maybe this will require an upgraded version of LEP, to be called LEP240!

The present status of the excluded region is such that the lightest higgson of the minimal SUSY model has a mass $M_h \geq 40$ GeV at a 95 percent confidence level and that, similarly, $M_A \geq 40$ GeV.[2] It is still clear that we have a long way to go before excluding masses as large as 90 GeV.

MEASUREMENTS OF THE STRONG COUPLING CONSTANT

During the first year of LEP, the third positive surprise came from the relative easiness with which one could measure the quark–gluon coupling constant, usually expressed as α_s, the equivalent of the fine structure constant $\alpha = 1/137$ of electromagnetism. Unfortunately, the subject is involved and we can only sketch the arguments here.

FIGURE 7 describes the electroweak creation of a quark–antiquark pair in the decay of a Z-boson, followed by the radiation of hard gluons, and then—at a distance of less than 10^{-13} cm—by the fragmentation of the virtual partons (quarks and gluons) and hadronization, which lead to the hadrons observed in the detector. The cascade of processes is very complicated and it has to be well understood to obtain, from the measured final states, the coupling constant α_s that determines the initial radiation of gluons. A further complication is due to the fact that the strong coupling constant of quantum chromodynamics (QCD) α_s is not at all constant but "runs," that is, has a value that depends on the energy scale at which the process takes place[9]

$$\alpha_s = \alpha_s(\mu). \tag{9}$$

The larger the energy scale μ, the smaller the distance at which the coupling constant $\alpha_s(\mu)$ is measured, so that $\alpha_s(\mu)$ is roughly the coupling constant that is effective at a distance $\hbar/\mu$.

The inverse of the coupling constant $\alpha_s(\mu)^{-1}$ has a simple physical meaning: qualitatively it equals the number of times one has to observe a quark before finding, close to it, a gluon of energy larger than μ. Strong interactions are called "strong"

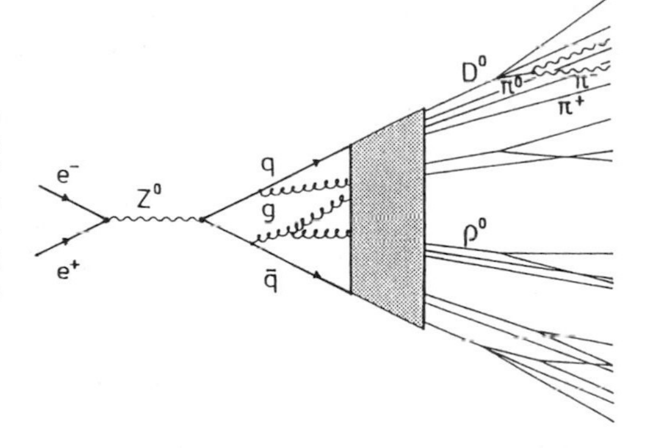

FIGURE 7. Representation of the creation of a quark–antiquark pair in the decay of a Z-boson. The virtual "hard" particles (q and $\bar{q}$) radiate virtual gluons (g) and quarks, which then fragment and hadronize giving eventual rise to the many hadrons observed in the detector.

because $\alpha_s^{-1} \approx \frac{1}{2}$ at $\mu \approx 0.5$ GeV, so that at distances of the order of 10^{-13} cm there is no such a thing as a "bare" quark.

The main property of QCD is "asymptotic freedom," which is the prediction that, due to virtual effects, $\alpha_s(\mu)$ decreases logarithmically when the scale μ increases, so that the radiation of gluons by quarks is less probable at small distances than at large distances. QCD is based on the gauge group $SU(3)$ of colored quarks and gluons and the running of the strong coupling constant can be computed with perturbative methods. For a given value of μ, the quarks that contribute to the running of the coupling are the "light" ones, that is, those that have masses much smaller than μ. The number f of relevant quarks is thus a function of μ. For scales $\mu \lesssim 5$ GeV (below the beauty quark mass $M_b = 5$ GeV), $f = 4$ because the only active quarks are u, d, s, and c. At LEP ($\mu \simeq 100$ GeV) the b quark also contributes and $f = 5$. In general, in this energy range the asymptotic first-order expression of the coupling constant is

$$\alpha_s(\mu) = \frac{12\pi}{(33 - 2f)\ln(\mu/\Lambda)^2}, \tag{10}$$

where $f = 5$ and Λ is the QCD parameter that has to be obtained from experiment. Its numerical value depends on the number of active quarks and on the scheme adopted in the definition of the coupling constant. The usual prescription[10] is the "modified minimal substraction scheme" ($\overline{MS}$) so that, with $f = 5$, a better symbol for the parameter Λ appearing in (**10**) is $\Lambda^{(5)}_{\overline{MS}}$.

Before the start-up of LEP, Guido Altarelli[9] summarized all the lower energy data on $\alpha_s(\mu)$, coming to the conclusion that $\Lambda^{(5)}_{\overline{MS}} = (140 \pm 60)$ MeV so that

> ... the prediction for α_s to be measured at LEP is very precise: $\alpha_s(\mu = M_Z) \simeq 0.11 \pm 0.01$. Establishing that this prediction is experimentally true would be a very quantitative and accurate test of QCD, conceptually equivalent but more reasonable than trying to see the running (of α_s) in a given experiment.

To obtain $\alpha_s(M_Z)$ the main methods are based on the measurement of a quantity that is proportional to α_s, and thus is zero in lowest order (corresponding to a quark–antiquark pair in the final state). Such a quantity is sensitive directly to the quark–antiquark–gluon coupling. To determine α_s it is necessary to also know the same quantity to the next order (α_s^2). The most widely used quantities at lower energies and at LEP are the fraction (R_3) of events containing three jets (the so-called *three-jet rate*) and the asymmetry of the energy–energy correlation (AEEC), which compares the width of a quark jet with the width of the combination of a quark and a gluon jet and, of course, it is zero if there is no gluon radiation.

The four LEP Collaborations have published by now a number of papers on the determination of $\alpha_s(M_Z)$ with these and other methods. The overall conclusion is that the value (0.11 ± 0.01) certainly covers the totality of the results. However, there are qualifications that require a long and detailed discussion and divide the experts. For instance, the final results of two recent reviews[11] can be summarized by the following two numbers:

$$\text{W. de Boer:} \quad \alpha_s(M_Z) = 0.107 \pm 0.007,$$

$$\text{S. Bethke:} \quad \alpha_s(M_Z) = 0.118 \pm 0.008. \tag{11}$$

Of the many ingredients that, according to the image of FIGURE 7, enter in the determination of α_s, the most uncertain one is the choice of the scale μ at which the hard radiation processes have to be computed. In fact, at LEP the number of events is large and statistics are not a problem. Even the nonperturbative fragmentation and hadronization processes taking place in the grey box of FIGURE 7 at these high energies can be kept under control, and give errors on $\alpha_s(M_Z)$, which are typically 3 percent.

As far as the choice of the scale μ, to be introduced in the second-order matrix element describing the hard radiation processes is concerned, because only α_s^2 calculations exists and there is no firm prescription, the rule of the game has been to vary it between $M_b \simeq 5$ GeV, the mass of the b-quark, and M_Z. A theoretical "scale" error is thus appended to the output value of $\alpha_s(M_Z)$, which turns out to be of the order of ± 0.010 for the three jet rates (R_3), but is smaller for the AEEC, because in that case the second-order corrections are less important. One of the main arguments of debate is whether this is an indication of the fact that the AEEC is a better quantity than R_3 from which to derive $\alpha_s(M_Z)$. The difference of opinion reflects itself, together with a somewhat different choice of data to average, in the difference between the central values of **(11)**. However, the quoted errors—which are a mixture of experimental and of (often maximal) theoretical errors and thus have *not* the usual meaning of a 68 percent confidence level—are large enough to cover it.

In conclusion, all measured values agree with the extrapolation from lower energies, thus confirming QCD predictions. Moreover, they also agree with the most recent analysis of *all* deep inelastic data[12]

$$\alpha_s(M_Z) = 0.109 \begin{cases} +0.004, \\ -0.005. \end{cases} \qquad \textbf{(12)}$$

A LEP VIEW OF GRAND UNIFIED THEORIES

LEP has been seen since its inception as a laboratory for very accurate measurements of electroweak parameters. It has been often called "*the (g-2) of electroweak interactions,*" and the accumulation of statistics will make this statement even more true: during 1992, the number of bunches will pass from 4 to 8 and, by the end of 1993, each Collaboration should have collected about five 10^6 Zs, with a potential decrease of the statistical errors by a factor of 5. Afterward LEP will run at higher energy (LEP 200), and with polarized beams it will give further accurate information on the electroweak parameters, such as $\sin^2 \Theta$, the mass M_W of the W-boson (to less than 100 MeV), and the never observed Z–W–W coupling. Toward the end of the century, the number of electron and positron bunches could be increased from 8 to 18, with a further increase of luminosity: a total of 25 10^6 Zs could be collected in a couple of years. This will open the way to, for instance, studies of $B\overline{B}$ oscillations that cannot be performed with the dedicated collider called "threshold beauty factories."

By combining the electroweak data with accurate measurements of the strong coupling constant $\alpha_s(M_Z)$, one can even glance at what happens at energies larger than that available at LEP, and perform meaningful consistency checks of the so-called grand unified theories (GUT). (The name is pompous: unified theories would be a better one.) These theories reduce the number of independent coupling constants by embedding the $SU(3) \otimes SU(2) \otimes U(1)$ group of the standard model into a group G, in which there is only one underlying gauge coupling constant at the unification scale. If at the scale M_{GUT} the group G is broken into the three independent gauge groups, the evolution of the corresponding coupling constants with the scale μ depends only on the particle content of the $SU(3) \otimes SU(2) \otimes U(1)$ model, but not on the structure of the group G. By evolving the coupling constants, as measured at low energies, toward high energies one can check which particle content is consistent with unification.

The starting point of the analysis of Amaldi *et al.*[13] is the work performed in 1987 in Amaldi *et al.*[14] Previous approaches to the same problem using LEP data have been done by Paul Langacker[15] and John Ellis *et al.*[16]

One of the results of Amaldi *et al.*[14] is summarized in FIGURE 8(a) [14, fig. 11], where the *inverses* of the standard model couplings are plotted versus the energy. The

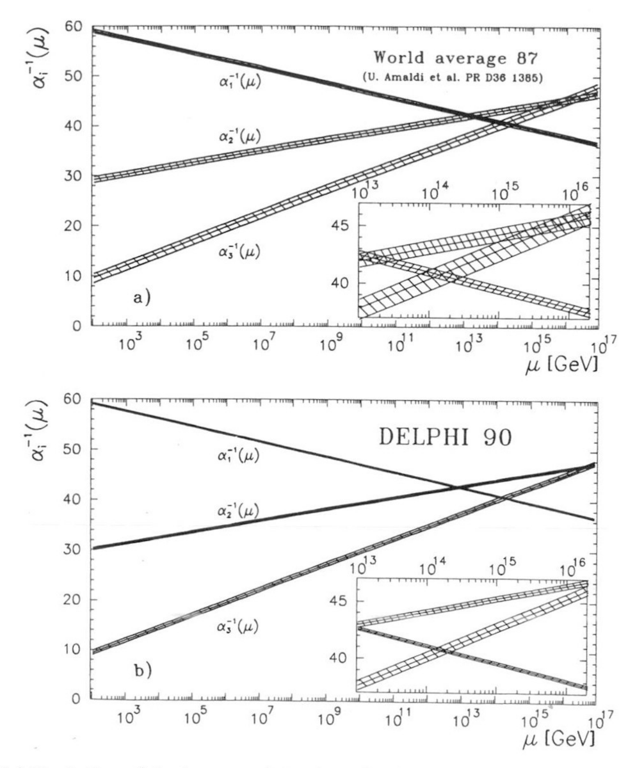

FIGURE 8. **(a)** Evolution of the inverse of the three fundamental coupling constants α_1, α_2, and α_3 in the minimal standard model. Input data are the world averages known in 1987.[14] **(b)** As in **(a)**, but using the data from one of the LEP experiments.

usual definitions of the coupling constants are

$$\alpha_1(\mu) = \frac{5\alpha(\mu)}{3\cos^2\Theta},$$

$$\alpha_2(\mu) = \frac{\alpha(\mu)}{\sin^2\Theta},$$

$$\alpha_3(\mu) = \alpha_s(\mu), \qquad \textbf{(13)}$$

where $\alpha(\mu)$ is the (running) fine structure constant, which for $\mu = 0$ has the well-known value $\alpha(0)^{-1} = 137$ and for $\mu = M_Z$ is $\alpha(M_Z)^{-1} = 127.9$.

According to the GUT hypothesis, there is an energy M_{GUT} such that $\alpha_1(M_{\text{GUT}}) = \alpha_2(M_{\text{GUT}}) = \alpha_3(M_{\text{GUT}})$. From **(13)** one obtains that $\sin^2\Theta(M_{\text{GUT}}) = 3/8 \simeq 0.375$ at $\mu = M_{\text{GUT}}$. The energy dependence of the coupling constant is completely determined by the particle content of each theory, and the evolution equations are known[17] up to second order in α_i. FIGURE 8(a) was computed in 1987 using for the values of α_1, α_2, and α_3 at $\sim$100 GeV the averages of *all* the available data, and making the assumptions of the minimal standard model with three families of matter and one Higgs doublet (which give rise to a single standard neutral higgson h^0). Clearly, around 10^{15} GeV all couplings become of the same order of magnitude, but they do not meet exactly in a single unification point. The deviation from perfect crossing, as expected in the simplest unified theories, was only two standard deviations in 1987. It is now possible to perform the same analysis with data from LEP.

Either the results from a single experiment or average values can be used. In FIGURE 8(b) the evolutions of the three coupling constants are shown with the published data of the DELPHI Collaboration; again the minimal standard model with three families and one Higgs doublet is assumed.[13] Compared to the results of 1987, the errors, indicated by the width of the lines, are considerably smaller. It is clear that a single unification point cannot be obtained within the present errors: the α_3 coupling constant misses the crossing point of the other two by more than 7 standard deviations. Even if one would neglect the nonunification and take the center of the triangle, there is a problem because the unification scale would be $M_{\text{GUT}} \simeq 10^{15}$ GeV. The proton lifetime is determined by M_{GUT}, since one expects the new heavy gauge bosons (X), which in GUT can transform a quark in a lepton, to have masses that are typically $M_X \simeq M_{\text{GUT}}$. If the proton decay is dominated by X-boson exchange, the proton lifetime for $M_X = 10^{15}$ GeV can be estimated to be

$$\tau_{\text{proton}} \approx \frac{1}{\alpha_{\text{GUT}}^2} \frac{M_X^4}{M_P^5} \approx 10^{31} \text{ y}, \tag{14}$$

where M_P is the proton mass. Such a proton lifetime does contradict the present experimental limits of $\tau_{\text{proton}} \geq 5.5\ 10^{32}$ years.[18]

Therefore, both the nonobservation of proton decay and the nonunification of the coupling constants independently rule out any minimal GUT, which breaks to the standard model below the unification point. In the framework of GUT this nonunification implies new physics. Such new physics could come from supersymmetry (SUSY), which introduces, as discussed in the third section, a symmetry between bosons and fermions. In SUSY GUTs the evolution equations are modified. Since one does not know at which scale the modifications start to be effective, in Amaldi *et al.*[13] both the unification scale M_{GUT} and the SUSY breaking scale M_{SUSY} were fitted. By definition, M_{SUSY} is a *phenomenological parameter* that determines the point at which the inverse of the coupling constants change slope. The fit was done by minimizing the χ^2 for a single unification point. The minimal SUSY contains two Higgs doublets. In the minimization the mass of the lightest Higgs doublet was chosen to be equal to M_Z, and the mass of the heavier doublet was taken to be equal to M_{SUSY}. These choices have practically no influence on the conclusions as long as the Higgs masses are less than a few times M_{SUSY}.

The second-order renormalization group equations give the results shown in FIGURE 9(a). Both the break due to the onset of SUSY (at $M_{\text{SUSY}} \simeq 1000$ GeV) and the fact that the three coupling constants meet in a single point around 10^{16} GeV are apparent from the figure. The resulting χ^2 distributions for M_{SUSY} and M_{GUT} are shown

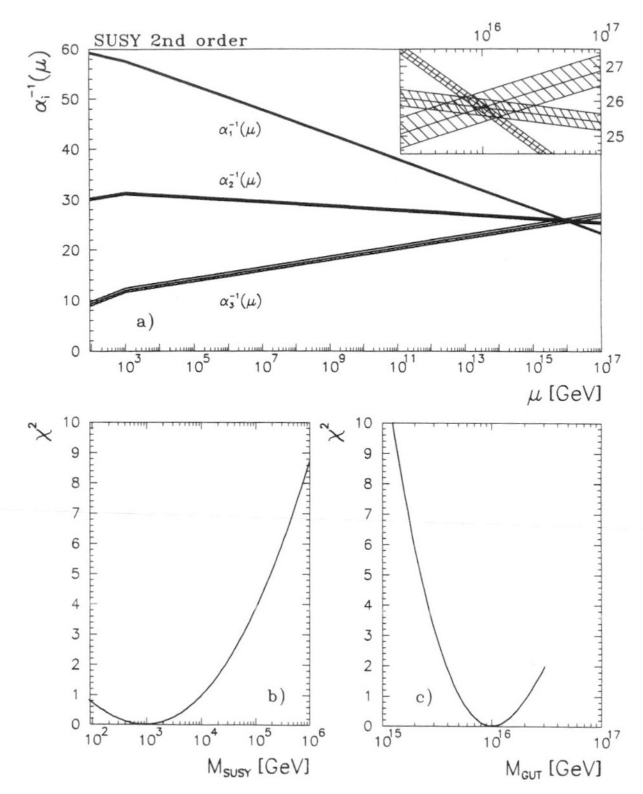

FIGURE 9. **(a)** Evolution of the inverse of the coupling constants in the minimal supersymmetric (SUSY) model width when it is required that they meet in a point at an energy M_{GUT}. The parameters of the fit are M_{GUT} and M_{SUSY}, the unknown energy at which the slopes of the curve change. **(b)** and **(c)** show the χ^2-distribution for M_{SUSY} and M_{GUT}.

in FIGURES 9(b) and (c). The minimum χ^2 is obtained for

$$M_{SUSY} = 10^{3.0\pm1.0}\ \text{GeV},$$

$$M_{GUT} = 10^{16.0\pm0.3}\ \text{GeV},$$

$$\alpha_{GUT}^{-1} = 25.7 \pm 1.7. \quad \textbf{(15)}$$

It is interesting that, by introducing these values in **(14)**, one obtains $\tau_{proton} \approx 10^{35\pm1.2}$ years, which is compatible with the present experimental limits *if* the decays, which are not dominated by X-boson exchange and are theoretically possible, are suppressed. It has to be added that, in standard and SUSY GUTs, the mass of the b-quark can be computed.[19] If the number of neutrinos is $N_\nu = 3$, the computed value $M_b = (5.2 \pm 0.3)$ GeV agrees very well with experiment, thus nicely closing the circle with the determination of N_ν from astrophysics[20] and the more accurate direct measurement of (5) provided by LEP.

In the framework of GUT, present LEP data favor the minimal SUSY (with two Higgs doublets) with respect to the minimal SM (which has one Higgs doublet). One can ask whether, by having more Higgs doublets, the situation changes. Further fits show that if, in the SM, one adds five additional Higgs doublets, unification can be obtained at a scale $M_{\rm GUT} = 4.10^{13}$ GeV. However, such a low scale is excluded by the limits on the proton lifetime. It has also been shown in Amaldi *et al.*[13] that nonminimal SUSY models with four or more Higgs doublets—having masses around or below the SUSY scale—yield unification. However, once more the unification scale is below the limit allowed by the proton decay experiments. Therefore only the minimal SUSY model gives a unification scale that may be compatible with the proton lifetime limit.

It has to be stressed that all these arguments are not a *proof* of SUSY, but tell us that SUSY GUTs are *consistent* with present LEP data at 100 GeV.

The best fit to the allowed minimal SUSY model, shown in FIGURE 9, is obtained for a SUSY scale around 1000 GeV or, more precisely, $M_{\rm SUSY} = 10^{3.0\pm1.0}$ GeV, where

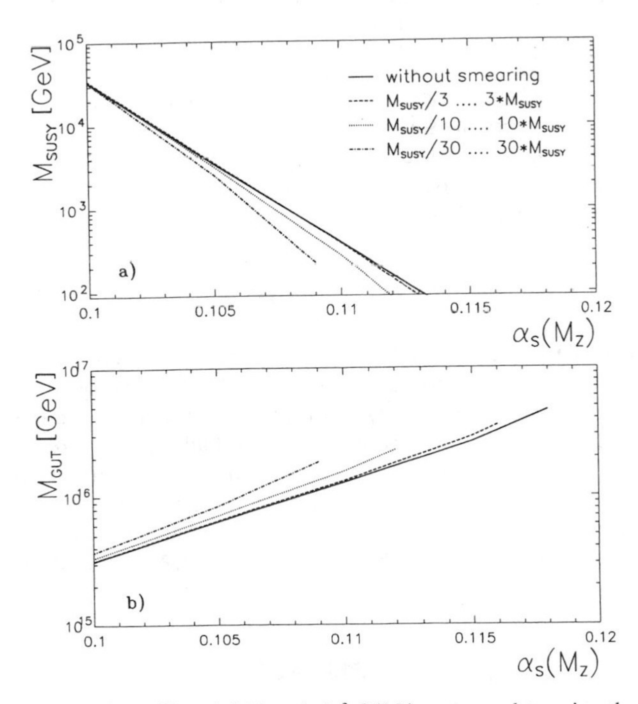

FIGURE 10. By knowing $\alpha(0) = 1/137$ and $\sin^2 \Theta(M_Z)$ one can determine the value of the phenomenological parameter $M_{\rm SUSY}$ as a function of α_Z (M_Z) (*full line*). The *other lines* correspond to varying the region of masses, around $M_{\rm SUSY}$, where the slopes change, so as to have a smoother transition between the regions below and above the break.

the error originates mainly from the uncertainty in the strong coupling constant. Averaging more experiments and better controlling the scale and fragmentation errors, the overall uncertainty of $\alpha_s(M_Z)$ will decrease with time below the 5 percent taken in Amaldi *et al.*,[13] where the input value was $\alpha_1(M_Z) = 0.108 \pm 0.005$ (65 percent confidence level). This will require a large but worthwhile concentrated effort from the side of both theorists and experimentalists.

The conclusion of this consistency check is that, *if* a supersymmetric GUT describes nature, SUSY particles, which are expected to have masses on the order of M_{SUSY}, could be within reach of the present or next generation of accelerators. Moreover, it is rewarding to find that the SUSY scale, which is obtained from the extrapolation of LEP data is, even with *large* errors, compatible with the value $M_{\mathrm{SUSY}} \leq 1$ TeV. This is indeed what is required by theorists to solve in a natural way the "hierarchy problem" of the incredibly small ratio $M_Z/M_{\mathrm{Planck}} \simeq 10^{-17}$, which was the main reason for introducing SUSY models in the first place.

Independently of the numerical values derived for M_{SUSY} and M_{GUT}, the main result of this type of analysis is summarized in FIGURE 10, where the relations between $\alpha_s(M_Z)$, M_{SUSY}, and M_{GUT} are plotted. To draw the lines it is necessary to use $\alpha(0) = 1/137$ and $\sin^2 \Theta$ from **(4)** and, moreover, to assume that:

1. the three coupling constants [defined as in **(13)** with a factor of 5/3 in α_1] pass through a single unification point;
2. the particle content of the theory is of the minimal SUSY model.

As underlined in the last section, the "best" central value for $\alpha_s(M_Z)$ and its 68 percent confidence level errors are subjects of debate among specialists. Eventually, the subject will be cleared and the final value of $\alpha_s(M_Z)$, with its reduced error, will be introduced in graphs similar to the ones in FIGURE 10 to obtain some informations on phenomena that take place at energies not yet available.

CONCLUSIONS

The year 1990 has been for LEP and its four detectors the year of the first harvest. The standard model predictions have been checked to an unprecedented level in the electroweak sector, and neutral higgsons of mass smaller than about 50 GeV have been excluded. The measurements performed on hadronic jets are in quantitative agreement with asymptotic freedom, a crucial property of quantum chromodynamics. Moreover, the combination of the measured electroweak and strong couplings have allowed a first glimpse into the energy domain that is not directly available to LEP, which in this way can justifiably be called the CERN Laboratory for Electrostrong Physics.

REFERENCES

1. DECAMP, D., *et al.* 1989. ALEPH Collab., Phys. Lett. **231B:** 519; AARNIO, P., *et al.* 1989. DELPHI Collab., Phys. Lett. **231B:** 539; ADEVA, B., *et al.* 1989. L3 Collab., Phys. Lett. **231B:** 509; AKRAWY, M. Z., *et al.* 1989. OPAL Collab., Phys. Lett. **231B:** 530.
2. DYDAK, F. 1991. Proc. 25th Int. Conf. on High Energy Physics, Vol. 1 (Singapore, Aug. 2–8, 1990), K. K. Phue and V. Venragouchi, Eds: 3–32. World Scientific. Singapore.

3. DECAMP, D., *et al.,* 1990. ALEPH Collab., Nucl. Instrum. Methods **A294:** 121; AARNIO, P., *et al.* 1990. DELPHI Collab., Nucl. Instrum. Methods. In print and CERN-PPE/90-128; ADEVA, B., *et al.* 1990. L3 Collab., Nucl. Instrum. Methods **A289:** 35; AHMET, K., *et al.* 1990. OPAL Collab., Nucl. Instrum. Methods. In print and CERN-PPE/90-114.
4. BOBBINK, G. & P. RATOFF. 1991. Invited talks given at the 26th Recontres de Moriond on Electroweak Interaction and Unified Theories, Mar. 10–17, 1991. To be published in the Proceedings.
5. DECAMP, D., *et al.* 1991. ALEPH Collab. Contribution to the Aspen, La Thuile, and Moriond Conferences, Winter 1991. CERN PPE/91-19; AARNIO, P., *et al.* 1991. DELPHI Collab. Contribution to the 26th Moriond Conference, Mar. 10–17, 1991; ADEVA, B., *et al.* 1990. L3 Collab., Search for neutral Higgs bosons. L3 Preprint No. 24; AKRAWY, M. Z., *et al.* 1991. OPAL Collab., Phys. Lett. **253B:** 511.
6. For reviews, see NILLES, H. P. 1984. Phys. Rep. **110C:** 1; HABER, H. E. & G. L. KANE. 1985. Phys. Rep. **117C:** 76; BARBIERI, R. 1988. Nuovo Cimento **11:** 1.
7. INOUE, K., *et al.* 1982. Prog. Theor. Phys. **67:** 1889; FLORES, R. & M. SHER. 1983. Ann. Phys. **148:** 95.
8. OKADA, T., M. YAMAGUCHI & T. YANAGIDA. 1991. Prog. Theor. Phys. Lett. **85:** 1 and Preprint TU-363/90; ELLIS, J., G. RIDOLFI & F. ZWIRNER. 1991. Phys. Lett. **B257:** 83 and CERN-TH 6002/91; HABER, H. E. & R. HEMPFLING. 1990. Preprint SCIPP-90/42; BARBIERI, R., M. FRIGENI & M. CARAVAGLIOS. 1990. Preprints IFUP-TH 46/90 and IFUP-TH2/91.
9. ALTARELLI, G. 1989. Ann. Rev. Nucl. Sci. **39:** 357.
10. T'HOOFT, G. 1973. Nucl. Phys. **B61:** 455; BARDEEN, W. A., *et al.* 1978. Phys. Rev. **D18:** 3998.
11. DE BOER, W. 1990. Invited Talk Given at the Topical Conference of the SLAC Summer Institute, Aug. 1990. CERN-PPE/90-161. To be published in the Proceedings; BETHKE, S. 1990. Invited Talk Given at the Workshop on Jet Physics at LEP and HERA, Dec. 9–15, 1990. CERN-PPE/91-36. To be published in the Proceedings.
12. MARTIN, A. D., R. G. ROBERTS & W. J. STIRLING. 1990. Precision analysis of the $\Lambda_{\overline{MS}}$ and the gluon distributions and its implication for jet and top quark cross-sections. RAL-90-084.
13. AMALDI, U., W. DE BOER & H. FURSTENAU. 1991. Phys. Lett. **B260:** 447.
14. AMALDI, U., A. BOEHM, L. S. DURKIN, P. LANGACKER, A. K. MANN, W. J. MARCIANO, A. SIRLIN & H. H. WILLIAMS. 1987. Phys. Rev. **D36:** 1385.
15. LANGACKER, P. & M. INO. 1991. Phys. Rev. **D44:** 817.
16. ELLIS, J., S. KELLEY & D. V. NANOPOULOS. 1990. Phys. Lett. **B249:** 441, and Phys. Lett. **B260:** 131.
17. EINHORN, M. B. & D. R. T. JONES. 1982. Nucl. Phys. **B196:** 475.
18. REVIEW OF PARTICLE PROPERTIES. 1990. Phys. Lett. **239B:** 1; BECKER-SZENDY, R., *et al.* 1990. IMB-3 Collab., Phys. Rev. **D42:** 2974.
19. CHANOWITZ, M. S., J. ELLIS & M. K. GAILLARD. 1977. Nucl. Phys. **B128:** 508; BURAS, A. J., J. ELLIS, M. K. GAILLARD & D. V. NANOPOULOS. 1978. Nucl. Phys. **B135:** 66; NANOPOULOS, D. V. & D. A. ROSS. 1982. Phys. Lett. **118B:** 99.
20. OLIVE, K. A., D. N. SCHRAMM, G. STEIGMAN & T. WALKER. 1990.

The Quark–Hadron Phase Transition

HELMUT SATZ

Theory Division
CERN
CH-1211 Geneva 23, Switzerland
and
Fakultät für Physik
Universität Bielefeld
D-4800 Bielefeld 1, Germany

RESULTS FROM STATISTICAL QCD

The quark structure of elementary particles leads rather directly to the concept of quark matter; at sufficiently high density, there should be a transition from hadronic matter to a plasma of deconfined quarks and gluons. We expect deconfinement in dense matter, because the presence of many color charges will screen the confining potential between the members of a given $q\bar{q}$ or qqq system. The density of the system can be increased either by "compression" (an increase in baryon number density or baryonic chemical potential μ_B), or by "heating" (an increase in the initial energy density ϵ). This leads to a critical transition curve in the phase diagram of strongly interacting matter, as shown in FIGURE 1. The crucial features for our present consideration are the values of the transition parameters (temperature, density, energy density, screening length). These questions have been addressed in statistical quantum chromodynamics (QCD), both in the lattice formulation and in various effective Lagrangian models.

In the computer simulation of statistical QCD on the lattice,[1] one tries to calculate the relevant quantities from first principles, without any simplifying physical assumptions. The quantitative reliability of the results is at present, however, still somewhat limited by technical restrictions (memory size, operating speed of available supercomputers). Nevertheless, the results of lattice QCD today do give us a reasonably good general understanding of the critical behavior of strongly interacting matter at vanishing baryon number density, and this appears to be the first time that basic dynamics leads directly to predictions for equilibrium thermodynamics.

In FIGURE 2, we show the behavior of energy density and pressure for QCD matter with light quarks of two flavors (u and d), as calculated on the lattice.[2] We note that at a critical temperature T_c, the energy density undergoes a rapid transition from low values, corresponding to a hadron gas, to much higher values, corresponding to a quark–gluon plasma. For ideal (i.e., noninteracting) systems, the ratio of the energy densities of pion gas to quark–gluon plasma is given simply by the corresponding degrees of freedom; for $N_f = 2$, this means $\epsilon_\pi/\epsilon_{QG} \simeq 1/10$. The energy density of the ideal plasma, including finite lattice corrections,[3] is also shown in FIGURE 2, and at high temperatures, the calculations appear to approach this value. In an ideal gas, however, energy density and pressure are related by $(\epsilon - 3P)/T^4 = 0$, and we see that this condition is certainly not fulfilled for $T \leq 1.5T_c$: below $1.5T_c$, the energy

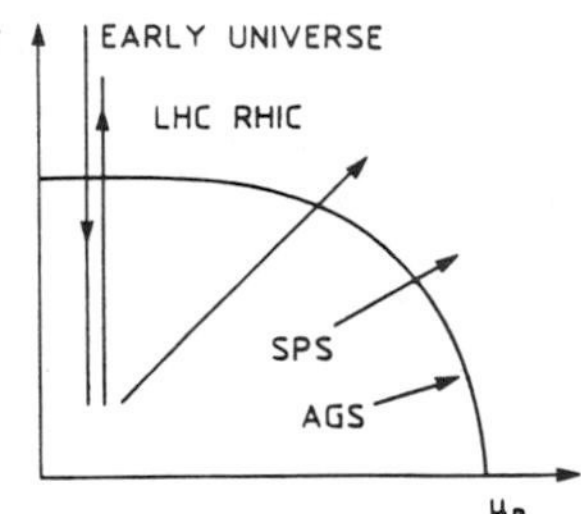

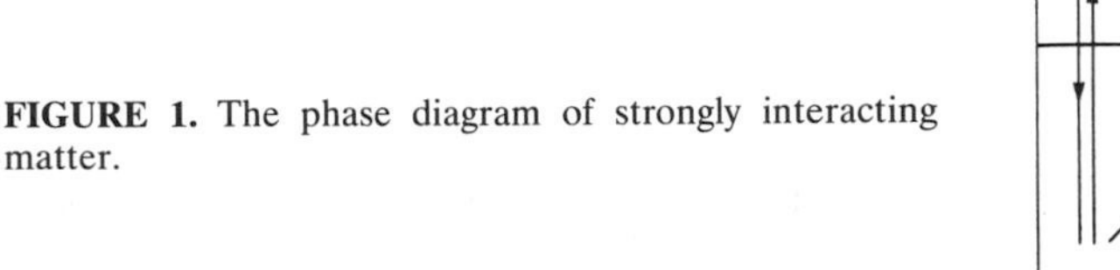

FIGURE 1. The phase diagram of strongly interacting matter.

density overshoots the Stefan–Boltzmann limit, the pressure falls much below it. For a pure $SU(3)$ gauge system (i.e., for $N_f = 0$), these deviations from ideal gas behavior were recently studied in detail;[4] the result is shown in FIGURE 3 and indicates that the system may not become ideal until even higher temperatures.

The order of the transition is at present under intense investigation by lattice studies.[1] One finds a first-order deconfinement transition for $N_f = 0$, and a first-order transition corresponding to chiral symmetry restoration and deconfinement for $N_f \geq 3$, in the limit of massless quarks. For $N_f = 2$, the transition appears to be continuous for present lattice sizes and quark masses; however, for two light and one heavy quark species (corresponding to the actual *u, d, s* quarks), it becomes first order when the strange quark mass reaches a certain value.[5] More detailed quantitative studies are needed here. All results at vanishing baryon number density agree, however, on the same transition point for deconfinement and chiral symmetry restoration.

The physical value of the transition temperature is for current, not yet asymptotic lattice sizes best determined by calculating both T_c and the hadron masses in units of the lattice spacing; the ratio then gives us T_c in terms of meson or baryon masses. In FIGURE 4, we show the result for T_c as determined from m_ρ, for different N_f. The most reasonable value for the actual physical case is $T_c \simeq 150$ MeV; we must keep in mind, however, that in present calculations the ratio of π to ρ mass has not yet reached its physical value, and hence quantitative results are not yet final. To be on the safe side, we shall consider the critical temperatures to lie in the range $T_c = 150$–200 MeV; the

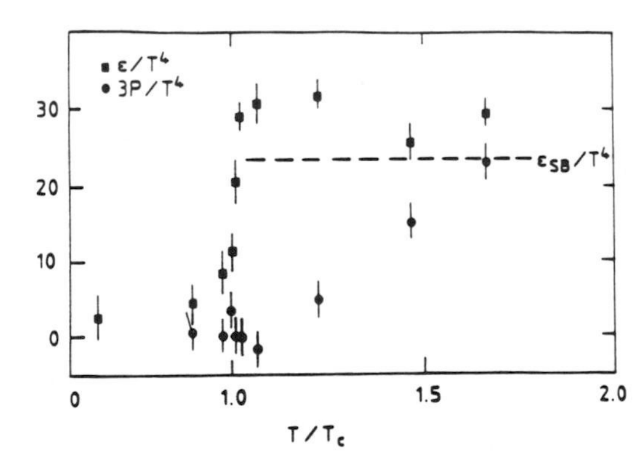

FIGURE 2. Energy density ϵ and pressure P for QCD matter with light quarks of two flavors, as a function of the temperature T. (From Gottlieb *et al.*[2] Reproduced by permission.)

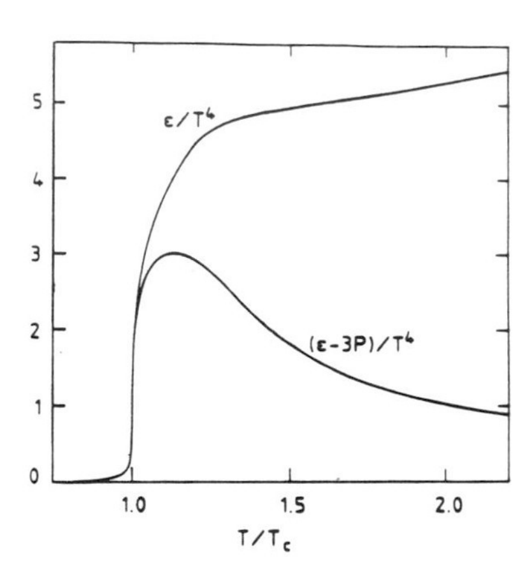

FIGURE 3. Energy density ϵ and interaction measure $(\epsilon - 3P)/T^4$ in pure $SU(3)$ gauge theory, as a function of the temperature T. (From Engels *et al.*[4] Reproduced by permission.)

corresponding critical values of the energy density necessary for deconfinement are $\epsilon_s \simeq$ 1–3 GeV/fm^3.

Finally we want to note the result of lattice calculations of the screening length. Since bound states will "melt" in dense matter when the screening radius becomes significantly smaller than the binding radius, the temperature dependence of the screening radius $r_D(T)$ gives us some idea of when specific bound states will disappear. From FIGURE 5 we see that above $T \simeq 1.2T_c$, even the tightly bound $c\bar{c}$ state J/ψ will become deconfined.

An alternative approach to statistical QCD is offered by the study of effective Lagrangian models. The basic idea here is to construct a model Lagrangian that incorporates as much as possible of the known low-density hadron physics, and then check what it predicts at higher densities. Rather detailed studies in the framework of chiral perturbation theory[6] reproduce the known pion physics at zero temperature, and predict chiral symmetry restoration at $T_c \simeq$ 190 MeV, in accord with lattice results. Another effective Lagrangian study[7] has also been extended to nonzero baryon number density; it leads to the interesting phase diagram shown in FIGURE 6,

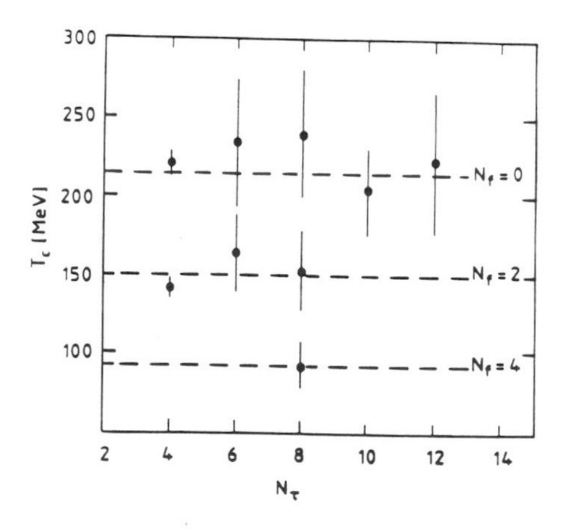

FIGURE 4. Deconfinement temperature T_c for QCD matter with N_f flavors of light quarks, as a function of the lattice size N_τ. (From Petersson.[1] Reproduced by permission.)

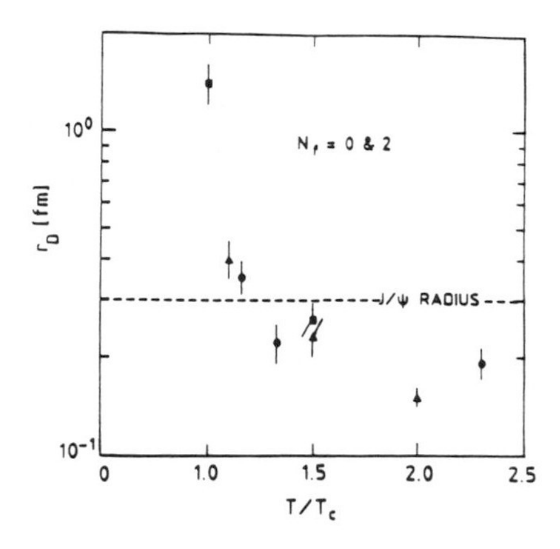

FIGURE 5. Color screening radius r_D as a function of the temperature T, for QCD matter with N_f light quarks. (From Petersson.[1] Reproduced by permission.)

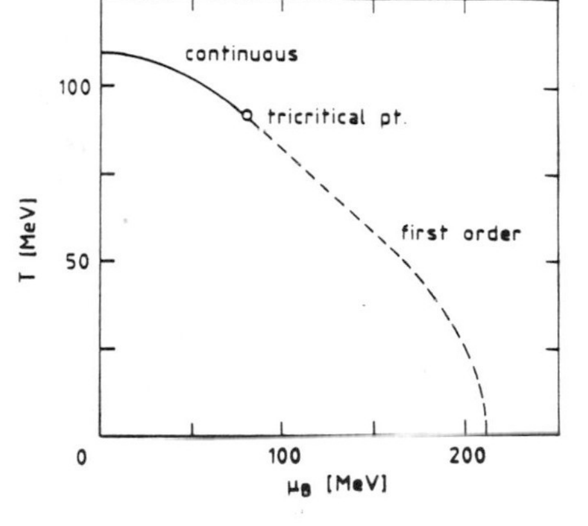

FIGURE 6. Phase diagram for chiral symmetry restoration in QCD matter. (From Barducci *et al.*[7] Reproduced by permission.)

with a continuous transition at zero baryonic chemical potential μ_B, which then turns into a first-order transition at some tricritical point for $\mu_B > 0$. This behavior illustrates that new features can still be expected in the region of the phase diagram not yet accessible to lattice studies.

CONDITIONS IN NUCLEAR COLLISIONS

In statistical QCD, we study the equilibrium thermodynamics of strongly interacting matter. We would like to test the results experimentally in high-energy heavy-ion collisions. This leads us immediately to the basic question that such studies have to face: Do heavy-ion collisions lead to systems dense enough, large enough, and long-lived enough to treat them by the equilibrium thermodynamics based on QCD? Let us first consider the density regime attainable.

The basic observable for an estimate of the initial energy density is a multiplicity of the secondary hadrons emitted in the collision. The total multiplicity (1.5 times the observed charged multiplicity) per unit central rapidity interval can be parametrized in high energy proton–proton collisions by

$$(dN/dy)_p = 0.8 \ln \sqrt{s}; \tag{1}$$

this form describes well all data from SPS to Tevatron energies,[8] with $(dN/dy)_p$ growing from 2.4 at 20 GeV to about 6 at 1.8 TeV. This is extrapolated to central A–A collisions by

$$(dN/dy)_A = A^{\alpha}(dN/dy)_p, \tag{2}$$

with $\alpha \geq 1$. For $\alpha = 1$, we simply have a superposition of A independent p–p collisions; if there is rescattering between the different nucleons and/or the produced secondaries, we will have $\alpha > 1$. Present data from nuclear collisions give $\alpha \geq 1.1$,[9] and this leads for the multiplicity per unit central rapidity in Pb–Pb interactions ($A = 208$) to the range of values:

SPS (17 GeV)	500–800	
RHIC (200 GeV)	900–1500	(3)
LHC (6300 GeV)	1500–2500.	

The lower number always corresponds to $\alpha = 1$, the higher to $\alpha = 1.1$. As large as these numbers may seem, it should be noted that fixed-target experiments at SPS and AGS, which cover several units of rapidity, have shown that multiplicities in the hundreds can indeed be handled.

If the observed secondaries have come from the initial interaction region in "free flow," then the initial energy density ϵ in a central A–A collision is given by[10]

$$\epsilon = (dN/dy)_A m_T/(\pi R_A^2 \tau), \tag{4}$$

where $m_T = (p_T^2 + m^2)^{1/2}$ denotes the transverse energy of the secondary, $R_A \simeq 1.2A^{1/3}$ the nuclear radius, and τ the formation time or longitudinal extension of the equilibrium system. With the estimates $m_T \simeq 0.5$ GeV and $\tau \simeq 1$ fm, we obtain from **(1)** and **(2)**

$$\epsilon = 0.09 A^{\alpha - 2/3} \ln \sqrt{s}, \tag{5}$$

and from this for central Pb–Pb collisions:

SPS	1.5–2.5 GeV/fm^3,	
RHIC	2.8–4.7 GeV/fm^3,	(6)
LHC	4.6–7.8 GeV/fm^3.	

The values we have used for m_T and τ are "conservative"—the transverse mass increases somewhat with $\sqrt{s}$, and enhanced nuclear stopping would reduce τ; there are models[11] that give $\tau \sim A^{-1/6}$. On the other hand, a diffuse edge of the interaction region would result in a larger value for R_A, which could offset these effects. Note that the values of **(6)** are 4–7 times higher than the average energy density in p–p collisions of the same $\sqrt{s}$.

In FIGURE 7 we show the variation of the average energy density in central Pb–Pb

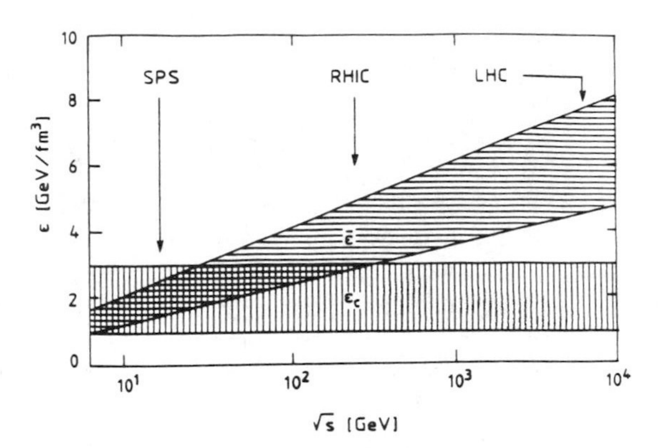

FIGURE 7. The expected range of the average initial energy density $\bar{\epsilon}$ vs. incident CMS energy $\sqrt{s}$; also shown is the predicted range of the critical energy density ϵ_c for deconfinement.

collisions, together with the critical energy density ϵ_c determined in the first section. We note that at the LHC we are well above this value and are in fact getting into the range in which ϵ becomes high enough to produce an ideal quark–gluon plasma.

We now want to consider briefly the question of how the energy density ϵ can in practice be varied for a given collider. In A–B collisions, with $A \ll B$, we have different values of ϵ for events at different impact parameter b, that is, for events of different multiplicity or different tranverse energy E_T. Going from peripheral to central collisions, we increase ϵ: the effective transverse overlap area remains constant, since the smaller projectile after complete "immersion" hits the larger target at smaller and smaller b; the effective number of participants increases, however, since the target contains more nucleons in the center, at $b = 0$, than at the edges. In "symmetric" A–A collisions, with large A, this is no longer the case, since over most of the impact parameter range, the number of participants and the overlap area are essentially proportional. The result[12] is illustrated in FIGURE 8, where we show ϵ as a function of $E_T(b)/E_T(0)$. For an infinite nucleus, ϵ becomes independent of $E_T(b)/E_T(0)$; deviations from constancy are thus of purely geometric origin. In particular, the drop of ϵ as $E_T(b)/E_T(0) \rightarrow 0$ is extremely dependent on the specific

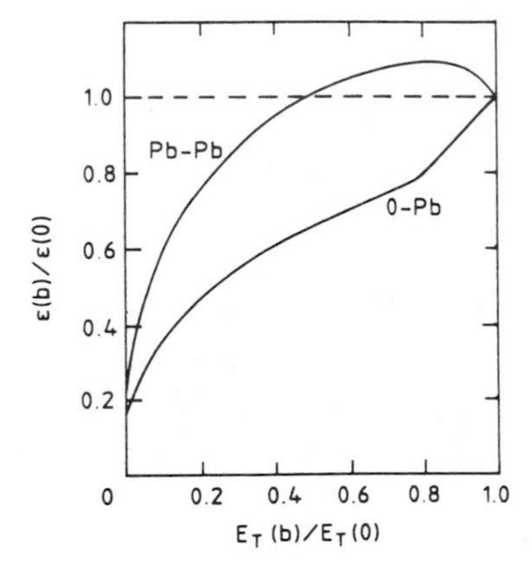

FIGURE 8. Energy density ϵ vs. transverse energy E_T as a function of the impact parameter b, for O–Pb and Pb–Pb collisions; the *dashed line* shows the behavior for A–A collisions in the limit of infinite nuclear size. (From Karsch and Satz.[12] Reproduced by permission.)

nuclear edge structure; it is more a surface than a volume effect and hence not useful to study the ϵ dependence of any observables. For A–A collisions, we thus have to find some other way to change ϵ.

In principle, the best way would be to vary the incident energy $\sqrt{s}$ for given A, since this would change ϵ at constant volume. In practice, this requires large variations of the beam energy, since $\epsilon \sim \ln \sqrt{s}$. Moreover, a reduction of $\sqrt{s}$ at a given collider is in general accompanied by a considerable luminosity drop,[13,14] and hence for many experiments not very useful.

This leaves us with A–A collisions at several A, and fixed $\sqrt{s}$ as the only viable road to different ϵ, even though this changes the associated volume as well. Going from U–U to S–S collisions reduces ϵ by a factor of 2 or more; to achieve such a change by varying $\sqrt{s}$ would require going from the peak LHC energy of 6.3 TeV down to one-hundredth of this value, which is expected to decrease the luminosity by almost two orders of magnitude.[13] In contrast, reducing A will generally *increase* the luminosity, so that most rates should not be too much affected. For these reasons, it is very important to include the capability to run at varying A from the beginning among the essential requirements in planning the heavy-ion mode of any collider.

Finally, we should note that in view of these considerations, the study of QCD thermodynamics at different accelerators, but for the same or similar A, plays an important complementary role.

What temperatures do the values of ϵ in **(6)** correspond to? For an ideal plasma with three flavors of massless quarks we have

$$\epsilon = (47.5\pi^2/30)T^4, \tag{7}$$

which leads to an initial temperature $T = \epsilon/(1953)^{1/4}$, with ϵ in GeV/fm^3 and T in GeV. We can now use the energy density estimates of **(6)** to get the corresponding temperatures. Before we list these, we want to note an alternative way to estimate T. If the initial bubble of matter undergoes longitudinal hydrodynamic expansion to attain the observed final state, rather than the free flow assumed in **(4)**, then the entropy is conserved, not the energy: part of the initial energy goes into work against the pressure of the vacuum on the system.[15,16] The initial entropy density s in a central A–A collision is obtained from

$$s = 3.6(dN/dy)_A/(\pi R_A^2 \pi), \tag{8}$$

which for an ideal plasma leads to $T = (s/2605)^{1/3}$, with T in GeV and s in fm^{-3}. As already noted, this leads to somewhat higher initial temperatures, and hence also to higher initial energy densities. The temperature values for Pb–Pb are (in MeV):

$$\begin{array}{lll} \text{SPS} & 170\text{–}190 & 160\text{–}190 \\ \text{RHIC} & 200\text{–}220 & 200\text{–}240 \\ \text{LHC} & 220\text{–}250 & 230\text{–}280. \end{array} \tag{9}$$

The first column corresponds to free flow, the second to isentropic expansion. With increasing collision energy, the difference between the two temperature estimates (and hence between the corresponding energy densities) increases. For Pb–Pb, we obtain for free flow $\epsilon_{Bj} \simeq 0.52 \ln \sqrt{s}$, for isentropic expansion $\epsilon_S \simeq 0.32(\ln \sqrt{s})^{4/3}$. The

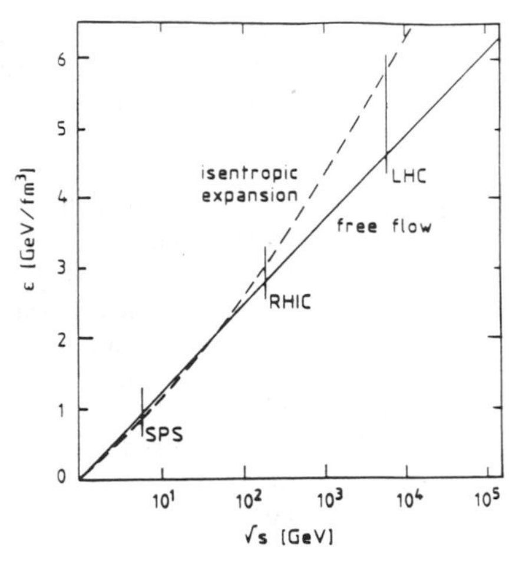

FIGURE 9. Initial energy density ϵ vs. incident CMS energy $\sqrt{s}$ for free flow[10] and hydrodynamic flow.[16]

resulting behavior of ϵ is illustrated in FIGURE 9. We note that the two have become really distinct only at LHC energies, so that then the effects of longitudinal hydrodynamic expansion should become evident even for average quantities.

In the first section, we had seen that so far the most reliable calculations are available for systems of vanishing baryon-number density; this is also the situation that presumably existed at the end of the quark–gluon phase of the early universe. When can we expect similar conditions in nuclear collisions? From *p–A* collisions at $A \simeq 200$, we know that in passing through the nuclear target, the projectile proton looses approximately two units in rapidity. The corresponding baryon-number distribution is then centered at $Y - \delta y$, where $Y \simeq \ln(\sqrt{s}/m_p)$ denotes the maximum rapidity and δy the rapidity shift of a nucleon passing a nuclear target; the distribution vanishes at Y and $Y - 2\delta y$. The overall baryon-free region in rapidity thus becomes a

$$(\Delta y)_0 \simeq 2(Y - 2\delta y) \qquad \textbf{(10)}$$

target. If δy is the same in *A–A* collisions as for *p–A*, then we get:

	$(\Delta y)_0$	$2Y$
SPS	0	5.8
RHIC	2.7	10.7
LHC	9.6	17.6.

(11)

Equation **(11)** is consistent with recent p–$\bar{p}$ data,[17] which show at $\sqrt{s} = 1800$ GeV, where $2Y \simeq 15$, a baryon-free region of at least 6.5 units; even with $\delta y = 2$ we expect seven units. It is not clear, however, if *A–A* collisions do not give enhanced stopping, that is, $\delta y > 2$, and hence a smaller baryon-free region. This has to be studied by event generators for nuclear collisions, and we shall return to such studies shortly. In

FIGURE 10 we show the baryon-free regions at the different accelerator energies. Even if there is more stopping than expected from *p–A* collisions, however, the LHC still provides an ample safety margin.

What can we say about the baryon-rich fragmentation region at very high energies? The kinematic compression experienced by target and projectile is essentially determined by the rapidity shift δy for nucleons;[18] if this is not dependent on $\sqrt{s}$, then the kinematic compression will not increase by going to higher incident energies. The energy deposited in target or fragmentation regions is also expected to depend only very weakly on the incident energy.[9] This leads to the conclusion that a baryon-rich state of strongly interacting matter is better studied at a fixed target machine with high *A* beams; there appears to be no particular reason to consider this regime at high-energy colliders, where it is not so easily accessible.

So far, we have addressed general aspects of nuclear collisions, extrapolating from known features of *p–p* and *p–A* interactions. This is done in more detail in specific event generators, which are obtained by assuming some particular form of

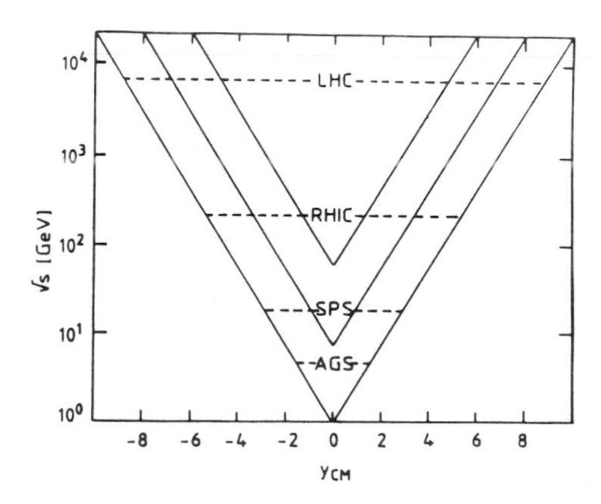

FIGURE 10. Baryon number density distribution in rapidity *y* vs. incident energy $\sqrt{s}$.

interaction in the course of the collision.[19] In PYTHIA, the interaction is simply taken to be a superposition of individual nucleon–nucleon collisions, without any secondary interactions; particular attention is paid, however, to obtain the correct high-energy behavior of the *p–p* interaction, including minijet effects. In VENUS, on the other hand, the interaction is assumed to be string fragmentation, with strings formed also between nucleons and secondaries, as well as between different secondaries. This model has been used so far more at lower energies, and it probably has to be complemented for high-energy features (minijets). It does already give us some idea of rescattering effects now. In addition, studies using FRITIOF and the dual parton model are under way. In all cases, the event generator should be tuned to account correctly for the general features of *p–p* data at all available energies. For *A–A* collisions, the different codes will then show us the role of different interaction schemes corresponding to different extrapolations.

Let us look at some first results.[19] In FIGURE 11 we see that both PYTHIA and VENUS describe correctly the energy dependence of the average multiplicity per

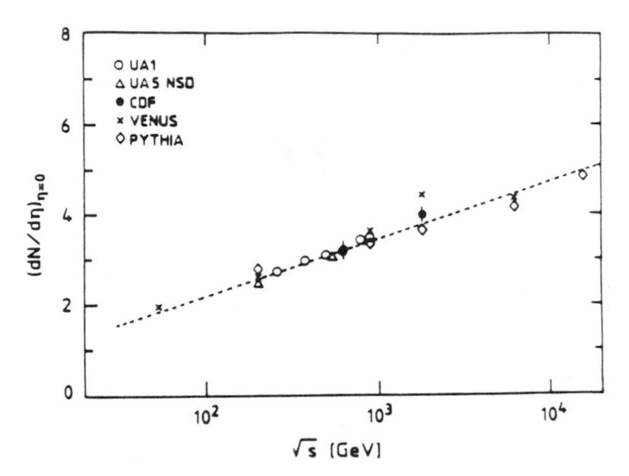

FIGURE 11. Multiplicity for unit central pseudorapidity η as a function of incident CMS energy $\sqrt{s}$ in p–p and p–$\bar{p}$ collisions, compared to calculations by PYTHIA and VENUS;[19] the *dashed line* is the fit from Abe *et al.*[8] used in the text.

unit central rapidity in p–p collisions, and that this dependence is in fact well reproduced by **(1)**. Going to Pb–Pb collisions at LHC energy, PYTHIA recovers the result listed in **(2)**, with $(dN/dy)_{\mathrm{Pb}} = 1500$. On the other hand, VENUS, because it includes rescattering, gives a considerably larger value, $(dN/dy)_{\mathrm{Pb}} = 2500$. First estimates from the dual parton model give much larger values still. It is thus possible that our "conservative" estimates of energy densities and temperatures for RHIC and LHC are in fact too conservative.

In FIGURE 12, we show the baryon-number distributions from PYTHIA and VENUS. For Pb–Pb at the LHC, PYTHIA gives about 14 units of baryon-free rapidity region. This is more than the 10 units we had found in **(11)** by extrapolating p–A data, since PYTHIA contains no nuclear stopping. VENUS, with more secondary interactions, gets only 8 units, which, however, is still only half of the total LHC rapidity region.

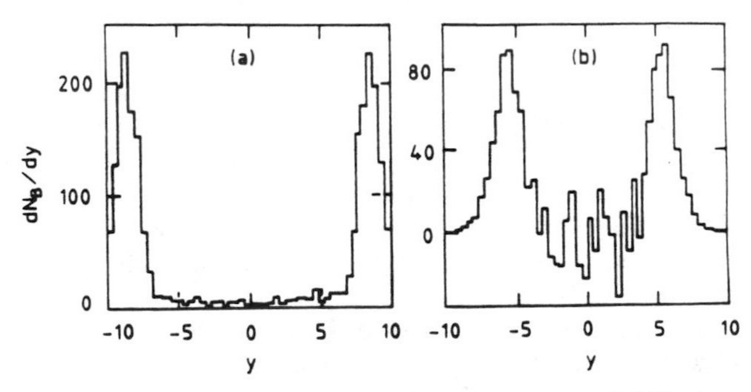

FIGURE 12. Net baryon number distribution for Pb–Pb collisions at LHC energy as predicted by PYTHIA **(a)** and VENUS **(b)**.[19]

These results are just the beginning of event-generator studies in this field. Much further work is in progress and should also give us information on particle ratios and momentum spectra.

VOLUMES AND LIFETIMES

Before we can use heavy-ion collisions to study equilibrium thermodynamics, we must make sure that the volume and the lifetime of the systems experimentally produced are sufficiently large. This becomes all the more crucial for the study of critical behavior, which becomes really "critical" only in the infinite volume limit. How can we check volumes and lifetimes?

The initial interaction volume has already been introduced in (**3**); for a central A–A collision, it is $V_0 \simeq \pi R_A^2 \tau$, where $R_A \simeq 1.2A^{1/3}$ denotes the effective transverse nuclear radius and $\tau \simeq 1$ fm the initial longitudinal extension of the system that will later thermalize, or, equivalently, the formation time. This system now expands, and at the time of transition to hadronic matter, it has attained the size $V_c = V_0(\epsilon_0/\epsilon_c)$, where we have used ϵ_0 to denote the initial energy density. This means that V_c increases with (dN/dy); at the LHC, the transition volume V_c is about 5–8 times larger than the initial volume. For Pb–Pb, this gives at the transition point a volume of some 800–1200 fm^3. The system now continues to expand until freeze-out. The freeze-out radius can be estimated by supposing that interactions stop when the energy density has dropped to that of an ideal pion gas at $T \simeq T_c = 150$ MeV. For lead beams, this gives us a freeze-out radius

$$R_F^{\epsilon} \simeq 1.24(dN/dy)^{1/3}. \qquad \textbf{(12)}$$

Another possible estimate for R_F is obtained if one supposes freeze-out to take place when the mean free path λ of pions has reached the size of the system.[20,21] With $\lambda = V_F/(dN/dy)\sigma_\pi$ and $\sigma_\pi \simeq 20$ mb, this leads to

$$R_F^{\lambda} \simeq 0.69(dN/dy)^{1/2} \qquad \textbf{(13)}$$

for the freeze-out radius. Introducing an energy dependence of σ_π[9,22] leads to yet another form, which grows as $(dN/dy)^{5/12}$ and thus falls between the two cases we had discussed.

Experimentally, the freeze-out size can be determined by particle interferometry, based on the Hanbury-Brown–Twiss method to measure star sizes. In FIGURE 13 we see that at present multiplicities, both forms (**12**) and (**13**) accommodate the data. It should be emphasized that these data lead to freeze-out radii that are almost a factor of 2 larger than the radii of the projectiles. This supports the idea that nuclear collisions indeed produce bubbles of expanding strongly interacting matter, and thus gives us a first hint that we are on the right track. From FIGURE 14 we conclude that already a lead beam in the SPS should allow us to distinguish the two forms (**12**) and (**13**), and that at the LHC lead–lead collisions should produce volumes of some 10^4 to 10^5 fm^3. Thus nuclear collisions can indeed be expected to lead to volumes that are very much larger than the few fm^3 typical of nucleon–nucleon interactions.

The use of like-particle interferometry to determine the freeze-out radii will, however, encounter some difficulties at very high energies. The linear size of the

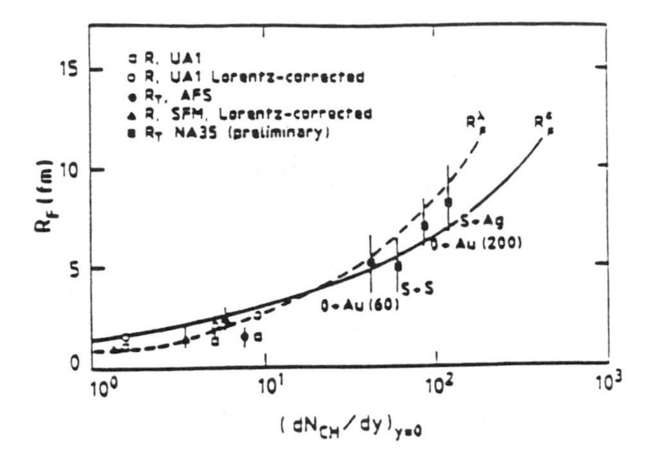

FIGURE 13. Freeze-out radii R_F vs. central charged multiplicity, determined from mean free path (λ) and ideal gas (ϵ), for hadron–hadron and nucleus–nucleus collisions. (From Stock.[21] Reproduced by permission.)

system is inversely proportional to the momentum resolution needed to study the interference between partners, and this may at LHC energy surpass the feasible. Moreover, at a radius of some 20 fm the Coulomb interaction between like-sign charged particles becomes very strong and may mask the Bose–Einstein interference. Hence the proposal[23,24] to study correlations in the longitudinal instead of the transverse dimension is very interesting. It would replace the transverse scale of nuclear size by the longitudinal scale of only one fermi, so that after expansion even by a factor of 5, the corresponding radii would still be less than 5–10 fermi.

The lifetime of the produced bubble in a possible plasma phase can be estimated if we assume longitudinal hydrodynamic expansion; from entropy conservation we

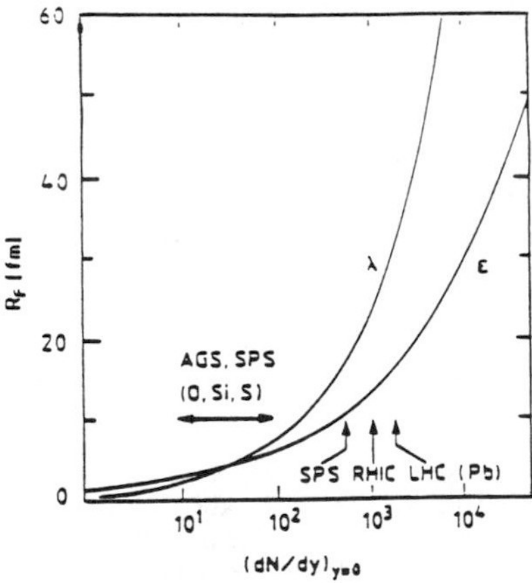

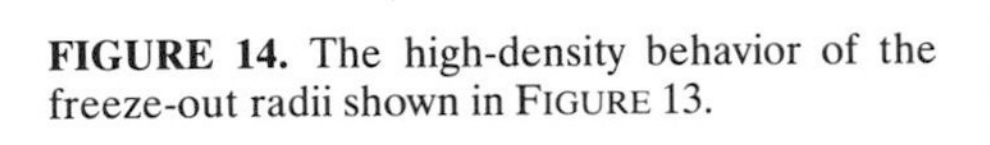

FIGURE 14. The high-density behavior of the freeze-out radii shown in FIGURE 13.

then get

$$\tau_c = \tau_0(T_0/T_c)^3, \tag{14}$$

with $\tau_0 \simeq 1$ fm for the initial state formation time. This gives

	SPS	RHIC	LHC
τ_c[fm]	1.2–2.0	2.4–3.2	3.6–6.5

(15)

for the plasma lifetime at the various energies. A higher initial temperature, and even more so a first-order transition, would prolong considerably the time until the system is completely hadronized.

THE ONSET OF THERMALIZATION

How can we test whether the system produced in a heavy-ion collision has reached thermal equilibrium? The starting point of the collision is evidently the interaction of individual nucleons in the projectile with those in the target. The extreme opposite to thermalization is thus a superposition of nucleon–nucleon collisions, with each projectile nucleon interacting with just one target nucleon. On the way to thermalization, we have "rescattering": a given nucleon will interact with more than one other nucleon, it will interact with secondaries produced in previous collisions, and these secondaries will interact with each other. Particle ratios provide us with an effective tool to check whether rescattering has occurred and brought us closer to thermal equilibrium. Consider as an example the production ratio K^+/π^+. In p–p interactions at $\sqrt{s} = 20$ GeV, it is found to be 0.07 ± 0.02.[25] Rescattering will

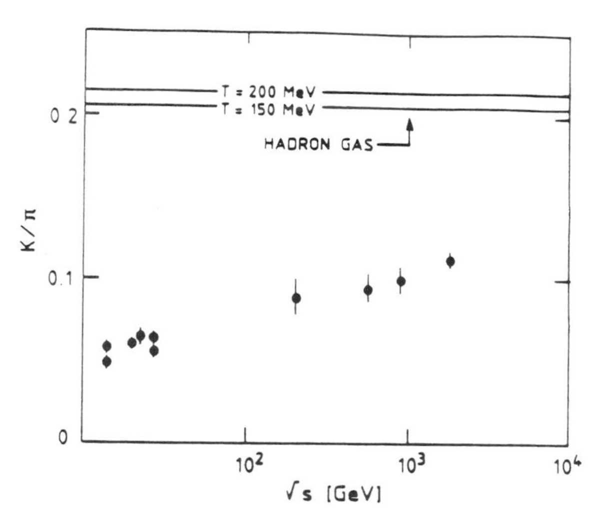

FIGURE 15. The K/π ratio in p–p and p–$\bar{p}$ collisions as a function of the incident energy $\sqrt{s}$. (From Ansorge *et al.*[25] Reproduced by permission.) Also shown are the hadron gas limits for two temperatures. (From Cleymans *et al.*[26] Reproduced by permission.)

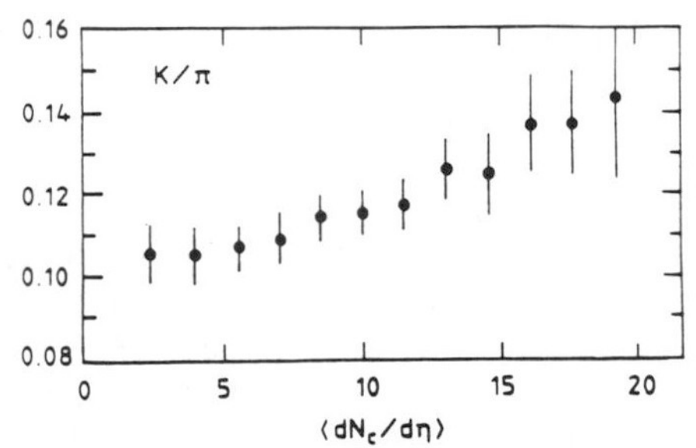

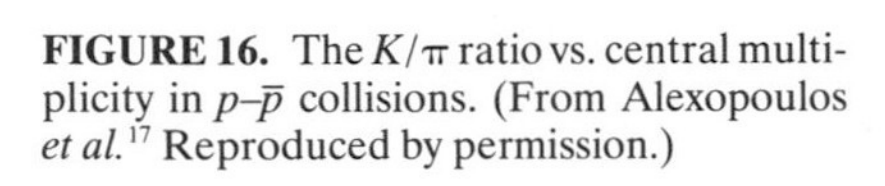

FIGURE 16. The K/π ratio vs. central multiplicity in p–$\bar{p}$ collisions. (From Alexopoulos *et al.*[17] Reproduced by permission.)

increase this ratio, and after "enough" rescattering, there will be equilibrium. The K^+/π^+ ratio attains a value of about 0.20 in an equilibrium hadron gas at $T = 150$ MeV and vanishing baryon number density.[26] We therefore want to check how far nuclear collisions bring us on the road to equilibrium, and the observation of $K^+/\pi^+ \simeq 0.19 \pm 0.03$ in Si–Au collisions at the AGS[27] certainly points toward an onset of thermal behavior. Data from the SPS are in agreement with this. At $\sqrt{s} = 20$ GeV, central S–S collisions[28] give 0.15 ± 0.03 at midrapidity for the K/π ratio, to be compared to the quoted p–p value; central S–W collisions give about 0.2 at this energy.[29] In general, we expect to move toward thermalization by increasing the effective A at fixed $\sqrt{s}$ and by increasing $\sqrt{s}$ at fixed A: in both cases, the number of secondaries, and hence the chance for rescattering increases. How far particle ratios such as the considered K/π ratio increase in event generator codes that include only one or two rescatterings and thus are probably not yet thermal still has to be checked, however. Furthermore, it should be noted that in p–p and p–$\bar{p}$ interactions, the K/π ratio increases with increasing $\sqrt{s}$ (FIG. 15)[25] as well as with increasing multiplicity (FIG. 16).[17] This could be due to rescattering among the produced secondaries, but as well to enhanced strangeness production in minijets. Both possibilities can again be checked by event generator studies.

The onset of thermalization can thus be tested by studying the evolution of strangeness away from the p–p level, for kaons as well as hyperons. We have here implied that a gas of noninteracting hadrons is the equilibrated end-phase; as well, one could consider the ratios obtained from an equilibrium quark–gluon plasma, which expands and subsequently hadronizes. It is found, however, that for most ratios the two scenarios agree,[30] even though it may take longer to attain equilibrium by hadron interactions alone.[31] There may be cases, however, for which hadron gas and quark–gluon plasma do lead to very different ratios, and hence the study of strangeness evolution has also been proposed as a way to obtain information about the primordial state;[31,32] we shall return to this point later on.

PRIMORDIAL FEATURES

There are two ways to test whether the system produced in a high-energy heavy-ion collision was in its early "primordial" history in a deconfined state.

We can look for signals that are produced at such early times and are not affected by the subsequent hadronization. Possible signals of this type are thermal dileptons or direct photons, which are emitted by the plasma and then escape.[33-35] In this spirit, one may also study the effect of the produced medium on the observed production rates of heavy quark bound states[36,37] or hard jets;[38] their initial production is nonthermal, presumably occurs very early in the collision, and can be understood reasonably well in terms of perturbative QCD.

Another approach is to look for primordial remnants in the observed hadronic features. Possible candidates considered in this vein are discontinuities in the momentum distribution of the secondaries, reflecting a first-order phase transition;[39] particle ratios that are significantly different for a hadron gas and a hadronizing quark–gluon plasma;[31,32] and droplets of strange matter, baryonic states of very low charge to mass ratio.[40] Here, however, we shall restrict ourselves to thermal dileptons and photons and to the study of heavy-quark bound states in dense matter.

Thermal Dileptons and Direct Photons

There are several sources for dilepton production in nuclear collisions. They are produced in the decay of the low-mass vector mesons ρ, ω, and ϕ. Thermal dileptons are emitted when $\pi^+\pi^-$ or $q\bar{q}$ pairs annihilate in a pion or quark gas. Finally, there are comparatively rare "hard" interactions between incident partons at a very early stage of the collision, leading to Drell–Yan production or to the production of heavy ($c\bar{c}$ or $b\bar{b}$) vector mesons that subsequently decay into lepton pairs. The distinction between dileptons from low-mass and high-mass vector mesons, as well as that between thermal dileptons from $q\bar{q}$ annihilation and Drell–Yan pairs is somewhat arbitrary; it is mainly motivated by the fact that the production mechanism of high-mass dileptons seems to be comparatively well understood in terms of perturbative QCD. In the typical "soft" hadronic regime around the ρ, ω, and ϕ this is not the case. However, it may well be that below the ρ region, for very soft dileptons, finite temperature perturbation theory may become applicable; this region has been the subject of much recent theoretical work.[41-43]

Thermal dileptons and Drell–Yan pairs moreover lead to different functional behavior in the dilepton mass M. We have

$$(d^2\sigma/dM^2\,dy)_{y=0} \sim \exp\,-\,(M/T) \tag{16}$$

for thermal pairs emitted from a system at temperature T, and

$$(d^2\sigma/dM^2\,dy)_{y=0} \sim M^{-4}f(\tau) \tag{17}$$

for Drell–Yan pairs, with $\tau \equiv M^2/s$. Equation (**17**) does not include contributions from scaling violation terms; these are, however, expected to contribute more at large P_T and not greatly affect the integrated cross section. Since the mass distribution of thermal dileptons contains directly the temperature of the emitting system, it has been proposed as a "thermometer" for strongly interacting matter.[34] Note here that if used in this way, it does not tell us whether the system is deconfined or not, since a pion gas as well as a quark–gluon plasma would emit thermal dileptons. In addition, however, we have to ask under what conditions thermal dileptons are at all observable: Is there some window between the "hadronic" region around the low-mass vector mesons and the high-mass Drell–Yan regime, in which thermal pairs should dominate? Theoretical studies[44] based on what we today consider extreme temperatures (T = 500–800 MeV) led to significant thermal production in the mass region above 2.5 GeV, dominating the Drell–Yan distribution. To check this more quantitatively, we need calculations of the thermal spectrum, based on LHC conditions as upper limit, as well as the corresponding Drell–Yan calculations for Pb–Pb collisions, with $3 \leq M \leq 10$ GeV, including higher order QCD contributions. For the latter task, we encounter some further problems: since at the LHC even M = 10 GeV leads to $\tau \simeq 10^{-6}$, we need quark and gluon structure functions at very small x, where they are not yet known. Moreover, it is known that at small x there is "shadowing" in nuclear collisions, that is, the structure function in a nucleus is at small x reduced in comparison to that in a nucleon. Detailed studies lead us to expect for Drell–Yan production at the LHC a reduction of up to 50 percent.[45]

Both thermal[46] and Drell–Yan[47] production have now been calculated for LHC conditions; for the latter, the DFLM[48] structure functions were used, without any nuclear shadowing. The result is shown in FIGURE 17 for two different initial-

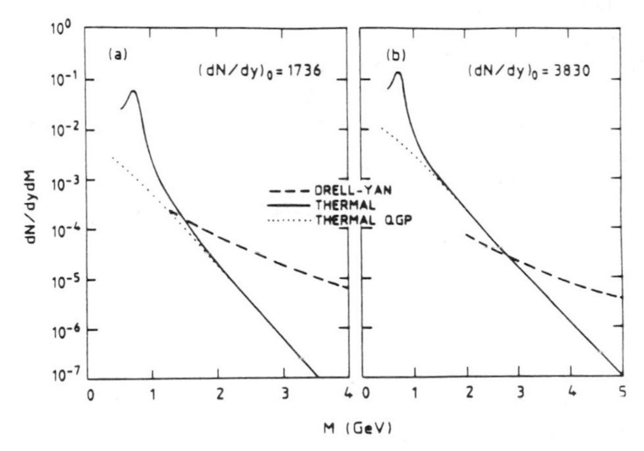

FIGURE 17. Production rates for thermal[46] and Drell–Yan dileptons,[47] as function of the dilepton mass M.

temperature values, corresponding to our conservative density estimate for the lower and to about twice that density for the higher value. We conclude that for the lower density, there does not seem to be a clear-cut window for the observation of thermal dileptons; this, incidentally, is in accord with the observation that the mass distribution for $1.7 \leq M \leq 3.5$ GeV in present dilepton data from nuclear collisions[49] is in accord with the functional form found in p–p Drell–Yan production. At the higher density, the situation improves somewhat. All in all, the usefulness of thermal dileptons as a thermometer for strongly interacting matter is not *a priori* clear; if we can get to densities at or above the upper end of our estimated average, it appears possible.

For direct photons, the main competition at low momenta comes from the decay of hadrons, mainly π^0 and η.[50] At high momenta, there are in addition direct photons from Compton scattering. The rates for thermal photons were compared[46] to those

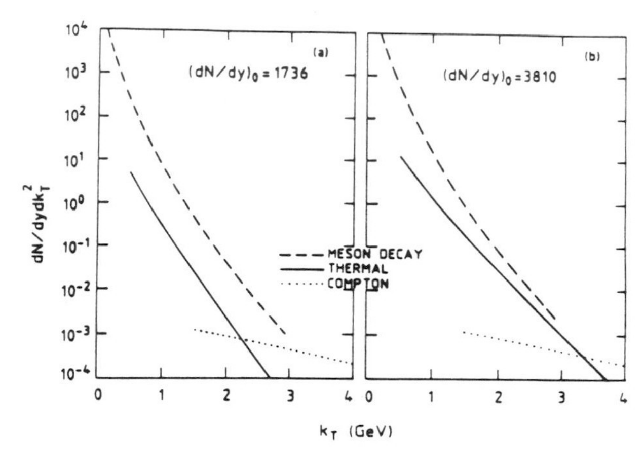

FIGURE 18. Production rates for direct (thermal) photons[46] compared to photons from hadron decay[50] and Compton scattering.

from the other two mechanisms; the result is shown in FIGURE 18, again for two possible densities. At low density, the thermal photon rate is a factor of 20 lower than that from hadron decays; at higher density, this improves, particularly at higher momenta. The feasibility thus depends very much on how well the hadron decays can be identified and eliminated. At low density, signal and background moreover have almost the same functional form; for higher densities (higher T and hence flatter thermal distributions) this again improves.

The Spectral Analysis of $c\bar{c}$ *and* $b\bar{b}$ *States*

In view of the problems encountered for thermal dileptons and photons as tools to probe the primordial features of strongly interacting matter, another means of analysis would certainly be very helpful. This may be provided by studying the spectra

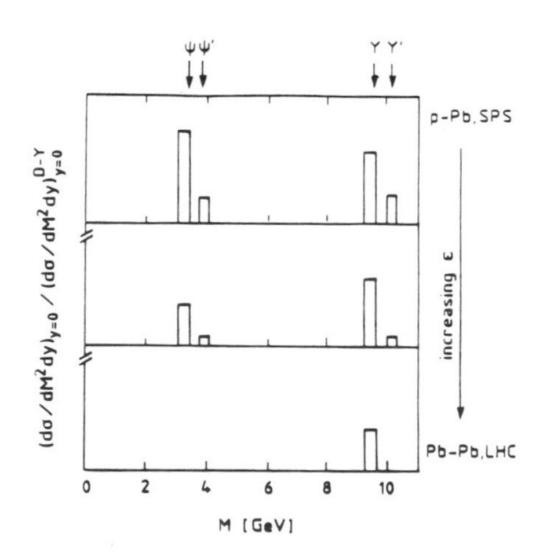

FIGURE 19. Schematic illustration of the spectral analysis of strongly interacting matter, using $c\bar{c}$ and $b\bar{b}$ bound states. (From Karsch and Satz.[12] Reproduced by permission.)

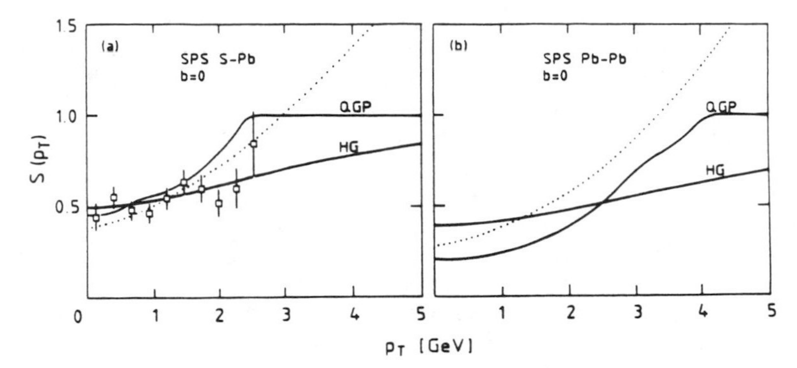

FIGURE 20. J/ψ suppression as a function of transverse momentum, in central S–U (**a**) and Pb–Pb (**b**) collisions at the SPS. The behavior by deconfinement (QGP), absorption (HG), and absorption after initial state scattering (IPS) is calculated with parameters tuned for S–U collisions.[12] (Data in (**a**) are from Abreu *et al.*[51] Reproduced by permission.)

of heavy-quark bound states ($c\bar{c}$ and $b\bar{b}$) produced in heavy-ion collisions. A suppression of the J/ψ signal relative to the Drell–Yan continuum had, in fact, been predicted as a signature for deconfinement,[36] and was subsequently observed by the NA38 collaboration at the CERN-SPS.[51] The observed features have in the meantime also been accounted for by absorption in dense hadronic matter, coupled with initial-state parton scattering.[52,53] The effect thus does seem to establish the production of dense, strongly interacting matter; to check whether this matter is already deconfined or still in a hadronic state, further experimental study is required. But we can use it as a starting point for a spectral analysis of QCD matter,[12] very similar to the spectral analysis of stellar matter used in astrophysics. Stellar matter emits radiation containing spectral lines from the excitation and ionization of various elements. The hotter the matter is, the lower the intensity of the spectral lines from low-lying atomic excitation/ionization states: with increasing temperature, these states become "suppressed" by thermal excitation. In much the same way, we expect

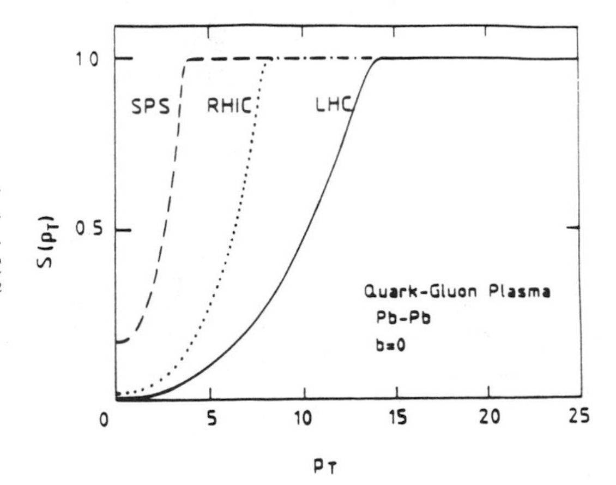

FIGURE 21. ψ suppression by deconfinement, as a function of transverse momentum, for central collisions at SPS, RHIC, and LHC energies. (From Karsch and Satz.[12] Reproduced by permission.)

"cold" QCD matter to show charmonium and bottonium signals at the same rate, relative to the Drell–Yan continuum, as in *p–p* collisions; an increase in temperature should then lead to J/ψ and ψ' suppression, and still further increase to stronger suppression and to that of higher mass bound states (FIG. 19). Different suppression mechanisms (deconfinement, absorption) will also lead to different suppression patterns in energy density and in the P_T of the bound states. In FIGURE 20, we show the predicted P_T distributions for J/ψ production in *S–U* and Pb–Pb collisions at the SPS, based on deconfinement, absorption, and absorption with initial state parton scattering as suppression mechanisms. We note that the present *S–U* data are in accord with all three schemes; considerable differences in the predictions arise, however, already for the large P_T behavior in Pb–Pb at the SPS. In FIGURE 21, the suppression patterns of the ψ at SPS, RHIC, and LHC energies are shown, as predicted by deconfinement. The difference between the suppression patterns by deconfinement and absorption is in FIGURE 22 illustrated for ψ and the Υ production at the LHC. The basis for the proposed analysis is certainly not yet complete,

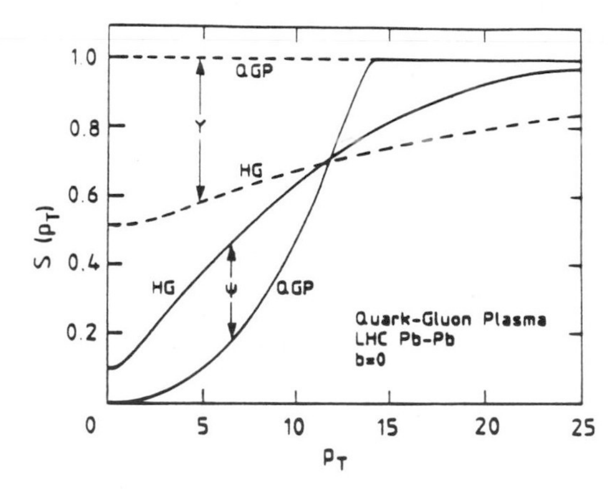

FIGURE 22. ψ and Υ suppression for central collisions at LHC energy, by deconfinement (QGP) and absorption (HG). (From Karsch and Satz.[12] Reproduced by permission.)

however: at TeV energies and large P_T, there is abundant J/ψ production from *B* decay,[54] and much of the observable Υ production will come from χ_b decay.[55] Both these effects have to be taken into account in a realistic analysis of charmonium and bottonium production in high-energy nuclear collisions.

CONCLUSIONS

We have seen that statistical QCD, in particular through the evaluation of the lattice formulation by means of computer simulation, predicts and describes the transition from hadronic matter to a quark–gluon plasma. On the experimental side, high-energy nucleus–nucleus collisions should provide us with an excellent tool for the experimental study of strongly interacting matter in general, and of the quark–gluon phase transition in particular. The results obtained so far, both theoretically and experimentally, are sufficiently promising to suggest that a study of the little bang in the laboratory is within reach.

REFERENCES

1. For a recent survey, see PETERSSON, B. 1991. Nucl. Phys. **525A:** 237c.
2. GOTTLIEB, S. *et al.* 1987. Phys. Rev. **D35:** 3972.
3. KARSCH, F. & H. W. WYLD. 1988. Phys. Lett. **B213:** 505.
4. ENGELS, J., *et al.* 1990. Non-perturbative thermodynamics of SU(N) gauge theories. Bielefeld Preprint BI-TP 29/90.
5. BROWN, F. R., *et al.* 1990. On the existence of a phase transition for QCD with three light quarks. Columbia Preprint CU-TP-470.
6. GERBER, P. & H. LEUTWYLER. 1989. Nucl. Phys. **B231:** 387.
7. BARDUCCI, A., *et al.* 1989. Chiral phase transitions in QCD for finite temperature and density. University of Geneva Preprint UGVA-DPT 1989/09-629.
8. ABE, F., *et al.* (CDF). 1990. Phys. Rev. **D41:** 2330.
9. HEINZ, U. 1991. *In* Proceedings of the Large Hadron Collider Workshop, Vol. II, CERN 90-10, 1990, p. 1079.
10. BJORKEN, J. D. 1983. Phys. Rev. **D27:** 140.
11. GYULASSY, M. & A. IWAZAKI. 1985. Phys. Lett. **B165:** 157; KERMAN, A., T. MATSUI & B. SVETITSKY. 1986. Phys. Rev. Lett. **56:** 219.
12. KARSCH, F. & H. SATZ. 1991. Z. Phys. C **51:** 209.
13. LUDLAM, T. & N. P. SAMIOS. 1988. Z. Phys. **C38:** 353.
14. BRANDT, D. 1990. Relativistic heavy ions in the LHC. CERN Note LHC/Note No. 87.
15. GYULASSY, M. 1984. *In* Multiparticle Dynamics 1983, P. Yager and J. F. Gunion, Eds. World Scientific. Singapore.
16. HWA, R. & K. KAJANTIE. 1985. Phys. Rev. **D32:** 1109.
17. ALEXOPOULOS, T., *et al.* (E 735). 1990. Phys. Rev. Lett. **64:** 991.
18. ANISHETTY, R., P. KOEHLER & L. MCLERRAN. 1980. Phys. Rev. **D22:** 2793.
19. See the reports of CERELLO, P., P. GIUBELLINO, U. GÖRLACH, M. MASERA, L. RAMELLO, J. RANFT & K. WERNER. 1991. *In* Proceedings of the Large Hadron Collider Workshop, Vol. II, CERN 90-10, 1990, p. 1091.
20. MEKJIAN, A. Z. 1978. Phys. Rev. **C17:** 1051.
21. For a recent survey, see STOCK, R. 1990. Pion interferometry in high energy hadron and nuclear collisions: Radial expansion and freeze-out. Frankfurt Preprint IKF 90-3.
22. GOITY, J. L. & H. LEUTWYLER. 1989. Phys. Lett. **B228:** 517.
23. MAKHLIN, A. N. & YU. M. SINYUKOV. 1988. Z. Phys. **C39:** 69.
24. FERENC, D. (NA35). 1990. Pion interferometry in S + AG collisions at 200 GeV/nucleon. Zagreb (Rudjer Boskovic). Preprint.
25. ANSORGE, R. E., *et al.* (UA5). 1988. Z. Phys. **C41:** 179.
26. CLEYMANS, J., *et al.* 1990. Phys. Lett. **B242:** 111.
27. ABBOTT, T., *et al.* (E802). 1990. Phys. Rev. Lett. **64:** 847.
28. STOCK, R. (NA35). 1991. Nucl. Phys. **525A:** 221c.
29. VAN HECKE, H. (HELIOS). 1991. Nucl. Phys. **525A:** 227c.
30. LEE, K. S., M. RHOADES-BROWN & U. HEINZ. 1988. Phys. Rev. **C37:** 1452.
31. For a survey, see KOCH, P., B. MÜLLER & J. RAFELSKI. 1986. Phys. Rep. **142:** 142.
32. For a recent survey, see EGGERS, H. C. & J. RAFELSKI. Strangeness and quark gluon plasma: Aspects of theory and experiment. 1990. GSI Preprint GSI-90-37.
33. FEINBERG, E. L. 1976. Nuovo Cimento **34A:** 39; SHURYAK, E. V. 1978. Sov. J. Nucl. Phys. **28:** 408.
34. For a survey, see GYULASSY, M.1984. Nucl. Phys. **A418:** 59c.
35. For a recent survey, see RUUSKANEN, P. V. 1990. *In* Quark Gluon Plasma, R. C. Hwa, Ed. World Scientific. Singapore.
36. MATSUI, T. & H. SATZ. 1986. Phys. Lett. **B178:** 416.
37. KARSCH, F. & R. PETRONZIO. 1987. Phys. Lett. **B193:** 105.
38. BJORKEN, J. D. 1982. Energy loss of energetic partons in quark-gluon plasma: Possible

extinction of high P_T jets in hadron-hadron collisions. Fermilab Preprint Pub-82/59-THY.
39. VAN HOVE, L. 1982. Phys. Lett. **B118:** 138.
40. GREINER, C., P. KOCH & H. STÖCKER. 1987. Phys. Rev. Lett. **58:** 1825.
41. BRAATEN, E., R. PISARSKI & T.-C. YUAN. 1990. Production of soft dileptons in the quark-gluon plasma. Brookhaven Preprint BNL 43882.
42. WELDON, H. A. 1990. Measuring the quark-gluon T_c with e^+e^- pairs. VIIIth International Conference on Ultrarelativistic Nucleus-Nucleus Collisions (Menton, France, May 1990).
43. For a recent survey, see SHURYAK, E. V. Physics of hot hadronic matter and quark-gluon plasma. Brookhaven Preprint BNL-44859.
44. KAJANTIE, K., *et al.* 1986. Phys. Rev. **D34:** 2746.
45. CASTORINA, P. & A. DONNACHIE. 1989. Z. Phys. **C45:** 141; CASTORINA, P. 1991. Report at this Workshop, Proceedings Vol. III.
46. RUUSKANEN, P. V. 1990. Emission of thermal dileptons in ultrarelativistic heavy-ion collisions. VIIIth International Conference on Ultrarelativistic Nucleus-Nucleus Collisions (Menton, France, May 1990); RUUSKANEN, P. V. 1990. Jyväskylä Preprint JYFL 12/90; RUUSKANEN, P. V. 1991.
47. GUPTA, S. 1991.
48. DIEMOZ, M., *et al.* 1988. Z. Phys. **C39:** 21.
49. DRAPIER, O. (NA38). 1990. Thesis, University of Lyon, Lyon, France, May 1990.
50. NEUBERT, M. 1989. Z. Phys. **C42:** 231.
51. ABREU, M. C., *et al.* (NA38). 1988. Z. Phys. **C38:** 117; BAGLIN, C., *et al.* (NA38). 1989. Phys. Lett. **B220:** 471.
52. For a recent survey, see SATZ, H. 1990. *In* Quark Gluon Plasma. R. C. Hwa, Ed. World Scientific. Singapore.
53. GAVIN, S. 1991.
54. GLOVER, E. W. N., A. D. MARTIN & W. J. STIRLING. 1988. Z. Phys. **C38:** 473.
55. BERGSTRÖM, L., R. W. ROBINETT & L. WEINKAUF. 1990. Aspects of ψ and Y production at supercollider energies. Pennsylvania State University Preprint PSU/TH/63.

The Epoch of Galaxy Formation

SIMON D. M. WHITE

Institute of Astronomy
Madingley Road
Cambridge CB3 0HA, England

INTRODUCTION

This paper combines a review of recent observational work on the state of the Universe at redshifts between 0.5 and 5, with a discussion of some current ideas about how galaxies may have formed. The two parts are related in that the first implicitly assumes that this time period includes most galaxy formation, as is indeed the case in the cold dark matter (CDM) theories on which most of the theoretical discussion is based. The epoch of galaxy formation was the subject of a very successful workshop held in 1988 in Durham, U.K. The proceedings of that meeting provide an excellent account of the state of the field two years ago, as well as a relatively complete source for references to earlier work.[1] As a result I will try to emphasize more recent developments in the present article, and I will not attempt a full review. Other relevant material will be found in this volume in the articles of Boyle, Cowie, Tyson, and Wright,[2] and in some of the contributed papers to the minisymposium on large-scale structure and galaxy formation.

It was clear from the Durham meeting that there is a wide spectrum of opinion about what is meant by galaxy formation, and about when and in what order it might have occurred. Possible definitions include the moment when the first stars were formed, when the first heavy elements were formed, when half the observed stars were formed, or when the present structure first came to equilibrium. The earliest, and most naive, formation models presupposed the rapid and monolithic collapse of a single protogalactic cloud. All these times are then more or less equivalent. However, at least as far back as the Yale meeting of 1977, both observational and theoretical arguments suggested that much more inhomogeneous models were required, that the epochs of star formation and of galaxy assembly could be spread out in time, and that they need not coincide (see the contributions by Searle, by Toomre, and by Rees in reference 3). Observations of extragalactic HII regions, of infrared-bright starburst galaxies, of high redshift radio galaxies, of merging systems, and of cooling flows all suggest possible paradigms for galaxy formation, but with a very wide range of formation sequences and of observable properties. I shall generally take galaxy formation to be the period during which star formation rates are highest and the bulk of the observed stars are formed. This period may extend to the present day for late-type galaxies, and it may substantially precede the equilibration of any elliptical galaxy produced by a merger.

For a long time the predominant, although by no means universal, view was that galaxy formation occurred at relatively high redshift. This was based mainly on the failure of very deep searches to detect line or continuum emission from "obvious" protogalaxies.[4] Further support came from evidence that the stellar populations in elliptical galaxies are all old, even in objects observed at redshifts up to 0.7. In addition, globular clusters in our own Galaxy seemed to show very little scatter in age, and to be only slightly younger than (or, perhaps, even older than!) the Universe. Recent work has pushed further on these points, but without resolving the fundamental issues. Searches for faint objects have revealed a very large population of blue, apparently star-forming systems that seem to be at redshifts of 1 to 3. However, the relationship of these systems to present-day galaxies is far from clear; they are almost certainly not the initial burst of star formation associated with a typical bright galaxy (see contributions by Tyson[2] and Cowie[2] to this volume). Red stellar populations have now been seen in radio galaxies out to redshifts beyond 3, but there is considerable argument about whether they are, in fact, "old" ($>10^9$ y in this context, see the third section); it is, in any case, very dangerous to treat such systems as representative of normal galaxies.

Some of the observational indications in favor of *recent* galaxy formation have also been appreciated for many years. A trivial point is that perhaps half the observed stars in the Universe are in disks, and that *all* the disk stars observed locally seem to be substantially younger than the globular clusters, and thus to have formed at low redshift. Star formation rates in many late-type disk galaxies are currently as high as they ever need to have been in the past. Thus models that predict spectacular early bursts of star formation, if they apply at all, probably apply to bulges of bright spirals and to luminous elliptical galaxies. It seems natural to associate the peak in the abundance of quasars at redshifts of 2 to 3 with the collapse of the central regions of such systems. However, the conditions for quasar formation are still poorly understood, and recent data have, if anything, reduced the evidence for a substantial fall-off beyond $z = 3$ (see Boyle's article[2] in this volume). The faint blue galaxies already referred to constitute a considerable stellar population forming at $z < 3$, probably more, and perhaps much more, than 10 percent of all observed stars; furthermore, enough HI is observed directly at $z \sim 2$ to make most of these stars (see the next section). Although both these facts seem to indicate substantial recent activity, the details of their relation to galaxy formation remain obscure.

The epoch since $z = 4$ may therefore include the formation of most stars in galaxies, but the case is still far from proved. I shall assume this to be so for much of the following discussion, because there are very few observational constraints at higher redshift, and because the models of the third and fourth sections predict such recent formation. I will not further discuss constraints from faint galaxy counts, colors, and clustering, nor will I discuss quasar abundances, nor possible constraints from COBE on the infrared background, and thus on galaxy formation; these questions are reviewed elsewhere in this volume. However, I will give brief reviews of recent work on quasar absorption lines (the following section) and on radio galaxies at high redshift (the third section). There has been considerable progress in these fields over the last two years, and they give us considerable information about the state of the Universe at moderate redshift. In the third section I present an attempt to understand how quasar absorption line clouds might be related to pregalactic and galactic structures in a cold dark matter universe. Finally, in the fourth section I give a more general discussion of galaxy formation in this and related hierarchical clustering models.

QSO ABSORPTION LINES

There are now about ten quasars known with redshifts exceeding 4. The spectra of these objects show the same variety of features as those of nearer bright systems, the only clear systematic difference being a considerable increase in the line density of the Ly-α forest longward of the quasi-stellar object (QSO) Ly-α emission.[5] Absorption by low column density neutral hydrogen is thus substantially more frequent at higher redshift. This may reflect a greater abundance of the absorbing clouds, or a decreased efficiency in their ionization by the general ultraviolet background. The quasar continua seem, nevertheless, to approach an unabsorbed level between the clouds, suggesting that any diffuse intergalactic medium is either absent or highly ionized at redshifts approaching 5. Recent very high-resolution observations of Ly-α forest clouds have led to a controversy about their physical state.[6] A number of the clouds appear to have linewidths that are too narrow to be consistent with the standard model, in which they are pressure-confined systems ionized by the UV background. In addition, their linewidth appears to correlate positively with column density, a result that is also unexpected in the standard model, and that suggests that the lines could be broadened by bulk motions rather than thermal effects. It may be possible to understand these results (which need independent confirmation) in terms of adiabatic cooling of expanding pressure-confined clouds.[7] Otherwise they appear to require the clouds to be very dense, very small (at least along the line-of-sight), and neutral. Another puzzling property is the fact that these systems are extremely weakly clustered along the line-of-sight; their correlations appear substantially weaker than would be expected for galaxies at the same redshift.[6] I discuss a possible physical model for the Ly-α clouds in the fourth section.

Clouds with larger neutral hydrogen column densities are less common, and usually also show absorption from heavier elements produced in stars. Such contamination suggests that these objects are related to galaxies. Their abundance variation with redshift may give direct information about galaxy formation, but is open to multiple interpretations.[8] Thus the abundance of systems with large enough column density to be opaque at the Lyman limit (912 Å: this corresponds to $N_{HI} > 10^{17}$ cm^{-2}) decreases from $z = 3$ to $z = 1$, whereas the abundance of systems with strong absorption in the C IV doublet (near 1550 Å) increases from $z = 3$ to $z = 1.5$. This may reflect the conversion of hydrogen into stars, with the consequent production of carbon, or it may be caused by changes in the amount of ionizing radiation, or it may be due to both. At lower redshifts, the abundance of clouds with strong absorption in the Mg II doublet (near 2800 Å) is found to *decrease* from $z = 1.5$ to $z = 0.3$. These absorption systems are known to be associated with the halos of normal bright galaxies,[9] and appear to indicate that heavy elements are mixed well beyond the regions in which stars are forming. In addition, in contrast with the Ly-α forest, such metal line systems are quite strongly clustered, and, indeed, there are some indications for superclustering of C IV systems that is substantially stronger than anything seen in the nearby galaxy distribution.[10]

Among the most intriguing systems seen in absorption against QSOs are those in which the column density of neutral hydrogen is so high ($n_{HI} \sim 10^{21}$ cm^{-2}) that the Ly-α profile is extremely broad and is dominated by its damping wings. Such column densities are typical for lines-of-sight through the disks of present-day spiral galaxies, and a number of properties of these absorption systems (for example, their size and their low turbulent velocities) suggest that they may be related to galaxy disks.[11] On the other hand, their metal and dust content is about a tenth of that seen in the interstellar medium of galaxies like our own,[12,13] and there are stringent upper limits on their possible molecular hydrogen abundance ($\sim 10^{-5}$ of the local value).[14] The total amount of neutral gas seen in such systems is surprisingly large. For $\Omega = 1$ and $H_0 = 50$ km/s/Mpc a recent estimate gives $\Omega_g = 0.003$ in gas at $z = 2$ after allowance for helium.[15] This is quite comparable to the amount of baryons seen as stars in the local universe, and greatly exceeds that seen as cold gas. This suggests that the damped Ly-α clouds may contain a significant fraction of the raw material needed for galaxy formation. The situation is somewhat less extreme if the universe is open, since the pathlength at high redshift is increased, and the inferred Ω_g is decreased. This effect is even greater in universes with a significant and positive cosmological constant. If these objects were really protogalaxies one might expect them to show the emission lines usually associated with star-formation regions. However, with one controversial exception,[16] Ly-α has never been detected in emission from damped systems, and the upper limits derived correspond to very low star-formation rates unless Ly-α is very efficiently absorbed by the small amount of dust these objects contain.[17] Such an explanation cannot be excluded at present, and it is important to search in the near-infrared for other emission lines, particularly [OII]3727, which is very strong in nearby star-forming regions.

In conclusion, the gas clouds detected as QSO absorption lines contain enough material to make galaxies, and show some signs of progressive metal enrichment. However, their clustering properties are complex and appear to vary with mass; where they are detected at modest redshift, they are associated with galaxy halos rather than with the regions containing stars; and they show evidence neither for the molecular content nor for the emission lines normally associated with star-forming regions. It seems likely that their properties provide valuable information about galaxy formation, but for the moment it is unclear how the various parts of the puzzle fall into place.

DISTANT RADIO GALAXIES

In the last few years a number of galaxies have been found with redshifts well beyond $z = 1$. With one exception (an object detected because of strong gravitational lensing by a foreground cluster[18]) all these objects were known first from their radio emission, were then identified on deep optical images, and finally had redshifts determined from optical emission lines. A discussion of the objects known in 1988 can be found in a number of articles in reference 1. Since that time a number of even more distant objects have been found by concentrating on faint radio sources with the steep radio spectra usually associated with intrinsically powerful radio galaxies. The current record holder is 4C41.17 at $z = 3.80$.[19] The spectrum of this galaxy has a flat and barely detected continuum. The single strong emission line (observed near 5800 Å) is identified as Ly-α, while emission from C IV (1550 Å rest) is very marginally detected. The lack of other strong lines, and a possible continuum jump across the strong line (due to the onset of Ly-α forest absorption?), are adduced as additional arguments in favor of the identification. The redshift is not definitive, but is probably correct.

The morphology of 4C41.17 in radio, optical, and infrared images is in many ways representative of these distant objects, although they are by no means all similar.[19] Its radio structure shows four strong lobes, an outer pair separated by about 12 arcsec., and an inner pair with a slightly different axis separated by about 2 arcsec. The optical and infrared emission is centered on one of the inner pair of lobes, but has considerable structure. Narrowband images show that the Ly-α emission, while concentrated to the galaxy "center," is considerably elongated along the axis of the inner radio lobes, and has a total extent exceeding 60 kpc (for $H_0 = 50$ km/s/Mpc and $q_0 = 1/2$). The emission in broad-band optical and infrared bands (at 7000, 9000, and 22,000 Å) is not detected to such large distances, but still shows some evidence for the same elongation. The evidence is rather marginal for the critical 2.2-μ K band. Recall that for $z = 3.8$, the Ly-α line is tracing a gas component, the broad optical bands are observing emission from the rest UV, and the K band corresponds approximately to the rest frame B band; thus K is the only band that is expected to contain significant emission from any but the very youngest stars.

The interpretation of such observations has given rise to substantial controversy. The Ly-α fluxes from many of these radio galaxies are very high, more than 100 times the upper limits deduced for the damped Ly-α absorption systems discussed in the last section. For standard models, even assuming negligible internal losses due to dust absorption, such fluxes imply star formation rates exceeding $100M_\odot$/y. However, the large extent of the emission, its large linewidth (often over 1000 km/s), and its alignment with the radio structure suggest explanations that link the exciting radiation directly to the radio source. At the very least, the star formation may be stimulated by the radio jets, rather than being an independent consequence of protogalactic collapse.[20,21] More radically the exciting radiation may be beamed directly from the central engine of the radio source and scattered in the observed gas. The UV continuum, could perhaps be similarly scattered, an interpretation that is supported by the detection of significant polarization.[22,23] Such models are currently rather sketchy, but certainly cannot yet be excluded as a major contribution to the observed light. However, the (rather marginal) detection of stellar absorption lines in the superposed spectra of two galaxies at $z \sim 1.8$ suggests that much of the UV emission in these systems does indeed come from stars.[24] The typical age of these stars is highly uncertain. The bulk of the stellar population is seen only in the K band, and alignment with the radio source (which is the major *prima facie* evidence in favor of a young population) is less convincing at this wavelength. On the other hand, stellar population models with ages between about 10^8 and a few times 10^9 years appear consistent with the rather sparse data currently available for these systems.[25] The relatively small scatter in K-band brightness (a factor of a few) of these radio galaxies argues against their 2.2-μ light being dominated by high mass supergiants from a young starburst,[26] but this point is disputed.[25]

Thus, although the detection of galaxies at high redshift is a major advance in the field, the evolutionary status of these objects remains unclear. They may be in the first phases of star formation, perhaps greatly enhanced by interaction with the radio jets, or they may possess a dominant "old" population on which the UV light of a relatively small starburst is superposed. In addition, the relation of powerful radio galaxies at high redshift to those seen nearby is quite unclear. Local objects show very little evidence for star formation, and the continuity in K band absolute magnitudes could be fortuitous. The formation of these rare, and perhaps extraordinary galaxies may be a dangerous paradigm to use for that of more typical systems.

FORMATION OF SMALL-SCALE STRUCTURE IN A CDM UNIVERSE

It is useful to supplement purely observational searches for galaxy formation with theoretical discussions of when and how this process may have occurred. In recent years one of the most popular cosmogonies has been the cold dark matter theory. This model is based on the hypotheses that the universe is flat, is dominated by some weakly interacting elementary particle that had small random motions at early times, and has the scale-free initial density fluctuations predicted by the simplest inflationary models for the origin of structure. Recent reviews can be found in Frenk's article in this volume[27] and in reference 28. Because this model has well-defined initial conditions from which structure grows gravitationally, it is relatively easy to study its nonlinear evolution using numerical simulations. There has been much investigation of its predictions for galaxy clustering and large-scale structure, but no realistic simulations of galaxy formation itself have so far been carried out. This is partly because a detailed treatment of dissipative hydrodynamics and of star formation is indispensable in such modeling, and is only just now becoming feasible; partly because the extremely rapid evolution of structure that the model predicts requires treatment of a very large range of scales even in dissipationless (N-body) simulations. I here report briefly on such simulations recently carried out by my own group (M. Davis, G. Efstathiou, C. S. Frenk, and S.D.M.W.) to study the growth of structure on scales that might be appropriate to dwarf galaxies or QSO absorption clouds.

FIGURE 1 shows the projected mass density in a cubic region with one side corresponding to 5 Mpc at $z = 0$ (for $H_0 = 50$ km/s/Mpc). At the time shown the fluctuation amplitude implies an rms mass fluctuation of 0.15 in a sphere of radius 16 Mpc; this corresponds to $z = 1.5$ in the highly biased CDM models of our original papers, and to $z \sim 3$ in the more weakly biased models that seem required by recent data on large-scale streaming motions. The simulation shown followed 262,144 particles of mass $3 \times 10^7 M_\odot$ over an expansion factor of 8; it cannot be continued to later times because structures comparable in size to the entire simulated region are beginning to go nonlinear. FIGURE 1 shows a complex structure with clumps with a wide range of sizes as well as a noticeable sheetlike overdensity almost perpendicular to the line-of-sight. The potential wells of the largest dark matter lumps in FIGURE 1 have equivalent circular velocities ($V_c^2 = GM/r$) of about 100 km/s, corresponding to the halos of dwarf galaxies like M33. The volume simulated is at least an order of magnitude too small to expect even a single galaxy as large as our own.

This simulation only followed the evolution of the dark matter distribution. However, if we wish to make contact with observations of QSO absorption lines, we must clearly make some assumptions about the relative distributions of gas and CDM, and in practice there is little choice but to assume that the two density distributions parallel one another. Detailed simulations show that this is indeed the case if the gas is taken to be adiabatic.[29] However, in the present situation, cooling

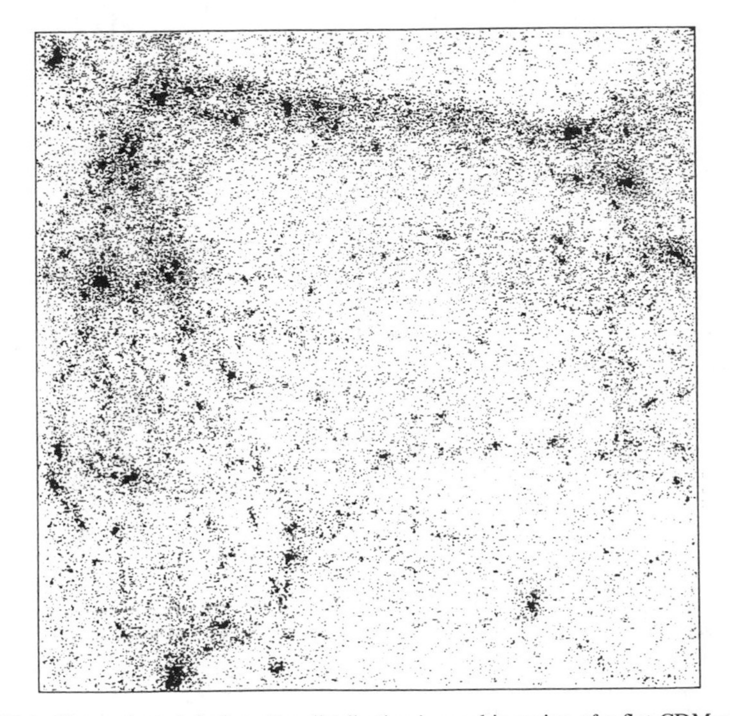

FIGURE 1. The projected dark matter distribution in a cubic region of a flat CDM universe. The side of the cube corresponds to 5 Mpc at the present day for our adopted Hubble constant of $H_0 = 50$ km/s/Mpc. The distribution is seen at redshift, $1 + z = 6.3/b$, where the "bias factor," b, is the usual parametrization of overall fluctuation amplitude in a CDM universe. The highly biased models of Davis *et al.*[39] correspond to $b = 2.5$, but a smaller value of b seems required by recent observations of large-scale flows in the universe.

processes (and, perhaps, also hydrodynamic heating due to star formation) are expected to be highly significant. Thus the assumption that gas follows dark matter may be seriously in error. With the additional assumption that the gas is in ionization equilibrium with a uniform UV background, we are able to calculate the local *neutral* hydrogen density at each point in our models. With standard ionization balance models we find;

$$n_{\rm HI} = 5.7 \times 10^{-13}\delta^2\Omega_b^2(1 + z)^6 T_4^{-3/4}/I_{21}\ {\rm cm}^{-3},$$

where δ is the local mass density in units of the critical density, Ω_b is the fraction of the universe in the form of gas, T_4 is the gas temperature in units of 10^4 K, and I_{21} is the UV background flux in units of 10^{-21} erg/s/cm^2/ster/Hz. The latter flux is approximately that inferred to be present at $z = 2$ if the observed lack of Ly-α forest clouds in the immediate neighborhood of quasars is assumed to reflect the local dominance of UV radiation from the quasar itself[30] (this is the so-called "inverse effect"). The required UV background is larger than can readily be accounted for by the observed population of QSOs at high redshift.[31]

When this model is applied to the simulation of FIGURE 1, we find that along almost all lines-of-sight with high neutral hydrogen column density, the dominant contribution comes from one, or occasionally two dense regions. Thus the simulation is indeed "cloudy." However, if we estimate the probability per unit path length of crossing a cloud with $N_{HI} > 10^{14}$ cm^{-2}, and we try to match the value from the simulation to that observed along real QSO lines-of-sight, we find that $\Omega_b^{-2} I_{21} \sim 50{,}000$ is required at $z \sim 2$. This value applies for our standard assumptions about biasing, for which FIGURE 1 corresponds to $z = 1.5$; for a larger fluctuation amplitude (i.e., weaker bias), an even larger value is implied. Thus if $I_{21} \sim 1$, as inferred from the inverse effect, our assumptions require $\Omega_b \sim 0.005$. Such a small value may be inconsistent with cosmological nucleosynthesis,[32] and is very difficult to reconcile with galaxy-formation models—the gas in large halos is never able to cool, and galaxies do not condense at their centers.[33] This estimate should probably be taken as referring to the amount of gas clustered in the same way as the dark matter. It is considerably below the upper limit on uniform intergalactic gas, $\Omega_b < 0.07$, inferred for the same value of I_{21} from the absence of any detectable continuum depression *between* the Ly-α absorption lines in quasar spectra.[34] In a CDM universe with a mean $\Omega_b \sim 0.1$ the gas either must avoid clustering with the dark matter, or must turn into stars in dark halos with high efficiency, or must be expelled at least partially from halos, presumably as a result of energy input from supernovae and stellar winds. The second possibility seems implausible, since less than 1 percent of the critical density is currently observed in the form of stars.

Assuming that some such effect can reduce the effective gas content of halos substantially, we can ask whether the other properties of the gas distribution in the simulations can agree with the observed properties of quasar absorption clouds. If the reduction in effective neutral density is taken to be the same everywhere, the probability of intersecting a cloud in the model is found to depend on the cloud's column density in the same way as in the observations ($dP/dN_{HI} \propto N_{HI}^{-1.6}$). This can be understood using an argument due to Rees.[35] If the gas is distributed in isothermal halos with $\rho \propto r^{-2}$, which is a fair approximation to the dark matter distribution in halos for a CDM universe, then it is easy to show that the cross section of each halo above column density, N_{HI}, is proportional to $N_{HI}^{-2/3}$. This gives $dP/dN_{HI} \propto N_{HI}^{-5/3}$ for an ensemble of such halos. A linewidth for "clouds" in the simulation can also be estimated from the velocity dispersion of the relevant dark matter particles. The linewidth distribution peaks at a width of about 20 km/s, but has a long tail extending to beyond 100 km/s. There is only a weak correlation of linewidth with column density. These properties are in good qualitative agreement with older observations,[36] but may conflict with the stronger correlation seen in one recent data set.[6] In addition, the simulations predict more clouds of large linewidth than are observed. This may reflect the action of the process responsible for reducing the gas content of halos; for example, it may reflect additional ionization by starlight in the more massive halos. Finally, line-of-sight correlations between clouds can be measured in the models for comparison with the very weak clustering observed in the Ly-α forest. The results turn out to be very sensitive to the assumptions made about line-blending effects. The model correlations are indeed weak, and for some blending models they are as weak as the clustering measurements of Webb.[37]

Thus from these experiments it becomes clear that at $z = 2$, before the formation of a major fraction of at least the disk stars in galaxies, the gas content of the Universe had been substantially affected by hydrodynamic and radiative processes in addition to gravity. It seems that unless the total gas content was very low at that time, the gas must have been distribution significantly more uniformly than the dark matter. If this is indeed the case, the distribution of halo properties in a CDM universe is qualitatively consistent with the linewidths, column densities, and the clustering of quasar absorption line clouds. Possible disagreements could arise from an overabundance of large linewidth clouds ($b \sim 100$ km/s) or from overly strong clustering, but the conflicts appear marginal at present given the major uncertainties in our modeling of the gas.

GALAXY FORMATION IN A CDM UNIVERSE

Studies of the large-scale structure of the universe have tended to ignore the details of galaxy formation. The simplest assumption to make is that the galaxy and mass distributions are identical on large scales. The meaning of "on large scales" is quite vague in this context; measures of the mass-to-light ratio of galaxy groups probe the relative distributions in objects of scale a few hundred kiloparsecs, while comparisons of large-scale motions to the galaxy distribution in the "Great Attractor" refer to a scale about two orders of magnitude larger. The approximate constancy of the derived values of mass per galaxy over this range is often taken as evidence that the global mean has been reached, and thus that the Universe is open ($\Omega \sim 0.1$). In fact, the uncertainties would allow quite a strong trend with scale, and a globally flat universe certainly cannot be ruled out on the basis of current dynamical data. In fact, the assumption that the mass "follows" the observed galaxy distribution is one of convenience, and masks uncertainties about the true relation between the two. This must depend on the nature of the dark matter and on the galaxy-formation process. The assumption can have at most limited validity, since the distributions of different types of galaxy (e.g., ellipticals and spirals) are observed to differ substantially.

A somewhat more sophisticated scheme has become popular in the last seven or eight years. The sites of galaxy formation are identified as high peaks of the smoothed *linear* density fluctuation field at high redshift. This seems physically reasonable because such peaks are closely related to the first objects to collapse with a mass comparable to that of the smoothing scale. Moreover, it results in a prescription for relating the galaxy and mass distributions that is well defined, is simple enough that many properties can be obtained analytically,[38] and is easily implemented in nonlinear simulations that can be directly compared with the observed galaxy distribution.[39] This scheme is usually referred to as "biased galaxy formation," because it predicts enhanced galaxy formation in regions where the mean mass density is high ("protoclusters") and suppressed galaxy formation in regions where it is low ("protovoids"). In such a universe galaxy clustering exaggerates the clustering of the underlying mass, and measurements of the mass of galaxy groups, clusters, and superclusters would produce masses per galaxy that systematically underestimate that of the universe as a whole. The disadvantage of the scheme is that it is highly idealized, and it is unclear how to relate the smoothing scale and the height of the peak to the properties of the resulting galaxy. As a result it is hard to obtain unambiguous predictions for observables such as the relative distributions of bright and faint galaxies. Nevertheless, most versions of the theory predict that faint galaxies should be less clustered than bright ones, while observational evidence suggests that any such effect is quite weak.[40]

With numerical simulations of clustering it is possible to treat galaxy formation somewhat more realistically. Dark matter "halos" can be identified as they form, and galaxies can be assumed to condense within them with properties determined in a plausible way by those of the halo. The major difficulty in such studies is the very large dynamical range needed if the formation of individual galaxies is to be resolved while simulating a large enough volume to get a reasonable measurement of galaxy clustering. Results so far show that the inferred galaxy distribution can differ from the underlying mass distribution both by a positional bias that enhances clustering as before,[41] and by a velocity bias[41] that reduces galaxy velocities below those of representative dark matter particles. Both effects result in the global mass per galaxy being underestimated, and together they may be sufficient to reconcile the observed kinematics of clustering with a flat universe. These results remain controversial because of the crude way in which they represent galaxy formation, and because a number of aspects of the simulations need further investigation to arrive at a complete understanding of the results.

Galaxy formation itself has been relatively little studied in the CDM model since its initial exploration.[43] Certain aspects of the process are not at all specific to the model. For example, the explanation of the fundamental distinction between disk and bulge subsystems is likely to be the same in any theory. A centrifugally supported thin stellar disk can form only from a centrifugally supported thin gaseous disk. Furthermore, such a stellar disk cannot be perturbed violently subsequent to the formation of its stars, since this would irreversibly thicken it. Thus observed disks must have formed in their current dynamical context from gas that had already settled onto coplanar, near circular orbits. Their observed angular momenta can be understood if this gas contracted from a much larger radius as a result of radiative cooling within a dynamically dominant dark halo.[44] The formation process would then be analogous to the cooling flows that appear to be present in the centers of many rich galaxy clusters, although differing characteristic parameters in the two cases are likely to lead to flows with significantly different structure. Numerical experiments show that ellipsoidal systems resembling bulges or elliptical galaxies are almost always formed in the contrary situation where star formation is largely complete *before* final equilibration of the system.[45] Thus the predominance of galaxies without disks, or without gaseous disks, in regions of high density can be understood as resulting from the combination of two factors: galaxies in such regions are unable to retain extended gaseous halos from which continuing disk formation can occur, and they are likely to have suffered violent perturbations or mergers that would destroy any preexisting disk.

Other properties of galaxies, such as their luminosity functions and their clustering, are much more sensitive to the details of the cosmological context in which galaxies are assumed to form. Recently, Carlos Frenk and I have made detailed analytic models for galaxy formation in a CDM universe.[33] These are based on the properties of the dark halos formed in numerical simulations, together with simple models for cooling flows and for star formation and evolution, and they allow the prediction of formation rates, luminosity functions, and relations between galaxy luminosity, metallicity, characteristic velocity, and environment. For the $\Omega = 1$ models that we studied, the characteristic luminosities of galaxies and of galaxy clusters can be approximately reproduced provided the density in gas is high $\Omega_g > 0.1$. This is in apparent conflict with the baryon density required by cosmological nucleosynthesis,[32] but a significantly smaller Ω_g results in too little cooling gas in massive halos to make a bright galaxy. The models also produce luminosity functions with too many faint galaxies, and they form most of their stars late, typically more than half after $z = 1$. In these, as in all hierarchical-formation models, bright galaxies form later than faint ones, a result that is contrary to the standard lore that the stellar populations in bright galaxies are older, in the mean, than those in faint galaxies.

The results of this kind of modeling are sensitive to the treatment of many poorly understood processes—gas cooling and shocking, the evolution of multiphase gaseous media, star formation and the resulting stellar initial mass function, hydrodynamic and radiative feedback from the stellar population into the surrounding gas, the evolution of the dust component of the interstellar medium—thus the resulting uncertainties are enormous. Only by tying the results to as great a variety of observational data as possible can real progress be made. For this reason new observational insights into the properties of cooling flows, of merging galaxies, of gas halos around our own and nearby galaxies, of quasar absorption line clouds, of known galaxies at high redshift, and of faint galaxies of as yet unknown redshift, can all play a major role in arriving at a proper understanding of galaxy formation. There are many indications that galaxies evolve substantially over the observationally accessible range of redshifts, and that a large fraction of their stars may form during this period. It is possible that some variant of these CDM models, for example $\Omega < 1$ models or models with different gas evolution or star-formation hypotheses, can remedy the shortcomings of our first attempts. Alternatively, an entirely different paradigm for galaxy formation may be required. It is clear that this is a field where theoretical advances will be strongly driven by observation over the next few years.

REFERENCES

1. FRENK, C. S., *et al.,* Eds. 1989. The Epoch of Galaxy Formation. Kluwer. Dordrecht, the Netherlands.
2. BOYLE, B. J. 1992. The quasar population at high redshifts. This issue; COWIE, L. L. 1992. Young galaxies. This issue; TYSON, J. A. 1992. Faint galaxies and dark matter. This issue; WRIGHT, E. L. 1992. COBE. This issue.
3. SEARLE, L. 1977. *In* The Evolution of Galaxies and Stellar Populations, B. M. Tinsley and R. B. Larson, Eds.: 219. Yale University Observatory, New Haven, Conn.; TOOMRE, 1977. *In* The Evolution of Galaxies and Stellar Populations, B. M. Tinsley and R. B. Larson, Eds.: 401. Yale University Observatory. New Haven, Conn.; REES, M. J. 1977. *In* The Evolution of Galaxies and Stellar Populations, B. M. Tinsley and R. B. Larson, Eds.: 339. Yale University Observatory. New Haven, Conn.
4. KOO, D. 1986. *In* Spectral Evolution of Galaxies, C. Chiosi and A. Renzini, Eds.: 419. Reidel, Dordrecht, the Netherlands.
5. SCHNEIDER, D. P., M. SCHMIDT & J. E. GUNN. 1989. Astrophys. J. **98:** 1507.
6. PETTINI, M., *et al.* 1990. Mon. Not. R. Astron. Soc. **246:** 544.
7. DUNCAN, R. C., E. VISHNIAC & J. P. OSTRIKER. 1991. Astrophys. J., Lett. **368:** L1.
8. SARGENT, W. L. W., A. BOKSENBERG & C. C. STEIDEL. 1988. Astrophys. J., Suppl. **68:** 609;

SARGENT, W. L. W., A. BOKSENBERG & C. C. STEIDEL. 1989. Astrophys. J., Suppl. **69:** 703.
9. BERGERON, J. 1988. *In* The Post-Recombination Universe, N. Kaiser and A. Lasenby, Eds.: 201. Kluwer. Dordrecht, the Netherlands.
10. HEISLER, J., C. HOGAN & S. D. M. WHITE. 1989. Astrophys. J. **347:** 52.
11. WOLFE, A. 1989. *In* The Epoch of Galaxy Formation, C. S. Frenk *et al.,* Eds. Kluwer. Dordrecht, the Netherlands.
12. PETTINI, M., A. BOKSENBERG & R. HUNSTEAD. 1990. Astrophys. J. **348:** 48.
13. FALL, S. M., Y. C. PEI & R. MCMAHON. 1989. Astrophys. J., Lett. **341:** L5.
14. BLACK, J. H., F. CHAFFEE & C. FOLTZ. 1987. Astrophys. J. **317:** 442.
15. LANZETTA, K., *et al.* 1991. University of California at San Diego Preprint.
16. HUNSTEAD, R., M. PETTINI & A. B. FLETCHER. 1990. Astrophys. J. **356:** 23.
17. SMITH, H. E., *et al.* 1989. Astrophys. J. **347:** 87.
18. MELLIER, Y., *et al.* 1991. Astrophys. J. In press.
19. CHAMBERS, K. C., G. K. MILEY & W. J. M. VAN BREUGEL. 1990. Astrophys. J. **363:** 32.
20. REES, M. J. 1989. Mon. Not. R. Astron. Soc. **239:** 1P.
21. BEGELMAN, M. C. & D. CIOFFI. 1989. Astrophys. J., Lett. **345:** L21.
22. FABIAN, A. C. 1989. Mon. Not. R. Astron. Soc. **238:** 41P.
23. DI SEREGO ALIGHIERI, S., *et al.* 1989. Nature **341:** 307.
24. CHAMBERS, K. C. & P. MCCARTHY. 1990. Astrophys. J., Lett. **354:** L9.
25. CHAMBERS, K. C. & S. CHARLOT. 1990. Astrophys. J., Lett. **348:** L1.
26. LILLY, S. 1989. Astrophys. J. **340:** 77.
27. FRENK, C. S. 1992. The cold dark matter model: Does it work? This issue.
28. FRENK, C. S. 1991. *In* Birth and Early Evolution of Our Universe, B. Gustafsson, J. Nilsson, and B. Skagerstam, Eds. World Scientific. Singapore. In press.
29. EVRARD, A. 1990. Astrophys. J. **363:** 349.
30. BAJTLIK, S., R. C. DUNCAN & J. P. OSTRIKER. 1988. Astrophys. J. **327:** 570.
31. SHAPIRO, P. R. & M. L. GIROUX. 1987. Astrophys. J., Lett. **321:** L107.
32. OLIVE, K. A., *et al.* 1990. Phys. Lett. B **236:** 454.
33. WHITE, S. D. M. & C. S. FRENK. 1991. Astrophys. J. In press.
34. STEIDEL, C. & W. L. W. SARGENT. 1987. Astrophys. J., Lett **318:** L11.
35. REES, M. J. 1986. Mon. Not. R. Astron. Soc. **218:** 25P.
36. CARSWELL, R. F. 1988. *In* QSO Absorption Lines: Probing the Universe, J. C. Blades, D. Turnshek, and C. A. Norman, Eds.: 91. Cambridge University Press. Cambridge, England.
37. WEBB, J. K. 1987. *In* Observational Cosmology, A. Hewitt, G. Burbidge, and L. Z. Fang, Eds.: 803. Reidel. Dordrecht, the Netherlands.
38. BOND, J. R., *et al.* 1986. Astrophys. J. **304:** 15.
39. DAVIS, M., *et al.* 1985. Astrophys. J. **292:** 371.
40. THUAN, T. X., *et al.* 1991. Astrophys. J. **370:** 25.
41. WHITE, S. D. M., *et al.* 1987. Nature **330:** 451.
42. CARLBERG, R. G., H. M. P. COUCHMAN & P. A. THOMAS. 1990. Astrophys. J., Lett. **352:** L29.
43. BLUMENTHAL, G. R., *et al.* 1984. Nature **311:** 527.
44. WHITE, S. D. M. 1990. *In* Dynamics of Galaxies and Their Molecular Cloud Distributions, F. Combes and F. Casoli, Eds.: 383. Kluwer. Dordrecht, the Netherlands.
45. WHITE, S. D. M. 1987. *In* Structure and Dynamics of Elliptical Galaxies, P. T. de Zeeuw, Eds.: 339. Reidel. Dordrecht, the Netherlands.

The Density and Clustering of Mass in the Universe

NICK KAISER

CIAR Cosmology Program
CITA
University of Toronto
Toronto, Ontario
M5S 1A1 Canada

1. INTRODUCTION

It has been known since the pioneering studies of clusters of galaxies by Zwicky[1] that there are copious amounts of dark matter in the universe, and that if the mass-to-light ratios of these systems are representative of the universe, then the value of the density parameter $\Omega = 8\pi G\rho/3H^2$ is on the order of one-fifth (for a comprehensive review, see reference 2). Most cosmologists are impressed by how close this value is to unity—this surely indicates that our understanding of the dynamics of the universe is not entirely wrong—and, by itself, this might well seem to be compatible with a true value of unity, since as one proceeds from the scale of the solar system through galaxies, galaxy halos, groups, and clusters one sees a systematic increase in mass-to-light ratios (M/L) as one casts one's net wider. However, studies of the deceleration of the Hubble expansion around the Local Supercluster in the early 1980s (see the review by Davis and Peebles[3]) seemed to show that as one went to the next scale of objects there was no further increase in M/L.

This empirical result stands in contrast to the prejudice that many people bring to this subject that the universe should have a density close to the closure value. The case for $\Omega = 1$ was championed by Dicke and Peebles,[4] who argued that if Ω was close to but not equal to unity, this would require a fine tuning of the conditions in the early universe, so that Ω was just departing from unity today. This argument (that for us to live just at the "special" time that Ω was peeling away from unity requires an unpalatable coincidence) is very persuasive, but should perhaps be tempered by the fact that the universe did very recently (on a logarithmic timescale at least) pass through not one but two special epochs—matter and radiation equality and recombination—and that current cosmological theory provides no firm explanation of why this should be.

A further boost to the prejudice that Ω should be equal to unity was given by the theory of inflation. This theory, originated by Alan Guth[5] in response to the monopole problem, offers a solution to a number of long-standing problems (horizon, flatness, etc.). While there is still a wide-ranging debate as to the extend to which inflation really solves these problems or rids us of the need to specify initial conditions, it is certainly the case that if the universe passed through an extended phase of accelerated expansion as predicted in inflation, then this at least allows the *possibility* that the large-scale homogeneity we see from the microwave background is the result of some physical process, rather than simply being a whim of the creator as in conventional (i.e., decelerating) Friedmann cosmology, and that this would also tend to prepare the universe in a state with Ω close to unity. A further result of the inflationary paradigm is that by adding a couple of additional assumptions (that the mass is dominated by some "cold" particle, and that the initial density fluctuations were Gaussian) one arrives at the "standard cold dark matter" model. This is an hypothesis of enormous predictive power, and one whose predictions for the formation of structure seem for the most part to be quite compatible with the observed properties of galaxies, clusters, and (qualitatively at least) the texture of large-scale structure.

Ranged against this theoretical (theological?) prejudice there is not only the empirical evidence for $\Omega \simeq 0.2$, but the indication for a low *baryon* density from big bang nucleosynthesis (BBN). As described by Dave Schramm,[6] this theory is very successful at accounting for the abundance of the light elements, but only for a low baryon density. The dynamics of clusters already indicates a value for Ω greater than Ω_{BBN}, and if, as I shall be arguing, the weight of the evidence now favors an even higher value of Ω, this must be primarily in a nonbaryonic form.

In the first part of this paper, I will review the dynamical estimates of Ω mentioned before. Concerning the dynamics of virialized systems I will attempt to review the work that has been done to explore the kinds of biases (statistical, dynamical, velocity, etc.) that might affect this. I then consider estimates of Ω obtained from supercluster dynamics, and show that recent developments here now reveal that there is a good correlation between mass and light, and that the best estimate of Ω from these studies is close to unity. Finally, I will consider the clustering of mass (revealed in part by the same studies), and the problems this poses for the "standard" cold-dark-matter (CDM) model.

2. DYNAMICAL ESTIMATES OF Ω ON SCALES $\sim 1\ H^{-1}$ MPC

2.1. Virial Analysis of Individual Objects

Just as observations of the motions of the planets allow one to calculate the mass of the sun, the motion of stars in globular clusters and galaxies give an estimate of the mass of these luminous systems; the motion of neutral hydrogen clouds and companion galaxies show that galaxies have extended massive halos; and the motion of galaxies in groups and clusters probe the mass distribution out to $\sim 1h^{-1}$ Mpc. The fundamental assumption used here is that what we are seeing are systems that are roughly in equilibrium (i.e., that the clusters are not all expanding or contracting), and that the mass is just given by $M \simeq v^2 r/G$. It is useful to express the results of these studies as mass-to-light ratios, and, in rough terms, one finds (in units such that the sun has $M/L = 1$) that the luminous central parts of galaxies have $M/L \simeq$ a few (times h, the Hubble parameter in units of 100 km/s/Mpc); that the halo M/L's increase roughly linearly with scale as far as they can be measured; and that clusters have M/L's of a few hundred h (see, e.g., Kent and Gunn[7]). Redshift surveys allow an estimate of the "luminosity density of the universe" (e.g., Efstathiou, Ellis, and Peterson[8]), from which one finds that a critical density corresponds to a mass-to-light ratio for the universe as a whole of $\simeq 1600h$. Thus, if mass-to-light ratios of clusters are representative of the universal value—and this is a big if—then Ω is about 0.2.

2.2. Cosmic Virial Theorem

In addition to studies of individual clusters, there is an alternative technique for "weighing" all the clumps in a redshift survey called the *cosmic virial theorem* (CVT). This was pioneered by Davis and Peebles,[9] and provides a relation between the pairwise velocity dispersion $\sigma^2(r_p)$, as a function of projected separation r_p, and the 2- and 3-point correlation functions of the galaxies. All these quantities can be estimated empirically, and this then determines the one free parameter that is the net mean density. The main assumption here is that the clustering is "statistically stable"; that is, that the correlation functions are effectively constant over a dynamical time (though they may change considerably over a cosmological time), and therefore that the distribution function $f(\mathbf{r}, \mathbf{v})$ for the separation and relative velocity of pairs of galaxies is independent of time. The consequence of this is Davis and Peebles' equation 36:

$$\frac{1}{\xi(r)} \frac{\partial(\xi(r)\langle v_i v_j \rangle)}{\partial r_j} = -2Gm(r)r_i/r^3 - \frac{2G}{\xi(r)} \int d^3z m(z) \zeta z_i/z^3. \quad (1)$$

This equation says that the "pressure gradient" on the left is balanced by the mean gravitational attraction between a pair of particles, and this in turn is composed of a direct attraction between the pair plus the last term that involves the 3-point function ζ, and that is the mean attraction between a pair due to the statistical enhancement of galaxies in their neighborhood. Davis and Peebles dealt with this by (1) neglecting the direct interaction, (2) replacing ζ by a constant Q times products of pairs of 2-point functions (a model that seems to fit the data quite well), (3) assuming that the velocity dispersion tensor is approximately isotropic: $\langle v_i v_j \rangle = A(r)\delta_{ij}$, (4) modeling $\xi(r)$ and $A(r)$ as power laws, and (5) assuming that the mass is concentrated in and around the galaxies, so $m(r)$ converges at a small radius. Most of these assumptions are very reasonable, though in Section 2.3.2 I will discuss the effect of relaxing the first and last of these assumptions. The upshot of all this is that the pairwise velocity dispersion [which is proportional to $A(r)$] is then given by

$$\sigma^2(r_p) = 6CQ(Hr)^2\xi(r)\Omega, \quad (2)$$

where C is a constant determined by the power law indices of $\xi(r)$ and $A(r)$. This result is physically very reasonable, since if galaxies trace the mass, the mean mass excess within distance r of a galaxy is $M \propto r^3\xi(r)$, so one expects $\sigma^2 \simeq GM/r \propto r^2\xi(r)$. Now what is remarkable is that this actually seems to work! The 2-point function has a slope $\gamma \simeq -1.8$, so $\sigma^2(r)$ should increase roughly as $r^{0.2}$, and this is just what is observed. From the constant of proportionality between σ^2 and $r^2\xi$ one infers $\Omega \simeq 0.2$, which is very similar to the value obtained from virial analysis of individual clusters.

2.3. Biases

The value of Ω inferred this way relies on the assumption that in the systems studied we are seeing a representative mix of dark matter and galaxies. To reconcile the inflationary prediction with observations requires that the number of galaxies per unit mass in the groups and clusters we use as laboratories for virial analysis is greater that for the universe as a whole by a factor 5 or so.

While our current understanding of how galaxies form is very limited, it is strongly suspected (see, e.g., reference 10 for a discussion of galaxy formation in the context of the standard CDM model) that galaxies are largely the result of collisional cooling that will allow the baryons to segregate and settle into the centers of the confining dark halos of an appropriate size and binding energy. Thus, when we find that in the luminous parts of galaxies a dynamical mass-to-light ratio of ~ 3 in solar units, we do not consider this damning evidence against the inflationary prediction, just that we need to cast our net wider to measure the total matter content.

I will now describe a number of effects that might plausibly have acted to bias the dynamical estimates of Ω from groups and clusters.

2.3.1. Statistical Bias

One way to segregate mass and light in clusters, etc., is if galaxy formation was more efficient than average in protoclusters. In hierarchical scenarios like the CDM model, while the amplitude of density fluctuations is decreasing with increasing scale, it does so rather slowly (the CDM spectrum being rather "red" on the scales between galaxies and clusters), and there may be significant "cross talk" between cluster scales (going nonlinear today) and galaxy scales (which went nonlinear at $z \simeq 3$ or so). One clue to how this might come about comes from the clustering properties of high peaks in Gaussian random fields. It is easy to show that if one takes such a field and clips off the regions above some high threshold (density contrast $> \nu$ in units of the rms σ), then these high peaks will be more strongly clustered than the underlying field. For large scales where the correlations are weak one finds that the 2-point function is enhanced by a factor $b^2 = (\nu/\sigma)^2$. This process can easily be visualized[11] if one thinks of a field with small-scale Gaussian fluctuations superposed on larger scale, "background" fluctuations $\Delta_b(\mathbf{r})$. One finds that for a high threshold, the space density of peaks is modulated roughly as $n_{\rm pk} \propto \exp(\nu\Delta_b/\sigma)$, and the long wavelength limit mentioned previously just comes from taking a first-order expansion in Δ_b, since the correlation function goes as the square of the density fluctuation.

The process can be applied with some justification to rich clusters today—the threshold being the overdensity required for collapse by the present—and qualitatively explains why clusters today are more strongly clustered than the mass,[12] and it is very appealing to imagine that the formation of galaxies at $z \simeq 2 - 3$ is analogous to cluster formation today. For galaxies, however, there is no obvious physical cause for the threshold that should be invoked. Perhaps a better way to think about biasing of galaxy formation is to think of how a "background" perturbation Δ_b would affect an individual galactic-scale condensation. For a positive Δ_b, the fluctuation will turn around and collapse sooner, and have a deeper potential well. If one hypothesizes that galaxies generate a luminosity determined mainly by the depth of the confining potential well, then this can introduce a bias. Calculations based on Press–Schechter[13] theory[14] suggest that this may produce a significant bias on the scale of clusters of galaxies (the Press–Schechter calculations are well described by an exponential modulation $n \propto \exp(\beta\Delta_b)$ at given L, and if one assumes that the luminosity is exactly proportional to the fourth power of the halo velocity dispersion, one finds $\beta \simeq 0.4$, giving perhaps a factor of 2 suppression of M/L in clusters from this process). These calculations are only meant as an illustrative model for what might really be happening. While assuming $L \propto v_{\rm halo}^4$ has some empirical appeal, it is by no means mandatory. It may be possible to believe that galaxies are not biased in this way,[15] or perhaps that the effect is somewhat stronger than this crude model would suggest. It seems quite likely that this "statistical bias" may be augmented in clusters and groups by other effects described below. The situation on small scales is therefore likely to be quite complicated, but on large scales the dominant effect should be the statistical one, and this should give a constant bias: $\delta n/n = b\delta\rho/\rho$, with $b = 1 + \beta$, and this is the model I will adopt in interpreting supercluster dynamics in Section 3.

In the absence of a complete theory for galaxy formation the situation is necessarily somewhat confused. In an attempt to circumvent the unacceptably large pairwise velocities predicted if galaxies trace the mass (and $\Omega = 1$), simulations[16] were made where a set of particles tagged as "galaxies" were laid down as expected for such peaks, with the peak threshold set to a rather high value and giving a strongly biased distribution $b \simeq 2.4$. While these simulations gave encouraging results, it subsequently became apparent that there is some indication that galaxies may not be as strongly biased as this on large scales, and that one should at least consider a much milder statistical bias (as expected from the simple model previously described), perhaps augmented on scales of clusters and below by other effects. Even if one were to back the specific $L \propto v_{\rm halo}^4$ model, or some variant thereof, the predicted large-scale bias factor b is very dependent of the form of the spectrum, and since, as we shall see, there is now some indication that the CDM spectrum does not work very well on large scales, there is no very strong reason to believe that it works in detail on scales between galaxies and clusters. If so, one should probably consider b to be a constant best determined by observation. The expectation that, for sufficiently large scales, one would expect $(\delta n/n)/(\delta\rho/\rho)$ to tend to a constant is a generic feature of this type of biasing scheme described before (though one would most naturally expect b to be a function of the luminosity or type of galaxy). Of course, it may be that there is some other effect that can more radically decouple mass and light, even on the rather large scales we will consider here. Again, it is probably best to test for this empirically rather than to simply speculate.

2.3.2. Dynamical Bias

I will now describe a quite different kind of bias that can arise in the process of nonlinear clustering. This bias might apply even if galaxies were initially (at their formation at a redshift of a few) distributed just like the dark matter on scales larger than individual galaxy halos, or it may act to augment the statistical bias previously described. The essential feature here is that if galaxies are created with a strongly segregated baryonic core in a dark matter halo, then the subsequent clustering of the luminous cores may be quite different from that of the typical halo particle.

Numerical simulations[17–20] show that if several core-halo structures merge to form a group, the resulting group inherits, to a large degree, the segregation of its subunits—the high phase-space density material in the inner parts of the subunits finding itself after merging in the inner parts of the newly formed group—and this could explain why, when we measure the mass-to-light ratio in the luminous parts of a group, we get a value much smaller than that of the entire system, or of the universe as a whole. As several groups come together to form a cluster, the process will be repeated, and the simulations suggest that this effect could plausibly be strong enough to account for the observed M/L's, even if the universe has $\Omega = 1$. This effect has been described as resulting from dynamical friction, though perhaps a more apt description would be *inherited segregation.*

The simulations show that the effect can be very strong, and if correct, would suggest that the "missing mass" is lurking just outside the regions where we use virial analysis to estimate the mass. Note that while the effect relies on the fact that the motion of a massive galaxy falling into a group or a massive group falling into a cluster is very different from a low-mass "test particle," once the group or cluster has settled down the luminous particles do behave as test particles and so their v^2r should correctly estimate the mass interior to their orbits.

There are several potential problems with this idea. The first is that the effect may be too efficient; Barnes's recent simulations actually show the formation of a single galaxy by merging rather than a group, so I have been rather liberal in my interpretation here. A second problem is that we see no sign of segregation of galaxies by luminosity in clusters.[21,22] The way I have sketched the process, one would have expected that a low-luminosity galaxy (with, I presume, a low-mass halo) should behave much as a typical dark-mass particle in falling into a group, and would therefore be less segregated. The expectation, I would have thought, would be that faint galaxies would exist as a more extended (and therefore probably hotter; see below) population in clusters, and it remains to be seen whether the scenario can survive this problem.

A further potential problem is that one would expect to see some continuous increase in M/L with increasing distance from the cluster center, yet the conventional wisdom is that M/L's are constant. If it were the case that one could show convincingly that M/L's are independent of scale, then this would be a strong argument against any of the biasing schemes previously discussed, and would strongly suggest either that Ω is small or that the "missing mass" is in some quite distinct form (e.g., in some very hot particle that does not partake at all of the clustering we see in galaxies). This is, however, I think, far from being the case. It is certainly true that models of individual clusters (such as that of Kent and Gunn[7] for Coma) with constant assumed M/L are *compatible* with the observed surface density and velocity dispersion profiles, and, similarly, that the relation between the pairwise velocity dispersion $\sigma^2(r_p)$ and the 2-point correlation function $\xi(r)$ is *consistent* with the idea that galaxies cluster just like the mass, but this does not necessarily rule out alternatives. For the case of the Coma cluster, The and White[23] have shown that the data are also consistent with a startlingly wide range of mass profiles—including models in which M/L increases strongly with scale—and, as I will now explain, there is also a viable interpretation of the CVT statistics with M/L increasing with scale.

Following the line of argument that led to (**2**), one might think that if M/L were an increasing function of scale, then one would see $\sigma^2(r_p)$ rising faster than predicted. To test this, I have calculated $\sigma^2(r_p)$ predicted using the CVT formulas of Davis and Peebles,[9] but for a model in which "galaxies" (with an assumed number density of

$n = 10^{-2}h^3$ Mpc^{-3}, as appropriate for typical bright galaxies, and with 2- and 3-point correlation functions, as observed) are surrounded by extended power-law halos. The result, and the observed $\sigma(r_p)$ values are shown in FIGURE 1, where I have given the "galaxies" nearly flat rotation curve halos with $\rho \propto r^{-2.1}$ and normalized to $v_{\rm rot}$ = 200 km/s at $r = 10h^{-1}$ kpc. The halos extend to $\simeq 5h^{-1}$ Mpc (which gives $\Omega = 1$ in total), and the predicted velocities look quite reasonable.

In this model, the clustering of mass and "galaxies" is identical on scales larger than the outer radius of the individual halos ($\simeq 5h^{-1}$ Mpc), but on smaller scales the clustering of galaxies is relatively enhanced. For instance, this model implies that for a cluster in which the number density of galaxies falls as $r^{-\alpha}$, the density profile is $\propto r^{0.9-\alpha}$. I should say that this is only a phenomenological model of the mass distribu-

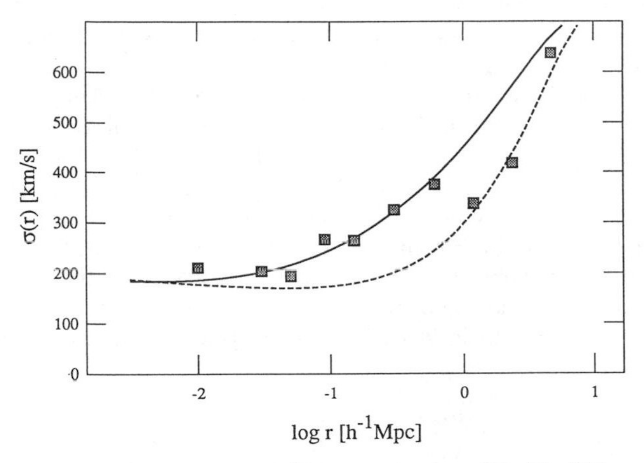

FIGURE 1. The data are the estimates of $\sigma^2(r_p)$ from Davis and Peebles.[9] The *solid line* is the predicted σ^2 for a model in which "galaxies" are distributed with a mean number density and low-order correlation functions as appropriate for bright galaxies, and where each galaxy is assumed to have $\rho \propto r^{-2.1}$ halos with parameters as described in the text. The calculation follows the steps described by Davis and Peebles, but modified to incorporate the extended halos (this feature was included in their analysis, but not used), and also to include the "direct interaction" term in (**1**) that Davis and Peebles dropped.

tion: I am not assuming that galaxies really carry around with them halos, which must overlap in groups and clusters, but which are individually rigidly locked to their galaxy owner. I am simply asking the question: What if the dark mass distribution happened to have that relation to the galaxy distribution; how would one expect the galaxies to be moving? The answer I find is that the prediction seems quite compatible with the observations, even though in this model, M/L increases roughly linearly with the scale of the system. This differs from the conventional interpretation for the following reasons: First, I have let $m(r)$ increase with r rather than assume that $m(r)$ converges to a constant value at small radius, as did Davis and Peebles. With this modification alone, one would expect $\sigma(r_p)$ to rise too fast with radius. However, the problem then is not that the predicted velocities would be too large at say $1h^{-1}$ Mpc, but that the velocities would be too small at small radius. This is cured by including the "direct interaction" term in **(1),** which Davis and Peebles dropped. This raises the small-scale velocities to an acceptable, nearly constant, level.

A further bonus of including an extended halo in the CVT is that this removes a technical problem found by Davis and Peebles. In calculating the acceleration arising from the 3-point function in **(1),** one finds three terms from the combination of pairs of products of ξ in the model for ζ. As discussed by Davis and Peebles, one of these terms vanishes by symmetry; the second represents the attraction of "galaxy A" toward "galaxy B" because of the mass clustered around galaxy B; and the third term (which with the model for ζ is actually somewhat larger than the second term) represents the attraction of A toward B due to the asymmetric distribution of mass close to galaxy A itself. Davis and Peebles argued that this was an unreasonable feature of the model for ζ, and chose to replace the third term with the second (note that dropping this term entirely would double their estimate of Ω). With an extended halo there is no large contribution of this kind, and the resulting prediction for $\sigma^2(r_p)$ is less sensitive to such vagaries of the model.

Finally, I should say that the results here are not in any fundamental conflict with Davis and Peebles' basic conclusion that the net mean density of mass clustered around galaxies on a scale $\simeq 1h^{-1}$ is substantially smaller than unity, since all I have done in this $\Omega = 1$ model is to "hide" the mass at a somewhat larger radius ($\simeq 5h^{-1}$ Mpc). What this does show, however, is that the conventional interpretation of these observations in terms of constant M/L is not unique, and that the data are quite compatible with a roughly linear increase in M/L with scale. If this latter interpretation is correct, we should find evidence for the "missing mass" in supercluster dynamics (Section 3) and in weak lensing of distant galaxies from the extended massive halos.

2.3.3. Velocity Bias

An alternative to, and a useful check on the validity of, the CVT is to compute the pairwise velocity dispersion of particles in an evolved *N*-body simulation. Davis and Peebles did this using a simulation by Efstathiou and Eastwood[24] and found, not surprisingly, that the small-scale velocities in this $\Omega = 1$ simulation were much larger than those actually observed. This is as expected from the CVT. The velocities here are those of the dark matter particles. A more recent development has been to attempt to identify "galaxies" in the simulations as dense clumps of particles, and a large number of experiments[25-30] have now shown that the pairwise velocity dispersion for these "galaxies" tends to be considerably smaller than those of the dark particles, and this phenomenon has been dubbed *velocity bias.* A number of possible causes for this effect have been discussed, and since the effect is seen in simulations with and without a dissipational component and with very different algorithms for identifying "galaxies," it is not inconceivable that there are several effects at work here. Since any velocity bias will also tend to bias one's estimate of Ω, it seems worthwhile to consider the possible causes, and to see how they relate to the biases discussed before.

One source of bias arises from the internal motions of the dark particles in the halos that is removed if one uses the center of momentum of the halo. While this effect must be present at some level, most discussions have concluded that this is a rather minor effect. This is certainly true for galaxies in clusters, where the relative galaxy velocities are much larger than any plausible internal motion, and the same is probably true for the general pairwise velocity dispersion.

Many of the previously mentioned workers have referred to the role of dynamical friction in "cooling" the massive galaxy halos and thus reducing their velocity dispersion. Whether or not friction will actually reduce the velocity dispersion depends, of course, on the shape of the potential. In a Keplerian potential well, for instance, friction will cause the velocity to rise, just as it does for a satellite suffering drag from the earth's atmosphere, but in a cluster with a shallow (shallower than r^{-2}) density profile, one would expect friction to put the galaxies on lower velocity orbits. Even in a pure logarithmic potential there will be an effect in this direction since there is a nice result (due, I am told, to Steve Kent) that for any finite population of particles in a singular isothermal potential well the mean square line of sight velocity of the particles is just two-thirds that of the dark-matter particles providing the confining potential. A reduction of this size agrees very well with the mild velocity bias found by Carlberg and Dubinski[28] in their cluster simulation, and the steeper profile that they found for the "galaxies" relative to the dark particles supports this interpretation. This kind of velocity bias is a natural consequence of the "clustering biases" described before: If dynamical evolution (or statistical bias for that matter) results in galaxy orbits that are centrally concentrated in clusters, etc., then their velocities will be biased relative to that of the dark particles that probe the potential further out. This effect will be particularly strong if the mass density profile in the cluster is shallower than r^{-2}.

While this explains what is happening in some of the simulations, this is almost certainly not the whole story. Carlberg[27] finds that the "galaxies" are no more concentrated than the dark matter, and yet he finds a stronger velocity bias than seen by Carlberg and Dubinski. In the recent simulation of Couchman and Carlberg the velocity bias is stronger still, and, while they state that the galaxies are more centrally concentrated than the dark matter in the clusters, they find that the correlation function of the galaxies on small ($\sim 1h^{-1}$ Mpc) scales is weaker than that of the dark matter. An alternative explanation for velocity bias has been offered by Couchman and Carlberg, and by Bertschinger and Gelb,[30] both groups suggesting that galaxies are less common in high-velocity dispersion systems due to increased merging. It is very plausible that this is what is happening in the simulations—culling galaxies in high-velocity regions will, of course, suppress σ^2, and, to the extent that high-velocity regions also tend to be high overdensity regions, this will also suppress ξ—but there is, I think, some reason to doubt that the properties of clusters produced in this way would be very realistic. The problem is that in real clusters we see a very large *enhancement* in the mass-to-light ratio (a factor of ~ 5 if Ω is unity), and since there is no noticeable difference in the characteristic luminosity of galaxies in dense regions, a similar enhancement in the number of galaxies per unit mass, rather than a deficit. This problem is illustrated by the results of Couchman and Carlberg who find that the 14 most massive groups in their simulation contain 6 percent of the galaxies and 8 percent of the mass. This is a mild antibias, and means that an observer in this simulation would infer $\Omega \simeq 4/3$ from virial analysis of these clusters. This is qualitatively compatible with the antibias they see in the correlation function of galaxies, but seems to be rather different to what we see in our universe.

In this discussion, I have assumed that, while dynamical or other effects may have caused the orbits of massive galaxies to be centrally concentrated relative to the mass, when we study their kinematics, these galaxies do behave essentially as test particles and therefore correctly tell us the mass interior to their radii. If this assumption is violated it would have a profound effect on estimates of Ω, and if, for instance, massive "galaxies" with their associated halos move more slowly than one would expect for a test particle, it might be possible to reconcile the observed velocities with a closed universe in which the mass-to-light ratios of clusters are really unbiased or antibiased. Luckily we can check on this by comparing the mass inferred from galaxy motions with that inferred from the temperature of the X-ray-emitting gas. To a reasonable approximation these mass estimates agree, and to the extent that there is any discrepancy, it is in the opposite sense to that needed to reconcile this puzzle. Incidentally, this check also provides a powerful constraint on schemes to reconcile $\Omega = 1$ with virial estimates by invoking long-range nongravitational forces acting on the dark matter.[31]

To summarize, there are (at least) two kinds of velocity bias seen in the numerical simulations. The first arises if the "galaxies" are relatively more concentrated than the dark matter within individual clusters. This bias would be caused by either the "statistical" or "dynamical" biases described before. This type of velocity bias will be particularly strong if the mass-density profile in the clusters is shallower than r^{-2}, and will generally be associated with an enhanced clustering strength of the "galaxies" relative to the mass. The second kind of bias arises if merging substantially depletes the number of "galaxies" per unit mass in dense, high-velocity dispersion regions relative to the "field." This also reduces the pairwise velocity dispersion, but results in a galaxy clustering strength that is suppressed with respect to that of the mass. While this is no doubt a real effect in some simulations, an antibias in clusters would be rather difficult to reconcile with observations.

3. Ω FROM SUPERCLUSTER DYNAMICS

As previously discussed, for dense systems ($\rho \gg \bar{\rho}$, $t_{dyn} \ll H^{-1}$) we can estimate $L \sim lz^2H^{-2}$ and $M \sim \sigma^2\theta/H$ from observations of the apparent luminosity l, the redshift z, the angular size θ, and the internal velocity dispersion σ, and form M/L. Multiplying by the ratio of the known (or at least estimated) luminosity density of the universe to the critical density gives an estimate of Ω, which one can think of as a partial accounting of the mass in the universe. When we go to supercluster scales we are dealing with systems that are still expanding ($\rho - \bar{\rho} \lesssim \bar{\rho}$, $t_{dyn} \sim H^{-1}$), and one must proceed differently. For linear perturbations (and there is good reason to think that most of the Local Group motion at least is produced by fluctuations in the linear regime), the velocity of a particle (or galaxy) should be proportional to the acceleration times an effective time over which the acceleration has been operating: $\mathbf{v}_{pec} = \mathbf{g}t$. The acceleration depends on Ω and on the bias factor b, and the effective age also depends on Ω, and the result is

$$\mathbf{v}_{pec} = \frac{\Omega^{0.6}}{b} H \int d^3r \frac{\Delta n}{n} \frac{\hat{\mathbf{r}}}{r^2}. \tag{3}$$

An equivalent statement is that

$$\nabla \cdot \mathbf{v}_{\text{pec}} \propto \frac{\Omega^{0.6}}{b} \frac{\Delta n}{n}, \tag{4}$$

which makes it more explicit that all one is really testing is the continuity equation.

An immediate problem is that one cannot determine Ω and b separately from observations in the linear regime. It might be thought that by going into the weakly nonlinear regime that one might be able to determine the two separately. This would be true if, for instance, one were prepared to assume that the bias factor was constant even when the density perturbations start to become nonlinear, but this is highly unlikely to be the case; certainly in the models discussed in Section 2.3.1 this would not be the case, the statistical contribution to the bias (defined as $(\delta n/n)/(\delta\rho/\rho)$) varying roughly exponentially with the *initial* density contrast, and in these models the departures from constant bias set in at about the same point that any departures from dynamical linearity set in. Given the imperfections in our understanding of dynamical nonlinearity and the imperfections in the velocity data any hope of determining Ω directly from this type of measurements is a forlorn one, and one must live with the ambiguity. In the past, it was popular to assume that $b = 1$—that galaxies fairly trace the mass on very large-scales—and the ratio of the velocity to the apparent acceleration was then interpreted as an estimate of $\Omega^{0.6}$, often expressed as a mass-to-light ratio of the universe. Nowadays, it is more common to assume that $\Omega = 1$, and express the result as a value for b. No matter how the result is expressed, the test is still the same. The important questions are: Does the apparent acceleration correlate with the observed motion? and, if so, What is the constant of proportionality?

To measure b (or Ω) one needs to measure some appropriate combination of velocities and galaxy density contrast. There is a range of possibilities. One natural division is between those tests that exploit the velocity of the Local Group inferred from the dipole anisotropy of the microwave background and those that use velocities of other galaxies (usually relative to the frame of the Local Group) as determined from Tully–Fisher or $D_n - \sigma$ techniques. In the former class one must of course estimate the acceleration of the Local Group, and one can distinguish two subclasses of this type of test: those that use only angular information (and exploit the property that since both gravity and luminosity obey inverse square laws, the dipole moment of the extragalactic sky should be proportional to the acceleration), and those that use redshift information. The latter allow one, in principle, to check whether the acceleration appears to have converged within the survey volume, whereas with the former one must simply assume that the survey is deep enough to encompass the source of our motion (though one has some indication of whether this is true by looking at the discrepancy in angle between the dipole and the actual motion). In the second class of test (using the motions of other galaxies), there are also two subclasses that can be usefully distinguished: One can either start from peculiar velocities, calculate $\nabla \cdot \mathbf{v}$, and compare this with the distribution of galaxies $\delta n/n$ or work in the other direction: measure $\delta n/n$, calculate $\mathbf{g}(\mathbf{r})$, and then compare with the observed motion. The main advantage of the former is that the relation between $\nabla \cdot \mathbf{v}$ and $\delta\, n/n$ is a local one; perturbations outside the observed region do not contribute to $\nabla \cdot \mathbf{v}$, whereas with the latter one needs a deep survey with nearly full sky coverage. Probably the biggest disadvantage with the former is that one starts with the less reliable data—there is always the worry that the small offsets in the Tully–Fisher or $D_n - \sigma$ relations may be influenced by environment—and any errors (systematic or otherwise) will propagate through the analysis. I do not think that there is any "best" solution; the different approaches can be used in a complementary manner, and it is best to apply all of these, and use as many diverse techniques as possible.

Historically, this kind of dynamical test was applied to the Local Supercluster. The Local Supercluster is certainly a prominent density enhancement, and dominates the view of the world obtained from surveys such as the RSA catalog.[32] It is also very clear that the direction to the Local Supercluster does not agree very well with the Local Group motion, so one is really forced to resort to tests using the motion of other galaxies with concomitant uncertainties. The situation as of the early 1980s was reviewed by Davis and Peebles.[3] Various groups had worked on this problem using a variety of data sets. The estimated density contrast (within our radius) was taken variously from the RSA and CfA surveys; the consensus value being a density contrast of 2–3. For the most part, the perturbation was modeled as an isolated spherical perturbation with parameters describing the radial profile, etc. At the time of the review by Davis and Peebles, there was quite a range of estimates of the slowing of the Hubble expansion by the Local Supercluster, ranging from about 150 km/s up to 300 km/s or more, and this gave a correspondingly wide range of estimates of Ω (but all below about 0.3). Since then, the estimates have tended to reduce the estimated infall velocity, and, using the same estimate for the density contrast one would infer $\Omega \simeq 0.1$ or even lower. At face value, this is a big problem for a biased $\Omega = 1$ model (at least with what I would consider a plausible—i.e., mild—bias parameter) since one would surely have expected to find a value of Ω larger than that obtained from virial analysis of clusters, whereas, if anything, the indication is for a smaller Ω. As I will now describe, there has since been a significant increase in the size and quality of available data, and there is now a broad consensus that the best estimate of Ω is a lot larger, and if Infrared Astronomical Satellite (IRAS) galaxies trace the mass at least, then the current best estimate of Ω is very close to unity. I do not want to go into a detailed *postmortem* of what might have gone wrong with the old Virgo infall calculations save to mention that (1) there was always some considerable doubt as to whether the adopted density contrast was correct; (2) it was always dubious to assume a single isolated spherical perturbation when the direction to the LSC was so discrepant from the motion of the Local Group, and it is now amply apparent that the LSC is really a rather minor perturbation in the whole picture; and (3) the state of the art has now progressed far beyond the need for assumptions of spherical infall, and all of the applications I will mention below make no restrictive assumptions of this kind.

3.1. Tests Using the Local Group Motion

The big observational development here has been the advent of the IRAS point source catalog and the galaxy catalog that can be extracted from it by a suitable color selection. The remarkable result is that the flux weighted dipole moment agrees in direction with the motion of the Local Group to a precision of about 10 degrees.[33]

The most natural explanation of this is that in this catalog we are seeing deep enough and a large enough solid angle to encompass the source of our acceleration, and that this survey might give a useful estimate of b for IRAS galaxies. Given an estimate of the physical depth of the survey from a redshift survey of a limited patch of sky,[34] it is possible to obtain an estimate of b, and Kaiser and Lahav[35] found $b \simeq 1.6$. The indication from IRAS galaxies then is that the large-scale dynamics may be compatible with a high value for Ω. The only important assumption about the geometry of the source of our acceleration is that it is not so distant as to be beyond the scope of the IRAS survey and that there is not a very large component of acceleration hiding in the plane of the galaxy. To the extent that the two vectors agree quite well there is empirical indication that the effect of such unseen superclusters is quite small. It seems from this result that if one wanted to believe that Ω is really much smaller than unity, then one would have to say that a good deal of our acceleration comes from a much greater distance, and happens to align in direction coincidentally with the acceleration from the stuff that is seen by IRAS. As I will shortly describe, one can make tests of this idea to some extent, and the results so far seem not to support this.

A further observational development made possible by IRAS are the all-sky redshift surveys. Strauss and collaborators[36] have made a complete survey to a flux limit of about 2 Jy, and the QDOT survey[37] is a 1-in-6 random sample to 0.6 Jy. The former survey has a higher space density nearby, but the crossover is at about 70 Mpc/h, beyond which the QDOT survey has the higher sampling rate. Both of these surveys give a 3-dimensional view of the structures around us (though of course in redshift space) and can be used to see if the acceleration does converge. Both of these surveys show that a major source of our acceleration is the supercluster surrounding the clusters of Hydra and Centaurus (coinciding roughly with the Great Attractor concentration inferred from peculiar motions). In the 2-Jy survey, the Perseus–Pisces supercluster appears to be almost as prominent, but in the QDOT survey it is much less so, for reasons that are not well understood (by me at least). Both of these surveys have been used to explore the convergence of our acceleration with distance. The 2-Jy survey seems to show a reasonable convergence (to a direction of 20 to 30 degrees away from the Local Group motions) within about 5000 km/s. In the QDOT survey there is a hint that there may be some additional acceleration (out to perhaps 8000 km/s), but the problem is that in both surveys the shot noise diverges rapidly at large distances. Both studies gave estimates of b (or Ω) very close to unity, in rough agreement with the results from the flux-weighted dipole. It is not at all clear that the precision of this test is really any better than that from the flux dipole alone, despite the extra redshift information. What these 3-dimensional studies do, however, is give confidence that the results are not severely affected by uncertainty in the acceleration from very nearby. This is reassuring since the uncertainty in the flux-weighted dipole is dominated by the brightest galaxies.

3.2. Tests Using Motions of Other Galaxies

As mentioned before, there are two possible routes here: one can either estimate $\nabla \cdot \mathbf{v}$ from the velocity data and compare with $\delta n/n$ or estimate $\mathbf{g}$ from $\delta n/n$ and compare with $\mathbf{v}$ (or the line-of-sight component thereof). The former route has been pioneered by Bertschinger and Dekel (BD), and their results are described by

Dekel.[38] One problem with this method is that the data are inhomogeneous in spatial coverage, and the noise becomes large at large distance. Bertschinger and Dekel have tried to circumvent this by smoothing the velocity field derived, but it turns out that a rather large smoothing radius is required. Albert Stebbins and I[39,40] have tried an alternative approach; what we did was to make a regularized fit to a "nonparametric" model for the velocity field (i.e., one with an effectively infinite number of degrees of freedom). Our model is a set of Fourier modes for the density field with velocity derived from that assuming linear theory. Thus, just as in the BD technique, the assumption that the velocity field is curl-free is built in from scratch. What we did was to calculate the most probable velocity field given the data (we used the compilation of Burstein[41]) and an assumed power spectrum for the density field. This is a Bayesian approach—the power spectrum defines our prior probability distribution for velocity field configurations—and the technique has similarities to maximum entropy modeling: for example, the solution is the "smoothest" map compatible with the data and therefore tends to relax to zero in the absence of data. Our initial results were very encouraging indeed; several prominent (and real!) features could be seen in the reconstructed density map, and, because in our approach the resolution is automatically adjusted to extract the most information at each point, the Local Supercluster, for instance, appears quite prominently. In order to explore the range of allowed results we vary both the slope and amplitude of the assumed spectrum and this gives a reasonable indication of which results can be believed. We find that the Great Attractor, for instance, appears quite robust. What we have subsequently discovered is that there is a bias in this kind of technique as we initially implemented it that makes the interpretation rather difficult. The problem is that if one starts with a map of the peculiar velocity as a function of *estimated* distance, then simply the effect of distance estimate errors will mimic the effect of gravity. What happens is that if one has a supercluster of galaxies, these tend all to lie at a similar redshift. In estimated distance space, the errors (on the order of 20 percent) will scatter the galaxies front and back, and so one will see an apparent infall pattern (which is, of course, spurious) just due to the effect alone. The overall effect is then a complicated combination of this Malmquist-type effect and real infall. We have tried an alternative solution working from a map of peculiar velocity in *redshift* space (in linear theory, at least, this is justifiable); here the effect of distance errors is to look like random thermal velocities rather than gravitational infall, and we have found that the resulting maps are unfortunately quite sensitive to what technique is used. We consider that this type of technique is certainly promising, and with better data and a fuller understanding of the type of bias described previously this may yield a reliable estimate of the bias factor. The results obtained so far with both techniques indicate a rather mild bias (high Ω) that is in accord with other indications.

The IRAS QDOT survey allows one to exploit the second approach.[42] What we have done here is to bin the galaxies on a grid with cells whose size increases linearly with distance from us and, after allowing for the radial selection function, incompleteness in the plane of the galaxy, etc., construct a map of the density contrast in redshift space. We then correct this for peculiar velocities in an iterative manner using linear theory to make a map of the density field in real space. From the resulting 3-dimensional density contrast map we can calculate the acceleration for each of the

1000 or so galaxies in the Burstein compilation (we actually binned these in the same cells used to make the density map), and thereby compare the predicted and observed motions. In making this test it is best to work in the Local Group frame. This is simply because there is a divergent uncertainty in the acceleration due to distant matter, but if we predict peculiar velocities relative to the Local Group any spurious acceleration from distant regions tends to cancel out. The result is shown in FIGURE 2. At first sight this may not look like a very wonderful correlation, but perhaps the best way to read this figure is to divide the points into two halves

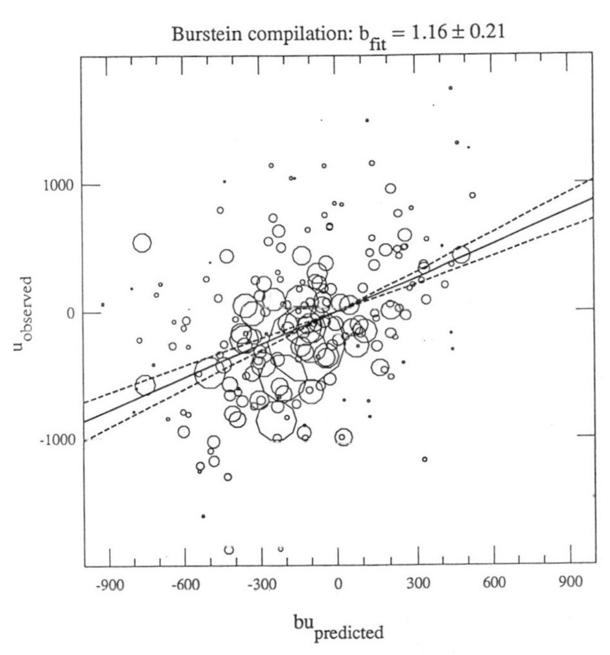

FIGURE 2. Scatter plot of observed peculiar velocities against those predicted from the spatial distribution of IRAS galaxies assuming gravitational instability. The galaxies with velocities have been binned into cells, and the area of the symbol is proportional to the statistical weight. With no peculiar velocity the χ^2 is about 1500 and is reduced to about 800 by the IRAS predictions. This is still bigger than the number of cells (about 200), so there is some excess scatter here over and above that from measurement errors. The *solid line* is the best linear fit, and the *dashed lines* correspond to one-sigma errors in *b*.

according to predicted velocity. It is then quite apparent that the mean velocity of the galaxies that IRAS predicts to be moving away from us is indeed positive and very significantly (at about the four-sigma level) larger than the velocity of the other set of galaxies. The uncertainty in the estimate of *b* obtained here (the slope of the correlation) is similar to that obtained using the Local Group motion, and the estimate here independently confirms the result obtained there that (if IRAS galaxies trace the mass), the best estimate of the density parameter is close to unity.

ratios (i.e., the number of galaxies per unit mass in the simulated clusters is *lower* than for the universe as a whole, rather than being a factor of 5 or so higher as we see in the real world). One could, of course, invoke some other algorithm for selecting "galaxies" that would avoid this problem, but I would then imagine that the predicted galaxy correlation function would become unacceptably large at small scales, and the pairwise velocity dispersion (which, in this simulation, is already uncomfortably large) would become larger still. However, more work is needed to see if this is really an insurmountable problem.

5. SUMMARY

In Section 2, I have reviewed methods for estimating the mass clustered around galaxies on scales of $\sim 1h^{-1}$ Mpc. At face value, these suggest $\Omega \simeq 0.2$, but, as I have described, there are various biases that may well have acted to cause this estimate to be too low; these include statistical effects imprinted at the time of galaxy formation and dynamical effects that may have further segregated mass from light on these scales. I have also tried to review the phenomenon of "velocity bias" found in numerical simulations, and to show how this may relate to clustering bias and to galaxies and clusters in the real world. In the biasing schemes I have discussed, the M/L's we infer from cluster dynamics are low simply because the galaxies we see are more concentrated than the mass; the mass is "hiding" just outside the luminous regions. This idea is quite testable, and one interesting test will be to see whether the predicted mass is detectable through the distortion of background galaxy images. In Section 3 I have shown that a wide variety of tests and techniques now give a consistent picture that shows that (1) the acceleration predicted by IRAS agrees quite well with the motions both of the Local Group and of other galaxies, and (2) if IRAS galaxies trace the mass, then $\Omega \simeq 0.8 \pm 0.3$. This is quite a change from earlier studies using the Virgo supercluster (and assuming spherical symmetry) that gave a much lower estimate of Ω. The estimate of Ω obtained by comparing predicted and observed motions of other galaxies relative to the Local Group gives a result that is largely independent from but that confirms earlier estimates obtained by comparing with the Local Group motion.

Problems and questions for the future are: What is happening in the Great Attractor region? Does the rather large discrepancy found there indicate mass hiding in the plane? A more distant attractor? A problem for $D_n - \sigma$? Other questions are: What about ordinary galaxies—are they more strongly biased than IRAS galaxies? Does the bias factor depend on luminosity? What will be found if these dynamical tests are pushed to even larger scales?

Concerning the statistics of large-scale structures: The success of the dynamical tests give some confidence that the fluctuations in galaxy counts do reflect underlying mass fluctuations and that the IRAS QDOT counts in cell analysis and the more recent power spectrum analysis do indeed suggest a serious problem for the standard CDM model. The main *caveat* here is that the result relies strongly on the influence of the Hercules supercluster and other superclusters at large distance (>100 Mpc/h), and it would be nice to have direct confirmation from distortion of the Hubble flow that these are "real" density fluctuations.

The weight of evidence from the dynamics of superclusters now seems to confirm the central prediction of the inflation scenario, and it is particularly exciting that the dynamically inferred Ω, while no doubt still compatible with a range of true values for Ω around unity, is certainly incompatible with the low baryon density required for big bang nucleosynthesis. The challenge now is to detect in the laboratory the mysterious dark matter that makes up most of our universe. It is ironic that the very same observations now point to serious problems for the "standard" cold dark matter model that for long has been held up as the paradigm of a successful inflationary cosmogonical theory. (Note that I refer here to the combination of assumptions of Gaussian fluctuations, etc., not simply the hypothesis that the universe is dominated by some initially cold particle.) Whether the salvation lies in a 17-keV neutrino,[46] texture,[47] or some other radical alternative remains to be seen.

REFERENCES

1. ZWICKY, F. 1933. Helv. Phys. Acta **6:** 110.
2. TRIMBLE, V. 1987. Ann. Rev. Astron. Astrophys. **25:** 425.
3. DAVIS, M. & P. J. E. PEEBLES. 1983. Ann. Rev. Astron. Astrophys. **21:** 109.
4. DICKE, R. H. & P. J. E. PEEBLES. 1979. *In* General Relativity: An Einstein Centenary Review, S. W. Hawking and W. Isreal, Eds. Cambridge University Press. Cambridge, England.
5. GUTH, A. 1981. Phys. Rev. D **23:** 347.
6. SCHRAMM, D. 1990. Presented at the 15th Texas/ESO–CERN Symposium on Relativistic Astrophysics, Cosmology and Fundamental Physics, Brighton, England, Dec. 16–21, 1990.
7. KENT, S. M. & J. E. GUNN. 1982. Astron. J. **87:** 945.
8. EFSTATHIOU, G., R. S. ELLIS & B. A. PETERSON. 1988. Mon. Not. R. Astron. Soc. **232:** 431.
9. DAVIS, M. & P. J. E. PEEBLES. 1983. Astrophys. J. **267:** 465.
10. BLUMENTHAL, G. R., S. M. FABER, J. R. PRIMACK & M. J. REES. 1984. Nature **311:** 517.
11. BARDEEN, J. M., J. R. BOND, N. KAISER & A. SZALAY. 1986. Astrophys. J. **304:** 15.
12. KAISER, N. 1984. Astrophys. J., Lett. **284:** L9.
13. PRESS, W. H. & P. SCHECHTER. 1974. Astrophys. J. **187:** 425.
14. COLE, S. & N. KAISER. 1989. Mon. Not. R. Astron. Soc. **237:** 1127.
15. PEACOCK, J. A. 1990. Mon. Not. R. Astron. Soc. **243:** 517.
16. DAVIS, M., G. EFSTATHIOU, C. S. FRENK & S. D. M. WHITE. 1985. Astrophys. J. **292:** 371.
17. BARNES, J. 1983. Mon. Not. R. Astron. Soc. **203:** 223.
18. BARNES, J. 1984. Mon. Not. R. Astron. Soc. **208:** 873.
19. EVRARD, A. E. 1987. Astrophys. J. **316:** 36.
20. WEST, M. J. & D. O. RICHSTONE. 1988. Astrophys. J. **335:** 532.
21. WHITE, S. D. M. 1977. Mon. Not. R. Astron. Soc. **179:** 33.
22. MERRITT, D. 1984. Astrophys. J. **276:** 26.
23. THE, L. S. & S. D. M. WHITE. 1986. Astron. J. **92:** 1248.
24. EFSTATHIOU, G. P. & J. W. EASTWOOD. 1981. Mon. Not. R. Astron. Soc. **194:** 503.
25. CARLBERG, R. & H. COUCHMAN. 1989. Astrophys. J. **340:** 47.
26. CARLBERG, R., H. COUCHMAN & P. THOMAS. 1990. Astrophys. J., Lett. **352:** L29.
27. CARLBERG, R. 1991. Astrophys. J. **367:** 385.
28. CARLBERG, R. & J. DUBINSKI. 1991. Astrophys. J. **369:** 13.
29. COUCHMAN, H. & R. CARLBERG. 1991. Preprint.
30. BERTSCHINGER, E. & J. GELB. 1991. Comput. Phys. **5:** 164.
31. FRIEMAN, J. A. & B. GRADWOHL. 1991. Preprint.
32. YAHIL, A., A. SANDAGE & G. A. TAMMANN. 1980. Astrophys. J. **242:** 448.
33. HARMON, R. T., O. LAHAV & E. J. A. MEURS. 1987. Mon. Not. R. Astron. Soc. **228:** 5p.
34. LAWRENCE, A., D. WALKER, M. ROWAN-ROBINSON, K. J. LEECH & M. V. PENSTON. 1986. Mon. Not. R. Astron. Soc. **219:** 687.

35. KAISER, N. & O. LAHAV. 1989. Mon. Not. R. Astron. Soc. **237:** 129.
36. STRAUSS, M. A. & M. DAVIS. 1988. *In* Large Scale Structure of the Universe, IAU Symp. 130, J. Audouze and A. Szalay, Eds. Reidel. Dordrecht, the Netherlands.
37. LAWRENCE, A., *et al.* 1991. In preparation.
38. DEKEL, A. 1991. Streaming velocities and the formation of large-scale structure. This issue.
39. STEBBINS, A. & N. KAISER. 1991. In preparation.
40. KAISER, N. 1990. Contemp. Phys. **31:** 149.
41. BURSTEIN, D. 1990. Compilation of distances to ca. 1000 galaxies. Available on request from burstein@asucps.
42. KAISER, N., G. EFSTATHIOU, W. SAUNDERS, A. LAWRENCE, M. ROWAN-ROBINSON, R. S. ELLIS & C. S. FRENK. 1991. Mon. Not. R. Astron. Soc. In press.
43. EFSTATHIOU, G., N. KAISER, W. SAUNDERS, A. LAWRENCE, M. ROWAN-ROBINSON, R. S. ELLIS & C. S. FRENK. 1990. Mon. Not. R. Astron. Soc. **247:** 10p.
44. MADDOX, S. J., G. EFSTATHIOU, W. SUTHERLAND & J. LOVEDAY. 1990. Mon. Not. R. Astron. Soc. **242:** 43p.
45. PEACOCK, J. & D. NICHOLSON. 1991. Preprint.
46. BOND, J. R. & G. P. EFSTATHIOU. 1991. Preprint.
47. PARK, C., D. N. SPERGEL & N. TUROK. 1991. Preprint.

The Beginning of the Universe

S. W. HAWKING
Department of Applied Mathematics and Theoretical Physics
Silver Street
Cambridge University
Cambridge, England

In this paper, I would like to discuss whether time itself has a beginning and whether it will have an end. All the evidence seems to indicate that the universe has not existed forever, but that it had a beginning about 15 billion years ago. This is probably the most remarkable discovery of modern cosmology. We are not yet certain whether the universe will have an end. When I gave a lecture in Japan, I was asked not to mention the possible recollapse of the universe, because it might affect the stock market. However, I can reassure anyone who is nervous about their investments that it is a bit early to sell: the end will not come for at least 10 billion years. By that time, maybe Britain will have joined the European Monetary Union.

The timescale of the universe is very long compared to a human life time. It was therefore not surprising that until recently the universe was thought to be essentially static and unchanging in time. On the other hand, it must have been obvious that society is evolving in culture and technology. This indicates that the present phase of human history cannot have been going for more than a few thousand years. It was therefore natural to believe that the human race, and maybe the whole universe, had a beginning in the fairly recent past. However, many people were unhappy with the idea that the universe had a beginning, because it seemed to imply the existence of a supernatural being, who created the universe. They prefered to believe that the universe, and the human race, had existed forever. Their explanation for human progress was that there had been periodic floods, or other natural disasters, that repeatedly set back the human race to a primitive state.

This argument about whether or not the universe had a beginning persisted into the nineteenth and twentieth centuries. It was conducted mainly on the basis of theology and philosophy, with little consideration of observational evidence. This may have been reasonable given the notoriously unreliable character of cosmological observations until fairly recently. As Eddington once said: "Don't worry if your theory doesn't agree with the observations, because they are probably wrong. But if your theory disagrees with the Second Law of Thermodynamics, it is in bad trouble." In fact, the theory that the universe has existed forever is in serious difficulty with the Second Law of Thermodynamics. This states that entropy, or disorder, always increases with time. Like the argument about human progress, it indicates that there must have been a beginning. Otherwise, the universe would have reached thermal equilibrium by now and everything would be at the same temperature. In an infinite and everlasting universe, every line of sight would end on the surface of a star. This would mean that the night sky would have been as bright as the surface of the sun. The only way of avoiding this problem would be if for some reason the stars did not shine before a certain time.

In a universe that was essentially static there would not have been any dynamical reason why the stars should have suddenly turned on at some time. Any such "lighting up time" would have to be imposed by an intervention from outside the universe. The situation was different, however, when it was realized that the universe is not static, but expanding. Galaxies are moving steadily apart from each other, which means that they were closer together in the past. One can plot the separation of two galaxies as a function of time. If there were no acceleration due to gravity, the graph would be a straight line. It would have gone down to zero separation about 20 billion years ago. One would expect gravity to cause the galaxies to accelerate toward each other. This will mean that the graph of the separation of two galaxies will bend downward below the straight line. So the time of zero separation would have been less than 20 billion years ago.

At this time, the big bang, all the matter in the universe would have been on top of itself. The density would have been infinite. It would have been what is called a singularity. At a singularity, all the laws of physics would have broken down. This means that the state of the universe, after the big bang, would not depend on anything that may have happened before, because the deterministic laws that govern the universe would break down in the big bang. The universe would evolve from the big bang, completely independently of what it was like before. Even the amount of matter in the universe can be different to what it was before the big bang, as the Law of Conservation of Matter would break down at the big bang.

Since events before the big bang have no observational consequences, one may as well cut them out of the theory and say that time began at the big bang. Events before the big bang are simply not defined, because there's no way one could measure what happened at them. This kind of beginning to the universe, and of time itself, is very different to the beginnings that had been considered earlier. These had to be imposed on the universe, by some external agency: there is no dynamical reason why the motion of bodies in the solar system cannot be extrapolated back in time, far beyond 4004 B.C., the date for the creation of the universe according to the book of Genesis. Thus it would require the direct intervention of God if the universe began at that date. By contrast, the big bang is a beginning that is required by the dynamical laws that govern the universe. It is therefore intrinsic to the universe, and is not imposed on it from outside.

Although the laws of science seemed to predict that the universe had a beginning, they also seemed to predict that they could not determine how the universe would have begun. This was obviously very unsatisfactory, so there were a number of attempts to get round the conclusion that there was a singularity of infinite density in the past. One suggestion was to modify the law of gravity so that it became repulsive. This could lead to the graph of the separation between two galaxies being an exponential curve that did not pass through zero at any finite time in the past. Instead, the idea was that, as the galaxies moved apart, new galaxies were formed in between from matter that was supposed to be continually created. This was the Steady State theory, proposed by Bondi, Gold, and Hoyle.

The Steady State theory was what Karl Popper would call "a good scientific theory": it made definite predictions that could be tested by observation, and possibly falsified. Unfortunately for the theory, they were falsified. The first trouble came with the Cambridge observations of the number of radio sources of different strengths. On average, one would expect that the fainter sources would also be the more distant. One would therefore expect them to be more numerous than bright sources, which would tend to be near to us. However, the graph of the number of radio sources, against their strength, went up much more sharply at low source strengths than the Steady State theory predicted.

There were attempts to explain away this number count graph by claiming that some of the faint radio sources were within our own galaxy, and so did not tell us anything about cosmology. This argument did not really stand up to further observations. But the final nail in the coffin of the Steady State theory came with the discovery of the microwave background radiation in 1965. This radiation is the same in all directions. It has the spectrum of radiation in thermal equilibrium at a temperature of 2.7 degrees above the absolute zero of temperature. There does not seem to be any way to explain this radiation in the Steady State theory. Hoyle still claims that it could be generated by iron needles distributed throughout intergalactic space and heated by ultraviolet light. However, the recent microwave background observations by the Cosmic Background Explorer sattelite show that it has such a perfectly thermal spectrum that I think even Hoyle should now abandon the Steady State theory. Knowing Hoyle, however, I expect he will still claim there is an alternative explanation.

Another attempt to avoid a beginning to time was the suggestion that maybe all the galaxies did not meet up at a single point in the past. Although on average the galaxies are moving apart from each other at a steady rate, they also have small additional velocities relative to the uniform expansion. These so-called "peculiar velocities" of the galaxies may be directed sideways to the main expansion. It was argued that as you plotted the position of the galaxies back in time, the sideways peculiar velocities would have meant that the galaxies would not have all met up. Instead, there could have been a previous contracting phase of the universe, in which galaxies were moving toward each other. The sideways velocities could have meant that the galaxies did not collide, but rushed past each other and then started to move apart. There would not have been any singularity of infinite density, or any breakdown of the laws of physics. Thus there would be no necessity for the universe, and time itself, to have a beginning. Indeed, one might suppose that the universe had oscillated, though that still would not solve the problem with the Second Law of Thermodynamics: one would expect that the universe would become more disordered each oscillation. It is therefore difficult to see how the universe could have been oscillating for an infinite time.

This possibility that the galaxies would have missed each other was supported by a paper by Lifshitz and Khalatnikov in 1963. They claimed that there would be no singularities in a solution of the field equations of general relativity, which was fully general in the sense that it did not have any exact symmetry. However, their claim was proved wrong by a number of theorems by Roger Penrose and myself. These showed that general relativity predicted singularities whenever more than a certain amount of mass was present in a region. The first theorems were designed to show that time came to an end inside a black hole formed by the collapse of a star. However, the expansion of the universe is like the time reverse of the collapse of a star. I therefore want to show you that observational evidence indicates the universe contains sufficient matter that it is like the time reverse of a black hole, and so contains a singularity.

In order to discuss observations in cosmology, it is helpful to draw a diagram of events in space and time, with time going upwards and the space directions horizontal. To show this diagram properly, I would really need a four-dimensional screen. However, because of government cuts, we could manage to provide only a two-dimensional screen. I shall therefore be able to show only one of the space directions.

As we look out at the universe, we are looking back in time, because light had to leave distant objects a long time ago to reach us at the present time. This means that the events we observe lie on what is called our past light cone. The point of the cone is at our position at the present time. As one goes back in time on the diagram, the light cone spreads out to greater distances, and its area increases. However, if there is sufficient matter on our past light cone, it will bend the rays of light toward each other. This will mean that, as one goes back into the past, the area of our past light cone will reach a maximum, and then start to decrease. It is this focusing of our past light cone by the gravitational effect of the matter in the universe that is the signal that the universe is within its horizon, like the time reverse of a black hole. If one can determine that there is enough matter in the universe to focus our past light cone, one can then apply the singularity theorems to show that time must have a beginning.

How can we tell from the observations whether there is enough matter on our past light cone to focus it. We observe a number of galaxies, but we cannot measure directly how much matter they contain. Nor can we be sure that every line of sight from us will pass through a galaxy. So I will give a different argument to show that the universe contains enough matter to focus our past light cone. The argument is based on the spectrum of the microwave background radiation. This is characteristic of radiation that has been in thermal equilibrium with matter at the same temperature. To achieve such an equilibrium, it is necessary for the radiation to be scattered by matter many times. For example, the light that we receive from the sun has a characteristically thermal spectrum. This is not because the nuclear reactions that go on in the center of the sun produce radiation with a thermal spectrum. Rather, it is because the radiation has been scattered by the matter in the sun many times on its way from the center.

In the case of the universe, the fact that the microwave background has such an exactly thermal spectrum indicates that it must have been scattered many times. The universe must therefore contain enough matter to make it opaque in every direction we look, because the microwave background is the same in every direction we look. Moreover, this opacity must occur a long way away from us, because we can see galaxies and quasars at great distances. Thus there must be a lot of matter at a great distance from us. The greatest opacity over a broad wave band for a given density comes from ionized hydrogen. It then follows that if there is enough matter to make the universe opaque, there is also enough matter to focus our past light cone. One can then apply the theorem of Penrose and myself to show that time must have a beginning if the General Theory of Relativity is correct.

The focusing of our past light cone implied that time must have a beginning if the General Theory of relativity is correct. But one might raise the question of whether General Relativity is correct. It certainly agrees with all the observational tests that have been carried out. However, these test General Relativity only over fairly large distances. We know that General Relativity cannot be quite correct on very small distances, because it is a classical theory. This means it does not take into account the Uncertainty Principle of Quantum Mechanics, which says that an object cannot have both a well-defined position and a well-defined speed: the more accurately one measures the position, the less accurately one can measure the speed, and vice versa. Therefore, to understand the very high-density stage when the universe was very small, one needs a quantum theory of gravity that will combine General Relativity with the Uncertainty Principle.

Many people hoped that quantum effects would somehow smooth out the singularity of infinite density, and allow the universe to bounce and continue back to a previous contracting phase. This would be rather like the earlier idea of galaxies missing each other, but the bounce would occur at a much higher density. However, I think that this is not what happens: quantum effects do not remove the singularity and allow time to be continued back indefinitely. But it seems that quantum effects can remove the most objectionable feature of singularities in classical General Relativity. This is that the classical theory does not enable one to calculate what would come out of a singularity, because all the Laws of Physics would break down there. This would mean that science could not predict how the universe would have begun. Instead, one would have to appeal to an agency outside the universe. This may be why many religious leaders were ready to accept the big bang and the singularity theorems.

Quantum effects do not seem to remove the singularity at the beginning of time. But quantum theory is based on complex numbers in an essential way, in contrast to classical theory, which is based on real numbers. In particular, quantum theory introduces the idea of complex time. That is, time measured with complex numbers, instead of just the ordinary real numbers.

One can draw a diagram with the ordinary, real direction of time horizontal, and the imaginary direction of time vertical. Points on this diagram that are on the same horizontal line represent events that are separated by real intervals of time. But points on a vertical line are separated by an imaginary interval of time. Imaginary intervals of time behave just like a fourth direction of space at right angles to the three normal directions of space. So the three space directions and imaginary time together make up a spacetime that is like Euclidean space, in that all directions are on the same footing. On the other hand, the three space directions and real time make up a spacetime that is like Minkowski space, in which the time direction is different from the space directions.

In the case of real time, there are only two possible behaviors: either time continues back into the past indefinitely, or time had a beginning at a singularity. However, in the imaginary direction of time, there is a third possibility: because imaginary time behaves like another direction in space, it is possible for space and imaginary time to form a spacetime that is finite in extent, but does not have a boundary or edge. It would be like the surface of the earth, but with two more dimensions. The surface of the earth is finite in extent, but it does not have any boundary or edge. I have tested this by experiment; I have been round the world, and I have not fallen off.

Jim Hartle and I have suggested that space and imaginary time together are indeed finite in extent, but without boundary. If this is the case, there would not be any singularities in the imaginary time direction at which the laws of physics would break down. And there would not be any boundaries to the imaginary time space-time, just as there are not any boundaries to the surface of the earth. This absence of boundaries means that the laws of physics determine the state of the universe uniquely in imaginary time. But if one knows the state of the universe in imaginary time, one can calculate the state of the universe in real time by analytical continuation. One would still expect some sort of big bang singularity in real time. After all, an analytic function that is not constant must have a singularity somewhere, so real time would still have a beginning. But one wouldn't have to appeal to something outside the universe, like God, to determine how the universe began. Instead, the way the universe started out at the big bang would be determined by the state of the universe in imaginary time. Thus, the universe would be a completely self-contained system. It would not require us to postulate the existence of anything, outside the physical universe that we observe.

The no-boundary condition is the statement that the laws of physics hold everywhere. Clearly, this is something that one would like to believe, but it is an hypothesis. One has to test it by comparing the state of the universe that it would predict with observations of what the universe is actually like. If the observations disagreed with the predictions of the no-boundary hypothesis, we would have to conclude the hypothesis was false. There would have to be something outside the universe to wind up the clockwork and set the universe going. Of course, even if the observations do agree with the predictions, that does not prove that the no-boundary proposal is correct. But one's confidence in it would be increased, particularly because there does not seem to be any other natural proposal for the quantum state of the universe.

What does the no-boundary proposal predict for how the universe began? In a quantum theory, the universe does not have just a single history. Rather, it can be thought of as having every possible history. With each history is associated a complex number called the *amplitude.* This determines how probable that history is. It can be thought of as a little arrow with a length and a phase, or angle, with the horizontal direction. The amplitude of a history depends on what happens in that history. For most histories, the phase of the amplitude varies very rapidly if the history is changed slightly. The amplitudes of such histories will be very nearly canceled, therefore, by the amplitudes of slightly different histories. It will be like adding a number of different arrows with different directions. The arrows will cancel each other out, and leave nothing. However, there will be certain histories for which the amplitude does not change if the history is changed slightly. In these cases, the amplitudes of slightly different histories will all add up. It will be like adding arrows that all point in the same direction. The arrows will reinforce each other. This will mean that these histories are by far the most probable.

These histories, for which the amplitude does not change when the history is changed slightly are solutions of the field equations of General Relativity. But they are solutions in which both time and space can be complex quantities rather than just real quantities, as in classical General Relativity. Of course, the time and space that one observes are always real. But the use of complex solutions of the field equations enables one to calculate the probabilities of measuring different real quantities.

The universe is the same at every point in space, and in every direction, to a first approximation. So I will consider closed Friedmann solutions, which contain a massive scalar field, phi. In order that the solutions satisfy the no-boundary condition, they must close up in the imaginary time direction like the surface of the earth closes up at the North Pole. One can think of the imaginary time direction as being like the distance from the North Pole, and the size of the universe as being like the length of a circle of latitude around the North Pole. In this analogy, the North Pole of the earth would be like the beginning of the universe in imaginary time. The universe would start out with zero size, just as the length of a circle of latitude is zero at the North Pole. As imaginary time increases, the universe would expand in size, reaching a maximum in imaginary time at the equator, and then decreasing again.

One can consider the behavior of the solutions on a diagram in which real time is shown in the horizontal direction and imaginary time in the vertical direction. One can choose to measure time from the point like the North Pole, where the size of the universe is zero. One imposes the condition that this is a regular point of spacetime, just as the North Pole is a regular point on the surface of the earth. Then one can integrate the field equations and determine the solution for all values of the complex time, given the value of the scalar field at the origin of time. In other words, there is a probable history for the universe for each value of the scalar field at the North Pole.

In the solutions, the size of the universe will increase up the positive imaginary time axis to a maximum value, and will then decrease to zero. This maximum value will be small, and will occur close to the origin if the initial value of the scalar field is large. From the point on the imaginary axis where the size of the universe is a maximum, one can integrate the solution in the real direction of time. Work by one of my students, Glenn Lyons, has shown that the size of the universe can remain almost real as one follows the solution in this direction. For this to happen, the initial value of the scalar field has to be chosen correctly. Of course, the size of the universe and the value of the scalar field that one observes are always real, but they are given by these solutions to a good approximation. In a sense, one could say that the universe expanded from zero to a certain size in the imaginary direction of time, and then changed to expanding in real time. But this is just one way of looking at what is really a sum over all complex histories.

One can not observe the absolute value of time, but only the time interval between events. Thus, along this horizontal line from the maximum on the imaginary axis, time will appear real. The universe will expand at first in an inflationary, or exponential manner. This is like the behavior in the chaotic inflation model proposed by Lindey. It would then go over to the normal hot big bang model.

In this simple model, the no-boundary condition thus predicts inflation. However, one can also consider more general models in which the universe is not exactly homogeneous and isotropic. If the departures from homogeneity are large enough, they will prevent the universe from inflating. However, preliminary calculations indicate that the probabilities of very inhomogeneous universes are low. Thus the no-boundary proposal predicts inflation with high probability. It also predicts the spectrum of small departures that would be expected from an exact Friedmann model. This has almost the scale-free form that is thought to be required for the standard model of galaxy formation. Thus the predictions of the no-boundary proposal seem to be consistent with observation. This does not prove that the no-boundary hypothesis is correct, but it does give one more confidence in it.

What does the no-boundary proposal predict for the future of the universe? Because it requires that the universe is finite in space, as well as in imaginary time, it implies that the universe will recollapse eventually. However, it will not recollapse for a very long time, much longer than the 15 billion years it has already been expanding. So, you will have time to sell your electricity shares before the end of the universe is nigh.

Originally, I thought that the collapse would be the time reverse of the expansion. This would have meant that the arrow of time would have pointed the other way in the contracting phase. People would have gotten younger as the universe got smaller. Eventually, they would have disappeared back into the womb.

However, I now realize I was wrong, as these solutions show. The collapse is not the time reverse of the expansion. The expansion will start with an inflationary phase, but the collapse will not in general end with an anti-inflationary phase. Moreover, the small departures from uniform density will continue to grow in the contracting phase. The universe will get more and more lumpy and irregular as it gets smaller, and disorder will increase. This means that the arrow of time will not reverse. People will continue to get older, even after the universe has begun to contract. So it is no good waiting until the universe recollapses to return to your youth. You would be a bit past it, anyway, by then.

The conclusion of this lecture is that the universe has not existed forever. Rather, the universe, and time itself, had a beginning in the big bang, about 15 billion years ago. The beginning of real time would have been a singularity at which the laws of physics would have broken down. Nevertheless, the way the universe began would have been determined by the laws of physics if the universe satisfied the no-boundary condition. This says that in the imaginary time direction, spacetime is finite in extent, but does not have any boundary or edge. The predictions of the no-boundary proposal seem to agree with observation. The no-boundary hypothesis also predicts that the universe will eventually collapse again. However, the contracting phase will not have the opposite arrow of time to the expanding phase. So we will keep on getting older, and we will not return to our youth. Because time is not going to go backwards, I think I better stop now. Thank you very much.

Particle Astronomy Minisymposium: Introductory Remarks

P. F. SMITH[a] AND B. SADOULET[b]

[a]*Rutherford Appleton Laboratory*
Chilton, Oxfordshire, England

[b]*Center for Particle Astrophysics*
University of California at Berkeley
Berkeley, California 94720

The objective of the papers in this Part of this volume is to review the status and future prospects for the detection of all types of particles incident (or hypothetically incident) upon the earth.

A summary of the types of particle discussed is shown pictorially in FIGURE 1. Photons are included under this heading if their wavelengths are less than nuclear dimensions, and hence classified as gamma rays. The other particles are cosmic rays, neutrinos from the sun, supernovae or hypothetical high-energy point sources, light- or heavy-particle candidates for galactic dark matter, and the as yet undetectable low-energy relic neutrinos. These require a wide range of surface and underground detectors, with a future need for very much larger detector areas and volumes, at present out of reach of present funding levels. Other speculative future directions include the siting of detectors on the moon to reduce background from atmospheric interactions, and possible twenty-first century zero-gravity experiments to detect the relic neutrino background.

This group of topics has evolved into an established interdisciplinary experimental subject during the past 10 years, as a result of its joint interest and importance to both astronomy and particle physics. A longstanding example of this joint interest is that of neutrino physics, where the question of neutrino mass, vital to the understanding of the lepton spectrum in particle physics, is also crucial to the questions of the apparent deficit of solar neutrinos, the nature of dark matter, the value of omega, and the evolution of structure. Further joint studies arose from subsequent conjectures that galaxies evolved with the aid of "cold dark matter" in the form of either very light axions or new heavy weakly interacting massive particles, both of which were predicted by specific particle models. This in turn has led to joint funding of experimental programs in both the United States and the United Kingdom, in addition to programs in the United States and Europe funded solely from particle physics. Several experiments are now planned that could observe or exclude specific types of dark matter particle.

Cosmic ray, gamma ray, and solar neutrino physics also now attract interest and activity from both particle physicists and astronomers. Detectors specifically for supernova neutrinos have been unattractive because of the infrequency of supernova events in our galaxy, but interest was greatly stimulated by the observation of supernova 1987A by several underground proton decay detectors, and this has led to speculation regarding very large future neutrino detectors with sensitivity to supernova events outside the Galaxy, providing more frequent data and also offering the prospect of estimating neutrino mass differences from arrival time differences. Large detectors are also of interest for the observation of neutrinos and gamma rays from hypothetical new types of ultrahigh-energy point sources, although it is perhaps unfortunate that the point source and supernova detectors do involve rather dif-

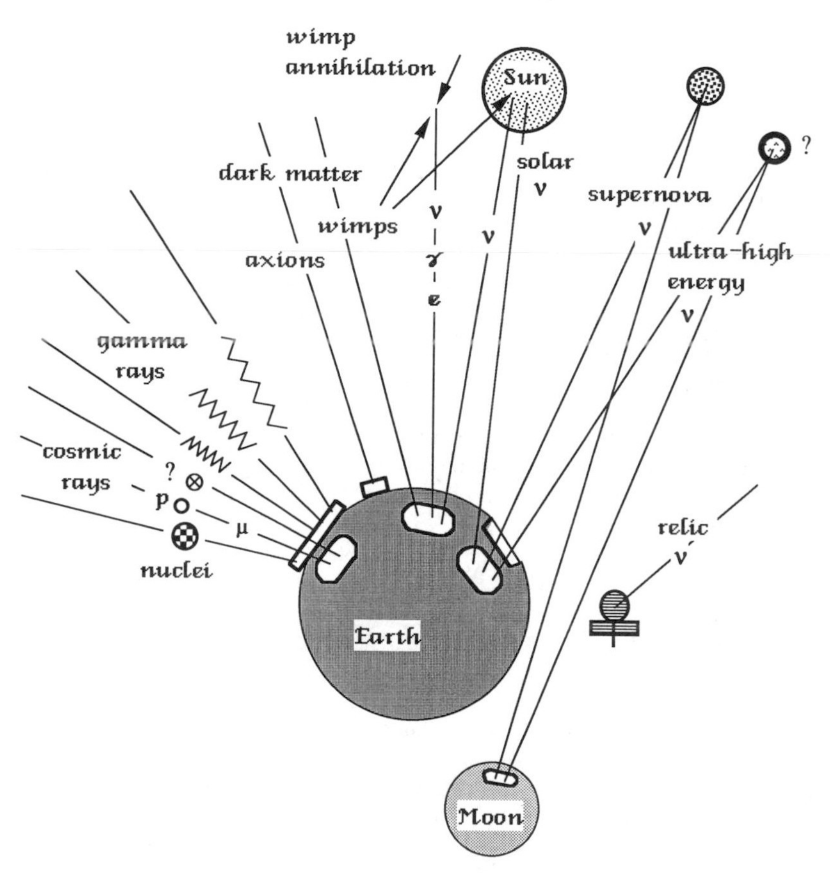

FIGURE 1. Pictorial summary of known and hypothetical particles incident on current and future detectors.

ferent energy ranges, so the objectives cannot easily be combined into a single detector.

The more established and noncontroversial topic of solar neutrino detection is now entering a critical and exciting phase, with the original chlorine rates and the more recent Kamioka rates both suggesting a deficit from the standard solar model.

There are also tantalizing preliminary indications of a deficit from the Soviet/American Gallium experiment SAGE, and with the other Gallium experiment GALLEX close to operation, these two experiments will have in the coming years the critical task of arriving at conclusive results, a difficult enterprise that will certainly necessitate calibration with isotopically enriched neutrino sources. Several years in the future, the large Sudbury Neutrino Observatory (SNO) will be able to study in detail the neutrino spectrum and may confirm the existence of oscillations between different neutrino flavors.

The first three sessions in this minisymposium were devoted to review papers aimed at summarizing the present status of high- and ultrahigh-energy gamma rays, high-energy muons, direct and radiochemical solar neutrino experiments, new large-detector projects, and the various techniques planned for detection of dark matter particle candidates. The final session was devoted to more speculative talks, covering the future of cosmic-ray physics, ultrahigh-energy point source detection, and twenty-first century prospects for extragalactic supernova detection and relic neutrino detection. This collection of papers thus provides an excellent summary of both the state of the art and future prospects for the still-expanding experimental and theoretical interface between astronomy and particle physics.

Very-high-energy Gamma-ray Astronomy: Overview and Prospects[a]

TREVOR C. WEEKES

Whipple Observatory
Harvard-Smithsonian Center for Astrophysics
P.O. Box 97
Amado, AZ 85645-0097

OVERVIEW: SOURCES

From several recent reviews (Weekes 1988; Fegan 1990; Vacanti 1990) one can compile a "source catalog" of some 22 objects that have been reported as TeV or PeV sources (TABLE 1). At first sight such a long list is encouraging and would suggest that the field is a rich one with great potential. A closer examination of the list shows that many of the reports have not been substantiated and that many have only marginal statistical significance. A broad division of gamma-ray sources can be made into (a) classical and (b) serendipitous. The former are generally steady, predicted, and verifiable; the latter are episodic, largely unexpected, and difficult to verify. The former category is sparsely populated, whereas there is an abundance of entries in the latter. It is these latter sources that are the most controversial aspect of the discipline. Here I will concentrate on sources in category (a), since these must constitute the foundation for very-high-energy (VHE) gamma-ray astronomy.

Diffuse Background

A diffuse background of cosmic rays inevitably results in a flux of gamma rays at some level. A particularly interesting source of background is the pileup that occurs at energies of order 100 TeV due to the cosmological interaction of primary protons with the microwave background (Halzen *et al.* 1990). The measurement of the diffuse background is difficult to make, since it presupposes the ability to absolutely identify an individual gamma-ray shower. From ground-based experiments it has only been possible to derive upper limits; these are listed in TABLE 2 in terms of the ratio I_g/I_p. It is generally believed that the flux measured by the SAS-2 satellite at high latitudes is a true measurement of the diffuse flux. The predicted ratio at 10–100 TeV is of order 10^{-5} (Halzen *et al.* 1990).

Galactic Plane

The strongest feature in the 100-MeV gamma-ray sky is the galactic plane; at some level it must also be detectable at TeV and PeV energies. At TeV energies the

[a]The research in VHE gamma-ray astronomy at the Whipple Observatory is supported by the U.S. Department of Energy, NASA, and the Smithsonian Scholarly Studies Fund.

TABLE 1. Source Catalog

Pulsars	X-ray Binaries	Supernova Remnants
PSR0355+54	Her X-1	Crab Nebula
PSR0531+21	Cyg X-3	
PSR0833−45	Vel X-1	
PSR1509−08	Sco X-1	Extragalactic
PSR1953+29	2A1822-37	Cen A
PSR1937+21	LMC X-4	M31
PSR0950+08	Cen X-3	
PSR1855+09	4U0115+63	Cataclysmic Variable
PSR1957+20	1E2259+58	AE Aqr
		AM Her

observational situation is confusing; there is some evidence for the detection of an antisource close to the plane, but no evidence for the broader source on which the antisource is supposed to be superimposed (Weekes 1988). At energies of 100 TeV and greater there are good upper limits from the Utah–Michigan experiment (Matthews *et al.* 1991); these limits assume that gamma-ray event selection can be made on the basis of their low muon content. Theoretical predictions (Berezinsky and Kudryavtsev 1990) for the emission from the galactic center region of the plane (which is not optimally visible from Utah) give $I_g/I_p = 7 \times 10^{-5}$. Hence the predicted and measured limits are close to confrontation.

Giant Molecular Clouds

The maps of the galactic plane at 100-MeV energies include evidence for emission from a number of Giant Molecular Clouds, including Taurus and Rho Ophiuchis (Issa and Wolfendale 1981). Estimates of emission from these objects at higher energies have been made by a number of authors. These estimates vary widely because the cosmic-ray spectrum within the clouds is unknown. Gurevich *et al.* (1985) assume that the higher energy cosmic rays are not trapped within the clouds, and hence the gamma-ray spectrum cuts off sharply. Aharonian (1990) predicts containment and possible enhancement of the cosmic-ray flux, and hence predicts the spectrum from the Taurus cloud shown in FIGURE 1. There are no upper limits from TeV observations, but the detection at 100-MeV energies and the upper limit from the Utah–Michigan experiment (Matthews *et al.* 1991) are shown in FIGURE 1.

TABLE 2.

	Ratio I_g/I_p		Flux
Energy	Diffuse	Galactic Plane	GMC (Taurus) Photons-cm^{2}-s^{-1}
1 TeV	$<10^{-2}$	1.5×10^{-2}	No measurements
200 TeV	$<4 \times 10^{-3}$	$<8 \times 10^{-5}$	$<3.8 \times 10^{-14}$

Supernova Explosions

The supernova outburst is one of the most violent events on a stellar scale, and it would not be surprising if there was a burst of gamma rays within the first few seconds of the explosion. As the outburst continues it is thought that cosmic rays could be accelerated in the expanding shock wave; as the surrounding shell thins, gamma rays produced by the collision of these cosmic rays with the gas shell can escape. The supernova in the Large Magellanic Cloud (1987A) is a classic laboratory for testing these hypotheses. To date no VHE emission has been detected; during the actual outburst the source was fortuitously within the field of view of the Durham telescope, but no signal was seen (Brazier *et al.* 1990).

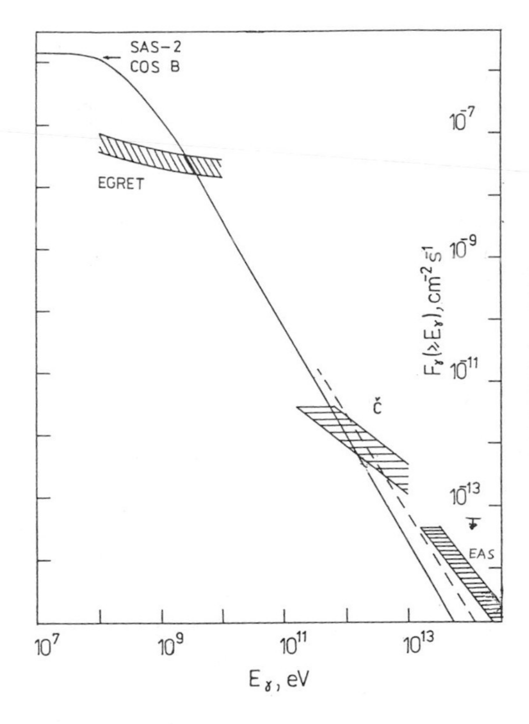

FIGURE 1. The flux predicted from a typical Grant Molecular Cloud for two different models of cosmic-ray intensity within the cloud (Aharonian 1989). The UM upper limit from the Taurus Giant Molecular Cloud (Matthews *et al.* 1991) is shown in the bottom right-hand corner.

Supernova Remnants

The aftermath of the supernova outburst is often a radio pulsar with surrounding expanding nebula that contains the cosmic rays accelerated by the pulsar. Such objects are called plerions, and are potential cosmic-ray sources. The observation of synchrotron emission from such objects confirms the acceleration of electrons to TeV energies by the pulsar, and a simple Compton-synchrotron model predicts an extended gamma-ray spectrum (Gould 1965). The relativistic electrons lose their periodic character as they move away from the pulsar, and gamma-ray emission from the nebula will be unpulsed and probably steady on timescales similar to the dimensions of the nebula.

The most prominent object in the plerion category is the Crab Nebula; this supernova remnant played a key role in the early history of radio and X-ray astronomy, and it would not be surprising if it played a similar role in very-high-energy (VHE) astronomy. Because of numerous predictions of possible emission it qualifies as a classical gamma-ray source. To date it must be regarded as the only truly well-established gamma-ray source at TeV energies.

The Crab Nebula has been detected by a number of experiments over a period of 20 years. There is no strong evidence for variability making it suitable as a standard candle for referencing the strengths of other sources and telescope sensitivities. The most significant detection is that reported by the Whipple collaboration (Vacanti *et al.* 1991). This group uses the Whipple Observatory optical reflector as an imaging device so that they can select gamma-ray showers from amongst the much more numerous cosmic-ray background (Cawley *et al.* 1990). It is estimated that the technique has a rejection factor of about 98 percent with an improvement in flux sensitivity over conventional techniques of a factor of 10.

The significant features of this detection (Vacanti *et al.* 1991) are the following.

(a) The statistical significance is at the twenty-sigma level.
(b) Previous reports of emission are verified (Weekes *et al.* 1989; Akerlof *et al.* 1989), and there are no conflicting limits.
(c) There is no evidence for variability on any timescale, and no evidence for periodicity at the pulsar period.
(d) The showers that constitute the signal are entirely consistent with those expected from gamma-ray primaries.
(e) The energy spectrum between 0.4 and 4 TeV is measured to be $N(E) \cdot dE = 2.5 \times 10^{-10} (E/0.4 \text{ TeV})^{-0.2.4\pm0.3}$ photons cm^{-2} s^{-1} TeV^{-1}.

The most significant VHE Crab observations are plotted in FIGURE 2 together with the COS-B observations. It is clear that the Crab pulsar spectrum dips sharply between 1 GeV and 1 TeV, and that the steady emission also steepens.

Radio Pulsars

Although there is no generally accepted model for pulsars, it is agreed that the rapidly rotating highly magnetized neutron star is a certain source for relativistic electrons, and that it is radiation from these electrons that gives the characteristic radio pulses. Such objects are natural sources of gamma rays with energies up to the maximum energy of the accelerated electrons. However, while the electrons will radiate gamma rays close to the neutron star surface, it is difficult to invent a scenario whereby the gamma rays can escape the pulsar without pair production in the strong magnetic field. Only if the electrons radiate well away from the neutron star surface can the gamma rays emerge; in general such gamma rays will retain the periodic character of the parent electrons.

Two of the strongest 100-MeV sources are associated with the Crab and Vela pulsars. The emission from both of these sources is variable in amplitude on timescales of months. No emission is seen from any other pulsar. The observational situation at TeV energies is somewhat confused. Several detections have been reported for TeV gamma-ray emission from the Crab and Vela pulsars (and a few others), but these do not appear to be reproducible. Observations of episodic emission have also been reported that are reminiscent of the reports of similar emission from the binary sources (next section). The most convincing evidence for periodic emission is reported by the Durham group (Dowthwaite *et al.* 1984). This emission has not been detected by more recent experiments by other groups (FIG. 2).

Emission at TeV energies from several other pulsars have also been reported, but usually only by one group. The most noteworthy of these is PSR1833 (Chadwick 1990). The quasi-steady signal from this millisecond pulsar is in agreement with the level predicted by Usov (1983). However, the observation of pulsars deserves many of the caveats associated with the serendipitous sources discussed below. In particular,

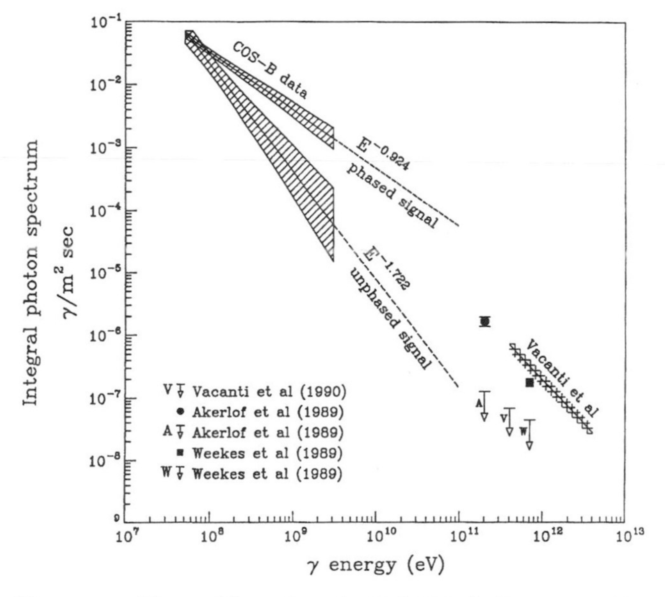

FIGURE 2. The measured integral fluxes from the Crab Nebula. Some upper limits for pulsed emission are also indicated.

the reported emission from PSR1802-23 by the Potchefstroom group (Raubenheimer *et al.* 1986) serves as a cautionary tale. The radio pulsar was one of three reported from radio observations in the error box associated with the COS-B gamma-ray source CG006-00. The Potchefstroom group reported a detection at the 99.8 percent confidence level. However, later, more extended observations by the same group (Nel *et al.* 1991) did not confirm the emission. Furthermore it was found that the original statistical significance had been overstated. Subsequent analysis of the COS-B database did not confirm CG006-00 as a discrete source, and further radio observations did not confirm the existence of PSR1802-23.

Extragalactic: AGNs

Although HE and VHE gamma-ray astronomy has been preoccupied with galactic sources, there are good reasons to believe that extragalactic sources may be detectable with a small increase in detection sensitivity. At 100-MeV energies there is good evidence for the detection of 3C273; it appears that there is more energy emitted in this decade of the electromagnetic spectrum than any other. The best candidates for detection at high energies appear to be AGNs (although at one stage it appeared that M31 had been detected). Detailed models are not generally available, but there are some predictions (Sikora and Shlosman 1989). An apparent TeV detection of Centaurus A led to the development of a Compton–synchrotron model for the source; however, the detection has not been confirmed. There are upper limits for 3C273 and several other potential sources at TeV energies (Vacanti *et al.* 1990).

Serendipitous Sources

This category includes all sources that do not have any firm *a priori* basis for their selection as candidate VHE gamma-ray sources, that is, there were no observations at other wavelengths that make a compelling case for the presence of nonthermal processes, nor are there detailed models to suggest VHE emission. Every new band of the electromagnetic spectrum has had its share of serendipitous sources, and the VHE band has been no exception. What has made this field particularly difficult to interpret has been that until the discovery of VHE emission from the Crab Nebula, *all* the sources reported appeared to be serendipitous sources. The fact that these sources were X-ray binaries that had not been expected to be sources of relativistic particles, that all the emission appeared to be episodic, and that the observations were based on the detection of a periodicity that could not always be predicted precisely, has made the credibility of the sources a matter of some debate. The issue is made urgent by reason of the large number of such sources, which suggests that they are major contributors to the cosmic radiation and, even stranger, that the properties of the detected signal are not that expected of gamma rays.

No attempt will be made to review the evidence for all of these sources; they are listed in TABLE 1. Several of these sources will be discussed by others at this symposium.

PROSPECTS

The current controversy in the field comes from the serendipitous sources whose reality is disputed. Despite the seemingly strong statistical evidence for the detection of a number of episodes of emission from some of these sources, there are some disturbing features of the whole phenomenon. These include the following.

1. Uncertainty as to the real statistical significance to apply to episodic events that are found *a posteriori*.

2. The failure of the statistical significance to increase as more and better experiments are applied to the study of these sources.
3. The apparent fading of some of the strongest source candidates; in particular Cyg X-3, the original and strongest source, is no longer detectable.
4. The failure of the putative signal to display the characteristics of gamma-ray showers (in those experiments that can characterize the primaries). This failure is now established for Her X-1 at both TeV and PeV energies; more disturbing is the absence of any shower parameter that is sufficiently anomalous to suggest that the signal was other than a fluctuation of the background.

Although there have been some experiments in this field for thirty years, the overall investment has been small, and it is only in recent years that there has been such a concentration of effort that real progress can be expected. It is at TeV energies that the first definite detection of a source has emerged, and it is at TeV energies that the existence of the episodic sources must be resolved. In TeV astronomy the integrated investment has been about one-hundredth that of the effort at MeV–GeV energies (which must be pursued from space) and only one-tenth that devoted to PeV gamma-ray astronomy. There is a similar discrepancy between the cost of TeV gamma-ray and neutrino astronomy.

Despite the meager investment, TeV gamma-ray astronomy appears to be on a firm footing. With the positive detection of at least one gamma-ray source, there is now a standard for calibrating experiments and for optimizing detection techniques. With existing telescopes it is possible to detect sources that are less than one-tenth the intensity of the Crab Nebula. With the completion of GRANITE, the extension of the Whipple Observatory experiment (involving the Smithsonian Astrophysical Observatory, the University of Michigan, and Iowa State University), this factor will become one-hundredth. The angular resolution with these techniques is now comparable to that which can be achieved in space gamma-ray telescopes, so that one can anticipate a very fruitful overlap with the Gamma Ray Observatory mission scheduled for launch in 1991.

At PeV energies the experimental situation is good also with several new major experiments coming on-line. The most sensitive of these employ large-area muon detectors in addition to the surface electron detectors. Of particular note is the Chicago–Michigan–Utah experiment that should have a flux sensitivity in excess of a hundred times better than the original Kiel experiment that detected Cygnus X-3. One note of caution here is that it has yet to be established that the muon-poor selection does have the efficiency claimed by simulations.

The present confused state of the field as regards the detection of episodic sources is to be expected given the rapid growth of the field and the paucity of sensitive telescopes, but the situation should clarify as more groups enter the field, as the more sensitive telescopes now being built come on-line, and as more sensitive analysis techniques are developed. It remains to be seen whether these new experiments will lead to the discipline becoming an established branch of astronomy . . . or whether it continues its controversial role on the borderline between high-energy astrophysics and particle physics.

ACKNOWLEDGMENTS

I am grateful for helpful comments from M. F. Cawley, M. A. Lawrence and P. T. Reynolds.

REFERENCES

AHARONIAN, F. A. 1990. Yerevan preprint, YERPHI 1254(40)-90.
AKERLOF, C., *et al.* 1989. *In* Proceedings of the Gamma Ray Observatory Science Workshop (Goddard Space Flight Center): 4–49.
BRAZIER, K. T. S., *et al.* 1990. *In* Proceedings of the 21st International Cosmic Ray Conference (Adelaide) **2:** 263.
BEREZINSKY, V. S. & V. A. KUDRYAVTSEV. 1990. Astrophys. J. **349:** 620.
CAWLEY, M. F., *et al.* 1990. Exper. Astron. **1:** 173.
CHADWICK, P. 1990. *In* Proceedings of the International Workshop on Gamma Rays (Ann Arbor, Mich.). In press.
DOWTHWAITE, J. C., *et al.* 1984. Astrophys. J., Lett. **286:** L35.
FEGAN, D. J. 1990. *In* Proceedings of the 21st International Cosmic Ray Conference (Adelaide) **11:** 23.
GOULD, R. J. 1965. Phys. Rev. Lett. **15:** 511.
GUREVICH, A. V., *et al.* 1985. Sov. Astron. Lett. **11:** 284.
HALZEN, F., *et al.* 1990. Phys. Rev. **D41:** 342.
ISSA, M. R. & A. W. WOLFENDALE. 1981. Nature **292:** 430.
MATTHEWS, J., *et al.* 1991. Astrophys. J. In press.
NEL, H. I., *et al.* 1991. Astrophys. J. In press.
RAUBENHEIMER, B. C., *et al.* 1986. Astrophys. J., Lett. **307:** L43.
SIKORA, M. & I. SHLOSMAN. 1989. Astrophys. J. **336:** 593.
USOY, V. V. 1983. Nature **305:** 409.
VACANTI, G. 1990. *In* Proceedings of the Relativistic Hadrons Workshop (Suhora, Poland). In press.
VACANTI, G., *et al.* 1990. *In* Proceedings of the 21st International Cosmic Ray Conference (Adelaide) **2:** 329.
VACANTI, G., *et al.* 1991. Astrophys. J. In press.
WEEKES, T. C. 1988. Phys. Rev. Rep. **160:** 1.
WEEKES, T. C., *et al.* 1989. Astrophys. J. **342:** 379.

Teraelectronvolt Gamma-ray Astronomy

K. E. TURVER

Department of Physics
University of Durham
Durham, England

INTRODUCTION

The purpose of this paper, following the overview of ground-based gamma-ray astronomy by Weekes (1990), is to describe the evidence for emission of teraelectronvolt (TeV) gamma rays from a limited number of objects. In so doing I wish to demonstrate:

(a) the wide range of astrophysical objects for which there have been claims for the emission of TeV gamma rays,
(b) the breadth of evidence that supports the genuine nature of many of these claims.

The Cerenkov light technique has been employed to study TeV gamma rays since the mid-1950s following the discovery of the atmospheric Cerenkov effect. Numerous developments took place during the 1960s and 1970s (e.g., the pioneering work done at the Whipple Observatory and the Crimean Astrophysical Observatory), but it was not until the 1980s that the numerous large-scale experiments necessary to establish the field were commenced. By the end of that decade about 10 gamma-ray observatories using many variations of the Cerenkov technique and covering the northern and southern skies were in operation. These provided, for the first time, comprehensive datasets and the possibility of independent confirmation of claims.

This paper will review in some detail the evidence for TeV emission from a selection of objects as diverse as a supernova remnant (the Crab nebula and pulsar), X-ray binary pulsars (4U0115 + 63 and Centarurus X-3), and millisecond pulsars (PSR 1855 + 09 and Cygnus X-3). Finally, an update on the flux limit at TeV energies from SN 1987A will be given. More extensive reviews of the evidence for TeV gamma-ray sources can be found in, for example, Weekes (1988), Turver (1990), and Chadwick, McComb, and Turver (1990).

EVIDENCE FOR TERAELECTRONVOLT GAMMA-RAY SOURCES

The Crab Nebula

The statistically significant and reproducible data showing emission of a persistent, weak but apparently unpulsed emission from the Crab nebula obtained with the Whipple telescope has been discussed by Weekes (1990). This is the first (and so far only) example of the successful application of an exciting new technique developed at the Whipple Observatory to provide a strong enhancement of the gamma-ray signal in the background of cosmic-ray nucleons. The successful application of this technique to other objects and an independent confirmation of the Crab nebula result will mark a major advance in the field.

PSR 0532 + 21

The Crab pulsar is perhaps the most extensively observed object at TeV energies. The results have been mixed, ranging from flux limits to statistically weak evidence for 33-ms periodicity. In many cases, due to the nonavailability of a simultaneous radio or optical observation, there existed uncertainty in the period, and it was not possible to compare the phase of the peak of the TeV light curve with that of the radio or 100-MeV gamma-ray measurement.

The resumption of the monitoring of PSR 0532 at radio wavelengths by the NRAL team at Jodrell Bank in 1981 meant that data at TeV energies could be

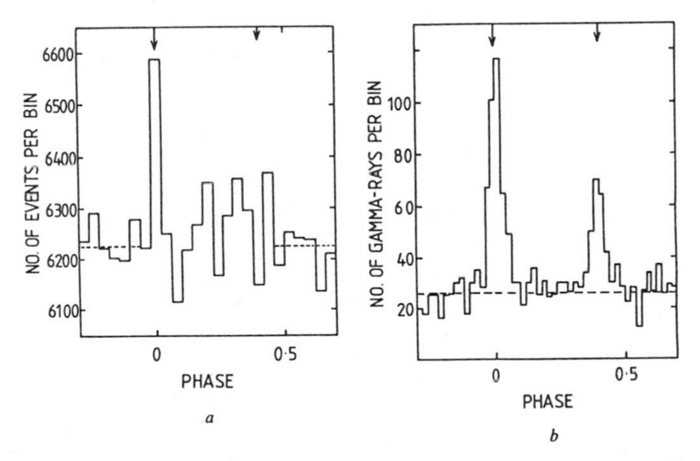

FIGURE 1. **(a)** The light curve for all events recorded from PSR 0532 from September 1982 to November 1983. The radio main and interpulses are indicated by *arrows*. **(b)** The average light curve for 100-MeV γ-rays emitted by PSR 0532 according to Wills *et al.* (1982).

"epoch folded" to provide a light curve that could be directly compared with that at, for example, 100-MeV gamma-ray energies. This ability to test the Cerenkov light data for periodicity at a well-measured contemporary period for which, as a bonus, the epoch of the radio main pulse was well known marked an advance in TeV measurements. The continuation over 10 years of the provision of radio ephemerides every month by Lyne and Pritchard has been of great assistance to TeV observers.

In 1981/1982 the Durham group, working with a quartet of TeV gamma-ray telescopes at Dugway, Utah, made a study of > 100 hr of data from the Crab pulsar spread over more than a year (Dowthwaite *et al.* 1984). The TeV light curve folded at the appropriate radio period is shown in FIGURE 1, where it is compared with that measured at 100 MeV by the COS-B experiment. (The phases of the two gamma-ray results are linked via the radio measurements.)

The advantages of the quartet of telescopes, in addition to the obvious increase in count rate and improved sensitivity and the ability to demonstrate the gamma-ray signal in each of the four telescopes, lies in the interpretation of those Cerenkov light events when multiple telescopes respond to the same light pool. (This occurs about 15–20 percent of the time.) For example, when two telescopes spaced by 60 m respond to a common light pool, it may be expected from such considerations (and it is confirmed by simulations) that the effective aperture will be less than that of a single telescope. This should lead to a reduced cosmic-ray background in the sample of such two-telescope events and a constant gamma-ray signal. The light curve of the "two telescope responses" from Dugway is shown in FIGURE 2. The strength of the gamma-ray peak coincident with the radio main pulse is doubled from 5 to 10 percent, as expected if the aperture [full width at half maximum (FWHM)] has decreased by the anticipated 40–50 percent.

One can proceed further in this direction; in cases when two or more telescopes respond to the same light pool, and the time of arrival at the various telescopes is accurately measured, the arrival direction within the field of view can be determined using the well-established light front timing techniques. It is then possible to

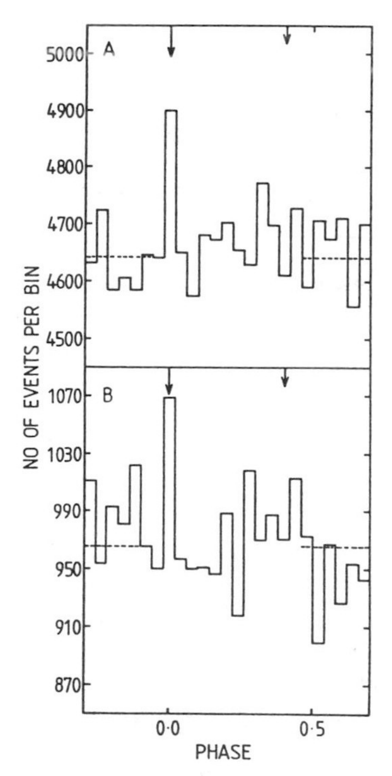

FIGURE 2. **(a)** The light curve for all events recorded from PSR 0532 in November 1982 that initiated single-telescope responses. **(b)** The light curve for all events recorded from PSR 0532 in November 1982 that initiated double-telescope responses.

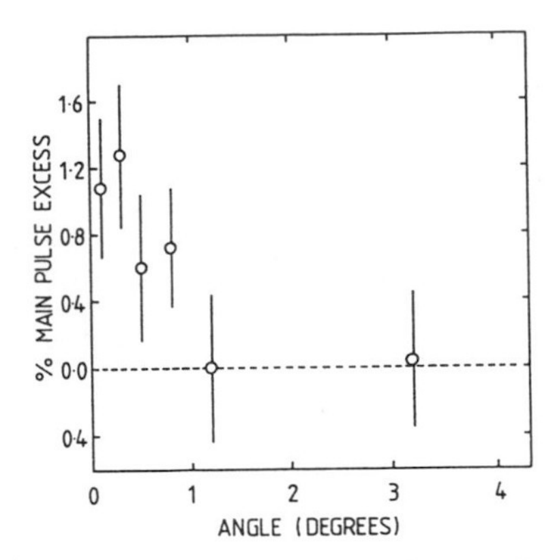

FIGURE 3. The signal strength of the main pulse emission as a function of angular distance from the center of the field of view.

categorize events according to their arrival direction within the aperture and to plot the strength of the "gamma-ray main pulse" as a function of angle from the center of the field. The results of this analysis are shown in FIGURE 3.

There is thus substantial evidence that supports the basic result of the period analysis shown in FIGURE 1. Unfortunately, the Dugway experiment had no independent simultaneous "off source" or background control data to confirm that the effects claimed were confined to the source direction, and so to add further to the evidence for the genuine nature of the claimed effects.

4U0115 + 63

This was the second X-ray binary (XRB) pulsar found to produce TeV gamma rays (the first was Her X-1), and it was observed with the Dugway telescopes in 1984 (Chadwick *et al.* 1985). The observations took place on the nights of September 21–29, 1984, lasted for 25 hr and yielded 31,000 Cerenkov flashes recorded from the direction of the XRB. Here the period analysis is more complicated; we do not have the benefit of contemporary period measurements at X-ray wavelengths, and the source is in a binary system. In such cases the period analysis of the individual observations (after reducing the time of the light flashes to the solar system barycenter, and allowing for the Doppler shifting due to the orbital motion of the source) is made using the Rayleigh test for uniformity in phase. The results of a recent analysis are shown in FIGURE 4. The result of an "incoherent" analysis—the distributions in chance probability for various test periods in the individual observations have been combined with no maintenance of phase between observations—are shown in FIGURE 4(a). The average signal strength is 2.4 percent of the cosmic-ray background count rate and the chance probability is 1.5×10^{-4}. When all the data are taken as a single gapped dataset (i.e., with phase maintained between observations), and a "coherent" analysis is made, the result is as shown in FIGURE 4(b). The aliases

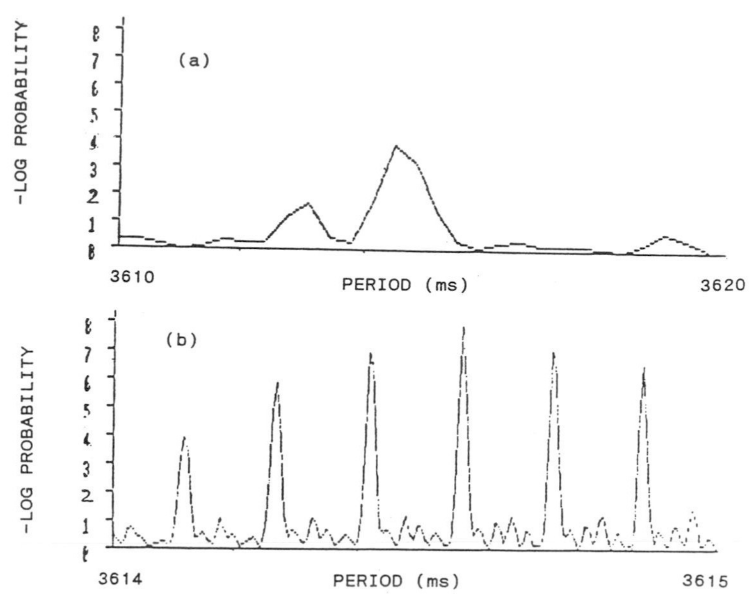

FIGURE 4. **(a)** The periodogram for all events recorded from 4U0115 when the data are considered to be an incoherent set (phase is not preserved between the separate observations). **(b)** The periodogram for all events from 4U0115 analyzed as a coherent, gapped dataset.

from the 24-hr gapped data are obvious; the chance probability is 1.7×10^{-8}. Note that the signal strength suggested by the "incoherent" analysis is retained in the "coherent" signal. This demonstrates a TeV gamma-ray signal that is periodic at the X-ray period and is coherent over an 8-day interval.

The sensitivity of the 4U0115 + 63 data to the corrections for the motion of the production region in the binary system (and hence the ability to measure the orbital parameters from TeV data) is demonstrated in FIGURE 5. Here the contours for the chance probability for periodicity at the X-ray period are plotted as a function of the values of the semimajor axis of the orbit (asin i) and the epoch for X-ray eclipse that are assumed for the orbital parameters. The minimum probability is at values of orbital parameters similar to those from previous X-ray observations.

Teraelectronvolt emission from this object has been noted by other observers. In 1985 the Halaekala collaboration recorded 39 hr of data and analyzed the events in independent 1000-s segments (Resvanis *et al.* 1987). Three of the 129 segments showed 3.6-s periodicity, significant at <0.001 chance level.

The Whipple telescope also observed 4U0115 for 38 hr and analyzed the data as nine segments each of data recorded over 3 days or less (Lamb *et al.* 1987). Data recorded during three consecutive nights showed periodicity at 3.617 s with chance probability of 0.01.

However, the significance of this result does not improve when the new enhancement technique is applied (see the earlier section on the Crab nebula).

Centaurus X-3

Centaurus X-3 (Cen X-3) is a 4.8-s X-ray pulsar in a 2.1-d orbit. The X-ray emission is well studied with recent observations by the Ginga and Mir-Kvant spacecraft. Teraelectronvolt gamma rays were reported from Cen X-3 by the Durham group following 207 hr of observations that sampled the whole orbit. Evidence was found, significant at the 10^{-6} level, for emission at the fundamental X-ray period, but the emission was confined to a 5 percent wide interval of orbital phase close to the ascending node. See FIGURE 6. The benefits of the extensive control data available at Narrabri showing that the clear association with the source direction are demonstrated. During these TeV gamma-ray observations lasting 3 years little evidence was found for a spin up of the pulsar of a magnitude seen previously, a fact confirmed by the contemporaneous Ginga and Mir-Kvant X-ray observations.

Independent confirmation of TeV emission has come from the Potchesfstroom group (North *et al.* 1990), who recorded 71 hr of data, and also found evidence for emission confined to the region around the ascending node (phase 0.7–0.8).

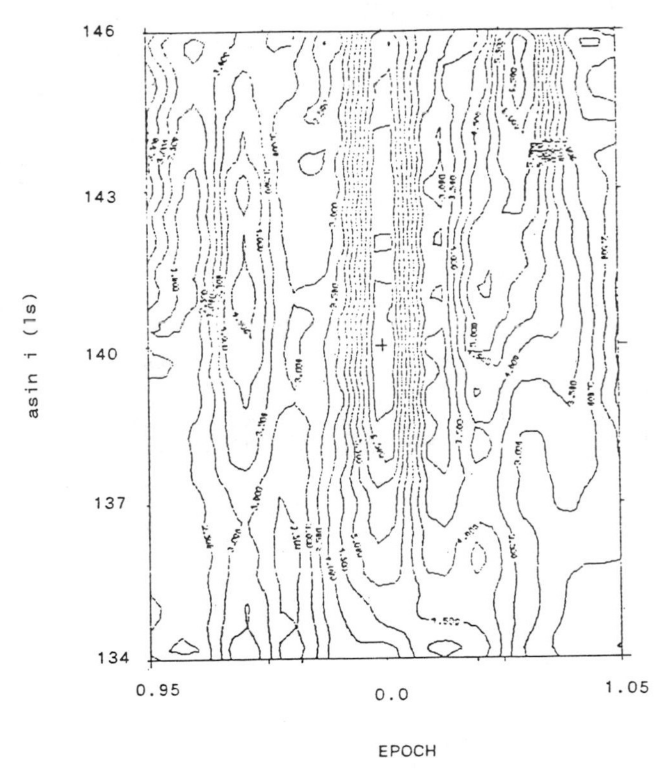

FIGURE 5. The contours of chance probability for periodicity at the X-ray period as a function of the values of the semimajor axis (asin i) and the epoch of the X-ray eclipse (advancing and retarding by 0.05 of the orbital period).

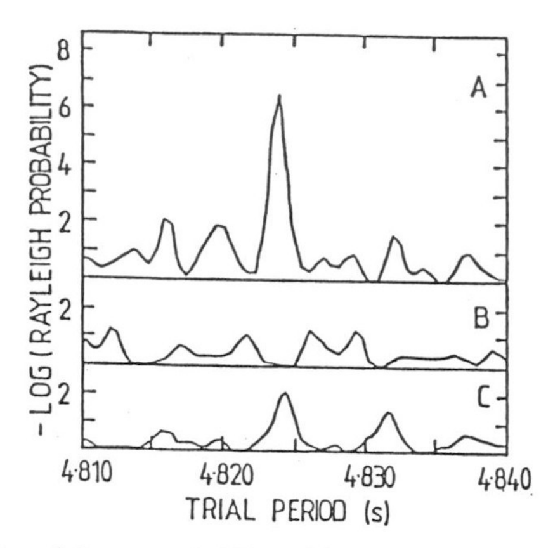

FIGURE 6. The variation of chance probability with period for data taken in the orbital phase 0.77–0.82 from Cen X-3. **(a)** Uses events that triggered the detector only. **(b)** Uses events that triggered both the central detector and an outer guard-ring detector. **(c)** Comprises events that triggered an outer guard-ring detector only. The data are from the University of Durham telescope at Narrabri.

PSR 1855 + 09

This is a 5-ms pulsar in a 12.3-d binary orbit for which a good radio ephemeris is available. Extensive TeV gamma-ray observations have been made at Potchefstroom and Narrabri since 1987. The Potchefstroom result (de Jager *et al.* 1989) indicated that periodicity was detected at a chance level of 0.01 (with power in the second harmonic as in the radio signal).

The Narrabri observations were made over 143 hr on 55 separate occasions in 1987–1990. The data recorded during each of the separate observations were analyzed using the Rayleigh test after each event time was adjusted to the solar system barycenter and focus of the binary system. An analysis of the total dataset (with no requirement for linking the observations in phase) shows some evidence that is significant at the 0.01 chance level for power in the second harmonic. One of the TeV observations at Narrabri (October 11, 1988) showed strong evidence for power in the second harmonic at precisely the radio period. See FIGURE 7(a). This observation was at phase 0.685 in the orbit. The very extensive off-source and control data that are a feature of the Narrabri experiment show no indication of similar periodicity. The chance probabilities for power in the second harmonic of the radio period for all the individual observations are plotted FIGURE 7(b). These results suggest that there is evidence for additional TeV emission clustered around the descending node in the orbit, in addition to the single strong occurrence shown in FIGURE 7(a).

Cygnus X-3

The claims for the detection of TeV and petaelectronvolt (PeV) gamma rays from Cygnus X-3 have been one of the most interesting developments in high-energy astrophysics during the 1980s. A summary of the various claims for TeV emission, including those for the detection of a 12-ms pulsar as the central engine of this unusual source has been given recently (Brazier *et al.* 1990). The Durham group on the basis of data from 1981–1985 suggested the phase in the orbit (close to X-ray maximum) at which the short (400-s) outburst of pulsed (12.5953-ms) TeV gamma rays was to be expected in 1988. As reported by Brazier *et al.,* this prediction was fulfilled during observations in 1988 at La Palma, Canary Islands, leading in turn to an even firmer prediction for later observations.

During additional observations at La Palma in 1989 made to test the predictions, no evidence for a 12-ms periodicity was obtained.

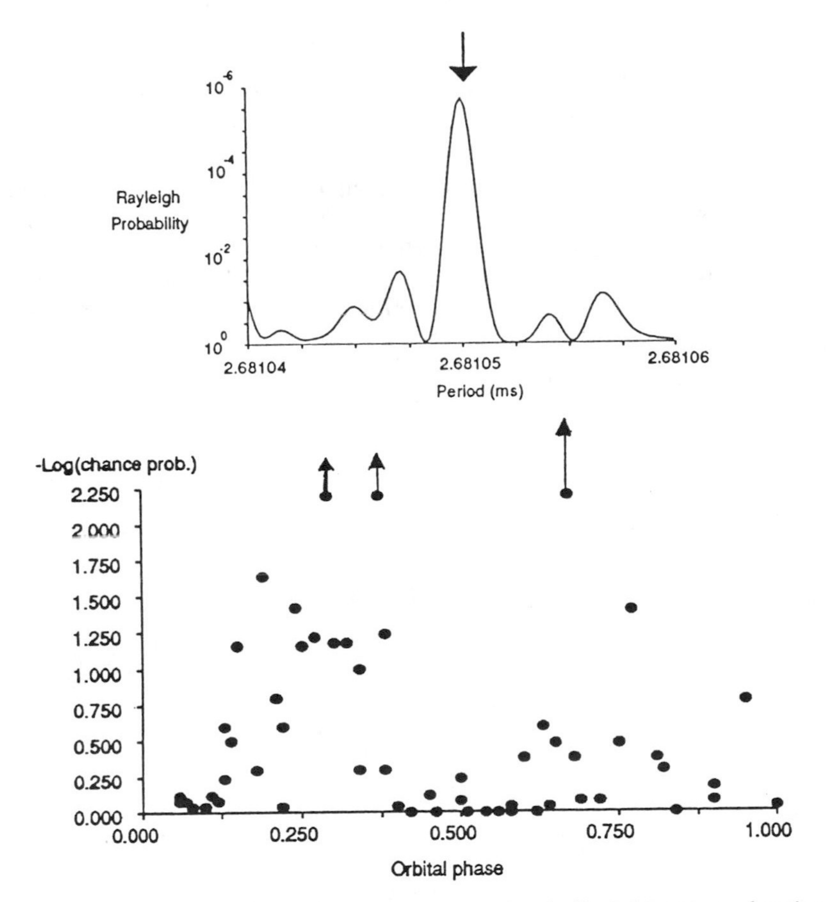

FIGURE 7. **(a)** The probability of chance occurrence using the Rayleigh test as a function of period from the observation of PSR 1855 + 09 on October 11, 1988. The *arrow* indicates the second harmonic of the expected radio period. **(b)** The chance probability for the periodicity at the radio period in individual observations plotted as a function of the orbital phase when the observation was made. The significances are based on repeated simulations of the analysis routine. The lower limits refer to the three observations with the smallest chance probabilities that require the most simulations to ascertain the true chance probabilities and for which the simulation is incomplete.

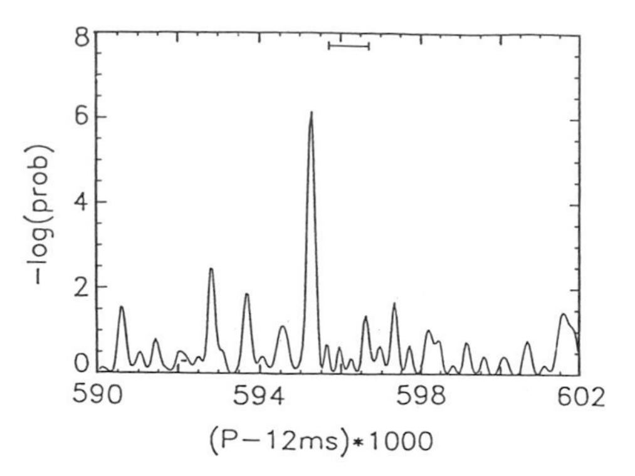

FIGURE 8. The periodogram for the data during two 600-s intervals on September 1 and 2, 1989 recorded by the Woomera telescope.

Independently the Adelaide, Australia, group using their telescope at Woomera came to a similar conclusion after a test of the Durham prediction (Gregory *et al.* 1990). This Southern Hemisphere observation is of interest, being the first attempt to use the technique of observing 100-TeV gamma rays using the atmospheric Cerenkov technique at large zenith angles (>70 deg), as suggested by Sommers and Elbert

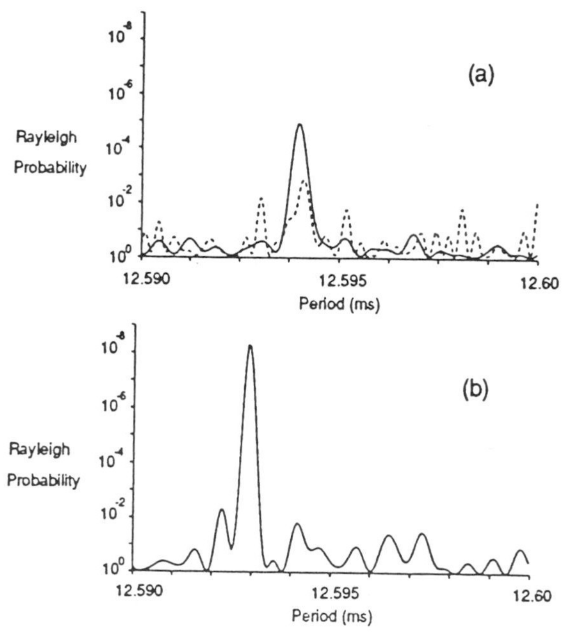

FIGURE 9. **(a)** The periodogram for the 600-s dataset identified by the Woomera group (see FIG. 8) obtained by the Durham group at La Palma on September 1, 1984. The *broken line* represents the chance probability for the whole 600-s dataset; the *solid line* represents that for a 300-s subset. **(b)** The periodogram for the data from La Palma during 300 s of the Woomera 600-s window on September 7, 1989.

(1985). However, the Adelaide group noted that during their observations in late August–September 1989, they had evidence for pulsed emission at a period of 12.5953 ms (0.0007 ms less than predicted) at a phase in the 4.8-hr cycle that was 1000 s earlier than predicted. Their strongest signal was during two intervals of 600 s (at the same phase in the 4.8-hr cycle) during observations from approximately 1000 to 1200 hours universal time coordinated (UTC) on September 1 and 2, 1989. See

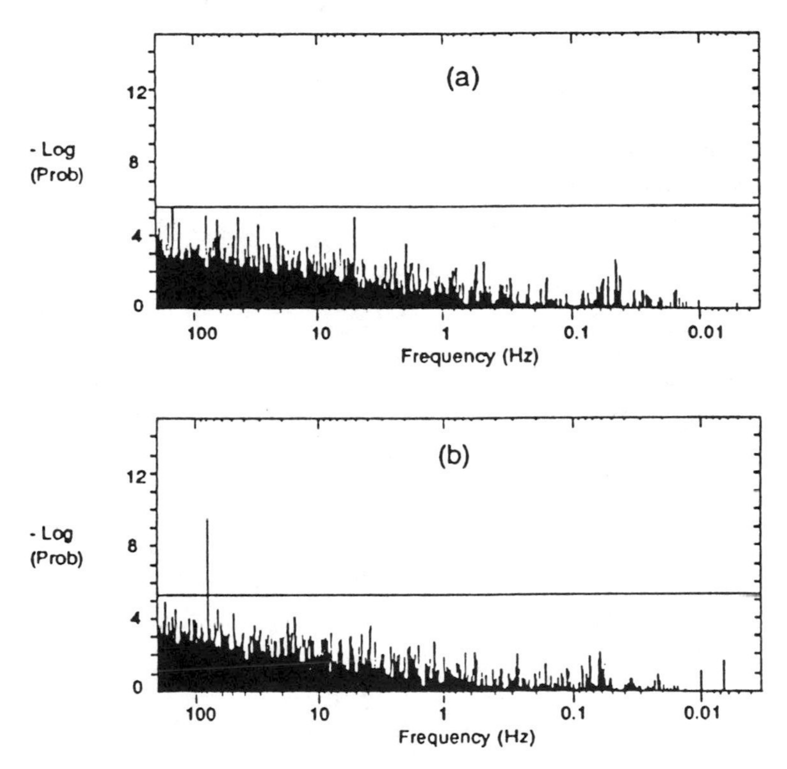

FIGURE 10. Fourier transforms for the 300-s segments in the 4.8-hr cycle shown to contain 12.5-ms pulsar activity. **(a)** From the dataset taken from La Palma on September 1, 1989. **(b)** From the dataset taken on September 7, 1989. The *horizontal lines* indicate the minimum probability expected by chance for each dataset.

FIGURE 8. The overall significance of the Woomera result is about 0.001, allowing for the choice of 600-s datasets and the trial periods.

Observations were made by the Durham group at La Palma from 2200–2400 hours UTC on September 1–7, 1989. (These were the observations that yielded data that failed to substantiate the predictions referred to previously.) The observation in La Palma on September 1, 1989 was thus between those made at Woomera that suggested periodicity (although at an energy 100 times lower). An investigation of the precise 600-s data segment in the La Palma data at exactly the same phase in the 4.8-hr cycle as the Woomera result is shown in FIGURE 9(a). Some evidence for periodicity at 12.5940 ms was found that was strongest in the second half of the 600-s segment. In addition, there was a 300-s segment within the "Woomera 600-s window" during the La Palma observation on September 7, 1989 that was strongly periodic (12.5930 ms) [see FIGURE 9(b)], and that showed a corresponding count rate excess. The Fourier transform of the two datasets for September 1 and 7, 1989 over the wide range of trial frequencies 200 Hz–0.003 Hz are shown in FIGURE 10.

There is thus increasing evidence for Cygnus X-3 containing a 12-ms pulsar as the central engine.

Supernova 1987A

To date no evidence for TeV emission from SN 1987A has been obtained. A typical value (from the Narrabri experiment) of the time averaged (3-sd) flux limit at an energy of about 400 GeV is about $1 \times 10^{-10}\ cm^{-2}\ s^{-1}$.

CONCLUSIONS

Evidence for the emission of TeV gamma rays from a range of astrophysical objects with small chance probabilities exists. In many cases there is a spectrum of additional observational facts that suggest that the effects are genuine together with control data that reinforce the association of the claimed effects with the astrophysical objects.

ACKNOWLEDGMENTS

It is a pleasure to acknowledge the contribution made by all colleagues in the Durham University high-energy gamma-ray astronomy group to many of the measurements and interpretations described here.

REFERENCES

BRAZIER, K. T. S., *et al.* 1990. Astrophys. J. **350:** 745.
CHADWICK, P. M., *et al.* 1985. Astron. Astrophys. **151:** L1.
CHADWICK, P. M., T. J. L. MCCOMB & K. E. TURVER. 1990. J. Phys. G **16:** 1773.
DE JAGER, O. C., *et al.* 1990. Proceedings of the International Conference on High Energy Gamma Ray Astronomy (Little Rock, Ark.). Nucl. Phys. B (Proc. Suppl.) **14B:** 117.
DOWTHWAITE, J. C., *et al.* 1984. Astrophys. J., Lett. **286:** L35.
GREGORY, A. A., *et al.* 1990. Astron. Astrophys. **237:** L5.
LAMB, R. C., *et al.* 1987. *In* Very High Energy Gamma Ray Astronomy, K. E. Turver, Ed.: 139. Reidel. Dordrecht, the Netherlands.
NORTH, A. R., *et al.* 1990. Proceedings of the 21st International Cosmic Ray Conference (Adelaide) **2:** 275.
RESVANIS, L. K., *et al.* 1987. *In* Very High Energy Gamma Ray Astronomy, K. E. Turver, Ed.: 131. Reidel. Dordrecht, the Netherlands.
SOMMERS, P. & J. ELBERT. 1987. J. Phys. G **13:** 553.
TURVER, K. E. 1990. Proc. R. Neth. Acad. In press.
WEEKES, T. C. 1988. Phys. Rep. **160:** 1.
WEEKES, T. C. 1990. Very-high-energy gamma-ray astronomy: Overview and prospects. This issue.
WILLS, R. D., *et al.* 1982. Nature **296:** 723.

Ultrahigh Energies

PIERRE SOKOLSKY

Physics Department
University of Utah
Salt Lake City, Utah 84112

INTRODUCTION

The struggle to discover the nature of the accelerating mechanism that generates the observed charged particle cosmic-ray spectrum is made more difficult by the sparcity of interesting features in the spectrum. The spectrum obeys a power law with index −2.6 from gigaelectronvolt to petaelectronvolt (10^{15} eV) energies, and then steepens to an index of between −3.0 and −3.1. This feature is known as the *knee*. The new power law continues up to an energy of 10 EeV (10^{19} eV). Beyond that the data suggest another flattening (known as the *ankle*). The spectrum must eventually terminate, either by the exhaustion of the acceleration mechanism or due to propagation effects. K. Greisen[1] pointed out as far back as 1966 that if the spectrum in the region of the ankle is composed of extragalactic protons and the sources of these protons are sufficiently distant, then we should observe a spectral cutoff between 50 and 100 EeV. This cutoff occurs due to the interaction of protons with the (then) recently discovered 2.7 deg black body radiation. This observation was made independently by Zatsepin and Kuzmin[2] in the same year.

THE GREISEN–ZATSEPIN–KUZMIN CUTOFF

The physical process underlying this Greisen–Zatsepin–Kuzmin (GZK) cutoff is quite simple. Protons with energies of 50 EeV or greater will interact inelastically with the 2.7 deg black body photons and produce secondary pions, protons, and neutrons. The secondary protons will have lower energies and generate an apparent steepening or cutoff in the observed protonic spectrum. There will also be associated gamma-ray and neutrino fluxes from the subsequent decays of pions and muons. Since the effective interaction length for protons of this energy is about 6 Mpc, relatively close extragalactic sources will not exhibit such a cutoff.

There have been a number[3] of recent calculations of the GZK effect incorporating such refinements as cosmological evolution of sources and the black body radiation, following the reinteraction of secondary particles in a full transport calculation, etc. These calculations are in qualitative agreement. Roughly speaking they predict a well-defined GZK cutoff below 100 EeV if cosmic-ray sources extend beyond 100 Mpc (as in the universal cosmic-ray origin hypothesis) and a less well-defined steepening above 100 EeV if sources are quasi local (such as sources in the Virgo supercluster). Observation of a GZK cutoff below 100 EeV would be strong evidence for the universal origin theory of ultrahigh-energy (UHE) cosmic rays, as well as direct confirmation of the universality of the black body radiation.

EXPERIMENTAL TECHNIQUES

There are two basic techniques for studying the ultrahigh-energy cosmic-ray spectrum: ground arrays and the Fly's Eye detector. We discuss how the different techniques work and the evidence to date for the presence or absence of the GZK cutoff.

Ground Arrays

At petaelectronvolt and exaelectronvolt energies, primary cosmic rays entering the atmosphere produce extensive air showers (EAS) of secondary particles, many of which reach the ground. Ground arrays sample these surviving secondaries, typically using large scintillation counters spaced by hundreds of meters to kilometers over areas of tens of square kilometers (such large areas are necessary because of the low flux at these energies). The core of the EAS can be found by a fit to the lateral particle density (the lateral distribution function) and the zenith angle of the incident primary can be determined by measuring the angle of the (approximately) plane shower front as it manifests itself in the relative delays between adjacent scintillation counters. Determining the energy of the primary particle is more complicated. Integrating the lateral distribution function is not a good measure of the primary energy, since large fluctuations can be induced by fluctuations in the interaction depth of the primary and subsequent shower development fluctuations. It has been found through Monte Carlo studies that the best measure of the primary energy is the charged-particle density at about 600 meters from the shower core. It is claimed that this parameter, $\rho(600)$, is approximately linearly related to the incident energy and is not strongly dependent on the composition of the primary particle or the interaction model used.[4]

The Yakutsk ground array also has Cherenkov light detectors. Since the total Cherenkov light observed at the earth's surface integrates over the shower development in the atmosphere, one can use such measurements to cross-calibrate the $\rho(600)$ energy scale with the more direct Cherenkov energy scale. Good agreement between the two techniques has been found by the Yakutsk group up to energies of 10 EeV.[5] The Haverah Park group has also done a careful study of the Yakutsk and Haverah Park energy analysis and claim that the two are quite compatible.[6]

Useful data are now available from three arrays: Haverah Park in the United Kingdom (which was turned off in 1988),[7] Yakutsk in the Soviet Union,[8] and Akeno in Japan (which is now being enlarged into a new array known as AGASA).[9] All these arrays have effective collecting apertures of 10 to 20 km^2sr and an approximately 100 percent duty cycle. They have been accumulating data for between 6 and 25 years.

The Fly's Eye

The Fly's Eye[10] is a unique detector that utilizes the scintillation properties of atmospheric N_2 molecules to detect the EAS produced by cosmic-ray primaries. As the EAS develops in the atmosphere, the secondary charged particles excite N_2 molecules, which subsequently emit scintillation light in the near UV. This light is

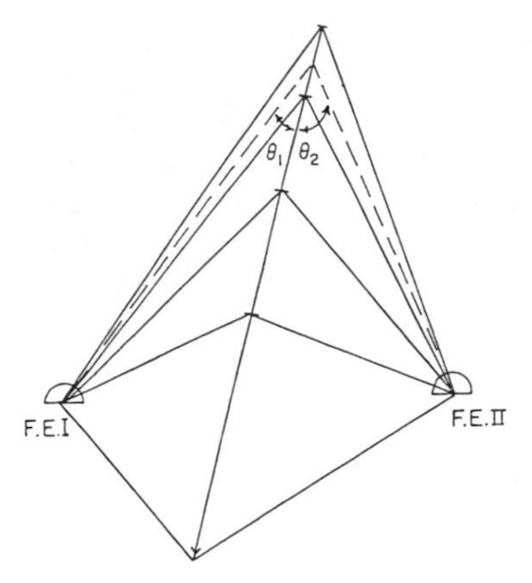

FIGURE 1. Schematic illustration of Fly's Eye operation. *Arrow* indicates the direction of the core of the EAS. Each phototube views the light from the EAS emitted at an angle θ in an atmospheric depth bin defined by the tube's direction and solid angle.

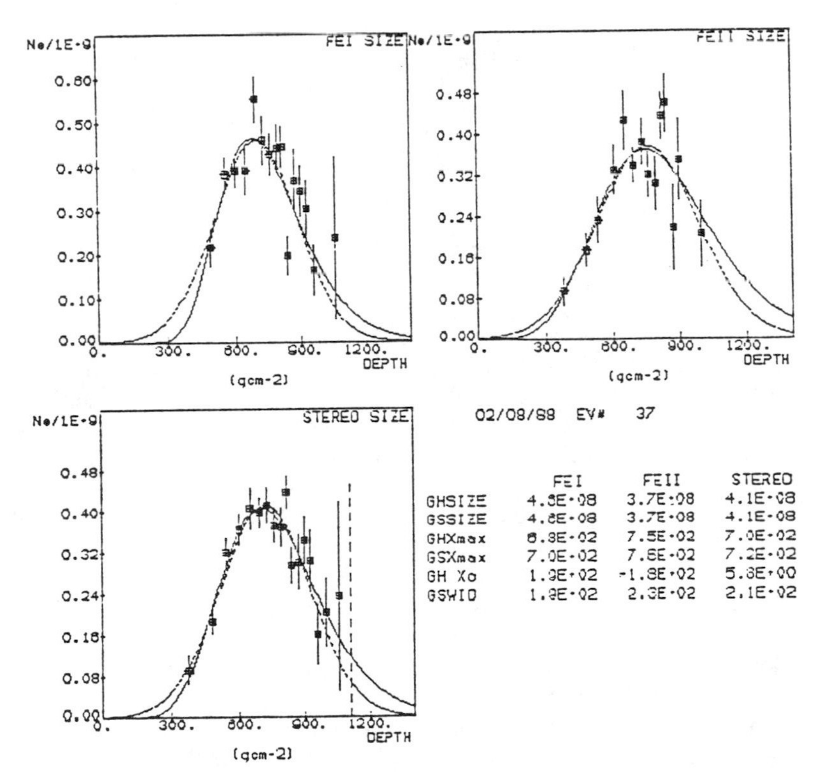

FIGURE 2. Longitudinal development curve of typical event seen by Fly's Eye I and Fly's Eye II.

collected by the Fly's Eye mirrors (67 1.6-m-diameter mirrors at one site and 36 at another site 3.5 km distant) and focused on photomultiplier (PM) tubes that give pulse height and relative timing information. FIGURE 1 shows how such a detector works schematically. The EAS can be thought of as a point source of light moving with the speed of light through the atmosphere and growing first brighter and then dimmer as the shower comes to its maximum, and then dissipates. Relative tube timing and stereo information allow the impact parameter and zenith and azimuth angles of the the track of the shower to be reconstructed. Once the geometry is known, pulse height information can be used to reconstruct the shower development curve (the longitudinal distribution) as a function of depth in the atmosphere. FIGURE 2 shows a typical event as seen from the first and second Fly's Eye and the mean longitudinal development curve. Once the shower longitudinal profile is measured, it can be integrated, and the primary energy is then just the integral times the mean charged-particle energy loss in the atmosphere. The energy determination is much more direct than in the case of ground arrays, the energy scale being set by the N_2 scintillation efficiency (a measured number) and the detector optical efficiency. The geometrical reconstruction, however, is more complex and the detector aperture is an increasing function of the particle energy, going from tens of km^2sr at 0.1 EeV to 1000 km^2sr at 50 to 100 EeV. In this case, it is the aperture that must be carefully Monte Carloed. Since this is an optical technique, the duty factor is 10 percent, leading to an effective aperture at 100 EeV of a few hundred km^2sr.

EXPERIMENTAL RESULTS

FIGURE 3 shows the results for the ground arrays and the Fly's Eye. Note that the spectrum is multiplied by E^3 to more easily see deviations from an E^{-3} power law. We can say in general that all the spectra are qualitatively similar, showing a near E^{-3} dependence up to 10 EeV, and then some evidence of flattening. In no case is this evidence for flattening stronger than a three-sigma effect, however. There is no strong evidence for a GZK cutoff at 50 EeV. With the possible exception of Haverah Park, which sees five events above 100 EeV, the combined data are not inconsistent with a GZK cutoff between 50 and 100 EeV. Given that the ground array and Fly's Eye technique are almost completely orthogonal, it is remarkable that the data agree as well as they do.

When interpreting spectral data plotted in this way it is important to bear in mind that resolution effects can also produce apparent deviations from a simple power-law behavior, as well as affect the shape near a cutoff. A constant fractional energy resolution of 0.3 (typical for these experiments) will not distort the measured spectral slope or produce significant spilldown beyond the GZK cutoff, but a resolution that worsens as a function of energy, or one with energy-dependent non-Gaussian tails, can give rise to apparent flattening of the spectra before the cutoff and a masking of the true cutoff energy. All the experiments claim that their estimated energy resolution is sufficiently well understood that this is not a problem, but it remains questionable how well tails of resolution functions can be understood in the greater than 50-EeV region where statistics are quite sparse.

In the case of the Fly's Eye, a significant fraction of events seen by one eye is also seen by the second one. These stereo events have more precise reconstruction, and the energy resolution can be checked, since two independent measurements are performed. The resultant spectrum, although poorer in statistics, has well-controlled

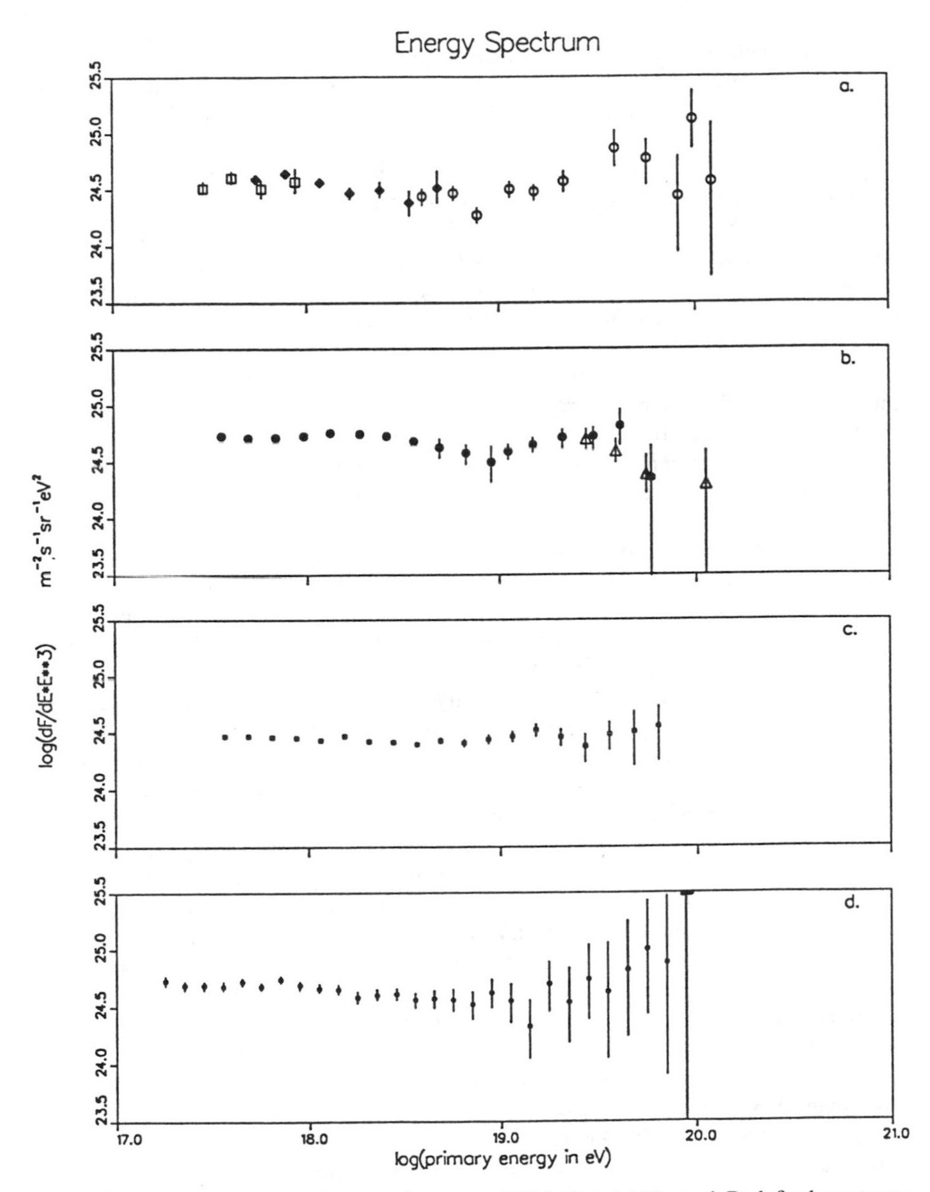

FIGURE 3. Spectra measured by ground arrays and Fly's Eye. **(a)** Haverah Park final spectrum; **(b)** Yakutsk spectrum; **(c)** Fly's Eye spectrum; **(d)** Akeno spectrum.

errors. It is in good agreement with the monocular Fly's Eye data, but has insuficient statistics to explore the GZK cutoff region.

IS THERE A SPECTRUM BEYOND THE GZK REGION?

Putting aside for the moment issues of energy resolution and energy scale (which may be systematically different between the different experiments), one may ask if it is possible for the GZK cutoff, which should lie between 50 and 100 EeV in the case of the universal cosmic-ray source hypothesis, to be masked by an increase in the flux of particles other than protons. There are several possibilities: an increase in the gamma-ray flux beyond the cutoff, and a dominance by heavy nuclei produced in the local supercluster. It is, of course, also possible that the spectrum in the region of the ankle is still galactic and made up of heavy nuclei (as has been proposed by Jokipii and Morfill[11]). In this case, any observed cutoff would be due to the termination of the injection spectrum and not due to propagation effects through the 2.7 deg black body radiation.

Most interesting is the observation, recently proposed by Aharonian *et al.*,[12] that the gamma-ray flux from pi-zero decay and inverse Compton scattering of decay electrons produced by the GZK mechanism on the relic black body radiation will, in fact, dominate over the resulting equilibrium proton flux above the GZK cutoff, thus effectively masking it if the detectors cannot distinguish between protons and gamma rays.

This is due to two reasons. First, the gamma-ray mean free path (with respect to interactions with the 2.7 deg black body background) increases with energy and reaches almost 20 Mpc at 100 EeV (at least for the pair-production energy-loss mechanism). The proton mean free path, on the other hand, is rapidly decreasing in this region. Hence one might be able to observe gamma rays produced by the GZK effect at larger distances than the secondary protons. Second, the gamma ray will produce an electromagnetic cascade as it traverses the relic black body radiation. This cascade is unusual in that the mean free path of secondary particles grows with energy. The properties of such cascades have been calculated, and under certain conditions, the net effect is to increase the transparency of the universe to UHE gamma rays beyond the single interaction case. The requirements are that the extragalactic magnetic field be less than 10^{-9} G (otherwise, intermediate electrons in the cascade will lose too much energy to syncroton radiation) and that the longwave radio background be sufficiently small. If these conditions are met, Aharonian *et al.*[12] calculate a large gamma-ray flux above the GZK cutoff (see FIG. 4). The upper and lower lines in the figure represent different assumptions about the slope of the equilibrium proton spectrum (−2.6. and −3.0, respectively). On the same figure are the results of a different calculation by Halzen *et al.*[12] Here the upper and lower spectra depend on assumptions about the trident production cross section, and no cosmological evolution has been included.

Although these calculations require significant assumptions (the nature of the trident and four-electron final state cross sections, intergalactic magnetic field and radio wave strength, etc.) it appears to be possible that significant gamma-ray fluxes could be found near 100 EeV. Another way in which this could come about has been

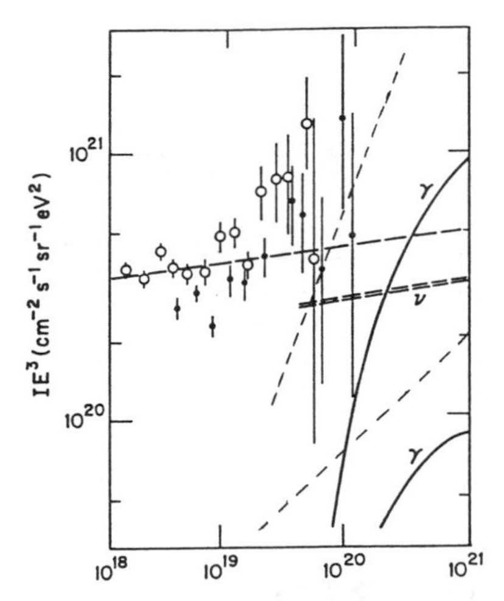

FIGURE 4. Continuation of gamma-ray spectrum beyond GZK cutoff. *Dashed lines:* Aharonian *et al.*;[12] *solid lines:* Halzen *et al.*;[12] data: Fly's Eye and Haverah Park.

suggested by Wolfendale.[13] In this scenario, the source of greater than 10-EeV extragalactic cosmic rays lies in the Virgo supercluster. Wolfendale argues that the extragalactic magnetic field in the supercluster is of order 3×10^{-9} G (based on equipartition with the cosmic-ray-flux energy density). Such a strong field would cause proton trajectories to be largely confined in this region, increasing the age of the particles, and allow many of these protons to experience the GZK effect even though the accelerating sources are quasi local. The resultant gamma rays would have only 10 Mpc or so to traverse to get to the earth, and hence could be directly detected. Here again, the GZK effect would be modified if gamma rays and protons could not be separately identified.

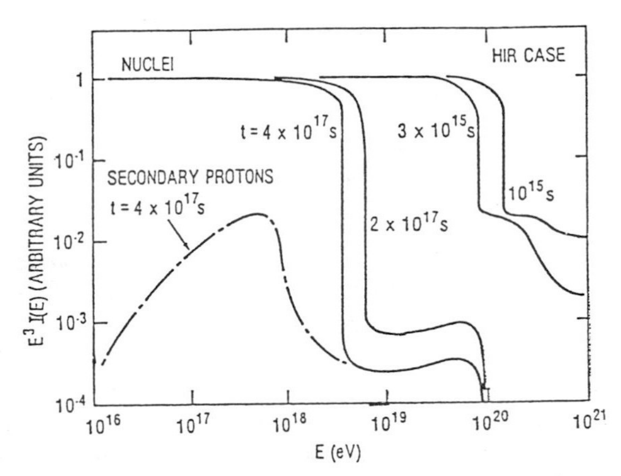

FIGURE 5. Effect of propagation through electromagnetic spectrum on pure Fe spectrum. (From Puget *et al.*[14] Reproduced by permission.)

Another possible reason why the GZK cutoff has not been observed is that the extragalactic flux is not protonic but dominated by heavy nuclei such as Fe. It is commonly argued that extragalactic fluxes cannot be heavy because of photospallation reactions on the 2.7 deg black body background. This is certainly true for sources greater than 100 Mpc away (i.e., it would apply to the universal source hypothesis). If sources are within the local supercluster, the situation is more complicated, the amount of stripping to light nuclei and protons being dependent on the total number of interaction lengths that the particles traverse in their lifetime. This may be significantly different from the line of sight distance to the source, depending on the nature of the intergalactic magnetic field. FIGURE 5 shows, for instance, the distortion in a pure Fe flux[14] due to interactions with the black body radiation and other electromagnetic radiation for two cosmic-ray lifetime assumptions. It is thus important to be able to measure the hadronic cosmic-ray composition as well as separating gamma rays from hadrons.

WHAT IS KNOWN ABOUT COMPOSITION BELOW 10 EeV

In this context it is clearly important to establish what the cosmic-ray composition is below 10 EeV where the flux is most likely galactic in origin and to contrast it to the

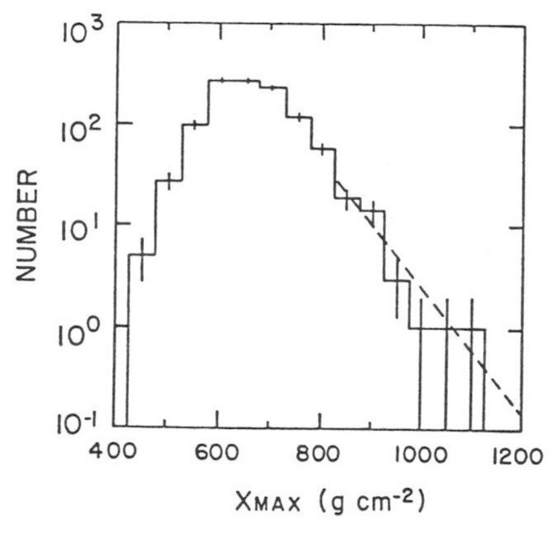

FIGURE 6. Measured Fly's Eye Xmax distribution.

composition in the ankle. The Fly's Eye group[15] has recently published results on the cosmic-ray composition in the region 0.1 to 10 EeV based on high-quality stereo data. The technique for establishing composition is to measure the distribution as a function of depth in the atmosphere of the EAS shower maxima. This so-called Xmax distribution should reflect the cosmic-ray composition, because heavy nuclei such as Fe will interact earlier, and thus have Xmax's higher in the atmosphere than protons. Heavy nucleus EAS are essentially superpositions of tens of lower energy showers initiated by the individual nucleons. Thus we expect that the width of the Xmax distribution (due to development fluctuations) will be significantly narrower

TABLE 1. Mean Values and Distribution Widths Expected for Pure Fe and Proton Fluxes

	Average Xmax (gm/cm^2)	Sigma (Xmax) (gm/cm^2)
Fe	705 ± 3	66 ± 2
Proton	803 ± 2	80 ± 1
Data	690 ± 3 ± 20	85 ± 2

for Fe-induced EAS than for protons. The overall Xmax distribution integrated from 0.1 to 10 EeV is shown in FIGURE 6. TABLE 1 shows the mean values and distribution widths expected for pure Fe and proton fluxes. These are generated using a standard interaction model that is in good agreement with lower energy accelerator data. One immediately notices that the width of the experimental distribution is consistent with a pure proton flux, while the mean value is consistent with a pure Fe flux. The only way out of this apparent quandary is to assume a mixed composition. Reasonably good agreement between the measured shape and Monte Carlo predictions based on the previously mentioned model can be had for Fe/proton ratios in the range of 1.0 to 0.5. The details of this estimate of composition will vary somewhat depending on the hadronic model used, but it is very difficult to salvage either a pure Fe or pure proton spectrum in this energy region.

FIGURE 7 shows a scatter plot of Xmax vs. energy. The solid line is the mean value of Xmax as a function of energy. The slope appears to be constant at 70 gm/cm^2 per decade of energy. This constancy is an indication that both the cosmic-ray composition and the hadronic interaction model are not changing very much in the energy region of interest. Similar results are claimed by the Yakutsk group,[16] using Cherenkov light techniques to determine Xmax, and by the Haverah Park array,[17] which measure the rise time of the pulses in their water Cherenkov detectors to estimate Xmax.

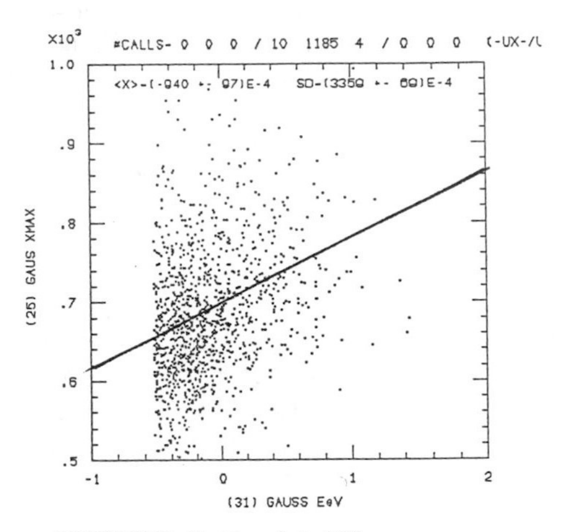

FIGURE 7. Scatter plot of Xmax vs. energy.

Above 10 EeV, there is very little statistics. However, examination of FIGURE 7 shows that the few highest energy events are more consistent with being Fe than protons. Since the stereo Fly's Eye may be biased toward detecting Fe showers above 10 EeV (this is still being studied), one cannot rule out the presence of protons, but the Fe component is clearly not going away (at least up to 30 EeV). It is amusing that the Haverah Park group claimed weak evidence for a change in the elongation rate above 10 EeV as well.[17] Further work on this extremely important question awaits the completion of the High Resolution Fly's Eye, which will have much higher statistics in this energy region.

PROSPECTS FOR DETECTING GAMMA RAYS ABOVE 10 EeV

Until recently, prospects for being able to separate hadrons from gamma rays in the region of the ankle by using EAS development measurements were not very good. The basic problem was that electromagnetic cascades initiated by greater than 10-EeV gamma rays and hadronic cascades initiated by greater than 10-EeV protons looked very similar both in fluctuation (width of Xmax distribution) and in mean Xmax. However, Dedenko[18] has recently pointed out that the Landau–Pomeranchuk–Migdal (LPM) effect begins to be important in air at about 10 EeV, and hence predictions of shower shapes for gamma rays would have to be recalculated. The LPM effect comes into play in a medium when the coherence length becomes longer than the spacing between nuclear scatterers. In this case, the scattering amplitudes must be added coherently and the Bethe–Heitler cross sections are modified from their standard form. Preliminary results indicate that inclusion of the LPM effect makes gamma-ray showers much deeper (by about 100 gm/cm^2) and much wider than comparable energy protons. Since the total energy of the primary must be conserved, this also means that the shower size at shower maximum is smaller (by about a factor of 2) than a comparable hadronic shower. Since the LPM effect threshold depends on the density of the medium, gamma-ray shower shapes will also depend on the primary particle zenith angle, since the atmosphere is exponential in density. This means that surface arrays will not be able to discriminate between gamma rays and hadrons, and may even reach incorrect conclusions about gamma-ray energies, since there is no way they can correct for the LPM effect on an event-by-event basis. A detector like the Fly's Eye, on the other hand, can use this shower profile elongation and deeper penetration as a signature of the presence of gamma rays. Since the energy determination in this case comes from an integral over the shower shape, there should not be a problem in estimating gamma-ray energies correctly (as long as enough of the shower profile can be seen).

One might think that the classic technique of distinguishing gammas from hadrons by their muon multiplicity might be more sensitive. Unfortunately, recent calculations by Aharonian *et al.*[19] have shown that standard shower models that predict that gamma-induced showers will have a few percent of the muon content of hadronic showers at petaelectronvolt energies come to quite different conclusions at exaelectronvolt energies. In fact, the calculations show that there is no significant difference between the muon content of exaelectronvolt gamma rays and hadrons. This is because the energy dependence of the gamma and hadronic EAS muon multiplicity is different, and the energy spectrum of gamma-induced muons is significantly softer than for hadronic showers. The net result is that there are about the same number of less than 1-GeV muons at sea level for exaelectronvolt hadrons and gamma rays. This conclusion changes if greater than 100-GeV muons are counted or if the experiment is done at much higher altitudes. If this calculation is correct, it does not appear that a large muon array will be sensitive to the gamma content of greater than 10-EeV showers.

Dedenko and others[19,20] have also recently pointed out a significant complication with the LPM effect picture. Indeed, it was pointed out as far back as 1981 by McBreen and Lambert,[21] that greater than 10-EeV gamma rays will interact with the earth's magnetic field and lose their energy by emitting lower energy electrons and gamma rays. In effect, the incoming gamma ray develops an electromagnetic shower in the magnetic field of the earth. The threshold for this effect is about 30 EeV. The products of this showering will appear in the troposphere as a superposition of lower energy electrons and gamma rays, and hence will appear as an EAS with anomalously early development and shallow Xmax! Sommers, Aharonian,[19,20] and others have calculated this effect and point out that it is only important if the trajectory of the incoming particle is perpendicular to the field lines of the earth's magnetic field. Particles arriving parallel to the field lines will not shower in the magnetosphere, and will appear as LPM lengthened and deepened showers. In effect, the angle between the particle direction and the field line direction switches the Xmax of the particles' EAS between shallow (900 gm/cm^2 at 100 EeV) and deep (1100 gm/cm^2 at 100 EeV) compared to standard expectations (1000 gm/cm^2 at the same energy). This rather bizzare effect can be turned into a sensitive test of the presence of gamma rays above 30 EeV.

FUTURE PROSPECTS

It is clear that the present detectors will not be able to resolve the issue of events above 100 EeV claimed by the Haverah Park array. Furthermore, a real search for the GZK cutoff requires much higher statistics in the 10- to 100-EeV region than the few tens of events per year currently achievable. A concomitant increase in the understanding of systematic errors and non-Gaussian tails is also necessary. Fly's Eye type of measurement of composition and search for gamma rays also requires much higher statistics and improvements in shower profile resolution. Fortunately, steps to significantly improve this situation are underway. The Akeno group has already enlarged their array, renaming it AGASA, to 100 km^{2}sr and plan to increase it to 200 km^{2}sr by next year. The Soviet group is in the process of constructing a new Soviet array, which initially will be 150 km^{2}sr in size and will eventually grow to 1000 km^{2}sr. The Fly's Eye group is in the prototyping stage of a new detector, the High Resolution Fly's Eye (HiRes), which will consist of three detector stations 15 km apart. The aperture at 100 EeV will be 10,000 km^{2}sr, and all events will be seen in stereo by at least two eyes. The resolution of the EAS development curve will be also significantly improved.

The HiRes and Soviet detectors will eventually accumulate about 200 events/year between 10 and 100 EeV, and about 20 events/year above 100 EeV (if the spectrum really does continue). These statistics, combined with the possibility of doing composition measurements, should allow us to finally answer the question raised in Greisen's paper in the coming decade.

REFERENCES

1. GREISEN, K. 1966. Phys. Rev. Lett. **16:** 748.
2. ZATSEPIN, G. T. & V. A. KUZMIN. 1966. Zh. Eksp. Teor. Fiz. Lett. **4:** 78.
3. HILL, G. T. & D. N. SCHRAMM. 1985. Phys. Rev. D. **31:** 564; BEREZINSKY, V. S. & S. I. GRIGOR'EVA. 1988. Astron. Astrophys. **199:** 1; AHARONIAN, F. A., B. L. KANEVSKY & V. V. VARDANIAN. 1990. Astrophys. Space Sci. **167:** 93; YOSHIDA, S. & M. TESHIMA. 1990. ICCR Rep. 221-90-12.
4. HILLAS, A. M., *et al.* 1971. Proceedings of the 12th ICRC (Hobart, Tasmania), Vol. 3, pp. 1001 and 1007.
5. CHRISTIANSEN, G. B. 1985. Proceedings of the 19th ICRC (La Jolla, Calif.), Vol. 9, p. 487.
6. WATSON, A. A. 1990. *In* Proceedings of ICRR International Symposium on Astrophysical Aspects of the Most Energetic Cosmic Rays (Kofu, Japan). In press.
7. LAWRENCE, M. A., R. J. O. REID & A. A. WATSON. 1990. Proceedings of the 21st ICRC (Adelaide, Australia), Vol. 3, p. 159.
8. EFIMOV, N. N. 1990. *In* Proceedings of ICRR International Symposium on Astrophysical Aspects of the Most Energetic Cosmic Rays (Kofu, Japan). In press.
9. TESHIMA, M. 1990. *In* Proceedings of ICRR International Symposium on Astrophysical Aspects of the Most Energetic Cosmic Rays (Kofu, Japan). In press.
10. BALTRUSAITIS, R. M., *et al.* 1985. Nucl. Instrum. Methods **A240:** 410.
11. JOKIPII, J. R. & G. MORFILL. 1987. Astrophys. J. **312:** 170.
12. AHARONIAN, F. A., V. V. VARDANIAN & B. L. KANEVSKY. 1990. Astrophys. Space Sci. **167:** 111; HALZEN, F., R. J. PROTHEROE, T. STANEV & H. P. VANKOV. 1990. Phys. Rev. D **41:** 342.
13. WOLFENDALE, A. W. & J. WDOWCZYK. 1990. *In* Proceedings of the ICRR International Symposium on Astrophysical Aspects of the Most Energetic Cosmic Rays (Kofu, Japan). In press.
14. PUGET, J. L., F. W. STECKER & J. H. BREDEKEMP. 1976. Astrophys. J. **204:** 638.
15. CASSIDAY, G. L., *et al.* 1990. Astrophys. J. **356:** 669.
16. EFIMOV, N. N., *et al.* 1987. *In* Proceedings of the 20th ICRC (Moscow, USSR), Vol. 5, p. 490.
17. WATSON, A. A. 1990. *In* Proceedings of the ICRR International Symposium on Astrophysical Aspects of the Most Energetic Cosmic Rays (Kofu, Japan). In press.
18. DEDENKO, L. Personal communication.
19. AHARONIAN, F. A., B. L. KANEVSKY & V. A. SAHAKIAN. 1990. Yerevan Physics Institute preprint.
20. DEDENKO, L. & P. SOMMERS. Personal communication.
21. MCBREEN, B. & C. J. LAMBERT. 1981. *In* Proceedings of the 17th ICRC, Vol. 6, p. 70.

Past and Future Dark Matter Experiments

J. RICH

DPhPE
CEN Saclay
91191 Gif-sur-Yvette, France

INTRODUCTION

Six years ago, Goodman and Witten[1] suggested that hypothetical weakly interacting massive particles (wimps) in our galactic halo could be detected via wimp-nucleus elastic scattering.[2,3] Wimps would orbit through the galaxy with velocities of order $10^{-3}c$, so gigaelectronvolt-mass wimps would produce nuclear recoils in the kiloelectronvolt range. Such events could be detected if the nucleus is contained in a "calorimetric" detector of sufficiently low energy threshold. The rate would depend on the wimp's cross section and local wimp density, estimated to be 0.3 GeV-cm^{-3}.

While early discussion of experimental possibilities centered around exotic cryogenic detectors, the only existing limits come from experiments using well-established semiconductor technology, based on either germanium[4,5] or silicon.[6,7] Conventional scintillators or gas proportional chambers are also sensitive to dark matter. The primary purpose of this paper is to discuss the various possible wimp detectors and to evaluate their ability to improve on the existing limits.

EXISTING LIMITS

The lowest limits on wimp cross sections come from experiments using kilogram-size germanium detectors originally designed to detect double-beta decay of ^{76}Ge.[4,5] A representative spectrum from one experiment is shown in FIGURE 1. The observed event rate, of order 1 (kg-day-keV)$^{-1}$, is presumably due to the ambiant radioactivity and is an upper limit on the rate of wimp scattering in germanium. The zone of wimp mass–cross-section space that is thus excluded is shown in FIGURE 2. Wimps with cross sections above the zone's lower diagnonal boundary would give rates in excess of that observed. The upper diagnonal boundary at high cross section is due to the fact that the experiments are underground (600 meters-water-equivalent for reference 5) so that for sufficiently high cross sections the wimps are thermalized before reaching the detector. The vertical boundary at low mass is due to the fact that low-mass wimps do not produce germanium recoils above the experimental threshold (about 10 keV for reference 5).

The exclusion zone includes a large mass range for a once popular dark matter candidate, a heavy Dirac neutrino. Such neutrinos with masses below 10 GeV are not excluded by the dark matter experiments, but are excluded by the recent LEP measurement of the number of unseen modes of Z^0 decay.[8]

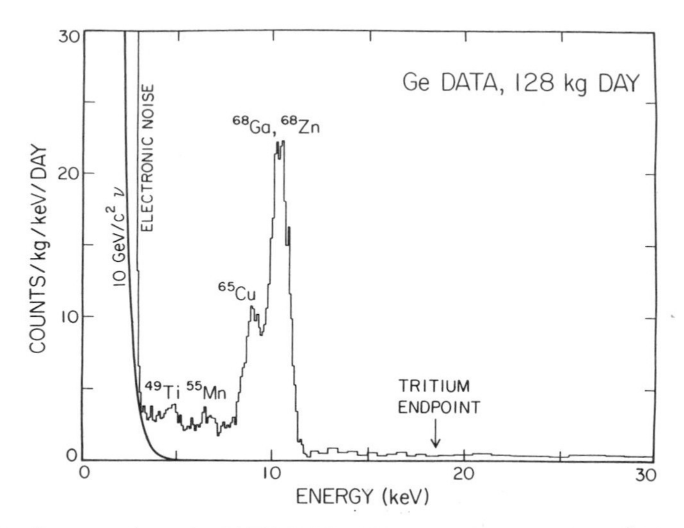

FIGURE 1. Spectrum from the UCSB/UCB/LBL germanium experiment.[5] The spectrum exhibits X-ray peaks from radionucleides produced by cosmic rays while the detector was at the earth's surface. The background above 20 keV is of uncertain origin. The experimental threshold, determined by the electronic noise, is about 3-keV photon energy or 10 keV for the recoil of a germanium nucleus.[10] The curve labeled "10 GeV/c^2 ν" is the recoil spectrum expected of a Dirac neutrino of that mass.

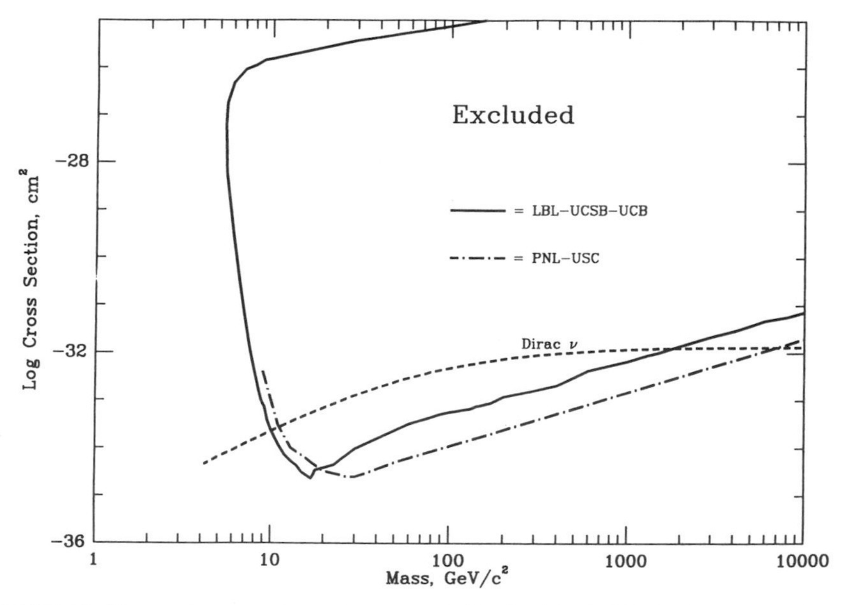

FIGURE 2. Mass and cross-section regions excluded by the LBL-UCSB-UCB experiment[5] and the PNL-USC experiment.[4] Cross-section–mass combinations inside the excluded zone would give rates that are incompatible with the experimental spectrum (e.g., FIG. 1).

The Dirac neutrino is a typical candidate having large "spin-independent" interactions. The cross section due to such interactions is proportional to the square of the number of nucleons in the target nucleus, and is thus large for heavy nuclei like germanium. The present germanium experiments are not yet sensitive to candidates that have primarily "spin-dependent" interactions. Such particles generally have much smaller cross sections and only scatter off nuclei with nonzero spin. An example of such a wimp is the photino. Photinos that close the universe would have a cross section of order 10^{-38} cm^2 on ^{73}Ge (the only germanium isotope with spin–7 percent isotopic abundance). The lightest supersymmetric particle would, in general, be a mixture of the photino and other particles leading to both spin-dependent and spin-independent interactions.[10] However, with its cross section that is about three orders of magnitude below the present limits, the pure photino represents a good target for future experiments.

The exclusion zone in FIGURE 2 has been extended to higher cross sections by a silicon experiment flown to high altitude in a balloon.[6] The results from this experiment constrain various types of charged or strongly interacting dark matter.[11,12]

The exclusion zone has recently been extended to smaller masses ($m_x = 3$ GeV) by an underground experiment using silicon.[7] This experiment essentially rules out all "cosmions" with spin-independent couplings. (Cosmions are dark matter particles that also serve to solve the solar neutrino problem.[13]

We should note that silicon and germanium detectors have been calibrated for dark matter detection via elastic neutron-nucleus scattering.[14,15] This procedure is necessary because recoil nuclei generally produce less signal than X-rays of the same energy. Hence, the standard calibration procedures using radioactive sources are not applicable.

CRITERIA FOR FUTURE EXPERIMENTS

In order to determine whether a given detector can improve on the limits shown in FIGURE 2, we need to determine the recoil spectrum expected for scattering due to a hypothetical wimp. Let the wimp have a mass m_x and differential scattering cross section $d\sigma/dE$ for a nuclear recoil energy E. The scattering rate per target nucleus is given by

$$\frac{dN}{dt\,dE} = \frac{\rho}{m_x} \int d^3v f(v) v \frac{d\sigma}{dE}, \tag{1}$$

where ρ is the local wimp mass density and $f(v)$ is the distribution of wimp velocities in the target frame. The form of the recoil energy spectrum is then determined by the form of the two functions $f(v)$ and $d\sigma/dE$. For the velocity distribution, one generally assumes a boosted Maxwellian:

$$f(v) = \frac{(3/2\pi)^{3/2}}{v_{\rm rms}^3} \exp\left[-\frac{3|v - v_e|^2}{2v_{\rm rms}^2}\right], \tag{2}$$

where v_e is the earth velocity estimated to be 220 km/sec and v_{rms} is estimated to be 270 km/sec.[2] The differential cross section is expected to have the form

$$\frac{d\sigma}{dE}(v, E) = \frac{\sigma}{E_{max}(v)} g(E),$$

where σ is the total cross section at $v = 0$, and $E_{max}(v)$ is the maximum recoil energy. The function $g(E)$ is a form factor that takes into account the loss of coherence at large momentum transfers. Without this factor, the scattering would be isotropic in the center of mass. For spin-independent interactions, it is often assumed to be given by $g(E) = \exp[-2m_N ER^2/3\hbar]$, where R is the rms radius of the target nucleus ($R \sim 0.9A^{1/3}$ fm).[16] A similar function will apply in the case of spin-dependent interactions.

The distribution of E calculated according to **(1)** falls with increasing E in a roughly exponential manner. The average value of E, $\langle E \rangle$, is shown in FIGURE 3 as a function of m_x and m_N. (We assumed the specific forms previously mentioned for the functions $f(v)$ and $g(E)$.) For $m_x < 20$ GeV or $m_N < 20$ GeV, the form factor is unimportant and $\langle E \rangle$ is uncertain to about 20 percent due to the uncertainty in the average velocity. For much larger nuclear and wimp masses, $\langle E \rangle$ is essentially determined by the form factor. For spin-dependent interactions, there are no reliable estimates of the uncertainty in the form factor, so the estimates of $\langle E \rangle$ in FIGURE 3 should be used with caution.

FIGURE 3 allows us to determine how one can extend the limits of FIGURE 2 to lower masses. For fixed m_N, $\langle E \rangle$ increases with increasing m_x. The value of m_x below which a given experiment starts to lose sensitivity can be calculated by equating the experiment's energy threshold with $\langle E \rangle$ in FIGURE 3. For the germanium experiment of reference 5, the threshold is about 10 keV and the experiment loses sensitivity below $m_x = 30$ GeV. For the silicon experiment of reference 5, the threshold is about 4 keV and the experiment loses sensitivity below $m_x = 10$ GeV. In order to extend the exclusion zone to smaller masses, any proposed experiment must do better than this. This can be done either by lowering the threshold energy or by lowering m_N.

In order to extend the limits of FIGURE 2 to lower cross sections, experimenters must either reduce the background below 1 (kg-day-keV)$^{-1}$, or use a nucleus that gives a higher scattering rate than germanium (for a given type of wimp). Assuming the experimental threshold is low enough to make the experiment sensitive to the average recoil, the signal is nearly the differential rate at zero recoil energy. Using **(1)**, this is:

$$m_N^{-1} \frac{dN}{dt\, dE}(E = 0) = \frac{1}{2} \frac{\sigma}{m_{red}^2} \frac{\rho}{m_x} \langle v^{-1} \rangle, \tag{3}$$

where m_{red} is the reduced mass of the nucleus-wimp system and $\langle v^{-1} \rangle$ is the average inverse velocity. If the background is fixed in (kg-day-keV)$^{-1}$, then the "optimal" target is the one that maximizes σ/m_{red}^2. Since σ is always proportional to m_{red}^2, their ratio is a simple function of the wimp-nucleus coupling. In general, the ratio is the sum of two terms:[10] a "spin-independent" term that grows like m_N^2, and a "spin-dependent" term that does not depend directly on m_N and is zero for nuclei with zero spin.

Improvements in the wimp signal over what one has in the germanium experiments can be made by increasing either of the two terms. The spin-independent term can be increased by using a nucleus heavier than germanium. However, this will lower $\langle E \rangle$ (see FIG. 3), so in order to profit from the increase in signal, the experimental threshold would have to be lowered below the 10 keV of present germanium detectors.

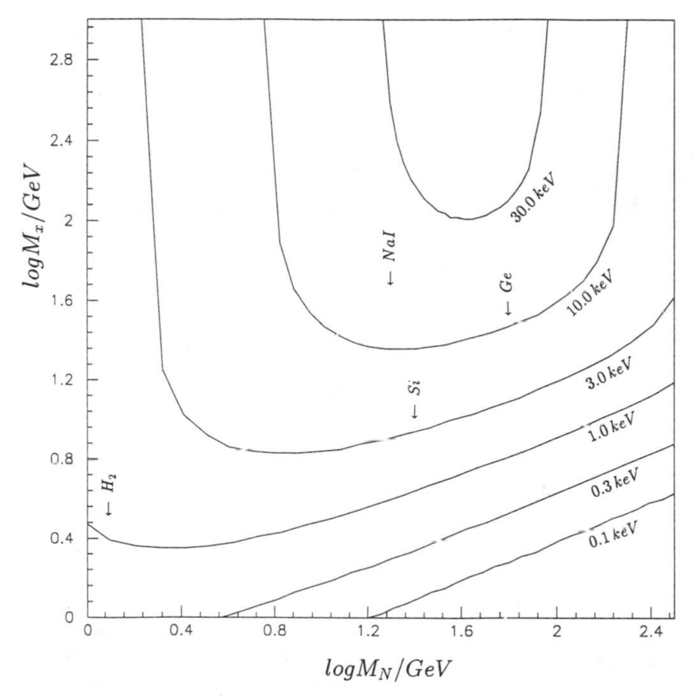

FIGURE 3. The average recoil energy, $\langle E \rangle$, for wimp-nucleus scattering as a function of nuclear mass, m_N, and wimp mass, m_χ. The hypotheses involved in this calculation are explained in the text. The *arrows* point to the nuclear mass and threshold of four detectors that have been calibrated for nuclear recoils: Ge,[5] Si,[7] NaI,[22] and H_2.[25]

The spin-dependent term can be increased by using a detector made from germanium enriched in ^{73}Ge. It can also be increased by using an element that has a stronger spin-dependent coupling than germanium. To first approximation, this coupling is due to the characteristics of the "odd" nucleon of an odd-Z or odd-N nucleus. Its calculation is then a two-step process. First the fundamental couplings of quarks are used to determine the couplings to protons or neutrons.[17] Second, some variant of the nuclear shell model is used to calculate the coupling of the nucleus in question.[18,19] Each of these steps has uncertainties of factors of 2 or more. There are odd-N nuclei that have a calculated σ/m_{red}^2 somewhat larger than that of ^{73}Ge. (^{3}He is more than a factor of 10 better.) For candidate wimps that couple primarily to protons (e.g., the photino), it would be useful to use an odd-Z nucleus. Good candidate nuclei are ^{7}Li and ^{19}F.

Since the possible target modifications mentioned earlier will probably not be sufficient to make possible the observation of supersymmetric dark matter, the most important thing for the future of wimp searches is to reduce the radioactive background. The guilty nuclei are impurities either in the detectors or surrounding material and are of two types. The first type consists of "natural" radioisotopes, primarily ^{40}K, ^{115}In, or nuclei from the uranium–thorium chains. The second type consists of long-lived radioisotopes produced in cosmic-ray interactions while material was at the earth's surface. This last source appears to dominate in the present generation of germanium detectors.[20] Methods for reducing radioactivity are discussed in references 21 and 22.

A second way to reduce background is to develop systems that can distinguish nuclear recoils from beta, gamma, and X-rays. One possibility is to use high-purity active shields to veto multiple Compton scatters or beta–gamma cascades.[3] Another is to develop calorimetric techniques that give signals whose character distinguishes between nuclear recoils and normal radioactivity where the detected particle is an electron. Possibilities for this will be discussed below.

POSSIBLE DETECTION TECHNIQUES

As remarked previously, there are many types of detectors sensitive to nuclear recoils in the kiloelectronvolt region. Using the preceding considerations, we will try to evaluate the possibilities for each to extend the wimp limits to lower masses or lower cross sections.

Semiconductor detectors in their germanium and silicon forms have set the standards for dark matter detection. This is because they have appropriate thresholds and because they have very low concentrations of uranium–thorium–potassium. Improvements on the limits at low mass may be coming soon from an improved low-threshold version of the silicon experiment of reference 7. Improvements on the limits at low cross sections could come about by further reducing surface cosmic-ray exposure of the detectors.[21,22] Improvements of order ten in the cross-section limits for wimps with only spin-dependent interactions could be made by using detectors made of germanium enriched in ^{73}Ge or of silicon enriched in ^{29}Si. The possibility of using more exotic odd-Z semiconductors like GaAs has also been discussed,[3] but these detectors are, for the moment, too small to yield useful dark matter limits.

Scintillators like NaI[23] or liquid xenon[24] have been suggested as possible dark matter detectors. Both of these detectors have high proportions of isotopes with spin, so they might be useful in extending the limits on cross sections for wimp's with primarily spin-dependent interactions.

Sodium iodide has the advantage of being inexpensive and easy to use. This could permit the construction of detectors that are sufficiently massive to use the anticipated annual modulation in event rate as a way to reject radioactive background.[16] Sodium iodide has recently been calibrated for sodium recoils using a neutron beam.[23] The light output is about one-quarter that for X-rays of the same energy down to recoils of 10 keV. This amount of light should make it possible to trigger on recoils of order 20 keV, making the low-mass sensitivity of NaI similar to that of germanium (see FIG. 3). It appears that some nuclear recoil-electron discrimination may be possible through pulse-shape analysis. The primary problem with the use of NaI is that commercially available sodium iodide has uranium–thorium impurities at the 10^{-9} level. This needs to be lowered by two orders of magnitude to be competitive with existing germanium experiments.[3]

Liquid xenon would have the advantage that it can be purified underground, thus reducing the problem of cosmogenic radioactivity. A big disadvantage is that the form factor for this large nucleus reduces the average recoil to below 20 keV (see FIG. 3). Since the scintillation light is in the UV range where light collection is difficult, it may be difficult to have a threshold this low if the light output for nuclear recoils is similar to that of NaI.

The final class of conventional dark matter detectors are gas proportional chambers. With a threshold near 1 keV, such chambers filled with H_2 are the standard detector for kiloelectronvolt-range neutrons.[25] They are thus sensitive to wimps with masses down to 1 GeV, and are the best proven detector in this regard (see FIG. 3). However, their low density would make them primarily useful for searching for wimps with relatively large cross sections like cosmions with purely spin-dependent interactions. The possibility of using range and energy information in low-pressure TPCs as a method to reject background has also been discussed.[26]

Finally, we discuss the cryogenic detectors that were discussed in the original Goodman and Witten article.[1] Bolometric detectors are capable of very low thresholds, making them sensitive to low-mass wimps. Unfortunately, this possibility has only been realized in microgram-size detectors. On the gram level, the detectors are still not competitive with ordinary silicon detectors, or, on the kilogram level with germanium detectors.

Instead of low thresholds, present wimp cryogenic detector development is directed at hybrid devices that measure both ionization and thermal pulses. Because the ratio of the thermal signal to the ionization signal is expected to be different for nuclear recoils and electrons, simultaneous detection of the two may give a big reduction in background. Two groups, one at Berkeley[27] and one in the United Kingdom,[28] have recently reported observing simultaneous thermal and ionization pulses induced by gamma rays. The 60-g germanium Berkeley detector has a threshold below 18 keV, in the range necessary for dark matter detection.

It has also been suggested that simultaneous measurement of a thermal signal and a scintillation signal could be used to reject background in dark matter experiments.[29] For the ultimate in background rejection, it has been suggested that crystal damage due to nuclear recoils may be observable in mica track etch detectors.[30] Such detectors would not be sensitive to normal radioactivity.

A detector with good electron–nuclear recoil rejection would still be sensitive to nuclear recoils caused by fast neutrons from spontaneous fission and (α, n) reactions. The fast neutron flux at the Gran Sasso laboratory is about 10^{-6} $cm^{-2}sec^{-1}$.[31] If unshielded, this flux would lead to an event rate of a few $kg^{-1}day^{-1}$. Appropriate shielding would reduce the rate to tolerable levels.

CONCLUSION

We have emphasized that a number of conventional and exotic techniques are capable of detecting galactic wimps. In the future, they can be used to search for wimps with masses or cross sections outside the present exclusion zone. While it is important to continue searches for low-mass wimps ($m_x < 10$ GeV), accellerator limits will continue to make such candidates less and less plausible.[32] The priority should then be on developing methods to combat the radioactive background that limits the sensitivity of present experiments. More than two orders of magnitude must be gained in order to reach the cross sections predicted for popular supersymmetric wimps, but we know of no fundamental reason why this sensitivity cannot be achieved.

ACKNOWLEDGMENTS

It is a pleasure to thank the following people for conversations concerning various aspects of this review: Steve Barwick, Gilles Gerbier, Yannick Giraud-Heraud, Kim Griest, Bernard Sadoulet, Dave Seckel, Peter Smith, Michel Spiro, Leo Stodolsky, and Charling Tao. I would especially like to thank Jon Engel for extensive correspondence concerning nuclear form factors.

REFERENCES

1. Goodman, M. W. & E. Witten. 1985. Phys. Rev. **D31:** 3059.
2. A general review of dark matter is given by Primack, J. R., D. Seckel & B. Sadoulet. 1988. Ann. Rev. Nucl. Sci **38:** 751–807.
3. A review of dark matter detection schemes is given by Smith, P. F. & J. D. Lewin. 1990. Phys. Rep. **187:** 203.
4. Ahlen, S. P., F. T. Avignone III, R. L. Brodzinski, A. K. Drukier, G. Gelmini & D. N. Spergel. 1987. Phys. Lett. **B195:** 603.
5. Caldwell, D. O., *et al.* 1988. Phys. Rev. Lett. **61:** 510.
6. Rich, J., R. Rocchia & M. Spiro. 1987. Phys. Lett. B **194:** 173.
7. Caldwell, D. O., *et al.* 1990. Phys. Rev. Lett. **67:** 1305.
8. Griest, K. & J. Silk. 1990. Nature **343:** 26.
9. We have used the nuclear matrix elements for ^{73}Ge given by Engel and Vogel[17] and the photino-squark mass combinations given by Gaisser, T. K., G. Steigman & S. Tilav. 1986. Phys. Rev. **D34:** 2206, table VII.
10. Griest, K. 1988. Phys. Rev. Lett. **61:** 666; also, Griest, K. 1988. Phys. Rev. **D38:** 2357.
11. Basdevant, J. L., *et al.* 1990. Phys. Lett. B **234:** 395.
12. Starkman, G., *et al.* 1990. Phys. Rev. D **41:** 3594.
13. Faulkner, J. & R. L. Gilliland. 1985. Astrophys. J. **299:** 994; Spergel, D. N. & W. H. Press. 1985. Astrophys. J. **294:** 663; Press, W. H. & D. N. Spergel. 1985. Astrophys. J. **296:** 679; Gilliland, R. L., *et al.* 1986. Astrophys. J. **306:** 703.
14. Gerbier, G., *et al.* 1990. Phys. Rev. **D42:** 3211.
15. Chasman, C., K. W. Jones & R. A. Ristinen. 1965. Phys. Rev. Lett. **15:** 245.
16. Freese, K., J. Freedman & A. Gould. 1988. Phys. Rev. **D37:** 3388.
17. Ellis, J. & R. A. Flores. 1988. Nucl. Phys. **B307:** 883.
18. Engel, J. & P. Vogel. 1989. Phys. Rev. **D40:** 3132.
19. Iachello, F., L. M. Krauss & G. Maino. 1991. Phys. Lett. B **254:** 220.
20. Martoff, C. J. 1987. Science **237:** 507.
21. Brodzinski, R. L., *et al.* 1990. Nucl. Instrum. Methods. *In* Phys. Res. **A292:** 337.

22. BARNES, P. D., *et al.* 1990. University of California at Santa Barbara Preprint UCSB-HEP-90-15. To be published in the proceedings of the Snowmass Workshop.
23. GERBIER, G. 1990. Saclay Preprint DPhPE 90-13.
24. ZIOUTAS, K. 1989. CERN Preprint CERN/LAA/PC/89-010; BELLI, P., *et al.* 1990. Nuovo Cimento **A103:** 767.
25. VERBINSKI, V. V. & R. GIOVANNINI. 1974. Nucl. Instrum. Methods **114:** 205.
26. MASEK, G., K. BUCKLAND & M. MOJAVER. 1989. *In* Paricle Astrophysics: Forefront Experimental Issues, Eric B. Norman, Ed. World Scientific. Singapore; also, GERBIER, G., J. RICH, M. SPIRO & C. TAO. 1989. *In* Particle Astrophysics: Forefront Experimental Issues, Eric B. Norman, Ed. World Scientific. Singapore.
27. WANG, N., *et al. In* After the First Three Minutes, V. Trimble *et al.,* Eds. To be published; SMITH, P. F. & B. SADOULET. 1992. Particle Astronomy Minisymposium: Introductory remarks. This issue.
28. SMITH, P. F. Private communication.
29. GONZALEZ-MESTRES, L. & D. PERRET-GALLIX. 1989. Nucl. Instrum Methods. *In* Phys. Res. **A279:** 382.
30. SNOWDEN, D., *et al.* 1990. *In* Proceedings of Conference on Particle Astrophysics (UCLA); BARWICK, S. Private communication; also PRICE, P. B. & M. H. SALAMON. 1986. Phys. Rev. Lett. **56:** 1226.
31. BELLI, P., *et al.* 1989. Nuovo Cimento **A101:** 959.
32. For example, KRAUSS, L. M. 1990. Phys. Rev. Lett. **64:** 999; GIUDICE, G. F. & S. RABY. 1990. Phys. Lett. **B247:** 423.

Axion Searches[a]

P. SIKIVIE

Physics Department
University of Florida
Gainesville, Florida 32611

The axion was postulated about thirteen years ago to guarantee that the strong interactions conserve P and CP even though the weak interactions do not.[1] It is a light pseudoscalar particle whose couplings are all proportional to its mass m_a. The mass is not known *a priori*, but high-energy physics laboratory searches and astrophysical and cosmological considerations have ruled out all values of m_a except those inside two windows:[1] 10^{-3} eV $< m_a \lesssim 10^{-6}$ eV and a rather narrow window (at most a factor of 2 or 3 in width) near $m_a = 3$ eV. The latter window is only allowed for axions whose coupling to the electron is much suppressed.

The $m_a = 0$ (3 eV) window can be searched by looking for the conversion of solar axions to photons in a laboratory magnetic field,[2,3] or by looking for monochromatic photons from the $a \rightarrow 2\gamma$ decay of axions that are gravitationally bound to galactic clusters.[4] Recently, results from a search of the latter type have been reported.[5] It excludes the range 3 eV $\lesssim m_a \lesssim$ 8 eV.

For a value of m_a of order 10^{-5} eV, one may obtain $\Omega = 1$ in the form of cold dark matter axions.[6] However, precisely which value of m_a yields $\Omega = 1$ is rather uncertain. One source of uncertainty is the contribution to the cosmological axion energy density from cosmic axion strings.[7,8] The problem is that it has proved impossible thus far to derive analytically the energy spectrum of axions radiated by decaying axion strings. However, Hagmann and I have recently obtained the spectrum by means of computer simulation.[8] Our results support the view that cosmic axion strings contribute to the cosmological axion energy density an amount of the same order of magnitude as the contribution due to initial vacuum misalignment.

Very recently, Turner and Wilczek have discussed the cosmological limits on the axion due to primordial quantum fluctuations in inflationary models.[9] Their results constrain a scenario in which m_a smaller than 10^{-6} eV is allowed because the initial vacuum misalignment is accidentally small.

Dark matter axions constituting the halo of our galaxy can be searched for by stimulating their conversion to microwave photons in an electromagnetic cavity permeated by a large magnetic field.[2,10] Two collaborations have reported results from such searches.[11,12] Their limits are summarized in FIGURE 1, which shows that the sensitivity of the existing detectors is approximately a factor of 500 short of that required to detect axions at the galactic halo density of one-half 10^{-24} g/cm^3. The goal of the University of Florida effort was, in fact, to build a pilot detector to study the feasibility of a full-scale experiment and to develop the necessary techniques. There is no doubt left in my mind that a full-scale experiment is feasible. In fact, a proposal has recently been put forth by the University of Florida and Lawrence Livermore

[a]This work was supported in part by the U.S. Department of Energy under Contract DE-FG05-86ER40272.

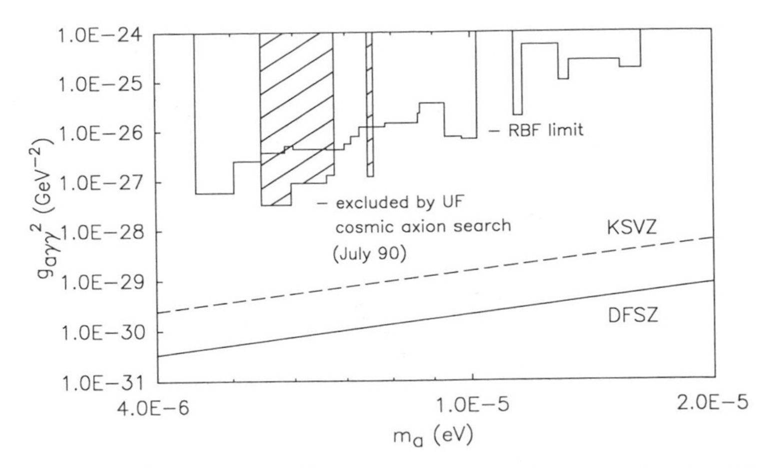

FIGURE 1. Upper limits on the coupling of the axion to two photons as a function of the axion mass, under the assumption that the galactic halo is made of axions. (From the cavity detectors of cosmic axions at Brookhaven National Laboratory (RBF) and the University of Florida (UF))

Laboratory (LLNL) to use two of the "axicell" magnets from the decommissioned mirror fusion test facility at LLNL for a cosmic axion search. The proposed detector would test the hypothesis that our galactic halo is made of axions over the mass range $0.6 \lesssim m_a \lesssim 16\ \mu$eV.

REFERENCES

1. Recent axion reviews include, Kim, J. E. 1987. Phys. Rep. **150:** 1; Cheng, H. Y. 1988. Phys. Rep. **158:** 1; Peccei, R. D. *In* the review volume CP Violation, C. Jarlskog, Ed.; Turner, M. Windows on the axion, Phys. Rep. In press.
2. Sikivie, P. 1983. Phys. Rev. Lett. **51:** 1415; Sikivie, P. 1985. Phys. Rev. **D32:** 2988.
3. van Bibber, K., P. M. McIntyre, D. E. Morris & G. G. Raffelt. 1989. Phys. Rev. **D39:** 2089.
4. Kephart, T. & T. Weiler. 1987. Phys. Rev. Lett. **58:** 171; Turner, M. 1987. Phys. Rev. Lett. **59:** 2489.
5. Bershady, M. A., M. T. Ressel & M. S. Turner. 1990. Telescope search for multi-eV axions. Fermilab Preprint Pub 90/244-A, December.
6. Abbott, L. & P. Sikivie. 1983. Phys. Lett. **120B:** 133; Preskill, J., M. Wise & F. Wilczek. 1983. Phys. Lett. **120B:** 127; Dine, M. & W. Fischler. 1983. Phys. Lett. **120B:** 137; Ipser, J. & P. Sikivie. 1983. Phys. Rev. Lett. **50:** 925.
7. Davis, R. 1985. Phys. Rev. **D32:** 3172; Davis, R. 1986. Phys. Lett. **180B:** 225; Davis, R. L. & E. P. S. Shellard. 1989. Nucl. Phys. **B324:** 167; Dabholkar, A. & J. Quashnock. 1990. Nucl. Phys. **B333:** 815.
8. Harari, D. & P. Sikivie. 1987. Phys. Lett. **B195:** 361; Hagmann, C. & P. Sikivie. 1990. Computer simulations of the motion and decay of global strings. Univ. of Florida Preprint UFIFT-HEP-90-30, November. Submitted to for publication in Nucl. Phys. B.
9. Turner, M. S. & F. Wildzek. 1991. Phys. Rev. Lett. **66:** 5.
10. Krauss, L., J. Moody, F. Wilczek & D. Morris. 1985. Phys. Rev. Lett. **55:** 1797.
11. De Panfilis, S., *et al.* 1987. Phys. Rev. Lett. **59:** 839; De Panfilis, S. 1989. Phys. Rev. **D40:** 3153.
12. Hagmann, C., P. Sikivie, N. Sullivan & D. Tanner. 1990. Phys. Rev. **D42:** 1297.

Indirect Detection of Dark Matter

KATHERINE FREESE

Physics Department
Massachusetts Institute of Technology
Cambridge, Massachusetts 02139

INTRODUCTION

The three favorite candidates for nonbaryonic dark matter are light massive neutrinos (mass ~ 30 eV), axions (mass $\sim 10^{-5}$ eV), and weakly interacting massive particles, or wimps (mass few GeV–few TeV). Light neutrinos are candidates for hot dark matter; in this scenario, structure forms first on very large scales, such as superclusters, and then fragments down to smaller scales. Axions and wimps are candidates for cold dark matter; here structure forms first on small scales and then clusters on ever larger scales. In this paper I will focus on wimp candidates for dark matter, which include supersymmetric (SUSY) particles in the gigaelectronvolt to teraelectronvolt mass range.

First, I will discuss the limits placed on dark matter particle masses by the results of recent experiments at LEP and FermiLab. The remaining interesting mass range for weakly interacting massive particles is 20 GeV–few TeV; if certain assumptions about parameters at the grand unified theory (GUT) scale are dropped, then the dark matter candidate may be as light as a few GeV.

Next, I will discuss the status of ideas for indirect detection of wimp particles. If these particles provided the dark matter in the halo of our Galaxy, they would annihilate with one another in the halo. The end products of the annihilation would be antiprotons, positrons, and photons, which could be detected by cosmic-ray experiments, and thus serve as signatures for the existence of wimps in the halo. Similarly, wimps could collect in the cores of the earth or the sun, annihilate with one another in the core, and give rise to a signal of massless neutrinos at the surface of the earth; these neutrinos could then be detected by proton-decay detectors.

THE REMAINING WIMP PARAMETER RANGE

FIGURE 1 plots the mass density of Dirac neutrinos $\Omega_\nu = \rho_\nu/\rho_c$ as a function of neutrino mass m_ν. Here, $\rho_c = 1.88 \times 10^{-29} h_0^2$ gm cm^{-3} is the critical density required to close the universe, and h_0 is the Hubble parameter in units of 100 km s^{-1} Mpc^{-1}. The observational constraint $\Omega \leq 2$ implies that the two cosmologically interesting mass ranges for Dirac neutrinos are $m_\nu \leq 100$ eV and $m_\nu \geq 2$ GeV. Note that at point A in the diagram ($m_\nu \simeq m_Z/2$), there is a pole in the propagator for annihilation, the annihilation cross section σ_A becomes very large, and $\Omega_\nu \propto (1/\sigma_A)$ is at a minimum. The mass density grows with increasing mass beyond this point, and can become $O(1)$ at $m_\nu \sim 100$ GeV (Enqvist, Kainulainen, and Maalampi 1989). A similar diagram can be drawn for many of the other wimp candidates. It is the mass range

$m \geq$ few GeV that defines the wimp category of dark matter, and it is this mass range that I will focus on for the rest of this paper.

The original wimp, the Dirac neutrino, is ruled out as the dark matter by a combination of experiments: (1) since neutrinos contribute to the width of the Z^0, the measurements of this width at LEP rule out any neutrino (other than the three light neutrinos ν_e, ν_μ, and τ_τ) lighter than 45 GeV (Krauss 1990); and (2) double β-decay experiments rule out the remaining mass range of Dirac neutrinos up to a mass of 4 TeV (Ahlen *et al.* 1987, Caldwell *et al.* 1988, Boehm *et al.* 1989).

Among the remaining wimp candidates are supersymmetric (SUSY) particles. The minimal supersymmetric standard model is the simplest phenomenologically acceptable supersymmetric version of the standard model of elementary particles (see, e.g., Haber and Kane 1985). There is one new particle, the SUSY partner, for each particle of the standard model. An additional difference from the standard model is that there are two Higgs doublets (instead of one) and their SUSY partners. In the neutral sector of the theory, there are four SUSY fermions that mix with one

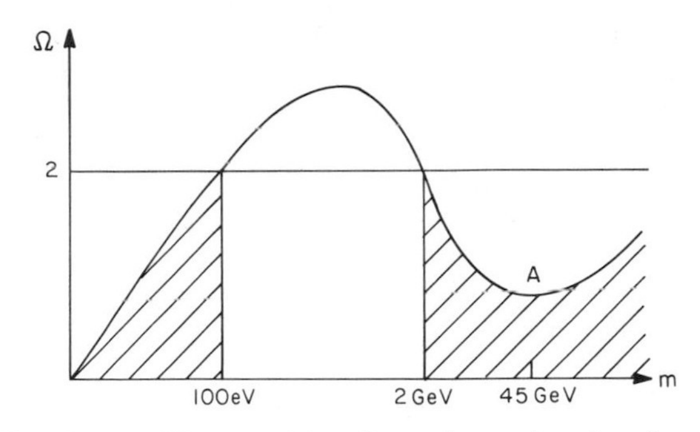

FIGURE 1. Mass density of Dirac neutrinos $\Omega_\nu = \rho_\nu/\rho_c$ as a function of neutrino mass m_ν.

another. These are known as neutralinos: $\tilde{B}$ is the partner of the $U(1)$ gauge boson B, $\tilde{W}_3$ is the partner of the neutral $SU(2)$ gauge boson W_3, and $\tilde{H}_1^0$ and $\tilde{H}_0^2$ are the neutral components of the SUSY partners of the Higgs. The mass mixing matrix is as follows:

$$(\tilde{B}\,\tilde{W}_3\tilde{H}_1^0\tilde{H}_2^0)\begin{pmatrix} M_1 & 0 & -M_z c_\beta s_w & M_z s_\beta s_w \\ 0 & M_2 & M_z c_\beta c_w & -M_z s_\beta c_w \\ -M_z c_\beta s_w & M_z c_\beta c_w & 0 & -\mu \\ M_z s_\beta s_w & -M_z c_\beta s_w & -\mu & 0 \end{pmatrix}\begin{pmatrix} \tilde{B} \\ \tilde{W}_3 \\ \tilde{H}_1^0 \\ \tilde{H}_2^0 \end{pmatrix}, \tag{1}$$

where $c_\beta = \cos\beta$, $s_\beta = \sin\beta$, $c_w = \cos\theta_w$, $s_w = \sin\theta_w$, θ_w is the Weinberg angle, and $\sin^2\theta_w = 0.23$. This mass mixing matrix can be diagonalized in order to obtain mass eigenstates, which are linear combinations of $\tilde{B}$, $\tilde{W}_3$, $\tilde{H}_0^1$, and $\tilde{H}_0^2$:

$$\chi_i = Z_{i1}\tilde{B} + Z_{i2}\tilde{W}_3 + Z_{i3}\tilde{H}_1^0 + Z_{i4}\tilde{H}_2^0 \tag{2}$$

where $(Z)_{ij}$ is a real orthogonal matrix that diagonalizes the neutralino mass matrix. Only the lightest neutralino χ_n is stable (if we assume a new conserved quantum number, R-parity) and can be the dark matter (see, e.g., Ellis *et al.* 1984).

The neutralino is a pure photino when we take the limit $M_2 \rightarrow 0$ in the preceding matrix. If we take $Z_{n1} = \cos\theta_w$, $Z_{n2} = \sin\theta_w$, and $Z_{n3} = Z_{n4} = 0$, then the lightest mass eigenstate is the photino, $\tilde{\gamma} = \cos\theta_w \tilde{B} + \sin\theta_w \tilde{W}_3$ (by analogy with the photon). The higgsino arises when we take the limit $\mu \rightarrow 0$. If we take $Z_{n1} = Z_{n2} = 0$, the higgsino is $h = \sin\beta \tilde{H}_1^0 + \cos\beta \tilde{H}_2^0$ with $m_{\tilde{h}} \propto \mu$. The zino arises if we take $Z_{n1} = -\sin\theta_w$, $Z_{n2} = \cos\theta_W$, and $Z_{n3} = Z_{n4} = 0$; the zino is $\tilde{Z}^0 = -\sin\theta_w \tilde{B} + \cos\theta_w \tilde{W}_3$ (by analogy with the Z^0).

The free parameters in the mixing matrix are (1) $\tan\beta = v_2/v_1 = \langle H_2^0 \rangle / \langle H_1^0 \rangle$ (note that v_2 gives mass to the up quark and v_1 gives mass to the down quark). The experimental limit on the top quark mass, $m_t > 50$ GeV implies that $v_2/v_1 > 1$ is favored; (2) μ, the higgsino mass parameter; and (3) three gaugino masses, M_1, M_2, and M_3 for $U(1)$, $SU(2)$, and $SU(3)$, respectively. If an addition simplifying assumption is made: $M_1 = M_2 = M_3$ at an epoch of grand unification (the GUT scale), then the number of free parameters is reduced. Scaling to lower energies with renormalization group equations, one finds $M_1 = 5/3\, M_2 \tan^2\theta_w$ today.

Recent measurements of the width of the Z^0 at LEP rule out any neutralinos that could have been decay products of the Z^0, that is, any neutralinos that satisfy the following two criteria: (1) neutralino mass $m_\chi \leq 45$ GeV, and (2) the neutralino couples to the Z^0. Since the coupling of the neutralino to the Z^0 is proportional to $Z_{n3}^2 - Z_{n4}^2$, not all neutralinos are constrained by the LEP measurements. For example, the photino, which has no Z_{n3} or Z_{n4} components, is unconstrained by LEP. Higgsino particles, on the other hand, are ruled out unless their Z_{n3} and Z_{n4} components happen to cancel. With the additional assumption $M_1 = M_2 = M_3$ at the GUT scale, then the limits placed by the CDF detector (at Fermilab) on the gluino mass $M_3 > 120$ GeV translate into limits on M_1 and M_2, and thus on SUSY particles. This combination of LEP and CDF constraints rules out neutralino masses below 20 GeV (Ellis *et al.* 1990). However, Nature may not be as simple as we would like, and it may be that the gaugino masses are not equal at the GUT scale. If the assumption of equality of gaugino masses at the GUT scale is relaxed (Drees 1985, 1986), then the available parameter space is much larger, and a neutralino can exist with a mass m_χ = few GeV with $\Omega_\chi = 1$ that satisfies all experimental constraints (Griest and Roszkowski 1991). In fact, the minimal SUSY model may be too simple to adequately describe nature; there may be additional unknown parameters and thus light dark matter candidates.

To reiterate, the interesting parameter ranges for wimps are as follows: (1) if $M_1 = M_2 = M_3$ at the GUT scale, then a combination of constraints from LEP and CDF imply that neutralino masses $m_\chi \leq 20$ GeV are ruled out, and the interesting mass range for SUSY wimps is then 20 GeV–few TeV; (2) if this assumption is relaxed, then one can have m_χ = few GeV with $\Omega = 1$; and (3) interesting new physics takes place if $m_\chi \geq 90$ GeV ($\simeq m_W \simeq m_Z$). New annihilation channels open up: $\chi\chi \rightarrow WW$, ZZ, HW, and HZ. Although I do not have time to discuss this regime, it has been investigated by Olive and Srednicki (1989), and Griest, Kamionkowski, and Turner (1990).

INDIRECT DETECTION OF DARK MATTER

WIMP Annihilation in the Galactic Halo

If wimps are the halo dark matter, then they can annihilate with one another to produce observable signatures of antiprotons, photons, and positrons.

$$\chi\bar{\chi} \overset{(i)}{\rightarrow} q\bar{q} \overset{(ii)}{\rightarrow} \begin{Bmatrix} \pi^{\circ} \\ \eta \\ X \end{Bmatrix} \rightarrow \gamma + Y \tag{3}$$

where χ is the wimp, $\bar{\chi}$ its antiparticle (note that Majorana particles such as Majorana neutrinos, photinos, and higgsinos are their own antipartners), $q\bar{q}$ is a quark/antiquark pair, X refers to hadronic end products of the annihilation, and Y refers to final decay products of the hadrons. The cross sections for the first part of the process (i) can be calculated for specific dark matter candidates. The second step (ii), hadronization, is not well understood. One can attempt to model this process by using e^+e^- accelerator data; however, $e^+e^- \rightarrow q\bar{q}$ has different branching ratios to various quarks, and the spectrum will not be correct. Instead, one uses a Monte Carlo such as Lund or Herwig, semiphenomenological models that match the e^+e^- data and handle the branching ratios and the decay of particles in flight correctly. In the process $\chi\bar{\chi} \rightarrow q\bar{q} \rightarrow 2\,jets$, the whole chain of particles in a given jet depends on which initial quark it emerged from. The jets from different quarks have different spectra of final particle energies; for example, light quarks give a harder spectrum of antiprotons and photons (more $\bar{p}$ and γ at high energies) than heavy quarks. Thus, since higgsinos have large branching fractions to heavy quarks, the higgsino annihilation spectrum would drop more rapidly as a function of antiproton energy than would, for example, the photino annihilation spectrum. The total yields of antiprotons or other particles also depends on which quark initiated the jet. For these (and other) reasons, correct accounting for the branching ratios to different quarks is important, and one uses Monte Carlos rather than the e^+e^- data directly.

Antiprotons

One of the signatures for wimp annihilation in the galactic halo is the production of antiprotons, $\chi\bar{\chi} \rightarrow q\bar{q} \rightarrow \bar{p} + \cdots$; cosmic-ray experiments can then look for these antiprotons. The flux of antiprotons from wimp annihilation would be given by

$$\Phi_{\bar{p}}(E_{\bar{p}}) = \frac{1}{4\pi} n_{\chi}^{2}(\sigma_A v) \frac{dN}{dE_{\bar{p}}} v_{\bar{p}} \tau_{\bar{p}} \tag{4}$$

where $(\sigma_A v)$ is the annihilation cross section of the particular wimp, $n_\chi = \rho_x/m_x$ is the number density of wimps, $v_{\bar{p}}$ is the antiproton velocity,

$$\frac{dN}{dE_{\bar{p}}} = \frac{1}{\sigma_A} \frac{d\sigma(\chi\chi \rightarrow \bar{p} + \cdots)}{dE_{\bar{p}}}$$

is the differential cross section for inclusive $\bar{p}$ production (the result of the Monte Carlo) suitably modulated by the solar wind, and $\tau_{\bar{p}}$ is the confinement time of antiprotons in the galactic halo.

In 1981, Buffington, Schindler, and Pennypacker reported a $\bar{p}$ detection significantly above what would be expected from ordinary astrophysical processes. Silk and Srednicki (1984) proposed that this might be evidence of a photino signal (see also Stecker, Rudaz, and Walsh 1985). However, new upper limits by Ahlen *et al.* (1988) and Salamon *et al.* (1990) placed the antiproton flux several orders of magnitude below the Buffington *et al.* claim. FIGURE 2 (taken from Ellis *et al.* 1988) plots these new experimental limits, as well as theoretical plots of $\bar{p}$ fluxes from wimp annihila-

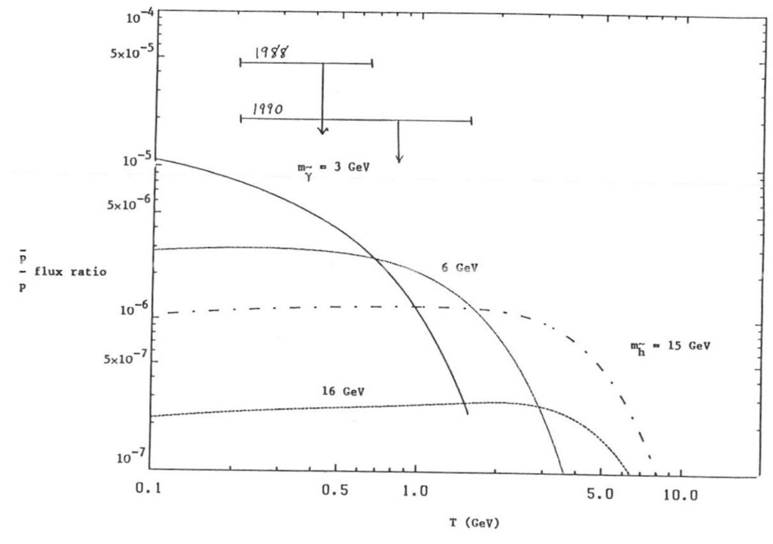

FIGURE 2. Experimental upper limits on the $\bar{p}$ flux from Ahlen *et al.* (1988) and Salamon *et al.* (1990), as well as theoretical plots from Ellis *et al.* (1988) of $\bar{p}$ fluxes from wimp annihilation as a function of $\bar{p}$ energy for wimp masses as labeled. Curves are calculated with $\rho_x = 0.3$ GeV cm^{-3} and $\tau_{\bar{p}} = 2 \times 10^{15}$ sec. Present upper limits on cosmic ray $\bar{p}$ do not exclude any range of SUSY particle masses.

tion as a function of $\bar{p}$ energy. The theoretical plots assume the wimps provide the halo density, $\rho_x = 0.3$ GeV cm^{-3}, and use a confinement time, $\tau_{\bar{p}} = 2 \times 10^{15}$ sec. One can see that present upper limits on cosmic-ray $\bar{p}$ do not exclude any range of SUSY particle masses. There is still a $\bar{p}$ problem at higher energies; the data from Golden *et al.* (1979) is still 4 × the predictions from production of cosmic-ray secondaries.

The most recent $\bar{p}$ experiment of Salamon *et al.* (1990) was from a balloon experiment that flew for 12 hours. Unlike previous experiments, this experiment was able to explicitly identify particles: protons, muons, pions, and some kaons. In 1.24×10^5 proton events, not a single $\bar{p}$ was seen in the energy interval 100–1580 MeV. To extent $\bar{p}$ measurements in this energy range to much greater sensitivity will require longer exposures. This could be accomplished with long-duration (10-day) balloon flights over the South Pole or by putting a detector on a satellite for about five years outside the earth's magnetosphere. Such experiments are being contemplated but will not be likely in the near future. In the next few years, the High-energy Antimatter Telescope (HEAT) balloon program intends to measure the positron spectrum to about 100 GeV and the high-energy $\bar{p}$ spectrum from 2–50 GeV. This experiment will check and extend the earlier results of Golden *et al.* (1979), which exceed theoretical predictions.

A problem with using antiprotons or any other end product of wimp annihilation as a test for halo dark matter is that there are astrophysical uncertainties in the predicted fluxes that are perhaps as large as a factor of $O(100)$. If a clear signal is seen, with count rates above what is predicted by astrophysical processes, then one can interpret this signal as a possible detection of dark matter; in this case, the astrophysical uncertainties are not critical. However, because of these uncertainties, it is difficult to use a null result, that is, the lack of an observed signal, to place limits on dark matter candidates. For example, there is large uncertainty in the value of $\tau_{\bar{p}}$, depending on whether one assumes disk or halo confinement for the $\bar{p}$'s. In addition, one cannot rule anything out because of a possible galactic wind (Ahlen *et al.* 1982). The dark matter could be there, and yet we might not be able to see the signal. There are also uncertainties in the local matter density in the halo, perhaps a factor of $O(10)$. Recent work by Long and Ostriker (1989) and Blitz and Spergel (1989) indicate that the halo may be flattened as in the models of Binney, May, and Ostriker (1987); in this case, the density of the dark halo near the sun, and thus $\bar{p}$, γ, and e^+ fluxes at the earth may be higher (ρ_0 may be as large as 2 GeV cm^{-3} in flattened halo models). The hope in these experiments has been that one would see a signal clearly identifiable as what one would expect from wimp annihilation in the halo.

Photons

The annihilation of wimps in the halo of our Galaxy can also lead to a signature of photons (Rudaz and Stecker 1988; Stecker 1988; Ellis *et al.* 1988; Freese and Silk 1989; Stecker and Tylka 1989). The photon flux at Earth would be

$$\Phi_\gamma(E_\gamma) = \frac{1}{4\pi} n_\chi^2 (\sigma_A v) \frac{dN}{dE_\gamma} aJ(b,l)$$

$$= 3.1 \times 10^{-6} \left(\frac{\rho_0}{0.4\ \mathrm{GeV\ cm^{-3}}}\right)^2 \left(\frac{a}{8\ \mathrm{kpc}}\right) \left[\frac{\frac{dN}{dE_\gamma}}{\mathrm{GeV^{-1}}}\right] \left[\frac{\sigma_A v}{10^{-26}\ \mathrm{cm^3 s^{-1}}}\right]$$

$$\times \left[\frac{M_\chi}{\mathrm{GeV}}\right]^{-2} J(b,l)\ \mathrm{cm^{-2} s^{-1} sr^{-1} GeV^{-1}}, \qquad (5)$$

where $(\sigma_A v)$ is the annihilation cross section of the particular wimp, the halo density, assumed spherically symmetric, is modeled as $\rho_x = \rho_0(a^2 + r_0^2)(a^2 + r^2)^{-1}$, ρ_0 is the local dark halo density, a is the halo core radius, r_0 is the distance of the sun from the galactic center, and $J(b, l)$ is a function of galactic longitude l and latitude b defined by

$$J(b,l) = \frac{(1+\alpha^2)^2}{2\beta^3}\left\{\frac{\pi}{2} + \frac{\alpha\beta}{1+\alpha^2}\cos b\cos l + \tan^{-1}\left(\frac{\alpha}{\beta}\cos b\cos l\right)\right\} \tag{6}$$

with $\alpha = r_0/a$ and $\beta^2 = 1 + \alpha^2 - \alpha^2\cos^2 b\cos^2 l$ (Gunn *et al.* 1978; Turner 1986). The factors $aJ(b, l)$ depend somewhat on the choice of parameters for modeling the galactic halo. Taking $a = 7$ kpc and $r_0 = 8.5$ kpc, the factor $J(b, l)$ varies between 1.24 ($b = 90°$; the galactic halo), 9.01 ($b = 0°, l = 0°$; the galactic center), and 0.61 ($b = 0°$, $l = 180°$; the galactic anticenter).

The diffuse γ-ray flux observed over 35–200 MeV (Fichtel *et al.* 1978) is the principal source of background. It consists of two components: an isotropic component, modeled by

$$i_{\mathrm{iso}}(E_\gamma) = 2.8 \times 10^{-8}\, E_{\gamma,\mathrm{GeV}}^{-3.4}\ \mathrm{cm^{-2}s^{-1}sr^{-1}GeV^{-1}}, \tag{7a}$$

and a galactic disk component, modeled by

$$I_{\mathrm{gal}}(E_\gamma) = 2.8 \times 10^{-6}(\sin|b|)^{-1} E_{\gamma,\mathrm{GeV}}^{-1.6}\mathrm{cm^{-2}s^{-1}sr^{-1}GeV^{-1}}. \tag{7b}$$

The observations extend out to photon energies of 200 MeV and, if power-law spectra are assumed, can be fit as earlier. Extending these power-law fits out to larger energies may be overestimates of the background; Dermer (1986) argues that if the primary contribution to the galactic γ-ray component is due to decay of π^0 produced by cosmic-ray collisions with particles in the interstellar medium, the high-energy asymptotic form of the γ-ray spectrum goes as $E_\gamma^{-2.75}$. To be conservative we use the fits in (**7**) as estimates of the backgrounds. The galactic component dominates above 70 (80) MeV at an average galactic latitude of 60° (90°).

FIGURE 3 (taken from Ellis *et al.* 1988) plots predicted photon fluxes as a function of photon energy for various wimp masses. The ratio of signal to background is ≤ 10 percent; however, the distinctive angular signature of the halo component means that it may be possible to distinguish it from the other diffuse backgrounds in a future experiment such as Gamma Ray Observatory (GRO).

A more readily identifiable signal could be obtained from monochromatic γ-ray line signals (Srednicki, Theissen, and Silk 1986; Rudaz 1986, 1989; Bergstrom and Snellman 1988; Giudice and Griest 1989; Bouquet, Salati, and Silk 1989): $\chi\bar{\chi} \to \gamma\gamma$ or $\chi\bar{\chi} \to V_\gamma$, where V is a $c\bar{c}$ or $b\bar{b}$ bound state. The final photon energy would be exactly equal to the initial wimp mass (Doppler broadening would spread out the line by an amount $(\Delta E_\gamma/E_\gamma) = \beta = 10^{-3}$). Such a narrow signal would be a "smoking gun" for a mechanism such as wimp annihilation. However, the problem turns out to be that the count rates are extremely small. The most favorable case is that of photino annihilation, $\tilde{\gamma}\tilde{\gamma} \to \gamma\gamma$ via fermion and sfermion box diagrams. The cross section for annihilation to photons is calculated to be $\sigma_A v \sim 4 \times 10^{-31}\ (m_{\tilde{\gamma}}/4\ \mathrm{GeV})^2$. A proposed detector such as ASTROMAG abroad a space station would have energy resolution of about 1 percent. At high galactic latitude, the expected signal to noise with this detector would then be $\sim 2.5 \times 10^{-2}\ (m_{\tilde{\gamma}}/4\ \mathrm{GeV})^{1.7}\ (\rho_0/0.4\ \mathrm{GeV\ cm^{-3}})^2\ (a/8\ \mathrm{kpc})$. With "area × efficiency factor" for ASTROGAM $\sim 7000\ \mathrm{cm^2\ sr}$, one would expect a few events per year. However, NASA has postponed the ASTROGAM mission. A Russian collaboration is currently expecting to build a similar detector (Fradkin, private communication; Dogiel 1988). Again, one has the problem that the backgrounds are not that well known; however, the advantage of a line signal is that it serves as a "smoking gun" for wimp annihilation.

Positrons

The analysis of continuum positrons from wimp annihilation is similar to the case of $\bar{p}$'s, but, unlike $\bar{p}$'s, e^+ lose significant energy during propagation through the interstellar medium (ISM) (by synchrotron and inverse Compton effects). The bounds on $\rho_\chi\sqrt{\tau}$ from e^+ are less stringent than from $\bar{p}$ (Ellis *et al.* 1988).

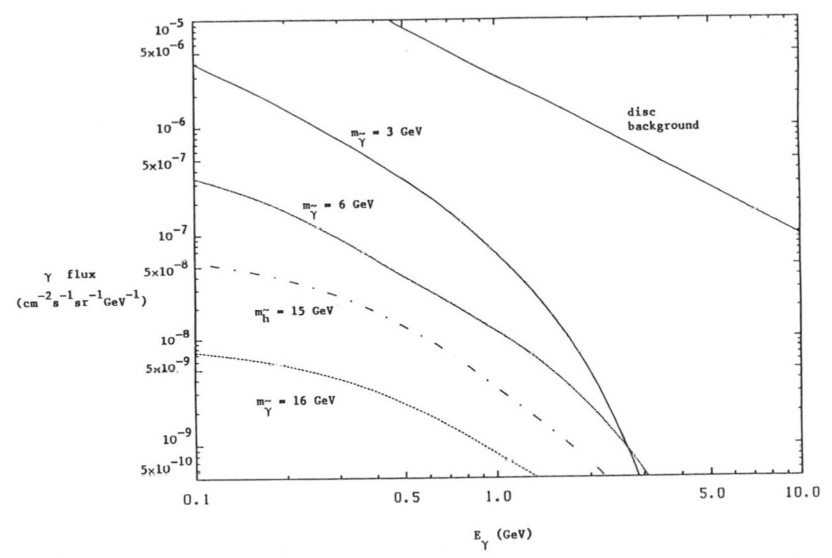

FIGURE 3. Predicted photon fluxes from wimp halo annihilation as a function of photon energy for various wimp masses for $\rho_\chi = 0.3$ GeV cm^{-3} and $a = 7$ kpc. Plot taken from Ellis *et al.* (1988). The disk γ-ray background of (**7b**) is also shown.

However, the line radiation signal from $\chi\bar{\chi} \rightarrow e^+e^-$ may be interesting (Tylka 1989; Turner and Wilczek 1990). The conventional explanation for cosmic-ray positrons is that primary cosmic rays (protons and He^4) interact with nuclei in the ISM to produce kaons, pions, and muons that decay to positrons. The theoretical expectations for positron production are $e^+/(e^+ + e^-) \sim 10$ percent at 0.3–1.0 GeV, as is observed, dropping to 3–5 percent at 10 GeV. Instead, the observed e^+ flux is seen to rise to 25 percent at 30 GeV (see, e.g., Mueller and Wang 1987). It has been proposed that wimp annihilation to a line signal may be responsible for this rise. The energy losses of positrons as they travel through the ISM would broaden the line. The distinctive signature in this case for wimp annihilation would be a sharp drop off in the number of observed positrons for $E_{e^+} \geq m_\chi$.

For the simplest neutralino, the wimp annihilation channel to e^+e^- is suppressed; the branching ratio is $BR \sim (m_e/m_\chi)^2 \leq 10^{-5}$. However, in models where the left- and right-handed SUSY electron masses are not the same, $m_{\tilde{e}_L} \neq m_{\tilde{e}_R}$, the neutralino can have a substantial branching ratio to e^+e^-, and the preceding scenario can take place.

FIGURE 4 is a plot of positron fraction $e^+/(e^+ + e^-)$ (taken from Turner and Wilczek 1990) as a function of positron energy. The solid curve is a theoretical curve with parameters chosen ($m_\chi = 25$ GeV) to best fit the data points. The HEAT experiment will look at the positron fraction up to energies of 70 GeV, thus testing

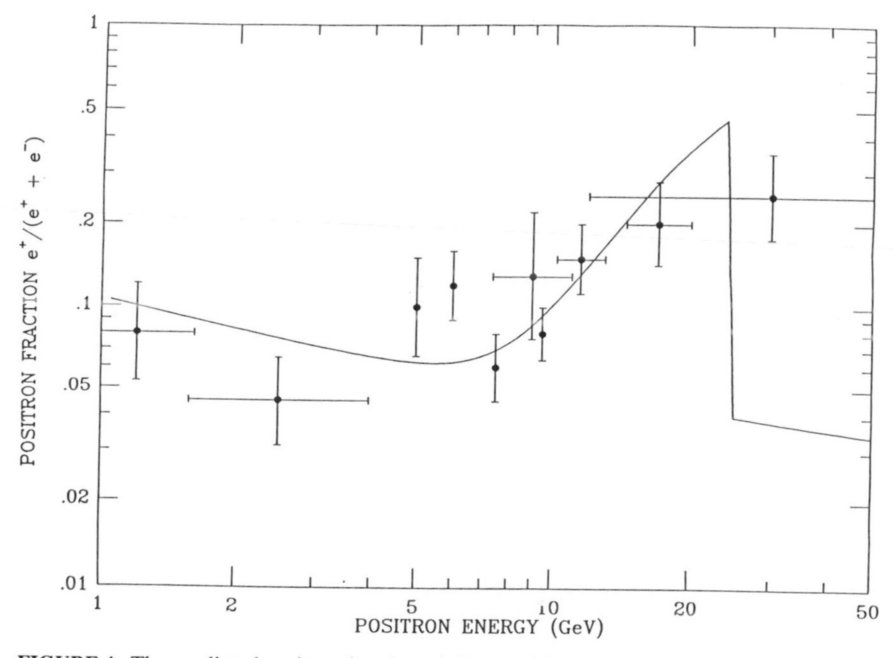

FIGURE 4. The predicted positron fraction, $e^+/(e^+ + e^-)$ from wimp line annihilation, and the existing experimental data (from Mueller and Wang 1987). The theoretical curves use $m_\chi = 25$ GeV and other parameters chosen to match the data. Conventional sources have been assumed to contribute a positron fraction $0.02 + 0.10\ (E/\text{GeV})^{-0.5}$, consistent with models of Protheroe (1982).

the preceding ideas. Again. ASTROMAG aboard the space station would be ideal for this experiment, but has been postponed; a similar Russian experiment is planned.

As an alternative explanation for the observed anomalous positron component at 10–20 GeV, Eichler (1989) proposed teraelectronvolt dark matter particles that are unstable at the GUT scale.

If the wimp mass $m_\chi > m_W$, then there is another mechanism for making high-energy e^+: $\chi\chi \rightarrow W^+W^-$ or $\chi\chi \rightarrow Z^0Z^0$ (Kamionkowski and Turner 1991). The signal would be narrow, with $E_{e^+} \sim m_\chi/2$. The best case for an observable feature would be nearly pure higgsino. An observation of this signal would be a "smoking gun" for wimp annihilation. The many astrophysical uncertainties make it difficult for a nonobservation to constrain wimp properties.

To reiterate, the signatures of halo annihilation include (1) antiprotons, (2) photons—continuum signal and monochromatic line signals, and (3) positrons—continuum and line signals. The problems with all these signals are low count rates and unknown astrophysics; one hope is that a line signal or a sharp cutoff in signal (for $E_{e^+} > m_\chi$) would serve as a "smoking gun" for wimp annihilation.

Annihilation in the Earth and Sun

Wimps traveling through the earth or sun can interact with nuclei, and some fraction of the wimps will be captured. The captured wimps sink to the core of the earth or sun gravitationally. There they annihilate with one another; among the annihilation products will be ordinary massless neutrinos. These neutrinos can then be detected in proton-decay detectors at the surface of the earth.

There are several advantages of such a neutrino signature over cosmic-ray signatures for heavy wimps. (1) The number density of neutralinos in the halo drops as a function of increasing wimp mass (assuming the wimps supply the halo density), $n_{\chi=\rho_\chi}/m_\chi \propto 1/m_\chi$. As we have seen, the annihilation rate in the halo $\propto n_\chi^2$, whereas the annihilation rate in the earth or sun $\propto n_\chi$ (in equilibrium, the annihilation rate is just half the capture rate and depends on only one power of the density). (2) The uncertainties in predicted event rates are smaller in the earth or sun than in cosmic rays. The propagation of neutrinos in the earth or sun is better understood than the propagation of cosmic rays in the galaxy, and the local halo density is known better than the dark matter distribution throughout the galaxy. (3) The neutrino signal is more easily distinguished from background ($\sim E_\nu^{-3}$) than are cosmic ray signals.

The original paper considering annihilation of wimps in the sun was by Silk, Olive, and Srednicki 1985; the first papers considering annihilation in the earth were by Freese (1986) and by Krauss, Srednicki, and Wilczek (1986). Additional work on this subject has been done by many authors, including Srednicki, Olive, and Silk (1987), Hagelin, Ng, and Olive (1987), Gaisser, Steigman, and Tilav (1986), Griest and Seckel 1987, Olive and Srednicki (1988), Ritz and Seckel (1988) (who used a Lund Monte Carlo to obtain branching ratios to neutrinos), Gelmini, Gondolo, and Roulet (1991) (who explored parameter space for $m_\chi < m_W$), and Kamionkowski (1991) (who explored parameter space for $m_\chi > m_W$).

Wimps traveling through the earth or sun can lose energy by interacting with several different elements i, for example, Fe, O, and H. Summing over the various elements, one can obtain the capture rate for halo neutralinos by the sun or earth:

$$C = c\,\frac{\rho_{0.3}}{v_{300}} f_i \sum_i X_i \frac{(\sigma_i/10^{-36}\ \mathrm{cm}^2)}{m_x m_i/\mathrm{GeV}^2}, \tag{8}$$

where $c = 5.88 \times 10^{28}\ \mathrm{sec}^{-1}$ for the sun and $c = 5.80 \times 10^{19}\ \mathrm{sec}^{-1}$ for the earth, σ_i is the cross section for wimp scattering with nucleus of type i, $\rho_{0.3} = \rho_\chi/0.3\ \mathrm{GeV\ cm}^{-3}$, $v_{300} =$

$v^{\chi}_{\rm rms}/[300\ {\rm km/s}]$, f_i is the mass fraction of element i, and X_i incorporates various factors including gravitational focusing and effects due to mass differences of m_i vs. m_χ.

There are two types of scattering contributions for wimp-nucleus scattering (see, e.g., Goodman and Witten 1985; Freese 1989): (1) coherent scattering, $\sigma \propto N^2$, where N is approximately the number of nucleons in the nucleus, and (2) spin-dependent scattering, $\sigma \propto \mathbf{S}_\chi \cdot \mathbf{J}_{\rm nuclear\ spin}$. The scattering cross section in (**8**) consists of coherent terms only in the earth (the earth is made up primarily of spinless nuclei; the core is primarily Fe and the mantle SiO_2, MgO, and FeO), and both coherent and spin-dependent terms in the sun.

Once the wimps are captured, they sink to the core of the earth or sun, where they annihilate with one another to produce an observable signal of ordinary massless neutrinos. In equilibrium, the annihilation rate is equal to half the capture rate (two wimps must be captured to achieve a single annihilation). The differential rate of neutrino production is

$$\frac{dR}{dE_v} = \frac{C}{2} \sum_Y B_{\chi Y} \frac{dN_{Yv}}{dE_v} \tag{9}$$

where $B_{\chi Y}$ is the branching ratio for the channel $\chi\chi \rightarrow Y$, $(dN_{Y\nu}/dE_\nu)$ is the differential neutrino yield from $Y \rightarrow \nu +$ "anything" (calculated with a Lund Monte Carlo by Ritz and Seckel 1988).

There are two distinct types of neutrino signals that can be identified by proton-decay detectors. (1) Contained events. Here, the neutrino interacts with a nucleus in the detector. This is useful for detecting ν_e and ν_μ. (2) Throughgoing events. This is useful for detecting ν_μ. Here the neutrino interacts with rock outside the detector, and a muon then travels into the detector where it is identified.

Proton-decay detectors that can detect such massless neutrino signatures of wimp annihilation in the earth or sun include IMB (Irvine, Michigan, and Brookhaven), Frejus, and Kamiokande. The event rate limits on ν_e and ν_μ with $E_\nu \geq 2$ GeV at the 90 percent confidence level from Frejus are: 4.1 events kton^{-1} yr^{-1} from the sun, and 6.4 events kton^{-1} yr^{-1} from the earth, from charged-current-contained and vertex-contained events. The event-rate limits on ν_μ with $E_\nu \geq 2$ GeV from IMB are: 8.4×10^{-14} cm^{-2} sec^{-1} from the sun, and 2.7×10^{-13} cm^{-2} sec^{-1} from the earth, obtained by considering upward going muons. The latter results from the earth are conservative, as these counts include all muons with zenith angles $>98°$, while the angular resolution of the detector is 3.5° and wimps are expected only in the earth's core. The experimental bound on the signal should be improved by a factor of 10 or even 100 by further analysis of the data (this is planned by John Learned).

I will report on the recent work of Gelmini, Gondolo, and Roulet (1991), who examined neutralino parameter space from $m_\chi < m_W$. FIGURE 5 shows contour maps of the mass of the lightest neutralino χ in gigaelectronvolts and its dominant component (in regions separated by dotted lines) in the $\mu - M_2$ plane, for $v_2/v_1 = 2$. FIGURE 6 shows what regions in this plane are excluded (for $m_{H_2} = 50$ GeV) by accelerator bounds, indirect searches from annihilation in the earth or sun, and direct searches with Ge spectrometers (designed to look for double β-decay).

I have not reported on indirect limits due to wimp annihilation in the earth or sun for $m_\chi > m_W$; for a discussion of this subject, see Kamionkowski (1991).

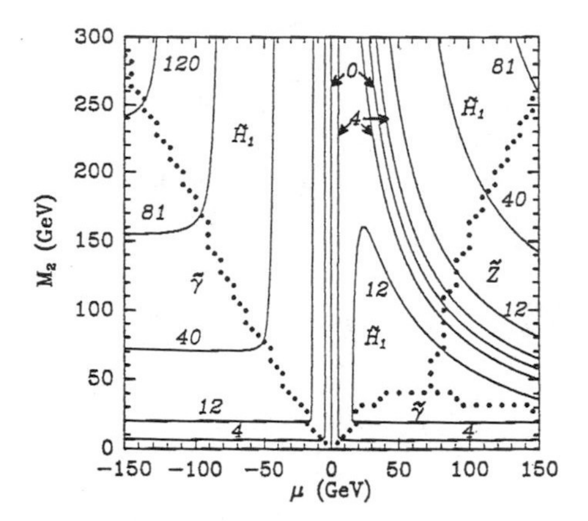

FIGURE 5. Contour maps of the mass of the lightest neutralino in GeV and its dominant component (in regions separated by dotted lines) in the $\mu - M_2$ plane for $v_2/v_1 = 2$. Plot taken from Gelmini, Gondolo, and Roulet (1991).

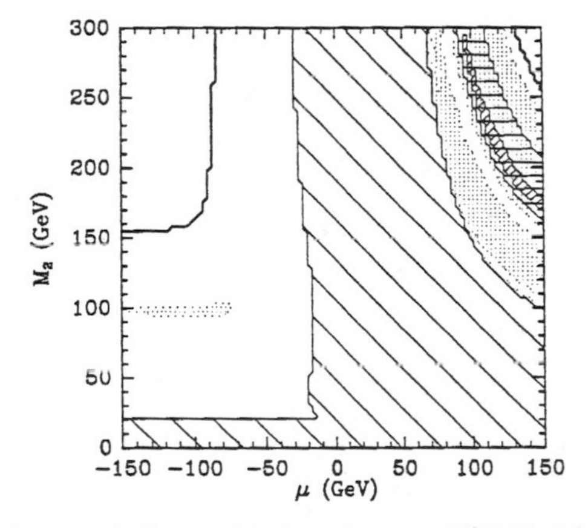

FIGURE 6. Excluded regions in the $\mu - M_2$ plane, for $m_{H_2} = 50$ GeV and $v_2/v_1 = 2$. The *central hatched area* is excluded by accelerator bounds. Indirect searches due to annihilation in earth or sun under the assumption $\rho_\chi = \rho_{\text{halo}}$ exclude the areas hatched with positive slope. Direct searches in Ge spectrometers exclude those with negative slope. Improvements by fators of 10 and 100 in the limit from upgoing μ's from the earth should exclude the horizontally hatched and dotted regions, respectively. Plot taken from Gelmini, Gondolo, and Roulet (1991).

CONCLUSIONS

If $M_1 = M_2 = M_3$ at the GUT scale, then a combination of accelerator bounds from LEP and CDF rule out wimps lighter than 20 GeV. Moving out in parameter space to examine the loophole of relaxing this equality of masses, one finds m_χ = few GeV can satisfy all experimental bounds even with $\Omega_\chi = 1$ (Griest and Roszkowski 1991).

Indirect detection of wimps includes looking for end products of wimp annihilation in the halo of our Galaxy: antiprotons, photons, and positrons. The predicted fluxes of these particles would be low, and the astrophysical uncertainties are high; thus, it is hard to use halo annihilation to rule out candidates. However, line radiation could be a "smoking gun" for wimps. In particular, the HEAT detector is going to look for a cutoff in the positron fraction spectrum at $E_{e^+} = m_\chi/2$.

Wimps can also collect in the earth or sun, and annihilate to a signal of ordinary massless neutrinos. A study of this signal already rules out regions of parameter space; better data analysis of upward going muons from the earth could improve the limits.

ACKNOWLEDGMENTS

I would like to thank Greg Tarle, Jack Vander Welde, Mark Kamionkowski, Graciella Gelmini, and especially Kim Griest and Dave Seckel for extremely helpful discussions.

REFERENCES

AHLEN, S. P., F. AVIGNONE, R. L. BRODZINSKI, A. K. DRUKIER, G. GELMINI & D. N. SPERGEL. 1987. Phys. Lett. B **195:** 603.

AHLEN, S. P., P. B. PRICE, M. H. SALAMON & G. TARLE. 1982. Astrophys. J. **260:** 20.

AHLEN, S. P., *et al.* 1988. Phys. Rev. Lett. **61:** 145.

BERGSTROM, L. & H. SNELLMAN. 1988. Phys. Rev. D **37:** 3737.

BINNEY, J., A. MAY & J. P. OSTRIKER. 1987. Mon. Not. R. Astron. Soc. **226:** 149.

BLITZ, L. & D. N. SPERGEL. 1989. Princeton University Preprint.

BOEHM, F., *et al.* 1989. *In* Theoretical and Phenomenological Aspects of Underground Physics. L'Aquila, Italy. To be published.

BOUQUET, A., P. SALATI & J. SILK. 1989. Phys. Rev. D **40:** 3168.

BUFFINGTON, A., S. M. SCHINDLER & C. R. PENNYPACKER. 1981. Astrophys. J. **248:** 1179.

CALDWELL, D. O., *et al.* 1988. Phys. Rev. Lett. **61:** 510.

DERMER, C. D. 1986. Astron. Astrophys. **157:** 223.

DOGIEL, V. A., *et al.* 1988. Space Sci. Rev. **49:** 215.

DREES, M. 1985. Phys. Lett. B **158:** 409.

DREES, M. 1986. Phys. Rev. D **33:** 1468.

EICHLER, D. 1989. Phys. Rev. Lett. **63:** 2440.

ELLIS, J., J. HAGELIN, D. V. NANOPOULOS, K. A. OLIVE & M. SREDNICKI. 1984. Nucl. Phys. B. **238:** 453.

ELLIS, J., R. FLORES, K. FREESE, S. RITZ, D. SECKEL & J. SILK. 1988. Phys. Lett. B **214:** 403.

ELLIS, J., D. V. NANOPOULOS, L. ROSZKOWSKI & D. N. SCHRAMM. 1990. Phys. Lett. B **245:** 251.

ENQVIST, K., K. KAINULAINEN & J. MAALAMPI. 1989. Nucl. Phys. B **316:** 456.

FICHTEL, C. E., G. A. SIMPSON & D. J. THOMPSOM. 1978. Astrophys. J. **222:** 833.

FREESE, K. 1986. Phys. Lett. B **167:** 295.

FREESE, K. 1989. WEIN Conference Proceedings (Montreal, P. Q., Canada).

FREESE, K. & J. SILK. 1989. Phys. Rev. D **40:** 3828.

GAISSER, T., G. STEIGMAN & S. TILAV. 1986. Phys. Rev. D **334:** 2206.

GELMINI, G., P. GONDOLO & E. ROULET. 1991. UCLA Preprint.

GIUDICE, G. F. & K. GRIEST. 1989. Phys. Rev. D **40:** 2549.

GOLDEN, R. L., *et al.* 1979. Phys. Rev. Lett. **43:** 1196.

GOODMAN, M. & E. WITTEN. 1985. Phys. Rev. D **31:** 3059.

GRIEST, K., M. KAMIONKOWSKI & M. S. TURNER. 1990. Phys. Rev. D **41:** 3565.

GRIEST, K. & D. SECKEL. 1987. Nucl. Phys. B. **283:** 681.

GRIEST, K. & L. ROSZKOWSKI. 1991. In preparation.
GUNN, J. E., B. W. LEE, I. LERCHE, D. N. SCHRAMM & G. STEIGMAN. 1978. Astrophys. J. **223:** 1015.
HABER, H. & G. KANE. 1985. Phys. Rep. **117:** 75.
HAGELIN, J., K. NG & K. OLIVE. 1987. Phys. Lett. B **180:** 375.
KAMIONKOWSKI, M. 1991. FermiLab Preprint.
KAMIONKOWSKI, M. & M. S. TURNER. 1991. Fermilab Preprint.
KRAUSS, L. 1990. Phys. Rev. Lett. **64:** 999.
KRAUSS, L., M. SREDNICKI & F. WILCZEK. 1986. Phys. Rev. D **33:** 2079.
LONG, & J. P. OSTRIKER. 1989. Princeton University Preprint.
MUELLER, D. & K. K. WANG. 1987. Astrophys. J. **312:** 183.
OLIVE, K. & M. SREDNICKI. 1988. Phys. Lett. B **205:** 553.
OLIVE, K. & M. SREDNICKI. 1989. Phys. Lett. B **230:** 78.
PROTHEROE, R. J. 1982. Astrophys. J. **254:** 391.
RITZ, S. & D. SECKEL. 1988. Nucl. Phys. B **304:** 877.
RUDAZ, S. 1986. Phys. Rev. Lett. **56:** 2128.
RUDAZ, S. 1989. Phys. Rev. D **39:** 3549.
RUDAZ, S. & F. W. STECKER. 1988. Astrophys. J. **325:** 16.
SALAMON, M. H., *et al.* 1990. Astrophys. J. **349:** 78.
SILK, J., K. A. OLIVE & M. SREDNICKI. 1985. Phys. Rev. Lett. **55:** 257.
SILK, J. & M. SREDNICKI. 1984. Phys. Rev. Lett. **53:** 624.
SREDNICKI, M., K. A. OLIVE & J. SILK. 1987. Nucl. Phys. B **279:** 804.
SREDNICKI, M., S. THEISSEN & J. SILK. 1986. Phys. Rev. Lett. **56:** 263.
STECKER, F. W. 1988. Phys. Lett. B. **201B:** 529.
STECKER, F. W., S. RUDAZ & T. WALSH.1985. Phys. Rev. Lett. **55:** 2622.
STECKER, F. W. & A. TYLKA. 1989. Astrophys. J., Lett. **336:** L51.
TURNER, M. 1986. Phys. Rev. D **34:** 1921.
TURNER, M. & F. WILCZEK. 1990. Phys. Rev. D **42:** 1001.
TYLKA, A. J. 1989. Phys. Rev. Lett. **63:** 840.

Direct Detection of Solar Neutrinos

D. SINCLAIR

Centre for Research in Particle Physics
Carleton University
Colonel By Drive
Ottawa, K1S 5B6, Canada

INTRODUCTION

The original motivation for searching for solar neutrinos was to exploit the great penetrating power of the neutrino to learn about the central, energy-generating regions of the sun. When the pioneering radio-chemical experiment of Ray Davis *et al.*[1] showed that the flux of high-energy neutrinos was much lower than predicted by the standard solar model[2] it was suggested[3] that the discrepancy might be due to the properties of the neutrino rather than some lack of completeness in the model. A second motivation for pursuing solar neutrino experiments then became the measurement of neutrino properties using the sun as an intense, distant source.

Direct-detection experiments offer a number of attractions for meeting these physics motivations. These include the possibility of measuring the neutrino spectrum directly, the use of directional response to identify the sun as the neutrino source, the measurement of the flux in real time to identify any temporal fluctuations, and the sensitivity to all neutrino flavors through neutral-current reactions.

There are two principal reasons for measuring the neutrino spectrum. The first is to test the predicted spectrum of the solar model, shown in FIGURE 1. The different groups carry complementary information about the solar interior. For example, Bahcall and Ulrich[4] have shown that the different groups have different radial sources within the sun. Thus they are sensitive to the solar conditions at different locations. Some groups carry rather specific information. Bahcall[5] has discussed the sensitivity of the hep (^{3}He(p, $e^+\nu$)^{4}He) reaction to mixing of the solar core. This group should be measurable in the next generation of experiments. Many authors have discussed the extreme temperature dependence of the ^{8}B group. Recently, Kim and Mann[6] have discussed measuring the ^{7}Be/^{8}B ratio to determine the central solar temperature in a largely model independent way. The second motivation in making a spectral measurement is to search for the distortions of the spectra due to possible Mikheyev–Smirrov–Wolfenstein (MSW) -type[7,8] neutrino oscillations.

The data from the chlorine experiment suggest that the neutrino flux is not constant, but varies with the 11-year sunspot cycle. The MSW effect could give temporal variation on a day–night, or seasonal timescale. To clarify this situation, a high-statistics real-time experiment must be carried out.

The most direct and unambiguous test for neutrino oscillations will come from a comparison of the total neutrino flux with the ν_e flux. Because it is assumed that all solar neutrinos are created as ν_e, any excess of total neutrino flux would confirm the oscillation hypothesis.

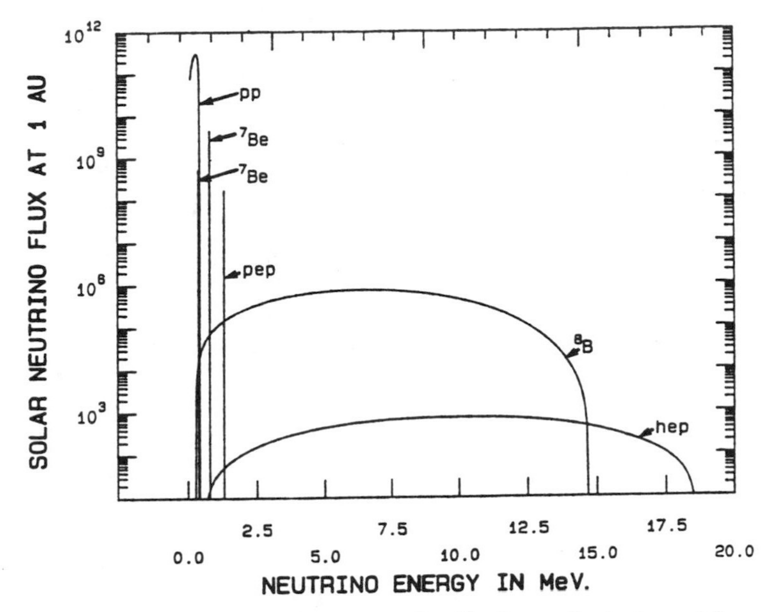

FIGURE 1. Spectrum of neutrinos predicted by the standard solar model.

THE KAMIOKANDE II EXPERIMENT

The first successful direct detection experiment was the light water Cerenkov detector at Kamiokande. The data from this project have recently been published.[9] Neutrinos interact in the light water by scattering from electrons. The recoiling electrons can have relativistic energies, and hence produce Cerenkov light in the water that is sensed by photomultiplier tubes that surround the water. Because the electrons are constrained to recoil in the forward direction, the signal can be distinguished from background by noting the direction of the light cone with respect to the sun. The data are shown in FIGURE 2.

The experiment confirmed that the flux is well below that of the solar models, especially

$$\Phi = 0.46 \pm 0.05\,(\text{stat}) \pm 0.06\,(\text{syst}) \times \text{solar model} \qquad (\text{Bahcall}^5).$$

The measured spectrum was consistent with that expected for neutrinos from the decay of ^8B, and this has a number of important consequences. First, it represents the first spectroscopic information about the solar neutrino spectrum. Second, it rules out some of the neutrino oscillation parameter space that would have led to a measureable distortion of the spectrum.

The data have been searched for possible time dependence, but no significant deviation from constant flux has been identified. However, the low statistical accuracy of these data preclude any definitive conclusions on time dependence.

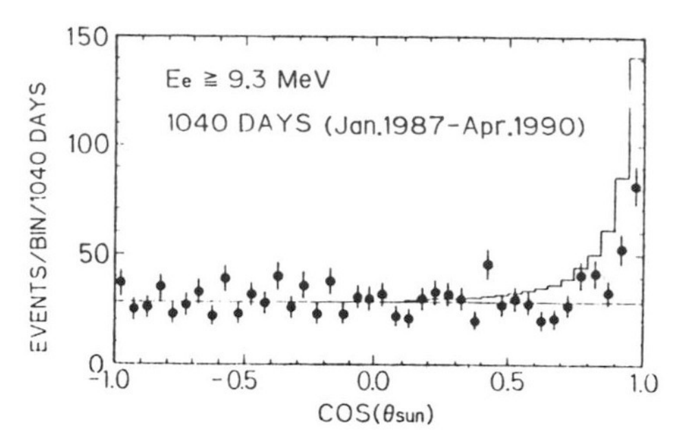

FIGURE 2. Events observed in the Kamiokande detector plotted as a function of the direction of the electron with respect to the direction of the sun. The neutrino events are contained in the peak at COS(Θ) = 1.

The most significant conclusion to be drawn from the Kamiokande experience is that direct counting experiments are feasible for solar neutrinos.

THE NEW EXPERIMENTS

Four classes of live detectors are being built or studied for detection of the high-energy solar neutrinos. These are light-water Cerenkov detectors, heavy-water Cerenkov detectors, scintillation detectors, and ionization detectors. All of these detectors offer spectral information and precise temporal information. Some also give directional information on the neutrino flux that is useful for confirming the source of the signals.

Three light-water Cerenkov detectors for solar neutrinos, Sudbury Neutrino Observatory (SNO),[10] Sunlab,[11] and Super Kamiokande,[12] are under consideration. The principle of detection is to observe the Cerenkov light from relativistic electrons created in the water by the elastic scattering of solar neutrinos. The light is detected using an array of photomultiplier tubes around the surface of the water. The energy of the electron is determined by the total number of photons detected, while the location of the electron track is determined by timing the arrival of the photons at the different phototubes and fitting to a point source. An important development over the past few years has been the introduction by several manufacturers of large phototubes with very good timing properties. The Kamiokande detector has proved the effectiveness of this detection process and has also demonstrated the experimental difficulties.[9] The ideal next-generation detector would be very large (to allow good statistical accuracy), and be well shielded from external background. It would be very deep to avoid the muon-induced spallation. None of the proposed detectors meet all of these objectives, but they should allow major improvements in the quality of the present data.

The most impressive detector of this type is the 50,000 tonne proposal referred to as Super Kamiokande[12] shown in FIGURE 3. This detector is essentially a scaled-up version of Kamiokande II using faster phototubes with higher coverage. The proper-

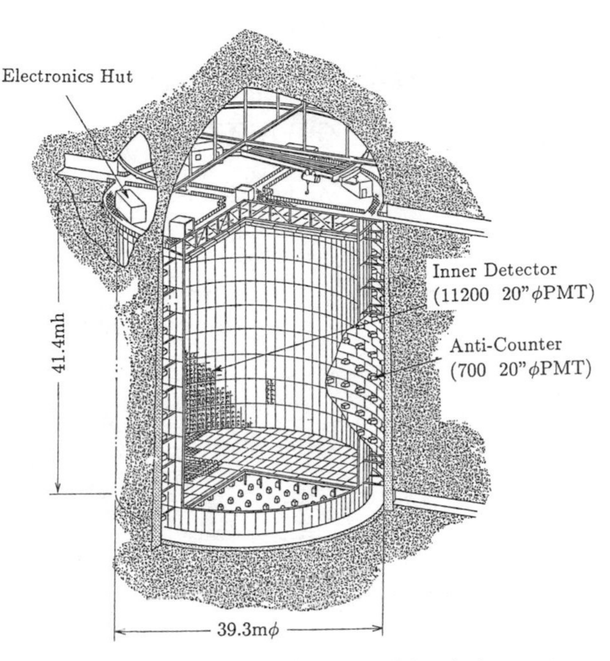

FIGURE 3. Outline of the Super-Kamiokande detector.

ties of the detectors are summarized in TABLE 1. The large enhancement in counting rate is clear. However, the background is still greater than the signal and is dominated by the muon-induced spallation.

The only proposal for a heavy-water Cerenkov detector is the Sudbury Neutrino Observatory. An outline of this detector is shown in FIGURE 4. The detector is similar in operation to the Kamiokande II detector, except that the sensitive volume is 1000 tonnes of heavy water (99.85 percent pure) contained in a transparent acrylic vessel. The heavy water is surrounded by a blanket of light water to shield against the

TABLE 1. Comparison of Kamiokande II and Super Kamiokande

Property	Kamiokande II	Super Kamiokande
Dimensions	$16m \times 19m\phi$	$41m \times 39m\phi$
Total mass	4500 t	50,000 t
Fiducial Mass	680 t	22,000 t
Number of PMTs	948	11,200
Vertex resolution	1.1–1.7 m	0.5 m
^{8}B signal	0.13/day	4.3/day
^{214}Bi decay	2/day	$\ll 1$/day
External γ's	0.7/day	0.4/day
Spallation products	1/day	6/day
U fission	0.02/day	0.6/day
Total background	3.4/day	8/day

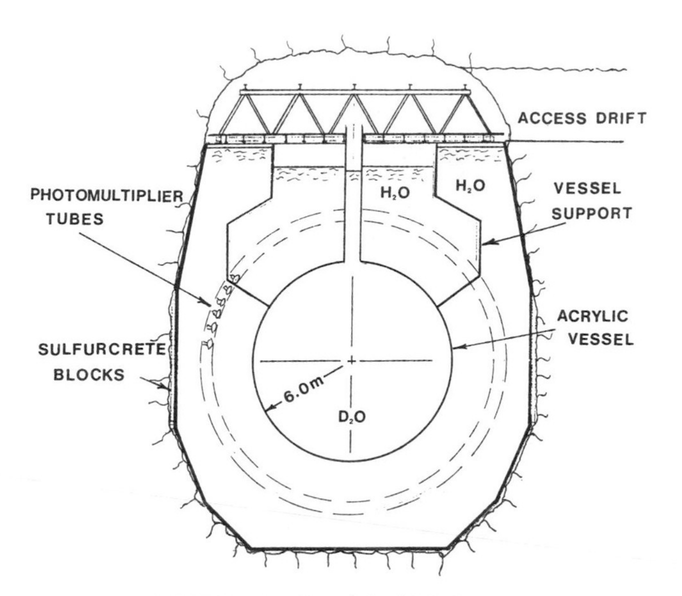

FIGURE 4. Outline of the SNO detector.

radioactivity in the rock, the phototubes, and the detector structure. Low-activity concrete blocks are used at the midpoint of the cavity to further attenuate the external background. Three reactions can be used to study the solar neutrinos. As in the light-water detectors, neutrino-electron scattering can be observed. This gives a very directional signal and conveys some spectral information. However, the rate is rather low. The presence of deuterium allows two additional reactions to be used. First, there is a charged-current process

$$\nu_e + d \rightarrow p + p + e^-.$$

As most of the free energy is given to the electron, this reaction is well suited to spectroscopic studies. The rate is one order of magnitude greater than the electron scattering reaction, so data of good statistical accuracy should result. The second reaction on deuterium is the neutral-current process

$$\nu_x + d \rightarrow n + p + \nu_x.$$

This reaction cannot be detected directly, since it produces no Cerenkov light. Instead the signature that is a free neutron in the heavy water must be used. Several methods for detecting this neutron are being considered. If the water is doped with NaCl, the capture on Cl will give rise to high-energy gamma rays that can subsequently be detected by the showers they produce. Alternatively, ^{3}He-filled proportional tubes may be used to detect the neutrons but not give a signal that can be confused with the charged-current signal. The main physics case for observing this reaction is that the total neutrino flux independent of neutrino flavor can be determined. This is the most model independent way of searching for neutrino flavor oscillations. There is a background problem that is unique to this reaction, however.

Any γ-ray of energy greater than 2.2 MeV can photodisintegrate a deuteron, creating a free neutron in the water. Because this cannot be distinguished from the neutral-current process, any such γ-rays must be carefully controlled and monitored. The most troublesome of these is the 2.614-MeV γ-ray in the decay of ^{208}Tl at the bottom of the ^{232}Th decay chain. Construction of the SNO detector has started and completion is scheduled for 1995.

Two scintillation experiments have been proposed to study the high-energy solar neutrinos. The scintillation process gives an intense light signal, allowing good energy and vertex resolution. These properties are particularly important for the rejection of backgrounds. If the backgrounds can be controlled, it will be possible to operate these detectors at lower thresholds than is possible for the Cerenkov detectors. The scintillation detectors lose the directional information, however. They are sensitive to all charged particles, including alphas. This can give rise to additional backgrounds, but the detectors can also be used to tag background events, and hence help in the rejection of internal activity. In some scintillators pulse-shape analysis will allow the alphas to be distinguished from electrons produced in the neutrino reactions.

Raghavan *et al.*[13] have proposed a boron-loaded scintillator experiment. BOREX, based on the many neutrino channels available with a ^{11}B target, as shown in FIGURE 5. The detector would contain 1760 tonnes of trimethoxyboroxine, which contains 18.7 percent boron by weight. The total fiducial boron content would be 197 tonnes. The optical coupling between the scintillator and the PMTs would be via light guides. The detector would be located in the Gran Sasso laboratory. As in SNO, three reactions are available for studying the solar neutrinos. The main signal for high-energy neutrinos is the charged-current reaction leading to states in ^{11}C. These reactions can be distinguished from background by the subsequent β signal from the decay of ^{11}C to ^{11}B. The high resolution of the detector should allow the gamma decay of the ^{11}B states at 4.4 and 5.0 MeV to be observed. This would allow a measure of the

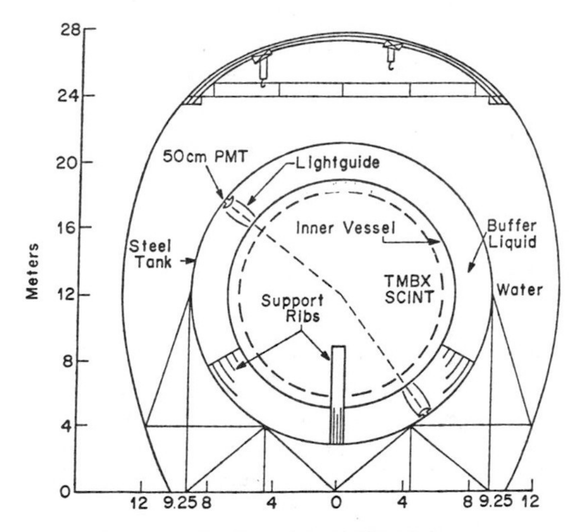

FIGURE 5. Outline of the BOREX detector.

total neutrino flux by giving the rate for neutral-current inelastic scattering of neutrinos from ^{11}B. However, these states are largely vibrational in structure, giving quite small M1 matrix elements, and hence low rates for the neutral-current excitation.

A dectector based on the C_6F_6 scintillator has been developed by Barabanov *et al.*[14] The main reaction to be employed is the charged-current reaction on ^{19}F leading to the mirror state in ^{19}N. The ^{19}N subsequently decays back to ^{19}F with a 17-second half-life, so that, as in the case of BOREX, the delayed coincidence can be used as a signature of the neutrino reaction to reduce the background. The proposed detector would comprise 1000 tonnes of this scintillator and be located at Baksan, USSR. An attraction of this material is that it contains 61 percent fluorine by weight and it is readily available.

The only proposal for an ionization detector for high-energy solar neutrinos is the ICARUS project.[15] The detector would consist of up to 3000 tonnes of liquid argon and would be located in the Gran Sasso Laboratory. Neutrinos would be detected with roughly equal probability through the elastic scattering from electrons or

TABLE 2. Expected Annual Counting Rates for Some of the High-Energy Neutrino Detectors

Experiment	ES	CC	NC	hep CC
S. Kamiokande	6000	—	—	—
SNO	300	3200	2700	20
BOREX	1000	800	120	4
Fluorine	300	1100	—	6
ICARUS	800	800	—	16
Assumed flux	^{8}B ν_e 2×10^6	^{8}B ν_e 2×10^6	^{8}B ν_x 6×10^6	1.6×10^4

charged-current reactions on the argon. The resulting electrons would produce ionization in the liquid that drifts under the influence of an electric field onto pads that allow the tracks to be reproduced. The advantages of such a detector are excellent energy, vertex, and directional resolution. It should also be possible to produce argon with very little radioactive impurity. At present, a 3-tonne prototype detector is undergoing tests at CERN.[16]

The counting rates for the new detectors discussed earlier are compared in TABLE 2. It can be seen that the physics objectives of high statistics data with spectral information and a determination of the total neutrino flux via neutral-current reactions for the ^{8}B neutrinos should be achieved by the next generation of experiments. It is also likely that an estimate of the hep flux will be possible. The hep neutrinos extend up to 19 MeV, and the counting rates shown are for neutrinos above the ^{8}B end point. This is a region between the radioactive background and the atmospheric neutrino signal, so the expected background counting rates are small compared with the predicted hep signal. The effect of mixing in the solar core is to greatly enhance these hep rates.

Low-Energy Live Detectors

The challenge of designing an experiment to measure the intense groups of solar neutrinos below 1 MeV has produced a wide range of very ingenious ideas. Feasible detectors require both a mechanism for generating an observable signal from the low energy available and a way of handling the very severe low-energy radioactive backgrounds. I will describe three proposals that appear to me to lead to feasible detectors in the near future.

The use of indium as a target for low-energy detection has been vigorously pursued since it was suggested by Raghavan.[17] The allowed charged-current reaction leads to the $7/2^+$ level of ^{115}Sn, which subsequently decays by emitting two gamma rays. Thus the signature of the reaction is very distinct. However, the beta decay of ^{115}Sn is a major source of background that requires some form of very fine-grained detector if the low-energy *pp* neutrinos are to be detected. At present no viable scheme exists to realize this, and the main activity is the design of a detector that could see the higher energy ^{7}Be group. The concept favored at present is a liquid scintillator loaded with indium. The recent development of such a scintillator loaded with 7 percent indium and having good light transmission is a major advance.[18] The detector might be made up of cells 2 m long and perhaps 12 cm $\times$ 12 cm, with phototubes at each end. About 3000 such cells would be required to make a detector with sufficient sensitivity.

A new proposal[19] to use a scintillation detector to look at the ^{7}Be neutrinos has come from the BOREX group. Known as BOREXINO, the detector would consist of a 200-tonne version of the BOREX scintillation detector. It offers much more than just a prototype for the full BOREX project, however, because it looks possible to measure the ^{7}Be neutrinos directly. Neutrinos would be detected mainly by observing the electron recoils following neutrino-electron elastic scattering. For a detection threshold of 250 keV, 50 events per day would be observed for the standard solar model ^{7}Be flux. Such a high counting rate offers many physics possibilities. First, it will test the standard solar model predictions for the ^{7}Be flux. Second, it would provide a very important low-energy measurement for pinning down possible MSW distortion of the neutrino spectrum. The detector could be tested with the GALLEX ^{51}Cr neutrino source. The statistical accuracy would allow the 7 percent annual variation in the flux due to the eccentricity of the earth's orbit to be clearly seen to establish the solar origin of the signal. Any other temporal variation of the signal (day–night or seasonal MSW-type effects, correlations with solar flares, sun spot activity, etc.) could be detected. The key to achieving this physics will be the elimination of radioactive material from the scintillator. This appears to be achievable. Indeed, because the mineral oil is not as aggressive a solvent as pure water, this material may be easier to purify and keep clean than water.

A number of cryogenic detectors for solar neutrinos have been suggested. One attractive proposal[20] is to look for excitations for superfluid ^{4}He. Neutrinos scatter elastically from electrons in the helium and the recoil energy produces rotons in the liquid. On reaching the surface the rotons vaporize the helium. Silicon wafers are placed above the liquid and, when the helium recondenses on these wafers, the energy is determined calorimetrically. A detector containing 10 tonnes of helium would see 20 events per day for the standard solar model flux of neutrinos. A key advantage of this detector concept is that helium is the purest material we can make, and hence the intrinsic background should be lower than in any other detector. A prototype detector is under construction at Brown University.

CONCLUSIONS

The coming decade promises to be a very exciting one for solar neutrino experiments. In addition to gallium experiments, several new projects are now demonstrated to be viable and some are under construction, so there remain no technical reasons why the objectives outlined in the Introduction should not be achieved. This will allow the questions posed by the last two decades of solar neutrino research to be answered. We should know if the solar neutrino problem is caused by deficiencies in the solar models or by flavor oscillations of neutrinos. We should know the intensities of several of the neutrino groups to allow more stringent tests of the models. Some aspects of these experiments remain unchanged, however. In 1949 L. Alverez wrote, "the most important experimental problems lie in the elimination of the various types of background." This is still very true today. The success of the various experiments will be determined largely by the extent to which the experimenters achieve the backgrounds estimated in their proposals.

ACKNOWLEDGMENTS

No review of the future of a field like this can be made by one person in isolation. Many people have contributed to this review through their letters, telephone conversations, discussions, and preprints. I would specially like to thank the following for their input: K. Lande (radiochemical experiments); E. W. Beier, A. Suzuki (Super Kamiokande); L. Peak (SUNLAB); A. Mann, N. Booth (Indium); S. P. Mikheyev, G. Domogatsky (Fluorine); R. S. Raghavan (BOREX, BOREXINO); D. Cline (ICARUS); R. Lanou (Helium).

REFERENCES

1. Davis, R., Jr. *et al.,* 1968. Phys. Rev. Lett. **20:** 205.
2. Bahcall, J. N. 1978. Rev. Mod. Phys. **50:** 881.
3. Pontecorvo, B. 1967. Zh. Eksp. Teor. Fiz. **53:** 717; Pontecorvo, B. 1968. Sov. Phys.—JETP **26:** 984.
4. Bahcall, J. N. & R. K. Ulrich. 1988. Rev. Mod. Phys. **60:** 297.
5. Bahcall, J. N. 1987. Rev. Mod. Phys. **59:** 505.
6. Kim, S. B. & A. K. Mann. 1990. Comments Nucl. Part. Phys. **19:** 173.
7. Mikheyev, S. P. & A. Yu. Smirnov. 1986. Nuovo Cimento C **9:** 17; Mikheyev, S. P. & A. Yu. Smirnov. 1985. Yad. Fiz. **42:** 1441.
8. Wolfenstein, L. 1978. Phys. Rev. **D17:** 2369.
9. Hirata, K. S., *et al.* 1990. Phys. Rev. Lett. **65:** 1297.
10. Ewan, G. T., *et al.* 1985. Sudbury Neutrino Observatory, SNO-85-3.
11. Bakish, A. M., *et al.* 1988. Nucl. Instrum. Methods **A273:** 853.
12. Kajita. 1989. Physics with the Super-Kamiokande Detector. ICR Report-185-89-2.

13. RAGHAVAN, R. S., *et al.* 1988. Design Concepts for Borex. Private communication.
14. BARABANOV, I. R., G. V. DOMOGATSKY & G. T. ZATSEPIN. 1988. ^{8}B solar neutrino experiment with fluorine scintillator. *In* Proceedings of the 13th International Conference on Neutrino Physics and Astrophysics (Boston), J. Schneps, T. Kafka, W. A. Mann, and P. Nath, Eds.: 331–336. World Scientific. Singapore.
15. ICARUS PROPOSAL. 1985. INFN/AE-85/7 Frascati, Italy.
16. BONETTI, S., *et al.* 1990. Nucl. Instrum. Methods A286, 135.
17. RAGHAVAN, R. S. 1976. Phys. Rev. Lett. **37:** 259.
18. SUZUKI, Y., K. INOUE, Y. NAGASHIMA, S. HASHIMOTO & T. INAGAKI. 1990. Nucl. Instrum. Methods **A293:** 615.
19. RAGHAVAN, R. S. BOREXINO. Private Communication.
20. LANOU, R. E., H. J. MARIS, F. S. PORTER, G. M. SEIDEL, C. A. SMITH & R. H. TORII. 1988. Particle detection via rotons in helium-4 superfluid. *In* Particle Astrophysics Workshop: Forefront Experimental Issues (Berkeley). World Scientific. Singapore.

Radiochemical Solar Neutrino Experiments

T. KIRSTEN
Max-Planck-Institut für Kernphysik
P.O. Box 103 980
D-6900 Heidelberg, Germany

Solar neutrino experiments serve to test the theory of stellar structure and, by virtue of the Mikheyev–Smirnov–Wolfenstein (MSW) effect, to test neutrino properties. A short account was given of the present experimental situation concerning the *radiochemical* experiments. For more detail, see, for example, references 1, 2, 3, 4. The Homestake Cl experiment,[5] sensitive mainly to ^{8}B neutrinos, persists to yield average results 2 to 4 times below the standard solar model predictions (2.19 ± 0.24 SNU vs. 7.9 ± 2.6 SNU). Moreover, there is an apparent anticorrelation between the measured neutrino flux and solar activity[5–7] that could be linked to a magnetic moment of the neutrinos,[8] yet the statistical proof is not free of inconsistencies, and also the effect is not seen in the (real-time) Kamiokande experiment.[9]

Radiochemical gallium experiments are pursued by the Soviet/American Gallium experiment (SAGE) collaboration[10] (using 60 t of metallic Ga) and by the Gallium experiment (GALLEX) (using 30 t of Ga in the form of aqueous gallium chloride solution). These experiments are sensitive also to the "*pp*" solar neutrinos from the primary proton–proton fusion reaction. SAGE has announced a preliminary result of <138 SNU (2σ).[10] GALLEX, in the Gran Sasso Underground Laboratory,[11,12] has been fully operating since August 1990,[13,14] but only in May 1991 has it completed the quantitative removal of the cosmogenic ^{68}Ge, which was produced in the target material before it was brought underground. This $T_{1/2} = 288$ d nuclide must be removed, since it compromises the detection of 11.4 d ^{71}Ge.

Other proposed radiochemical experiments include iodine,[15] bromine,[15] and lithium, but they are currently not actively pushed. $I^{127}(\nu, e^-)^{127}Xe$ resembles in many respects $Cl^{37}(\nu, e)^{37}Ar$ with the advantage of *a* $\sim 5\times$ higher production rate.

"Geochemical" experiments are based on the detection of long-lived (or even stable) neutrino capture products accumulated in natural minerals over geological time.[2] The most advanced and promising experiment of this type ($^{98}Mo(\nu, e)^{98}Tc$)[17] has, unfortunately, come to an end without success. The LOREX-collaboration[18] (from lorandite, a thallium ore) continues work on $^{205}Tl(\nu, e)^{205}Pb$,[19] but difficulties are large and the ore is not sufficiently shielded. Proposals involving bromine and tellurium[19] seem also not feasible.

REFERENCES

1. VIGNAUD, D. 1991. Some facts and some dreams about solar models and solar neutrinos. Proceedings of the Moriond Workshop. In press.
2. KIRSTEN, T. 1989. Upcoming experiments and plans in low energy neutrino physics. *In*

Neutrino '88 (Boston), J. Schneps, T. Kafka, and W. A. Mann, Eds.: 742–764. World Scientific. Singapore.
3. BAHCALL, J. 1989. Neutrino Astrophysics. Cambridge University Press. Cambridge, England.
4. POMANSKY, A. A. 1990. Current and proposed solar neutrino experiments. Presented at the Symposium on Solar Neutrino Detection with Thallium (Dubrovnik, Yugoslavia), October 1990.
5. DAVIS, R., *et al.* 1990. *In* Proceedings of the 21 ICRC (Adelaide), Vol. 12, R. J. Protheroe, Ed.: 143.
6. LANDE, K. 1990. Presented at Neutrino '90 (Geneva, Switzerland). June 1990.
7. FILLIPONE, B. W. & P. VOGEL. 1990. Phys. Lett. **B246:** 546.
8. AKHMEDOV, E. K. 1991. Phys. Lett. **B257:** 163.
9. HIRATA, K. S., *et al.* 1991. Phys. Rev. Lett. **66:** 9.
10. ABAZOV, A. I., *et al.* 1991. First results from the Soviet-American gallium experiment. *In* Neutrino '90. Nucl. Phys. B (Proc. Suppl.) **19:** 84.
11. KIRSTEN, T. 1991. The status of Gallex. *In* Neutrino '90. Nucl. Phys. B (Proc. Suppl.) **19:** 77.
12. CRIBIER, M. 1990. The status of Gallex. *In* Proceedings of the 25th International Conference on High Energy Physics (Singapore). In press.
13. VIGNAUD, D. 1991. Report on the Gallex experiment. Presented at the Moriond Workshop 1991.
14. KIRSTEN, T. 1990. The Gallex project. *In* Inside the Sun, G. Berthomieu and M. Cribier, Eds.: 187–199. Kluwer. Dordrecht, the Netherlands.
15. HAXTON, W. C. 1988. Los Alamos 40048-46-IVZ.
16. HURST, G. S., *et al.* 1984. Phys. Rev. Lett. **53:** 1116.
17. COWAN, G. A. & W. C. HAXTON. 1982. Science **216:** 51.
18. POMANSKY, M. K., Ed. 1990. Proceedings of the Symposium on Solar Neutrino Detection with Thallium (Dubrovnik, Yugoslavia).
19. HAXTON, W. C. 1990. Phys. Rev. Lett. **65:** 809.

First Measurement of the Integral Solar Neutrino Flux by the Soviet/American Gallium Experiment[a]

V. N. GAVRIN[b]

Institute for Nuclear Research of the Academy of Sciences of the USSR Moscow 117312, USSR

INTRODUCTION

About forty years ago the inverse beta-decay of ^{37}Cl was proposed to measure the solar neutrino flux.[1,2] Such a measurement was initiated in the 1960s in the Homestake gold mine in the United States. The measured value in the chlorine experiment averaged over 1970–1988 was 2.2 ± 0.3 (1σ) solar neutrino units (SNU)[3] (1 SNU = 10^{-36} captures per target atom per second). Recent calculated values of the flux are 7.9 ± 2.6 (3σ) SNU in the Bahcall–Ulrich SSM.[4] This deficit has now been corroborated by the Kamiokande 2 Water Cherenkov experiment[5] in Japan, which observed only 0.46 ± 0.05 ± 0.06 of the flux predicted by the Bahcall–Ulrich SSM, in fair agreement with the chlorine result of recent years.

The low-energy neutrinos produced in the proton–proton (p–p) fusion in the sun, which account for more than 90 percent of solar neutrinos, are far below the threshold of the chlorine and Kamiokande experiments, which are primarily sensitive to the high-energy ^{8}B solar neutrinos whose production rate depends critically on the core temperature of the sun. The p–p neutrino production rate in the sun is fundamentally linked to the observed solar luminocity, and is insensitive to alteration in the solar models.

A radiochemical experiment using ^{71}Ga as the capture material[6] provides a feasible means of measuring the p–p neutrino flux.

THE BAKSAN GALLIUM EXPERIMENT

The Soviet/American Gallium Solar Neutrino experiment (SAGE) is situated in an underground laboratory specially built in the Baksan Valley of the Northern Caucasus, USSR. The laboratory is 60 m long, 10 m wide, and 12 m high. It is located 3.5 km from the entrance of a horizontal adit driven into the side of Mount Andyrchi, and has an overhead shielding of about 4700 meters of water equivalent (MWE).

[a] This work was supported by the Academy of Sciences of the USSR, the Institute for Nuclear Research of Academy of Sciences of the USSR, the Division of Nuclear Physics of the U.S. Department of Energy, the National Science Foundation, and the Los Alamos National Laboratory.

[b] Spokesman of the SAGE Collaboration.

Chemistry

The experiment exploits a radiochemical procedure. About 30 tons of the Ga metal is kept molten (~30°C) in four chemical reactors, each with an internal volume of 2 m^3 and lined with Teflon. The process for extracting the ^{71}Ge has been described in detail elsewhere,[7] and will be only briefly described here. At the beginning of each run, approximately 120 microgrammes of natural Ge carrier is added to each reactor in the form of solid Ga–Ge alloy. After a suitable exposure interval (typically 3 to 4 weeks), the Ge carrier and any ^{71}Ge atoms that have been produced by neutrino capture are chemically extracted from the gallium. The extraction procedure concentrates the carrier and ^{71}Ge atoms from the Ga metal to 100 ml of final solution. From this final solution the counting gas GeH_4 (germane) is synthesized. The efficiency of extraction of Ge from the reactors is measured in two stages of the extraction procedure by atomic absorption analysis. A final determination of the quantity of Ge is made by measuring the volume of sythesized GeH_4. The overall extraction efficiency is typically 80 percent. The uncertainty in the extraction efficiency is typically ±6 percent.

The GeH_4 is mixed with a measured quantity of Xe, and is inserted into a low-background proportional counter with an internal volume of about 0.75 cm^3.

Counting

The proportional counter is placed in the well of a NaI detector inside a large passive shield and counted for 2 to 3 months. The counting of the ^{71}Ge decays has been described in detail elsewhere.[7] The number of ^{71}Ge atoms is determined by detecting the Auger electrons and/or X-rays in the K- and L-peaks of Ge (at 10.4 and 1.2 KeV, respectively) that are produced by ^{71}Ge decay. Due to considerably higher background in the L-peak, only the K-peak can be used in the analysis presented here.

Pulse-shape discrimination based on rise-time measurements is used to separate the ^{71}Ge decays from the background. In contrast to the localized ionization produced by Auger electrons or X-rays from ^{71}Ge decay, background radioactivity primarily produced fast electrons in the counter, which produced extended ionization. Pulses from the counter are differentiated with a time constant of 10 ns. The amplitude of the differentiated pulse is proportional to the product of the amplitude and the inverse of the rise time of the pulse. For every event in the counter, the energy, the amplitude of the differentiated pulse, the time, and any associated NaI signal are recorded. The total background rate of selected counters from 0.7 to 13.0 keV is approximately 1.5 counts per day (cts/day). The counter is calibrated at one month intervals using an external ^{55}Fe source, which illuminates the central part of the counter through a thin side window. The ^{55}Fe calibration is used to generate an acceptance window in a two-dimensional plot of inverse rise time versus energy. The acceptance cut in energy is 98 percent, and in inverse rise time it is 95 percent. The measured total counting efficiency is 35 percent. This efficiency includes geometrical effects inside the counter. The position of the acceptance window for the ^{71}Ge K-peak is calculated by extrapolating from the ^{55}Fe-peak. The extrapolation procedure was checked by filling a counter with $^{71}GeH_4$ together with the standard counter gas. These data clearly show that when the extrapolation method is used correctly it includes the ^{71}Ge events in the window.

The analysis searches for events that are within the ^{71}Ge K-peak acceptance window and that have no NaI activity in coincidence. A maximum likelihood analysis[8] is then carried out on these events by fitting the time distribution to an 11.4-day half-life exponentional decay plus a constant ray background.

The SSM predicts a production rate of 1.2 ^{71}Ge atoms/day in 30 tons of Ga. The mean number of detected ^{71}Ge atoms expected in each run is 4.0.

Backgrounds

The main source of ^{71}Ge in the gallium other than from solar neutrinos is from protons arising as secondary particles produced by external neutrons, internal radioactivity, and cosmic-ray muons. The measurement and calculation of these background channels indicate that the total background production rate of ^{71}Ge is less than 2.5 percent of the SSM-predicted rate.

The main task at the beginning of the experiment was to remove from the Ga long-lived ^{68}Ge (half-life 288 days) produced by cosmic rays while the Ga was on the surface. The ^{68}Ge activity in the first extraction from 30 tons of Ga was 7700 cts/day in the Ge K-peak. ^{68}Ge decays by electron capture, so its decays cannot be differentiated from those of ^{71}Ge.

The experiment began operating in May of 1988, when the removal of the ^{68}Ge from 30 tons of Ga commenced. Data from extractions made during 1988–1989 are not included in the analysis here because of the high background in earlier extractions due to ^{68}Ge and Rn.

Additional purification procedures were implemented beginning with the January 1990 extraction that resulted in the reduction of the radioactive backgrounds of the extraction samples.

RESULTS

Extractions from 30 tons of Ga were carried out in January, February, March, April, and July of 1990.

The results of the maximum likelihood analysis are shown in TABLE 1.

Systematic Effects

Systematic effects fall into three categories: uncertainties in efficiencies, a possible variation in time of the detector background causing an incorrect background subtraction, and uncertainties in the extrapolation of the ^{71}Ge K-peak acceptance window from the ^{55}Fe calibration.

The overall uncertainty in the detection efficiency due to the systematic uncertainties in the chemical extraction and counting efficiency produced a change in the ^{71}Ge

TABLE 1. Statistical Analysis

Run	Best Fit (SNU)	68% CL (SNU)	95% CL (SNU)
January	0	60	118
February	39	83	142
March	90	175	276
April	0	94	174
July	0	149	275
Sum	20	35	60

TABLE 2. Extraction Efficiency of the Ge Carrier and ^{71}Ge

Run	Carrier (mg)	^{71}Ge Atoms	Carrier Efficiency (%)	^{71}Ge Efficiency (%)
1	410 ± 10	4900 ± 500	77 ± 2	81 ± 8
2	97 ± 2	980 ± 220	80 ± 2	84 ± 19
3	21 ± 1	130 ± 45	84 ± 4	71 ± 25
Sum	528 ± 10	6010 ± 550		
Amount added	525 ± 26	6050 ± 600		

rate of not more than 5 SNU [68 percent coincidence level (CL)] and 14 SNU (90 percent CL).

The uncertainty in the background subtraction under ^{71}Ge decay curves due to possible time variations in the counter background produced a maximum change in the ^{71}Ge rate of 30 SNU (68 percent CL) and 35 SNU (90 percent CL).

The uncertainty in the position of the ^{71}Ge acceptance window increased the upper limit by 10 SNU (68 percent CL) and 23 SNU (90 percent CL).

The results of the analysis are, assuming that the extraction efficiency for ^{71}Ge atoms produced by solar neutrinos is the same as that measured using natural Ge carrier,

$$^{71}\text{Ga capture rate} = 20 + 15/-20\ (\text{stat}) + 32\ (\text{syst})\quad \text{SNU}.$$

Upper limits were determined by adding the statistical and systematic errors in quadrature and then adding this linearly to the best fit value. The upper limits are

$$^{71}\text{Ga capture rate} < 55\ \text{SNU (68\% CL)},\qquad <79\ \text{SNU (90\% CL)}.$$

While all available information leads one to expect that the extraction efficiency for ^{71}Ge atoms produced by solar neutrinos should be the same as for the carrier, it is important to test this assumption. As an initial test of the extraction process, we doped one of the reactors holding 7 tons of Ga with a known number of ^{71}Ge atoms, along with the natural Ge carrier. Three extractions were carried out and the number of ^{71}Ge atoms was counted.

TABLE 2 shows the results of this measurement, and indicates that the extraction efficiency of the natural carrier and the ^{71}Ge track very closely.

FUTURE PLANS

Monthly extraction from 30 tons of Ga will continue. At the same time, the detector will be extended to include the 60 tons of Ga. The additional 30 tons is now stored underground and will be installed into reactors shortly. A final calibration of the detector is planned using the ^{51}Cr neutrino source.[7]

SUMMARY

The first data from the Baksan Gallium Solar Neutrino experiment are consistent with no solar-neutino-induced events being observed. The initial data indicate that the flux may be less than that expected from p–p neutrinos alone, thus indicating that the solar neutrino problem also applies to the low-energy p–p neutrinos.

ACKNOWLEDGMENTS

The SAGE Collaboration wishes to thank A. E. Chudakov, G. T. Garvey, M. A. Markov, V. A. Matveev, J. M. Moss, V. A. Rubakov, and A. N. Tavkhelidze for their continued interest in our work and for stimulating discussions. We are also grateful to J. N. Bahcall and A. Yu. Smirnov for useful discussions.

REFERENCES

1. PONTECORVO, B. 1946. Inverse beta decay. Chalk River Report PD-205; PONTECORVO, B. 1983. Usp. Fiz. Nauk. **141:** 675 (Sov. Phys. Usp. **26:** 1087).
2. ALVAREZ, L. W. 1949. A proposed experimental test of the neutrino theory. Univ. California Radiation Laboratory Report UCRL-328, Berkeley, Calif.
3. DAVIS, R., *et al.* 1988. Proceedings of the Neutrino '88 International Conference: 518.
4. BAHCALL, J. N. & R. K. ULRICH. 1988. Rev. Mod. Phys. **60:** 297.
5. TOTSUKA, Y., *et al.* 1990. Presented at the 15th Texas/ESO-CERN Symosium on Relativistic Astrophysics, Cosmology and Fundamental Physics, Brighton, England, Dec. 16–21.
6. KUZ'MIN, A. 1964. Lebedev Physical Institute Preprint A-62; KUZ'MIN, A. 1965 Sov. Phys.—JETP **49:** 1532.
7. GAVRIN, V. N., *et al.* (SAGE). 1989. Proceedings of the "Inside the Sun" Conference (Versailles France), G. Berthomieu and M. Cribier, Eds.: 201. Kluwer. Dordrecht, the Netherlands.
8. CLEVELAND, B. T. 1983. Nucl. Instrum. Methods **214:** 451.

The Origin of Ultrahigh Energy Cosmic Rays

X. CHI,[a] J. SZABELSKI,[a,b] M. N. VAHIA,[c]
AND A. W. WOLFENDALE[a,d]

[a]*Department of Physics*
University of Durham
Durham DH1 3LE, England

[c]*Tata Institute of Fundamental Research*
Homi Bhabha Road
Bombay 400 005, India

INTRODUCTION

We have studied cosmic rays (CR) of energy above 10^{19} eV using data published by the World Data Centre C2 for Cosmic Rays (see references). Three catalogues are available that contain information from four large CR experiments at Volcano Ranch, Haverah Park, SUGAR (Sydney), and Yakutsk, respectively.

The data have been summed to study the sky as seen in the most energetic particles. If these particles are mostly protons, then their trajectory Larmor radii will be large, and there is therefore some hope for the detection of their sources if they are in the Galaxy. However, it is well known that the arrival directions of these particles are almost entirely isotropic, with only a hint of some small anisotropies, but with no clear evidence of identified point sources put forward so far. The small anisotropies have already been investigated assuming specific models, for example, galactic plane enhancement or avoidance[1] and supercluster (Virgo) origin of the highest energy cosmic rays.[2,3]

Here we present a different statistical approach to the same problem. Analyzing arrival directions of the extensive air showers (EAS), we have used statistical models that are not astrophysical model dependent: the n-point correlation function and the searches for the grouping of the arrival directions of the EAS for the most energetic events (which might be from different experiments).

We have started by examining the zenith angle distributions of arrival directions in each experiment in the energy ranges $E > 10^{19}$ eV and $E > 3 \cdot 10^{19}$ eV. Then, assuming equal coverage of the sidereal day by working time for each experiment, and azimuthal symmetry, we have calculated the predicted declination distribution in celestial coordinates. The predicted and actual distributions agree well (perhaps with the exception of the south celestial pole direction), giving us some confidence in the analysis.

[b]On leave from the Institute of Nuclear Studies, PO Box 477, 90-950 Łódź, Poland.
[d]To whom correspondence should be addressed.

n-POINT CORRELATION FUNCTIONS

The n-point correlation function has been derived in the usual way, by studying the number of events as a function of their angular separation.

The calculations have been performed up to $n = 5$, and general agreement has been found between the observed and the predicted values in different declination regions and energy bands, with two exceptions, both of which are in the energy band $E > 3 \cdot 10^{19}$ eV:

- for $-90° < \delta < -60°$ there is one event with four other events within 6 degrees of it, whereas 0.01 is expected (14 events in the region and a chance probability of $6.6 \cdot 10^{-4}$)
- for $-60° < \delta < -30°$ there are three events with three other associated events within 6 degrees, whereas 0.067 is expected. There are also three other events in this region, with two associated events within 6 degrees, whereas 0.57 is expected

All these events are in the south celestial hemisphere, where only the Sydney (SUGAR) data are available. The positions of the most energetic events in each group are given in TABLE 1.

TABLE 1.

Central CR Event Energy (10^{19} eV)	Number of Associated Events + 1	α	δ	l	b
4.06	5	254°	−82°	311°	−23°
7.20	4	340°	−42°	355°	−60°
6.61	4	281°	−55°	341°	−21°
4.33	4	93°	−45°	253°	−25°
5.44	3	123°	−37°	255°	− 1°
4.45	3	116°	−35°	250°	− 5°
4.65	3	144°	−32°	262°	15°

GALACTIC PLANE ENHANCEMENT IN THE NORTH CELESTIAL HEMISPHERE

A further selection has also been made. For each event with $E > 3 \cdot 10^{19}$ eV we calculate the number of events with $E > 10^{19}$ eV being within 6 degrees of it. All events with $E > 3 \cdot 10^{19}$ eV having at least five associated events are plotted in FIGURE 1. Starting with the groups near to the north celestial pole, $\delta > 60°$, we expect roughly the number observed by chance. However, it is surprising that those observed are so close together (the high chance probability is because the received particles density is the largest there due to exposure). The three groups in the region $\alpha = 180° - 190°$ and $\delta \approx 10° - 20°$ are in the direction of the Virgo cluster ($\alpha = 187°$, $\delta = 13°$), and one group coincides with the first group listed in TABLE 1. The two groups far from the galactic plane are near to the peak of CR energy intensity of

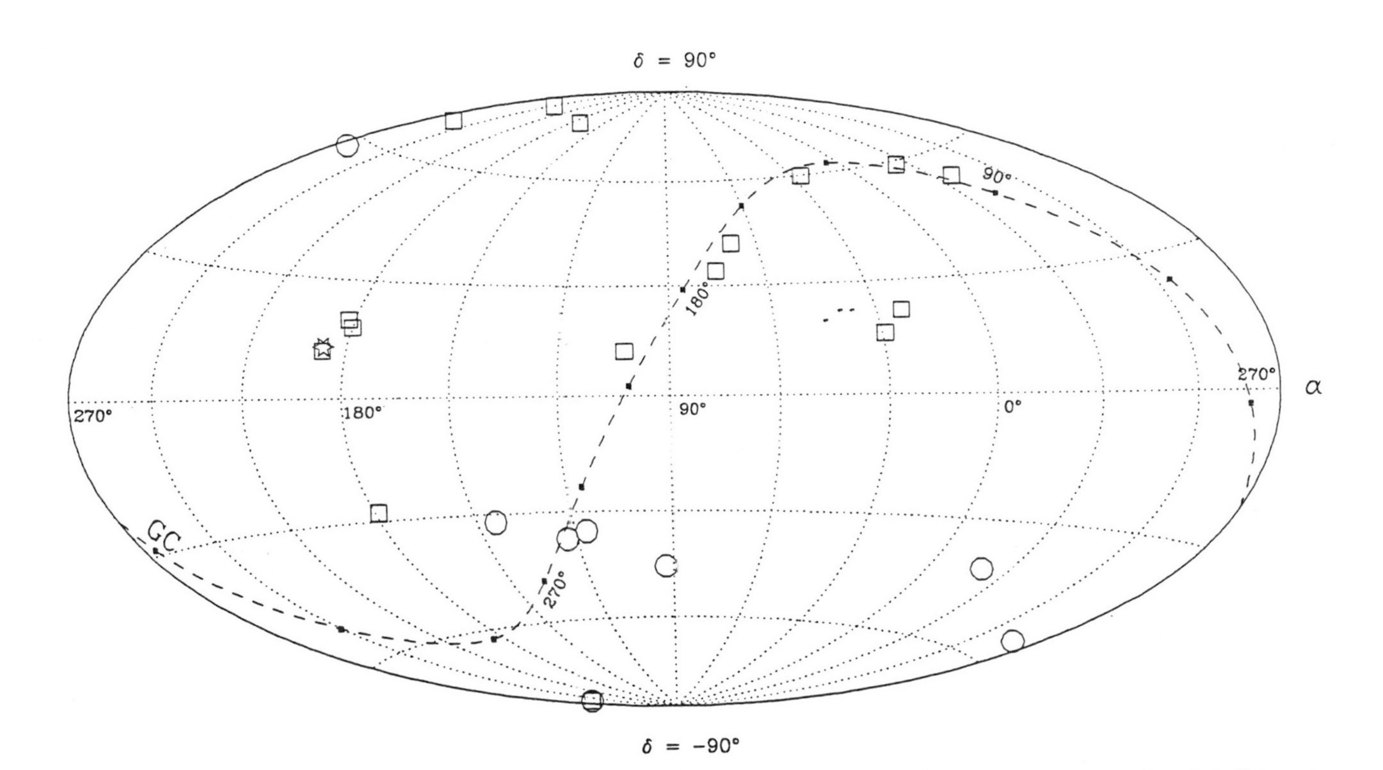

FIGURE 1. Aitoff plot of "groups" of more then four CR events ($E > 10^{19}$ eV) within 6° of an event (plotted as *squares*) with $E > 3 \cdot 10^{19}$ eV and groups listed in the TABLE 1 (plotted as *circles*). The Galactic plane is indicated. The *star* indicates the Virgo cluster (M87) direction.

about 10^{18} eV observed by the Akeno array (Kofu Conference, 1990): $\alpha = 32°$, $\delta = 32°$, and six of the remaining seven events are within 6 degrees of the galactic plane direction.

There are five such groups in the $\delta = 30° - 60°$ region. In this region there are 212 events with $E > 10^{19}$ eV, with 33 events among them with $E > 3 \cdot 10^{19}$ eV. Events of number 207.3 and 35, respectively, are expected from the zenith angle distribution, so there is no excess of DC signal in this declination region. FIGURE 2 presents the expected and the observed galactic latitude distributions of these events. It will be noticed that there is a clear excess of events close to the galactic plane, particularly for those particles above $3 \cdot 10^{19}$ eV. This fact, taken together with the detection of groups, makes the case for a galactic plane enhancement overhelming. (We expect 4.67 events within 6° of the galactic plane in this declination band, and 12 were observed; see FIG. 2.)

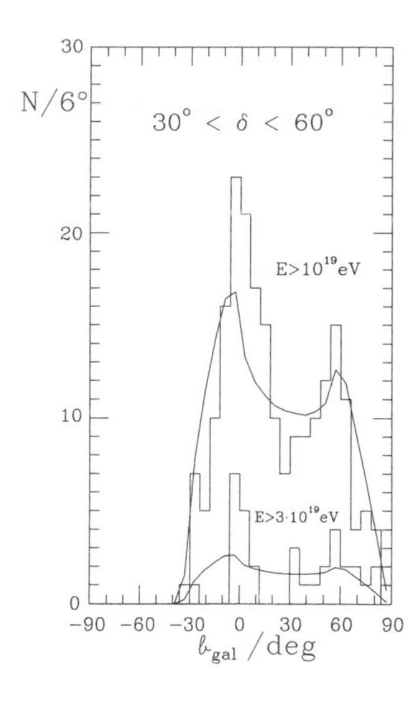

FIGURE 2. Galactic plane enhancement in the northern hemisphere: the predicted and observed galactic latitude distributions.

The derivation of the chance probability that these groups are due to discrete and separate sources is model dependent. For example, it can be determined by the spatial distribution of the proposed sources and their luminosity distribution. Alternatively, if the groups are due not to a few sources but to statistical fluctuations associated with numerous sources, or even some form of distributed acceleration, then again we would need to know the details. This point is explored further in the next section.

DISCUSSION

The two methods of analysis given previously give us a new picture of the origin and propagation of the highest energy cosmic rays. We have found statistically significant groups of events in both the southern celestial hemisphere and in the inner galactic hemisphere. Most of these groups are within about 25° of the galactic plane.

In the northern celestial hemisphere (and particularly in the outer Galaxy) there are concentrations in the arrival directions of the highest energy particles near to the galactic plane, and some groupings in the direction of the Virgo cluster. These findings are consistent with our previous work.[4,5] The fact that the groups are largely located in the galactic plane calls for serious consideration to be given to the idea that many (and perhaps all) of the highest energy CR are of galactic origin. The effective distance from the sources that preserves directional information would be of the order of 1 kpc or less. Presumably, the source density would be higher in the galactic spiral arms, but because of galactic "diffusion" in the tangled magnetic field, we can see our nearby Orion arm only, in the galactic longitudinal range l = 100°–200°. Information on more distant sources would be largely erased by the diffusion process, and the lack of detected galactic plane sources in the inner galaxy is therefore easily explained.

We have not seen a significant energy dependence of particle deflections for EAS's in groups (apart from the case for the Virgo cluster direction as described earlier[4,5]). Such a relationship would be expected if there was only one source responsible for each group. Therefore the emission is probably not coherent, and the grouping might be explained by assuming relatively low efficiency of the single sources and some grouping of these objects (e.g., substructure within spiral arms, to account for the groups in the galactic plane anticenter direction, and some small associations of nearby objects, to account for the high-latitude groups).

Proceeding with the assumption that the groups are in some way associated with discrete sources that are bright in other wavelengths (despite the arguments in the previous section), an attempt has been made to identify them with objects from several astronomical catalogues by comparing directions and luminosity. The High Energy Astrophysical Observatory (HEAO) X-ray point source data and the Wood's Catalogue of Interacting Binary Stars (both available on Starlink) have been processed and we have looked in the direction of nearby bright galaxies. The gamma-ray burst (GRB) catalogues[6,7] have also been processed, and we conclude that there is no significant correlation between the arrival directions of GRBs (localized within one degree) and the highest energy CRs. It must be remarked, however, that from 77 GRB events, three are in the directions of the groups listed in TABLE 1: $\alpha = 255°$ $\delta = -82°$, $\alpha = 335°$ $\delta = -42°$, and $\alpha = 94°$ $\delta = -46°$. These GRBs do not have particular properties from which we could extract them from the bulk of data, however, and it cannot be claimed that some of the ultrahigh energy CR came from the GRBs that have been detected thus far (one can invoke other arguments against this hypothesis, too, of course).

The observed properties of the highest energy CR can be explained better by assuming that they contain a mixture of iron (and similar "heavy" elements) and hydrogen nuclei. Because of their larger electric charge, the arrival directions of iron nuclei will be very nearly isotropic at these energies, whereas the proton component should preserve some directional information. The idea seems to be very attractive, and we are currently checking the hypothesis by searching for differences between the EAS characteristics of the particles in the groups and the rest. Such differences would also exist in this model between all galactic plane events and those elsewhere. We would expect a different EAS detector response to different parent particle showers and probably different energy determination, too. It will be interesting to relate the respective responses of different EAS arrays to such a mixed mass hypothesis from the point of view of the difference in reported energy spectra, and to see whether such an assumption would help to obtain a consistent picture of the high-energy end of the CR spectrum.

ACKNOWLEDGMENTS

Two of the authors (J. S. and M. N. V.) thank the Royal Society and the University of Durham for support and hospitality.

REFERENCES

1. Wdowczyk, J. & A. W. Wolfendale. 1984. J. Phys. G **10:** 1453–1463.
2. Wdowczyk, J. & A. W. Wolfendale. 1979. Nature **281:** 356–357.
3. Giler, M., J. Wdowczyk & A. W. Wolfendale. 1980. J. Phys. G **10:** 1561–1573.
4. Szabelski, J. 1988. Genesis and Propagation of Cosmic Rays (NATO ASI C220), M. M. Shapiro and J. P. Wefel, Eds.: 97–104. Reidel. Dordrecht, the Netherlands.
5. Szabelski, J., J. Wdowczyk & A. W. Wolfendale. 1986. J. Phys. G **12:** 1433–1442.
6. Atteia, J. L., *et al.* 1987. Astrophys. J., Suppl. Ser. **64:** 305–382.
7. Golenetskii, S. V., Yu. A. Guryan, G. B. Dumov, A. V. Dyatchkov, V. N. Panov, N. G. Khavenson & L. O. Sheshin. 1986. A. F. Ioffe Physical-Technical Institute. Preprint No. 1026.
8. Cunningham, G., *et al.* 1980. Catalogue of the Highest Energy Cosmic Rays, Giant EAS, No. 1: 61–97. Haverah Park. Published by the World Data Centre C2 for Cosmic Rays, Institute of Physical and Chemical Research, Itabashi, Tokyo, Japan.
9. Efimov, N. N., T. A. Egorov, D. D. Krasilnikov, M. I. Pravdin & I. Ye. Sleptsov. 1988. Catalogue of the Highest Energy Cosmic Rays, Giant EAS, No. 3. Yakutsk. Published by the World Data Centre C2 for Cosmic Rays, Institute of Physical and Chemical Research, Itabashi, Tokyo, Japan.
10. Linsley, J. 1980. Catalogue of the Highest Energy Cosmic Rays, Giant EAS, No. 1: 3–59. Volcano Ranch. Published by the World Data Centre C2 for Cosmic Rays, Institute of Physical and Chemical Research, Itabashi, Tokyo, Japan.
11. Winn, M. M., J. Ulrichs, L. S. Peak, C. B. A. McCusker & L. Horton. 1986. Catalogue of the Highest Energy Cosmic Rays, Giant EAS, No. 2. SUGAR. Published by the World Data Centre C2 for Cosmic Rays, Institute of Physical and Chemical Research, Itabashi, Tokyo, Japan.

Low-temperature Extragalactic Supernova Neutrino Detector

L. STODOLSKY

Max-Planck Institut für Physik und Astrophysik
Föhringer Ring 6
8000 Munich 40 Germany

INTRODUCTION

It is a pleasure to talk about one of my favorite crazy subjects, a cold fantasy for the twenty-first century: systematic extragalactic supernova neutrino studies, operating on the nuclear coherent scattering reaction. When we first began to think about detecting neutrinos by using the large cross section of neutrino–nuclear coherent scattering in combination with cryogenic methods,[1,a] this was one of the ideas that naturally came up. It seemed hopelessly utopian then, and perhaps it still is. There has been, however, a lot of work on low-temperature detectors since then,[2] Supernova 1987a has happened and the twenty-first century has drawn closer. So perhaps its not totally out of place to give the subject another look. We shall find that many of the technical tasks are on the ragged edge of feasibility or thinkability; the major problem appears to be that old nuisance, background.

MOTIVATION

The main motivation for extragalactic detection is to be able to look at more galaxies and thus see supernovas in neutrinos more regularly. We might guess the rates to be expected from the Milky Way, a starburst group of galaxies at 4 Mpc, and the Virgo cluster at 10 Mpc as follows, where 1 ~ few:

More Solar Neutrinos:

10 kpc	4 Mpc	10 Mpc
1/100 yrs	1/1 yr	1/wk

We see that if we could get out to some megaparsecs, supernova neutrinos could be a rather regular affair instead of a very occasional fling.

[a] Nuclear coherent scattering in the electroweak standard model was first calculated by D. Z. Freedman, Phys. Rev., Vol. D9, 1974, p. 1389.

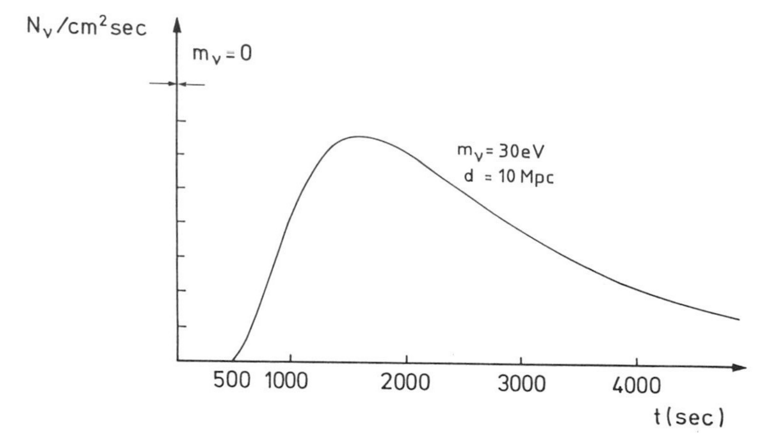

FIGURE 1. Time distribution of neutrino pulses arriving from 10 Mpc, with a first mass state with approximately zero mass and a second with $m^2 = 10^3$ eV^2. The presumed third pulse is not shown. For details see Reinartz and Stodolsky.[3]

Longer Flight Time

Also, the great distances involved means that time-of-flight effects[3] due to possible nonzero neutrino masses will be much magnified compared to Milky Way supernovas. The time delay between two neutrino states whose mass difference squared is given by δm^2, and coming from a distance d ($c = 1$ units), is

$$\delta t = d\delta m^2/2E^2,$$

or for $\delta m^2 = (10 \text{ eV})^2$ and an energy of 20 MeV (remember the muon neutrinos come from deeper in the star and have a higher energy),

$$\approx 12 \text{ msec/kpc}.$$

FIGURE 1 shows what the delay effect looks like for a 30-eV neutrino,[1,a] that is, $\delta m^2 = 10^3$ eV^2, arriving from $d = 10$ Mpc. Time zero is taken at the arrival of the first neutrino type, the second one arrives about 20 minutes later, the third is not shown. Note from the formula that the rule is to scale with mass squared to get delay.

Neutrino Mixing

An interesting point here is the role of neutrino mixing. The three pulses do not necessarily correspond to e-mu-tau types. As opposed to reactor or accelerator experiments where mixing leads to oscillation effects, here the great distances mean that the mass eigenstates separate spatially[3] (into the analogues of the K_l and K_s of the neutral K-meson system). A supernova probably presents the only occassion we will ever have to see this separation into mass eigenstates. A further amusing point is that there should be "echoes." A given neutrino flavor type is contained, in general, in all three mass eigenstates, and thus in all three pulses it should reappear with the same properties—although not in the same amounts—three times. Unfortunately, neither the detector type I shall discuss here nor any other I know of for this purpose, such as Dave Cline's "magic mountain," can distinguish flavor. For more on echoes and similar points such as CPT or CP tests for ν's from supernovas, see Reinartz and Stodolsky.[3]

Spreading

Another aspect of time-of-flight effects for nonzero masses is that the pulse spreads out; this was, in fact, Zatsepin's[4] original point in starting this subject. This spreading, which is very evident in FIGURE 1, is due to the dipsersion of energies and therefore of velocities of the neutrinos when they are produced. For detection purposes this otherwise interesting effect might turn out to be too much of a good thing. If the pulse is too spread out, it may sink into the background and get lost. This is a very real problem and should be kept in mind, especially when relatively substantial neutrino masses are considered.

COHERENT SCATTERING AND RATES

With low temperature we should be able to see coherent scattering on the nucleus. This process has the largest known cross section for neutrinos in this energy range, for example, with $E_\nu \approx 10$ MeV and Pb as the target material:

$$\sigma \approx 10^{-38}\ \mathrm{cm}^2.$$

This is a huge cross section for neutrinos, one or two orders of magnitude bigger than for any other process, a point that has stimulated some activity.

The nuclear coherent scattering reaction is simply

$$\nu + A \rightarrow \nu + A,$$

and is described by the standard model formula[1,a]

$$\sigma \approx \frac{G^2}{4\pi} N^2 E^2,$$

where N = neutron number of the nucleus. The process has the following particularities:

- high rate, hence low mass for the detector
- all ν types are detected equal

and on the minus side:

- no directionality
- no distinction between neutrino types
- there is only the small nuclear recoil energy, here keV's, to look at for detection purposes; this is the main difficulty

This is where the low temperature comes in; it should make it be possible to detect such small energies in massive detectors. For the principles and ramifications of the low-temperature developments, see references 1 and 2.

Taking the preceding cross section, a neutrino pulse as would be expected from the center of our galaxy, $10^{12}/\text{cm}^2$, and a ton as $10^{27} - 10^{28}$ atoms, we have as a naive estimate

$$\begin{aligned}\text{Scatters} &= (\text{No. } \nu/\text{cm}^2)(\text{No. atoms})(\text{cross section})\\ &\approx (10^{12}/\text{cm}^2)(10^{27} - 10^{28})(10^{-38} - 10^{-37}\ \text{cm}^2)\\ &\approx (100 - 10)/\text{ton}.\end{aligned}$$

To go a bit beyond this very crude estimate, we note that there will naturally be various losses due to small recoils, incomplete thermalization, and so forth. TABLE 1 shows some estimates[1] where we attempted to take some of these effects into account

TABLE 1. Some Sample Configurations of a Superconducting Grain Detector*

Supernova Burst					
T (K)	Material	R (μm)	E_{Th} (KeV)	Rate 1 per 100 kg	Rate 2 per 100 kg
0.4	Sn	7.6	5	2.3	1.8
0.4	Ge (La)	7.9	3	2.1	1.4
0.4	Pb	8.0	10	1.9	0.9
1.0	Pb	5.1	10	1.9	0.9
1.0	Ge (La)	7.4	10	0.5	0.3

*R is the radius of the grain, E_{Th} the energy deposit to flip the grain, assuming a 10-mK temperature jump is necessary. In Rate 1 it is assumed that all the deposited energy goes into heating, while in Rate 2 it is taken that 30 percent is lost.

for various configurations of a detector based on superconducting grains.[5] In view of these numbers, let us take 1/100 kg for the number of scatters per unit mass for a 10-kpc supernova.

If we now define 100 scatters as a good event, we arrive at the following table for the mass of detector necessary to get a good event at our different distances:

10 kpc	4 Mpc	10 Mpc
10 tons	1.6 MT	10 MT

These masses are wild, especially for a technology that has not been proved yet, but perhaps not totally unthinkable. A mere megaton of Sn, after all, would fit in your local cinema, and maybe also improve the quality of the showings.

LOW TEMPERATURE

Low-temperature work is currently being actively pursued, particularly for dark matter detection, as well as for neutrinos. In this context the following technological points particular to the present problem might be noted:

- With the kiloelectronvolts resulting from 10-MeV neutrinos we have a relatively high recoil energy, on the order of that sought for in the currently planned dark matter detectors, and much more than we would have from solar or reactor neutrinos
- Therefore the superconducting grains, bolometers, or other low-temperature devices do not have to be *very* perfect
- Also this means that some of the potential devices, such as the grains, can work at He^4 ($\sim$K) temperatures, and that fancy refridgerators are not necessary. This is nothing to sneeze at when we are talking of megatons.

Cooling Down

One of the objections that immediately comes up is that just the amount of liquid He and cooling power involved seems overwhelming. This does not look as bad as might appear at first glance. Assume that with liquid nitrogen, of which there is plenty, we have gotten ourselves down to the liquid He regime. Then we still have a quantity of heat on the order of

$$\sim 1 \text{ J/Mole} \sim 10^4 \text{ J/T}$$

to get rid of with liquid helium.

Let us now take the cooling plant of the HERA accelerator at DESY. It is described (DESY Journal 3-90) as having a cooling plant performance of 14 kW at 4.5 K. Stating the previously given requirement in these units, we get:

$$10^4 \text{ J/T} = 10^4 \frac{\text{Watt-sec}}{\text{ton}} = \frac{1 \text{ HERA-sec}}{\text{ton}}$$

or

$$1 \text{ megaton} \rightarrow 10^6 \text{ sec} \sim \text{few months.}$$

So very crudely, the cooling power seems within range, and of course the SSC will be cooling even an order of magnitude more. We now turn to the bad news.

BACKGROUND

We defined a good event as $\equiv$ 100 scatters in 1 sec. With such large amounts of mass, just random activity in our detector could frequently produce such "good events." Now, of course, the coherent scattering on a nuclei produces a different kind of signal than most backgrounds, which are ionizing, and perhaps the supernova pulse has certain distinguishing characteristics we can use. However, for a first look at background let us pose the question this way: What $\bar{n}$ (average background rate) can we tolerate so that fluctuations by Poisson statistics produce only one fake supernova a year? In other words, the chance of a 100 count fluctuation of the background above its average level in one second should be 10^{-7} or less. Since this is about 7σ, we want a background level, $\bar{n}$, of about 200/sec or less in our detector.

Naturally there are dangers in assuming naive Poisson statistics, since some rare process producing a lot of counts all at once may be less rare than independent fluctuations of small multiplicity processes; however, this kind of argument will at least give us some kind of orientation.

We then find the following requirements for the extragalactic cases:

4 Mpc	12 Mpc
1.6 MT 10^{-2}/kg-day	10 MT 10^{-3}/kg-day

For the Milky Way detector we won't be getting one supernova a year, so we have to be more stringent, say by a factor of 10:

10 kpc
10 T 100/kg-day

where we give the tolerable $\bar{n}$ in counts/kg-day.

For the Milky Way detector the purity requirements look "easy," at least by the elevated standards of present-day low counting techniques. On the other hand, for our main interest, the extragalactic detectors, the numbers resemble those just being achieved for small, highly sophisticated state-of-the-art experiments of the double beta decay type.

We can also use this kind of argument to see how dangerous the "spreading" discussed previously can be: Let the pulse last an hour instead of a second. Then with 10^4 hours in a year, the probability of the fluctuation should be 10^{-4}. This is about 4σ, so $\bar{n}$ must be less than 625/hour instead of 200/sec, implying considerably more stringent requirements.

In placing our emphasis on the detector material we take it there is no problem with background coming from outside. Indeed, one of the advantages of anything this massive is that except for neutrons, which will have to be moderated and absorbed, and cosmic ray muons, which should be identifiable, is that everything will be stopped near the surface. The prospect of having to produce megatons of such high purity material suggests that if our fantasy is to come true, then the twenty-first century will have to learn new material preparation techniques. While this does not seem impossible, background looks like our major obstacle.

Optical Confirmation

In connection with background it is interesting to consider the role of optical confirmation. That is to say, if we think we have seen a supernova in neutrinos, and then only accept it as such if it is also seen by an optical telescope, we will get some help on our purity requirements. Now we can say there should be no fake supernova within an hour instead of within a year, as previously assumed. This can ease our purity requirements by perhaps a factor of 10. On the other hand, by insisting on optical confirmation we would lose some of the most interesting aspects of the subject, such as catching supernovas obscured optically by dust, or even—another fantasy—"black supernovas," those that collapse and give neutrinos but no optical outburst. This seems to happen often on the computer in supernova simulations, so maybe it has a counterpart in reality. The most plausible strategy will probably be to rely on optical confirmation at first, then having learned what a good signal looks like, to become independent of it.

Dark and Other Matters

Finally, there is another form of background that could kill us, and this would be a lovely way to die—namely, the dark matter. By the time we get well into the twenty-first century we should know if particle dark matter is to be discovered. If it exists even with the low rates of around 1/kg-day, as currently discussed, this is substantial at the level of our background figures. Furthermore, dark matter particles with mass in the gigaelectronvolt range or above are expected to yield recoil energies in the kilovolt range, like that from our supernova neutrinos. But perhaps we will understand the dark matter so well that we can separate it. One way out might be that the dark matter particle couples only to the spin of the nucleus. In that case, it could be avoided be using a spin-zero nucleus for the detector.

In addition to dark matter, solar and terrestial neutrinos are also present, of course. These lead to a much lower recoil energy (the recoil energy goes as the square of the incident particle momentum) and can in principle be distinguished. The best and most beautiful, naturally, would be for our cool megaton observatory to be sensitive to small (few electronvolt) recoils, too, with on-line energy resolution. We then have a combined solar and supernova observatory. However, I'm not sure I would know how to do this at the moment, even in principle.

CONCLUSIONS

In conclusion, our salient points for the future would be:

- To push on with the low-temperature developments currently underway, to see where we get to
- In this development not to lose sight of the relatively low-tech path of using small detection elements, like grains, that can operate at ordinary liquid helium temperatures

- There is a challenge to technology: the mass production of high purity materials
- Last, but not least, to understand the dark matter and solar neutrinos so they can be put in their place as background.

ACKNOWLEDGMENT

I am grateful to Peter F. Smith for discussions and suggestions on these matters.

REFERENCES

1. DRUKIER, A. K. & L. STODOLSKY. 1984. Phys. Rev. **D30:** 2295.
2. VON FEILITZSCH, F. 1992. Recent results on cryogenic detector developments. This issue; PRETZL, K., N. SCHMITZ & L. STODOLSKY, Eds. 1987. Proceedings of the Workshop on Low Temperature Detectors for Neutrinos and Dark Matter. Springer-Verlag. New York/Berlin; GONZALEZ-MESTRES, L. & D. PERRET-GALLIX, Eds. 1989. Proceedings of the Workshop on Low Temperature Detectors for Neutrinos and Dark Matter II. Editions Frontieres. Gif-sur-Yvette, France; BROGIATO, L., D. V. CAMIN & E. FIORINI, Eds. 1990. Proceedings of the Workshop on Low Temperature Detectors for Neutrinos and Dark Matter III. Editions Frontieres. Gif-sur-Yvette, France.
3. REINARTZ, P. & L. STODOLSKY. 1985. Z. Phys. **C27:** 507.
4. ZATSEPIN, G. T. 1968. Zh. TEF Pis. Red. **8:** 333.
5. For some recent work on grains, through which other references may be traced, see FRANK, M., A. SINGSAAS, L. STODOLSKY & S. COOPER. 1991. Phys. Rev. **B43:** 5321.

Extragalactic Supernova Detector, Neutrino Mass, and the Supernova Watch

DAVID B. CLINE

Departments of Physics and Astronomy
University of California at Los Angeles
405 Hilgard Avenue
Los Angeles, California 90024

THE SUPERNOVA WATCH

The detection of neutrinos from SN 1987A by the Kamiokande II, IMB, and other detectors demonstrated that the final stages of a type II supernova (SN) was more or less as had been expected by theory and modeling studies. However, there were many questions left open about the details of the collapse and the properties of the neutrinos that are emitted. In the future the detection of the neutrino burst by several terrestrial detectors would have a profound effect on may diverse fields of physics, such as elementary particles (neutrino mass) and nucleon physics (equation of state). Recently, a workshop was held at UCLA to discuss the SuperNova Watch and to discuss a possible real time network that could link the different detectors. In TABLE 1 we list some of the supernova detectors operating, being constructed, or being planned around the world. This is an impressive array of detectors. The basic idea of a supernova watch network is shown schematically in FIGURE 1. A real-time supernova watch discussion group has been formed following the UCLA meeting, and progress in the development of this concept is expected during the next year or so.

SUPERNOVA NEUTRINO BURSTS

FIGURE 2 shows the expected luminosity function and mean energy of the ν_e, $\bar{\nu}_e$, $\overset{(-)}{\nu}_\mu$, $\overset{(-)}{\nu}_\tau$ neutrinos from the collapse. The important times are:[a]

(i)	Prompt ν_e burst	~(3–6) ms
(ii)	Rise time and time internal to acretion pulse	~(100–200) ms
(iii)	Width of the acretion pulse	~400 ms
(iv)	Explosion starts	~(300–600) ms
(v)	Start of neutrino cooling	~(300–600) ms
(vi)	Full width of cooldown	~(10–20) sec

These characteristic times of the supernova neutrino emission process are set by scale of the supernova collapse dynamics.[1] In turn the detailed study of these

[a] I have been coached in this subject by Jim Wilson.

TABLE 1. Supernova Detectors in Operation, Construcion, or Design

For Supernova Watch	Mass	Rate for GSN
IMB (UCIrvine–Michigan–Brookhaven) (H_2O)	6800	1300
Kamiokande (Kamioka Neutrino Detector Experiment) (H_2O)	2140	500
Baksan (Scintillator)	200	50
SNO (Sudbury Neutrino Observatory) ($H_2O + D_2O$)	1600(H_{20}) 1000 (CO_{20})	1200
LVD (Large Volume Detector) (Scintillator) (Gran Sasso)	1800	6000
LSD (Large Scintillation Detector) (Scintillation) (Mont Blanc)	~100	30–50
MACRO (Monopole, Astrophysics, and Cosmic Ray Observatory) (Scintillator)	1000	350
Caltech Scintillator (Scintillator)	1000	400
LSND (Liquid Scintillating Neutrino Detector) (CH_4)	200	90
Homestake (^{37}Cl, ^{137}I)	1000	420
ICARUS (Imaging of Cosmic And Rare Underground) (^{137}I)	1000–3000	60–180
Baksan (^{37}Cl) Signals (^{40}Ar)	3000	130
SNBO (Supernova Neutrino Burst Observatory) ($CaCO_2$)	100,000	10,000
JULIA (Joint Underwater Laboratory and Institute for Astrophysics) (H_2O)	40,000	10,000
Low-temperature Detectors (?) (?)	~4	~150
$\geq$ 15 Operating/Planned SN Detectors in World		

characteristic time structures in a future supernova event would provide real detailed information about the dynamics of the supernova process. In addition, these time structures are important if we are to use the supernova neutrinos' flight time to determine supernova mass (see the next section for more details).

These times also set the scale of the techniques that may be used to detect a finite neutrino mass by the time-of-flight method. In FIGURE 3 we show the difference

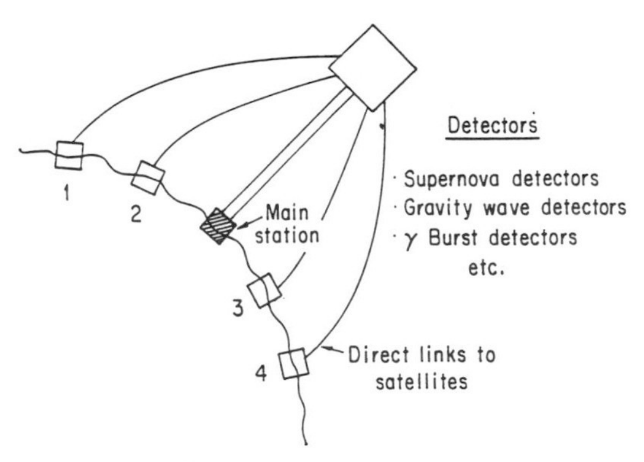

FIGURE 1. A schematic of a real-time supernova watch network connecting several detectors. Purpose of network: (1) To provide a real-time detection of a neutrino burst; possible to provide real-time information to various detectors; (2) to detect very faint signals that are not observable in a single detector (or between gravity wave and SN, etc.).

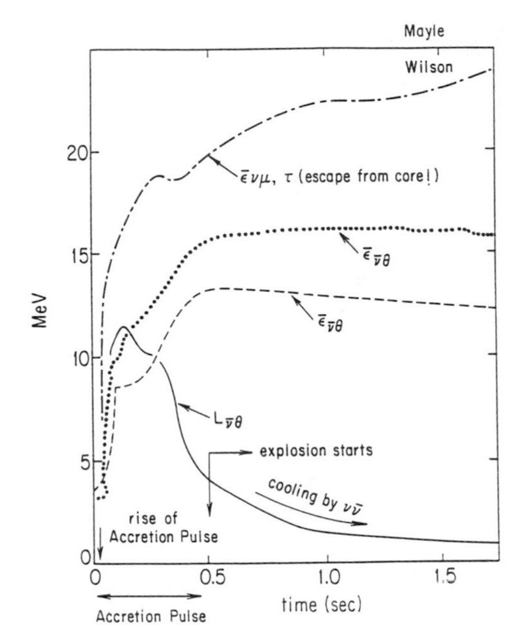

FIGURE 2. The neutrino luminosity function and average energies of the various neutrinos vs. time.

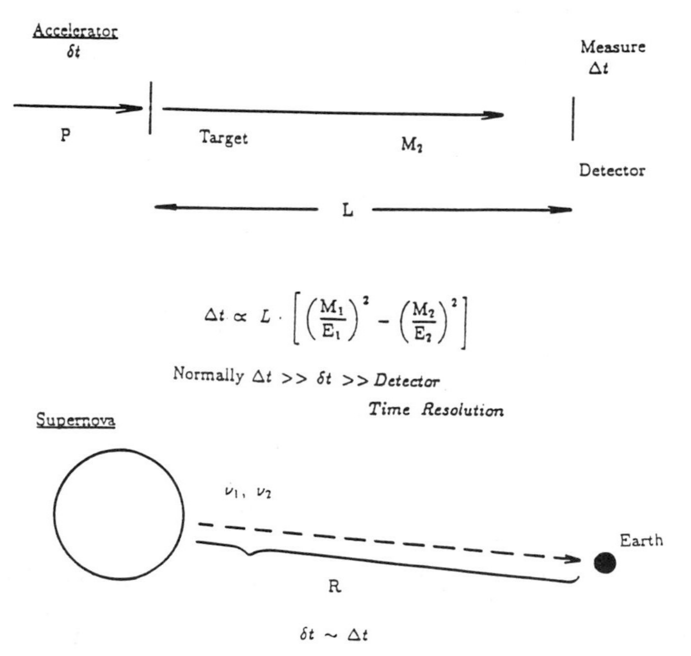

FIGURE 3. Schematic of the technique to measure neutrino mass using time of flight from a supernova explosion compared to a similar accelerator experiment.

between a "time-of-flight" mass measurement at an accelerator and using a supernova.

DETERMINATION OF NEUTRINO MASS OF COSMOLOGICAL SIGNIFICANCE

One or more stable neutrinos, with a mass in the region of 30 eV, could supply the missing or dark matter. The mass relationship is $\Omega_\nu = 1$ for $M_{\nu_x} = 92h^2$ eV, where H is the Hubble constant ($h = 1$ for $H = 100$ km/s^{-1}Mpc^{-1}; for $h = 1/2$, $M_{\nu_x} = 23$ eV to give closure of the universe, and for $h = 0.6$, $M_{\nu_x} = 33$ eV). We thus consider the neutrino mass range of 10–40 eV to be of cosmological significance. There are no known laboratory techniques to uniquely detect such a mass directly. It is possible that some form of neutrino oscillation experiment could be used to infer a mass in this range; however, this will depend on the uncertain level of neutrino mixing. The only technique that is known to provide a unique mass measurement is to use the difference in flight times for neutrinos from a distant supernova that goes as

$$\Delta t = 51.4\, R_{\text{Mpc}} \left[\left(\frac{M_{\nu_x}}{E_{\nu_x}} \right)^2 - \left(\frac{M_{\nu_e}}{E_{\nu_e}} \right)^2 \right] \text{sec},$$

where M_{ν_x} and M_{ν_e} are measured in electronvolts, E_{ν_x}, E_{ν_e} are measured in megaelectronvolts, and R (the distance to the supernova) is measured in megaparsecs. The μ and τ neutrinos are expected to have higher average energies since they escape from deep inside the supernova core.[1]

We assume an instantaneous source of ν_x neutrinos with a distribution of the form[2]

$$E_{\nu_x}^2 e^{-E_{\nu_x}/T}$$

and an assumed detection efficiency that scales as $E_{\nu_x}^3$ (i.e., the cross-section scales like $E_{\nu_x}^2$ and detection of secondary products like E_{ν_x}) gives

$$\delta t = 51.4 R_{\text{Mpc}} \frac{\int E_{\nu_x}^5 (M_{\nu_x}/E_{\nu_x})^2 e^{-E_{\nu_x}/T}\, dE_{\nu_x}}{\int E_{\nu_x}^2 e^{E_{\nu_x}/T}\, dE_\nu},$$

and for the case of μ, τ neutrinos we expect[2-5] $T \simeq 25/3$, giving

$$\delta t = 0.037 M_{\nu_x}^2 R_{\text{Mpc}} \text{ sec}.$$

For a galactic supernova $R_{\text{Mpc}} = 0.01$ Mpc and for $M_{\nu_x} = 30$ eV, we find

$$\delta t = 330 \text{ ms}.$$

Note that the mean time separation and shape of the time distribution are altered in a characteristic manner by the different neutrino masses. It is this characteristic that must be used to extract a cosmologically significant mass from a galactic supernova. The detailed time distribution for a galactic supernova is shown in FIGURE 4. TABLE 2 lists the reactions that can be used for supernova burst detection

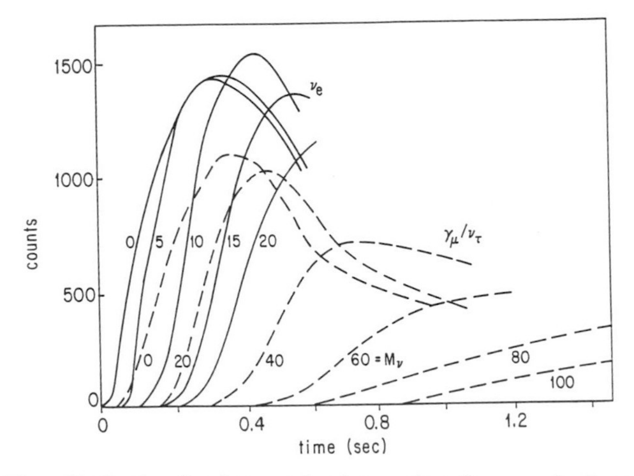

FIGURE 4. Time distribution for the neutrino interactions from a galactic supernova as a function of the neutrino mass. The *solid curves* refer to ν_e, and the *dashed curves* refer to ν_μ/ν_τ.

and the experimental techniques that can be employed to detect them. Note from TABLE 2 that no detector will provide information on all the possible channels. In this paper we are mainly concerned with the prospects for extracting a neutrino mass of the μ and τ neutrino if the mass value is in the 10–40 eV mass range of cosmological

TABLE 2. Supernova Neutrino Detection in the 1990s

Reactions	$\bar{\nu}_e p \rightarrow e^+ n$	$\nu_x e \rightarrow \nu_x e$	$\nu_x N \rightarrow \nu_x N$	$\nu_x N \rightarrow \nu_x N^*$ $\hookrightarrow n$
Parameters				
Cross section	Large (KII, SK, IMB, LVD)	Small $\sim E_\nu^2$ (ICARUS)	Large For coherent process	Large At high E_{ν_x} SNO/SNBO
Neutrino Energy Estimate	Yes $\sim E_e$	Partial $E_{\nu_e} \sim f(E_e)$	No	No But a threshold may set E_{ν_x}
ν direction	No	Yes	No	No
Time information	Yes	Yes	Yes	Yes
Down time (guess)	$\gtrsim 10\%$	$\sim 30\%$	?	(Could be small)
Maximum Detector Size	2×10^5 T (H_2O) LENA $\lesssim 10^4$ tons liquid scintillator (LVD)	$\sim 2 \times 10^5$ T (H_2O) or $\sim 10^3$ T Cryogenic (ICARUS)	? Kilograms No detector proposed so far	$\sim 10^5 - 10^7$ T of $CaCO_3$ (SNBO) or $\sim 10^3$ T D_2O(SNO)
Backgrounds	Small If e^+ and n capture detected –OK for H_2O galactic signal	Small If directionality used to reject background	?	Depends on radioactivity of material

TABLE 3. Comparison of Future Supernova ν Detectors

Process	$\bar{\nu}_e p \to e^+ n$ $\bar{\nu}_e d \to ppe^-$	$\bar{\nu}_x e \to \bar{\nu}_x e$ $\bar{\nu}_e e \to \bar{\nu}_e e$	$x = \mu, \tau$	$\nu_e N \to N^* \nu_e$ $\hookrightarrow n$	$\nu_\mu N \to N^* \nu_x$ $\hookrightarrow n$	ν_e prompt
Detectors						
ICARUS (3 kT)	—	~140	25	—	—	4*
SNO (1 kT)	~500	60	20	~200	~400	5*
($D_2O + H_2O$)	(H_2O shield + D_2O)			—	—	5–20*
LVD/MACRO (3 kT) scent	~1000	—	—	—	—	—
Kamiokande II/IMB	(~480)	(~60)	(~20)	—	—	—
SUPER Kamiokande (30 kT) H_2O	~4000	~600	200	—	—	~5*
SNBO (100 kT)	(100s)?			~100s	10,000	—
Comments	Measure t_ν $E_\nu \sim E_e$ No direction	t_ν E_ν estimated from E_e Θ_ν measured		t_ν only No E_ν! No Θ_ν		$\Delta t \simeq$ 10 ms

*Depends on energy spectrum of prompt ν_e and detector threshold.

significance. Note also that the mean time distribution for these cases are

$$\delta t = 35 \text{ ms} \qquad M_{\nu_x} = 10 \text{ eV}$$

$$\delta t = 590 \text{ ms} \qquad M_{\nu_x} = 40 \text{ eV}.$$

In a sense the galaxy is simply too small to obtain large time differences for a cosmologically interesting neutrino mass. From FIGURE 1 it is clear that the shape of the time pulse changes with M_{ν_x} and that the mean width of the initial pulse is

$$\delta t \simeq 400 \text{ ms} \quad \text{for} \quad M_{\nu_x} \simeq 10 \text{ eV}$$

$$\delta t \simeq 1 \text{ sec} \quad \text{for} \quad M_{\nu_x} \simeq 40 \text{ eV}$$

and thus

$$(\delta t)_{\text{pulsewidth}} > (\delta t)_{\text{time difference }(-)\text{ between } \overset{(-)}{\nu_x} \text{ and } \overset{(-)}{\nu_x} \text{ neutrino arrival}}.$$

It is clear that a very large number of events and very good time resolution is required to resolve the effects in the lower neutrino mass range due to the shape of the time pulse near the origin. We propose to use the derivative of the pulse to give a zero crossing estimate of the arrival times to obtain the required accuracy.

We now consider the expected event ratio for various channels for planned or proposed detectors for supernova detection in the 1990s and beyond. TABLE 3 lists the approximate event rates for several detectors in the construction or planning stage (ICARUS, SNO, LVD, MACRO), and for two newly proposed detectors:[2]

1. Super Kamiokande
2. SNBO (Supernova Neutrino Burst Observatory) as well as the existing IMB and Kamiokande II detectors (other detectors are likely too small to give additional information).

We refer to reference 3 for discussions of the proposed Super Kamiokande detector. The SNBO detector would have the active mass of 100,000 tons of $CaCO_3$, and would be instrumented with a large number of neutron detectors. The detector concept has been described in references 2 and 4. This detector is mainly sensitive to ν_μ and ν_τ neutrinos, due to the dynamics of the neutral current process that strongly discriminates against lower energy ν_e and $\bar{\nu}_e$ events; thus, it is a ν_μ, ν_τ detector. Therefore, this zero crossing technique could be applied to obtain the desired time resolution.

A SUPERNOVA NEUTRINO BURST OBSERVATORY

A new concept in supernova neutrino burst detection has recently been proposed. The detector uses high-energy neutrinos from stellar collapse to drive neutral-current inelastic-scattering excitation of nuclei in the detector. The excited nuclei can decay by emitting neutrons. These neutrons would be detected by inexpensive counters. There are four important points regarding this proposed detector.

1. Large, relatively pure deposits of SNBO detector material exist in nature in well-shielded sites. Preparation of the detector material should therefore be minimal. In addition, detectors are relatively simple, inexpensive BF_3 neutron counters. The technology of these detectors has been available since the 1950s and is well understood. The SNBO potentially is an inexpensive, easily prepared, and easily maintained neutrino-burst detector that would run for decades with little maintenance. This is important because the stellar collapse rate in our galaxy is only one event every ten to one hundred years, so that neutrino detectors will probably have to run for several decades before the next galactic supernova is seen.

2. Because the basic design of the SNBO is to place neutron counters into holes drilled into the detector medium, the mass of the SNBO is easily increased by drilling new holes and adding new counters. The advantage of the SNBO over water detectors is clear. The feasibility of the detector may be tested by placing a neutron counter into a chunk of detector medium that could then be transported to a neutrino source (such as the Los Alamos Meson Physics Facility) and tested. Should the detector indeed prove feasible, a small-scale SNBO could then be started with only a few neutron counters in place. As experience with the detector grows, more holes could be drilled and more counters added. This is the only proposed cosmic neutrino detector that could be scaled up to very large size. The ultimate mass of the SNBO could reach 10^5 tons or more, which would give tens of thousands of counts for a galactic supernova. This large number of counts could yield detailed information on the supernova mechanism as well as provide a good limit on the μ and τ neutrino masses.

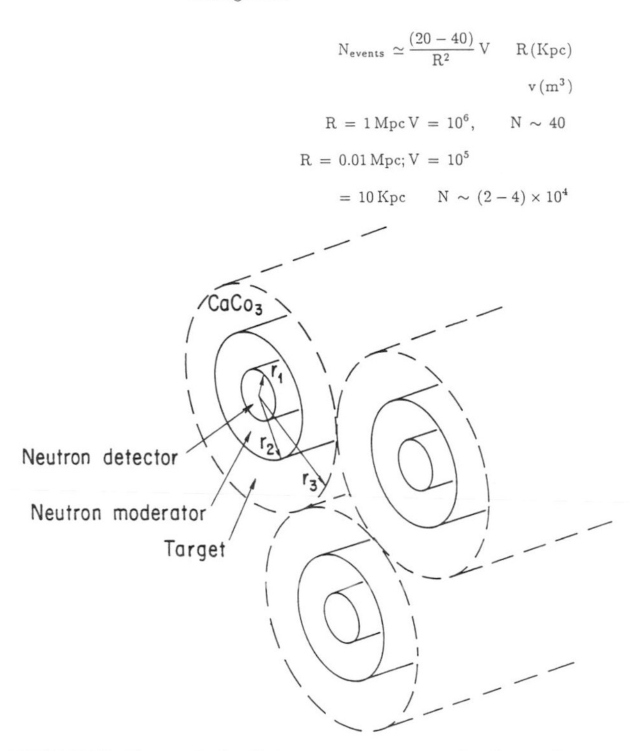

FIGURE 5. Concept of a distant supernova neutrino burst detector.

3. Because the detector medium produces neutrons almost exclusively by neutral current reactions, the SNBO would be able to count neutrinos of all three flavors coming from a supernova. In addition, the threshold for the detector, which is set by the neutron separation energy of the nuclei of the detector medium, is high ($\gtrsim 12$ MeV) so that only high-energy neutrinos would be counted. In the case that there are no oscillations between neutrino flavors, the average energy of electron neutrinos from a supernova is roughly 15 MeV, while that of the μ and τ neutrinos are more like 20–25 MeV. As a consequence, the SNBO would essentially count only μ and τ neutrinos.

Neutrinos with a mass in the range 10–50 eV are again becoming attractive as candidates for the dark matter of the universe. The upper limit on the mass of the electron neutrino is currently 18 eV. Improvement in ^{3}H endpoint experiments could reduce this upper limit below the cosmologically interesting range. Any Mikheyev–Smirnov–Wolfenstein (MSW) resonant neutrino oscillation explanation for the solar neutrino puzzle would suggest that, at best, $\Delta m^2 \approx 10^{-6}$–$10^{-4}$ eV^2 so that m_{ν_e} would be very small, but conceivably $m_{\nu_\mu} \simeq 10^{-2}$ eV and "see-saw" schemes for neutrino masses would then suggest $m_{\nu_\tau} \sim 10$–40 eV. This would leave the τ neutrino as the

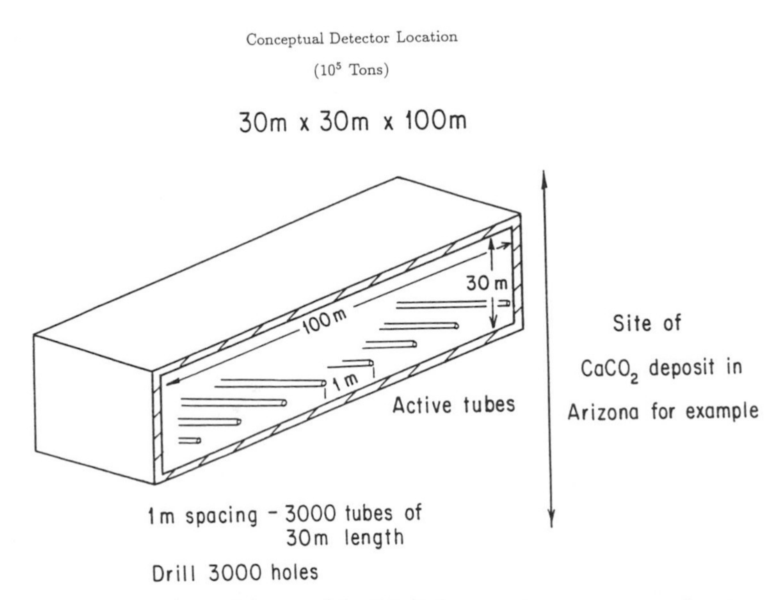

FIGURE 6. A schematic layout of the SNBO detector that uses neutron detectors.

potential closure neutrino. The only way to measure the mass of μ and τ neutrinos in this mass range is, at present, by time-of-flight measurements from supernovae, a task for which the SNBO is ideally suited because of its ability to operate for a long time, its large number of counts, and its particular sensitivity to μ and τ neutrinos.

4. If vacuum mixing angles and mass difference between ν_e and ν_μ or ν_τ are as expected in any of the MSW oscillation solutions to the solar neutrino problem, then neutrino flavors might be mixed on their way down the density gradient in the supernova. This would complicate the clear identification of high-energy neutrinos as corresponding to ν_μ and ν_τ. We point out, however, that in a mass hierarchy in which $m_{\nu_\tau} > m_{\nu_\mu} > m_{\nu_e}$ the antineutrinos have no resonant transitions, so that a $\overline{\nu}_e$ signal in a large-volume detector (LVD) or a water detector serves to give a "fiducial mark" to calibrate the ν_e ($\overline{\nu}_e$) flux. This, coupled with a detailed time history of events from SNBO, could give interesting constraints on neutrino masses and mixing angles in a region of parameter space potentially very different from solar neutrino experiments. We note that since our experiment involves neutral currents, we can still identify high-energy peaks from $\overline{\nu}_\mu$ and $\overline{\nu}_\tau$, and place mass limits as discussed in point.

Note that the SNBO detector could also provide a higher mass limit on one of the neutrinos if the rate of events is considerably below the prediction. This would imply that one of the neutrinos has a mass of > 100 eV or is unstable. The two newly proposed detectors provide additional information that could be used to uniquely detect a ν_μ or ν_τ mass even in the 10-eV range. This is due to the fact that the Super Kamiokande detector gives adequate numbers of $\nu_x e \rightarrow \nu_x e$ events to possibly make the separation in the high mass case, and the SNBO detector (when combined with

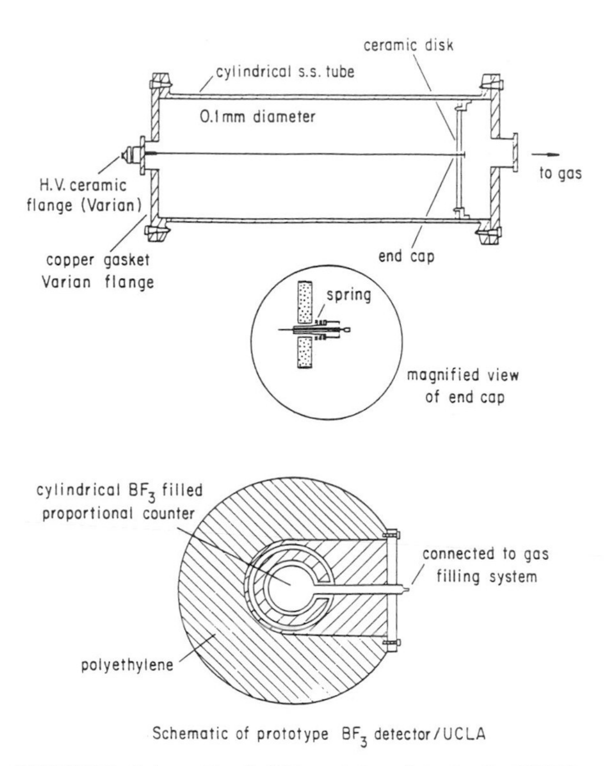

FIGURE 7. Schematic of a BF_3 prototype detector for SNBO.

the other detector results) provides a very large number of pure ν_μ and ν_τ events. With 5000 ν_μ and 5000 ν_τ events it should be possible to make a unique separation of low-mass neutrinos (i.e., $M_{\nu_\mu} \simeq 0$; $M_{\nu_\tau} \sim 30$ eV) and mixed cases such as ($M_{\nu_\mu} \simeq 10$ eV, $M_{\nu_\tau} \sim 30$ eV). (Note that the zero crossing technique is difficult to use in the case of a mixed, overlapping ν_e, ν_x sample.)

The construction and long-term operation of such detectors in a self-triggered "supernova watch" detection mode is essential in determining whether the μ and τ (and possibly ν_e) neutrinos have mass values that are important for the cosmology of the universe. At this time there appears to be no other viable proposal of techniques to carry out the important measurement. In addition to the neutrino mass determination the detection of thousands of $\bar{\nu}_e$, ν_e, and ν_μ/ν_τ events from a future supernova will provide crucial information about other properties of neutrinos (such as a magnetic moment in the range of 10^{-11}–10^{-14} μ_B) and exotic $\nu\nu$ interactions, as well as the dynamics of stellar collapse and explosion.

Detailed response calculations have been carried out at LLNL for the configuration shown in FIGURE 5.[2] The SNBO could be constructed in a limestone deposit in places like Arizona. FIGURE 6 shows a conceptual design of such a detector. Prototype neutron detectors are being developed by the collaboration (see FIG. 7).[2]

POSSIBILITY OF EXTRAGALACTIC SUPERNOVA DETECTION

The search for supernova neutrino bursts poses a serious problem: the rate of galactic supernova formation is expected to be small (~ 1/50 years), and the distance to the next large galaxy is large. At the SuperNova Watch meeting a lively discussion of the distance to a galaxy was held (see FIG. 8).[5] It was concluded that at a distance of ~ (1–2) Mpc the SN II rate would be ~ 1 per year. In order to detect such a supernova, a very massive SNBO-type detector is required (~ 10^7–10^8 tons). This might be accomplished if a technique can be found to construct such a detector for the cost of ~ 5 dollars per ton.

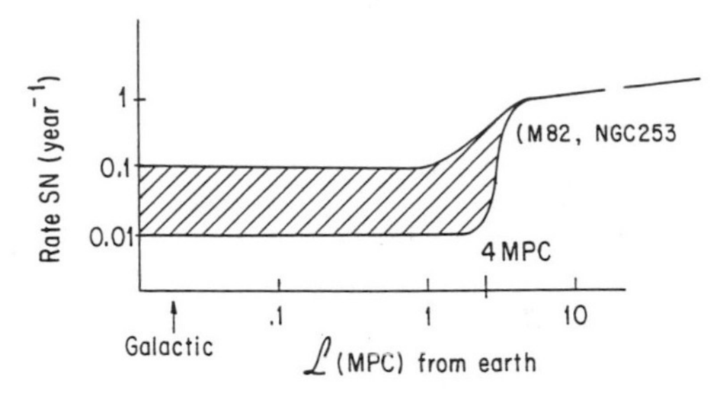

FIGURE 8. Estimated SN II rate as a function of distance from earth from the recent UCLA workshop.

TABLE 4.

	Mass	Tate (ν_x)	Distance	M_ν
H_2O	3 × 10^4 T (Super Kamio-kande)	~ 200	10 kpc	20 eV
	(10^8) T	~ 200	0.6 Mpc	2 eV
$CaCo_2$	10^7 T (SNBO)	2 × 10^4	10 kpc	20 eV
	(10^7) T	300	0.6 Mpc	2 eV
Cryogenic	10 T*	30*	10 kpc*	20 eV*
(T < 300 MK)	(10^4) T*	30*	0.6 Mpc*	2 eV*

NOTE: To measure M_ν ~ 2 eV, we must use an extragalactic supernova.

Now we can compare the prospects for detecting a finite mass neutrino from the use of galactic and extragalactic supernovas as given in TABLE 4. While the lowest mass detector is one using very low temperature sensors (such as 300 mK superconducting grams), the most easily constructed is the SNBO.[6]

ACKNOWLEDGMENT

I wish to thank George Fuller, Jim Wilson, Peter Smith, and Adam Burrows, as well as other members of the SNBO Collaboration for the help they have given me.

REFERENCES

1. WILSON, J. & R. MAYLE. 1988. Livermore Preprint (1988) and submitted for publication in Astrophys. J.
2. BOYD, R., *et al.* 1991. Proposal to study a new type of neutrino burst detector: The SuperNova Neutrino Burst Observatory (SNBO), University of California at Los Angeles.
3. COWSIK, R. & J. MCCLELLAND. 1972. Phys. Rev. Lett. **29:** 660.
4. CLINE, D., *et al.* 1989. A new method for detection of distant supernova neutrino bursts. Astro. Phys. Lett. Commun.
5. SMITH, P. F. Private communication.
6. PROCEEDINGS OF THE UCLA SUPERNOVA WATCH WORKSHOP. 1991. UCLA preprint. To be published.
7. PIRAN, T. & J. WILSON (unpublished) and CLINE, D., *et al.* 1990. Proposal to study a new type of neutrino burst detector: The supernova neutrino burst observatory (SNBO). UCLA preprint.
8. CLINE, D. 1989. Neutrino astronomy. *In* Proceedings of the 14th Texas Symposium on Relativistic Astrophysics (Dallas, Texas, 1988). Ann. N. Y. Acad. Sci. **571.**
9. BURROWS, A., M. TURNER & R. E. BRICKMAN. 1989. Phys. Rev. D.
10. The event rate estimates for the SNO detector come from DOE, P. 1988. The SNO detector. *In* Observational Neutrino Astronomy, D. Cline, Ed.: 92. World Scientific. Singapore.

Prospects for Relic Neutrino Detection

P. F. SMITH

Rutherford Appleton Laboratory
Chilton, Didcot, Oxon,
OX11 OSU, U.K.

1. INTRODUCTION

The standard big bang model predicts a universal background of relic neutrinos, comparable in number density to the background microwave photons.[1] Since the latter are observed, one can be confident that the neutrino background will also be present, and observations on this background could, in principle, provide information on neutrino masses and mixings (since all neutrino types would have been produced). In addition, if the heaviest neutrino has a mass ~ 25–50 eV, it could both explain the galactic dark matter and produce a closed universe.

This neutrino background is undetectable at the present time, firstly because the neutrino energy is very low (10^{-4}–10^{-5} eV), resulting in a very low-energy transfer to any conceivable detector, and secondly, the low energy gives a low interaction cross section, and hence a very low event rate per unit mass. These obstacles have so far precluded any realistic proposal for relic neutrino detection.[1]

However, it is also difficult to accept that these neutrinos will never be detected, and it is therefore of interest to ask what technical developments would be needed to make this feasible, and whether any realistic experimental possibilities can be foreseen for the future. The aim of this paper is to illustrate the difficulties by summarizing six detection ideas that have been previously considered, indicating in each case the problems that have prevented the idea from being developed into an experimental proposal. The most promising direction for further study would appear to be that of coherent interactions, from which a considerably increased cross section results for scattering from bulk matter, producing small macroscopic forces. So far, no investigations of this idea have resulted in a practical detection scheme, but in this paper one new variation is suggested that could in principle give an observable effect, if the necessary stringent experimental conditions could be created. It is suggested that this may become possible with the aid of foreseeable twenty-first century developments in nanotechnology.

2. BASIC NEUTRINO NUMBERS AND ENERGIES

For the case of zero neutrino mass, the number density ρ_ν of relic neutrinos (which decouple from matter at $T \sim 1$ MeV, ~ 1 s after the big bang) can be related[2] to the number density of relic photons (~ 400 cm^{-3}) giving $\rho\nu \sim 100$ cm^{-3} $(\nu + \overline{\nu})$ for each of the three generations. The energy distribution would be spatially uniform and would have a red-shifted Fermi distribution with a present-day mean momentum $\sim 5.10^{-4}$ eV, and a flux $\sim 10^{12}$ cm^{-2}s^{-1}.

If one neutrino (for example, the tau neutrino) has a mass $m_\nu \sim$ 20–50 eV, then it could be clustered in galaxies with a density similar to that estimated for the dark matter in our own galaxy.[3] From considerations of phase-space constraints, bound velocity, and dark matter density ~ 0.4 GeV cm^{-3}, one obtains a "best fit" with $m_\nu \sim$ 25 eV, characteristic momentum $p_\nu \sim 0.02$ eV, number density $\rho_\nu \sim 2.10^7$ cm^{-3}, and flux $\sim 5.10^{14}$ cm^{-2} s^{-1}.[3-5]

For completeness a third possibility should be noted, in which the relic neutrinos have a nonzero mass, but have not clustered in galaxies. In this case, they would be accelerated into and through our galaxy by its gravitational potential giving, in the region of the sun, a reduced number density $\rho_\nu \sim 1$ cm^{-3}, increased momentum $p_\nu \sim$ 0.1 eV, and a relatively low momentum spread $\Delta p_\nu/p_\nu \sim 10^{-2}$.[6]

For studies of relic neutrino detection ideas, it is usual to assume the case of galactic clustering, giving the set of numbers in the second paragraph of this section.

3. DISCUSSION OF DETECTION IDEAS

3.1. *Single-Particle Cross Section. Electron Scattering*

The lower curves in FIGURE 1 show the momentum dependence of the neutrino cross section for elastic neutral current scattering on neutrons. They become constant at low momentum if the neutrino has a nonzero mass. Charged-current interactions are a factor of 4 higher (including the elastic scattering of an electron neutrino from an electron, which can also proceed via the W). The elastic scattering amplitude for a ν_μ or ν_τ on an electron, or any ν from a proton, is lower by the standard factor $(1 - 4\sin^2\theta_w)/2 \sim 0.04$.

Suppose we were to consider experiments based on a target of quasi–free electrons, for example, in a metallic foil. The typical recoil energy from a collision with a galactic neutrino of momentum 2.10^{-2} eV would be $\sim 10^{-9}$ eV. At first sight this appears somewhat encouraging, since stabilization and measurement of potentials <1 nV is already achievable, and it might be thought that with sufficient technological effort an ultracold electronic system might be developed that would be sensitive to single-electron recoils in the nanovolt range. The problem, however, lies in the very low expected rate that, for an incoming neutrino mass ~ 30 eV, would be ~ 1 day^{-1} ktonne^{-1}. Thus the target mass required for a reasonable event rate appears totally incompatible with any envisaged ultralow noise experimental environment.

3.2. *Bremsstrahlung from Free Electrons*

One method of detecting neutrino + electron collisions in a large target might be to look for the accompanying bremsstrahlung photons. Loeb and Starkman[7] have estimated the photon production from the interaction of relic neutrinos with conduction electrons in a multilayer metallic target (FIG. 2) and find a rate of only $\sim 10^{-3}$ day^{-1} ktonne^{-1}. If, however, the neutrino has a magnetic moment μ_ν, the rate becomes $\propto(\mu_\nu/\mu_B)^2$, where μ_B is the Bohr magneton. The proportionality constant was estimated in reference 7 as $\sim 10^8$, but a more accurate estimate of the number of

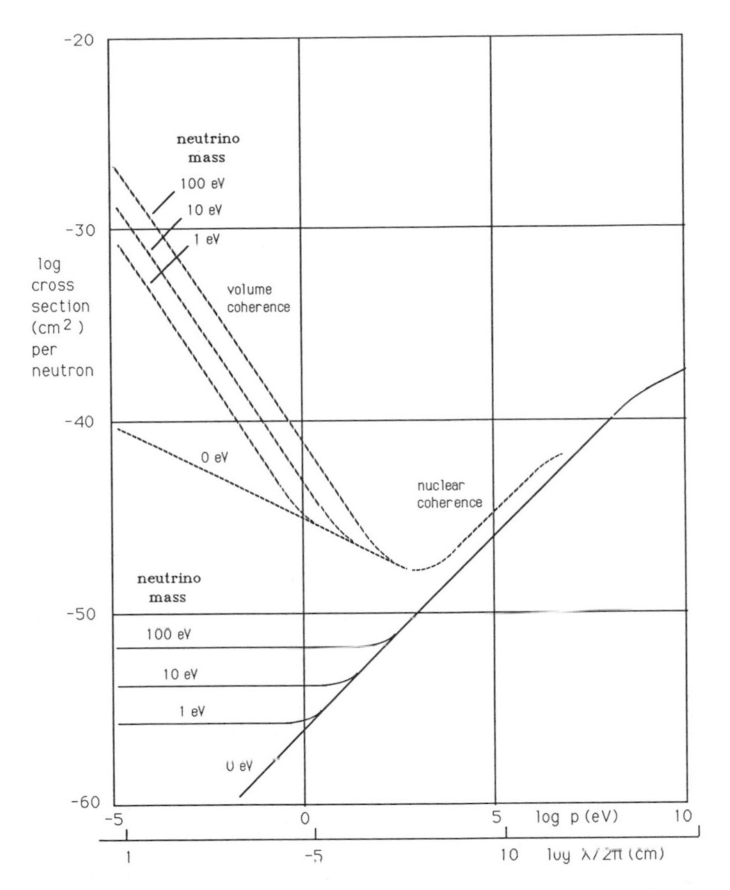

FIGURE 1. Neutrino elastic scattering cross section vs. momentum p and corresponding wavelength λ. *Full lines:* single-particle scattering. *Dashed lines:* coherent scattering.

participating free electrons has indicated[8] a much lower coefficient $\sim 10^{-2}$. However, this is a relatively minor issue compared with that of detecting the photons themselves, which have typical energies $< 10^{-5}$ eV. There is at present no method foreseen for the detection of single photons of this energy, and indeed any such device would have many other experimental applications, for example, in connection with possible

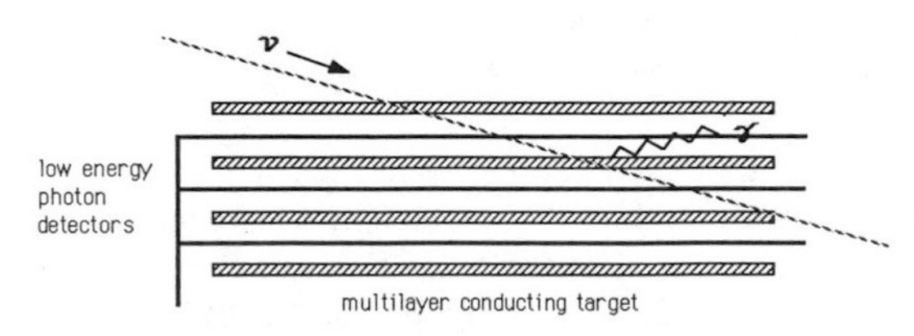

FIGURE 2. Suggested detection of bremsstrahlung photons from scattering of relic neutrinos by free electrons.

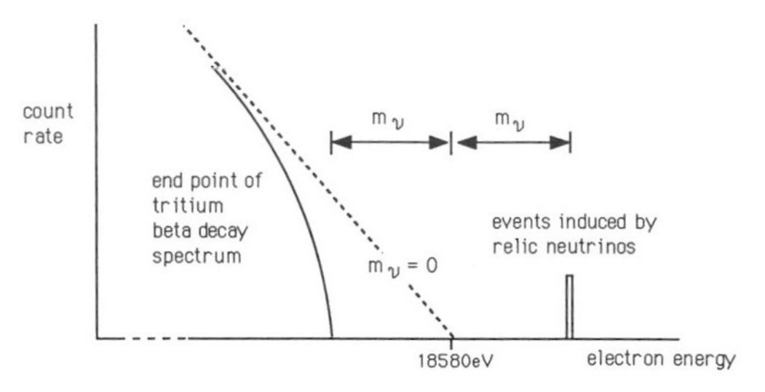

FIGURE 3. Suggested detection of galactic electron neutrinos by induced beta decay in tritium.

axion searches [see reference 1, p. 217]. Thus this does not appear to be a promising approach.

3.3. *Tritium Beta Decay Spectrum*

The tritium beta decay process, $^3H \rightarrow {}^3He + e + \nu_e$ would, in the presence of an electron–neutrino background, be supplemented by events of the form $\nu_e + {}^3H \rightarrow {}^3He + e$, detected by searching for events displaced beyond the normal endpoint of the spectrum (by a distance equal to the neutrino rest mass) as indicated in FIGURE 3. This idea was originally discussed informally by Weinberg and Turner,[9] and subsequently examined in more detail by Irvine and Humphries.[10] The cross section for this inverse reaction is unexpectedly high ($\sim 10^{-43}$ cm^2, compared with 10^{-53} cm^2 for elastic scattering), being enhanced firstly by the factor m_e/m_ν, and secondly by the high ratio of final to initial center-of-mass momentum for interaction with a decaying system. A further enhancement ~ 100 was suggested in Irvine and Humphries[10] due to Fermi momentum, but this is now believed to be incorrect. The outcome is a rate ~ 1 day^{-1} g^{-1}, at first sight encouraging, but in practice not so, since existing tritium beta decay experiments operate with only ~ 1 μg of tritium, and have considerable difficulty in obtaining a resolution ~ 10 eV near the endpoint. In addition, the fact that this experiment would rely on the electron–neutrino providing the dominant component of the galactic dark matter discourages further study of its feasibility.

3.4. *Coherent Interactions: Electrons in Superconductors*

The major hope for relic neutrino detection lies in the use of coherent scattering to increase the low-energy cross section. The principle of coherent scattering is that if a number N of scattering centers are rigidly connected, that is, the energy transfer is insufficient to excite the structure so that the group recoils as a whole, then the scattering amplitudes A must be added prior to squaring to obtain the cross section. If the size of the group is less than the wavelength of the momentum transfer, then the scattering amplitudes add in phase to give a total amplitude $\sim NA$ and a cross section $\propto (NA)^2$, compared with NA^2 for independent scatterers. Thus there is a gain N in cross section, which in the case of the long wavelengths associated with relic neutrino interactions, can be extremely large. This is shown by the upper curves in

FIGURE 1. There are two stages to the coherent increase. For momentum transfers in the range 10^4–10^6 eV there is a significant gain arising from coherence over the atomic nucleus, which could be used to decrease the target size required for the detection of solar or supernova neutrinos.[11] As the momentum decreases further the coherent volume reaches atomic dimensions, and for the relic neutrino region 10^{-2}–10^{-4} eV the coherence extends to volumes $\sim 10^{-8}$–10^{-2} cm^3, containing 10^{15}–10^{21} atoms and giving the very large cross-section gains shown. The problem of applying this experimentally lies in the fact that the cross section now refers the recoil of a large assembly of particles, so that for a given momentum transfer the energy transfer ($p^2/2m$) becomes extremely small and difficult to observe.

The first application we consider is that of improving on the single-electron scattering process considered in Section 3.1 by using the coherent interaction with electrons in a superconducting circuit, to produce a small but measurable current.[12,13] A detailed numerical analysis of this scheme[13] shows that it involves practical constraints that are at present unattainable. As an illustration of the problem, a typical requirement would be a 1-m^3 target comprising $\sim 10^6$ layers of superconductor each 10 m × 1 m, connected to a superconducting quantum interference device (SQUID) to measure the small changes in current (FIG. 4). When oriented parallel to the galactic motion (resulting in a "neutrino wind"), the coherent neutral current interaction would be equivalent to an electric field $\sim 10^{-36}$ V cm^{-1} and a total circuit voltage 10^{-27} V, giving a linear rise in current amounting to 10^{-18} A after 10 days. Thus unprecedented levels of noise and mechanical stability would be required, and to prevent spurious induced voltages from external sources it would be necessary to shield the system magnetically to changes $< 10^{-8}$ flux quanta ($\pm 10^{-19}$ G over 1 m^2) in 10 days.

Even if it is believed that these stringent requirements might be met with sufficient future technological effort, there remains a further problem of principle that has not been resolved—that a superconducting circuit is quantized, with a level spacing many orders of magnitude greater than the energy transfer from each neutrino scattering, so that it is not clear that the electrons will in fact recoil independently of the lattice material.[12] In addition, the idea suffers from the same disadvantage as that of Section 3.3—that it is applicable only to electron neutrinos, the cross section for muon or tau neutrinos being lower (see Section 3.1) by a factor of $\sim 10^{-3}$.

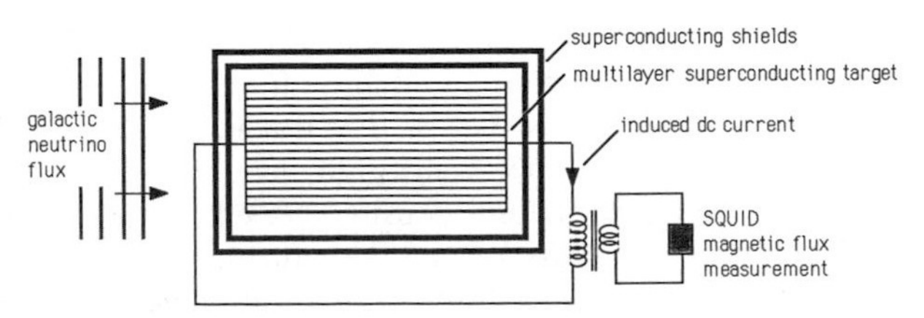

FIGURE 4. Hypothetical galactic electron–neutrino detector based on coherent momentum transfer to superconducting electrons.

3.5. *Coherent Interactions: Macroscopic First-order Forces*

The momentum transfer from a flux of low-energy neutrinos by coherent scattering from a material target is equivalent to a small force on that target, and we now consider the feasibility of measuring such forces. The easiest route to studying this is via neutrino optics, and a significant amount of work on this has been carried out in the past 10 years. There is a close analogy with the more familiar case of neutron optics: for a low-energy neutral particle passing through a region of matter the coherent interaction is equivalent to an average potential U (positive or negative) that changes the momentum p of the particle while in the material. This defines an effective refractive index $n = p_2/p_1$, which can then be used to calculate reflection and refraction effects in the medium, and any consequent forces on the material. The value of U is proportional to the weak coupling constant G_F, the number of atoms/cubic centimeter, and a factor that depends on the neutrino type and the values of A and Z for the target material.[4,15] A typical value for U in a high-density material is 2.10^{-13} eV (compared with 2.10^{-7} eV for neutrons). The refractive index n is related to U (and hence to G_F) by $n^2 = 1 - 2Um_\nu/p^2$, giving $(n - 1)$ typically $\sim 10^{-8}$ for hypothetical galactic dark matter neutrinos. This increases as p^{-2} until for that portion of the neutrino spectrum with $p < 10^{-6}$ eV total reflection would occur from a surface, as in the case of photons and sufficiently low-energy neutrons.

An immediate point of interest is that one could in principle construct optical elements (from high-density materials), for example, Fresnel lenses and Bragg mirrors, which could provide a means of modifying or modulating the neutrino flux. One cannot, however, obtain a significant overall focusing or enhancement of the flux, in view of its largely isotropic nature (but with a superimposed directional component added by the galactic motion of the solar system). The use of Bragg mirrors could provide selective reflection of a portion of the neutrino spectrum, but the total reflectivity cannot be increased by this means.[6] It is of interest that the size of neutrino optical systems would not be as great as the small value of $(n - 1)$ would suggest. This is because the linear scale of an optical system increases only as $(n - 1)^{-0.5}$,[6] indicating a typical size (e.g., focal length) $\sim$100–1000 m for neutrino optical elements. Thus the latter would be very substantial structures, but not unreasonable in comparison with the scale of other modern particle physics or astrophysics projects.

Turning now to detection ideas, it was originally suggested that, since a prism deflects an incident beam through an angle $\propto(n - 1)$, corresponding mechanical forces $\propto(n - 1)$ could be produced in appropriately shaped targets [FIG. 5(a)]. Subsequent studies showed that for targets immersed in a uniform neutrino flux [FIG. 5(b)] these first-order forces must always cancel, leaving only effects $\propto(n - 1)^2$.[16] However, it would be incorrect to conclude that no ideas can be created based on first-order forces. What these proofs show is that the first-order force is $\propto(n - 1)$ grad ρ_ν, where grad ρ_ν is the number density gradient. However, although relic neutrinos would have negligible natural density gradients (e.g., arising from the gravitational potential gradient, or from random spatial fluctuations), it is nevertheless possible to envisage that significant distortions in ρ_ν could be produced by the previously mentioned neutrino lenses or mirrors, so that over a small spatial region (e.g., 10–100 cm) grad ρ_ν would be sufficient to produce a force $\propto(n - 1)$ that could be detected by a sensitive torsion balance [FIG. 5(c)]. Moreover, since alternating forces would be generally easier to detect than constant forces, one could envisage oscillating or rotating structures (perhaps in orbit) to produce the time-varying gradients in ρ_ν. Although this general point was made many years ago,[4] no attempt has been made to estimate the maximum value of grad ρ_ν that might be achievable

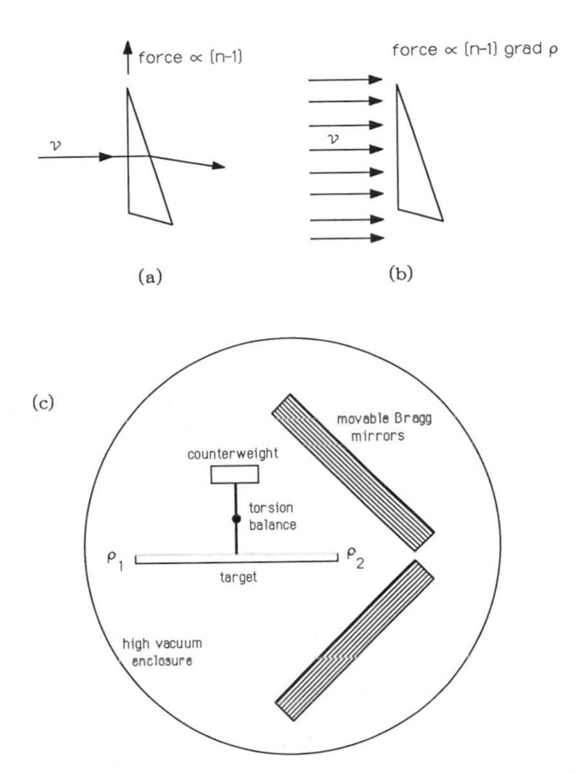

FIGURE 5. (**a**) Force proportional to $n - 1$ resulting from deflection of a neutrino by a material prism. (**b**) Dependence of first-order force on neutrino density gradient for a prism immersed in neutrino flux. (**c**) Conceptual production of neutrino-density gradient by Bragg mirror system, and detection with torsion balance.

with neutrino optics. The magnitude of a first-order force that might be produced artificially is thus still unknown, so although approximate (unpublished) calculations are somewhat discouraging, this idea cannot yet be excluded as a practical possibility.

3.6. *Coherent Interactions: Macroscopic Second-order Forces*

Coherent scattering of a uniform flux of neutrinos produces a force on the target $\propto(n - 1)^2$ that is simply due to the optical reflection of the incident neutrinos. For example, at normal incidence the reflection coefficient from a single surface is $C =$

$(n-1)^2/(n+1)^2$, with more complicated formulas for other angles and for a finite-thickness slab.[a] By periodically varying the material density one obtains the equivalent of a Bragg mirror, which will increase the reflectivity for a restricted wavelength range, while the momentum transfer remains approximately constant [FIG. 6(a)]. The outcome of our detailed analysis of these effects[4–6] was that the magnitude of the force on a target due to reflection of a galactic neutrino flux would

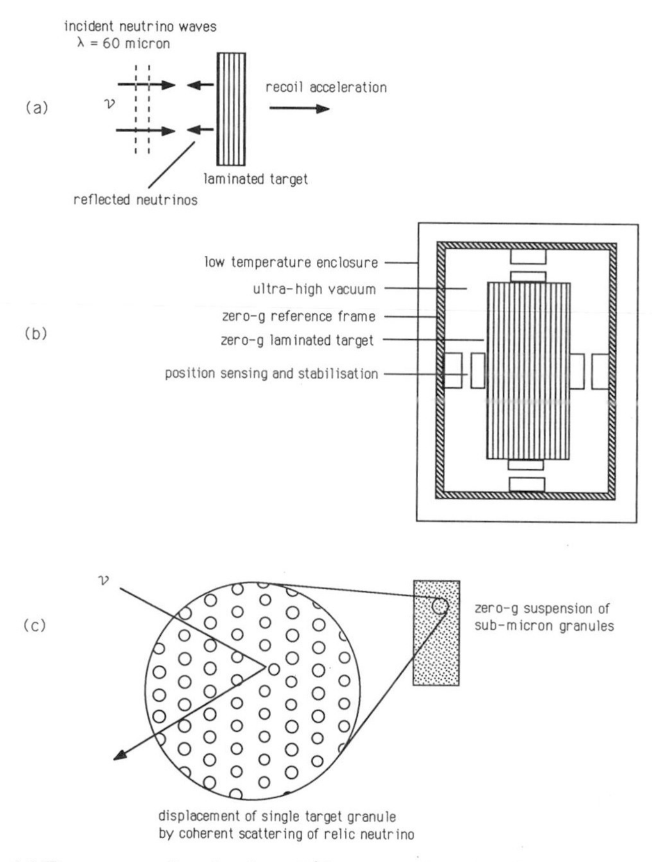

FIGURE 6. **(a)** Force proportional to $(n-1)^2$ from neutrino reflection. **(b)** Conceptual zero-*g* experiment to detect second-order force. **(c)** Conceptual zero-*g* experiment with multiple-granule target.

not exceed $\sim 10^{-23}$ dyne g^{-1} (for a material target density $\sim 10^{-20}$ g cm^{-3}). Thus this case is more straightforward than that of Section 3.5 in that an upper limit for the force per unit target mass due to second-order forces can be rigorously established.

It is of some interest that measurement of a force of this order, although not technically feasible at the present time, appears possible in principle. For a freely

[a] The neutrino reflection formulas given in reference 4 contain a misprint in eq. 13b, where the plus sign should be a minus sign as in eq. 13a.

suspended target (e.g., using a "drag-free" satellite enclosure to achieve the equivalent of zero-gravity[17]) the displacement would be $\sim 10^{-15}$ cm in a few hours, and $\sim 10^{-13}$ cm in a day, easily exceeding the displacements nowadays measurable in gravitational radiation detectors.[17,18] To reduce noise from gas collisions below this level requires a pressure $\sim 10^{-16}$ torr around the target, achievable by cooling to 50 mK (at which temperature all vapor pressures fall below this level). Photon noise would be smaller than the gas collision noise, and noise from cosmic-ray collisions could be reduced sufficiently by measuring the differential movement between the target and a surrounding (also freely floating) reference enclosure of lower density and hence reduced neutrino sensitivity [FIG. 6(b)]. There remains the problem of the constancy of the neutrino flux, making signal identification possible only through the correlation between the force direction and the galactic orbit, but as previously indicated, it is also possible in principle to add a modulating component by an array of movable Bragg mirrors around the satellite.

Having established that a "physics design" is possible, it is nevertheless equally clear that such a system could present insurmountable difficulties in construction and operation, and this scheme is manifestly unrealistic both technologically and economically in the foreseeable future.

4. COHERENT INTERACTION WITH MULTIPLE TARGETS

We now consider a variation of the method in Section 3.6, not previously studied, that is in principle capable of giving a larger signal. The principal problem with the single-target scheme just described is that the coherent interaction results in $\sim 10^{7}$ interactions per tonne per second, but produces only a very small acceleration of the whole target. It is evident that this is wasteful, since only a small fraction of the target participates in the collisions, while the whole target mass limits the recoil. If instead each individual coherent region (<10 μm diameter) were free to recoil independently, then an individual neutrino scattering event would produce a larger recoil velocity, and hence a larger displacement (of that small volume) in a given time. As an example, for a target granule of diameter 0.1 μ a typical single-neutrino scattering would produce a recoil velocity $\sim 10^{-9}$ cm s^{-1} and a displacement $\sim 10^{-4}$ cm after one day. For a 100-Å granule the recoil distance would be 1 mm after one day. Sufficient total target mass would be required to give an adequate event rate (e.g., 1/day), but would in this case be in the form of a zero-g suspension of very small granules [FIG. 6(c)].

The choice of granule size could range from the lower limit of single atoms or molecules, to particles of the maximum coherent size ~ 10 μm (and larger sizes could also be considered in conjunction with a lower momentum subset of the neutrino spectrum). There is no immediate optimum diameter, since smaller granules would produce larger displacements, while larger granule sizes would improve the coherent cross section and reduce the total mass required. The displacement must also be sufficient to satisfy the quantum measurement limit (uncertainty principle).

As an example, for a single galactic neutrino interaction in tungsten the displacement d(μm) of a granule of mass m(pg) in time t(days) would be $d \approx 10^{-3}\, t/m$, and the total target mass M(kg) required for a single event in time t(days) would be $M \approx 10^{-3}/mt$. The quantum limit imposes the restriction $d > 10$ μm. Some examples of target masses and displacements satisfying these conditions for 1- and 10-day observation times are as follows:

Observation Time (days)	1			10		
Mass of single granule (pg)	10^{-4}	10^{-5}	10^{-6}	10^{-4}	10^{-5}	10^{-6}
Total target mass (kg for one event)	10	10^{2}	10^{3}	1	10	10^{2}
Recoil distance (μm)	10	10^{2}	10^{3}	10^{2}	10^{3}	10^{4}

It can be seen that this scheme is of interest in allowing a reasonably small total target mass, and giving much larger (and in principle more easily observable) mechanical displacements than is possible in the case of a single target. It is equally clear that this type of scheme would be technically impossible at the present time since, in addition to the zero-*g* requirement, the system would require methods firstly for dissipating residual motion of the granules to provide an ultracold state with negligible initial velocity, and secondly for identifying individual displacements in the array (without using photons, which would cause larger recoils than the neutrinos). Other refinements could include the suspension of each granule within an outer shell as a means of discriminating against cosmic-ray collisions.

It can, however, be envisaged that schemes based on this principle could become practicable with foreseeable twenty-first century technology. Today's technology, in which microengineering has reached the submicron level, could place substantial instrumentation and computing power within devices <1 cm in diameter. If in the future engineering becomes possible on the molecular scale, it will then be similarly possible to place instrumentation within the submicron granules, allowing velocity changes relative to neighboring members of the array to be recognized and recorded.

It might even be argued that this could already be the case, using selected molecules (or even biological cells) that would behave in a characteristic way on displacement and collision with neighbors. However, highly controlled engineering on the molecular scale will certainly begin to be possible within the next 20 years. The ability to lift and reposition single atoms and portions of molecules using the scanning tunneling microscope has already demonstrated[b] the beginnings of nanotechnology, which will lead to the design and construction of molecular machines in turn capable of assembling other molecular machines (including self-replication) from atomic and molecular subconstituents.[19] It is envisaged that some forms of these machines will act as programmed assemblers for the construction of microrobots and macroscopic objects from a reservoir of chemical constituents,[c] with revolutionary consequences for medicine, engineering, and artificial intelligence.[20] Of course, "conventional" microlithography and microelectronics have even today reached the 10–100-nm level, but it is likely that this "top-down" micromachining approach will eventually be superseded by the "bottom-up" approach of atomic assembly.

These developments are still at an uncertain distance into the future, but (remembering the development speed of microelectronics and computer technology)

[b] For example, nanometer-size words and patterns have been formed with single-atom linewidth by positioning Xe atoms, S atoms, and CO molecules on a substrate (see New Scientist, No. 1712, 1990, p. 22; No. 1753, 1991, p. 31; No. 1757, 1991, p. 20).

[c] Just as plant seeds can be regarded as "programmed nanomachines" that contain all of the information and molecular machinery necessary to extract and assemble C, H, O, and N atoms into macroscopic edible objects or vertical cylinders of wood.

are likely to happen faster than is generally appreciated, with major advances in the next 30 years. Thus, although not immediately applicable to neutrino physics, these developments are certainly relevant to questions of the form "Will it ever be possible?" The conclusions from the preceding discussion are thus (1) that coherent interaction with submicron targets could in principle give observable signals in the form of mechanical motion, and (2) that complex structures and instrumentation within and around such submicron targets will eventually become possible, making it at least conceivable that suitable experiments could be proposed. It is also hoped that the preceding summary will stimulate further thought, and perhaps new ideas for relic neutrino detection.

ACKNOWLEDGMENTS

The earlier studies of coherent detection were carried out in collaboration with J. D. Lewin, to whom I am grateful for continuing helpful discussions. I would also like to thank R. W. P. Drever for discussions on the detectability of small forces.

REFERENCES

1. For an earlier summary and additional references see Secs. 3 and 4 of SMITH, P. F. & J. D. LEWIN. 1990. Phys. Rep. **187:** 203.
2. See, for example, COLLINS, P. D. P., A. D. MARTIN & E. J. SQUIRES. 1989. Particle Physics and Cosmology: 389. Wiley, New York; BARROW, J. 1983. Fundam. Cosmic Phys. **8:** 83; MARCIANO, W. J. 1981. Comments Nucl. Part. Phys. **9:** 169.
3. TREMAINE, S. & J. E. GUNN. 1979. Phys. Rev. Lett. **42:** 407.
4. SMITH, P. F. & J. D. LEWIN. 1983. Phys. Lett. **127B:** 185.
5. SMITH, P. F. & J. D. LEWIN. 1985. Acta Phys. Pol. **B16:** 837.
6. SMITH, P. F. & J. D. LEWIN. 1984. Acta Phys. Pol. **B15:** 1201.
7. LOEB, A. & G. D. STARKMAN. 1990. Princeton Preprint IASSNS-AST 90/10.
8. HONG, W. UCLA. Personal communication.
9. See ORAL DISCUSSION. 1981. Proc. Am. Inst. Phys. **72:** 353.
10. IRVINE, J. M. & R. HUMPHRIES. 1983. J. Phys. **G9:** 847.
11. See, for example, DRUKIER, A. & L. STODOLSKY. 1984. Phys. Rev. **D30:** 2295; SMITH, P. F. 1986. Z. Phys. **C31:** 265.
12. LEWIN, J. D. & P. F. SMITH. 1984. Astrophys. Lett. **24:** 59.
13. OPHER, R. 1984. Astrophys. J. **282:** 398.
14. SMITH, P. F. & J. D. LEWIN. 1987. Astrophys. J. **318:** 738.
15. SMITH, P. F. 1984. Nuovo Cimento **83A:** 263 (note misprint in eq. 4.3 for which the square brackets should be positioned as in the corresponding eq. 3.2).
16. CABIBBO, N. & L. MAIANI. 1982. Phys. Lett. **B114:** 115; LANGACKER, P., J. P. LEVEILLE & J. SHEIMAN. 1983. Phys. Rev. **D27:** 1228.
17. See, for example, BRAGINSKY, V. B. & A. B. MANUKIN. 1977. Measurements of Weak Forces in Physics Experiments. University of Chicago Press. Chicago.
18. See, for example, BRAGINSKI, V. 1988. Sov. Phys. Usp. **31:** 836; and reviews by AMALDI, E. 1982. Proceedings of the 2nd Marcel Grossman Meeting on General Relativity, Vol. 1, R. Ruffini and N. Holland, Eds.: 21; AMALDI, E. 1989. Proceedings of the 4th Winter School Hadronic Physics (Folgaria, Italy), R. Cherubini, Ed.: 423. World Scientific. Singapore.
19. See, for example, review by SCHNEIKER, C. 1987. Artificial life. *In* Proceedings of the Interdisciplinary Workshop on Synthesis of Living Systems, C. Langton, Ed.: 443. Los Alamos, N. Mex.
20. For an account of future possibilities see reference 19 and DREXLER, K. E. 1986. Engines of Creation, The Coming Era of Nanotechnology. Anchor. New York.

Gamma-ray Astronomy above 10^{13} eV

A. A. WATSON

Department of Physics
University of Leeds
Leeds LS2 9JT, England

Observations of candidate gamma-ray sources at about 1 TeV are made using the air–Cerenkov technique, which has the advantage that a single detecting telescope has an effective collecting area of $\sim 10^4$ m^2. Relatively high count rates can be achieved on clear moonless nights (see accompanying papers by Weekes and Turver for details), and the yield of possible sources obtained thus far is impressive. At higher energies the rate of events is too low to make this technique feasible, and advantage is taken of the fact that above $\sim 10^{13}$ eV particles in the cascades generated by cosmic nuclei and cosmic gamma-rays when they hit the atmosphere reach ground level and are spread out over a large area. Arrays of detectors, usually scintillation counters, are deployed over areas comparable to or greater than 10^4 m^2. The direction of the cascade [or extensive air shower (EAS)], and hence of the particle initiating it, can be measured to within about 1° or perhaps somewhat better in the best experiments. Unlike the air–Cerenkov technique the particle arrays operate in all weathers and monitor a large solid angle, typically about 1.5 sr, continuously. However, the cosmic-ray and gamma-ray spectrum fall steeply, and to optimize the detection it is desirable to have some method for discriminating against charged cosmic rays that are highly isotropic in their arrival direction distribution and are very much more numerous.

It was recognized in the early 1960s that one powerful discriminatory tool would be to select events in which the muon content of the extensive air shower was much lower than normal. This led to the operation of a 60-m^2 muon detector at the center of the Bolivian Air Shower Joint Experiment (BASJE) in Chacaltaya. Despite the height of this laboratory (5200 m), technical limitations restricted the shower threshold to above 10^{14} eV. No convincing gamma-ray signals were detected, and the work was wound down following the discovery of the 3 K black body radiation in 1964 and the subsequent recognition that gamma rays of 10^{14}–10^{16} eV would be quite severely attenuated by the reaction $\gamma + \gamma \rightarrow e^+ + e^-$.

Interest in possible point sources of gamma rays was revived in 1983 when Samorski and Stamm (1983a), working with data taken at the University of Kiel shower array in the late 1970s, reported the detection of gamma rays from the enigmatic radio object Cygnus X-3. The detection consisted of a strong DC signal (16.6 events above a background of 14.4 ± 0.4 events) and of evidence of emission modulated with the well-established 4.8-hr periodicity of the source noted from X-ray observations. This object had been observed at TeV energies in the early 1970s by the Crimean group, and again in the latter part of that decade by groups from the Smithsonian Institute, working at Mount Hopkins, and the University of Durham, working at Dugway in the United States. The first confirmation of the Kiel result came from the Haverah Park group (Lloyd-Evans *et al.* 1983) who analyzed data from water–Cerenkov detector arrays and when using a common ephemeris, found supporting evidence for emission at the same phase in the 4.8-hr cycle as that noted by the Kiel group. The angular resolution of the Haverah Park array allowed only a weak (1.7σ) DC detection. Subsequently the Kiel group (Samorski and Stamm 1983b) reported the unexpected presence of muons in the events defining the Cygnus X-3 signal, but because of the relatively simple nature of the muon detector and supportive nature of the Haverah Park data, this report was largely discounted.

The Kiel result was totally unexpected, but led rapidly to a plethora of interpretative papers linking the detection of gamma rays from Cygnus X-3 to the problem of the origin of cosmic rays (e.g., Hillas 1984). Most EAS groups worldwide searched data bases for evidence of signals from Cygnus X-3 or from such candidate sources as were visible from their latitudes. By the time of the International Cosmic Ray Conference in La Jolla (1985) there had been reports of gamma-ray emission at about 10^{15} eV from the Crab pulsar, Her X-1, Vela X-1, LMC X-4, and Cen A, in addition to Cygnus X-3. While the rapporteur report (Watson 1985) at that meeting

TABLE 1. Characteristics of a Sample of the Shower Arrays Now Operating as Ultrahigh Energy Gamma-Ray Telescopes

Site and Name	Atmospheric Depth (g-cm^{-2})	Number of Detectors	Sensitive Area (10^4 m^2)	Energy Threshold (100 TeV)	Muon Counter Area (m^2)
Los Alamos, U.S. (Cygnus)	800	108	> 4	0.5	244
Dugway, U.S. (CASA-MUMA)	850	1089	25	0.5	1280*
Haverah Park, U.K. (GREX)	1020	36	~ 4	3	40
La Palma (HEGRA)	800	~ 100*	4	1	—
Black Birch Mt., New Zealand (JANZOS)	930	70	> 0.3	2	—
South Pole (SPASE)	695	24	0.7	1	—

*To be expanded.

was relatively upbeat, it was pointed out that even areas of 10^4 m^2 and an angular resolution of 1° were likely to be insufficient to establish the ultrahigh-energy gamma-ray astronomy field. It was predicted that by the following biennial meeting in 1987 (Moscow) Cygnus X-3 would be clearly established as a PeV source, but the rest of the sky would be strewn with doubtful "three-sigma" detections where confirmation had been difficult to get because of poor statistics and time variability. Sadly, only the second of these predictions has turned out to be true.

It was also clear in 1985 that arrays having areas of 10^5–10^6 m^2 should be constructed to study candidate sources in sufficient detail to pin down firmly time variability. With a 1-km^2 array about 10 gamma rays might be expected per 4.8-hr cycle from Cygnus X-3 about 3×10^{14} eV. So far this difficult target has not been achieved, but arrays of significant collecting area have been constructed and have begun to report data. TABLE 1 lists the parameters for six of the largest of these, which are also "paired" in the sense of being at closely similar longitudes, an important factor when anticipating the detection of short bursts of radiation at low statistical significance.

The arrays listed in TABLE 1 all have comparable angular resolution, somewhat better than 1°, which, in the case of Los Alamos, Dugway, La Palma, and Black Birch Mountain, has been verified by the seeing of the shadow of the sun and the moon. The latitudes of the Haverah Park and South Pole instruments prevent such a test, but other methods have been devised (e.g., Walker *et al.* 1991). The results of observations with all telescopes except the CASA-MUMA instrument have been reported. (For recent reviews, see Protheroe 1987, Nagle, Gaisser, and Protheroe 1988, Weekes 1988, Fegan 1990.) No convincing evidence of emission from Cygnus X-3 has been offered above 10^{13} eV since the spate of measurements summarized at the La Jolla meeting (Watson 1985). The Los Alamos and Haverah Park exposures are now such that had the object been emitting at the efficiency reported by the Kiel group, then signals of ~ 9 sigma DC should have been seen in both telescopes. The Los Alamos group (Dingus *et al.* 1988a) have reported a period of episodic emission in April/June 1986, soon after the instrument was turned on. However, this signal is not confirmed at Haverah Park, although there is some support from the Baksan experiment (Alexeenko *et al.* 1987). The muon content of the events claimed to be associated with Cygnus X-3 is not entirely consistent with gamma-ray emission. Although the situation is clearly far from satisfactory with regard to Cygnus X-3, the evidence from many experiments in the early 1980s is very hard to dismiss, particularly in view of the alignment of the phase in the 4.8-hr cycle at which emission is reported, and it seems to me that a probable explanation is that Cygnus X-3 is currently in a quiescent phase as regards its gamma-ray emission at the highest energies. It should be recalled that there was a period of almost seven years after its first detection at 1 TeV before it was again seen, and that the intense radio bursts that distinguish this source are also erratic.

By comparison with the other arrays listed in TABLE 1, that at the South Pole, which is operated jointly by groups from the Bartol Research Institute and Leeds University, is not particularly large. However, the location of the array on the rotation axis of the earth and its high elevation mean that the threshold energy is low and the counting rate is high. Additionally, source candidates are continuously monitored. Analysis of the first six months of data taken during 1988 has been reported (Smith *et al.* 1990). No emission has been detected from candidate sources (SMC X-1, LMC X-4, Cen X-3, and Vela X-1), claimed as TeV emitters, nor from SN1987a. However, an all-sky survey has revealed an excess emission from a region centred on $\alpha = 174.5°$ $\delta = -62.5°$. This lies close to the galactic plane and to two X-ray binaries, 1E1145 − 6141 and 4U1145 − 619, the latter of which was independently shown to be an emitter of TeV gamma rays. This "source" has a strength that is about 5 percent of cosmic-ray background, but no periodic emission has been detected. Confirmatory searches are under way using data from Black Birch Mountain and further data from the SPASE array.

At present the results from the very high-energy (VHE) gamma-ray scene are relatively discouraging, and much of the enthusiasm of five years ago has been replaced by hard-nosed scepticism. This may be no bad thing, since the field has presented so many puzzles. The most noteworthy of these is the summer of 1986 behavior of Hercules X-1. This object, first identified as a TeV emitter by the Durham group in 1984 (Douthwaite *et al.* 1984), was thought to emit gamma rays modulated at exactly the X-ray pulsar period. However, on May 13 and June 11, 1986 the Haleakala (Resvanis *et al.* 1988) and Whipple (Lamb *et al.* 1988) groups reported episodic TeV emission at periods of 1.23593 ± 0.00018 s and 1.23579 ± 0.00020 s, respectively, lasting for 15 and 25 minutes. These results would be difficult to dismiss, but have become even harder to ignore since, at higher energy, the Los Alamos (Dingus *et al.* 1988b) group has reported episodic emission on July 24, 1986, the emission being modulated with a period of 1.23568 ± 0.00020 s. All three blue-shifted periods are remarkably similar. However, the Los Alamos events contained approximately the number of muons expected if they had been initiated by hadrons. Such a remarkable result throws doubt on the explanation of the events as being produced by gamma rays, but it would require radically new physics to explain them, either in the form of a new particle or of very anomolous photon behavior. It is perhaps more conservative to believe that we have been victims of freak statistical fluctuations.

The subject of gamma-ray astronomy above 10^{13} eV will surely be established or become dormant within the next three to five years: the telescopes are labor-intensive and costly to run, and one must hope that the effort will be rewarded by convincing source signals. However, there will certainly be a diffuse gamma-ray signal from the galactic plane resulting from interactions of cosmic rays and gas in the intersteller medium (ISM). The predicted level is only $\lesssim 10^{-4}$ of the cosmic-ray flux, but some of the larger experiments, with good muon discrimination, are probably capable of detecting this. This detection will be a significant achievement that would have been unlikely to have been stimulated without the drive to see point sources. Perhaps both aims will be rewarded.

REFERENCES

ALEXEENKO, V. V., *et al.* 1987. Proceedings of the 20th International Cosmic Ray Conference (Moscow) **1:** 219.

DINGUS, B. L., *et al.* 1988a. Phys. Rev. Lett. **60:** 1785.

DINGUS, B. L., *et al.* 1988b. Phys. Rev. Lett. **61:** 1906.

DOUTHWAITE, J. C., *et al.* 1984. Nature **309:** 691.

FEGAN, D. J. 1990. Proceedings of the 21st International Cosmic Ray Conference (Adelaide, Australia) **11:** 23.

HILLAS, A. M. 1984. Nature **312:** 50.

LAMB, R. C., *et al.* 1988. Astrophys. J., Lett. **328:** L13.

LLOYD-EVANS, J., *et al.* 1983. Nature **305:** 784.

NAGLE, D. E., T. K. GAISSER & R. J. PROTHEROE. 1988. Annu. Rev. Nucl. Part. Sci. **38:** 609.

PROTHEROE, R. J. 1987. Proceedings of the 20th International Cosmic Ray Conference (Moscow) **8:** 21.

RESVANIS, L. K., *et al.* 1988. Astrophys. J., Lett. **328:** L9.

SAMORSKI, M. & W. STAMM. 1983a. Astrophys. J., Lett. **268:** L17.

SAMORSKI, M. & W. STAMM. 1983b. Proceedings of the 18th International Cosmic Ray Conference (Bangalore, India) **11:** 244.

SMITH, N. J. T., *et al.* 1990. Submitted for publication in the Proceedings of the 1990 Vulcano, Italy, Workshop on "Frontier Objects in Astrophysics and Particle Physics."

TURVER, K. E. 1992. Teraelectronvolt gamma-ray astronomy. This issue.

WALKER, A., *et al.* 1991. Nucl. Instrum. Methods Phys. Res. **A301:** 574.

WATSON, A. A. 1985. Proceedings of the 19th International Cosmic Ray Conference (La Jolla, Calif.) **9:** 111.

WEEKES, T. C. 1988. Phys. Rev. Rep. **160:** 1.

WEEKES, T. C. 1992. Very-high-energy gamma-ray astronomy: Overview and prospects. This issue.

The Deep Underground Sky: Astrophysical Sources and High-energy Muons

MARVIN L. MARSHAK

School of Physics and Astronomy
University of Minnesota
Minneapolis, Minnesota 55455

INTRODUCTION

For the past several decades, physicists have recognized that nonthermal stars may provide unique elementary particle physics laboratories, complementary to those man-made laboratories available on earth. Neutron stars, which power many of these objects, often possess large magnetic fields, as a result of magnetic flux conservation during their formative supernova collapse. The combination of these magnetic fields and rapid rotation rates (which result from angular momentum conservation during collapse) may produce large electric fields, which are capable of accelerating charged ions to extremely high energies.

These possible high-energy particle beams, similarly possible "beam dump" geometries (with a companion star or an accretion disk as a target), and the long propagation path to earth permit a number of interesting particle physics experiments. However, it is important to note the considerable difference between the experimental context of conventional particle physics and the observational nature of investigations of these distant systems. Experiments can be repeated at will with rigorously predictable results. Observational results are in part stochastic; extraction of repeatable data may be difficult or even impossible. Indeed, accelerator-based particle physicists often have a problem believing which cannot be repeated at will.

The standard experimental technique in very-high-energy (VHE) and ultrahigh-energy (UHE) physics is the observation of air showers initiated by primary quanta incident on the earth's atmosphere.[1] Previous measurements clearly indicate that the number of quanta incident on the earth *directly* from a particular source is, at best, small compared to the background of protons and heavy nuclei cosmic rays, given the intrinsic air shower angular resolution. These direct quanta must be electrically neutral (because of galactic magnetic fields) and stable enough to reach the earth. Indeed, these quanta are presumed as photons, because the other known neutral, stable particle (the neutrino) does not interact at a sufficient rate for straightforward detection.

Several calculations[2] show that photon-induced, and therefore electromagnetic (EM) air showers should be distinguishable from proton or heavy-nucleus-induced (that is, hadronic) air showers. Possibly differentiating characteristics include (1) shower shape—the EM shower is narrower and has fewer fluctuations; (2) arrival time dispersion of charged particles—the EM shower front is more distinctly defined in time; and (3) muon multiplicity—the EM shower has a μ/e ratio less than 0.1 that of a hadronic shower. However, with the possible exception of atmospheric Cerenkov observations of the Crab nebula enhanced by shower imaging,[3] no detector has yet produced clear evidence for the EM nature of source-related air showers.

TIME MODULATION

This combination of a poor signal-to-background ratio and an historic inability to enhance that ratio on the basis of internal shower characteristics has led observers to use other techniques. The most common is a search for a time modulation in the VHE and UHE flux that is synchronized with an intrinsic source time modulation. This source time modulation is presumed known from X-ray, optical, radio, or infrared measurements. The advantages of this technique are both significant improvement in the signal-to-noise ratio and unique identification of flux from a certain direction with a particular source in that direction.

Unfortunately, the pitfalls in this technique are numerous and often subtle. Past experience indicates that the ambiguity of the observational context stimulates all manner of exploratory data analysis. However, confidence levels, the quotation of which are expected by the standards of the field, cannot be reliably estimated from *a posteriori* tests. The possible problems somewhat depend on the length of the period. Orbital periods ranging from hours to days are generally well enough known from other measurements, such as X-ray or megaelectronvolt gamma-ray observations. However, some detectors—atmospheric Cerenkov detectors, for example—have some difficulty making flux measurements that are systematically reliable over long intervals and are thus disadvantaged in measuring flux modulation with long periods. Short periods—such as millisecond to second neutron star rotation periods—minimize systematic detector uncertainties, but they are generally not well enough known for an unambiguous comparison between expectation and observation.

A typical procedure (especially for short periods) is to search for a period at which the arrival time series exhibits maximal modulation. In reality, however, such a period is no more "correct" than a nearby period with somewhat less modulation, because modulation power differences at nearby periods are influenced mostly by random background fluctuations. An estimation of other parameters such as flux using the "most powerful" period will produce biased results. A further complication is that most detectors (other than those located at the South Pole) experience severe sidereal flux modulation. The rotation of the earth manifests itself as beats or nonindependence of nearby frequencies, which further complicates the interpretation of the event arrival time series.[4]

Despite these difficulties, considerable progress has been made in understanding how to do a proper event arrival-time analysis. One constraint is that the apparent modulated signal must be consistent with the time-invariant excess of events from the source direction. That is, evidence for a signal must both contradict the uniform, random background hypothesis and support the signal hypothesis. However, even the time-invariant excess analysis has some difficulties. There is no clear evidence at the 1 percent level that the VHE and UHE fluxes are indeed isotropic, when viewed with fine-grained angular resolution. In particular, most likely sources lie in the galactic plane. Although by no means conclusive, some data suggest a several percent flux deficit associated with the plane.[a] If such a deficit exists, it would invalidate any "D.C." analysis that does not take such a deficit into account.

I close this discussion of time modulation with a comment on the nature of confidence levels quoted in the VHE and the UHE literature. In many cases, these "confidence levels" seem considerably overstated. The effects are clearly small. While VHE fluxes are adequate, atmospheric Cerenkov observing time is very limited. UHE fluxes are just very small. The ambiguities of signal period, possible "burst" time structure, and source Doppler shifts are sufficiently large that highly confident rejection of the no-signal hypothesis is not currently attainable. This situation is not likely to change without a technique for a clean discrimination on a shower-by-shower basis between signal and background. In my opinion, the field would be better served by honest evaluations of confidence levels. A 10^{-2} probability for the no-signal hypothesis should be more than adequate to suggest a signal, if the stated confidence level is real. The current situation seems unfortunately cyclic; extremely strong confidence levels are required by the community and are routinely produced by observers, without much consideration of what is actually possible or credible.

ALTERNATE VIEWS OF THE VHE AND UHE WORLD

I believe that the considerable progress of the past 25 years of VHE and UHE observations can be summarized as a choice among three possibilities.

1. *There are no VHE or UHE sources.* This possibility should not be discounted. There is little doubt of gamma-ray emission by nonthermal stars up to energies of ≈ 100 MeV. However, the spectrum must cut off somewhere. A few hundred megaelectronvolt cutoff is as possible as any other cutoff point. In this view, all source-related VHE and UHE observations are just incorrectly interpreted, statistical fluctuations of the background.
2. *There is only one confirmed VHE source—the Crab nebula; there are no UHE sources.* The Crab nebula is a theoretically likely source of synchrotron radiation up to energies of order 1 TeV. Thus, the Whipple observations of a steady, nonmodulated flux from the Crab nebula are consistent with our general understanding of the physics of that system. Of course, if the Crab nebula is a VHE source, there must be other synchrotron-radiation VHE sources; they just have not yet been observed. The evidence for this hypothesis is primarily the Whipple atmospheric Cerenkov imaging data, which, in principle, permit considerable background rejection on a shower-by-shower basis. Apparently, when this imaging cut is applied to all previous source-suggesting Whipple observations, the evidence for the source goes away.[5]
3. *There exist a variety of sources (isolated pulsars, binaries, etc.), but in at least some cases, the primaries or the interactions of the primaries are unconventional.* The most likely such sources (based on the number of claimed observations) are Cygnus X-3 and Hercules X-1. We discuss this possibility further in the section below.

[a] These data are summarized and referenced in the review by Weekes in reference 1.

Of course, I have left out here the possibility that there exist multiple, already observed, absolutely conventional sources of VHE and UHE photons. As discussed in the next section, I find the evidence for this alternative both sparse and unconvincing. If this alternative were indeed correct, the entire field of VHE and UHE source observations would be in much better shape than it actually is today.

THE EVIDENCE FOR UNCONVENTIONAL PRIMARIES OR PRIMARY INTERACTIONS

If hypotheses (1) or (2) are correct, the past 25 years of VHE and UHE source literature is almost totally nonsense and the efforts of many people have been in vain. Some positive observations are likely wrong, but hypotheses (1) and (2) require almost all of them to be wrong. This situation seems unlikely. Thus, I presume that sources other than the Crab nebula exist. If so, there seems ample reason to believe that some or all of the presumably photon primaries are either unconventional or that they have unconventional interactions. Three bodies of evidence suggest to me such a view.

First, with the possible exception of the Whipple imaging technique on the Crab nebula data, considerable detector optimization work over the past 25 years has generally failed to improve signal-to-background ratios. For example, the modern Haleakala detector seems to function no better (and perhaps worse) than the fairly simple detector deployed by the University of Durham at Dugway about a decade ago. The improvement program on the Whipple detector has resulted in a marked *decrease* in the number of sources observed. This lack of progress can be contrasted with other areas, such as elementary particle physics, in which the sophistication and background rejection capabilities of detectors have improved by more than an order of magnitude over the past decade. A possible explanation for this lack of progress is that the hypotheses underlying detector optimization are flawed. Source-related (photon-induced) air showers are presumed to be more compact and to have fewer spatial, angular, and temporal fluctuations than background air showers. If air showers associated with sources do not have these properties, that is, if they behave as if they are hadron-induced, detector optimization will not work.

The second body of evidence is the repeated tendency for detectors with such capabilities to find anomalous hadronlike features in apparently source-related showers. The initial measurement of this kind was the high muon content reported by the Kiel group[6] in showers associated with Cygnus X-3. Excess muons were also reported by Moscow State University[7] in showers associated with Cygnus X-1 and by the CYGNUS detector[8] at Los Alamos in showers associated with Cygnus X-3 and Hercules X-1. The Hercules X-1 data, taken in the spring of 1986, showed a consistent blue-shifted periodicity, similar to that reported by two atmospheric Cerenkov detectors during that same season.[9] Another example is the result reported by the Whipple detector from analysis of four years of data from Hercules X-1, where reasonably convincing evidence for periodicity was destroyed by application of the same imaging cut that enhances the Crab nebula signal.[10]

A final body of evidence are the observations of underground muons associated with Cygnus X-3 by the Soudan 1 and NUSEX detectors.[11] These data have few systematic problems; there is little ambiguity that the observed tracks are muons. The statistical interpretation of these data have been correctly criticized for the use of *a posteriori* hypotheses; however, both groups repeated their observations during the interval 1985–1990 with arguably similar, although less dramatic results.[12] Some investigators in the field seem willing to accept excess muons at the surface but not underground. In my opinion, this distinction is implausible. The evidence for anomalous showers—surface and underground—stands together. It seems unlikely to me that surface measurements of anomalous showers are correct and underground measurements incorrect or vice versa. One should either accept all the evidence of this type or find some explanation to argue it all away.

Of course, any consideration of such arguments must deal with data that contradict this picture. There have been arguments that measurements of shower age parameters indicate whether showers are of photonic or hadronic origin. I believe a reasonable view is that age parameter determinations are not a reliable indicator of the nature of the primary. Many age cuts described in the literature seem determined on an *ad hoc* basis with no presented justification for either the level or the direction of the cut.

More serious, in my view, is the contradictory observational evidence on underground muons related to sources. Both the Frejus and Kamiokande detectors[13] have published persuasive evidence against observation of underground muons related to Cygnus X-3. Because their data at least in part overlap with the Soudan 1 and NUSEX data, it is difficult to assign this discrepancy to temporal variation. On the other hand, both detectors may have some angular resolution problems, and neither group has pursued the question of underground muons as vigorously or for as long a time interval as Soudan 1 and NUSEX. Overall, I regard the observational situation as confusing, but in the end, one should weigh all the evidence that is available at any given time. Because new underground muon data are still being recorded, the Cygnus X-3 observational situation is fluid and subject to change.

CONCLUSIONS

It is easy in this field to oscillate between optimism and pessimism. From time to time a serendipitous source appears. One always hopes for a "golden event"—one instance that would clarify all of VHE and UHE physics. Unfortunately, serendipitous events by definition almost never repeat and the confusion continues to develop. However, in recent years, real progress has been made. A number of very capable detectors—both surface and underground—have now recorded extensive data sets on a variety of sources. As previously indicated, the issues are more clearly defined than they were several years ago. The answer is not yet at hand, but it is possible to conceive that a generally accepted answer will exist before the end of this decade (century and millenium).

REFERENCES

1. Recent reviews include WEEKES, T. C. 1988. Phys. Rep. **160:** 1; NAGLE, D. E., T. K. GAISSER & R. J. PROTHEROE. 1988. Ann. Rev. Nucl. Part. Sci. **38:** 609; and FEGAN, D. J. 1990. Proceedings of the 21st International Cosmic Ray Conference **11:** 23.

2. For example, HILLAS, A. M. 1985. Proceedings of the 19th International Cosmic Ray Conference **3:** 445.
3. LANG, M. J., *et al.* 1990. Proceedings of the 21st International Cosmic Ray Conference **2:** 139.
4. LEWIS, D. A., R. C. LAMB & S. D. BILLER. 1991. Astrophys. J. In press.
5. WEEKES, T. C. 1992. Very-high-energy gamma-ray astronomy: Overview and prospects. This issue.
6. SAMORSKI, M. & W. STAMM. 1983. Astrophys. J., Lett. **268:** L17.
7. FOMIN, YU. A., *et al.* 1988. Proceedings of the 20th International Cosmic Ray Conference **1:** 397.
8. DINGUS, B. L., *et al.* 1988. Phys. Rev. Lett. **60:** 1785 and **61:** 1906.
9. LAMB, R. C., *et al.* 1988. Astrophys. J., Lett. **328:** L13; and RESVANIS, L., *et al.* 1988. Astrophys. J., Lett. **328:** L9.
10. LEWIS, D. A., *et al.* 1988. Presented at the Vulcano Workshop—Frontiers in Astrophysics and Particle Physics.
11. MARSHAK, M. L., *et al.* 1985. Phys. Rev. Lett. **54:** 2079 and **55:** 1965; BATTISTONI, G., *et al.* 1985. Phys. Lett. **155B:** 465.
12. AGLIETTA, M., *et al.* 1990. Proceedings of the 21st International Cosmic Ray Conference **9:** 388; JOHNS, K., *et al.* 1990. Proceedings of the 21st International Cosmic Ray Conference **9:** 402.
13. BERGER, C., *et al.* 1987. Nucl. Instrum. Methods **A264:** 24; OYAMA, Y., *et al.* 1986. Phys. Rev. Lett. **56:** 991.

Thermal Detection of Dark Matter

ETTORE FIORINI

Dipartimento di Fisica
Universitá di Milano and Istituto
Nazionale di Fisica Nucleare
Milano 20133, Italy

INTRODUCTION

There is a general consensus, based on experimental observations,[1,2] that visible matter accounts for less than 10 percent of the universe, the rest being in the form of dark matter (DM). Conversely, visible matter (or better observed baryons) accounts for only ~0.7 percent of the average density needed to "close" the universe. This missing mass could be accounted for by nonvisible barionic matter, like Jupiters of black holes, but the theory of nucleosynthesis predicts that all baryonic dark matter cannot account for more than about 10 percent of the closure density.[1-3] There is therefore a strong theoretical prejudice, supported by some experimental indications, that baryons account only partially for the density of the universe.

Nonbaryonic DM could be made by *hot* particles like light neutrinos, whose masses have been proved to be below 10–20 eV for the electron flavor.[4] Their detection seems very difficult, at least at present. *Cold* particles could be axions with a mass around 10^{-5} eV, which have already been searched for by looking to their interactions with magnetic fields.[1] I will concentrate on the *direct*[a] detection of *cold* particles with masses ranging from a fractions of a gigaelectronvolt to a few teraelectronvolts, like the heavy neutrino or supersymmetric particles like the photinos, sneutrinos, or Higgsinos.[5-7] Particles of much larger masses, like grand unified theory (GUT) magnetic monopoles, will not be considered here. All of the following considerations will be based on an average DM density in the solar system of $\sim 0.007 M_{\text{sun}}\ \text{pc}^{-3} = 5.2 \times 10^{-25}\ \text{g cm}^{-3} = 0.36\ \text{GeV cm}^{-3}$, and on a velocity of these particles with respect to the detector of 10^{-3} the speed of light.

I will assume that these particles will interact weakly (wimps), coherently, or incoherently. Particles with coherent vector coupling, like Dirac or scalar neutrinos, will have a cross section:

$$\sigma = 10^{-38} \frac{m^2 M^2}{(m + M)^2} K_N^2, \tag{1}$$

where m and M are the wimp and target mass, respectively, and the "coherence factor" K_N is roughly equal to half of the number of the neutrons. The maximum recoiling energy for various targets and wimp mass are reported in TABLE I.

Detection of particles with incoherent axial coupling (like photinos or Majorana neutrinos) looks considerably more difficult since the cross section, not enhanced by coherence, is expected to be of at least two orders of magnitudes less. The rate

[a] Indirect detection was discussed at this conference by K. Freese.

TABLE 1. Maximum Recoiling Energies Induced by Massive Particles (keV)

	m_{wimp}(GeV)				
	0.1	1	10	100	1000
Hydrogen	0.0174	0.499	1.57	1.84	1.87
Iron	0.0004	0.037	2.70	45.0	(94.2)
Lead	0.0001	0.010	0.937	(44.9)	(271)

NOTE: The figures in brackets refers to events where the momentum transfer to the nucleus is so large to produce incoherence.

should, however, be proportional to $J(J + 1)$, where J is the total nuclear spin. This condition could allow searches based on the type of target and also on its isotopic aboundance.

Experimental difficulties in detecting DM are connected with the low recoiling energy and the low event rate. In iron we would expect about 160 coherent interactions per kilogram if the mass of wimp is around 10 GeV. In lead this rate would be six times larger. The rates for incoherent interactions are expected to be lower by two to three orders of magnitude, depending mainly on the spin. As a consequence it is essential to adopt a detector, or an array of detectors, that can reveal small transfers of energy, to place it underground to get rid of cosmic rays, and to shield it against local radioactivity.[8-10] A positive effect could be unambiguously attributed to DM from its seasonal variation. The velocity of the earth with respect to the galactic coordinates varies from 219 to 249 km/s with a maximum between June second and third.

Most of the present limits for direct observation of dark matter come from nondedicated experiments that are mainly intended to search for double beta decay of ^{76}Ge carried out with Ge diodes.[2,11-15] These detectors are already very well shielded against radioactivity and cosmic rays, but double beta decay is expected to occur at much higher energies than predicted for DM. Most of the detectors were,

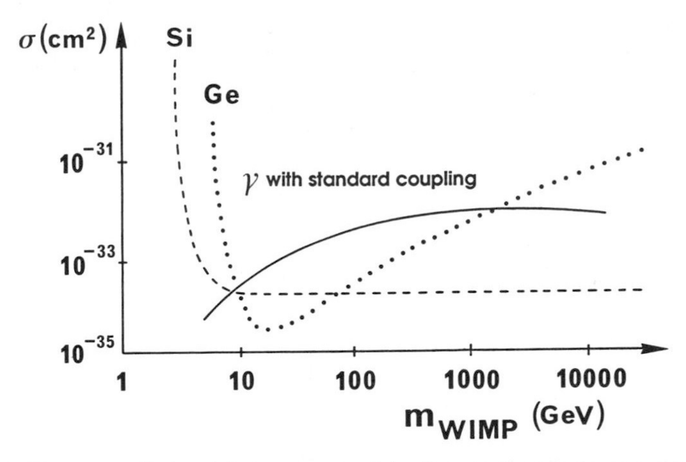

FIGURE 1. The present limits of dark matter particles from semiconductor experiments. The standard prediction for heavy neutrino interactions in Ge is also shown.

however, already sensitive to the low-energy spectrum, eventually with some modification in the readout. An experiment[16] specially dedicated to DM searches on DM with a lower mass nucleus, has been installed by the UCSB–LBL–UCB–Saclay Collaboration, by replacing two of their eight Ge diodes with four Si detectors of 17 g each. The results are shown in FIGURE 1 together with those of the Caltech–Neuchatel–PSI Collaboration,[15] which at present gives the most stringent limit for germanium and are discussed in the paper by Caldwell *et al.*[16] It can be seen that Si excludes lower masses than Ge, which was expected. It can also be seen from the figure that a massive neutrino with a standard weak coupling is also excluded in the mass region 10 to 2000 GeV.

THERMAL DETECTORS

In addition to the previously mentioned semiconductors, other ionization detectors are been considered.[1,17,b] One has to note, however, that only about one-third of the energy of a charged relativistic particle is given in the form of ionization (ion pairs or electron-hole pairs) in all these detectors. For a recoiling nucleus the situation is even worse: for energy from 1 to 10 keV the percentage of energy delivered in the form of ionization ranges from 5 to 10 percent.[13,18] Since most, if not all, of the energy is delivered in the form of heat, thermal detectors are, at least in principle, more appropriate to directly detect heavy DM particles.[8,9,19,20] One could consider a large array of superheated superconducting granules (SSG),[19–21] made by grains of a type I superconductor in a magnetic field. These grains are in a metastable superconducting state just below the critical curve of the magnetic field versus the temperature. The heat delivered by a particle to the grain could increase its temperature, thus inducing its flip to the normal state. The corresponding change in magnetization can be detected and measured. This technique has been found to be capable of detecting charged particles of various types, and can also be used to reveal DM.[19,21] One has to point out, however, that the energy delivered by the wimp to the granule can induce its flip to the normal state only if the diameter of the grain is of a micron or slightly more. One has therefore to construct a very large array of grains of small and reasonably uniform radii, and to monitor all of them, a formidable task indeed.

A thermal system more appropriate to search for DM particles is the one suggested about six years ago[8,22] especially for detection of rare events.[8] I will call it "bolometric," even if this definition strictly applies also to the SSG. Let us consider an "absorber" of heat capacity C to which a particle delivers an energy ΔE. The consequent increase in temperature will be

$$\Delta T = \frac{\Delta E}{C}. \quad (2)$$

[b] An experiment based on scintillation of xenon and NaI is planned by the Rome–Scalay–Bejing Collaboration (R. Bernabei, private communication).

For a pure crystal of diamagnetic and dielectric material, the heat capacity is given at low temperature by

$$C = 1944(V/V_m)\,(T/T_D)^3 \text{ J/K}, \quad (3)$$

where V and T are the volume and temperature of the crystal, and V_m and T_D its molar volume and Debye temperature. At very low temperatures the heat capacity can be so small as to allow a measurable increase of temperature even for the tiny energy delivered by a particle. The energy resolution could "theoretically" be extremely good, as shown by the expression:[22]

$$\Delta E_{\text{FWHM}} = \xi\sqrt{kCT^2}, \quad (4)$$

where the resolution is at full width at the half maximum (FWHM), k is the Boltzmann constant, and ξ is a numerical factor normally of a few units. No experiment has yet reached this resolution, due to mechanical or electrical noise or to effects affecting the thermal response of the detector.

The increase of temperature in (2) can be measured by a thermistor (doped Ge or Si, etc.) glued in thermal contact with the absorber or even "integrated" in it by implanting donors or acceptors into Si. It can also be measured by means of a superconducting edge detector[9,19,20,23] or a magnetic sensor.[24]

The results obtained so far are impressive, even if the resolutions are still quite worse than expected "theoretically." Many small detectors have been developed[19] with materials ranging from carbon to rhenium,[25] and a FWHM resolution of 7.2 eV has been obtained by the NASA–Wisconsin Collaboration.[26] Searches for dark matter can obviously be performed with large-volume detectors. From the thermal point of view, detectors with atoms of low atomic weight are normally easier to be constructed because one can employ materials with large Debye temperatures like diamond, lithium, calcium fluoride, silicon, and sapphire. In particular, the Imperial College–Holloway–Bedford groups (the U.K. Dark Matter Collaboration) have successfully developed various Si and LiF absorbers with germanium thermistors[27] for experiments to be carried out in the Boulby mine. A similar activity is being performed by the Aarus–CERN–Gotheburg–Paris–New York Collaboration,[28] which, after a pioneering work on diamond bolometers, is now concentrating on large sapphire crystals. A record sapphire detector of 280 g has been constructed by the group of TUM in Munich[23] and exposed to α-particles.

These detectors are obviously essential to explore the existence of wimps in the low-energy mass region, but for the large mass region, detectors made by large A material should also be considered, despite their less attractive thermal properties. Large A is also favored in experiments on double beta decay. The Milano group has constructed bolometers of masses of up to 190 g of germanium,[29] and more recently of masses up to 40 g of Te and TeO_2.[30] These last ones are mainly intended for detection of γ-rays and of rare events like double beta decay of ^{130}Te, but they could also be employed in searches for very massive wimps. Despite the unfavorable thermal properties (Debye temperature from 120 to 150 K), a resolution of 10 to 20 keV for energies ranging from 0.5 to 2.5 MeV was obtained with a 6-g crystal of TeO_2. Their experimental setup, the first thermal one operating in low background conditions, is shown in FIGURE 2. It is installed in Hall A of the Gran Sasso

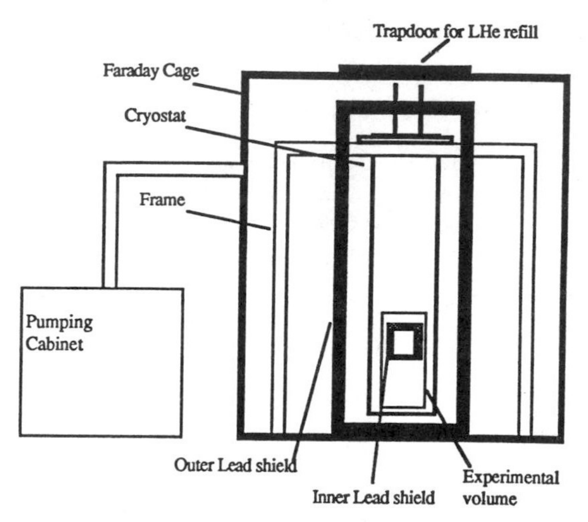

FIGURE 2. The underground cryogenic setup of the Milano group.

Laboratory under about the equivalent of 3300 hectogram cm^{-2} of standard rock, and the entire dilution refrigerator is shielded with a minimum thickness of 10-cm lead against local radioactivity and by a Faraday cage against electromagnetic interferences. A spectrum of this detector exposed to a source of ^{232}Th is shown in FIGURE 3.

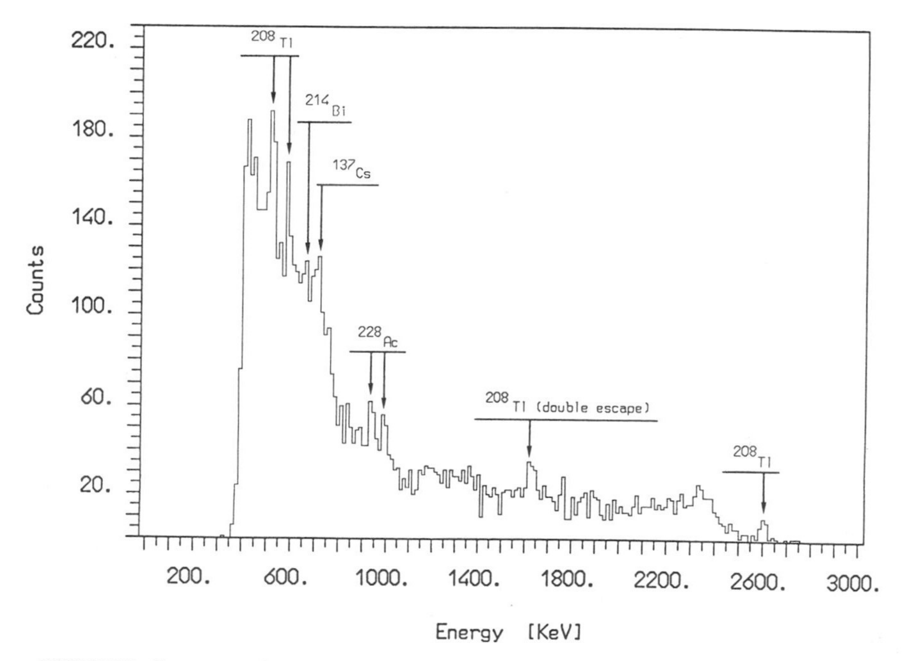

FIGURE 3. Spectrum of environmental radioactivity obtained with a 6-g TeO_2 bolometer.

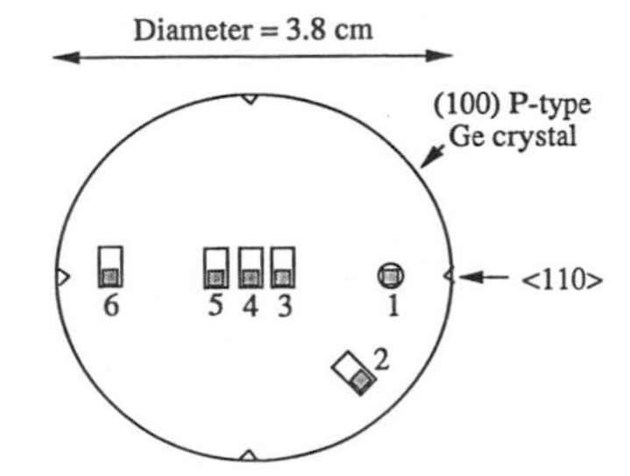

FIGURE 4. The thermal detector of the Center for Particle Astrophysics (Berkeley).

A 40-g TeO_2 crystal is now being tested in Milano, and will be installed in the Gran Sasso Laboratory soon.

While the Milano group is mainly interested in measurements on double decay or in the determination of the neutrino mass, intense activity especially devoted to direct detection of dark matter is going on in the Center for Particle Astrophysics in Berkeley.[31] The group has recently constructed and operated a detector, operated at a temperature of about 30 mK, that is capable of measuring contemporarily ionization and heat. The absorber is a disk of germanium of 60 g (FIG. 4) in thermal contact with six neutron transmutation doped (NTD) germanium thermistors.[32] The top and bottom of this disk are metallized by implantation and kept at a moderate electric tension in order to measure ionization. The standard deviations of the Gaussians (not the FWHM!) reported by the authors for the 18- and 60-keV photons from ^{241}Am are 4 and 5 keV for ionization and 4 keV at both energies for thermal pulses (FIG. 5). The contemporary measurement of ionization and heat is obviously going to play an essential role in searches for dark matter for the suppression of the background. The contemporary measurement of heat and scintillation has also been suggested,[21] and is being investigated by the Milano group in collaboration with the Institute of Physics and Engineering of Moscow, with CaF crystals, in view of an

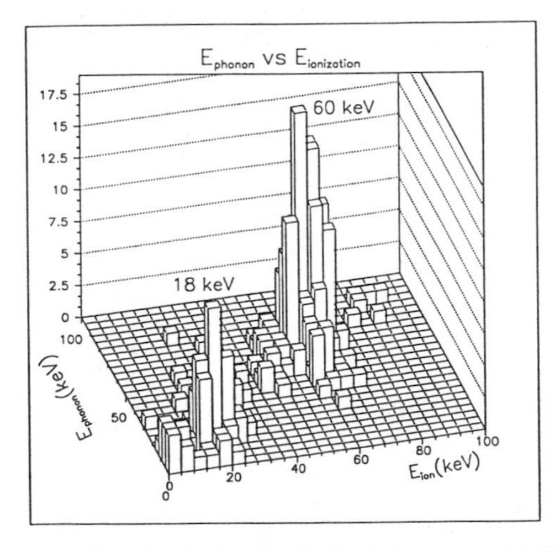

FIGURE 5. Results obtained by the Berkeley group with ^{241}Am.

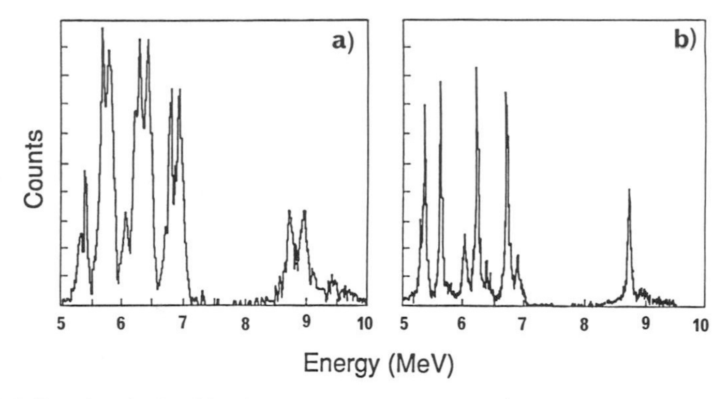

FIGURE 6. Results obtained by the Milano group on detection of nuclear recoils. (**a**) Alpha particles and implanted nuclei. (**b**) The implanted nuclei have mostly decayed.

experiment on double beta decay of ^{48}Ca. Its use can obviously be extended to searches for dark matter.

Both the Berkeley and Milano groups have profited much from the use of NTD germanium sensors developed by the group of Prof. Haller,[33] which have been found

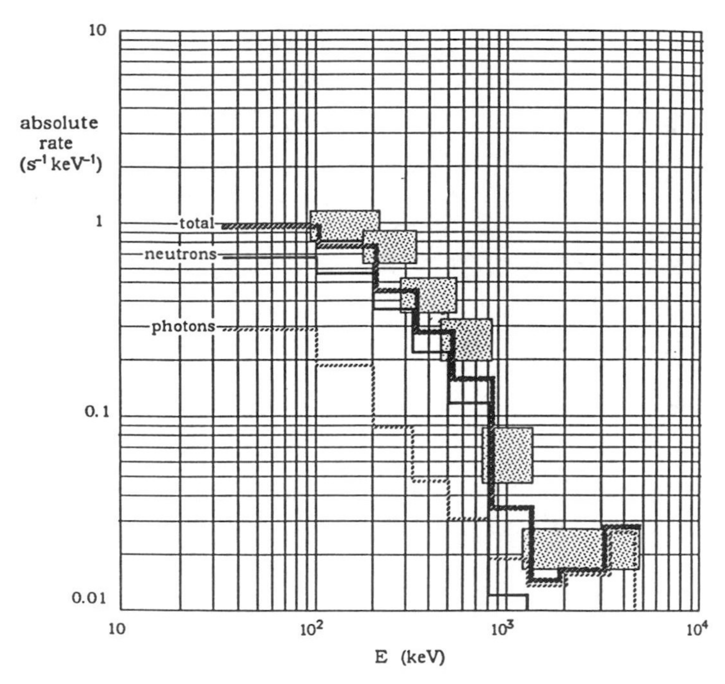

FIGURE 7. Counting rates for 4-g Si targets exposed to neutrons. *Histograms* show Monte Carlo estimate. *Shaded boxes* show observed counting rates.

to be much more reproducible and dependable than commercial chips. Recent measurements carried out by the Milano group on freshly irradiated chips have, however, shown a considerable residual radioactivity, due in particular to ^{68}Ge and ^{65}Zn. These isotopes have rather large half-lives (271 and 243.8 days, respectively), and could be a problem in very low background experiments like those on dark matter.

The first indication of detection of nuclear recoils in thermal detectors is due to the Bonus Collaboration,[34] and has been proved by the Milano group,[35] which has also measured the thermal efficiency. A germanium detector has been implanted with ^{224}Ra from a ^{228}Th source that is also used to bombard the detector. As a consequence, only the peaks at 5.34 and 5.42 due to ^{228}Th appear as single [FIG. 6(a)]. All the others, coming from the ^{224}Ra chain, are doublets with the low-energy line due to the α-particle alone, and the other due to the sum of the energies of the α-particle and recoil energy of the implanted nucleus. In order to prove this interpretation, the nuclei have been left to decay for 15 days, about four half-lives of ^{224}Ra. As shown in FIGURE 6(b), the doublets have almost disappeared. The split of the doublets, which is due to the recoil, indicate that the thermal efficiency for the recoil itself is 1.20 ± 0.15 larger than for an α-particle, that is, it is 100 percent compatible. Nuclear recoils have been detected thermally[36] by exposing Si and LiF crystals of 4 and 32 g, respectively, to neutrons from an Am/Be source. As shown by FIGURE 7, their spectrum is in excellent agreement with the prediction of a Monte Carlo calculation that also takes into account the 4.4-MeV γ-rays emitted by the source.

UNDERGROUND CRYOGENICS

The contemporary desire to reach low temperatures and a low background of spurious events is leading us to a new interdisciplinary field which I would like to call "underground cryogenics." Experiments on rare low-energy events carried out with "standard" techniques have already achieved an extraordinary reduction of this background by (1) installing the detector underground in order to reduce cosmic rays; (2) shielding it against local radioactivity; and (3) using previously tested low-activity materials in the construction of the detector. The Milano group has constructed, in collaboration with Oxford Instruments, a low background dilution refrigerator where all materials present in substantial amounts had a radioactive contamination of less then 10^{-2} Bq/kg. This already mentioned installation (FIG. 2) is currently operating in the Gran Sasso Laboratory. The present counting rate in the region of neutrinoless double beta decay (2–3 MeV) is of about 10^{-3} counts keV^{-1} h^{-1} kg^{-1}, which is lower by more that 1000 then on the surface, and is not far from the rate in the present germanium experiments. Similar underground facilities to search specifically for dark matter are planned by the United Kingdom and French Dark Matter Collaboration, by the Max Planck Institute–TUM Collaboration in Munich, and by the Center for Particle Astrophysics in Berkeley.

Underground cryogenics could also allow indirect detection of DM.[c] As shown in

[c] I am greatly indebted to O. V. Lounasmaa, T. O. Niinikoski, and F. Pobell for very interesting discussions on this subject.

TABLE 1, DM particles of mass below 1 GeV produce recoils that are nondetectable by ionization and sometimes even by thermal detectors of single particles.[37] They could, however, heat a solid, or simply "heat" the nuclei by producing a spin-flip without imparting a visible nuclear recoil. In the field of ultralow temperature (tens of microkelvin), where nuclear demagnetization refrigerators are used in series with a dilution refrigerator, the ultimate limit is represented by the *heat leak.* There is a contribution to this leak[8] from cosmic rays and radioactivity, which is reported in TABLE 2. Recently, the Bayreuth refrigerator[38] has reached 12 μK with a heat leak of 4×10^{-14} W/g^{-1}, which is only three times larger than expected for radioactivity and cosmic rays. We are not far from the need to go underground, where the "environmental" heat leak can be of 10^{-18} W/g^{-1} or less. This figure, easy to be reached in a large bloc of copper, would correspond by about 20,000 coherent interactions of 100-GeV DM particles per kilogram per day.

It is sometimes surprising for those working in fundamental physics to realize that a thermal experiment of DM has already been carried out and published.[39] This experiment, on the nuclear magnetism of silver nuclei, is performed with a cascade refrigerator similar to the one operating in Bayreuth. The ultimate "nuclear temperature" has been measured by nuclear orientation of silver and found to be ~17 nK. The warmup rate is slow (~3 nK/h) due to the long spin-lattice relaxation

TABLE 2. Predicted Environmental Heat from Measurements with Ge Diodes (W/g)

	Cosmic Rays	Radioactivity	Total
Milano (unshielded)			1.4×10^{-14}
Milano (shielded)	0.8×10^{-14}		0.8×10^{-14}
Mont Blanc (unshielded)	—	2.8×10^{-14}	2.8×10^{-14}
Mont Blanc (shielded)		2×10^{-18}	2×10^{18}

time in silver and to a "nuclear" heat leak of 5×10^{-16} W/g. In fact, at least 80 percent of the warmup rate can be accounted for by the usual relaxation, which leaves ~10^{-16} W/g as a limit for the heat produced by DM. We would like to note, however, that this limit is model dependent, since it refers only to DM particles interacting by spin-flip. It is, in fact, already two orders of magnitude lower than the "atomic heat" produced by cosmic rays.

CONCLUSIONS

If gravitation is not the only iteration of wimps, their most detectable product should be heat. Nuclear recoils produced by them can deliver almost all their energy in the form of heat, about ten times more than in the form of ionization. The possibility of measuring heat together with ionization or scintillation is very promising and should be pursued strongly.

It is essential to install the experiments on heavy DM particles underground, and to devote all efforts to reducing not only radioactive contaminations but also all electromagnetic interference, electronic noise, microphonics, etc., that could "heat" the detector. This could reduce the heat leak to such an extent as to allow it to reveal DM from the heat it produces in the detector!

The glorious history of the contributions of cryogenics to fundamental physics is going to enrich itself. One cryogenic detector is already operating underground and four new ones will join it soon. We are at the beginning of a new technique that could bring unexpected new discoveries. The detection of elusive particles like wimps could be one of the most important of them.

DEDICATION

I would like to dedicate this paper to the memory of Edoardo Amaldi as an inadequate expression of my gratitude. Like all those he honored with his friendship, advice, and help, I know it was a unique priviledge to have met a scientist and a man like him.

REFERENCES

1. SMITH, P. F. & J. D. LEWIN. 1990. Phys. Rep. **187:** 204; also for previous references.
2. CALDWELL, D. O. 1990. The search for dark matter. Mod. Phys. A. In press; also for previous references.
3. PAGEL, B. E. J. 1992. Primordial nucleosynthesis and observed light element aboundances. This issue.
4. See, for instance, WILKERSON, J. F. 1990. Neutrino mass limit from tritium beta decay. Presented at the Neutrino 90 Conference, CERN, June 1990. (Nucl. Phys.)
5. BOTTINO, A., *et al.* 1991. Indirect search for neutralinos at the neutrino telescopes. Preprint.
6. ELLIS, J. 1990. Phys. Lett. **245B:** 251.
7. ELLIS, J. & R. FLORES. 1991. Elastic supersymmetric relic-nucleus scattering revisited. Preprint.
8. FIORINI, E. & T. O. NIINIKOSKI. 1984. Nucl. Instrum. Methods **224:** 83.
9. FIORINI, E. 1991. Physica **B167:** 191.
10. BARNES, P. D., *et al.* 1990. Low-background underground facilities for the direct detection of dark matter. UCSB-HEP-90-15. Preprint.
11. FIORINI, E. 1960. Suppl. Nuovo Cimento **17:** 132.
12. MOE, M. Experimental review of double beta decay. Presented at the Neutrino 90 Conference, CERN, June 1990. [Nucl Phys. B (Proc. Suppl.) 19: 158 (1991).]
13. RICH, J. 1990. Ionization detectors for heavy particles. Presented at the 15th Texas/ESO-CERN Symposium on Relativistic Astrophysics, Cosmology and Fundamental Physics, Brighton, England, Dec. 16–21, 1990.
14. AHLEN, S. P., F. T. AVIGNONE, R. L. BROZINSKI, A. K. DRUKIER, G. GELMINI & D. N. SPERGEL. 1987. Phys. Lett. **195:** 603.
15. REUSSER, D., *et al.* 1991. Phys. Lett. B **255:** 143.
16. CALDWELL, D. O., *et al.* 1990. Phys. Rev. Lett. **65:** 1967.
17. RICH, J. & M. SPIRO. 1988. Saclay preprint DPhPe 88-4.
18. GERBIER, G., *et al.* 1990. Measurement of the ionization of slow silicon nuclei in silicon for the calibration of a silicon dark matter detector. DPhE 90-02, Feb. 1990.
19. For details on thermal detectors, see PRETZL, K., Ed. 1988. Proceedings of the Conference on Low Temperature Detectors for Neutrinos and Dark Matter, Vol. I. Springer-Verlag. New York/Berlin; GONZALES-MESTRES, V. & L. PERRET-GALLIX, Eds. 1989. Proceedings of the Conference on Low Temperature Detectors for Neutrinos and Dark Matter, Vol. II. Editions Frontieres. Gif-sur-Yvette, France; BROGIATO, L., D. CAMIN & E. FIORINI, Eds. 1990. Proceedings of the Conference on Low Temperature Detec-

tors for Neutrinos and Dark Matter, Vol. II. Editions Frontieres. Gif-sur-Yvette, France; WAYSAND, G. & G. CHARDIN Eds. Superconducting and Low Temperature Particle Detectors. Elsevier. Amsterdam, the Netherlands.
20. VON FEILITZSCH, F. 1990. Phonon detectors for heavy particles and neutrinos. Presented at the 15th Texas/ESO-CERN Symposium on Relativistic Astrophysics, Cosmology and Fundamental Physics, Brighton, England, Dec. 16–21, 1990.
21. GONZALES-MESTRES, L. & D. PERRET-GALLIX. 1989. New results on detector developments for low energy neutrinos and dark matter. LAPP-EXP-89-07, June 1989.
22. MOSELEY, S. H., J. C. MATHER & D. MCCAMMON. 1984. J. Appl. Phys. **56:** 1257.
23. SEIDEL, W., W. CHRISTEN, F. VON FEILITZSCH, G. FORSTER, H. GOBEL, F. PROEBST & R. MOESSBAUER. 1990. Phys. Lett. **B483:** 483.
24. UMLAUF, E. & M. BUHLER. 1989. Development of a magnetic calorimeter for neutrino detection. *In* Superconducting and Low Temperature Particle Detectors, G. Waysand and G. Chardin, Eds. Elsevier. North-Holland, the Netherlands.
25. GALLINARO, G., F. GATTI & S. VITALE. 1991. Europhys. Lett. **14:** 225.
26. MCCAMMON, D. 1991. Private communication.
27. SMITH, P. F., G. J. HOMER, S. F. J. READ, D. J. WHITE, J. D. LEWIN & N. J. C. SPOONER. 1990. Phys. Lett. B **245:** 265.
28. CORON, N. *et al.* 1989. Thermal spectrometry of particles and γ-rays with cooled composite bolometers of mass up to 25 grams. *In* Superconducting and Low temperature Particle Detectors, G. Waysand and G. Chardin, Eds., Elsevier. Amsterdam, the Netherlands.
29. ALESSANDRELLO, A., *et al.* Cryogenic detectors for rare decays. Nucl. Instrum. Methods. In press.
30. ALESSANDRELLO, A., *et al.* 1990. Phys. Lett. **247:** 442.
31. SADOULET, B. 1990. Summary talk on Dark Matter. Presented at the 15th Texas/ESO-CERN Symposium on Relativistic Astrophysics, Cosmology and Fundamental Physics, Brighton, England, Dec. 16–21, 1990.
32. WANG, M. 1991. Private communication.
33. SADOULET, B., N. WANG, F. C. WELLSTOOD, E. E. HALLER & J. BEEMAN. 1990. Phys. Rev. B **41:** 3761.
34. STROKE, H. H., *et al.* 1986. IEEE Trans. Nucl. Sci. **NS-33:** 759.
35. ALESSANDRELLO, A., D. V. CAMIN, E. FIORINI & A. GIULIANI. 1988. Phys. Lett. **202:** 611.
36. SMITH, P. F., G. J. HOMER, S. F. J. READ, D. J. WHITE, M. J. LEWIN & N. SPOONER. 1991. Phys. Lett. B **255:** 454.
37. NIINIKOSKI, T. O. 1988. Ann. Phys. (Paris) Suppl. **13:** 143.
38. GLOOS, K., P. SMEIBIDL, C. KENNEDY, A. SINGSAAS, P. SEKOWSKI, R. M. MULLER & F. POBELL. 1988. J. Low Temp. Phys. **73:** 101.
39. HAKONEN, P. J., S. YIN & O. V. LOUNASMAA. 1990. Phys. Rev. Lett. **64:** 2707.

Recent Results on Cryogenic Detector Developments[a]

FRANZ VON FEILITZSCH

Physik-Department
Technische Universität München
D-8046 Garching, Germany

INTRODUCTION

Among many other experimental developments in other fields of physics numerous attempts to explore the limits of the standard theory of electroweak interaction as well as new generation experiments in X-ray astrophysics and cosmology require detectors with considerably improved energy resolution and very low-energy threshold. Currently used detection techniques based upon semiconductors or gas ionization have been developed for several decades. They are scarcely expected to be further improved considerably. It therefore seems advisable to search for new detection principles.

Generalizing, one can understand a detector as a system consisting of a material in which the absorption of energy leads to detectable excitations. The amount of these excitations is a measure of the energy deposited. The minimal energy required for one excitation determines the detection threshold energy and the statistics of the excitations. This statistics for a given amount of energy absorbed finally limits the energy resolution if the detector operates in an idealized way.

For the operation of a detector it is important that no other radiation than the one that is supposed to be detected leads to an excitation in the detector. First of all, to avoid thermal excitations this requires

$$KT \ll \Delta,$$

where Δ is the energy gap for the excitations mentioned, and K and T are the Boltzmann constant and the operating temperature, respectively.

Basically, any solid-state or atomic property showing a level structure with small energy gaps can serve as a basis for such a detector, provided an adequate readout technique is available. There is, in fact, an increasing number of experimeters working on such new detection techniques. In this paper I shall discuss briefly developments based upon phonon excitations and the detection of quasi particles in superconductors.

DETECTION OF PHONONS

Phonons are lattice vibrations in a solid-state crystal, and can be excited by the absorption of radiation. Typical phonon excitation energies are less than 10^{-3} eV,

[a]This work was supported by the Bundesministerium für Ferschung und Technologie (BMFT) of the Federal Republic of Germany.

and thus well below the ionization energy in semiconductors. The primary frequency distribution of the phonons depends upon the energy density at the point of absorption. High energetic phonons decay into low energetic phonons with a lifetime that strongly depends on the phonon frequency. Acoustic phonons, for example, decay spontaneously through the effect of anharmonic lattice potentials with a rate given by

$$\Gamma_{\text{dec}} = \gamma\omega_D \left(\frac{\omega}{\omega_D}\right)^5,$$

where $\gamma = 3.3 \times 10^{-4}$ for longitudinal phonons in Si and ω_D is the Debeye frequency.

During their lifetime the phonons propagate through the crystal and may be scattered on crystal defects of any kind. These may be impurities, lattice defects, or isotopic atoms.

In a pure crystal low energetic phonons may propagate ballisticly with the speed of sound ($v \sim 5000$ m/s) over long distances. Due to the anisotropy of the lattice structure of a crystal the ballistic phonons show preferential propagation in certain lattice directions (phonon focusing).

There are two possibilities for phonon detection. The detection shortly after their creation by radiation absorption in the detector consisting of an absorber and phonon sensors on top of it. In this case the phonons may not yet have reached a homogeneous spatial distribution nor a thermal frequency distribution. The second possibility is to measure the phonons at a later time when they have reached thermal equilibrium, manifesting themselves as an increase in the temperature of the absorber.

In the first case, they propagate as nonequilibrium phonons undergoing decay, scattering, and ballistic propagation. During this time period, using several phonon sensors on the surface of the absorber crystal, the point of radiation absorption in the crystal can be determined. This can be achieved by means of the time differences between the arrival of the phonon wavefront at the different phonon sensors.

An experiment demonstrating this possibility was performed by Th. Peterreins *et al.*[1] Three $Al/Al_2O_3/Al$ tunnel diodes were evaporated on one surface of Si monocrystals with the dimensions $10 \times 20 \times 3$ mm^3 and $10 \times 20 \times 10$ mm^3. The sensitivity to phonons of these tunnel diodes has an energy threshold due to the Cooper pair binding energy Δ in the Al films. The minimal energy of phonons being detected is

$$\hbar\omega_{\text{Ph}} = 2\Delta_{\text{Al}},$$

which is for aluminum 0.17 meV.

In FIGURE 1 the geometry of this experiment is shown. The Si crystal was irradiated with 5.5 MeV α-particles at five different points opposite to the surface onto which the Al-tunnel diodes where evaporated. This way the time differences between the arrival of the α-induced phonon pulses as well as their pulse hights could be measured for each α-particle absorbed by the crystal. The measured effective speed of the phonon wavefront was 520 m/sec, which is approximately ten times slower than what is expected for purely ballistic phonons. This indicates that scattering dominates the propagation of the phonons observed in this case.

FIGURE 2 shows the time correlation of the signals between diode 1 and diode 3

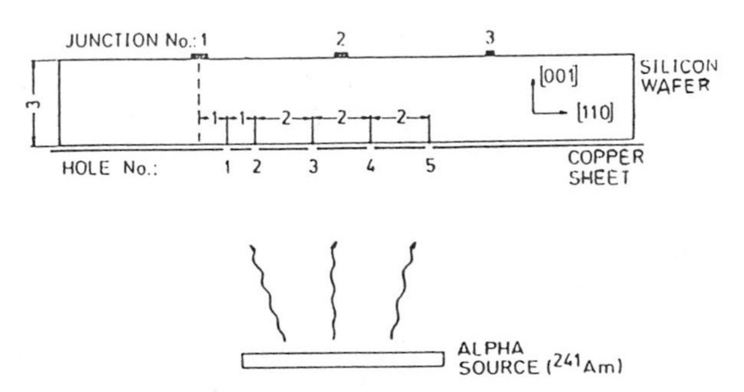

FIGURE 1. Geometry of the experiment to measure nonequilibrium phonons with three superconducting Al-tunnel junctions.[1]

(see FIG. 1) for the five points of irradiation leading to a position sensitivity of 200 μm full width at half maximum (FWHM).

After an integration time of about 1 ms the phonons are approximately homogeneously spread over the whole crystal, even though they still have not reached thermal equilibrium. This was measured by comparing the pulse hights between the different tunnel diodes for different integration times. A measurement of the rise times of the phonon-induced pulses in the tunnel diodes allows for a distinction between focused phonons and phonons coming from other directions. This is depicted in FIGURE 3 where the inverse rise time versus the pulse hight is plotted.

The signals stemming from phonons propagating along focusing directions are clearly separated from those stemming from other directions.[2]

For the detection of phonons in thermal equilibrium superconducting tunnel junctions are not adequate. Here a sensor with an energy gap Δ being smaller than the energy of thermal phonons is required. For this sensor essentially two systems

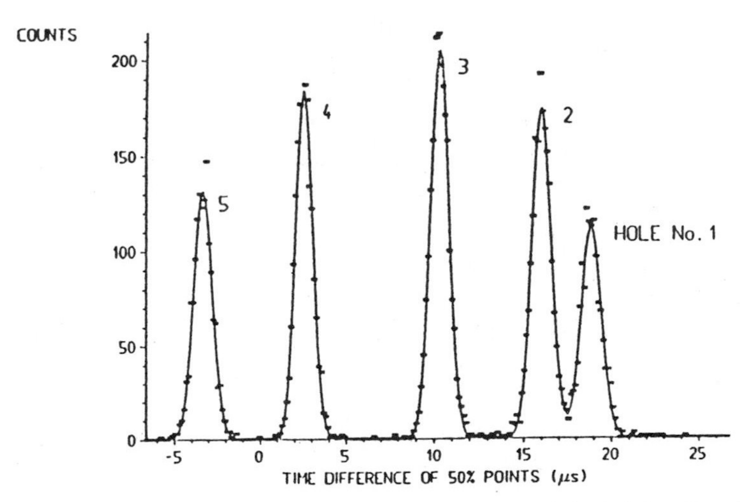

FIGURE 2. Time correlations of the signals between diode 1 and diode 3 (see FIG. 1) for 5 points of irradiation.

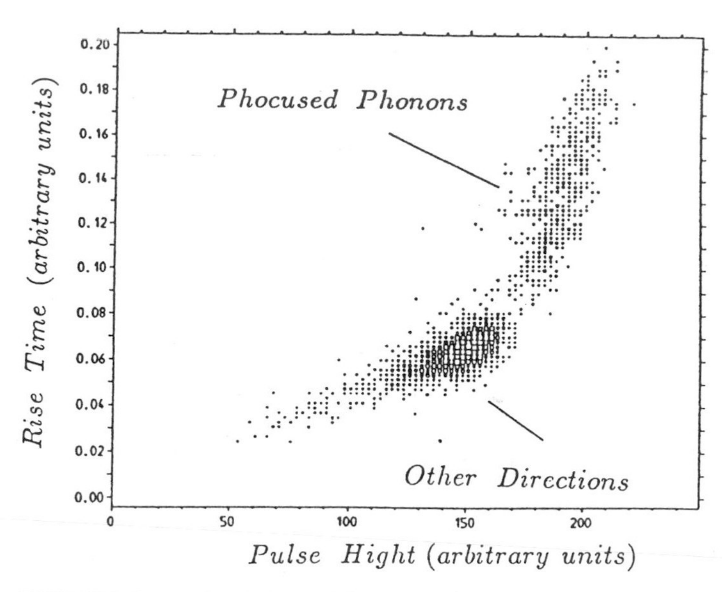

FIGURE 3. Separation of focused phonons and those from other directions.

have been explored recently, superconducting phase-transition thermometers and specially doped semiconducting thermistors.

With a superconducting Ir phase-transition thermometer the α-induced rise in temperature of a 280-g sapphire was measured with an accuracy of 1.2 percent and an integral nonlinearity of less than 0.3 percent.[3] This accuracy corresponds to 54 nK if the pulse is interpreted as being purely thermal.

In FIGURE 4 the principle of detection with a phase-transition thermometer is shown. The thermometer consisting of a thin superconducting stripe is thermally stabilized at its transition temperature where the electric resistivity changes from the superconducting to the normal conducting state.

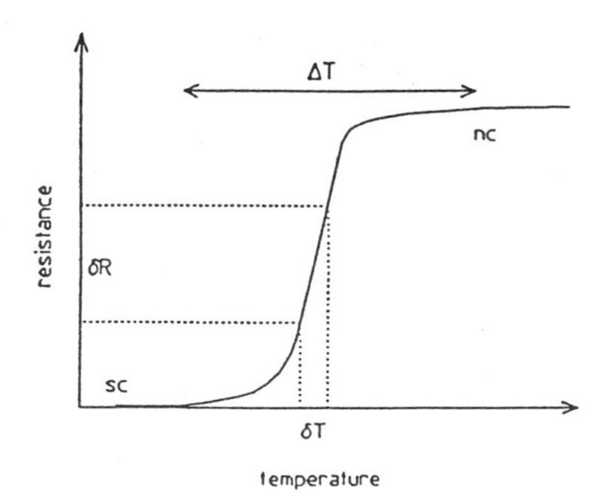

FIGURE 4. Principle of temperature measurement with a phase-transition thermometer.

A measurement of its resistivity in this transition range is very sensitive to small changes in temperature. In the experiment previously mentioned the resistivity was measured by means of a DC-SQUID system.

Another purely thermal measurement of radiation absorption was performed using a specially doped silicon thermistor with a small-gap semiconducting absorber glued on top of the thermistor.[4] The signal was measured as a change in resistivity of the semiconducting thermistor by means of conventional electronics. The absorber with a mass of ~ 10 μg was irradiated with ^{55}Mn X-rays and showed an energy resolution of 7 eV for the 5.9 keV ^{55}Mn$K_{\alpha_{1,2}}$ line, which is the best energy resolution obtained with a calorimeter up to now.

In FIGURE 5 the measured spectrum is shown indicating already a separation of the K_{α_1} and K_{α_2} lines.

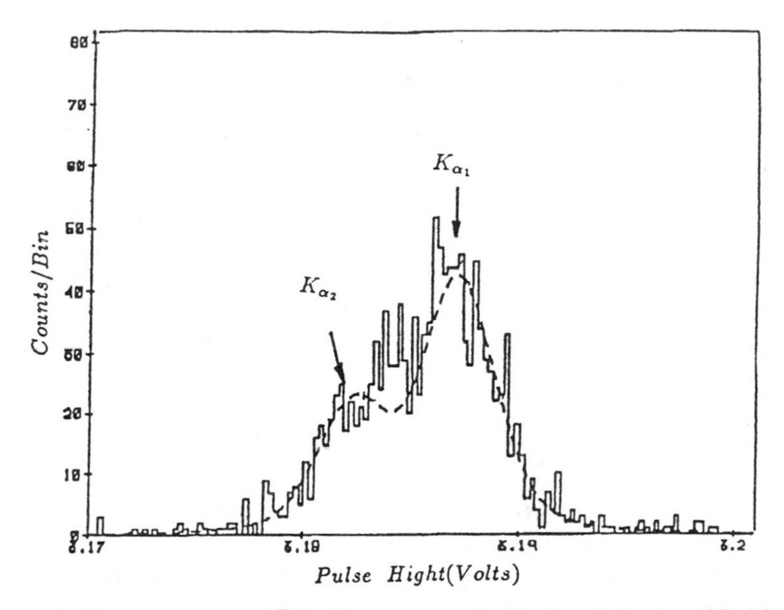

FIGURE 5. Energy spectrum of ^{55}Mn X-rays measured with the calorimeter of D. McCammon.[4]

In a resistive thermometer, due to the readout current, heat is dissipated and a thermal noise is added to the signal. An increasing current leads to an increasing signal, but also to an increasing thermal noise. In principle a capacitive or inductive readout might be more favorable from this point of few if they have sufficient sensitivity.

DETECTION OF QUASI PARTICLES IN A SUPERCONDUCTOR

Instead of the detection of phonons the breaking of Cooper pairs in a superconductor can be detected directly if the radiation is absorbed in the superconductor.

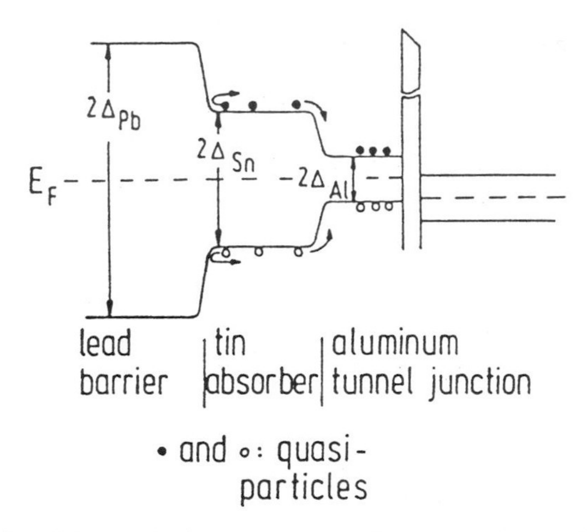

FIGURE 6. Quasi particle trapping for a sequence of the superconductors Pb, Sn, and Al. The decreasing energy gaps, Δ, representing the Cooper pair binding energies, are drawn. Quasi particle diffusion in the direction of decreasing Δ is possible; diffusion in the opposite direction is energetically forbidden.

The minimal energy for this process, depending on the critical temperature of the superconductor, is given by the Cooper pair binding energy being of the order of millielectronvolts or less. The detection principle was demonstrated in experiments where superconducting tunnel diodes were irradiated with α-particles and X-rays.[5] The radiation-induced signal, represented by an increase of the quasi particle tunnel current in the diode, is proportional to the quasi particle density in the tunnel diode. Therefore large tunnel diodes are not favorable, since they lead to a reduced quasi particle density for a given deposited energy and to the large capacity of the diode. On the other hand, a detector may require a certain size for adequate detection efficiency and solid angle. To overcome this problem quasi particle trapping has been independently suggested and demonstrated by N. Booth and H. Kraus.[6]

Quasi particle trapping occurs from a superconductor 1 with Δ_1 into a supercon-

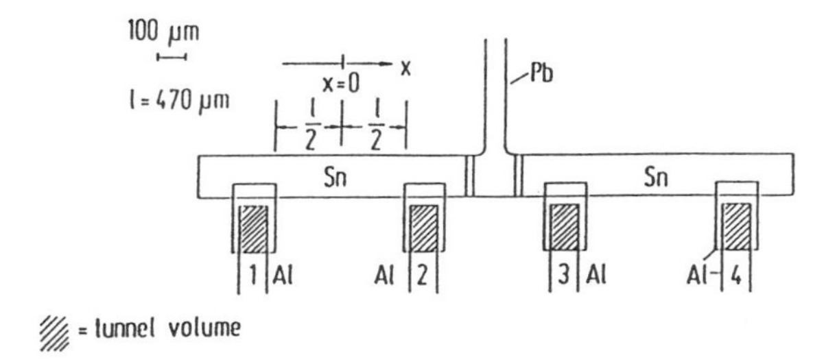

FIGURE 7. Geometry of the detector to demonstrate quasi particle trapping. The Pb serves as electrical contact and as barrier for quasi particles created in the Sn absorber. The quasi particles are trapped and detected in the $Al/Al_2O_3/Al$ tunnel diodes numbered with 1 to 4.

ductor 2 with Δ_2 if $\Delta_1 > \Delta_2$. This is schematically depicted in FIGURE 6 for a sequence of Pb, Sn, and Al.

H. Kraus used a superconducting Sn film as an absorber for radiation, Al/Al_2O_3/Al tunnel diodes into which quasi particles were trapped for detection, and Pb films for electrical contacts. The geometry of this detector is shown in FIGURE 7.

Quasi particles created in the Sn stripe cannot diffuse into the Pb contact line, but are trapped into the Al tunnel diodes. A comparison of the pulse heights in diode 1 and diode 2, as well as for diode 3 and diode 4, gives information on the position along the Sn stripe where the radiation was absorbed. The sum of the signals is proportional to the total amount of quasi particles created in the Sn absorber, and hence proportional to the energy absorbed. The Sn absorber was irradiated with ^{55}Mn X-rays. This way a position resolution of better than 5 μm and an energy resolution of 50 eV were achieved.

REFERENCES

1. PETERREINS, TH., *et al.* 1988. Phys. Lett. **B202:** 161; STRICKER, D. A., *et al.* Submitted for publication in IEEE Transactions on Magnetics; YOUNG, B. A., B. CABRERA & A. T. LEE. Submitted for publication in Phys. Rev. Lett.
2. JOCHUM, J. 1989. Diplom Theses, TUM.
3. SEIDEL, W., *et al.* 1990. Phys. Lett. **B236:** 483.
4. MCCAMMON, D., *et al.* 1990. *In* Low Temperature Detectors for Neutrinos and Dark Matter III, L. Brogiato, D. V. Camin, and E. Fiorini, Eds. Editions Frontières. Gif-sur-Yvette, France. Also private communications.
5. KRAUS, H., *et al.* 1986. Europhys. Lett. **1:** 161; TWERENBOLD, D. 1986. Europhys. Lett. **1:** 209.
6. BOOTH, N. 1987. Appl. Phys. Lett. **50:** 293; GOLDIE, D. J., *et al.* 1990. Phys. Rev. Lett. **64:** 954; KRAUS, H. 1985. Diplom Theses, TUM; KRAUS, H., *et al.* 1989. Phys. Lett. **B231:** 195.

Neutrino Astronomy with Large Cerenkov Detectors

JOHN G. LEARNED

Department of Physics and Astronomy
University of Hawaii, Manoa
2505 Correa Road
Honolulu, Hawaii 96822

INTRODUCTION

In this paper we summarize the present status of experiments and prospects for future high-energy neutrino astrophysics endeavors employing Cerenkov radiation detection. Economics almost surely dictate the employment of massive water (or ice) Cerenkov detectors in the next decades in beginning neutrino astronomy, since present and near-future underground experiments may not be large enough to detect point sources of neutrinos. Roughly six third-generation detectors (in the $>10{,}000\text{-m}^2$ class) are in various stages of proposal, test, or construction, some of which will come to operation by the mid-1990s. We describe the DUMAND II detector, now in construction for deployment in Hawaii for operation beginning in late 1993, in some detail. We also briefly discuss several novel detection techniques, particularly for application in the Antarctic, and examine prospects for the future. For the present it seems that water (or ice) Cerenkov detectors employing photomultipliers remain the most cost effective means to reach the 1 km^2 sizes needed for neutrino astronomy in the next generation.

NEUTRINO ASTRONOMY AND CERENKOV DETECTORS

Physicists have dreamed of starting neutrino astronomy since the mid 1960s.[a] Nobody questions the importance of neutrinos in cosmology, in that neutrinos outnumber protons by 10^9 or so, and in that if neutrinos have only a small mass (>30 eV, summed over flavors), they may dominate the mass of the universe. Unfortunately, no practical plans have been put forward to detect these relic neutrinos, so we must begin with the much more easily detected and surprisingly abundant higher energy neutrinos. For example, more cosmic-ray neutrinos exist at the earth's surface and below, at all energies, than any other (free) particle of similar kinetic energy (e.g., high-energy muons). The miniscule interaction cross section for neutrinos, down by about 14 orders of magnitude from gamma rays at 1 GeV, drives one to think about detecting higher energies.

One might attempt to "see" the universe in neutrinos produced by nuclear processes, in the 10-MeV range. Unfortunately, though stars produce huge fluxes of

[a] It is a little hard to say just who thought of the idea first, but early dreamers were certainly K. Greisen, M. A. Markov, and F. Reines.

neutrinos due to nuclear burning, we have had a very hard time even detecting our own sun (see the various papers summarizing the "solar neutrino problem" at this meeting[1]). Detecting stellar fusion neutrinos from point sources beyond our solar system (or even the aggregate of many sources, as from the whole galactic nucleus) seems unlikely (and terrestrial reactors would provide a significant background).

However, the tremendous burst of neutrinos from gravitational stellar collapse (GSC) should be fairly easily detectable from throughout our galaxy. The dramatic observation of supernova SN1987A by Irvine, Michigan, and Brookhaven (IMB)[2] and Kamiokande,[3] yielded about 10 events above 10 MeV in 1000 tons of Cherenkov detector from a distance of 50 kpc [the Large Magellanic Cloud (LMC)]. The rate of Type II supernovae in our galaxy being no more frequent than one in 10 to 50 years[b] (and quiet GSC not much more frequent than once every few years), one would like to observe neutrinos from our galactic cluster. Recently, there has been renewed interest in the possibility for a detector of a few megatons effective volume, massive enough to sense events out to a few megaparsecs, detecting roughly one supernova per year.[4] We will not discuss that initiative further here, but will focus upon the detection of higher energies where beginning neutrino astronomy may be easier, and probably will be realized earlier.

The attraction of utilizing natural bodies of water for achieving the necessary megatonnage needed to begin neutrino astronomy has long been obvious. The least expensive of bulk materials, say concrete, cost on the order of US$25–$50/ton. Even (reverse osmosis or micropore) filtered water costs this much if one includes the cost of the filtration plant, plumbing, and container. While the costs of deploying detectors in the deep ocean, deep lakes, or polar ice may be higher than in a laboratory situation, these differences tend to become smaller as the scale of the detector grows.

In general the surface arrays must pay a penalty of high civil engineering costs and greater photocathode coverage per unit area (in order to beat the fierce downward flux of muons, dominating the upward muon flux by 10^{12}), while receiving the benefit of accessibility. The net photocathode penalty of working at the surface is about a factor of 2 compared to deep detectors, but when further accounting for effective solid angle of the array, surface Cerenkov detectors require about six times the photocathode area per unit solid angle area for neutrino detection.

The advantage of a Cerenkov detector in contrast with a scintillation counter or tracking counter is simply that the expensive part of the apparatus need not intercept the track and only occupies a small fraction of the track sensing area. In Monte Carlo simulations for many different geometries of deep ocean muon detectors we have found that one needs about one (40 cm diameter) photomultiplier per 100 m^2 of total array effective area, or about 0.0012 fractional projected photocathode area, largely independently of geometry of the array. At a typical large photomultiplier cost of roughly \$2.5/cm^2, this translates to \$18/m^2 of array effective area. One may compare this cost with a cost of a few hundred dollars per square meter for any kind of ordinary counter that intercepts the track ($X = Y$, at least two planes).

[b] S. T. Dye, University of Hawaii doctoral dissertation, Dec. 1988 (Preprint UH-511-667-89), contains a good summary of various means of inferring the supernova rate in the Milky Way. See also reference 4.

One may wonder if photomultiplier tubes (PMTs) can be improved upon for detecting the Cerenkov light. As yet, solid-state detectors remain too expensive and noise limited. Gas photocathodes, as employed in ring-imaging Cerenkov detectors at accelerators, seem quite possibly useful when coupled to grid multiplier structures. So far these chambers have only been developed with gases that have sensitivity in the UV below about 200 nm; for use in water the sensitivity would need to be in the 300–500-nm range.

Another approach, improving traditional PMTs by coupling them to wavelength shifters may yield collection gains in the range of a factor of 2. However, this improvement was explored and rejected by the DUMAND group (mainly for mechanical reasons), and was not very successful for the IMB-2 detector (increases noise and smears timing). Reflectors may help enclosed detectors, and have recently been added to the Kamiokande detector. (However, a cylindrical reflecting surface, in the short-lived HPW detector in the Silver King mine in Utah, made event reconstruction nearly impossible.)

One may also consider radiation in other parts of the electromagnetic spectrum. Only the radio frequencies have reasonable transmission distances in solid media. Because Cerenkov radiation depends upon sign, most of the radiation from a particle cascade cancels out. The coherent signal power (depending upon the square of the net radiating charge excess) arises mainly from electrons around the periphery of the cascade by Compton scattering. Recent calculations[5] indicate that the energy threshold for detection of a cascade at 1 km is discouragingly high, probably above 5 PeV in the best of conditions (good geometry, thermal noise limited).

The only other (identified) potentially useful long-range natural radiation that might be utilized to detect muons or cascades at a distance arises from the acoustic pulse produced by rapid heating of the medium traversed by ionizing particles.[6] The kilometer attenuation lengths of sound in water make tempting the idea of using hydrophones to achieve gigaton detector sizes, except for the catch that the practical detection threshold appears to be in the range of 10^{16} eV.

Thus one is hard-pressed to beat the use of natural water (or ice) and the employment of Cerenkov detection by PMTs for construction of the enormous detectors needed to begin neutrino astronomy. Yet one has no guarantee that this will always be the most cost-effective technique, and particularly not beyond the extremes in high ($>10^{16}$ eV) and low energy (<10 MeV) considered herein.

The DUMAND group has long recognized that the best beginning for neutrino astronomy is likely to be in the teraelectronvolt neutrino regime, via the use of Cerenkov detection of muons from ν_μ charged current interactions.[7] Since the νP cross section rises with energy, the muon range increases with energy, the angle between the neutrino and muon decreases with energy, and the cosmic-ray neutrino background is more steep ($E_\nu^{-3.8}$) than expected sources (E_ν^{-2}), we expect that the signal to noise gains strongly with energy. The rationale for anticipating flat neutrino spectra is fairly broadly based,[8] coming independently from observations in gamma rays, acceleration models, and the relation to the cosmic-ray spectrum (and we have no example model for a steep neutrino spectrum). Folding these factors together, the time required to detect a given source falls with increasing muon threshold energy, out to an energy in the range of 100 GeV to 1 TeV.[9] Depending upon the flux level, one soon runs out of signal, so further raising of the threshold is not useful. Hence a detector sensitive to >100 GeV muons is desirable, which corresponds to neutrino detection in the teraelectronvolt range (for the hypothesized flat spectrum neutrino sources).

Given that teraelectronvolt neutrino detection may be the best for beginning neutrino astronomy, one must face the crucial design question of how large a muon-detection area is needed to get into business. There have been calculations of neutrino flux based upon energetics of celestial objects,[8] but all they really demonstrate is that detectable fluxes, even with existing underground detectors, are allowed (but not required). We do have one way to calculate lower limits on possible neutrino fluxes with reasonable reliability, and that is via the observations of very-high-energy (VHE, teraelectronvolt energy range) and ultrahigh energy (UHE, petaelectronvolt energy range) gamma rays. It is generally believed that at least the UHE gamma rays are beam-dump products (that is, γ's from π° decay), and not electromagnetic in origin. The somewhat delicate requirement of a target thick enough to make gamma rays, but not so thick as to absorb them (10–100 $\mathrm{gm/cm^2}$), makes it seem *a priori* unlikely that one would see any gamma rays.

Nevertheless, if one accepts the (now disputed) VHE and UHE γ observations, one can predict lower limits on the associated neutrino fluxes. Typically, the ν/γ flux ratio is expected to be 1–50 considering time variation. Existing underground neutrino detectors in the range of 100 to 1000 m^2 of muon counting area could potentially detect such point sources. They have not yet done so, and since flux sensitivity will only increase as the square root of time, it seems unlikely that present generation instruments will discover point neutrino sources unless a burst should be observed.

Many possible sources of neutrinos do not have accompanying gamma rays,[8] however. Neutrinos may come from the decay or annihilation of relic particles, associated with the "missing mass" problem. Neutrinos may come from beam-dump situations where the target matter is sufficiently thick to kill the associated gamma rays. And, one of the most interesting prospects are neutrinos from γP interactions wherein the photons from subsequent π° decay are thermalized by $\gamma\gamma \rightarrow e^+e^-$ in the dense photon fields surrounding a compact object. A new model by Stecker *et al.*[10] makes predictions for neutrino fluxes that exceed the cosmic-ray background above about 20 TeV from the integral over all active galactic nuclei (AGNs). This flux would result in thousands of >10 TeV muons/year in DUMAND II, which events would be distinguishable from the cosmic-ray neutrino background by energy alone.[11] In fact, there should be an equivalent ν_e flux such that even direct resonant (6.4 PeV) $\nu_e e^- \rightarrow W^-$ production may be observable (order of 10 events/year contained in DUMAND II at the nominal Stecker *et al.* model flux).

The best current data set for point source neutrino searches, via upcoming muons observed in the IMB detector,[12] shows tempting visual correlations with the highest peaks with the galactic plane that, however, are not statistically compelling.

Hence it seems that detectors one or two orders of magnitude larger than present instruments in area, and with an order of magnitude higher muon energy threshold, may be necessary to detect the first neutrino point sources. However, it has been long recognized that probably a full 10^6 m^2 will be needed to really begin regular neutrino astronomy.[c]

[c] See Proceedings of the 1976 DUMAND Summer Workshop, A. Roberts, Ed., University of Hawaii, Honolulu, 1976.

SURVEY OF HIGH-ENERGY NEUTRINO CERENKOV DETECTORS

The first natural neutrinos were observed in scintillation and flash-tube detectors located in deep mines in South Africa[13] and in India[14] in the mid-1960s, and the larger of them, CWI, collected about 100 events. Another experiment in Utah collected a few events.[15] The first experiments did not have very good directionality, and for most events one could only deduce the projected direction. The energy threshold was low as well, in the 100-MeV range, so that one could not hope to do much with point source astronomy. Nevertheless, these experiments made the first atmospheric neutrino flux measurements, and did set the first limits on extraterrestrial neutrino fluxes.[13,14]

Little activity in the field took place for about a decade, until the search for proton decay became fashionable, in the late 1970s. Since the early 1980s eight large detectors have operated, six of them continuing, and several more are in various stages of proposal, feasibility testing, or construction (see TABLE 1). The two biggest detectors were, and still are, the IMB and Kamiokande water Cerenkov instruments, which have been spectacularly successful in applications from a few megaelectronvolts (e.g., solar neutrinos in Kamiokande) to searches for all manner of higher energy exotica, most prominently nucleon decay, for which these collaborations have reported most of the strongest limits.

The 1990s will see the completion of several new underground detectors, such as MACRO[16] and LVD[17] in Italy, SNO in Canada,[d] and SuperKamiokande[18] in Japan,

TABLE 1. Summary of Large Underground Instruments with High-energy Neutrino Detection Capability, 1960s through mid-1990s

Detector, Location	Status	μ Area (m^2)	Direction of Sens	Technique	Primary Purpose
KGF, South India	X	110	N	LS + FT	Observe ν's
CWI, South Africa	X	10	N	PS + FT + Fe	Observe ν's
Silver King, Utah	X	30	Y	WC + Ctrs + Fe	Observe ν's
KGF, South India	R	20	N	St tubes	PDK
Baksan, Caucasus	R	250	Y	LS tanks	ν's
IMB, Ohio	R	400	Y	WC	PDK
HPW, Utah	X	100	Y	WC	PDK
Kamioka, Japan	R	120	Y	WC	PDK
NUSEX, Mt. Blanc	R	10	N	ST + Fe	PDK
Frejus	X	90	N	ST + Fe	PDK
Soudan I	R	10	N	ST + Concrete	PDK
Soudan II	C/R	100	N	DT + Concrete	PDK
MACRO	C/R	1100	Y	LS + ST +	Monopoles
LVD	C	800	Y	LS tanks + ST	SN ν's
SNO	C 1996	300	Y	D_2O	Solar ν's
SuperKamiokande	C > 1996	740	Y	WC	PDK
Borex	T	<100	Y	LS	Solar ν's

KEY FOR TABLE: P = proposal; T = testing and development; C = construction; R = operating; X = shut down; WC = water Cerenkov; ST = streamer tubes; LS = liquid scintillator; PS = plastic scintillator; FT = flash tubes.

[d] See the paper by Sinclair cited in reference 1.

TABLE 2. Summary of New Initiatives in High-energy Neutrino Astronomy

Detector	Location	Status	μ Area (m^2)	Depth (mwe)	Technique	Threshold
DUMAND II	Hawaii	C 1993	20,000	4,800	WC	20 GeV
Baykal NT-200	Siberia	C 1993	3,000	1,000	WC	10 GeV
SINGAO	S Italy	T	15,000	10–0	RPC	2 GeV
LENA	Japan	T/P	30,000	0–30	WC	6 GeV
GRANDE	Arkansas	P	30,800	0–50	WC	6 GeV
NET	Gran Sass	P	90,000	0–70	WC	11 GeV
NESTOR	SW Greece	D/T	?	4,000	WC	?
?? Sov. DUM, East	Pacific?	D/T	?	?	WC	?
?? GRANDE type	USSR	D	?	0–?	WC	?
?? GRANDE type	Australia	D	?	0–?	WC	?
AMANDA	South Pole	T	?	> 1,000	WC in ice	?
RAMAND	Antarctic	T	10^6	0–1,000	ice μwv	> 100 TeV
?? World Detector	?	D	10^6	> 4,000	WC	> 100 GeV

KEY FOR TABLE: D = discussion; P = proposal (possible operational date); T = testing and development; C = construction (operational date); WC = water Cerenkov detector; RPC = resistive plate chamber; μwv = microwave detection.

so that there will be about nine ongoing experiments underground through the middle of the decade, as summarized in TABLE I. Three of them employ the water Cerenkov technique (IMB, SNO, and SuperKamiokande). Because these detectors approach the maximum stable mine cavity size, much larger underground detectors appear unlikely.

The prospects for other large neutrino detectors, as indicated in TABLE 2, are less clear. We can divide the new $> 10^4$-m^2 initiatives into two classes: surface and underwater (or ice). Detectors located on the earth's surface, such as GRANDE,[19] LENA,[20] and NET,[21] aim at using a covered pond for studying upcoming muons from neutrinos. The detection, as in the deep water detectors, is via the Cerenkov radiation of particles in water, the light being sensed by large downlooking photomultipliers. Surface detectors may also study extensive air showers (EAS) with a layer of upward-facing PMTs in the same detector (though curiously NET does not include uplooking PMTs). An exception to the use of Cerenkov radiation in the surface neutrino detector category is the SINGAO proposal,[22] which would employ novel resistive plate chambers. The second category is characterized by deep water Cerenkov detectors employing open natural bodies of water, generically of the DUMAND type, with strings of PMTs floating upward from bottom moorings.

The underwater approach is being pursued by the international DUMAND collaboration in Hawaii,[e] about which more below, and also by other groups in the USSR[f] and Europe.[g,h] The Soviets have a substantial ongoing program in Lake

[e] The DUMAND Collaboration consists of groups from Aachen, Bern, Boston, Hawaii, KEK, Kiel, Kinki, Kobe, Okayama, Scripps, Tohoku, Tokyo, Vanderbilt, Washington, and Wisconsin.

[f] There appear to be at least three separate groups working on DUMAND-type neutrino detectors in the Soviet Union. One, headed by V. I. Domogatsky and L. Bezrukov is in Baykal (see reference 23), another by I. M. Zheleznykh and N. M. Surin aims toward a detector in the Mediterranean or Atlantic, and a third group under A. A. Petrukhin is directed at a device in

Baykal in Siberia.[23] Another group has been carrying out tests for an ocean-based DUMAND-style instrument, possibly for emplacement in the Mediterranean.[i]

In looking at TABLE I one should be aware that deep underearth detectors (which can look for neutrinos arriving from slightly above the horizon) have a solid angle advantage of about a factor of 3 over flat surface arrays (which must restrict their viewing region to below 20° below the horizon), so that area comparison alone is misleading.

DUMAND HAWAII

The DUMAND organization got started with a series of workshops in the mid-to-late 1970s, with the goal of building a very large under-ocean detector. This stimulated the first serious considerations of types of neutrino experiments, venues, energy ranges, and techniques. The best location was soon realized to be in the abyssal deep off Hawaii. Since the early 1980s the collaboration has been engaged in studying the environment and backgrounds, developing the necessary technology, and carrying out system design studies. A prototype experimental demonstration was carried out in 1988, measuring muon fluxes in the open ocean from depths of 2 to 4 km. The group received U.S. Department of the Environment approval in April 1990, to proceed with a long-term ocean-bottom-moored array, which is scheduled for full operation in late-1993, at a total project cost of about US$10 million.

The design goal for DUMAND II[7] was for a deep-ocean-moored instrument with 20,000 m^2 of muon area and an angular resolution of 1°. The configuration arrived at is a 100-m-diameter octagon of strings, with a ninth string in the center. These strings will float upward from a 4.8-km-deep ocean bottom, about 30 km off the Island of Hawaii, at 19° 44′ N, 156° 19′ W. Each string consists of 24 optical modules, plus laser calibration units, hydrophones (5 per string), and environmental monitoring instruments. The strings will have instrumentation beginning 100 m off the bottom, optical detectors every 10 m for 230 m above, and a float package at the string top, some 350 m off the bottom, to provide tension to keep the strings near vertical in the small ocean bottom currents.

Signals from the PMTs will be digitized to 1-ns accuracy at the base of each string and multiplexed onto a single 625-MBd fiber-optic link for each string. The strings attach to a junction box, which has other instrumentation including TV and lights, and links to shore via a 12-fiber cable that also delivers 5.5 kW to the array. On-shore digital signal processors filter out interesting events for on-line reconstruction. Fast

---

the Pacific. The former two groups are from the Institute of Nuclear Research in Moscow, the latter from the Moscow Engineering Physics Institute.

[g]NESTOR is a project proposal headed by L. K. Resvanis of the University of Athens, Greece, with a detector proposed for Mediterranean waters off southwest Greece.

[h]JULIA is a similar project proposal to NESTOR, headed by P. Bosetti of Aachen, directed toward low-energy neutrinos.

[i]Personal communication from I. M. Zheleznykh and N. M. Surin, report on muon counting tests from the R/V D. Mendeleev during the research cruise of November–December 1989 in the Mediterranean and Atlantic.

data links to collaborating institutions will permit remote monitoring and control, as well as simultaneous data distribution.

An experiment has been proposed to employ the intense neutrino beam from the planned Fermilab main injector to study neutrino oscillations via muon neutrino disappearance enroute to DUMAND.[24] While convincing Fermilab to invest in the 30° downward neutrino beamline will certainly be difficult, the physics to be done is quite unique because the typically 20-GeV ν_μ will have the possibility to experience significant matter oscillations in traversing the 6000-km distance to DUMAND.[25] For an optimal $\delta m^2 \approx 0.01$ eV^2, DUMAND II would observe a measurable deficit out to a (surprisingly small, due to resonance) mixing angle of $\sin^2(2\theta) \approx 0.02$,[25] if the oscillation goes from $\nu_\mu \rightarrow \nu_e$.

NEW APPROACHES AND FUTURE PROSPECTS

Beyond the third-generation high-energy neutrino detectors previously discussed, several prospects have appeared for the farther future, as listed in TABLE II (which should be regarded with due caution). These include employment of microwave radiation from UHE showers in ice (RAMAND), acoustic detection in the ocean, and the possibility of employing Cerenkov radiation in the deep clear antarctic ice.

One must go to the antarctic in order to find ice with a long enough attenuation length to make microwave neutrino detection possible, because the attenuation in ice at the relevant frequencies ($>$300 MHz) only approaches kilometer distances for temperatures below -60°C, as found near the Soviet Vostok Station. The Soviet team has conducted tests for several years, and some U.S. groups have now taken an interest in the possibility as well. However, it remains to be demonstrated that the noise background is low enough and that the technique is practical.

Acoustic detection had a short-lived period of activity about a decade ago, but was largely dropped when it was realized that there was probably no way to get the threshold down from 10 PeV to the teraelectronvolt region where at least atmospheric neutrino signals are known to exist. Some Soviet workers have continued work on the idea, and reportedly will make an experiment in the Atlantic in 1991.[26] The DUMAND group has kept the idea alive as a background operation to the practical purpose of acoustic surveying of the array geometry, and they are now considering the prospects for detecting resonant W^- in the neighborhood of the DUMAND II array. The flash of Cerenkov light from a 6.4-PeV cascade may be seen for several hundred meters, and the arrival time of the acoustic pulse could give the vertex location, thereby permitting event energy determination. Whether this actually will work remains to be determined.

Another novel method for attempting neutrino astronomy in antarctica involves the use of Cerenkov radiation in the ice. An initiative named AMANDA,[27] now getting underway in active field study, plans to place PMTs in three 1-km-deep holes at the South Pole in the winter of 1991–1992. Beyond the obvious attraction of working from a solid surface and simply melting holes in which to place the PMTs, albeit in a difficult environment and inaccessible location, a most intriguing aspect of the proposal is that there should be no optical background in deep clear ice. If one neglects the cost of infrastructure at the South Pole (fuel costs to melt the holes, for example), the technique may be economically attractive as well, though the difference between deep ocean and deep ice may not be much in technical difficulty in the end. This author believes that the major advantage of a deep ice detector may be in the (hopefully) negligible optical background, and thus an advantage in building a major supernova detector with few megaelectronvolt sensitivity and multimegaton effective volume.

All of the previously mentioned new detection techniques are, unfortunately, probably a few years from practical application. Even if realized, the threshold energy may be very high for the microwave technique and the acoustic techniques. Active exploration of the optical technique in ice has just begun, and we need to understand the environment (optical characteristics of ice, depth for bubble-free ice, verification of lack of optical background, etc.) before realistic plans for a large detector can be put forward, which could take place within several years.[27] Hence, it appears that the competitors for beginning very-high-energy neutrino astronomy through middecade will be (some of) the Soviet and Hawaii DUMAND detectors, the GRANDE- and SINGAO-style detectors, and perhaps an AMANDA detector at the South Pole. It is a healthy situation for there to be several such instruments, and with various techniques employed. The first signals are not likely to be large, so independent confirmation will probably be necessary (and usually teaches one something).

Beyond the mid-1990s we need to begin to contemplate the next step, which on a logarithmic scale suggests a 1-km^2 detector. While such a device will almost surely not be realized before the turn of the millennium, it is certainly not too soon to begin to work on the means to achieve such an instrument. The author currently favors an extension of the DUMAND approach, which does scale well to great size, requiring about 10^4 modules (PMTs) and roughly US$100 million. While this represents a reasonable cost scale by present-day accelerator standards, it will probably require a substantial international collaboration to obtain the necessary resources.

However, in order to proceed with such grand visions we must have success in detecting the first astrophysical point sources of high-energy neutrinos in the shorter term. Indeed the physics might point in other directions, which we cannot now know. Perhaps also, new technology will come along that will make other techniques more attractive. We must continue exploring detection technology, and most important, make the upcoming generation of instruments work as well as planned. For now, it appears to this author that the simple Cerenkov detection technique, employing natural bodies of deep ice or ocean and photomultiplier tubes, will be hard to beat.

ACKNOWLEDGMENTS

The author would first like to thank the Brighton meeting organizers for a stimulating meeting. I want to thank all the DUMAND collaborators for the work of theirs upon which I have drawn so heavily for this report. I must say also that the material about other experiments reported upon herein may well be wrong, particularly in reference to those detectors in the proposal stage, about which I have used preliminary and in some instances guessed numbers. Nevertheless, the scale of things presented herein (in TABLE II in particular) are about right, but one should be careful about inferring anything other than overall trend.

REFERENCES

1. SINCLAIR, D. 1992. Direct detection of solar neutrinos. This issue; KIRSTEN, T. 1992. Radiochemical solar neutrino experiments. This issue; STODOLSKY, L. 1992. Low-temperature extragalactic supernova neutrino detector. This issue; ZATSEPIN. 1990. Presented at the 15th Texas/ESO-CERN Symposium on Relativistic Astronomy, Cosmology and Fundamental Physics, Brighton, England, Dec. 16–21, 1990.
2. BIONTA, R. M., *et al.* 1987. Phys. Rev. Lett. **58:** 1494.
3. HIRATA, K., *et al.* 1987. Phys. Rev. Lett. **58:** 1490.
4. TAMMANN, G. A. 1976. *In* Proceedings of the 1976 DUMAND Summer Workshop (hereafter D76), A. Roberts, Ed., University of Hawaii, Honolulu, Sept. 6–19, 1976.
5. HALZEN, F., E. ZAS & T. STANEV. 1990. Preprint MAD/PH/606.
6. LEARNED, J. G. 1979. Phys. Rev. D **19:** 3293, and reference therein.
7. BOSETTI, P., *et al.* 1988. DUMAND II Proposal, HDC-2-88, Aug. 1988.
8. BEREZINSKY, V. S., S. V. BULANOV, V. A. DOGIEL, V. L. GINSBURG & V. S. PTUSKIN. 1990. Astrophysics of Cosmic Rays, Chap. VIII. Elsevier. Amsterdam, the Netherlands.
9. LEARNED, J. G. 1990. Hawaii preprint HDC-7-90.
10. STECKER, F. W., C. DONE, M. H. SALAMON & P. SOMMERS. 1991. NASA-LHEAPTH-91-007. Submitted for publication in Phys. Rev. Lett.
11. LEARNED, J. G. & T. STANEV. 1991. HDC-1-91.
12. BECKER-SZENDY, R., *et al.* 1990. Proceedings of the 25th ICHEP, Singapore, Aug. 1990; SVOBODA, R., *et al.* 1987. Astrophys. J. **315:** 420; and BECKER-SZENDY, R. 1991. Doctoral dissertation. In preparation.
13. REINES, F. & M. F. CROUCH. 1974. Phys. Rev. Lett. **32:** 493.
14. KRISHNASWAMY, M. R., *et al.* 1979. Proceedings of the 16th ICRC. Kyoto, Japan, **13:** 14, 24.
15. BERGESON, H. E., *et al.* 1967. Phys. Rev. Lett. **19:** 1487.
16. CALICCHIO, M., *et al.* (MACRO Collaboration). 1988. Nucl. Instrum. Methods **A264:** 18.
17. PLESS, I. A. 1988. Proceedings of the 13th International Conference on Neutrino Physics and Astrophysics (Boston, June 5–11, 1988), J. Schneps *et al.,* Eds: 297. World Scientific. Singapore.
18. TOTSUKA, Y. 1990. Proceedings of the International Symposium Underground Physics Experiments, at Science Council of Japan, April 1990.
19. ADAMS, A., *et al.* 1990. Proposal to construct the first stage of the GRANDE facility . . . GRANDE Report 90-005. University of California at Irvine, Mar. 1990.
20. SASAKI, M., *et al.* 1988. Proceedings of the 2d Workshop on Elementary Particle Picture of the Universe, KEK, M. Yoshimura *et al.,* Eds.: 181.
21. GENONI, M., *et al.* 1990. Proceedings of the 2d International Workshop on Neutrino Telescopes, M. Baldo-Ceolin, Ed.: 243. Venice.
22. PISTILLI, P. 1988. Proceedings of the 1st International Workshop on Neutrino Telescopes, M. Baldo-Ceolin, Ed.: 344. Venice.
23. BELOLAPATIKOV, I. A., *et al.* 1990. Nucl. Phys. B **14B:** 51.
24. LEARNED, J. G., & V. Z. PETERSON. 1989. Proceedings of the Workshop on Physics at the Main Injector, Fermilab, May 18, 1989. Hawaii preprint HDC-5-89. Fermilab Proposal number P-824, M. Webster spokesperson.
25. PANTALEONE, J. 1990. Hawaii preprint UH-511-699-90.
26. ZHELEZNYKH, I. M. Personal communication.
27. BARWICK, S., F. HALZEN, D. LOWDER, D. MILLER, R. MORSE, B. PRICE & A. WESTPHAL. 1991. Preprint UCI-PA-91-1, University of California at Irvine.

Detection of High-energy Astronomical Point Sources of Neutrinos

P. J. LITCHFIELD

Rutherford Appleton Laboratory
Chilton, Didcot, Oxon
OX11 OQX, England

1. PRODUCTION OF HIGH-ENERGY NEUTRINOS

High energy in this review means above 100 MeV. This means that the neutrinos are the end result of an acceleration process and are nonthermal (i.e., not the outcome of element burning in stars or of supernova collapse). I became interested in this subject from working on the Soudan 2 proton-decay experiment where we are in the business of detecting neutrinos as the main background to proton decay. Also, if previous results are correct, we may be able to detect astronomical point sources using the high-energy cosmic-ray muons passing through the detector. The combination of the two subjects to detect neutrino point sources is obvious. However, the conclusion of this paper is that although it looks quite possible in principle to detect such sources, much bigger, dedicated experiments are probably required. Despite this disappointment the investigation was interesting to me and I hope that this summary of the expectations and problems, garnered from the work of many previous authors, will be interesting to others.

The existence of high-energy neutrino point sources follows directly from the detection of high-energy γ-ray point sources.[1] Some doubts are probably justified on individual claims of observation of Cygnus X-3, Hercules X-1, the Crab nebula, Geminga, Vela X-1, LMC X-4, and others at teraelectronvolt and petaelectronvolt energies, but I take as the starting point for this article that some at least of these observations are correct. Many of the preceding sources are known, from their radiation at lower energies, to be variable binary stars (this could be an experimental bias due to the fact that the extra information of the known time period helps to discriminate the signal from the large backgrounds). This has given rise to the standard model for the production of high-energy γ-rays in which an accelerating mechanism, probably associated with the high magnetic fields present around compact stars, produces a beam of high-energy protons. This beam interacts in a dense medium, producing π°'s, which decay to γ-rays. Possible configurations that could give rise to γ-ray beams are proton beams from the remnant of a supernova interacting in the blown off outer portions of the star, a binary system of a compact star and massive companion in which the beam interacts in the atmosphere of the companion, or a similar system where the beam interacts in an accretion disk around the compact object. Whatever the production mechanism of the π°'s, charged pions must be produced at the same time, and these will decay to produce beams of electron and muon neutrinos.

Despite the similarity in the production mechanism there are important differences in the conditions required to produce a γ and a ν beam. For the γ beam the conditions are quite restrictive. First, a target mass of greater than about 50 gm cm^{-2} is required to efficiently produce π° mesons, but second the total mass of the column between the target and earth must be less than about 200 gm cm^{-2}; otherwise, the γ's are absorbed. For a ν beam, again a target mass of greater than 50 gm cm^{-2} is required to produce $\pi^\pm$ mesons, but now only column masses of greater than about 10^{10} gm cm^{-2} will attenuate the beam. An additional requirement for a neutrino beam is that the $\pi^\pm$ and their decay $\mu^\pm$ should not interact or be significantly slowed in their flight path before decaying. The path length before decay is proportional to the meson energy. Thus, the mass present in the meson flight path produces an upper energy cutoff that is proportional to this mass density. Above this cutoff the ν come mainly from heavy particle (charm and beauty) decays, but the flux from these is down by factors of 10^2–10^3 on the flux from light mesons.

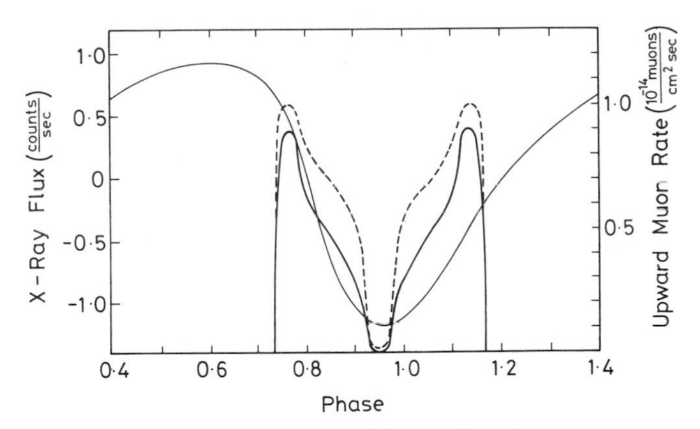

FIGURE 1. The light curve of Cygnus X-3 for X-rays (*thin solid line,* measured) and for ν calculated by Gaisser and Stanev. The *thick solid line* is for a 10^8-GeV beam and the *dashed line* for a power-law spectrum.

A number of authors[2] have made detailed calculations of the neutrino fluxes from typical sources, such as Cygnus X-3 based on the measured γ-ray fluxes. FIGURE 1 shows the ν light curve for Cygnus X-3 calculated by Gaisser and Stanev on the assumption that a proton beam from the compact star interacts in the atmosphere of its companion. The thin solid line, which covers the whole width of the plot, is the measured X-ray light curve, and the remaining lines are the calculated ν light curve for two assumptions on the proton beam, either a monoenergetic 10^8-GeV beam (thick solid line), as in the original Hillas model, or an E^{-2} power-law spectrum (dashed line). It can be seen that the ν's are 180° out of phase with the X-rays, since the X-rays are only visible when the compact star is between us and the companion, while the reverse is the case for ν. The flux begins just as the compact object is eclipsed by the companion and rises rapidly. When the compact star is on the far side of the companion, the flux is reduced due to the absorption of ν's passing through the core of the companion. Gamma-rays would be produced only at the wings of the distribution where the mass between us and the proton source is between about 50 and 200 gm cm^{-2}. There is thus likely to be a large enhancement (typically 20:1) in the flux of ν relative to γ-rays from such a source.

We can now list some of the exciting astrophysical results that would come from the detection of ν as well as γ-rays from high-energy sources.

1. It would confirm the hadronic production mechanism of high-energy photons, and possibly the production mechanism of all high-energy cosmic rays.
2. Comparison of the ν and γ light curves and energy spectra would enable detailed measurements to be made of the density distribution in the target region. These should distinguish between the stellar atmosphere, accretion disk, and other production models.
3. The central dip in the ν light curve would provide unprecedented information on the internal structure of the companion star.
4. In addition ν detection would open up the possibility of finding sources that are currently invisible in γ-rays, either because they are shrouded in dense matter or because the tight γ beams that may be produced miss the earth.

2. DETECTION OF ASTRONOMICAL NEUTRINOS

There are two crucial ingredients in the successful detection of any signal; first, the actual level of the signal, second, the level of the background. In this section I will describe calculations that have been made for both, and give my estimates of the prospect of successfully detecting point sources.

2.1. ν *Flux*

The estimated interaction rates of neutrinos from astronomical sources in today's biggest detectors (proton decay detectors having a mass of several kilotons) is of the order of 1 event every 1000 years. Much bigger detectors have been proposed with masses in the 10^7-ton range. However, bigger detectors in general have higher energy thresholds (100 GeV against 100 MeV in the smaller detectors), and thus, because of the expected falling power-law spectrum, the calculated event rate inside even the largest projected detector is discouragingly low. Only one trick has been devised to overcome this problem. This is to use the material surrounding the detector as the target and detect the muons produced in ν_μ charged-current events. High-energy muons have very long ranges (about 5 km in water at 1 TeV), and the effective target volume is increased to the detector area (A) times the muon range (R_μ). Note that the critical detector parameter is now not its volume, but is area, and that relatively thin detectors are sufficient, at a considerable cost saving. At 1-TeV effective target masses of 100 Mtons are obtained with feasible detector areas of 20,000 m^2. In such a detector the muon range and the ν cross sections increase with ν energy, whereas, of course, the flux decreases with energy. Using an expected E^{-2} flux distribution, this results in a broad maximum in the detection rate at around 10 TeV.

Cherry[3] has calculated the expected event rate for Cygnus X-3, scaled from the measured high-energy γ-ray fluxes. The expected number of observed muons N_μ is given by

$$N_\mu = 0.3NA \int_{\text{threshold}}^{\infty} \lambda(E) \frac{dS_\gamma}{dE} \sigma_\nu R_\mu(E)\, dE,$$

where N is Avogadro's number, A is the detector area, $\lambda(E)$ is the enhancement factor of the ν flux relative to the γ flux, (dS_γ/dE) is the γ-ray flux (parameterized as $E^{-\alpha}$, where α is the spectral function), σ_ν is the neutrino cross section, and R_μ is the muon range. Note that α is a critical parameter because typically the γ-ray fluxes are measured at teraelectronvolt or petaelectronvolt energies, whereas the thresholds for ν detectors go down to the gigaelectronvolt level. Thus small differences in the assumed α give very large differences in the expected numbers of events.

Normalizing to some typical expected values, Cherry obtained

$$N_\mu = \frac{\epsilon A}{0.4 \times 10^4\ \text{m}^2} \cdot \frac{S_\gamma(E_\gamma \geq 3 \times 10^{15}\ \text{eV})}{1.5 \times 10^{-14}\ \text{cm}^{-2}\ \text{sec}^{-1}} \cdot \frac{\lambda}{20} \cdot \frac{1/3}{f_{\mu-\text{wave}}} \cdot \Gamma(\alpha),$$

where N_μ is the number of muons expected/year, ϵ is the source visibility factor (typically 0.4), the detector area is taken as 10^4 m^2, the flux is normalized to the Haverah Park Cygnus X-3 measurement,[4] the ν:γ enhancement factor discussed previously is taken as 20:1, $f_{\mu\text{-wave}}$ is the absorption factor of the γ-rays on the microwave background between the source and earth, and $\Gamma(\alpha)$ is the result of the integrations over E. The following table shows some values of N_μ as a function of α:

N_μ(year^{-1})	α
0.5	1.5
8	2.0
300	2.5
3000	2.75

Note the rapid increase of N_μ with α, and that α's as large as 2.75 are already ruled out by present-day detectors. The models of neutrino production favor α's of around 2.0 that would give measurable numbers of muons/year. It should be noted, however, that recent measurements of the γ flux from Cygnus X-3 have given somewhat lower fluxes; indeed, some experiments have detected nothing at all. This could be due to variability of the source or to favorable statistical fluctuations of the earlier measurements. Obviously an order of magnitude decrease in the measured γ flux would require an order of magnitude increase in detector area to compensate. However, the other quantities in the expression (α, λ, and $f_{\mu\text{-wave}}$) also have large uncertainties so, as for all new physics measurements, one should not place too much reliance on theoretical calculations but go out and do the experiment.

2.2. *Backgrounds*

There are two main backgrounds to muons produced by astronomical ν in underground detectors. The first is muons produced in cosmic-ray interactions in the atmosphere above the detector, and the second is muons produced around the detector by ν formed in cosmic-ray interactions in the atmosphere all around the earth.

The first background is trivially removed by going to sufficiently large absorption lengths of material. FIGURE 2 shows the rate of muons observed in the Kolar gold fields proton-decay experiment (the deepest of all underground experiments) as a function of zenith angle.[5] The transition from atmospherically produced muons to ν produced underground, around the detector, is clearly seen at zenith angles around 60°. For shallower detectors, including detectors close to the surface, atmospherically produced muons can be removed by only accepting muons traveling upward from below 90° zenith angles.

The background due to μ produced in the atmosphere is indistinguishable from the signal and is the ultimate limit of the sensitivity of experiments done on earth. The flux of upward-going muons has been measured in a number of underground detectors, particularly the large proton-decay detectors. FIGURE 3 shows a compilation of data and calculation due to Gaisser and Stanev.[6] The calculation and data agree to within 20 percent. The atmospheric flux is 2–3 orders of magnitude greater than the fluxes calculated in Section 2.1.

Two features rescue the signal. First, the atmospheric neutron spectrum is suppressed at high energies because high-energy muons hit the earth and are ranged out before they decay, producing the neutrinos. This approximately adds an extra factor of E to the spectral function at high energies. Second, point sources are localized, whereas the atmospheric neutrino background is diffuse. The enhancement of signal to noise is then dependent on the resolution of the detector.

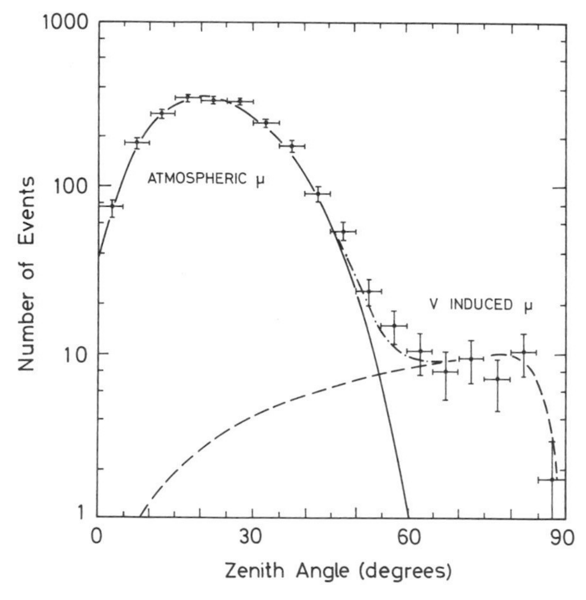

FIGURE 2. The rate of muons as a function of zenith angle observed in the Kolar gold fields proton-decay experiment.

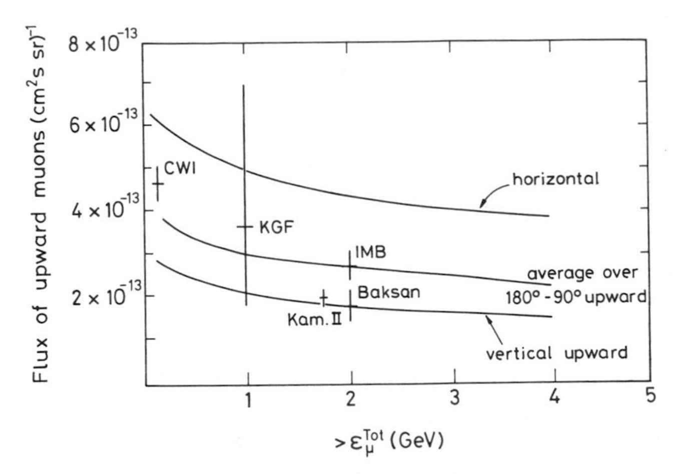

FIGURE 3. Comparison of the fluxes of upward neutrino-induced muons as measured in various experiments with the calculation of Gaisser and Stanev.[6] The fluxes are integrated from the threshold that is plotted on the horizontal scale.

The typical intrinsic angular resolution of detectors considered as neutrino telescopes is less than 1°. There are two physics contributions to the resolution that limit the overall resolution to about this value, making it unnecessary to go to better detector resolutions. These are the multiple scattering of the muons in the rock that

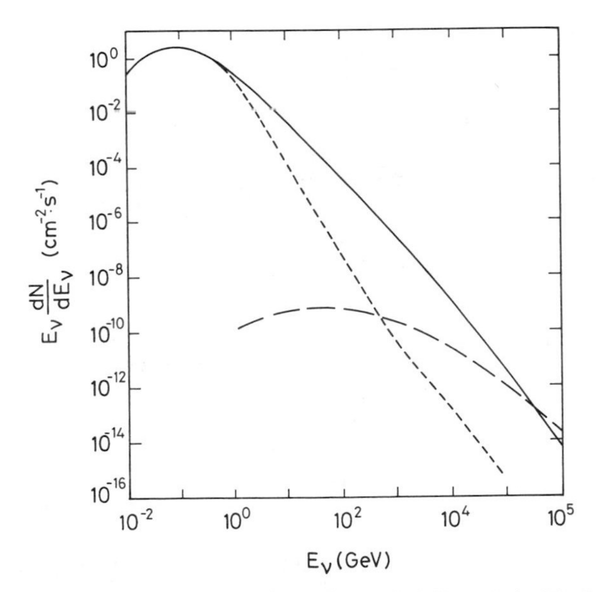

FIGURE 4. Calculation of the neutrino flux from Cygnus X-3 (*long-dashed line*) compared with the neutrino flux from atmospheric cosmic-ray interactions (*solid line*). The flux within 1° of the source is shown as the *short-dashed line.*

varies with energy, but is about 0.5° at 1 TeV, and the spread in the angle between the muon and the incident neutrino that is given approximately by 2.6 [100/E(GeV)], which is 0.8° at 1 TeV. Thus, at 1 TeV the pointing angle of the muon will have a resolution of about 1°. If the muon energy is not measured, as will usually be the case, the spectrum must be integrated from threshold to the highest energies. For an experiment such as GRANDE with a threshold of 10 GeV and an assumed spectral function (α) of 2.1, 90 percent of the signal would be contained within a 1° cut. This would rise to 90 percent in a 2° cut if the spectral function was 2.3.

FIGURE 4 shows a calculation of the flux from Cygnus X-3 and atmospheric backgrounds with and without an angular cut due to Gaisser and Stanev.[7] It can be seen that above 1 TeV backgrounds are not expected to be a problem, provided the signal is as high as expected. It can also be concluded that low-energy events are confusing and dilute the signal to noise. Neutrino telescopes are high-energy devices and should have high thresholds.

3. FUTURE PROSPECTS

The article by John Learned[8] in these proceedings gives an excellent account of the experiments currently running and on the drawing board. The current crop were almost all initiated as proton-decay detectors. The much larger detectors to arrive in the near future, designed specifically with neutrino astronomy in mind, are all water Cherenkov detectors, this being the least expensive way of instrumenting a large volume. Two distinct branches have evolved: the deep underwater detectors, such as DUMAND, where shielding from atmospheric muons is provided by an overblanket of water, and surface water detectors, such as the GRANDE proposal, when the rejection of atmosphere muons is completely active, depending on directionality. To my knowledge only DUMAND stage 2 and its sister experiment at Lake Baikal have reached the stage of being funded and under construction. It would surely be a pity if at least one surface detector was not constructed in the world.

Hopefully these new detectors will observe point neutrino sources. However, it is reasonable to ask, What if they don't? The reason would have to be either that the signal is too low or the background too high.

Too low a signal would presumably mean that the γ-ray point source claims upon which rates are based are overoptimistic, maybe entirely false. However, we do know that high-energy cosmic rays exist, and these have to have been accelerated by some reasonably compact mechanism that would probably produce γ-rays and neutrinos as well. Thus a search using even larger detectors, on the scale of kilometers, would be interesting and scientifically justified. An upgrade of the present DUMAND II to something similar to the original DUMAND proposal would seem to be feasible. Other, more exotic methods of observing large masses, such as optical or microwave emission in the antarctic ice and acoustic detectors, are described in John Learned's article.[8]

Too high a background could be overcome in a number of ways. The most obvious, given that these sources are probably episodic, is to search for neutrinos in coincidence with γ-ray bursts. Unfortunately, current γ-ray and ν telescopes are mutually exclusive, one looking upward and the other downward, and one detector cannot observe one source in both modes at the same time. However, two well-connected detectors in the northern and southern hemispheres could carry out a time-correlated search. A second possibility is to raise the detector threshold, since as we saw earlier, it is expected that atmospheric background will be reduced at high energies. The rate of production of high-energy showers from bremsstrahlung and pair production rises rapidly with muon energy. Selecting muons accompanied by large showers would provide an enhanced sample of high-energy events. High energy can be imposed by size and range. A full-sized DUMAND detector would have a threshold of 100–200 GeV. Some thought should be given to planar detectors, where instead of instrumenting a large volume, one instruments only two relatively narrow planes separated by a 0.5-TeV absorption length. Solid angle is an obvious problem, but may be overcome by some ingenuity. If backgrounds cannot be beaten down in muon–neutrino interactions, then electron neutrino detectors could be successful if high-energy fluxes were substantial. The background of atmospheric neutrinos at high energies is expected to be low because of the ranging out of high-energy muons by hitting the earth, whereas the rate in astrophysical sources is likely to be near the production ratio of two muons to one electron neutrino. Unfortunately, the range of electron showers is much less than muon ranges, so the trick of enhancing the production volume by using the rock in front of the detector is much less effective. However, the rate of contained or semicontained electron showers in a next-generation DUMAND detector might be sufficient to see a signal.

4. CONCLUSIONS

High-energy neutrino astronomy is very interesting as a probe of stellar structure and the mechanisms responsible for very-high-energy cosmic rays. Reasonably accurate predictions of the expected rates for the detection of astrophysical neutrinos can be made if we accept the claims for the detection of sources of high-energy γ-rays. Present-day experiments are well below the sensitivity required, but the next generation of deep underwater or surface water Cherenkov detectors have a good chance of observing a few sources. A number of ideas ranging from grand to wild exist for very-large-scale detectors that would open up this entirely new and fascinating window on the universe.

REFERENCES

1. There have been many claims to observe sources in high-energy γ-rays. A useful introduction to the subject is SOKOLSKY P. 1988. Introduction to Ultrahigh Energy Cosmic Ray Physics. Frontiers in Physics Series. Addison-Wesley. Reading, Mass.; also, the review article by WEEKES, T. C. 1988. Phys. Rep. **160:** 1.
2. BEREZINSKY, V. S., C. CASTAGROLI & P. GALEOTI. 1985. Nuovo Cimento **8C:** 185; GAISSER, T. K. & T. STANEV. 1985. Phys. Rev. Lett. **54:** 2265; KOLB, E. W., M. S. TURNER & T. P. WALKER. 1985. Phys. Rev. **D22:** 1145.
3. CHERRY, M. L. 1990. Nucl. Phys. B **14A:** 38.

4. LLOYD-EVANS, J., *et al.* 1983. Nature **305:** 784.
5. ADARKAS, H., *et al.* 1989. Proceedings of the 21st International Cosmic Ray Conference, Adelaide, Australia.
6. GAISSER, T. K. 1990. Nucl. Phys. B **14A:** 381.
7. GAISSER, T. K. & T. STANEV. 1985. Proceedings of the 19th International Cosmic Ray Conference, Vol. 8, F. C. Jones, J. Adams, and G. M. Maxin, Eds. NASA Conference Publication 2376. Washington, D.C.
8. LEARNED, J. G. 1992. Neutrino astronomy with large Cherenkov detectors. This issue.

Solar Neutrino Flux Variations

V. GAVRYUSEV[a] AND E. GAVRYUSEVA[b]

[a]*Nuclear Physics Institute*
Moscow State University
Moscow, USSR

[b]*Institute for Nuclear Research of the*
Academy of Sciences of the USSR
Moscow, USSR

INTRODUCTION

Almost uninterrupted measurements of the ^{37}Ar production rate using 610 tons of perchlorethylene, which have been conducted over the past 20 years by R. Davis and his collaborators, offers a unique possibility for the investigation of the processes in the deep interior of the sun.[1] The first results obtained by this group have already shown an intriguing discrepancy between the high-energy neutrino fluxes predicted for the sun by the stellar evolution theory and the measured values of the ^{37}Ar production rate. Approximately at the end of the first ten years of measurements, a number of investigators asked whether there were time variations in the data of the counting rate of neutrinos in the Cl–Ar detector.[2–19] The investigations showed that there is a noticeable anticorrelation between the ^{37}Ar production rate and the solar activity. Time variations with other periods were also revealed. However, the basic question of whether these effects are real or stipulated by a great inaccuracy of the measurements of the ^{37}Ar production rate remains unanswered.

SPECTRUM ANALYSIS

In order to reveal hidden periodicities we computed the Davis data spectrum with the help of the Fourier transform. Within an interval of periods T from 4 to 200 months the values of amplitudes A and phases φ of corresponding harmonics were calculated. The power spectrum of Davis's experimental data obtained with the help of this program is shown in FIGURE 1.

DISTRIBUTION OF PEAKS IN SPECTRA OF RANDOM DATA SERIES

The Fourier analysis reveals possible hidden periodicities, but it is not directly indicative of which peaks shall be regarded as an actual "signal" and which originate from the "noise" component of the initial data.

We conducted a special investigation of the statistical properties of peaks in the Fourier spectra of the "noise" component of the experimental data. Several sets of artificial data series have been simulated. They are random samples of normally distributed random values around a predetermined most probable value R_0 and with

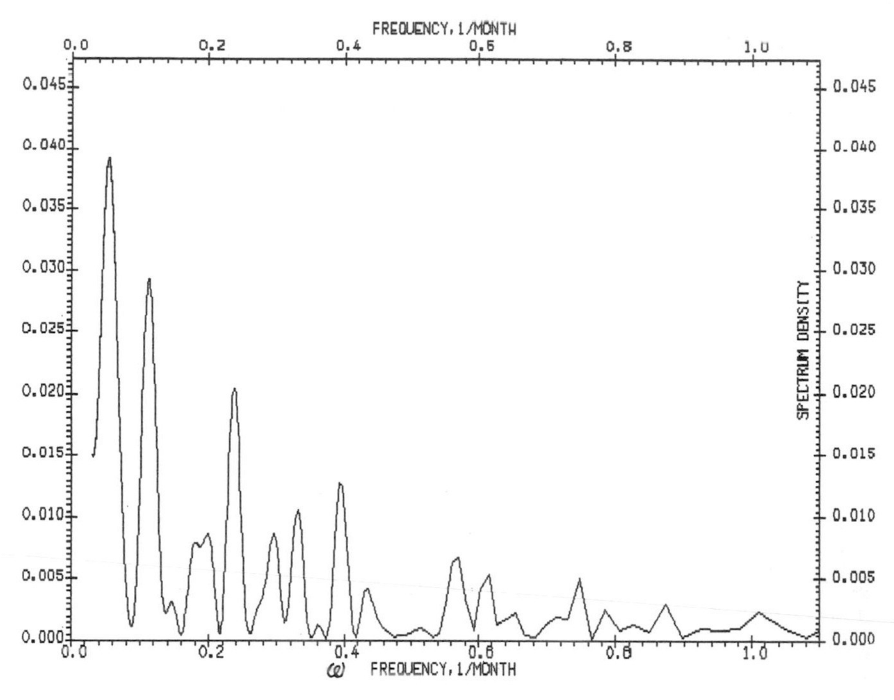

FIGURE 1. Spectral density of ^{37}Ar production rate in Cl–Ar detector.

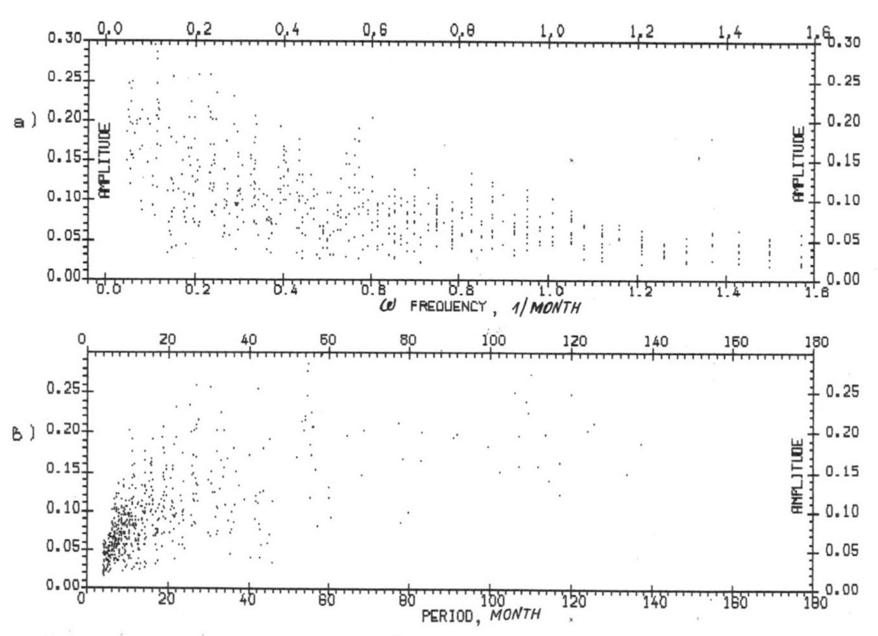

FIGURE 2. Distribution of peaks in the spectra of first simulated data set in the coordinates "amplitude frequency" and "amplitude period." The most probable value is equal to experimental values of ^{37}Ar production rate. The root-mean-square deviation was assumed to be equal to the experimental values for each run.

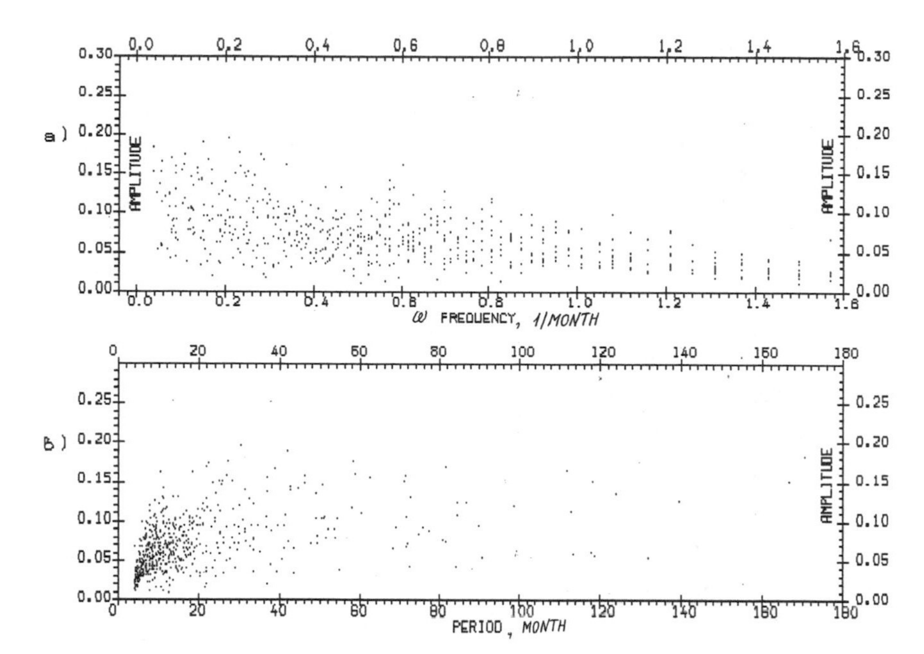

FIGURE 3. Distribution of peaks in the spectra of second simulated data set in the coordinates "amplitude frequency" and "amplitude period." The most probable value is equal to mean experimental value of ^{37}Ar production rate. The root-mean-square deviation was assumed to be equal to the experimental values for each run.

a root-mean-square deviation σ (both constant and time-dependent). The spectrum was calculated and all peaks available in it were distinguished for each series of data incorporating as many points as Davis's series has. Every set consists of 25 such series of data, and a total number of spectrum peaks in a set varies from 500 to 600.

Though we had no data from several installations similar to the Davis detector at our disposal, there is the possibility of simulating a data series set where the Davis series is one of them. The other series were obtained in a way similar to the one previously described. The experimental values of ^{37}Ar production rates for each run are taken as the most probable values, as is each corresponding root-mean-square deviation. Then we assume that the difference in results obtained at different installations is caused only by a variable random component with the same statistical characteristics. These conditions are used to build a simulated data set. We then test to see how the analysis of peak distribution in the spectra of such data sets can distinguish nonrandom components, even with so unfavorable a "signal"/"noise" ratio. The calculation results for this version are shown in FIGURE 2. In a comparison with the second version, where the most probable value is constant and σ varies with the experiment (FIG. 3), it is obvious that these distributions are different even in outward appearance. This difference is still more easily seen while plotting the histograms for the frequency distribution of the peaks (FIG. 4).

The probability of the appearance of a peak with definite properties caused by "noise" (in the present interval of frequencies, i.e., in bin, and with an amplitude exceeding some value) can be evaluated according to the histogram by a number of

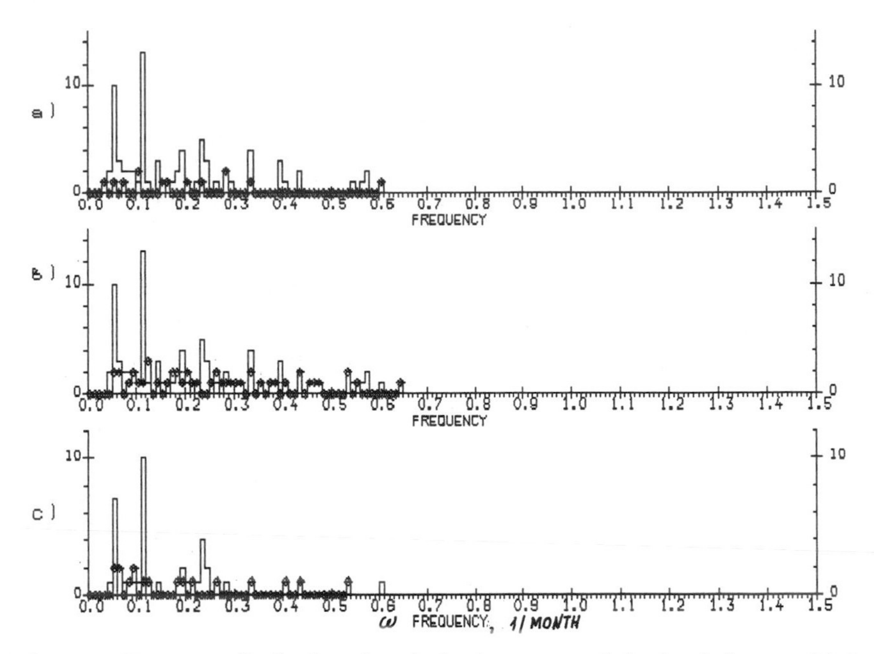

FIGURE 4. Frequency distribution of peaks in the spectra of simulated data set (**a**) in comparison with second data set (R_0 = const., σ – experimental values for each run, ◇), amplitudes >0.16; (**b**) in comparison with third data set (the values R_0 are of random realization of constant, ◇), amplitudes >0.16; (**c**) the same as (**b**), but amplitudes >0.2.

events in bin divided by *N,* which is the total number of peaks with any value of amplitude in bin. In our case, it is ~25. This number depends on the bin width, and we had a choice between the two conflicting arguments: maximum resolution or maximum statistics. As seen in FIGURES 2 and 3, one of the symptoms of the "noise" component is a "spreading" of signal amplitude on the amplitude-frequency diagram. There are also such spectrum realizations in which amplitudes of corresponding peaks are either less or larger than the actual one. Thus, if only peaks with the amplitude higher than some others are included in the histogram for a set of a purely random series, then with a sufficiently high threshold the number of hits into the bin can be reduced to a single hit (or completely excluded). If for the data set presumably containing the signal and with the same threshold in the same range of bins, a sufficient number of hits appears inside some bin (on some frequency), it can be said that a signal can be seen at this frequency with a significance level of $P = 1 - 1/N$. In our case $P \sim 96$ percent. Certainly, the number of hits into the same bin (or into a peak in a histogram occupying several bins) is an important factor. The larger it is, the more reliable the determination of the signal.

TABLE 1 contains the characteristics of the peaks in the Davis data spectrum that "survive" in terms of the described method of analysis, when *P* is a lower bound of their significance levels.

In the previously described analysis we made a comparison [FIG. 4(a)] of the characteristics of the set based on the Davis data with the characteristics of the set

TABLE 1. Characteristics of the Peaks in the Spectrum Density of the Davis Data Series with Significance Level $P > 96$ Percent

	T (months)	ω (1/month)	A (atoms/day)	φ (rad)	P (%)	P_1 (%)
1	113.8	0.552E − 01	0.198E + 00	2.410	99	99
2	55.6	0.113E + 00	0.171E + 00	−2.940	99	99
3	34.8	0.181E + 00	0.889E − 01	2.406	96	
4	31.6	0.199E + 00	0.926E − 01	1.579	96	
5	26.6	0.236E + 00	0.143E + 00	0.679	99	99
6	19.0	0.331E + 00	0.103E + 00	−0.769	99	
7	16.0	0.393E + 00	0.113E + 00	−1.612	99	

based on a constant (in terms of the most probable value). However, Davis's data themselves can be regarded as some random realization of measurements of the constant. It is obvious that during simulation of the set on such a base a periodicity accidentally inserted into an initial base series will also reveal itself. Therefore, a comparative analysis was also performed with a third data set where some random realization of the data series from the set with R_0 = const., σ − variable, was used as the most probable value. FIGURE 4(b) is identical to FIGURE 4(a) (the histogram incorporates all peaks with an amplitude exceeding 0.16), but the comparison has been conducted with a new set. It is evident that the significance level of determination of nonrandom peaks decreases. If the threshold has risen, we proceed to the previous significance level (96 percent); in this case, the number of "survived" peaks decreases [FIG. 4(c)]. The significance levels for these peaks are given in TABLE 1, column P_1.

χ^2- AND t- (STUDENT) CRITERIA

The primary hypothesis, which we adhere to, lies in the fact that a series of ^{37}Ar production rates contains a regular component ("signal") and a random component ("noise") with a normal distribution.

The preceding analysis found several harmonics that are very probably regular. There are two sets of such harmonics represented in TABLE 1. The harmonics marked by values P_1 were chosen according to more strict criteria than these marked by P. In every set the harmonics also differ in their significance levels. While constructing the "signal" from different combinations of these harmonics we can determine with the help of the χ^2-criterion how much the distribution of deviations from the "signal" complies with the assumption of normality. At the same time the Student criterion allows us to understand how our evaluation of the most probable value is displaced. The "good" hypothesis is the one whose significance level is high enough by both criteria. High significance levels by both criteria can be obtained if the hypothesis incorporates a sufficiently fast changing regular component with a small amplitude against a background of great "noise." Column γ in TABLE 2 contains such a characteristic of the hypothesis as a ratio of root-mean-square deviation from the "signal" to a mean value of the "signal" amplitude. In the case of

TABLE 2. χ^2-, t-Values and Corresponding Significance Levels P for Some Combinations of Harmonics Available from TABLE 1

	χ^2	P (%)	t	P (%)	γ	1	2	3	4	5	6	7
0	3.06	39.35	0.00	100.00								
1	8.34	4.21	0.13	89.92	1.65	+	−	−	−	−	−	−
2	4.55	20.92	0.17	86.62	2.06	−	+	−	−	−	−	−
3	0.76	85.85	0.05	96.24	3.99	−	−	+	−	−	−	−
4	0.07	99.40	0.04	97.12	3.81	−	−	−	+	−	−	−
5	3.98	26.81	0.11	90.88	2.73	−	−	−	−	+	−	−
6	2.71	44.66	0.13	89.97	3.43	−	−	−	−	−	+	−
7	1.68	64.60	0.12	90.18	3.18	−	−	−	−	−	−	+
8	0.07	99.40	0.30	76.75	1.01	+	+	−	−	−	−	−
9	0.18	98.02	0.29	76.91	0.91	+	+	+	−	+	−	+
10	0.99	80.39	0.16	87.54	0.81	+	+	−	−	+	−	−
11	0.87	83.12	0.35	72.58	0.80	+	+	−	−	+	+	+
12	1.33	72.16	0.22	82.41	0.92	+	+	+	+	+	−	−
13	1.22	74.91	0.36	71.83	0.84	+	+	+	+	−	−	−
14	1.68	64.60	0.30	76.64	0.79	+	+	+	−	+	+	−
15	2.25	52.40	0.20	84.40	0.93	+	+	+	−	+	−	−
16	2.71	44.66	0.28	77.83	0.75	+	+	−	+	+	−	+
17	3.06	39.35	0.26	79.57	0.87	+	+	−	−	+	−	+
18	3.06	39.35	0.34	73.67	1.04	+	+	+	−	−	−	−
19	3.52	32.27	0.28	77.64	0.73	+	+	−	+	+	+	−
20	3.75	29.16	0.18	85.36	0.80	+	+	−	+	+	−	−
21	3.75	29.16	0.40	68.85	0.81	+	+	+	+	+	+	+
22	4.44	22.10	0.26	79.42	0.71	+	+	−	−	+	+	−
23	4.67	19.85	0.32	75.02	0.81	+	+	+	+	+	+	−

NOTE: The numbers of harmonics available in the combination are marked by (+) in the right part of the table, everywhere a constant component equal to a mean value of ^{37}Ar production rate is included in "signal."

large γ the hypothesis should be regarded as a badly discriminated one by both criteria.

A combination of the two harmonics with longest periods, which are the most reliably determined by FIGURE 4(c), is a sufficiently good one by both criteria. The addition of a third harmonic with a period of about 27 months makes the significance of evaluating the normality of the distribution of deviations somewhat worse, and it reduces the displacement of the most probable value. In total, it can be evaluated as a good significance level for the given hypothesis.

It is interesting to note that the ratios of these periods are near to two:

$$113.8/55.6 = 2.05, \qquad 55.6/26.6 = 2.09.$$

Considering that there is some inaccuracy in determing the periods, it is highly probable that there is a deep physical relationship between these harmonics.

The next hypothesis according to its significance level by both criteria is obtained by adding the harmonics with periods of about 19 or 16 months to the previously mentioned combination. It can be seen easily that $113.8/19 = 5.99$, $113.8/16 = 7.11$,

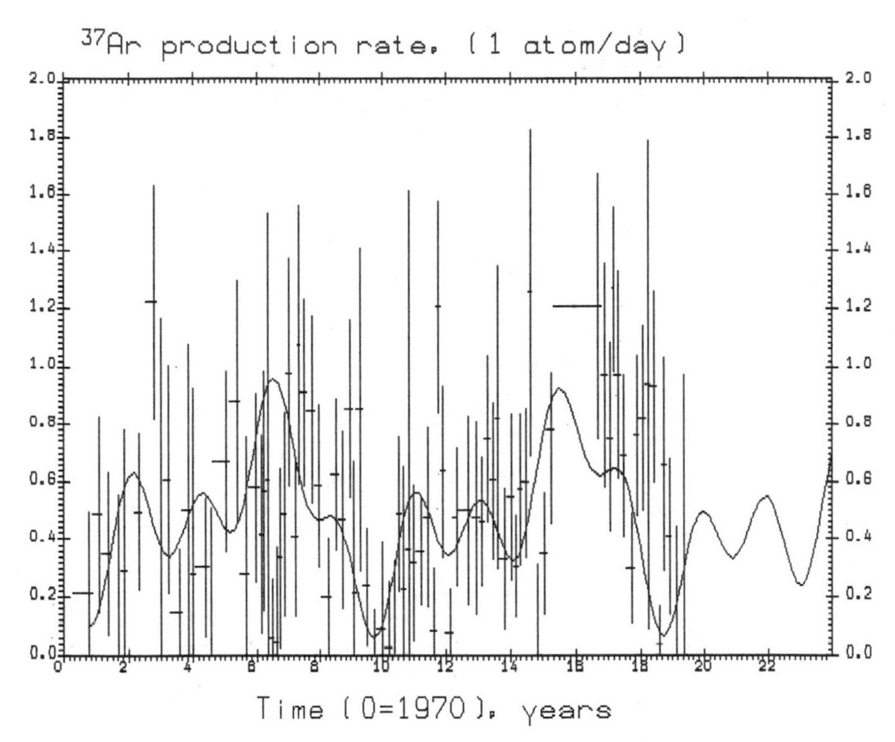

FIGURE 5. Diagram of "signal" corresponding to hypothesis 10 from TABLE 2 (harmonics with periods of 113.8, 55.6, and 26.6 months) in comparison with data of the Davis group.

that is, these harmonics are rather the overtones of basic period too. However, they do not pass by criterion P_1.

Thus the analysis of TABLE 2 results in the conclusion that from the standpoint of χ^2- and t-criteria the best combinations are three combinations of harmonics from TABLE 1, the combination of which incorporates all of them being of a sufficiently high significance level. The ratios between periods of these harmonics are such that they allow us to speak about the presence of a primary tone with a period of ~ 113.8 months, the second, fourth, and with less confidance, the sixth and seventh overtones. The combination of harmonics with periods of 113.8, 55.6, and 26.6 months is compared with the Davis data in FIGURE 5.

CORRELATION ANALYSIS

The anticorrelation of Davis's data series with such well-known manifestations of solar activity as the monthly averaged numbers of sunspots and solar flares is a very strong argument in favor of the presence of a periodic phenomenon in the ^{37}Ar production rate.

The Davis data correlation coefficient with the previously mentioned characteristics,[20–22] as well as the measurements of the sun's diameter,[23,24] are shown in FIGURE 6. An argument in this figure is a time shift between the data. The negative shifts signify that in computation the correlation coefficient (K_{corr}), a value of the ^{37}Ar production rate was taken at earlier moments than that of another characteristic. The anticorrelation significance level of the Davis data related to sunspots and solar flares is very high. In less than 0.1 percent of the cases two random noncorrelated series of such a volume will give K_{corr} exceeding 0.35. In this case, the value of K_{corr} depends on the method used for its calculation. Since the data of the neutrino counting rate and solar flares (sunspots, etc.) are related to various moments, they are required to be interpolated. However, the different methods of calculating K_{corr} do not reduce the significance level to less than 98 percent. The correlation of the ^{37}Ar production rate with the variation in solar diameter must be regarded only as preliminary, since we did not have enough measurement data of the diameter.

Because the periodic component in the Davis data is rather a combination of primary tone and overtones, sufficiently high values of the correlation coefficient can be expected between individual subseries of data obtained while dividing the total measurement time into intervals equal to a period of the primary tone. Only two such

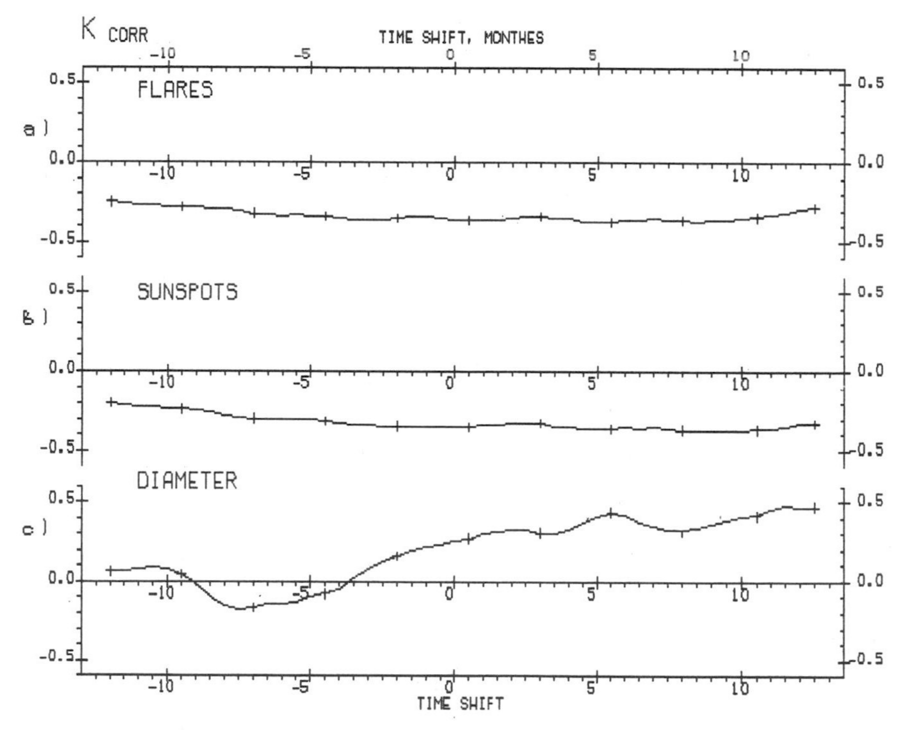

FIGURE 6. Correlation coefficients of the Davis data and (**a**) monthly averaged numbers of solar flares; (**b**) monthly averaged numbers of sunspots; (**c**) variations of the sun diameter as functions of time shift between these data.

intervals can be seen in the available data, so it is practically necessary to calculate the K_{corr} of the first half of the Davis data series with the second one. The result is 0.43 (with a significance level of 99.4 percent), and it can be seen as additional confirmation of the reality of the 1:2:4 ratio between the periods of harmonics that were most reliably revealed in the spectrum of the ^{37}Ar production rate.

VACUUM OSCILLATIONS AND HALF-YEAR VARIATIONS OF THE NEUTRINO CAPTURE RATE

The results of measuring the neutrino fluxes by Cl, Ga, and Kamiokande detectors can be explained by the matter-enhanced neutrino oscillations.[25,26] But time variations in the counting rate in Cl do not have a good physical description. The low capture rate of neutrinos in the Cl and Ga experiments could be caused by neutrino vacuum oscillations.[27–29] At 10^{-10} eV^2 < Δm^2 < 10^{-9} eV^2 the oscillations wavelength becomes comparable with the variation of the distance between the sun and the earth: $\Delta R \approx 5 \cdot 10^6$ km. In this case, the oscillation effect would be expressed in the periodic changes of the neutrino capture rate.[28,30] Periodicities of half a year and less are possible. The neutrinos of the Be branch are mainly responsible for this kind of capture rate variation. The best fit of Davis's data mapping of the half-year interval corresponds to $\Delta m^2 = 4.9 \cdot 10^{-10}$ eV^2. Obviously, the hypothesis about the description of the experimental data doubling Q by a theoretical curve $V(t)$ with a random superposition of the errors normally distributed is most probable when the correlation coefficient is close to 1 and the values of the t-criterion of Student and the χ^2-criterion are small. The correlation coefficient equals 0.83, $t = 0.07$, $\chi^2 = 1.0$. This means that the reliability of the approximation of the Q range by $V(t)$ at $\Delta m^2 = 4.9 \cdot 10^{-10}$ eV^2 is 96 percent by means of the t-criterion and 80 percent by means of the χ^2-criterion. The represented results are obtained by using Davis's data from the 18 to 70 run for the solar model with $T = 15.27 \cdot 10^6$ K (mixed core equals 0.4 $M_\odot$), a mixing angle of neutrino 45°. The doubling of Davis's data on the half-year interval Q and the time variation of the neutrino capture rate in the Cl detector at $\Delta m^2 = 4.9 \cdot 10^{-10}$ eV^2 are represented in FIGURE 7. It should be mentioned here that the expected

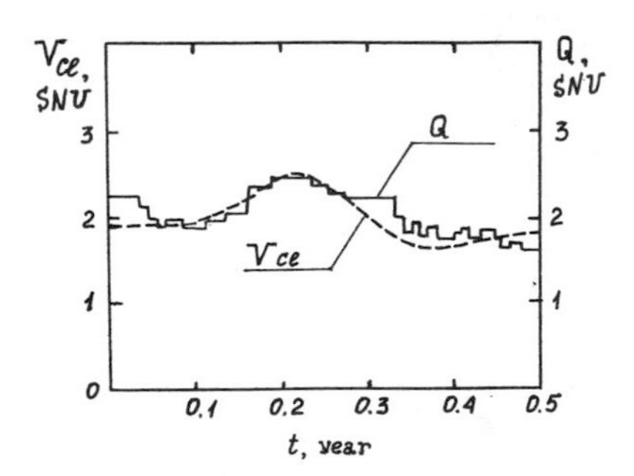

FIGURE 7. The doubling of the Davis data on a half-year interval Q and time variation of neutrino capture rate in the Cl detector at $\Delta m^2 = 4.9 \cdot 10^{-10}$ eV^2 – V_{Cl}.

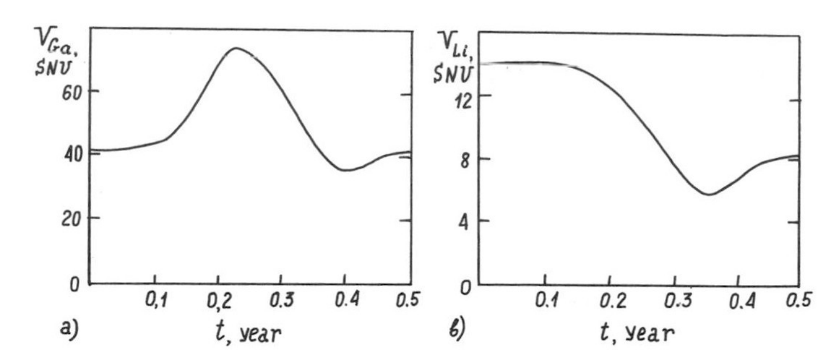

FIGURE 8. Time variations of neutrino capture rate in Ga and Li detectors during half year at $\Delta m^2 = 4.9 \cdot 10^{-10}$ eV.

amplitude of the time variation in the Ga detector is much larger. If this effect is real, it could be visible in the neutrino measurements by the Ga and Li detectors (FIG. 8). We hope to get an answer soon, because the Ga detector, which will provide us with new information about the main flux of the solar neutrino, has begun operation.

CONCLUSION

The highly significant anticorrelation of the neutrino flux and the monthly averaged number of flares and sunspots confirms the reality of the harmonic with longest period (quasi–eleven-year) harmonic in the neutrino fluxes. Its existence emphasized the importance of a question about the possibility that active phenomena on the surface of the sun are caused by processes in its core. To explain this anticorrelation it was assumed that the neutrino has an anomalously great magnetic moment.[16] Then the movement of neutrinos through the layers of the sun with a strong enough magnetic field causes a decrease in the ^{37}Ar production rate in perchlorethylene. Such interaction should lead to the anticorrelation of neutrino fluxes with solar activity, and to appreciable half-yearly variations of the capture rate of neutrinos, because of changes in the location of the earth in relation to the plane of the sun's equator. There are peaks with periods of about six months in the ^{37}Ar production rate, but with a large degree of probability that they can be caused by the existence of random noise.

Substantial differences should be emphasized in the spectra of the neutrino fluxes and the monthly averaged numbers of flares and spots on the sun (FIG. 9). As was mentioned earlier,[10] in the latter spectra the ratio of the amplitude of the quasi–eleven-year component to the other peak amplitudes is overwhelmingly large while the amplitudes of all three harmonics in the neutrino spectrum are comparable among them. If the time variations of the neutrinos are caused by the migration of a strong magnetic field from the high latitudes of the sun to its equator, then the reason for the appearance of the other harmonics would not be quite clear. Integer-valued relations between the periods also will not be understood.

An assumption about a primary of processes occurring in the solar core looks preferable. In this case, the observed time variations in the ^{37}Ar production rate

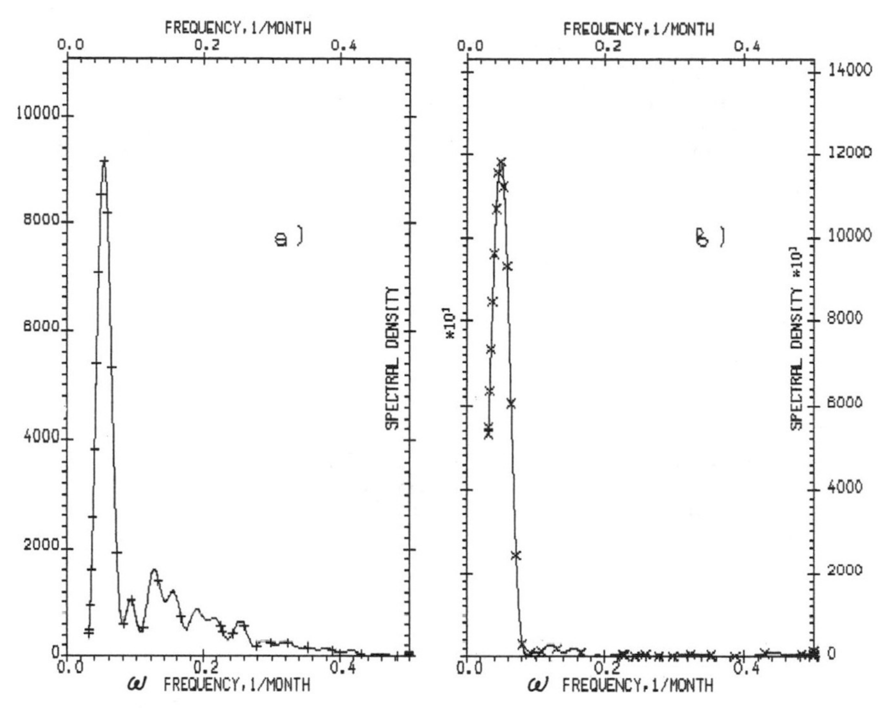

FIGURE 9. Spectral density of monthly averaged numbers of (**a**) sunspots, (**b**) solar flares.

should be caused almost completely by variations in the ^{8}B (and hence ^{7}Be) concentrations in the interior of the sun. Since a part of the energy generated by the beryllium branch of the pp chain is rather small, its time change in regard to a possible dissipation in the process of diffusion from the center to the surface might very slightly affect the integral luminosity of the sun. However, this energy could be enough to influence the processes in the shell.

ACKNOWLEDGMENTS

The authors express their sincere gratitude to academician G. T. Zatsepin for his permanent and stimulating interest in this work and for the useful discussions, and to Professor R. Davis for kindly providing unpublished data. We also wish to thank Professor E. Fossat for helpful discussion.

REFERENCES

1. DAVIS, R. 1990. Private communication; DAVIS, R., K. LANDE, C. LEE, B. CLEVELAND & J. ULLMAN. 1990. *In* Inside the Sun. Astrophysical and Space Science Library, Vol. 159: 171. Kluwer. Dordrecht, the Netherlands.
2. ZATSEPIN, G. & V. KUZ'MIN. 1964. Vestn. Akad. Nauk. USSR (2): 50.

3. SUBRAMANIAN, A. 1979. Curr. Sci. **48:** 705.
4. SAKURAI, K. 1980. Publ. Astron. Soc. Japan **32:** 547.
5. SAKURAI, K. 1981. Sol. Phys. **74:** 35.
6. LANZEROTTI, L. & R. RAGHAVAN. 1981. Nature **293:** 122.
7. GAVRIN, V., YU. KOPYSOV & N. MAKEEV. 1982. Pis'ma Zh. Eksp. Teor. Fiz. **35**(11): 491.
8. BAZILEVSKAYA, G., YU. STOZHKOV & T. CHARAHCH'YAN. 1982. Pis'ma Zh. Eksp. Teor. Fiz. **35**(11): 237.
9. GAVRYUSEVA, E. & V. GAVRYUSEV. 1983. Sov. Phys.—Lebedev Inst. Rep. **9:** 33.
10. RIVIN, YU., V. GAVRYUSEV, E. GAVRYUSEVA & L. KOSHELEVA. 1983. Proceedings of the 11th Conference on Muon and Neutrino Investigations in Large Water Volumes. (USSR, Alma-Ata), p. 33.
11. RIVIN, YU., V. GAVRYUSEV, E. GAVRYUSEVA & L. KOSHELEVA. 1983. *In* Geomagnetic Variations, Electric Fields and Currents: 153. Izmiran. Moscow.
12. GAVRIN, V. & A. KOPYLOV. 1984. Pis'ma Astron. Zh. **10**(2): 154.
13. RAYCHAUDHURI, P. 1984. Sol. Phys. **93:** 397.
14. ROWLEY, J., *et al.* 1985. Proceedings of the AIP Conference on Solar Neutrinos and Neutrino Astronomy (New York).
15. HAUBOLD, H. & E. GERTH. 1985. Astrophys. Space Sci. **112:** 397.
16. VOLOSHIN, M., M. VISOTSKI & L. OKUN. 1986. Zh. Eksp. Teor. Fiz. **91:** 754.
17. DELACHE, P., F. LACLARE & H. SADSAOUD. 1988. IAU Symp. **123**.
18. BAHCALL, J., G. FIELD & W. PRESS. 1987. Astrophys. J., Lett. **320:** L69; BAHCALL, J. 1989. Neutrino Astrophysics:326. Cambridge University Press. Cambridge, England.
19. RIVIN, YU. 1989. Astron. Tsirk.: 1539.
20. SOLAR GEOPHYSICAL DATA. 1987. **516:** Part 1, August.
21. SOLAR GEOPHYSICAL DATA. 1987. **519:** Part 1, November.
22. SOLAR GEOPHYSICAL DATA. 1987. **515:** Part 1, July.
23. LACLARE, F. 1983. Astron. Astrophys. **125:** 200.
24. DELACHE, P. & F. LACLARE. 1985. Nature **6036:** 416.
25. MIKHEYEV, S. P. & A. YU. SMIRNOV. 1985. Yad. Fiz., **42:** 1441; MIKHEYEV, S. P. & A. YU. SMIRNOV. 1988. Phys. Lett. **200B:** 560; MIKHEYEV, S. P. & A. YU. SMIRNOV. 1989. Progress in Particle and Nuclear Physics, Vol. 23: 41.
26. BAHCALL, J. N. & W. C. HAXTON. 1989. IASSNS-AST 89/14, Princeton, N.J. Preprint.
27. PONTECORVO, B. M. 1969. Zh. Eksp. Teor. Fiz. **53:** 1717; PONTECORVO, B. M. 1957. Zh. Eksp. Teor. Fiz. **33:** 549; PONTECORVO, B. M. 1958. Zh. Eksp. Teor. Fiz. **34:** 247.
28. PONTECORVO, B. M. 1977. Usp. Fiz. Nauk **123**(2): 181.
29. BILENKY, S. M. & B. M. PONTECORVO. 1984. Lett. Nuovo Cimento **40**(6): 161.
30. GAVRYUSEV, V., E. GAVRYUSEVA & I. GOLUBEV. 1985. INR AN USSR, P-0423. Preprint.

Black Holes in the Galaxy

JEFFREY McCLINTOCK

Harvard-Smithsonian
Center for Astrophysics
Cambridge, Massachusetts 02138

INTRODUCTION

There is no decisive evidence that black holes exist. However, it is widely believed that they power active galactic nuclei (AGN) and some X-ray binaries, and that they inhabit the nuclei of many normal galaxies. The compulsion to believe that black holes are physical objects is driven by our faith in general relativity, the idea that a black hole is a logical end point of stellar and galactic evolution,[1] and the knowledge that the familiar neutron star is a small step away from a black hole. The clearest evidence for black holes comes from dynamical studies of four X-ray binaries, which are described and compared in the following section. Topics in subsequent sections include nondynamical signatures of black holes, alternative models for Cyg X-1, and recent results obtained for Cyg X-1 and A0620-00.

THE FOUR LEADING BLACK HOLE CANDIDATES

The best evidence for the existence of black holes comes from studies of four X-ray binaries: Cyg X-1,[2,3] LMC X-1,[4] LMC X-3,[5] and A0620-00.[6,7] Three lines of evidence make the case that these binaries contain black holes: (1) a luminous ($\sim 10^{38}$ erg s^{-1}) and rapidly variable X-ray source establishes beyond doubt that each binary contains a compact object at least as dense and massive as a neutron star;[8] (2) the mass function of the optical star and some additional data imply that the mass of the compact object exceeds $3M_{\odot}$; (3) according to theory, a neutron star more massive than $3M_{\odot}$ is unstable and will collapse to form a black hole.[9] In this paper we discuss only the second point: the dynamical evidence for the existence of massive, compact stars.

The nominal characteristics of the four binary systems are summarized in TABLE 1. The X-ray luminosities are the maximum that have been observed; the visual magnitude of A0620-00 is for the quiescent state (see below). For all four systems, the orbital solutions are consistent with circular orbits, and X-ray eclipses are not observed.

The most important and reliable observable is the optical mass function:

$$f(M) \equiv \frac{(M_x \sin i)^3}{(M_x + M_c)^2} = \frac{PK^3}{2\pi G},$$

where M_x and M_c are the masses of the X-ray source and the optical companion star,

TABLE 1. Properties of Four Black Hole Binaries

	Cyg X-1	LMC X-1	LMC X-3	A0620-00
L_x(erg s^{-1})	2×10^{37}	2×10^{38}	3×10^{38}	1×10^{38}
MK type	O9.7I$_{ab}$	late O	B3V	K5V
d(kpc)	2.5(?)	55	55	1(?)
m_v	9	14	17	18
e*	0.00 ± 0.01†		0.13 ± 0.05	0.01 ± 0.01
K(km s^{-1})	74.7 ± 1.0†	68 ± 8	235 ± 11	443 ± 4
P(days)	5.6	4.2	1.7	0.32
$f(M/M_\odot)$	0.25 ± 0.01	0.14 ± 0.05	2.3 ± 0.3	2.91 ± 0.08

*Orbital eccentricity.
†The values of e and K for Cyg X-1 were derived from the high-excitation line data.

respectively, and i is the orbital inclination angle.[a] The value of the mass function (TABLE 1, bottom line) is given by the term on the right, which depends only on the orbital period, P, and the velocity semiamplitude, K. Thus the mass function is a cubic equation in M_x, with M_c and i as parameters.

For the discussion at hand, the following fact is crucial: M_x cannot be less than the value of the mass function, that is, $M_x \geq f(M)$. This rock-bottom limit on M_x corresponds to a system with a zero-mass companion ($M_c = 0$) viewed at the maximum inclination angle ($i = 90°$). Thus, based solely on the observed values of the mass function (TABLE 1), one obtains the following 3σ lower limits on the masses of the compact X-ray sources: M_x (Cyg X-1) $> 0.22 M_\odot$, M_x (LMC X-1) > 0, M_x (LMC X-3) $> 1.4 M_\odot$, and M_x (A0620) $> 2.7 M_\odot$. Therefore, A0620-00 is the only candidate that qualifies as a probable black hole by virtue of having a large mass function. The other three candidates require additional supporting evidence (e.g., large values of M_c and/or small values of i). The weakest candidate is plainly LMC X-1.

How reliable are the values of the mass function ($PK^3/2\pi G$) given in TABLE 1? With the possible exception of LMC X-1,[4] uncertainties in the orbital period, P, are negligible (less than the uncertainty in G!). The K velocity, however, may be affected by several sources of systematic error including X-ray heating, tidal distortion, and nonsynchronous rotation of the secondary, and by spectral contamination due to blending of emission lines from an accretion disk, gas streams, or circumstellar material. Several factors[b] suggest strongly that the mass functions measured for Cyg X-1 and A0620-00 are free of significant systematic effects. On the other hand, less is known about the spectra of the LMC sources, and their K velocities may be subject to systematic errors. It has been speculated, for example, that the true K velocity of LMC X-3 may be $\lesssim 4\sigma$ less than the quoted value and the mass of the compact object as low as $\sim 2.5 M_\odot$.[11]

A comparison of the four candidates, based on published models, is shown in FIGURE 1. In each case the model depends on the mass function and additionally on two less reliable parameters, the orbital inclination angle, i, and the mass of the companion star, M_c. The value of M_c is determined primarily from the MK spectral

[a] For $i = 90°$ the orbital plane is perpendicular to the plane of the sky.
[b] For example, the absence of contaminating emission in a large collection of high signal-to-noise spectra of Cyg X-1,[3,10] the absence of eccentricity in the orbits of Cyg X-1 and A0620-00 (TABLE 1), and the large K velocity and narrow lines of A0620-00.[6]

class of the companion star, a method that is reasonably reliable if a star is normal. However, the companion star in an X-ray binary system has had a violent and uncertain history and cannot be presumed normal. Moreover, in pulsating X-ray binaries, the companion is often found to be undermassive by a factor of 2 or more.[12] Just possibly, the companions of black holes are severely undermassive.[13] For example, in the extreme and speculative model of LMC X-3 mentioned earlier, $M_c \sim 0.7 M_\odot$ (and $M_x \sim 2.5 M_\odot$).[11]

These gross uncertainties in the value of M_c motivated a formulation of a firm lower limit on M_x for Cyg X-1, which is independent of spectral type and evolutionary history.[14,15] This limit depends only on the value of the mass function, the absence of an X-ray eclipse, and the dereddened spectral flux of the companion star:

$$\frac{M_x}{M_\odot} > 3.4\,(d/2\text{ kpc})^2.$$

The distance, d, to Cyg X-1 probably exceeds 2 kpc,[16,17] which implies $M_x > 3M_\odot$, although this distance limit cannot be considered certain.[18] An attempt to firm up the

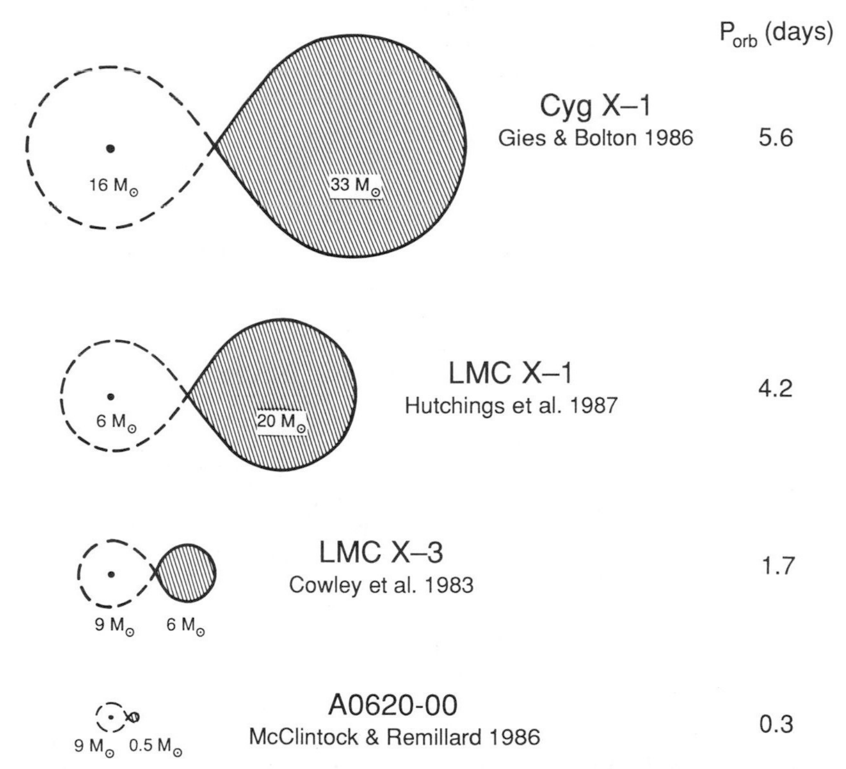

FIGURE 1. Schematic sketch, to scale, of plausible models for four dynamical black hole candidates. The optical companions (*shaded regions*) are shown filling their critical Roche equipotential lobes. The models shown here are favored by the teams named in the figure and they are described in the following references: Cyg X-1,[3] LMC X-1,[4] LMC X-3,[5] and A0620-00.[6,7]

distance by making very long baseline interferometry (VLBI) radio observations of Cyg X-1 is underway (PI: N. Bartel).

An analogous limit for LMC X-3, which has a known distance, is given by Paczynski:[19]

$$\frac{M_x}{M_\odot} > 2.3 + 0.32\left(\frac{R_c}{R_\odot}\right)^{1.8},$$

where R_c is the radius of the companion star. Paczynski's limit assumes that the companion star is the sole source of optical flux. This assumption is true for the supergiant system Cyg X-1 but, unfortunately, it is false for LMC X-3, which undergoes sizable light variations.[11] The variable light source, which is probably an accretion disk, invalidates Paczynski's limit and opens the door to the possibility that $M_x < 3M_\odot$.[11]

The orbital inclination angle, i, is the final parameter in the mass function. The absence of X-ray eclipses, for an assumed system geometry, provides upper limits on i. Additional constraints on i are sometimes achieved by modeling the light curve of the tidally distorted companion star. This is not straightforward because the ellipsoidal model depends on additional parameters (the mass ratio and the extent to which the companion star fills its Roche lobe), and especially because the light from the companion star is often mixed with light from an accretion disk or from gas streams. Occasionally, supplementary data (such as the rotational velocity of the companion[3] or the K velocity of the X-ray star[20]) constrain the mass ratio. The best determination of i has been obtained for Cyg X-1: $i = 33°$ with a formal uncertainty of 1.4°.[3]

In brief, the specific models shown in FIGURE 1 should be taken with a grain of salt, given the uncertainties in M_c and i. In my opinion, however, it is likely that $M_x > 3M_\odot$ for all four candidates, and I consider it out of the question that all four have $M_x < 3M_\odot$.

BLACK HOLE SIGNATURES: RELIABLE AND UNRELIABLE

Several X-ray signatures of an accreting black hole have been proposed (e.g., millisecond flickering and a bimodal spectrum). However, as indicated in TABLE 2, all of them have been forged by neutron stars and therefore are unreliable, with the possible exception of the "power-law tail." Plainly these signatures cannot be used to establish that a particular X-ray star is a black hole, although they may be useful for preselecting black hole candidates. (The X-ray properties of black hole candidates have been reviewed by Tanaka.[29])

Today, the only sure signature of the presence of a black hole in an X-ray binary is dynamical proof that $M_x > 3M_\odot$. Unfortunately, this evidence for a black hole is indirect; it does not depend on the unique, strong-field properties of a black hole. Phenomena that eventually may provide a decisive signature of a black hole include the following: identifiable spectral lines that have suffered extreme gravitational redshift ($z > 2$);[30] a gradual, energy-dependent rotation of the plane of linear polarization;[31] distinctive Fe line profiles emitted from the inner accretion disk;[32] and gravitational radiation from collisions involving black holes.

TABLE 2. X-ray Properties of Black Hole Candidates*

Source Name	Bimodal Spectrum	Ultrasoft High State	Power-law Tail	Flickering ($\tau \lesssim 10$ ms)	Comments	References
Cyg X-1†	X	X	X	X		
A0620-00†	X	X	X			
LMC X-3†		X	X			
LMC X-1†		X	X			
GX339 − 4	X	X	X	X		21
GS2023 + 33			X	X	Transient	22
GS2000 + 25		X	X		Transient	23
CAL 87		X			$K_x \sim 40$ km s^{-1}	24
4U1543 − 47		X			Transient	25
4U1630 − 47		X			Transient	26
4U1755 − 33		X			Persistent	26
4U1957 + 11		X			Persistent	26
Neutron Stars that Mimic Black Holes						
Cir X-1	X	X		X	Type-1 burster	27
X0331 + 53				X	X-ray pulsar	28

*SS433, which has been regarded as a black hole candidate, is not considered here because of its unique properties.

†Dynamical black hole candidate (see text).

CYGNUS X-1

For a review on Cyg X-1 see Liang and Nolan.[33] Cyg X-1 was identified as a dynamical black hole candidate in 1972 and, until the 1983 discovery of LMC X-3's massive primary, it was the only such candidate. During those eleven years, several non–black hole models for Cyg X-1 with $M_x < 3M_\odot$ were proposed. Consider the following three such models in order of decreasing importance. *Triple-star models* were proposed[34] and considered seriously (e.g., an O9.7 supergiant, a $\sim 1.4M_\odot$ neutron star, and an $8M_\odot$ BV star, which contributes ~ 1 percent of the total light). The predicted effects of these models have not been observed; however, they remain marginally viable. A *severely undermassive companion* (e.g., $M_c \sim 0.5M_\odot$) would imply $M_x < 3M_\odot$. Such a model was proposed in which the O9.7 supergiant was reduced to a hot subdwarf with $d \sim 1$ kpc.[35] The model was ruled out by reddening studies that gave $d \gtrsim 2$ kpc.[16,17] A *massive accretion disk* ($M_x \sim 10M_\odot$) encircling a $\sim 1.4M_\odot$ neutron star is the substitute for a black hole; the outer portion of the disk is assumed to be in near-rigid rotation around the relatively minor neutron star.[36] Such a disk is almost certainly (but not definitely) unstable. Nevertheless, in the eleven years since the model was introduced, neither its creators nor anyone else has examined the zeroth-order problem of Maclaurin instabilities. I conclude, therefore, that the model is not taken seriously.

The black hole model of Cyg X-1 prevailed alone against the models just mentioned and others for eleven years. The situation was summarized by E. E. Salpeter, who said, "A black hole in Cyg X-1 is the most *conservative* hypothesis."[37] Today, Cyg X-1 is joined by three additional black hole candidates, which makes the non–black hole models appear even less probable.

A reanalysis of HEAO-3 γ-ray data for Cyg X-1 has revealed a complex, 3-state spectrum that extends to $\sim$1.5 MeV.[38] Most remarkable is the spectrum of the low (γ_1) state that contains a strong, broad "bump" centered at $\sim$1 MeV. The bump has been interpreted as emission from a $\approx$400-keV pair-dominated plasma.[39] If true, this result provides the first clear connection to a large body of published theoretical work on accretion-powered pair plasmas.[40]

Additionally, the HEAO-3 discovery may provide an explanation for the unique 511-keV positron-annihilation source, which is located in the vicinity of the galactic center. In the past, a relatively massive black hole ($\sim 10^4$–$10^6\ M_\odot$) was favored as the positron source. Now, in light of the γ-ray behavior of Cyg X-1, an "ordinary" black hole X-ray binary with $M_x \sim 10M_\odot$ is considered by some to be the most probable model.[41] In this context, it is worth asking if Cyg X-1 emits a narrow 511-keV line, similar to the one observed from the galactic center. Unfortunately, the answer is only "perhaps." The HEAO-3 team reports a 1.9σ feature at 511 keV.[42]

A report of a broad Fe emission feature at 6.2 keV in the spectrum of Cyg X-1[43] has been modeled as fluorescent emission produced by hard X-ray irradiation of cold gas that is orbiting close to a black hole.[32] The modelers assert that future higher resolution studies of the Fe feature will allow a determination of the mass of the compact object and the properties of the accretion flow. The energy resolution of existing data are inadequate to rule for or against the model, and there is an alternative model.[43] Nevertheless, the Fe-line disk model, which is similar to an H_α disk model for AGN,[44] is a stimulating possibility.

A0620-00

A0620-00, an X-ray nova (1917, 1975), is fundamentally different from the other candidate black hole systems. For two months in the fall of 1975, A0620-00 was the brightest celestial X-ray source, and was then identified with a blue star of twelfth magnitude. Over the course of a year the star faded to $V \sim 18.3$,[6] which is approximately its present magnitude. By 1978 its X-ray luminosity had fallen by more than six orders of magnitude—to undetectable levels.[6] Of special interest is an X-ray precursor, which was observed for about one day during the rise to maximum; it had a hard spectrum ($\sim$30 keV), whereas the spectrum during outburst was soft ($kT \sim 1$ keV).[45]

A0620-00 has remained quiescent since its 1975 eruption. About half the V-band light comes from the K dwarf companion and the other half from an accretion disk. In the following we discuss the results of some recent studies. Johnston *et al.*[46] modeled the optical emission line profiles (principally H_α) arising from the accretion disk. Using their determination of K_c (468 $\pm$ 44 km s^{-1}), they concluded that $M_x/M_c \gtrsim 13.3$, which implies that $M_x > 6.6M_\odot$ for a plausible, assumed value of $M_c = 0.5M_\odot$ (limits are 3σ). Their result strengthens the case for a black hole; however, the result should be viewed with reserve for two reasons. First, if one uses a more refined value for K_c (TABLE 1), then one finds an extraordinarily large value of the mass ratio, $M_x/M_c \gtrsim 40$ (3σ), which disagrees with the value quoted below. Second, the model for the profile of disk emission lines is known to have shortcomings.[47]

Haswell and Shafter[20] inferred the orbital radial velocity variations of the compact star from observations of the H_α disk emission line. They report $K = 43 \pm 8$ km s^{-1}, which implies $M_x/M_c = 10.3 \pm 1.9$ (for $K_c = 443$ km s^{-1}). They conclude that the *minimum* masses (for $i = 90°$) are $M_x = 3.19 \pm 0.15 M_\odot$ and $M_c = 0.31 \pm 0.06 M_\odot$. This result is less model dependent than the result discussed earlier; nevertheless, the extraction of reliable orbital parameters from observations of the broad, multi-component H_α line is by no means straightforward.

The rotational broadening of the K dwarf's absorption lines also gives a measure of $M_x/M_c(\equiv q)^3$ if one makes the reasonable assumption that the secondary is in corotation:

$$\frac{V \sin i}{K} \leq \frac{0.462(1 + q)^{2/3}}{q}.$$

The [in] equality holds if the K star [under] fills its Roche lobe. McClintock and Remillard[7] measured $V \sin i$ and found $M_x/M_c > 9$ (preliminary), which corresponds to $M_x > 4.5 M_\odot$ for an assumed value of $M_c = 0.5 M_\odot$ (limits are 2σ).

The preceding results, which constrain or fix the mass ratio, support a black hole model for A0620-00. It is hoped that work now in progress on modeling light curve data for 1981–1990 will lead to clear constraints on i and further define the model.

CONCLUSION

Do black holes exist? The evidence is very strong that the masses of some compact X-ray sources exceed $3 M_\odot$. Perhaps the greatest uncertainty about black holes lies with Einstein's field equations, which are not unique and have been tested only in weak gravity ($GM/Rc^2 \lesssim 2 \times 10^{-6}$). The goal must be to test general relativity in fields with $GM/Rc^2 \sim 1$ by observing the unique properties of black holes in distilled form. Our standard of performance should be the quality of evidence that now defines for us a neutron star.

REFERENCES

1. REES, M. 1984. Ann. Rev. Astron. Astrophys. **22:** 471.
2. GIES, D. R. & C. T. BOLTON. 1982. Astrophys. J. **260:** 240.
3. GIES, D. R. & C. T. BOLTON. 1986. Astrophys. J. **304:** 371.
4. HUTCHINGS, J. B., D. CRAMPTON, A. P. COWLEY, L. BIANCHI & I. B. THOMPSON. 1987. Astron. J. **94:** 340.
5. COWLEY, A. P., D. CRAMPTON, J. B. HUTCHINGS, R. REMILLARD & J. E. PENFOLD. 1983. Astrophys. J. **272:** 118.
6. McCLINTOCK, J. E. & R. A. REMILLARD. 1986. Astrophys. J. **308:** 110.
7. McCLINTOCK, J. E. & R. A. REMILLARD. 1990. Bull. Am. Astron. Soc. **21**(4): 1206.
8. JOSS, P. C. & S. A. RAPPAPORT. 1984. Ann. Rev. Astron. Astrophys. **22:** 537.
9. CHITRE, D. M. & J. B. HARTLE. 1976. Astrophys. J. **207:** 592.
10. NINKOV, Z., G. A. H. WALKER & S. YANG. 1987. Astrophys. J. **321:** 425.
11. MAZEH, T., J. VAN PARADIJS, E. P. J. VAN DEN HEUVEL & G. J. SAVONIJE. 1986. Astron. Astrophys. **157:** 113.
12. HUTCHINGS, J. B., A. P. COWLEY, D. CRAMPTON, J. VAN PARADIJS & N. E. WHITE. 1979. Astrophys. J. **229:** 1079.

13. Van den Heuvel, E. P. J. 1983. *In* Accretion-Driven Stellar X-ray Sources, W. H. G. Lewin and E. P. J. van den Heuvel, Eds.: 303. Cambridge University Press. Cambridge, England.
14. Paczynski, B. 1974. Astron. Astrophys. **34:** 161.
15. Bahcall, J. N. 1978. Ann. Rev. Astron. Astrophys. **16:** 241.
16. Bregman, J., D. Butler, E. Kemper, A. Koski, R. P. Kraft & R. P. S. Stone. 1973. Astrophys. J., Lett. **185:** L117.
17. Margon, B., S. Bowyer & R. P. S. Stone. 1973. Astrophys. J., Lett. **185:** L113.
18. McClintock, J. E. 1986. *In* Physics of Accretion onto Compact Objects, K. Mason *et al.*, Eds.: 211. Springer-Verlag. Heidelberg.
19. Paczynski, B. 1983. Astrophys. J., Lett. **272:** L81.
20. Haswell, C. A. & A. W. Shafter. 1990. Astrophys. J., Lett. **359:** L47.
21. Ilovaisky, S. A., C. Chevalier, C. Motch & L. Chiappetti. 1986. Astron. Astrophys. **164:** 67.
22. Kitamoto, S., *et al.* 1989. Nature **342:** 518.
23. Tsunemi, H., S. Kitamoto, S. Okamura & D. Roussell-Dupre. 1989. Astrophys. J., Lett. **337:** L81.
24. Cowley, A. P., P. C. Schmidtke, D. Crampton & J. B. Hutchings. 1990. Astrophys. J. **350:** 288.
25. Van der Woerd, H., N. E. White & S. M. Kahn. 1989. Astrophys. J. **344:** 320.
26. White, N. E. & F. E. Marshall. 1984. Astrophys. J. **281:** 354.
27. Tennant, A. F., A. C. Fabian & R. A. Shafer. 1986. Mon. Not. R. Astron. Soc. **221:** 27P.
28. Makishimi, K., *et al.*, 1990. Publ. Astron. Soc. Japan. **42:** 295.
29. Tanaka, Y. 1989. Proceedings of the 23rd ESLAB Symposium, Vol. I, J. Hunt and B. Battrick, Eds.: 3. European Space Agency. Paris.
30. Weinberg, S. 1972. Gravitation and Cosmology: 334. Wiley. New York.
31. Connors, P., T. Piran & R. Stark. 1980. Astrophys. J. **235:** 224.
32. Fabian, A. C., M. J. Rees, L. Stella & N. E. White. 1989. Mon. Not. R. Astron. Soc. **238:** 729.
33. Liang, E. P. & P. L. Nolan. 1984. Space Sci. Rev. **38:** 353.
34. Bahcall, J. N., F. J. Dyson, J. I. Katz & B. Paczynski. 1974. Astrophys. J., Lett. **189:** L17.
35. Trimble, V., W. K. Rose & J. Weber. 1973. Mon. Not. R. Astron. Soc. **162:** 1P.
36. Kundt, W. & D. Fischer. 1989. J. Astrophys. Astron. **10:** 119.
37. Shapiro, S. L. & S. A. Teukolsky. 1983. Black Holes, White Dwarfs & Neutron Stars: 388. Wiley. New York.
38. Ling, J. C., W. A. Mahoney, W. A. Wheaton & A. S. Jacobson. 1987. Astrophys. J., Lett. **321:** L117.
39. Liang, E. P. & C. D. Dermer. 1988. Astrophys. J., Lett. **325:** L39.
40. Zdziarski, A. A., G. Ghisellini, I. A. George, R. Svensson, A. C. Fabian & C. Done. 1990. Astrophys. J., Lett. **363:** L1.
41. Lingenfelter, R. E. & R. Ramaty. 1989. Astrophys. J. **343:** 686.
42. Ling, J. C. & W. A. Wheaton. 1989. Astrophys. J., Lett. **343:** L57.
43. Barr, P., N. E. White & C. G. Page. 1985. Mon. Not. R. Astron. Soc. **216:** 65P.
44. Chen, K. & J. P. Halpern. 1989. Astrophys. J. **334:** 115.
45. Ricketts, M. J., K. A. Pounds & M. J. L. Turner. 1975. Nature **257:** 657.
46. Johnston, H. M., S. Kulkarni & J. B. Oke. 1989. Astrophys. J. **345:** 492.
47. Horne, K. & T. R. Marsh. 1986. Mon. Not. R. Astron. Soc. **218:** 716.

Nuclear Physics of Dense Matter[a]

C. J. PETHICK AND D. G. RAVENHALL

Department of Physics
University of Illinois at Urbana-Champaign
1110 West Green Street
Urbana, Illinois 61801
and
NORDITA
Blegdamsvej 17
DK-2100 Copenhagen Ø, Denmark

INTRODUCTION

Over the past few years there have been a number of important developments in applications of nuclear physics to astrophysical dense matter. Because of the limited space available we shall focus on these aspects of the subject in this article. For more general surveys we refer to recent conference proceedings.[1-3]

In dense matter the conditions encountered are so different from those under which laboratory measurements are made that one cannot simply take over the tabulated properties of stable and unstable nuclei and apply them directly. For example, neutron star matter is typically much more neutron rich than terrestrial matter, and one cannot *measure* the properties of nuclei that are expected to exist in neutron stars. Rather one must first distill the available data on stable and marginally unstable nuclei into a model that accounts for why nuclei work the way they do, and then apply the model to the conditions encountered in neutron stars. The examples we shall consider illustrate this general theme.

In what follows we shall discuss only the nuclear and hadronic constituents assumed in the nuclear models. In dense matter there are also electrons, muons, and neutrinos as the thermodynamic conditions call for them. They interact electromagnetically and/or via weak interactions, and their thermodynamic properties may be calculated by the standard methods for almost ideal relativistic Fermi gases.

NUCLEI BELOW NEUTRON DRIP

With increasing density, the electron chemical potential, μ_e, rises. Thus, as a consequence of the beta equilibrium condition, $\mu_n = \mu_p + \mu_e$, matter becomes increasingly neutron rich, and the chemical potentials, μ_n and μ_p, of neutrons and protons are pushed increasingly further apart. Eventually the neutron chemical potential reaches the neutron rest mass, its value at the bottom of the continuum of unbound levels, and neutrons "drip" out of nuclei. The properties of nuclei below neutron drip have been estimated previously[4] using a semiempirical mass formula[5] in

[a]This work was supported in part by National Science Foundation Grants PHY86-00377, PHY91-00283, and PHY89-21025, and in part by NASA Grant NAGW-1583.

which nuclei are treated as liquid drops, with additional corrections to take into account shell structure. The parameters of the model are determined by fits to observed nuclei. The results of these calculations reflect strongly the shell structure observed in terrestrial nuclei, as is indicated by the large number of equilibrium nuclei expected to have 82 neutrons, the well-known shell closing.

Recently Haensel, Zdunik, and Dobaczewski[6] have performed a microscopic calculation of nuclei below neutron drip using the Hartree–Fock–Bogoliubov approximation, a method particularly well suited to obtaining shell-closing effects. The nucleon–nucleon interaction employed is of the Skyrme type, chosen to reproduce nuclear energy levels. In these calculations, shell effects are not put in *a priori,* but are a result of the calculation. Haensel *et al.*[6] examine the most stable nuclei in the density range $\rho = 8 \times 10^6$ g cm^{-3} to 4.5×10^{11} g cm^{-3}, or baryon number density range $n = 4 \times 10^{-8}$ fm^{-3} to 2.7×10^{-4} fm^{-3}. They find that the favored nuclei tend to have closed proton, rather than neutron, shells, with proton numbers of 28 and 40. (See [6, tables 1 and 2].)

The reason that neutron closed-shell effects are relatively unimportant in neutron-rich nuclei is undoubtedly that these effects are weakened, and dominated, by the proximity of the neutron Fermi level to the continuum. It is interesting to note that the $Z = 40$ proton closed shell plays an important role, since that is also found for nuclei above neutron drip.[7] That $Z = 40$ is favored, rather than $Z = 50$ as in terrestrial nuclei, implies a reduction in the spin–orbit splitting in neutron-rich nuclei relative to the spacing between major shells: the $Z = 40$ shell corresponds to completing the $2p$, $1f$ shell without the 10 extra protons from the $1g_{9/2}$ orbital. The spin–orbit interaction is proportional to the derivative of the particle densities (the Thomas form). Presumably the neutron excess causes a spatial separation of the proton and neutron surfaces, and the protons are moved away from the neutron surface somewhat, thereby reducing their spin–orbit splitting.

The weakening of the neutron shell effects prompts the question whether there could be significant changes in nucleosynthesis predictions from the *r*-process, which depends sensitively on the position of neutron closed shells.[b]

The preceding results demonstrate the necessity of calculating properties of dense matter with microscopic models that capture the physics of observed nuclei, rather than making direct extrapolations from empirical data.

BULK MATTER AT NUCLEAR DENSITIES AND ABOVE

The basis for the calculations of reference 6 and many other discussions of the properties of matter in the density range below nuclear saturation ($n_s \simeq 0.16$ fm^{-3}, $\rho_s \simeq 2.6 \times 10^{14}$ g cm^{-3}) and below the critical temperature T_c, is usually an effective-interaction Hamiltonian of the Skyrme type. These approximate in a crude, simple way the basic properties of the in-medium nucleon–nucleon interaction. The several Skyrme interactions are usually designed to give correctly some of the properties of stable nuclei and of nuclear vibrational dynamics. They include the gradient terms needed to calculate surfaces and finite nuclei. Since the stable forms of matter in the

[b] This point was raised by W. Hillebrandt.

density range below n_s consist of mixtures of nuclei (in various shapes) and a low-density, weakly interacting vapor of neutrons (with protons also, at sufficiently high temperatures), the only extrapolation required from laboratory nuclei involves the large neutron excess. That must necessarily be theoretical.

The problem of calculating the properties of matter at densities greater than that of nuclear matter is more difficult, because there is very little direct information available. It is therefore necessary to use what one knows about interactions between nucleons, and from it calculate the properties of dense matter.

Two-body Interactions

Until temperatures or Fermi energies comparable to the pion rest mass are reached, the properties of the two-nucleon system (deuteron and scattering) can be represented by nonrelativistic nucleons interacting through a two-body potential. The potentials are as complicated as is allowed by the various invariance properties (exchange of space, spin, isospin; spin–orbit, tensor, in all combinations), and their spatial dependence cannot be determined unambiguously. (This is why there is not just one such potential, but several.) But their properties and parameters may be determined by comparison with the experimental energies and phase shifts. We take as a recent, well-explored example the $V14$ potential described by Pandharipande and coworkers,[8] and used by Wiringa, Fiks, and Fabrocini,[9] which we refer to as WFF. Techniques of many-body physics (basically two: the Brueckner–Bethe–Goldstone G-matrix method; and development of the variational approach, with correlations treated by the hypernetted chain method) permit the calculation of properties of the uniform-density systems of nuclear and neutron matter. With these two-body interactions, and including up to four-body correlations, WFF systematically find agreement between the two methods as to the deduced nuclear properties (binding energies and sizes of the light nuclei, saturation density, and energy of nuclear matter). For both methods, the results *disagree* with the experimental values.

Three- or More-body Forces

The same physics (meson exchange) that induces two-body forces also gives three-body forces. They are necessarily much more varied in their possible form, and are thus much less well determined by the relatively small amount of available data. The procedure adopted by WFF[8,9] is to choose a form suggested by theory, and to fit the parameters so that agreement is obtained with the experimental values of the binding energies of few-body nuclei and of nuclear matter. The variational Monte Carlo method is used for the fits to the three- and four-body nuclei. The results obtained are almost independent of the calculational technique used, and the nuclear matter saturation density is close to the experimental value.

What these calculations demonstrate is that three-body interactions are essential in calculations of matter at densities expected to occur in the interiors of neutron stars. A second important observation is that central densities of neutron stars with a mass of $1.4M_{\odot}$ are about $4n_s$ for the WFF equations of state. At such a density the average nucleon kinetic energy is only about 90 MeV, and it is therefore a reasonable approximation to treat the nucleons as nonrelativistic particles interacting via a potential that can be deduced from scattering data for lab energies less than about that value.

Despite the fact that WFF's two different choices for the internucleon interaction lead to rather similar predictions for the equation of state of dense matter, they do give significantly different values for the proton concentration for matter in beta equilibrium, a fact that will be important for our discussions of neutrino emission in the following section. This reflects, among other things, the fact that the isospin dependence of the three-body interation is not at all well determined. Clearly, it is important in the future to better pin down the character of three- and higher-body interactions.

NEUTRON-STAR COOLING AND THE DIRECT URCA PROCESS

Young neutron stars cool primarily by emitting neutrinos. For standard neutron star matter consisting of neutrons, protons, electrons (and possibly neutrinos), such as was considered earlier, the simplest weak-interaction processes one can imagine are $n \rightarrow p + e^- + \bar{\nu}_e$ and $e^- + p \rightarrow n + \nu_e$, which we shall refer to collectively as the direct Urca process. At temperatures well below typical Fermi temperatures ($\sim 10^{12}$ K), fermions participating in such a reaction must have momenta close to the respective Fermi momenta, and since the neutrino and antineutrino momenta are of order k_BT/c, which is small compared to the Fermi momenta, a necessary condition for momentum to be conserved, and therefore for the process to occur, is that $p_{Fe} + p_{Fp} \geq p_{Fn}$.

If matter consists only of neutrons, protons, and electrons, charge neutrality requires that the electron and proton densities be equal, and therefore, since $n_i = p_{Fi}^3/(3\pi^2\hbar^3)$, p_{Fe} is equal to p_{Fp}. The momentum condition is therefore $p_{Fp} \geq (1/2)p_{Fn}$, or $n_p \geq (1/2)^3 n_n$. Thus the proton fraction $x_p = n_p/(n_p + n_n)$ must exceed $(1/2)^3/[(1/2)^3 + 1] = 1/9 \approx 11.1$ percent. If the electron chemical potential exceeds the muon rest energy, muons as well as electrons will be present. Under those conditions $p_{Fe} \neq p_{Fp}$, and the critical proton fraction for the Urca process to proceed is somewhat higher, but always less than $1/[(1 + 2^{-1/3})^3 + 1] \approx 14.8$ percent, the value it attains for $\mu_e \gg m_\mu c^2$. If the proton fraction is high enough to allow the direct Urca process, the luminosity per unit volume in neutrinos and antineutrinos is given by[10]

$$\epsilon_{\text{Urca}} = \frac{457\pi c}{10{,}080\hbar^4} G_F^2 \cos^2\theta_C(1 + 3g_A^2)m_n m_p \mu_e \left(\frac{k_BT}{\hbar c}\right)^6$$

$$= 4 \times 10^{27}\left(Y_e \frac{n}{n_s}\right)^{1/3} T_9^6 \text{ erg cm}^{-3}\text{s}^{-1}, \qquad (1)$$

where G_F is the weak-interaction coupling constant, and θ_C is the Cabibbo angle. The temperature dependence may be understood as being due to a single power of T for each degenerate particle in the initial and final states, since degeneracy restricts participating particles to energies within k_BT of the Fermi surface, and a factor T^3 coming from the 3-dimensional phase space available for final neutrinos and antineutrinos. If the proton fraction is less than the one for the threshold, the dominant process is then the so-called modified Urca process.[11,12] It differs by having an additional nucleon bystander in the interaction whose role is to permit momentum conservation. Its luminosity per unit volume is given by[13]

$$\epsilon_{\text{modUrca}} = 9 \times 10^{21} \left(Y_e \frac{n}{n_s} \right)^{1/3} T_9^8 \text{ erg cm}^{-3}\text{s}^{-1}. \quad \textbf{(2)}$$

This process varies as T^8, the two additional powers of T coming from the additional nucleon in the initial and final states. Roughly speaking, the modified Urca rate is $\sim (T/T_F)^2$ times that for the direct process, where T_F is ~ 100 MeV, and therefore at $T \sim 10^8$ K, it is of order 10^8 times slower than the direct process. Thus, if the proton fraction is sufficiently high, the direct Urca process permits rapid cooling rates of this simple n, p and e^-, μ^- mixture of a magnitude usually associated with more exotic forms of matter, such as quarks.

For two forces of comparable physical plausibility, the proton fractions obtained by WFF[9] (see [9, fig. 11]) extend almost to the interesting value needed. (The AV14 + UVII potential gives smaller x_p than the UV14 + UVII one, presumably because the peak values occur in a density range where there are significant short-range spin-isospin correlations corresponding to an incipient neutral pion condensate, making it energetically less favorable to have protons.) In this density range, $n \sim 3n_s$, the three-body force that contributes strongly is assumed to have, as we observed earlier, a very restricted isospin dependence, so that the values of x_p are rather uncertain, and it is possible that they exceed the threshold value, or are far below. The source of this uncertainty is lack of knowledge of the symmetry energy above nuclear matter density, which deserves further study.

SUPERFLUID AND SUPERCONDUCTING GAPS

Soon after the BCS theory of superconductivity was developed, Migdal suggested that neutrons in neutron stars could be superfluid and the protons superconducting. Since the identification of pulsars with neutron stars, many people have estimated the neutron and proton gaps, and the associated transition temperatures. One of the main problems in such calculations is to calculate the effective two-body interaction. While methods for evaluating total energies of many-body systems are very highly developed, as we described in the third section, there has been somewhat less work devoted to the effective interactions for particles close to the Fermi surface, which determine superfluid and superconducting properties, as well as transport properties of the normal state.

Physically the effective interaction consists of two parts, a direct interaction that may be thought of as taking into account the short-range correlations produced by the nucleon–nucleon interaction, and an induced part that reflects the modification of the interaction by the response of the nuclear medium through which the two nucleons move.[14–17] The induced interaction is analogous to the phonon-induced interaction between electrons that gives rise to superconductivity in terrestrial metals. If one adopts a simple model for the nucleon–nucleon interaction in which the energy density is independent of momentum, one finds the induced interaction for a pair of neutrons in a spin-singlet state is given by[15]

$$V_{nn}^{\text{singlet}} = -\frac{(f_{nn}^{s})^2\chi(q)}{1+f_{nn}^{s}\chi(q)} + 3\frac{(f_{nn}^{a})^2\chi(q)}{1+f_{nn}^{a}\chi(q)}, \tag{3}$$

where $\chi(q)$ is the static Lindhard function (which is positive) as a function of momentum transfer, q, and the effective interaction between nucleons of species i and j is given by $f_{ij}^{s} + f_{ij}^{a}\sigma_i \cdot \sigma_j$, where σ_i and σ_j are Pauli matrices. (For simplicity we neglect small concentrations of protons in estimating the induced interaction.) The first term in **(3)** is the interaction induced by density fluctuations, and the second is due to spin fluctuations. For the usual singlet superconductivity, the effective pairing interaction is V^{singlet} averaged over the Fermi surface. Density fluctuations therefore enhance the pairing interaction for such a state, but spin fluctuations depress it, an effect well known in metals. Detailed calculations[16,17] show that for singlet nucleon pairs the induced interaction depresses the superconducting gap, indicating that the repulsion due to exchange of spin fluctuations, with its associated spin-degeneracy factor of 3, overwhelms the attraction due to density fluctuations. While both these calculations predict suppression of the gap, there are significant differences between them, even at densities somewhat less than that of nuclear matter.

For protons moving in a medium of neutrons, the induced interaction is given by the analog of **(3)**, but with f_{np} instead of f_{nn}, and therefore exchanges of fluctuations have effects qualitatively similar to those for n–n pairs, but with different coupling constants. Recently, Ainsworth *et al.*[17] have calculated proton gaps and find that, as in the case of neutrons, the induced interaction suppresses s-wave pairing, which again indicates that spin fluctuation exchanges dominate.

Above nuclear matter density, the n–n interaction in free space is most attractive in the $^3P_2$ channel, and if one uses this interaction to estimate gaps, one is led to expect neutrons in this density range to be paired in a $^3P_2$ state. To see how the induced interaction affects $^3P_2$ pairing, we observe that the pairs are in a spin-triplet state, and the relevant interaction is

$$V_{nn}^{\text{triplet}} = -\frac{(f_{nn}^{s})^2\chi(q)}{1+f_{nn}^{s}\chi(q)} - \frac{(f_{nn}^{a})^2\chi(q)}{1+f_{nn}^{a}\chi(q)}, \tag{4}$$

which is always negative. To find the effective pairing interaction for a P state ($l = 1$), one averages the interaction over the Fermi surface with a factor $P_1(\mu) = \mu = 1 - q^2/(2_{pF})^2$, where $\mu = \hat{p} \cdot \hat{p}'$. Here $\mathbf{p}$ is the initial momentum of one of the neutrons, $\mathbf{p}'$ is its final momentum, and $\mathbf{q} = \mathbf{p} - \mathbf{p}'$. Since $\chi(q)$ is a monotonically decreasing function of q, V^{triplet} is an increasing function of q, and thus the contribution to the $l =$ 1 pairing interaction is negative. Spin and density fluctuations both tend to enhance $^3P_2$ pairing.

To date, calculations of pairing have been based on the assumption that the basic interactions between nucleons are two-body ones. As we observed earlier, there are strong reasons to believe[9] that three-body forces play a significant role at nuclear densities and above. In the nuclear medium, three-body forces give rise to an effective two-body interaction that will affect the effective pairing interaction. They will generally be more important in the pairing interaction than in the energy density, because the effective two-body interaction is, crudely speaking, the second derivative of the energy density with respect to density. In a simple model, one can expect two- and three-body interactions to give rise to contributions to the energy density of the form $c_2n^3 + c_3n^3$. The effective interaction would be of order $2c_2n^2 + 6c_3n^3$, enhancing the three-body pairing contributions by a factor of 3. This argument is suggestive, but it is not rigorous, because the interactions have both negative and positive contributions, and for realistic potentials the density dependence is more complicated than this. It is therefore important to estimate the pairing interaction, including contributions from the three-body force.

Over the years, nuclear physics has provided important data for neutron star studies. In turn, the study of neutron stars has forced nuclear physicists to consider conditions quite different from those in the laboratory, and this has stimulated them to strive for a deeper understanding of nuclear phenomena. The examples discussed here illustrate the continuing vitality of this interplay between nuclear physics and astrophysics, which contributes to progress in both areas.

ACKNOWLEDGMENT

We thank our colleagues G. Baym, J. W. Clark, F. K. Lamb, J. M. Lattimer, M. Prakash, and particularly V. R. Pandharipande, for helpful conversations and advice.

REFERENCES

1. GREINER, W. & H. STÖCKER, Eds. 1989. The Nuclear Equation of State, Part A. Plenum. New York.
2. SOYEUR, M., H. FLOCARD, B. TAMAIN & M. PORNEUF, Eds. 1989. Nuclear Matter and Heavy Ion Collisions. Plenum. New York.
3. PROCEEDINGS OF THE 3D INTERNATIONAL CONFERENCE ON NUCLEUS-NUCLEUS COLLISIONS. 1988. Nucl. Phys. **A488.**
4. For example, BAYM, G., C. J. PETHICK & P. SUTHERLAND. 1971. Astrophys. J. **170:** 299.
5. As in, for example, MYERS, W. D. 1977. Droplet Model of Atomic Nuclei. Plenum. New York.
6. HAENSEL, P., J. L. ZDUNIK & J. DOBACZEWSKI. 1989. Astron. Astrophys. **222:** 353.
7. NEGELE, J. W. & D. VAUTHERIN. 1973. Nucl. Phys. **A207:** 298.
8. CARLSON, J., V. R. PANDHARIPANDE & R. SCHIAVILLA. 1991. *In* Electron Nucleus Scattering, B. Frois and I. Sick, Eds. World Scientific. Singapore. In press.
9. WIRINGA, R. B., V. FIKS & A. FABROCINI. 1988. Phys. Rev. **C38:** 1010.
10. In this section our discussion follows LATTIMER, J. M., C. J. PETHICK, M. PRAKASH & P. HAENSEL. 1991. Phys. Rev. Lett. **66:** 2701.
11. CHIU, H.-Y. & E. E. SALPETER. 1964. Phys. Rev. Lett. **12:** 413.
12. BAHCALL, J. N. & R. A. WOLF. 1965. Phys. Rev. **B140:** 1452.
13. FRIMAN, B. L. & O. V. MAXWELL. 1979. Astrophys. J. **232:** 541.
14. PINES, D. 1971. *In* Proceedings of the XIIth International Conference on Low Temperature Physics, E. Kanda, Ed.: 7 (Keigaku, Tokyo).
15. CLARK, J. W., C.-G. KÄLLMAN, C.-H. YANG & D. A. CHAKKALAKAL. 1976. Phys. Lett. **61B:** 331.
16. CHEN, J. M. C., J. W. CLARK, E. KROTSCHECK & R. A. SMITH. 1986. Nucl. Phys. **A451:** 509.
17. AINSWORTH, T. L., J. WAMBACH & D. PINES. 1987. Phys. Lett. **B228:** 73; WAMBACH, J., T. L. AINSWORTH & D. PINES. 1991. *In* Neutron Stars: Theory and Observations, J. Ventura and D. Pines, Eds.: 37. Kluwer, Dordrecht, the Netherlands.

Neutron Star Crusts[a]

LARS HERNQUIST[b]

Lick Observatory
University of California at Santa Cruz
Santa Cruz, California 95064

INTRODUCTION

Broadly speaking, the crusts of neutron stars can be divided into two regimes. The outer crust, which lies at densities lower than neutron drip, $\rho \lesssim 4.3 \times 10^{11}$ gm-cm^{-3}, is composed of nuclei whose atomic weight and nuclear charge increase with depth.[1] At slightly higher densities, free neutrons appear, and the nuclei eventually dissolve at densities $\sim 2 \times 10^{14}$ gm-cm^{-3}, which demarcates the boundary of the inner crust. For typical equations of state the total mass of the crust is negligible compared with neutron star masses. Nevertheless, a proper understanding of crusts is important, since they mediate many observed properties of neutron stars, including, possibly, radio emission from pulsars, thermal X-ray fluxes, glitches, and γ-ray bursts. From a purely theoretical point of view, neutron star crusts are of interest because of the wide range of physical processes they support. The largest temperature and density gradients in neutron stars are found in their crusts, magnetic fields can affect their surface structure, and crusts are thought to consist of both solid and liquid components.

The following is a summary of recent developments in our understanding of the physics of neutron star crusts. Of necessity, this review focuses on a limited number of topics; mainly those related to isolated neutron stars, that is, those not in binaries, and which involve effects occurring at low densities in the outer crust.

PHYSICAL STATE OF THE CRUST

At low densities the structure of neutron star crusts can be influenced by a variety of effects. For example, in sufficiently strong magnetic fields the orbits of free electrons are quantized and have discrete energies:

$$E_n = (p_z^2c^2 + m^2c^4 + 2n\hbar\omega_B mc^2)^{1/2} \sim mc^2 + \frac{p_z^2}{2m} + n\hbar\omega_B\,, \tag{1}$$

where $\omega_B = |e|B/mc$, $n = 0, 1, 2, \ldots$ labels these Landau levels, and the second equality obtains in the nonrelativistic limit. For a degenerate, nonrelativistic gas the index of the highest occupied level is $n_{\max} \sim \mu_{NR}/\hbar\omega_B$, where $\mu_{NR} = p_F^2/2m$ is the chemical potential of the electrons in the nonrelativistic domain. Thus, the limiting

[a]This work was supported in part by the Alfred P. Sloan Foundation, in part by NASA Theory Grant NAGW-2422, and in part by National Science Foundation Grant AST 90-18526.
[b]Alfred P. Sloan Foundation Fellow.

density for a given number of levels is roughly

$$\rho < 10^5 B_{12}^{3/2} \left(\frac{n_{\max}}{3}\right) \text{gm/cm}^3, \tag{2}$$

where B_{12} is the field strength in units of 10^{12} Gauss. Typically, magnetic effects on the structure of the matter are small unless only a few Landau levels are occupied, and so, according to (2), are mainly important near the surface.

MAGNETICALLY CONDENSED MATTER

In strong magnetic fields, that is, when *free* electrons occupy only the lowest Landau levels, *atoms* acquire large quadrupole moments. Early in the study of the properties of neutron star crusts, it was recognized that this distortion can lead to a quantum mechanical analog of covalent bonding and that atoms would bind into linear chains.[2] These chains attract one another, resulting in a latticelike structure. If chains form, then the neutron star surface lies at a relatively high density, $\rho \sim 4 \times 10^3 B_{12}^{6.5}$ gm-cm^{-3}, and there is no atmosphere. The question of whether or not this "magnetically condensed matter" exists rests on the value of the cohesion energy, E_c, which is simply the difference of the binding energy of isolated atoms relative to that of atoms in linear chains. The first estimates of E_c were in the range 10–20 keV; sufficiently large to inhibit the extraction of charged particles from the surfaces of neutron stars, suggesting that some pulsar emission mechanisms might be inoperable. However, it now appears that these calculations were in error: E_c is much smaller than either of the binding energies used to compute it, implying that higher numerical precision is required than in most previous analyses.

The properties of matter in quantizing magnetic fields have been considered recently by a number of groups.[3–5] In particular, Neuhauser *et al.* obtain cohesion energies for atomic species up to iron in fields $B = 1 - 5 \times 10^{12}$ G using a Hartree–Fock variational approach. They conclude that bonding will not occur for iron and that helium will be only weakly bound. For helium, $E_c = 25$ eV at $B = 10^{12}$ G and $E_c = 150$ eV at $B = 5 \times 10^{12}$ G; far too low to affect the extraction of charged particles. In fact, it is likely that such weakly bound structures would be disrupted by thermal effects, suggesting that linear chains are probably not important for most observable properties of neutron stars. Note, however, that some work remains on this problem, since Neuhauser *et al.* ignore interactions between chains. It is possible that chain–chain interactions may lead to weak bonding, even if individual chains are not energetically favored.[4,5]

Surface Structure

For an unmagnetized, degenerate gas of free electrons the relationship between the electron density, n_e, and the chemical potential can be written

$$n_e = \frac{1}{3\pi^2}\left(\frac{2m}{\hbar^2}\right)^{3/2} \mu_{\rm NR}^{3/2}. \tag{3}$$

In the presence of a strong magnetic field, when only the lowest Landau level is occupied, this relation becomes instead

$$n_e = \frac{1}{\sqrt{2}\pi^2}\frac{m^{3/2}}{\hbar}\omega_B\,\mu_{NR}^{1/2}. \tag{4}$$

For a given value of μ,

$$\frac{n_e(B)}{n_e(0)} = \frac{3}{4}\frac{\hbar\omega_B}{\mu_{NR}} \gg 1, \tag{5}$$

where the inequality follows from the fact that $n_{max} = 0$. To lowest order, the chemical potential is determined by the equation of hydrostatic equilibrium ignoring magnetic fields, and increases linearly with depth. Equation **(5)** implies that the density at a given depth will be significantly higher in the presence of a strong magnetic field.

Fushiki *et al.*[6] have computed the surface structure of neutron stars, including magnetic effects on atomic structure. They determine the equation of state using the Thomas–Fermi and Thomas–Fermi–Dirac approximations. Although these more exact models give results that differ significantly from that of a free electron gas, the density at a given depth is higher in the presence of a magnetic field, in accord with **(5)**. In some cases the pressure goes to zero at finite density, so neutron stars may have high density surfaces even if magnetically condensed matter does not form. Fushiki *et al.* suggest that magnetic corrections to the equation of state may affect the thermal structure of neutron stars. However, this does not appear likely, since the temperature distribution in neutron star crusts rapidly converges to the radiative zero solution.[7] A more reasonable possibility is that magnetic fields will modify the X-ray spectra of cooling neutron stars. An optical depth of order unity in X-rays requires $\sim 10^{14}$ gm of material spread over the entire surface, extending to a maximum density $\rho \sim$ 10–20 gm-cm^{-3} and a depth $\sim$1 cm, where magnetic effects can be quite important.[7]

RADIATIVE TRANSFER AND COOLING

Departures from Blackbody

Past comparisons between theoretical cooling curves of isolated neutron stars and observations generally assumed that the surface emits as a blackbody. Romani[8] has constructed LTE model atmospheres for neutron stars using a temperature correction approach. These calculations make use of Los Alamos opacity tables, and were performed for a variety of surface compositions. His synthetic spectra show that departures from blackbody are possible, especially if the surface contains helium. For low effective temperatures $T_{eff} \sim 10^5$–$10^{5.5}$ the soft X-ray flux as measured by an instrument like the Einstein IPC is much greater than the blackbody value, for helium composition. In the case of iron surfaces, the effective temperature is roughly equal to the blackbody value, but absorption edges appear at the energies relevant to current and planned X-ray detectors and may be detected spectroscopically.

Miller[9] has computed LTE atmospheres also using a temperature correction approach, but including the effects of magnetic fields on atomic structure and opacities. Some departure from a blackbody spectrum is noted, but mainly at high energies. Contrary to Romani's findings for an unmagnetized neutron star, Miller predicts that the blackbody temperature, T_{bb}, should be less than the effective temperature, as judged by Einstein. Two effects increase the opacity in the relevant energy range. First, strong magnetic fields shift the atomic energy levels higher, placing ionization edges nearer observed energies. Second, in a strong magnetic field the bound-free cross section declines with frequency as ν^{-2}, rather than as ν^{-3} in an unmagnetized atmosphere. Owing to the increased opacity, the levels of the atmosphere contributing to the spectrum are higher up and cooler. Consequently, the flux is closer to the blackbody value at the observed energies.

The inference that magnetic fields act to push the flux nearer its blackbody value is significant since calculations implying that $T_{eff} < T_{bb}$ would require neutron stars to cool more rapidly than standard models. It should be noted, however, that Miller's calculations were performed for a restricted set of effective temperatures and assuming that, that the ideal gas law obtains in the presence of a quantizing magnetic field. As indicated earlier, significant departures from the ideal gas law are expected near the surface,[6] so the validity of this assumption is unclear.

Heat Flow with Magnetic Fields

The presence of a magnetic field can influence the flow of heat through neutron star crusts in two ways. First, it can modify the total flux via its quantum effect on conductive opacities. Second, since the opacity is no longer a scalar, the heat flow will be anisotropic and can produce a nonuniform distribution of temperature on the stellar surface. Hernquist[10] computed the heat flux through a magnetized neutron star crust assuming a purely radial field. In that limit the surface temperature is higher for a given core temperature than in the absence of a magnetic field. Essentially, the field enhances heat conduction along the field by quantizing the density of states. However, the effect appears to be rather modest; even in the most extreme case the total heat flux is increased by only a factor of ~ 3 relative to an unmagnetized crust, a finding supported by more detailed calculations.[11] Moreover, simple geometric arguments suggest that this enhancement is mostly canceled for a more realistic magnetic topology, since heat flow perpendicular to the field is strongly inhibited.[10]

A serious flaw of these earlier works was their neglect of anisotropic conductivity. Since heat can flow more easily along field lines than perpendicular to them, one expects the poles of a magnetized neutron star to be hotter than its equatorial regions. A proper treatment of this effect requires the solution of the heat transport equation in multidimensions. Significant progress on this problem has been made by Schaaf,[12] who computes the flow of heat through a neutron star crust in a uniform, axisymmetric field. To maintain numerical stability in his iterative scheme, Schaaf restricts his analysis to weak fields by linearizing the relevant equations in the field strength. For weak fields $B \approx 10^{11}$ G it appears that geometrical effects are more important in determining the total heat flux than electron quantization, and the effective temperature is lower than the surface temperature of an unmagnetized neutron star. The surface distribution of temperature is nonuniform, as expected, with a maximum pole-to-equator variation of roughly a factor of 2 for low effective temperatures. This implies that the surface X-ray emission will be modulated by rotation, though at a much lower level than the temperature variation when averaged over the visible surface. However, these results must be regarded as preliminary, since field strengths $B \gtrsim 10^{12}$ G are probably more relevant to young cooling neutron stars.

EVOLUTION OF MAGNETIC FIELDS

Thermomagnetic Effects

The origin of magnetic field in neutron stars is poorly understood. In conventional models, it is assumed that the field arises from flux freezing during the core collapse of the progenitor star. A number of observational difficulties with this viewpoint motivated Blandford *et al.*[13] to consider an alternate model in which the magnetic field is generated by thermal processes in the crusts of neutron stars long after they form. Crudely speaking, a horizontal seed field in the presence of a vertical heat flux will give rise to a horizontal heat flux through the Nernst effect. Associated with this horizontal heat flux are horizontal temperature perturbations that induce horizontal electric fields by the Seebeck effect. It is straightforward to show that these horizontal electric fields have nonvanishing curl, and hence can amplify the horizontal magnetic field. Rigorously, the induction equation is

$$\frac{\partial \mathbf{B}}{\partial t} = \nabla \times (\mathbf{v} \times \mathbf{B}) - \nabla Q_0 \times \nabla T - \nabla \times \left[\frac{c^2 \nabla \times \mathbf{B}}{4\pi\sigma}\right], \qquad (6)$$

where Q_0 is the thermopower and the three terms on the right-hand side of (6) represent field convection, an effective battery, and ohmic decay. To linear order, $\nabla Q_0 \times \nabla T \propto \nabla\rho \times \nabla T$. If the condition of hydrostatic equilibrium obtains, then $\nabla\rho \times \nabla T \equiv 0$ and field amplification can occur only in the solid crust.[13]

A linear analysis shows the field can grow by these thermomagnetic effects and that the fastest growth occurs near the melt surface, where ∇T is largest. However, for the field to grow to sufficiently large values to account for radio pulsars, either the heat flux must be increased by a factor of ~ 3 compared to standard cooling calculations or some field must be generated in the liquid crust. In addition, the nonlinear development of a thermally produced magnetic field is uncertain. In particular, it has yet to be demonstrated that this mechanism can produce fields having long-range order.

This problem was subsequently examined by Urpin *et al.*,[14] who do not enforce hydrostatic equilibrium in the liquid crust and argue that $\nabla\rho \times \nabla T$ can be nonzero there owing to Coriolis forces. If thermomagnetic processes operate in the liquid crust, growth occurs much more rapidly than in the solid, since the largest temperature gradients are supported at densities lower than the melt surface. In fact, Urpin *et al.* obtain an *e*-folding time for field growth of roughly 70 days if the surface temperature is 5×10^6 K. In principle, this would obviate most of the difficulties encountered by Blandford *et al.*, and has the potentially desirable consequence of linking field growth to the rotation period.

However, it has not been rigorously shown that Coriolis forces can indeed allow departures from hydrostatic equilibrium in the liquid crust. The time scale to achieve hydrostatic equilibrium is $\tau_{hyd} \sim z/c_s$, where z is the depth and c_s is the sound speed. For a degenerate, relativistic gas

$$\tau_{hyd} \approx 3 \times 10^{-6} g_{14}^{-1} \rho^{1/6}\ \mathrm{s}, \tag{7}$$

where g_{14} is the surface gravity in units of 10^{14} cm-s^{-2}. At a density $\rho = 10^7$ gm-cm^{-3}, $\tau_{hyd} \sim 4 \times 10^{-5}$ s. The ratio of this time scale to that obtained by Urpin *et al.* for field growth is $\tau_{hyd}/\tau_{gr} \approx 7 \times 10^{-12}$. Moreover, $\tau_{hyd} \ll P$, where P is the period of a rapidly rotating neutron star. On this basis it is somewhat difficult to see how Coriolis forces could act to allow nonzero $\nabla\rho \times \nabla T$ in the liquid crust. Even if the field could grow as envisaged by Urpin *et al.,* the magnetic dipole moment would not be sufficient to account for radio pulsars. The dipole moment is given roughly by

$$M \sim \frac{3\pi R^2}{8} B_{max} z_{max}, \tag{8}$$

where B_{max} is the field strength at the maximum depth, z_{max}, to which it penetrates, and R is the stellar radius.[13] Numerically, $M \sim 5 \times 10^{27} R_6^2 B_{max,12} z_{max,40}$ G-cm^3, where $z_{max,40}$ is the depth in units of 40 meters. In the calculations by Urpin *et al.* the field peaks at $z \sim 40$ meters. Field strengths $B_{max} \sim 2 \times 10^{14}$ G would then be needed to produce the dipole moments of radio pulsars, which are typically $M \sim 10^{30}$ G-cm^3. This does not seem reasonable, since a variety of effects act to saturate field growth in the liquid at much lower field strengths.[13] Thus, it is likely that significant field growth must occur in the solid crust if thermal processes are to explain pulsed emission from young neutron stars.

Field Decay

A variety of observations suggest that magnetic fields in neutron stars are not static. In the case of radio pulsars it appears that fields decay ohmically.[15] (A gradual alignment of the rotation and magnetic axes has a similar effect, but studies of polarization variations as a function of radio pulse phase indicate that torque decay occurs mainly by field decay.[16]) This interpretation is complicated by observations implying that some old neutron stars remain strongly magnetized while others do not. Millisecond pulsars have only weak fields $B \sim 10^8 - 10^9$ G, and their statistics suggest that the field decays at most slowly for $B \lesssim 3 \times 10^9$ G.[17] Binary radio pulsars have fields that are on average weaker than isolated pulsars, yet often appear to be older than $\sim 10^9$ years.[18] X-ray pulsars are strongly magnetized, $B \gtrsim 10^{12}$ G, but are older than typical radio pulsars. X-ray bursters do not have measurable fields, while γ-bursters may still be strongly magnetized.

A generic difficulty with interpreting these data is that magnetic fields at high densities in neutron stars decay only slowly, if at all. In the solid crust the conductive opacity is dominated by electron–phonon scattering with corresponding ohmic decay time

$$\tau_{ohm} \sim \frac{4\pi\sigma L^2}{c^2} \sim 2 \times 10^3 \rho_9^{4/3} T_8^{-1} g_{14}^{-2}\ \mathrm{yr}. \tag{9}$$

If one requires an ohmic decay time shorter than $10^{6.5}$ years, the field must be confined to densities $\rho < 3 \times 10^{11} T_8^{3/4} g_{14}^{3/2}$ gm-cm^{-3}; that is, it must reside primarily in the outer crust.

The ohmic decay of crustal magnetic fields has been considered by Sang and Chanmugam.[19] They find that ohmic decay is nonexponential and is effective only at rather low densities. Two effects act to retard field decay with time. First, even if the field lies entirely at very low densities initially, it can diffuse ohmically to higher densities where the conductivity is large. Second, as the star cools, the interior temperature drops and the conductivity increases. A detailed study shows that the nonexponential decay found by Sang and Chanmugam is consistent with pulsar statistics if the field is confined to low densities in the outer crust.[20] In fact, some residual field may be long-lived, possibly explaining the weak fields associated with millisecond pulsars.

Thus, if the field is initially confined mainly to the outer crust, as in models of thermal field growth, its subsequent ohmic decay may provide a consistent model of neutron star magnetization in most contexts. If, instead, the magnetic field was generated by flux freezing at the birth of the neutron star, then one expects some field to thread through the core where ohmic decay times are much longer than a Hubble time.

Several proposals have appeared to reconcile this difficulty with observations of neutron stars. In one class of models, it is assumed that magnetic fields in pulsars do not decay significantly and that weak-field objects have different evolutionary tracks from strongly magnetized ones. One possibility is that objects such as X-ray bursters and millisecond pulsars do not have the same progenitors as radio pulsars and result from the accretion-induced collapse of white dwarfs in cataclysmic variables.[21] It is unclear, however, if this suggestion is consistent with the statistics of progenitors, especially in globular clusters. A more elaborate scenario suggests that magnetic field decays only as the result of external influences. A long-standing hypothesis, apparently first made by Bisnovatyi-Kogan and Komberg,[22] but curiously overlooked by subsequent workers, is that mass accretion may "screen" or bury the field. Calculations of evolutionary tracks suggest that field screening is effective if the field is confined to the crust,[23,24] but it is not known if core field will be affected. (The suggestion that accretion can drive an inverse battery effect to destroy the field[25] is probably not tenable owing to significant departures from spherical symmetry expected when accreting onto a strongly magnetized star.)

Other models that reconcile the existence of core field with indications of decay rely on nonohmic processes to expel flux from the core on sufficiently short time scales to account for torque decay in radio pulsars. Muslimov and Tsygan[26] note that quantized flux tubes in the superconducting interior would be subject to buoyancy forces and would simply float out of the core. The first estimates of the time scale for the flux to be expelled in this way suggested that it could not account for field decay in pulsars;[26,27] however, a more recent treatment by Jones predicts that the flux expulsion time is roughly 10^7 years.[28] Thus, it may be possible to expel the flux to densities $\sim 2 \times 10^{14}$ gm-cm^{-3} on interestingly short time scales. Still, the conductivity in the inner crust is far too large for field there to decay significantly in a Hubble time. In a subsequent refinement to this model, Jones[29] argues that Hall drift in the inner crust will push the field to densities below neutron drip where it can decay. However, the efficiency of this mechanism is unknown.

A novel proposal has been put forward by Wang and Eichler,[30] who note that poloidal field trapped in the core may be subject to overturning instability that would reduce the large-scale dipole moment of a neutron star. In their analysis they examine the stability of field in an infinite conducting cylinder surrounded by a thin insulating layer with finite conductivity, σ. The overturning instability develops on a time scale determined by the ohmic decay time of the insulating layer. So, if σ is sufficiently small, the field can undergo such an instability on a time scale similar to that needed to explain pulsar deaths. That is, although significant field may still be present, the magnetic dipole moment will be reduced to the extent that pulsed emission is no longer possible. Although this model is appealing in its simplicity, its relation to field decay in actual neutron stars is problematic. The viability of the scenario will rest to a large extent on which layers in the crust determine the time scale for the onset of the instability, since the ohmic decay time at high densities in the inner crust is longer than a Hubble time.

ACKNOWLEDGMENTS

I thank R. Blandford and S. Phinney for much enlightening discussion.

REFERENCES

1. Baym, G., C. Pethick & P. Sutherland. 1971. Astrophys. J. **170:** 299.
2. Chen, H.-H., M. A. Ruderman & P. G. Sutherland. 1974. Astrophys. J. **191:** 473.
3. Neuhauser, D., S. E. Koonin & K. Langanke. 1987. Phys. Rev. A **36:** 4163.
4. Jones, P. B. 1986. Mon. Not. R. Astron. Soc. **218:** 477.
5. Kössel, D., R. G. Wolff, E. Müller & W. Hillebrandt. 1988. Astron. Astrophys. **205:** 347.
6. Fushiki, I., E. H. Gudmundsson & C. J. Pethick. 1989. Astrophys. J. **342:** 958.
7. Hernquist, L. & J. H. Applegate. 1984. Astrophys. J. **287:** 244.
8. Romani, R. W. 1987. Astrophys. J. **313:** 718.
9. Miller, M. C. 1991. Preprint.
10. Hernquist, L. 1985. Mon. Not. R. Astron. Soc. **213:** 313.
11. van Riper, K. A. 1988. Astrophys. J. **329:** 339.
12. Schaaf, M. E. 1990. Astron. Astrophys. **235:** 499.
13. Blandford, R. D., J. H. Applegate & L. Hernquist. 1983. Mon. Not. R. Astron. Soc. **204:** 1025.
14. Urpin, V. A., S. A. Levshakov & D. G. Yakovlev. 1986. Mon. Not. R. Astron. Soc. **219:** 703.
15. Gunn, J. E. & J. P. Ostriker. 1970. Astrophys. J. **160:** 979.
16. Lyne, A. G. & R. N. Manchester. 1988. Mon. Not. R. Astron. Soc. **234:** 477.
17. van den Heuval, E. P. J., J. A. van Paradijs & R. E. Taam. 1988. Nature **322:** 153.
18. Kulkarni, S. R. 1986. Astrophys. J. **306:** L85.
19. Sang, Y. & G. Chanmugam. 1987. Astrophys. J., Lett. **323:** L61.
20. Sang, Y. & G. Chanmugam. 1987. Astrophys. J. **363:** 597.
21. Chanmugam, G. & K. Brecher. 1987. Nature **329:** 696.
22. Bisnovatyi-Kogan, G. S. & B. V. Komberg. 1974. Astron. Zh. **51:** 373.
23. Shibazaki, N., T. Murakami, J. Shaham & K. Nomoto. 1989. Nature **342:** 656.
24. Romani, R. W. 1990. Nature **347:** 741.
25. Blondin, J. M. & K. Freese. 1986. Nature **323:** 786.

26. MUSLIMOV, A. G. & A. I. TSYGAN. 1985. Astrophys. Space Sci. **115:** 43.
27. HARVEY, J. A., M. A. RUDERMAN & J. SHAHAM. 1987. Phys. Rev. D. **33:** 2084.
28. JONES, P. B. 1987. Mon. Not. R. Astron. Soc. **228:** 513.
29. JONES, P. B. 1988. Mon. Not. R. Astron. Soc. **233:** 875.
30. WANG, Z. & D. EICHLER. 1988. Astrophys. J. **324:** 966.

The Pulsar Radio-Frequency Emission Problem[a]

JOANNA M. RANKIN[b]

Physics Department
University of Vermont
Burlington, Vermont 05405

HIGHLIGHTS OF IAU COLLOQUIUM NO. 128

I have been asked in this paper to do two very different things: On the one hand, to review the results of IAU Colloquium No. 128 "The Magnetospheric Structure and Emission Mechanisms of Radio Pulsars," held in Łagów, Poland, June 17–23, 1990, and on the other, to summarize my own work that attempts to construct an empirical model of the pulsar RF emission process.

The format of IAU Colloquium No. 128 very much represented the perspectives (and prejudices) of observers in approaching the pulsar emission problem. Sessions were organized around questions that were empirical in nature: What is the structure and evolution of the pulsar magnetic field? How do average profile properties constrain RF emission models? What can we learn from X-ray, γ-ray, and millisecond pulsars about the RF emission problem? What processes operate in the polar-cap region? How is the coherent RF radiation generated, and how is it (de)polarized?

Some of the highlights of the colloquium were as follows:

- New theoretical results suggest that magnetic field decay is associated with the particular rotational history of a pulsar in a binary system. This appears to explain the paucity of evidence for field decay and alignment in the slow pulsar population and the smaller fields of millisecond pulsars (Bhattacharya *et al.* 1992). A review of these results by Dr. Srinivasan was presented during this symposium (Srinivasan 1991).
- Recent evidence indicates that pulsar magnetic fields are surprisingly dipolar—at least in the emission region. Profile observations made over a 200-to-1 frequency range show no systematic low frequency misalignment that might be attributed to the effects of poloidal field components. Neither do most pulsars exhibit high frequency alignment offsets associated with quadrupole field components (Hankins *et al.* 1992).
- The two somewhat different methods of interpreting the polarization of pulsar average profiles by Lyne and Manchester (1992) and myself (Rankin 1992) were thoroughly discussed, and I will say more about this below.

[a]This work was supported in part by grants from the National Science Foundation (AST 89-17722) and the Vermont EPSCoR Program. Arecibo Observatory is operated by Cornell University under contract to the National Science Foundation.

[b]Current address: Raman Research Institute, C. V. Raman Avenue, Sadashivanagar, Bangalore 560080, India.

- Three groups presented pulsar emission theories: The Lebedev group (Beskin, Gurevich, and Istomin 1992), the Georgian group (Kazbegi, Machabeli, and Melichidze 1992), and the Paris group (Asseo, Pelletier, and Sol 1992). Each theory is based on more sophisticated plasma theory than previous efforts. As yet only the Lebedev model makes detailed observational predictions and many, but not all, aspects of this model appear promising as a framework for understanding the observations. A number of theorists regarded the Georgian and French work as better grounded physically.
- The work function for ions at the stellar surface is now generally regarded to be very small, undermining one of the principal tenets of some existing pulsar models.
- General relativistic effects near the stellar surface have been shown to contribute importantly to the potentials in the polar cap region and help explain how pair production avalanches can be sustained in the slower pulsars (Muslimov and Tsygan 1992).

THEORETICAL AND OBSERVATIONAL APPROACHES

The problem of pulsar emission has presented great challenges for both theorists and observers. In part because of preoccupation with technical difficulties, efforts have tended to fragment into distinct observational and theoretical areas that have not always been complementary.

Some theorists have complained that too much emphasis was being given to emission-region models that could then be compared with the observations. Certainly the primary work, they would say, is in building global pulsar models on the solid foundation of basic physics. Curt Michel (1992), in particular, has expressed strong cautions about attempts to "verify" emission-region theories that are not grounded on global models. He compares these to "snake-oil cures"—that is, attempts at treatment of a disease that are not grounded immediately on an epidemiology of the disease and ultimately on basic biochemical research.

Observers, however, have had a very different problem: Not even the most credible emission-region theories have been able to make meaningful predictions about specific individual pulsars. While certainly quantitative, their level of detail can only begin to account for the emission characteristics of pulsars as a class. Of course, we observers have no option but to observe specific individual pulsars. This severely limits what can be learned through comparing the observations and theories, because these theories cannot be falsified empirically owing to their lack of specificity.

This latter concern has motivated the analytical work that is described below. In a number of instances historically, a phenomenological approach has proved a useful means of gaining physical insight in situations that were too complex to approach from theoretical first principles. The work on stellar classification and its physical interpretation by Hertzsprung and Russell represent one such example.

A system of classification of pulsar emission phenomena has been developed that serves both as an overall description, or "empirical theory," and as a source of insight into how individual pulsars differ from others. In the sections that follow, we shall see that, for radio pulsars, this insight has primarily been of benefit in helping to understand the basic emission geometry of individual stars.

PULSAR PROFILES: PATTERNS OF FREQUENCY EVOLUTION

The system of classification that is the fundamental empirical foundation for this study has been described in a series of published papers (Rankin 1983a, b; Rankin 1986; Rankin *et al.* 1989). It considers the morphological characteristics of polarized average profiles with particular attention to their formal evolution with radio frequency as well as certain pulse-sequence properties, mode changing, drifting subpulses, and pulse nulling.

Classification of pulsar characteristics has long been recognized as a potential source of physical insight into the emission process. Inspired by Radhakrishnan and Cooke's (1969; hereafter RC) hollow-cone emission model, Huguenin *et al.* (1971) first proposed a classification scheme for average profiles, and Backer (1976) then greatly elaborated it. The latest system, which is summarized in TABLE 1, builds directly upon these early ideas.

On the basis of their profile properties, five major categories of pulsar are delineated, the core-single ($\mathbf{S}_t$), conal single ($\mathbf{S}_d$), double (**D**), triple (**T**), and five-component (**M**). A sixth class, the conal triple (*c***T**) is also known, which is closely related to the **M** stars. These species in turn are found to entail two distinct types or mechanisms of radiation, *core* emission and *conal* emission. Depending upon (a) the relative strengths of the core and conal contributions to the overall emission, and (b) the orientation of our sight line to the emission beam, one of the several characteristic profiles is observed. The possibilities for an entirely conal geometry are depicted in FIGURE 1.

One of the species, the five-component (**M**) profile, deserves special mention. More than 15 pulsars have been identified that appear to be members of the **M** profile species. These profiles all have a single, central core component and two pairs of conal outriders. The outer component pairs (I and V, and II and IV, respectively) apparently result from a double-conal emission beam, and the central core component (III), from a smaller pencil beam within it. This geometry is diagrammed in FIGURE 2. They emit both core and conal radiation in roughly comparable amounts. Pulsars of the **M** class exhibit the highest degree of profile complexity known. Indeed, no convincing examples of pulsars with more than five components have been identified.

Summarizing the material in TABLE 1 and anticipating some of the conclusions below, let us compare the characteristics of the core and conal radiation. Core emission is primary in some 70 percent of all pulsars. It is prominent in core-single ($\mathbf{S}_t$), triple (**T**), and five-component (**M**) pulsars, associated with younger pulsars, and emitted at low altitude. Its modulation tends to be steady, and it is frequently marked by sense-reversing (antisymmetric) circular polarization. Finally, it typically has a softer spectrum, and thus tends to diminish at high frequency.

Conal emission, by contrast, is primary in about 30 percent of all pulsars. It is most prominent in conal single ($\mathbf{S}_d$), double (**D**), and conal triple (*c***T**) pulsars, associated with the older stars, and emitted at altitudes of 10 to 20 stellar radii. Conal

TABLE 1. Pulsar Classification System

Profile Species	Type	%	Emission Type	Primary Identification Criteria	
				Spectral Evolution	Polarization
Core Single	$\mathbf{S}_t$	34	Core (and conal at high frequencies)	• Profile has a *single component* near 400 MHz and usually develops a *conal outrider pair* above 1400 MHz. $\mathbf{S}_t$ profiles thus evolve to triple or even double forms at very high frequencies, given the steeper spectrum of core emission.	• *Circular:* moderate in degree, often with an antisymmetric change of hand; moderate symmetric (nonreversing) also observed • *Linear:* nearly complete to unpolarized • *PA traverse:* unsystematic; does not appear to follow the single-vector (RC) model.
Triple	**T**	25	Central core component plus conal component pair(s)	• Profile consits of both a *central core component* and one *conal outrider pair.* **T** profiles tend to evolve into double forms at very high frequencies, given the steeper spectrum of core emission.	• *Circular:* moderate in degree, often with an antisymmetric change of hand; moderate symmetric (nonreversing) also observed; 50%–70% circular is observed in some stars •*Linear:* moderate to virtually unpolarized; polarization-mode depolarization at wings • *PA traverse:* S-shaped, steep, usually near 180° for **M**, less for **T**; closely following the single-vector (RC) model • *Correlation:* if circular sense change is LH to RH, linear traverse is usually cw (neg.), and vice versa
Five Component	**M**	9	As above	• Profile consits of both a *central core component* and *two conal outrider pairs.* **M** profiles tend to evolve into featureless "boxy" forms at very high frequencies, as the components merge and the core emission subsides.	
Conal Single	$\mathbf{S}_d$	15	Conal	• Profile has a *single component* near 400 MHz, which *broadens and bifurcates* progressively at lower frequencies, due to the high altitude spreading of the conal emission beam. ↕ *closely connected* ↕	• *PA traverse:* shallow (≤90°), closely following the single-vector (RC) model • *Linear:* moderate to virtually unpolarized; polarization-mode depolarization at wings • *Circular:* small and unsystematic ↕ *closely connected* ↕
Conal Double	**D**	10	As above	• Profile has a *two components* virtually throughout spectrum, which slowly broaden at lower frequencies, due to the high altitude spreading of the conal emission beam.	• *PA traverse:* S-shaped, ~180°, closely following the single-vector (RC) model • *Linear:* moderate to virtually unpolarized; polarization-mode depolarization at wings • *Circular:* small and unsystematic
One-sided Triple	$\mathbf{T}_{1/2}$	3	As above	• One-sided "triple" pulsars are similar to triple (*T*) pulsars in all outrider is apparent. It is not clear whether the outrider is actually or overlain with the core component.	
Conal Triple	*c***T**	3	Conal	• Pulsars with *c***T** profiles are closely related to those with *M* profiles, the polar cap region at the periphery of the conal emission appears as a central conal component.	
One-sided Double	$\mathbf{D}_{1/2}$	>1	Conal	*No pulsars belonging to this species have been identified, although there is no excitation of the conal emission beam is "patchy"—and there are many is at least nonuniform.*	
Others?	?	1?	?	?	?

TABLE 1. Continued

Geometry	Beam size (°)	Acceler. Potential B_{12}/P^2	Profile Mode Changes	Subpulse Modulation	Nulling	Age $\langle \log_{10} \gamma \rangle$ (10^6 years) $\langle \lvert z \rvert \rangle$
Uncertain: the linear angle traverse apparently provides little reliable information about where the sight line crosses the emission beam	$\frac{2.45°}{P^{1/2} \sin \alpha}$ (FWHM)	Large ≥ 2.5; typically 10	Not observed	Little; steady, flat fluctuation spectrum	None?	Young 6.1 160 pc
The sight line crosses the polar cap region somewhat peripherally; i.e., away from the center of the conal emission beam	• Core beam diam. 2.45° • Conal beam radii 4.3° and 5.9° • Both scale as $P^{-1/2}$ and $\sin^{-1} \alpha$	Full range; large to small	Yes (always?)	Stationary subpulse modulation, associated with the conal components	Yes	Middle age to old 6.6 210 pc
The sight line crosses the polar cap region near the magnetic axis; i.e., near the center of the conal emission beam		Small, generally less than 2.5	As above	As above	As above	Old 7.3 $\lesssim$400 pc
The sight line crosses the polar cap region near the periphery of the conal emission beam ↕ *closely connected* ↕	Not well known	Small $\lesssim 2.5$; lowest known value 0.16	Maybe; difficult to discern	Yes, orderly drifting subpulses	Yes	Old 7.6 335 pc
The sight line crosses the polar cap region near the magnetic axis; i.e., near the center of the conal emission beam	As above	As above	As above	Little or no drifting; stationary subpulse modulation	As above	Old 7.7 260 pc
respects, except that only one conal missing or whether it is simply merged		...As with triple...				
except that the sight line crosses beams; thus, the inner conal beam		$\lesssim 2.5$	Yes	*Drifting* subpulses	Yes	Old 7.7 $\lesssim$400 pc
a priori *reason why they should not if the indications that the illumination*		...*Presumably very much like double (D) pulsars*...				
?	?	?	?	?	?	?

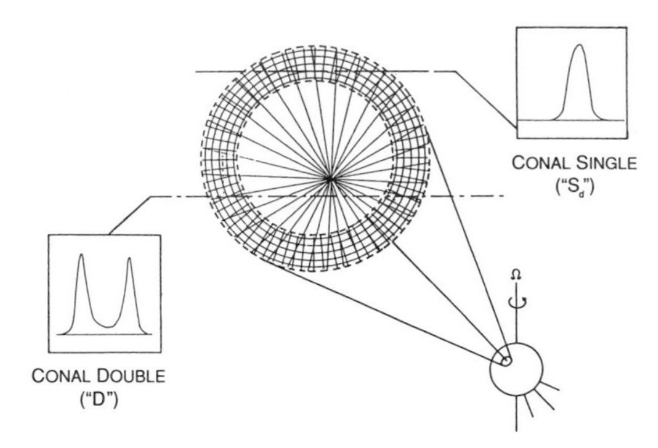

FIGURE 1. Schematic diagram showing the geometry of conal beaming. Single or double profiles result, depending upon the centrality of the sight-line traverse.

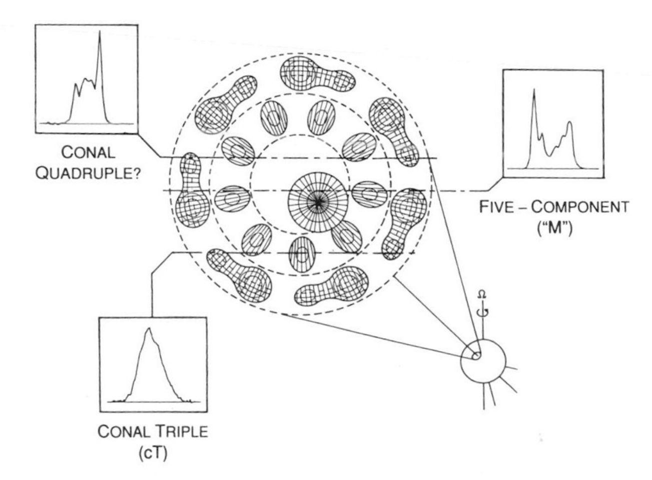

FIGURE 2. Schematic diagram showing the geometry of both the core beam and the double-conal beam. Central traverses of the sight line give a five-component (**M**) profile; more oblique traverses provide other, entirely conal possibilities. Several examples of conal triple (*c***T**) profiles have been identified, but no conal quadruple (**Q**) stars are yet known. Furthermore, core emission together with a single conal zone produces a triple (**T**) profile, and core emission by itself a core-single (**S_t**) profile.

emission exhibits complex, quasi-periodic modulation patterns, characteristic linear polarization, and a harder spectrum.

PHYSICAL BASIS OF SPECIATION

The foregoing classification scheme has clear physical consequences, and these circumstances provide clues to the evolutionary significance of the various species. FIGURE 3 gives the magnetic field as a function of spindown age for some 150 pulsars

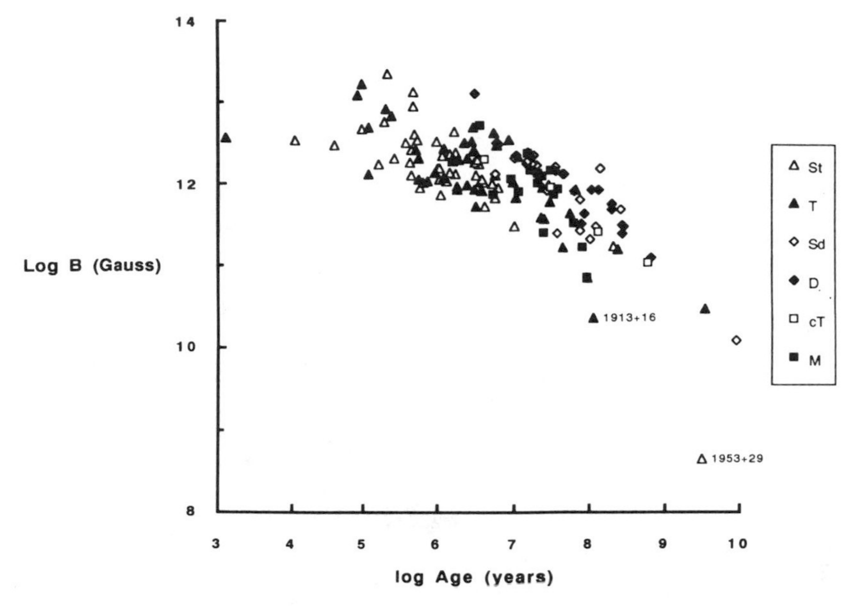

FIGURE 3. Inferred magnetic field vs. spindown age for some 150 pulsars. Note the segregation of the $\mathbf{S}_t$ pulsars at the left of the diagram and the $\mathbf{S}_d$, **D**, *c***T**, and **M** stars at the right. The triple (**T**) pulsars are found throughout the entire range of values.

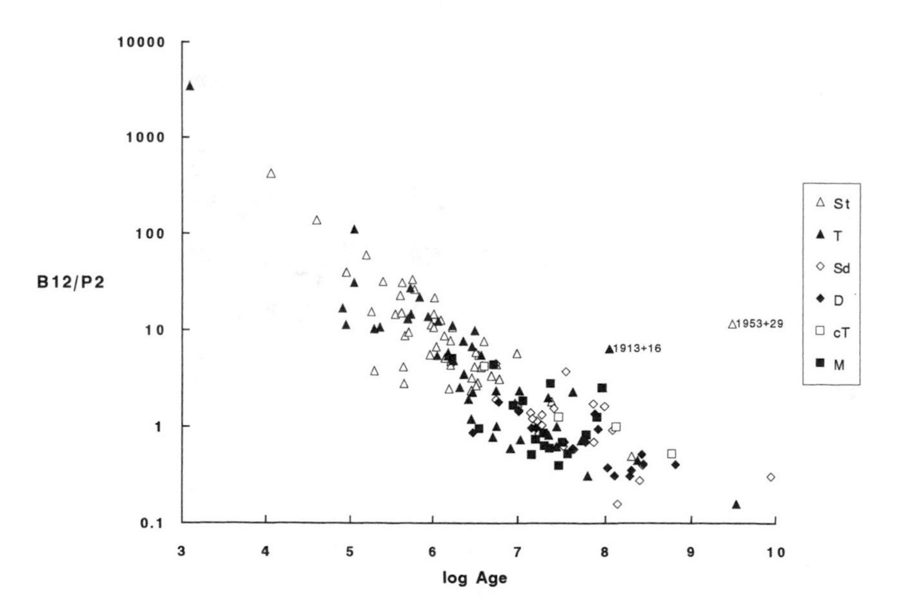

FIGURE 4. "Acceleration parameter" B_{12}/P^2 vs. spindown age for some 150 pulsars (see text).

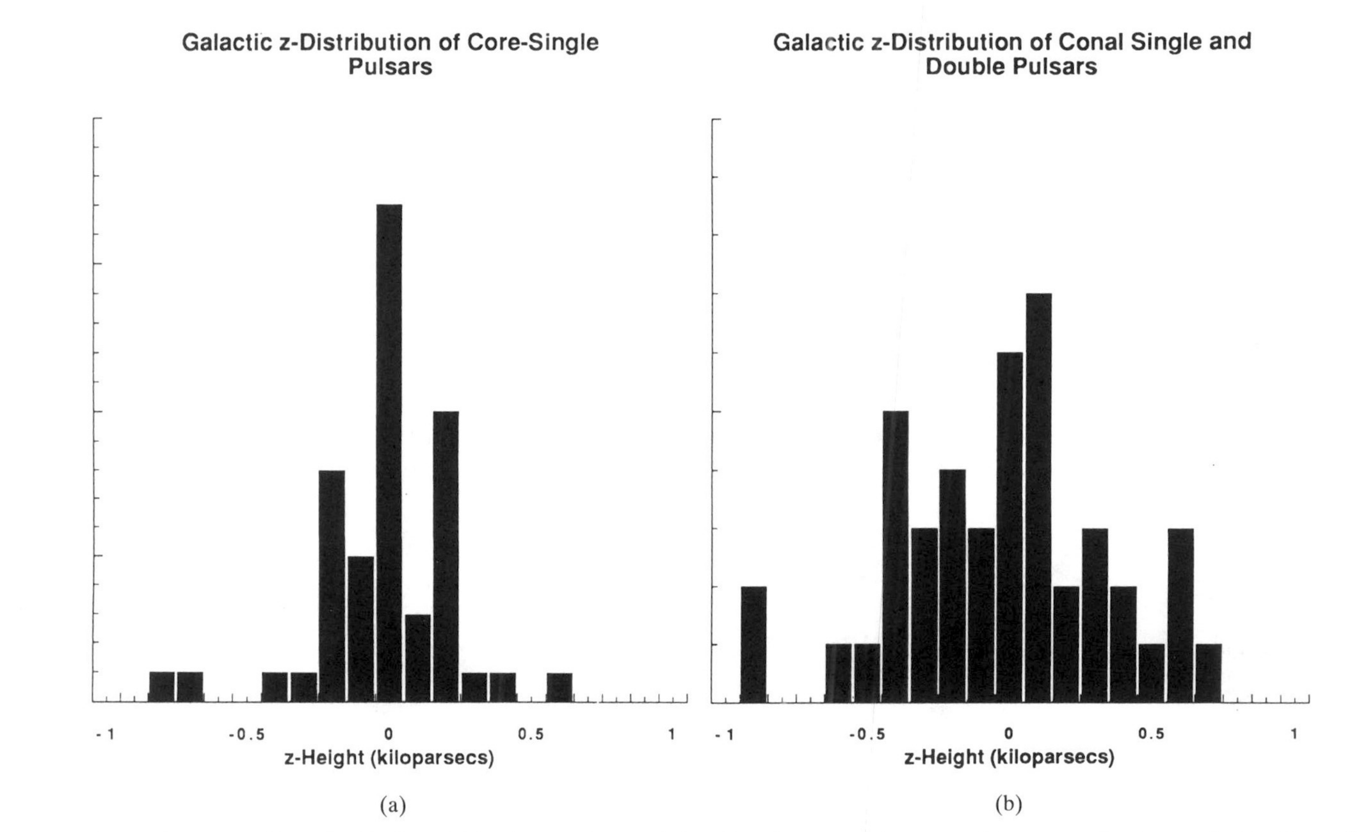

FIGURE 5. Histogram of the z-distribution of **(a)** pulsars with core-single (**S**$_t$) profiles, and **(b)** pulsars with conal single (**S**$_d$) and double (**D**) profiles.

with well-determined classifications; note the tendency of the various species to "clump" in certain regions of the diagram. This tendency is even more marked in FIGURE 4, where the "acceleration potential," proportional to B_{12}/P^2, is plotted against spindown age. The "linear" form of the plot is expected in that the two quantities are correlated. Note, however, how the $\mathbf{S}_t$ pulsars have B_{12}/P^2 values greater than about 2.5, whereas the $\mathbf{S}_d$, **D**, *c***T**, and **M** pulsars almost always fall below 2.5. The triple (**T**) pulsars, by contrast, seem to populate the entire range of B_{12}/P^2 values.

Furthermore, the various species differ significantly in age. The $\mathbf{S}_t$ group is by far the youngest, with a mean log age of 6.1, whereas the $\mathbf{S}_d$, **D**, *c***T**, and **M** pulsars are all old, with mean log age values of about 7.5. The **T** pulsars again fall in between with a mean value of about 6.6. These differences in age are reiterated in FIGURE 5, which gives the galactic z-height distribution for $\mathbf{S}_t$ pulsars in contrast with the combined group of $\mathbf{S}_d$ and **D** stars. Note that the $\mathbf{S}_t$ pulsars have a scale height $\langle |z| \rangle$ of only about 160 parsecs, whereas the other group has a value almost twice as great.

INTERPULSARS

Twelve pulsars are now known that exhibit core emission in their main-pulse and/or interpulse profiles. These pulsars are very important because only for these few stars do we have any direct information about the angle α between their rotation and magnetic axes. Study of these interpulsars having core emission has resulted in the conclusion that most, but not all, have magnetic axes that are nearly orthogonal to their rotation axes (Rankin 1990).

Six of the stars have core components whose width (i.e., full width at half maximum) can be measured with reasonable accuracy and interpolated to 1 GHz. When fitted against period, these values exhibit a surprisingly accurate power-law relation. A least-squares fit to these values yields the result that

$$W = 2.45^\circ P^{-0.50}.$$

The core components of most other interpulsars also have widths comparable with those given by the preceding relation, but cannot be so accurately determined.

This core-width–period relationship has a very simple interpretation in terms of the magnetic field structure of the pulsar. Assuming a dipolar magnetic field, the angle ρ between the field-line tangent and the magnetic axis can easily be evaluated. Taking the 1-GHz width of the core component as twice this value

$$W_{\text{core}} = 2\rho \approx 2.49^\circ (r/R)^{1/2} P^{1/2},$$

where r is the emission height, measured from the center of the star, and R the stellar radius, assumed to be 10 km.

Comparing the empirical relationship for W with the geometrically derived expression for W_{core} just given, both equations have a $P^{-1/2}$ term, and thus it appears that the period dependence of the core width is geometrical in origin. However, the two expressions can only be reconciled numerically if the ratio of the emission height r to the stellar radius R is about unity. This in turn suggests that the core emission comes from very near the stellar surface.

Most pulsars, of course, do not have interpulses, and in general, we must consider how an emission beam of angular radius ρ about the magnetic axis projects onto the sight-line direction. Simple geometrical arguments (see Rankin 1990) suggest that the preceding relation can be generalized as follows to any pulsar with a core component

$$W_{\mathrm{core}} = \frac{2.45° P^{-1/2}}{\sin \alpha}. \tag{1}$$

Apparently, this simple relationship describes the angular width of core emission beams at 1 GHz. The relationship depends only on the pulsar period, which determines the height of the velocity-of-light cylinder and thus the polar cap radius, and the angle α, which enters in considering how much of the angular rotation cycle the core beam occupies.

GEOMETRY OF CORE EMISSION

The 1-GHz half-power width of pulsars with core-single ($\mathbf{S}_t$) profiles are plotted as a function of period in FIGURE 6. The points corresponding to the interpulsars (filled symbols) as well as several other stars are labeled, and a line showing the fitted widths of the interpulsars indicated. Note the minimum width defined by the four interpulsars; several other pulsars—most notably the 6-msec pulsar 1953+29—have core widths just in excess of the interpulsar minimum, but none have smaller widths.

The widths of the central core components of pulsars with triple (**T**) and five-component (**M**) profiles are plotted in FIGURE 7. Pulsars with triple profiles are

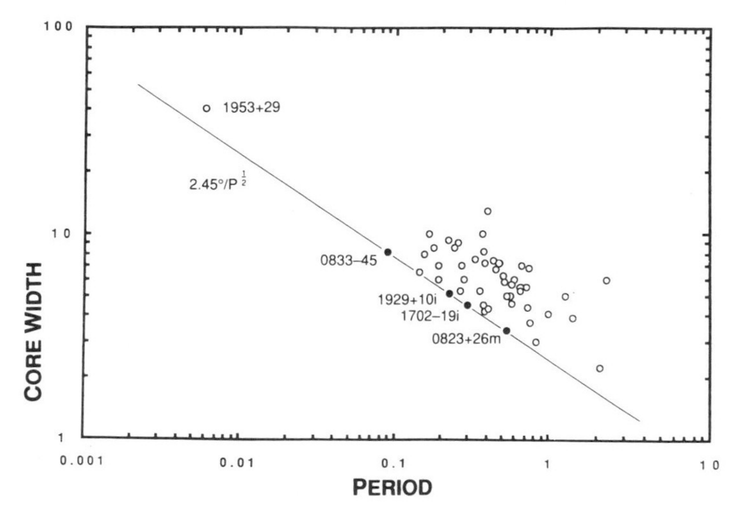

FIGURE 6. The half-power profile width of core-single ($\mathbf{S}_t$) pulsars as a function of period. The symbols of a few prominent pulsars are labeled, and those with interpulses indicated by a *filled symbol.* (The suffixes *i*, *m*, and *p* indicate interpulse, main pulse, and postcursor, respectively.) The indicated curve is $2.45° P^{-1/2}$ (see text).

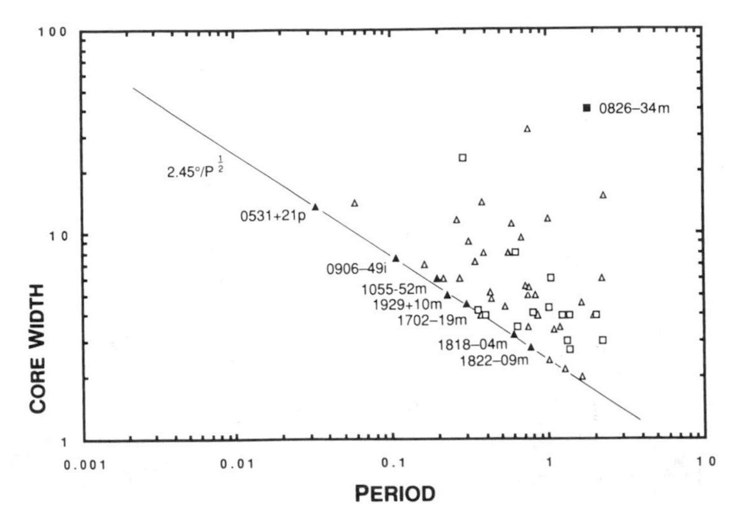

FIGURE 7. The half-power core-component widths of triple (**T**) pulsars (*open triangle symbol*) and five-component (**M**) pulsars (*open square symbol*) as a function of period as in FIGURE 6.

indicated by triangular symbols and the **M** stars with squares. Filled symbols are used to indicate the interpulsars which, along with several other prominent stars, are explicitly identified. A curve indicating the fitted widths of the interpulsars is again superposed. The core-width values of the seven triple interpulsars (filled triangles) all fall accurately on this curve within their measurement or estimation errors. (By contrast, the core-width value for 0826−34*m*, which has an aligned geometry, falls furthermost from the curve.)

We can interpret the core-width–period relationship in FIGURES 6 and 7 as deriving primarily from the polar cap geometry, and as such it should be well described by (**1**). Only two factors apparently determine the 1-GHz width of a core component: (a) the angular radius of the polar cap emission region at the stellar surface, which goes as $2.45°P^{-1/2}$, and (b) the angle between the magnetic axis and the rotation axis α.

ESTIMATION OF α VALUES

Assuming that the angular width of all core emission is determined by the angular extent of the open field lines at the stellar surface, then (**1**) can be inverted to estimate the magnetic orientation angle α. Histograms of the α values are then given in FIGURE 8: (a) The angles for the $\mathbf{S}_t$ stars range between some 15° and 90° and exhibit a median value of about 35°. (b) The α values for the **T** (filled bars) and **M** (striped bars) stars range more widely than for the $\mathbf{S}_t$ stars. The median value is still about 35°, but the distribution is flatter and a larger fraction are very nearly aligned.

The population of $\mathbf{S}_t$ pulsars has been divided into two subgroups according to their spindown age. The younger group (solid bars) has a mean log age of 5.6, whereas the older one (striped bars) has a mean log spindown age of 6.7. The older

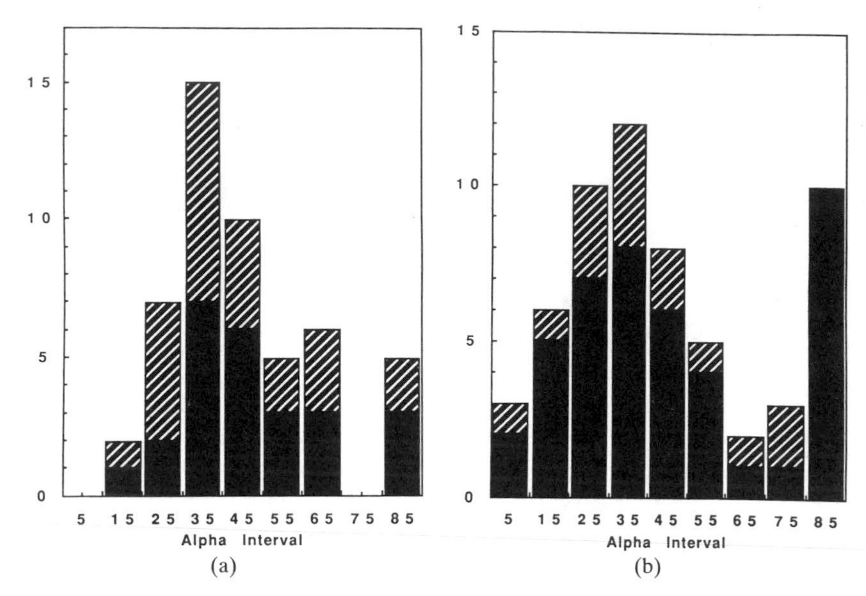

FIGURE 8. Histograms of inferred α values for (**a**) core-single (S_t) pulsars, and (**b**) triple (**T**) and five-component (**M**) pulsars. The S_t population is divided into two equal groups, a younger group (*solid bars*) whose mean log spindown age is 5.6, and an older group (*striped bars*) whose mean log age is 6.7. Similarly, the **T** (*solid bars*) and **M** (*striped bars*) pulsars have mean log ages of 6.6 and 7.3, respectively.

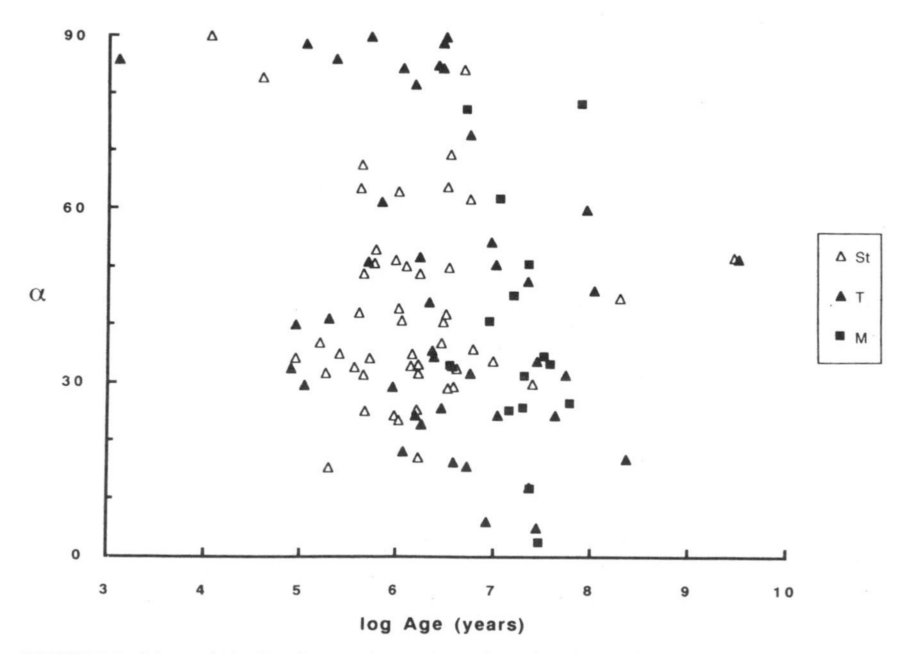

FIGURE 9. Magnetic inclination angle α values plotted against spindown age for populations of S_t, **T**, and **M** pulsars.

group shows a weak tendency toward alignment, which is probably not significant at this level of analysis. The α distributions for the **T** (filled bars) and **M** pulsars are very similar despite differences in mean log age (6.6 and 7.3, respectively). These results are summarized in FIGURE 9; clearly, we find no support here for the proposition that the magnetic axes of pulsars tend to align with age.

CIRCULAR POLARIZATION AND THE FIGURE OF THE POLAR CAP REGION

A phenomenological study of the circular polarization associated with pulsar emission has shown that most is observed in core components—that is, in core-single ($\mathbf{S}_t$) profiles and in the central components of triple (**T**) and five-component (**M**) profiles (Radhakrishnan and Rankin 1990). Two extreme types of circular signature are identified in the observations: (a) an antisymmetric type wherein the circular polarization changes sense in midpulse, and (b) a symmetric type wherein it is predominantly of one sense.

In pulsars with triple (**T**) and five-component (**M**) profiles, the antisymmetric type is usually correlated with the sense of rotation of the linear position angle. Transitions from positive (LH) to negative (RH) are found to accompany negative (clockwise) rotations of the position angle, and vice versa.

In the general framework of models in which the radio emission is produced by curvature radiation from charge bunches constrained to follow field lines, the linear polarization is intrinsic to the emission mechanism, and is, furthermore, a purely geometric property independent of the polarity of the magnetic field or of the sign of the charges. The correlation we find then requires that the antisymmetric circular polarization be also a purely geometric property of the emission process. Curvature radiation will have significant net circular polarization if there are gradients in the emissivity over angular scales comparable with the emission cone of a single charge (i.e., $\gamma \sim 1$, where γ is the Lorentz factor of a charge bunch). The observation of significant circular polarization therefore implies that $\gamma \lesssim 20$ for the core emission. Furthermore, no net circular polarization is produced if the emissivity is circularly symmetric about the magnetic dipole axis. The sign of the correlation we have observed is consistent with an emission region more extended in longitude than in latitude (referred to the rotation axis).

GEOMETRY OF CONAL EMISSION

We saw in (**1**) that the 1-GHz width of core components can be described in terms of the rotation period P and the angle between the rotation and magnetic axes α. The angle α can then be determined in any pulsar with a core component.

On the basis of this knowledge of α, the radius of the conal emission zone can be determined from the width of the conal component pair and the sweep rate of the linear polarization angle. The geometry is depicted in FIGURE 10.

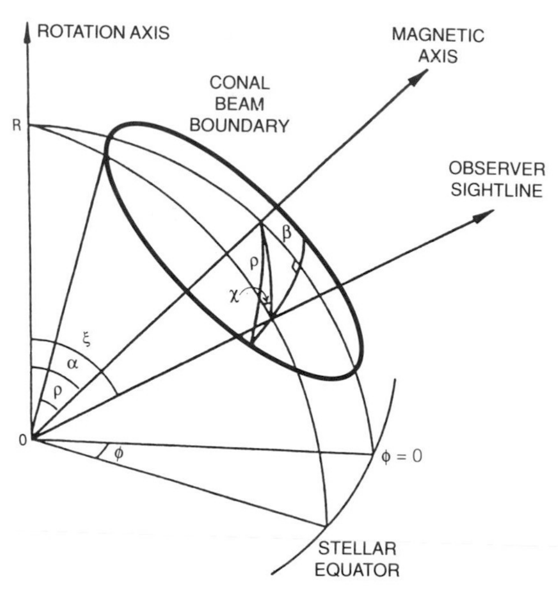

FIGURE 10. Geometry of the conal emission region.

The full 1-GHz width of the conal profile pair (to the outside half-power points) $\Delta\psi$ is

$$\Delta\psi = 4 \sin^{-1} \left\{ \left(\frac{\sin(\rho/2 + \beta/2)\sin(\rho/2 - \beta/2)}{\sin\alpha \sin\zeta} \right) \right\}^{1/2}$$

where ρ is the radius of the conal emission zone and β the "impact angle" of the sight line with the magnetic axis. Solving for ρ, we have

$$\rho = \cos^{-1}\left\{\cos\beta - 2\sin\alpha\sin\zeta\sin^2\frac{\Delta\psi}{4}\right\}. \quad \textbf{(2)}$$

The angle β can then be calculated via the relation that

$$|d\chi/d\varphi|_{\varphi_0} = \frac{\sin\alpha}{\sin\beta},$$

where χ is the polarization angle and φ is the longitude.

On the premise that the core emission is produced close to the ($R \simeq 10$ km) surface of the neutron star, the 1-GHz emission height of the conal radiation can be estimated via the following relation, which assumes a purely dipolar geometry in the emitting region

$$h_{1\,\mathrm{GHz}} = 10\ \mathrm{km}[2\rho/W_{\mathrm{core}}]^2 = 10\ \mathrm{km}[2\rho/(2.45°P^{-1/2}/\sin\alpha)]^2.$$

Using the preceding relationships, the conal emission geometry of five-component (**M**) pulsars can be calculated with some confidence, owing to the

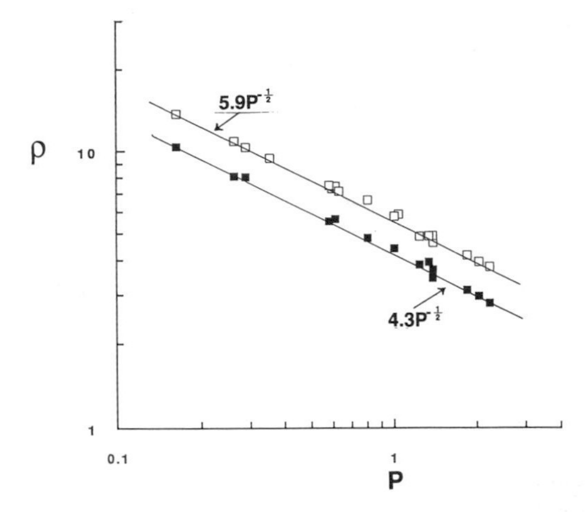

FIGURE 11. Conal radius ρ as a function of period P for the inner and outer conal zones of five-component (**M**) pulsars. The fitted curves $4.3°P^{-1/2}$ and $5.9°P^{-1/2}$ are indicated.

presence of both a core component and two pairs of conal components. The results of the calculations are shown in FIGURE 11. The 1-GHz angular radii of the inner and outer conal zones exhibit a very regular behavior and indeed also scale as $P^{-1/2}$ as follows:

$$\rho_{\text{inner}} = 4.3°P^{-1/2} \qquad \text{and} \qquad \rho_{\text{outer}} = 5.9°P^{-1/2}.$$

Because both the core and conal widths scale as $P^{-1/2}$, the emission height is

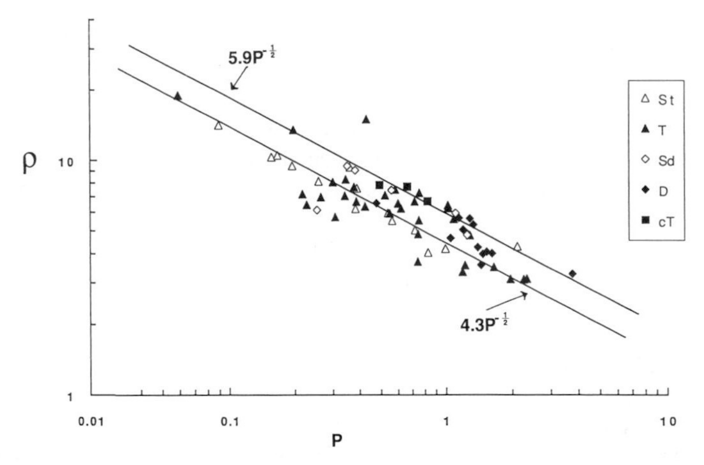

FIGURE 12. Conal emission radius ρ as a function of period for populations of $\mathbf{S}_t$ and **T** pulsars. These values are computed from the full conal analysis as described in the text; ρ values are also shown for a small group of $\mathbf{S}_d$, **D,** and c**T** pulsar stars. Here the α values and other observational data are taken from the analysis of Lyne and Manchester (1988). Curves corresponding to the inner and outer conal zones of the **M** stars, $4.3°P^{-1/2}$ and $5.9°P^{-1/2}$, are indicated.

found to be independent of the period. The curves fitted to the 1-GHz inner and outer conal emission heights then correspond to values of

$$h_{\text{inner}} = \sim 120 \text{ km} \quad \text{or} \quad \sim 12 \text{ stellar radii}$$

and

$$h_{\text{outer}} = \sim 210 \text{ km} \quad \text{or} \quad \sim 21 \text{ stellar radii,}$$

where we have assumed, for the sake of calculation, that the inner and outer conal zones are emitted along the same peripheral field lines at different heights. Clearly, other relationships between the two emission zones are possible.

In any case, let us here compare these results for **M** stars with other pulsars having conal emission, looking first at those pulsars with core-single ($\mathbf{S}_t$) and triple (**T**) profiles. Exactly the same procedure can be followed: first compute the orientation of the magnetic axis α by evaluating the width of the core component, and then calculate ρ, the radius of the conal emission beam, using (**2**).

The results for $\mathbf{S}_t$ (open triangle) and **T** (solid triangle) pulsars are given in FIGURE 12. Notice the overall $P^{-1/2}$ trend in the curve, with significant departures. Note, however, that the $\mathbf{S}_t$ stars generally lie close to the lower curve; it is the **T** stars that have the least orderly behavior.

Finally, a few points representing $\mathbf{S}_d$, **D**, and *c***T** pulsars are also plotted on the diagram. None of these pulsars have core components and so the technique described earlier is not applicable. Instead, α values (as well as the other observational data) have been taken from the analysis of Lyne and Manchester (1988), as it is for just such cone-dominated profiles that their analysis is best motivated. Note that the conal single and double pulsars have ρ values that generally lie between the two solid curves, and that the conal triple values fall essentially on the outer curve as expected.

DISCUSSION

We reemphasize that pulsars exhibit two modes or mechanisms of pulsar emission. Core emission is found in a majority (60–70 percent) of mostly younger pulsars, whereas conal emission predominates in the profiles of the remaining older pulsars.

The width of core components have a regular dependence on rotation period and on the magnetic inclination angle α. The magnitude of these widths suggests that the core radiation is emitted close to the neutron-star surface throughout the entire polar cap region. Furthermore, its circular polarization properties seem compatible only with a radiating population of low γ particles in an asymmetric emission region, that is, one that is longitudinally extended.

The conal emission, by contrast, seems to be emitted at heights of $\sim$100–200 km. This value results from consideration of 1-GHz profiles; the lower frequency emission will then come from larger heights and vice versa. There is no reason to doubt the long-standing association of conal emission with relatively high γ particles, and this is further supported by its linear polarization, which (apart from mode changes) closely follows the single-vector (RC) model.

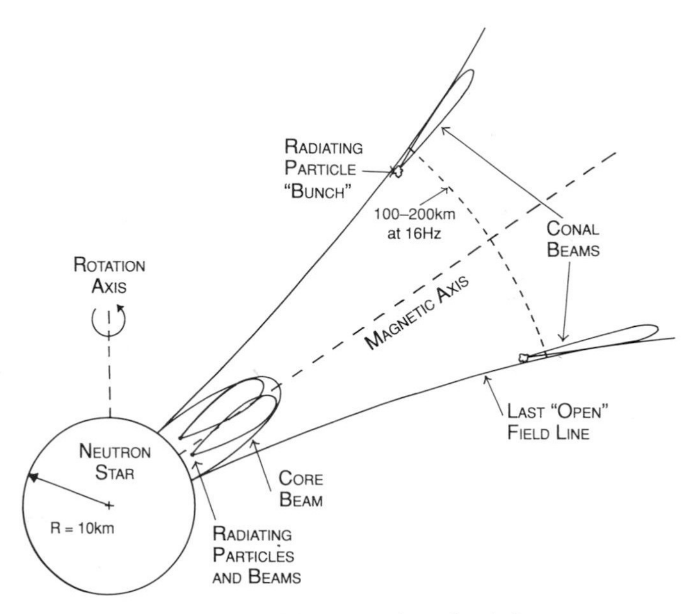

FIGURE 13. Cartoon of the core and conal emission geometry.

FIGURE 13 illustrates the core and conal emission geometry in the form of a cartoon.

The classification of pulsar profiles appears to give useful information about the geometry of the emission region. The orderly sequence of profile types as well as their generally orderly evolution with frequency suggests, among other things, that the pulsar magnetic field is usually highly dipolar. Similarly, the observational evidence suggests that the "patchy" or incomplete cones of emission suggested by Lyne and Manchester (1988) are rather unusual.

Finally, the five-component **M** pulsars exhibit the highest degree of profile complexity, and appear to express the full potentialities of the pulsar emission process. The inner conal zone has geometrical properties that are very similar to the type of conal emission exhibited by core-single ($\mathbf{S}_t$) pulsars at high frequency, and the the outer conal zone seems to have much in common with the conal radiation pattern seen in conal triple (*c***T**), conal single ($\mathbf{S}_d$), and double (**D**) pulsars. Triple (**T**) pulsars are apparently less regular in their behavior and warrant further detailed study.

In their 1988 paper Lyne and Manchester have considered a large number of average polarization measurements and have used the RC model to calculate the basic emission geometry (angles α and β) for a large number of pulsars. Despite their *ad hoc* assumption that the conal radius ρ scales as $P^{-1/3}$, their analysis gives reasonable agreement with the preceding results for a majority of the pulsars common to the two studies.

More significantly, Lyne and Manchester come to *very* different conclusions about the physical nature of the emission region. They argue that there is no fundamental distinction between the core and conal components, but that they assume a range of properties depending upon their centrality with respect to the magnetic axis. In order to account for the variety of average-profile forms, they appeal to "patchy" conal beams and nonuniform emission in the polar cap region near the stellar surface.

I disagree strongly with this interpretation. It seems to me unphysical and at variance with the observational evidence. If, however, one takes as an unalterable rule that the linear polarization-angle behavior of pulse components *always* conforms to the single-vector (RC) model, then I agree that one might be tempted to draw such conclusions. The shallow linear angle traverses of many core-single ($\mathbf{S}_t$) pulsars might reasonably be interpreted as one component of a conal double profile (the other of which is missing) under the rule of this assumption. I have argued that the RC model is not reliable for core components, and indeed it *should* not be if the core emission is in fact a composite of low γ emission from particles flowing along many open field lines near the star (Radhakrishnan and Rankin 1990).

Both Lyne and Manchester's interpretation and the one outlined in this paper were presented at the IAU Colloquium No. 128 and discussed extensively, but no clear position emerged that permitted the participants to achieve a consensus regarding which (if either) was the more correct. Several different lines of work, however, can usefully be pursued in an effort to amplify the differences between the two interpretations and assess their implications.

- Consider the frequency evolution of $\mathbf{S}_t$ profiles between meter wavelengths where they are core dominated and centimeter wavelengths where they develop conal outriders and become conal dominated. The question is then whether the linear angle signature of the isolated core component bears any relation to that of the conal outrider pair at high frequency.
- Look for changes in the spacings between components in multicomponent profiles as a function of frequency. If all the components are emitted at about the same height above the pulsar, then there will be much less reason to expect that such changes will occur.
- Assess whether there is any natural explanation within Lyne and Manchester's interpretation for the $P^{-1/2}/\sin\alpha$ width relationship exhibited by the core components.
- Further explore the five-component (**M**) class of pulsars, and question whether any convincing examples of profiles with six or more components exist.
- Independently repeat Lyne and Manchester's analysis on the entire body of available polarimetric observations (once Lyne and Manchester's survey is published and available in the literature) and reassess their conclusions.

REFERENCES

ASSEO, E., G. PELLETIER & H. SOL. 1992. Proceedings of IAU Colloquium #128 on the Magnetospheric Structure and Emission Mechanisms of Radio Pulsars, T. H. Hankins *et al.,* Eds. University of Zielona Gora Press, Poland.

BACKER, D. C. 1976. Astrophys. J. **209:** 895.

BESKIN, V. S., A. V. GUREVICH & YA. N. ISTOMIN. 1992. Proceedings of IAU Colloquium #128 on the Magnetospheric Structure and Emission Mechanisms of Radio Pulsars, T. H. Hankins *et al.,* Eds. University of Zielona Gora Press, Poland.

BHATTACHARYA, D., G. SRINIVASAN, A. G. MUSLIMOV & A. I. TSYGAN. 1992. Proceedings of IAU Colloquium #128 on the Magnetospheric Structure and Emission Mechanisms of Radio Pulsars, T. H. Hankins *et al.,* Eds. University of Zielona Gora Press, Poland. See also SRINIVASAN, G., D. BHATTACHARYA, A. G. MUSLIMOV & A. I. TSYGAN. 1990. Curr. Sci. **59:** 1.

HANKINS, T. H., V. A. IZVEKOVA, V. M. MALOFEEV, J. M. RANKIN, YU. P. SHITOV & D. R. STINEBRING. 1992. Proceedings of IAU Colloquium #128 on the Magnetospheric Structure and Emission Mechanisms of Radio Pulsars, T. H. Hankins *et al.,* Eds. University of Zielona Gora Press, Poland.

HUGUENIN, G. R., R. N. MANCHESTER & J. H. TAYLOR. 1971. Astrophys. J. **169:** 97.

KAZBEGI, A. Z., G. Z. MACHABELI & G. I. MELIKIDZE. 1992. Proceedings of IAU Colloquium #128 on the Magnetospheric Structure and Emission Mechanisms of Radio Pulsars, T. H. Hankins *et al.,* Eds. University of Zielona Gora Press, Poland.

LYNE, A. G. & R. N. MANCHESTER. 1992. Proceedings of IAU Colloquium #128 on the Magnetospheric Structure and Emission Mechanisms of Radio Pulsars, T. H. Hankins *et al.,* Eds. University of Zielona Gora Press, Poland.

LYNE, A. G. & R. N. MANCHESTER. 1988. Mon. Not. R. Astron. Soc. **234:** 477.

MICHEL, F. C. 1992. Proceedings of IAU Colloquium #128 on the Magnetospheric Structure and Emission Mechanisms of Radio Pulsars, T. H. Hankins *et al.,* Eds. University of Zielona Gora Press, Poland.

MUSLIMOV, A. G. & A. I. TSYGAN. 1992. Proceedings of IAU Colloquium #128 on the Magnetospheric Structure and Emission Mechanisms of Radio Pulsars, T. H. Hankins *et al.,* Eds. University of Zielona Gora Press, Poland.

RADHAKRISHNAN, V. & D. J. COOKE. 1969. Astrophys. Lett. **3:** 225.

RADHAKRISHNAN, V. & J. M. RANKIN. 1990. Astrophys. J. **352:** 258.

RANKIN, J. M. 1983a. Astrophys. J. **274:** 333.

RANKIN, J. M. 1983b. Astrophys. J. **274:** 359.

RANKIN, J. M. 1986. Astrophys. J. **301:** 901.

RANKIN, J. M. 1990. Astrophys. J. **352:** 247.

RANKIN, J. M. 1992. Proceedings of IAU Colloquium #128 on the Magnetospheric Structure and Emission Mechanisms of Radio Pulsars, T. H. Hankins *et al.,* Eds. University of Zielona Gora Press, Poland.

RANKIN, J. M., D. R. STINEBRING & J. M. WEISBERG. 1989. Astrophys. J. **346:** 869.

SRINIVASAN, G. 1991. Evolution of the magnetic fields of neutron stars. This issue.

Evolution of the Magnetic Fields of Neutron Stars

G. SRINIVASAN

Raman Research Institute
Bangalore 560 080, India

DO NEUTRON STAR MAGNETIC FIELDS DECAY?

Soon after pulsars were discovered, Gunn and Ostriker (1970) conjectured that their magnetic fields may be decaying in relatively short timescales ~ a few million years. They invoked field decay to reconcile the discrepancy between the *characteristic ages* of pulsars and their estimated *dynamical ages.* Although there was no identified mechanism for such a field decay, during the last 20 years this hypothesis has become an essential ingredient in most evolutionary scenarios for pulsars. But more recent analyses of the statistics of observed pulsars—whose number has now grown to nearly 500—suggest that if the magnetic fields of neutron stars do decay, then the timescale for such a decay cannot be as short as envisaged earlier (~ a few million years), but must be much greater than the typical lifetime of radio pulsars. A qualitative way to see this result is through a plot of the *pulsar current* in the *B–P* plane. What is plotted in FIGURE 1 is the "true current" estimated from the observed population allowing for various selection effects. The current is defined in the usual way as

$$J(B, P) = \frac{1}{\Delta P} \sum_i \frac{S(P_i, \dot{P}_i, Z_i)}{f(P_i)} \dot{P}_i, \tag{1}$$

where $S(P, \dot{P}, z)$ is the scaling factor to account for various selection effects and $f(P)$ is the "beaming factor" for pulsars (for details we refer to Narayan 1987; Deshpande and Srinivasan 1992). As may be seen from FIGURE 1, in the region of the *B–P* plane where the number of pulsars is maximum—and therefore the statistical errors minimum—the *constant current lines* are *horzontal* right up to the death line, implying that there is essentially no field decay. A more quantitative estimate can be obtained by computing the *current-weighted average field* in various characteristic age bins and deriving the best fit decay timescale (see FIG. 2). Such a procedure gives $\tau_d \gtrsim 20$ Myr (Deshpande and Srinivasan 1992), in an agreement with the estimates of Krishnamohan (1987) and Narayan and Ostriker (1990).

Thus, it appears that no conclusive inference can be drawn about the decay of magnetic fields of neutron stars by studying radio pulsars alone. But this conclusion does not preclude the possibility of field decay over much longer timescales. In this case, one has to ask if there is any evidence for field decay among the "dead pulsars." During the last couple of years the existence of strong magnetic fields $\geq 10^{12}$ G in very old neutron stars $\sim 10^{10}$ yr have been inferred. We are referring to the gamma-ray burst sources and the discovery of cyclotron absorption lines in their

X-ray spectrum (Murakami *et al.* 1988). If one accepts the standard argument (based on the isotropic distribution of gamma-ray bursters) that these must be very old neutron stars, then one is forced to conclude that the magnetic fields of some of the neutron stars at least do *not* decay significantly even in timescales comparable to the age of the galaxy itself.

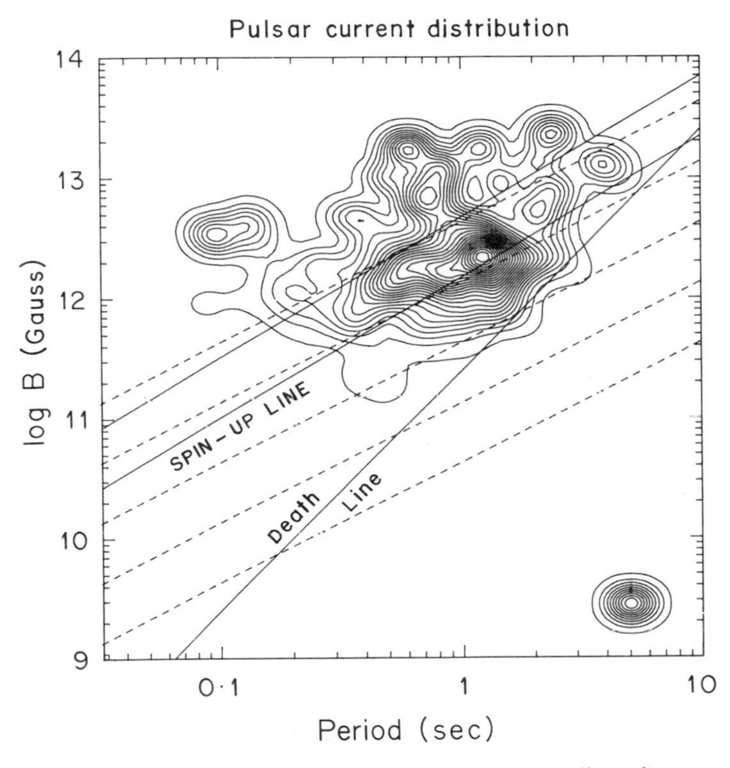

FIGURE 1. The distribution of pulsar current. The contours represent lines of constant current defined through (**1**). (From Deshpande and Srinivasan 1992. Reproduced by permission.)

THE LOW MAGNETIC FIELD BINARY PULSARS

Nevertheless, there is strong circumstantial evidence that the magnetic fields of some neutron stars do decay. Invariably, these happen to be neutron stars with binary companions. The first clue to this came with the discovery of the Hulse–Taylor pulsar PSR 1913 + 16. The most satisfactory explanation for its anomalous combination of short period of rotation and low magnetic field has been in terms of the "recycling" or "spin-up" scenario: its magnetic field is low because it had decayed during the time between the first supernova and the onset of mass transfer, and it is spinning rapidly because it has been spun-up during the mass accretion phase (for a

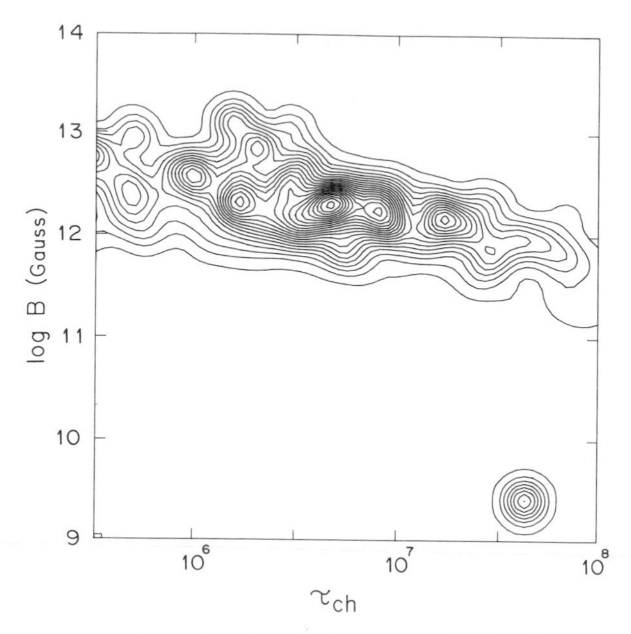

FIGURE 2. Pulsar current as a function of the magnetic field and characteristic age. A current-weighted average of the magnetic field as a function of the characteristic age suggests a long decay timescale $\gtrsim 20$ Myr. (From Deshpande and Srinivasan 1992. Reproduced by permission.)

detailed discussion, as well as references, see Srinivasan 1989). This hypothesis has been substantially strengthened by the discovery of the low field millisecond pulsars that are also believed to be spun-up pulsars.

That the decay of the magnetic field is somehow related to evolution in a binary is consistent with the following argument. Let us concentrate on the subset of low field pulsars ($B \lesssim 10^{11.5}$ G) inside the dashed circle in FIGURE 3. The paucity of such low field pulsars with periods $\lesssim 100$ ms was previously cited as evidence for rapid field decay (Radhakrishnan 1982). But the preceding discussion shows that this conclusion is no longer tenable. As shown in FIGURE 4, these pulsars form a *distinct population.* It has been recently argued that these pulsars did not evolve from the *left* of the diagram, but have been *injected* close to the spin-up line at a characteristic age $\sim 10^{7.5}$ yr (Deshpande and Srinivasan 1992). Further, as shown in FIGURE 5, these pulsars are born at substantial distances from the galactic plane. The most reasonable explanation for this is that these are recycled pulsars from binary systems that had "floated up" from the galactic plane during the time interval between the first and second supernova explosions, and released from them during the second explosion. If this hypothesis is correct, then one should find the second-born pulsars—presumably with short periods and high fields—*also* injected at large z-distances; this is indeed the case (Deshpande and Srinivasan 1992).

To sum up, our main point is that the magnetic fields of neutron stars with a binary history do decay, whereas the fields of solitary pulsars do not!

ASYMPTOTIC FIELDS

If the fields of the first-born neutron stars in binary systems decay, then for some reason this decay seems to stop. The fact that pulsars like PSR 0655 + 44 and the millisecond pulsars (which are almost certainly older than $\sim 10^9$ yr) retain fairly large fields is evidence for this concept of *asymptotic fields* (Kulkarni 1986; Bhattacharya and Srinivasan 1986; van den Heuvel *et al.* 1986).

FIELD DECAY DUE TO MASS ACCRETION

Clearly, any attempt to explain the evolution of neutron star magnetic fields must keep in mind the various observational clues mentioned earlier.

Even before the discovery of the first binary pulsar, Bisnovatyi-Kogan and Komberg (1974) had anticipated that pulsars in binaries may have low fields because the accreted matter may "screen" or "bury" the field. Several authors have recently pursued this scenario (see, e.g., Romani 1990, and references cited therein). In our opinion, various mechanisms invoked so far, while they can destroy or modify the crustal field, are unlikely to have any significant influence on the magnetic field that resides in the *core* of the neutron star. It is to a discussion of how to destroy this field that we shall now turn.

THE NATURE OF THE CORE FIELD

In a paper published soon after the formulation of the microscopic theory of superconductivity by Bardeen, Cooper, and Schrieffer, and nearly 10 years before

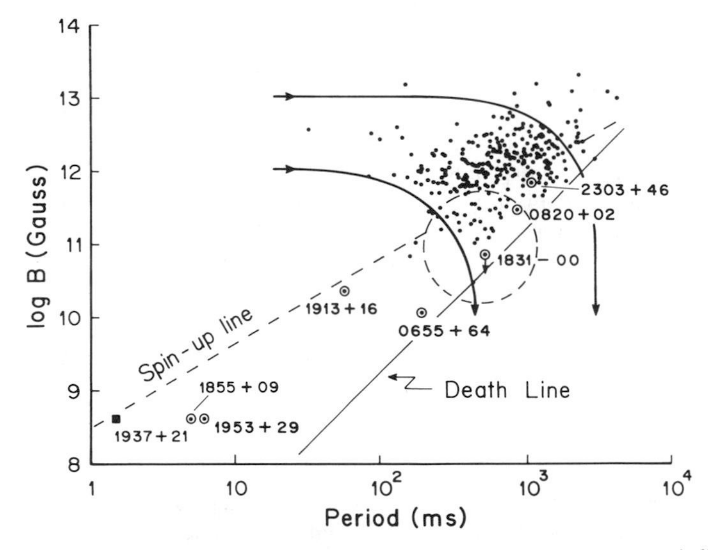

FIGURE 3. The distribution of the observed pulsars. There are strong reasons to believe that the low field pulsars inside the *dashed circle* did not evolve from the left of the diagram, but were "injected" close to the spin-up line.

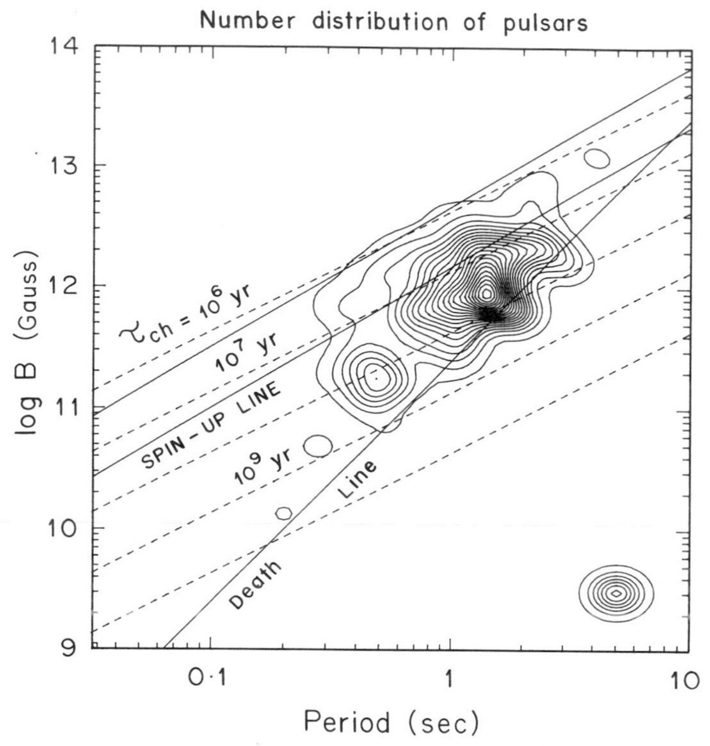

FIGURE 4. A contour representation of the number distribution of the true pulsar population clearly shows that the low field pulsars form a distinct population.

pulsars were discovered, A. B. Migdal (1959) argued that superfluid states should be expected inside neutron stars. A few years later Ginzburg and Kirzhnits (1964) showed that the protons in the interior will become superconducting, while the electrons will remain normal, highly degenerate, relativistic plasma. Soon after pulsars were discovered Baym, Pethick, and Pines (1969) went on to argue that the protons will become a *type II superconductor,* which, because of the magnetic field, will be in a *vortex state.* The number of these *Abrikosov vortices* (or *fluxoids*) is given by

$$N_f = \pi R^2 B/\phi_0 \sim 10^{31} B_{12}, \tag{2}$$

where $\phi_0 = hc/2e$ is the quantum of flux. (For an excellent discussion of superfluidity in neutron stars, see Sauls 1989.) The point of interest to us is that the timescale for flux expulsion from a vortex state is very large and comparable to the ohmic dissipation timescale in the normal state, which in the present case will be larger than the age of the universe.

How, then, does the magnetic field decay? In order for the flux trapped in the superconducting interior to decay, the fluxoids must somehow migrate to the crust where the field can decay due to ohmic dissipation. Muslimov and Tsygan (1984) have suggested that this may happen due to buoyancy forces acting on the fluxoids.

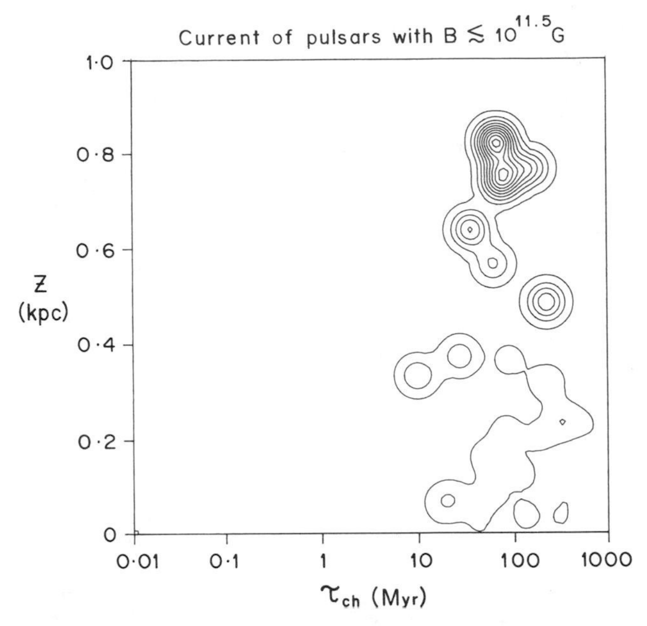

FIGURE 5. The current of the low field pulsars. The injection of these pulsars at a characteristic age $\sim 10^{7.5}$ yr, and at z-distances all the way up to ~ 800 pc from the galactic plane strongly suggests that these are recycled pulsars from binary systems. (From Deshpande and Srinivasan 1992. Reproduced by permission.)

While this is attractive, it is hard to understand why there is a residual field and why low-field pulsars always occur in binaries. Also, it does not explain why some neutron stars show no field decay!

THE FISHING RODS

So one is left speculating about a mechanism that will transport the fluxoids to the crust. What we need to do is essentially "fish out" the fluxoids from the interior. We shall now outline a simple mechanism for doing this (Srinivasan *et al.* 1990). *The underlying postulate is that the mechanism of field expulsion from the interior is related to the slowing down of the neutron star.*

Before elaborating on this let us first recall the nature of the neutron fluid in the core of the star. Following Migdal's original suggestion, Ginzburg and Kirzhnits showed that the neutrons will become superfluid, and because of the *rotation* of the star the superfluid will be in a *vortex state,* with each vortex having a quantum of circulation $h/2m$. The total number of these *Onsager–Feynman vortices,* N_v, will be related to the angular velocity Ω of the star by the relation

$$N_v \cdot \frac{h}{2m} = 2\pi R(\Omega R), \qquad N_v \sim 10^{16} P^{-1}, \tag{3}$$

where R is the radius of the neutron star, and P the period in seconds. Ginzburg and Kirzhnitz argued that because the neutron superfluid is threaded by these vortices parallel to the rotation axis, on a macroscopic scale it will mimic rigid body rotation, and thus have its classical moment of inertia.

Thus there are *two* families of vortices in the neutron star. The Abrikosov fluxoids parallel to the magnetic axis and the Onsager–Feynman vortices parallel to the rotation axis (FIG. 6). Although the superfluid vortices and the fluxoids have in the past been invoked to understand some aspects of the rotational dynamics and the decay of the magnetic field, respectively, the consequences of the *interaction* between these two families of vortices have not been seriously explored so far. Recently, Srinivasan *et al.* (1990), and Sauls (1989), have argued that the fluxoids and vortices may be strongly *interpinned.* We will now argue that if such interpinning is important,

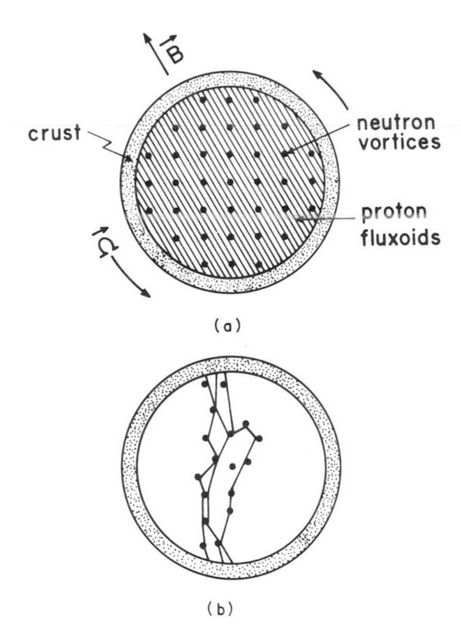

FIGURE 6. (**a**) An idealized geometry showing the quantized fluxoids in the proton superconductor and the Onsager–Feynman vortices in the neutron superfluid. The former are parallel to the magnetic axis, and the latter to the rotation axis. (**b**) A more realistic geometry taking into account the strong interaction between the two families of vortices.

then one can construct an elegant model for the field decay based upon flux expulsion from the interior as the neutron star slows down.

As Jim Sauls and his collaborators have shown (Sauls 1989), the solid crust and the core superfluid are strongly coupled. Because of this, as the crust slows down the superfluid will respond by destroying the required number of vortices. This slowing will happen by a radial *outward flow* of the vortices and their annihilation at the interface between the inner crust and the superfluid core. If the pinning of the vortices and the Abrikosov fluxoids is sufficiently strong, as various estimates suggest (Sauls 1989; Srinivasan *et al.* 1990), then as the vortices move out radially, the fluxoids will also *be dragged* to the crust and the field will eventually decay there due to Ohmic dissipation. As will be seen, this very simple idea of using the Onsager–

Feynman vortices as "fishing rods" to pull the fluxoids out to the crust is able to explain, in one stroke, all the observational features that one set out to explain.

FIELD EVOLUTION OF SOLITARY PULSARS

We will now argue that this mechanism will not result in any significant field decay for *isolated* neutron stars. Since the energy of the electromagnetic radiation and the relativistic wind emitted by a pulsar comes at the expense of its stored rotational energy, the equation governing the slowing of the neutron star is the following:

$$-\frac{d}{dt}\left(\frac{1}{2}I\Omega^2\right) = \frac{2}{3c^3}B^2R^6\Omega^4. \tag{4}$$

If the evolution of the magnetic field is directly related to the secular slowing down, as assumed in the model, then one can substitute $B(t) \propto \Omega((t)$ in the preceding formula. Integrating this one finds that at late times

$$B(t) \propto \left(\frac{t}{\tau_{sd}}\right)^{-1/4} \tag{5}$$

Here τ_{sd} is the characteristic slowdown timescale of the pulsar at late times (typically $\sim 10^6$–10^7 yr). This is a very slow rate of decay indeed. In other words, if the mechanism suggested by us is correct, then one is forced to conclude that the magnetic fields of isolated neutron stars cannot decay significantly, even over a timescale comparable to the age of the universe! The point is simply the following. An isolated neutron star that slows down spontaneously by virtue of its electromagnetic luminosity will never slow down significantly for a reasonable fraction of the flux to be dragged out from its interior.

NEUTRON STARS IN BINARIES

The difficulty mentioned at the end of the last section is overcome if the neutron star is in a binary system. For, in that case, the neutron star can be slowed down much more than if it were left to itself. The main evidence for this comes from the 20 or so binary X-ray pulsars whose rotation periods range from a few seconds to ~ 900 seconds. Indeed, such large periods were predicted by Illarianov and Sunyaev even before the slow X-ray pulsars were discovered. To recall briefly, when the companion star is still in its main sequence, the weak stellar wind emanating from it will not be able to accrete onto the neutron star, but will be expelled. In the process the neutron star will be slowed down until it attains an equilibrium period at which the corotation radius and the Alfvén radius match.

Earlier, while discussing isolated neutron stars, we concluded that they will not slow down significantly due to their electromagnetic luminosity. Consequently, there was not much field expulsion. In the present case, a substantial fraction of the field trapped in the superconducting interior can be expelled during this slowdown phase. Whether or not the expelled field actually decays *even* as the star slows down will depend upon whether the slowdown timescale is short or long compared to the ohmic dissipation timescale in the crust. Both circumstances can obtain; but in either case, given enough time the field will decay in the crust.

THE RESIDUAL FIELD

In this simple picture it is easy to understand why there is an asymptotic or residual field, and what determines its value.

When the companion star evolves, and accretion sets in, the neutron star will, in fact, be *spun up.* Consequently, the degree to which flux has been expelled from the interior will be determined by the *maximum period to which the neutron star was slowed down* prior to this spin-up.

Regarding why there is no further decay, there are two simple reasons. First, during the spin-up to ultrashort periods, the superfluid core will respond by creating new vortices that will now move *radially inward,* thus pushing the remaining core field deeper into the superconducting core. It is worth pointing out that if the neutron star is spun up to a few milliseconds, then during this spin-up the number of Onsager–Feynman vortices will increase a *millionfold*(!), thus making the entanglement between the fluxoids and vortices more effective than ever before! Second, after the spin-up phase, the slow-down rate of the recycled pulsar will be determined by its electromagnetic luminosity appropriate to the *reduced* value of the field. Hence, there will be essentially no further decay.

To summarize, the simple model based on the effective pinning of the Abrikosov fluxoids in the proton superconductor and the Onsager–Feynman vortices in the neutron superfluid is able to give a natural explanation for the following facts that we set out to explain.

- How the flux can be expelled from the superconducting interior.
- Why most low field pulsars are in binaries.
- Why solitary neutron stars show no field decay.
- Why there is an asymptotic field.
- What determines the value of this residual field.

ACKNOWLEDGMENTS

The work outlined here was done mainly in collaboration with A. A. Deshpande and D. Bhattacharya of the Raman Research Institute. I wish to thank both of them for numerous enlightening discussions.

REFERENCES

BAYM, G., C. PETHICK & D. PINES. 1969. Nature **224:** 673.
BHATTACHARYA, D. & G. SRINIVASAN. 1986. Curr. Sci. **55:** 327.
BISNOVATYI-KOGAN, G. S. & B. V. KOMBERG. 1974. Sov. Astron. **18:** 217.
DESHPANDE, A. A. & G. SRINIVASAN. 1992. In preparation.
GINZBURG, V. L. & D. A. KIRZHNITS. 1964. Zh. Eskp. Teor. Fiz. **47:** 2006.

GUNN, J. E. & J. P. OSTRIKER. 1970. Astrophys. J. **160:** 979.
KRISHNAMOHAN, S. 1987. *In* IAU Symposium 125, the Origin and Evolution of Neutron Stars, D. J. Helfand and J. H. Huang, Eds.: 377. Reidel. Dordrecht, the Netherlands.
KULKARNI, S. R. 1986. Astrophys. J., Lett. **306:** L85.
MIGDAL, A. B. 1959. Zh. Eskp. Teor. Fiz. **37:** 249.
MURAKAMI, T., *et al.* 1988. Nature **335:** 235.
MUSLIMOV, A. G. & A. I. TSYGAN. 1985. Sov. Astron. Lett. **11:** 80.
NARAYAN, R. 1987. Astrophys. J. **319:** 162.
NARAYAN, R. & J. P. OSTRIKER. 1990. Astrophys. J. **352:** 222.
RADHAKRISHNAN, V. 1982. Contemp. Phys. **23:** 207.
ROMANI, R. W. 1990. Astrophys. J., Lett. **339:** L29.
SAULS, J. A. 1989. *In* Timing Neutron Stars, H. Ogelman and E. P. J. van den Heuvel, Eds.: 457. Kluwer. Dordrecht, the Netherlands.
SRINIVASAN, G. 1989. Astron. Astrophys. Rev. **1:** 209.
VAN DEN HEUVEL, E. P. J., J. A. VAN PARADIJS & R. E. TAAM. 1986. Nature **332:** 153.

Pulsars in Globular Clusters[a]

SHRINIVAS R. KULKARNI

Division of Physics, Mathematics and Astronomy, 105-24
California Institute of Technology
Pasadena, California 91125

INTRODUCTION

In the late 1970s, astronomers had come to appreciate the connection between massive X-ray binaries and binary radio pulsars (such as 1913+16). Mass transfer in the X-ray binaries, observed via the intensity of the X-ray emission, was seen to affect the rotation rate of the accreting neutron star. Such observations gave rise to the notion that mass transfer led to rapidly rotating pulsars. However, the maximum rotation is limited by the inner edge of the accreting disk, which is the Alfvén radius and hence inversely proportional to the dipole magnetic field strength of the accreting neutron star. Given the large magnetic field strengths of the massive X-ray binaries, the resulting spun-up pulsars were expected to have periods probably no faster than tens of milliseconds, as was the case for 1913+16 (Bisnovatyi-Kogan and Komberg 1975; Smarr and Blandford 1975).

The brightest X-ray binaries are the low-mass X-ray binaries (LMXBs), of which the Galaxy has about 100 with about 10 in the globular cluster system (Bradt and McClintock 1983). The absence of pulsations and the presence of bursts (inferred to be unstable thermonuclear burning on the surface of neutron stars) gradually led to the idea that the neutron stars in these systems had low magnetic field strengths (or no magnetic field strengths). This raised the possibility of spinning up neutron stars to their breakup rates, around a millisecond, and consequently pulsed millisecond X-ray emission was sought, but with no success.

Independently (and in blissful ignorance) radio astronomers stumbled on a millisecond pulsar in 1982 (Backer *et al.*). The discovery reinforced the evolutionary connection between LMXBs and millisecond pulsars. Specifically, Alpar *et al.* (1982), noting the high specific incidence of LMXBs in the cluster system, suggested that millisecond pulsars should be found in globular clusters. Prompted by this suggestion, Hamilton, Helfand, and Becker (1985) surveyed a dozen nearby and rich clusters and found an intriguing source with high linear polarization and steep spectral index, features peculiar only to pulsars in the cluster M28. Eventually this was shown to be a 3-ms single pulsar (Lyne *et al.* 1987).

The current score, a result of intensive efforts at Arecibo, Jodrell Bank, Parkes, and the very large array (VLA), stands at 28 pulsars (TABLES 1 and 2). In TABLE 1, we give the observed parameters, and in TABLE 2, the derived parameters. The references (marked numerically) to the discoveries and related observations are given following the reference section for the paper. As can be seen from TABLE 1, the parameters for most cluster pulsars need to be determined. This will take some time

[a] This work is supported by a grant from the Perkin Fund.

TABLE 1.

Cluster	Pulsar	RA (1950) Dec hr	min	sec	°	′	″	P_p (ms)	$\dot{P}$ (ss^{-1})	DM cm^{-3} pc	$f(M)$ $M_\odot$	S_{430} (mJy)	S_{610} (mJy)	S_{1400} (mJy)	Reference
47 Tuc	0021 − 72C	00	21	53.0*	−72	21	00*	5.76	$(-4 \pm 5) \times 10^{-17}$	24.5			3	0.6	18
	0021 − 72D	00	21	53.0*	−72	21	00*	5.36		24.5					13
	0021 − 72E	00	21	53.0*	−72	21	00*	3.54		24.2					13
	0021 − 72F	00	21	53.0*	−72	21	00*	2.62		24.5					13
	0021 − 72G	00	21	53.0*	−72	21	00*	4.04		24.5					13
	0021 − 72H	00	21	53.0*	−72	21	00*	3.21		24.5					13
	0021 − 72I	00	21	53.0*	−72	21	00*	3.48		23.7					13
	0021 − 72J	00	21	53.0*	−72	21	00*	2.10		24.5					13
	0021 − 72K	00	21	53.0*	−72	21	00*	1.79		24.9					13
	0021 − 72L	00	21	53.0*	−72	21	00*	4.35		24.0					13
M53	1310 + 18	13	10	28.3*	18	26	02*	33.16		24	9.8×10^{-3}	1			11
M5	1516 + 02A	15	16	1.9*	2	15	51*	5.5		29.5		0.5			24
	1516 + 02B	15	16	1.9*	2	15	51*	7.9		29.5†		0.5			24, 25
M4	1620 − 26	16	20	31.1	−26	24	58	11.08	8.2×10^{-19}	62.9	0.008	15		1.4	8, 10, 14, 20
M13	1639 − 36A	16	39	53.6	36	32	55	10.38	$< 4.5 \times 10^{-23}$	30.4		3			11
	1639 − 36B	16	39	54.2*	36	33	16*	3.5		30.4†	1.8×10^{-3}				5
Trz 5	1744 − 24A	17	44	57.7	−24	45	38	11.56	$(-5 \pm 9) \times 10^{-20}$	242	3.215×10^{-4}		5	2.5	16, 21
NGC 6440	1745 − 20	17	45	54.1*	−20	20	39*	288.60		210				1.3	19
NGC 6539	1802 − 07	18	2	7.1*	−7	35	24*	23.10		187	9.7×10^{-3}			0.8	7
NGC 6624	1820 − 30A	18	20	27.9*	−30	23	15*	5.44		86			2		6
	1820 − 30B	18	20	27.9*	−30	23	15*	378.59		86			2		6
M28	1821 − 24	18	21	27.4	−24	53	51	3.05	1.55×10^{-18}	120		27	12	1.1	8, 9, 12, 15, 17
NGC 6760	1908 + 00	19	8	39.4*	0	56	49*	3.6		200					4
M15	2127 + 11A	21	27	33.2	11	56	49	110.66	-2×10^{-17}	67.2		1.7		0.2	12, 26
	2127 + 11B	21	27	33.6	11	56	48	56.13	8.8×10^{-18}	67.2		1.1			3
	2127 + 11C	21	27	36.2	11	57	26	30.53	4.99×10^{-18}	67.1	0.1528	0.6			3, 22
	2127 + 11D	21	27	33.2	11	56	48	4.8	-1.1×10^{-18}	67.2†					22
	2127 + 11E	21	27	33.2	11	56	57	4.65	1.8×10^{-19}	67.2					22

*Position of cluster core from Shawl and White (1986), assumed pulsar position.
†DM assumed to be the same as for A pulsar.

since most of the cluster pulsars are faint and require long observations and tedious data reduction. [Excluded from the tables are the first two pulsars reported in 47 Tuc (references 1 and 2), since there has been no independent confirmation of these two pulsars. More worrisome is that the dispersion measure of these two pulsars is inconsistent with that of the nearly dozen pulsars found in the same cluster.]

TABLE 2.

Cluster	Pulsar	P_p (ms)	$P/2\dot{P}$ (y)	B (G)	P_b (d)	e	M_2 ($M_\odot$)	Reference
47 Tuc	0021 − 72C	5.76			S			18
	0021 − 72D	5.36			S			13
	0021 − 72E	3.54			~2			13
	0021 − 72F	2.62			S			13
	0021 − 72G	4.04			S			13
	0021 − 72H	3.21			~1?			13
	0021 − 72I	3.48			~1?			13
	0021 − 72J	2.10			0.12			13
	0021 − 72K	1.79			?			13
	0021 − 72L	4.35			?			13
M53	1310 + 18	33.16			255.84	<0.01	0.3	11
M5	1516 + 02A	5.5			S			24
	1516 + 02B	7.9			6.8	0.13		24, 25
M4	1620 − 26	11.08	2.2×10^8	3×10^9	191.4	0.025	0.4	8, 10, 14, 20
M13	1639 + 36A	10.38	$>3.6 \times 10^9$	5×10^8	S			11
	1639 + 36B	3.5			1.26	~0	0.4	5
Trz 5	1744 − 24A	11.56			0.0756	$<10^{-3}$	0.09	16, 21
NGC 6440	1745 − 20	288.60			S			19
NGC 6539	1802 − 07	23.10			2.62	0.22		7
NGC 6624	1820 − 30A	5.44			S			6
	1820 − 30B	378.59			S			6
M28	1821 − 24	3.05	3.1×10^7	2×10^9	S			8, 9, 12, 15, 17
NGC 6760	1908 + 00	3.6			?			4
M15	2127 + 11A	110.66			S			12, 26
	2127 + 11B	56.13			S			3
	2127 + 11C	30.53	1×10^8	1×10^{10}	0.335	0.68	1.4	3, 22
	2127 + 11D	4.8			S			22
	2127 + 11E	4.65			S			22

APOLOGIA

This is by no means a complete review. I have attempted to summarize the latest happenings in this field along with some perspective. The single most important contribution of this review paper are the two tables and the associated reference list. The list is complete as of February 1991.

The reader is referred to the following recent reviews. Verbunt (1990) gives an excellent summary of the stellar evolutionary models for LMXBs, millisecond pulsars both in the disk and the globular clusters. Bhattacharya and van den Heuvel (1991) present a very complete review of the formation and evolution of binary and radio pulsars. Phinney and Kulkarni (1991) review the origin and evolution of cluster pulsars. Finally, there were several talks specifically on globular cluster pulsars at a recent workshop (a NATO-ITP workshop, Santa Barbara, January 1991); the proceedings is expected to be published soon.

FORMATION SCENARIOS

There are two models by which cluster pulsars are supposed to have been formed: by spin-up of old neutron stars (as outlined earlier), and by accretion-induced collapse of massive white dwarfs (see Verbunt 1990 for a review). In both cases, a degenerate star is assumed to tidally capture a nondegenerate star. Until the discovery of cluster pulsars, only 2-body captures were considered. The presence of cluster pulsars in low-density clusters like M53 *require* far more efficient capture mechanisms, the leading candidate for which is 3-body captures (degenerate star and primordial binaries; Phinney and Kulkarni 1991).

The advantage of the accretion-induced collapse over the old neutron star hypothesis is that, for any reasonable initial mass function (IMF), the white dwarfs will outnumber the neutron stars. In addition, if cluster neutron stars were born with large velocity kicks (as in the disk of the Galaxy), then only about a fifth of the original population would have been retained in the feeble gravitational potential well of protoclusters. The principal problem with the accretion-induced collapse model is the uncertain physics of the collapse itself. Indeed, the standard model for a Type Ia supernova requires the disruption of an accreting white dwarf. Apart from this, there are some quantitative issues. It appears that many conditions need to be satisfied in order for the accretion to result in a net growth of the mass of the white dwarf (as opposed to novae ejection), and in addition, an *ad hoc* assumption needs to be made regarding the period and magnetic field strength of the newly formed pulsar (both need to be low). It is unclear what the final efficiency of this channel is.

PULSAR CONTENT

The abundance of pulsars in clusters has continued to astonish everyone—theorists and observers. At one end, in clusters like 47 Tuc, 10 pulsars have already been reported. Equally astonishing is the presence of pulsars in low density clusters like M53, since in such clusters tidal capture is supposedly inefficient. As we shall see below, such clusters require a large abundance of neutron stars as well as efficient tidal capture mechanisms.

Typical radio pulsar surveys with sensitivity to millisecond periods are complete to a distance of 1 kpc. Nearby millisecond pulsars have radio luminosities (at 400 MHz), Sd^2 of about 5 mJy kpc^2. In contrast, the typical cluster is 5 kpc away, and searches for cluster pulsars reach flux densities, $S \lesssim 1$ mJy. Thus only the most luminous cluster pulsars are observable. This means that any estimate of the pulsar content of globular cluster necessarily involves large extrapolation.

The most model-free estimate is to use the observed birthrates. Multiplying the birthrate by the duration of the current phase (after all, clusters are not static structures) then yields the pulsar content. Unfortunately, this estimate is highly uncertain since the $\dot{P}$ for most cluster pulsars are not known. However, in due course there is much hope for this method. Phinney and Kulkarni (1991) estimate about $2000(1 + \beta) f^{-1}$, where f is the beaming fraction. Estimates for the beaming factor of pulsars vary from 0.2 to 1 (for fast pulsars). The term β is the ratio of pulsars in tight binaries to those in wide systems (including single pulsars). Pulsars in tight binaries escape detection in pulsar searches because of the changing Doppler velocity over the course of observations. Note that imaging observations are impervious to this effect.

The most model-dependent method presupposes an underlying luminosity function, and estimates the rest of the iceberg by looking at the tip. Kulkarni, Narayan, and Romani (1990) estimate a number similar to that given previously.

Fruchter and Goss (1990) estimate between $500f^{-1}$ and $2000f^{-1}$ based on observations of integrated flux from clusters. Note the absence of β since the results are based on VLA imaging observations.

High-resolution imaging observations, such as those from the VLA, are useful in placing constraints on β, especially toward those clusters with known pulsars. Kulkarni *et al.* (1990) find $\beta \ll 1$.

Wijers and van Paradijs (1991) compare the radio luminosities of known pulsars with the integrated emission from the rest of the clusters and conclude that the total number of pulsars is between $300f^{-1}$ and $700f^{-1}$. In each of the three clusters that Wijeres and van Paradijs investigated, they find that the one detected pulsar accounts for most of the flux. Thus the rest of the pulsars must be either very faint or few. Assuming a realistic luminosity cutoff, they find that the latter hypothesis is favored. This test, while powerful (in the sense that the observations exist), is prone to measurement errors. Pulsar flux densities are usually badly measured. Observers usually measure them when the pulsar is bright (due to scintillation). In addition, recent observations by Fruchter and coworkers of one of the three clusters (Terzan 5) reveal the presence of diffuse radio emission with relatively steep spectral index in the central region. The simplest inference is that this cluster contains many pulsars. Careful series of observations (both imaging and especially pulsed) are urgently needed in order to usefully apply this test.

The issue of the relative weight of a cluster in respect to its propensity for pulsars is important in determining the global number of cluster pulsars. The two-body capture mechanism, popular before the discovery of cluster pulsars, would result in the weights proportional to $M_c\rho_c$, where M_c and ρ_c are the core mass and the core density, and hence heavily favor dense clusters. On the other hand, 3-body capture mechanisms favor lower density clusters because the primordial binaries may be hardened or broken up in dense clusters (Phinney and Sigurdsson 1991).

Johnston, Kulkarni, and Phinney (1991) have combined all the measurements (pulsed, integrated, and high-resolution imaging) and find that the current data are consistent with a luminosity function similar to disk millisecond pulsars ($dN/dL \propto L^{\gamma}$, $\gamma \sim -2$) and a cluster weighting function based on both 2-body and 3-body capture mechanisms. The total number of pulsars inferred in this study is about $1500f^{-1}$.

Thus the total number of pulsars in clusters is still a controversial issue. We need more $\dot{P}$ measurements, careful flux density measurement, and high angular resolution observations to firmly pin down this number. Finally, it is possible that some time-dependent effects may well be operational. To wit, Liller 1 is expected to have a large number of pulsars, but the integrated flux density from this cluster is very low (Johnston, Kulkarni, and Goss 1991). Clusters like M28 and M4 have rather young pulsars, age $\sim 10^8$ y! Such time-dependent behavior may make most of the analyses previously presented somewhat suspect.

RELATION TO LMXBS

The original motivation in searching for cluster pulsars was the high specific incidence of LMXBs in clusters. It is natural to suppose that cluster pulsars descend from cluster LMXBs, in which case the birthrates of the two species must match. However, most of the cluster LMXBs are found in dense, metal-rich clusters, whereas pulsars have been found in all kinds of clusters: metal-rich, metal-poor, dense, and not so dense. Thus it is incorrect to compare the birthrates of LMXBs with those of cluster pulsars (Phinney and Kulkarni 1991).

The preceding dichotomy, the observed large birthrate of pulsars and the small value of β, can be reconciled by invoking a two-channel model: a *fast* channel in which the accretion phase is short-lived, and a *slow* channel in which the accretion phase is long-lived (Ray and Kluzniak 1990; Phinney and Kulkarni 1991). The latter authors identify the fast channel with the destructive interaction of field stars (binary or otherwise) with an existing neutron star, and the slow channel with long-lived tidal captures. Thus in this model, most of the resulting pulsars are single and produced in the fast channel, neatly explaining the small value of β. The long-lived tidal captures produce the small number of observed LMXBs. Ray and Kluzniak (1990) make a similar identification, replacing the neutron star with a massive white dwarf that eventually is converted to a pulsar.

USES OF GLOBULAR CLUSTER PULSARS

Cluster pulsars offer a variety of uses ranging from interstellar medium to gravitational radiation.

When multiple pulsars are found in a single cluster, the observed dispersion measures will provide the best constraints or perhaps even detection of intracluster gas. The sensitivity of the observations will exceed current upper limits (obtained from HI 21-cm or Hα emission) by several orders of magnitude.

Structural parameters can be deduced from timing observations, especially of multiple pulsars in a single cluster. Pulsars undergo acceleration due to the gravitational attraction of the mean gravitational field (affecting $\dot{P}$), and jerks (change of acceleration) due to encounters of nearby and distant stars (affecting $\ddot{P}$; Blandford, Romani, and Applegate 1987). The changes in these two parameters can be many orders of magnitude larger than those caused by magnetic braking, especially for dense clusters. Indeed, in M15, two of the measured period derivatives are negative (references 22 and 26). Phinney (1991), using all the available data (optical photometry, radial velocity, and pulsar $\dot{P}$'s), gets a beautiful fit to a multicomponent mass model of M15. The advantage of the pulsars over optical measurement is that the pulsars are located close to center of the cluster where most of the degenerate stars are located.

Tidal captures result in exotic binaries. The pulsar in Terzan 5 (references 16 and 21) has an orbital period of about 2 h, and the pulsar appears to be ablating its low-mass companion, similar to PSR 1957+20. In M15, the pulsar 2127+11C (reference 3) appears to consist of two neutron stars and is rather similar to 1913+16. Already gravitational-radiation (GR) effects have been observed (reference 22). Unlike in the disk, the spin and the orbital angular momenta are not expected to be aligned, and hence there is a possibility of observing geodetic precession (Phinney and Kulkarni 1991). Like 1913+16, the binary is expected to coalesce due to GR radiation in about 10^8 y, contributing to the list of sources detectable by terrestrial gravity wave interferometers.

IMPLICATIONS AND THE FUTURE

It is now clear that pulsars in clusters are offering a new and relatively unexpected window into the massive star content of these ancient subunits of our galaxy. How protoclusters survived the formation of a few thousand neutron stars without being blown apart requires some finely tuned birth process. (Perhaps we see the few that survived the process.) Moving to the current time, highly accurate timing observations will tell us much about the structure of clusters and their interstellar medium. The orbital period distribution, the birthrates, and the number of pulsars in different clusters has reawakened interest in the physics of 2- and 3-body tidal captures. Exotic binaries allow us to investigate the effects of pulsar winds on normal stars, conduct tests of GR, and finally act as sources of GR.

The future is clearly bright. Deep integrations will yield faint pulsars that will improve the statistics of the incidence of pulsars in clusters. Timing observations of the bright pulsars will enable us to carry out the many applications listed earlier. Above all, there may well be new unexpected discoveries that will be revealed by long-term timing observations.

ACKNOWLEDGMENT

I am grateful to Helen Johnston for compiling the tables and the associated reference list and for a careful reading of the manuscript.

REFERENCES

ALPAR, M. A., A. F. CHENG, M. A. RUDERMAN & J. SHAHAM. 1982. Nature **300:** 728.
BACKER, D. C., S. R. KULKARNI, C. E. HEILES, M. M. DAVIS & W. M. GOSS. 1982. Nature **300:** 615.
BHATTACHARYA, D. & E. J. P. VAN DEN HEUVEL. 1991. Phys. Rep. In press.
BISNOVATYI-KOGAN, G. S. & B. V. KOMBERG. 1975. Sov. Astron. **18:** 217.
BLANDFORD, R. D., R. W. ROMANI & J. H. APPLEGATE. 1987. Mon. Not. R. Astron. Soc. **225:** 51P.
BRADT, H. V. D. & J. E. MCCLINTOCK. 1983. Annu. Rev. Astron. Astrophys. **21:** 13.
FRUCHTER, A. S. & W. M. GOSS. 1990. Astrophys. J., Lett. **365:** L63.
HAMILTON, T. T., D. J. HELFAND & R. H. BECKER. 1985. Astron. J. **90:** 606.
JOHNSTON, H. M., S. R. KULKARNI & E. S. PHINNEY. 1991. In preparation.
JOHNSTON, H. M., S. R. KULKARNI & W. M. GOSS. 1991. Astrophys. J., Lett. **382:** L89.
KULKARNI, S. R., R. NARAYAN & R. W. ROMANI. 1990. Astrophys. J. **356:** 174.
KULKARNI, S. R., W. M. GOSS, J. M. MIDDLEDITCH & A. WOLSZCZAN. 1990. Astrophys. J., Lett. **363:** L5.
LYNE, A. G., A. BRINKLOW, J. MIDDLEDITCH, S. R. KULKARNI, D. C. BACKER & T. R. CLIFTON. 1987. Nature **328:** 399.

PHINNEY, E. S. 1991. Mon. Not. R. Astron. Soc. In press.
PHINNEY, E. S. & S. R. KULKARNI. 1991. Nature. In press.
PHINNEY, E. S. & S. SIGURDSSON. 1991. Submitted for publication in Astrophys. J.
RAY, A. & W. KLUZNIAK. 1990. Nature **344:** 415.
SMARR, L. L. & R. D. BLANDFORD. 1975. Astrophys. J. **207:** 546.
WIJERS, R. A. M. J. & J. VAN PARADIJS. 1991. Astron. Astrophys. In press.
VERBUNT, F. M. 1990. *In* Neutron Stars and Their Birth Events, W. Kundt, Ed.: 179.

REFERENCES CITED IN TABLES

1. ABLES, J. G., C. E. JACKA, P. A. HAMILTON, P. M. MCCULLOCH & P. J. HALL. 1988. I.A.U. Circ. #4602.
2. ABLES, J. G., D. MCCONNELL, C. E. JACKA, P. M. MCCULLOCH, P. J. HALL & P. A. HAMILTON. 1989. Nature **342:** 158–161.
3. ANDERSON, S., P. GORHAM, S. R. KULKARNI, T. PRINCE & A. WOLSZCZAN. 1990. Nature **346:** 42–44.
4. ANDERSON, S., S. R. KULKARNI, T. PRINCE & A. WOLSZCZAN. 1990. I.A.U. Circ. #5013.
5. ANDERSON, S., S. R. KULKARNI, T. PRINCE & A. WOLSZCZAN. 1991. Personal communication.
6. BIGGS, J. D., A. G. LYNE, R. N. MANCHESTER & M. ASHWORTH. 1990. I.A.U. Circ. #4988.
7. D'AMICO, N., A. G. LYNE, M. BAILES, S. JOHNSTON, R. N. MANCHESTER, L. STAVELEY-SMITH, J. LIM, A. S. FRUCHTER & W. M. GOSS. 1990. I.A.U. Circ. #5013.
8. FOSTER, R. 1991. Personal communication.
9. FOSTER, R. S., D. C. BACKER, J. H. TAYLOR & W. M. GOSS. 1988. Astrophys. J., Lett. **326:** L13–L15.
10. GOSS, W. M., S. R. KULKARNI & A. G. LYNE. 1987. Nature **332:** 47–48.
11. KULKARNI, S. R., S. B. ANDERSON, T. A. PRINCE & A. WOLSZCZAN. 1991. Nature **349:** 47–49.
12. KULKARNI, S. R., W. M. GOSS, A. WOLSZCZAN & J. MIDDLEDITCH. 1990. Astrophys. J., Lett. **363:** L5–L8.
13. LYNE, A. G. 1991. Presented at the NATO Workshop on X-ray Binaries and the Formation of Binary and Millisecond Pulsars, Santa Barbara, Calif., Jan. 21–25, 1991.
14. LYNE, A. G., J. D. BIGGS, A. BRINKLOW, M. ASHWORTH & J. MCKENNA. 1988. Nature **332:** 45–47.
15. LYNE, A. G., A. BRINKLOW, J. MIDDLEDITCH, S. R. KULKARNI, D. C. BACKER & T. R. CLIFTON. 1987. Nature **328:** 399–400.
16. LYNE, A. G., R. N. MANCHESTER, N. D'AMICO, L. STAVELY-SMITH, S. JOHNSTON, J. LIM, A. S. FRUCHTER, W. M. GOSS & D. FRAIL. 1990. Nature **347:** 650–652.
17. MAHONEY, M. J. & W. C. ERICKSON. 1985. Nature **317:** 154–155.
18. MANCHESTER, R. N., A. G. LYNE, S. JOHNSTON, N. D'AMICO, J. LIM & D. A. KNIFFEN. 1990. Nature **345:** 598–600.
19. MANCHESTER, R. N., A. G. LYNE, S. JOHNSTON, N. D'AMICO, J. LIM, D. A. KNIFFEN, A. S. FRUCHTER & W. M. GOSS. 1989. I.A.U. Circ. #4905.
20. MCKENNA, J. & A. G. LYNE. 1988. Nature **336:** 226–227.
21. NICE, D. J., S. E. THORSETT, J. H. TAYLOR & A. S. FRUCHTER. 1990. Astrophys. J., Lett. **361:** L61–L63.
22. PRINCE, T., S. B. ANDERSON, S. R. KULKARNI & A. WOLSZCZAN. 1991. Submitted for publication in Astrophys. J.
23. SHAWL, S. J. & R. E. WHITE. 1986. Astrophys. J. **91:** 312–316.
24. WOLSZCZAN, A., S. ANDERSON, S. KULKARNI & T. PRINCE. 1989. I.A.U. Circ. #4880.
25. WOLSZCZAN, A., S. ANDERSON, S. KULKARNI & T. PRINCE. 1991. Personal communication.
26. WOLSZCZAN, A., S. R. KULKARNI, J. MIDDLEDITCH, D. C. BACKER, A. S. FRUCHTER & R. J. DEWEY. 1989. Nature **337:** 531–533.

Gamma-ray Burst Observations: Overview of Recent Results

G. VEDRENNE

Centre d'Etude Spatiale des Rayonnements
Centre National de la Recherche Scientifique
Universite Paul Sabatier
BP 4346-31029 Toulouse cedex, France

INTRODUCTION

We present a brief overview of the main observational characteristics of gamma-ray bursts (hereafter GRB), taking into account the latest results obtained by three missions: GINGA, PHOBOS, and SMM. For an extensive analysis of the results accumulated on GRB during the last 20 years, excellent complete reviews have recently appeared.[1–3] No discussion of the various models proposed to explain the GRB mystery will appear here; several comprehensive articles have been published over these last few years.[3–6]

After a short summary of the main temporal characteristics of GRB, two aspects will be discussed: the spectral characteristics of GRB, and their spatial distribution, taking into account excellent models recently developed, mainly by Hartmann and colleagues, and by Paczynsky. These models allow us to reconcile many of the divergences reported in the past when the spatial distribution was confronted with the log N–log S curve.

TIME HISTORY

FIGURE 1 illustrates the diversity of GRB time profiles. Very short well-structured profiles (≤ 100 ms) are shown in FIGURE 2; they have been observed on the GRANAT mission by two GRB detectors: the SIGMA anticoincidence[7] and the PHEBUS experiment. The attempts to classify GRB according to duration have produced few conclusive results, even if two main classes were suggested (below 1 sec and above a few seconds), the minimum being near 1 or 2 seconds.[2,8,9] As far as the rise and decay times of bursts are concerned, here too the observations do not lead to a special classification.[10] The search for periodicities in GRB has not been very successful. The results are in general negative, except for the exceptional GRB of March 5, 1979b with a possible counterpart in the direction of N49 in the Large Magellanic Cloud (LMC), which displays a clear 8-sec periodicity.[11] In the region of few seconds some results have been reported,[12–15] but they are at the limit of statistical significance.

Another property of GRB sources that has been searched for is their possible recurrence; we will see in the fourth section that the GRB distribution allows us to set limits on the recurrence time of burst sources. This time is not precisely determined, but a generally accepted value is ≥ 10 years. This means that most of the bursters that have been observed up to now do not repeat. The exceptions are given in TABLE 1.[1] One of them, SGR 1806−20, recently detected by the PROGNOZ 9 Franco-Soviet experiment,[16] and confirmed by the Los Alamos group (ICE results),[17] is situated in the general direction of the galactic center. The bursts have time profiles and spectra quite similar from one burst to the next. They are short (≈ 10 ms) and they have soft spectra ($kT \approx 20$ keV, FIG. 3). These properties allowed an identification of more than 100 recurrences in the ICE data.[18] These repetitive bursts appear as another class of GRB, not at all typical of the properties of common GRB.

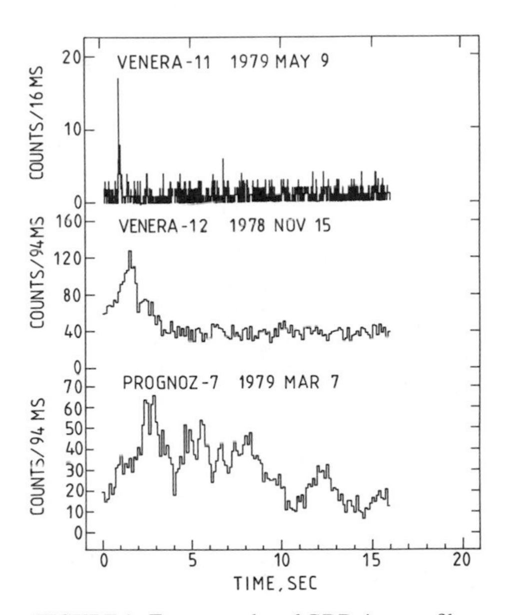

FIGURE 1. Few examples of GRB time profiles.

One last property of GRB time histories is the difference between profiles at low energies (≤ 20 keV) and above. The P78-1 and Hakucho satellites, and more recently GINGA and PHOBOS, have reported many results on this property. The conclusions are as follows: the soft X-ray emission (< 10 keV) may appear at the same time as or before the GRB onset; possible precursors have been detected.[19,20] There is a clear X-ray tail after the end of the gamma-ray emission (lasting up to ~ 100 sec) (FIG. 7). Thus the soft emission may exhibit temporal variation that seems independent of the gamma-ray emission. The energy content is a few percent of the GRB energy.[21]

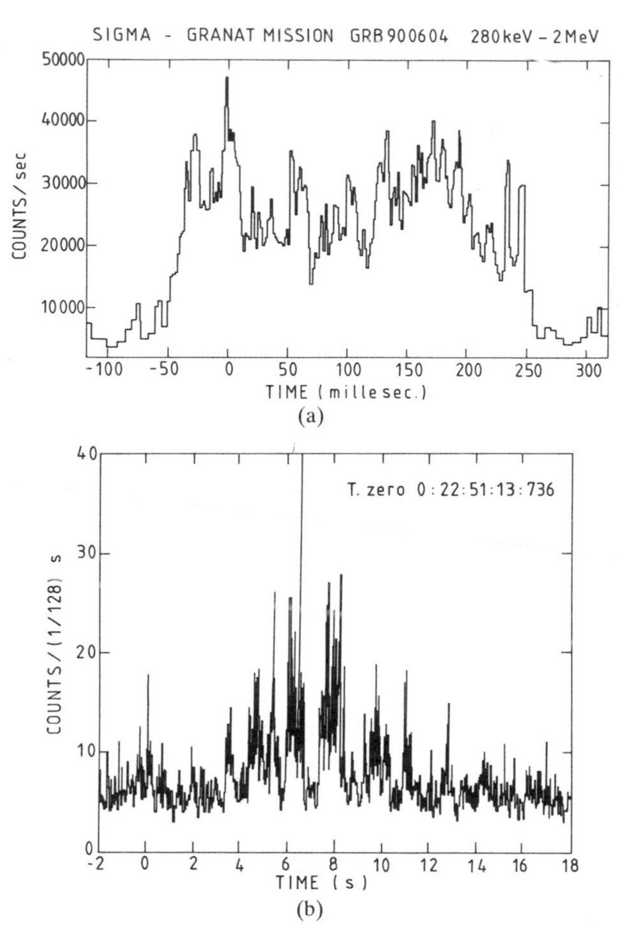

FIGURE 2. Two examples of short GRB observed with the French (**a**) SIGMA and (**b**) PHEBUS experiments on the GRANAT satellite.

TABLE 1. The Three Known Repeaters

	March 5, 1979b	B1900+14	SGR 1806−20
Number of bursts	16	3	110
Inverval between bursts	0.6–100 d	1 d	1 s–2 y
Time histories	Short	Short	Short
Spectrum kT, keV	35	35	40
Range of fluences, S	44(3000):1	4:1	30:1
Slope of log N–log S curve	−0.5	—	$-1 \rightarrow -2$
Association	N49?	Galactic plane?	Galactic bulge?

*SOURCE: From Hurley.[1] Used with permission.

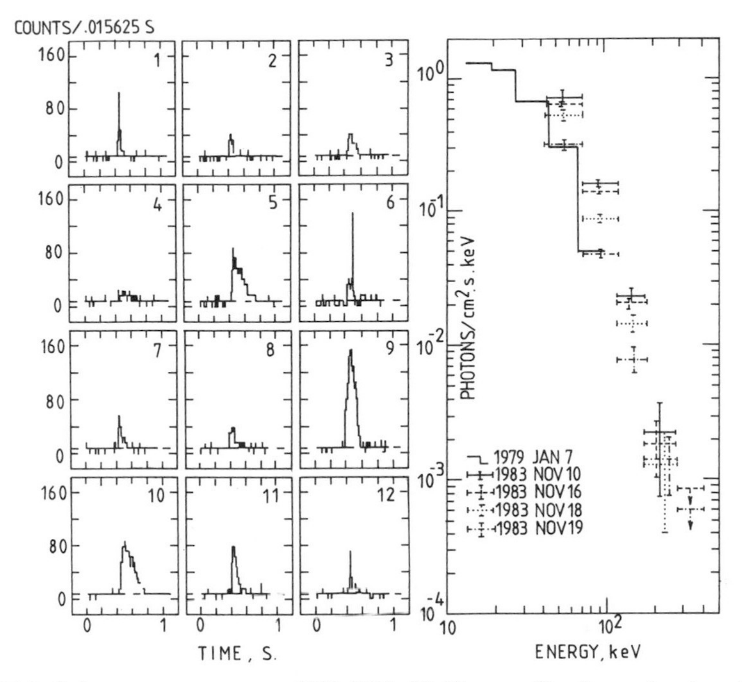

FIGURE 3. Soft gamma-ray repeater: SGR 1806−20. Time profile of some bursts and their characteristic spectra. (From Atteia *et al.*[49] Reproduced by permission.)

SPECTRA OF GRB: THE MAIN CHARACTERISTICS

We will analyze successively the complete burst spectrum (integrated over the total duration of the burst), the possible features at low and high energy, the behavior of the high-energy part of the GRB with respect to the low-energy part, and a related property, the hardness ratio.

Complete GRB Spectrum

FIGURE 4 shows examples of spectra obtained by SMM.[22] They are best fitted above 300 keV by a power law with an index between 2 and 3 for most events. At least equally important, an extension of the spectrum is currently above 1 MeV in more than 60 percent of the events.[23] With the APEX experiment on PHOBOS, which used a large CsI detector ($\phi = 10$ cm, $h = 10$ cm), as for SMM, the detector efficiency above 1 MeV was high, and most of the detected GRB also displayed an extension of their spectra above 1 MeV. Moreover, where measurements extended down to 100 keV, a break around 500 keV was very often observed (FIG. 5). This is a common feature that is also observed for some events reported by SMM.

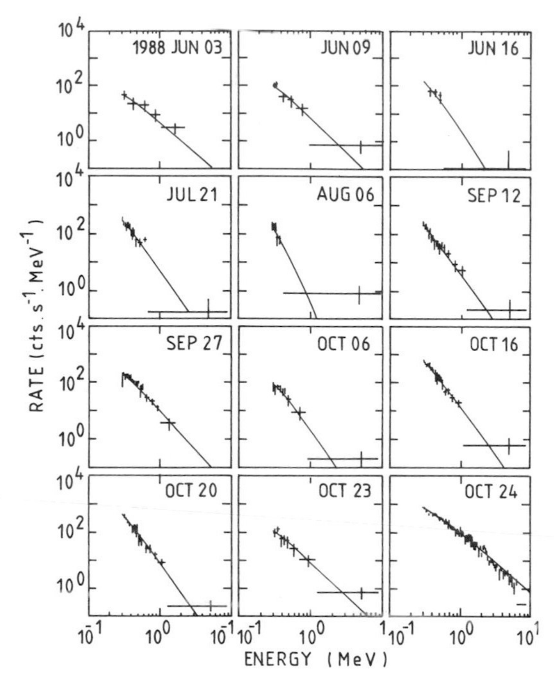

FIGURE 4. Examples of GRB spectra observed by SMM. (From Share *et al.*[22] Reproduced by permission.)

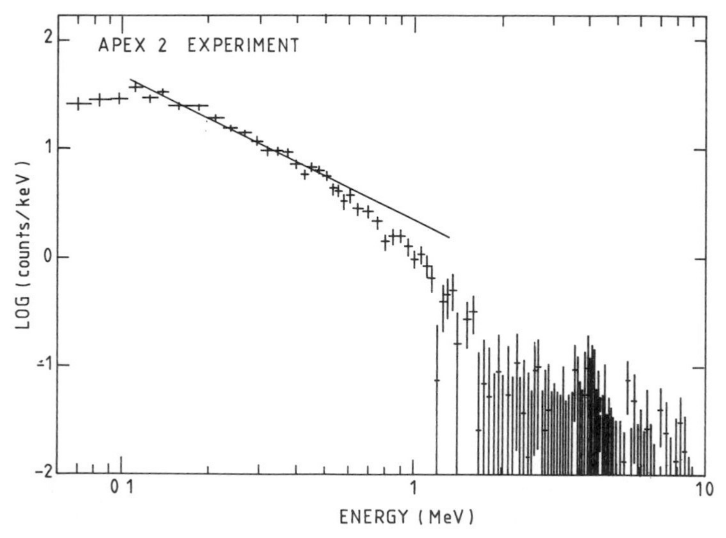

FIGURE 5. Example of GRB spectra observed by the APEX experiment and corresponding to the total GRB duration.

Of course, the raw APEX spectra have to be deconvolved into photon spectra to determine the change in the shape and to determine whether the best fit is given by two power laws with a break around 500 keV, or if, as was reported long ago by Desai and Cline,[24] the best fit is an exponential at low energy and a power law above ≈500 keV. The deconvolution at low energy is difficult because very often we do not know the arrival direction with respect to the axis of the detector for each burst. In any case, a flattening below 100 keV is a common feature of the APEX results, and finally, even if the fluences of the bursts are quite variable, many of them present the same general shape. We will not review the various fits that have been proposed. Thermal and nonthermal models, as well as two-component models, have been discussed extensively.[5]

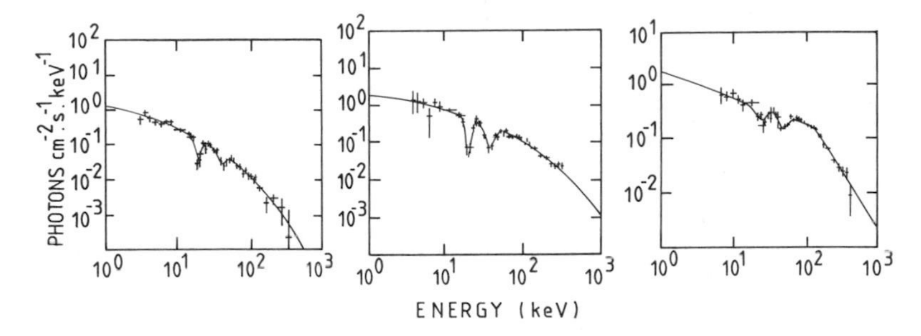

FIGURE 6. Absorption cyclotron features observed with GINGA. (From Murakami.[14] Reproduced by permission.)

Features in GRB Spectra

The first reports of features in GRB spectra were due to the KONUS experiment on VENERA 11 and 12.[25] Two kinds of features were observed one at low energy, below 40 keV, and one around 400–500 keV. We discuss each of these very important observations.

Low-energy Features

At low energies the features were, in general, absorption features, or very often breaks in the continuum spectra fitted by an optically thin thermal bremsstrahlung (OTTB) continuum. Much controversy surrounded these results[26] because the break at low energy naturally depends strongly on the choice of the continuum. Without reviewing this debate, it is clear that if one chooses an OTTB continuum as opposed to a simple exponential fit, the absorption feature may be present or it may disappear.[27] This simply illustrates one difficulty with this energy domain. Nevertheless, Mazets and his group reported features that were not only breaks at low energy, but also real absorption features compatible with the resolution of the detectors. Of course, due to the small area of these detectors, the significance level was generally low. Nevertheless, the HEAO A 4 experiment[28] confirmed this low-energy behavior of GRB spectra. But the discussion of the reality of this very important characteristic was not closed. We had to await the GINGA results to clearly, or more clearly, demonstrate the evidence for absorption features. The great advantage of the GINGA detectors is to be able to measure the spectra below 20 keV (the limits of the KONUS experiments) to a few kiloelectronvolts. The results reported by Murakami *et al.*[29] (FIG. 6) and analyzed in details by Fenimore *et al.*[30] conclusively demonstrated the presence of features, at two energies at least, that can be considered to be the first and second cyclotron absorption harmonics in the presence of a strong magnetic field ($\approx 2.10^{12}$ Gauss) that can exist at the surface of neutron stars. This is of primary importance because, since we have not identified any counterpart to GRB sources (see the fourth section), this is the first evidence that GRBs (although perhaps not all) are produced at the surface of neutron stars.

The LILAS experiment on the PHOBOS mission used two cleaved NaI detectors (ϕ = 5.3 cm, h = 3 cm) at the end of the solar panels with a sky coverage around 4π and a 5-keV–1-MeV energy range. This experiment also observed absorption features. One example was analyzed recently by C. Barat *et al.*;[31] the counting rate and photon spectra are given in FIGURE 7.

All of these recent observations confirm, at least for some bursts, the presence of cyclotron absorption features with two harmonics and perhaps even three. However, the detailed analysis, and particularly the choice of the continuum spectrum, have a decisive influence on the strengths of the reported lines and their significance level.

Annihilation Features

Annihilation features were first reported from the KONUS experiments.[25] Very often these features are bumps with full width at half maxima of 200–250 keV. These were subject to controversy,[32,33] the statistical significance level very often being low; but in some cases features were reported by two different detectors. With the APEX experiment aboard the PHOBOS probe, seventy bursts were detected by the CsI detectors (ϕ = 10 cm, h = 10 cm). The large dimensions and the spectral analysis (108 channels, with a time-to-spill method for accumulating the spectra) allowed a search for features. Eight of the strongest bursts have been examined in detail, with the time resolution for the spectra varying between tens of milliseconds for the strongest events and half a second for the weakest one. A systematic analysis of the excesses at the 3σ level has been done after fitting the continuum with a four-parameter polynomial function. Looking at the excesses that were observed (some of them of statistical origin), it appears that most cluster in the regions around 400 keV, 500 keV, and 800 keV,[34] although none has a high confidence level. FIGURE 8 shows an example of a feature at 475 keV with a $>3\sigma$ confidence level that was observed for 2.4 sec (corresponding to the first peak of the GRB).

Another example is particularly interesting because it corresponds to a very strong burst observed on October 24, 1988. The main peak of this strong event was analyzed with a time resolution of about 30 ms. Several features are quite visible: a strong absorption at low energy in spectrum b that disappears after 30 ms and two lines in spectra b and a, respectively, at 520 $\pm$ 30 keV and 820 $\pm$ 60 keV (3.2σ and 3.9σ excesses).[35,36] These lines are also quite transient—their duration is around 30

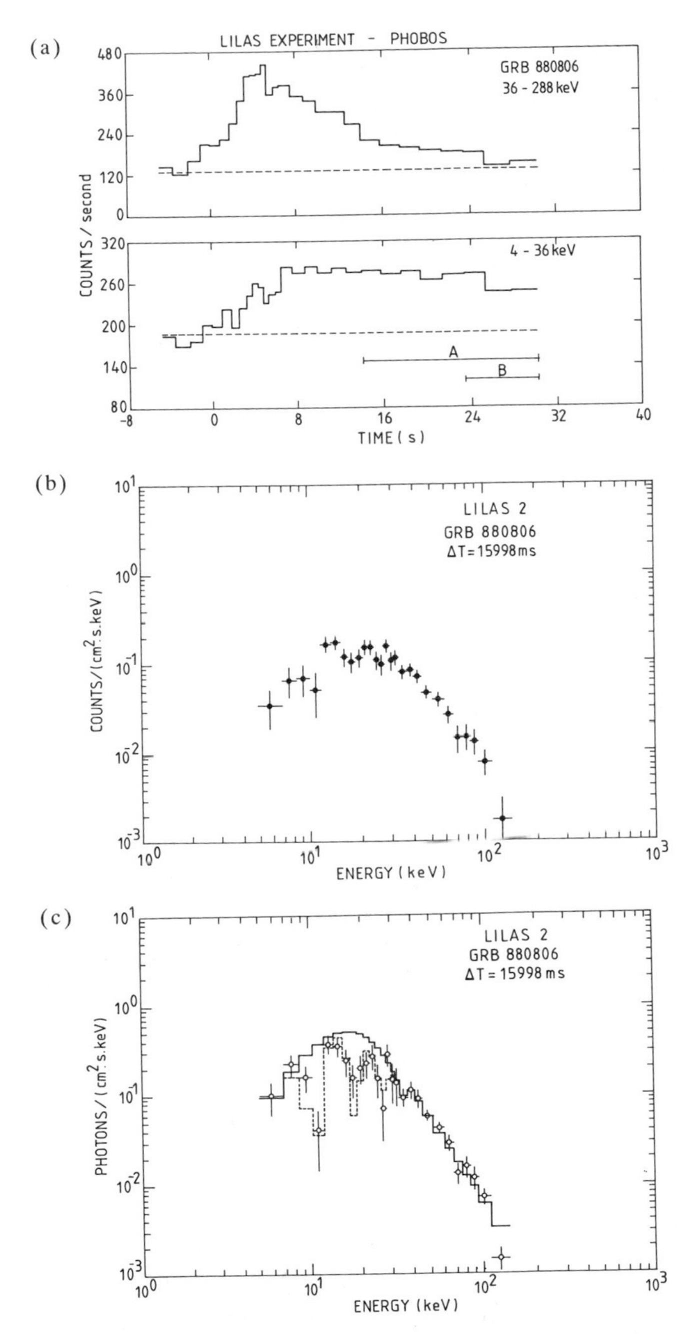

FIGURE 7. Low-energy absorption features observed with the LILAS experiment aboard the PHOBOS mission. (**a**) Time profile. (**b**) Counting rate vs. energy. (**c**) Photon spectrum.

ms (FIG. 9). This example demonstrates the difficulties of the spectral analysis when such features have durations as short as a few tens of ms. It means that only if the detector is very large or the burst very intense, which is quite rare, can the search for features be positive. And if this is the case, it also means that many spectra within a particular burst have to be treated before a feature appears; under these conditions, the statistical significance of the feature will be strongly decreased due to the large number of trials needed to reveal something. For the APEX spectra, this analysis has

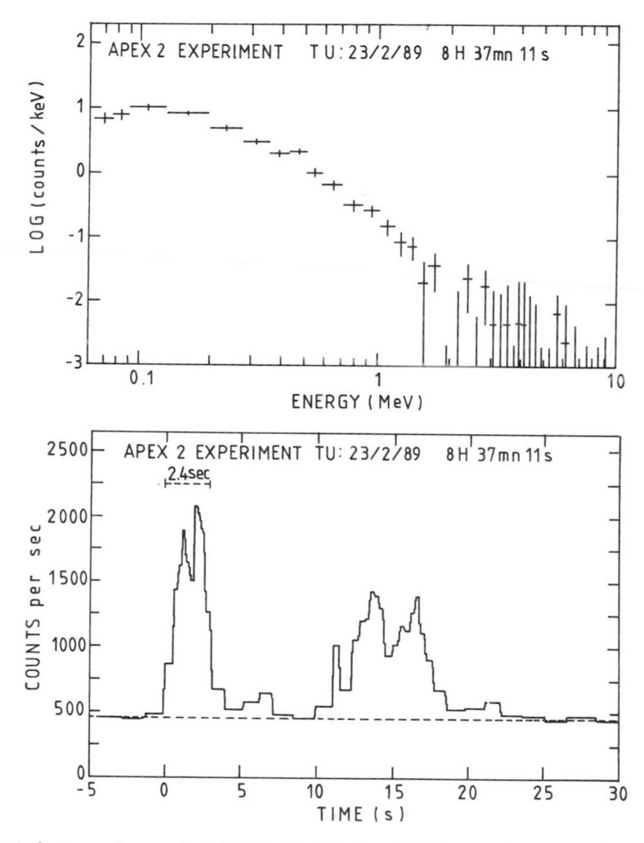

FIGURE 8. A feature observed at 475 keV with the APEX experiment during the first peak (2.4-sec duration) of the GRB: GB890223.

been carried out to look for features whose energy widths are compatible with the detector resolution (40–50 keV).

Another, more difficult task has to be performed to search for broader features; in this case, to reach a definite conclusion the photon spectrum must be obtained, which requires knowledge of the direction of the burst with respect to the detector axis. Moreover, a continuum fit must be chosen, and this is not easy. Depending on this choice, the presence or absence of the broad feature can be debated; see, for instance, the discussion of GB781104.[32] Such a discussion can be pursued for other examples; for instance, the strong GB781111 detected by the KONUS and SIGNE experiments[37] (FIG. 10). The continuum chosen by Mazets *et al.* is the traditional optically thin thermal bremsstrahlung, and the figure indicates the presence of a broad excess above the continuum, which has been attributed to a broad annihilation feature. If we look at the SIGNE results for the same period (FIG. 10), it appears that during the few seconds of the first peak the spectra were changing very rapidly; in the first time interval there was a very flat spectrum, below 500 keV, and after that, a rapid evolution. In this case, to explain the spectra it is not necessary to introduce a broad annihilation feature. The shape can be explained by the addition of different spectra changing very quickly over <0.5 sec. Many other examples in the SIGNE data might also be considered in the same way as GB791111, demonstrating only the fast variability of GRB with very often a break around 500 keV that appears as a pivot point in the fast temporal variation of the spectrum.

The preceding analysis finds another kind of confirmation if we consider the GRB spectra integrated for the total duration of the burst. In this case no features are seen either by SMM with an integration time of 16 sec for the GRB spectra or on the PHOBOS mission with APEX. Therefore GRB spectra have to be studied on short timescales. Only large detectors, such as the BGO detectors on GRANAT, and BATSE on GRO, will be able to define the fast variability of GRB spectra and provide possible evidence for transient annihilation features.

Time Variation of the GRB Spectra at Low and High Energies

Here, high energy means energies higher than 1 MeV. With the APEX CsI detector, it has been possible to detect the high-energy portions of GRB spectra and to show that in many cases the time variations of the high- and low-energy parts of these spectra are quite independent. FIGURE 11 gives two examples of this behavior. Moreover, there seems to be an anticorrelation between high- and low-energy variation (see FIG. 11).[34,38] Another way to study this point is to consider the hardness ratio (HR). It was introduced by Mazets[39] to show a clear correlation between the HR and the time profiles of GRB. Later SMM results[40] demonstrated, for other bursts and in a different energy range, a different behavior, with an HR decreasing systematically from the beginning of a GRB peak to its end.

With APEX, the strong GB891024 was studied, taking for the hardness ratio the ratio of the 64-keV–200-keV and 1030-keV–9218-keV counting rates. For this burst a generally good correlation in the HR evolution with respect to the GRB time profile is observed. This behavior does not confirm the SMM results, but it might prove that GRBs do not obey the same rules. GRBs might have different origins and possibly different energy production mechanisms.

DISTRIBUTION IN SPACE OF GRB

Two types of information have to be considered:

- The precise localizations obtained for a dozen GRBs with the interplanetary network, which allow counterpart searches in other wavelength domains.
- The other localizations, too crude to expect any counterpart identification but that can be used as a whole to define the characteristics of the spatial

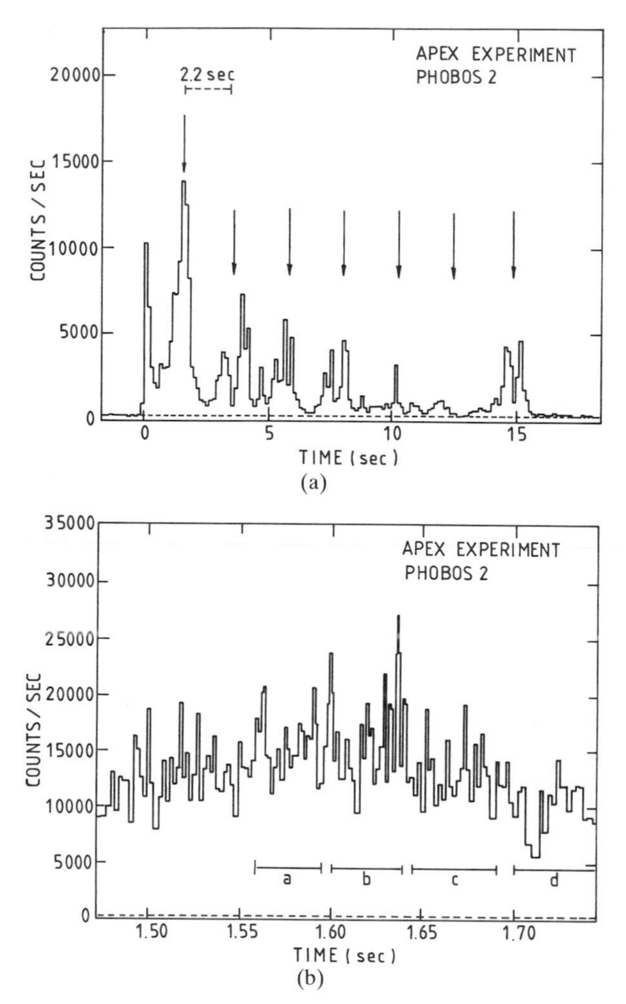

FIGURE 9. Short time spectral variability for a very strong GRB: GB881024. (**a**) Time profile of the GRB. (**b**) Time-to-spill analysis for the main peak. (**c**) Spectra corresponding to the four intervals of (**b**).

distribution and its isotropy or anisotropy. This point is crucial for determining whether we have a galactic or extragalactic GRB population.

Well-localized GRBs and the Possible Counterparts

Considerable work has been done to find counterparts at various wavelengths. In the optical domain, two approaches have been considered:

- The search for optical transients at the times of the bursts or on archival plates. No optical objects have been detected at the time of a burst. The search for

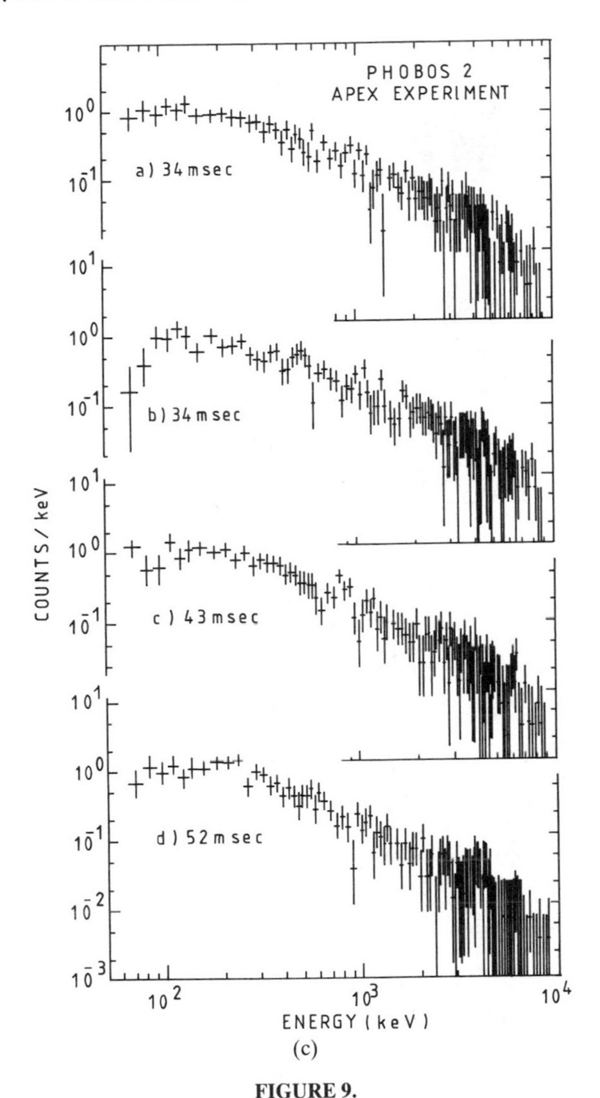

(c)

FIGURE 9.

transients on archival plates has yielded few candidates. The results are controversial due to many sources of errors. General reviews of the main results obtained in this field were given by G. Pizzichini *et al.* (1984),[41] B. Schaefer (1984),[42] and Hudec (1990).[43]

- The search for quiescent candidates. Here again, no confirmed quiescent objects that can be unambiguously associated with GRB have been detected down to magnitude 22–23 (excepting, of course, the GB790305 in the direction of N49 in the Large Magellanic Cloud). A complete analysis of this search has been published recently by Hudec.[43]

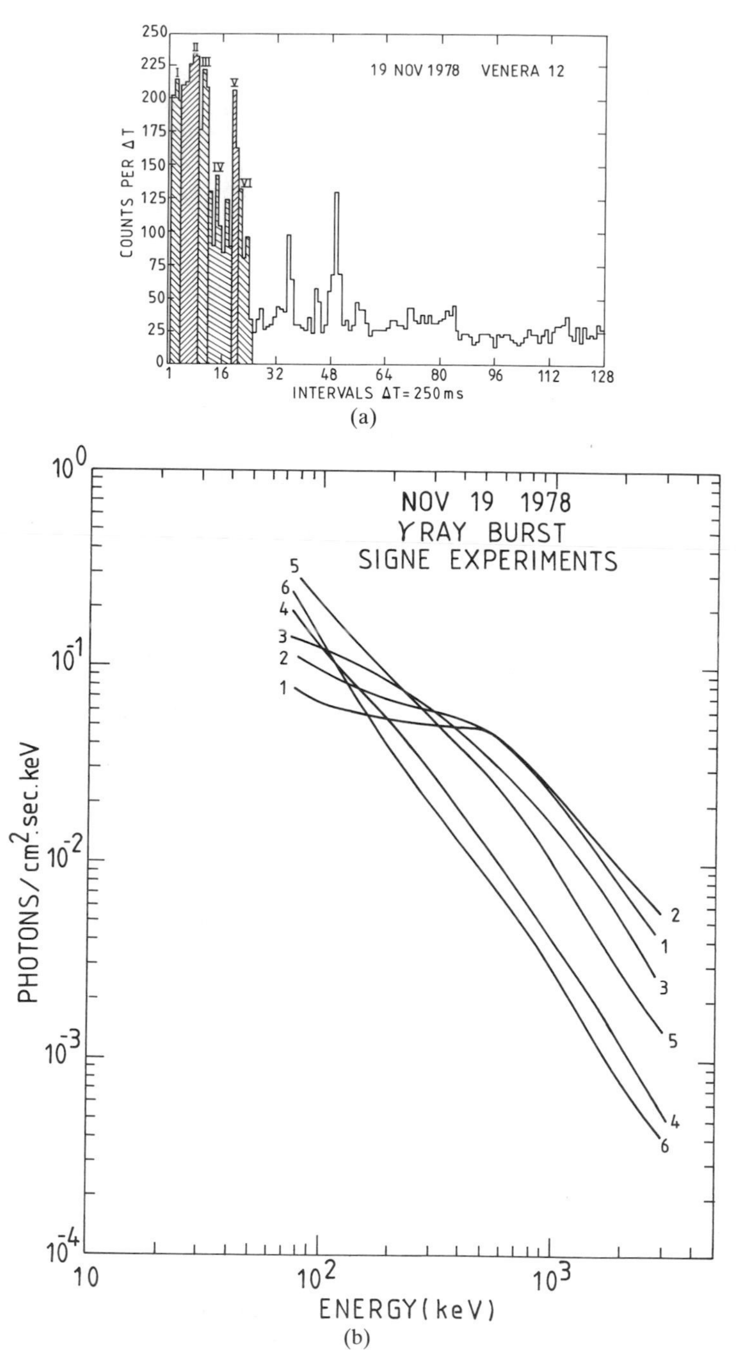

FIGURE 10. (**a**) GRB: GB781111 observed on a short timescale by SIGNE experiments. (**b**) An example of fast spectral variability.

Similar negative results were obtained in infrared, and also in X-rays from Einstein and Exosat.[44,45] A marginal (3.5σ signal) Einstein source was found in the error box of GB791119,[46] but not confirmed by Exosat.

To conclude, except for GB790305, there is no accepted evidence for any identification of the GRBs that have been localized at the few arcminute level by the interplanetary network.

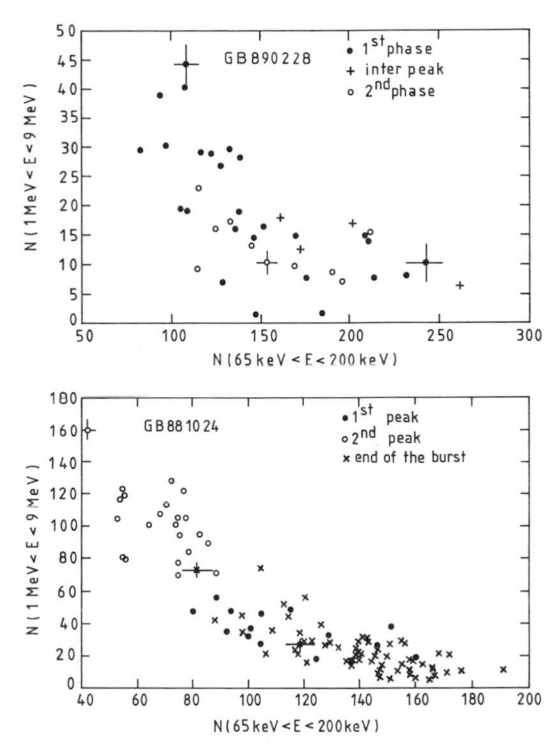

FIGURE 11. Two GRB observed by the APEX experiment showing the independence in the variation of the GRB spectra at high and low energy.

The Spatial Distribution of GRBs

The spatial distribution of GRBs involves about 170 sources now. FIGURE 12 gives their distribution on the sky using two catalogs that have been published: one from KONUS observations on the VENERA probes (see Mazets *et al.*,[47] Golenetskii *et al.*[48]), and the other obtained with the interplanetary network (see Atteia *et al.*[49]).

A first glance at the GRB distribution shows no evidence of concentration in the galactic disc. In fact, it has been shown that the distribution is uniform with no indication of significant clustering on scales between 1° and 100° (Hartmann and Blumenthal, 1989).[50] Moreover, using the catalog of the interplanetary network, the calculation of dipole and quadrupole moments of the GRB distribution on the celestial sphere shows that this distribution is quite consistent with isotropy (Hartmann and Epstein, 1989).[51] Thus, it is not possible to draw any conclusion about the origin of these objects. Nevertheless, this distribution has been used to give a limit (≥ 10 years) to the GRB recurrence time from a single source (Atteia *et al.*).[49]

Another piece of information that can be used to understand the origin of bursts is the well-known log N log S curve, N being the number of bursts with fluence above S. For the observed uniform distribution of the bursts, a power law with a 3/2 slope is expected. Therefore, the flattening that is generally observed for fluences below 10^{-5} erg/cm^2 has been the object of controversy, since it is apparently in contradiction with the uniform distribution of bursts on the sky. In fact, it has been shown that selection effects can explain the flattening at low fluences. Thus, it was proposed not to use S, but rather P, which is not the total fluence of the burst but the peak energy flux. Here again, as shown, for instance, by Higdon and Lingenfelter,[52] there exist selection effects associated with the varying energy spectra and durations of bursts. Using the KONUS size frequency distribution of sources, and taking into account spectral selection effects, they reconcile the apparent contradiction between the isotropic distribution of the observed source locations and the flattening of the size frequency distribution.

To avoid or minimize these selection effects, Paczynski and Long (1988)[53] suggest using simply C_{max}, the raw peak photon count rate: C_{max} has to be calculated for the same ΔE and Δt used to trigger the burst detectors. If N is now the number of bursts that have a maximum photon count rate in excess of C_{max}, a relation $N \propto C_{max}^{-1.5}$ is expected if the source distribution is isotropic in the sky; it does not matter how complicated the intrinsic distributions of burst strengths, durations, and spectra are. The application of this method to the KONUS catalog gives $N \propto C_{max}^{\alpha}$, with $\alpha = -1.2$, but $\alpha = -1.5$ can be accepted.

It is possible to go a little further and, for instance, take into account a possible time evolution of the detector background. This background variation can be caused by many things: position of the satellite along the orbit, gain shifts, changes in the

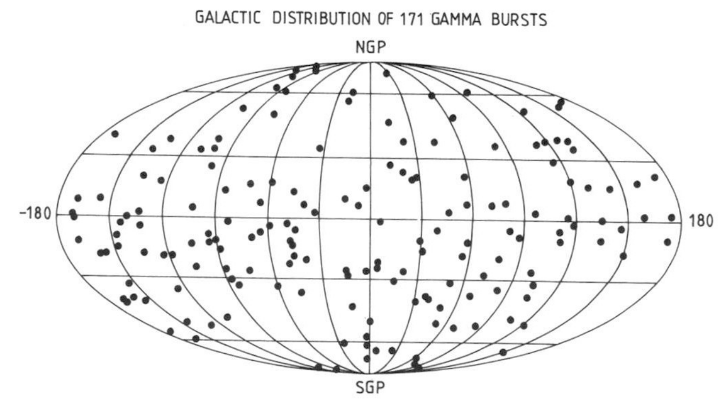

FIGURE 12. Spatial distribution of the GRB that have been localized by the KONUS experiments on VENERA probes (Mazets *et al.*, 1981)[47] (Golenetskii *et al.*, 1986),[48] and by the interplanetary network (Atteia *et al.*, 1987)[49]

TABLE 2. Principal Characteristics of the Three Burst Samples Obtained with the SIGNE Experiments on VENERA and with APEX and LILAS on PHOBOS Mission*

Experiment	VENERA-SIGNE	APEX	LILAS
Detector	NaI (cyl.)	CsI (cyl.)	NaI (cyl.)
Size [diam. × h (cm)]	9 × 3.7	10 × 10	5.3 × 3
Energy range	50–350 keV	120–700 keV	20–300 keV
Lifetime	11/81–4/83	8/88–3/89	8/88–3/89
Number of bursts	169	57	44
Bursts detection frequency (year^{-1})	~125	~115	~90
$\langle V/V_{max} \rangle$	0.435 ± 0.022	0.367 ± 0.038	0.472 ± 0.044

*SOURCE: From Atteia *et al.*[58] Used with permission.

detection thresholds, etc. In this case, it was suggested to use C_{max}/C_{min}; C_{max} has the same meaning as previously and C_{min} is the limiting counting rate (not the photon number) that would trigger the burst detector over the same time interval Δt and energy range ΔE. From this it is quite natural to introduce the V/V_{max} test (Schmidt, 1968);[54] V is the volume of the sphere whose radius is the burster distance, and V_{max} is the total volume in which the given source could have been detected. The relation $V/V_{max} = [C_{max}/C_{min}]^{-3/2}$ can easily be established.[55] If the source population is spatially uniform, the distribution of V/V_{max} will be uniform between 0 and 1. The mean value of V/V_{max} for such a sample is 0.5, with an rms error of $(12n)^{-1/2}$; here n is the number of bursts in the sample. The value V/V_{max} was calculated by Higdon and Schmidt (1990)[56] for the VENERA 11 and 12 KONUS catalog (Mazets and Golonetskii, 1981).[57] The mean value $\langle V/V_{max} \rangle = 0.45 \pm 0.03$ is consistent with 0.5. Hence, there is no contradiction with the spatial distribution of these bursts. Schmidt *et al.* (1988)[55] also used this test on a small sample of 13 gamma-ray bursts and they found $\langle V/V_{max} \rangle = 0.40 \pm 0.08$.

For our VENERA (SIGNE) experiments, as well as for the APEX and LILAS experiments on the Soviet PHOBOS mission, the mean values $\langle V/V_{max} \rangle$ have also been calculated. The results are given in TABLE 2.[58] It is clear that the LILAS and VENERA-SIGNE results are more compatible with an isotropic source distribution. For APEX results, $\langle V/V_{max} \rangle$ is significantly lower than 0.5 and, interestingly, the same tendency is observed by SMM: 0.4 ± 0.025.[59] The two experiments, with their higher energy thresholds, are more sensitive to bursts that extend to higher energies, and we have some reason to believe that the luminosity function at low-energy thresholds (a few tens of kiloelectronvolts) and high-energy thresholds (a few 100 keV) are not the same.[58]

CONCLUSION

Although much work has already been done on gamma-ray bursts, no definitive conclusions can be reached concerning the following:

- Their temporal characteristics: fine structure on a very short timescale exists, but there is no strong evidence of periodicities.

- Their spectral characteristics: there are no unique spectral characteristics, but there are time-dependent spectra with spectral features in some cases at low energy (cyclotron lines that now seem well established) or higher energy (annihilation lines or possibly strongly varying spectra very often with a break around 500 keV), and observed independent behavior of high- (> 1-MeV) and low-energy emissions as a function of time.

There is no noncontroversial identification with any object at other wavelengths (except for GB790305 in the direction of N49). There is no conclusion on the spatial distribution, except to say that it still seems uniform. Therefore, galactic and extragalactic origins are both possible, even though the recent APEX PHOBOS and SIGNE results are in favor of a galactic distribution.[58] These results are important in the light of recent models by Paczynski (1990)[60] and Hartmann *et al.* (1990)[61] showing that the distribution of population I neutron stars is quite consistent with current observations of gamma-ray bursts.

To solve the problem of the origin of GRBs there are two paths that have to be explored simultaneously, the first one being crucial.

1. Identification and spatial distribution.
   (a) Identification of GRBs with known objects. HETE[62] will go in this direction by allowing simultaneous detections of GRBs with UV and X-ray observations. The localizations, which can be expected at the level of at least arc minutes, will be transmitted quickly to the ground with the possibility of pointing large telescopes in the direction of the identified burst with a short delay (a few minutes or tens of minutes in some cases).
   (b) To pursue the triangulation of a large number of bursts with ULYSSES[63] and other probes: Mars Observer Mission, MARS 94, PVO, and near-earth satellites such as GRO and GRANAT.
   (c) To localize less precisely as many bursts as possible with the BATSE experiment[64] on GRO to see if the spatial distribution is changing at low fluences.
2. Temporal spectral analysis.

To progress in the knowledge of the spectral characteristics of GRBs (short-time variability and spectral features), several missions exist or will be launched very soon. The PHEBUS experiment on GRANAT, GINGA, and soon, BATSE on GRO, will certainly make decisive contributions to the study of the short time variability of GRB spectra. Concerning gamma-ray lines, as mentioned, BATSE and PHEBUS will be very important in the near future. The germanium spectrometers aboard MOM, WIND, and SPECTRUM X will certainly provide definitive conclusions on the existence of these lines, if they are not greatly broadened. The presence of several similar detectors in space at the same time would be extremely useful, as this would inspire confidence in the reality of the detections.

REFERENCES

1. HURLEY, K. 1988. NATO Advanced Study Institute, Erice, Sicily, Italy.
2. HURLEY, K. 1989. Proceedings of the 14th Texas Symposium on Relativistic Astrophysics. Ann. N.Y. Acad. Sci.: 571–442.
3. LIANG, E. P. 1989. Proceedings of the GRO Science Workshop, Greenbelt, Md.

4. HURLEY, K. & D. HARTMANN. 1988. *In* Particle Astrophysics: Forefront Theoretical Issues, E. Norman, Ed.: 248. World Scientific. Singapore.
5. LAMB, D. Q. 1988. Proceedings of the AIP Conference, No. 170, N. Gehrels and G. Share, Eds.: 265.
6. LAMB, D. Q. 1983. Proceedings of the 11th Texas Symposium on Relativistic Astrophysics, D. Evans Ed. Ann. N.Y. Acad. Sci: 422,237.
7. MANDROU, P. 1984. *In* Advances in Space Research, G. F. Bignami and R. A. Sunyaev, Eds.: 3,525. (High Energy Astrophysics and Cosmology).
8. MAZETS, E. P., S. V. GOLENETSKII, YU. GUZYAN & V. N. ILYINSKI. 1981. A. F. Ioffe. Phys. Tech. Inst. Rep. 738.
9. NORRIS, J. P., T. L. CLINE, U. D. DESAI & J. B. TEEGARDEN. 1984. Nature **308:** 434.
10. BARAT, C., R. I. HAYLES, K. HURLEY, M. NIEL, G. VEDRENNE, I. V. ESTULIN & V. M. ZENCHENKO. 1984. Astrophys. J. **285:** 791.
11. CLINE, T., *et al.* 1980. Astrophys. J., Lett. **237:** L1.
12. BARAT, C., *et al.* 1984. Astrophys. J., Lett. **286:** L5.
13. KOUVELIOTOU, C., U. D. DESAI, T. L. CLINE, B. R. DENNIS, E. E. FENIMORE, R. W. KLEBESADEL & J. G. LAROS. 1988. Astrophys. J. **330:** 101.
14. WOOD, K. S., E. T. BYRAM, T. A. CHUBB, H. FRIEDMAN, J. F. MEEKINGS, G. H. SHARE & D. J. YENTIS. 1981. Astrophys. J. **247:** 632.
15. OWENS, A., D. BHATTACHARYA & S. SEMBAY. 1990. Astrophys. J. **352:** 741.
16. ATTEIA, J. L., *et al.* 1987. Astrophys. J., Lett. **320:** L105.
17. LAROS, J. G., *et al.* 1986. Nature **322:** 152.
18. LAROS, J., *et al.* 1987. Astrophys. J., Lett. **320:** L111.
19. MURAKAMI, T. 1990. *In* Proceedings of the XXVIII Cospar Meeting, G. Palumbo, L. Bassani, an G. Vedzenne, Eds. The Hague, the Netherlands.
20. LAROS, J. G., W. D. EVANS, E. E. FENIMORE, R. W. KLEBESADEL, S. SHULMAN & G. FRITZ. 1984. Astrophys. J. **286:** 681.
21. HARDING, A. K., V. PETROSIAN & B. J. TEEGARDEN. 1984. Proceedings of the AIP Conference, No. 141. P. Edison, G. Liang and P. Petrosian, Eds.: 77.
22. SHARE, G. H., D. C. MESSINA, A. JADICCO, S. M. MATZ, E. RIEGER & D. J. FORREST. 1990. Workshop on Cosmic Gamma-Ray Bursts, Taos, N. Mex., August.
23. MATZ, S. M., D. J. FORREST, W. I. VESTRAND, E. L. CHUPP, G. H. SHARE & E. RIEGER. 1985. Astrophys. J., Lett. **288:** L37.
24. CLINE, T. L. & U. DESAI. 1975. Astrophys. J., Lett. **196:** L43.
25. MAZETS, E. P., S. V. GOLENETSKII, R. L. APTEKAR, YU. A. GUR'YAN & V. N. IL'INSKII, 1981. Nature **290:** 378.
26. FENIMORE, E. E., J. G. LAROS, R. W. KLEBESADEL, R. E. STOCKDALE & S. KANE. 1981. Proceedings of the Workshop on Gamma Ray Transients and Related Astrophysical Phenomena, University of California San Diego, La Jolla, Calif.
27. FENIMORE, E. E., R. W. KLEBESADEL & J. G. LAROS. 1982. Proceedings of the XXIV COSPAR Meeting, Ottawa, Canada, May.
28. HUETER, G. J. & D. E. GRUBER. 1982. Accreting Neutron Stars MPE Rep. 177, N. Brinkmann and J. Trumper, Eds.
29. MURAKAMI, T., *et al.* 1988. Nature **335:** 234.
30. FENIMORE, E. E., *et al.* 1988. Astrophys. J., Lett. **335:** L71.
31. BARAT, C., R. TALON, R. SUNYAEV, N. BLINOV, A. KUZNETSOV & O. TEREKHOV. 1990. Workshop on Cosmic Gamma-Ray Bursts, Taos, N. Mex., August.
32. FENIMORE, E. E., R. W. KLEBESADEL, J. G. LAROS, R. E. STOCKDALE & S. R. KANE. 1982. Nature **297:** 665.
33. NOLAN, P. L., G. H. SHARE, E. L. CHUPP, D. J. FORREST & S. M. MATZ. 1984. Nature **311:** 360.
34. JOURDAIN, E. 1990. Thèse Doctorat, Toulouse, France.
35. ATTEIA, J. L., *et al.* 1991. Astron. Astrophys. **244:** 363–366.
36. MITROFANOV, I., *et al.* 1991. Planet. Space Sci. **39:** 23–37.
37. BARAT, C., G. CHAMBON, K. HURLEY, M. NIEL, G. VEDRENNE, I. V. ESTULIN, A. V. KUZNETSOV & V. M. ZENCHENKO. 1981. Astrophys. Space Sci. **75:** 83.
38. JOURDAIN, E., *et al.* 1990. Workshop on Cosmic Gamma-Ray Bursts, Taos, N. Mex., August.
39. GOLONETSKI, S., V. ILYINSKII & E. MAZETS. 1984. Nature **307:** 41.

40. NORRIS, J. P., G. H. SHARE, D. C. MESSINA, B. R. DENNIS, U. D. DESAI, T. L. CLINE, S. M. MATZ & E. L. CHUPP. 1986. Astrophys. J. **301:** 213.
41. PIZZICHINI, G., B. SCHAEFER & K. HURLEY. 1984. Proceedings of the AIP Conference, No. 141, P. Edison, G. Liang, and V. Petrosian, Eds.: 39.
42. SCHAEFER, B. 1984. Proceedings of the AIP Conference, No. 141, P. Edison, G. Liang, and V. Petrosian, Eds.: 47.
43. HUDEC, R. 1990. Workshop on Cosmic Gamma-Ray Bursts, Taos, N. Mex., August.
44. PIZZICHINI, G., *et al.* 1986. Astrophys. J. **301:** 641.
45. BOER, M., *et al.* 1988. Astron. Astrophys. **202:** 117.
46. GRINDLAY, J., *et al.* 1982. Nature **300:** 730.
47. MAZETS, E. P., *et al.* 1981. Astrophys. Space Sci. **80:** 3.
48. GOLENETSKII, S. V., Y. A. GURYAN, G. B. DUMOV, A. V. DYATCHOV, V. N. PANOV, N. G. KHAVENSON & L. O. SHESHIN. 1986. Preprint 1026.
49. ATTEIA, J. L., *et al.* 1987. Astrophys. J., Suppl. Ser. **64:** 305.
50. HARTMANN, D. & G. R. BLUMENTHAL. 1989. Astrophys. J. **342:** 521.
51. HARTMANN, D. & R. I. EPSTEIN. 1989. Astrophys. J. **346:** 960.
52. HIGDON, J. C. & R. E. LINGENFELTER. 1986. Astrophys. J. **307:** 197.
53. PACZYNSKI, B. & K. LONG. 1988. Astrophys. J. **333:** 694.
54. SCHMIDT, M. 1968. Astrophys. J. **151:** 353.
55. SCHMIDT, M., J. C. HIGDON & G. HUETER. 1988. Astrophys. J., Lett. **329:** L85.
56. HIGDON, J. C., & M. SCHMIDT. 1990. Astrophys. J. **355:** 13.
57. MAZETS, E. P. & S. V. GOLONETSKII. 1981. Appl. Space Sci. Rev. **1:** 205.
58. ATTEIA, J. L., *et al.* 1991. Nature **351:** 296.
59. HIGDON, J. C., S. M. MATZ, G. H. SHARE, D. MESSINA & A. IADICCIO. 1990. Workshop on Cosmic Gamma-Ray Bursts, Taos, N. Mex., August.
60. PACZINSKI, B. 1990. Astrophys. J. **348:** 485.
61. HARTMANN, D., R. I. EPSTEIN & S. E. WOOSLEY. 1990. Astrophys. J. **348:** 625.
62. RICKER, G., *et al.* 1988. Proceedings of the AIP Conference, No. 170, N. Gehrels and G. Share, Eds.: 407.
63. COTIN, F., *et al.* 1983. ESA Special Publication SP 1050.
64. FISHMAN, G., C. MEEGAN, T. PARNELL, R. WILSON & W. PACIESAS. 1984. Proceedings of the AIP Conference, No. 115: 651.

Gamma-ray Burst Theory

DIETER HARTMANN
Department of Physics and Astronomy
Clemson University
Clemson, South Carolina 29634

CASTLES OF KHASHABRIZ

Since the discovery of the first cosmic γ-ray burst in 1967 (Klebesadel, Strong, and Olson 1973; Klebesadel 1988), much work has gone into detailed simulations of various theoretical aspects of these bursts, but a generally accepted theory has yet to emerge. At times, it seems as if our attempts to unravel this mystery are doomed to failure just like those of scientists in the city of Khashabriz, where the nature of the castles and pilasters that surround and confine the city cannot be revealed (Lightman 1984). In our community, as in theirs, some have ventured toward that outer fortress, only to find the castles receding in step, discouraging further exploration. Perhaps some of us have been left content with isolation and lack of progress. Over the years, a truly impressive number of scenarios has been developed to explain γ-ray bursts. The number of theories almost equals the number of people who discuss them at frequent society meetings. A small group of Khashabriz scientists, known for their detachment, have proposed that the castles on the horizon are merely a mirage. They say that irregularities in the atmosphere cause light rays to bend, that the air can act as a misshapen lens, distorting some images and creating others. Could this be true for γ-ray bursts as well? Indeed, all efforts to identify γ-ray burst counterparts in other wavelength bands, quiescent or transient, have not been successful. Perhaps some misshapen cosmic lens has led us to search in the wrong place? Lacking counterparts, we turn to global statistical properties of γ-ray bursts to determine their distance scale. The observed sky distribution is isotropic, which does not reveal any obvious length scale. We either sample a very local part of a galactic population, or the events are of extragalactic or even cosmological origin. In the latter case, gravitational lensing might, in fact, have caused enough distortion to mislead us in the counterpart search. Few believe this explanation, but the data do not allow us to exclude this possibility. Other groups have quietly abandoned their unpopular theories, without proof or disproof. Others have become philosophical about the quest and decided to wait for new data to shed light on the burster mystery. In fact, recent observations with the Japanese Ginga satellite have led to a growing party of believers in the Galactic Magnetic Neutron Star paradigm.

REVIEW REVIEW

Soon after the discovery of γ-ray bursts Ruderman (1975) presented the first comprehensive review of the observations and the already large number of theoretical models proposed at that time. Most of the scenarios discussed in that first review are still with us today, but many more have been developed since then and discussions of them can be found in a series of reviews. Here we mention a few recent review articles that should provide the reader with an up-to-date overview of most aspects of the γ-ray burst phenomenon (Lamb 1984; Liang and Petrosian 1986; Taam 1987; Hartmann and Woosley 1988; Hurley 1989; Higdon and Lingenfelter 1990; Hurley and Lamb 1991; Harding 1991). In this review we do not consider the possible new class of soft repeaters, but focus our attention on *classical bursts* that apparently do not repeat on timescales shorter than about 10 years. Classical bursts account for the majority of observed events. We also do not discuss theoretical models designed to describe particular bursts, such as the famous March 5 event, or those scenarios for which insufficient detail does not allow a meaningful comparison with data.

NEUTRON STAR PARADIGM

In the conclusion of his review, Ruderman (1975) points out that all but one of the very many proposed models have in common that they will not be an explanation of γ-ray bursts. He did not identify the exception, but suggested an accretion-ridden black hole as his favorite in the race. Less than a decade later the fashion had shifted to neutron stars (e.g., Verter 1982) based on the observational facts of rapid variability, lack of counterparts, and spectral line features at energies below 100 keV (interpreted as cyclotron resonances in strong magnetic fields) and at energies around 450 keV (interpreted as redshifted $e^{\pm}$ annihilation lines). The arguably strongest observational support for the neutron star hypothesis has recently been obtained from the Ginga detection of double cyclotron lines in the spectra of three bursts (Murakami *et al.* 1988; Fenimore *et al.* 1988; Yoshida *et al.* 1991) and the possible identification of triple harmonic structure in the low-energy spectrum of GB880806 observed with the LILAS experiment aboard the Phobos mission (Barat *et al.* 1991). The observed harmonic structure strongly suggests that these features are caused by resonant scattering in fields of a few times 10^{12} gauss. Some potential difficulties with strongly magnetized neutron stars were recognized later. In particular, high-energy photon degradation by pair creation in these strong fields appears to be in conflict with the substantial emission above several megaelectronvolts observed by SMM (Matz *et al.* 1985). Another strong constraint on the emission mechanism and site stems from the observed paucity of X-ray emission, because it is hard to find γ-ray emission models that do not simultaneously produce X-rays, or to devise geometries that avoid the reprocessing of γ-rays into lower energy bands (Imamura and Epstein 1987). Also, the general lack of periodic modulation and the apparent lack of magnetic field decay appear to argue against the paradigm. However, for most of these "problems" remedies have been proposed, which has allowed neutron star models to maintain their lead in the race. However, it may well happen that more than one jockey riding the neutron star horse will cross the finish line.

Assuming that γ-ray bursts indeed originate on or near galactic neutron stars, any viable model attempting to represent the bulk of the observed events must allow for repetition. A rather optimistic estimate of the total number of neutron stars in our galaxy is $N_* \sim 10^{10}$, so that the observed event rate, $R_\gamma \sim 100$/yr, and the mean age of galactic neutron stars ($\sim 5 \times 10^9$ yr) imply that each star produced ~ 50 bursts over its lifetime. Of course, there are probably fewer neutron stars in the Galaxy, and we are definitely observing only a small fraction of the whole population. Also, not every neutron star will evolve into a burster source. Thus, the total number of events is probably a much larger number. If we sample about 0.1 percent of the neutron star volume and assume a total of 10^9 stars, about 10^6 events per star are needed for the current event rate. There are many alternative γ-ray burst models that invoke neutron stars and the large dynamic range observed for most burst properties seems to suggest that in fact several of these could be responsible for the observed variety of bursts. However, even with multiple mechanisms at our disposal, recurrence is an essential feature of models for classical bursts. If the duration of the burst activity phase is comparable to the mean age of galactic neutron stars ($\sim 5\ 10^9$ yr), the question arises whether the observed cyclotron lines are in conflict with the standard assertion of rapid magnetic field decay in neutron stars. Neutron stars residing in external galaxies could explain the observations even if a γ-ray burst is a rare or even singular event in a neutron star's life. In that case, the implied large distances require burst mechanisms that are rather different from those invoked for galactic scenarios. Extragalactic or even cosmological models that do not involve neutron stars (e.g., Babul, Paczynski, and Spergel 1987; Paczynski 1988; Zdziarski, Svensson, and Paczynski 1991) remain strong contenders in the race, although none of these models provide a satisfactory explanation of the observed low-energy lines. However, direct observational evidence for strong magnetic fields through these features is only available for a small fraction of all recorded events, so that it is reasonable to consider the possibility that perhaps *some* γ-ray bursts originate at large cosmological redshifts (e.g., Zdziarski, Svensson, and Paczynski 1991).

THE GREAT DEBATE

The distances to individual γ-ray bursts are not known, because no source has yet been unambiguously associated with any quiescent object that would have allowed a distance determination by standard astronomical techniques. The possible association of the March 5 transient with a supernova remnant in the Large Magellanic Cloud (LMC) is ignored here. Searches for counterparts are hampered by the lack of sufficiently accurate localizations and by a lack of criteria that uniquely identify the expected appearance of counterparts. The task is far worse than the needle in the haystack problem, because at least we know all about the appearance of needles. The detection of archival optical transients (OTs) inside γ-ray burst error boxes generates hope for a breakthrough in the γ-ray burst counterpart search (Schaefer 1990, and references therein), but no counterpart of these OTs has been identified either. Recently, the reality of archival OT images has been questioned (Zytkow 1990; Greiner 1991a; but see also Schaefer 1990). Recent work by Greiner (1991b), who analyzed the original plates of the Zytkow–Schaefer debate, seems to support the view that most, if not all, of the OT images are due to plate faults.

Lacking reliable counterpart identifications, we thus turn to statistical properties of γ-ray bursts to derive a rough distance scale (Epstein and Hurley 1988). The angular distribution of bursts on the sky has been analyzed by using either a low-order multipole expansion (Hartmann and Epstein 1989) or the two-point correlation function (Hartmann and Blumenthal 1989; Hartmann, Linder, and Blumenthal 1991). No anisotropies or angular clustering is evident in the data. If we are dealing with a galactic distribution of neutron stars, then this lack of deviations from isotropy constrains the sampled volume to a small fraction of the Galaxy. Detailed simulations of the kinematics and spatial distribution of old neutron stars in our galaxy (Hartmann, Epstein, and Woosley 1990; Paczynski 1990a), using initial conditions obtained from radio pulsar observations, suggest that the sampling depth of current detectors should not exceed ~2 kpc. An extended halo population with scale of ~100 kpc might still be consistent with the data, but no calculations have been carried out. A minimum sampling depth of ~150 pc is required to satisfy the observed burst frequency together with a limit of ~10 yr on the event recurrence timescale. If γ-ray bursts trace the large-scale matter distribution in the universe, then the observed lack of angular correlation can also be used to derive a minimum sampling depth for such cosmological models (Hartmann and Blumenthal 1989). If the underlying spatial clustering resembles that of galaxies or galaxy clusters, we must observe events to distances in excess of ~100 Mpc. This scale might not be appropriate if bursts trace dark matter with unknown clustering properties. A recent optical constraint on extragalactic burst models (Schaefer 1991), based on the absence of bright galaxies inside γ-ray burst error boxes, implies even larger distances of ~1 Gpc, but is confined to models in which bursts would occur in M31-like galaxies (the majority of galaxies is much fainter than M31, and the preceding constraint is thus not too realistic). From the angular distribution statistics we conclude that bursts are either very close and galactic (kiloparsec scale), reside in an extended halo (100 kpc), or are extragalactic (100 Mpc), or possibly cosmological. Only a few intermediate-distance scales, such as those comparable to the distance to the galactic center, appear to be ruled out by the observed sky distribution.

In addition to the angular tests, the burster brightness distribution can be used to constrain certain spatial distribution models. Recent progress in this analysis is due to the introduction of the so-called V/V_{max} test (Schmidt, Higdon, and Hueter 1988). The statistic of γ-ray burst light curves obtained with this tool removes all selection effects due to uncertainties in the spectral deconvolution process and allows an individual treatment of events that takes variations in detector backgrounds properly into account. The V/V_{max} test has now been applied to the data sets of HEAO-A1 (Schmidt, Higdon, and Hueter 1988), KONUS on Venera 11/12 (Higdon and Schmidt 1990), SMM (Matz *et al.* 1991), SIGNE on Venera 11/12 (Atteia *et al.* 1991), the Lilas/Apex experiments on the Soviet Phobos mission (Atteia *et al.* 1991), and the PVO burst detector (Hartmann *et al.* 1991). If the observed bursts are uniformly distributed in Euklidean space, then a uniform V/V_{max} distribution with mean value 1/2 should be obtained. The data suggest that γ-ray burst space distribution is slightly skewed toward higher space densities near the solar position, that is, a mean V/V_{max} less than 1/2. This is consistent with expectations for galactic neutron stars (Hartmann, Epstein, and Woosley 1990; Paczynski 1990a), but it certainly does not prove the neutron star paradigm or a galactic origin of γ-ray bursts.

ENERGY BUDGET

If γ-ray bursts originate from nearby sources, a moderate flux of 10^{-4} erg/cm^2 s observed at earth requires a γ-ray luminosity of 10^{40} D^2(kpc) erg/s. A source at 100 pc, emitting isotropically, would thus exceed the Eddington luminosity for a standard neutron star. From the statistical estimate of maximum sampling depth for galactic neutron stars, one infers a maximum recurrence time of $\sim 10^3 fN_8$ yr (Hartmann, Epstein, and Woosley 1990), where $fN_8 10^8$ is the total number of neutron stars active as γ-ray burst sources. Thus, a neutron star produces typically more than $\sim 10^6$ events during its lifetime. The corresponding requirement for the total energy budget is thus $\sim 10^{46}$ erg. It is not too difficult to imagine scenarios that generate the right amount of energy for a single event, but it is much harder to find a burst mechanism that allows storage of 10^{46} erg over sufficiently long times and the occasional release of just the right energy installments. This point was recently emphasized by Blaes *et al.* (1989), who reassessed neutron starquake models for γ-ray bursts. This global energy crisis argues in favor of multiple mechanisms for γ-ray bursts, consistent with the observed variety of their various properties. The energy for a single galactic event ($\sim 10^{40}$ erg) may be generated by one of several mechanisms.

1. Thermonuclear explosions of matter slowly accreted from a companion star (or from the interstellar medium (ISM) if the neutron star velocity is small and/or the medium density is high). This model has recently been reviewed by Taam (1987) and Hameury and Lasota (1986). There might be a problem for this model related to a partial or complete suppression of the accretion rate due to radiation pressure from the neutron star (Kluzniak 1991). With an energy release efficiency of $Q_{nuc} \sim 1$ MeV per particle one must accrete about 10^{22} g of matter to explain a typical burst fluence at a distance of 1 kpc. Over the neutron star lifetime about 5×10^{-6} solar masses are accumulated (little of that amount should be ejected because $Q_{nuc} \ll Q_{grav} \sim 100$ MeV per particle). The recurrence time would thus be about 5×10^3 yr and the overall average accretion rate is 10^{-15} solar masses per year. Current limits on quiescent X-ray emission from γ-ray error boxes (Boer *et al.* 1991, and references therein) are beginning to constrain these rates for nearby sources (~ 200 pc) if the accretion flow is confined by strong magnetic fields to small polar-cap areas of ~ 1 km^2. These parameters resemble those of the Type II "detonation model" of Woosley and Wallace (1982), which considers burning of low-metallicity gas due to either accretion from a Population II companion star or due to gravitational settling of heavy elements (Wallace, Woosley, and Weaver 1982). Alternatively, the accretion process itself could reduce the metallicity of accreted matter due to the spallation breakup of CNO isotopes by fast protons that pass through CNO-rich layers that have already settled due to their larger coulomb stopping power (Bildsten 1991).

2. A Rayleigh Tayler instability occurring after the buildup of a density inversion due to pycnonuclear burning (cold fusion) of a very slowly accreting neutron star (Blaes *et al.* 1990). The calculated long recurrence time for this model ($\sim 10^9$ yr) argues against this scenario as a model for the bulk of γ-ray bursts, even if one considers some modifications due to improved reaction rate estimates or the possibility of localized instabilities. Furthermore, on such long timescales neutron stars have orbited the Galaxy many times, and have thus gone through phases of much larger accretion rates than those considered in this model.

3. Sudden accretion models (comet or asteroid impacts or accretion instabilities) have recently received enhanced attention due to new developments in our understanding of comet clouds (Mitrofanov and Sagdeev 1990; Pineault 1990, 1991; Colgate 1991). Observations of comet Halley by the Vega spacecraft (Sagdeev *et al.*

1986) have shown that the albedo of cometary material is smaller than commonly assumed, which implies that there must be a larger number of comets than previously estimated. This in turn implies a larger event rate or shorter recurrence timescale (Mitrofanov and Sagdeev 1990). The low probability of comet accretion has always been a strong argument against this class of models (e.g., Liang and Petrosian 1986). This model also depends sensitively on the poorly known distribution of comet masses (Bailey 1990). Again the observations of comet Halley seem to suggest more favorable values ($\sim 10^{18}$ g) than previously assumed (Sagdeev *et al.* 1986). With these new values comet impact on neutron stars can explain the energetics of γ-ray bursts for distances less than ~ 1 kpc, but whether the observed rates can be obtained solely from these interactions remains very uncertain (Pineault 1990). The possibility that comets may act as "matches" to ignite abundant latent pulsars (Ruderman and Cheng 1988) could provide the missing boost. Another unsolved question for this class of models is whether the hard spectra of γ-ray bursts can be obtained with comet impacts (Katz 1986). However, including white dwarfs in this scenario, the model might more naturally apply to soft repeaters and OT sources.

4. Star quake models have been significantly refined in recent years (Muslimov and Tsygan 1986; Epstein 1988; Blaes *et al.* 1989) due to an improved understanding of the physics of neutron star interiors as well as the coupling between the superfluid core and the neutron star crust. The development of starquake models has greatly benefited from observations of energetic period transients ("glitches") in the Vela radio pulsar that have led to a new theoretical understanding of core–crust coupling involving angular momentum transfer from the free neutron gas in the inner parts of the crust to the outer crust (Pines *et al.* 1980; Sauls 1991). Sudden angular momentum loss of the superfluid neutrons might involve a catastrophic unpinning of vortex lines if the velocity shear between crust and superfluid reaches a critical value or a displacement of vortex lines (without unpinning) caused by cracking of the crust due to the stresses on the crust from the pinned vortex lines. Due to large uncertainties in the microphysics, it is not clear whether stress release occurs through brittle fracture or plastic flow (Ruderman 1991). The maximum amount of elastic energy that the crust can store is $\sim 10^{44}$ erg, clearly enough for a single burst but insufficient to satisfy the global energy requirements (Blaes *et al.* 1989). In this model, the formation of a γ-ray burst is believed to be related to the seismic excitation of magnetic field oscillations that transfer the energy to the magnetosphere where particle acceleration can occur, which in turn radiate γ-rays. Blaes *et al.* (1989) estimate that about 1 percent of the total energy released in the quake can be transferred to γ-rays on timescales consistent with typical burst durations. It is not clear whether the remaining energy can be prevented from producing too many X-rays. All models based on transient magnetospheric instabilities, whether caused by interior disturbances (glitches) or exterior triggers (comet collisions) have as an important ingredient the acceleration of relativistic electrons by electric fields parallel to the magnetic field. After long periods of dormancy a neutron star might thus temporarily act like a rotation-powered pulsar, where the braking of the neutron star rotation provides the energy for steady emission. Spectral and other properties of such pulsarlike models have recently been discussed by a number of authors (Ruderman and Cheng 1988; Mitrofanov 1989; Harding, Sturrock, and Daugherty 1990; Melia 1990).

Sources of γ-ray bursts at cosmological distances were thought to be ruled out by their rapid time variability and large fluxes (e.g., Ruderman 1975; Taam 1987). The energy required for bursts at cosmological distances ($\sim 10^{51}$ erg for 100 Mpc) will, if released in a compact volume with radius r_i, lead to an optically thick pair dominated plasma-radiating blackbody radiation at a temperature determined from $L \sim 4\pi r_i^2 \sigma T^4$. Without confinement, this fireball will expand at relativistic speeds until it becomes optically thin when T drops below the pair creation limit $T_p \sim 2 \times 10^8$ K (Paczynski 1986; Goodman 1986). Considering the effects of relativistic motion on the received versus emitted radiation, Zdziarski *et al.* (1991) have shown that the observed variability timescales and fluxes are in fact compatible with cosmological source models as long as their redshifts are less than $\sim 10^2$–10^3 so that the opacity of the universe does not exceed unity (Babul, Paczynski, and Spergel 1987; Paczynski 1988; Zdziarski and Svensson 1989). The revival of cosmological models has led to renewed interest in the physics of lepton and/or photon "fireballs" that were originally proposed as a model for γ-ray bursts by Cavallo and Rees (1978). Sudden release of large amounts of radiative energy in a compact region could be accomplished in a number of different ways, including hypothetical neutron star deflagration (Michel 1988), coalescence of neutron star binaries (Eichler *et al.* 1989), and accretion-induced collapse (AIC) of a white dwarf to a "naked" neutron star (Dar and Ramaty 1990). The latter two mechanisms are accompanied by neutrino bursts. Annihilation of these neutrinos just outside the neutrinosphere creates a lepton pair fireball (Goodman, Dar, and Nussinov 1987) that might produce the γ-ray burst. The electromagnetic signal of this fireball is sensitive to the baryonic pollution in the resulting super-Eddington wind. Even when a small amount of baryonic material is present, most of the energy of the fireball will be converted to matter kinetic energy, greatly reducing the γ-ray appearance of cosmic fireballs (Shemi and Piran 1990). To generate cosmological γ-ray bursts, the intrinsic spectrum must be much more energetic than observed burst spectra to accommodate large redshifts. Paczynski (1990b) has shown that that this requires a relativistic wind. A detailed hydrodynamical AIC model by Woosley and Baron (Woosley 1991) has resolved a wind of about 0.002 solar mass per second. These authors find that the terminal wind velocity is only 0.1 c, so that the wind actually prevents the formation of sufficiently hard photons and reduces the overall luminosity to about 10^{40} erg. Cosmological AIC models of γ-ray bursts appear thus to be ruled out due to the accompanying mass loss. However, nonspherical fireballs (particle jets) could produce γ-ray bursts at large redshifts by Compton scattering of the cosmic background radiation even if the particle accelerator does not produce hard photons directly (Zdziarski, Svensson, and Paczynski 1991).

SPECTRAL CONSTRAINTS

In the previous section we have discussed the task of satisfying the global energy budget for various γ-ray burst models. It is an even greater challenge to explain why the burst energy is of so pure gamma-ray quality. Only ~ 2 percent of the total emitted energy emerges in the X-ray band, and the power spectrum peaks between 100 keV and 1 MeV. This paucity of X-rays places severe constraints on models that overproduce X-rays because of energy thermalization (Epstein 1986, 1989). This is of particular importance for those models in which the energy is derived from an "external" agent (cometary impacts, accretion instabilities, thermonuclear explosions, etc). However, the observed low values for L_x/L_γ can be accommodated if the interaction of these soft photons with the electrons in the neutron star "corona" acts as a filter. That Compton up-scattering in a coronal wind can very efficiently hide the X-rays was shown by Hameury and Lasota (1989) in the framework of the thermonuclear model. This mechanism could also alleviate the constraints on the emission site due to the reprocessing of γ-rays in the stellar surface (Imamura and Epstein 1987). The ability of inverse Compton (IC) scattering to suppress the soft photon flux has recently been used to construct spectral models based on the up-scattering of the thermal surface emission by relativistic beams of electrons (Ho and Epstein 1989; Dermer 1990).

A low-energy thermal afterglow (X-ray tail) following the main γ-ray emission for some time was identified as a signature of the thermonuclear model. Recent Ginga data have provided evidence for these tails (Yoshida *et al.* 1989, 1991; Murakami 1990). The X-ray/γ-ray flux ratios measured by Ginga at the peak of the γ-ray emission are consistent with the 2 percent ratios mentioned earlier. If the Ginga data indeed imply blackbody radiation, one infers a distance–size relationship of $R(\text{km}) \sim D(\text{kpc})$, consistent with polar-cap emission from galactic neutron stars at distances consistent with angular distribution constraints.

Another common spectral feature of γ-ray bursts appears to be high-energy power-law emission extending well beyond 1 MeV (Matz *et al.* 1985). If this spectral component originated close to the high field regions in which the cyclotron lines are formed, γB pair creation would prevent the emergence of these high-energy tails. Using a simplified geometry, Matz *et al.* (1985) derived a limit of $\sim 10^{11}$ G for the formation region of the high-energy emission. This seems to suggest a two-component model in which cyclotron lines and high-energy tails are produced in separate regions (e.g., Hartmann and Woosley 1988). Another solution would be to postulate a correlation between neutron star field strengths and the production of high-energy photons (only low field objects produce high-energy photon tails). However, a detailed treatment of radiation transport effects, including field-strength changes along a photon's path, show that the large percentage of SMM bursts with high-energy tails is actually consistent with neutron star surface fields as large as $\sim 10^{12}$ G (Meszaros, Bagoly, and Riffert 1989). Furthermore, beaming of γ-rays along field lines, a natural consequence of the electron jet models mentioned earlier, also solves this problem because of the strong dependence of the annihilation cross section on the angle between photon direction and the field vector.

The observation of cyclotron resonances in γ-ray burst spectra has led to a flurry of theoretical work on radiation processes and transport in strongly magnetized plasmas (see Harding 1991, and references therein). In particular, a detailed treatment of magnetic Compton scattering is essential for using the observed lines as a diagnostic tool. One of the perhaps surprising observational findings was the comparable strengths of the lines in light of a drastically reduced opacity with increasing harmonic number. The explanation of this effect is "spawning" of photons. Scattering of photons involving the excitation of electrons to higher Landau orbits leads to radiative de-excitation that (to first order) generates multiple photons at the fundamental frequency when the electron returns to the ground state in steps satisfying the selection rule $\Delta n = 1$. These photons "fill in" the absorption line caused by the $0 \rightarrow 0$ transition. Another surprise is the relatively small width of the lines in contrast to the temperature associated with the continuum. This reflects the fact that in the line-forming region the plasma temperature is driven to the equilibrium between heating and cooling as determined by the Compton resonant scattering of photons near the line center, where the optical depth is large compared to that of the continuum. This Compton equilibrium temperature is roughly given by $kT_c = 0.25\,\hbar\omega_c$ (Lamb *et al.* 1989; Lamb, Wang, and Wasserman 1990), which yields the observed FWHM of ~ 5 keV for a field of $\sim 2 \times 10^{12}$ G. Requiring that the line-dominated radiation force on the scattering layer be less than its gravitational binding constrains the distances of the Ginga events with line features to less than ~ 200 pc (Lamb, Wang, and Wasserman 1990). This limit is consistent with isotropy, but if the fraction of bursts with strong fields is closer to 100 percent, one might have a conflict with minimum distances required by burst statistics.

RELATIVISTIC BULK MOTION

In the previous paragraphs we have pointed out three fundamental "problems" of γ-ray burst theory: (1) the global energy crisis, (2) the lack of X-rays, and (3) the problem of getting high-energy photons out of strong magnetic field regions. Recently, Krolik and Pier (1991) have presented arguments in favor of relativistic bulk motion (RBM) in γ-ray bursts as a general feature of classical bursts that solves or eases all three problems simultaneously. The CUSP models for galactic neutron stars, as well as most of the proposed cosmological models, appear to be a special subclasses of the general models with RBM discussed by these authors. If RBM plays a major role in γ-ray burst physics, the energy requirements for a given burst fluence are roughly reduced by the square of the Lorentz factor of the flow. This effect could revive a number of models that have been abandoned because of insufficient energy supply. The effect of RBM to reduce the L_X/L_γ ratio and to decrease the γB pair creation opacity reduces the required γ-ray production efficiency and allows the origin of hard photons to be closer to the line-forming region. RBM might become yet another paradigm associated with the γ-ray burst phenomenon.

MAGNETIC FIELD DECAY

Direct measurements of neutron star field strengths have been obtained only for a few X-ray pulsars that exhibit cyclotron features in their spectra. Her X-1 is the prototype object of this kind. Typical values derived for these pulsars are $\sim 4 \times 10^{12}$ G. A larger set of indirect field measurements is derived from the observations of radio pulsars. Statistical analysis of the known ~ 450 pulsars seems to suggest a decay of the dipole surface fields of these pulsars on timescales as short as $\sim 10^7$ yr (Lyne 1991). A detailed analysis of the pulsar density contours in *B–P* space (Srinivasan 1991) possibly indicates two pulsar populations as an alternative in which rapid field decay need not occur. How do γ-ray bursts fit in this picture? The absence of a correlation between radio pulsars and burst error boxes implies a minimum delay of more than $\sim 10^7$ yr between pulsar formation and the start of burst activity. If burst activity develops shortly after the pulsar phase, the observed fraction of burst spectra with cyclotron features (~ 0.1) can be used to constrain field decay models. If the fields remain roughly constant, this fraction reflects the field distribution at birth. With decay, the age distribution of galactic neutron stars implies that exponential decay cannot occur on a timescale shorter than $\sim 10^9$ yr (Hartmann *et al.* 1991), unless birth fields are much larger than the critical field $B_{cr} = 4.414 \times 10^{13}$ G ($\hbar\omega_c = m_e c^2$). If decay proceeds more rapidly, no cyclotron features should have been observed in burst spectra. A plausible explanations for this inconsistency is that the field strength measured by cyclotron lines does not determine the dipole surface field, but only some localized high field region on the stellar surface. The field could be chaotic, concentrated in islands, or exhibiting high multipolar structure. In this scenario the average dipole component could have decayed to much smaller values. There are several proposals for the generation of such spots, ranging from the thermoelectric effect (Blandford, Applegate, and Hernquist 1983) to neutron star plate tectonics (Ruderman 1991). Alternatively, the burst active phase might be biased toward young neutron stars. In that case it is obviously easier to obtain a large fraction of bursts that exhibit cyclotron lines. However, to avoid the inevitable anisotropies associated with a young population, one would require a rather small sampling depth ($\ll 1$ kpc) of current detectors, which could conflict with the required minimum distance even if all neutron stars become burst sources.

A VERY IMPORTANT PROBLEM

After more than two decades of research no definite understanding of γ-ray bursts has emerged. A paradigm has been formed. At present, γ-ray bursts still share the label VIP (very interesting problem) with such phenomena as the solar neutrino problem, or the large-scale flow of galaxies (Trimble 1991). If the galactic neutron star origin is proved correct in the near future, our understanding of the γ-ray burst phenomenon will have followed an historic path similar to that of cosmic rays, Cepheid variables, and white dwarfs, where a basic understanding of the phenomenon did not follow immediately after the discovery, but was obtained after considerable delays (Trimble 1991). However, it is still possible that a final understanding of γ-ray bursts will require a change in the current paradigm. The BATSE experiment aboard NASA's forthcoming Gamma Ray Observatory (GRO) mission, as well as a number of additional recent and future spacecraft (Granat, Ulysses, Wind, Rosat, Phobos) will undoubtedly enhance our knowledge of γ-ray bursts and might even determine their spatial distribution. However, ultimately we must obtain unambiguous identification of burster counterparts to be able to fully investigate the nature of the objects responsible for γ-ray bursts. If burst localization techniques continue to yield error boxes devoid of obvious counterpart candidates, the multiwavelength transient experiment HETE (Ricker *et al.* 1988) appears to be our best bet for solving this VIP. The new decade of γ-ray astronomy might bring the long-sought explanation for cosmic γ-ray bursts, and if the paradigm turns out to be correct, a new era will begin in which the study of γ-ray bursts is used to advance our understanding of such exciting problems as quantum electrodynamics (QED) in strong magnetic fields and the evolution of neutron stars.

REFERENCES

ATTEIA, J.-L., *et al.* 1991. Nature **351:** 296.
BABUL, A., B. PACZYNSKI & D. SPERGEL. 1987. Astrophys. J., Lett. **316:** L49.
BAILEY, M. E. 1990. *In* Baryonic Dark Matter, D. Lynden-Bell and G. Gilmore, Eds.: 7. Kluwer. Dordrecht, the Netherlands.
BARAT, C., *et al.* 1991. *In* Gamma-Ray Bursts, C. Ho, Ed. Cambridge University Press. Cambridge, England.
BILDSTEN, L. 1991. *In* Gamma-Ray Line Astronomy, P. Durouchoux and N. Prantzos, Eds. AIP Conference Proceedings: **232:** 401.
BLANDFORD, R., J. H. APPLEGATE & L. HERNQUIST. 1983. Mon. Not. R. Astron. Soc. **204:** 1025.
BLAES, O., *et al.* 1989. Astrophys. J. **343:** 839.
BLAES, O., *et al.* 1990. Astrophys. J. **363:** 612.
BOER, M., *et al.* 1991. Appl. Space Sci. In press.
CAVALLO, G. & M. REES. 1978. Mon. Not. R. Astron. Soc. **183:** 359.
COLGATE, S. A. 1991. *In* Gamma-Ray Bursts, C. Ho, Ed. Cambridge University Press. Cambridge, England.
DAR, A. & R. RAMATY. 1990. Preprint.
DERMER, C. 1990. Astrophys. J. **360:** 197.
EICHLER, D., *et al.* 1989. Nature **340:** 126.
EPSTEIN, R. I. 1986. Lecture Notes in Physics **255:** 305.
EPSTEIN, R. I. 1988. Phys. Rep. **163:** 155.
EPSTEIN, R. I. 1989. *In* Cosmic Gamma Rays, Neutrinos, and Related Astrophysics, M. M. Shapiro and J. P. Wefel, Eds. In press.
EPSTEIN, R. I. & K. HURLEY. 1988. Astrophys. Lett. Commun. **27:** 229.
FENIMORE, E. E., *et al.* 1988. Astrophys. J., Lett. **335:** L71.
GOODMAN, J. 1986. Astrophys. J., Lett. **308:** L47.
GOODMAN, J., A. DAR & S. NUSSINOV. 1987. Astrophys. J., Lett. **314:** L7.
GREINER, J. 1991a. *In* Gamma-Ray Bursts, C. Ho, Ed. Cambridge University Press. Cambridge, England.
GREINER, J. 1991b. Preprint.
HAMEURY, J. M. & J. P. LASOTA. 1986. *In* Gamma-Ray Bursts, E. Liang and V. Petrosian, Eds.: 177. Am. Inst. Phys. Press. New York.
HAMEURY, J. M. & J. P. LASOTA. 1989. Astron. Astrophys. **211:** L15.
HARDING, A. K., P. A. STURROCK & J. K. DAUGHERTY. 1990. Preprint.
HARDING, A. K. 1991. Phys. Rep. **206:** 328.
HARTMANN, D. & S. E. WOOSLEY. 1988. *In* Multiwavelength Astrophysics, F. Cordova, Ed.: 189. Cambridge University Press. Cambridge, England.
HARTMANN, D. H. & R. I. EPSTEIN. 1989. Astrophys. J. **346:** 960.
HARTMANN, D. & G. R. BLUMENTHAL. 1989. Astrophys. J. **342:** 521.
HARTMANN, D., R. I. EPSTEIN & S. E. WOOSLEY. 1990. Astrophys. J. **348:** 625.
HARTMANN, D., E. V. LINDER & G. R. BLUMENTHAL. 1991. Astrophys. J. **367:** 186.
HARTMANN, D., *et al.* 1991. *In* Gamma-Ray Bursts, C. Ho, Ed. Cambridge University Press. Cambridge, England.
HIGDON, J. C. & M. SCHMIDT. 1990. Astrophys. J., Lett. **355:** L13.
HIGDON, J. C. & R. E. LINGENFELTER. 1990. ARA&A, **28:** 401.
HO, C. & R. I. EPSTEIN. 1989. Astrophys. J. **343:** 277.
HURLEY, K. 1989. Ann. N.Y. Acad. Sci. **571:** 442.
HURLEY, K. & D. Q. LAMB. 1991. Phys. Rep. In press.
IMAMURA, J. N. & R. I. EPSTEIN. 1987. Astrophys. J. **313:** 711.
KATZ, J. I. 1986. Astrophys. J. **309:** 253.
KLEBESADEL, R. W., I. B. STRONG & R. A. OLSON. 1973. Astrophys. J., Lett. **182:** L85.
KLEBESADEL, R. W. 1988. *In* Physics of Neutron Stars and Black Holes, Y. Tanaka, Ed.: 387. University Academy Press. Tokyo.
KLUZNIAK, W. 1991. X-ray binaries and their evolution. This issue.
KROLIK, J. H. & E. A. PIER. 1990. Submitted for publication in Astrophys. J.
LAMB, D. Q. 1984. Ann. N.Y. Acad. Sci. **422:** 237.
LAMB, D. Q., *et al.* 1989. Ann. N.Y. Acad. Sci. **571:** 460.
LAMB, D. Q., C. L. WANG & I. M. WASSERMAN. 1990. Astrophys. J. **363:** 670.

LIGHTMAN, A. P. 1984. *In* Time Travel Papa Joe's Pipe: 111. Penguin. Baltimore.
LIANG, E. P. & V. PETROSIAN. 1986. Gamma-Ray Bursts, Am. Inst. Phys. Publ. New York.
LYNE, A. 1991. Paper presented at the 15th Texas/ESO-CERN Symposium on Relativistic Astrophysics, Cosmology and Fundamental Physics, Brighton, England, December 16–21, 1990.
MATZ, S., *et al.* 1985. Astrophys. J., Lett. **288:** L37.
MATZ, S., *et al.* 1991. *In* Gamma-Ray Bursts, C. Ho, Ed. Cambridge University Press. Cambridge, England.
MELIA, F. 1990. Astrophys. J. **351:** 601.
MESZAROS, P., Z. BAGOLY & H. RIFFERT. 1989. Astrophys. J., Lett. **337:** L23.
MICHEL, C. 1988. Astrophys. J., Lett. **327:** L81.
MITROFANOV, I. G. 1989. Appl. Space Sci. **165:** 137.
MITROFANOV, I. G. & R. Z. SAGDEEV. 1990. Nature **344:** 313.
MURAKAMI, T., *et al.* 1988. Nature **335:** 234.
MURAKAMI, T. 1990. Adv. Space Res. **102**(2): 63.
MUSLIMOV, A. G. & A. I. TSYGAN. 1986. Appl. Space Sci. **120:** 27.
PACZYNSKI, B. 1986. Astrophys. J., Lett. **308:** L43.
PACZYNSKI, B. 1988. Astrophys. J. **335:** 525.
PACZYNSKI, B. 1990a. Astrophys. J. **348:** 485.
PACZYNSKI, B. 1990b. Astrophys. J. **363:** 218.
PINEAULT, S. 1990. Nature **345:** 233.
PINEAULT, S. 1991. *In* Gamma-Ray Bursts, C. Ho, Ed. Cambridge University Press. Cambridge, England.
PINES, D., J. SHAHAM, M. A. ALPAR & P. W. ANDERSON. 1980. Prog. Theor. Phys. Suppl. **69:** 376.
RICKER, G., *et al.* 1988. *In* Nuclear Spectroscopy of Astrophysical Sources, N. Gehrels and G. H. Share, Eds.: 407. Am. Inst. Phys. Press. New York.
RUDERMAN, M. 1975. Ann. N.Y. Acad. Sci. **262:** 165.
RUDERMAN, M. 1991. Astrophys. J. **366:** 261.
RUDERMAN, M. & K. S. CHENG. 1988. Astrophys. J. **335:** 306.
SAGDEEV, R. Z., *et al.* 1986. Nature **321:** 259.
SAULS, J. A. 1991. Paper presented at the 15th Texas/ESO-CERN Symposium on Relativistic Astrophysics, Cosmology and Fundamental Physics, Brighton, England, December 16–21, 1990.
SCHAEFER, B. E. 1990. Astrophys. J. **364:** 590.
SCHAEFER, B. E. 1991. *In* Gamma-Ray Bursts, C. Ho, Ed. Cambridge University Press. Cambridge, England.
SCHMIDT, M., J. C. HIGDON & G. HUETER. 1988. Astrophys. J., Lett. **329:** L85.
SHEMI, A. & T. PIRAN. 1990. Astrophys. J. **365:** L55.
SRINIVASAN, G. 1991. Evolution of the magnetic fields of neutron stars. This issue.
TAAM, R. E. 1987. 13th Texas Symposium, M. P. Ulmer, Ed.: 546. World Scientific. Singapore.
TRIMBLE, V. 1991. *In* Gamma-Ray Bursts, C. Ho, Ed. Cambridge University Press. Cambridge, England.
VERTER, F. 1982. Phys. Rep. **81:** 293.
WALLACE, R. K., S. E. WOOSLEY & T. A. WEAVER. 1982. Astrophys. J. **258:** 696.
WOOSLEY, S. E. & R. K. WALLACE. 1982. Astrophys. J. **258:** 716.
WOOSLEY, S. E. 1991. *In* Gamma Ray Line Astrophysics, N. Prantzos and Ph. Durouchoux, Eds. Am. Inst. Phys. Press. New York. In press.
YOSHIDA, A., *et al.* 1989. Publ. Astron. Soc. Japan **41:** 509.
YOSHIDA, A., *et al.* 1991. *In* Gamma-Ray Bursts, C. Ho, Ed. Cambridge University Press. Cambridge, England.
ZDZIARSKI, A. A. & R. SVENSSON. 1989. Astrophys. J. **344:** 551.
ZDZIARSKI, A. A., R. SVENSSON & B. PACZYNSKI. 1991. Astrophys. J. **366:** 343.
ZYTKOW, A. 1990. Astrophys. J. **359:** 138.

X-ray Binaries and Their Evolution[a]

WŁODZIMIERZ KLUŹNIAK[b]

Physics Department
Columbia University
New York, New York 10027

INTRODUCTION

Although existence of solitary neutron stars was anticipated by Landau and Baade, neutron stars were first discovered in binary systems transferring mass. The credit goes to R. Giacconi, H. Gursky, F. R. Paolini, and B. B. Rossi (1962) for gathering "Evidence for X-rays from sources outside the Solar System" and for refining in subsequent flights the position of the first source detected, Sco X-1; to Sandage *et al.* (1966) for identifying the optical counterpart of Sco X-1; and to Igor S. Shklovsky for deriving constraints on the physical parameters of the emitting region and for concluding that (Shklovsky 1967) "By all its characteristics this model, obtained . . . without any prior hypotheses about the nature of the source, corresponds to a neutron star in a state of accretion . . . [Sco X-1 is a] close binary . . . in which one of the components is a neutron star . . . [accreting at the rate of] $\sim 10^{-9}$ $M_\odot$/yr."

I will not have much to say about solitary neutron stars (whose radio signature was first seen by Jocelyn Bell in the year following Shklovsky's work) except to note that they cannot be accreting much matter. As pointed out recently by Blaes *et al.* (1990), solitary neutron stars cannot begin to accrete at a rate exceeding 10^{10} g/s until they spin down to periods greater than $\sim 10^3$ s. Work (in progress) by Ruderman, Kluźniak, and Bhattacharya confirms this conclusion: solitary neutron stars of canonical 10^{12}-G magnetic dipole field and spin period $P \sim 10$ s can undergo accretion at most at the rate of $\sim 10^6$ g/s regardless of the ambient density of the interstellar medium (placing the star in a dense molecular cloud does not help). It is not clear whether such stars can ever slow down to the periods $> 10^3$ s required for accretion to reach levels interesting for thermonuclear (or pycnonuclear) models of gamma-ray bursts discussed in this meeting by Hartmann (1991).

Today it is clear that bright galactic X-ray sources are binary (or triple?) systems in which mass transfer occurs from the companion to a compact star, usually a neutron star. Observations and theories of X-ray binaries were recently reviewed in Lewin and van den Heuvel 1983, White 1989, Canal, Isern, and Labay 1990.

Massive companions have a life of their own. For example, O stars have a strong wind, while Be stars probably have an excretion disk. A compact object orbiting such a companion may easily capture and accrete part of the outflowing matter. In other cases, the companion expands on a (fairly short) nuclear timescale, and may spill matter from its Roche lobe into the lobe of the companion. In these high-mass

[a]This work was supported in part by National Science Foundation Grant AST 86-02831.
[b]Visiting the Astrophysics Department, Oxford University, Oxford, England.

systems, the compact object is usually a strongly magnetized neutron star (X-ray pulsar) or a black hole candidate.

LOW MASS X-RAY BINARIES

I will concentrate on low-mass X-ray binaries (LMXBs)—systems composed of a neutron star accreting matter from a low-mass companion, most often a late-type main-sequence star or a degenerate dwarf. With the exception of a couple of sources the neutron stars in those systems never show regular pulsations (unlike the X-ray

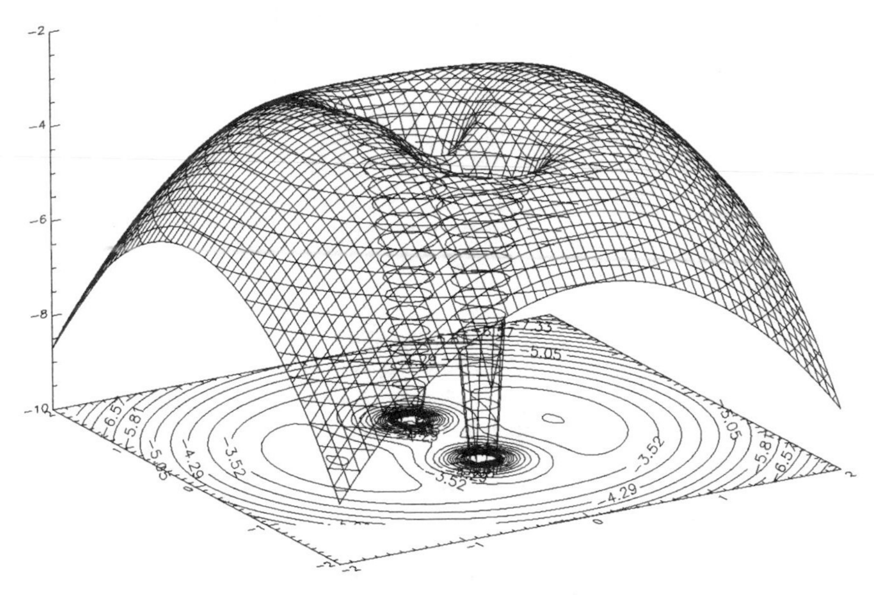

FIGURE 1. The "three-dimensional surface" is a mapping of the equatorial plane of a binary system into a vertical axis by the Jacobi integral of motion. The two stars have equal masses, as can be seen from the symmetry of the surface. Roche lobe overflow can be visualized by filling one of the wells with liquid up to its brim (known as the Lagrangian point L1).

pulsars). Many sources exhibit X-ray bursts, and *none* of these burst sources is an X-ray pulsar. It is believed that in all X-ray bursters and in the majority of the remaining LMXBs the strength of the dipole magnetic field of the neutron star does not exceed $\sim 10^9$ G or perhaps 10^{10} G. I stress that this is an upper limit, the actual value may be significantly lower (no lower limit exists to the field value in LMXBs).

FIGURE 1 and the following figures are graphic representations of the Jacobi integral of motion for the restricted three-body problem (e.g., Danby 1962). The *x-y* plane is the orbital plane in a (corotating) frame in which both components of the binary system are at rest on one of the two axes, separated by one (dimensionless) unit of distance. The origin (0, 0) is at the center of mass of the system.[c] Several contours of the Jacobi integral are shown in the orbital plane. The "three-dimensional" surface is the Jacobi integral restricted to the orbital plane, in the sense of a mapping of the orbital plane into the real axis (in short, each contour has been raised to an appropriate level).

In the hydrostatic case, the contours can be thought of as corresponding to equipotential surfaces (to be precise, the intersection of the latter with the orbital plane). The direction and magnitude of tidal forces can then be directly gleaned from the figures by relying on our everyday experience of motion under the influence of gravity on the slopes of hills and valleys known to us. This intuitive understanding is valid to some extent also for fluids or particles moving with respect to the two binary stars, but in that case—as is well known—the Coriolis force must be included. Furthermore, in this representation angular momentum has no simple interpretation. For rapid motion (velocities comparable to orbital velocity) all but one of the contours lose their utility. In such cases all that is known is that a certain zone is excluded; the motion in the allowed zone cannot then be obtained by looking at the remaining contours, but must instead be computed (Danby 1962).

FIGURE 1 illustrates a binary system in which both companions have equal masses. It is immediately apparent that no static configuration is possible when one and only one of the "wells" is filled with fluid above the level of the saddlepoint. The star extending past this point (the "inner Lagrangian point L1") must spill matter onto the other star (later I will discuss how this conclusion is modified when radiation pressure from the compact star is taken into account).

For many years it was thought that such Roche lobe overflow is the only avenue of mass transfer open to LMXBs. In those systems the timescale for evolution of the companion is longer than that for orbital period change due to angular momentum losses from the system, and no large *intrinsic* mass loss from the (late-type) companion is thought to occur. It seemed clear that mass transfer could occur in such systems only when the Roche lobe contracted to the size of the companion star. In our parlance, this would occur only when one of the "potential wells" was filled to the brim with fluid (FIG. 2).

A novel possibility was suggested by the Columbia group (Kluźniak *et al.*, 1988b) to explain the past evolutionary history[d] of the eclipsing pulsar system 1957+20, in which the $0.02M_\odot$ companion is deep within its Roche lobe. The answer to the question posed in the caption of FIGURE 2 is: use a siphon. Specifically, the idea (going back to Arons 1973 and Basko and Sunyaev 1973) of self-excited winds (a.k.a. bootstrapped evolution) in its Ruderman *et al.* 1989 incarnation was invoked. Radiation from an accreting neutron star impinging on the companion can drive an evaporative plume from its surface; part of that plume when captured and accreted by the neutron star gives rise to the radiation capable of raising the plume.

If it is indeed possible for mass transfer to proceed in LMXBs when the companion is underfilling its Roche lobe, many axioms of LMXB astrophysics should be reexamined. In particular, the probability of eclipses of the X-ray source would be

[c] The mass ratio of the two stars is therefore the inverse ratio of their distances to the origin.
[d] Other models have been advanced in which the companion had been ablated by the radio pulsar to its present mass (Phinney *et al.* 1988; van den Heuvel and van Paradijs 1988).

lower, limits on the inclination angle of any particular system derived from the lack of observed eclipses would be less stringent (as would therefore also be lower limits on the companion masses), the evolutionary timescales and the sign of the derivative of the orbital period could differ from those predicted by models assuming Roche lobe contact, etc. Regarding the last point, note that the orbital decay timescale has now been measured for a few systems and it disagrees with standard theory. For example, $P_{orb}/\dot{P}_{orb} = -1 \times 10^7$ y for the 11-minute binary 1820–30 (Tan *et al.* 1991) instead of the $+1 \times 10^7$ y predicted (Rappaport *et al.* 1987).

Even if one does not believe in the Columbia University siphon theory, effects of radiation from the neutron star on mass transfer are worth examining. To contrast the case already discussed with the findings below, the central zone of FIGURE 1 is given in FIGURE 3 (it has been rotated by 90°, and the vertical scale has been expanded to emphasize the shape). FIGURE 4(a), quite different to the eye, is for the mass ratio appropriate to the Terzan 5 eclipsing pulsar 1744−24A, $m_c = 0.089M_{\odot}$, $m_{ns} = 1.4M_{\odot}$. FIGURE 5(a) illustrates the same binary system, but now the gravitational pull exerted on the test particle by the neutron star has been reduced by one half. In an accreting system this is equivalent to including radiation pressure corresponding to one-half Eddington luminosity ($L_{Edd}/2$). In a radio pulsar system the same pressure can be exerted on ionized plasma at much lower luminosities of the neutron star. It is apparent from those figures, and the corresponding contours in FIGURES 4(b) and 5(b) that (optically thin) matter leaving the companion may find it easier to leave the system than to enter the "potential well" of the neutron star.

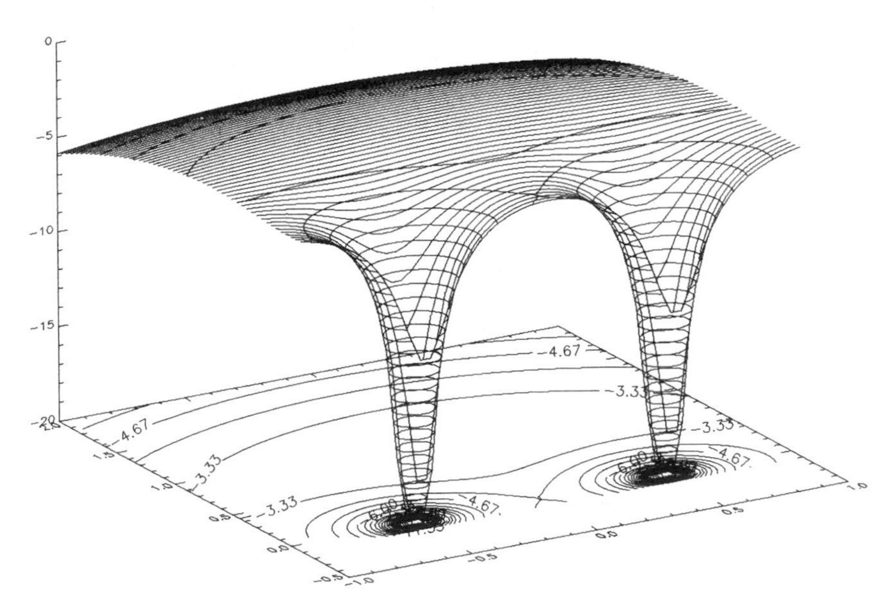

FIGURE 2. A cutaway view of the "equipotential wells" of FIGURE 1. To get hands-on experience in recent theories of mass transfer, the reader is invited to fill one of the wells with liquid up to a contour some distance *below* the saddle point (use a color pencil). How can this well be then emptied of its fluid? (Answer: .nohpis a esU)

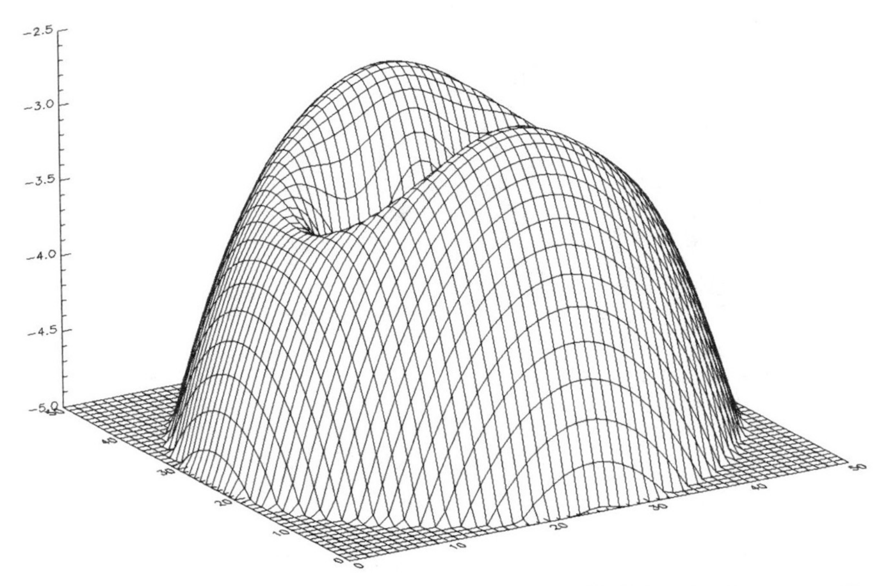

FIGURE 3. Detail of the central portion of FIGURE 1. The orbital plane has been rotated by 90°, and the vertical axis has been stretched to emphasize the maxima of the surface (Lagrangian points L4 and L5). (In this figure only, the x and y coordinates are arbitrary.)

Conservative evolution is therefore not very likely for LMXBs. For completeness, FIGURE 6 shows the Jacobi map when the compact object radiates X-rays at full Eddington luminosity. The neutron star is now located at the top of this helmet, and it is the location least likely to be visited by fluid gushing out from the companion's well. It is easy to imagine that the projection on the orbital plane of the stream of fluid flowing down the slope of this surface would resemble a cometary tail similar to the one envisioned for the eclipsing pulsar 1957+20 (Kluźniak *et al.* 1988a, 1988b; Phinney *et al.* 1988; Rasio, Shapiro, and Teukolsky 1989). However, the eclipses of 1744−24A are quite different (Lyne *et al.* 1990). It seems quite likely then that in that system the pulsar is quite feeble and motion is controlled by the surface depicted in FIGURE 4. The Lagrangian points L4 and L5 are shown in greater detail in FIGURE 4(c). Fluid particles of energy corresponding to the level of the grid plane shown would have to circumnavigate these twin peaks before leaving the system. If optically thick, this fluid would cause eclipses of length $\sim 1/2$ of orbital period, as observed.

NEUTRON STARS OR BLACK HOLES?

Two questions of great interest in this symposium is whether the compact object in bright X-ray binaries can be a black hole and how would we know it if it were. I think it is fairly clear that today the answer to the first question is an article of faith while the answer to the second is not at all straightforward. The best evidence for the presence of black holes in binary systems comes from the studies of the mass function

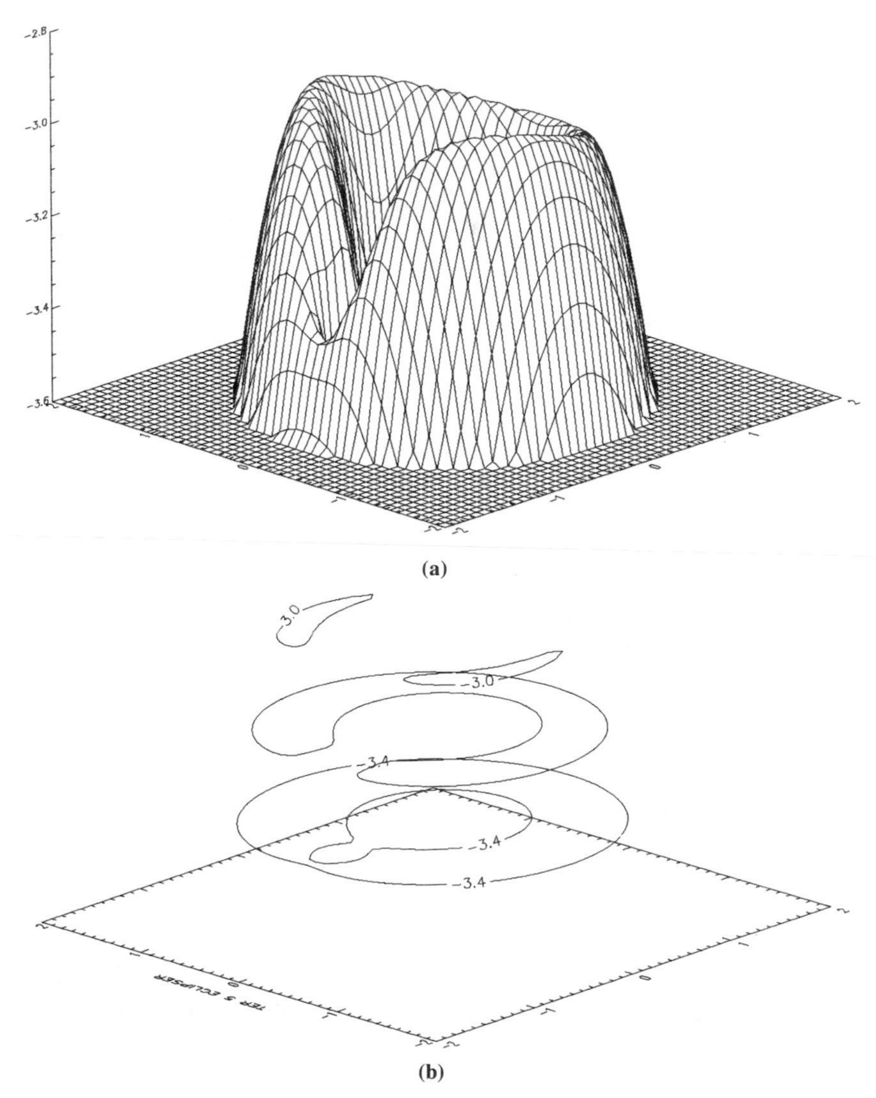

FIGURE 4. (**a**) The Jacobi integral in the orbital plane of a binary system with mass ratio of the two components equal to 0.063. This figure is the same as FIGURE 3, except for the mass ratio, which is 1.0 in that figure. Note the deep cleft in the cliffs surrounding the less massive star. (**b**) Some contours of the surface in part (**a**). Note that the inner contour at level -3.4 envelopes both stars. A particle of energy -3.4 can be captured by the more massive star, but cannot leave the system due to the potential barrier between the inner and outer contour at this level. (**c**) This is the same as part (**a**), but the vertical scale has been expanded to show Lagrangian points L4 and L5 in greater detail. As in part (**a**) and (**b**), the companion is located at $\approx(-1,0)$ at the mouth of Charybdis, while the neutron star is behind Scylla at $\approx(0,0)$. Fluid particles of energy -3.00 rolling on the horizontal grid from the companion toward the neutron star can leave the system only after they reach the opening on the other side of these two rocks. If the neutron star is a radio pulsar, it can be eclipsed for about one-half of the orbital period (as PSR1744−24A in the globular cluster Terzan 5 is) by fluid trying to cicumnavigate the tidal obstacles. (The jagged nature of the peaks is a numerical artifact added for greater effect.)

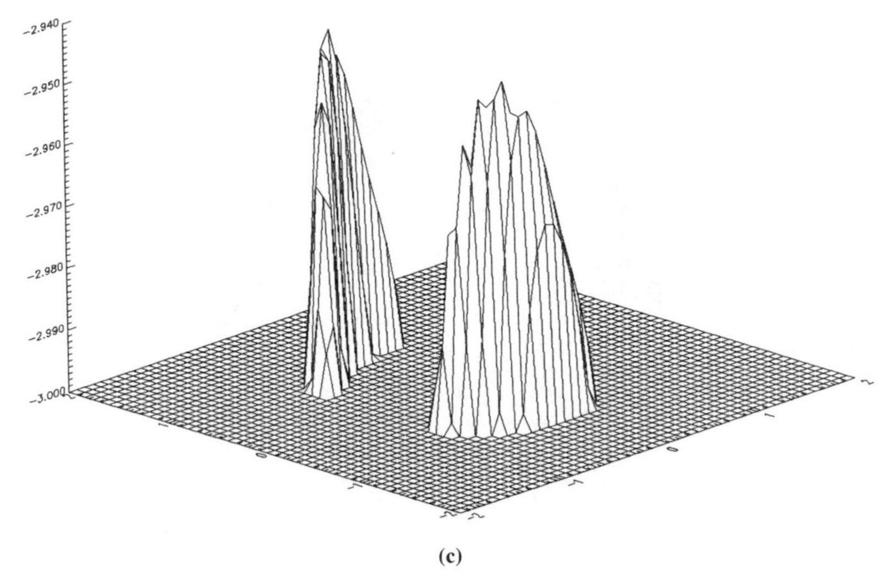

(c)

FIGURE 4.

and has been summarized in this session by J. McClintock (1991). The argument is that no stable stellar configurations above a certain mass are allowed by the theory of general relativity (GR) for a certain class of equations of state of nuclear matter. This means that even if one excludes stellar models based on nonconventional equations of state (such as those of Bahcall, Lynn, Selipsky 1990, for instance), one still needs to *assume* that GR is valid before one can conclude that a $4M_\odot$ or a $10M_\odot$ compact object is a black hole. Thus, we are at present unable to prove the existence of black holes. Instead, we are reduced to identifying black hole candidates (based on their mass and our belief in GR). This underscores the need for tests of GR predictions in the strong field limit (one such test has been proposed by Kluźniak, Michelson, and Wagoner 1990).

Temporal variability and spectral criteria are also commonly used for identifying candidate black hole candidates. At present, these seem to have no theoretical justification. The X-ray signatures are supposed to arise in an accretion disk around a black hole (more precisely, in a two-temperature disk or in a hot corona above a single-temperature disk). But no foolproof argument has been given to show that neutron star disks cannot have similar properties and signatures. To the contrary, it is known that many standard models allow neutron stars to fit comfortably within the smallest circular orbit allowed by GR, that is, inside standard black hole disks (Kluźniak and Wagoner 1985; Kluźniak, Michelson, and Wagoner 1990). X-rays from the neutron star will not necessarily Compton cool a postulated high-temperature corona. In fact, work by Kluźniak and Wilson (1989, 1991) shows that the expected surface emission from such a compact neutron star is quite hot; power-law spectra up to ~ 200 keV were obtained. These surface spectra are quite reminiscent of the hard component of spectra of transient black hole candidates

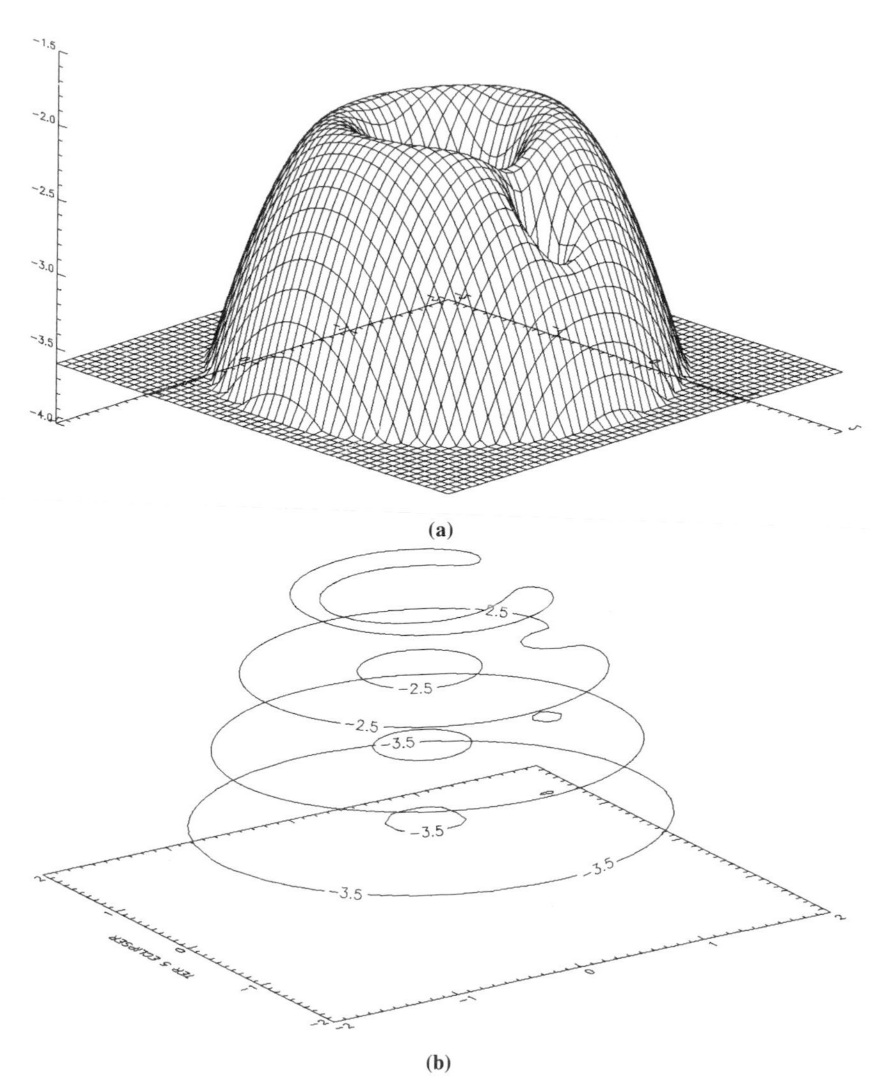

FIGURE 5. (**a**) This binary system has the same mass ratio, binary period, and binary separation as the one in FIGURE 4(a), but the effective gravity of the neutron star acting on the test particles has been reduced by one-half. This figure would be appropriate, for example, for optically thin ionized plasma in a binary system in which the more massive component is emitting X-rays at $\frac{1}{2}L_{\mathrm{Edd}}$. Note the potential barrier between the two stars. (**b**) Some contours for the Jacobi integral of part (**a**). (The orbital plane has been rotated by 180°.) Contrast this with FIGURE 4(b). The doughnut-shaped contour that looks as though a piece had been bitten off illustrates that a particle of energy −2.5 can either move near the neutron star or is free to go to infinity (with the option of resting near the companion star or circling the whole system), but not both. Clearly, it is is easier for matter in the companion star to leave the system than to be accreted.

(compare Tanaka 1989). Thus, it seems that not only a "black hole disk" can exist around a neutron star, the neutron star itself is capable of emitting very hard X-rays usually ascribed to such a disk.

Finally, I would like to comment on the very interesting results of the SIGMA experiment presented by Dr. Sunyaev (1991) (also Mandrou *et al.* 1990). It is not at all clear that the source of the hard X-rays and presumably of the famous narrow and variable e^+e^- annihilation line is a black hole. The original argument for a black hole origin was the presumed coincidence of the source with the galactic center. It is now known that Sgr A is not the source. That the source(s) would not be coincident with the galactic center has been predicted by a model (Kluźniak *et al.* 1988c) explaining

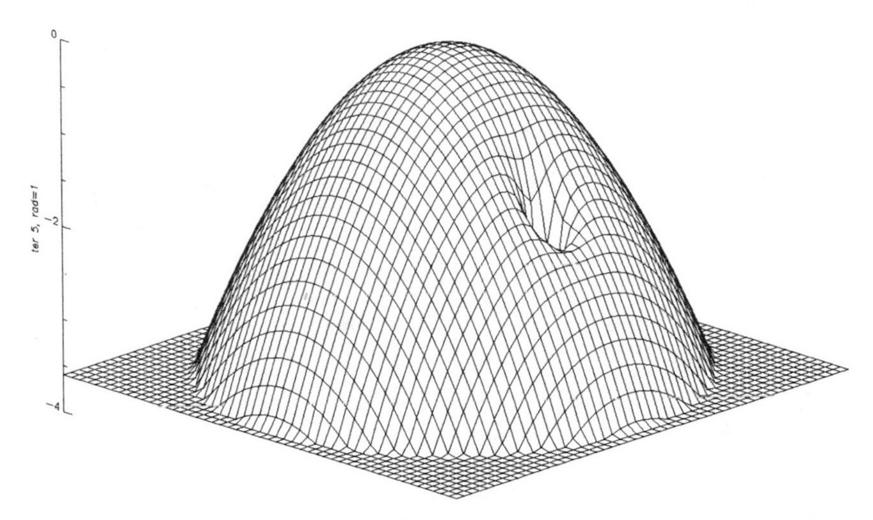

FIGURE 6. Again this is the same binary system as in FIGURES 4 and 5, but now the radiation pressure from the neutron star balances exactly its gravitational attraction exerted on test fluid particles (corresponding to X-ray luminosity of L_{Edd}). Note that the shape of the Jacobi integral function is far from being flat like a table top, which it would be for a solitary Eddington-luminosity star. This point may not have been sufficiently appreciated in the literature. Optically thin fluid is inhibited by centrifugal forces present in the binary system from visiting the neutron star, which commands the position at the top of the hill.

the spectrum and the copious production of positrons by processes occurring in an accreting neutron star of $\sim 10^9$-G field, that is, in a LMXB. Whether or not 1E1740.7−2942 is an accreting neutron star remains to be seen.

CONCLUSIONS

1. In low-mass X-ray binaries, mass loss from the system seems quite likely. It should not be surprising that predictions of standard "conservative evolution" scenarios are often not verified.
2. Anything a black hole can do, a neutron star can do better.

ACKNOWLEDGMENTS

It is a pleasure to thank Dr. George Efstathiou for hospitality at Oxford University. The illustrations were prepared using the computer facilities of the Astrophysics Department at Oxford; special thanks go to Ivan. I am also grateful to Dr. Josh Grindlay for reading the manuscript and for helpful comments.

REFERENCES

ARONS, J. A. 1973. Astrophys. J. **184:** 539.

BASKO, M. M. & R. A. SUNYAEV. 1973. Astrophys. Space Sci. **23:** 117.

BLAES, O., R. BLANDFORD, P. GOLDREICH & P. MADAU. 1989. Astrophys. J. **343:** 839.

BAHCALL, S., B. W. LYNN & S. B. SELIPSKY. 1990. Astrophys. J. **362:** 251.

CANAL, R., J. ISERN & J. LABAY. 1990. Annu. Rev. Astron. Astrophys. **28:** 183.

DANBY, J. M. A. 1962. Fundamentals of Celestial Mechanics. Macmillan. New York.

GIACCONI, R., H. GURSKY, F. R. PAOLINI & B. B. ROSSI. 1962. Phys. Rev. Lett. **9:** 439.

HARTMANN, D. 1991. Gamma-ray burst theory. This issue.

KLUŹNIAK, W., M. RUDERMAN, J. SHAHAM & M. TAVANI. 1988a. *In* Timing Neutron Stars, H. Ögelman, and E. P. J. van den Heuvel, Eds.: 641–645. Kluwer. Dordrecht, the Netherlands.

KLUŹNIAK, W., M. RUDERMAN, J. SHAHAM & M. TAVANI. 1988b. Nature **334:** 225.

KLUŹNIAK, W., M. RUDERMAN, J. SHAHAM & M. TAVANI. 1988c. Nature **336:** 558.

KLUŹNIAK, W. & R. V. WAGONER. 1985. Astrophys. J. **297:** 548.

KLUŹNIAK, W., P. MICHELSON & R. V. WAGONER. 1990. Astrophys. J. **358:** 538.

KLUŹNIAK, W. & J. R. WILSON. 1989. *In* 23rd ESLAB Symposium on Two Topics in X-ray Astronomy, J. Hunt and B. Battrick, Eds.: 477–478. European Space Agency. Paris.

KLUŹNIAK, W. & J. R. WILSON. 1991. Astrophys. J., Lett. **372:** L87.

LEWIN, W. H. G. & E. P. J. VAN DEN HEUVEL, EDS. 1983. Accretion-driven Stellar X-ray Sources. Cambridge University Press. Cambridge, England.

LYNE, A., *et al.* 1990. Nature **347:** 650.

MANDROU, P., *et al.* 1990. IAU Circ. 5140.

MCCLINTOCK, J. 1991. Black holes in the galaxy. This issue.

PHINNEY, E. S., C. R. EVANS, R. D. BLANDFORD & S. R. KULKARNI. 1988. Nature **333:** 832.

RAPPAPORT, S., L. A. NELSON, C. P. MA & P. C. JOSS. 1987. Astrophys. J. **322:** 842.

RASIO, F. A., S. L. SHAPIRO & S. A. TEUKOLSKY. 1989. Astrophys. J. **342:** 934.

RUDERMAN, M., J. SHAHAM, M. TAVANI & D. EICHLER. 1989. Astrophys. J. **343:** 292.

SANDAGE, A. R., *et al.* 1966. Astrophys. J. **146:** 316.

SHKLOVSKY, I. S. 1967. Astrophys. J., Lett. **148:** L1–L3.

SUNYAEV, R. 1991. Paper presented at the 15th Texas/ESO-CERN Symposium on Relativistic Astrophysics, Cosmology, and Fundamental Physics, Brighton, England, December 16–21, 1990.

TAN, *et al.* 1991.

TANAKA, Y. 1989. *In* 23rd ESLAB Symposium on Two Topics in X-ray Astronomy, J. Hunt and B. Battrick, Eds.: 3–13. European Space Agency. Paris.

VAN DEN HEUVEL, E. & J. VAN PARADIJS. 1988. Nature **334:** 227.

WHITE, N. 1989. Astron. Astrophys. Rev. **1:** 85.

Coalescing Binary Neutron Stars[a]

TAKASHI NAKAMURA[b] AND KEN-ICHI OOHARA[c]

[b]*Yukawa Institute for Theoretical Physics,*
Kyoto University
Kyoto, 606, Japan

[c]*National Laboratory for High Energy Physics*
Oho, Tsukuba-shi, Ibaraki-ken, 305, Japan

INTRODUCTION

There are several motivations for the study of coalescing binary neutron stars. One is the real existence of binary neutron stars. In particular, PSR1913 + 16 has been observed precisely and it is believed to consist of two neutron stars of mass $1.445M_{\odot}$ and $1.384M_{\odot}$.[1] Very recently it was found that two more binary neutron stars, PSR2127 + 11C and PSR1534 + 12, exist. The orbital parameters of these binary neutron stars are very similar to those of PSR1913 + 16. They look like twins. Moreover the total mass is almost the same in these three systems, that is, $\sim 2.8M_{\odot}$. It seems that six neutron stars in these three binaries have almost the same mass $\sim 1.4M_{\odot}$. Therefore two neutron stars in PSR1913 + 16, PSR2127 + 11C, and PSR1534 + 12 will coalesce in $\sim 10^8$ y because of the emission of gravitational waves. Using these new data we can follow the argument by Clarke, van den Heuvel, and W. Sutantyo[2] to estimate the event rate of coalescence of binary stars. Since the time before the coalescence is much less than the age of the universe, we can reasonably assume the steady state between the formation and coalescence of the binary neutron stars. Depending on the Type II supernova rate, the frequency in coalescence of binary neutron stars is estimated as $6 \sim 60$ events/year $(d/100 \text{ Mpc})^3$ within the distance of d Mpc. If we can observe coalescence events by gravitational waves and electromagnetic wave such as γ-rays, we may have important information on equations of state, the structure of neutron stars, and the physics in strong gravity, including black hole physics.

Another kind of coalescence of neutron stars is theoretically expected to occur.[3] If the core of the progenitor of Type II supernova has a large angular momentum, the centrifugal force will be important in some stage of the collapse into a final neutron star. The core radius where the centrifugal force is comparable to the gravitational force is proportional to the square of the angular momentum. When the size of the core decreases to this radius, the core contracts principally along the rotational axis, and then a thin disk will be formed. Such a thin disk is known to be gravitationally unstable irrespective of the equation of state,[4] and fragments into several pieces in a free-fall time scale. Each fragment looks like a neutron star and is called a protoneutron star. Protoneutron stars will coalesce again to form a single neutron star, owing to the emission of gravitational waves. If the number of fragments is two,

[a]This work was supported in part by the Grant-in-Aid for Scientific Research of the Ministry of Education, Science and Culture (01306006, 01652509).

the system is essentially the same as a binary neutron star like PSR1913 + 16 in the final coalescence stage. If a scenario like this applies to a largc fraction of Type II supernovae, the frequency in the events of burst emission of gravitational waves within a distance of 10 Mpc increases to ~ 30 events/year.

The third possibility is the accretion-induced collapse (AIC) of white dwarfs leading to the formation of neutron stars. AIC has been considered as the possible formation mechanism of neutron stars in low-mass X-ray binaries as well as millisecond pulsars in globular clusters. If this occurs, the collapsing white dwarfs should have angular momentum of the accreting matter. This situation is different from that in a Type II supernova. In a Type II supernova the collapse may be spherically symmetric. However, in AIC the collapse should be more or less nonspherical due to the angular momentum of the accreted matter. In reality, depending on the strength of the magnetic fields of the white dwarfs and the accretion rate, the angular momentum of AIC can be so large that the scenario discussed in the previous paragraph may occur.

Now to know the final destiny of the coalescing binary neutron stars we must use a fully general relativistic three-dimensional numerical code including the evolution of matter and metric. However, such a code does not exist at present. We therefore use a Newtonian three-dimensional hydrodynamics code to assess the final destiny of coalescing binary neutron stars as well as the wave pattern and the amplitude of the gravitational waves.

Since all the results in this paper are based on our four papers (Paper I,[5] Paper II,[6] Paper III,[7] Paper IV[8]), one can refer to the original paper for details.

BASIC EQUATIONS AND NUMERICAL METHODS

In the early version of our simulations, in Paper I, we used Burke and Thorne-type radiation reaction potential expressed by the fifth time derivative of the quadrupole moment. However, here we use a radiation reaction potential proposed by Blanchet, Damour, and Schäfer,[9] in which only the third time derivative of the quadrupole moment is used. The basic equations are

$$\frac{\partial\rho}{\partial t} + \frac{\partial\rho v^j}{\kappa x^j} = 0, \tag{1}$$

$$\frac{\partial\rho w^i}{\partial t} + \frac{\partial\rho w^i v^j}{\partial x^j} = -\frac{\partial P}{\partial x^i} - \rho\frac{\partial(\psi + \psi_{\mathrm{react}})}{\partial x^i}, \tag{2}$$

$$\frac{\partial\rho\epsilon}{\partial t} + \frac{\partial\rho\epsilon v^j}{\partial x^j} = -P\frac{\partial v^j}{\partial x^j}, \tag{3}$$

$$P = (\gamma - 1)\rho\epsilon, \tag{4}$$

$$\Delta\psi = 4\pi G\rho, \tag{5}$$

$$\Delta R = 4\pi G\left(\frac{d^3}{dt^3}D_{ij}\right)x^i\frac{\partial\rho}{\partial x^j}, \tag{6}$$

$$v^i = w^i + 0.8\frac{G}{c^5}\left(\frac{d^3}{dt^3}D_{ij}\right)w^j, \tag{7}$$

$$\psi_{\text{react}} = 0.4G\left(-R + \left(\frac{d^3}{dt^3}D_{ij}\right)x^i\frac{\partial\psi}{\partial x^j}\right), \tag{8}$$

and

$$D_{ij} = \int \rho(x^i x^j - \tfrac{1}{3}\delta_{ij} r^2)\, dV,$$

where all the variables except w^i have the usual meanings. In the original formalism by Blanchet, Damour, and Schäfer,[9] 1PN and 2.5PN quantities should be evaluated. However, for this we must solve nine Poisson equations, which is numerically highly time-consuming. Therefore, in this paper we evaluate only terms that are directly connected with the radiation reaction. In this case, we need to solve only three Poisson equations.

We take the units of

$$M = M_\odot, \quad L = \frac{GM_\odot}{c^2} = 1.5\text{ km}, \quad T = \frac{GM_\odot}{c^3} = 5 \times 10^{-6}\text{ sec}. \tag{9}$$

To express the hard equation of state, we use a polytropic equation of state with $\gamma = 2$. Then the pressure is expressed by

$$P = K\rho^2, \tag{10}$$

where K is a constant that is related to the radius of the spherical star as

$$K = \frac{2r_0^2 G}{\pi}. \tag{11}$$

We assume that each neutron is rigidly rotating around the z-axis with the angular velocity Ω. Then the equilibrium is determined by

$$\nabla(\psi + h - 0.5(x^2 + y^2)\Omega^2) = 0, \tag{12}$$

and

$$\Delta\psi = 4\pi G\rho, \tag{13}$$

where

$$h = 2K\rho.$$

The integral of **(12)** becomes

$$\psi + h - 0.5(x^2 + y^2)\Omega^2 = C. \tag{14}$$

The solution to the preceding equation can be obtained by the iteration as is shown in detail in Paper III.

We will estimate the amount of gravitational radiation emitted using the quadrupole formula. Details of how to evaluate the third time derivative of the quadrupole moment without using the numerical time difference is shown in Paper II.

A numerical scheme for the hydrodynamics equations is shown in Paper I. We solve three Poisson equations as **(5)** and **(6)** by the ICCG method described in Oohara and Nakamura[10] under appropriate boundary conditions.

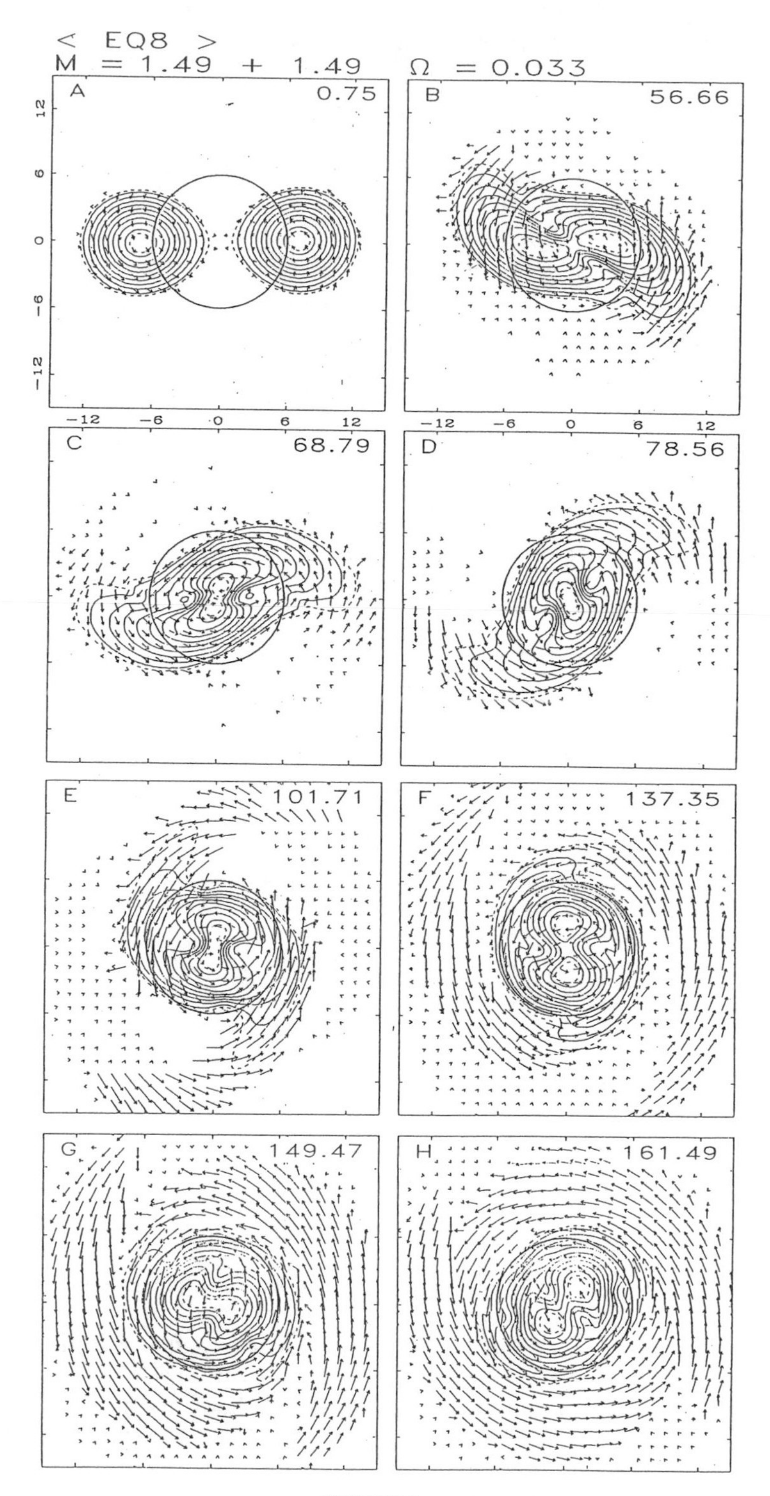
< EQ8 >
M = 1.49 + 1.49 Ω = 0.033
A 0.75
B 56.66
C 68.79
D 78.56
E 101.71
F 137.35
G 149.47
H 161.49
12
6
0
−6
−12
−12 −6 0 6 12
−12 −6 0 6 12

FIGURE 1.

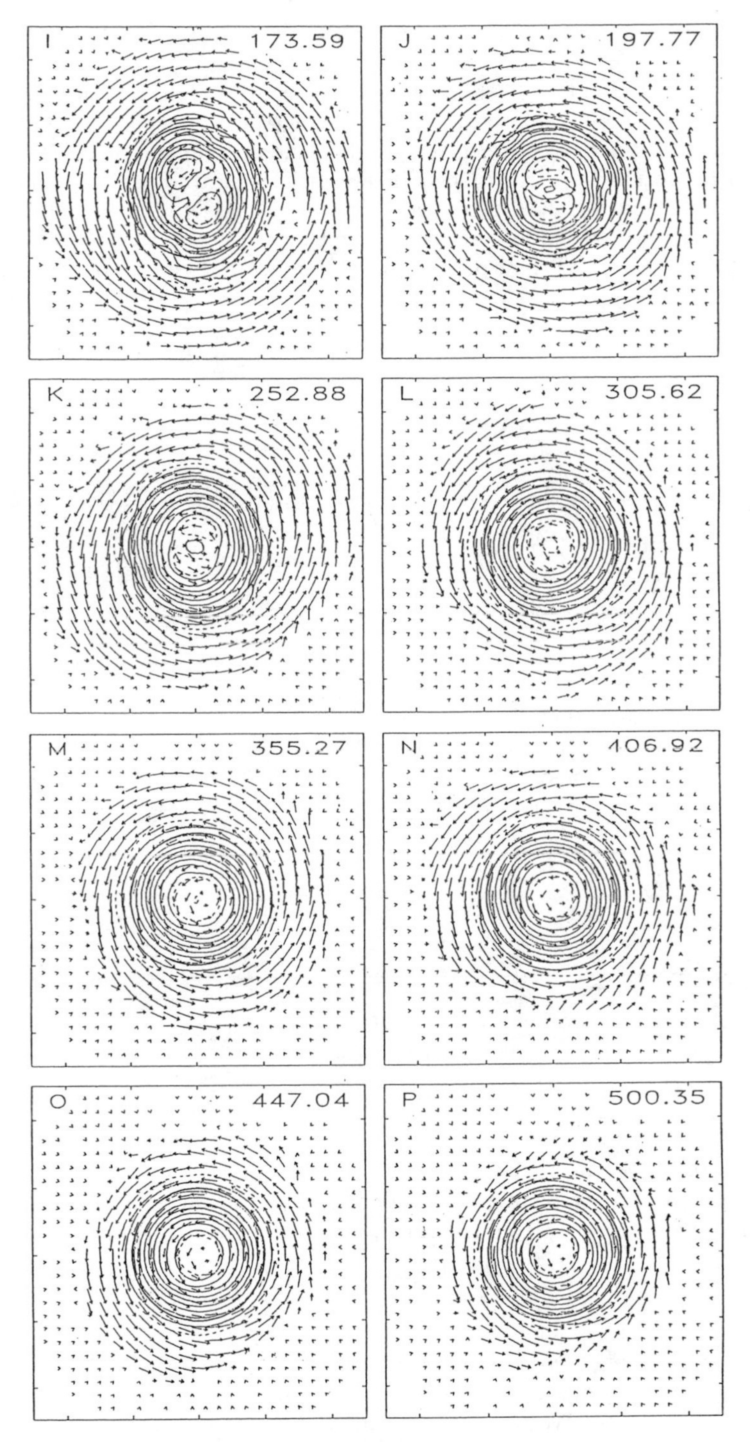

FIGURE 1.

RESULTS

We take a $141 \times 141 \times 130$ grid under the assumption of the reflection symmetry about $z = 0$ plane. The grids cover $[-21, 21]$ in the x and y directions, and $[0, 16.9]$ in the z direction. We have performed eight simulations so far. At the conference we showed the results by a video movie. However, it is hard to present them in the proceedings. Here we show two typical ones called model EQ8 and TIDAL2. A typical CPU time needed to perform one model is ~ 240 hours with $\sim 90{,}000$ time steps. Thus a typical CPU time per one step is about 10 sec by a HITAC S820/80 with

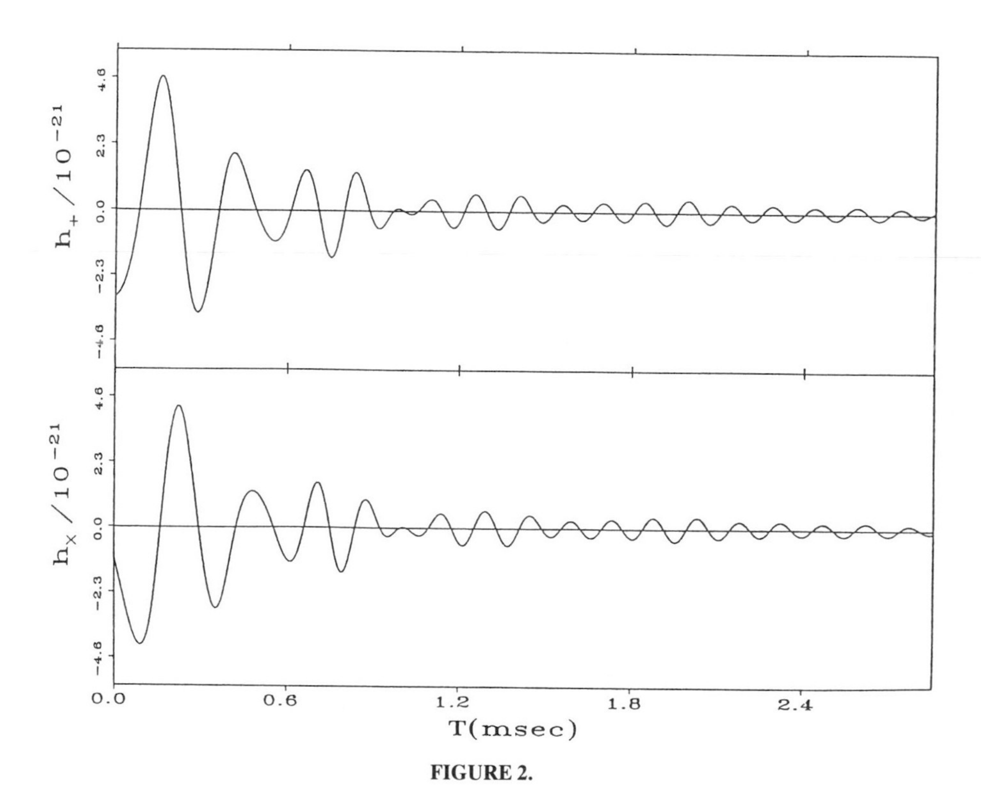

FIGURE 2.

3GFLOPS peak speed. We performed the numerical calculation for each model until $t = 600 \sim 1000$ in our units.

In the case of a two-point mass binary of mass m_1 and m_2, with separation r in a circular orbit, r decreases at a rate given by

$$\dot{r} = -\frac{64 m_1 m_2 (m_1 + m_2)}{5r^3}. \tag{15}$$

Inserting $r = 2r_0$, $m_1 = m_2 = 1.49$, we have $\dot{r} = -0.086$. In model EQ8, we put the approaching velocity of each neutron star at $t = 0$. Namely, for $x \geq 0$, $w_x = -0.08$, and for $x \leq 0$, $w_x = 0.08$. We show the evolution of the density contours for EQ8 in

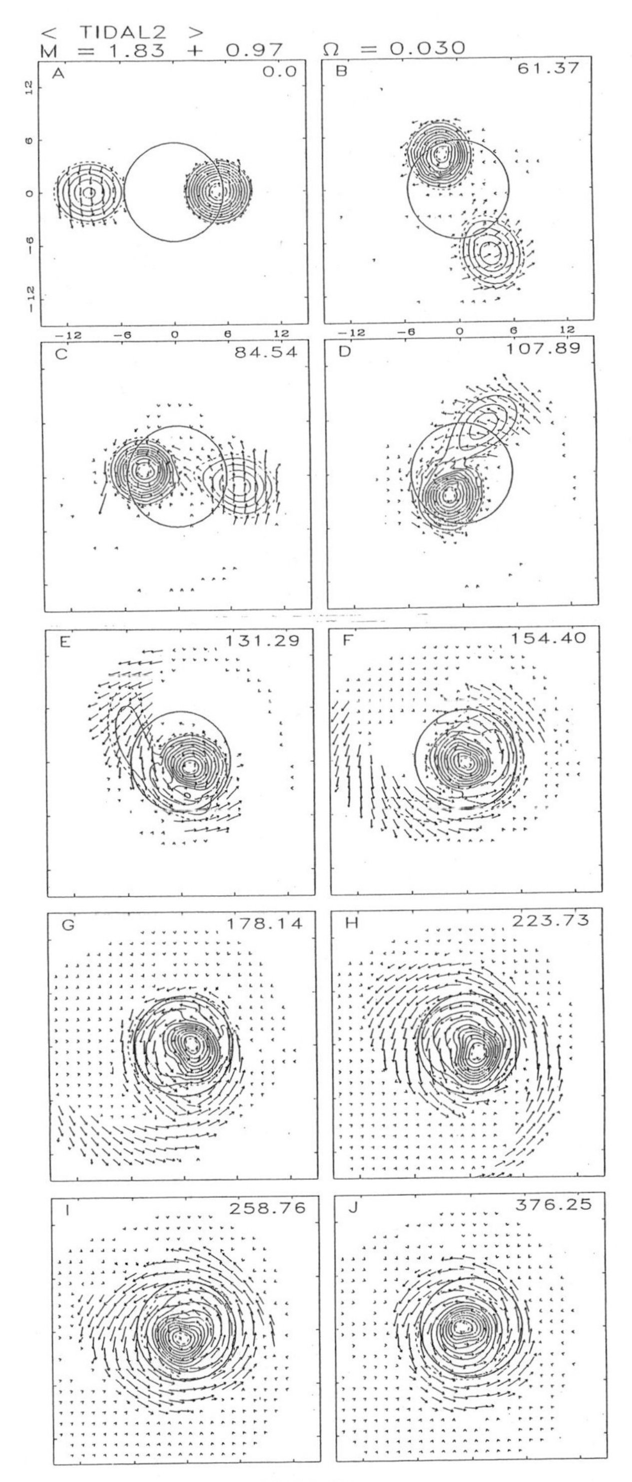

< TIDAL2 >
M = 1.83 + 0.97 Ω = 0.030
A 0.0
B 61.37
C 84.54
D 107.89
E 131.29
F 154.40
G 178.14
H 223.73
I 258.76
J 376.25
12
6
0
−6
−12
−12 −6 0 6 12

FIGURE 3.

FIGURE 1(a) to (p). In EQ8 the coalescence is initiated not only by the radiation reaction but also by the approaching velocity 0.08 at t = 0, since $\tau_{loss} \sim 100$ is comparable to r_0/w_x, where τ_{loss} is the timescale for a 10 percent loss of the total angular momentum. Therefore the luminosity is not constant even at $t \leq 100$. Spiral arms can be seen [FIG. 1-(c) to (h)]. Two neutron stars appear again after the coalescence [FIG. 1(f) and (h)]. The reason is as follows: When the coalescence proceeds and the core shrinks, the centrifugal force increases against the gravity. Then the core expands up to the reappearance of the original two neutron stars. However, in EQ8 after several appearances of the two neutron stars we have a quasi–steady-binary-like system in the central part [FIG. 1(i) and (j)]. Due to the angular momentum loss the binary coalesces very slowly in FIG. 1(k). Finally, the central binary becomes a ringlike system in FIG. 1(l). This ring evolves to the disk in the final stage [FIG. 1(m) to (p)]. In the final stage of the evolution the system becomes almost axisymmetric with the very low luminosity $\sim 3 \times 10^{-6}$ while it is $\sim 2 \times 10^{-4}$ at $t = 0$. We show the waveform observed on the z-axis at 10 Mpc in FIGURE 2. The maximum amplitude of the gravitational waves is $\sim 5 \times 10^{-21}$. Since the proposed sensitivity of large laser interferometric detectors of gravitational waves such as LIGO is $h \sim 10^{-21}$, we may observe the coalescence event up to 50 Mpc. Then the event rate will be more than 1 event/year from the argument in the Introduction. In FIGURE 3(a) to (j) we show the density contour for TIDAL2. In this case $m_1 = 1.8$ and $m_2 = 0.97$ with initial separation 21 km. One can see the tidal disruption of the smaller mass neutron star. The details of TIDAL2 will be shown in Paper IV.[8]

ACKNOWLEDGMENT

The numerical calculations were performed on a HITAC S820/80 at the Data Handling Center of National Laboratory for High Energy Physics (KEK).

REFERENCES

1. TAYLOR, J. H. 1987. *In* General Relativity and Gravitation 11, M. A. H. MacCallum, Ed.: 209. Cambridge University Press. Cambridge, England.
2. CLARKE, J. P. A., E. P. J. VAN DEN HEUVEL & W. SUTANTYO. 1978. Astron. Astrophys. **72:** 120.
3. NAKAMURA, T. & M. FUKUGITA. 1989. Astrophys. J. **337:** 466.
4. GOLDREICH, P. & D. LYNDEN-BELL. 1965. Mon. Not. R. Astron. Soc. **130:** 97.
5. OOHARA, K. & T. NAKAMURA. 1989. Prog. Theor. Phys. **82:** 535. (Paper I.)
6. NAKAMURA, T. & K. OOHARA. 1989. Prog. Theor. Phys. **82:** 1066. (Paper II.)
7. OOHARA, K. & T. NAKAMURA. 1990. Prog. Theor. Phys. **83:** 906. (Paper III.)
8. NAKAMURA, T. & K. OOHARA. 1991. Prog. Theor. Phys. **86:** 73. (Paper IV.)
9. BLANCHET, L., T. DAMOUR & G. SCHÄFER. 1990. Mon. Not. R. Astron. Soc. **241:** 289.
10. OOHARA, K. & T. NAKAMURA. 1989. Prog. Theor. Phys. **81:** 360; OOHARA, K. & T. NAKAMURA, T. 1989. *In* Frontiers of Numerical Relativity, D. Hobill, C. Evans, and S. Finn, Eds.: 74. Cambridge University Press. Cambridge, England.

Film of the Extension of Schwarzschild Space Through the $r = 0$ Singularity

D. LYNDEN-BELL,[a] J. KATZ,[a,b] I. H. REDMOUNT,[a,c] AND E. L. LYNDEN-BELL[a]

[a] *Institute of Astronomy*
Madingley Road
Cambridge CB3 OHA, England

[b] *Racah Institute of Physics*
Hebrew University of Jerusalem
Jerusalem, Israel

In 1934 Tolman[1] gave the general solution to Einstein's equations for the motion of a spherically symmetrical distribution of dust. In freely falling coordinates the dust moves with the coordinate mesh, and Tolman showed the metric is

$$ds^2 = -d\tau^2 + \frac{(\partial r/\partial R)^2\, dR^2}{[1 + 2E(R)]} + r^2(R, \tau)(d\theta^2 + \sin^2\theta\, d\phi^2). \tag{1}$$

Here the function $r(R, \tau)$ is the solution of

$$\frac{1}{2}\dot{r}^2 - \frac{GM(R)}{r} = E(R), \tag{2}$$

and $\dot{r}$ is the rate of change following the motion (i.e., with R constant). Equation **(2)** is identical to the Newtonian energy equation for a spherical distribution. The functions $E(R)$ and $M(R)$ are the "constants" of integration of the Einstein equations. The dust density $\rho(R, \tau)$ is related to $M(R)$ by

$$\frac{dM}{dR} = 4\pi r^2 \rho \frac{\partial r}{\partial R}. \tag{3}$$

From **(2)** we see that

$$\ddot{r} = -\frac{GM(R)}{r^2} \tag{4}$$

so that M is the gravitating mass within $r(R, \tau)$. However, M is not the sum of the rest masses within R because the volume between r and $r + dr$ is not $4\pi r^2 dr$, but is given by **(1)** as

$$dV = 4\pi r^2 dr[1 + 2E(R)]^{-1/2}. \tag{5}$$

Notice that $M(R)$ increases when r increases with R, but *decreases* when $\partial r/\partial R < 0$.

[c] Current address: Department of Physics, Washington University, St. Louis, Missouri 63130.

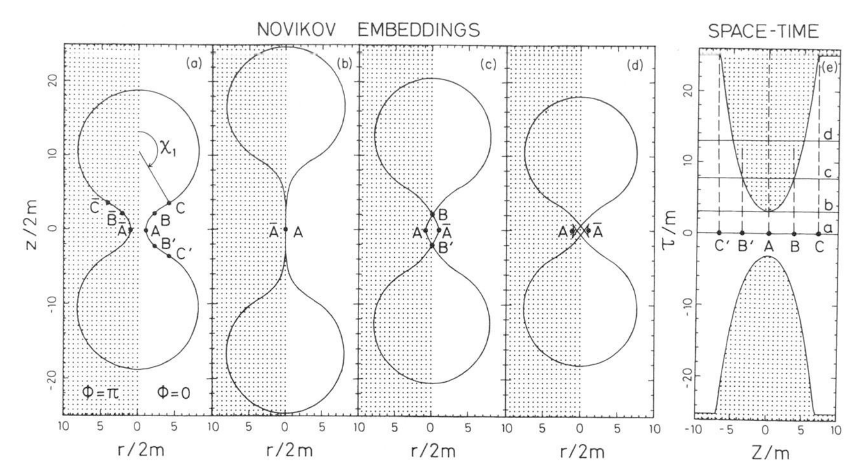

FIGURE 1. (**a**) Two Friedmann $\Omega > 1$ universes are here joined by a Schwarzschild neck that makes an Einstein–Rosen bridge between them. In this momentarily static configuration Novikov sets observers momentarily at rest. The subsequent diagrams (**b**) and (**c**) are isometric embeddings of the space at later times measured by these observers. The slope of the surface at A or B or C is the same at all times, so when B meets $\overline{B}$ at $r = 0$ in (**c**) an interesting conical point develops in the space. By moment (**d**) Novikov's coordinates have developed coordinate singularities above and below A and $\overline{A}$, so the space is no longer properly represented in the diagram. (**e**) The spacetime diagram in which the comoving coordinate Z is the initial height of the observer above the neck. (**a**)–(**d**) are the cross sections so labeled. $r = 0$ on the *two parabolas,* and the regions in the half-plane $\phi = \pi$ are *shaded.*

This happens, for example, when one considers regions more than halfway around a closed universe from the origin of r.

To picture metrics such as **(1)** we represent different moments of proper time τ by the different frames of a film. Constant τ reduces the metric **(1)** to three dimensions, but that is still too many for us to be able to visualize their curvature. We therefore set $\theta = \pi/2$ and leave only the symmetry about the ϕ axis to represent what is, in truth, full spherical symmetry. With this convention the metric **(1)** on each frame of the film takes the form

$$ds^2 = [1 + 2E(R)]^{-1}\, dr^2 + r^2\, d\phi^2. \qquad \textbf{(6)}$$

We have written dr to emphasize that this quantity is only dr when τ is held fixed. To show the curvature of these spaces we make isometric embeddings of them into cylindrical polar coordinates in flat space. In that flat space

$$ds^2 = dr^2 + dz^2 + r^2\, d\phi^2, \qquad \textbf{(7)}$$

and on the surface $z = z(r)$ given by the solution of

$$1 + \left(\frac{dz}{dr}\right)^2 = [1 + 2E(R)]^{-1} \qquad \textbf{(8)}$$

the metric is precisely **(6)**. This is because **(7)** restricted by **(8)** gives **(6)**. Our procedure is to solve **(2)** for $r(R, \tau)$, to substitute the result into **(8)**, and solve it for

$z(r, \tau)$. We then view the resulting coordinate system in perspective to make the different frames of the film.

Equation (**8**) already tells us a trivial but interesting theorem. That is, dz/dr, the gradient of the surface in the embedding, is a function of R alone. Thus, it depends on the freely falling observer considered, but each observer keeps the same gradient for all times τ. This has interesting implications when observers with opposite gradients meet as in FIGURE 1(c) at B.

Considerations of the geometry of forming the singularity in Schwarzschild space have been bedeviled by the difficulties of the physics of matter at ultrahigh density. When neither the physics nor the mathematics of the forming singularity is understood, one swims in treacle. To avoid this we have taken a specially simple case in

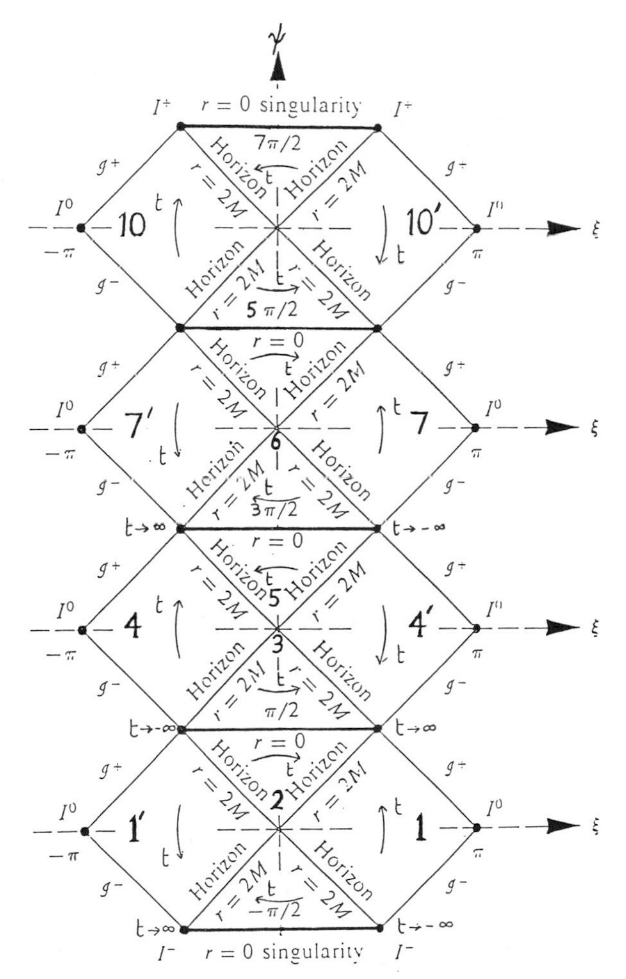

FIGURE 2. Schwarzschild spacetime is *periodic* in extended Penrose coordinates with period 2π. The sense of increase of Schwarzschild time t is illustrated.

which the singularity forms in empty space. We consider the system defined at a moment of time symmetry by an almost completely closed Friedmann universe at its moment of greatest radius, which is surmounted by an Einstein–Rosen bridge representing Schwarzschild space, which is in turn surmounted by another almost complete Friedmann universe, as in FIGURE 1. This whole construction is, of course, a particular example of a Tolman solution. All the mass lies in the two "Friedmann universes." There is no mass in the Schwarzschild bridge between them. Nevertheless it is this that collapses first and forms the singularity. In this region we take our observers and their coordinates to be initially at rest. Such coordinates were first used by Novikov.[2] Our gradient theorem allows us to continue through the singularity. Indeed, close to $r = 0$ the right-hand side of **(2)** is negligible, and **(2)** integrates to give $\frac{2}{3} r^{3/2} = \pm\sqrt{2GM}(\tau - \tau_0)$, and thence

$$r = \left(\frac{9}{2} GM\right)^{1/3} (\tau - \tau_0)^{2/3}. \qquad \textbf{(9)}$$

Notice that $(\tau - \tau_0)^{2/3}$ is positive for $\tau < \tau_0$ and for $\tau > \tau_0$ so that r remains *real*.

Detailed explorations of the resulting extension and of the reason why Kruskal and Penrose coordinates[3] fail beyond $r = 0$ may be found in Lynden-Bell and Katz.[4] A simple extension of Penrose's diagram leads to FIGURE 2 for the case of a pure Einstein–Rosen bridge with no Friedmann universes attached. This demonstrates that pure Schwarzschild space is periodic in Penrose's conformal time.

FIGURE 3 shows a sequence of stills from the film at different values of the time of Novikov's freely falling observers. Notice the crossing through $r = 0$ and the conical

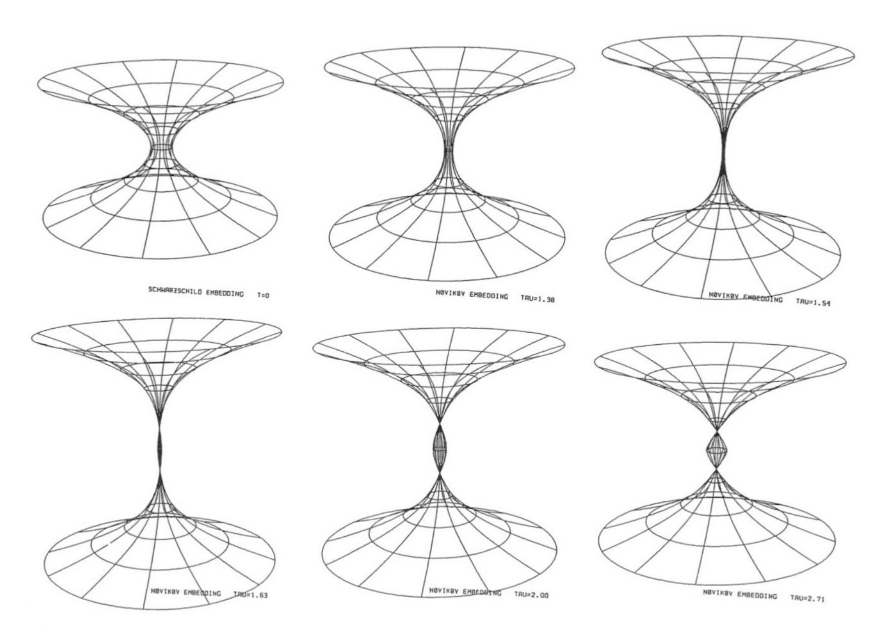

FIGURE 3. A sequence of stills from the film of the development of Schwarzschild space in Novikov time.

form of the embedding surface where it crosses $r = 0$. It is this topological change rather than any mass point that gives the mathematical singularity.

NOTE ADDED IN PROOF: There is now evidence that the extension of Schwarzschild space proposed in the preceding discussion has a "flash" of tension along $r = 0$ and is therefore *not* empty.

REFERENCES

1. TOLMAN, R. C. 1934. Proc. Nat. Acad. Sci. USA **20:** 169.
2. NOVIKOV, I. D. 1966. Sov. Astron. **10:** 731.
3. PENROSE, R. 1965. *In* Relativity Groups & Topology, C. de Witt and B. de Witt, Eds: 565. Blackie. London.
4. LYNDEN-BELL, D. & J. KATZ. 1990. Mon. Not. R. Astron. Soc. **247:** 651.

Magnetohydrodynamics of Black Holes and the Origin of Jets

M. CAMENZIND

Landessternwarte Königstuhl
D-6900 Heidelberg, Germany

INTRODUCTION

The standard model for active galactic nuclei (AGNs)[1,2] assumes the existence of a supermassive rotating black hole in the center of a galaxy surrounded by a Keplerian accretion disk with an extension of a few hundred Schwarzschild radii. This inner disk is fueled by material present in the form of molecular clouds on the scale of a few parsecs to a few hundred parsecs. In the innermost few parsecs, galactic matter is assembled in the form of a molecular torus, quite similar to structures found in our galactic center. Direct evidence for the existence of the inner accretion disks follows from the presence of a strong optical-UVX excess in the continuum of AGNs.[3] Cold molecular gas appears as a thermal component in the far-infrared, while hot dust is the origin of the near-infrared emission in radio-weak objects.[2]

Galactic material builds up large-scale magnetic fields with field strengths having energy densities of the order of the energy density in the turbulence. On the parsec scale we therefore find field strengths of the order of a few milligauss. These fields are also advected inwards by the accretion flow and redistributed in the Keplerian disk by strong differential rotation and turbulent diffusion. As a consequence, the central black hole appears immersed in a rotating magnetosphere built up by the inner accretion disk. The global structure of magnetic fields generated by accretion disks cannot yet be studied in terms of magnetohydrodynamics (MHD) simulations, since relativity and turbulent conductivity are essential ingredients for the transport of magnetic fields in disks in the background of a Kerr geometry. Instead of this we treat the problem in a two-step process: first we study the evolution of magnetic fields in disks with a corona and then use the boundary distribution for the magnetic fields as input for the calculation of self-consistent wind magnetospheres.

The structure of stationary and axisymmetric magnetospheres follows from solutions of the relativistic Grad–Schlüter–Shafranov equation, which can now be formulated for any rapidly rotating compact object including millisecond pulsars, supermassive rotators, and rapidly rotating black holes.[4] The basic elements of this theory are discussed shortly and applied to the case of Kerr black holes for the accretion process and the structure of magnetized winds.

THE MAGNETIC STRUCTURE OF AGN DISKS

Fully turbulent disks are ideal setups for magnetic dynamos.[5–8] A small magnetic field in the accreting matter will be amplified in a Keplerian disk by the strong differential rotation. At the same time, the turbulence in the disk will generate a turbulent magnetic field and a turbulent velocity field. These turbulent fields and motion will also generate large-scale magnetic fields over the helicity $\alpha_D(R, z) = L_T^2 \Omega \, \overline{\alpha}(z)/H(R)$, quite similar to the dynamo action observed in the disks of spiral galaxies[9] (FIG. 1).

In such turbulent disks, the mean magnetic fields evolve according to the induction equation of mean field electrodynamics

$$\partial_t \mathbf{B} + (\mathbf{u} \cdot \nabla)\mathbf{B} = (\mathbf{B} \cdot \nabla)\mathbf{u} - (\nabla \cdot \mathbf{u})\mathbf{B} + \nabla \wedge \{\epsilon - \eta_T \nabla \wedge \mathbf{B}\}, \tag{1}$$

where $\mathbf{B}$ is the mean magnetic field; $\mathbf{u}$ the mean velocity field in the disk, $\mathbf{u} = (u_R, R\Omega, 0)$; η_T is the total magnetic diffusivity including the microscopic one and the turbulent diffusivity; and $\epsilon = \alpha_D \mathbf{B}$ represents the electromotive force induced by correlations between turbulent motions and turbulent magnetic fields.

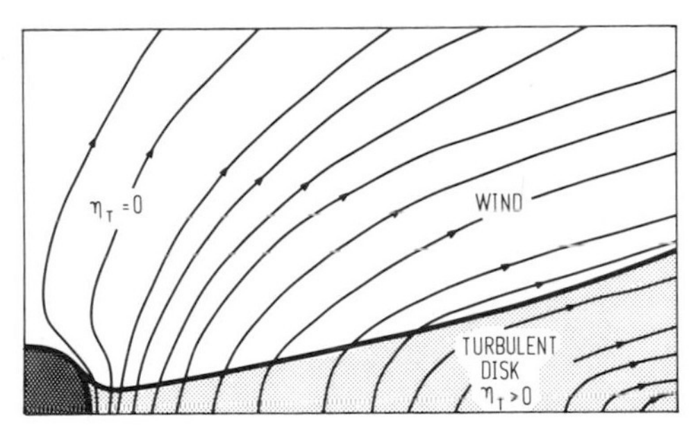

FIGURE 1. The general setup of a fully turbulent disk around a collapsed or noncollapsed object. Differential rotation and turbulence will generate large-scale magnetic fields that build up a rotating magnetosphere around the central object. Plasma from the disk can flow into this magnetosphere and initiate strong outflows.[7,10]

For axisymmetric fields, the preceding equation can be written as a coupled system for the poloidal magnetic flux $\psi(t, R, z)$ and the poloidal current flux $T(t, R, z) = RB_\phi(t, R, z)$[7]

$$B_R = -\frac{1}{R}\frac{\partial\psi}{\partial z}, \quad B_z = \frac{1}{R}\frac{\partial\psi}{\partial R}, \tag{2}$$

$$\frac{\partial\psi}{\partial t} + u_R \frac{\partial\psi}{\partial R} - \eta_T R^2 \nabla \cdot \left\{\frac{1}{R^2}\nabla\psi\right\} = \alpha_D T, \tag{3}$$

$$\frac{\partial T}{\partial t} + u_R \frac{\partial T}{\partial R} - R^2 \nabla \cdot \left\{\frac{\eta_T}{R^2}\nabla T\right\} + T\frac{R\partial(u_R/R)}{\partial R} = R^2 \mathbf{B}_p \cdot \nabla\Omega + R^2 \nabla \cdot \left\{\frac{\alpha_D}{R^2}\nabla\psi\right\}. \tag{4}$$

These equations show the redistribution of the fluxes ψ and T over the effect of spatial diffusion and radial advection.

Solutions of these equations have been studied mainly in the thin disk approximation for vanishing α-effect.[11–13] In this case accretion from the molecular torus advects a total magnetic flux inwards and builds up either a global dipolar or quadrupolar structure around the black hole.[7]

When the dynamo effect is included, dipolar fields are in general unstable and quadrupolar fields are the first modes to be excited in geometrically thin disks.[5,7,8] Whether this result is in general true is an unsolved problem and must be studied in the future with general dynamo codes, including the exterior wind magnetosphere and the backreaction of the magnetic fields on the turbulence.[14]

For accretion rates $\dot{m} = \dot{M}/\dot{M}_{\rm ED} > 0.001$ the disks around supermassive black holes are dominated by radiation pressure, and the growth of the toroidal magnetic

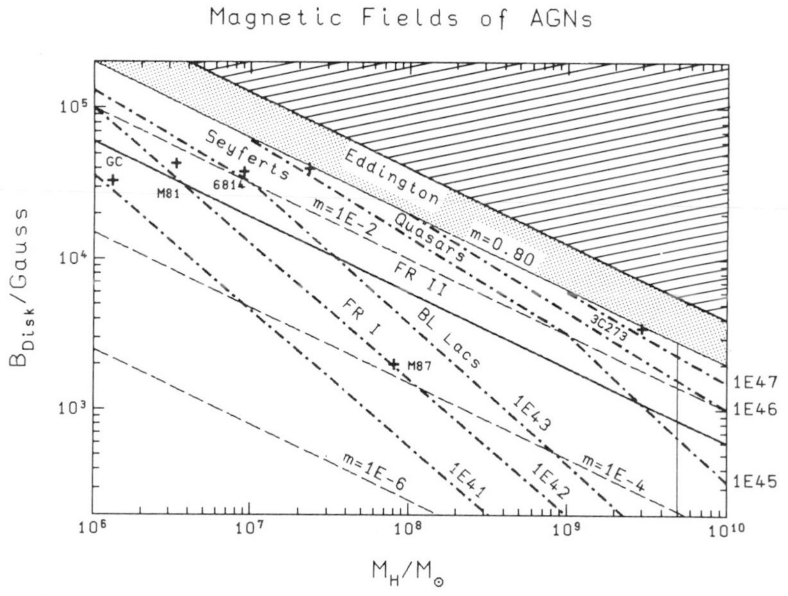

FIGURE 2. The classification of AGN in terms of central masses M_H and the typical toroidal magnetic field B_D of the accretion disk. The accretion rate is parametrized in terms of the Eddington accretion rate, $m = \dot{M}_{\rm acc}/\dot{M}_{\rm Ed} < 1.0$. Seyfert galaxies, QSOs, and quasars all have high accretion rates ($m \geq 0.01$), while FR I radio galaxies and BL Lac objects have rather low accretion rates ($m \simeq 0.001$). Also shown are lines of constant accretion power in units of ergs/second (*dot–dashed lines*).[15]

fields in the disk is limited by the turbulent pressure in the disk, which makes the magnetic field to be practically independent of the accretion rate $\dot{m}$[6,15]

$$B_D \simeq B_{\rm ED}\left(\frac{M_H}{R}\right)^{3/4}, \quad B_{\rm ED} = \sqrt{\frac{L_{\rm ED}}{M_H^2\,c}} = 4 \times 10^4\ {\rm G}\ M_{H,8}^{-1/2}. \tag{5}$$

For low accretion rates, gas pressure dominates, and the disk field scales as

$$B_D \simeq 6 \times 10^3\ {\rm G}\sqrt{\frac{\dot{m}}{M_{H,8}}}\sqrt{\frac{R}{H}}, \tag{6}$$

depending now on the relative scale height of the disk. The disk field (**5**) corresponds to a total poloidal current driven by the differential rotation

$$I_D \simeq 3 \times 10^{18}\ \mathrm{A}\, M_{H,8}^{1/2} \left(\frac{R}{M_H}\right)^{1/4}. \tag{7}$$

Magnetic flux $\psi_D = \pi R^2 B_p$ carried by the disk and the current I_D are the essential new elements of magnetized disks.

MAGNETOHYDRODYNAMICS OF RAPIDLY ROTATING COMPACT OBJECTS

The theory of stationary MHD on any rotating background is now quite complete. The basic elements are quite similar to the special relativistic theory, and they can be formulated with respect to FIDOs. In this sense, we discuss Maxwell's equations, the MHD relations, and the critical surfaces for plasma flows.

Fields and Plasma

The magnetospheres of rapidly rotating neutron stars and black holes are essentially 3-dimensional objects and therefore time-dependent. These magnetospheres cannot be handled as vacuum solutions of Maxwell's equations, since they are populated by plasma injected either from the central object or from the ambient medium, such as an accretion disk. In a first approximation, we handle this plasma with infinite conductivity, though there is in general no global justification for this assumption. It will break down in some special regions of the magnetosphere and give rise to effects beyond the ideal magnetohydrodynamics, such as resistive reconnection. In particular, the transport of magnetic fields in the accretion disk itself is heavily dominated by the finite resistivity of the disk plasma.

For the following considerations we neglect the finite resistivity and work in the ideal approximation. We also restrict ourselves to *axisymmetric configurations.* Many problems in the case of pulsar physics have only been discussed in the one-dimensional approximation (such as pulsar winds). The basic features of magnetohydrodynamic winds can be discussed already in the 2-dimensional approximation.

The metric of rapidly rotating compact objects is in general given as

$$ds^2 = \alpha^2\, dt^2 - R^2(d\phi - \omega\, dt)^2 - e^{2\lambda}\, dr^2 - e^{2\mu}\, d\theta^2, \tag{8}$$

where α is the lapse function, or the redshift factor between the local FIDOs and infinity,[16] and ω the gravitomagnetic potential due to the spin of the central object. Due to stationarity and axisymmetry, all the metric coefficients are only functions of r and θ. This form of the metric includes the Kerr solution for rotating black holes, but also the solutions for rapidly rotating neutron stars. In this latter case, no analytic solution is known. In the asymptotic domain, the angular momentum J_*, $J_* = J_H = M_H a_H$ for a black hole, is the source of the gravitomagnetic potential ω. The fiducial observers move in this space time with the 4-velocity $U_{\mathrm{FIDO}} = (\xi + \omega\mathbf{m})/\alpha(\xi = \partial_t, \mathbf{m} = \partial_\phi)$.

The structure of axisymmetric magnetospheres follows from the magnetic flux function $\psi(r, \theta)$ given by

$$\psi = \int_\Sigma \mathbf{B}_p \cdot d\mathbf{S}, \quad \mathbf{B}_p = \frac{\nabla\psi \wedge \mathbf{m}}{2\pi R}, \tag{9}$$

a surface integral over a surface Σ with radius R around the rotational axis. Axisymmetric magnetospheres consist of a family of nested magnetic surfaces, each of which is obtained by taking a single **B**-line and rotating it around the rotational axis. The toroidal field $\mathbf{B}_T$ is then responsible for the spiraling. Each magnetic surface is characterized by its total magnetic flux contained inside the surface.

Due to the high conductivity in the plasma, the electric fields as measured in the plasma frame vanish. The magnetic field lines are therefore rotating with velocity $R\Omega^F$ with respect to infinity, and consequently with a velocity $\mathbf{v}^F$ with respect to the FIDOs such that

$$\mathbf{E}_T = 0, \quad \mathbf{E}_p = -\mathbf{v}^F \wedge \mathbf{B}_p, \quad \mathbf{v}^F = \frac{\Omega^F - \omega}{\alpha} R\mathbf{m}. \tag{10}$$

Different field lines may rotate with different velocities, but Maxwell's equations (and in particular, Faraday's equation)

$$\nabla \cdot \mathbf{E} = 4\pi\rho_e \tag{11}$$

$$\nabla \cdot \mathbf{B} = 0 \tag{12}$$

$$\nabla \wedge (\alpha\mathbf{E}) = -(\partial_t - L_\beta)\mathbf{B} = (\mathbf{B} \cdot \nabla\omega)R\mathbf{m} \tag{13}$$

$$\nabla \wedge (\alpha\mathbf{B}) = (\partial_t - L_\beta)\mathbf{E} + 4\pi\alpha\mathbf{j} = -(\mathbf{E} \cdot \nabla\omega)R\mathbf{m} + 4\pi\alpha\mathbf{j} \tag{14}$$

imply then that the field line velocity Ω^F is constant along the flux surfaces $\psi = \text{const}$, that is, $\Omega^F = \Omega^F(\psi)$ (Ferraro's law of isorotation).

These magnetospheres are also filled up with plasma that is injected at various places and by various processes in the magnetosphere. In particular, the conservation of particle number along flux surfaces is described in terms of another constant of motion $\eta(\psi)$ related to the poloidal velocity $\mathbf{u}_p$ of the 4-velocity u^μ

$$\mathbf{u}_p = \frac{\eta(\psi)}{\alpha n}\mathbf{B}_p, \quad u^t(\Omega - \Omega^F) = \frac{\eta(\psi)}{\alpha n}B^\phi. \tag{15}$$

The redshift factor accounts for the transformation between the proper time of the FIDOs and the universal time t; n is the density in the plasma frame. These relations are well known from Newtonian MHD, and also special relativistic MHD. The great advantage of this scheme is that the structure of the magnetosphere already determines the flow properties of the plasma. Plasma is streaming along the magnetic surfaces. The kinematics of this plasma follows from two additional constants of motion,[17] the total energy E and total angular momentum L,

$$u_t = \frac{E}{\mu}\frac{1 - M_g^2 - \Omega^F L/E}{1 - M_g^2 - \Omega^F l(\Omega^F)} \tag{16}$$

$$l = \frac{l(\Omega^F)(1 - \Omega^F L/E) - M_g^2 L/E}{1 - M_g^2 - \Omega^F L/E}, \quad l(\Omega^F) = -\frac{g_{\phi\phi}(\Omega^F - \omega)}{g_{tt} + \Omega^F g_{t\phi}}. \tag{17}$$

Here, we introduced the general Mach number

$$M^2 = \frac{4\pi\mu\eta^2}{n} = \frac{4\pi\alpha^2\mu n u_p^2}{B_p^2}, \quad M_g^2 = \frac{M^2}{g_{tt} + \Omega^F g_{t\phi}}. \tag{18}$$

The poloidal velocities $\mathbf{u}_p$ diverge at the horizon. This fact is accounted for by the redshift factor α in the Mach number, so that M is a well-defined concept also near the horizon.

Relativistic Alfvén Surfaces

When plasma is streaming along rotating magnetic surfaces, it must decouple from the field geometry at special points; otherwise, it would rotate with superluminal velocity with respect to the FIDOs. In Newtonian MHD this point is known as the Alfvén point, implying that at this point $M_A^2 = 1$ and $L = R_A^2\Omega^F$. Similarly, the relativistic expression **(16)** tells us that there are critical points in the flow, whenever

$$M_g^2 = 1 - \Omega^F \frac{L}{E}, \quad l_A(\Omega^F) = \frac{L}{E}. \tag{19}$$

The relativistic Alfvén point is shifted by the amount of electromagnetic energy carried in the flow, so that for $\Omega^F L/E \to 1$, the Alfvén point is located at the points with vanishing Mach number. This is a special relativistic effect. The condition for the Alfvén point can also be written in a different form, when we use

$$f = \xi + \Omega^F \mathbf{m}, \quad (f, f) = g_{tt} + 2\Omega^F g_{t\phi} + (\Omega^F)^2 g_{\phi\phi}. \tag{20}$$

The Alfvén point occurs at the position, where $M_A^2 = (f, f)_A$.

This expression shows that magnetically dominated flows, $M_A^2 \ll 1$, *must have their Alfvén points at the light surfaces of the magnetosphere, which are defined by* $(f, f) = 0$. For special relativistic flows, there is only one light surface given by its cylindrical radius at $R_L = c/\Omega^F$. A rotating magnetosphere of a Kerr black hole has two different light surfaces. The outer light surface is formed by the rapid rotation in the same manner as in pulsar models. It is, however, slightly deformed by the underlying geometry. The existence of the inner light surface is a consequence of the frame dragging effect. For given rotation law $\Omega^F = \Omega^F(\psi)$ we can determine the spatial shapes of the two light surfaces, $r_L^{\text{in}}(\theta, \Omega^F)$ and $r_L^{\text{out}}(\theta, \Omega^F)$. In general, Ω^F is limited to the range

$$\Omega_{\text{min}} \le \Omega^F \le \Omega_{\text{max}}, \quad \Omega_{\text{min}} = \omega - \frac{\alpha}{R}, \quad \Omega_{\text{max}} = \omega + \frac{\alpha}{R} \tag{21}$$

required by the existence of an inner Alfvén point with $M_A^2 > 0$. When $\Omega^F \to 0$, the outer light surface moves to infinity and the inner one toward the static limit. For $\Omega^F = \Omega_H$, the inner light surface moves toward the horizon. When plasma is sitting on rotating field lines, the corotation radius r_c decides about the fate of the plasma. When plasma is injected at $r_{\text{in}} < r_c$, it will accrete onto the hole; for $r_{\text{in}} > r_c$ it will flow away as a wind

$$r_c = (M_H(1 - a_H\Omega^F)^2)^{1/3}(\Omega^F)^{\mp 2/3} = M_H(1 - a_H\Omega^F)^{1/3}\left(\frac{R_L}{M_H}\right)^{2/3}. \tag{22}$$

In fact, corotation is always inside the outer light surface, but it touches the outer light surface, when $\Omega^F = \Omega_{max}$. The magnetosphere in a Kerr background also has two Alfvén surfaces, an outer and an inner one. They are always located between the two light surfaces.[18]

The Hot Wind Equation and the Magnetosonic Points

Since the redshift factor u_t and the specific angular momentum $l = -u_\phi/u_t$ of the plasma flow are determined by conservation laws, we can derive an algebraic equation for the poloidal plasma flow[17,18] u_p with $u_p^2 = -u^A u_A$,

$$u_p^2 + 1 = \left(\frac{E}{\mu}\right)^2 \frac{(f,f)x_A^4 - 2x_A^4 M^2 - KM^4}{((f,f) - M^2)^2}. \tag{23}$$

This is still an implicit equation for the poloidal velocity u_p, since the Mach number is a function of u_p itself

$$M^2 = \frac{\mu}{m_p} \alpha u_p \sqrt{-g} \frac{\overline{\Phi}(r;\psi)}{\sigma_D}, \tag{24}$$

where the function $\overline{\Phi}$ represents the flux function of the magnetic surfaces, $\overline{\Phi} = \Phi_D/\Phi(r, \psi)$; σ_D is the magnetization parameter for the flux surface, which denotes the ratio between the Poynting flux and the mass energy flux at the injection point

$$\sigma_D = \frac{\Phi_D^2}{4\pi m_p J_D R_L^2} \simeq \frac{\psi_D^2}{2\dot{M}_w R_L^2} \simeq \pi^2 \frac{\dot{m}}{\epsilon_H} \left(\frac{3M_H}{R_D}\right)^{3/2} \tag{25}$$

for radiation-pressure-dominated disks $0.001 < \dot{m} < 1.0$. The poloidal equation shows two additional critical points, the slow and fast magnetosonic points.[17,18] Physical outflows must pass at least through the fast magnetosonic point. This is also true for magnetized accretion flows onto black holes. The magnetization parameter is the crucial quantity for the question of the outflow speeds.[7,19] When $\sigma_D > 1$, the outflow is accelerated to relativistic speeds with Lorentz factors $\gamma \simeq \sigma_D$; for $\sigma_D < 1$ only transrelativistic speeds are achieved.

MAGNETIZED ACCRETION ONTO RAPIDLY ROTATING BLACK HOLES AND THE BLANDFORD–ZNAJEK PROCESS

Near a black hole, turbulence in the disk has a rather minor influence on the accretion process, since differential rotation is overtaken by the increasing accretion velocity. The ideal MHD approximation is therefore valid in the innermost part of the disk around black holes, but certainly the force-free approximation is not valid for the accreting disk plasma. This means that the poloidal magnetic structure is parallel to the flow structure and we can study the magnetized accretion process with the full MHD program. This approximation is justified for radii smaller than the marginally stable radius. The magnetic fields are frozen into the disk as long as the inflow is subsonic. This determines in a way the field line velocity $\Omega^F \simeq \Omega(r_{ms}) = M_H/(\sqrt{r_{ms}^3} + \sqrt{M_H}a_H)$.

Inflow solutions have been considered starting at the corotation radius with vanishing poloidal velocity for a monopole-type geometry in the equatorial plane.[18] For $a_H = 0$, that is, $\Omega_H = 0$, and therefore $\Omega_H < \Omega^F < \Omega_{max}$, the accreting plasma brings in a positive energy and angular momentum, $P^r = nu^r E$ and $J^r = nu^r L$, which will accelerate a slowly rotating black hole. Of special importance is the case $0 < \Omega^F < \Omega_H$, in which a negative energy flux is possible. This follows from the wind equation **(23)**, written as

$$E - \Omega^F L = \mu_{in}\sqrt{(f,f)_{in}} \tag{26}$$

at the injection point ($u_{p,in} = 0$). Together with the Alfvén condition **(19)** this implies for the total energy and angular momentum

$$E = \frac{\mu_{in}\sqrt{(f,f)_{in}}(g_{tt} + \Omega^F g_{t\phi})_A}{(f,f)_A}, \quad L = -\frac{\mu_{in}\sqrt{(f,f)_{in}}(g^A_{\phi\phi}(\Omega^F - \omega_A)}{(f,f)_A}. \tag{27}$$

The total energy becomes negative if the Alfvén point occurs in a region where $(g_{tt} + \Omega^F g_{t\phi})_A < 0$. This can only occur if the Alfvén point is located inside the ergosphere ($0 < \Omega^F < \Omega_H$). In this situation, $\Omega^F < \omega_A$, so that also negative angular momentum is carried into the hole. This particular example for accretion with finite inertia is a realization of the Blandford–Znajek process, originally discussed only for force-free conditions.[16,20,21]

AXISYMMETRIC MAGNETOSPHERES OF BLACK HOLES AND THE ORIGIN OF JETS IN AGNS

The structure of a rotating magnetosphere follows from Ampère's equation **(14)** with the toroidal current density given by force-balance perpendicular to the flux surfaces

$$\rho_e \mathbf{E}_\perp + \mathbf{j}_T \wedge \mathbf{B}_P + \mathbf{j}_P \wedge \mathbf{B}_T = \mathbf{f}_\perp, \tag{28}$$

where $\mathbf{f}_\perp$ is the inertial force due to the streaming plasma. Plasma streaming on AGN magnetospheres is in general quite thin, so that the force-free approximation is a reliable one, $\mathbf{f}_\perp \simeq 0$, except for certain critical regions in the magnetosphere. Under force-freeness, the condition $\mathbf{f}_T = 0$ also implies $\mathbf{j}_P \| \mathbf{B}_P$, and therefore the relativistic Grad–Schlüter–Shafranov (GSS) equation [20,22–24]

$$\nabla \cdot \left\{\frac{\alpha D}{R^2}\nabla\psi\right\} + \frac{\Omega^F - \omega}{\alpha}\frac{d\Omega^F}{d\psi}|\nabla\psi|^2 + \frac{8\pi^2}{R^2\alpha}\frac{dI^2}{d\psi} = 0, \tag{29}$$

$$D = 1 - \frac{R^2(\Omega^F - \omega)^2}{\alpha^2}. \tag{30}$$

This GSS equation has critical points at the light surfaces of the magnetosphere that are identical with the Alfvén surfaces for the force-free condition. This is reflected in the fact that charge density and toroidal current density also have critical points at these surfaces. Under force-free conditions, the inner fast magnetosonic point also moves toward the horizon so that $B_P(r_H)_\perp = 0$. If finite inertia is included, the fast point moves away from the horizon and the boundary conditions along the horizon are no longer clear.[25]

We solved the GSS equation with a given flux distribution along the disk,[24] $\psi(r, \theta = \pi/2) = \psi_D(r)$ for dipolar and quadrupolar configurations. When the GSS equation is normalized in geometrical units,

$$\nabla \cdot \left\{\frac{\alpha D}{R^2} \nabla\psi\right\} + g_I f_2(\psi) = 0, \tag{31}$$

$$g_I = 8\pi^2 \frac{I_D^2 M_H^2}{\psi_D^2} = 3 \times 10^{-3} \left(\frac{I_D}{10^{17}\mathrm{A}}\right)^2 \left(\frac{M_H}{10^8 M_\odot}\right)^{-2} \left(\frac{B_D}{1\ \mathrm{kG}}\right)^{-2}, \tag{32}$$

solutions can only be found for coupling constants g_I smaller than some critical value. The total current I_D sustained by the magnetosphere cannot be arbitrarily large. We observe, however, a clear bending of the magnetic surfaces toward the rotational axis for large currents in the case of dipolar fields, which favors the formation of collimated relativistic jets.

REFERENCES

1. REES, M. J. 1984. Ann. Rev. Astron. Astrophys. **22:** 471–506.
2. SANDERS, D. B., E. S. PHINNEY, G. NEUGEBAUER, B. T. SOIFER & K. MATTHEWS. 1989. Astrophys. J. **347:** 29.
3. LAOR, A. & H. NETZER. 1989. Mon. Not. R. Astron. Soc. **238:** 897–916; **242:** 560.
4. CAMENZIND, M. 1989. *In* Neutron Stars and Their Birth Events, W. Kundt, Ed.: 139–177. Kluwer. Dordrecht, the Netherlands.
5. PUDRITZ, R. E. 1981. Mon. Not. R. Astron. Soc. **195:** 881–896, 897.
6. KUPERUS, M. 1987. *In* Gravitation in Astrophysics, Vol. 156, B. Carter and J. B. Hartle, Eds.: 195–208. Plenum. New York.
7. CAMENZIND, M. 1990. *In* Reviews of Modern Astronomy, Vol. 3, G. Klare, Ed.: 234–265. Springer-Verlag. Heidelberg/New York.
8. STEPINSKI, T. F. & E. H. LEVY. 1990. Astrophys. J. **362:** 318–332.
9. RUZMAIKIN, A. A., A. M. SHUKUROV & D. D. SOKOLOFF. 1988. Magnetic Fields of Galaxies. Kluwer. Dordrecht, the Netherlands.
10. BLANDFORD, R. D. & D. G. PAYNE. 1982. Mon. Not. R. Astron. Soc. **199:** 833.
11. LOVELACE, R. V. E., J. C. L. WANG & M. E. SULKANEN. 1987. Astrophys. J. **315:** 504.
12. KÖNIGL, A. 1989. Astrophys. J. **342:** 208–223.
13. KHANNA, R. & M. CAMENZIND. 1992. Astron. Astrophys. In press.
14. NOZAKURA, T. 1991. Mon. Not. R. Astron. Soc. **248:** 389–397.
15. CAMENZIND, M. 1991. *In* Variability of Active Galactic Nuclei, S. Wagner, W. Duschl, and M. Camenzind, Eds. Springer-Verlag. Heidelberg/New York: 201–210.
16. THORNE, K. S., R. H. PRICE & D. M. MACDONALD. 1986. Black Holes: The Membrane Paradigm. Yale University Press. New Haven, Conn.
17. CAMENZIND, M. 1986. Astron. Astrophys. **162:** 32–44.
18. TAKAHASHI, M., S. NITTA, Y. TATEMATSU & A. TOMIMATSU. 1990. Astrophys. J. **363:** 206–217.
19. CAMENZIND, M. 1989. *In* Accretion Disks and Magnetic Fields in Astrophysics, G. Belvedere, Ed.: 129–143. Kluwer. Dordrecht, the Netherlands.
20. BLANDFORD, R. D. & R. L. ZNAJEK. 1977. Mon. Not. R. Astron. Soc. **179:** 433.
21. BLANDFORD, R. D. 1989. *In* Theory of Accretion Disks, F. Meyer, W. J. Duschl, J. Frank, and E. Meyer-Hofmeister, Eds.: 35–57. Kluwer. Dordrecht, the Netherlands.

22. MACDONALD, D. & K. S. THORNE. 1982. Mon. Not. R. Astron. Soc. **188:** 345.
23. CAMENZIND, M. 1987. Astron. Astrophys. **184:** 341–360.
24. HAEHNELT, M. & M. CAMENZIND. 1991. Submitted for publication in Mon. Not. R. Astron. Soc.
25. PUNSLY, B. & F. CORONITI. 1990. Astrophys. J. **354:** 583.

Is Physics Consistent with Closed Timelike Curves?[a]

JOHN L. FRIEDMAN

Department of Physics
University of Wisconsin-Milwaukee
Milwaukee, Wisconsin 53211

INTRODUCTION

An underlying motivation for considering closed timelike curves (CTCs) is the possibility that at scales where quantum fluctuations of the metric are of order unity, the notion of spacetime—of manifold and Lorentz metric—retains its meaning. Local fluctuations of the metric would be independent of fluctuations several Planck lengths away, and the resulting randomly oriented field of light cones would imply a sea of small CTCs. One would therefore expect to avoid microscopic CTCs only by adopting a theory in which causal structure is fundamental (e.g., a theory based on causal sets), or one in which the metric is not meaningful or not Lorentzian on small scales.

There is a second motivation, having to do with changes in spatial topology. When a black hole forms and evaporates, information is lost to external observers. If one does not know the initial state of the collapsing system, arbitrarily accurate measurements of the exterior spacetime cannot recover the state inside the horizon. If, however, the collapsing matter reexpands, forming a child universe, it would be possible to avoid loss of information in the full spacetime.

The usual picture of a trousers universe is a spacetime in which a single S^3 bifurcates to form two S^3, and it allows no Lorentz metric. If, however, one removes three balls from CP^2, the resulting manifold admits a Lorentz metric in which each of the three spherical boundaries is spacelike, representing one initial and two final universes. The metric is everywhere smooth and nonsingular, but the spatial manifold has changed, and the metric necessarily has CTCs.

More generally, a spacetime with CTCs is the price of Lorentzian topology change. Any two 3-manifolds, S_1 and S_2, can be joined by a 4-manifold with a smooth Lorentz metric for which S_1 and S_2 are spacelike. Geroch[1] showed that if S_1 and S_2 are not homeomorphic, or if M is not a product of the form $S_1 \times R$, then the metric must have CTCs. As stated in Geroch's paper, the spacetime must either have closed timelike curves or be time nonorientable, but by going to the covering space, one can easily prove that there must be CTCs in all but one case: the creation of a universe from nothing. (A simple extension of Geroch's theorem from closed to asymptotically flat spacetimes is given in reference 2.)

Then if one allows spacetimes with CTCs in a sum-over-histories framework, topology change is implied by a Lorentzian path integral—there are nonzero

[a]Time spent in preparing this brief overview was supported in part by National Science Foundation Grant PHY 8603173.

amplitudes for continuous paths that start at any 3-manifold and end at any other. Within the framework of a nonsingular Lorentzian metric, the converse is also true: Topology change requires CTCs.

CTCs have commonly been regarded as pathologies of general relativity that do not arise in the real world, and that lead to an inconsistent theory, allowing almost no consistent initial data at the classical level and no consistent interpretation at the quantum level. This paper serves as a brief introduction to recent work on whether the laws of physics might permit CTCs. A paper by Kip Thorne[3] and reference 4 are more detailed reviews, and this paper spends somewhat more time on the lesser issues with which I have been directly involved.

MODEL SPACETIMES WITH CTCS

A 3-dimensional wormhole can be constructed by removing two balls from Euclidean space and identifying their spherical boundaries Σ_1 and Σ_2, as shown in FIGURE 1. To describe a corresponding spacetime with a smooth, time-independent, traversable wormhole one can choose a metric of the form

$$ds^2 = -dt^2 + dl^2 + r^2(l)\, d\Omega^2;$$

for example,

$$ds^2 = -dt^2 + dl^2 + (l^2 + a^2)\, d\Omega^2.$$

Here l is a radial coordinate that runs from $-\infty$ to ∞, and has the value zero halfway

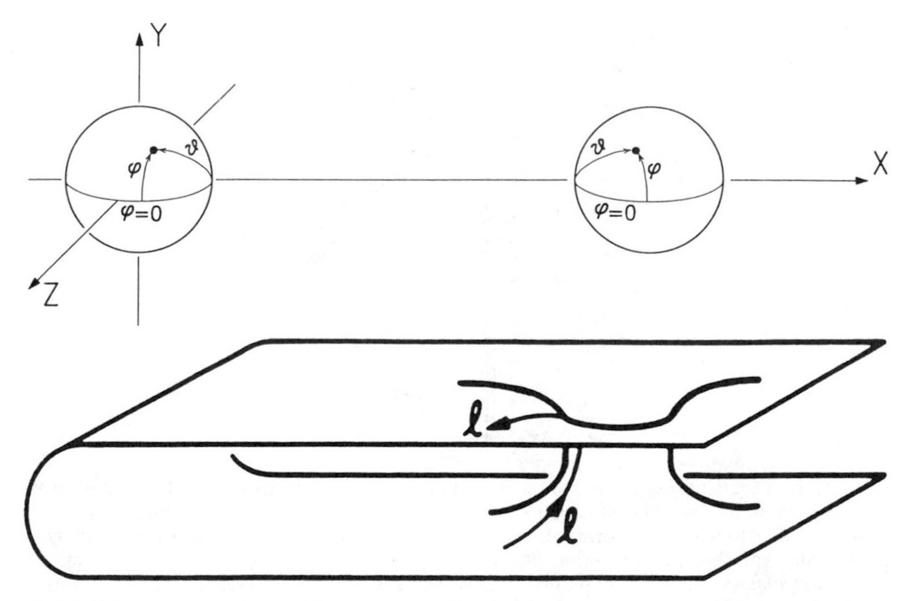

FIGURE 1. A spatial diagram showing how points on the two wormhole mouths are identified. Points with the same values of θ and ϕ are the same.

through the wormhole throat, and $2\pi r(l)$ is the circumference of a sphere surrounding the hole at coordinate l.

Morris and Thorne[5,6] considered spacetimes in which the two mouths of a wormhole move toward one another, as seen by the external spacetime. The relative boost alters the separation between identified points so that an initially spacelike separation becomes null and then timelike (FIG. 2). In an alternative spacetime of Frolov and Novikov,[7] one surrounds one wormhole mouth by a cloud of matter. With no matter, the timelike Killing vector has the same length everywhere, but the presence of the cloud means that the Killing vector field defined outside the two mouths meets at the identified spheres with a discontinuity in its length, due to the gravitational redshift of the cloud. Again identified points with initial spacelike separation become timelike separated, joined by CTCs.

Gott[8] has recently considered two-dimensional spactimes whose only source is two infinitely long cosmic strings that pass each other at relativistic speeds. These do not violate the weak energy condition, but they are not asymptotically flat. In a detailed description of Gott's solutions, Cutler[9] shows that each of the spacetimes has a spacelike surface prior to which there are no CTCs, although the region containing CTCs extends to spatial infinity. There is a natural way to associate a total 4-momentum with a system of infinite strings, and (as Deser *et al.*[10] and Carroll *et al.*[11] show) the total 4-momentum associated with the strings is negative for Gott's spacetimes.

There are several ways in which physics might prevent such spacetimes on a macroscopic scale. First, matter satisfying the weak energy condition appears not to allow asymptotically flat spacetimes with CTCs;[12,13] wormholes, for example, would collapse.[5] Second, the Cauchy horizon might be unstable, with the energy density of

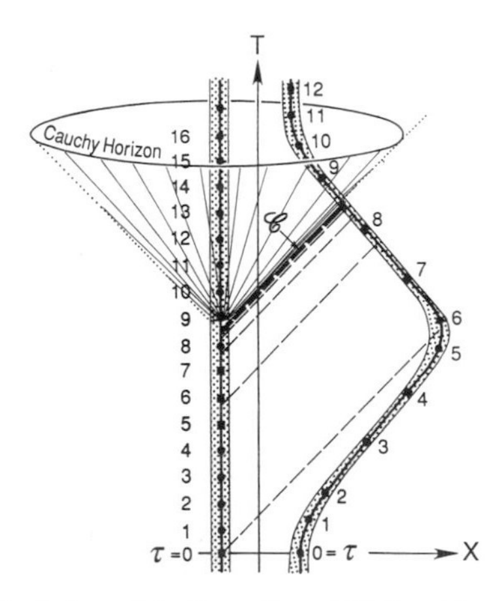

FIGURE 2. A spacetime depiction of the formation of CTCs arising from the relative motion of wormhole mouths. The *stippled region* is excised from the spacetime and its outer edges are joined together to make the wormhole throat. *Numbered dots* represent proper time read by a clock in the wormhole, and *two dots* at the same proper time represent the same spacetime point. *Straight lines* joining these points are closed curves; they are timelike in the future of the Cauchy horizon. (From Friedman *et al.*[4] Reproduced by permission.)

initial fields growing without bound before the region of CTCs is reached. Third, the Cauchy problem itself might not be well defined: for generic initial data there might be no solutions to its field equations, or there might be more than one solution. Finally, there might be no consistent quantum field theory in a spacetime with CTCs, one that allows no decomposition into space × time. The recent work has been primarily concerned with these questions.

DOES PHYSICS MAKE SENSE ON SPACETIMES WITH CTCS?

Classical Free Fields

It has commonly been thought that in spacetimes with closed timelike curves, one cannot find consistent time evolutions of classical fields for generic initial data—that the Cauchy problem is not well defined. A simple example is the temporally compact spacetime obtained from the piece of Minkowski space between the $t = 0$ and $t = T$ planes by identifying the points $(t = 0, x)$ and $(t = T, x)$. A well-defined solution exists only if the Minkowski evolution is periodic in time with period T—there is no solution for generic data. For the wormhole spacetimes mentioned previously, however, the set of closed timelike and null geodesics has measure zero. Morris, Thorne, and Yurtsever[6] noted that the evolution of free fields is well defined in the limit of geometrical optics; and this in turn makes it seem likely that a multiple scattering series converges to a solution for arbitrary initial data.[4,5,7]

Morris and I[14] have recently obtained a formal proof that the Cauchy problem is well defined on a class of static wormhole spacetimes with CTCs. Because there are no complete spacelike hypersurfaces, one must pose data at past null infinity. Standard existence theorems for hyperbolic equations are based on energy inequalities or the spectral theorem for a self-adjoint operator, and they fail on spacetimes with CTCs. It nevertheless turns out that one can build a solution as a convergent sum of nonorthogonal modes. The formal proof is technical, and a brief discussion of the heuristic argument is more illuminating.

An incoming wave from $\mathscr{I}^-$ may be thought of as initially scattering off of the wormhole mouths, producing at each mouth some purely outgoing scattered wave composed of a reflected part that comes from the initial scattering at that mouth and a transmitted part that comes from the initial scattering at the other mouth. These first-scattered waves in turn scatter again, producing new reflected and transmitted waves, and so on. In each scattering, in the geometrical optics approximation, the amplitude of the wave decreases by a factor $a/2d$ in case (a) or $a/[2d(1 - v)]$ in case (b) in traversing the wormhole and then traveling from one mouth to the other. Again the series converges when $a/[2d(1 - v)]$ is small compared to one. We have explicitly constructed this series for the static spacetime, (a), when the spacetime external to the wormhole throat is flat. A proof of existence can then be secured by a proof that this multiple-scattering series is convergent. For small mouths, one can prove convergence for wavelengths short enough that geometrical optics must be valid ($\lambda \ll a$); and one already had convergence in the geometrical optics limit. We therefore have little doubt that, at least for handles with small mouths, the multiple scattering series converges.[15] We thus claim an alternate, partly numerical, construction of the modes whose existence we have previously established.

Interacting Systems: Classical Billiard Balls

For a free scalar field, the grandfather paradox is easily resolved for the wormhole spacetimes. An initial field that propagates along a CTC returns to its starting point with a different value; but because the field is linear, a unique consistent solution is given by summing the result of such traversals, by a multiple scattering series of waves that have propagated through the wormhole an arbitrary number of times. For self-interacting fields, the grandfather paradox is much more interesting. Echeverria, Klinkhammer, and Thorne[16] have examined a simple nonlinear problem in which a billiard ball traverses a time-tunnel, and on its return, collides with its earlier self. The collision is violent enough to prevent the original ball from entering the wormhole, and it appears at first that one has no self-consistent solution. Remarkably, however, it appears that solutions always exist: The incident ball is struck a glancing blow on its way to the wormhole, deflecting it in such a way that when it emerges, it is aimed correctly to give its earlier self precisely that glancing blow! Solutions of this type are not unique—in fact, many initial trajectories allow an infinite number of solutions with the ball traversing the wormhole an arbitrarily large number of times. Although they have no formal proof of existence, Echeverria *et al.* searched for initial trajectories in which a solution fails to exist and they found none.

Novikov[17] discusses a class of more complicated classical interactions, with bombs whose debris returns to prevent the initial explosion. It appears that by adding an extra degree of freedom to the billiard balls one *can* find a classical system for which some initial data has no classical evolution.[18] Suppose that one has a variable—call it spin—that can take two values—up and down—and that the interaction energy of the switch is small compared to the collision energy. One can choose a Hamiltonian so that when two down balls collide, they change to two up balls, but when up collides with down, the spins do not change. If one sends an up ball through the wormhole, and if it collides with an up ball (its future self) before traversing the wormhole, its spin will change to down; this contradicts the fact that it emerged from the wormhole with spin up. If it collides with a down ball, it remains up, contradicting the fact that its future self was down. So there appears to be no self-consistent solution. If the billiard ball trajectories of Echeverria *et al.* always allowed solutions with arbitrarily small glancing angle, one could get around the problem by choosing the glancing angle small enough that the collision energy was smaller than the spin-flip energy. Then the spin flip could disrupt the collision. But the glancing angles for a given initial trajectory appear to be bounded above some minimum value, and it therefore appears that one cannot find a consistent classical solution.

Although the classical theory of interacting billiard balls is not well-defined, the evolution of a quantum mechanical wave function has a better chance. Because the Schrödinger equation and its relativistic counterpart are linear, it is plausible that a solution exists and is unique. Before considering this question, let us return to the problem of maintaining traversable wormholes, and turning them into time machines.

QUANTUM PHYSICS

Can Generic Properties of Quantum Fields Forbid Macroscopic CTCs?

To prevent collapse of a wormhole, one must violate the the averaged weak energy condition:[5] If k^a is tangent to a null geodesic with affine parameter λ, the energy-momentum tensor must satisfy,

$$\int T_{ab} k^a k^b \, d\lambda < 0.$$

More generally, Tipler proved[12] that if a spacetime is asymptotically flat, satisfies the weak energy condition, and is geodesically complete, then it cannot have CTCs. This has recently been substantially strengthened by Hawking,[13] who shows that the assumption of geodesic completeness is not needed. (The wormhole spacetimes have an incomplete geodesic that winds to the Cauchy horizon, but the spacetimes are smooth and they allow well-defined free fields.)

The local weak energy condition is violated by particle creation, by the Casimir effect, and by squeezed vacuum states.[19] The averaged weak energy condition is more difficult to violate. It holds for the scalar field in 4-dimensional Minkowski space and for asymptotically flat 2-dimensional spacetimes,[20–23] but not in a generic four-dimensional spacetime. Consequently, the question of whether one can maintain traversable, macroscopic wormholes remains open.

Quantum Instability of the Cauchy Horizon

The fact that the Cauchy problem is well defined for free classical fields means that the Cauchy horizon is classically stable to linear perturbations. In the quantum theory, however, Kim and Thorne[24] find that the renormalized stress tensor diverges at points joined by closed null geodesics. Their calculation uses the Hadamard form of the Green's function to compute $\langle T_{ab}\rangle$ by point splitting. In the geometrical optics approximation, the Green's function, $G^{(1)}(x',x)$, normally has two terms that diverge when the (squared) geodesic distance, $\sigma(x,x')$, goes to zero. In a normal neighborhood, σ is unique, and one regularizes $G^{(1)}$ by subtracting the divergent part. When there are CTCs, however, the remaining finite part of the Green's function contains a sum of large terms corresponding to null geodesics that are nearly closed, and that join x to x'. That is, $G^{(1)}$ has the form,

$$\begin{aligned} G^{(1)}(x',x) &= \langle\psi|\phi(x)\phi(x') + \phi(x')\phi(x)|\psi\rangle \\ &= \sum_N \frac{\Delta_N^{1/2}}{4\pi^2}\left(\frac{1}{\sigma_N} + u_N \ln|\sigma_N| + w_N\right). \end{aligned}$$

To regularize, one removes only the $N = 0$ term. Then $\langle T_{ab}\rangle$, constucted by taking two derivatives of G, diverges as one approaches the Cauchy horizon. When the distance Δt to the Cauchy horizon is of the order of the Planck length, as measured by an observer moving along one of the wormhole mouths, T_{ab} is very small. For $\Delta t > l_{\text{Planck}}$, and the maximum fractional strain is less than 10^{-35}, when the wormhole separation is macroscopic (1 m). This may mean that the semiclassical approximation cannot show more than a negligible disruption of the spacetime in its regime of validity. The smallness of the strain, however, is observer dependent, and Hawking[18] argues that the divergence is serious: An observer moving along a path joining the point x to the intersection of the Cauchy horizon with the wormhole mouth sees a fractional strain of order unity when she is within a Planck time of the Cauchy horizon. Quantum fields may satisfy a version of the weak energy condition that prevents one from maintaining spacetimes with CTCs. If not, the quantum instability of Kim and Thorne may prevent one from ever forming them.

Consistent Quantum Theory on a Background Spacetime with CTCs

While we have encountered several possible ways in which one might rule out macroscopic CTCs, it is at least plausible that CTCs pervade a small-scale Lorentzian foam. A first question is whether one can generalize quantum field theory to accommodate a background spacetime with CTCs.[25] An obvious candidate for such a generalization is a sum-over-histories framework. If the CTCs are confined to a compact region, one could look at histories that end at a spacelike hypersurface S to the future of the region containing CTCs. Probabilities for different classes of histories in the CTC-region would have the same formal expression as probabilities for histories when there are no CTCs. That is, if $A(\Phi)$ is the amplitude for a field Φ with value $\Phi|_s$ on S, then the probability for a class C of histories has the form[26]

$$P(C) = \frac{\sum_{\Phi|_S} |\sum_{\Phi \in C} A(\Phi)|^2}{\sum_{\Phi|_S} |\sum_{\Phi} A(\Phi)|^2}.$$

For a static time-tunnel spacetime, or one in which the relative velocity of wormhole mouths is slow, one can consider a nonrelativistic limit of the theory. Thorne and Klinkhammer have looked at the quantum mechanics of a billiard ball on such a spacetime, using a Wentzel–Kramer–Brillouin (WKB) approximation to construct an explicit path-integral. They find that all classical histories that are short enough so that the wave packet stays together are equally likely, while paths that are long enough so that the wave packet has spread (so that the WKB approximation is invalid) are expected to have almost zero probability. This is a reassuring picture: The lack of uniqueness present for the classical billiard balls has been replaced by a well-defined probability for each of those classical paths, so quantum theory would seem to have overcome one of the classical difficulties.

However, one in general violates unitarity—probabilities can differ if one places the final hypersurfaces before the region of CTCs rather than after. At least for the case discussed earlier, where there appear to be no classical solutions, this leads to a loss of causality in regions that do not interact with the CTCs.[12] In some cases, Thorne and Klinkhammer show that one can avoid a loss of causality of this type by a prescription that discontinuously changes the surface S from a location in the past of the CTC-region to a location in its future. That is, to determine the probability of events before the Cauchy horizon, one uses a surface in the past of the CTCs, while for events after the Cauchy horizon one takes S in the future. However, one cannot, in general, avoid the problem in this way for spacetimes containing disjoint regions with CTCs.

ACKNOWLEDGMENT

I am grateful to Kip Thorne for a rapidly transmitted copy of his review (reference 3).

REFERENCES

1. GEROCH, R. P. 1967. J. Math. Phys. **8:** 782–786.
2. FRIEDMAN, J. L. & A. HIGUCHI. 1990. Nucl. Phys. **B339:** 491–515.
3. THORNE, K. S. 1991. Do the laws of physics permit closed timelike curves? *In* Sixth Florida Workshop in Nonlinear Astronomy: Nonlinear Problems in Relativity and Cosmology. N.Y. Acad. Sci. **631:** 182–193.
4. FRIEDMAN, J. L., M. S. MORRIS, I. D. NOVIKOV, F. ECHEVERRIA, G. KLINKHAMMER, K. S. THORNE & U. YURTSEVER. 1990. Phys. Rev. D **42:** 1915–1930.
5. MORRIS, M. S. & K. S. THORNE. 1988. Am. J. Phys. **56:** 395–412.
6. MORRIS, M. S., K. S. THORNE & U. YURTSEVER. 1988. Phys. Rev. Lett. **61:** 1446–1449.
7. FROLOV, V. & I. NOVIKOV. 1990. Phys. Rev. **D42:** 1057; NOVIKOV, I. 1989. Zh. Eksp. Teor. Fiz. **95:** 769. [Sov. Phys. JETP **68:** 439].
8. GOTT, J. R. 1991. Phys. Rev. Lett. **66:** 1126–1129.
9. CUTLER, C. 1991. Caltech preprint GRP-282; see also ORI, A. 1991. Phys. Rev. **D44:** 2214–2215.
10. DESER, S., R. JACKIW & G. H. 'T HOOFT. 1991. MIT preprint CTP #2011.
11. CARROLL, S. M., E. FARHI & A. H. GUTH. 1991. MIT preprint CTP #2009.
12. TIPLER, F. J. 1976. Ann. Phys. **108:** 1–36.
13. HAWKING, S. 1991. The chronology protection conjecture. Submitted for publication in Phys. Rev. Lett.
14. FRIEDMAN, J. L. & M. S. MORRIS. 1991. Phys. Rev. Lett. **66:** 401–404.
15. FRIEDMAN, J. L. & M. S. MORRIS. 1991. Paper in preparation.
16. ECHEVERRIA, F., G. KLINKHAMMER & K. S. THORNE. 1991. Phys. Rev. **D44:** 1077–1099.
17. NOVIKOV, I. D. 1991. Phys. Rev. D. In press.
18. FRIEDMAN, J. L. & J. SIMON. Paper in preparation.
19. BRAUNSTEIN, S. & M. MORRIS. Paper in preparation.
20. YURTSEVER, U. 1990. C. Quantum Gravity **7:** L251.
21. KLINKHAMMER, G. 1991. Phys. Rev. **D43:** 2542–2548.
22. FORD, L. D. 1991. Phys. Rev. **D43:**3972.
23. WALD, R. M. & U. YURTSEVER. 1991. Phys. Rev. **D44:** 403–416.
24. KIM, S.-W. & K. S. THORNE. 1991. Phys. Rev. **D43:** 3929.
25. HARTLE, J. B. 1991. *In* Quantum Cosmology and Baby Universes (Proceedings of the December 27, 1989–January 4, 1990 Winter School in Jerusalem, Israel).
26. SINHA, S. & R. D. SORKIN. 1991. Foundations of Physics. In press.

Summary: Miniworkshops on Space-based Astrophysics

HARLAN J. SMITH

Astronomy Department
University of Texas at Austin
Austin, Texas 78712
and
Lunar and Planetary Institute
Houston, Texas

INTRODUCTION

On the twentieth anniversary of the first Apollo landing on the moon, President Bush defined major strategic goals for the U.S. space program. These goals became subsumed under the heading of the "Human Exploration Initiative," later renamed "Space Exploration Initiative," or SEI. While recognizing the importance of robotic explorations, the SEI accepts the concept of major human participation in the next two or three decades of large-scale developments in space—successively with *space stations,* then *permanent bases on the moon,* and finally *human presence on Mars.* Each of these phases involves science, including especially some very important new observational opportunities for astronomy, astrophysics, and physics.

With one exception, the scientific developments or prospects associated with human activities in space are much less well known than some of the spectacular dedicated robotic missions such as Einstein, IRAS, COBE, and Hipparchos. Accordingly, in responding to the Meeting Organizers' call to pull together a couple of miniworkshops on observational astrophysics off the earth, I decided to concentrate on the prospects for manned or man-tended activity, dealing first with currently or (hopefully) soon to be operational earth-orbiting systems, and to use this as a springboard for the second workshop on the remarkable prospects that lunar-based observatories should represent. These meetings offered something of a change of pace from the rather intense and concentrated topics of the full-day sessions, and also contain material of which theoreticians should be aware in order to help build the consensus needed to accomplish great future space missions upon which in turn much of twenty-first century relativistic astrophysics will depend.

WORKSHOP #1: SPACE-BASED ASTROPHYSICS

This first workshop dealt essentially with earth-orbiting observatories: Hubble Space Telescope, the Soviet space station MIR, and the prospective U.S. space station Freedom.

Hubble Space Telescope

The Hubble Space Telescope (HST) is the first space observatory to be both man-launched and to require occasional human tending. Its two flaws, discovered subsequent to the May 1990 launch, are well known indeed. What is not so well known, even among the wider scientific community, is the remarkable success that this great instrument already represents.

Alec Boksenberg, Principal Scientist for the HST European Wide-field Camera Instrument Team, reviewed the current state and prospects of the telescope. He made a number of principal points.

- On brighter objects, HST already achieves essentially its theoretical diffraction-limited performance of about 0.07 arcsecond. This is nearly a factor of 10 better than all but the tiniest fraction even of high-quality images obtained by ground-based telescopes over the past century, and is about a factor of 5 better than the best. It is a far cry indeed from the irresponsible newspaper reports of total disaster, complete incompetence, and inexcusable waste that accompanied the first disappointing and somewhat out-of-focus images to be returned. Boksenberg showed many stunning images, including, for example, novel clouds on Saturn with resolution comparable to images returned by the Pioneer and Voyager probes, and rich details never before seen in the active core of the Seyfert galaxy NGC1068.
- However, the magnificent resolution obtainable on bright objects comes from the 15 percent of light in the sharp, diffraction-limited core of the spherically aberrated image resulting from Perkin-Elmer Corporation's delivery of an imperfect primary mirror. The remaining 85 percent of the light, smeared over broad wings surrounding each brighter point in the images, must be taken care of by eliminating all but the brighter portions of images, and/or by computerized image restoration using the extremely well-known modulation transfer function of the aberrated system. As a practical matter, it is not possible to recover all of the information that would be present in a perfect image; also exposures must be longer than would otherwise be required to achieve a given level of information content in the final image. Fortunately, at only a few percent of the original cost of HST it will be possible to restore the originally designed performance. All of the instruments were designed to be replaced by astronauts, and—because they were designed and built so long ago—work had already been under way for several years to build the second-generation instruments. These will now be provided with simple optical correctors to eliminate the spherical aberration. The precise mechanism for doing this will almost certainly be through replacing the fast photometer with a "COSTAR" device that will ensure that each of the other instruments has its appropriate corrector when required. So, by 1994, HST should be restored to essentially its full design optical performance.
- The second problem with HST stems from the European-provided solar arrays that flex upon experiencing the thermal shock of moving from light to shade or back again. Since the telescope has no inertial base, this motion causes the entire system to vibrate for some minutes in a manner difficult to control with

the moment gyros. The principal consequence of this is to reduce the exposure times available for the other instruments. Careful programming of the control gyros during the intervals of thermal shock substantially reduces the problem, but the final solution may lie in replacing the solar arrays with new ones having much less thermal flexure. This could also be done when the main instruments are replaced.

- Despite these problems, every instrument on HST is itself performing as designed, and several of them are rather little affected. As a result there remains at least a fivefold oversubscription of outstanding science to be done by HST even in its present condition. To give only one example, the telescope is already busy with UV studies of relatively nearby galaxies, quasi-stellar objects (QSOs), and intergalactic absorptions for which the redshift is insufficient to bring the Lyman-alpha forest into accessibility for ground-based telescopes.

MIR, and Other Related Soviet Space Missions

Until recently, relatively little was known (or at least widely recognized) concerning the growth in amount and quality of Soviet space science. With glasnost this is rapidly changing. A high spot of the evening minisymposia was the spirited presentation by Rashid Sunyaev of some of the properties of the large Soviet space station MIR, the high-energy astrophysical observatories attached to it now or in the near future, and some work by similar Soviet high-energy satellites now operating or under construction.

Although the habitable volume of the huge U.S. Skylab space station sent up some 20 years ago (before the United States inexplicably trashed its Saturn 5 rocket capability) has not yet been exceeded, the MIR concept allows almost indefinite growth by relatively simple accretion of units onto central station docking ports. Some of these new units in turn offer more ports for further expansion. The original 20-ton central living and working unit of MIR has now been orbiting for about five years. This unit will continue to be used until 1994, then be replaced with a similar one scheduled for use until 2000; it is not likely that a contemplated MIR II will be built.

Three scientific units—*Kvant I* (1987), *Kvant II* (1989), and *Kristall* (1990)—totaling nearly an additional 50 tons have been added to the central MIR module, and two more units each of 20 tons will be added in 1991 and 1992. Even the first Kvant gave X-ray imaging over the range 2 to 30 keV with 2-arcminute resolution in a 15 × 15 degree field, and spectrometry of sources in the 15 to 150-keV range. It, and its companion instruments, have been timing X-ray pulsars, studying bursters, observing Supernova 1987A over the range 2 to 800 keV, and surveying black hole candidates. (One concern with science from manned space stations, that the movements of astronauts would disturb the pointing of instruments, has not proved to be a problem on MIR because of its great inertia and because its present instruments require only arc-minute stability.)

While there have been some good results from the attached MIR payloads, the most exciting Soviet high-energy data have come from the *Granat* spacecraft. It is blessed with a high-apogee 4-day orbit, and carries a coded mask for good positional sensitivity. The 2200-kg scientific instrument has system efficiency of 59 percent, and the remarkable spectral range of 4 to 1500 keV. It has proved especially good for imaging X-rays and soft gamma rays, for broad-band spectroscopy, for timing sources and bursters, and even for monitoring of solar X-rays. Its most exciting result to date is the discovery of what appears to be the principal and highly variable positronium source near the galactic center (actually 40 arc-minutes away), which when most active puts out up to 40 percent of its energy in the positronium annihilation band. Major variations range on timescales down to as short as days, while the general spectrum is consistent with a comptonized disk at several hundred kiloelectronvolts—the hottest source so far identified. These properties are consistent with a black hole of perhaps 10 solar masses. Curiously, the possible black hole of roughly 10^6 solar masses at the galactic center, presumably corresponding to Sgr A, is barely detectable to the Granat instruments.

Granat should continue to function into 1992. *Astro D,* in cooperation with Japan and NASA, is scheduled for January 1993, to work out to 10 keV with very high-quality optics for imaging, timing, and spectroscopy. *Spectra x-gamma,* supported by a dozen countries, will also go up in 1993 or 1994, in a similar orbit and with the same-size payload as Granat, to do a 10^5 source sky survey and virtually all kinds of X-ray astronomy over the range 0.1 to 100 keV. Several other scientific satellites using the same general spacecraft design as Spectra x-gamma—particularly *Radioastron* and *UFT* (a 170-cm HST-like UV telescope)—are scheduled for the 1994 to 1997 time frame. Apart from political instability, the prospects for these and subsequent instruments are good, since at least at present the Soviet military-industrial complex, like that in the United States, is very motivated to convert at least in part to scientific research.

U.S. Space Station Plans

When this part of the session was scheduled in late summer of 1990, the plans for U.S. Space Station Freedom were still intact, complete with hopes, if not even expectation, for those attached science payloads that had emerged from a rigorous competitive selection process. By the time of the meeting in Brighton, considerable retrenchment had occurred, and essentially during the meeting itself the report of the influential Augustine Commission emerged, recommending among many other things decoupling of most if not all science from the station.

Faced with this problem, William Taylor, Head of the Space Station Science Office, made an effective presentation of science still likely to be accomplished on some version of a space station, probably with international cooperation. He described the two projects that are most appropriate and least likely to be dropped from the long list of excellent science instruments that have been recommended for the space station. *LAMAR* is a very-large-aperture X-ray system covering the range 0.15 to 2 keV, and giving 30-arcsecond resolution of a wide field. The *X-ray Background Survey Spectrometer* would concentrate on the low-energy diffuse background over wide fields in the sky. However, as of early 1991, funding for these instruments—indeed for the entire U.S./international space station, whatever form it may take—is in limbo.

WORKSHOP #2: LUNAR-BASED ASTROPHYSICS AND PHYSICS

As noted in more detail in my longer article on moon-based astrophysics in this volume, several factors caused the shortening of this minisymposium to an hour; also the concentration of the first two originally scheduled papers into a single one, leaving only two presentations to be given at the minisymposium. Because full articles covering both of these appear in this volume, the summary below is less extensive than the one for the minisymposium on space-based astrophysics.

Astrophysics from the Moon

Much astronomical effort throughout the past century has gone into finding and developing ever better sites for observation. The progression led first to remote and towering peaks on earth, then to systems orbiting above the atmosphere. This process has one more major step to take—to the moon.

The moon offers an extraordinary constellation of factors conducive to observational astronomy and astrophysics. In the main moon-base article referred to before these were summarized under three primary categories: *ultrahigh vacuum, lunar mass and size,* and *location and motions of the moon.* As a consequence of the many detailed factors, the virtues of orbiting telescopes can be achieved on the moon with much simpler construction, more versatility, and greatly simplified operation. The initial lunar telescopes will be relatively small, and probably robotically landed and operated. However, manned lunar base(s) seem virtually certain to come into existence in the first decade of the coming century. This will make possible the gradual development of lunar telescope systems having size and performance vastly exceeding anything that can be done on earth, and probably far beyond what would be feasible to do in free space.

Much of the material covered in the Brighton presentation came from the large February 1990 Annapolis Workshop on *Astrophysics from the Moon* (Mumma and Smith, *American Institute of Physics Conference Proceedings,* Vol. 207). In particular that workshop raised the question of vital astronomical problems that will probably still be of the highest interest to the next generation, and that will probably require lunar-based telescopes for their resolution. While a very large number of problems will ultimately fit these categories, the Annapolis workshop centered on those connected with ultrahigh resolution (range of 100 to 1000 km) on the surfaces of stars, with the details of development of solar systems, with detection and characterization of earthlike planets, with the search for life in the universe, with directly observing at least some of the accretion disks around the black holes presumed to lie at the nuclei of many if not most galaxies and quasars, and with probing to the very beginning of the crystallization of large discrete luminous objects in the universe.

Three principal types of giant instruments seem most likely to emerge. Most important will be UV/optical/IR imaging interferometers that may ultimately achieve resolutions in the microarcsecond range. Almost equally important will be immense diffraction-limited filled-aperture telescopes again spanning the wavelength range from far UV out to submillimeter. A third class of instruments will be immense radio telescopes and interferometers including the little-known VLF wavelength region, especially on the radio-quiet lunar backside.

We should see the beginnings of lunar-based astronomy before the end of this decade, and with increasing international cooperation in the admittedly expensive and difficult task of setting up the first permanently inhabited lunar base, the beginnings of major moon-based astronomy and astrophysics programs in the early years of the coming century. Given continuance of civilization of earth, such developments are certain to occur on some timescale. Given the difficulties of beginning and sustaining major new thrusts in science, large-scale support from the community will be needed to ensure that they begin to happen on a timescale of interest to us.

Particle Astrophysics from the Moon

For a number of years scientists have occasionally noted possible advantages of using the moon for various kinds of physical research, including plasmas, general relativity, gravitational radiation, X-ray, gamma-ray, and particle physics. Many of these topics were considered at some length at the 1989 First NASA Workshop on *Physics and Astrophysics from a Lunar Base* (A. Potter and T. Wilson, Eds., *American Institute of Physics Conference Proceedings,* Vol. 202, 1990).

In the short time available to him on our program Kenneth Lande chose to concentrate on particle astrophysics on the moon, the subject of his paper in this volume. It is also true that many of the proposed physics experiments that would profit from lunar basing will require a very large lunar-base infrastructure that is unlikely to be in place for some decades to come. Accordingly Lande also focused on the several experimental approaches that might be practicable in the relatively early years of lunar development, especially those involving neutrino and cosmic-ray detectors. Research programs that he suggested included the following.

High-energy Neutrinos

Until now it has not been possible to locate astronomical sources of high-energy neutrinos ($E > 100$ MeV), because of the noise background from atmospheric neutrinos. On the moon, with need to average only the angular difference between the direction of the neutrino incident on the detector and that of the secondary charged particle that is actually detected, there may be some prospect of finding celestial sources. Given an adequately large and thick detector, the lack of lunar atmosphere also would allow detection of astronomical neutrinos (if any exist) with energies up even into the range of 10^{14} eV. Conversely, however, low-energy neutrinos are best observed from the earth, because geomagnetic shielding deflects cosmic-ray primaries below 15 GeV, and these are the major source of low-energy neutrinos.

Nucleon Decay

For the neutrinos of interest in the search for proton or other nucleon decay, the absence of atmosphere gives the moon an advantage of about 10^3 over the earth. A solid detector with mass of about 10 tons/m^2 could be build with active particle-sensitive layers interspersed with passive ranging layers, beginning with a relatively small area, but continuing to increase in size as resources permitted. This instrument should also serve as the neutrino astronomical-source detector.

Earth–Moon Neutrino Vacuum Oscillations

By aiming a beam of neutrinos generated by a terrestrial accelerator toward a far-side lunar detector, the flight path between earth and moon would serve as the oscillation region, while transversal through the moon would serve as the neutrino conversion region. This would sample the only flight path available to experiment other than the earth–sun or the terrestrial diameter. Also, by adjusting the energy of the primary beam of the accelerator, the energy spectrum of the emitted neutrino beam can be varied, allowing the measurement of the probability of either muon neutrino disappearance or appearance as a function of neutrino energy. In turn this measurement determines whether neutrino vacuum oscillations occur, and if so the mass difference involved.

Cosmic-ray Mass Spectroscopy

The large lunar neutrino detector outlined earlier would also be extremely well suited to study directly high-energy cosmic rays, especially above 1 TeV per nucleon.

Tests of General Relativity

Lande's talk was followed by a brief interpolated presentation by Arnold Rosenblum, who reported continued progress in work by Herbert Walther at the Max Planck Institute for Quantum Optics in Munich toward developing clocks with long-term stability and accuracies of the order of 1 part in 10^{18}, and who pointed out that such clocks in orbit around the moon would permit an accurate test of dragging of inertial frames. (I regret to report Rosenblum's sudden and totally unexpected death only a few days after the Brighton meeting, so no manuscript of his talk is available; however, he presented an earlier version of this idea in the *AIP Conference Proceeding,* Vol. 202, referred to previously.)

Particle Astrophysics on the Moon[a]

KENNETH LANDE

University of Pennsylvania
Philadelphia, Pennsylvania 19104

INTRODUCTION

The moon provides a site with rather special characteristics that permit us to view the particle emission of the astronomical world and study properties of particles in an environment not accessible on the earth. However, the expense of installing and operating instrumentation on the moon is enormous, and maintaining that apparatus is extremely difficult. Thus, only very special problems should even be considered for investigation on the moon, and even such investigations can only be considered as part of a larger lunar program. In this review, I will describe some of the questions that will benefit from observatories on the moon, indicate why these cannot be carried out on either the earth or in earth orbit, and consider characteristics of a multipurpose detector with which some of these questions could be attacked. Many of these ideas have been discussed in detail in three previous conferences.[1-3]

The two features of the moon that distinguish it from the earth are the lack of an atmosphere and the absence of a magnetic field. Although both of these features also exist in various earth orbits, the moon provides large, stable surfaces on which apparatus can be mounted together with a large mass that can serve either as a particle filter or as a interaction target, for example, for neutrinos.

An additional feature that is not commonly considered is the possibility of using the earth–moon system as a long lever arm or long flight path laboratory. The earth–moon separation, 3.8×10^8 m, is about 30 times the diameter of the earth and about one-half the radius of the sun. These parameters suggest that the earth–moon path has the correct dimensions to investigate vacuum flavor oscillations of neutrinos for mass ranges of interest in understanding the questions raised by the solar neutrino investigations.

LIMITATIONS OF TERRESTRIAL OBSERVATORIES

High-energy Neutrinos

In the past three decades, we have watched an enormous development in the investigations of astronomical neutrinos. During the 1960s three large underground detectors were constructed: one in the East Rand Proprietary Mine in South Africa, by F. Reines and collaborators, another in the Kolar Gold Field by the Tata Institute group, and a third in the Homestake Gold Mine in Lead, South Dakota, by Raymond Davis.

[a]This work was supported by grants from the Astronomy Division of the National Science Foundation, and the University of Pennsylvania Research Foundation.

The first two groups focused their attention on high-energy neutrinos, mainly muonic neutrinos, with $E > 100$ MeV. These neutrinos arose primarily from the decay of pions produced in the upper atmosphere of the earth by the interaction of cosmic-ray primaries. The muonic neutrinos produced in these decays interacted with the rock surrounding the underground detectors, producing muon secondaries. The detectors then observed the passage of these muon secondaries. The flux of the neutrino produced muons was about 4×10^{-13} per cm^2 sec. There were small variations of flux associated with the zenith angle giving rise to a slightly higher rate for zenith angles near 90°. There were also geomagnetic latitude effects associated with the deflection of lower energy cosmic-ray primaries by the earth's magnetic field.

These first underground neutrino detectors were used to look for localized astronomical sources of neutrinos. No signal above that of the atmospheric neutrinos was seen. Since neutrinos from astronomical sources have localized (point) origins, the signal-to-noise ratio should improve as the angular resolution of the detection process is improved. Unfortunately, neutrinos are only detectable as the result of another interaction, either an elastic scattering or the production of a lepton secondary. In both cases there is an angular difference between the direction of the incident neutrino and the secondary charged particle. That angular difference limits the angular resolution of the apparatus and thus the sensitivity of any search that must be conducted in the presence of atmospheric neutrinos.

Search for Nucleon Decay

In the 1970s a fourth large underground detector, a 300-ton water Cerenkov detector, was established at the Homestake Mine with the prime goals of looking for neutrinos from stellar collapse (supernovae), and improving the limits on nucleon stability previously established by Reines in the South African detector. In the 1980s, the specific prediction of GUTS that the nucleon had a lifetime of about 10^{30} years, led to the construction of additional large underground detectors. Among these were two additional water Cerenkov detectors, one at the Kamioka Mine in Japan, and a second at the Morton Salt Mine in Ohio, the Irvine, Michigan, and Brookhaven (IMB) detector. Three ranging hodoscopes were also built, one at the Kolar Gold Fields, and two under the Alps, one in the Mt. Blanc tunnel and another in the Frejus tunnel. These detectors had as their primary intent the search for nucleon decay. Cosmic-ray neutrino interactions inside these detectors and in the rock surrounding them became the limiting factor in searching for nucleon decay. Lifetime limits of the order of 10^{32} yr were established. It is easy to see that this limit is closely related to the interaction rate of neutrinos produced in the atmosphere, since

$$R(\nu) = \phi(\nu) \times \sigma(\nu).$$

At 1 GeV, this interaction rate is about 10^{-38} per sec. The inverse of this rate gives an effective lifetime limit of about $1/2 \times 10^{31}$ yr. An improvement of about an order of magnitude in sensitivity over this limit was obtained by requiring that the sum of the visible momenta be zero. However, the atmospheric neutrino background was still the limiting factor. In order to search for nucleon decays with longer lifetimes, it will be necessary to eliminate the background from atmospheric neutrino interactions. It is unlikely that this can be done on the earth.

Underwater Neutrino Detectors

The underground detectors previously described have typical linear dimensions of 10 to 15 m, and thus surface areas of about 200 m^2. A new class of much larger astronomical neutrino detectors is now under construction. The new detectors, DUMAND and BAIKAL, utilize cubic kilometers of deep ocean (DUMAND) and deep lake (BAIKAL) water to serve as both neutrino target and secondary detector. Their goals are to search for extremely energetic astronomical neutrinos, $E > 1$ TeV. At these energies, the probability of decay of an atmospherically produced pion is greatly inhibited. The mean free path for decay of a pion is $7.5E/m_0c^2$, where $m_0c^2 =$ 140 MeV. For a pion of 10^{12} eV, for example, the decay length is 50 km. But the absorption length for such a pion in the atmosphere is only a few kilometers, so that the decay probability is greatly reduced. The higher the neutrino (and pion) energy, the greater the background suppression. At energies of 10^{14} eV the background suppression is about a factor of 10^3.

In order to obtain the preceding suppression factor it is necessary to have a very thick detector, so that very energetic neutrino secondaries can be selected, and a very large area detector, so that a reasonable flux can be detected. It remains to be seen if indeed there are any astronomical sources of neutrinos in the 10^{14}-eV range.

Low-energy Neutrinos

Low-energy astronomical neutrinos, $E < 50$ MeV, have proved rather interesting recently. Two astronomical sources of this type of neutrino that have been observed are the sun, with neutrino emission below 15 MeV, and a supernova in a nearby galaxy, SN87a in the Large Magellanic Cloud (LMC). The observation of neutrinos from SN87a confirmed our assumptions about the mechanism involved in the collapse of a massive star into a neutron star. From the dispersion of the neutrino burst we could put a significant limit on neutrino mass, $m_\nu < 4$ eV$/c^2$, and from the difference in arrival time of the neutrinos and the first increase in light, it was possible to demonstrate that differences in neutrino and light velocities are less than 1 m/sec. The detection of a source 50 kpc away indicated that detectors in the earth are not background limited for nearby sources. Thus, it appears that supernova neutrino detection can be carried out on earth and does not require a lunar detector.

A similar comment can be made for solar neutrino detection. Three solar neutrino detectors are now in operation, the Homestake Chlorine Detector, which has been observing the sun for over two decades, the Kamiokande water Cerenkov detector, which has been observing solar neutrinos since 1987, and the Baksan gallium detector, which began observations in 1990. The preceding observations have led to very interesting conclusions, namely, that the observed electron neutrino flux is considerably less than that predicted to be produced in the sun. One intriguing suggestion is that a neutrino flavor resonant transition converts electron neutrinos to another neutrino flavor [Mikheyev–Smirnov–Wolfenstein (MSW) theory] during the passage of the neutrinos through the sun. The proposed interaction has never been seen on earth and violates our assumption of lepton flavor conservation. If verified, the MSW mechanism could have a major impact on our understanding of weak interactions, might suggest possible neutrino contributions to dark matter, and could provide an additional probe of the internal structure of the sun. Not only have rather big strides been made in studying the neutrino emission of the sun, but in the process, it has been clearly demonstrated that terrestrial sites seem very well suited for solar neutrino observations.

Indeed, for low-energy neutrinos, an underground terrestrial site should have a lower background than one on the moon. Cosmic-ray primaries below 15 GeV, the major source of low-energy neutrinos, are magnetically deflected away from the earth. But, these low-energy cosmic-ray primaries hit the moon with impunity. Thus, the moon has a much higher flux of secondaries from these particles. Low-energy pions, below 100 MeV, have a very short, several meter, decay path and so even on the moon low-energy pions have an appreciable probability of decaying before interaction. Thus, the earth may have a lower background from these low-energy neutrinos than does the moon.

LUNAR ASTRONOMICAL AND ELEMENTARY PARTICLE DETECTORS

Neutrino and Nucleon Decay Detectors

On the moon, the absence of an atmosphere means that cosmic-ray primaries interact directly with the lunar rock, whose density is about 3, compared to the density of the earth's atmosphere, 10^{-3}. Now the pion absorption length is only a few meters, and thus the pion decay probability and therefore the cosmic-ray-produced neutrino background is a factor of 3×10^3 lower than it is on the earth. For nucleon decay searches, this implies extending the limit from 10^{32} yr to 10^{35} yr. Of course, very-large-volume detectors, about 10^{35} nucleons, will be required in order to have a reasonable decay rate. The construction of such large-volume detectors will require utilization of local lunar materials and so must await the development of a lunar construction industry.

The suppression of cosmic-ray-produced neutrinos on the moon relative to that on the earth, previously discussed, which increased the sensitivity for nucleon decay searches, will also apply to searches for localized astronomical sources of neutrinos. This search is particularly interesting at lower energies, below 1 TeV, where terrestrial detectors have very limited search sensitivity.

Given the enormous expense of detector construction on the moon, it is very desirable that a single detector serve as both the neutrino source detector and the nucleon decay detector. The nucleon decay detector depends on detector volume, while the detector that searches for astronomical neutrinos depends on detector surface area. A compromise can be reached by making a detector whose thickness is about 10^3 gm/cm^2. Such a detector has sufficient thickness to measure the energy and momentum deposited by the nucleon decay secondaries, and thus characterize nucleon decays, and still have a large surface area to mass ratio, the figure of merit for a neutrino source detector. Such a detector can also be layered, with active, particle-sensitive layers interspersed with passive ranging layers. A modular approach will permit the initial use of a smaller detector to begin the neutrino source search and serve as the initial element of a larger, multipurpose detector.

Earth–Moon Neutrino Vacuum Oscillations

In order for the MSW mechanism to operate within the density profile of the sun, it is necessary that the mass difference between the electron neutrino and at least one of the other neutrino species be between 10^{-2} and 10^{-4} eV/c^2. For large values of the lepton flavor violating interaction, an appreciable vacuum lepton flavor transition should also occur. Such a transition has been looked for in atmospheric neutrinos traversing the earth without any clear positive signal. Longer flight paths with controlled beams are most desirable. The only such flight path that is likely to be available to us in the foreseeable future, other than that from the sun to the earth, is that between the earth and the moon.

One can thus visualize accelerator-generated neutrino beams directed toward a neutrino detector on the far side of the moon. The flight path between the earth and the moon would serve as the oscillation region, while the traversal through the moon would serve as the neutrino conversion region. The most readily detectable neutrino is the muon neutrino, since it produces a long-range secondary, the muon upon interaction. The muon neutrino is also the easiest neutrino to produce at accelerators, as it arises from the decay of the pion.

Thus, the first experiment is likely to be a disappearance experiment, one in which a beam of muon neutrinos is aimed at the lunar neutrino detector and the ratio of observed to expected flux is measured. Such experiments are not very satisfying since many other factors could reduce the detected intensity—poor beam aiming, miscalculation of flux, poor detector response, etc. A more sophisticated approach would be to direct beams of either electron or tau neutrinos toward the lunar detector and look for the appearance of muons whose flight paths point back toward the terrestrial source. The challenge now lies in the generation of directed, intense beams of electron or tau neutrinos.

Measurement of the Mass of the Tau Neutrino

The earth–moon neutrino flight path also provides for the possibility of the measurement of neutrino masses by their time of flight over this path. For a neutrino of mass m and energy E, the time delay relative to a massless neutrino is

$$\Delta\tau = \frac{L}{c}\left(\frac{m}{E}\right)^2$$

or

$$m < (\Delta\tau)^{1/2} \times E(0.9)$$

where $L/c = 1.3$ sec. The neutrino with the largest upper limit in mass is the tau neutrino with an upper limit of 35 MeV/c^2. If we assume that the combination of terrestrial-accelerator-beam time structure and neutrino-detector time resolution is 10 nsec, and $E = 10$ GeV, then we can limit M to less than 1 MeV/c^2. Better time resolutions or lower neutrino energies can obviously establish lower mass limits. However, it seems very unlikely that we will ever be able to use such a short flight path to obtain cosmologically interesting limits on the tau neutrino mass.

Naturally occurring neutrinos, such as those produced in the atmosphere or at stellar sources, generally have a spectrum of energies and occur at uncontrolled times. Neutrinos produced at accelerators can be controlled both in energy spectrum and time of emission. By adjusting the energy of the primary beam of the accelerator, we can vary the energy spectrum of the emitted neutrino beam and thus measure the probability of either muon neutrino disappearance or appearance as a function of neutrino energy. This is precisely the measurement required to determine whether neutrino vacuum oscillations occur and, if they do, the mass difference involved in the oscillation. Considerable experience has been acquired in establishing neutrino beams from accelerators, but not with the aiming precision and collimation required for a beam aimed at a target on the moon.

Cosmic-ray Mass Spectroscopy

There is considerable interest in understanding the processes that generate high-energy cosmic rays and the conditions in the intra- and intergalactic medium that determine the propagation of these particles. The interest is particularly intense at the high-energy end of the spectrum, above 1 TeV per nucleon, where production in supernovae might terminate and other, as yet unknown, processes might be detectable. In order to see these primaries directly it is necessary to operate detectors above the atmosphere, thus requiring satellite-based detectors.

Since the flux of cosmic rays decreases as a high power of the energy, large detector areas are required and long observation times are necessary. These conditions can, in principle, be met in satellites. However, the lunar neutrino detector, previously described, will also have these characteristics and can be used jointly for these studies while also serving the functions described earlier. The availability of thick absorbers to measure energy and a stable platform in the lunar detector will produce far better information than can be expected from a detector with similar goals in a satellite.

CONCLUSIONS

We are only now beginning to think about the expansion of our experimental domain beyond the confines of the earth and the blanket of gas that envelopes us. Our present ideas and suggestions are based on extrapolations of observations made within the confines of our terrestrial world. The suggestions in this paper are of that character. The real excitement lies in those new unanticipated discoveries that will be made as we expand our operational horizons beyond the earth. We have, historicaliy, experienced the opening of a sufficient number of frontiers to know that the real payoff lies in the unexpected and the unanticipated. We can only envy the next generation as they step off the earth and expand their accessible world into the solar system. Decades will pass by as dreams turn into reality and fantasy turns into apparatus and data and a clearer picture of the universe within we live.

REFERENCES

1. MENDELL, W. W., ED. 1985. Lunar Bases and Space Activities of the 21st Century. Lunar and Planetary Institute.
2. POTTER, A. E. & T. L. WILSON, EDS. 1989. Physics and Astrophysics from a Lunar Base. AIP Conference Proceedings **202**. New York.
3. MUMMA, M. J. & H. J. SMITH, EDS. 1990. Astrophysics from the moon. AIP Conference Proceedings **207:** 656 pp. New York.

Astrophysics from the Moon

HARLAN J. SMITH

Astronomy Department
University of Texas at Austin
Austin, Texas 78712
and
Lunar and Planetary Institute
Houston, Texas

INTRODUCTION

Why should this topic, as an entire minisymposium, impinge on a conference in relativistic astrophysics?

The answer is simple. A speaker at this meeting pointed out that Eddington once commented that whenever theory and observation disagree, one should check the data! Nevertheless, as we all know, without data much theory would never be born and none of it would be tested. This is conspicuously true in astronomy where very few indeed of the remarkable discoveries of the past century have been predicted, and where theories tend to grow like weeds—in large part because it is so hard to test them against the particular universe in which we happen to find ourselves. Accordingly, theoreticians should be informed about and should become strong supporters of activities having major new observational potential. The use of the moon as a base for much of the frontier observational astronomy of the twenty-first century is an outstanding example.

As indicated in earlier drafts of programs for this Texas/ESO-CERN meeting, the lunar-base minisymposium was to feature (in addition to Ken Lande representing lunar-based physics) Roger Bonnet, Scientific Director of the European Space Agency, who has written about and was to have introduced the extraordinary range of advantages that the moon offers for observational astronomy and astrophysics. His presentation was to have been followed by Mike Mumma of the NASA Goddard Space Flight Center, a leader in infrared astronomy and one of the principal U.S. proponents of telescopes on the moon, who would have treated some of the major lunar instruments and scientific breakthroughs to be expected when these great instruments can finally be put into operation. Unfortunately, each of these speakers encountered irreconcilable schedule conflicts very shortly before the meeting. However, since I have also been working in this field for half a dozen years, it was possible for me to combine their two topics into a single presentation at the minisymposium and as the subject of this paper.

Recently, Mumma and I were able to organize a large workshop on the subject of astrophysics from the moon. The papers from that meeting, held in February 1990, at Annapolis, Maryland, have now been published (Mumma and Smith 1990). Nearly all of the material in this present brief review appears in far greater detail in that volume.

WHY IS THE MOON SO GOOD FOR ASTRONOMY?

I have reviewed many of the principal advantages of the moon at some length in a paper in the book just noted. However, being a conference proceedings, this rather large workshop volume may not be as widely available as the regular astrophysics journals. It may thus be useful to begin by summarizing the general factors bearing on the utility of the moon for observers. These fall into the following three major categories.

1. Ultrahigh vacuum. The lunar atmosphere has a density typically only 10^{-13} that of the earth at sea level. This is so high a vacuum that the atmospheric particles are noncollisional in ballistic trajectories after each encounter with the surface. As a consequence:

- Transmittance is effectively perfect at all wavelengths
- There are no "seeing" effects—instruments may be used at their full diffraction limits
- Interferometers need only their separate fundamental optical elements (no vacuum lines or wavefront corrections required)
- There is no appreciable scattering of light even in the daytime sky, permitting day as well as night observing and work at very small solar elongation
- The sky is cold, allowing telescopes to radiate down to temperatures probably well below 100 K, in turn permitting noncryogenically cooled telescopes to work in the thermal IR out to beyond 10 μm, limited only by zodiacal emission
- The total lack of wind greatly simplifies telescope design (only static loads need be considered), and requires domes to serve only as ultralightweight sunshades or micrometeorite shields

2. Lunar mass and size. The moon is virtually a planet in its own right (only slightly smaller than Mercury), with mass 1/81 that of earth, and diameter of nearly 3500 km. As a consequence:

- Just as on earth, the moon's effectively infinite inertia permits the simplest types of telescope mounting and drive, with completely smooth tracking
- The moon is very rigid and stable, its seismic activity being only about 10^{-8} that of the earth—an invaluable property for ultralarge and powerful interferometers
- Lunar gravity (one-sixth terrestrial) is small enough to greatly simplify telescope design, but large enough to be biologically much more friendly than orbital free-fall; also to cause dust, debris, and outgassing to sink out of the field of view and to allow dropped tools and parts to fall gently to a surface for easy retrieval
- Unlimited material is available to shield humans and sensors from solar storms and galactic cosmic rays; also for an increasing range of construction purposes as lunar development proceeds
- Enormous surface area is available for ultimate astronomical development, especially for large interferometers

3. Location and motions of the moon. The mean distance of nearly 400,000 km, and the moon's synchronous rotation and revolution, offer significant advantages for astronomy:

- The distance is great enough to quarantine the moon from nearly all terrestrial noise and interference (lower frequency radio, on the near side, is the only significant exception), yet
- The moon is close enough (3-second round-trip communication time) to allow real-time control of telescopes and equipment from earth, and direct transmission of data back to earth—in other words, no astronomers really need to be on the moon
- Any observatory located on the near side can be in continuous communication with the earth, yet if located near the lunar limb can have the earth always tucked away near the horizon, shielded from telescopes and not significantly interfering with their access to the sky
- The slow sidereal rotation of nearly 30 days gives full-dark nights of two weeks duration, permitting very long uninterrupted integrations on ultrafaint or variable sources, and reduces the incidence of thermal shocks to only twice a month

This constellation of factors makes the moon the site of choice for most kinds of observational astronomy, the more so as instruments—especially interferometers—become very large.

DRAWBACKS OF THE MOON FOR ASTRONOMY

Nothing in this world is perfect—even the moon has several significant disadvantages for astronomy.

- While lunar telescopes can be very simple and cost effective compared with their terrestrial counterparts and with most free-fliers, at present the cost of placing them on the moon is very high (though this should go down sharply as lunar development proceeds).
- Solar power is unavailable during lunar night, requiring either power storage or some type of power plant (proximity to a manned lunar base should solve this problem).
- The temperature of the lunar surface changes by about 300 K between night to day, requiring careful thermal design and shielding.
- As in space away from the earth's magnetic field, the lack of lunar magnetic field allows direct penetration of the solar component of cosmic rays, and speaks for shielding humans and many detectors.
- Any single observatory can study at most only half the sky at any one time (if at a pole, half the sky is forever invisible), thus requiring antipodal observatories if essentially all the sky must be observable at all times.

The preceding problems all appear to be either tractable or acceptable in view of the great advantages of lunar siting and the potential of some of the spectacular instruments and observations that can be achieved perhaps only with lunar observatories, as outlined in the next section.

SOME LUNAR-SPECIFIC TELESCOPES AND OBSERVATIONAL GOALS

While the Annapolis workshop alluded to earlier considered a wide range of astronomical and astrophysical topics, it was primarily directed toward bringing out scientific observational goals requiring instruments that may only be practicable on the moon. Half a dozen particularly important problem areas emerged from many different areas of the papers and discussions at this meeting. Each of these is likely still to be of the highest interest to our astronomical children and grandchildren.

(a) Can we observe, and come to understand, other stars with the same resolution and detail that is becoming available for the sun?

Our general understanding of the sun is marred by several especially conspicuous gaps, such as the neutrino deficit. But it is likely that the physics of surface and near-surface phenomena will prove the most intractable because of the complexity of these phenomena and the extraordinary difficulty of the magnetohydrodynamic theory involved.

A major—perhaps indispensable—key to tackling these problems lies in exquisitely detailed observations down to the hundreds of kilometer scale. Such observations should become routinely available for the sun over the next decade or two. But detailed physical understanding will surely profit from a comparable understanding of the range of similar or different phenomena in most of the menagerie of kinds of stars.

Can we imagine instruments able to observe at least the nearest stars with something approaching this kind of resolution? The answer is yes. Consider the following rough order-of-magnitude arguments. At the sun's distance of $\sim 10^8$ km, surface resolution of better than 1000 km requires about 1-arcsec telescopic resolution. The nearest stars are about 10^6 times more distant than the sun; hence, resolution of about 10^{-6} arcsec will be required to study their surfaces in roughly comparable detail (resolution on more distant stars will of course decline in proportion to their distance, but will still be of great interest even out to the order of hundreds of parsecs).

To achieve optical resolution (say at 500 nm) of 10^{-6} arcsec, the diffraction equation indicates that imaging interferometers having extreme baselines of the order of 100 km will be required. Even our scientific children may be stretched to produce such systems, but given the expected progress of science and technology, this should not present very serious problems to our grandchildren. The power of such an instrument for so many purposes in astronomy is so great that I like to think of it as the "Nirvana Telescope."

(b) How, in precise detail, do planetary systems form?

To answer this question, we will need to be able to examine in rather close detail some solar systems actually being born. Unless our present general conceptions are completely off base, this means resolving structures and large condensations in accretion disks around stars undergoing birth from relatively dense clouds of material in the interstellar medium.

The nearest star-forming region with likely candidate T Tauri stars is the Taurus-Auriga complex, about 140 parsec distant. The inner parts of accretion disks, within several tens of astronomical units of their central stars, are presumably the planet-forming regions in which we are most interested. These are optically thick at wavelengths shorter than about 100 microns, but should be thin enough to permit observation in the range 100 to 600 microns. A principal goal is to measure the chemical composition, densities, temperatures, and internal dynamics with enough accuracy to permit understanding of the dissipational and viscous processes that cause transfer of mass and angular momentum.

Such close examination of the accretion disk requires resolution at least of the order of an astronomical unit. To achieve this with a submillimeter array would involve several dozen 4- to 6-m elements spaced to a diameter of about 30 km. This is a substantially less challenging task than the "ultimate" optical/IR interferometer noted earlier.

Another interesting possibility is to look for gaps in the accretion disk caused by the sweeping action of large planets in the process of formation. Such observations could probably be made at around 10 microns, and would be straightforward for an IR interferometer only a tenth the size of the one eventually desired for solar-stellar studies, but still in the 5- to 10-km-diameter range.

(c) What is the distribution of solar systems in type and number? In particular, are there earth-like planets around other stars, and if so what is their distribution in terms of type of and distance from the parent star?

The first part of this question, as to the number and types of other solar systems, will in part be answered by earth- and space-based instruments over the next decade or two. But complete and accurate statistics, along with the answer to the second question will almost certainly require careful direct-imaging detection of the planets (if any—none are yet known) around thousands of nearby stars. In turn this can be done either using a very large (16-m class or bigger) diffraction-limited filled-aperture telescope, having hypersmooth optical surfaces to reduce scattered light from the central stars, or using a close-spaced optical/IR interferometer with enough well-spaced elements to cover the UV plane unambiguously. In principle such telescopes could perhaps be built and operated in free space, but problems including assembly, pointing, tracking, momentum dumping, station-keeping, and maintenance appear to speak for the construction of these on the moon, in the form of simplified terrestrial-type telescopes.

(d) Do life and intelligence exist outside our Solar System?

Very large filled-aperture telescopes, such as those referred to in (c), have an interesting additional power. Assuming earth-like planets are found around nearby stars, then their spectroscopic examination by the giant telescopes can test for nonequilibrium gas concentrations, especially of oxygen, ozone, and methane. Such detections would virtually assure the presence of life.

At present only the SETI (search for extra-terrestrial intelligence) programs using sensitive multiband radio receivers offer even the slightest hope of detecting the existence of another civilization in space. Tuning in on their communications seems a truly remote prospect, because of such factors as their presumed tight beaming, low power, and modulation that might look to us like noise. But even highly advanced civilizations might well still use RF power beams to send energy from great radiation collectors around their stars, or require radars, especially to assure the safety of space travel. With proper equipment such broad-beamed, very-high-power, conspicuously nonnatural signals should be easy to pick up over distances of the order of the size of our galaxy.

To answer the SETI question relatively definitively we need to search the entire sky over the entire accessible RF spectrum with equipment limited in sensitivity only by natural backgrounds. Today, because of the rapid increase in human uses of radio, only tiny slivers of relatively unpolluted RF spectrum remain anywhere within a considerable distance of the earth. The sole exception is, of course, the back side of the moon, which has the distinction of being the only place in the universe that never gets to look at the earth. Using robotic labor it should be possible one day to build on the lunar backside filled-aperture telescopes kilometers in diameter, or interferometers with baselines of up to thousands of kilometers if desired, opening the entire useful part of the RF spectrum to observations of ultimate sensitivity.

(e) Can we observe directly the inner core activity in galactic nuclei and quasars?

Each of the great interferometers referred to earlier could and would be used to study the monsters at the hearts of galaxies, each with their accessible spatial resolutions and spectral coverages. Consider the Nirvana Telescope studying the nearest superquasar, 3C273. Since this object is about 10^8 times as distant as the nearest stars, the telescope could thus resolve features of the order of size of 10^8 times larger than the 1000-km features visible on nearby stars, or about 10^{11} km. This is only $\sim 10^3$ astronomical units, well within the spectroscopically detected broad-line region. Nuclei of galaxies up to a thousand times closer could be observed with resolutions down to an astronomical unit, the expected order of size of the innermost parts of the accretion disk around the large black holes believed to lie at the centers of at least some of the galaxies.

(f) What happened in the early universe?

To get closer to the answers we will probably need to search substantial sky areas for the most distant objects, from the earliest epoch of formation of discrete self-luminous objects, that are bright enough to be seen at that great distance and through whatever obscurations may intervene. In turn the study of these objects in any detail will surely require the largest telescopes we can build on the moon, working over the widest accessible spectral range.

HOW LIKELY IS ALL THIS TO HAPPEN?

Assuming the civilized world holds together, there can be no significant doubt that the human race—to be sure helped greatly by its robotic surrogates—will gradually begin to establish presence off the earth. In fact, the current U.S. strategic plan, as laid out in July 1989 by President Bush, contemplates completion of some form of station in this decade, establishment of a permanent lunar base in the first decade of the twenty-first century, and the beginnings of human visits to Mars in the following decade. This is a perfectly reasonable timetable, especially if effective large-scale international cooperation can be established toward these goals.

There are actually many reasons to begin human development of the moon. These include:

- The very real human imperative to explore and move out from the familiar
- The rapid progress of facilitating technologies
- The eventual retooling of some of the vast military budgets toward civilian spending, along with specific need to constructively employ skilled people and specialized machinery to "beat missiles into spaceships"
- The competition and/or cooperation with strong and still rapidly developing space programs not only of the United States and Russia, but now being joined by a uniting Europe, as well as by China, Japan, India, Brazil, Israel, and others
- The fact that the proximity of the moon makes it so clearly the first target for a permanent space base
- The need for much manned space experience at this nearby base before undertaking the much more difficult and dangerous step to Mars
- The prospect in at least the foreseeable future for economic return and even tourism resulting from lunar development
- Many scientific returns, including astronomy

In fact, I believe the moon would be developed by humans if there were no stars in the sky. But the prospect of this development offers to astronomy opportunities that in the end may make—even exceed—the visions noted in this paper, and it is already time for astronomers to begin to plan and work to achieve them.

REFERENCES

MUMMA, M. J. & H. J. SMITH, EDS. 1990. Astrophysics from the Moon. American Institute of Physics Conference Proceedings **207:** 656 pp. New York.

The Cold Dark Matter Model: Does It Work?

CARLOS S. FRENK

Physics Department
University of Durham
Durham, England DH1 3LE

1. INTRODUCTION

When the organizers of this symposium made up the title of my paper, they no doubt felt that the question posed should not be answered with a simple "yes" or "no." Indeed, in the past few months new studies of large-scale structure have produced results that simultaneously challenge and support different basic aspects of the cold dark matter theory. In this article, I discuss some of these developments and attempt to assess the status and prospects for the theory.

The *standard* cold dark matter (SCDM) model (see White 1988 and Frenk 1991a for recent reviews) is a theory of the formation of large-scale structure based on the following assumptions.

1. The dark matter consists of collisionless elementary particles that, like photinos or axions, had a low velocity dispersion at early times.
2. The cosmological density parameter, $\Omega = 1$.
3. The primordial seeds for the formation of cosmic structures are adiabatic density fluctuations with random phases and a scale invariant power spectrum (i.e., with $n = 1$ in $|\delta_k|^2 \propto k^n$).
4. Structure grows by gravitational clustering at least on scales larger than a typical galaxy.
5. The distribution of galaxies is related to the distribution of mass via the "linear biasing model":

$$\sigma_{\text{gal}} = b\sigma_\rho, \tag{1}$$

where σ_{gal} and σ_ρ are the rms fluctuations in the galaxy and density fields, respectively, filtered with a "top hat" window function. For bright, optically selected galaxies σ_{gal} is measured to be about unity at 8 h^{-1} Mpc. (Here and below the Hubble constant is taken to be $H_0 = 100\ \text{h}^{-1}$ Mpc.)

The motivation for assumptions (1), (2), and (3) comes primarily from particle physics. The standard theory of big bang nucleosynthesis constrains the fraction of baryons to be in the range $0.01 \leq \Omega_b\, \text{h}^2 \leq 0.02$ (Olive *et al.* 1990). The upper limit is already smaller than values often obtained from virial analyses of galaxy clusters or other dynamical determinations (e.g., White 1991). Certainly, if $\Omega = 1$, as predicted by the inflationary model of the early universe (see Guth 1991 for a review), the dark matter is likely to be one of the nonbaryonic candidates suggested by particle physicists. The standard model of inflation also provides the motivation for the type of primordial fluctuations postulated in (3). Assumption (4) is suggested by the well-known instability of a Friedmann universe to the growth of gravitational perturbations (e.g., Peebles 1980). Finally, assumption (5) is the least motivated of all. Although some form of biasing is required to reconcile dynamical estimates with the theoretical imperative $\Omega = 1$, the particular form of **(1)** is little more than a matter of mathematical convenience (Bardeen *et al.* 1986).

The beauty of the SCDM model lies in its simplicity, in its connection with fundamental physics and, above all, in the fact that it is well specified and therefore falsifiable. The basic framework is as follows. Assumptions (1), (2), and (3) determine the *linear* density field at early times such as the (re)combination epoch. Together with assumption (4) this determines, in turn, the *nonlinear* density field at late times, including the present. Finally, assumption (5) leads to a prediction for the observable galaxy distribution. In the SCDM model, there is only one free parameter, the amplitude of the primordial density fluctuations, or, equivalently, the value of b in **(1)**. Traditionally, b is fixed by matching some dynamical measure of galaxy clustering such as the amplitude of small-scale peculiar velocities or large-scale streaming motions. Most estimates use bright, optically selected galaxies and produce values in the range $1.5 \lesssim b \lesssim 2.5$ (Davis *et al.* 1985; Kaiser and Lahav 1988). In principle, though, different galaxy types might be biased by different amounts.

Some of the basic tenets of the theory are directly verifiable. A *prima facie* test is, of course, to search for cold dark matter particles. As discussed elsewhere in this volume, it is fortunate that some candidates are liable to experimental detection with current technology. Other fundamental elements of the theory, also within grasp, are the value of the mean cosmic density, the nature of the density field on large scales, and the presumption that structure grows by gravity. In this article I concentrate on these last three issues. Much of the discussion relies on results from the "QDOT" survey described in Section 2.1. The estimate of Ω from this survey, the test for gravity, and the large-scale field are discussed in Sections 2.2 and 2.3. A summary and discussion of future prospects is given in Section 3.

2. TESTS OF THE COLD DARK MATTER MODEL USING THE QDOT SURVEY

2.1. The QDOT Survey

The QDOT redshift survey of galaxies detected by the Infrared Astronomical Satellite (IRAS) is named after the initials of the participating institutions: QMWC (A. Lawrence and M. Rowan-Robinson), Durham (J. R. Allington-Smith, R. S. Ellis, C. S. Frenk, I. R. Parry, and X. Xiaoyang), Oxford (G. Efsthatiou and W. Saunders), and Toronto (N. Kaiser). (Guest stars include J. Crawford, B. Moore, and D. Weinberg.) This survey is ideal for large-scale studies because it provides almost full-sky coverage by galaxies that are selected uniformly and are virtually unaffected by inclination effects or galactic extinction. Since the survey was designed for studies of scales larger that $\sim 10\ h^{-1}$ Mpc, whereas the characteristic clustering length of galaxies is $\sim 5\ h^{-1}$ Mpc, the optimal strategy was to randomly sample the galaxy distribution (Kaiser 1986). We adopted a sampling rate of one-in-six and a flux limit at 60 μm of 0.6 Jy. This gave a sample of 2163 galaxies at galactic latitude $|b| > 10°$, covering about 9 sr, with a median velocity of ~ 9000 km s^{-1} and a limiting velocity of $\sim 30{,}000$ km s^{-1}. Of these, 1211 galaxies did not have measured redshifts and we obtained those during 1986–1988 using the Isaac Newton Telescope (INT), William Herschel Telescope (WHT), and Anglo-Australian telescope (AAT).

An impression of the three-dimensional distribution of the QDOT survey is given in FIGURE 1. To make this plot we place each galaxy at its redshift distance, weight it with the inverse of the selection function (computed from the Saunders *et al.* (1990) luminosity function), smooth the resulting distribution with a Gaussian of dispersion λ, carve out a sphere of radius R around us, and plot those regions where the smooth overdensity exceeds ν times its rms dispersion. In FIGURE 1(a), $\lambda = 15$ h^{-1} Mpc, $R =$ 100 h^{-1} Mpc, and $\nu = 0.7$. This plot shows several well-known nearby northern structures such as the Local Supercluster (LS), the Abell 1367/Coma complex, and the "Great Attractor" region. In the south, there are several less well-known

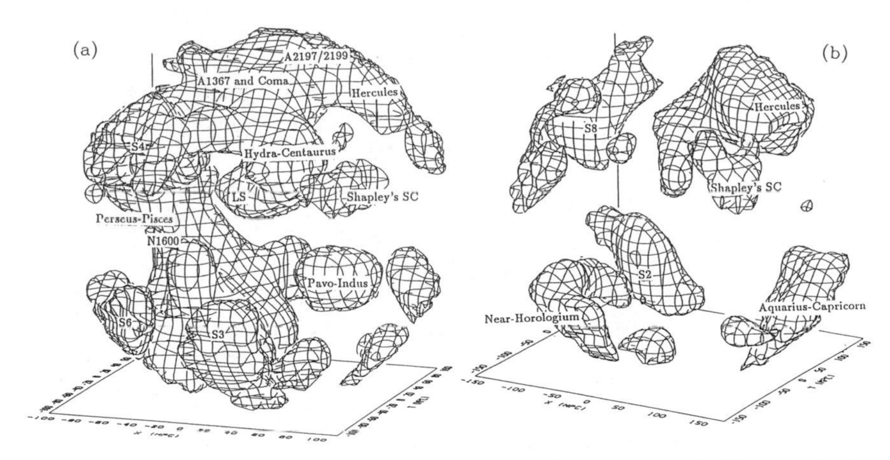

FIGURE 1. Isodensity contours of the smooth galaxy distribution in the QDOT survey for two different choices of parameters as discussed in the text. The observer is located at the center and the north galactic pole is along the z-axis. (Adapted from Moore *et al.* 1991.)

superclusters, some of which (e.g., S3, S6) are labeled according to the notation introduced by Saunders *et al.* (1991). FIGURE 1(b) shows the largest superclusters in the survey. Many are located in a shell at ~ 100 h^{-1} Mpc and, as discussed in Section 2.3, dominate the number density fluctuations in the survey. The parameters in FIGURE 1(b) are $\lambda = 24$ h^{-1} Mpc, $R = 150$ h^{-1} Mpc, and $\nu = 1.3$.

2.2. The Value of Ω

The QDOT survey provides a map of the nearby universe of galaxies. Assuming the "linear biasing model" of **(1)**, this is also a map of the nearby density field. Let us imagine smoothing the galaxy distribution on some scale; define b_I as the ratio of galaxy number density to mass fluctuations on this scale, $b_I = (\delta N/N)/(\delta\rho/\rho)$. (Note that b_I is related but not identical to the usual biasing parameter, b, of **(1)**; see, e.g., Frenk *et al.* 1990.) Let us assume further that gravity is the dominant interaction on the scales of interest. Then, at each point in space, we can predict the peculiar acceleration generated by this mass. This prediction depends on Ω and on the assumed connection between galaxies and mass. In linear theory it is easily computed to be (Peebles 1980)

$$\mathbf{v}(\mathbf{r}) = \frac{\Omega^{0.6} H_0}{4\pi b_I} \int d^3r' \, \frac{\delta N}{N} [\mathbf{r}' - \mathbf{r}] \frac{(\hat{\mathbf{r}}' - \hat{\mathbf{r}})}{(\mathbf{r}' - \mathbf{r})^2}. \tag{2}$$

The integral is assumed to be carried out over the whole of space. Thus, all-sky surveys like the QDOT survey, offer the first opportunity to use this equation reliably. (Note that the predicted $\mathbf{v}(\mathbf{r})$ is independent of H_0.)

We now use the smooth galaxy field and **(2)** to predict the radial component of $\mathbf{v}(\mathbf{r})$ at the position of the ~1000 galaxies in D. Burstein's compilation of (independently determined) peculiar velocities. (Details of the procedure, including a prescription to correct for redshift space distortions, are given in Kaiser *et al.* 1991.) The result is a reasonable correlation between predicted and observed peculiar velocities, and a best fit gives the estimate $\Omega^{0.6}/b_I = 0.86 \pm 0.16$.

We can also compare the predicted $\mathbf{v}(\mathbf{0})$ at the position of the Milky Way with the velocity inferred from the measured dipole in the microwave background radiation (Lubin and Villela 1986). For this, we replace the integral in **(2)** by $n^{-1} \Sigma \, \Phi(r')^{-1} \hat{\mathbf{r}}' / \mathbf{r}'^2$, where n is the mean density of the sample and $\Phi(r)$ is the selection function of the survey. (See Rowan-Robinson *et al.* 1990 for further details.) An encouraging result, originally found by Strauss and Davis (1988), is that the predicted vector points within only 20° of the dipole, suggesting that the mass traced by the QDOT survey is responsible for the Milky Way's motion. This identification requires $\Omega^{0.6}/b_I = 0.81 \pm 0.15$, very close to our previous estimate. Note that our two estimates are not redundant since they weight different parts of the survey in different ways; Burstein's galaxies are all within ~4000 km s^{-1}, whereas the predicted peculiar acceleration of the Milky Way converges only at ~10,000 km s^{-1}.

The correlation between predicted and measured peculiar velocities, including the agreement for the Milky Way's motion, tells us something rather fundamental. It implies, first, that galaxies have something to do with the bulk of the mass and, second, that velocities have something to do with gravity. Although by ensuring that the continuity equation is satisfied, it may be possible to arrange for nongravitational forces to mimic the velocity field produced by gravity, this seems rather contrived. The straightforward conclusion is that we have evidence that large-scale structure is primarily driven by gravity. This vindicates an assumption often taken for granted (and included explicitly as assumption (4) of the SCDM model in Section 1). The fact that we can use galaxies to infer the effects of the mass is equally significant. Our simple biasing prescription is simplistic and almost certainly wrong in detail. Nevertheless we are entitled to the hope of learning about the dynamical properties of our universe from studies of the galaxy distribution.

As with all dynamical tests, **(2)** constrains Ω only up to uncertainties introduced by biasing. (The assumption "galaxies trace the mass" is simply a particular biasing model.) Even with this uncertainty, our results strongly favor a density close to critical. Indeed, to have $\Omega < 0.2$, as suggested by optical studies in which galaxies are *assumed* to trace the mass, would require $b_I < 0.46$, implying that IRAS galaxies are much *more weakly* clustered than the mass. Although some degree of antibiasing might plausibly be defended, such a large effect seems ruled out. As FIGURE 2 shows, IRAS galaxies do seem somewhat less clustered than optical galaxies, but not nearly by as much as the factor of $\sim b_I^{-2} = 5$ required if optical galaxies traced the mass and $\Omega = 0.2$ (Moore *et al.* 1991; Frenk 1991b).

The obvious interpretation of our results is that $\Omega = 1$ and $b_I = 1.2 \pm 0.2$. Our estimate of $\Omega^{0.6}/b_I$ is larger than estimates based on optically selected galaxies, partly because IRAS galaxies are rather less biased, and partly because our extensive sample allows us to reliably bypass complications associated with nonlinear processes. At the very least, our results imply that, for once, it is the supporters of an open universe who must go through some contorsions to defend their viewpoint.

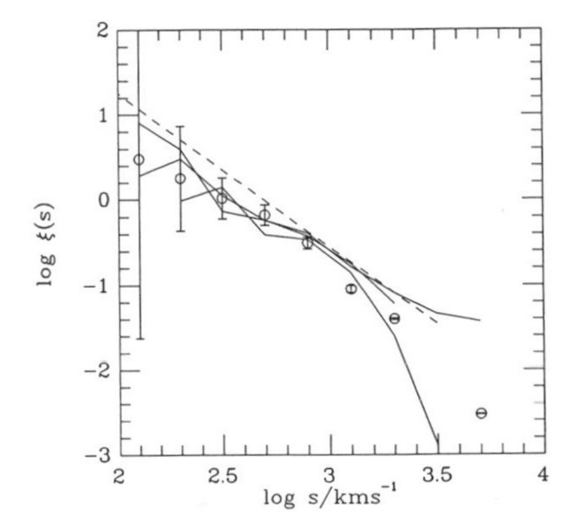

FIGURE 2. Autocorrelation function of galaxies in the QDOT survey as a function of velocity separation. The *open circles* give the estimate for the sample as a whole, while the *three lines* correspond to samples volume-limited at $s_{max} = 4000$, 6000, and 8000 km s^{-1}, respectively. The *dashed line* is $\xi(s) = (s/500 \text{ km s}^{-1})^{-1.8}$, which approximates the data for optically selected galaxies. (Adapted from Moore *et al.* 1991.)

2.3. Large-scale Power

The QDOT survey contains information on the *amplitude* and *phases* of primordial density fluctuations. The former is measured by the autocorrelation function discussed earlier and, more reliably, by the counts-in-cells described here. The latter are constrained by topological considerations.

Let us assume that the distribution of galaxies is a stationary point process. Then the variance of counts-in-cells of volume V is simply $\langle (N - nV)^2 \rangle = nV + N^2V^2\sigma^2$, where n is the mean density and σ^2 is given by a double volume integral of the autocorrelation function (Peebles 1980). Assuming each galaxy is at its redshift distance, we have computed this variance in two ways.

1. *Shell method.* The survey is divided into radial shells, and the moments of counts-in-cells in each shell are computed. For any given cell size, the variance for the sample as a whole is obtained using a maximum-likelihood estimator. This method has two advantages: it does not require an accurate knowledge of the selection function of the survey and it provides some information on the

radial dependence of the variance. The disadvantage is that it assumes Gaussian statistics, and so the uncertainties are model dependent. The results (Efstathiou *et al.* 1990) are plotted as solid squares in FIGURE 2.

2. *Box method.* We weight each galaxy by the inverse of the selection function, smooth the distribution in a manner analogous to that of FIGURE 1, and compute the moments of counts-in-cells. This method has the advantage that it does not assume Gaussian statistics, but it is sensitive to errors in the selection function. The results (Saunders *et al.* 1991) are plotted as solid circles in FIGURE 3.

In FIGURE 3 we also plot the variance of the counts expected in the SCDM model normalized to the data points on the smallest scales we have considered ($\sim 7\ h^{-1}$

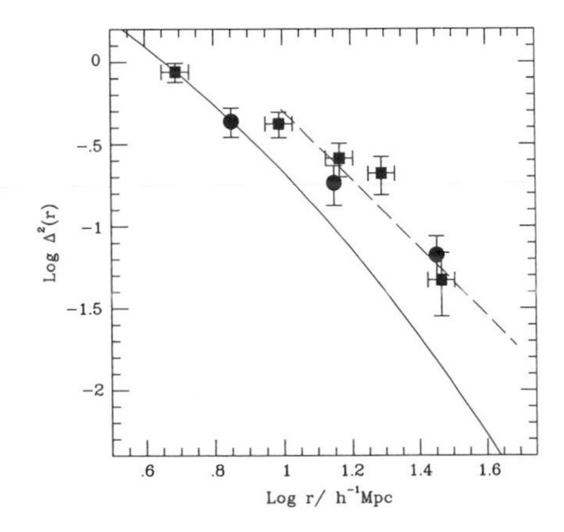

FIGURE 3. Variance of counts-in-cells. *Solid squares* and *solid circles* give results for the QDOT survey using the "shell" method (Efstathiou *et al.* 1990), and the "box" method (Saunders *et al.* 1991), respectively. The *vertical error bars* are 1σ errors; the *horizontal error bars* reflect uncertainties in scaling between different cell shapes and smoothing functions. The *solid line* gives the expected variance in a cold dark matter universe, normalized to the observations at $\sim 7\ h^{-1}$ Mpc. The *dashed line* gives the *slope* of the power spectrum inferred from a topological analysis. (Adapted from Moore *et al.* 1991.)

Mpc). This normalization requires $b_I \sim 1.2$, consistent with the results of Section 2.2. The QDOT data show more large-scale power than predicted in the model. The maximum discrepancy is a factor of ~ 1.8 in the standard deviation of the counts at a scale $r \sim 20\ h^{-1}$ Mpc. This result is consistent with the projected autocorrelation function of galaxies measured in the automated plate measuring machine (APM) survey, the first study to argue convincingly in favor of large-scale power (Maddox *et al.* 1990). A puzzling feature of the QDOT data is that most of the "excess" variance comes from a shell between $\sim 80 - 120\ h^{-1}$ Mpc (see Efstathiou *et al.* 1990, fig. 1). As FIGURE 1(b) shows, the largest superclusters are located there and are well distributed over angle on the sky. A deeper survey is required to investigate this peculiarity in more detail.

The *slope* of the power spectrum (but not its amplitude) is constrained by the topology of the galaxy distribution. This provides, in addition, a simple test of deviations from a Gaussian distribution of fluctuations. The main ideas in this area stem from work by Bardeen *et al.* (1986), Gott *et al.* (1986), and Hamilton *et al.* (1986), and have been applied to the QDOT data by Moore *et al.* (1991). Briefly, one computes the genus of the smooth galaxy distribution, a quantity related to the difference between the number of holes and the number of isolated regions. This is calculated from data such as the contours of FIGURE 1, using the algorithm given by Weinberg (1988). Evidently, the genus can be computed for different overdensity levels, ν, and a plot of the genus per unit volume as a function of ν is called a genus curve. A very useful result is that, for Gaussian fields, the genus curve is proportional to $(1 - \nu^2) \exp(-\nu^2/2)$, and the proportionality constant gives the logarithmic derivative of the power spectrum on the smoothing scale. In practice, difficulties arise in real surveys that, like the QDOT survey, are affected by shot noise on large scales. These difficulties can be overcome using *N*-body and Monte-Carlo simulations as described by Moore *et al.*

The genus curve for the QDOT survey, smoothed on scales between 10 and 50 h^{-1} Mpc, is consistent with that expected from a Gaussian distribution. (This is often referred to as the distribution having a "spongelike topology.") Although certain non-Gaussian distributions could have a genus curve of similar shape, this agreement lends support to the inflationary prediction of quantum primordial fluctuations, emodied in assumption (3) of the SCDM model. Further evidence for Gaussianity comes from the fact that the amplitude of the genus curve implies a power spectrum slope similar to that derived from the counts-in-cells approach. The topology analysis gives $n \simeq -1$ (in $|\delta_k|^2 \propto k^n$), roughly constant in the region considered, whereas the effective n in SCDM slowly rolls over from $n \simeq -1$ at 10 h^{-1} Mpc to $n \simeq 0.5$ at 50 h^{-1} Mpc. The maximum discrepancy in the QDOT data is $\sim 2\sigma$ at ~ 30 h^{-1} Mpc.

3. DISCUSSION

The QDOT survey of IRAS galaxies has proved to be extremely useful as a test of the cold dark matter cosmogony. A stringent test is possible only because the theory is so well specified—its one free parameter is fixed once and for all by one single diagnostic of clustering on small scales. From analysis of the QDOT data we must conclude that the standard model is falsified. This has been interpreted in some quarters as meaning that the entire CDM paradigm is falsified. Such extrapolation is, in my view, premature. Indeed the very same data set that contradicts the SCDM model also supports one of its most important (and controversial) tenets, that $\Omega = 1$. If we accept this result and standard big bang nucleosynthesis, we are forced to accept nonbaryonic dark matter as well. Unfortunately, the most obvious alternative to cold dark matter, light ($\sim$30-eV) neutrinos were shown long ago to lead to inconsistencies with observations of galaxy properties including large-scale clustering (see White 1988 and Frenk 1991a for reviews).

On balance, the evidence at our disposal today seems to me to indicate that cold dark matter is still the most attractive nonbaryonic candidate. We must then naturally ask whether other tenets of the SCDM model might be wrong. There are two obvious possibilities. The first is assumption (3) concerning the nature of primordial fluctuations. A neat way around this would be to invoke cosmic strings or textures (e.g., Brandenberger 1991; Turok 1991) as non-Gaussian seeds for galaxy formation. One argument against this, however, is that topological analyses of the QDOT and other surveys show no evidence for deviations from a Gaussian distribution of fluctuations on large scales (Gott *et al.* 1986; Moore *et al.* 1991). Even if the fluctuations were Gaussian, they might not be scale invariant as predicted by the simplest model of inflation. It is, in fact, entirely feasible to obtain different fluctuation spectra by altering the details of inflation (Guth 1991; Liddle 1991), but it is difficult to assess how plausible these alternatives are.

In my view, the SCDM model has one particularly feeble assumption. This is the linear biasing model for which there is no physical justification. Some form of biasing of the galaxy distribution is essential in any theory that assumes $\Omega = 1$, but there is no reason why the bias factor should be constant on all scales. The one physical mechanism that has been shown to lead to a segregation of galaxies from mass on large scales, the "natural bias" discussed by White *et al.* (1987), Efstathiou *et al.* (1988), and Frenk *et al.* (1988), produces a galaxy correlation function whose amplitude, relative to the mass correlation function, varies with scale. This deviation, however, does account for the excess large-scale power demonstrated in the QDOT and APM surveys. More recently, Carlberg and Couchman (1991) have argued that an N-body simulation of a CDM model with b *less* than unity gives an excellent fit to the APM projected galaxy autocorrelation function. Perhaps what is happening in this case is that mergers of clumps depress the correlation function on small scales. Since the normalization is carried out on intermediate scales, this has the effect of boosting the large-scale power. Further work is needed to check this interesting result.

Tampering with the bias model has been likened to a Ptolomeic attempt at saving a theory that is clearly doomed. Although it is true that the beauty of the simple model is under threat if a complicated bias is required, this criticism is simplistic. Biasing lies at the root of galaxy formation, the details of which are likely to remain poorly understood for some time. It would, of course, be encouraging to find some direct evidence for a scale-dependent bias. One such piece of evidence might be a variation of the luminosity function of galaxies with environment that could give rise to apparent superclustering in the galaxy distribution, not necessarily related to superclustering in the mass distribution. One can begin to look for effects of this sort in large data sets like the QDOT survey. The clearest signature of biasing would be a discrepancy between inhomogeneities in the *galaxy* and in the *mass* distributions. The latter give rise to fluctuations in the temperature of the microwave background radiation that are almost within reach of present detector sensitivity (Bond *et al.* 1991). Another handle on mass fluctuations is provided by analyses of large-scale streaming motions. Although these are more difficult to interpret than the microwave background experiments, the rapid increase in the amount and quality of the data gives reasons for optimism.

I will end by placing the disagreement between the SCDM model and measures of superclustering in context. The SCDM model has been singularly successful in accounting for a number of detailed properties of structures on small and intermediate scales (see White 1988; Frenk 1991a, and references therein). Among the successes, one may single out the following features of the model: approximately flat rotation curves in galactic halos, about the right abundance of galaxies and galaxy clusters, the correct mass-to-light ratios and internal structure of galaxy clusters, an explanation for the origin of the Hubble sequence, the existence of a "natural bias," the correct galaxy autocorrelation function, and other measures of clustering on scales $\sim 1 - 10\ h^{-1}$ Mpc, including the presence of filaments, voids, and "bubbles." Some predictions that are still to be convincingly tested concern the universe at high redshift. Galaxy formation is predicted to have occurred recently and to have involved a great deal of merging and other violent dynamical activity at redshifts $z \lesssim 1$ (White and Frenk 1991). Similarily, the abundance and properties of galaxy clusters are predicted to have evolved significantly at recent epochs (Frenk *et al.* 1990).

It remains to be seen how radical a departure from the standard model will be required. It may be that new fundamental physics will be needed to account, for example, for a nonstandard primordial fluctuation spectrum, for a nonzero cosmological constant (Efstathiou *et al.* 1991), or for an altogether different type of dark matter. Alternatively, it may be that new fundamental astrophysics are required to understand the connection between galaxies and mass. Ultimately, however, the most direct test of the CDM paradigm is to search for the hypothetical particles it postulates.

ACKNOWLEDGMENTS

I would like to thank my QDOT collaborators for allowing me to present some results prior to joint publication.

REFERENCES

Bardeen, J. M., J. R. Bond, N. Kaiser & Szalay. 1986. Astrophys. J. **304:** 15.

Bond, J. R., G. Efstathiou, P. M. Lubin & P. R. Meinhold. 1991. Phys. Rev. Lett. In press.

Brandenberger, R. 1991. *In* The Birth and Early Evolution of Our Universe, Nobel Symposium No. 79, J. S. Nilsson, B. Gustafsson, and B.-S. Skagerstam, Eds. World Scientific, Physica Scripta **T36:** 114.

Carlberg, R. & H. Couchman. 1991. CITA preprint.

Davis, M., G. Efstathiou, C. S. Frenk & S. D. M. White. 1985. Astrophys. J. **292:** 371.

Efstathiou, G., C. S. Frenk, S. D. M. White & M. Davis. 1988. Mon. Not. R. Astron. Soc. **235:** 715.

Efstathiou, G., N. Kaiser, W. Saunders, A. Lawrence, M. Rowan-Robinson, R. Ellis & C. S. Frenk. 1990. Mon. Not. R. Astron. Soc. **247:** 10 pp.

Efstathiou, G., W. J. Sutherland & S. J. Maddox. 1990. Nature **348:** 705.

Frenk, C. S. 1991a. *In* The Birth and Early Evolution of Our Universe, Nobel Symposium No. 79, J. S. Nilsson, B. Gustafsson, and B.-S. Skagerstam, Eds. World Scientific, Physica Scripta **T36:** 70.

Frenk, C. S. 1991b. *In* Observational Tests of Inflation, T. Shanks, A. Banday, C. S. Frenk, R. Ellis, and A. Wolfendale, Eds.: 355. Kluwer. Dordrecht, the Netherlands.

Frenk, C. S., S. D. M. White, G. Efstathiou & M. Davis. 1988. Astrophys. J. **327:** 507.

Frenk, C. S., S. D. M. White, G. Efstathiou & M. Davis. 1990. Astrophys. J. **351:** 10.

Gott, J. R., A. L. Melott & M. Dickinson. 1986. Astrophys. J. **306:** 341.

GUTH, A. 1991. *In* Observational Tests of Inflation, T. Shanks, A. Banday, C. S. Frenk, R. Ellis, and A. Wolfendale, Eds.: 1. Kluwer. Dordrecht, the Netherlands.
HAMILTON, A. J. S., J. R. GOTT & D. WEINBERG. 1986. Astrophys. J. **309:** 1.
KAISER, N. 1986. Mon. Not. R. Astron. Soc. **219:** 785.
KAISER, N. & O. LAHAV. 1989. Mon. Not. R. Astron. Soc. **237:** 129.
KAISER, N., G. EFSTATHIOU, R. ELLIS, C. S. FRENK, A. LAWRENCE, M. ROWAN-ROBINSON & W. SAUNDERS. 1991. Mon. Not. R. Astron. Soc. **252:** 1.
LIDDLE, A. R. 1991. *In* Observational Tests of Inflation, T. Shanks, A. Banday, C. S. Frenk, R. Ellis, and A. Wolfendale, Eds.: 23. Kluwer. Dordrecht, the Netherlands.
LUBIN, P. & T. VILLELA. 1986. *In* Galaxy Distances and Deviations Form Universal Expansion, B. F. Madore and R. B. Tully, Eds.: 161. Reidel. Dordrecht, the Netherlands.
MADDOX, S. J., G. EFSTATHIOU, W. SUTHERLAND & J. LOVEDAY. 1990. Mon. Not. R. Astron. Soc. **242:** 43 pp.
MOORE, B., C. S. FRENK, D. H. WEINBERG, W. SAUNDERS, A. LAWRENCE, M. ROWAN-ROBINSON, N. KAISER, G. EFSTATHIOU & R. ELLIS. 1991. Durham preprint.
OLIVE, K. A., D. N. SCHRAMM, G. STEIGMAN & T. P. WALKER. 1990. Phys. Lett. B **236:** 454.
PEEBLES, J. P. E. 1980. The Large-Scale Structure of the Universe. Princeton University Press. Princeton, N.J.
ROWAN-ROBINSON, M., *et al.* 1990. Mon. Not. R. Astron. Soc. **247:** 1.
SAUNDERS, W., *et al.* 1991. Nature **349:** 32.
SAUNDERS, W., M. ROWAN-ROBINSON, A. LAWRENCE, G. EFSTATHIOU, N. KAISER, R. ELLIS & C. S. FRENK. 1990. Mon. Not. R. Astron. Soc. **242:** 318.
STRAUSS, M. & M. DAVIS. 1988. *In* Large Scale Motions in the Universe: A Vatican Study Week, V. C. Rubin and G. V. Coyne, Eds.: 256. Princeton University Press. Princeton, N.J.
TUROK, N. 1991. *In* The Birth and Early Evolution of Our Universe, Nobel Symposium No. 79, J. S. Nilsson, B. Gustafsson, and B.-S. Skagerstam, Eds. World Scientific, Physica Scripta **T36:** 135.
WEINBERG, D. H. 1988. Publ. Astron. Soc. Pac. **100:** 1373.
WHITE, S. D. M. 1988. *In* Minnesota Lectures on Clusters of Galaxies and Large-scale Structure, J. M. Dickey, Ed.: 197. Astronomical Society of the Pacific Conference Series, **5:** 197.
WHITE, S. D. M. 1991. *In* Observational Tests of Inflation, T. Shanks, A. Banday, C. S. Frenk, R. Ellis, and A. Wolfendale, Eds.: 279. Kluwer. Dordrecht, the Netherlands.
WHITE, S. D. M. & C. S. FRENK. 1991. Astrophys. J. **379:** 52.
WHITE, S. D. M., C. S. FRENK, M. DAVIS & G. EFSTATHIOU. 1987. Nature **330:** 451.

The Great Attractor: A Challenge to Theory

V. N. LUKASH

Astro Space Centre
Academy of Sciences
Profsoyuznaya 84/32, 117810 Moscow, Russia

INTRODUCTION

There are two groups of experiments nowadays that, when completed, give us the direct detection of the spectrum of the primordial density perturbations on scales of $l \sim 3\ (10\text{–}10^3)\ h^{-1}$ Mpc. They are large-scale $\Delta T/T$ ($\Theta > 1°$) observations and the Great Attractor (GA) phenomenon. If the latter is a standard feature of the Universe structure, then the model independent estimate for $\Delta T/T$ on a few degrees of arc is $\sim (1\text{–}2)\ 10^{-5}$, which is quite amenable for COBE detection. Both optimistic predictions are true if the cosmological perturbation were Gaussian.

We see the following reasons for these conclusions.

The two experiment groups confront each other in theory. Indeed, if $\Delta T/T$ upper limits that we have up to now make us reduce the perturbation amplitudes on large scales, then the GA existence needs high enough primordial power on scales $l \sim (10\text{–}10^2)\ h^{-1}$ Mpc. Now the gap between these requirements is a factor of 2 for Gaussian perturbation theories. More of this, COBE and RELIC II sensitivity levels, $\Delta T/T \sim (3\text{–}5)\ 10^{-6}$, will leave practically no room for any theoretical speculations if GAs are representative and standard features in the Universe.

Both groups of observations are close to being completed, and the "dubbing" satelite experiments at a similar angle, frequency, and time intervals are fantastically lucky for the current cosmology.

Below, some results of the GA simulation are presented. We conclude that the standard cold dark matter (CDM) model cannot explain the GA appearance. New models of the early Universe accounting for the required primordial spectrum are proposed.

GA MODELING

GA is a large-scale phenomenon in the galactic peculiar velocity field (Dessler *et al.* 1987; Linden-Bell *et al.* 1988; Methewson *et al.* 1990; Dressler and Faber 1990; Burstein *et al.* 1990). Since more than half of the GA volume is still to be measured, the interpretation is not unambigious yet. To make the confrontation with theory more constructive, one has to determine the GA model first.

From GA we will understand a large-scale coherent flow of galaxies and clusters with the characteristic correlation length $l_c \geq 40\ h^{-1}$ Mpc, which is larger than the averaging scale ($l_f \sim 10\ h^{-1}$ Mpc) to be used for getting this velocity field. If so, then the comparison should be done in terms of the realization probability of the given velocity field expected in this or that theory of galaxy formation. This test is much more crucial and predictive than just the analysis of the bulk velocities.

Indeed, if rms velocity is a result of the contribution of the positive and negative fluctuations, then coherent flow requires the peak of a certain sign and shape. For example, the probability of finding the large-scale profile with velocities ≤ 600 km s^{-1} for Virgo and ≥ 800 km s^{-1} for Centaurus is extremely small in the CDM model, while the rms velocity $\sim 300\ h^{-1}\ b^{-1}$ km s^{-1} for scale $\sim 40\ h^{-1}$ (where h and b are the Hubble and biasing parameters, respectively) is close to the bulk observable motion of the volume (~ 400 km s^{-1}; see Dekel 1990).

For the simplest GA model the region between two spheres with the Centaurus and local group radia can be approximated by the spherically symmetric infall. Under this hypothesis, Hnatyk *et al.* (1991) considered the GA origin as a result of the evolution of a high-density peak of the adiabatic Gaussian perturbations in the two-component medium: dark matter (N-body approach) and hydrodynamic primordial plasma. Different perturbation spectra for $\Omega_{tot} = 1$ were tried. They found that some basic parameters of the linear initial peak required for the GA formation are model independent (e.g., the overdensity in the peak center $\delta(0) = (1 - 2)(1 + z)^{-1}$). The nonlinear evolution of the peak best fitted to Faber and Burstein's observational data (1988) at redshift $z = 0$, results in the core-radius $r_{GA} = (18–25)\ h^{-1}$ Mpc and the overdensity $\delta(0) \geq 3$, which corresponds to $\delta \leq 2$ in the central part of GA when averaged with filter $R_f = 12\ h^{-1}$ Mpc.

However, the realization probability of the GA peaks is drastically different in different models. The former depends on many parameters, the most important of which is the symmetry of GA. Say, under spherical symmetry GA can form in the CDM, hot dark matter (HDM), and hybrid models in density peaks of the levels $\delta(0) \geq 5\,\sigma$, $\sim (1–2)\sigma$, and $\sim (2–3)\sigma$, respectively, depending on b, h, and r_{GA}.

Taking this result into account, the CDM standard model appears to be in a similar situation in which the HDM model was a few years ago: CDM is good for galaxy formation, while only a few GAs can be found within the horizon; on the contrary, GAs are standard objects for HDM while the latter is helpless in the galactic origin. In fact, both models present two limit spectra for the Gaussian theories of the large-scale structure formation. The goal is to find models that can provide for the postrecombination primordial spectrum with high enough power (more than that in CDM) on GA scales as well as sufficient amplitude (higher than that in HDM) on galactic scales.

HYBRID MODELS

One of the ways to construct the desired spectrum is to change the transfer function assuming the standard Harrison–Zeldovich postinflationary spectrum. It is a two-component dark matter model that could solve the problem. The model is supposed to contain the hot (like massive neutrino) and cold (heavy particles or axion) components that are keeping the critical density together now. We have one free parameter here, for example, the total density of the massive neutrinos $\Omega_m \in$

(0, 1). For $\Omega_m = 0$ the model is the standard CDM, and for $\Omega_m > 0$ the extra power on large scales appears (see FIG. 1).

We can shift the position of the extra power "peak" with the help of another free parameter—the total amount of all weakly interacting relativistic particles. It directly affects the equality horizon scale and therefore the characteristic scale of the transfer function. However, this second dark matter parameter cannot change the spectrum shape (Lukash 1989).

Let us look at two interesting properties of the spectra presented at FIGURE 1. With Ω_m growing from zero to small positive values, the spectrum slope from galactic to GA scales becomes more and more flat. For the spectra with $\Omega_m > 0$ there is no problem in accounting for the difference between the ξ_{cc} and ξ_{gg} correlation functions since the extra power on suppercluster scales favors a larger ζ_{cc} amplitude.

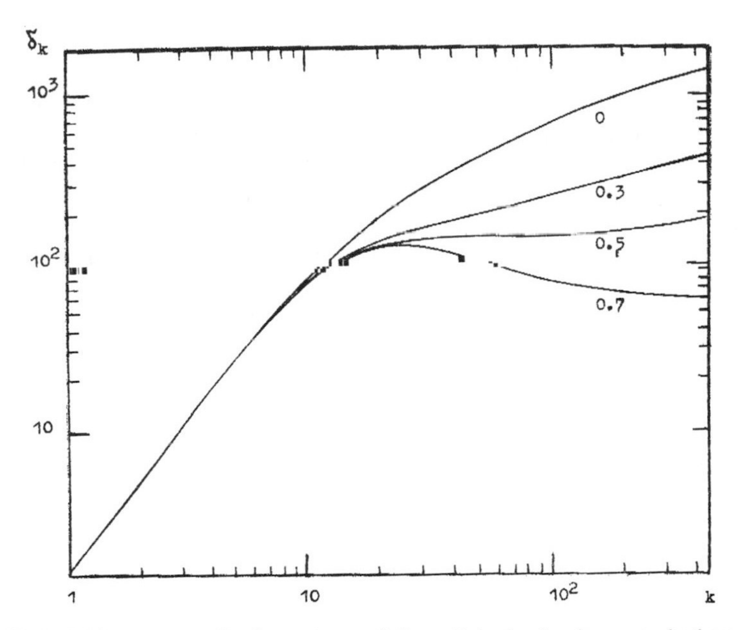

FIGURE 1. Arbitrary normalized spectrum of the adiabatic density perturbations at the postrecombination era for the two-component dark matter model with $\Omega_m = 0, 0.3, 0.5, 0.7$; $\delta_k^2 = k^3P(k)/2\pi^2$, $P(k)$ is the power spectrum, $\lambda = 450/kh^2$ [Mpc] is the perturbation wavelength.

CONCLUSIONS

There are a few important predictions produced by the GA simulation.

The GA symmetry itself could clarify the principal assumption the comparison with theory is based on: that about the GAs abundance. If GA deviates from spherical symmetry, then such objects are more or less typical ones independently of the model.

The absence of galactic contraflows and thermal activity in the GA-center testifies for the small overdensity in the initial peak center $\delta(0) < 1.5(1 + z)^{-1}$, and thus for the broad distribution of the GA mass, $r_{GA} \sim 20\ h^{-1}$ Mpc.

GA as a large-scale object can form only by gravitational instability. If GAs are additionally the representative phenomena in the Universe, then there is the need for the extra power in the primordial perturbations on scales of (10–100) h^{-1} Mpc relative to the standard CDM spectrum. Such an extra power appears naturally in the hot–cold dark matter models with the Harrison–Zeldovich post inflationary spectrum. This might prove to be the crucial test regarding inflation and dark matter problems.

The GA primordial perturbations, when required on the level $\sim 2\sigma$, bring about the model independent prediction $\Delta T/T$ ($\Theta \geq 1°$) $\sim 10^{-5}$, which reaches the prediction level of the standard neutrino HDM model. Therefore, both GA and $\Delta T/T$ observations may play a key role in the direct detection of the postrecombination primordial perturbation spectrum.

REFERENCES

BURSTEIN, D., S. M. FABER & A. DRESSLER. 1990. Astrophys. J. **354:** 18.
DEKEL, A. 1991. Streaming velocities and the formation of large-scale structure. This issue.
DRESSLER, A. & S. M. FABER. 1990. Astrophys. J. **354:** 13.
DRESSLER, A., S. M. FABER, D. BURSTEIN, R. L. DAVIES, D. LYNDEN-BELL, R. J. TERLEVICH & G. WEGNER. 1987. Astrophys. J., Lett. **313:** L37.
FABER, S. M. & D. BURSTEIN. 1988. Large scale motion in universe. Proceedings of the Pontific. Academy Study Week, No. 27.
HNATYK, B. I., V. N. LUKASH & B. S. NOVOSYADLY. 1991. Astron. Zh. Lett. **17:** 217, 659.
LUKASH, V. N. 1989. *In* Large Scale Structure and Motions in the Universe, M. Mezzetti *et al.*, Eds.: 139. Kluwer. Dordrecht, the Netherlands.
LYNDEN-BELL, D., S. M. FABER, D. BURSTEIN, R. L. DAVIES, A. DRESSLER, R. J. TERLEVICH & G. WEGNER. 1988. Astrophys. J. **326:** 19.
MATHEWSON, D. S., *et al.* 1990. Aust. J. Phys. **100:** 100.

The Origin of the X-ray Background

A. C. FABIAN

Institute of Astronomy
Madingley Road
Cambridge CB3 0HA, England

INTRODUCTION

The origin of the first cosmic background to be discovered (Giacconi *et al.* 1962) is still uncertain. The detection of many individual X-ray sources shows that some fraction of the X-ray background (XRB) must be due to unresolved sources, but so far no class of known sources has clearly been shown to contribute more than 50 percent of the total intensity.

The isotropy of the XRB strongly suggests that it is extragalactic. A weak galactic anisotropy is observed in the 2–10-keV band (Warwick *et al.* 1980; Iwan *et al.* 1982), but otherwise the X-ray sky is isotropic to within a few percent on large scales. On the largest scale, a dipole is observed that is consistent with the Compton–Getting effect due to the motion of our galaxy in the direction measured from the dipole in the microwave background (Shafer 1982; Shafer and Fabian 1983; Boldt 1988). Unless the photons now observed as X-rays originated as γ-rays, with consequent energy problems, it is most likely that the XRB originates from within the redshift range $0 < z < 10$. It therefore represents the integrated X-ray emission from all objects and processes since the Universe was about a billion years old. Its intensity and spectrum measure the total X-ray power of the Universe since then, and its anisotropies reflect the spatial distribution of the sources (Bagoly *et al.* 1988; Barcons and Fabian 1988; Mešzáros and Mešzáros 1988). This is of importance to studies of large-scale structure of the Universe and galaxy formation, since, unlike the microwave background, the XRB probes regions where the formation of structure is nonlinear.

After a brief review of the spectrum and isotropy of the XRB, I discuss the problems that arise with either a diffuse origin or an origin in known classes of discrete source. I then make the case for some new classes of source, or rather old classes with new faces, dominating the XRB. After discussing a possible signal found in the autocorrelation function (ACF) of *Ginga* data of the XRB and its interpretation, I outline some prospects for future studies of superclusters, clumps, and voids.

THE SPECTRUM AND ISOTROPY OF THE XRB

The XRB is detected over a wide energy band from soft X-rays of energies about 0.1 keV to gamma rays above an MeV. This is four decades of frequency or wavelength and so is a factor of 10 wider than the band stretching from 100 μ to 1000 Å. In practice, most of the soft XRB below 1 keV (see McCammon and Sanders 1990) appears to be galactic and due to the local hot bubble, and much of the hard XRB above 100 keV is probably due to the unresolved emission from active galactic nuclei (AGN) (see, e.g., Bignami *et al.* 1979). Consequently, when I talk about the XRB I shall concentrate on the band between 0.3 and 300 keV, which is comparable in relative size to the UV–optical–IR one mentioned earlier. The large bandwidth should be borne in mind whenever considering models for the XRB based on data from only a small part of the whole XRB band. The spectrum of an object detected at, say, 100 μ needs to be known with great accuracy before confident predictions can be made about its contribution in the UV. Similarly, we cannot yet be sure of the contribution of a quasars detected below 1 keV to the XRB at 40 keV.

The spectrum of the XRB closely resembles that of 40-keV thermal bremsstrahlung (Marshall *et al.* 1980). Most of the energy density ($\sim 5 \times 10^{-17}$ erg cm^{-3}) occurs between 10 and 100 keV. An excess is measured below about 2 keV (Garmire and Nousek 1980; Wu *et al.* 1990) and also above 100 keV. This last excess turns over around an MeV to form the "MeV Bump." Carbon-band observations of the soft XRB, in which absorption by galactic hydrogen clouds was searched for, limit the total extragalactic contribution at around 0.25 keV to an intensity of about twice that obtained by a power-law extrapolation of the spectrum above 3 keV (McCammon and Sanders 1990).

The large-scale isotropy of the XRB has already been mentioned. On small angular scales, unresolved sources dominate the observed granularity. These fluctuations can be modeled as "$P(D)$ noise" (see, e.g., Scheuer 1974; Warwick and Stewart 1989) based on predictions of the shape of the X-ray source counts, and any excess fluctuations, attributable to say clumping of the sources, can be measured or constrained. Generally, a limit of less than about 2 percent has been found on scales of 3 to 5 deg (Shafer and Fabian 1983) and also on about 1 to 2 deg (Butcher *et al.* 1991). Any excess fluctuations on a scale of a few arcminutes must be less than a factor of 2 in the 1 to 3-keV band (Hamilton and Helfand 1987; Barcons and Fabian 1990). These last two studies found such small fluctuations in the *Einstein Observatory* Imaging Proportional Counter (IPC) data they conclude that the X-ray source counts must flatten in the 1 to 3-keV band below a limit of 2.5×10^{-14} erg cm^{-2} s^{-1} in the 1 to 3-keV band (the deep survey limit).

Correlation analyses of the XRB are a powerful tool for testing for clumping of sources. Only limits have been published so far; Persic *et al.* (1989) measure $W = \Delta I^{-}_{1}\Delta I^{-}_{2}/I^2 < 4 \times 10^{-4}$ on the scale of 3 deg and Carrera *et al.* (1991a) and Martin-Mirones *et al.* (1991) find values of $W < 10^{-4}$ on scales of $\sim$2–3 deg. Analysis of IPC data by Barcons and Fabian (1989) shows a signal on a scale of about 5 arcmin of $W \sim 0.1$, which is treated as an upper limit since the IPC is known to have some intrinsic irregularities on that scale. Soltan (1991) has reduced this limit considerably. New results on W are presented later in this paper and in the paper by Martin-Mirones *et al.* (1992). Finally, an interesting result is emerging from work by Jahoda *et al.* (1991), in which fluctuations in the HEAO-1 data are cross-correlated with galaxy counts. A significant correlation is found, which is probably due to the sources that are just unresolved (mostly AGN) lying in regions of higher than average galaxy density.

A DIFFUSE ORIGIN?

Diffuse models for the XRB were among the first to be put forward. Felten and Morrison (1966) suggested that it might be due to inverse Compton scattering of

MWB photons by intergalactic electrons. Little has since been done with this idea, except to note that the break in the spectrum at 30–40 keV is difficult to produce. Hoyle (1962) proposed that the XRB was thermal bremsstrahlung from the hot intergalactic medium (IGM) created by decaying neutrons in the steady-state cosmology. This was soon seen to overproduce the XRB (Gould and Burbidge 1963; see also Friedman 1990).

The basic idea of thermal bremsstrahlung from a hot IGM in a big bang cosmology was taken up again by Cowsik and Kobetich (1972) and Field and Perrenod (1977). The temperature required corresponds to $kT_{IGM} \sim 40(1 + z_{heat})$ keV, where z_{heat} is the redshift at which the gas is heated. Later work by Guilbert and Fabian (1986) and by Barcons (1987), including relativistic corrections to the bremsstrahlung formulas at these high temperatures, showed that the baryonic density, Ω_{baryon}, needed to be at least 23 percent of the critical density. This density exceeds that allowed by standard cosmological nucleosynthesis. Plausibility problems were also raised about the high-energy density needed in hot gas (comparable to the energy density in the MWB). The model is now ruled out since it predicts a Compton distortion in the spectrum of the MWB, which is not observed in the spectrum obtained from the COBE satellite (Mather *et al.* 1990).

Clumping of the IGM can reduce the required value of Ω_{baryon}, but this is also ruled out since at high redshifts the clumps produce observable fluctuations in the MWB by the Sunyaev–Zeldovitch effect (Barcons and Fabian 1988; Barcons *et al.* 1991), and at low redshifts the clumps would be detectable as individual X-ray sources (the source cannot then be called the IGM). There are also serious confinement problems for any such clumps.

The IGM may still contribute at some level to the spectrum of the XRB, possibly in the soft X-ray band. The formation of galaxies, groups, and clusters in which some or all of the baryons are heated to the virial temperature can also lead to an observable soft XRB. Conversely, the observed soft XRB can limit Ω_{baryon} involved in dissipational structure formation (Thomas and Fabian 1990).

THE CONTRIBUTION OF KNOWN CLASSES OF DISCRETE SOURCE

The major classes of known extragalactic X-ray source are clusters of galaxies and AGN. We can be certain that clusters are not a major contributor to the XRB since their luminosity function is well-determined and its evolution with redshift known. The sense of the evolution of the X-ray luminosity of clusters is negative in the sense that the most luminous clusters are most numerous at the present epoch (Edge *et al.* 1990; Gioia *et al.* 1990). An integration of their contribution to the XRB in the 4–12-keV band gives a result of about 3 percent.

Estimating the contribution of AGN is more difficult. Splitting them into the two major classes of Seyfert galaxies and quasars, we can then make fair estimates of each class separately. The Seyfert luminosity function breaks at a luminosity, $L_x \sim 10^{43}$ erg s^{-1}, which dominates the resulting contribution of 10–20 percent. Their evolution is uncertain (see Danese *et al.* 1986 for estimates). The quasar contribution (and perhaps Seyferts evolve from quasars) is more difficult to assess since there is no reliable luminosity function. Many attempts have been made, however, by using some parameterization of the X-ray to optical luminosity ratio of quasars and their optical source counts. The results vary between a few tens to 100 percent. More recently a number of isotropy results, to be discussed later, require that the quasar contribution is less than about 50 percent.

A major problem with AGN as the source of the XRB is their spectra. Most of them are well-fitted by a power law of spectral index $\alpha > 0.5$, in the 2–10-keV band at least, whereas the XRB spectral index in that band is ~ 0.4. Radio-quiet quasars, the most numerous class, appear to have soft X-ray spectra with $\alpha > 1$ (Wilkes and Elvis 1987; Canizares and White 1989). However, it is possible that the spectra of AGN are complex, perhaps with $\alpha \sim 0.7$ at lower energies, flattening to an index of 0.4 above 10 keV, and then steepening to 1.5 or greater above about 100 keV (Schwartz and Tucker 1988).

Others have suggested that the intrinsic spectra of many AGN are steep, but are flattened at energies below about 20 keV or so by intrinsic photoelectric absorption (Setti and Woltjer 1989; Morisawa *et al.* 1990; Grindlay and Luke 1989). Seyfert 2 galaxies may be the dominant contributor. Although some reasonable fits have been made to the spectrum of the XRB, this model provides no simple explanation for the 40-keV break in the XRB spectrum.

Observations now show that the spectra of Seyfert galaxies do flatten above about 10 keV (Pounds *et al.* 1990; Matsuoka *et al.* 1990), the reason for which will be discussed later. Intrinsic absorption also appears to be common in low luminosity AGN, as indicated by the paucity of nearby AGN as serendipitous sources in soft X-ray images (Fabian *et al.* 1981; see also the comparison of source counts from different X-ray bands by Warwick and Stewart 1989, and references therein).

As mentioned already, the observed isotropy of the XRB provides some strong constraints on the contribution of AGN. The first limit came from an analysis of fluctuations in deep images from the Einstein Observatory IPC made by Hamilton and Helfand (1987). This showed that the (1–3-keV) X-ray sources counts must flatten just below the so-called deep survey level at which the faintest sources could be directly resolved. An independent analysis by Barcons and Fabian (1990) of the same images obtained a similar result. Basically, background fluctuations are expected from the unresolved sources in the field, and the data showed that these fluctuations are smaller than expected if the source counts continued to much fainter flux levels. Such a flattening of the source counts is expected (Barcons and Fabian 1990) since most of the deep survey sources are identified with faint quasars (Griffiths *et al.* 1987) and optical quasar number counts flatten off below twenty-second magnitude. The residual background must then be very smooth on the arcminute scale, requiring more than 1000 sources per square degree.

Very recently, Soltan (1991) has questioned the significance of the X-ray result, carrying out yet another (different) analysis of the IPC deep fields using more of the data and accounting for intrinsic gain variations in the detector. However, Hasinger (private communication) confirms from Röntgenstrahlen Satellit (ROSAT) images that a flattening of the source counts does occur. This flattening, at a flux level too low to account for all the XRB intensity from brighter sources, emphasises that sources contributing to imaging surveys are not the origin of all the XRB.

These limits apply to a range that is softer than the main 3–100-keV XRB band over which the spectrum is well determined. The fraction of the XRB estimated from sources contributing to the observed IPC source counts above the level at which the counts flatten is about 50 percent of the intensity obtained by extrapolating the preceding 3-keV spectrum to lower energies with a power law index of 0.4. Since an excess is measured in the spectrum in the 0.3–3-keV band, it is not clear whether the IPC sources contribute at all to the higher energy background, perhaps only constituting the soft extragalactic excess background.

Some limits have been obtained at higher energies (3–12 keV) from estimates of "excess" fluctuations. These are fluctuations in excess of those expected from the sources that are just not resolved and require that the sources contributing most of the XRB do not cluster on scales $\gtrsim 10$ Mpc. Stronger limits have been obtained from ACF analyses of the XRB in the HEAO-1 (Persic *et al.* 1989; de Zotti *et al.* 1990) and GINGA data (Carrera *et al.* 1991a). This last work allows only 50 percent of the XRB to originate in sources at redshifts of 1 to 2 with a correlation length exceeding 8 Mpc. Boyle (private communication) finds that 400 optically selected quasars (with redshifts out to 2.50) have a constant comoving correlation scale of ~ 8 Mpc.

The recent result of Shanks *et al.* (1991), who find about 70 quasars per square degree in a fairly deep ROSAT image, has confirmed that most quasars are X-ray sources (although the Einstein Observatory detected X-ray emission from many quasars, most of them are at $z < 1$). The quasars identified by Shanks *et al.* have a similar redshift distribution to that of faint optically identified quasars. Direct integration of the observed quasar emission accounts for about half of the soft XRB observed in the same image. Quite how important this is for the 3–300-keV XRB is not yet clear, since what they may have done is to clearly resolve the soft excess in the XRB spectrum. Again, the fluctuation analysis of Hasinger shows that quasars do not produce *all* the XRB, even at soft X-ray energies. We know from previous observations that the soft X-ray spectra of quasars are steep (they must rise in the soft X-ray/EUV band to match their optical spectra). The high density of quasars observed with ROSAT may then be a consequence of the higher soft X-ray (i.e., < 0.28-keV) sensitivity of ROSAT when compared with the IPC.

The probable picture that is forming from current observations is that the observed spectra and number density of quasars and known AGN cannot account for the 3–300-keV XRB. At most they can account for 50 percent of it. The most precise results have been obtained from deep imaging studies in the soft X-ray band where an excess intensity has been measured in the XRB. The straightforward interpretation is that the observed quasars (down to, say, $m_B \sim 24$) produce this excess intensity and contribute little to the 3–300-keV XRB, which remains a puzzle. Although it is possible that quasar spectra are complex and flatten above a few kiloelectron volts, their number density appears to be too low to be consistent with the fluctuation analyses.

A NEW CLASS OF SOURCE

The simplest interpretation of the data obtained so far is that the XRB is due to some new class of source, or at least a new face on an already known class. A strong candidate has been the starburst galaxy (Bookbinder *et al.* 1979; Stewart *et al.* 1982; Fabbiano *et al.* 1982; Weedman 1987; Griffiths and Padovani 1990; Helfand private communication). Supernovae or X-ray binaries in young, star-forming galaxies might produce large amounts of X-ray emission. Support for this view has been obtained from the high ratio of X-ray to optical luminosity of some nearby starburst galaxies, particularly where the metal abundance is low (e.g., Stewart *et al.* 1982).

The main problems with the starburst hypothesis have been the uncertain evolution (see de Zotti *et al.* 1989 for estimates; Infrared Astronomical Satellite (IRAS) counts suggest steep evolution) and the apparent softness of the spectrum of at least one promising starburst galaxy (NGC 5408; Stewart, private communication). A further problem might be the lack of any obvious population of starburst candidates in the deep ROSAT images (this statement is based on the high proportion of the objects identified by Shanks *et al.* (1991) with quasars; they identify very few sources with galaxies). Miyaji and Boldt (1990) obtained a strong upper limit on the local X-ray emissivity of the Universe from a study of source contributions to the dipole anisotropy, which does not allow a significant component of the XRB to arise from starbursts (or anything else) at the current epoch.

Perhaps the best candidate has been an earlier phase in the life of AGN. Leiter and Boldt (1982) and Boldt and Leiter (1984) suggested that a high accretion phase when the central engines were rapidly gaining mass might lead to a qualitatively different X-ray spectrum than that observed now from AGN that might be past their prime (in the sense of the ratio of the actual accretion rate to that corresponding to the Eddington luminosity). Zdziarski (1988) and Wandel (1990) have produced specific models that account for the overall spectrum in a physical way.

I have been convinced that the 40-keV break is an important feature that any successful model should generate in a very basic manner. It should not be due to some arbitrary combination of, say, optical depth and temperature, but should involve some basic physics. Only then will many sources, presumably with different luminosities, redshifts, etc., create a relatively sharp break in the observed XRB spectrum. Some energy above 40 keV (when the source redshifts are taken into account) must therefore be identified. The electron rest mass is probably the relevant one, but not through the production of electron–positron pairs unless a redshift of 20 for the sources is acceptable (Fabian *et al.* 1988). A simpler way of involving the electron mass is through Compton recoil. A continuous, hard X-ray spectrum incident onto a thick slab of matter gives rise to a "reflection" spectrum (i.e., the spectrum scattered back) that has a break at about 150 keV due to Compton recoil (Lightman and White 1988). A hard spectrum is produced in many AGN, and there is a large slab of matter there in the form of the accretion disc. Consequently, a reflection spectrum of the right shape is expected from AGN, if at redshifts of about 2 (Fabian *et al.* 1990).

The reflection spectrum is now observed in the X-ray spectra of many Seyfert galaxies and the geometry is confirmed by the accompanying iron fluorescence line (Pounds *et al.* 1990; Matsuoka *et al.* 1990). It appears to be the cause of the flattening of the spectra of AGN above 10 keV. Photoelectric absorption in the reflection slab causes there to be few photons reflected at lower energies. (To some extent, the reflection model for the XRB is here similar to the absorption models discussed in the last section.) The main problem with the model is that it requires only about 7 percent of the direct hard spectrum to be visible. This may be due to the geometry of the central engine or to anisotropic emission of the hard X-ray source. It does predict a large EUV background, since much of the hard X-ray emission is absorbed by the surrounding gas and is reradiated as quasi-thermal radiation. There are indications from other wavebands of the existence of such a large background (e.g., Steidel and Sargent 1989).

The 3–100-keV XRB in this model is predominately due to young AGN at $z \sim$ 1–3. The IPC fluctuation limit requires a high surface density of $\gtrsim 1000$ sources per square degree, which is comparable to the integrated surface density of all AGN (about 1 percent of all galaxies). The X-ray luminosity of the individual sources is then $\sim 10^{44}$ erg s^{-1}.

A SIGNAL IN THE ACF?

Major progress in our understanding of the origin of the XRB will occur when we have some better handles on it. The break in the spectrum is one handle, discussed previously. Fluctuations identifiable with the bulk of the XRB have remained elusive. When observed we can use them to both observe the distribution of candidate sources at other wavelengths and measure the large-scale structure of the (X-ray) Universe.

Preliminary analysis of GINGA scan data suggest that a signal is emerging. The first sign of this occurred in the ACF from pointed GINGA observations (Carrera *et al.* 1991a), where a strong limit on any positive signal was obtained ($\lesssim 10^{-4}$). Curiously, the ACF appeared at low ($\sim 2\sigma$ significance) to be *negative* at 2 and 10 deg. The scan data now analyzed by Carrera *et al.* (1991b) appear to show a significant negative signal at angles between 3 and 6 deg and also between 9 and 11 deg. A positive signal is found at around 7 and 14 deg. The scans are about 25 deg in length and parallel to each other. The fluctuations in them are repeatable over a 6 month interval, and so associated with the real sky. The amplitude is small and corresponds to fluctuations of only a few percent on a scale of a few degrees.

This observation of an anticorrelation in the distribution of the XRB, if significant and not due to a chance alignment of sources (simulations are being carried out to test this), suggests that voids may be important in the XRB. If some of the sources contributing to the XRB lie in "walls" either side of voids, then when we see a maximum in intensity due to a wall lying along the line of sight, we can expect to see a dip either side. Identifying the relevant redshift as about 0.3, from the amplitude of the fluctuations and the luminosity function of AGN (a source at the peak of the luminosity function, about 10^{43} erg s^{-1}, should give a flux of about 10^{-12} erg cm^{-2}s^{-1} keV^{-1} at $z = 0.3$) the comoving size of the peaks in the ACF corresponds to about 250 Mpc ($H_0 = 50$ km s^{-1} Mpc^{-1}), close to the large-scale structures observed in the redshift distribution of galaxies by Broadhurst *et al.* (1990).

Further work is clearly needed before this preliminary result can be taken any further. The current studies are, however, reaching the levels where signals should emerge. ROSAT data will define the soft X-ray spectrum of the XRB, the mean shape of quasar spectra, and their source counts. The high fraction of quasars detected in ROSAT images offers the exciting possibility of mapping the spatial distribution of complete (i.e., flux-limited) samples of them on scales of many tens of megaparseconds, and presumably thereby mapping the underlying mass distribution on those scales. Fluctuation and ACF analyses will measure the clustering of all sources, including quasars, and hopefully reveal other structures in the soft XRB (it is to be hoped that absorption by galactic "cirrus" does not seriously complicate such studies). The spectrum of the XRB above 3 keV will be very well-determined by the detectors on the recently flown BBXRT, which will also provide us with a taste of the results expected from ASTRO-D, Spectrum-X, AXAF, and XMM. These telescopes should make X-ray studies of both young AGN and the large-scale structure of the Universe an important research field.

ACKNOWLEDGMENTS

I thank my collaborators in the GINGA studies, F. Carrera, X. Barcons, J. Butcher, G. Stewart, and R. Warwick for much help, and B. Boyle and O. Lahav for discussions.

REFERENCES

BAGOLY, Z., A. MÉSZÁROS, & P. MÉSZÁROS. 1988. Astrophys. J. **335:** 54.
BARCONS, X. 1987. Astrophys. J. **313:** 547.
BARCONS, X. & A. C. FABIAN. 1988. Mon. Not. R. Astron. Soc. **230:** 189.
BARCONS, X. & A. C. FABIAN. 1989. Mon. Not. R. Astron. Soc. **237:** 119.
BARCONS, X. & A. C. FABIAN. 1990. Mon. Not. R. Astron. Soc. **243:** 366.
BARCONS, X., A. C. FABIAN & M. J. REES. 1991. Nature. In press.
BIGNAMI, G. F., C. E. FICHTEL, R. C. HARTMAN & D. J. THOMPSON. 1979. Astrophys. J. **232:** 649.
BOLDT, E. 1988. Phys. Rep. **146:** 215.
BOLDT, E. A. 1989. *In* X-ray Astronomy, 2. AGN and the X-ray Background. ESA SP-296:797. Noordwijk, the Netherlands.
BOLDT, E. A. & D. LEITER. 1984. Astrophys. J. **276:** 427.
BOOKBINDER, J., *et al.* 1980. Astrophys. J. **237:** 647.
BROADHURST, T. J., R. S. ELLIS, D. C. KOO & A. S. SZALAY. 1990. Nature **343:** 726.
BUTCHER, J., *et al.* 1990. In preparation.
CANIZARES, C. R. & J. L. WHITE. 1989. Astrophys. J. **339:** 27.
CARRERA, F. J., X. BARCONS, J. BUTCHER, A. C. FABIAN, G. C. STEWART, R. S. WARWICK, K. HAYASHIDA & T. KII. 1991a. Mon. Not. R. Astron. Soc. In press.
CARRERA, F. J., X. BARCONS, J. BUTCHER, G. C. STEWART, R. S. WARWICK & A. C. FABIAN. 1991b. In preparation.
COWSIK, R. & E. J. KOBETICH. 1972. Astrophys. J. **177:** 585.
DANESE, L., G. DE ZOTTI, G. FOSANO & A. FRANCESCHINI. 1986. Astron. Astrophys. **161:** 1.
DE ZOTTI, G., L. DANESE, A. FRANCESCHINI, M. PERSIC & L. TOFFOLATTI. 1989. *In* X-ray Astronomy, 2. AGN and the X-ray Background: 737. ESA SP-296. Noordwijk, the Netherlands.
DE ZOTTI, G., M. PERSIC, A. FRANCESCHINI, L. DANESE, G. G. C. PALUMBO, E. A. BOLDT & F. E. MARSHALL. 1990. Astrophys. J. **351:** 22.
EDGE, A. C., G. C. STEWART, A. C. FABIAN & K. ARNAUD. 1990. Mon. Not. R. Astron. Soc. **245:** 559.
FABBIANO, G., *et al.* 1982. Astrophys. J. **256:** 297.
FABIAN, A. C., C. DONE & G. GHISELLINI. 1988. Mon. Not. R. Astron. Soc. **232:** 21 pp.
FABIAN, A. C., A. K. KEMBHAVI & M. J. WARD. 1981. Space Sci. Rev. **30:** 113.
FABIAN, A. C., I. M. GEORGE, S. MIYOSHI & M. J. REES. 1990. Mon. Not. R. Astron. Soc. **242:** 14 pp.
FELTEN, J. E. & P. MORRISON. 1966. Astrophys. J. **146:** 686.
FIELD, G. B. & S. C. PERRENOD. 1977. Astrophys. J. **215:** 717.
FRIEDMAN, H. 1990. The Astronomer's Universe: 284. Norton. New York.
GARMIRE, G. & J. NOUSEK. 1981. Bull. Am. Astron. Soc. **12:** 853.

GIACCONI, R., H. GURSKY, F. PAOLINI & B. ROSSI. 1962. Phys. Rev. Lett. **9:** 439.
GIOIA, I. M., J. P. HENRY, T. MACCACARO, S. L. MORRIS, J. T. STOCKE & A. WOLTER. 1990. Astrophys. J. Lett. **356:** L35.
GOULD, R. & G. BURBIDGE. 1963. Astrophys. J. **138:** 969.
GRIFFITHS, R. E., I. R. TUOHY, R. J. V. BRISSENDEN, M. WARD, S. S. MURRAY & R. BURG. 1987. *In* The Post Recombination Universe, N. Kaiser and A. N. Lasenby, Eds.: 91. Kluwer. Dordrecht, the Netherlands.
GRIFFITHS, R. E. & P. PADOVANI. 1990. Astrophys. J. **360:** 483.
GRINDLAY, J. & LUKE. 1989. Preprint.
GUILBERT, P. W. & A. C. FABIAN. 1986. Mon. Not. R. Astron. Soc. **230:** 439.
HAMILTON, T. T. & D. J. HELFAND. 1987. Astrophys. J. **318:** 93.
HOYLE, F. 1963. Astrophys. J. **137:** 993.
IWAN, D., F. E. MARSHALL, E. A. BOLDT, R. F. MUSHOTZKY, R. A. SHAFER & A. STOTTLEMEYER. 1982. Astrophys. J. **260:** 111.
JAHODA, K., O. LAHAV, R. F. MUSHOTZKY & E. A. BOLDT. 1991. In preparation.
LEITER, D. & E. A. BOLDT. 1982. Astrophys. J. **260:** 1.
LIGHTMAN, A. P. & T. R. WHITE. 1988. Astrophys. J. **335:** 57.
MARSHALL, F. E., E. A. BOLDT, S. S. HOLT, R. B. MILLER, R. F. MUSHOTZKY, L. A. ROSE, R. E. ROTHSCHILD & P. J. SERLEMITSOS. 1980. Astrophys. J. **235:** 4.
MARTIN-MIRONES, J. M., *et al.* 1990. Submitted for publication in Astrophys. J.
MARTIN-MIRONES, J. M., G. DE ZOTTI, A. FRANCESCHINI & L. DANESE. 1991. Fluctuations in the X-ray background. This issue.
MATHER, J. C., *et al.* 1990. Astrophys. J. Lett. **354:** L37.
MATSUOKA, M., L. PIRO, M. YAMAUCHI & T. MURAKAMI. 1990. Astrophys. J. **361:** 440.
MCCAMMON, D. & W. T. SANDERS. 1990. Annu. Rev. Astron. Astrophys. **28:** 657.
MÉSZÁROS, A. & P. MÉSZÁROS. 1988. Astrophys. J. **327:** 25.
MIYAJI, T. & E. A. BOLDT. 1990. Astrophys. J. Lett. **353:** L3.
MORISAWA, K., M. MATSUOKA, F. TAKAHARA & L. PIRO. 1990. Astron. Astrophys. **236:** 299.
PERSIC, M., G. DE ZOTTI, E. A. BOLDT, F. E. MARSHALL, L. DANESE, A. FRANCESCHINI & G. G. C. PALUMBO. 1989. Astrophys. J. Lett. **336:** L47.
POUNDS, K. A., K. NANDRA, G. C. STEWART, I. M. GEORGE & A. C. FABIAN. 1990. Nature **344:** 1132.
SCHEUER, P. G. 1974. Mon. Not. R. Astron. Soc. **167:** 419.
SCHWARTZ, D. A. & W. H. TUCKER. 1988. Astrophys. J. **332:** 157.
SETTI, G. & L. WOLTJER. 1989. Astron. Astrophys. **224:** L21.
SHAFER, R. A. 1982. Ph. D. Thesis. University of Maryland, College Park.
SHAFER, R. A. & A. C. FABIAN. 1983. *In* Early Evolution of the Universe and Its Present Structure. G. O. Abell and G. Chincarini, Eds. Reidel. Dordrecht, the Netherlands.
SHANKS, T., *et al.* 1991. In preparation.
SOLTAN, A. 1991. Mon. Not. R. Astron. Soc. In press.
STEIDEL, C. C. & W. L. W. SARGENT. 1989. Astrophys. J. Lett. **343:** L43.
STEWART, G. C., A. C. FABIAN, R. J. TERLEVICH & C. HAZARD. 1982. Mon. Not. R. Astron. Soc. **200:** 61 pp.
THOMAS, P. A. & A. C. FABIAN. 1990. Mon. Not. R. Astron. Soc. **246:** 156.
WANDEL, 1990. Paper presented at the 15th Texas/ESO-CERN Symposium on Relativistic Astrophysics, Cosmology and Fundamental Physics, Brighton, England, December 16–21, 1990.
WARWICK, R. S., J. P. PYE & A. C. FABIAN. 1980. Mon. Not. R. Astron. Soc. **190:** 243.
WARWICK, R. S. & G. C. STEWART. 1989. *In* X-ray Astronomy, 2. AGN and the X-ray Background: 727. ESA SP-296. Noordwijk, the Netherlands.
WEEDMAN, D. 1987. *In* Proceedings of the Conference on Star Formation in Galaxies. NASA Conf. Publ. **2466:** 351.
WILKES, B. J. & M. ELVIS. 1987. Astrophys. J. **323:** 243.
WU, X., T. HAMILTON, D. J. HELFAND & Q. WANG. 1990. Astrophys. J. In press.
ZDZIARSKI, A. 1988. Mon. Not. R. Astron. Soc. **233:** 739.

The Void Spectrum in Numerical Model Universes[a]

ADRIAN L. MELOTT[b] AND GUINEVERE KAUFFMANN[c]

[b]*Department of Physics and Astronomy*
University of Kansas
Lawrence, Kansas 66045

[c]*Institute of Astronomy*
Cambridge University
Madingley Road
Cambridge CB3 0HA, England

Recently an algorithm that rapidly evaluates the spectrum of sizes of voids was used to examine galaxy data.[1] We believe this method may present a simple, rapid way of characterizing the spectrum of sizes of voids that is rather insensitive to errors resulting from discreteness or incompleteness. The program fits voids to a cubical gridded data set. Voids are defined as *maximal-sized* groups of grid cubes that are empty of or below some threshold density. The groups of grid cubes that constitute a void are constrained to be approximately spherical or ellipsoidal in shape. The void spectrum is presented as a histogram of the fraction of volume filled by cubes of various diameters. For more details, please see the original publication.[1]

Recently, a letter[2] based on the QMWC–Durham–Oxford–Toronto (QDOT) survey announced the discovery of several new voids. The work on which our present one is based[1] found all of these voids in a galaxy catalog with an inhomogeneous selection function, except those voids lying in the "no-man's-land" between northern and southern optical catalogs. This is strong evidence of the usefulness of the method. We would like to better understand its relation to dynamics.

In order to understand the operation of this algorithm better, we have applied it to an extensive series of two-dimensional numerical simulations, which are being studied in a number of complementary ways.[3–6] Of course, we have substituted squares for cubes and area for volume in the void spectrum. Here we report on work in progress at the time of this Symposium.

The simulations have nine different types of power spectra for their Gaussian initial conditions. Four different realizations of each type were done, for a total of 36 simulations. Working on a Cray-2 permits production and analysis of relatively larger numbers of simulations with 512^2 particles on an equivalent mesh with a photomultiplier (PM) code. In FIGURE 1 we show a subset (1/64) of the particles in such a simulation group. It is impossible to plot all the particles, or the picture would be mostly black. For this reason color plots are preferred.[3,7] All the simulations shown in

[a]This work was supported in part by NASA under Grants NAGW-1288 and NAGW-1793, and in part by the National Science Foundation under Grant AST-8911327. Travel to the symposium was supported by the National Science Foundation under Grant AST-8721484 at Princeton University.

this picture had identical phases for the same Fourier coefficients in their initial conditions.

We note that for Gaussian initial conditions, and apparently in the actual large-scale structure, the mass distribution is spongelike, with interconnecting tunnels rather than isolated voids.[8–10] We wish to verify that the VOIDSEARCH algorithm can pick out the proper lengthscale in either situation. Our test of this consists of analysis of the Gaussian initial conditions for a dynamical model with a truncated power spectrum of perturbations, rather like hot dark matter. The dynamically evolved field acquires a bubble topology.[9]

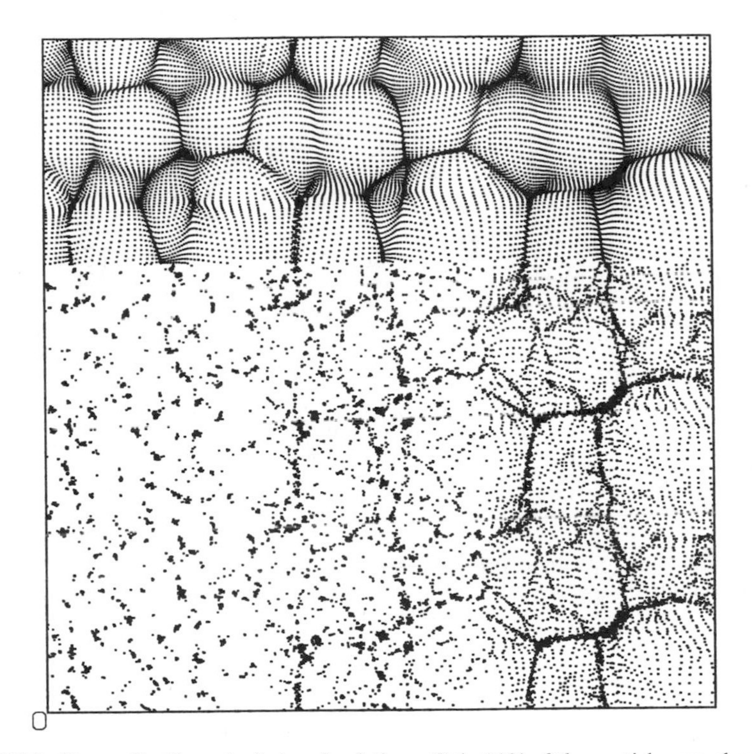

FIGURE 1. One realization set of nine simulations. Only 1/64 of the particles are shown, to avoid overload. The models had different initial power spectra, but the same phases for the same Fourier components.

We must define our voids. The simulations are designed to model a smooth density field to the extent possible. Particles are not the fundamental entities; they are merely markers in phase space. Defining a void as the absence of particles would give a different answer for two simulations that modeled the same initial conditions with different particle densities. A similar criticism of past work on redshift surveys is possible. Defining a void as a completely empty region means that a large void can disappear if a single galaxy is found inside. We recommend adopting an alternative

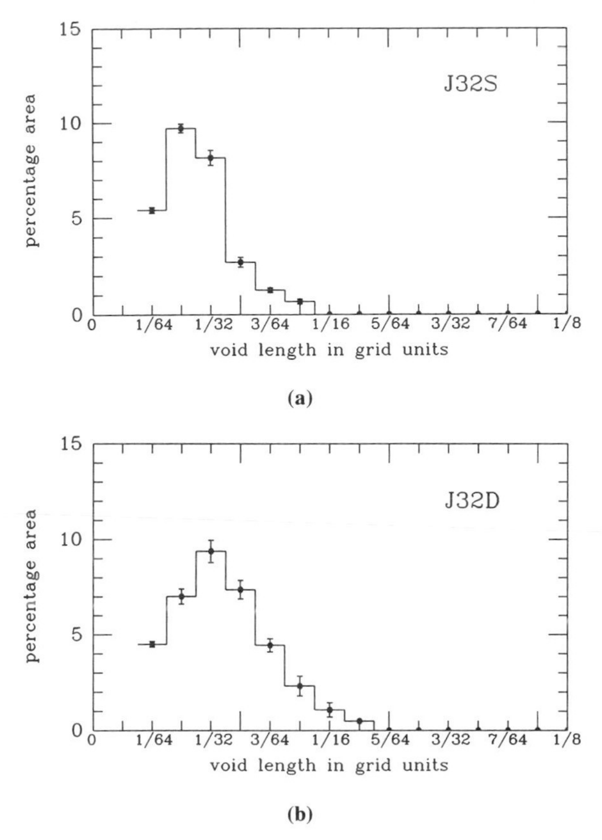

FIGURE 2. (**a**) The void spectrum of fractional area vs. diameter for initial conditions with shortest wavelength perturbation 1/32 of the box size. The peak appears in the appropriate position, considering that walls are thicker in the initial conditions. (**b**) The same as (**a**) except for the density field being evolved to pancaking ($\delta_\rho/\rho = 1$). With thin walls, the peak moves to a void diameter equal to the cutoff wavelength, as expected.

approach based on smoothed density fields, in the spirit of the topology studies.[8-10] For now, we are defining a void as the maximal region that contains no cells at or above the mean density. Given the use of Cloud-In-Cell smoothing, only clumps and not individual particles will appear as enhancements above the mean. The results presented here are preliminary, based on binning the particles into a 128^2 density mesh. This introduces further smoothing. We plan to explore exactly how much smoothing is needed for this method to work.

With this definition, the void spectra of our model appear in FIGURE 2(a) and (b) for the initial conditions (spongelike) and the evolved model (bubbly). Both indicate that this algorithm correctly picked out the right scale. A similar good fit was found for a model with a different cutoff scale in its power spectrum.

We have also conducted tests of models with pure power-law density fluctuation power spectra $P(k) \propto k^n$ in their initial conditions for $n = -2$, 0, and +2. We have made four realizations of each. Due to space limitations, we will not present full results here; we will describe them and present examples.

The model with an $n = -2$ power spectrum does not scale properly. This is in some sense a pathological model, with $\delta_\rho/\rho \propto (\ln R)^{-1/2}$. Fluctuations should exist, but cannot exist at and above the scale of the box. It could be said that in some sense, even the whole universe would not be a fair sample of the universe.

The model with $n = +2$ scales nicely. We have taken data in all models at each doubling of the lengthscale on which fluctuations are nonlinear, from the Nyquist frequency to nearly the box size. A very nice peak void size arises and scales properly from the moment it can be resolved until it approaches the size of the box. We

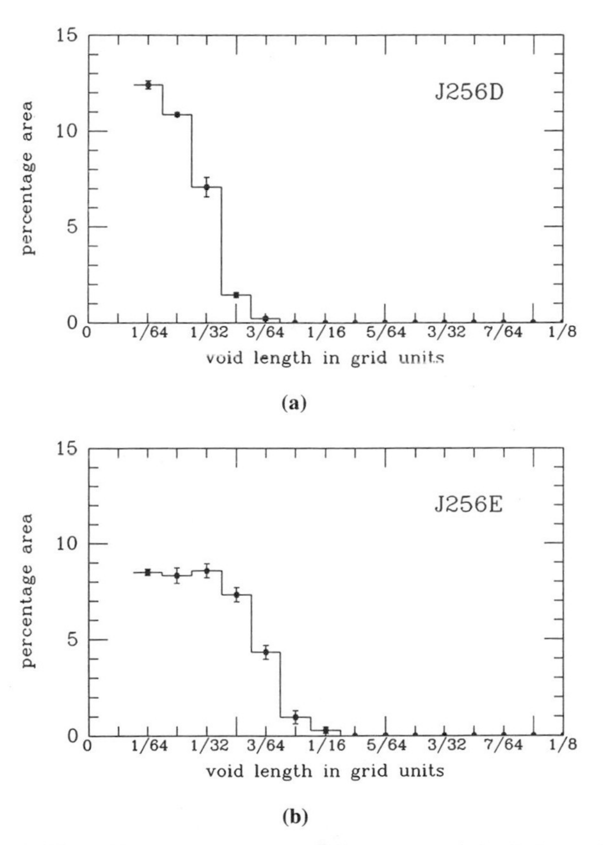

FIGURE 3. (a–e) The void spectrum of an evolving group of simulations with an $n = 0$ power-law initial power spectrum. Each successive histogram is taken when the nonlinear lengthscale has doubled from the previous one, so for perfect scaling the maximum and "characteristic" void sizes would double at each step. The maximum fails, then the large-void wing is suppressed until the "characteristic" fails also (see text for definition of "characteristic" void).

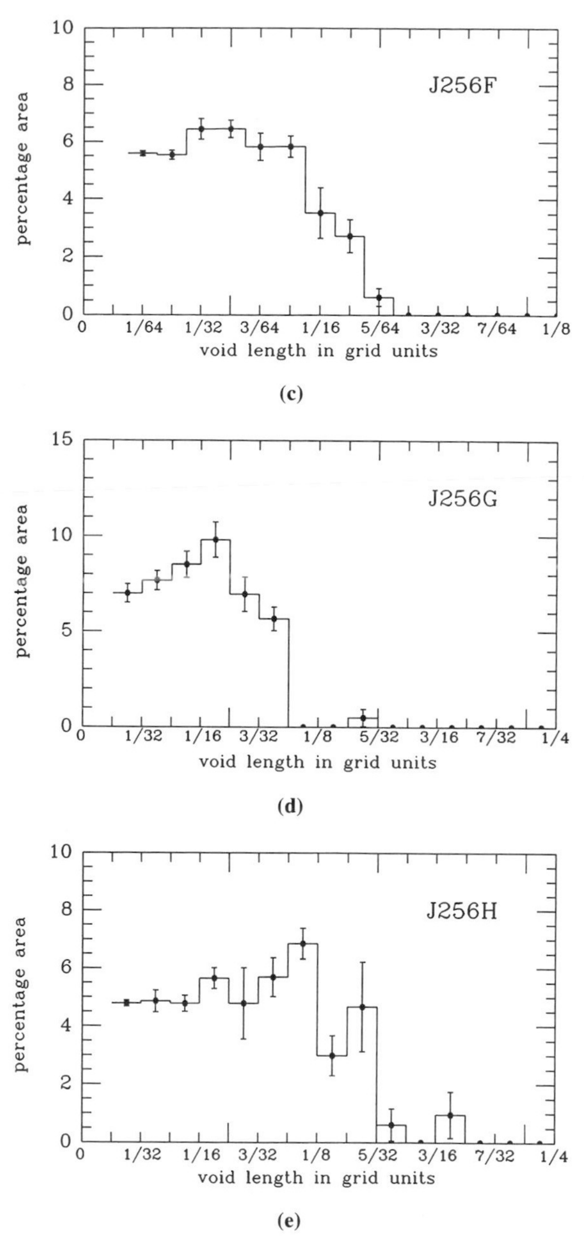

(c)

(d)

(e)

FIGURE 3. Continued.

observe that both the peak (mode) and maximum void size scale well, although the maximum begins to show deterioration when it is one-eighth or so of the box size.

An intermediate model with $n = 0$ is more troublesome. In FIGURE 3, we show the evolution of the void spectrum in this model. The maximum void size scales for a few steps, then is quickly suppressed. There is no clear peak, but we can find a characteristic void scale definition that works, at least for this binning: the diameter at which the fractional area first drops by more than one standard deviation (in the difference between adjacent values). This definition works as well as any we have formulated so far for this model and the preceding one; we will use it here and test it with varying resolutions in future work.

Even this "characteristic" definition fails when the nonlinear wavelength is one-quarter the box diameter. However, the size of the maximum void found fails to scale when it is only one-sixth the box size. We believe that this makes it extremely dangerous to use criteria based on maximum void size in numerical simulations. Even in the $n = +2$ model the maximum failed first and when it was rather small compared with the box.

We have developed a rule of thumb that seems to work to describe the scaling failures in these simulations. It is based on a hypothesis about the cause being the "missing modes." There can be no power between wavenumber $k = 0$ and the fundamental mode of the box. Since δ_ρ/ρ is determined by an integral over k, one can estimate how much of the density contrast on the void scale is missing due to the absence of these modes. When it is about 10 percent, the maximum-size void fails to scale. When it reaches 25 percent, our characteristic fails.

CONCLUSIONS

We studied the evolution of the size of voids in numerical models of clustering in the expanding universe. The void algorithm used was shown to correctly pick out the proper length scale whether the voids are bubbly or interconnected (as in a sponge topology). We found a characteristic void scale based on a smoothed distribution that is a more reliable indicator than the maximum void size. We found a good scaling behavior in simulations with little power in large scales. However, we also found evidence that the size of voids may be systematically suppressed for simulations with moderate large-scale power, so that previous results on voids must be reexamined.

This obviously requires considerably more study. We would like to answer these and other questions:

1. Would other density threshold choices provide better definitions?
2. What are the effects of varying resolution?
3. Is the absence of large-scale power in the initial conditions the cause of the scaling failure, or is it an effect of the constraint that it cannot develop as the simulation evolves, both, or something else unanticipated? (We are now beginning numerical experiments to test this.)
4. Are the scaling failures similar in three dimensions?
5. Can this algorithm reliably pick out the preferred void scale in a density field traced by a magnitude-limited sample of galaxies?

ACKNOWLEDGMENTS

We are grateful to the organizers of this meeting where some of our planning and analysis could take place. We appreciate a grant of Cray-2 time at the National Center for Supercomputing Applications, Urbana, Ill.

REFERENCES

1. KAUFFMANN, G. & A. FAIRALL. 1991. Mon. Not. R. Astron. Soc. **248:** 313.
2. SAUNDERS, W., *et al.* 1991. Nature **349:** 32.
3. BEACOM, J. F., K. G. DOMINIK, A. L. MELOTT, S. P. PERKINS & S. F. SHANDARIN. 1991. Astrophys. J. **372:** 351.
4. SCHERRER, R. J., A. L. MELOTT & S. F. SHANDARIN. 1991. Astrophys. J. **377:** 29.
5. FRY, J. N., A. L. MELOTT & S. F. SHANDARIN. 1991. Accepted for publication in Astrophys. J.
6. SHANDARIN, S. F. & K. G. DOMINIK. 1991. Accepted for publication in Astrophys. J.
7. MELOTT, A. L. & S. F. SHANDARIN. 1990. Nature **346:** 633.
8. GOTT, J. R., A. L. MELOTT & M. DICKINSON. 1986. Astrophys. J. **306:** 341.
9. MELOTT, A. L. 1990. Phys. Rep. **193:** 1.
10. COLES, P. & M. PLIONIS. 1991. Mon. Not. R. Astron. Soc. In press.

New Limits on the Cosmic Microwave Background Fluctuations on a 5° Angular Scale

R. D. DAVIES,[a] R. A. WATSON,[a] J. HOPKINS,[a]
R. REBOLO,[b] C. GUTIERREZ,[b] J. E. BECKMAN,[b] AND
A. N. LASENBY[c]

[a]*Nuffield Radio Astronomy Laboratories*
Jodrell Bank
Macclesfield, Cheshire SK11 9DL, England

[b]*Instituto de Astrofisica de Canarias*
38200 La Laguna, Tenerife, Spain

[c]*Mullard Radio Astronomy Observatory*
Cambridge CB3 OHE, England

INTRODUCTION

All current scenarios of galaxy formation require there to be fluctuations in the angular distribution of matter and radiation at the epoch of recombination ($z \sim 1000$). The mass perturbations on an angular scale of arcminutes give rise to galaxy clusters, while those on degree scales correspond to the largest structures (voids, filaments of galaxies, etc.) seen in the nearby Universe. The amplitude of the predicted angular fluctuations depends on the galaxy-formation scenario assumed and the cosmological parameters H_0, Ω, Λ, and the biassing factor b; predictions of the various scenarios are in the range $\Delta T/T = 3 \times 10^{-6}$ to 10^{-3} as outlined below.

Baryon-dominated adiabatic cosmologies with densities in the range $\Omega = 0.1$ to 0.4 as derived from the dynamics of galaxy clusters (Davis and Peebles 1983) are inconsistent with cosmic microwave background (CMB) fluctuation measurements at arcminute and degree scales. As a consequence cold dark matter (CDM) scenarios were introduced (e.g., Bond and Efstathiou 1984; Vittorio and Silk 1985), since they predicted lower amplitude CMB fluctuations; these scenarios were compatible with inflation, which required $\Omega = 1$. Isocurvature scenarios (Efstathiou and Bond 1987; Peebles 1987) have also been considered, but tend to produce slightly more power in perturbations on degree scales than CDM, because of an enhanced Sachs–Wolfe effect. Cosmic string theories for the origin of fluctuations (Brandenberger *et al.* 1986) involve distinctive steplike discontinuities on the CMB surface. As the observations pushed to higher degrees of isotropy (e.g., Uson and Wilkinson 1984; Davies *et al.* 1987; Readhead *et al.* 1989) further scenarios were investigated that gave lower fluctuations. These included the effects of a nonzero cosmological term Λ (Sugiyama *et al.* 1989, 1990) and the effects of biassing.

We will describe our new observations on scales of 5°–12° and comment on their significance for various scenarios of galaxy formation.

CHOICE OF OBSERVING PARAMETERS

The considerations leading to the adoption of the angular scale, frequency, site, and the required integration time are outlined below.

Angular Scale

A pivotal angular scale in cosmology is the horizon scale θ_H at the redshift of interest that is $z = 1000$ for the CMB, assuming $\Omega = 1$, $\theta_H = z^{-1/2}$ rad $\simeq 2°$. Regions on scales greater than this cannot communicate and are independent in all respects. Another reason for choosing larger angular scales is that any mechanism that caused a reionization of the intergalactic medium subsequent to $z = 1000$ would lead to a smearing out of primordial CMB structure that originated at $z = 1000$ on scales up to $\sim 1°$; larger scales would be unaffected. A further reason for choosing degree scales has been the popularity of the standard CDM model that predicts larger amplitude fluctuations on these scales. Finally, recent measurements of the distribution of individual galaxies and galaxy clusters in the nearby Universe have demonstrated the existence of structures up to scales of at least 100 Mpc, which is equivalent to 1° at recombination (Postman *et al.* 1989; Huchra *et al.* 1990; Saunders *et al.* 1991). Our observing systems were chosen to be sensitive to angular structure in the range 3° to 15° with maximum sensitivity in the range 5° to 10°.

Frequency and Site

Long integrations covering many beam areas are required to make a significant contribution to the study of fluctuations. A choice had to be made between operation on an accessible sea-level site at lower frequency where galactic contamination would be large or working at higher frequencies on a less accessible high dry site. We have used a 5-GHz interferometer at Jodrell Bank to measure galactic emission on a scale of 2°. Our most sensitive measurements are made at 10.4 and 15 GHz at the Teide Observatory on Tenerife at an altitude of 2300 m. Work is in progress on a 30-GHz system for Tenerife. It should be noted that galactic contamination from Infrared Astronomical Satellite (IRAS) cirrus becomes as important at frequencies above 100 GHz as does the contamination from synchrotron emission below 10 GHz. The Teide Observatory site has proved to be clear of cloud and mist for 70 to 80 percent of the year, during which times data are limited by receiver noise.

Integration Time Required

In an integration time of 10^5 sec per beamwidth an rms sensitivity of 15 μK is achieved with an observing bandwidth of 1.5 GHz using a beamswitching receiver. One hundred beamwidths will take one year of observation, yielding an rms sensitivity of $\Delta T/T = 5 \times 10^{-6}$ per beamwidth. Repeated observations at several frequencies will take several years and require substantial logistical support; operation at a well-equipped site such as the Teide Observatory fulfills these needs.

REALIZATION

The observing strategy that eventually proved to be most successful consisted of switching between a central beam and two adjacent beams, one each side of the central beam (Davies *et al.* 1987). This configuration effectively removed the emission from large-scale structure in the atmosphere and residual drifts in the receiver systems to such a level that the output noise was limited by the receiver. By using a four-port switched circulator two independent receiver systems could be run simultaneously, thereby increasing the sensitivity by a factor of $\sqrt{2}$. The first observations at 10.4 GHz were made using beams of 8° half-power beamwidth (HPBW) switched 8° (Davies *et al.* 1987). In a subsequent series of observations at 10.4 GHz and 15 GHz the beamwidth was reduced to 5° HPBW. The latter configuration was responsive to structure on scales from 3° to 15°.

A further important contribution to the success of the experiment was the use of corrugated horns that have extremely low spillover in a configuration with no moving parts except a wagging mirror. The triple beams were scanned daily through 24 hr of right ascension (RA) at fixed declination by earth rotation. There was accordingly no variable pickup during the day, which in the past has been found to be the major source of systematic error in observations with parabolic telescopes (Lasenby and Davies 1983).

These systems are operated on the Teide Observatory, Tenerife, at an altitude of 2300 m in a collaboration between Jodrell Bank and the Instituto de Astrofisica de Canarias.

THE RESULTS AT 10.4 GHZ

Observations at 10.4 GHz using the triple-beam configuration previously described have been made in two series of experiments with resolutions [full width half-power (FWHP)] of 8° and 5°. We have mapped an area of the sky approximately one steradian in total extent, comprising two 15°-wide (declination) strips centered at Dec = 0° and 40° covering the full 24 hr of RA. Both these areas cover high galactic latitudes where galactic emission is at a minimum. They are both observed within 30° of the zenith at Tenerife (latitude = 28°N). The 8° resolution survey is complete, while the 5° survey requires 20 percent more observation to give the same coverage.

FIGURE 1 shows the independent 10.4-GHz maps made of the RA = 6^h to 18^h region of the sky centered on Dec = 40° with 8° and 5° beamwidths. The region from 9^h to 18^h has been well mapped at both resolutions. Similar structure is evident on both maps indicating the presence of a real microwave sky contribution. Data were taken at different times over an interval of four years; observations made when the sun or moon were within three hours of transit were removed from the data set.

It is not clear from these data alone whether the emission detected is from the Galaxy or is structure in the CMB. This question can only be resolved by making observations at higher frequencies. In any case the 10.4-GHz observations provide the basic data set from which corrections for galactic emission can be deduced.

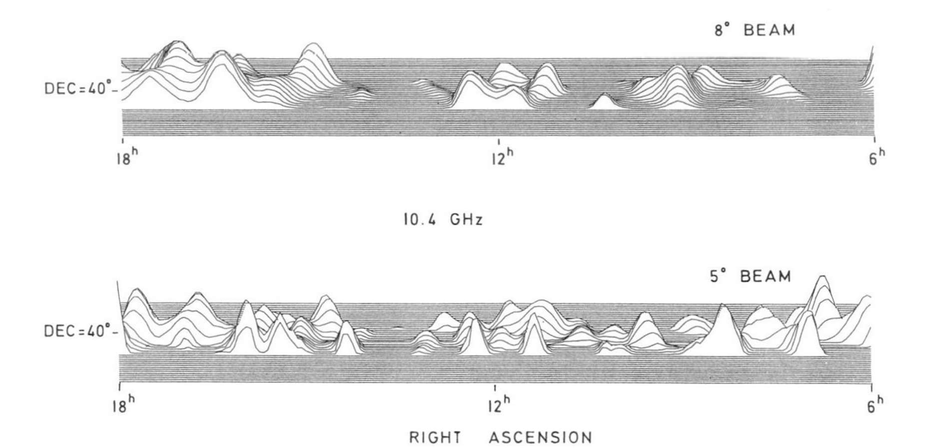

FIGURE 1. 10.4-GHz maps (MEM reconstruction) centered at Dec = 40° taken with a beamwidth of 8° (*upper plot*) and 5° (*lower plot*). At right ascension greater than ~09 hr the two maps have comparable sensitivities. Between 06 and 09 hr the 5° observations are not yet complete and, as a consequence, have a higher level of noise.

GALACTIC AND EXTRAGALACTIC EMISSION

Estimates of the galactic contribution to the microwave background can be made from all-sky surveys at low frequencies (Banday and Wolfendale 1990). It is necessary to calculate the spectral index point-by-point at these low frequencies before making the extrapolation to the higher frequencies where the CMB anisotropy searches are made. Such a calculation requires a knowledge of the zero levels of the surveys. Lawson *et al.* (1987) have used the published 408- and 1420-MHz surveys and found brightness temperature spectral indices $\beta = 2.6$ to 3.0 at high northern galactic latitudes. They find that the spectral index increases with frequency. FIGURE 2 gives a comparison of our 10.4-GHz scan at Dec = 40° reconstructed by a maximum entropy method using a 5° beam and calculated from the 408-MHz data extrapolated to 10.4 GHz with an assumed spectral index of 2.8. It can be seen that there is no detailed match between observation and prediction, particularly in connection with the feature at RA $\approx 15^h$. We believe that there are several reasons for this. First, the spectral index over the range 408 MHz to 10.4 GHz may vary with position; the 408- to 1420-MHz spectral index is not a predictor for the 10.4-GHz brightness temperature. Second, there are substantial instrumental (baseline) effects in all the published surveys at 408 MHz and above that make the calculation of spectral index uncertain, quite apart from the effect of the uncertainty in the zero level. The direct measurement of the background contribution by our experiment at 10.4 GHz provides the best way forward in estimating the galactic contribution at higher frequencies.

The second extraterrestrial component of background emission is the integrated contribution of the radio sources. This amounts to 3K at 408 MHz, compared with the minimum galactic contribution of 15K. Franceschini *et al.* (1989) predict from source counts at various frequencies that the extragalactic source contribution is

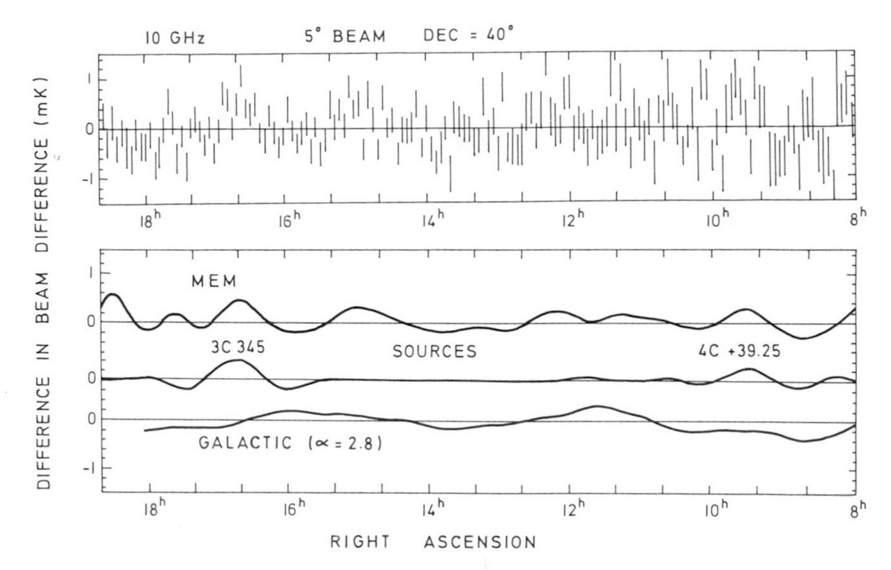

FIGURE 2. The 10.4-GHz 5° beamwidth scan at higher galactic latitudes. Points are plotted every 1°. An attempted noise-free reconstruction of the data set is shown separately as a *continuous curve* (MEM). This curve is formed by reconvolving the maximum entropy reconstruction of the sky given by the stacked data with a 5° beam switched through 8°. The expected point-source and galactic contributions are also plotted. The two strongest point sources are clearly evident.

$\Delta T/T \lesssim 10^{-5}$ on a 1° scale in the frequency range 20 to 200 GHz. The contribution will be considerably less on 5°–8° scales; we estimate $\Delta T = 20\ \mu K$ at 10 GHz. Strong sources such as 3C345 and 4C + 39.25 in the Dec = 40° scan in FIGURE 2 can be allowed for in the subsequent analysis.

15-GHZ RESULTS

A new high-sensitivity receiver system incorporating cryogenically operating high-electron mobility transistor (HEMT) amplifiers has been constructed and commissioned at 15 GHz with the same beamswitching characteristics as the 10.4-GHz 5° beamwidth system. This enables a direct comparison to be made between 10.4- and 15-GHz observations. The sensitivity of the 15-GHz system is twice that at 10.4 GHz for the same integration. FIGURE 3 shows the addition of eight days of data at Dec = 40°.

The added data after low-order baseline removal for one month of observations is shown in FIGURE 4 for the high-latitude part of the scan. It can be seen immediately that the repeatable signals at a level of several hundred μK at 10.4 GHz are not present at 15 GHz. It is therefore evident that the 10.4-GHz signals are galactic emission; with a spectral index of ~3.0 between 10.4 and 15 GHz they will be a factor of ~3.5 less at the higher frequency and will consequently be below the detection level. Similarly, the point sources will be a factor of ~3 less at the higher

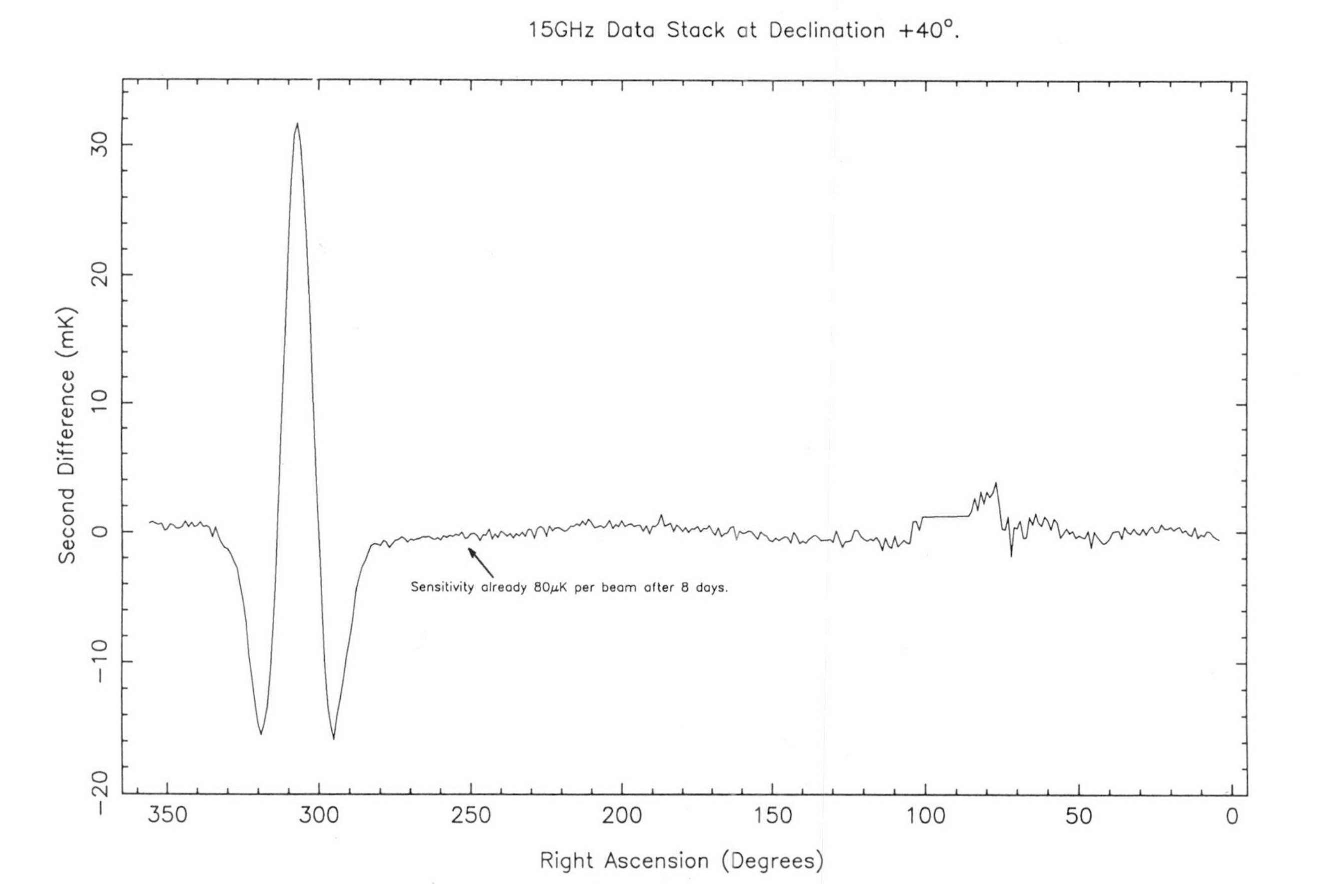

FIGURE 3. 15-GHz stacked data from 8 days of observation with the triple beam system. Only limited observations were available on these days in the 60° to 120° region, which includes the weak galactic plane crossing.

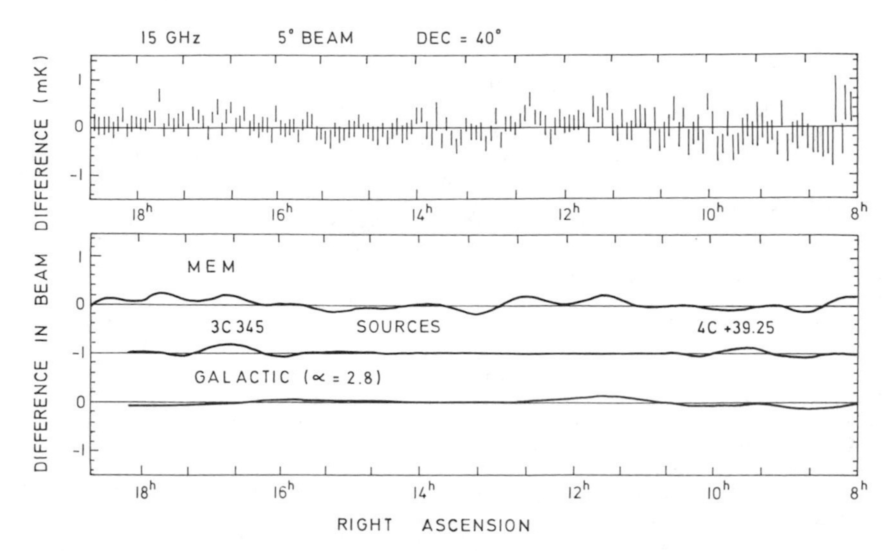

FIGURE 4. The 15-GHz 5° beamwidth scan at high galactic latitudes. Points are plotted every 1° in RA. The MEM reconstruction as described for FIGURE 2 is shown as a separate curve. The expected galactic and point-source contributions are also plotted.

frequency because of the reduced collecting area and the spectral index; this reduction is confirmed by the data.

A Bayesian analysis of the high latitude 15-GHz data gives a 95 percent upper limit to fluctuations on scales of 5° to 12° of $\Delta T/T \lesssim 2 \times 10^{-5}$. This limit is a factor of 2 less than that actually detected in the 10.4-GHz data. It represents a stringent new limit on intrinsic CMB fluctuations.

DISCUSSION AND CONCLUSIONS

The new measurements have shown clearly that our earlier detection of fluctuations on 5° and 8° scales was galactic emission rather than the intrinsic CMB. The new upper limit for CMB fluctuations on scales of 5°–12° at 15 GHz of $\Delta T/T < 2 \times 10^{-5}$ provides interesting constraints on most scenarios of galaxy formation in the early Universe. Many of these constraints are well-illustrated in the plots of Sugiyama *et al.* (1990), who include the effects of a nonzero cosmological constant.

Purely baryonic scenarios are ruled out by this result for $\Omega < 0.2$ and $h = 0.5$ to 1.0, both with and without a cosmological constant; the small angular scale result of Readhead *et al.* (1989) eliminates the whole range of Ω.

Isocurvature models are all ruled out by factors between 1.5 and 10 for all values Ω and cosmological constant. These models include those with baryons, CDM, hot dark matter (HDM), and the Peebles (1987) reionization scenario, and are best tested on the angular scale of the present experiments.

It is planned to map the declination range 35° to 45° at all right ascensions with this new 15-GHz system, thereby covering a steradian of solid angle.

REFERENCES

BANDAY, A. J. & A. W. WOLFENDALE. 1990. Mon. Not. R. Astron. Soc. **245:** 182.
BOND, J. R. & G. EFSTATHIOU. 1984. Astrophys. J., Lett. **285:** L45.
BRANDENBERGER, R., A. ALBRECHT & N. TUROK. 1986. Nucl. Phys. B. **277:** 605.
DAVIES, R. D., A. N. LASENBY, R. A. WATSON, E. J. DAINTREE, J. HOPKINS, J. BECKMAN, J. SANCHEZ & R. REBOLO. 1987. Nature **326:** 462.
DAVIS, M. & P. J. E. PEEBLES. 1983. Astrophys. J. **267:** 465.
EFSTATHIOU, G. & J. R. BOND. 1987. Mon. Not. R. Astron. Soc. **227:** 33P.
FRANCESCHINI, A., L. TOFFOLATTI, L. DANESE & G. DE ZOTTI. 1989. Astrophys. J. **344:** 35.
HUCHRA, J. P., J. P. HENRY, M. POSTMAN & M. G. GELLER. 1990. Astrophys. J. **365:** 66.
LASENBY, A. N. & R. D. DAVIES. 1983. Mon. Not. R. Astron. Soc. **203:** 1137.
LAWSON, K. D., C. J. MAYER, J. L. OSBORNE & M. L. PARKINSON. 1987. Mon. Not. R. Astron. Soc. **225:** 307.
PEEBLES, P. J. E. 1987. Astrophys. J., Lett. **315:** L73.
POSTMAN, M., D. N. SPERGEL, B. SUTIN & R. JUSKIEWICZ. 1989. Astrophys. J. **346:** 588.
READHEAD, A. C. S., C. R. LAWRENCE, S. T. MYERS, W. L. W. SARGENT, H. E. HARDEBECK & A. T. MOFFET. 1989. Astrophys. J. **346:** 566.
SAUNDERS, W., *et al.* 1991. Nature **349:** 32.
SUGIYAMA, N., M. SASAKI & K. TOMITA. 1989. Astrophys. J., Lett. **338:** L45.
SUGIYAMA, N., N. GOUDA & M. SASAKI. 1990. Astrophys. J. **365:** 432.
USON, J. M. & D. T. WILKINSON. 1984. Astrophys. J., Lett. **277:** L1.
VITTORIO, N. & J. SILK. 1985. Astrophys. J., Lett. **297:** L1.

The Angular Large-scale Structure

Y. HOFFMAN,[a] C. SCHARF,[b] AND O. LAHAV[b]

[a] *Racah Institute of Physics*
The Hebrew University
Jerusalem, Israel

[b] *Institute of Astronomy*
Madingley Road
Cambridge, CB3 0HA, England

INTRODUCTION

The Infrared Astronomical Satellite (IRAS) survey allows the construction of an almost full-sky, flux-limited catalogue of the angular distribution of galaxies that can serve as a good data base for the statistical analysis of the large-scale structure of the universe. Although a two-dimensional sample, the large number of objects gives a smaller shot noise compared with the IRAS redshift surveys, and the depth information can still be incorporated via a radial selection function. The angular distribution of galaxies has been traditionally analyzed by measuring the angular two-point correlation function, following its pioneering determination by Peebles and Hauser (1974) from the Lick catalogue. Indeed such an analysis for the small scales has been applied to the IRAS samples (e.g., Lahav *et al.* 1990; Strauss *et al.* 1991). Here we suggest a complementary approach to explore the large scales by expending the observed distribution in spherical harmonics. This is first used to reconstruct maps of the galaxy distribution that are produced by applying a cutoff at some maximal harmonic number. Then the angular power spectrum is evaluated and is used as a probe of the primordial density perturbation field. This analysis is a generalization for any harmonic l of the analysis of the much-discussed dipole anisotropy, $l = 1$ (e.g., Kaiser and Lahav 1989). Here we focus on the cold dark matter (CDM) model, and in particular on two variants of it, both of which obey the inflation paradigm of a flat universe with a scale invariant perturbation field. The first is the "standard" CDM model with $\Omega_0 = 1.0$ and $h = 0.5$, and the other is a model of an $\Omega_0 = 0.2$ and $h = 1$ universe dominated by a cosmological constant. The calculations given here are based on the formalism developed by Peebles (1974) for calculating the angular correlation function. A spherical harmonics analysis has also been applied to IRAS galaxies by Fabbri and Natale (1990).

COSMOGRAPHY

We have expanded the angular distribution of the IRAS color-selected sample (Meurs and Harmon 1989) in spherical harmonics and plotted the reconstructed surface density by using only components up to some l_{max}. This provides a very "natural" smoothing of the density on a scale $\sim\pi/l_{max}$ and a visual interpretation of the information in particular components. FIGURE 1 shows a comparison of the

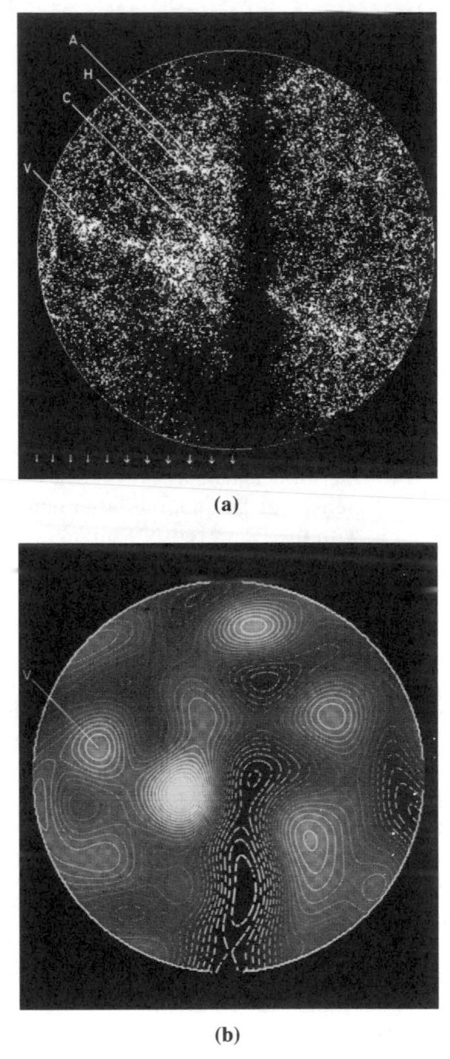

FIGURE 1. The projected distribution of optical and IRAS galaxies. (**a**) distribution of optical galaxies from the UGC, ESO, and MCG catalogues in the direction ($l = 307°$; $b = 9°$) of streaming motion of the elliptical galaxies of the "7 Samurai." *A*, *H*, *C*, and *V* indicate the Antila, Hydra, Centaurus, and Virgo clusters, respectively. This figure (by Lahav, see Lynden-Bell *et al.* 1988) illustrates the existence of a major concentration of galaxies in the direction of Centaurus, the so-called "Great Attractor." The *dark band* is due to galactic obscuration. (**b**) Contours of number density of IRAS galaxies deduced by spherical harmonics analysis out to order $l = 10$. The most prominent feature is again the Great Attractor, which is probably even more enhanced by contribution from the more distant Shapley supercluster. Another very prominent feature appears at the top of the figure, in the direction of Puppis, although it is not apparent in the optical map. The peak of this enhancement (using spherical harmonics out to order $l = 30$) is at galactic coordinates ($l = 243°$; $b = -3°$).

IRAS reconstruction with data from optical surveys. FIGURE 1(a) (by O. Lahav) shows the distribution of optical galaxies from the optical UGC, European Southern Observatory (ESO), and MCG catalogues in the direction of the "Great Attractor" (Lynden-Bell *et al.* 1988). FIGURE 1(b), in identical coordinates and projection, is the *raw* IRAS, $l = 1$ to $l = 10$ reconstruction. The IRAS plot has a set of equally spaced contours placed on top of an equivalent gray scale. The most prominent feature is again the Great Attractor (which is probably even more enhanced by contribution from the more distant Shapley supercluster). The topmost strong feature in the IRAS picture lies in the direction of Puppis, very close to the galactic plane. This feature, although not apparent in the optical map, is indeed a cluster of galaxies at velocity distance $800 < cz < 2700$ km/sec at galactic coordinates $l \approx 240°$; $|b| < 8°$ (Kraan-Korteweg and Huchtmeier 1991).

A contour map constructed from the dipole and quadrupole terms is presented in FIGURE 2. Here the map is centered on the north galactic pole. The (near) alignment

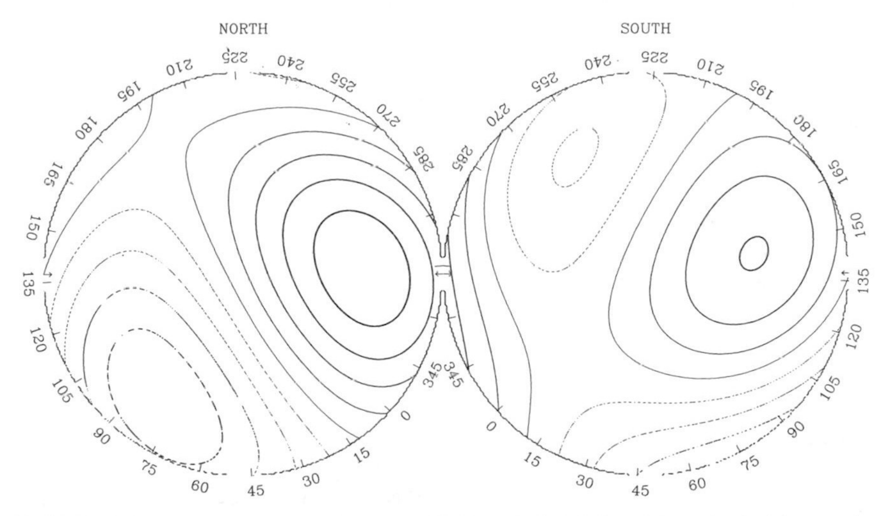

FIGURE 2. Contours of number density of IRAS galaxies deduced by spherical harmonics analysis out to order $l = 2$ (i.e., dipole and quadrupole), plotted in the galactic northern and southern hemispheres.

of the dipole and the eigenvector corresponding to the largest eigenvalue of the quadrupole tensor produces a structure of two maxima and two minima. The absolute maximum of the reconstructed projected density coincides with the Great Attractor/Centaurus complex at the northern hemisphere and the other with the Perseus/Pisces supercluster in the south.

THEORY

Consider now a given angular distribution of galaxies obtained from an incomplete survey of the whole sky. The incompleteness is represented by a mask that allows galaxies to be selected only outside the masked region. The spherical harmonics transform of the masked sky is

$$b^m_{l\,\mathrm{obs}} = \frac{1}{N_{\mathrm{obs}}} \sum_{i \notin \mathrm{mask}} Y^m_l(i), \tag{1}$$

where $Y^m_l(i)$ is the spherical harmonics evaluated at the angular position of the ith galaxy and b^m_l is normalized by the total number of galaxies outside the mask, N_{obs}. In the theoretical domain one considers a given density perturbation field $\delta(\mathbf{r})$. To relate the matter and galaxy distribution we adopt the convention of constant biasing factor and normalize the power spectrum by the empirical determination of the variance of IRAS galaxy counts in spheres of radius $R = 8h^{-1}$ Mpc, defined as σ_8^2. Assuming a full-sky coverage, the spherical harmonics transform yields

$$a^m_{l\,(\mathrm{th})} = \int d^2\,\Omega Y^m_l(\Omega) \int r^2\,dr\,\Phi(r)\,\delta(\mathbf{r}). \tag{2}$$

Here, the mean field is subtracted, which makes the $l = 0$ to vanish and the selection function is normalized such that $1 = 4\pi\bar{n} \int_0^\infty dr\, r^2\Phi(r)$ where $\bar{n}$ is the mean number density of galaxies.

For an isotropic and stationary Gaussian field the ensemble average $a^m_{l\,(\mathrm{th})}$ vanishes, there is no mode–mode coupling, and the angular power spectrum is

$$A^2_{l\,(\mathrm{th})} = \langle |a^m_{l\,(\mathrm{th})}|^2 \rangle = \frac{2}{\pi} \int k^2\,dk\,P(k) \left| \int r^2\,dr\,\Phi(r)\,j_l(kr) \right|^2, \tag{3}$$

where $P(k)$ is the power spectrum and $j_l(x)$ is the spherical Bessel function of order l. In the absence of an intrinsic clustering the distribution of galaxies is Poisson, which is statistically independent of the δ field, and the total power is given by

$$A^2_l = A^2_{l\,(\mathrm{th})} + \frac{1}{4\pi N}. \tag{4}$$

Next, the incomplete sky coverage is to be added. For a given mask we define

$$c^m_{l\,(\mathrm{th})} = \frac{4\pi}{\Omega_{\mathrm{obs}}} \int d^2\,\Omega Y^m_l W(\Omega)(\Omega) \int r^2\,dr\,\Phi(r)\delta(\mathbf{r}), \tag{5}$$

where $W(\Omega) = 1$ for Ω outside the mask, and it vanishes otherwise, and $\Omega_{\mathrm{obs}} = \int d^2\,\Omega W(\Omega)$. Note that c^m_l is related to the directly observable b^m_l by

$$c^m_l = b^m_l - \frac{4\pi}{\Omega_{\mathrm{obs}}} I^m_l, \tag{6}$$

where I^m_l is the spherical harmonics transform of $W(\Omega)$. An ensemble average of c^m_l is zero, but the mask introduces mode–mode coupling such that

$$\langle c^m_l c^{*m'}_{l'} \rangle = \left(\frac{4\pi}{\Omega_{\mathrm{obs}}}\right)^2 \sum_{l'',m''} W^{mm''}_{ll''} W^{*m'm''}_{l'l''} A^2_{l''}. \tag{7}$$

The transformation tensor is defined by

$$W^{mm'}_{ll'} = \int d^2\,\Omega Y^m_l(\Omega) Y^{m'}_{l'}(\Omega) W(\Omega). \tag{8}$$

The angular power spectrum of the masked data is

$$C_l^2 = \left(\frac{4\pi}{\Omega_{obs}}\right) \frac{1}{2l+1} \sum_{l',m,m'} |W_{ll'}^{mm'}|^2 A_{l'}^2. \tag{9}$$

It is convenient to represent the angular power spectrum by $O_l^2 = (4\pi)^2 \mathcal{N} C_l^2$, where $\mathcal{N}$ is the mean number of galaxies per solid angle. To normalize the power spectrum of the various models we adopt here $\sigma_8 = 0.7$ as obtained from the IRAS correlation function $(r/4h^{-1}\ \text{Mpc})^{-1.7}$, derived by Lahav *et al.* (1990). A Gaussian filter of a smoothing length of $3h^{-1}$ Mpc has been applied for all models.

DATA VS. MODELS

To compare the models with the data we correct the observable b_l^m by subtracting the mean sky. For a flux limit of $S_{lim} = 0.7$ Jy the volume-weighted selection function peaks at about $R = 90h^{-1}$ Mpc. The lth multipole is determined by the structure on a linear scale of $r \sim R/l$. The mask used here is a cut in galactic latitude, $|b| < 20°$. This mask leaves about 7000 IRAS galaxies, which corresponds to a shot noise level of $\sim 10^{-5}$, which dominates A_l at about $l = 15$. Note that the mask affects only the magnitude of the shot noise [as (**9**) is valid for any mask], but the convergence of the sum over l' depends on it. To calculate C_l up to $l = 10$ this sum converges at $l' = 15$.

The predictions of the "standard" and open CDM models are given in FIGURE 3 where O_l is plotted against l up to $l = 10$. In each figure part the IRAS results are plotted, and the theoretical models are represented by the rms values and the lower and upper 95 percent confidence levels. It is clear that the open CDM model fits the observation best, with all the IRAS data points except the $l = 10$ lying within the 95 percent confidence levels, as opposed to the standard CDM model, where about half the data points lie above the upper confidence level. The combined effect of most of the C_l being at or above the upper 95 percent confidence level makes the standard CDM to be incompatible with the data.

DISCUSSION

The results reported here are in agreement with other recent studies of the large-scale galaxy distribution. The galaxy distribution on the scale of a few tens of megaparseconds has been probed recently by measuring the angular correlation function calculated from the APM optical survey (Efstathiou *et al.* 1990) and the count in cells analysis of the "one in six" redshift survey of the IRAS galaxies (Saunders *et al.* 1991). In both cases an excess of power as compared to the "standard" CDM model has been found that is of the same magnitude as reported here. The observed angular correlation function is better fitted by an open, $\Omega_0 h = 0.2$ model, in agreement with our results. Note that the "one-in-six" study and the angular power spectrum are based on the IRAS galaxies, while the APM survey uses optical galaxies, which seems to cluster differently and to have a different biasing parameter. Still, all of them seem to find the same trend of excess of power.

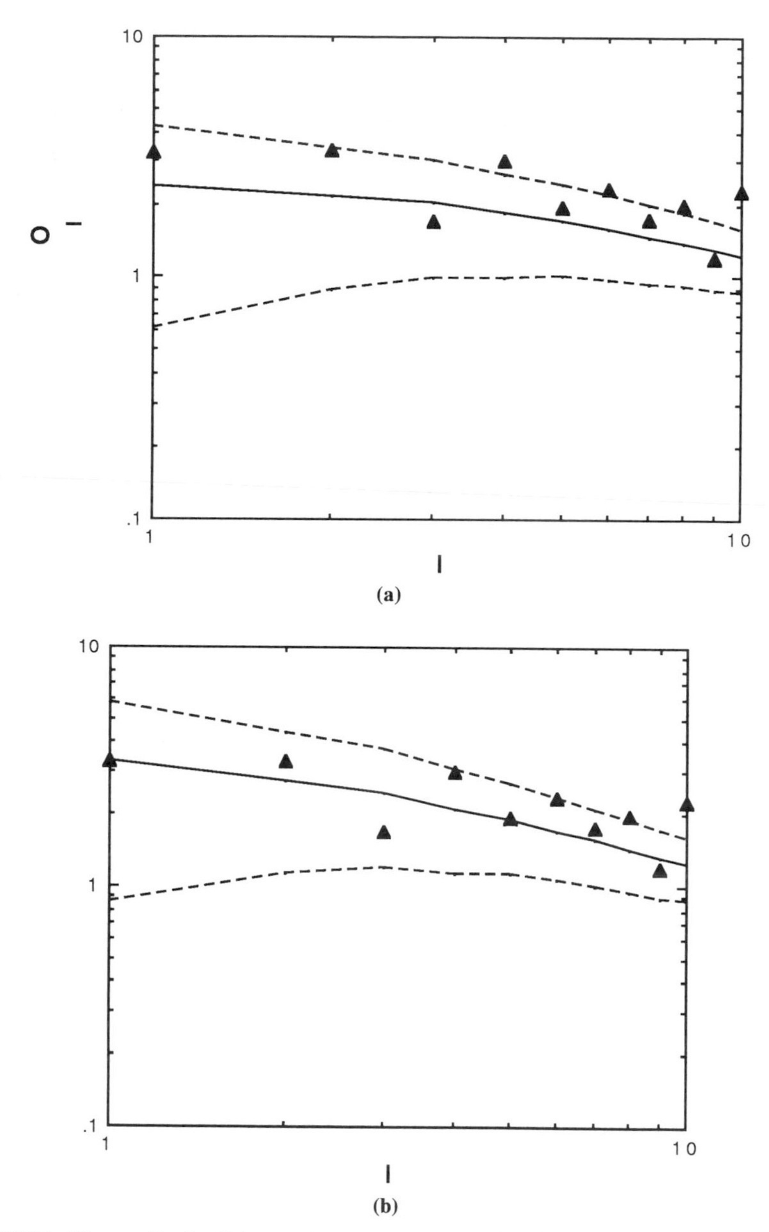

FIGURE 3. The amplitude of the angular power spectrum O_l vs. l. The rms value is represented by the *solid line,* and the two *dashed lines* are the lower and upper 95 percent confidence limits. The solid triangles are the IRAS data. **(a)** The "standard" CDM model. **(b)** Corresponds to the $\Omega_0 = 0.2$ open CDM model.

ACKNOWLEDGMENTS

Professor D. Lynden-Bell is gratefully acknowledged for his continued encouragement and many stimulating discussions.

REFERENCES

EFSTATHIOU, G., W. J. SUTHERLAND & S. J. MADDOX. 1990. Nature **348:** 705.
FABBRI, R. & V. NATALE. 1990. Astrophys. J. **363:** 3.
KAISER, N. & O. LAHAV. 1989. Mon. Not. R. Astron. Soc. **237:** 129.
KRAAN-KORTEWEG, R. C. & W. K. HUCHTMEIER. 1991. In preparation.
LAHAV, O., R. J. NEMIROFF & T. PIRAN. 1990. Astrophys. J. **350:** 119.
LYNDEN-BELL, D., S. M. FABER, D. BURSTEIN, R. L. DAVIES, A. DRESSLER, R. J. TERLEVICH & G. WEGNER. 1988. Astrophys. J. **326:** 19.
MEURS, E. J. A. & R. T. HARMON. 1989. Astron. Astrophys. **206:** 53.
PEEBLES, P. J. E. 1974. Astrophys. J. **185:** 413.
PEEBLES, P. J. E. & M. G. HAUSER. 1974. Astrophys. J., Suppl. Ser. **196:** 1.
SAUNDERS, W., *et al.* 1991. Nature **349:** 32.
STRAUSS, M. A., M. DAVIS, A. YAHIL & J. P. HUCHRA. 1991. Preprint.

Large-scale Anisotropy of Cosmic Background Radiation in Open Cosmological Models

NAOTERU GOUDA,[a] NAOSHI SUGIYAMA,[a]
AND MISAO SASAKI[b]

[a] *Department of Physics*
Kyoto University
Kyoto 606, Japan

[b] *Uji Research Center*
Yukawa Institute for Theoretical Physics
Kyoto University
Uji 611, Japan

INTRODUCTION

Up until quite recently, the cold dark matter scenario, with an initially Zeldovich spectrum and biased galaxy formation in a spatially flat universe, was regarded as the most successful scenario for our universe. It was based on the inflationary universe paradigm and, if not cold dark matter, little doubt was thrown on the validity of the Zeldovich spectrum and the flatness.

As the number of observations on the large-scale structure increased, however, the data began to indicate various inconsistencies of the standard cold dark matter scenario. Among others, many recent observations seem to favor a low-density universe, which conflicts with the prediction of the inflationary scenario unless a cosmological constant is introduced. In fact, it has been pointed out that a flat universe with cosmological constant is not only less constrained by cosmic background radiation (CBR) anisotropies (Vittorio and Silk 1985; Holtzman 1989; Sugiyama, Gouda, and Sasaki 1990) but also is most favored observationally (Fukugita *et al.* 1990). Yet many people, particularly theorists, seem to disfavor the introduction of cosmological constant, but would rather resort to an open universe model. Under this kind of a confused situation, it is important to investigate various possibilities without any prejudice. As one such attempt, in this paper we carefully evaluate the CBR quadrupole anisotropy in a number of open universe models for both adiabatic and isocurvature perturbations, and derive constraints on them from the observational upper limit of the quadrupole. The main motivation to consider the quadrupole is that it would have never been influenced by local phenomena after decoupling, such as reionization, and serves as a very direct probe of primordial inhomogeneities of the universe.

The CBR quadrupole moment has been calculated by many authors in various cosmological models (Wilson and Silk 1981; Wilson 1983; Tomita and Tanabe 1983; Peebles 1982; Kofman and Starobinskii 1985; Gouda and Sasaki 1986; Vitorio *et al.* 1988; Górski and Silk 1989). But all of these were unsatisfactory in the sense that they were restricted either to the spatially flat background or to an incomplete evaluation of the quadrupole moment. In other words, there has not been any work that considers an open universe model and includes both the generalized Sachs–Wolfe effect [i.e., that which includes the effect of time variation of the gravitational potential (Wilson 1983)] and intrinsic photon fluctuations at complete decoupling, in evaluating the quadrupole moment. This unsatisfactory situation is mainly due to a technical difficulty in estimating the present quadrupole and/or higher multipoles of the CBR anisotropy in an open universe. However, we have recently succeeded in deriving a formula by which one may calculate any multipole moment of the CBR anisotropy with practically arbitrary precision (Gouda, Sugiyama, and Sasaki 1990; hereafter Paper I). This formula is fully utilized in this paper.

BASIC TOOLS

We consider linear fluctuations around a Robertson–Walker spacetime with curvature $K \leq 0$:

$$ds^2 = -dt^2 + a(t)^2\left(d\chi^2 + \frac{1}{(-K)}\sinh^2(\sqrt{-K}\chi)(d\theta^2 + \sin^2\theta d\varphi^2)\right). \qquad (1)$$

To evaluate the present-day quadrupole anisotropy, the data at decoupling time have been prepared by numerically solving the evolution equations for density perturbations, for which we have adopted the gauge-invariant formalism (Bardeen 1980; Kodama and Sasaki 1984). The numerical method is the same as the one used in Gouda, Sasaki, and Suto (1989) and Sugiyama (1989). The initial spectrum is assumed to be a power law:

$$|\Delta(k)|^2 \propto \tilde{k}^n,$$

where $\tilde{k}^2 \equiv k^2 + K$ and Δ is the gauge-invariant *total* density perturbation. Note that our power-law index differs by 4 from that used in, for example, Wilson (1983) for isocurvature perturbations (i.e., $n = n_{\text{Wilson}} + 4$). The evolution is numerically solved until the universe becomes sufficiently matter dominated and optically thin.

The observable temperature anisotropy Θ_m, defined with respect to the matter restframe (Kodama and Sasaki 1986, 1987), is given by

$$\Theta_m(\eta_0, x_0, \gamma) = \Theta_{\text{int}} + \Theta_{\text{sac}} + \Theta_{\text{dif}};$$

$$\Theta_{\text{int}} \equiv \Theta_m(\eta, x(\eta), \gamma) - \frac{1}{k}V(\eta)Y(x(\eta))_{|i}\gamma^i + \frac{1}{k}V(\eta_0)Y(x_0)_{|i}\gamma^i,$$

$$\Theta_{\text{sac}} \equiv \left(\Psi(\eta)Y(x(\eta)) - \frac{1}{k}\frac{a'}{a}V(\eta)Y(x(\eta))\right) - \left(\Psi(\eta_0)Y(x_0) - \frac{1}{k}\frac{a'}{a}V(\eta_0)Y(x_0)\right),$$

$$\Theta_{\text{dif}} \equiv 2\int_{\eta}^{\eta_0}\left(\frac{d}{d\eta'}\Psi(\eta')\right)Y(x(\eta'))\,d\eta', \qquad (2)$$

where η is the conformal time at an arbitrary epoch after decoupling; Ψ and V are the gauge-invariant gravitational potential and velocity perturbations, respectively; Y is an harmonic on the negative curvature 3-space, a prime ($'$) means $d/d\eta$, and we focus on perturbations belonging to a single eigenvalue k^2; $\Theta_{\rm int}$ describes intrinsic photon fluctuations at decoupling; $\Theta_{\rm sac}$ the Sachs–Wolfe effect (Sachs and Wolfe 1967); and $\Theta_{\rm dif}$ is the effect solely due to nonvanishing curvature. We call the term $\Theta_{\rm GS} \equiv \Theta_{\rm sac} + \Theta_{\rm dif}$ the generalized Sachs–Wolfe effect.

The multipole expansion of Θ_m is given by

$$\Theta_m(\eta, x, \gamma) = \sum_{l=0}^{\infty} i^{-l}\theta_{m(l)}(\eta)G_l[Y(x)], \tag{3}$$

where

$$G_l[Y(x)] = (ik)^{-l}Y_{|i_1...i_l}(x)P_{(l)}^{i_1...i_l}. \tag{4}$$

The tensor $P_{(l)}^{i_1\cdots i_l}$ is defined in Wilson (1983). Provided that G_l acts on a harmonic function along a photon trajectory (i.e., a null geodesic), $Y(x(\eta))$, it can be regarded as a polynomial of degree l with respect to a differential operator $\hat{p} = (ik)^{-1}\, d/d\eta$; $G_l(\hat{p})$. In the flat case, $Y = e^{ik\cdot x}$ and G_l reduces to the Legendre polynomial P_l.

After some algebra, one can show that the quadrupole moment of the CBR anisotropy is given by (Paper I)

$$\begin{aligned}\theta_{m(2)}(\eta_0) = -\,5\Bigg[&\sum_{n=0}^{\infty} i^{-n}\theta_{\rm int(n)}(\eta)G_n(\hat{p})Z_\nu^2(\eta - \eta_0) \\ &+ \left(\Psi(\eta) - \frac{1}{k}\frac{a'}{a}V(\eta)\right)Z_\nu^2(\eta - \eta_0) \\ &+ 2\int_{\eta}^{\eta_0}\left(\frac{d\Psi}{d\eta}\right)Z_\nu^2(\eta - \eta_0)\,d\eta\Bigg],\end{aligned} \tag{5}$$

where $Z_\nu^l(\eta - \eta_0) = i^{-l}X_\nu^l\,(\eta_0 - \eta)$ $(l = 2)$ with X_ν^l being the radial function of a harmonic function:

$$X_\nu^l(\chi) = \left(\frac{\pi}{2\sinh(\sqrt{-K}\chi)}\right)^{1/2}(\nu^2 + 1)^{l/2}P_{i\nu-1/2}^{-(l+1/2)}(\cosh(\sqrt{-K}\chi)),$$

and

$$\theta_{\rm int(n)} = \begin{cases}\theta_{m(1)} + V & \text{for } n = 1,\\ \theta_{m(n)} & \text{for } n \neq 1.\end{cases}$$

The eigenvalue k^2 is related to ν as $k^2/(-K) = \nu^2 + 1$.

In Paper I, it is shown that the following formula holds for arbitrary nonnegative integers n and l:

$$G_nZ_\nu^l = \frac{c_n}{c_l}G_lZ_\nu^n; \qquad c_n = \begin{cases}1 & \text{for } n = 0,\\ \displaystyle\prod_{j=1}^{n}\frac{\nu^2 + j^2}{\nu^2 + 1} & \text{for } n \geq 1.\end{cases} \tag{6}$$

Then in particular for $l = 2$, we have

$$G_n Z_\nu^2 = \frac{c_n}{c_2} G_2 Z_\nu^n = \frac{c_n}{c_2}\left(\frac{3}{2}\hat{p}^2 - \frac{1}{2}\right) Z_\nu^n$$

$$= \frac{3}{2}\frac{(n-1)n}{(2n-1)(2n+1)} q_n Z_\nu^{n-2}$$

$$+ \left[\frac{3}{2}\frac{n^2}{(2n-1)(2n+1)}\frac{\nu^2+n^2}{\nu^2+1} q_n + \frac{3}{2}\frac{(n+1)^2}{(2n+1)(2n+3)} q_{n+1} - \frac{1}{2}\right] Z_\nu^n$$

$$+ \frac{3}{2}\frac{(n+1)(n+2)}{(2n+1)(2n+3)} q_{n+2} Z_\nu^{n+2}, \tag{7}$$

where

$$q_n = \begin{cases} 1 & \text{for } n < 3, \\ \prod_{j=3}^{n} \dfrac{\nu^2 + j^2}{\nu^2 + 1} & \text{for } n \geq 3. \end{cases}$$

Equation (5), with $G_n Z_\nu^2$ given by (7), has been used to evaluate the quadrupole anisotropy.

RESULTS

Models we have examined in this paper are baryon-dominated universe models (BDMs) and cold dark matter models (CDMs) with initially adiabatic and isocurvature perturbations, hot dark matter models (HDMs) with initially adiabatic perturbations, and Peebles' reionized universe models (Peebles 1987).

Following Wilson and Silk (1981) and Wilson (1983), the quadrupole moment Q^{ij} of the CBR anisotropy is defined to be

$$Q^{ij} = \frac{5}{4\pi}\int d\Omega\,(\gamma)\Theta_m P^{ij}_{(2)}, \tag{8}$$

and the rms quadrupole anisotropy Q is given by

$$Q^2 = \langle Q_{ij} Q^{ij}\rangle = \frac{3}{2}\frac{1}{2\pi^2}\int_0^\infty \tilde{k}^2\, d\tilde{k}\,|\theta_{m(2)}(\eta_0)|^2 \frac{(\tilde{k}^2 - 4K)}{(\tilde{k}^2 - K)}. \tag{9}$$

Since our analysis is in the scope of linear theory, the amplitude of perturbations should be normalized appropriately, based on a reliable quantitative measure of the present cosmological structures that reflects the amplitude in the linear regime. Lacking in such a measure, we have used empirically (probably) the most conservative normalization method among others (Davis and Peebles 1983):

$$J_3(25\ \mathrm{Mpc}h^{-1}) = 780\ \mathrm{Mpc}^3 h^{-3},$$

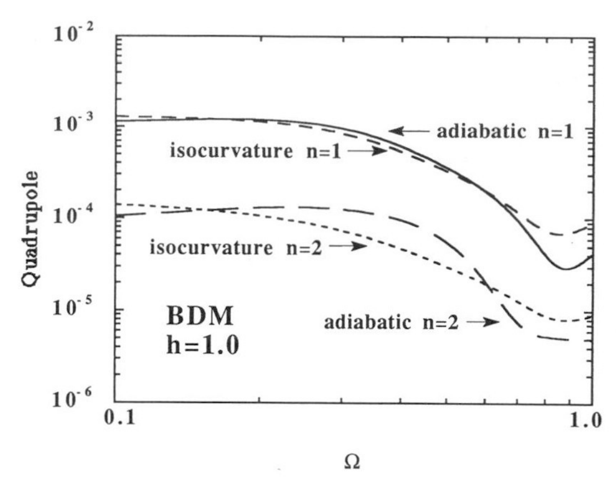

FIGURE 1. The rms quadrupole moment in adiabatic and isocurvature BDMs for the initial power-law indices $n = 1$ and 2.

where h is the Hubble constant normalized by 100 km/s/Mpc. This normalization usually gives the amplitude that is roughly a half of those with other normalization methods for a low-density universe. Correction factors for several other normalization schemes can be found in Gouda, Sasaki, and Suto (1989) and Sugiyama (1989) for BDMs and CDMs, respectively.

The results for some characteristic cases are shown in FIGURES 1–3. The rms quadrupole moments Q for BDMs are shown in FIGURE 1, for CDMs in FIGURE 2, and for HDMs and Peebles' models in FIGURE 3. We mention that, for CDMs and HDMs, the value of baryon mass density Ω_b is set fixed at 0.03 in the present calculations. Because of length considerations, only the results for $h = 1.0$ are shown. We have found that a lower h case generally gives a larger Q.

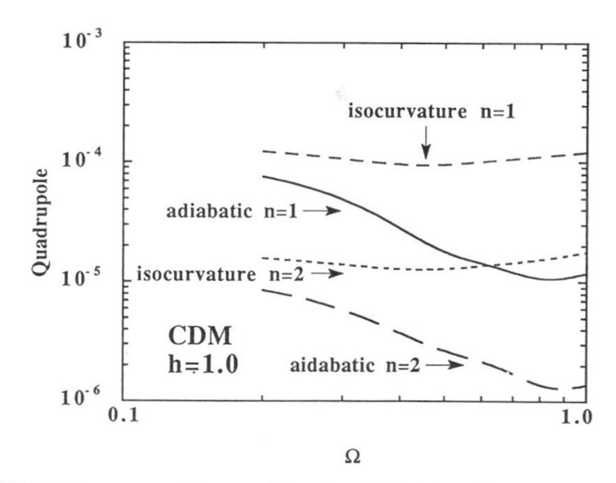

FIGURE 2. The same as FIGURE 1, but in adiabatic and isocurvature CDMs.

Observations of the CBR quadrupole moment show that $Q < 7 \times 10^{-5}$ (Lubin *et al.* 1985) and $Q < 3 \times 10^{-5}$ (Klypin *et al.* 1987). The expected rms quadrupole moment shown in the figures is not directly comparable with the observed upper limit of the quadrupole moment, since what we observe is the quadrupole moment at our location but not its rms value, not to mention the ambiguity of normalization. So, one has to keep in mind that the curves in FIGURES 1–3 are only of a qualitative nature and are not to be taken too literally. In deriving constraints on cosmological models, we have used the upper limit $Q < 10^{-4}$, but allowed violation of the limit by a factor of 2 or so because of the reasons noted earlier.

For adiabatic models, we find that low Ω_0 BDMs, CDMs, and HDMs, all with $n \leq 1$, are excluded. This conclusion is similar to the one drawn from the small-angle CBR anisotropies. As for isocurvature models, we find that all models with $n \leq 1$ are excluded. This exclusion is more stringent than in the case of small-angle anisotropies. As for Peebles' model, we find that models with $n \geq 3$ are allowed.

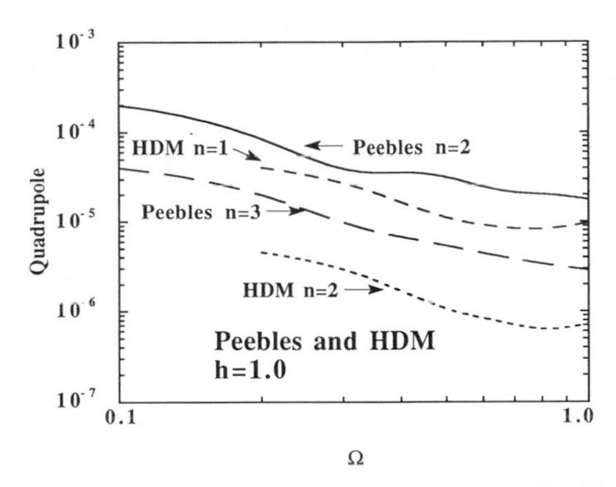

FIGURE 3. The same as FIGURE 1, but in adiabatic HDMs and Peebles' models.

In summary, we have calculated the rms quadrupole anisotropy in a variety of open universe models. Comparing the results with the observed upper limit of the quadrupole anisotropy, we have found that except for high Ω_0 adiabatic BDMs, CDMs, and HDMs, all conceivable models with $n \leq 1$ are excluded. Although the present conclusion is similar to the one obtained from the upper limit of small-angle CBR anisotropies, it must be taken more seriously. The reason for this is that the quadrupole anisotropy is a faithful carrier of primordial features of density fluctuations, while small-angle CBR anisotropies could be affected by phenomena that might have occurred after decoupling, such as reionization or gravitational lensing.

REFERENCES

BARDEEN, J. M. 1980. Phys. Rev. **D22:** 1882.
DAVIS, M. & P. J. E. PEEBLES. 1983. Astrophys. Lett. **267:** 465.

FUKUGITA, M., F. TAKAHARA, K. YAMASHITA & Y. YOSHII. 1990. Astrophys. J., Lett. **361:** L1.
GOUDA, N., M. SASAKI & Y. SUTO. 1989. Astrophys. J. **341:** 557.
GOUDA, N., N. SUGIYAMA & M. SASAKI. 1991. Prog. Theor. Phys. **85:** 1023. (Paper I)
GOUDA, N. & M. SASAKI. 1986. Prog. Theor. Phys. **76:** 1016.
GÓRSKI, K. M. & J. SILK. 1989. Astrophys. J., Lett. **346:** L1.
HOLTZMAN, J. A. 1989. Astrophys. J., Suppl. Ser. **71:** 1.
KLYPIN, A. A., M. V. SAZHIN, I. A. STRUKOV & D. P. SKULACHEV. 1987. Sov. Astron. Lett. **13:** 104.
KODAMA, H. & M. SASAKI. 1984. Prog. Theor. Phys. Suppl. **78:** 1.
KODAMA, H. & M. SASAKI. 1986. Int. J. Mod. Phys. **A1:** 265.
KODAMA, H. & M. SASAKI. 1987. Int. J. Mod. Phys. **A2:** 491.
KOFMAN, L. A. & A. A. STAROBINSKII. 1985. Sov. Astron. Lett. **11:** 271.
LUBIN, P., T. VILLELA, G. EPSTEIN & G. SMOOT. 1985. Astrophys. J., Lett. **298:** L1.
PEEBLES, P. J. E. 1982. Astrophys. J., Lett. **263:** L1.
PEEBLES, P. J. E. 1987. Astrophys. J., Lett. **315:** L73.
SACHS, R. K. & A. M. WOLFE. 1967. Astrophys. J. **147:** 73.
SUGIYAMA, N. 1989. Prog. Theor. Phys. **81:** 1021.
SUGIYAMA, N., N. GOUDA & M. SASAKI. 1990. Astrophys. J. In press.
TOMITA, K. & K. TANABE. 1983. Prog. Theor. Phys. **69:** 828.
VITTORIO, N., S. MATARRESE & F. LUCCHIN. 1988. Astrophys. J. **328:** 69.
VITTORIO, N. & J. SILK. 1985. Astrophys. J., Lett. **297:** L1.
WILSON, M. L. 1983. Astrophys. J. **273:** 2.
WILSON, M. L. & J. SILK. 1981. Astrophys. J. **243:** 14.

Analyzing the Cosmological Velocity Potential

ALAN HEAVENS

Department of Astronomy
University of Edinburgh
Royal Observatory
Edinburgh EH9 3HJ
Scotland

PROLOGUE

The flow of galaxies from all directions into the Great Attractor is used as a simple but powerful test for the spectrum of density fluctuations and the value of the density parameter, Ω. The most straightforward interpretation of the depth of the potential well at the Great Attractor is that the density parameter is rather high, and there is some strong observational evidence supporting the theoretical prejudice that $\Omega = 1$. This test is particularly effective because it probes the regime in which the perturbations are still linear, very few assumptions are required, and there are few of the uncertainties that beset other tests. Provided only that the perturbations are gravitational and Gaussian, and that galaxies trace the velocity field, one can use the depth of the potential at the Great Attractor, plus the degree of nonlinearity of the density field to put tight constraints on models of galaxy formation. For example, limiting the nonlinearity of the density field by current microwave background limits constrains $\Omega \gtrsim 0.5$ for cold dark matter (CDM) models. Using the Infrared Astronomical Satellite (IRAS) density field to fix the level of nonlinearity as well argues that, on large scales, IRAS galaxies trace the mass density. Consideration of a wider range of power-law fluctuation spectra allows Ω to be reduced to around 0.2 to 0.3, but values as low as required by nucleosynthesis for a baryonic Universe are strongly excluded.

INTRODUCTION

Deviations from uniform expansion of the Universe can be used as a useful diagnostic for the density parameter, since linear theory predicts that the magnitude of the peculiar velocities should scale with $\Omega^{0.6}\,\delta\rho/\rho$ (e.g., Peebles 1980). In practice, of course, one does not know $\delta\rho/\rho$ directly. Rather, one has to assume some relationship between the number density of a plausible population of luminous objects that can be detected, and the mass density, that cannot. A popular method is to take, for example, the density N of IRAS galaxies, and assume they are "biased" tracers of the mass distribution: $\delta N/N = b\delta\rho/\rho$, where the "bias" parameter is not directly known. The net result is that comparison between the velocity field and the number density of galaxies gives information on the quantity $\Omega^{0.6}/b$.

There are two principal ways of making the comparison between the number density and velocity fields. One can start with the distribution of objects and calculate the acceleration at the Local Group. Comparison with the observed dipole motion then constrains $\Omega^{0.6}/b$ (Strauss and Davis 1988; Rowan-Robinson *et al.* 1990). Alternatively, one can start from the velocity field, and take $\nabla \cdot \mathbf{v}$ to estimate a mass overdensity (e.g., Yahil 1990). The latter method requires well-sampled knowledge of the velocity field, and so involves considerable work, and the field must be smoothed over quite a large scale (a Gaussian filter of $12h^{-1}$ Mpc is used by Bertschinger *et al.* 1990). However, it has some advantages over the first method, which has uncertainties in it caused by incomplete sky coverage, uncertain luminosity function, and the possible effect of sources outside the survey volume. The luminosity function needs to be known accurately, since it affects how many low-luminosity sources are assumed to exist but are not seen.

This report describes a third method, which uses the properties of "attractors"—points in the Universe into which there is infall from all directions. The flow into the Great Attractor probably falls into this category, although the infall from the far side is not firmly established (Dressler and Faber 1990). The analysis uses the fact that large-scale flows are expected theoretically to be irrotational, so the peculiar velocity field is expressible in terms of a peculiar velocity potential: $\mathbf{v} = \nabla\Phi$. The analysis is thus concerned with local minima of the velocity potential. It is useful to state here the assumptions involved in this analysis:

1. On large scales, galaxies trace the velocity field.
2. The peculiar velocities are gravitational in origin.
3. The velocity potential on large scales is a random Gaussian field.

The meaning of the first assumption is that, if there is dark matter in the Universe, its peculiar velocity is the same as that of the galaxies. The peculiar velocity field is smoothed on a large scale, for a number of reasons. First, the smoothed density and velocity potential fields are still reasonably linear; second, the random errors in the distance indicators are reduced; third, in order to construct the potential, the velocity is required at all points in space. In addition, on small scales, the flow need not be irrotational, dissipation having violated the conditions of Kelvin's circulation theorem. Finally, the fields are only expected to be Gaussian on large scales. Some observational support comes from topology studies (Gott *et al.* 1989) and number density distributions (Rowan-Robinson *et al.* 1990).

MINIMA IN THE PECULIAR VELOCITY POTENTIAL

The growing mode of gravitational perturbations is irrotational, so the velocity field can be expressed as the gradient of a potential field. Using galaxies as tracers of the velocity field, and smoothing on a large scale ($\gtrsim 12h^{-1}$ Mpc), Bertschinger *et al.* (1990) have constructed this potential within a distance of about $60h^{-1}$ Mpc. The Great Attractor appears as a local minimum of the potential, indicative of infall from all directions. Assuming the field is Gaussian, it is possible (Heavens 1991) to analyze the properties of local minima, using an extension of the theory of density maxima (Peacock and Heavens 1985; Bardeen *et al.* 1986). The main result is that the joint number density distribution of potential drop and overdensity has a relatively simple form

$$n_{ga}(\varphi, \nu; r)\, d\varphi\, d\nu = \frac{f(\nu)}{(2\pi)^{5/2} R_\Phi^3 \beta_1} \exp\left[-\frac{1}{2\beta_1^2} (\varphi^2 - 2\varphi\nu\beta_2 + (\beta_1^2 + \beta_2^2)\nu^2) \right] d\varphi\, d\nu,$$

where $\nu \equiv \delta/\sigma_0$, with $\delta \equiv \delta\rho/\rho$ and $\sigma_0 \equiv \langle(\delta\rho/\rho)^2\rangle^{1/2}$; $f(\nu)$ is defined in Bardeen *et al.*:

$$f(\nu) \equiv \frac{(\nu^3 - 3\nu)}{2} \left\{ \mathrm{erf}\left[\left(\frac{5}{2}\right)^{1/2} \nu \right] + \mathrm{erf}\left[\left(\frac{5}{2}\right)^{1/2} \frac{\nu}{2} \right] \right\}$$
$$+ \left(\frac{2}{5\pi}\right)^{1/2} \left[\left(\frac{31\nu^2}{4} + \frac{8}{5}\right) \exp\left(-\frac{5\nu^2}{8}\right) + \left(\frac{\nu^2}{2} - \frac{8}{5}\right) \exp\left(-\frac{5\nu^2}{2}\right) \right].$$

The remaining variables are all calculated from the power spectrum of the density fluctuations, $|\delta_k|^2$, which includes a smoothing function $\exp(-k^2 R_f^2)$. The $\beta(r)$ come from the autocorrelation function of the field:

$$\beta_1^2(r) \equiv \hat{\sigma}^2 + (1 - A)^2 (1 - \alpha^2); \qquad \beta_2(r) \equiv \frac{\Box}{\alpha} + \alpha\,.$$

$R_\Phi \equiv \sqrt{3}\sigma_{-1}/\sigma_0$, where the quantities $\sigma_j^2 \propto \int_0^\infty |\delta_k|^2 k^{2j+2}\, dk$ are moments of the smoothed power spectrum of the density fluctuations; $\psi \equiv \int_0^\infty |\delta_k|^2 k^{-2} \sin(kr)/(kr\sigma_{-2})\, dk$ is the autocorrelation of the field normalized to unity at $r = 0$; $A \equiv (\psi + \Box)/(1 - \alpha^2)$ and $\alpha = \sigma_{-1}^2/(\sigma_0\sigma_{-2})$;

$$\hat{\sigma}^2 \equiv 1 - \frac{\psi^2}{(1 - \alpha^2)} - \frac{(2\alpha^2\psi + \Box)\Box}{\alpha^2(1 - \alpha^2)} - \frac{5}{\alpha^2}\left(\frac{R_\Phi^2 \partial\psi/\partial r}{r} - \Box\right)^2 - \left(\frac{R_\Phi \partial\psi/\partial r}{\alpha}\right)^2,$$

and $\Box \equiv R_\Phi^2 \nabla^2 \psi/3$. Full details may be found in Heavens (1991).

The variable φ is related to the potential drop $\Delta\Phi$ by $\varphi \equiv \Delta\Phi/(\Omega^{0.6}\sigma_{-2}) = 3\alpha\Delta\Phi/(R_\Phi^2\Omega^{0.6}\sigma_0)$, so increasing Ω or σ_0 has the effect of allowing deeper potential wells. This distribution, combined with the known depth and distance of the Great Attractor, gives likelihood estimates for the parameter(s) Ω (and ν) if σ_0 can be found by other methods. For example, using the microwave background to constrain σ_0 for particular fluctuation spectra (Vittorio *et al.* 1990) gives a likelihood function for CDM shown in FIGURE 1. Alternatively, σ_0 may be estimated using one of the IRAS surveys (Saunders *et al.* 1990). In this case, we estimate the overdensity from the IRAS counts (Yahil 1990) to get $\nu \simeq 2$, and obtain the likelihood for $\Omega/b^{5/3}$ (FIG. 2). Again, we see that, if $b \sim 1$, $\Omega \gtrsim 0.5$ is indicated. We may put these two methods together to constrain $0.7 \lesssim b \lesssim 1.4$, for the bias parameter on a scale of $12h^{-1}$ Mpc.

Potential Minimum?

If the infall from the far side of the Great Attractor is not confirmed, then we may still use the potential drop between the Local Group and the Great Attractor to constrain the density parameter and Ω. The essential point is that *for a given power spectrum* the constraint on Ω is little changed (Heavens 1991). However, in order to remove the minimum, one may wish to introduce more large-scale power, and

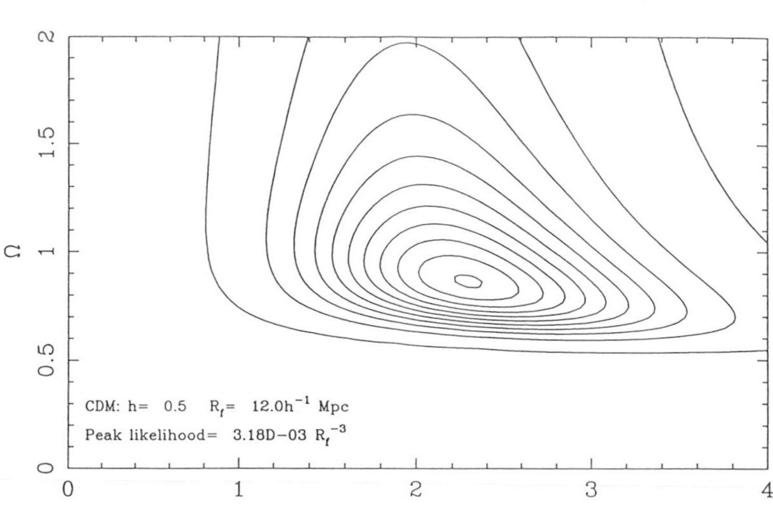

FIGURE 1. The likelihood as a function of Ω, using the UCSB south pole microwave background experiment to constrain σ_0 for CDM with $h = 0.5$. This assumes that σ_0 is as large as possible, with marginal consistency with the 95 percent upper limit of $\Delta T/T$. Contours at 1 percent, 10 percent, 20 percent . . . 90 percent and 99 percent of peak.

change the spectrum. In this case, the constraints on high Ω are less strong. With complete freedom to specify the power spectrum it is not possible to exclude any value of Ω, but investigations of power-law spectra indicate that values less than 0.2–0.3 are unlikely to be acceptable.

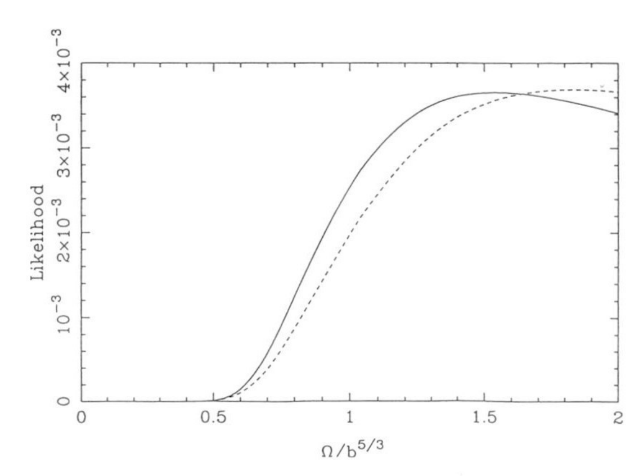

FIGURE 2. The likelihood distribution for CDM, using IRAS galaxies to obtain σ_0, for two values of $hb^{5/3}$, 0.5 and 1.0, shown *full* and *dashed.*

THE DENSITY AT THE GREAT ATTRACTOR

The distribution of attractors may be integrated over depths to give the distribution of density at attractors. FIGURE 3 shows the overdensity distribution (in units of the rms σ_0) for all attractors. This distribution is

$$p(\nu) = \frac{5^{3/2}2}{29 - 6\sqrt{6}} f(\nu) \exp\left(-\frac{1}{2}\nu^2\right),$$

which is independent of the spectrum of fluctuations. Note that attractors need not be particularly extreme: the mean overdensity is about $2\sigma_0$, and δ may be as low as σ_0. Thus, attractors need not be associated with particularly extreme density enhancements.

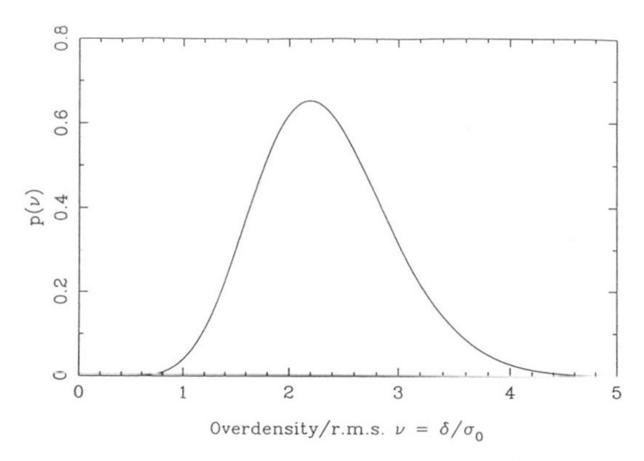

FIGURE 3. The overdensity distribution at *all* attractors, independent of the power spectrum. Note that attractors need not be exceptionally overdense.

If we calculate the conditional density distribution for attractors that are *as deep as* the Great Attractor, then the story changes. This distribution depends on the spectrum and the value of Ω. FIGURE 4 shows conditional distributions for CDM with $\Omega/b^{5/3} = 0.1, 0.3$, and 1.0. Here we see that, if Ω is low, then we *do* require an extreme density enhancement to account for as deep a potential well as the Great Attractor. Such an object would be expected to be fairly spherical (Peacock and Heavens 1985; Bardeen *et al.* 1986), but exponentially rare. This graph is essentially making the same point as FIGURE 2: the IRAS data suggest that $\nu \sim 2$, which is near the peak of the distribution for $\Omega/b^{5/3} = 1$, but which is clearly incompatible with $\Omega/b^{5/3} = 0.1$.

CONCLUSIONS

Assuming that galaxies trace the peculiar velocity field, and that the peculiar velocity potential arises from Gaussian gravitational perturbations, the depth of the potential at the Great Attractor is a useful and clean test of fluctuation spectra and

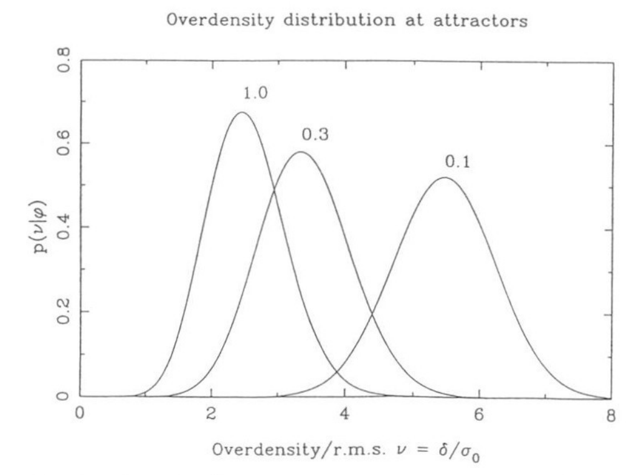

FIGURE 4. The overdensity distribution at attractors as deep as the Great Attractor, for CDM with different values of $\Omega/b^{5/3}$. If Ω is low, then an extreme overdensity is required, which would be extremely rare.

the value of the density parameter. It is sufficiently deep to put a firm constraint $\Omega \gtrsim 0.5$ for CDM models. If $\Omega = 1$, then the microwave background limits and the depth of the Great Attractor constrain the bias parameter for IRAS galaxies on large scales to be in the range 0.7–1.4.

REFERENCES

BARDEEN, J. M., J. R. BOND, N. KAISER & A. S. SZALAY. 1986. Astrophys. J. **304:** 15.

BERTSCHINGER, E., A. DEKEL, S. M. FABER, A. DRESSLER & D. BURSTEIN. 1990. Astrophys. J. **364:** 370.

DRESSLER, A. & S. M. FABER. 1990. Astrophys. J. **354:** 13.

GOTT, J. R. III, *et al.* 1989. Astrophys. J. **340:** 625.

HEAVENS, A. F. 1991. Mon. Not. R. Astron. Soc. **251:** 267.

PEACOCK, J. A. & A. F. HEAVENS. 1985. Mon. Not. R. Astron. Soc. **217:** 805.

PEEBLES, P. J. E. 1980. The Large Scale Structure of the Universe. Princeton University Press. Princeton, N.J.

ROWAN-ROBINSON, M., *et al.* 1990. Mon. Not. R. Astron. Soc. **247:** 1.

SAUNDERS, W. S., *et al.* 1990. Nature **349:** 32.

STRAUSS, M. A. & M. DAVIS. 1988. *In* Large Scale Motions in the Universe, V. C. Rubin and G. V. Coyne, Eds. Princeton University Press. Princeton, N.J.

VITTORIO, N., P. MEINHOLD, R. MUCIACCIA, P. LUBIN & J. SILK. 1990. Preprint.

YAHIL, A. 1990. *In* The Early Universe and Cosmic Structures, A. Blanchard, Ed. Editions Frontières. Gif sur Yvette Cedex, France.

Fluctuation of Luminosity Distance and the Peculiar Velocity Field

NAOSHI SUGIYAMA,[a,d] MISAO SASAKI,[b]
AND MASUMI KASAI[c]

[a]*Department of Physics*
Kyoto University
Kyoto 606, Japan

[b]*Department of Physics*
Kyoto University
Kyoto 606, Japan

[c]*Department of Physics*
Hirosaki University
Hirosaki 036 Japan

INTRODUCTION

Recently, various observations about the large-scale structure of the Universe have been done and it has been found that the universe has very complicated features, for example, the great wall, the Great Attractor, and the 128-Mpc periodic structure. When one compares such structures with cosmological models, the propagation of photons through these structures must be considered. We could not know true observed quantities if this propagation is not taken into account. For example, some authors calculated the velocity field caused by density perturbations on some cosmological models and directly compared them with the observed bulk flow (Vittorio *et al.* 1986; Vittorio and Turner 1987; Sugiyama 1989). However, these calculations should not be compared with observations before considering the effect of the photon propagation through density fluctuations.

Sasaki (1987) studied the propagation of photons from a point source in the perturbed Friedmann universe and obtained linear fluctuations of the luminosity distance that are caused by the existence of density anisotropies. *In this paper,* we numerically calculate the dipole component of these fluctuations and get the observed apparent peculiar velocity field expected for some cosmological models, that is, cold dark matter models with adiabatic and isocurvature perturbations and hot dark matter models with adiabatic perturbations. We also calculate the quadrupole component of these fluctuations used by Sasaki's formula.

FLUCTUATIONS OF THE LUMINOSITY DISTANCE

The metric of the dust-dominated, perturbed Friedmann universe on the flat spacetime is assumed to have the form

$$ds^2 = a(\eta)^2(-(1 + 2\Psi)d\eta^2 + (1 - 2\Psi)dx^2), \tag{1}$$

[d]Current address: University of Tokyo, Tokyo 113, Japan.

where η is a conformal time, $a(\eta)$ is a scale factor, and Ψ is a gravitational potential that is satisfied the Poisson equation

$$\frac{1}{a^2}\nabla^2\Psi = 4\pi G\rho\Delta, \tag{2}$$

with Δ being the density perturbation. Then the fluctuations of the luminosity distance are obtained by Sasaki (1987)

$$\begin{aligned}\frac{\Delta d_L}{d_L} = &\frac{1}{\lambda_s}\int_0^{\lambda_s} d\lambda[(\lambda - \lambda_s)\lambda\nabla^2\Psi + 2(\Psi - \Psi_0)] : \text{expansion of light ray}\\ &+ \frac{1}{2}\left(\frac{\eta_0}{\lambda_s} - 3\right)(\Psi_s - \Psi_0) : \text{Sachs–Wolfe effect}\\ &+ \frac{1}{6}\left(\frac{\eta_0}{\lambda_s} - 3\right)[\eta_s(\Psi_{|i}\gamma^i)_s - \eta_0(\Psi_{|i}\gamma^i)_0] : \text{peculiar velocity}\\ &- \frac{1}{3}\eta_0(\Psi_{|i}\gamma^i)_0 + \frac{\eta_0^2}{18}\nabla^2\Psi_0,\end{aligned} \tag{3}$$

where λ is an affine parameter along the ray and can be identified as the coordinate distance r on the flat space, γ^i is a unit vector, and the suffixes s and 0 represent the source and observer, respectively. It should be noted that the expansion term is caused by the gravitational lensing effect.

We decompose the preceding equation into multipole components as is usually done to analyze the anisotropies of a physical quantity

$$\frac{\Delta d_L}{d_L} = \sum_{l,m} C_{l,m}(z)Y_{lm}(\Omega), \tag{4}$$

with $Y_{lm}(\Omega)$ being the conventional spherical harmonic and

$$\begin{aligned}C_{l,m}(z) = 4\pi i^l \int d^3\mathbf{k}\Bigg[&\frac{1}{y_s}\int_0^{y_s} dy\,(y(y_s - y)j_l(y) + 2[j_l(y) - j_l(0)])\\ &+ \frac{1}{2}\left(\frac{\eta_0}{\lambda_s} - 3\right)[j_l(y_s) - j_l(0)]\\ &+ \frac{1}{6}\left(\frac{\eta_0}{\lambda_s} - 3\right)((y_0 - y_s)j_l'(y_s) - y_0 j_l'(0))\\ &- \frac{1}{3}y_0 j_l'(0) + \frac{y_0^2}{18}j_l(0)\ \Psi_k Y_{lm}^*(\Omega_k)e^{i\mathbf{k}\cdot\mathbf{x}_0}\end{aligned} \tag{5}$$

where $y_s = k\lambda_s$, $y_0 = k\eta_0$, $j_l(y)$ is the spherical Bessel function of order l, $j'_l(y)$ is its derivative, Ψ_k is a Fourier component of Ψ, Ω_k denotes the solid angle of $\mathbf{k}$, and $\mathbf{x}_0$ is the spatial coordinate of the observer.

PECULIAR VELOCITY FIELD

In this section, we obtain the observed apparent peculiar velocity field from the dipole component of **(4)**. The dipole component of fluctuations of the luminosity

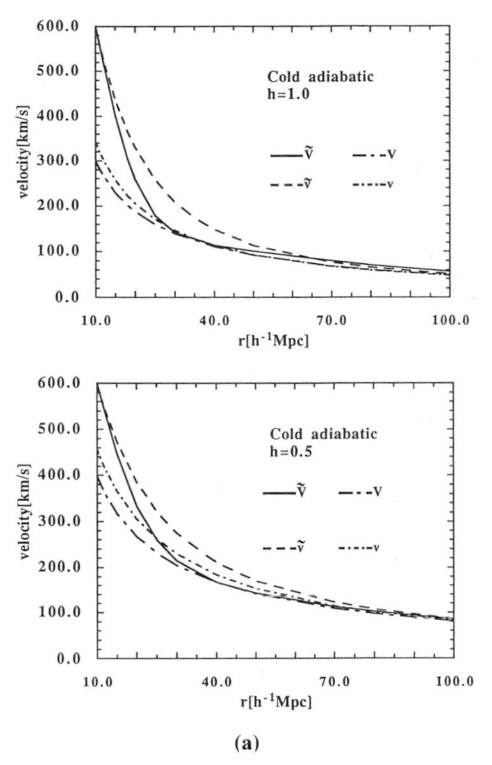

(a)

FIGURE 1. Peculiar velocity fields for various cosmological models are shown. $\tilde{V}$ is the expected value taking into account the condition that the observer moves 600 km/s and V is the rms peculiar velocity. Both $\tilde{V}$ and V are obtained from $\Delta d_L/d_L$. $\tilde{\upsilon}$ and υ are expected and rms peculiar velocities are obtained by using the ordinary method, for example, $\upsilon^2 = \int d^3\mathbf{k}|4\pi \int dr\, r^2 j_0(kr)W(r, r_s)\upsilon_k(r)|^2$, where $\mathbf{v}_k = -2i\mathbf{k}\Psi_k/3(Ha)_0$.

distance is expressed by using the dipole vector **D** as

$$\left(\frac{\Delta d_L}{d_L}\right)_{l=1} \equiv D_i \lambda^i, \tag{6}$$

where

$$\mathbf{D} = \sqrt{\frac{3}{4\pi}}\left(\frac{C_{1,1} + C_{1,-1}}{\sqrt{2}}, i\,\frac{C_{1,1} - C_{1,-1}}{\sqrt{2}}, C_{1,0}\right). \tag{7}$$

Since this dipole vector corresponds to the apparent velocity on the observer rest frame, the observed apparent velocity $\mathbf{V}(\lambda_s)$ on the matter rest frame is

$$\mathbf{V}(\lambda_s) = -\frac{2\lambda_s}{\eta_s}\mathbf{D} + \mathbf{v}(0). \tag{8}$$

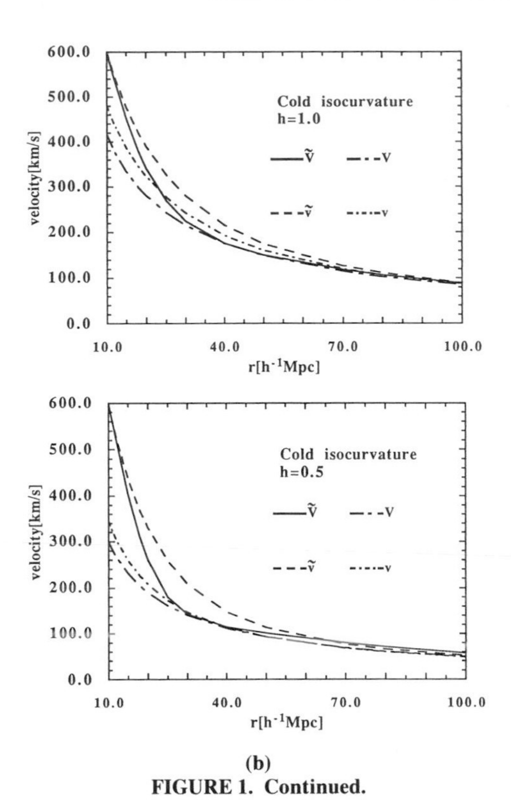

(b)
FIGURE 1. Continued.

The peculiar velocity of the observer $\mathbf{v}(0)$ is generated by density fluctuations as

$$\mathbf{v}(0) = -\frac{2}{3}\frac{1}{(Ha)_0}\nabla\Psi_0. \tag{9}$$

As a result, the apparent velocity $\mathbf{V}(\lambda_s)$ is obtained as

$$\begin{aligned}\mathbf{V}(\lambda_s) = \int d^3\mathbf{k}\Bigg[& -\frac{6}{k\eta_s}\int_0^{y_s} dy\,(y(y_s - y)j_1(y) - 2j_1(y)) \\ & - 3\left(1 - \frac{2\lambda_s}{\eta_s}\right)j_1(y_s) \\ & - \frac{1}{3}\left(1 - \frac{2\lambda_s}{\eta_s}\right)(y_0 - y_s)(j_0(y_s) - 2j_2(y_s))\Bigg] i\Psi_k\frac{\mathbf{k}}{k} \\ \equiv \int d^3\mathbf{k}\mathbf{V}_\mathbf{k}. \end{aligned} \tag{10}$$

In order to compare this formula with actual observations, we introduce the Gaussian window function $W(r, \mathbf{r}) = \exp(-r^2/2\bar{r}^2)/(2\pi)^{3/2}$. Then the apparent rms peculiar velocity of the source object at the distance r_s from the observer is obtained as

$$\langle V(r_s)^2 \rangle = 4\pi \int dk\, k^2 \left| 4\pi \int dr\, r^2 W(r, r_s) V_k(r) \right|^2. \tag{11}$$

Since the observer moves relative to the cosmic background radiation, we consider two measurements of flow field used by Vittorio *et al.* (1986). We take the velocity of the local group whose radius is r_0 as $\mathbf{v}_0$. The conditional probability becomes

$$dP(\mathbf{V}(r_s)|\mathbf{v}_0) = p(\mathbf{V}(r_s), \mathbf{v}_0) d^3\mathbf{V}(r_s)/p(\mathbf{v}_0), \tag{12}$$

where

$$p(\mathbf{v}_0) = \frac{1}{(2\pi\sigma_{00})^{3/2}} \exp\left(-\frac{v_0^2}{2\sigma_{00}^2} \right), \tag{13}$$

$$p(\mathbf{V}(r_s), \mathbf{v}_0) = \frac{1}{(2\pi)^3 |A|^{3/2}} \exp\left(-\frac{1}{2} (\mathbf{V}(r_s), \mathbf{v}_0) A^{-1} \begin{pmatrix} \mathbf{V}(r_s) \\ \mathbf{v}_0 \end{pmatrix} \right), \tag{14}$$

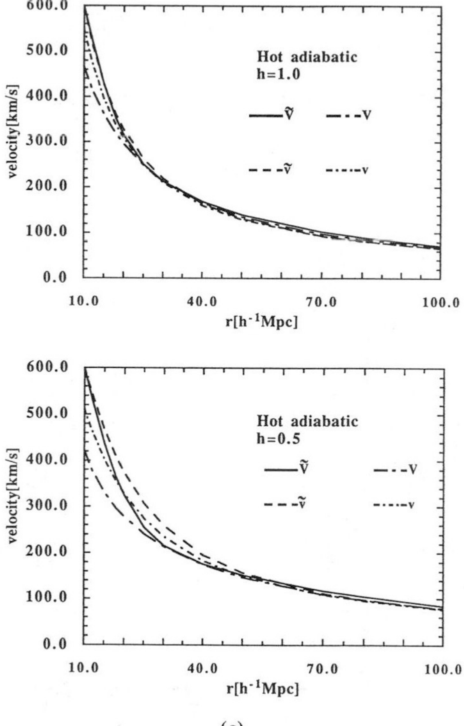

(c)
FIGURE 1. Continued.

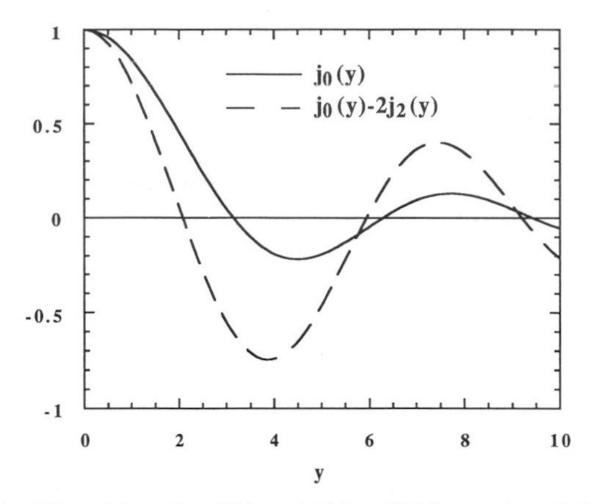

FIGURE 2. Spherical Bessel function $j_0(x)$ and $j_0(x) - 2j_2(x)$ are plotted. The former and the latter make a main contribution to the peculiar velocity field obtained by using the ordinary method, and are obtained from $\Delta d_L/d_L$, respectively.

with

$$A = \begin{pmatrix} \sigma_{ss}^2 & \sigma_{0s}^2 \\ \sigma_{0s}^2 & \sigma_{00}^2 \end{pmatrix}, \qquad \sigma_{ij}^2 = \frac{1}{3}\langle V(r_i)V(r_j)\rangle\,. \tag{15}$$

Then the expected value of the peculiar velocity field is obtained as

$$\begin{aligned} \tilde{V}(r_s)^2 &= \int |\mathbf{V}(r_s)|^2\, dP(\mathbf{V}(r_s)\,|\,\mathbf{v}_0) \\ &= \frac{1}{\sigma_{00}^2}\left(3\sigma_{00}^2\sigma_{ss}^2 - 3\sigma_{0s}^4 + \frac{\sigma_{0s}^4}{\sigma_{00}^2}v_a^2\right). \end{aligned} \tag{16}$$

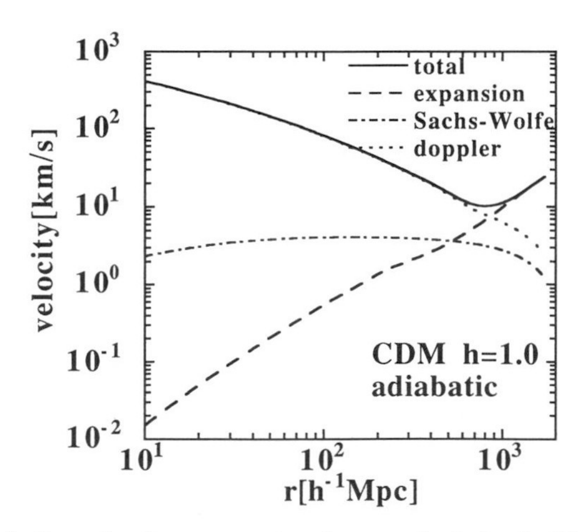

FIGURE 3. Contributions of each component on the rms velocity for the CDM model.

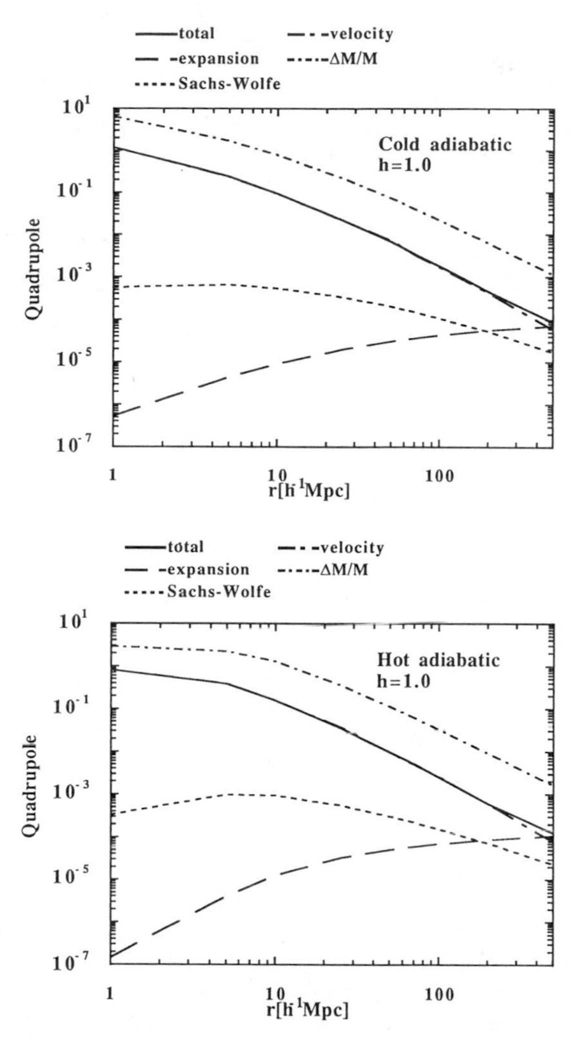

FIGURE 4. The rms quadrupole component of $\Delta d_L/d_L$ for some cosmological models is shown. The contribution of the expansion term, Sachs–Wolfe term, and peculiar velocity term are shown in these figures. $\delta M/M$ is also plotted because the behavior is expected to be very similar to the quadrupole component of $\Delta d_L/d_L$.

In FIGURE 1, we show the expected values $\tilde{V}$ for some cosmological models together with the rms velocity $\sqrt{\langle V(r_s^2\rangle} \equiv V$ when we set $r_0 = 10h^{-1}$ Mpc and $\upsilon_0 = 600$ kms^{-1}. We also show the expected ($\tilde{\upsilon}$) and rms values (υ) of the peculiar velocity field that are obtained by using the ordinary method in these figures, for comparison. We use J_3 normalization at $10h^{-1}$ Mpc and $25h^{-1}$ Mpc for cold-dark-matter-dominated models and hot-dark-matter-dominated models, respectively. It is very interesting that $\tilde{V} < \tilde{\upsilon}$ and $V < \upsilon$ for all models on a small scale (<50 Mpc). The observed apparent peculiar velocity field would become a little smaller than that expected from ordinary velocity perturbations! It is easy to understand this result. The main contribution of the dipole moment is the peculiar velocity term $j_0(y_s) - 2j_2(y_s)$ ($y_s = kr_s$). In the ordinary method, this term is replaced with $j_0(y_s)$. On a small scale (small r), the contribution of $j_0(y_s)$ is greater than that of $j_0(y_s) - 2j_2(y_s)$ as shown in FIGURE 2. Then the observed apparent velocity field becomes smaller.

We show the contributions of each component on the apparent rms velocity for the cold dark matter model with initial adiabatic perturbations in FIGURE 3. On small scales, the peculiar velocity term dominates. However, this term gets smaller on larger scales and the expansion term begins to dominate at $r \gtrsim 500$ Mpc. It is remarkable that we do not approach the "CMB rest frame" even at 1000 Mpc because of the expansion term.

QUADRUPOLE MOMENT

The quadrupole component of the fluctuations of the luminosity distance with Gaussian window function is described as

$$\left\langle \left(\frac{\Delta d_L}{d_L}(r_s) \right)^2_{l=2} \right\rangle = 20\pi \int dk\, k^2 |\Psi_k|^2 |4\pi \int dr\, r^2 W(r, r_s)$$

$$\left[\frac{1}{y_s} \int_0^{y_s} dy\, (y\,(y_s - y) j_2(y) + 2j_2(y)) \text{ (expansion)} \right.$$

$$+ \frac{1}{2}\left(\frac{\eta_0}{\lambda_s} - 3 \right) j_2(y_s) \text{ (Sachs–Wolfe)}$$

$$+ \frac{1}{6}\left(\frac{\eta_0}{\lambda_s} - 3 \right) (y_0 - y_s) j_2'(y_s)\, |^2 \text{ (peculiar velocity).} \qquad \textbf{(17)}$$

In FIGURE 4, we show each component and total quadrupole moments for cold-dark-matter- and hot-dark-matter-dominated universe models with adiabatic perturbations. It should be noted that the expansion term dominates at $r \gtrsim 500$ Mpc. We also show $\delta M/M$ in FIGURE 4, and find it behaves like a quadrupole component except for total normalization. In the limit of small $y_s(=kr_s)$, it is easy to see that the distance dependence of the dipole component is similar to that of $\delta M/M$.

In the future, it will become a very interesting and important problem to measure the quadrupole component of fluctuations of the distance.

REFERENCES

SASAKI, M. 1987. Mon. Not. R. Astron. Soc. **228:** 653.
SUGIYAMA, N. 1989. Prog. Theor. Phys. **81:** 1021.
VITTORIO, N., R. JUSZKIEWICZ & M. DAVIS. 1987. Nature **323:** 132.
VITTORIO, N. & M. S. TURNER. 1987. Astrophys. J. **316:** 475.

Natural Inflation[a]

KATHERINE FREESE

Physics Department
Massachusetts Institute of Technology
Cambridge, Massachusetts 02139

INTRODUCTION

Here we present results from three papers:

1. The requirements of sufficient inflation and microwave background anisotropy limits on the generation of density fluctuations place important constraints on inflationary models. The fine-tuning of parameters required in inflationary models can be quantified (Adams, Freese, and Guth 1991). Rolling fields need flat potentials: the ratio of height to $(\text{width})^4$ of the potential must be $\chi = [\Delta V/(\Delta\psi)^4] \leq O(10^{-8})$; for extended inflation, $\chi \leq O(10^{-15})$.
2. Double field inflation (Adams and Freese 1991) couples two scalar fields to obtain a time-dependent nucleation rate of true vacuum bubbles and thus a successful end to (old) inflation. Cosmic strings responsible for large-scale structure may form at the end of the inflationary phase transition, and bubble collisions may give rise to interesting structure.
3. We show (Freese, Frieman, and Olinto 1990) that a pseudo-Nambu–Goldstone boson, with a potential of the form $V(\phi) = \Lambda^4[1 + \cos(\phi/f)]$, can naturally give rise to an epoch of inflation in the early universe. The potential is flat, as required in Adams *et al.* (1991) without any fine-tuning. Successful inflation can be achieved if $f \sim m_{\text{pl}}$ and $\Lambda \sim m_{\text{GUT}}$. Such mass scales arise in particle physics models with a large gauge group that becomes strongly interacting at a scale $\sim\Lambda$, for example, as can happen in the hidden sector of superstring theories. The density fluctuation spectrum is non-scale-invariant, with more power on large length scales.

The inflationary universe model was proposed (Guth 1981) to solve several cosmological puzzles, notably the horizon, flatness, and monopole problems. During the inflationary epoch, the energy density of the universe is dominated by a (nearly constant) false vacuum energy term $\rho \simeq \rho_{\text{vac}}$, and the scale factor $R(t)$ of the Universe expands exponentially: $R(t) = R(t_1)e^{H(t-t_1)}$, where $H = \dot{R}/R$ is the Hubble parameter, $H^2 = 8\pi G\rho/3 - k/R^2$ $(\simeq 8\pi G\rho_{\text{vac}}/3$ during inflation), and t_1 is the time at the beginning of inflation. If the interval of exponential expansion satisfies $t_{\text{end}} - t_1 \gtrsim 65H^{-1}$, a small causally connected region of the universe grows to a sufficiently large size to explain the observed homogeneity and isotropy of the universe today. In the process, any overdensity of magnetic monopoles produced at an epoch of grand unification is diluted to acceptable levels. The predicted grand unified theory (GUT)

[a]This work was supported in part by a National Science Foundation Presidential Young Investigator Fellowship, in part by Sloan Foundation Grant 26722 and a Sloan Foundation Fellowship, and in part by NASA Grant NAGW-1320.

abundance of monopoles is $\Omega_{mon} \simeq 10^{12}$, whereas the energy density of our universe is observed to be within an order of magnitude of $\Omega = 1$; here the excess monopoles are simply "inflated away" beyond our visible horizon. Inflation predicts a geometrically flat universe ($k = 0$), $\Omega \equiv 8\pi G\rho/3H^2 \rightarrow 1$.

I begin with a discussion of the proposals and problems of the earliest inflationary models, now known as "old" inflation (Guth 1981) and "new" inflation (Linde 1982; Albrecht and Steinhardt 1982). The problems with these original models then lead into a discussion of new ideas proposed to circumvent some of these problems.

OLD INFLATION

In the model of old inflation, the universe passes through a first-order phase transition at a critical temperature T_c, for example, at the GUT epoch $T_c \simeq 10^{14}$ GeV. FIGURE 1 plots the effective potential for the scalar field at three different temperatures. Above the critical temperature, that is, for $T > T_c$, the potential has only one minimum, at point A. The (expectation value of the) scalar field is situated at this minimum. At $T = T_c$, there are two equally energetic minima. Once the temperature drops below T_c, however, there is only one true minimum, at point B. The old minimum at point A is now called the false vacuum, and the new minimum at point B is now called the true vacuum. The universe wants to go from A to B, since B has lower energy, but is prevented from doing so by an energy barrier. The universe can get from the false vacuum to the true vacuum by either thermal fluctuations or vacuum tunneling. As illustrated in FIGURE 2, bubbles of true vacuum nucleate in a sea of false vacuum. In the inflationary model, the nucleation rate for vacuum tunneling (Voloshin, Kobzarev, and Okun 1975; Coleman 1977) is arranged to be very slow, so that the universe is trapped in the metastable false vacuum for a long time. The difference in energy density between points A and B is the vacuum energy density ρ_{vac}. While the universe is trapped in the false vacuum, this energy density dominates over matter and radiation energy, and the universe expands exponentially: $H^2 = 8\pi G\rho_{vac}/3 \simeq$ const has the solution $R \propto e^{Ht}$. Sufficient inflation to solve the flatness, horizon, and monopole problems requires that the scale factor at the

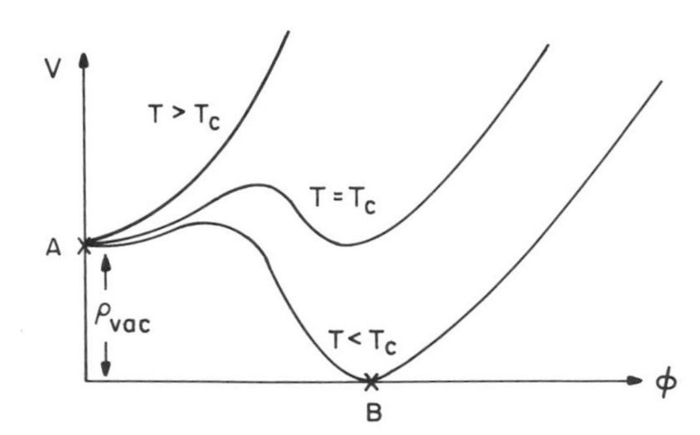

FIGURE 1. Effective potential for scalar field at first-order phase transition for $T > T_c$, $T = T_c$, and $T < T_c$.

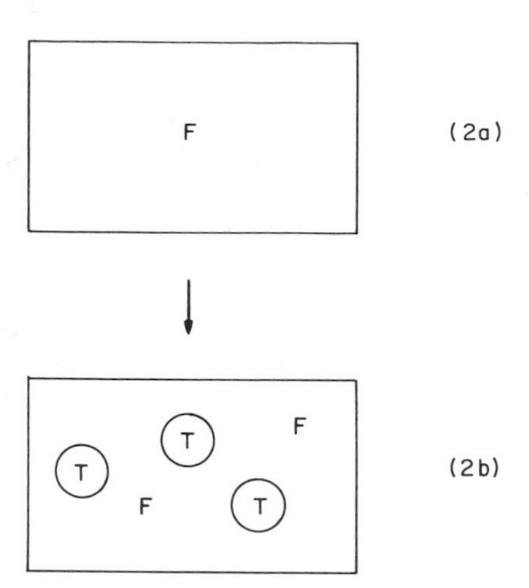

FIGURE 2. Nucleation of true vacuum bubbles at first-order phase transition. (**a**) For $T > T_c$, the Universe is entirely in the false vacuum phase. (**b**) For $T < T_c$, bubbles of true vacuum start to nucleate in the sea of false vacuum.

end of inflation satisfy $R_{\text{end}} = 10^{27} R_{\text{begin}} = e^{65} R_{\text{begin}}$, where $R_{\text{begin}} = R(t_1)$ is the scale factor at the beginning of inflation; that is, approximately 65 e-foldings of inflation are required.

Although old inflation can successfully obtain these 65 e-foldings of expansion, the original model has the problem that it cannot reheat, that is, after the inflationary phase, the energy in the vacuum cannot be converted to ordinary radiation (Guth and Weinberg 1983). In a successful inflationary model, the period of exponential expansion must be followed by a return to a radiation-dominated epoch, so that the universe can subsequently evolve according to the standard model.

In old inflation, all of the latent heat of the phase transition is converted into kinetic energy of the bubble walls. In order to end up with a universe like ours, the energy somehow has to be extracted from the bubble walls into thermalizing the interior. One might hope that bubble collisions could achieve this thermalization. However, although the bubbles expand with the speed of light, the false vacuum background expands exponentially, and the bubbles cannot find one another and do not percolate and thermalize. Occasionally finite clusters of a few bubbles form; in this case, the largest bubble dominates the dynamics and again thermalization does not take place. Thus the final result of old inflation is the "Swiss cheese universe" of FIGURE 2(b), with many empty bubbles of true vacuum in a false vacuum sea. This is not our universe.

NEW INFLATION

Shortly after the realization that old inflation fails to reheat, "new" inflation was proposed (Linde 1982; Albrecht and Steinhardt 1982). In this model, it is assumed that the diagram of the effective potential (or free energy) of the inflation field ψ has a very flat plateau, and the field evolves by "slowly rolling" off the plateau. FIGURE 3 illustrates a typical new inflationary potential. The phase transition can be second order or only weakly first order. This model is conceptually different in that a coherent region of the universe can be described as rolling down the hill described by the potential. While the scalar field ϕ rolls along the flat portion of the potential, the value of the potential is almost constant and vacuum energy density dominates the energy density of the universe. As before, $R \propto e^{Ht}$, where $H^2 = 8\pi G\rho_{vac}/3$; the universe expands exponentially. Then, once the field approaches the minimum of the potential, the ϕ field decays into particles and radiation, and reheating takes place to an ordinary radiation-dominated epoch. In this way a "graceful exit" from inflation is achieved.

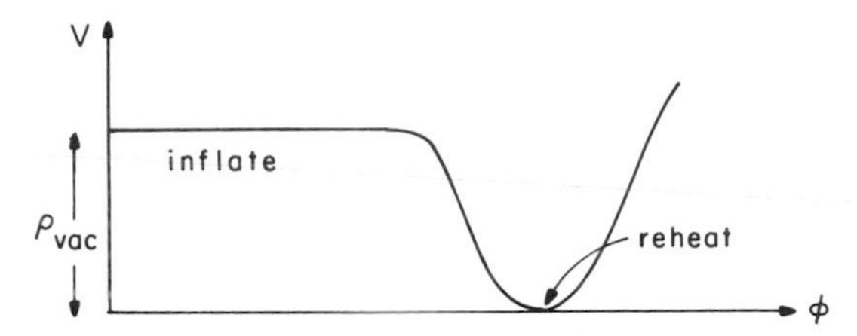

FIGURE 3. A typical new inflationary potential. The universe inflates while ϕ "rolls down" the flat portion of the potential and reheats while ϕ oscillates around the minimum.

CONSTRAINTS ON INFLATIONARY POTENTIALS

There are several requirements on potentials in inflationary models (Steinhardt and Turner 1984). The two most restrictive are:

1. Sufficient inflation: the scale-factor must grow by at least 65 e-foldings

$$N_e(\phi_1, \phi_2, f) \equiv \ln (R_{\text{end}}/R_{\text{begin}}) = \int_{t_1}^{t_2} H\, dt = \frac{-8\pi}{m_{\text{pl}}^2} \int_{\phi_1}^{\phi_2} \frac{V(\phi)}{V'(\phi)} d\phi \geq 65. \tag{1}$$

   Here, N_e is the number of e-foldings of the scale factor, m_{pl} is the Planck mass, ϕ_1 and t_1 are the value of the scalar field and the time at the beginning of inflation, ϕ_2 and t_2 are the value of the scalar field and the time at the end of inflation, and prime denotes derivative with respect to ϕ.

2. Amplitude of density fluctuations: quantum fluctuations in the scalar field give rise to density perturbations of amplitude (Guth and Pi 1982; Hawking 1982; Starobinskii 1982; Bardeen, Steinhardt, and Turner 1983)

$$\frac{\delta\rho}{\rho} \simeq \left.\frac{H^2}{\dot{\phi}}\right|_{\text{horizon}}. \tag{2a}$$

   The amplitude of these fluctuations is constrained not to exceed the observed limits on the anisotropy of the microwave background (see, e.g., Meinhold and

Lubin 1991; Readhead *et al.* 1989; Uson and Wilkinson 1985)

$$\frac{\delta\rho}{\rho} \simeq \frac{\delta T}{T} \leq O(10^{-5}). \quad \textbf{(2b)}$$

There are two nice features of inflationary density perturbations: (1) causal processes can explain their origin, and (2) during inflation, H and $\dot{\phi}$ vary slowly, and thus, from **(2a)**, we can see that the amplitude of the fluctuations at the time they enter the horizon is nearly the same on all scales (for many inflationary models), that is, one obtains a Harrison–Zel'dovich spectrum. The problem, however, is that inflationary models typically overproduce the amplitude of the fluctuations unless one fine-tunes the parameters of the potential.

In order to obtain sufficient inflation and to satisfy microwave background anisotropy limits on density fluctuations, the potential of the field responsible for inflation (the *inflaton*) must be very flat. Paper I (Adams, Freese, and Guth 1991) shows that, for a general class of inflation models involving a single slowly rolling field [including new, chaotic (Linde 1983), and double field inflation (Adams and Freese 1991; Linde 1990)], the ratio of the height to the (width)^4 of the potential must satisfy the constraint (Adams, Freese, and Guth 1991):

$$\chi \equiv \Delta V/(\Delta\phi)^4 \leq O(10^{-6} - 10^{-8}), \quad (3)$$

where ΔV is the change in the potential $V(\phi)$ and $\Delta\phi$ is the change in the field ϕ during the slowly rolling portion of the inflationary epoch. [For extended inflation, a model described below (La and Steinhardt 1989), $\chi \leq O(10^{-15})$ (Adams, Freese, and Guth 1991).] Thus, the inflaton must be extremely weakly self-coupled, with effective quartic self-coupling constant $\lambda_\phi < O(\chi)$ (e.g., $V(\phi) \simeq \frac{1}{4}\lambda_\phi\phi^4$; in realistic models, $\lambda_\phi < 10^{-12}$). Naturally, if there is only one mass scale in the problem, one expects $\chi = O(1)$.

NEW IDEAS WITH FIRST-ORDER PHASE TRANSITIONS

In order to circumvent the reheating problems of old inflation and the fine-tuning in new inflation, several new ideas have been proposed. Some of these involve returning to first-order phase transitions (as in old inflation), where bubbles of true vacuum (T) nucleate in a sea of false vacuum (F).

Consider the nucleation efficiency $\beta \equiv \Gamma_N/H^4$ (during a period of exponential expansion, the probability of remaining in the false vacuum is approximately $p(t) \simeq e^{-(4\pi/3)\beta Ht}$). Here Γ_N is the nucleation rate for true vacuum bubbles and the Hubble parameter H characterizes the expansion rate of the universe. A successful inflationary model must have β small initially, so that the universe is trapped in the metastable false vacuum for a long time and there is a sufficiently long period of inflation. Then, β must become very large, that is, nucleation must abruptly become very efficient. In this case, all of the universe can go from false to true vacuum at once. The bubbles of true vacuum are all of the same size, they can find one another and percolate, and presumably bubble collisions can thermalize the interiors. Thus one is not left with a Swiss cheese universe as in old inflation; instead, a thermal

Friedmann–Robertson–Walker universe results. This abrupt change of nucleation efficiency from very small to very large is illustrated in FIGURE 4. Guth first pointed out that a time-dependent nucleation rate would solve the reheating problems of old inflation in 1981.

One can achieve a time-dependent nucleation efficiency $\beta \equiv \Gamma_N/H^4$ by varying either the numerator or the denominator. Extended inflation (La and Steinhardt 1989), which uses Brans–Dicke gravity, has a time-dependent denominator $H = H(t)$. The scale factor initially expands exponentially, but quickly develops a power-law dependence on time. True vacuum bubbles can percolate in the sea of false vacuum, which is only expanding as power law (not exponentially as in old inflation). However, χ as defined in (**3**) must be small (Adams, Freese, and Guth 1991): $\chi \leq O(10^{-15})$. Hyperextended inflation (Steinhardt and Accetta 1990) uses more complicated couplings to gravity.

Another way to achieve a time-dependent nucleation efficiency β is to modify the numerator, that is, to have a time-dependent nucleation rate. Double-field inflation is the model described in Paper II (Adams and Freese 1991) to implement a time-dependent nucleation rate (a similar idea has been proposed by Linde 1990). In this model, there are two coupled scalar fields. The potential $V_1(\phi)$ of one field, in the absence of coupling, is much like that of old inflation (see FIG. 5). The universe passes through a first-order phase transition, and expands exponentially while it is trapped in the false vacuum. This first field ϕ is the inflaton. The potential of the second field $V_2(\psi)$, in the absence of coupling, describes a rolling field as in new inflation. The sole purpose of this second ψ field is to catalyze a change in the nucleation rate for the inflaton ϕ field. The total potential for the system is

$$V_{\text{tot}} = V_1(\phi) + V_2(\psi) + V_{\text{int}}(\phi, \psi), \tag{4}$$

where, for calculational simplicity, we take $V_{\text{int}}(\phi, \psi) = -\gamma(\phi - a)\psi^3$ for the interaction term between the two fields, although the results should hold for more generic interaction terms as well. Here, a is the minimum of the potential $V_1(\phi)$ and γ is the coupling constant between the two fields. It can be shown that the nucleation rate $\Gamma_N \propto e^{-S_E} \propto e^{-\text{const}/\epsilon_{\text{eff}}^3}$, where S_E is the Euclidean action and $\epsilon_{\text{eff}} = \rho_{\text{vac}} + 2a\gamma\psi^3$. In other words, the nucleation rate of the ϕ field has a term proportional to the value of the ψ field. When the ψ field is near the top of its potential and has a value close to

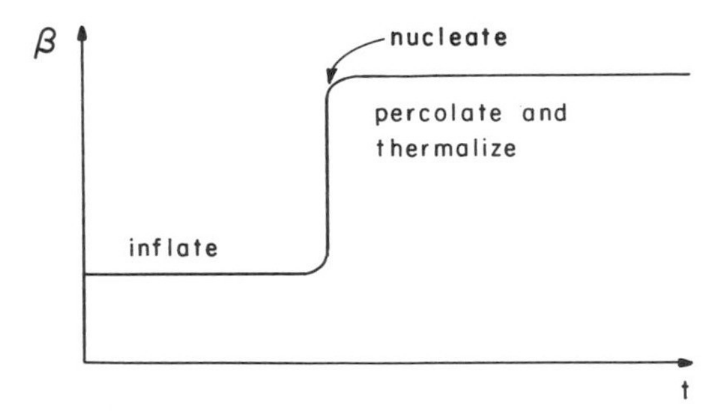

FIGURE 4. Required time dependence of nucleation efficiency β of true vacuum bubbles at first-order phase transition (for successful inflation).

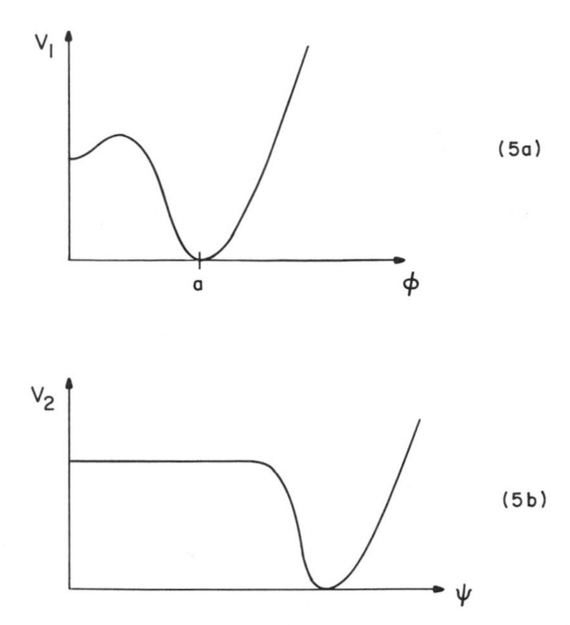

FIGURE 5. Potentials required for double field model. (**a**) The potential $V_1(\phi)$, in the absence of coupling, for the inflaton field ϕ, is much like that of old inflation. (**b**) The potential $V_2(\psi)$, in the absence of coupling, describes a rolling field as in new inflation. This purpose of this ψ field is to catalyze a change in the nucleation rate for the inflaton ϕ field.

zero, the nucleation rate of the inflation field ϕ is very small. When the ψ field nears the minimum of the potential and has a large value, then the nucleation rate suddenly gets very large and the phase transition completes. Thus one has the required β as in FIGURE 4, and the universe successfully reheats. Double field inflation obtains a "graceful exit" from inflation at a first-order phase transition with bubble nucleation.

There are several implications for large-scale structure. The same field that produces inflation can give rise to cosmic strings at the end of inflation. Inflation dilutes any preexisting cosmic strings. It is a nice feature of the double field model that cosmic strings, which may be important for structure formation, may be formed at the end of the inflationary phase transition. In addition, the collisions of true vacuum bubbles may give rise to interesting large-scale structure. (Both of these features, strings and bubble structures, also arise from extended inflation.)

However, the rolling fields in the double field model must have flat potentials [small values of χ in (**3**)] to avoid overproduction of density fluctuations.

NATURAL INFLATION WITH PSEUDO-NAMBU–GOLDSTONE BOSONS

Paper III (Freese, Frieman, and Olinto 1990) describes a model that naturally provides the flat potentials that are required for rolling fields in inflation [for either an inflationary model with only a single scalar field, which is a rolling field; or for the rolling field in the context of a double field model]. This paper takes advantage of pseudo-Nambu–Goldstone bosons [particles such as axions (Weinberg 1978; Wilczek 1978) and schizons (Hill and Ross 1988)] in particle physics that can naturally have potentials with two widely disparate mass scales for height and width.

For the past ten years, people have realized that rolling fields in inflation require flat potentials and that the parameters of the potentials must typically be fine-tuned. To reiterate, Paper I quantifies this statement, $\chi = [\Delta V/(\Delta\psi)^4] \leq 10^{-8}$. Most particle physics models require $\chi = O(1)$. But we know of a particle with a small ratio of scales: the "invisible" axion has self-coupling $\lambda_a \simeq [\Lambda_{QCD}/f_{PQ}]^4 \simeq 10^{-64}$ (Dine, Fischler, and Srednicki 1981; Wise, Georgi, and Glashow 1981). Paper III uses a potential similar to that for axions in inflation; we obtain "natural" inflation, without any fine-tuning.

The potential we obtain is of the form

$$V(\phi) = \Lambda^4[1 + \cos(\phi/f)], \tag{5}$$

as in FIGURE 6 (the potential takes this form for temperatures $T \leq \Lambda$). The height of the potential is $2\Lambda^4$, and the width of the potential is πf. Thus, the height and the width are given by two different mass scales, Λ and f. As explained below, f is the scale of spontaneous symmetry breaking of some global symmetry, and Λ is the scale at which a gauge group becomes strong. For example, for the quantum chromadynamics (QCD) axion, f is given by the Peccei–Quinn scale, $f_{PQ} \sim 10^{15}$ GeV, and Λ by the QCD scale, $\Lambda_{QCD} \sim 100$ MeV. The ratio of these two scales to the fourth power is a very small number, $\chi \sim 10^{-64}$. For inflation, we do not need a ratio quite this small, but, in order to satisfy various constraints on the model, we will need the mass scales to be higher. We can use a particle similar to the QCD axion (but not the QCD axion itself). Inflation needs $\Lambda \sim m_{GUT}$ and $f \sim m_{pl}$.

I will now briefly illustrate in what sense these potentials satisfy the criterion of

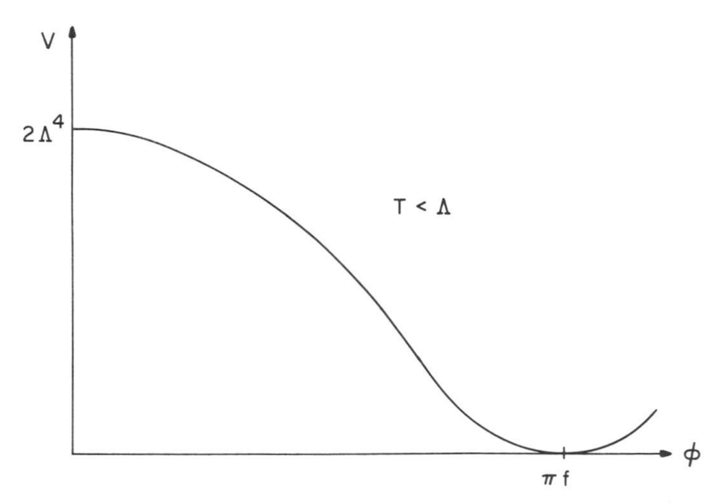

FIGURE 6. Potential of (**5**) for natural PNGB inflation; also, axion potential $V(\phi)$ for temperatures $T \leq \Lambda$.

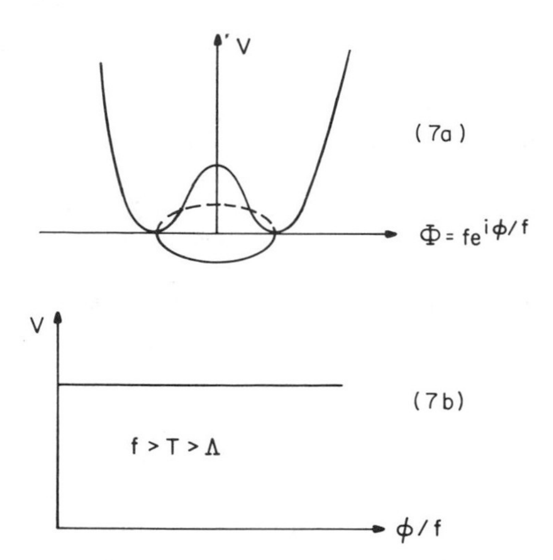

FIGURE 7. (**a**) Potential of Peccei–Quinn field for $T \leq f$. The angular degree of freedom around the bottom is the axion field ϕ. (**b**) Axion potential $V(\phi)$ for temperatures $f \geq T \gtrsim \Lambda$.

naturalness. I will use the definition of naturalness proposed by 't Hooft (1979): a small parameter α is natural if, in the limit $\alpha \to 0$, the symmetry of the system increases. I will show how the axion satisfies this criterion. [For references on axion cosmology, see Preskill, Wise, and Wilczek 1983; Abbott and Sikivie 1983; and Dine and Fischler 1983]. In order to solve the strong CP problem of QCD, Quinn and Peccei (1977) introduced a global U(1) symmetry that is broken at a scale f, that is, for $T \leq f$, the potential of the PQ (Peccei–Quinn) field is as illustrated in FIGURE 7(a). For $T \ll f$, the radial modes are frozen out because they are very massive, and the only remaining degree of freedom is the angle ϕ/f around the bottom of the Mexican hat-shaped potential. This angular degree of freedom ϕ is the axion field (Weinberg 1978; Wilczek 1978).

One can plot the value of the potential around the bottom of the Mexican hat, that is, V as a function of the angular variable ϕ/f. For temperatures $f \geq T \gtrsim \Lambda$, the potential has the same value for any choice of ϕ. This flat potential is plotted in FIGURE 7(b). The Lagrangian for the ϕ degree of freedom is just $L_\phi = \frac{1}{2}(\partial_\mu \phi)^2$, which is invariant under the transformation $\phi \to \phi + \text{const}$. In this sense, the U(1) symmetry is said to be nonlinearly realized (any point around the bottom of the Mexican hat is equivalent).

For the axion, however, this is not the entire story because of the chiral anomaly in QCD. The axion part of the QCD Lagrangian is

$$L_\phi = \frac{1}{2}(\partial_\mu \phi)^2 + \frac{g^2}{32\pi^2}\frac{\phi}{f}\, tr\,(F\tilde{F}).$$

At finite temperature, the free energy density path integral is dominated by instanton configurations.

As the temperature drops below $T \leq \Lambda$, instanton effects turn on, and the bottom of the Mexican hat develops ripples in it. Not all points in the bottom are equivalent any more. One can think of this as placing a block under one side (or under several points) of the hat, so that the potential as a function of the angular variable ϕ is as in FIGURE 6. This is the cosine potential given in (**5**). This cosine potential is invariant under the transformation $\phi \rightarrow \phi + 2\pi fN$, where N is the number of minima (ripples) of the potential [in (**5**) I have set $N = 1$]. These ripples around the bottom of the Mexican hat break the nonlinearly realized symmetry from continuous to discrete. If one were to set $\Lambda = 0$, then the potential would always retain the flat form of FIGURE 7(b) for all temperatures, rather than picking up the cosine shape from instantons. Thus, taking the limit $\Lambda \rightarrow 0$ restores a continuous symmetry. By the definition of naturalness given earlier, small Λ is therefore natural. As mentioned before, for the case of QCD axions, $(\Lambda_{QCD}/f)^4 \sim 10^{-64}$.

For the inflation, we used a scalar field with small self-coupling χ, similar to the case of axions, although the ratio of parameters does not need to be as small. Thus, we considered an axionlike model with scales Λ and f as free parameters; we found that inflation is successful for $f \sim m_{pl}$ and $\Lambda \sim m_{GUT} \sim 10^{15}$ GeV. Here, f is the scale at which a global symmetry is spontaneously broken, and Λ is the scale at which a gauge group becomes strongly interacting. These mass scales can arise naturally in particle physics models. For example, in the hidden sector of superstring theories, if a large non-Abelian group remains unbroken, the running gauge coupling can become strong at the GUT scale, that is, $\Lambda \sim m_{GUT}$; a hidden sector group that becomes strong at the GUT scale is suggested as a way to break supersymmetry in a phenomenologically viable way. In this case, the role of the PNGB inflaton might be played, for example, by the model-independent axion (Witten 1984).

Paper III considered several requirements on inflationary models with naturally small couplings provided by PNGBs. Since the potential as a function of ϕ is flat for temperatures $T \gtrsim \Lambda$, we assumed that the value of the ϕ field is initially laid down at random anywhere in the range $0 \leq \phi \leq 2\pi f$ in different causally connected regions. Once the cosine potential appears, the value of ϕ has an equal probability of starting anywhere on the potential; subsequently, ϕ rolls down the hill. We estimated the probability of success of our scenario by calculating (as a function of f) the probability of ϕ being close enough to the top of the potential to have sufficient inflation. The requirement that sufficient inflation occurred with a reasonable probability drives the value of f to be near the Planck mass m_{pl}.

We considered only the overdamped (slowly rolling) regime, ($\ddot{\phi} \ll 3H\dot{\phi}$). We required at least 65 e-foldings of inflation; we checked that we did not need to specify the initial value of the ϕ field to greater accuracy than allowed by quantum fluctuations; we checked that reheating to radiation can be successful; and we checked that spatial gradients and topological defects do not prevent the onset of inflation.

In addition, we ensured that density fluctuations are not overproduced; this requires $\Lambda \leq m_{GUT}$ [as expected, since then χ as defined in (**3**) is $\leq 10^{-12}$]. We found that the density fluctuation spectrum is non-scale-invariant, with extra power on large length scales: in Fourier space, $|\delta_k|^2 \sim k^n$, where $n \simeq 1 - (m_{pl}^2/8\pi f^2)$, for $f \leq 3m_{pl}/4$.

CONCLUSIONS

Inflation can explain the homogeneity, isotropy, flatness, oldness, and low flux of monopoles of our universe by having a period of exponential expansion in the early universe. Old inflation, which involves nucleation of true vacuum bubbles at a first-order phase transition, suffers from the problem that there is no thermalization of bubble interiors, and thus no return to the ordinary radiation-dominated universe. New inflation suffers from fine-tuning of parameters. The quest for a successful inflationary model without these problems had led to new ideas in the past few years.

In this paper, I reported on three papers:

1. The fine-tuning of parameters required in inflationary models can be quantified (Adams, Freese, and Guth 1991). Rolling fields need flat potentials; the ratio of height to $(\text{width})^4$ of the potential must be $\chi = [\Delta V/(\Delta\psi)^4] \leq O(10^{-8})$. For extended inflation, the constraint is somewhat more restrictive: $\chi \leq O(10^{-15})$.
2. Double field inflation (Adams and Freese 1991) couples two scalar fields to obtain a time-dependent nucleation rate of true vacuum bubbles, and thus a successful end to inflation with a first-order nucleating transition (as in old inflation).
3. Natural inflation with PNGBs (Freese, Frieman, and Olinto 1990) provides flat potentials (as required by Paper I) without any fine-tuning. A pseudo-Nambu–Goldstone boson (e.g., an axion), with a potential [(**5**)] that arises naturally from particle physics models, can lead to successful inflation if the global symmetry breaking scale $f \simeq m_{\text{pl}}$ and $\Lambda \simeq m_{\text{GUT}}$.

ACKNOWLEDGMENTS

I wish to acknowledge my collaborators, Fred Adams, Angela Olinto, Josh Frieman, and Alan Guth.

REFERENCES

Abbott, L. & P. Sikivie. 1983. Phys. Lett. **120B:** 33.
Adams, F. C. & K. Freese. 1991. Double field inflation. Phys. Rev. D. Jan.
Adams, F. C., K. Freese & A. H. Guth. 1991. Constraints on the scalar field potential in inflationary models. Phys. Rev. D. Feb.
Albrecht A. & P. J. Steinhardt. 1982. Phys. Rev. Lett. **48:** 1220.
Bardeen, J., P. Steinhardt & M. S. Turner. 1983. Phys. Rev. D **28:** 679.
Coleman, S. 1977. Phys. Rev. D **15:** 2929.
Dine, M. & W. Fischler. 1983. Phys. Lett. **120B:** 137.
Dine, M., W. Fischler & M. Srednicki. 1981. Phys. Lett. **104 B:** 199.
Freese, K., J. Frieman & A. Olinto. 1990. Natural inflation with pseudo-Nambu-Goldstone bosons. Phys. Rev. Lett. **65:** 3233.
Guth, A. H. 1981. Phys. Rev. D **23:** 347.
Guth, A. H. & S.-Y. Pi. 1982. Phys. Rev. Lett. **49:** 1110.
Guth, A. H. & E. Weinberg. 1983. Nucl. Phys. B **212:** 321.
Hawking, S. W. 1982. Phys. Lett. **115B:** 295.

HILL, C. T. & G. G. ROSS. 1988. Phys. Lett. **203B:** 125. HILL, C. T. & G. G. ROSS. 1988. Nucl. Phys. **B311:** 253.
LA, D. & P. J. STEINHARDT. 1989. Phys. Rev. Lett. **376:** 62; LA, D. & P. J. STEINHARDT. 1989. Phys. Lett. **220 B:** 375.
LINDE, A. D. 1982. Phys. Lett. **108 B:** 389.
LINDE, A. D. 1983. Phys. Lett. **129 B:** 177.
LINDE, A. D. 1990. Phys. Lett. **249 B:** 18.
MEINHOLD, P. & P. LUBIN. 1991. Unpublished.
PRESKILL, J., M. WISE & F. WILCZEK. 1983. Phys. Lett. **120 B:** 127.
QUINN, H. & R. PECCEI. 1977. Phys. Rev. Lett. **38:** 1440.
READHEAD, A. C. S., *et al.* 1989. Astrophys. J. **346:** 566.
STAROBINSKII, A. A. 1982. Phys. Lett. **117B:** 175.
STEINHARDT, P. J. & F. ACCETTA. 1990. Phys. Rev. Lett. **64:** 2740.
STEINHARDT, P. J. & M. S. TURNER. 1984. Phys. Rev. D **29:** 2162.
'T HOOFT, G. 1979. *In* Recent Developments in Gauge Theories, G. 't Hooft, *et al.,* Eds.: 135. Plenum. New York and London.
USON, J. M. & D. T. WILKINSON. 1985. Nature **312:** 427.
VOLOSHIN, M. B., I. YU. KOBZAREV & L. B. OKUN. 1975. Sov. J. Nucl. Phys. **20:** 644.
WEINBERG, S. 1978. Phys. Rev. Lett. **40:** 223.
WILCZEK, F. 1978. Phys. Rev. Lett. **40:** 279.
WISE, M., H. GEORGI & S. L. GLASHOW. 1981. Phys. Rev. Lett. **47:** 402.
WITTEN, E. 1984. Phys. Lett. **149B:** 351.

Relic Cosmological H II Regions and the Origin of the Lyman α Forest

AVERY MEIKSIN AND PIERO MADAU

Space Telescope Science Institute
and
Department of Physics and Astronomy
The Johns Hopkins University
Baltimore, Maryland 21218

INTRODUCTION

At epochs corresponding to $z \lesssim 1000$, the intergalactic medium (IGM) is expected to recombine and remain neutral until objects form that are capable of reionizing it. This appears to have been accomplished by $z \sim 5$ (Schneider, Schmidt, and Gunn 1989), in order to satisfy the Gunn–Peterson constraint on the amount of neutral hydrogen in the IGM. We consider the reionization of the IGM by discrete radiation sources at $z > 5$, and discuss the possibility that the Ly-α forest is a byproduct of the reionization. We suggest that several generations of sources with finite lifetimes have been active in photoionizing the IGM since the Compton era ($z \sim$ 10–15), achieving complete reionization by $z \sim 5$. Plausible candidate sources we examine are primeval galaxies and quasars. It is expected that these sources are short-lived: galaxies are likely to undergo an early burst of star formation, and quasars may dim with time. These sources would then not be able to maintain the advance of their ionization fronts indefinitely. The high pressure of the resulting relic H II regions will drive shocks into the cold, neutral ambient medium. We discuss the hydrodynamical evolution of relic cosmological H II regions generated at redshifts $z \sim$ 5–15. The shock wave surrounding each H II region will sweep the surrounding material into a thin shell. The peculiar velocity of the shell is typically 10–20 km s^{-1}. The dense shell may cool and fragment into objects with baryonic masses as high as 10^6–$10^8\,M_\odot$ and column densities of order 10^{19} cm^{-2}. These values are characteristic of the photoionization process, and are independent of the nature of the photoionizing source. They are set only by the density of the IGM, the sound speed of 10^4 K gas, and the age of the shell. The clouds formed may account for the Ly-α absorption systems seen in the spectra of high redshift quasi-stellar objects (QSOs). The shock propagation ceases once the filling factor of the H II regions reaches unity and the IGM is fully photoionized. For a filling factor of the H II regions near unity, the cloud autocorrelation function will be small if the ionizing sources are randomly distributed, in agreement with the observed absence of correlations in the Ly-α forest. A fuller account is provided in Madau and Meiksin (1991).

COSMOLOGICAL H II REGIONS

The evolution of an expanding cosmological ionization front generated by a point source of ionizing radiation is governed by the equation

$$4\pi r_I^2 n_{\rm H}\left(\frac{dr_I}{dt} - Hr_I\right) = S - \frac{4\pi}{3} r_I^3 n_{\rm H}^2 \alpha_B (1 + \chi) \tag{1}$$

(Shapiro 1986), where r_I is the proper radius of the I-front, H is the Hubble constant, S is the number of ionizing photons emitted per second from the central source, n_H is the intergalactic number density of hydrogen, α_B is the total recombination coefficientto the excited states of hydrogen, and $\chi = 0.2$ corrects for helium. The right-hand side of (1) vanishes at the Strömgren radius $r_{St} = \{3S/[4\pi n_H^2 \alpha_B(1 + \chi)]\}^{1/3}$.

We consider two sources of ionizing radiation: QSOs and primeval galaxies. In the case of QSOs, we generalize the constant luminosity result of Shapiro and Giroux (1987) and Donahue and Shull (1987) by taking the ionizing flux to decrease with redshift z as a power-law of the form $S_Q = S_{in}(x/x_{in})^q$, where $x = 1 + z$, $x_{in} = 1 + z_{in}$, z_{in} is the redshift at which the source turns on, $\alpha = 3S_{in}/(4\pi n_0 H_0)$, and $\beta = n_0\alpha_B(1 + \chi)/H_0$). Here $n_0 = (8.08 \times 10^{-6}\ \mathrm{cm}^{-3})\Omega_{IGM}h^2$ is the current number density of hydrogen, H_0 is the present value of H, $h = H_0/100\ \mathrm{km\ s^{-1}\ Mpc^{-1}}$, and Ω_{IGM} is the ratio of the IGM density to the closure density. The cosmological generalization of the Strömgren radius is $r_{St} = (\alpha/\beta)^{1/3} x^{q/3-2}/x_{in}^{q/3}$. For $r_I \lesssim 0.5 r_{St}$, (1) can be integrated directly with good accuracy to obtain ($K = Ht$ for $\Omega = 0$ or 1),

$$r_I \simeq \left(\frac{\alpha}{q + 3K/2 - 5/2}\right)^{1/3} x_{in}^{K/2-5/6} x^{-1} \left[1 - \left(\frac{x}{x_{in}}\right)^{q+3K/2-5/2}\right]^{1/3}. \tag{2}$$

The Strömgren sphere of a rapidly dimming quasar ($q > 3$) shrinks in comoving coordinates: the density decrease associated with the Hubble expansion cannot balance the decrease in ionizing photons. Once the peculiar velocity of the ionization front slows to $\sim 40\ \mathrm{km\ s^{-1}}$, corresponding to a near crossing of the front and the Strömgren sphere, a shock is driven into the neutral gas of the IGM. For the case $\Omega = 1$ and $q = 4$, the comoving radius of the ionized bubble is $R_0 = (1 + z_s)r_I(z_s) \simeq$ (23–29 Mpc)$h_{0.5}^{-1}(\Omega_{IGM}/0.1)^{-1/3}(S_{in}/10^{56}\ \mathrm{s}^{-1})^{1/3}$, where $h_{0.5} \equiv h/0.5$, z_s is the redshift at which the shock occurs, and the coefficient has been adjusted to match numerical integration including recombinations for x_{in} in the range 9–13. For $\Omega = 0$, R_0 is 20 percent larger. In FIGURE 1(a) the evolution of r_{St} and r_I is displayed for the case $q = 4$, $\Omega = 0$ (including recombinations).

The ionization history of a young galaxy undergoing a burst of star formation can also be determined, since the observed metallicity of a typical galaxy can be related to the total energy that has been emitted in ionizing radiation. The efficiency of mass conversion into ionizing radiation can then be estimated to be $\epsilon \simeq 2 \times 10^{-5}$ (Miralda-Escudé and Ostriker 1990). In the case of an exponential star formation rate with decay timescale τ, a galactic spheroid (disk halo or elliptical) of mass M_G will provide a time-dependent ionizing photon luminosity equal to

$$S_G \simeq (10^{54}\ \mathrm{s}^{-1}) \left(\frac{M_G}{10^{11} M_\odot}\right) \left(\frac{\epsilon}{2 \times 10^{-5}}\right) \left(\frac{\tau}{1\ \mathrm{Gyr}}\right)^{-1} \left(\frac{f}{0.2}\right) \exp\left(\frac{-\Delta t}{\tau}\right), \tag{3}$$

where f is an absorption factor, that is, the fraction of radiation that escapes from the Galaxy. In FIGURE 1(b) we show the evolution of the ionization front and Strömgren radius of the HII region associated with such a galaxy. We present two cases, one with $\tau = 0.3$ Gyr and the other with $\tau = 0.8$ Gyr. We take $\Omega = 0$, and turn the galaxies on at $z = 10$ in both cases. The resulting HII regions shock at $z = 7.7$ and $z = 5.3$, respectively. The total energy radiated is held fixed for both cases and corresponds to an HII region with a comoving radius of $8h_{0.5}^{-1}$ Mpc.

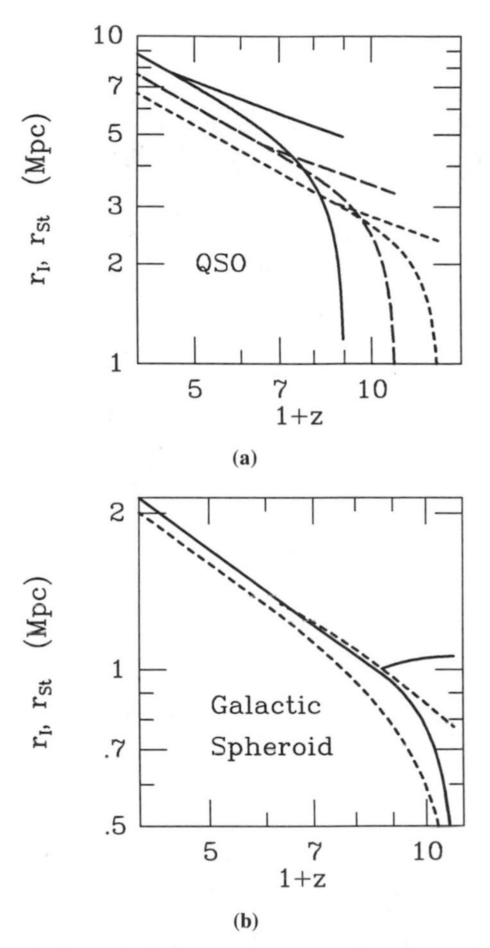

FIGURE 1. (**a**) The evolution of the ionization front r_I and Strömgren radius r_{St} as a function of redshift for a QSO with $S_{in} = 10^{56}$ s^{-1} and $q = 4$. ($\Omega = 0$, $\Omega_{IGM} = 0.1$, $h = 0.5$.) The curves of slope $-2/3$ represent r_{St}, while the curves of varying slope represent r_I. When these radii cross, the central source can no longer maintain the advance of the I-front and a shock is driven into the IGM. The three families of curves correspond to initial redshifts of $z_{in} = 8$ (*solid line*), 10 (*long-dashed line*), and 12 (*short-dashed line*). (**b**) Same as (**a**), except for a primeval galactic spheroid evolving according to (**3**) of the text. The two families of curves correspond to the star formation timescales $\tau = 0.3$ (*solid line*) and 0.8×10^9 yr (*short-dashed line*). The curves converge to the same comoving radius (8 Mpc), since they correspond to the same total energy input from the Galaxy (see text).

IMPACT OF THE RELIC HII REGION

Flow Structure

We wish to obtain an estimate for the mass that will be swept up by the shock driven into the IGM by the relic HII region, as well as estimates for the thickness and column densities of the swept-up material. We do so by approximating the HII region as a uniform sphere of hot gas that expands into the surrounding medium due to its high overpressure. We treat the separation of the hot gas and the IGM as a discontinuity. The expanding gas occupies only a thin layer compared with the radius of the HII region. In this approximation, the expansion behaves as a one-dimensional cosmological shock tube, and can be solved analytically.

Two types of behavior for the evolution of the discontinuity are possible. For a sufficiently low flow speed within the hot driver gas, the discontinuity will split into a shock wave that propagates into the low-pressure external medium, and a rarefaction wave that moves into the hot driver gas. The flow is composed of four regions: (1) the preshock IGM, (2) the postshock gas, (3) the rarefaction wave, and (4) the relic HII

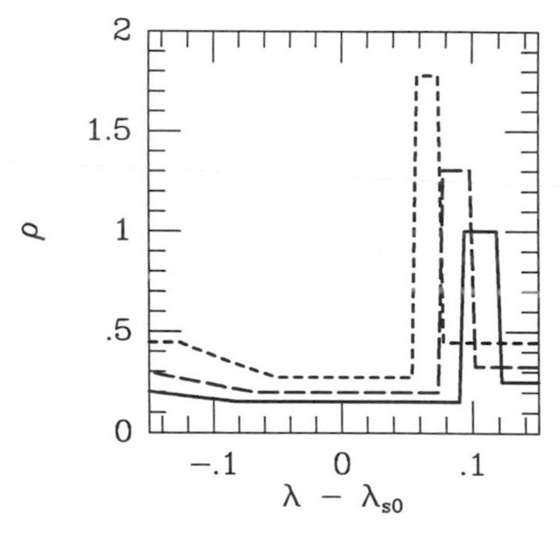

FIGURE 2. Evolution of physical gas density ρ as a function of comoving coordinate $\lambda - \lambda_{s0}$, where λ_{s0} is the comoving coordinate of the initial pressure discontinuity. The density is shown at the times $t/t_s = 1.5, 1.75$, and 2 for the case $\Omega = 1$, where t_s is the time of shock formation. A rarefaction wave moves into the driver gas while a shock front propagates into the IGM. Note that the density decreases uniformly throughout the flow structure like t^{-2}. The density normalization is arbitrary.

region driving the shock. The spatial behavior of the fluid variables for this case is shown in FIGURE 2. The shock wave and rarefaction wave propagate in opposite directions from the initial (pressure) discontinuity, and are separated by a region of uniform flow (within regions 2 and 3). The contact discontinuity is stationary with respect to the gas. Behind the strong shock (region 2) the neutral gas density rises to a value that permits exact pressure balance at the discontinuity, despite the lower temperature in that region.

A second type of flow occurs when the outward velocity of the undisturbed driver gas is sufficiently large, and a wind-driven type of solution occurs. In this case, the pressure decline in the rarefaction wave cannot be maintained. Instead, the original discontinuity splits into two oppositely moving shock waves, which again are separated by a region of uniform flow. The pressure of the gas across both postshock regions is constant. The flow divides into the same regions as in the previous case, except that region 3 now refers to the postshock gas behind the shock front propagating into the HII region.

The simplicity of the flow structures is intimately related to the adiabatic decrease of the sound speed inside and outside the relic HII region. The $(1 + z)$ cosmological dependence of the sound speed allows the solution to contain regions of uniform flow, where pressure gradients are absent. This can be readily demonstrated as follows. The velocity of the material in the shell with respect to the center of the HII region is given by

$$\frac{dr}{dt} = \frac{Kr}{t} + v_s \left(\frac{t}{t_s}\right)^{-K}, \tag{4}$$

where t_s is the time of shock formation, and v_s is the peculiar velocity of the gas at $t = t_s$, constant through the region of uniform flow. Integrating with respect to time we have

$$r = r_s \left(\frac{t}{t_s}\right)^K + \frac{v_s t}{2K - 1}\left(\frac{t}{t_s}\right)^{-K}\left[\left(\frac{t}{t_s}\right)^{2K-1} - 1\right]. \tag{5}$$

The effective gravitational field in the frame of the shock is

$$g_{\rm eff} = -\frac{d^2r}{dt^2} - \frac{\Omega}{2}H^2 r = \frac{r}{t^2}\left[(1 - K)K - \frac{\Omega}{2}K^2\right] = 0. \tag{6}$$

The absence of pressure gradients in the shell implies zero net acceleration: the shell propagates with constant velocity in a $\Omega = 0$ universe, and decelerates due to the gravitational pull of the matter interior to it in a $\Omega = 1$ universe. *The shell is in free-fall.* This behavior is peculiar to a solution in which the temperature declines due to adiabatic expansion alone, and affects the stability properties of the flow pattern. More generally, we expect different cooling mechanisms to set up pressure gradients in the shocked gas.

The Structure of the Thin Shell

The amount of shocked material in the shell depends on the initial sound speed in the HII region at the time of the shock. We determine the temperature assuming an IGM with cosmological abundances (He/H = 0.1), and including hydrogen and helium recombination radiation, thermal bremsstrahlung, and Compton cooling of electrons off the cosmic background radiation photons. We find that for $\Omega_{\rm IGM}h^2 = 0.03$ today, at $z = 10$ the temperature of the HII region is 1.6×10^4 K, corresponding to an adiabatic sound speed of 20 km s^{-1}. For $\Omega = 0$, we find that the mass swept up by the shock,

$$M_s \simeq (1.8 \times 10^{13}\, M_\odot) h_{0.5} \left(\frac{\Omega_{\rm IGM}}{0.1}\right)\left(\frac{R_0}{30\ {\rm Mpc}}\right)^2 \left(\frac{c_{4s}^{ad}}{20\ {\rm km\ s^{-1}}}\right)\left(1 - \frac{x}{x_s}\right), \tag{7}$$

is concentrated in a thin shell of thickness

$$\Delta r_s \simeq (9.8\ \text{kpc})\, h_{0.5}^{-1} \left(\frac{c_{4s}^{ad}}{20\ \text{km s}^{-1}}\right)\left(\frac{x}{6}\right)^{-1}\left(1 - \frac{x}{x_s}\right), \tag{8}$$

and hydrogen column density, $N_H = 4n_H \Delta r_s$, given by

$$N_H \simeq (5.3 \times 10^{18}\ \text{cm}^{-2}) h_{0.5} \left(\frac{\Omega_{IGM}}{0.1}\right)\left(\frac{c_{4s}^{ad}}{20\ \text{km s}^{-1}}\right)\left(\frac{x}{6}\right)^2\left(1 - \frac{x}{x_s}\right). \tag{9}$$

The results are comparable for $\Omega = 1$ and $5 < z < 10$.

THE FORMATION OF THE LYMAN α FOREST

The overdense shell will cool as a result of the formation of H_2. The density in the shell may then become sufficiently high for self-gravitating unstable modes to grow. The characteristic unstable fragment mass is

$$M_c \simeq (1.7 \times 10^8\, M_\odot) h_{0.5}^{-1} \left(\frac{\Omega_{IGM}}{0.1}\right)^3\left(\frac{c_{4s}^{ad}}{20\ \text{km s}^{-1}}\right)^3\left(\frac{x}{6}\right)^2\left(1 - \frac{x}{x_s}\right)^3, \tag{10}$$

for $\Omega = 0$. It is a factor of 100 smaller for $\Omega = 1$. Irrespective of whether the shocked gas is gravitationally unstable, the cooling shell may be subject to hydrodynamical instabilities. We estimate a characteristic fragment mass of $\sim 10^7\, M_\odot$. In either case, the sizes of the fragments will be 5–10 kpc, and their total hydrogen column densities will be $\sim 10^{19}$ cm^{-2}. These properties are intrinsic to the fragmentation of the shell and are independent of the properties of the radiation source that generated the HII region.

The cloud parameters are consistent with those estimated by Foltz, Weymann, and Roser (1984) from the observations of the lensed double quasar QSO2345 + 007A,B, assuming the clouds to be in photoionization equilibrium with the background UV radiation. It should be noted, though, that the clouds may be considerably modified from the time of formation at $z > 5$ to the time of observation at $z \sim$ 2–4; for instance, as a result of reionization at $z \lesssim 5$. It is possible that only clouds in which self-gravity is important will survive to $z < 4$, where they are observed, those remaining expanding until their column densities are below the threshold of detection.

We can estimate the fraction of the IGM processed into Ly-α clouds by considering the comoving volume that could be ionized by a source that puts out a total energy ΔE in ionizing radiation. Whenever the timescale for full recombination is longer than the cosmological timescale, a simple relation can be derived between the (comoving) volume V of the HII bubble and ΔE

$$\frac{V}{\Delta E} = \frac{56\pi G m_p}{15 H_0^2 \Omega_{IGM} e_I} = (2.3 \times 10^{17}\ \text{cm}^3 \text{erg}^{-1}) h_{0.5}^{-2} \left(\frac{\Omega_{IGM}}{0.1}\right)^{-1}, \tag{11}$$

where $e_I = 13.6$ eV, and we assumed a monochromatic ionizing emissivity. Giant HII regions with comoving radii of 30 Mpc will be generated with an input of 1.5×10^{61}

erg, an energy requirement that can easily be met by quasars. From the metallicity argument made previously, the "sphere of influence" of the spheroid of a young star-forming galaxy will be

$$R_G = (11\ \text{Mpc})h_{0.5}^{-1}\left(\frac{M_G}{10^{11}M_\odot}\right)^{1/3}\left(\frac{\epsilon}{2\times10^{-5}}\right)^{1/3}\left(\frac{f}{0.2}\right)^{1/3}. \qquad \textbf{(12)}$$

The fraction of the IGM that could be processed by the shocks is given by

$$\frac{\Omega_s}{\Omega_{\text{IGM}}} = 12\frac{\Delta r_s}{r_B}Q_{\text{HII}}, \qquad \textbf{(13)}$$

where Ω_s pertains to the shocked material and $Q_{\text{HII}} = (4/3)\pi r_B^3 n_B$ is the porosity parameter of the relic HII bubbles of proper radius r_B and number density n_B. For $z_s \sim 9$, $\Delta r_s \sim 4$ kpc at $z = 5$. Then $Q_{\text{HII}}^{-1}\Omega_s/\Omega_{\text{IGM}} \sim 4 \times 10^{-2}$ and 10^{-2} for shells with comoving radii of 8 and 30 Mpc, respectively.

More generally, galaxies will form continuously with redshift. By choosing a specific galaxy formation rate (GFR) $\psi(z)$, we can compute the evolution of $\Omega_s/\Omega_{\text{IGM}}$ with redshift. Following Miralda-Escudé and Ostriker (1990), we model the GFR as a Gaussian in redshift

$$\psi(z)\,dz \propto \exp\left[-\frac{1}{2}\left(\frac{z-z_f}{w}\right)^2\right]dz. \qquad \textbf{(14)}$$

We consider three cases: $z_f = 8, w = 4, \Omega = 0$; $z_f = 4, w = 2, \Omega = 0$; and $z_f = 4.5, w = 2, \Omega = 1$. Miralda-Escudé and Ostriker show that the last case is a very good fit to the star formation rate expected in the cold dark matter (CDM) scenario, as computed by Carlberg and Couchman (1989). The shocked fraction of the IGM for these cases is shown in FIGURE 3 (ignoring overlapping of the HII regions). Current Gunn–Peterson limits require the IGM to be ionized by $z \lesssim 5$. The shocks will then dissipate into sound waves by this epoch, so we do not plot the shocked fraction for $z < 5$. At this time, we find $\Omega_s \sim 4 \times 10^{-2}\Omega_{\text{IGM}}$ for the case with $z_f = 8$, $\Omega_s \sim 2 \times 10^{-3}\Omega_{\text{IGM}}$ for $z_f = 4$, and $\Omega_s \sim 9 \times 10^{-4}\Omega_{\text{IGM}}$ for the CDM case.

The observed Ly-α cloud distribution in redshift along a given line of sight, with rest-frame equivalent width greater than 0.36 Å, can be expressed as $dN_c/dz \simeq 3(1+z)^{\gamma_0}$, where $\gamma_0 \approx 2.3$ (Murdoch *et al.* 1986). If the clouds have a typical diameter of $D = 10D_{10}$ kpc, and are photoionized by a metagalactic flux at the Lyman limit of $J = J_{-21}10^{-21}$ erg cm^{-2} s^{-1} sr^{-1} Hz^{-1} to a temperature of $T = 3 \times 10^4 T_{30}$ K, we estimate the fraction of the critical density contained in Ly-α absorbers as

$$\Omega_{\text{Ly-}\alpha} \sim 5\times10^{-4}\left(\frac{1+z}{4.5}\right)^{1.8} D_{10}^{1/2}J_{-21}^{1/2}T_{30}^{3/8}h_{0.5}^{-1} \qquad \textbf{(15)}$$

in an $\Omega = 0$ universe, and twice as large for the $\Omega = 1$ case. For $\Omega_{\text{IGM}} \sim 0.1$, this is comparable to the amount predicted for the $z_f = 4$, $w = 2$ case. For the $z_f = 8$, $w = 4$ model, $\Omega_{\text{IGM}} \sim 0.01$ would give an Ω_s that matches $\Omega_{\text{Ly-}\alpha}$. For the CDM scenario, however, Ω_{IGM} would have to be unreasonably large. For QSO-created HII regions that shock at $z \sim 9$, it follows from **(13)** that the amount of the IGM that could be processed by the shocks is comparable to $\Omega_{\text{Ly-}\alpha}$ for shells with comoving radii $\lesssim 30$

Mpc. Thus the driving radiation sources could have been as powerful as quasars with $S_{in} \simeq 10^{56}\ s^{-1}$ and still have created a sufficient density of Ly-α clouds to be consistent with observations before the shells began to overlap.

No two-point line-of-sight correlations in the redshift distribution of the Ly-α forest have been detected on velocity scales >300 km s^{-1} (Sargent *et al.* 1980). On smaller scales, 50–300 km s^{-1}, weak clustering has been reported by Webb (1987). In our model, the clouds will be correlated because they form on shells, even if the seeds for the HII regions are uncorrelated. If the line of sight intersects a cloud at redshift z, then there will be an excess number of clouds within a range Δz of z resulting from

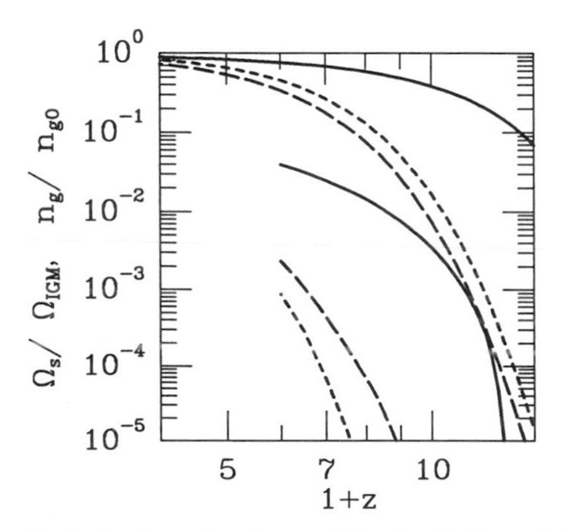

FIGURE 3. Fraction of galaxies formed, n_g/n_{g0}, and fraction Ω_s/Ω_{IGM} of the IGM shocked by relic HII regions generated by the galaxies, as a function of $1 + z$. Three galaxy formation rate cases are shown: $z_f = 8$, $w = 4$, $\Omega = 0$ (*solid lines*), $z_f = 4$, $w = 2$, $\Omega = 0$ (*long-dashed lines*), and $z_f = 4.5$, $w = 2$, $\Omega = 1$ (*short-dashed lines*) (see text). The last case corresponds to a universe dominated by cold dark matter. The curves for Ω_s/Ω_{IGM} are truncated at $z = 5$, since the IGM is ionized for $z < 5$ and the shocks cease to propagate. We assume a decay time for the galaxies of $\tau = 0.3$ Gyr and a luminosity corresponding to $8h_{0.5}^{-1}$ Mpc for the comoving radius R_0 of the relic HII regions, for the cases with $\Omega = 0$. For the $\Omega = 1$ CDM case, we assume $\tau = 0.1$ Gyr and $R_0 = 9h_{0.5}^{-1}$ Mpc.

the same bubble. Following Babul (1991), the line-of-sight autocorrelation function ζ of clouds distributed on comoving shells is, for $l < 2r_B$,

$$\zeta(l) = \frac{1}{6Q_{\rm HII}}\frac{r_B}{l}. \tag{16}$$

The relation between Δz and l is $\Delta z = (1 + z)Hl/c$. Although in this derivation the bubbles are assumed not to overlap ($Q_{\rm HII} < 1$), **(16)** suggests the correlations will be very weak for $Q_{\rm HII} \gtrsim 1$. We therefore expect the Ly-α clouds to be distributed very nearly like a Poisson process on scales $l > r_B/6$. The absence of correlations on scales greater than 300 km s^{-1} at redshifts as large as 3.5 suggests $(1 + z)r_B \lesssim 36h_{0.5}^{-1}$ Mpc for $\Omega = 0$, and $(1 + z)r_B \lesssim 17h_{0.5}^{-1}$ Mpc for $\Omega = 1$.

A porosity of order unity arises naturally in the galaxy scenarios described earlier. By $z = 5$, 30–75 percent of the present-day galaxies will have formed. For a present-day density of bright galaxies of $10^{-3}h_{0.5}^3$ Mpc^{-3}, and a comoving bubble radius of $8h_{0.5}^{-1}$ Mpc, the porosity parameter at this epoch is $Q_{\rm HII} \sim 1$–2. Although recombinations within the relic HII regions may be too numerous to permit satisfaction of the Gunn–Peterson limit for $z < 5$ (a topic beyond the scope of this work), the pressure of the IGM will have been raised to a level that the shocks will decay into sound waves: there no longer will be a low-pressure medium present for the shocks to travel in. No further Ly-α clouds will then form.

REFERENCES

BABUL, A. 1991. Mon. Not. R. Astron. Soc. **248:** 177.
CARLBERG, R. G. & H. M. P. COUCHMAN. 1989. Astrophys. J. **340:** 47.
DONAHUE, M. J. & J. M. SHULL. 1987. Astrophys. J., Lett. **323:** L13.
FOLTZ, C. B., R. J. WEYMANN & H. ROSER. 1984. Astrophys. J., Lett. **281:** L1.
MADAU, P. & A. MEIKSIN. 1991. Astrophys. J. **374:** In press.
MIRALDA-ESCUDÉ, J. & J. P. OSTRIKER. 1990. Astrophys. J. **350:** 1.
MURDOCH, H. S., R. W. HUNSTEAD, M. PETTINI & J. C. BLADES. 1986. Astrophys. J. **309:** 19.
SARGENT, W. L. W., P. J. YOUNG, A. BOKSENBERG & D. TYTLER. 1980. Astrophys. J. Suppl. **42:** 41.
SCHNEIDER, D. P., M. SCHMIDT & J. E. GUNN. 1989. Astron. J. **98:** 1507.
SHAPIRO, P. R. 1986. Publ. Astron. Soc. Pac. **98:** 1014.
SHAPIRO, P. R. & M. L. GIROUX. 1987. Astrophys. J., Lett. **321:** L107.
WEBB, J. K. 1987. *In* IAU Symposium 124: Observational Cosmology, A. Hewitt, G. Burbidge, and L. Fang, Eds.: 803. Reidel. Dordrecht, the Netherlands.

Inflationary Axion Cosmology

D. H. LYTH

Department of Physics
University of Lancaster
Lancaster LA1 4YB, England

INTRODUCTION

We have revisited inflationary axion cosmology, which was the subject of intense investigation in the early 1980s. The present treatment involves fewer assumptions than in the past, and we arrive at a scenario specified by the values of three parameters f_a/N, $N\bar{\theta}$, and $N\sigma_\theta$ ($\sqrt{2}f_a$ is the vacuum value of of the modulus of the Peccei–Quinn field, $\bar{\theta}$ and σ_θ are the mean and rms dispersion of its phase θ just before the axion mass switches on, and N is the number of distinct vacuum values of θ once the mass has turned on). From the observational bounds on the large-scale microwave background anisotropy we exclude part of the parameter space, and on this basis arrive at the following conclusions. First, axionic domain walls can be cosmologically interesting only if f_a/N is within an order of magnitude of its extreme astrophysical bound lower bound 2×10^8 GeV. Second, the axion density perturbation can be either Gaussian or of the χ^2 type, but the latter case is likely only if $f_a/N \lesssim 10^{10}$ GeV. Third, at least in the absence of walls the axion density perturbation can probably not become big enough to be the cause of the observed structure, though the non-Gaussian case requires further investigation. Finally, we make the additional assumption that interactions of the Peccei–Quinn field do not alter the effective value of f_a, while relevant scales leave the horizon during inflation. This leads to the strong bound on the inflationary energy density. It rules out the three currently favored mechanisms for structure formation, namely an adiabatic density perturbation originating as a quantum fluctuation of the inflation field, cosmic strings, and textures. In common with previous discussions, the present work assumes that while relevant scales leave the horizon during inflation, the Peccei–Quinn field is sitting in its vacuum. The conditions under which this is true require further investigation.

THE ASSUMPTIONS

The present treatment of axion cosmology is described in references 1–3, and was done partly in collaboration with A. D. Linde. By now there is a vast literature on the axion, which is reviewed, for example, in references 4 and 5. A crucial parameter is f_a/N, where $\sqrt{2}f_a$ is the vacuum value of the modulus of the canonically normalized Peccei–Quinn field, and N is the number of vacuum values of its phase θ. The astrophysics of red giants and of supernova 1987A, plus the fact that the axion is not seen in laboratory experiments, leads to a lower bound on f_a/N, which taking all the uncertainties into account[6,7] is $f_a/N > 2 \times 10^8$ GeV. As a result the axion has a very small mass, $m_a < 3 \times 10^{-2}$ eV. It should be emphasized that this bound can be saturated only if there is nonstandard astrophysics;[7] otherwise, the bound becomes $f_a/N > 2 \times 10^9$ GeV, and even the latter bound can be achieved only with particular assumptions about the axion couplings, generic assumptions leading to $f_a/N \gtrsim 10$ GeV. We shall see that the precise bound is crucial for axion cosmology. (There may be in addition a very narrow window at $f_a/N \sim 10^5$ GeV,[8] but we shall not consider that possibility.)

Our interest in this work is in the cosmology. There are two fundamentally different cases, depending on whether or not the temperature after inflation exceeds the value, of order f_a, that is necessary to restore the symmetry. If it does, we have the "thermal scenario," in which the axion field retains no memory of inflation. In order not to overproduce axions one definitely requires $f_a/N < 10^{12}$ GeV[9] (unless decaying dark matter generates extra entropy[10]), and some authors put the bound 100 times lower.[11] In the latter case there is at best a narrow window for the thermal scenario.

If the temperature after inflation is too low to restore the symmetry we have an "inflationary scenario." There are several possible scenarios, most of which have yet to be explored. Existing discussions make a large number of usually implicit assumptions. Here we will try to make all the assumptions explicit, and at the same time reduce their number somewhat.

We exclude a significant breaking of the Peccei–Quinn symmetry at the classical level.[12] We make the standard assumption that the density of the universe is critical, this being a natural consequence of inflation, though not mandatory,[13] and we assume that the cosmological constant is zero. This implies the existence of nonbaryonic dark matter;[14] we do not assume that it consists entirely, or even mostly, of axions, but we do assume that it is cold. We also assume that it is stable, which rules out variants of the axion scenario described in reference 10. We assume standard ("new") inflation, as opposed to extended inflation and its variants (for reviews of them, see reference 14). During inflation, nonzero modes of the inflation field and of the axion field are assumed to be in the conformal vacuum.

More assumptions are needed during inflation, while relevant scales leave the horizon. During inflation, f_a^2 may acquire a time-dependent effective value $f_a^2(t)$ through the coupling of the axion field to the inflation field or to the gravitational field.[15,2] We assume that $f_a^2(t) > 0$ so that the Peccei–Quinn symmetry is not restored. (If it is, one has a scenario similar to the thermal one, whose analogue for the case of global strings has been considered in references 16 and 17.) We assume also that the Peccei–Quinn field is sitting at the vacuum value $f_a(t)$; the conditions for this are under investigation[18] and within the chaotic paradigm, and it seems that a necessary condition is

$$\frac{H(t)}{f_a(t)} \lesssim 1, \tag{1}$$

where H is the Hubble parameter. This being so, the only degree of freedom is θ, whose nonzero modes are assumed to be in the conformal (Bunch–Davies) vacuum.[19]

This means that it is Gaussian,[a] with a spectrum given after horizon exit by,[19]

$$P_\theta^{1/2}(k) = \frac{1}{2\pi}\frac{H(t)}{f_a(t)}, \tag{2}$$

with t the epoch of horizon exit.

The final simplification to be made is that $f_a(t)$ changes only slowly while relevant scales leave the horizon, so that it has the practically scale independent value f_{a1}, where the subscript 1 denotes the epoch when the observable universe leaves the horizon. One also has $H(t) \simeq H_1$ because the inflation is exponential, so the spectrum is flat on all relevant scales. This is the case explored in some detail in the present work. The opposite case, in which the spectrum may vary sharply with scale, has not been explored in detail, though its existence has been recognized.[15]

After inflation, f_{a1} goes to its late time value f_a if it is not already there. The axion field is specified by θ and its mass is zero until quantum chromodynamics (QCD) effects begin to switch it on at $T \sim 1$ GeV. We denote the value of θ at this epoch by $\tilde{\theta}$, and similarly for other quantities. We are interested in the stochastic properties of $\tilde{\theta}$ within a comoving region that is now an order of magnitude bigger than the observable universe, so that we can evaluate its effect on the large-scale microwave background anisotropy. There is some mean value $\bar{\tilde{\theta}}$, and a Gaussian distribution with some rms dispersion[2,3] $\sigma_\theta \simeq 4P_\theta^{1/2}$. After the axion mass begins to switch on, θ rolls down to the minimum of its potential, except at those surfaces in space where $N\tilde{\theta} = \pi$ (modulo 2π) and domain walls form.[2] After it has fully switched on there are axions with some mass density, plus possibly domain walls, whose cosmological effect we need to determine.

THE RESULTS

Without loss of generality we take $0 < N\bar{\theta} < \pi$. Then the most numerous wall-forming surfaces are at $N\tilde{\theta} = \pi$, corresponding to ν standard deviations from the mean, where

$$\nu \equiv (1 - N\bar{\theta}/\pi)(N\sigma_\theta/\pi)^{-1}. \tag{3}$$

In FIGURE 1 is shown the line $\nu = 1$. The region above this line corresponds to $\nu < 1$, which means that walls are very common. Furthermore, the Euler number density is negative in this regime,[21] and there is typically a wall stretching across every region, no matter how large. Such a wall is a cosmological disaster, so we can rule out the regime $\nu < 1$.

In contrast, when ν exceeds some critical value ν^* the domain walls will occur as extremely rare bags and have no observable effect. This is the case dealt with in the present work. A quantitative estimate of ν^* and cosmological effects of the bags when $1 < \nu < \nu^*$ will be given in a future work, but since we are dealing with a Gaussian distribution we do not expect ν^* to be very large. In FIGURE 1 is shown the

[a] Allen, Grinstein, and Wise have considered non-Gaussian effects,[20] arising from a fourth-order derivative axion coupling, but they are small as long as (1) is satisfied.

line $\nu = 3$ corresponding to walls at three standard deviations, for which the volume enclosed by the bags is already only 0.0028 of the total, and it is probably safe to assume that ν^* will not be much bigger than 3.

We shall see in a moment that observational limits on the large-scale microwave background anisotropy give the bound on $N\sigma_\theta$ shown in FIGURE 1. Any effect of the walls could only tighten this bound, so we see that the walls can be significant only if $f_a/N \lesssim 10^9$ GeV.

In the absence of walls, we can calculate the "initial" axion mass density ρ_a^{in}, evaluated just after the mass has fully switched on. At each point in space ρ_a^{in} depends

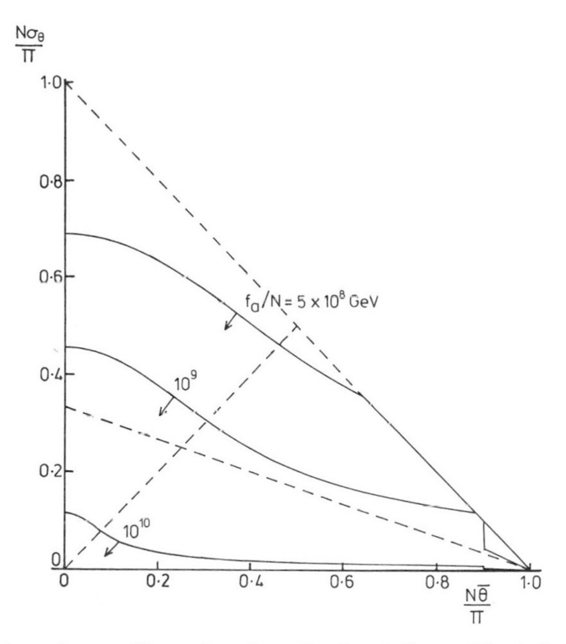

FIGURE 1. The bound on σ_θ. The region above the *dotted diagonal line* is forbidden because there would be a domain wall across the observable universe. Below the other *dotted line,* walls will be very rare because they correspond to fluctuations of $\tilde{\theta}$ by more than three standard deviations. Limits on the microwave background anisotropy exclude a region that depends on f_a/N as indicated.

only on $\tilde{\theta}$ at the same point, and is given by reference 22

$$a^3\rho_a^{in} = \tfrac{1}{2}\,(f_a/N)^2 m_a \tilde{m}_a \tilde{a}^3 N^2 \tilde{\theta}^2 y(N\tilde{\theta}), \tag{4}$$

where a is the scale factor of the universe and m_a is the axion mass. The formula applies in the interval $0 < N\tilde{\theta} < \pi$, with ρ_a^{in} having the same periodicity as the axion potential U outside this range, and its normalization is uncertain by a factor of a few. The function y accounts for deviations of the potential from the quadratic form, as one moves away from $\tilde{\theta} = 0$. It is equal to 1 at $\tilde{\theta} = 0$, and increases monotonically, rising to 2 at $N\tilde{\theta}/\pi \simeq 0.9$.[22]

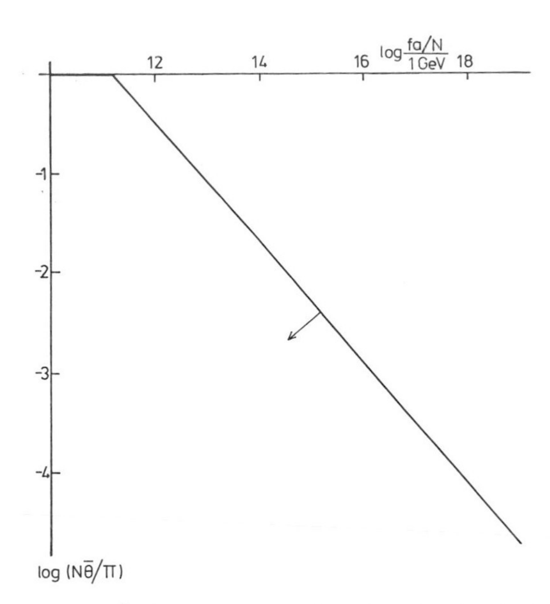

FIGURE 2. The bound on $\bar{\theta}$. The requirement that the axion density be less than critical excludes the region above the line shown.

This expression determines the mean $\bar{\rho}_a^{\text{in}}$ of ρ_a^{in} and also the stochastic properties of the inhomogeneity $\delta\rho_a^{\text{in}}/\bar{\rho}_a^{\text{in}}$. From the mean follows[22] the mean axion density parameter $\Omega_a = \bar{\rho}_a/\bar{\rho}_m$, which, setting $y = 1$, is

$$\Omega_a \simeq 10 \left(\frac{f_a/N}{10^{12}\text{ GeV}}\right)^{1.2} \left(\frac{N}{\pi}\right)^2 (\bar{\theta}^2 + \sigma_\theta^2). \tag{5}$$

We must have $\Omega_a < 1$, and if f_a/N is much bigger than 10^{12} GeV, this means that $N\bar{\theta}$ is very small, as shown in FIGURE 2. It has been argued[22,23] that nature may nevertheless have chosen a value $f_a/N \gg 10^{12}$ GeV, the anthropic principle ensuring a sufficiently small value of $\bar{\theta}$. The inhomogeneity is small and Gaussian if $\bar{\theta} \gtrsim \sigma_\theta$, but of order 1 and of the χ^2 type if $\bar{\theta} \lesssim \sigma_\theta$. The latter case is cosmologically acceptable only if $\Omega_a \ll 1$ is small, so as to make the initial total matter density perturbation $\Omega_a\delta\rho_a^{\text{in}}/\bar{\rho}_a^{\text{in}}$ small. The spectrum of the latter is constrained by observational limits on the microwave background anisotropy to be $P_m^{1/2} < 1 \times 10^{-4}$. The corresponding bound on $N\sigma_\theta$ is shown in FIGURE 1.

As a result of this bound, P_m cannot become of order 1 by the present treatment on scales of order 10 Mpc.[24] At least in the Gaussian case this means that it cannot explain the observed structure,[24] and the same is probably true in the non-Gaussian case, although further investigation is required along the lines of reference 25. It is perhaps possible that it could become of order 1 on smaller scales, so as to cause galaxy formation with something else responsible for the larger scale structure, and this too requires further investigation.

BOUNDS ON THE INFLATIONARY POTENTIAL

Finally, we make the extra assumption that $f_{a1} \simeq f_a$. Then the spectrum P_θ, given by **(2)**, is related to the inflationary Hubble parameter H_1, and hence to the inflationary energy density V_1. As a result there is the very strong bound on V_1 shown in FIGURE 3. We see that the bound on $V_1^{1/4}$ is about 10^{14} GeV if $f_a/N = 10^{10}$ GeV, rising to about 10^{16} GeV if f_a/N is at the Planck scale. This latter value is, however, only achieved if $N\bar{\theta}$ is much less than the already tiny value required to give $\Omega_a = 1$, and there seems to be no anthropic justification for such a choice. Thus it seems reasonable to assume that $\Omega_a = 1$ if $f_a/N \gg 10^{12}$ GeV, and in that case the bound drops to about 10^{15} GeV even if f_a/N is at the Planck scale.

This bound would practically rule out the three most currently favored mechanisms for structure formation, namely adiabatic perturbations arising from vacuum fluctuations of the inflaton field (popularly known as the cold dark matter or CDM hypothesis because it requires cold dark matter to have a chance of working), cosmic strings, and textures.[17] Like the whole of FIGURE 3, it can be evaded if $f_a \neq f_{a1}$, or if the Peccei–Quinn field is not sitting in its vacuum during inflation.[18]

Recently the bound on V_1 in reference 1 has been rediscovered,[26] and extended in a rough way to the case $\bar{\theta} \lesssim \sigma_\theta$ to produce roughly the bound shown in FIGURE 3. The authors suggest that the usually considered range of values $f_a/N \lesssim 10^{13}$ GeV is disfavored, because in that range the bound on V_1 is even tighter than in the range $f_a/N \gtrsim 10^{13}$ GeV, which would mean that existing searches for axionic dark matter

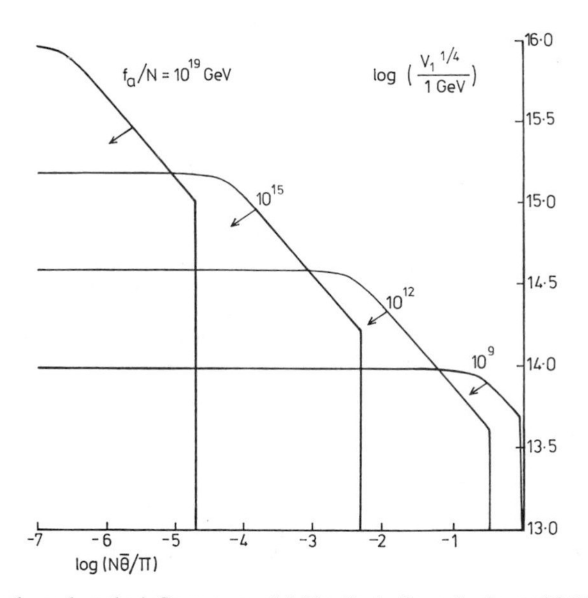

FIGURE 3. The bound on the inflaton potential. For the indicated values of f_a/N are shown the bounds on $V_1^{1/4}/m_P$, where V_1 is the inflaton potential evaluated when the observable universe leaves the horizon. The vertical part of the curves for $f_a/N = 10^{12}$ GeV upwards correspond to the bound in FIGURE 2, and the remainder corresponds to the bound in FIGURE 1.

are probably looking in the wrong mass range. The discussion we have just given, based on FIGURE 3, does not support this view.

ACKNOWLEDGMENT

The part of of this work published in reference 2 was done at CERN in collaboration with A. D. Linde, and much of the remainder has benefited from discussions with him.

REFERENCES

1. LYTH, D. H. 1990. Phys. Lett. **B236:** 408.
2. LINDE, A. D. & D. H. LYTH. 1990. Phys. Lett. **B246:** 353.
3. LYTH, D. H. 1991. Lancaster preprint TH8.
4. KIM, J. E. 1987. Phys. Rep. **150:** 1; CHENG, H. Y. 1988. Phys. Rep. **158:** 1.
5. TURNER, M. S. 1990. Phys. Rep. **197:** 67.
6. RAFFELT, G. 1990. Phys. Rep. **198:** 1.
7. ELLIS, J. & P. SALATI. 1990. Nuc. Phys. **B342:** 317.
8. ENGEL, J., D. SECKEL & A. C. HAYES. 1990. Phys. Rev. Lett. **65:** 960.
9. HARARI, D. & P. SIKIVIE. 1987. Phys. Lett. **B195:** 361.
10. LAZARIDES, G., R. K. SCHAEFFER, D. SECKEL & Q. SHAFI. 1990. Nuc. Phys. **B346:** 193; STEINHARDT, P. J. & M. S. TURNER. 1983. Phys. Lett. **B129:** 51; LAZARIDES, G., *et al.* 1987. Phys. Lett. **192B:** 323; DIMOPOULOS, S. & L. J. HALL. 1988. Phys. Rev. Lett. **60:** 1899.
11. DAVIS, R. L. 1986. Phys. Lett. **B180:** 225; DAVIS, R. L. & E. P. S. SHELLARD. 1989. Nuc. Phys. **B324:** 167; LINDE, A. D. 1988. Phys. Lett. **B201:** 437.
12. SIKIVIE, P. 1982. Phys. Rev. Lett. **48:** 1156.
13. LYTH, D. H. & E. D. STEWART. 1990. Phys. Lett. **B252:** 336; ELLIS, G. F. R., D. H. LYTH & M. B. MIJIC. 1990. SISSA preprint.
14. KOLB, E. W., D. S. SALOPEK & M. S. TURNER. 1990. Fermilab preprint 90/116-A; COPELAND, E. J., E. W. KOLB & A. R. LIDDLE. 1990. Phys. Rev. **D42:** 2911; KOLB, E. W. 1990. Preprint FNAL-CONF-90/195A, to appear in The Birth and Evolution of the Universe. 1991.
15. LINDE, A. D. 1985. JETP Lett. **40:** 1333; LINDE, A. D. 1985. Phys. Lett. **B158:** 375; SECKEL, D. & M. S. TURNER. 1985. Phys. Rev. **D32:** 3178; KOFMAN, L. A. 1986. Phys. Lett. **B173:** 400; KOFMAN, L. A. & A. D. LINDE. 1987. Nuc. Phys. **282:** 555; STAROBINSKY, A. A. 1985. JETP Lett. **42:** 124; SILK, J. & M. S. TURNER. 1987. Phys. Rev. **D35:** 419; KOFMAN, L. A. & V. F. MUKHANOV. 1987. *In* Quantum Gravity, M. Markov *et al.,* Eds. World Scientific. Singapore.
16. VISHNIAC, E. T., K. A. OLIVE & D. SECKEL. 1987. Nuc. Phys. **B289:** 717; YOKOYAMA, J. 1988. Phys. Lett. **B212:** 273; YOKOYAMA, J. 1989. Phys. Rev. Lett. **63:** 712.
17. LYTH, D. H. 1990. Phys. Lett. **B246:** 359.
18. LYTH, D. H. & E. D. STEWART. In preparation.
19. BIRREL, N. D. & P. C. W. DAVIES. 1982. Quantum Fields in Curved Spacetime. Cambridge University Press. Cambridge, England.
20. ALLEN, T. J., B. GRINSTEIN & M. B. WISE. 1987. Phys. Lett. **B197:** 66.
21. BARDEEN, J. M., J. R. BOND, N. KAISER & A. S. SZALAY. 1986. Astrophys. J. **304:** 15; MELOTT, A. L. 1990. Phys. Rep. **193:** 1.
22. TURNER, M. S. 1986. Phys. Rev. **D33:** 889.
23. PI, S.-Y. 1984. Phys. Rev. Lett. **52:** 1725.
24. EFSTATHIOU, G. & J. R. BOND. 1986. Mon. Not. R. Astron. Soc. **218:** 103.
25. MOLLERACH, S., S. MATARESSE, A. ORTOLAN & F. LUCCHIA. 1990. Padova preprint.
26. TURNER, M. S. & F. WILCJEK. 1991. Phys. Rev. Lett. **66:** 5.

Dissipationless Galaxy Formation?

MASSIMO STIAVELLI[a]

European Southern Observatory
Karl-Schwarzschild-Strasse 2
D8046 Garching bei München, Germany

INTRODUCTION

In principle, galaxy formation could be studied within the framework of cosmological simulations by specifying an initial spectrum of perturbations, the nature and amount of dark matter, and integrating forward in time until galaxies form. In practice, such a calculation is still beyond the present numerical capabilities, since cosmological *N*-body codes have limitations in both number of particles and spatial resolution that make them unable to simulate in detail the galaxies that may be formed. Although these two limitations may be overcome in the near future thanks to the availability of better computers and more powerful numerical techniques (see, e.g., Villumsen 1989), in the meantime one must resort to a different approach. A possibility is to study the formation of a single galaxy starting from a given initial state. By considering a large enough set of initial conditions and studying in detail the characteristics of the end-products, it is possible to determine which subset of initial states produces objects that more closely resemble real galaxies. These initial conditions are then the required final states of cosmological simulations meant to produce the objects currently observed in the universe.

Here, we discuss the results of such an approach in the context of the formation of elliptical galaxies. Four major scenarios have been proposed for the formation of ellipticals; namely, dissipationless collapse (van Albada 1982), dissipative collapse (Larson 1969, 1974), merging (Toomre and Toomre 1972), and cooling flows (Fabian, Nulsen, and Canizares 1982). Of these, the last seems to be restricted, at most, to brightest cluster members, where large amounts of hot gas are still observed. Merging is a more serious candidate, especially because merging events are seen to occur even at the present time (Schweizer 1982). Dissipationless collapse starting from rather general initial conditions shares the same ability to produce objects with photometric properties resembling those of ellipticals. In contrast, "fine-tuning" of initial parameters is required to produce ellipticals in the dissipative scenario. However, there are no studies where the properties of the end-products of these different mechanisms are compared in detail, not only to surface brightness profiles of real ellipticals, but also to velocity dispersion profiles and other observable quantities. Our aim is to perform such a study in the case of dissipationless collapse that is characterized by a smaller set of parameters defining the initial states. We consider the case of normal ellipticals, since the observational data on dwarf ellipticals are less clear.

[a] Current address: Classe di Scienze, Scuola Normale Superiore, Piazza dei Cavalieri 7, 56126 Pisa, Italy.

In the next section, we discuss the case where no dark matter is present that is especially relevant to the study of low-luminosity normal ellipticals. The case of bright elliptical galaxies, for which evidence for the presence of massive dark halos is available (Saglia, Bertin, and Stiavelli 1991), is discussed in the third section. Simple models where star formation and galaxy collapse occur simultaneously are described in the fourth section.

DISSIPATIONLESS COLLAPSE WITHOUT DARK MATTER

Since the pioneering study of van Albada (1982), several authors have addressed the problem of dissipationless collapse. These studies have shown that a luminosity profile well in agreement with the $R^{1/4}$ law (de Vaucouleurs 1948) for the surface brightness of ellipticals can be obtained whenever the initial virial, $2T/|W|$, is small enough. Aguilar and Merritt (1990) have analyzed the properties of the end-products of dissipationless collapse from a large set of initial states and shown that the radial-orbit instability (occurring during the collapse) is able to produce the observed range of ellipticity in elliptical galaxies. In addition, they have found that negative temperature models (Merritt, Tremaine, and Johnstone 1989) seem to provide a good fit to the phase space properties of the final systems. However, a detailed comparison with observations was missing. Also, the identification of the end-products of dissipationless collapse with negative temperature models was puzzling, especially in view of the results by Stiavelli and Sparke (1991), who find that, for the extreme anisotropies of these models, the radial-orbit instability modifies the density profile by increasing its core-radius, and thus renders the models less attractive for modeling real galaxies.

These considerations prompted Londrillo, Messina, and Stiavelli (1991) to perform an even larger set of collapse simulations with better dynamic range and spatial resolution. The major findings of this study were that spatial inhomogeneities can be important in determining the type of departures from the $R^{1/4}$ law that are produced. A better fit and more realistic departures are obtained when "long tails" are present in the initial density distribution. This translates, in terms of initial perturbations, to more power in long wavelengths and to a larger value of the initial density contrast $\delta\rho/\rho$, in agreement with the suggestion by Lake (1990). As far as the phase space structure of the end-products is concerned, Londrillo *et al.* (1991) find that the positive-temperature f_∞ models, introduced by Bertin and Stiavelli (1984), give a good description of the systems obtained, and especially of the central velocity distribution $N(v)$. These models are, however, characterized by insufficient anisotropy. In contrast, negative-temperature models have the correct anisotropy, but do not possess the correct central velocity distribution $N(v)$. Bertin and Stiavelli conclude that none of these models is able to represent in detail the properties of the systems produced by purely dissipationless collapse as modeled by the N-body simulations. Unfortunately, neither $N(v)$ nor the amount of anisotropy are directly observable quantities.

When various analytical and N-body collapse models are compared with real galaxies, it turns out that both the N-body systems and the f_∞ models (see Bertin, Saglia, and Stiavelli 1988) are able to fit adequately the photometric and kinematic properties of normal (especially low-luminosity) elliptical galaxies. In FIGURE 1 we compare the velocity dispersion profile of NGC 3379 (Davies and Birkinshaw 1988) with $f\infty$ and negative-temperature models and an N-body system. In order to discriminate between these systems, more extended kinematical data and additional diagnostics are required. Indeed, negative-temperature models have probably to be ruled out also on the basis of their double-peaked central line-profile (see, e.g., Merritt 1989).

Finally, some additional information could be gathered from the structure of isophotes. The models produced in collapses tend to have round isophotes in the center and a flattening increasing outward. We find that the radial-orbit instability in negative-temperature models produces, instead, isophotes of constant ellipticity. Both kinds of behavior are observed (see, e.g., Peletier *et al.* 1989).

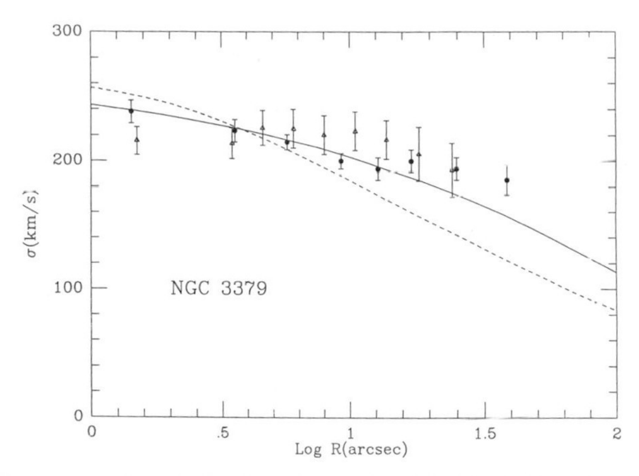

FIGURE 1. We compare the velocity dispersion profile of NGC 3379 (*filled circles,* Davies and Birkinshaw 1988), with an N-body system (*triangles*), an $f\infty$, $\Psi = 25$ model (*solid line*), and a negative-temperature $\Psi = -1, f_1$ model (*dashed line*). The *error bars* for the N-body system are due to the limited number of particles used (2×10^4). All these models give an acceptable fit to the photometry of this galaxy.

DISSIPATIONLESS COLLAPSE WITH DARK MATTER

From a detailed fitting to the observational properties of a set of bright objects Saglia, Bertin, and Stiavelli (1992) have recently given new evidence for dark matter in elliptical galaxies. Bright, giant ellipticals seem indeed to possess sizable amounts of dark matter, especially in the form of diffuse halos (see, e.g., the case of NGC 4472). It should be noted that comparatively large amounts of dark matter could be "hidden" by a radially anisotropic velocity dispersion. Since bright ellipticals seem to be characterized by rather flat profiles, the possibility that these are produced by adding very massive halos to a stellar component with a very radial velocity dispersion would imply very large M/L ratios within R_e.

Collapse simulations with dark matter are characterized by a larger number of initial parameters that has so far precluded an extensive exploration of a suitable set of initial conditions. We have only recently begun such a study, and thus our results are still preliminary. Within the small sample of initial states that we have explored, none was found characterized by a reasonably flat velocity dispersion profile and a realistic amount of dark matter. The failure of simulations with dark matter to produce realistic systems, might be due either to our assumptions in choosing the initial conditions (e.g., the total angular momentum is set to zero), or to an intrinsic sensitivity to the details of the initial states.

We would like to stress that Saglia *et al.* (1992) have shown that two-component models derived on the basis of f_∞ distribution function, are able to give very good fits to the observable properties of elliptical galaxies, but are apparently unable to fit the systems produced by dissipationless collapse with dark matter. Since it is known that dissipation must play some role in the formation of ellipticals, given the presence of color and metallicity gradients (Peletier *et al.* 1989; see also Matteucci 1990), it would be interesting to see whether the amount of dissipation required to produce such gradients is also able to reduce the amount of anisotropy to, for example, the level typical of f_∞ models. The next section is devoted to such a question.

COLLAPSE WITH A SIMPLE MODEL OF DISSIPATION

The initial systems that evolve into galaxies are made of gas and dark matter: stars are formed only due to some dissipative process. By assuming that the process of star formation is fast compared to the collapse time, one ends up with an essentially dissipation-free scenario. On the other hand, the two processes might well be concurrent. In this respect, the key effect of dissipation is to "slow down" the collapse. Gas and dark matter can collapse slowly while the newly formed stars are characterized by a small $2T/|W|$ and can then relax violently, producing the observed photometric profile. Thus, this scenario can be thought as a "delayed" collapse.

A complete study of dissipative collapse is still lacking. In the early work by Larson (1969, 1974), stars were unrealistically modeled by a fluid and dark matter was not included. Before embarking in full-fledged simulations where stars and dark matter are followed with a collisionless code and the gas with a fluid code, we plan to explore the main features of the "delayed" collapse scenario (Stiavelli and Matteucci 1991) with more approximate means. It turns out that its essence can be grasped even with a collisionless (multicomponent) code. In fact, by using particles to simulate the fluid, it is possible to have the overall gas dynamics correctly represented, especially if one thinks of the gas not as a uniform medium but rather as a collection of clouds internally cold, but moving at relatively large velocities with respect to each other. The major physics left outside this scheme is shocks, heat conduction, heating by supernovae, and chemical mixing. Shocks would appear in two major ways: at small scales, possibly inducing star formation, and at large scales, forming a disk. We can simulate the former phenomenology by forming stars according to the law

$$\dot{\rho}_{\mathrm{stars}} = K\rho_{\mathrm{gas}}^{\alpha}, \tag{1}$$

where we have so far taken $\alpha = 1$. The formation of a disk is here neglected. Indeed, in one of our simulations the system was, in the intermediate phases when only about half of the stars were formed, highly flattened. It is very likely that a disk would have been formed in this case, had the gas dynamics been properly simulated. While it is unclear what the effect of a more realistic (slower) heat conduction may be, it is easy to see that the heating by supernovae should not be critical for our studies, provided we stick to large ellipticals where outflows and winds are expected to occur only at relatively late stages (Matteucci and Tornambé 1987; D'Ercole *et al.* 1989). Finally, our instantaneous chemical mixing will tend to partly wash out any gradient that may be formed.

In this first cursory exploration, we have considered a system initially composed of clumps where one-quarter of the mass was gas and the remaining (collisionless) dark matter. The two are distributed equally and have the same velocity dispersion. The initial $2T/|W|$ considered were all larger than 0.15, which can be considered as the radial-orbit instability limit (Londrillo *et al.* 1991). According to **(1)** stars were formed by converting gas particles into stellar particles. Stars either die instantly or live forever, so that the stellar particle mass does not change throughout the simulation. The epoch of formation is recorded for each stellar particle. At the end of the simulation, a one-zone model with detailed stellar and chemical evolution, but no dynamics (Matteucci and Tornambé 1987), is run for a galaxy characterized by the same global parameters, so that the abundances of iron, magnesium, and the total metallicity are obtained as a function of time. These are in turn used to compute the average abundances as a function of radius starting from the average epoch of star formation as a function of radius derived from the simulation. In FIGURE 2 we show the run of [Fe/H], [Mg/H] with the projected radius for one of our models. We conclude that, even for a moderately short star-formation period ($\sim$400 Myr for a $10^{11}\ M_\odot$ galaxy), metallicity gradients comparable to those observed in real galaxies can be produced. At the same time the models turn out to be characterized by a small radial anisotropy, the ratio of radial to tangential kinetic energies, $2K_r/K_\rho$ being in the range 1.1 ÷ 1.4. When we compare models corresponding to different values of the total galactic mass, we have indications of gradients increasing with the mass of the galaxy.

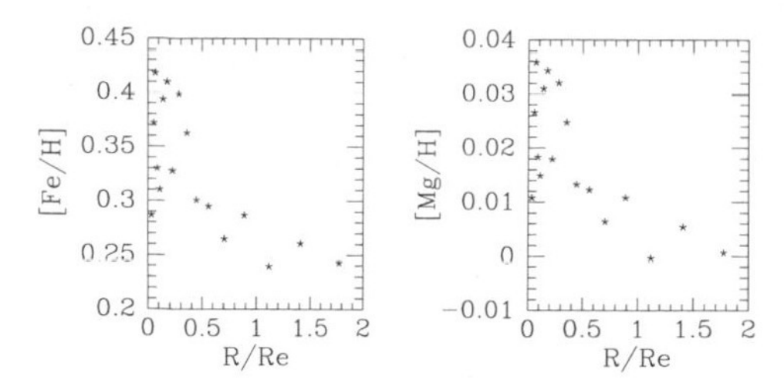

FIGURE 2. The abundance gradients [Fe/H] and [Mg/H] as a function of radius are shown for one of our simulations with dark matter ($M_L = 10^{11}\ M_\odot$, $M_L/M_D = 1/3$, $2T/|W|_{\text{initial}} \simeq 0.3$).

CONCLUSIONS

We have reviewed the constraints that the comparison of the observational properties of elliptical galaxies with the results of *N*-body simulations and the assumption that they are primary objects, that is, not formed at later times by, for example, merging of spirals, can set to cosmological simulations. It appears that in cases where no dark matter is required, pure dissipationless collapse is adequate to produce realistic systems. When dark matter is present, some fine-tuning of the initial parameters or the resort to a delayed star formation, occurring in the same time interval when the galaxy collapses, are necessary. The latter solution would also explain the observed metallicity gradients. In this respect, more work on both the observational and the theoretical side is called for. On the one hand, it would be important to measure metallicity gradients for a large set of ellipticals covering a wide range of luminosities. Work on this is in progress (Matteucci 1990). On the other hand, the cursory calculations described here should be followed by more realistic calculations, both in the case of collapse and in that of merging. The latter would be especially interesting, since in this case one would naively expect a trend with galactic luminosity opposite to the one expected for collapses, that is, gradients decreasing with increasing luminosity.

ACKNOWLEDGMENTS

Comments from G. Bertin, L. Lucy, F. Matteucci, and W. Zeilinger are gratefully acknowledged.

REFERENCES

AGUILAR, L. & D. MERRITT. 1990. Astrophys. J. **354:** 33.
BERTIN, G. & M. STIAVELLI. 1984. Astron. Astrophys. **137:** 26.
BERTIN, G., R. P. SAGLIA & M. STIAVELLI. 1988. Astrophys. J. **330:** 78.
DAVIES, R. L. & M. BIRKINSHAW. 1988. Astrophys. J., Suppl. Ser. **68:** 409.
D'ERCOLE, A., A. RENZINI, L. CIOTTI & S. PELLEGRINI. 1989. Astrophys. J., Lett. **341:** L9.
DE VAUCOULEURS, G. 1948. Astrophys. Ann. **11:** 247.
LAKE, G. 1990. Astrophys. J., Lett. **364:** L1.
LARSON, R. B. 1969. Mon. Not. R. Astron. Soc. **145:** 405.
LARSON, R. B. 1974. Mon. Not. R. Astron. Soc. **166:** 585.
LONDRILLO, P., A. MESSINA & M. STIAVELLI. 1991. Mon. Not. R. Astron. Soc. **250:** 54.
MATTEUCCI, F. 1990. *In* Morphological and Physical Classification of Galaxies, M. Capaccioli *et al.,* Eds. Kluwer. Dordrecht, the Netherlands. In press.
MATTEUCCI, F. & A. TORNAMBE. 1987. Astron. Astrophys. **185:** 51.
MERRITT, D. 1989. *In* Dynamics of Dense Stellar Systems, D. Merritt, Ed.: 75. Cambridge University Press. Cambridge, England.
MERRITT, D., S. TREMAINE & D. JOHNSTONE. 1989. Mon. Not. R. Astron. Soc. **236:** 829.
PELETIER, R. F., R. L. DAVIES, G. ILLINGWORTH, L. E. DAVIS & M. CAWSON. 1989. Astron. J. **100:** 1081.
SAGLIA, R. P., G. BERTIN & M. STIAVELLI. 1992. Astrophys. J. Jan. 1.
SCHWEIZER, F. 1982. Astrophys. J. **252:** 455.
STIAVELLI, M. & F. MATTEUCCI. 1991. Astrophys. J., Lett. **377:** L79.
STIAVELLI, M. & L. S. SPARKE. 1991. Astrophys. J. **382:** 466.
TOOMRE, A. & J. TOOMRE. 1972. Astrophys. J. **178:** 623.
VAN ALBADA, T. S. 1982. Mon. Not. R. Astron. Soc. **201:** 939.
VILLUMSEN, J. V. 1989. Astrophys. J., Suppl. Ser. **71:** 407.

The Adhesion Model: Some Results in Two and Three Dimensions

VARUN SAHNI

Inter University Center for Astronomy and Astrophysics
Poona University Campus
Ganeshkhind, Pune 411007, India

The Zeldovich approximation that so successfully describes the motion of dissipationless, self-gravitating matter until the formation of caustics, breaks down soon after and therefore cannot be used to study the evolution of large-scale structure at late nonlinear times. However, some useful insight into the formation and evolution of caustics at late times can be gained by means of the adhesion model that incorporates an effective viscosity into the Zeldovich approximation in order to mimic the effects of nonlinear gravity and thereby to stabilize the thickness of pancakes. The equations of motion for the velocity field in the framework of the adhesion approximation are equivalent to Burgers' equation:[1,2]

$$\left[\frac{\partial \vec{v}}{\partial b} + (\vec{v} \cdot \nabla)\vec{v} = \nu \Delta \vec{v}\right] \tag{1}$$

where $v = (1/a\dot{b})\,[\vec{V} - H\vec{r}]$, and $b(t)$ is the growing mode of the linear density contrast, and admit the exact solution[2]

$$\vec{v}(x, b) = \frac{\int d^3q[(\vec{x} - \vec{q})/(b(t))] \exp\left(-\{[\Phi_0(q) + [(\vec{x} - \vec{q})^2]/2b]/2\nu\}\right)}{\int d^3q \exp\left(-\{[\Phi_0(q) + [(\vec{x} - \vec{q})^2/2b]]/2\nu\}\right)}, \tag{2}$$

Φ_0 being the potential of the linear velocity field: $\vec{v}(\vec{x}, 0) = \nabla\Phi_0$. ($\Phi_0(q) \propto -\phi(q)$, where $\phi(q)$ is the linear gravitational potential[3,4]). For small values of the viscosity parameter $\nu \rightarrow 0$, the integrals in (2) can be evaluated using the method of steepest descents. The resulting solution has a geometrical interpretation.[1,4] Briefly, whether each particle can be located (or stuck) within a pancake is found by descending a paraboloid with radius of curvature $2b(t)$ onto the linear gravitational potential characterizing the particle distribution at very early times. The assignment of a particle having Lagrangian coordinate q^* at time parametrized by $b(t)$ to a pancake depends upon whether the paraboloid that is constructed to be tangential to the gravitational potential at q^*, actually touches or intersects the potential at any other point $q \neq q^*$. If it does, then the particle in question has already entered into a caustic [FIG. 1(b)]; otherwise, it has not [FIG 1(a)]. Thus one can divide the Lagrangian space at any given time into *stuck* and *free* regions. The distribution of caustics can be found by moving the border between these regions by means of the Zeldovich approximation. The resulting particle distribution for caustics is plotted superimposed on the pictures of *stuck* and *free* Lagrange regions in FIGURE 2(a), and superimposed on the results of the N-body simulations in FIGURE 3(b) for two distinct initial conditions. Comparison with the direct N-body simulations of Melott[5]

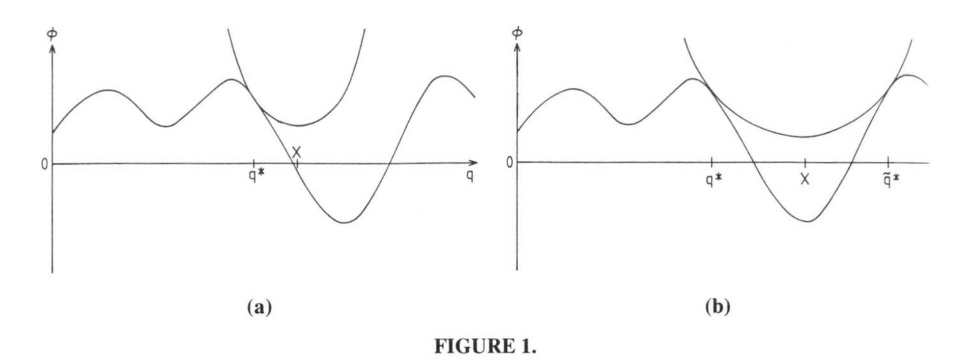

FIGURE 1.

[FIG. 2(b), 3(b)] shows very good agreement. *Note:* The pictures correspond to the time when $\delta_{\text{linear}} = \langle(\delta\rho/\rho)^2\rangle^{1/2} = 2$, that is, long after the Zeldovich approximation has broken down. FIGURES 2(a), (b) and 3(a), (b) show the results of the adhesion model for 256×256 particles and the corresponding N-body simulations for a linear density spectrum $\delta_k^2 = k^2$ with only the lowest 4 and 32 harmonics in the spectrum contributing, respectively. (See, however, references 6 and 7, where an alternative procedure has been adopted.)

Using the adhesion approximation it is possible to determine the fraction of dissipationless self-gravitating matter that is in caustics at any given time for an almost arbitrary initial density spectrum. FIGURE 4(a) shows the percentage of matter in caustics as a function of δ_{linear} for the three-dimensional *hot dark matter* model[8] obtained using 64^3 particles in a periodic box with side $\sim 50\ h_{100}^{-2}$ Mpc, and indicates that over 90 percent of the matter is in caustics by $\delta_{\text{linear}} \simeq 3$. A comparison of FIGURE 4(a) and results for the two-dimensional case, FIGURE 4(b), shows that the infall of matter into caustics is faster at late times in three dimensions than in two

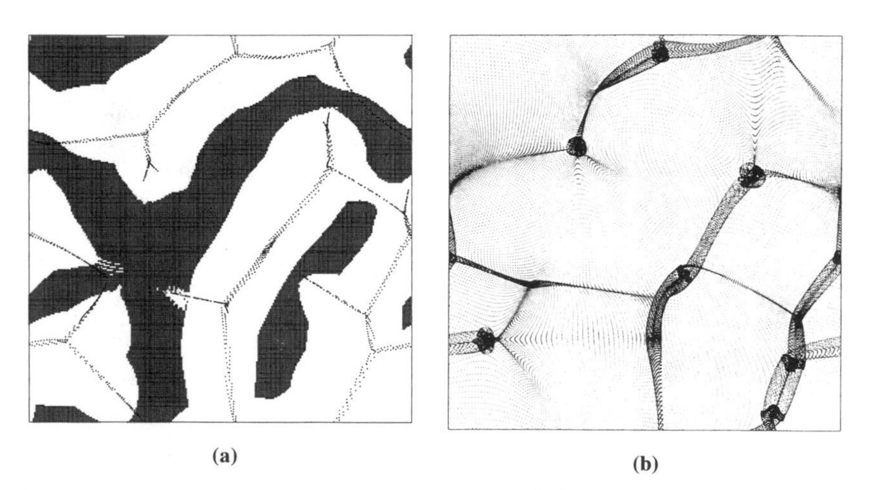

FIGURE 2. **(a)** The *shaded areas* represent free regions (voids) in Lagrange space; superimposed are shown caustics which form in Euler space. **(b)** N-body simulations[5]—Q4F4.

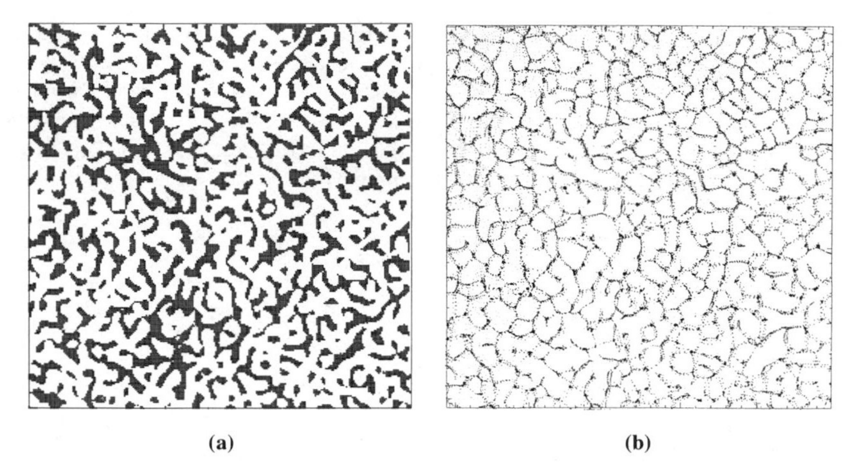

FIGURE 3. (**a**) The *shaded regions* represent stuck regions in Lagrange space. (**b**) The results of *N*-body simulations[5]—Q32F4, are shown superimposed on the adhesion model results for identical initial conditions.

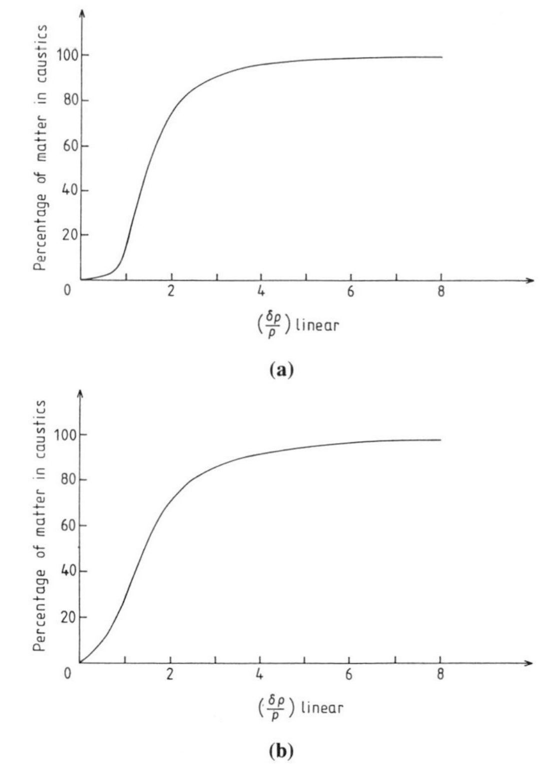

FIGURE 4. (**a**) Three dimensional: $\delta_k^2 = k^n\, e^{-(k\times R_f)2}$ $\quad n = \pm 1 \quad R_f = (4 \times k_{\min})^{-1}$. (**b**) Two dimensional: $\delta_k^2 = k^n\, e^{-(k\times R_f)2}$ $\quad n = 0{,}2 \quad R_f = (4 \times k_{\min})^{-1}$, where $k_{\min} = (2\pi/\lambda_{\text{box}})$.

dimensions, perhaps because as the radius of curvature of the paraboloid increases, its chances of intersecting the gravitational potential also increase significantly, being greater in three dimensions than in two dimensions at late times. The rate of infall of matter into caustics is not very sensitive to details of the initial density spectrum, being very similar for different spectra, as indicated in FIGURE 4(a), (b). This result will be explored in greater detail in a forthcoming paper.[9]

REFERENCES

1. GURBATOV, S. N., A. I. SAICHEV & S. F. SHANDARIN. 1985. Sov. Phys. Dokl. **30:** 921; GURBATOV, S. N., A. I. SAICHEV & S. F. SHANDARIN. 1989. Mon. Not. R. Astron. Soc. **236:** 385.
2. BURGERS, J. M. 1984. The Nonlinear Diffusion Equation. Reidel. Dordrecht, the Netherlands.
3. KOFMAN, L. A., D. POGOSIAN & S. F. SHANDARIN. 1990. Mon. Not. R. Astron. Soc. **242:** 200.
4. KOFMAN, L. A. & S. F. SHANDARIN. 1988. Nature **334:** 129.
5. MELOTT, A. L. & S. F. SHANDARIN. 1989. Astrophys. J. **343:** 26.
6. NUSSER, A. & A. DEKEL. 1990. Astrophys. J. In press.
7. WEINBERG, D. H. & J. E. GUNN. 1990. Mon. Not. R. Astron. Soc. **247:** 260.
8. BOND, J. R. & A. S. SZALAY. 1983. Astrophys. J. **274:** 443.
9. SAHNI, V. 1990. In preparation.

Large-scale Structure from Doppler Instability[a]

CRAIG J. HOGAN

Astronomy and Physics Departments, FM-20
University of Washington
Seattle, Washington 98195

INTRODUCTION

Historically, the account of the formation of large-scale structure has been split into two pieces. The first piece is the "initial conditions," the creation of protogalactic perturbations of appropriate magnitude in the very early universe.[1] Astronomers have been happy to leave this part of the problem to the specialists in inflation and quantum gravity. They have concentrated instead on the second piece of the problem, the evolution of these fluctuations in the relatively recent universe and the consequent formation and clustering of galaxies. Phenomenological models have been remarkably successful in accounting for the properties of galaxies and many of the statistical properties of their clustering and motions on large scales.[2] Standard theory connects the very early universe with these phenomenological models using the linear theory of perturbation growth by gravitational instability, modified by dissipative transport effects in the big bang plasma that depend on the nature of the cosmic dark matter.

There is, however, no compelling observational reason to believe that gravity has always been the dominant agent for structuring the universe on large scales. Indeed, problems arise if the purely gravitational evolution of perturbations is traced back to the earliest moments of the big bang. For example, the presence of fluctuations at the epoch of cosmic recombination, when the cosmic microwave background radiation last interacted with matter, tends to generate anisotropy in the radiation that has not been observed.[3] This difficulty is usually cited as evidence for a universe dominated by exotic primordial cold dark matter, but it could also be interpreted as a sign that the extrapolation of gravitational instability to very early times is incorrect. Moreover, the spectrum of fluctuations that works best empirically for observations of large-scale structure is not that naturally produced by processes in the very early universe, so it makes sense to investigate new processes that could modify the spectrum.

A particularly interesting new avenue to explore is whether nongravitational effects could have amplified fluctuations instead of just dissipating them. Here I outline a physical instability mechanism that may have intervened during the pregalactic era to greatly amplify the primordial perturbations.[4] This mechanism has several potentially attractive attributes for solving difficulties in the theory of cosmic structure. By providing a true instability in ordinary matter at very low density that

[a]This work was supported by NASA Grant NAGW-1703 and an Alfred P. Sloan Foundation Fellowship.

could spontaneously amplify primordial perturbations, it would effectively generate perturbations from scratch after cosmic recombination. Such an instability would play the central role in the early development of cosmic structure. It could circumvent the problem of excessive microwave background anisotropy even in open universes, because fluctuations would be generated after decoupling. For the same reason it would allow a wider variety of forms of dark matter, including hot massive neutrinos or ordinary baryons as well as cold dark matter. This paper outlines the conditions necessary for the instability to produce linear fluctuations with approximately the right amplitude and scale to agree with that required empirically for the gravitational formation of large-scale structure.

RESONANCE SCATTERING INSTABILITY

The basic principle behind the instability can be observed even in the behavior of a single atom immersed in an isotropic background radiation field that has a photon occupation number increasing with frequency in the vicinity of the ground state resonance.[b] If the spectral energy density $i(\nu_\alpha)$ has the corresponding logarithmic slope $\gamma \equiv d \ln i(\nu)/d \ln \nu > 3$, the radiation field at the particular rest frequency of the resonance in a moving frame of reference has its minimum in the forward direction. Any peculiar velocity of the atom gets amplified by resonance scattering of the radiation, an effect that translates into a dynamical instability of matter in the limit of small-amplitude perturbations. In an extended medium, the structures produced in the end by the instability depend on the nonlinear feedback of matter motions on the radiation spectrum. It is shown below that these effects tend to favor the amplification of motions on large scales.

Pure primordial blackbody radiation has a monotonically decreasing photon occupation number $(e^{h\nu/kT} - 1)^{-1}$ and can never cause this instability. To make it happen, some new source of photons, such as a decaying particle, needs to be added to the standard big bang model to produce the energy in the spectral feature.[6,7] In addition, appropriate opacity effects are required to get the increasing occupation number in the right part of the spectrum, near an electronic resonance of a neutral atom or molecule. For example, suppose a decaying neutrino produces $\simeq$ 22-eV photons. Around the time of cosmic recombination, neutral hydrogen Lyman continuum absorption preferentially removes redshifted lower energy photons and modifies the decay spectrum in the vicinity of the 584-Å HeI resonance to create an instability in the neutral helium. We will concern ourselves here with the general way the instability works rather than any such specific model for making the spectral distortion.

Suppose that some such process generates a steep spectral feature in the background radiation field that triggers an instability in the neutral gas. If the new radiation is sufficiently intense, the instability is able to create perturbations of the type required to generate galaxies and large-scale structure. Because the instability amplifies perturbations exponentially, the original fluctuations can have a negligible amplitude. Although it is not a prediction of the bare-bones standard big bang

[b] The reverse effect is used in laboratories, in the Doppler cooling of trapped atoms and ions by suitably tuned lasers;[5] unless the lasers are very stable, the atoms tend to run away.

model, and it is not clear whether there is a plausible decaying particle candidate for the energy source, the radiation intensity required to have all of these interesting consequences is surprisingly small by cosmic standards, and indeed may even be discouragingly difficult to detect with current techniques.

SATURATION

The development of the instability in an extended medium (as opposed to the acceleration of a single atom) is determined primarily by the backreaction on the radiation field as the matter moves around. The complement of the instability in the matter is that photons change their energy slightly when they scatter off of moving atoms, leading to a steady degradation of the spectral feature. When the feature is almost erased close to the line center, the instability saturates.

The velocities resulting from the instability can be estimated from the available photon energy, as the acceleration continues until the photon spectrum is changed. For each photon frequency, the photon energy losses must balance the kinetic energy gains of the matter *due to these same photons.* Let $\langle\beta^2\rangle$ denote the mean square velocity of the atoms; then the rate of change of kinetic energy density of the matter $(1/2)nm\langle\beta^2\rangle$ (where n denotes the number density of atoms and m their mass) approximately equals the energy extracted from the photons. The spectrum at offset δ changes due to the instability at a rate

$$\frac{d(\nu i(\nu))}{dt} \simeq -nc\sigma\nu i(\nu)\langle\beta^2\rangle,$$

where the classical cross section for unpolarized scattering from an atom with frequency ν_0 is given by

$$\sigma(\nu) = \sigma_T[\nu^4/(\nu_0^2 - \nu^2)^2],$$

where $\sigma_T \equiv 8\pi r_e^2/3$ is the Thomson cross section. (Expanding this expression about the line center ν_0 yields the classical damping profile, $\sigma = \sigma_T/4\delta^2 = (2\pi/3)r_e^2/\delta^2$, where the classical electron radius $r_e \equiv e^2/m_e c^2 = 2.8 \times 10^{-13}$ cm, and $\delta = (\nu - \nu_0)/\nu_0$ is the fractional displacement of the photon from line center.) Eventually, the velocities grow to the point where the instability saturates because of the photon reaction.

Photons with the smallest δ, although they initially contribute the fastest growth rate, saturate more quickly at smaller velocities, which is why they can be neglected in the final outcome. Saturated photons near the line center, for which $\gamma < 3$, tend to suppress the instability on small scales; thus the motions of the matter tend to be coherent, and most of the photon energy goes into large-scale motions rather than heat.

For situations of interest for large-scale structure, it is appropriate to describe the instability and its saturation in the limit of spatial photon trapping, viewing the photons as a gas. They form an unusual gas since as photons are redshifted they become more numerous; for $\gamma > 3$ the effective photon pressure increases as a region expands. Photons thus contribute a negative compressibility to the atomic gas, with a growth rate corresponding to the (imaginary) sound speed $i\beta_s = (\partial p_\gamma/\partial\rho c^2)^{1/2}$. Under such conditions, a medium spontaneously develops large-scale correlations.

This classical form of thermodynamic instability, in which a medium displays critical-point fluctuations, is how the instability manifests itself in the thick limit.

The effective pressure of trapped line radiation depends on scale.[8] For the thick limit to apply, the diffusion rate $t_d^{-1} = c/\lambda^2 n\sigma(\delta)$ for photons to escape from a region of size λ must be less than the growth rate for the instability on that scale, $\omega(\lambda) \simeq ((\gamma - 3)\delta(\lambda)\nu i(\nu)/nmc)^{1/2}/\lambda$. We can then solve for the critical offset $\delta(\lambda)$ of marginally trapped photons; to order of magnitude accuracy,

$$\delta(\lambda) = \left(\frac{\nu i(\nu)}{nmc^3}\right)^{1/3} (\lambda n r_e^2)^{2/3}$$

$$\omega(\lambda) = \left(\frac{\nu i(\nu)}{nmc^3}\right)^{2/3} (n r_e^2)^{1/3} \lambda^{-2/3}.$$

In the trapped limit, saturation occurs after the volume of a region has changed by a fraction $\delta(\lambda)$; after expanding by this amount the trapped photons have changed in energy by enough to escape faster than the perturbation is growing. The peculiar velocity associated with this volume change is $\lambda\omega(\lambda)\delta(\lambda)$, or

$$\beta_{\text{sat}} \simeq \left(\frac{\delta(\lambda)\nu i(\nu)}{nmc^3}\right)^{1/2}.$$

Since $\delta(\lambda)$ increases with scale, so does the saturated velocity. Therefore, the instability tends to generate peculiar velocities with a large coherence length; larger scale fluctuations tap the energy of photons far from line center, and therefore reach large velocities, while on small scales the relative velocities are damped.

In fact, the largest part of the energy is just lost as usual to the largest scale coherent motion of all, the general expansion of the universe. This sets an upper limit on the coherence scale of the peculiar motions. Since the growth rate $\omega(\lambda)$ decreases with scale, there is a maximum characteristic scale, given by $\omega(\lambda_{\text{max}}) = H$, below which fluctuations get amplified at a rate larger than the expansion rate H. The net efficiency of the instability (the fraction of each photon's energy injected into peculiar motions) is given by the offset of the photons that are trapped on this scale, that is, the δ for which $\lambda_{\text{diff}} = \lambda_{\text{max}}$:

$$\delta(\lambda_{\text{max}}) \simeq \frac{\nu i(\nu)}{nmc^3} \frac{r_e^2 nc}{H},$$

and the characteristic maximum scale is

$$\lambda_{\text{max}} \simeq \frac{c}{H} \frac{\nu i(\nu)}{nmc^3} \left(\frac{r_e^2 nc}{H}\right)^{1/2}.$$

The final peculiar velocity is obtained by integrating the net efficiency over the energy available in the entire width of the spectral feature $\epsilon\nu$ and equating this with the matter kinetic energy density,

$$\langle \beta^2 \rangle_{\text{sat}} \simeq \epsilon \left(\frac{\nu i(\nu)}{nmc^3}\right)^2 \frac{r_e^2 nc}{H}.$$

The factor $(r_e^2nc/H)^{1/2}$ emerges as an opacity figure of merit for the instability. For typical situations after recombination, this number is of the order of unity. The characteristic amplitude of the gas density perturbation on the scale λ_{max} is of the order of the width of the spectral feature ϵ. The other dimensionless parameter is the radiation flux in units of the mass density, $(\nu i(\nu)/nmc^3)$. These expressions show how the properties of the radiation spectral distortion determine both the amplitude of the perturbations and the characteristic scale where they appear.

For suitable choices of these parameters the instability produces fluctuations on scales large enough to match observations[9,10] of the large-scale galaxy distribution. Problems with the usual cold dark matter spectrum at small scales—for example, not forming quasi-stellar objects (QSOs) early enough[11]—are also easily solved. The fluctuations produced by the instability qualitatively resemble those one would prefer on purely empirical grounds.[10]

For an estimate of the radiation intensity needed to generate the observed structure, we can make an order of magnitude estimate of λ_{max}:[4]

$$[H_0\lambda_{max}]_0 \simeq 6000 \text{ km sec}^{-1}\left(\frac{\Omega_{feature}h^2}{10^{-6}}\right)\left(\frac{\epsilon}{0.1}\right)^{-1}\left(\frac{(\gamma-3)}{h}\right)^{1/2}\left(\frac{\Omega_b h^2}{0.025}\right)^{-1/2}\left[\frac{(1+z_i)}{1000}\right]^{5/4}.$$

For comparison in these units, the microwave background energy density is $\Omega_{bb}h^2 = 2.5 \times 10^{-5}$, so the required flux is small by cosmic standards. On the other hand, it is at a level potentially accessible to the experiments on COBE (Wright[3]).[12,13] At some level, one should also expect a Compton distortion of the blackbody spectrum, although the amplitude of this effect depends on the history of the electron heating and is very model-dependent.

REFERENCES

1. KOLB, E. & M. S. TURNER. 1990. The Early Universe. Addison-Wesley. Reading, Mass.
2. DEKEL, A. 1991. Streaming velocities and the formation of large-scale structure. This issue; WHITE, S. D. M. 1991. The epoch of galaxy formation. This issue; FRENK, C. S. 1991. The cold dark matter model: Does it work? This issue; KAISER, N. 1991. The density and clustering of mass in the universe. This issue.
3. WRIGHT, E. L. 1991. *COBE.* This issue; DAVIES, R. D., R. A. WATSON, J. HOPKINS, R. REBOLO, G. GUTIERREZ, J. E. BECKMAN & A. N. LASENBY. 1991. New limits on the cosmic microwave background fluctuations on a 5° angular scale. This issue.
4. HOGAN, C. J. 1991. Nature. **350:** 469; HOGAN, C. J. 1992. Submitted for publication in Astrophys. J.
5. COHEN-TANNOUDJI, C. N. & W. D. PHILLIPS. 1990. Phys. Today **43**(10): 33.
6. LYUBARSKY, Y. E. & R. A. SUNYAEV. 1983. Astron. Astrophys. **123:** 171.
7. FUKUGITA, M. & M. KAWASAKI. 1990. Astrophys. J. **353:** 384.
8. REES, M. J. 1985. Mon. Not. R. Astron. Soc. **213:** 75P.
9. MADDOX, S. J., G. EFSTATHIOU, W. J. SUTHERLAND & J. LOVEDAY. 1990. Mon. Not. R. Astron. Soc. **242:** 43P–47P.
10. PEEBLES, P. J. E. 1987. Nature **327:** 210–211.
11. EFSTATHIOU, G. & M. J. REES. 1988. Mon. Not. R. Astron. Soc. **230:** 5P–11P.
12. RUBIN, V. C. & G. V. COYNE. Eds. 1988. Large Scale Motions in the Universe: A Vatican Study Week. Princeton University Press. Princeton, N.J.
13. MATHER, J., *et al.* 1990. Ap. J. **354:** L37–L41.

Cosmological Relic Distribution of Conducting String Loops

B. CARTER

Département d'Astrophysique Relativiste et de Cosmologie
Observatoire de Paris
92 Meudon, France.

Previous notions of the potential cosmological significance of currents in cosmic strings have recently been shown to need radical revision. Earlier discussions, and in particular that of Ostriker, Thompson, and Witten,[1] tended to concentrate attention on the effects (such as cosmological void formation) that might have resulted from the very strong accompanying electromagnetic fields that were at first expected, on the basis of what can now, with hindsight, be seen to have been a rather too hasty analysis. However, the perspective was changed when Davis and Shellard[2] drew attention to considerations that greatly diminish the *a priori* likelihood of macroscopic superconducting cosmic string networks with such large currents as had been postulated. The new picture results from taking due account of the purely mechanical effects that had been overlooked or underestimated in previous work, but that should actually be considered generally to be of dominant importance for the dynamics of such strings on a macroscopic scale, the effects of electromagnetic coupling being describable as relatively secondary, or even completely absent if, as is easily conceivable, the currents were of neutral type.

The essential point is that (unless they subsequently decay by some quantum tunneling process whose timescale would have to be rather finely tuned to avoid having destroyed them in advance) such conducting string loops would not be able to radiate away all their energy, but would ultimately leave stationary centrifugally supported relics. The requirement that the mass density of the relic distribution should not exceed the cosmological closure density of the universe provides a severe upper limit on the Higgs mass scale of the vacuum symmetry breaking phase transition, producing the vortex defects that constitute the cores of the strings. Many details remain to be clarified, but it would seem that the Higgs mass scale of the grand unified theory (GUT) transition (that was assumed to be the relevant one in the earlier discussions) can be ruled out by a very wide margin on the basis of very general considerations.[2] Somewhat more specific assumptions lead to an estimate[3] that is compatible with a Higgs mass scale in the range (extending to a few teraelectron volts) that is commonly postulated for the electroweak symmetry-breaking phase transition, but only by a comparitively narrow margin, which means that the ensuing relic loops would be likely to constitute a significant fraction of the closure density of the universe as cold dark matter. Previous lower limits[4] on the masses of "chump" (charged ultramassive particle) dark matter candidates were marginally compatible with the predicted mass range of the relic loops, whose order of magnitude worked out as perhaps a few hundred teraelectron volts. However, more recent work may already have ruled this range out altogether, thereby placing what would be interpretable as experimental limits on the admissible string-forming generalizations of the standard (stringless) Glashow–Weinberg–Salem electroweak unification theory. Since there are no correspondingly severe limits for "nump" (neutral ultramassive particle) dark matter candidates, a cosmologically significant distribution of neutral (as opposed to electrically coupled) current-carrying string-loop relics would remain as a physically conceivable consequence of the electroweak phase transition.

Since the effects on which the foregoing conclusions depend are of an essentially local nature (long-range effects of gravitational and, when relevant, electromagnetic self-coupling being allowable for, since perturbations in a weak coupling limit treatment), it is appropriate as a lowest order approximation to analyze the behavior of the strings in question using a recently developed formalism[5,6] that provides a treatment of their macroscopic comportment in a manner that is independent of the detailed microscopic structure of the string, except in so far as it determines the equation of state giving its energy per unit length U, say, as a function of its tension T. An essential feature distinguishing a string loop of the generic conducting kind from the degenerate limit case of a nonconducting cosmic string, is that the conducting loop is characterized by two independent conserved numbers, of which the first, N say, is a topological phase-winding number that is the analogue of the conserved particle number in an ordinary elastic string loop, while the second, C say, is a charge number that is also interpretable as representing the Kelvin-type circulation round the loop. These numbers are expressible as integrals

$$2\pi N = \int dx^{\mu}\, \epsilon_{\mu\nu} b^{\mu}, \qquad C = \int dx^{\mu}\, \epsilon_{\mu\nu} c^{\mu} \tag{1}$$

of corresponding worldsheet currents b^{μ} and c^{μ}, say, that obey local two-surface current conservation laws that are expressible in terms of covariant differentiation with respect to the background spacetime metric $g_{\mu\nu}$ in the form

$$\eta^{\mu}_{\nu}\nabla_{\mu} b^{\nu} = 0, \qquad \eta^{\mu}_{\nu}\nabla_{\mu} c^{\nu} = 0, \tag{2}$$

where η^{μ}_{ν} is the fundamental tensor of projection onto the worldsheet, which is given in terms of the two-surface measure tensor ($\epsilon^{\mu\nu} = u^{\mu}v^{\nu} - u^{\nu}v^{\mu}$, where u^{μ} and v^{μ} are basis vectors of any orthonormal tangent frame in the worldsheet) by $\eta^{\mu}_{\nu} = \epsilon^{\mu\rho}\epsilon_{\rho\nu}$. Assuming there is no violation of the classical law of conservation of these quantities by quantum decay or processes or collisions, gravitational (and electromagnetic) radiation will not be able to take away all the energy of a generic string loop (as it can for an "ordinary" nonconducting cosmic string loop), so that instead of contracting to nothing, such a loop should presumably evolve toward a stationary relic state with the minimum energy allowed by its values of the conserved numbers C and N.

To complete the determination of the evolution of the system, the internal equations of motion (**2**) must be supplemented by the external dynamic equations governing the evolution of the worldsheet itself. The required extrinsic evolution equation can be shown under very general conditions,[7] covering the case not only of strings but also systems with higher dimensional worldsheets (of which the only nontrivial possibility in a four-dimensional spacetime is the case of a membrane such as that of a domain wall) to be expressible in terms of the relevant surface stress momentum energy density $T^{\mu\nu}$, say, in a form that reduces simply to

$$T^{\mu\nu}K_{\mu\nu}{}^{\rho} = 0 \tag{3}$$

whenever external forces (which would otherwise give a contribution on the right-hand side) can be neglected, where $K_{\mu\nu}{}^{\rho}$ denotes the kinematically defined second fundamental tensor of the worldsheet, which is definable in terms of the first fundamental tensor η^{μ}_{ν} by $K_{\mu\nu}{}^{\rho} = \eta^{\sigma}{}_{\mu}\eta^{\tau}_{\nu}\nabla_{\tau}\eta^{\rho}_{\sigma}$.

What distinguishes different kinds of string (or membrane) models is the prescription for the surface stress momentum energy tensor that is to be used in (3). In the two-dimensional case of a string this tensor will be genericly expressible in terms of a preferred choice of the, respectively, timelike and spacelike tangent vectors u^{μ} and v^{μ} of an orthonormal surface frame in the form

$$T^{\mu\nu} = Uu^{\mu}u^{\nu} - Tv^{\mu}v^{\nu}, \tag{4}$$

which the eigenvalues U and T are, respectively, interpretable as the energy density (per unit length) and the tension of the string. In terms of these quantities the general equation of extrinsic motion (3) reduces to the form

$$U\gamma^{\mu}{}_{\nu}u^{\rho}\nabla_{\rho}u^{\nu} = T\gamma^{\mu}{}_{\nu}v^{\rho}\nabla_{\rho}v^{\nu}, \tag{5}$$

in which the dual symmetry[5] of the string (as opposed to membrane or higher dimensional) case is made manifest.

To complete the determination of the evolution of the string it remains to specify the algebraic relationship between the quantities U and T and the surface currents b^{μ} and c^{μ}. The relevant information is expressible most concisely in terms of the appropriate string two-surface action density, L say, that (as recently discussed by Peter[8]) will be obtainable in the cosmological context with which we are concerned here as the cross-sectional integral of the spacetime action density of an underlying field-theoretical model over the core of the vacuum vortex defect that constitutes the microscopic internal structure of the string. In the simple case of the Nielsen–Olesen-type of vortices,[9] whose cosmological significance was first discussed by Kibble,[10] one obtains a simple Goto–Nambu-type action, the resulting expression for L being just a constant that is equal and opposite to the values of both of T and U, whose fixed common value T_0, say, given in order of magnitude (in Planck units) by the square of the relevant Higgs mass scale, m_x say, that is, one has

$$-L = T_0 \simeq m_x^2. \tag{6}$$

What we are interested in here, however, are models of the more general current-carrying kind, whose conceivable existence and cosmological relevance was originally discussed in terms of a toy field theoretical model of a more general kind by Witten.[11]

It was suggested by Witten that if the relevant current-carrier particles were endowed with an electromagnetic coupling constant, e say, presumably with the standard value, $e^2 \simeq 1/137$, then a particle number current, c^{μ} say, would imply a corresponding electromagnetic current

$$I^{\mu} = ec^{\mu}, \tag{7}$$

whose magnitude could be very large, being limited only by a current-saturation value for which he provided a dimensional estimate expressible in the form

$$c^{\mu}c_{\mu} \lesssim m_y^2 \tag{8}$$

where m_y is the relevant mass scale. Witten's original estimate was based on the implicit supposition that the scale m_y should be of the same order as the relevant Higgs mass scale m_x, but it was demonstrated by Hill, Hodges, and Turner[12] and by Babul, Piran, and Spergel[13] that a lower value is likely to be more appropriate, a better estimate according to Peter[8] being obtainable by identifying the quantity m_y in **(8)** with the mass of the relevant current-carrier particles, which can easily be conceived to be very much smaller than the relevant Higgs mass m_x. A possibility that seems to have been generally overlooked prior to the work of Davis and Shellard[2] is that of a timelike as opposed to spacelike current vector, which, if electromagnetically coupled, would imply the presence of an electrostatic charge density on the string. Such an electric-type (as opposed to magnetic-type) state would, of course, be subject in principle to charge loss by a quantum tunneling pair creation process, the condition for this to be negligible, implying that the positive upper limit given by **(8)** for $I^{\mu}I_{\mu}$ should be completed by an opposite limit on the negative side that can be expected on dimensional grounds to be given in order of magnitude by

$$-m_e^2 \lesssim e^2 I^{\mu}I_{\mu}, \tag{9}$$

where m_e is the mass of the lightest electromagnetically charged particle, which may be presumed to be the electron. Since it may safely be assumed that both $e^2m_y^2$ and m_e^2/e^2 are smaller than the relevant Higgs scale m_x^2, these limits imply that in both cases the absolute magnitude $|I^{\mu}I_{\mu}|$ will always be small compared with the value m_x^2 that by **(6)** gives an estimate of the energy density U and tension T in the null state limit, and hence presumably also in the neighboring electric-type and magnetic-type states, at least as a first, approximate order-of-magnitude guess. This implies that electromagnetic self-interaction effects within the string should be able to be considered only as small perturbations, and hence that the macroscopic string formalism described earlier should be applicable as a very good approximation, since internal effects of self-gravitation are even more obviously negligible provided m_x is small compared with (Planck) unit value, since it will be even in the GUT case, and hence *a fortiori* in the electroweak unification case that is of primary interest in the present context.

In the models with which we are concerned, the local state of the vacuum vortex will be determined by the gradient of a single real phase variable, φ say, so that the effective string action density will be a function just of the single scalar function

$$w = \eta^{\mu\nu}(\nabla_{\mu}\varphi)\nabla_{\nu}\varphi. \tag{10}$$

Starting from the functional relation between L and w we can obtain a derived function $\mathscr{K}$ and a dual Lagrangian function $\tilde{L}$ according to the rules

$$\mathscr{K} = -2\frac{dL}{dw}, \qquad \tilde{L} = L + w\mathscr{K}, \tag{11}$$

and one can then go on to construct independent phase-angle and charge-number currents

$$b^{\mu} = \epsilon^{\mu\nu}\nabla_{\nu}\varphi, \qquad c^{\mu} = \mathscr{K}\eta^{\mu\nu}\nabla_{\nu}\varphi, \tag{12}$$

which are both conserved, the relevant surface stress energy momentum tensor being given in terms of the latter by

$$T^{\mu\nu} = L\eta^{\mu\nu} - \mathscr{K}^{-1}c^{\mu}c^{\nu}. \tag{13}$$

In the magnetic case for which b^{μ} is timelike, and hence for which c^{μ} is spacelike, this system can be cast into the standard form as given earlier by making the identifications

$$U = -L, \qquad T = -\tilde{L}, \qquad \nu = \sqrt{w}, \qquad \mu = \mathscr{K}\nu, \qquad b^{\mu} = \nu u^{\mu}, \qquad c^{\mu} = \mu v^{\mu}. \tag{14}$$

In the electric case for which w is positive so that b^{μ} is spacelike, and hence for which c^{μ} is timelike, the system can be cast into the standard form as given previously by making the identifications

$$U = -\tilde{L}, \qquad T = -L, \qquad \mu = \sqrt{-w}, \qquad \nu = \mathscr{K}\mu, \qquad c^{\mu} = \nu u^{\mu}, \qquad b^{\mu} = \mu v^{\mu}, \tag{15}$$

(the only exception for which the standard form does not apply being the case that occurs where w vanishes so that b^{μ} and c^{μ} are both null). In each case, the number density ν and the corresponding effective mass (or relativistic chemical potential) μ defined in this way will be given by a well-defined equation of state in terms of a single independent variable that might be U or T according to convenience, via relationships of the standard form

$$\mu = \frac{dU}{d\nu}, \qquad \nu = -\frac{dT}{d\mu}, \qquad \mu\nu = U - T. \tag{16}$$

It therefore follows[6,7] that the characteristic propagation speeds c_T and c_L of extrinsic (transverse) perturbations of the worldsheet and of sonic-type (longitudinal) perturbations within the worldsheet will, respectively, be given in either case by

$$c_T^2 = \frac{T}{U}, \qquad c_L^2 = -\frac{dT}{dU}. \tag{17}$$

The simplest configuration that can be envisaged for a stationary equilibrium state of string loop of this kind is that of a centrifugally supported circular ring[2,4,14] of radius r, say, with angular momentum quantum number J given by

$$J^2 = (CN)^2 = 4\pi^2 r^4 UT, \tag{18}$$

its mass being given by

$$M^4 = (2\pi CN)^2 \frac{(U+T)^4}{UT}, \tag{19}$$

where the tension T and mass density U are determined[7,14] via the equation of state as functions of the ratio C/N. The ground state will, however, be degenerate, other "bent ring" configurations being possible with the *same* mass M but generically lower angular momentum, the value of J given by **(18)** being the maximum that is possible. It follows from a recent general study of stationary string configurations[15] that all (bent or circular) ground state configurations of a closed loop in a flat background must have in common (as the condition for centrifugal balance) the property of being transcharacteristic, meaning that the running velocity of the longitudinal motion must be equal to the characteristic speed c_T as given by **(17)** (whereas, a straight, but therefore necessarily open, configuration could have a freely chosen longitudinal running speed).

No rigorous comparative study of the stability of these equilibrium configurations has yet been carried out, but heuristic considerations suggest that the maximum angular momentum circular configurations will be at least marginally stabilized by the effect of (gravitational or electromagnetic) radiation backreaction, provided they are *subsonic* in the sense that the transverse characteristic speed c_T is smaller than the speed c_L, say, of longitudinal "sound" perturbations as given by **(17)**, the condition for this being expressible as

$$\frac{d^2L^2}{(dw)^2} > 0. \tag{20}$$

The basis for conjecturing such a stability requirement is the work of Friedman and Schutz[16] who showed (originally in the context of self-gravitating fluids, but using arguments of such a general nature that their applicability to the string case seems highly plausible) that a nonaxisymmetric perturbation mode of a rigidly rotating configuration will be damped by radiation unless it has relatively but not absolutely retrograde rotation. A generic rotating fluid system will have an extensive class of perturbation modes, including many in the dangerous (slow forward) angular velocity range, so that it will always tend to be destabilized by gravitational radiation reaction. However, in the string case there are only two kinds of mode. By the transcharacteristic property, relatively retrograde transverse modes are actually static and therefore marginally positively stable, while relatively retrograde longitudinal modes will be absolutely retrograde and therefore absolutely stable, provided the subsonicity condition **(20)** is satisfied.

The subsonicity condition $c_T < c_L$ will not only be satisfied for normal (violin-type) elastic strings (provided they are not too near breaking point), but also for the linear cosmic string model used in most early discussions, whose equation of state gives a constant sum $U + T$, for which it is well known[17] that longitudinal perturbations travel at the speed of light, that is, $c_L = 1$. According to this criterion the special integrable case[18] of the model with constant product UT characterized by the nondispersive transonic property $c_T = c_L$ would still have marginally stable ring configurations. However, for a nonlinear string model of the the kind that would result from the Witten mechanism in the generic case, one would expect the existence of supersonic states with $c_T > c_L$, and hence the possibility of actual instability, in regimes of high current (near the critical point for local longitudinal instability where dT/dU changes sign). This instability would presumably lead to phase bunching, a likely outcome in the long run being decay (by quantum tunneling) to a new state (with a modified value of N) in which the current norm $I^\mu I_\mu$ would have been reduced to a value compatible with stability.

Since the square of the relevant Higgs mass, m_x say, can be used as a rough dimensional estimate for the magnitudes of both U and T, it follows from **(19)** that a corresponding order-of-magnitude estimate for the equilibrium relic loop mass M will be given by

$$M \simeq m_x\sqrt{CN}. \quad \textbf{(21)}$$

Assuming (since $m_y < m_x$) that the random fluctuations in the phase of the current carrier are determined by the ordinary thermal correlation length that will be of the order of m_x^{-1} at the epoch of the relevant cosmological phase transition, random walk considerations lead me to guess[3] that the average values of the numbers N and C characterizing a string loop formed with initial length, ξ_x say, would be given by a formula of the form

$$C \simeq N \simeq \sqrt{m_x\xi_x}. \quad \textbf{(22)}$$

Since the dynamical effect of the current will become important only at a later stage after significant loop contraction has taken place, results from the study of "ordinary" nonconducting cosmic string formation should be applicable as a guide to what happens during and immediately after the phase transition. Although a certain number of larger loops are formed (in a "scaling" distribution) later on as the cosmological horizon expands, computer simulations[19] seem to confirm the conjecture that most of the string loops belong to the population that was formed at the outset, with a typical length scale ξ_x that is presumably to be identified with the Higgs field correlation length at the time of string creation. Assuming that this length scale also provides a (generous) dimensional estimate of the mean separation between the loops at the time of their formation, one can estimate the (approximately conserved) number density ratio of blackbody photons to string loops as of the order of $(m_x\xi_x)^3$. This means that ultimately when they have settled down to equilibrium with the typical final loop mass given by **(21)**, the mass per photon, m_c say, of the loop distribution will be given by

$$m_c \simeq M(m_x\xi_x)^{-3} \simeq m_x^{-3/2}\xi_x^{-5/2}, \quad \textbf{(23)}$$

which, of course, must not exceed the empirically measured cosmological closure value given roughly by $m_c \simeq 10^{-26}$ (i.e., about 10^2 eV in conventional as opposed to Planck units).

The estimation of the appropriate correlation length scale ξ_x is one of the most delicate questions in the theory, but it is obvious that it must lie in the range $m_x^{-1} < \xi_x < m_x^{-2}$, the lower limit being given by the thermal correlation length and the upper limit by the Hubble radius at the instant of the phase transition. This uncertainty range would only amount to a few powers of ten if m_x were to be identified with the GUT Higgs mass scale, $m_x \simeq 10^{-3}$ (as has been assumed in most numerical simulations), but it becomes very much greater if one is interested in the electroweak mass scale that is usually thought of as nearer to $m_x \simeq 10^{-15}$. A plausible estimate provided provisionally by the work of Kibble[20] would give ξ_x as the two-to-one weighted geometric mean between extreme values of the conceivable range, which amounts to taking

$$\xi_x \simeq m_x^{-4/3}, \tag{24}$$

which leads directly to

$$m_c \simeq m_x^{11/6}. \tag{25}$$

The striking (anthropic?) coincidence to which I wish to draw attention is that substitution of the electroweak Higgs mass scale $m_x \simeq 10^{-15}$ in **(25)** gives a result that is remarkably comparable with the cosmologically allowed upper limit $m_c \simeq 10^{-26}$ (a result that is not accutely sensitive to moderate adjustments of the relation **(24)** on which it is based). A shortfall with respect to the cosmological closure density, or a higher allowed Higgs mass scale, would be obtainable if a large fraction of the string loops failed to survive due to quantum decay processes or other mechanisms that might be imagined (as a subject for future work), but it would be very hard to get as far as the GUT mass scale. Substitution of **(24)** in **(21)** and **(22)** leads to the estimate

$$M \simeq m_x^{5/6} \tag{26}$$

giving perhaps a hundred to a thousand teraelectronvolts for the typical relic loop masses, and to the rather insensitive estimates

$$C \simeq N \simeq m_x^{1/6}, \tag{27}$$

for the corresponding quantum numbers, which would thus typically fall in the range from a few tens to a hundred, so that if they were electromagnetically coupled their charges would lie in the same range as those of ordinary heavy elements, the main difference being that negative as well as positive values would be allowed. Although large (by a factor of order N) compared with the microscopic string vortex radius, so that a classical description is justifiable, the dimensions of such relic loops would nevertheless be small compared with nuclear length scales, so that for most experimental purposes they could be treated as ultramassive charged or neutral-point particles.

ACKNOWLEDGMENTS

I am particularly indebted to Tsvi Piran for helping to clarify the ideas presented in this paper, and I should also like to thank Patrick Peter for recent discussions.

REFERENCES

1. OSTRIKER, J. P., C. THOMPSON & E. WITTEN. 1986. Phys. Lett. **B180:** 231.
2. DAVIS, R. L. & E. P. S. SHELLARD. 1988. Phys. Lett. **B209:** 485.
3. CARTER, B. 1990. *In* Early Universe and Cosmic Structures (Xth Moriand Astrophysic Meeting), A. Blanchard and J. Tran Than Van, Eds. Editions Frontieres. Gif sur Yvette, France.
4. DE RUJULA, A., S. L. GLASHOW & U. SARID. 1990. Nucl. Phys. **B333:** 173.
5. CARTER, B. 1989. Phys. Lett. **B224:** 61.
6. CARTER, B. 1989. Phys. Lett. **B228:** 446.

7. CARTER, B. 1990. *In* The Formation and Evolution of Cosmic Strings, G. Gibbons, S. Hawking, and T. Vachaspati, Eds. Cambridge University Press. Cambridge, England.
8. PETER, P. 1991. Preprint, D.A.R.C., Observatoire de Paris, Meudon.
9. NIELSEN, H. B. & P. OLESEN. 1973. Nucl. Phys. **B61:** 45.
10. KIBBLE, T. W. B. 1976. J. Phys. **A9:** 1387.
11. WITTEN, E. 1985. Nucl. Phys. **B249:** 557.
12. HILL, C. T., H. M. HODGES & M. S. TURNER. 1988. Phys. Rev. **D37:** 263.
13. BABUL, A., T. PIRAN & D. N. SPERGEL. 1988. Phys. Lett. **B202:** 307.
14. CARTER, B. 1990. Phys. Lett. **B238:** 166.
15. CARTER, B., V. P. FROLOY & O. HEINRICH. 1990. Preprint 90087, D.A.R.C., Observatoire de Paris, Meudon.
16. FRIEDMAN, J. L. & B. F. SCHUTZ. 1978. Astrophys. J. **22:** 281.
17. SPERGEL, D. N., T. PIRAN & J. GOODMAN. 1987. Nucl. Phys. **B291:** 847.
18. CARTER, B. 1990. Phys. Rev. **D41:** 3886.
19. BOUCHET, F. & D. BENNETT. 1990. Phys. Rev. **D41:** 2408.
20. KIBBLE, T. W. B. 1980. Phys. Rep. **67:** 183.

Cosmic Strings and Large-scale Structure: A Status Report[a]

ROBERT BRANDENBERGER

Physics Department
Brown University
Providence, Rhode Island 02912

INTRODUCTION

The cosmic string model[1] is a promising theory for the origin of structure in the Universe. It is a prototype for a model based on nonadiabatic seed perturbations and not on random phase adiabatic plane wave inhomogeneities.

The cosmic string model makes several distinctive predictions that contrast with those of other popular models such as the "standard" cold dark matter model[2] or the theory based on global textures.[3] The first distinctive feature relates to large-scale structure. In the cosmic string model it is dominated by planar perturbations forming in the wake of long rapidly moving strings[4] with a distinguished length scale given by the Hubble radius at the time of equal matter and radiation t_{eq}.[5] An interesting quantitative measure of the topology of large-scale structure is the genus curve[6] that shows a peculiar asymmetry.[7] Lensing produces a distinguished line jump in the temperature of the microwave background.[8]

In this review, I briefly summarize the main features of the cosmic string model. Then the distinguishing features of the theory will be explained and compared with those of their models. Emphasis is placed on the fact that in the context of the cosmic string theory, hot dark matter is a viable dark matter candidate.[9]

Units are used in which $c = k_B = \hbar = 1$. The scale factor of the Universe will be denoted by $a(t)$; $H(t)$ is the expansion rate, often expressed as $H(t_0) = h_{50}$ kms^{-1} Mpc^{-1}, where t_0 is the present time; G is Newton's constant; $z(t)$ is the redshift at time t.

BASIC POINTS

Cosmic strings are one-dimensional topological defects that arise in a large number of local quantum field theory models.[1] They may be superconducting[10] or not. Here, I consider only nonsuperconducting strings. Cosmic strings can also be viewed more simply as lines of trapped energy density.

As an example of a model that admits cosmic strings, consider the $U(1)$ gauge theory with a complex scalar field φ with potential

$$V(\varphi) = \lambda(|\varphi|^2 - \sigma^2)^2, \qquad (1)$$

[a]This work was supported in part by U.S. Department of Energy Grant DE-AC02-76ER03130 Task K & A, and by the Alfred P. Sloan Foundation.

where σ is the scale of symmetry breaking, and λ is the self-coupling constant. The space of ground states M (values of φ that minimize $V(\varphi)$) is a circle. Hence its first homotopy group $\pi_1(M)$ is nontrivial. This is the criterion for strings.

For comparison, the texture model[3] of structure formation requires a particle physics theory with a global symmetry. Theories with local symmetries and vacuum manifolds M with $\pi_0(M) \neq 1$ or $\pi_2(M) \neq 1$ give rise to domain walls and monopoles, respectively, and are ruled out on cosmological grounds if σ is large[11] (corresponding to the scale of grand unification).

A cosmic string model is characterized by a single free parameter, the mass per unit length $\mu \simeq \sigma^2$. If the theory is to be relevant for structure formation,[12] the dimensionless number $G\mu$ must be of the order 10^{-6}, which corresponds to the particle physics scale $\sigma \sim 10^{16}$ GeV of grand unification.

In any model that admits cosmic string solutions, a network of such strings inevitably will form during a phase transition in the very early Universe at a temperature $T_c \sim \sigma$. The mechanism by which this occurs was first discussed by Kibble.[13] At high temperatures $T > T_c$, $\varphi(\mathbf{x})$ will oscillate with a large amplitude $A > \sigma$. By causality, the phases of oscillation are uncorrelated on scales larger than the Hubble radius. As the Universe cools and $T < T_c$, $\varphi(\mathbf{x})$ will be constrained to lie in M. However, the orientation will be random on scales larger than H^{-1}, and thus, a network of strings with correlation length $\xi(T_c) < t_c$ will be formed at $T = T_c$.

The evolution of the cosmic string network has been studied extensively by numerical simulations.[14] It has now been well established that the network of infinite strings approaches a scaling solution in which the statistical properties are the same at all times when scaled to the Hubble radius. In particular, at all times there will be a couple of long string segments crossing each Hubble volume, and thus the energy density in these strings

$$\rho_\infty(t) = \nu\mu t^{-2} \tag{2}$$

will remain a constant and small fraction of the total energy density. The constant ν is of the order 1. The numerical uncertainties[14] in ν are moderate (about a factor of 5). The energy loss from long strings required to maintain **(2)** comes from loop production.[15] Loops with radius $\ll t$ are produced, for example, when a long string self-intersects. The spectrum of radii of loops is not yet well understood. See reference 14 for differing views. The consequences of the numerical uncertainties for the cosmic string structure formation scenario are discussed later.

The scaling solution for the cosmic string network is sketched in FIGURE 1. Thanks to the existence of a scaling solution we understand the network of cosmic strings at t_{eq}, the time of equal matter and radiation, when structure formation starts. There are two very different mechanisms by which cosmic strings give rise to structures. Loops at distances larger than their radius have the same time-averaged[16] gravitational field as a point mass and will therefore seed structures by gravitational accretion. Long moving strings, on the other hand, give rise to planar density enhancements ("wakes") in their wake.

To understand the origin of wakes, consider the gravitational field of an infinitely long straight string.[17] Locally, the Newtonian gravitational force vanishes. However, the global geometry is nontrivial. Space perpendicular to the string is a cone, that is, a

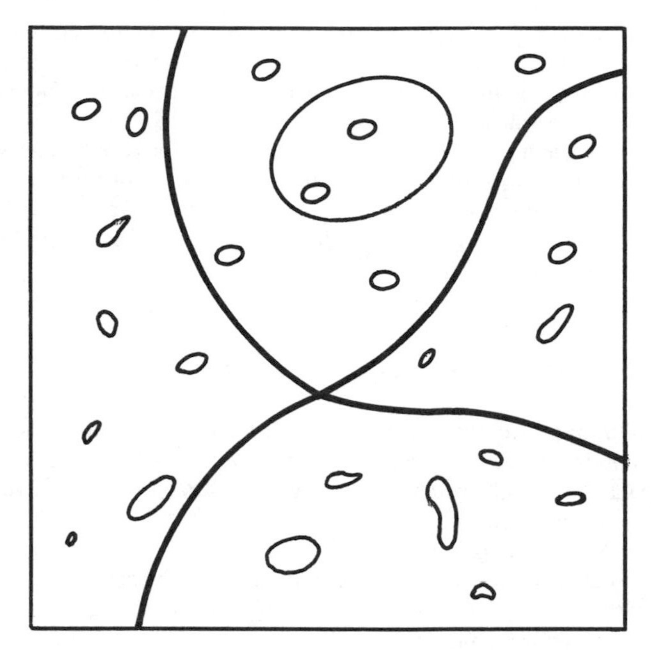

FIGURE 1. Sketch of the scaling solution. The box corresponds to one Hubble volume. At any time t, a small number of long string segments cross it, and it also contains a distribution of loops.

plane with deficit angle α

$$\alpha = 8\pi G\mu.$$

Consider now a string moving with velocity v. As illustrated in FIGURE 2 and best seen in the rest frame of the string, a velocity perturbation $\delta v = 4\pi G\mu v$ toward the plane in the wake of the string will be induced. This, in turn, leads to a planar density perturbation.

The relative importance of loops and wakes depends on the detailed form of the scaling solution. The initial quantitative cosmic string model of structure formation[12] was based on the hypothesis that loops seed individual clusters and galaxies. This hypothesis was only self-consistent under the assumption that many loops formed at time t with radius $R(t) \simeq t$. However, the recent simulations show that at the time of

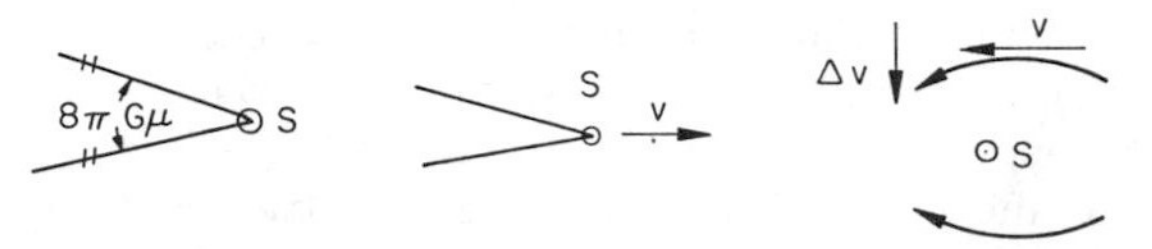

FIGURE 2. The origin of a cosmic string wake: space perpendicular to a straight string S is a plane with deficit angle $8\pi G\mu$ **(left).** If the string moves with velocity v through the plasma **(center),** then, as seen most clearly in the frame in which the string is at rest and the plasma is in motion, a velocity perturbation Δv of the plasma is induced **(right).**

formation $R \ll t$ (for other problems with the original model, see reference 18). Based on the recent simulations, it follows that wakes dominate. Wakes seed the large-scale structure of the Universe, and galaxies and clusters form mostly by fragmentation of the wakes.

The major problem with the current picture is that it has not been developed in nearly enough quantitative detail. There are two main reasons for this. First, the initial conditions are not yet known with sufficient accuracy. Second, to understand the small-scale structure in this model it will be necessary to use hydrodynamical and nonlinear gravitational techniques.

LARGE-SCALE STRUCTURE

In the following, I will discuss large-scale structure formation in a cosmic string model with hot dark matter.[5] With cold dark matter, the basic features would be the same, although there would be changes in the details. In particular, the relative role of loops would be greater.

First, let us recall the basic reason why hot dark matter is viable if perturbations are due to nonadiabatic seeds as opposed to random phase adiabatic perturbations.[9] Free streaming is the crucial issue. At time $t > t_{eq}$, hot dark matter particles typically move a distance (comoving units)

$$\lambda_J^c(t) = v(t) z(t) t, \tag{3}$$

in one Hubble expansion time, where $v(t)$ is the velocity of the dark matter particles. Thus, all primordial perturbations of dark matter on scales $\lambda < \lambda_J^c(t_{eq})$ are erased. This scale is larger than the mean separation of galaxies. Thus, there is insufficient power on small scales.

In contrast, nonadiabatic seed perturbations are not erased by free streaming. Growth of small-scale perturbations is delayed but not prevented; growth starts on a scale λ as soon as

$$\lambda = \lambda_J^c(t). \tag{4}$$

The time when this occurs will be denoted $t_J(\lambda)$. Thus, there is power on small scales to form galaxies.

Returning to large-scale structure formation, we have already mentioned wakes as the basic mechanism that creates planar density perturbations. In contrast to other models of structure formation, no nonlinear effects are needed to explain the observed sheetlike perturbations.[19] The cosmic string model also provides a distinguished scale for the sheets: the comoving Hubble radius at t_{eq}. This comes about as follows. At all times $t > t_{eq}$, the long strings seed wakes. The initial velocity perturbation is independent of time. Hence, the most prominent perturbations will be those that have the longest time to grow, namely those laid down at t_{eq}. They will also be most numerous, their length, width, and mean separation being of the order $t_{eq} z(t_{eq}) \simeq 40 h_{50}^{-1}$ Mpc.

These heuristic arguments can be confirmed by calculating the accretion of dark matter onto wakes.[5] Consider a dark matter particle with initial comoving distance q

from the wake. Its physical height is

$$h(q, t) = a(t)(q - \psi(q, t)), \quad (5)$$

where $\psi(q, t)$ is the comoving displacement. The thickness of the wake at time t can be defined by the value of q for which $\dot{h}(q, t) = 0$. Hot dark matter can be accommodated in this calculation by setting $\psi(q, t) = 0$ until $t_f(q)$. This is the "modified Zel'dovich approximation." The equation of motion for ψ follows by combining the Newton and Poisson equations. Solving the resulting equation in the case of hot dark matter, we find that no structures are bound to the wake until a time t_{cr} when $\dot{h}(q_{max}) = 0$, where

$$q_{max} = v(t_{eq})z(t_{eq})t_{eq}. \quad (6)$$

At t_{cr}, the entire sheet with thickness $2q_{max}$ becomes bound. With $v(t_{eq}) = 0.05$, q_{max} is about 2 Mpc. Requiring that $z(t_{cr}) > 1$ gives a lower bound on $G\mu$

$$G\mu > 5 \cdot 10^{-7}. \quad (7)$$

This lower bound is not yet in conflict with the upper bounds on $G\mu$ from the absence of detected anisotropies in the microwave background[8] and from millisecond pulsar timing data.[20]

A convenient way to quantify the predictions for large-scale structure is by using the genus curve.[6] Given a distribution of mass points (e.g., galaxies) and a smoothing length λ_s, we construct a smooth density distribution by replacing each mass point by a Gaussian mass distribution with width λ_s. For any density ρ, the surface $\rho(\mathbf{x}) = \rho$ can be characterized by its genus $\gamma(\rho)$:

$$\gamma(\rho) = \# \text{ of holes} - \# \text{ of disconnected pieces}. \quad (8)$$

Genus $\gamma(\rho)$ can then be plotted as a function of ρ. FIGURE 3 shows the result for a toy model of cosmic string wakes.[7] In the simulation a network of sheets was laid down at random (random positions and random orientations). Then, mass points (galaxies) were distributed on the sheets at random (with a number density chosen to give a consistent mass).

The distinctive feature in $\gamma(\rho)$ for the cosmic string model is the marked asymmetry between positive and negative overdensities. For high densities, the large negative value of $\gamma(\rho)$ reflects the large number of individual clumps within the wakes, while for underdensities, we are probing the small number of voids between the wakes [small negative $\gamma(\rho)$]. The cosmic string model predicts that in complete surveys of depth larger than $z(t_{eq})t_{eq}$, the genus curve should have a shape similar to that of FIGURE 3. The "standard" cold dark matter model predicts a very different shape.

There are other direct signatures for cosmic strings. Several are lensing effects due to the deficit angle of the surface perpendicular to the string. As illustrated in FIGURE 4, an observer sees two images of objects located behind long straight strings. A line of double quasars or double galaxies (the two images having equal redshift and spectrum) would be a "smoking gun" for cosmic strings. From FIGURE 4 we see that if the long string is moving with large velocity, the microwaves reaching the observer

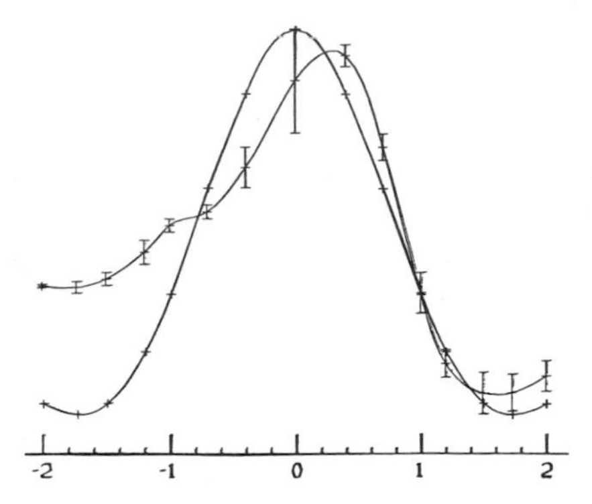

FIGURE 3. The *genus curve* (with *statistical error bars*) in a toy model of cosmic string wakes compared to the *corresponding curve* for a theory with random phase adiabatic perturbations (+). The *vertical axis* is the genus, the *horizontal axis* is overdensity (0 corresponds to the average density). Both axes are scaled. The specific parameters of the toy model were: wake size 40 × 40 × 3 Mpc^3, mean separation of the wakes 13 Mpc, smoothing length 3 Mpc, and galaxy mass $10^{12}M_\odot$, where $M_\odot$ is the solar mass.

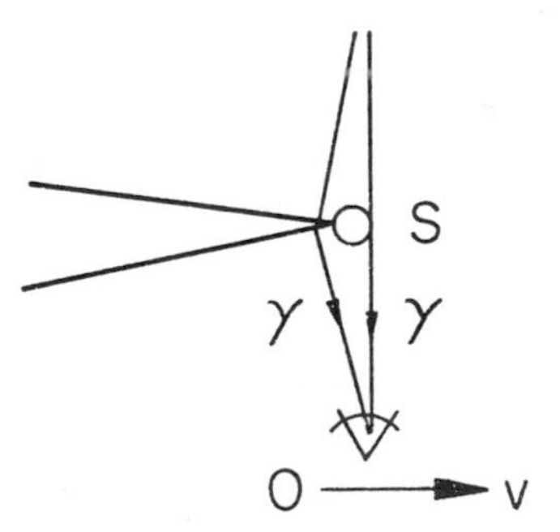

FIGURE 4. Gravitational lensing of light rays γ in the plane perpendicular to a long straight string *S* moving with velocity v. *O* is the observer.

will experience a Doppler shift[8]

$$\frac{\delta T}{T} \simeq 8\pi G\mu v < 3 \cdot 10^{-5} \qquad (9)$$

for $G\mu = 10^{-6}$. Thus, a line discontinuity in the temperature of the microwave background would be another clear signal for cosmic strings. This effect is very different from the signature in the global texture theory (hot and cold disks[21] on the microwave sky).

CONCLUSIONS

A structure formation model based on cosmic strings and hot dark matter has been presented. Its qualitative features look very promising, especially in the light of the new results presented at this meeting[22] that hint at more structure on scales of above $50h_{50}^{-1}$ Mpc than is compatible with the "standard" cold dark matter model.

The predicted large-scale structure is dominated by a network of wakes with length, width, and mean separation of the order the comoving Hubble radius $t_{eq}z(t_{eq})$ at equal matter and radiation. There is a prefactor ξ that must be determined from numerical simulations. With $\xi = 1$, $h_{50} = 1$, and $G\mu = 10^{-6}$, the dimensions of the wakes would be $(40 \times 40 \times 4)$ Mpc^3.

In the context of this model, hot dark matter is viable. This is particularly interesting in the light of the renewed speculations voiced at this meeting that neutrinos might have masses of the right magnitude to constitute the dark matter.[23]

The theory presented here does not have the usual biasing problem. A significant fraction of the dark matter remains homogeneously distributed,[24] thus providing an explanation of why dynamical measurements in clusters give $\Omega_{cl} < 0.2$ and allowing us to reconcile this with $\Omega_{total} = 1$.

The cosmic string model with hot dark matter also predicts a distinctive signal in the microwave background. For all these reasons, it is falsifiable.

The most significant problem is the possible mismatch of scales. Current cosmic string simulations give $\xi < 1$, and thus the scale of wakes would be too small. A possible solution of this problem involves baryon wakes,[25] in which case the distinguished scale corresponds to the Hubble radius at recombination.

Much more work is required to determine the predictions of this model for smaller scale structure. In particular, hydrodynamical and nonlinear gravitational effects will have to be considered.

ACKNOWLEDGMENTS

I am grateful to my collaborators E. P. S. Shellard, L. Perivolaropoulos, A. Stebbins, and J. Gerber for all their help.

REFERENCES

1. VILENKIN, A. 1985. Phys. Rep. **121:** 263.
2. BLUMENTHAL, G., S. FABER, J. PRIMACK & M. REES. 1984. Nature **311:** 517; DAVIS M., G. EFSTATHIOU, C. FRENK & S. WHITE. 1985. Astrophys. J. **292:** 371.
3. TUROK, N. 1989. Phys. Rev. Lett. **63:** 2625.
4. SILK, J. & A. VILENKIN. 1984. Phys. Rev. Lett. **53:** 1700; VACHASPATI, T. 1986. Phys. Rev. Lett. **57:** 1655; STEBBINS, A., S. VEERARAGHAVAN, R. BRANDENBERGER, J. SILK & N. TUROK. 1987. Astrophys. J. **322:** 1.
5. PERIVOLAROPOULOS, L., R. BRANDENBERGER & A. STEBBINS. 1990. Phys. Rev. **D41:** 1764.
6. GOTT, J. 1988. Publ. Astron. Soc. Pac. **100:** 1307.
7. GERBER, J. & R. BRANDENBERGER. 1991. In preparation.
8. KAISER, N. & A. STEBBINS. 1984. Nature **310:** 391; BOUCHET, F., D. BENNETT & A. STEBBINS. 1988. Nature **335:** 410.
9. BRANDENBERGER, R., N. KAISER, D. SCHRAMM & N. TUROK. 1987. Phys. Rev. Lett. **59:** 2371.
10. WITTEN, E. 1985. Nucl. Phys. **B249:** 557.
11. ZEL'DOVICH, YA. & M. KHLOPOV. 1978. Phys. Lett. **79B:** 239; PRESKILL, J. 1979. Phys. Rev. Lett. **43:** 1365.

12. TUROK, N. & R. BRANDENBERGER. 1986. Phys. Rev. **D33:** 2175; SATO, H. 1986. Prog. Theor. Phys. **75:** 1342; STEBBINS, A. 1986. Astrophys. J., Lett. **303:** L21.
13. KIBBLE, T. W. B. 1976. J. Phys. **A9:** 1387.
14. ALBRECHT, A. & N. TUROK. 1984. Phys. Rev. Lett. **54:** 1868; ALBRECHT, A. & N. TUROK. 1989. Phys. Rev. **D40:** 973; BENNETT, D. & F. BOUCHET. 1988. Phys. Rev. Lett. **60:** 257; ALLEN, B. & E. P. S. SHELLARD. 1990. Phys. Rev. Lett. **64:** 119.
15. SHELLARD, E. P. S. 1987. Nucl. Phys. **B283:** 624; MATZNER, R. 1988. Comput. Phys. **1:** 51; MORIARTY, K., E. MYERS & C. REBBI. 1988. Phys. Lett. **207B:** 411; SHELLARD, E. P. S. & P. RUBACK. 1988. Phys. Lett. **209B:** 262.
16. TUROK, N. 1984. Nucl. Phys. **B242:** 520.
17. VILENKIN, A. 1981. Phys. Rev. **D23:** 852; GOTT, J. 1985. Astrophys. J. **288:** 422; HISCOCK, W. 1985. Phys. Rev. **D31:** 3288; LINET, B. 1985. Gen. Relativ. Grav. **17:** 1109; GARFINKLE, D. 1986. Phys. Rev. **D32:** 1323; GREGORY, R. 1987. Phys. Rev. Lett. **59:** 740.
18. BRANDENBERGER, R. & E. P. S. SHELLARD. 1989. Phys. Rev. **D40:** 2542.
19. DE LAPPARENT, V., M. GELLER & J. HUCHRA. 1986. Astrophys. J., Lett. **302:** L1; DE LAPPARENT, V. 1990. Paper presented at the 15th Texas/ESO-CERN Symposium on Relativistic Astrophysics, Cosmology and Fundamental Physics, Brighton, England, December 16–21.
20. STINEBRING, D., M. RYBA, J. TAYLOR & R. ROMANI. 1990. Phys. Rev. Lett. **65:** 285.
21. TUROK, N. & D. SPERGEL. 1990. Phys. Rev. Lett. **64:** 2736.
22. SAUNDERS, W., *et al.* 1990. Paper presented at the 15th Texas/ESO-CERN Symposium on Relativistic Astrophysics, Cosmology and Fundamental Physics, Brighton, England, December 16–21.
23. BLUDMAN, S. & D. SCHRAMM. 1990. Papers presented at the 15th Texas/ESO-CERN Symposium on Relativistic Astrophysics, Cosmology and Fundamental Physics, Brighton, England, December 16–21.
24. BRANDENBERGER, R., L. PERIVOLAROPOULOS & A. STEBBINS. 1990. Int. J. Mod. Phys. **A5:** 1633.
25. REES, M. 1986. Mon. Not. R. Astron. Soc. **222:** 27.

Clustering in a Non-Gaussian Cold Dark Matter Cosmogony[a]

L. MOSCARDINI,[b] F. LUCCHIN,[c] A. MESSINA,[d] AND S. MATTARRESE[e]

[b]*Astronomy Centre*
University of Sussex
Falmer, Brighton BN1 9QH, England

[c]*Dipartimento di Astronomia*
Università di Padova
Vicolo dell'Osservatorio 5
I-35122 Padova, Italy

[d]*Dipartimento di Astronomia*
Università di Bologna
Via Zamboni 33
I-40126 Bologna, Italy

[e]*Dipartimento di Fisica "Galileo Galile"*
Università di Padova
Via Marzolo 8
I-35131 Padova, Italy

INTRODUCTION

The most thoroughly investigated model of galaxy formation is the standard cold dark matter (CDM) scenario, based on the assumption that the universe is flat ($\Omega_0 = 1$), the primordial power spectrum of adiabatic fluctuations is scale invariant, and the statistics of initial perturbations is Gaussian. The model proved to be quite satisfactory in explaining most of the basic features of galaxy clustering (see, e.g., references 1–3), without serious conflicts with the observed limits on the cosmic microwave background (CMB) anisotropies (see, e.g., reference 4). Its main difficulty comes from the lack of structure on scales larger than $\sim 10\,h^{-1}$ Mpc. In fact, the only characteristic length of the model is the horizon scale at the equivalence epoch $\lambda_{eq} \approx 13\,(\Omega_0 h^2)^{-1}$ Mpc, with h the Hubble constant in units of 100 km sec^{-1} Mpc^{-1}.

The observational evidence for large-scale power has recently been confirmed by the analysis of the galaxy angular correlation function in the APM survey,[5] by the quasi periodicity of the galaxy redshift distribution observed in deep pencil-beam surveys[6] and by the recent analysis of the density field as traced by the Infrared Astronomical Satellite (IRAS) galaxies.[7]

To overcome such a difficulty we considered primordial CDM perturbations with non-Gaussian statistics. In order to select suitable statistical distributions we fol-

[a] One of the authors (L.M.) acknowledges SERC financial support. This work was supported in part by the Italian MURST and in part by CNR (*Progetto Finalizzato: Sistemi Informatie e Calcolo Parallelo*).

lowed some simple criteria: some sort of statistical scale invariance can be invoked; we built up the perturbation process by performing nonlinear, multiplicative transformations on a Gaussian random field. At present there are no natural models able to generate these perturbations. Inflation-generated perturbations are generally expected to show this type of non-Gaussian behavior,[8] though on superhorizon scales.[9,10] Specific inflation models may provide adiabatic non-Gaussian fluctuations on relevant scales, but with a high level of fine-tuning (Bardeen, unpublished). Isocurvature axion perturbations with non-Gaussian statistics can also be obtained (see, e.g., reference 11).

CLUSTERING EVOLUTION

A class of noncorrelated non-Gaussian models was investigated by Messina *et al.*[12] Recently, we performed N-body simulations of the evolution of structures in non-Gaussian CDM models (Moscardini *et al.*[13]). In particular, three types of non-Gaussian statistics, chosen as distributions for the peculiar gravitational potential $\Phi(\mathbf{x})$ before the inclusion of the CDM transfer function, were considered:

- The *convolution* (hereafter C) model is obtained by the convolution of two independent Gaussian processes;[14]
- The *lognormal* model (hereafter LN), which is obtained by taking the exponential of a Gaussian random field, is the extreme case of multiplicative distribution;
- The *chi-squared* model (hereafter χ^2) with one degree of freedom is obtained by squaring a Gaussian field.

Statistical distributions related to the last two cases have already been considered in the cosmological context (see, e.g., references 15 and 16). Both the lognormal and the chi-squared statistics split in two different models: the *positive* models (LN_p and χ^2_p), characterized by positive skewness for the linear mass fluctuation (i.e., $\langle\delta^3_M\rangle > 0$), implying that overdense regions occur more frequently than underdense ones; and the *negative* models (LN_n and χ^2_n), characterized by negative skewness ($\langle\delta^3_M\rangle < 0$). A Poisson-like equation relates the local gravitational potential to the linear mass fluctuation, $\delta_M(\mathbf{x}, t) = 2\nabla^2\Phi(\mathbf{x})/3\dot{a}^2(t)$, where $a(t)$ denotes the scale factor and a dot a time derivative.

For all our distributions, Φ has power spectrum

$$\mathcal{P}_\Phi(k) = \tfrac{9}{4}\mathcal{P}_0 k^{-3}T^2(kl), \tag{1}$$

where $\mathcal{P}_0 k$ is the primordial Zel'dovich spectrum of density fluctuations, $T(kl)$ is the CDM transfer function (see, e.g., reference 2), and $l = (\Omega_0 h^2)^{-1}$ Mpc. We considered only models with $\Omega_0 = 1$; all lengths in our simulations scale as h^{-2}. We also compare with the evolution of the Gaussian CDM (hereafter G) model.

A particle-mesh code with $N_p = 64^3$ particles on $N_g = 64^3$ grid points has been used. Computations were performed at the CINECA Centre (Bologna) on a Cray YMP/432 running under UNICOS. The box size of our simulations was $L = 65h^{-2}$ Mpc, implying a particle mass $m = 2.94 \times 10^{11}h^{-4}M_\odot$.

The local gravitational potential $\Phi(\mathbf{x})$ was obtained by the convolution of the real function $\tau(\mathbf{x})$, representing the inverse Fourier transform of the transfer function T, with a zero-mean, homogeneous and isotropic, random field $\phi(\mathbf{x})$,

$$\Phi(\mathbf{x}) = \int d^3y\, \tau(\mathbf{y} - \mathbf{x})\, \varphi(\mathbf{y}), \tag{2}$$

the latter being related to a Gaussian process by some nonlinear transformation. The models listed earlier are indeed characterized by different choices of $\phi(\mathbf{x})$. The initial particle positions and velocities were assigned via the Zel'dovich algorithm. After evolving all our models starting from the same amplitude we analyzed their clustering properties from a reference *nonlinearity time,* t_{nl}, up to the *present time,* t_0, by different statistics. At the time t_{nl} nonlinear events (i.e., $\delta_M \sim 1$) first occur in the Gaussian model; at this time the non-Gaussian models still display most of their initial features, although some nonlinear evolution already occurred in moderately high peaks. The time t_0 is fixed when the particle two-point function is best fitted by a power law $\xi(r) = (r/r_0)^{-\gamma}$, with $\gamma = 1.8$, in a suitable interval. FIGURE 1 shows slices of thickness $1/16$ of the box size at t_0, all drawn from the same underlying Gaussian realization. (Note that all our models are obtained by performing nonlinear transformations on a *flicker-noise* Gaussian field.) The main result is that both the dynamics of clustering and the present texture are very sensitive to the sign of the initial skewness. Positive models, which cluster more rapidly, show a lumpy structure with small correlation length and small peculiar velocities. Large-scale structures in negative models form late: the merging of shells surrounding primordial underdense regions gives rise to a cellular structure with filaments, sheets, and large voids; both the correlation length and rms bulk motion are large. The structure and the dynamics of the C model, which has vanishing initial skewness, look similar to that of the G model: the multiplicative process here only mixes the phases without producing coherent structures on large scales. The presence of remarkably longer filaments and larger voids in negative models has been quantitatively confirmed by the application of suitable statistical tests: the quadrupole test[17] and the nearest neighbor analysis.[18]

TABLE 1 gives the value of the scale factor for each model at t_0 normalized to the value at t_{nl}. The slope of the correlation function and the corresponding correlation length r_0, in units of h^{-2} Mpc, are also reported. The errors represent one standard deviation. Models G and C get the right slope at $a_0 \approx 2.3$. Positive models evolve faster, $a_0 \approx 1.65$. Negative models need a long nonlinear evolution, $a_0 \approx 7.5$, before reaching the present. Due to the different rates of nonlinear evolution, the correlation length turns out to be quite different for the various models. Models G and C have small correlation length; the "classical" value $r_0 = 5h^{-1}$ Mpc can only be obtained with the uncomfortably low value $h \approx 0.3$. Even smaller is r_0 for positive models. In both cases a substantial amount of biasing would be required. Negative models have enough time to push the nonlinearity to larger scales; in this case the classical r_0 value is obtained with $h \approx 0.8$. With $h = 0.5$ the high correlation length of negative models can easily accommodate the recently favored value $r_0 \approx 7.5h^{-1}$ Mpc.[19]

All the models show a critical wavelength λ_{cr} below which the growth is more than linear, a signal of hierarchical clustering: the dynamics of small-scale fluctuations is dominated by those occurring in lumpy regions, corresponding to a locally closed universe. Such a critical wavelength is found at $\sim 13h^{-2}$ Mpc for the G and C models,

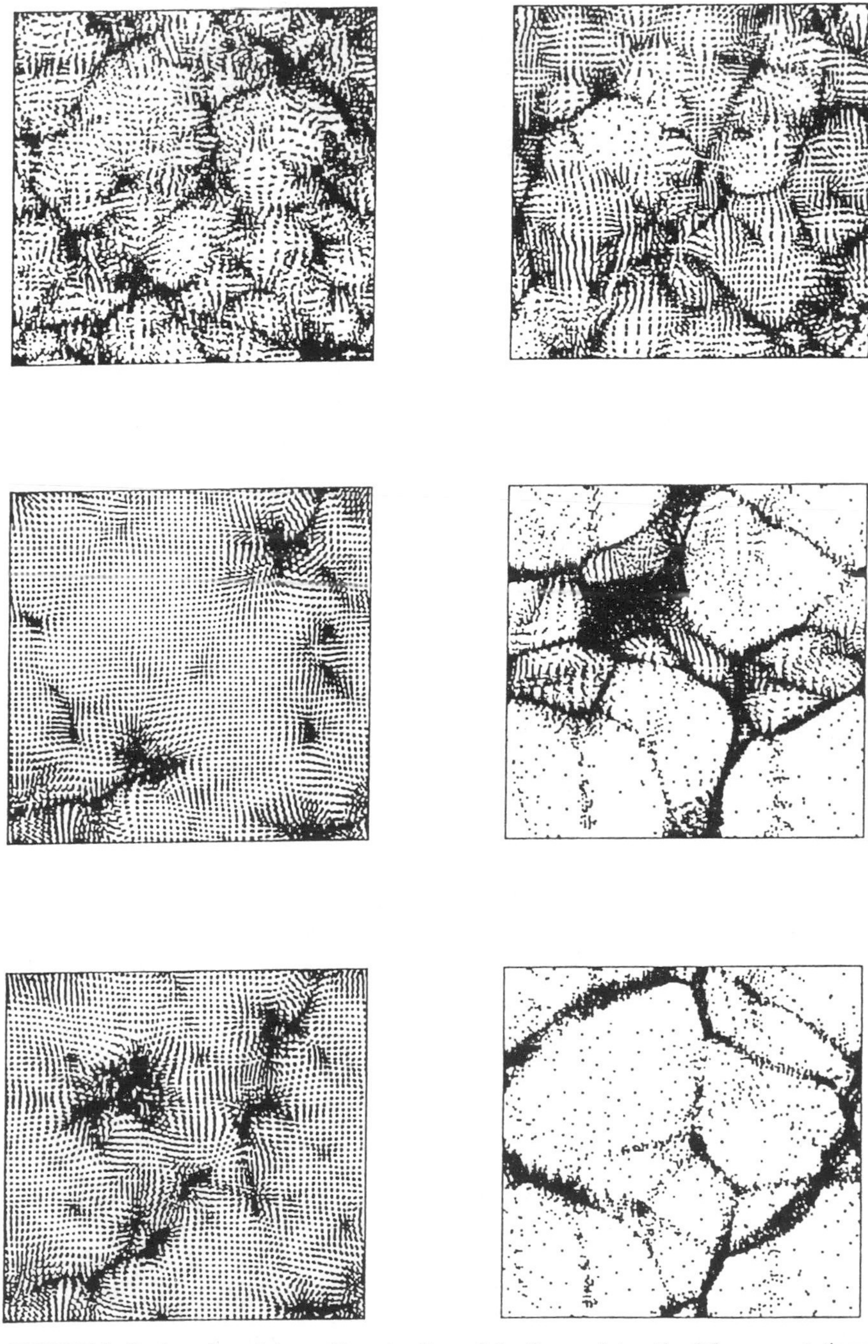

FIGURE 1. Projected particle positions in slices of depth one-sixteenth of the computational box-size at the *present time* t_0. The slices refer to different models: G (**top left**), C (**top right**), LN_p (**center left**), LN_n (**center right**), χ^2_p (**bottom left**), and χ^2_n (**bottom right**).

TABLE 1. Model Parameters

Model	a_0	γ	r_0
G	2.4	1.83 ± 0.02	1.84 ± 0.08
C	2.2	1.81 ± 0.03	1.80 ± 0.10
LN_p	1.8	1.90 ± 0.09	1.54 ± 0.14
LN_n	7.0	1.79 ± 0.01	4.14 ± 0.18
χ^2_p	1.5	1.88 ± 0.08	1.61 ± 0.15
χ^2_n	8.0	1.81 ± 0.02	4.04 ± 0.17

somewhat smaller for the negative and larger for the positive ones; λ_{cr} is of the same order as the CDM characteristic scale λ_{eq}. Above this scale the growth is essentially linear for all but the negative models that display a slowing down in their evolution at large scales: the dynamics of fluctuations in this case is monitored by the large underdense regions, corresponding to a locally open universe. The strong nonlinearity of negative models on small scales is to be ascribed to the long evolution after t_{nl}. In FIGURE 2 we plot the power spectrum at t_0 for G and LN models (χ^2 models behave very similarly to LN ones). It is clear that the long nonlinear evolution of negative models provides a remarkable excess of power at small wavenumbers compared to the Gaussian case.

The analysis of the *contour genus* (see, e.g., reference 20) of the mass distribution shows that positive models have a *meatball* topology, while negative models have a *cellular* one. Although the topology of negative models might appear in contrast with the one observed in the galaxy distribution,[21] one should realize that the topology can

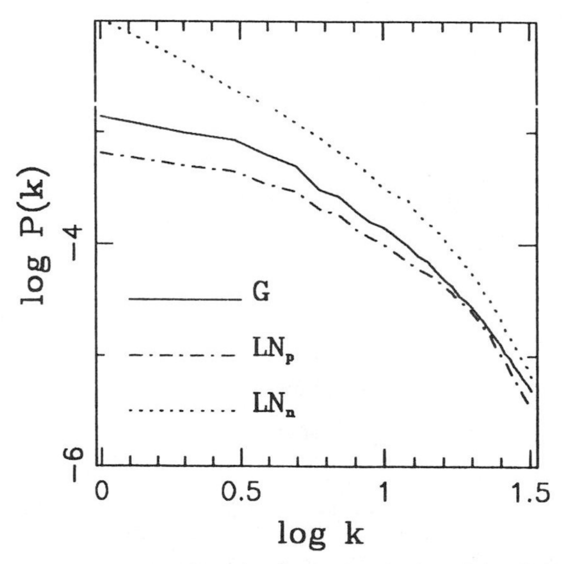

FIGURE 2. The power-spectrum $\mathscr{P}(k)$ at the present time is plotted, for G, LN_p, and LN_n, with $h = 0.5$. Wavenumbers are normalized in units of $2\pi/L$, L being the size of the computational box (13,000 km/sec).

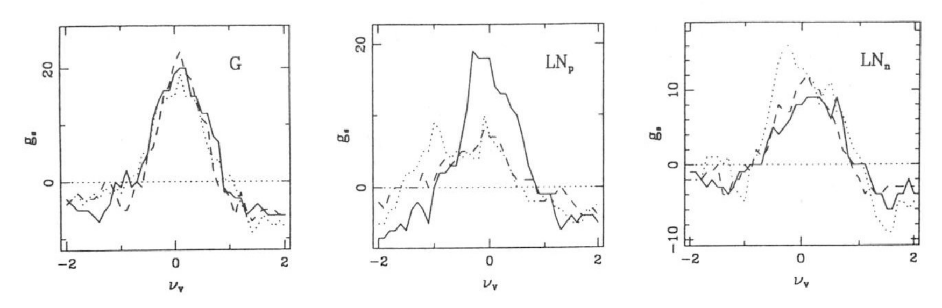

FIGURE 3. The genus curve for particles (*continuous curve*) and for groups with more than three (*dashed curve*) and five (*dotted curve*) particles.

be extremely sensitive to the possible presence of bias. This issue will be considered in a future paper. Some insight on this problem can be obtained by considering the effects of grouping on the topology. Using the percolation algorithm (with linking parameter $b = 0.25$, in units of the mean interparticle distance) systems formed of at least $N_{\min}$ particles have been selected in the simulations at t_0. Two group catalogs with $N_{\min} = 3$ and $N_{\min} = 5$ have been generated by placing a *group* in the center-of-mass of each system.

In FIGURE 3 we report the mean *contour genus* per unit volume g_S, for G, LN_p, and LN_n, as a function of the effective relative threshold ν_V (for details, see reference 13). The simulated data have been filtered with a Gaussian of radius $R_f = 7h^{-2}$ Mpc. Both the LN models display a shift of the peak toward the meatball topology, in qualitative agreement with the observational trend. The shift is more pronounced as $N_{\min}$ is increased. The G model, on the other hand, keeps its *spongelike* topology even when analyzed in terms of groups. The behavior of χ^2 models is quite similar to that of LN ones.

The simulated data can also be studied to derive information about the trend of the peculiar velocity field by analyzing the rms bulk velocity $V_{\text{bulk}}(R)$ and the rms

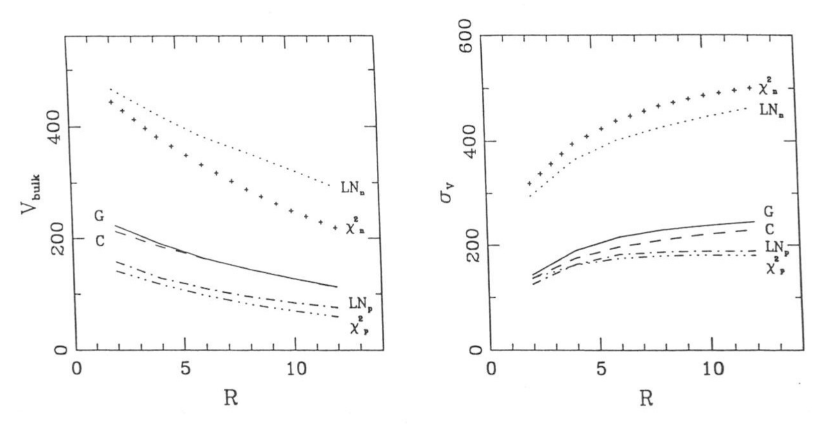

FIGURE 4. The bulk velocity $V_{\text{bulk}}(R)$ (**left**) and the velocity dispersion $\sigma_V(R)$ (**right**), in units of h^{-1} km/sec, versus the radius R, in units of h^{-2} Mpc, at t_0.

velocity dispersion $\sigma_V(R)$ of spherical patches of radius R. FIGURE 4 reports the preceding two quantities, evaluated at the end of the simulation, with R ranging from 2 to $12h^{-2}$ Mpc.

The most relevant result is that negative models are characterized by large bulk motions. We can consider a patch of radius 500 km/sec, the assumed scale of the *local group,* endowed with a velocity of 600 km/sec with respect to the CMB rest frame. For $h = 1$, LN_n has $V_{bulk} \approx 400$ km/sec, while χ_n^2 has $V_{bulk} \approx 360$ km/sec; *G, C,* and positive models all have $V_{bulk} < 200$ km/sec. For $h = 0.5$, negative models have $V_{bulk} \sim 900$ km/sec; all the other ones have $V_{bulk} < 400$ km/sec. Because of the high level of nonlinearity on small scales, negative models also have larger velocity dispersion.

CONCLUSIONS

The results reported here represent the first analysis, obtained by numerical simulations,[13] of the nonlinear clustering evolution in a CDM scenario, where the standard random-phase assumption has been removed. We found that the evolution strongly depends on the initial skewness of mass fluctuations. The main result is that *negative models,* that is, distributions with a substantial excess of primordial underdense regions, yield a possible solution to the large-scale problem of the standard CDM cosmogony: preserving the bottom-up hierarchical process they succeed in producing coherence on large scales and high bulk motions by the slow nonlinear process of merging of voids and disruption of low-density bridges. The resulting cellular structure displays interesting similarities with the recently proposed "Voronoi models" (see, e.g., reference 22). We can conclude that the lack of large-scale structure in the standard CDM model is not to be ascribed to the gravitational instability mechanism of structure formation. Still preserving the scale-invariant power spectrum, the assumption of suitable non-Gaussian primordial statistics could represent the key extra parameter of a CDM scenario.

ACKNOWLEDGMENTS

The staff and the management of the CINECA Computer Centre (Bologna) are warmly acknowledged for their assistance and for allowing the use of computational facilities.

REFERENCES

1. BLUMENTHAL, G. R., S. M. FABER, J. R. PRIMACK & M. J. REES. 1984. Nature **311:** 517–525.
2. DAVIS, M., G. EFSTATHIOU, C. S. FRENK & S. D. M. WHITE. 1985. Astrophys. J. **292:** 371–394.
3. WHITE, S. D. M., C. S. FRENK, M. DAVIS & G. EFSTATHIOU. 1987. Astrophys. J. **313:** 505–516.
4. VITTORIO, N., P. MEINHOLD, P. F. MUCIACCIA, P. LUBIN & J. SILK. 1991. Astrophys. J., Lett. In press.

5. MADDOX, S. J., G. EFSTATHIOU, W. J. SUTHERLAND & J. LOVEDAY. 1990. Mon. Not. R. Astron. Soc. **242:** 43P–47P.
6. BROADHURST, T. J., R. S. ELLIS, D. C. KOO & A. S. SZALAY. 1990. Nature **343:** 726–728.
7. SAUNDERS, W., *et al.* 1991. Nature **349:** 32–38.
8. MATARRESE, S., A. ORTOLAN & F. LUCCHIN. 1989. Phys. Rev. **D40:** 290–298.
9. SALOPEK, D. S. & J. R. BOND. 1991. Phys. Rev. **D43.** In press.
10. MOLLERACH, S., S. MATARRESE, A. ORTOLAN & F. LUCCHIN. 1990. Preprint DFPD/90/A/28.
11. KOFMAN, L. 1990. *In* Nobel Symposium No. 79: The Birth and Early Evolution of Our Universe, B. Skagerstam, Ed. In press.
12. MESSINA, A., L. MOSCARDINI, F. LUCCHIN & S. MATARRESE. 1990. Mon. Not. R. Astron. Soc. **245:** 244–254.
13. MOSCARDINI, L., S. MATARRESE, F. LUCCHIN & A. MESSINA. 1991. Mon. Not. R. Astron. Soc. **248:** 424–438.
14. PEEBLES, P. J. E. 1983. Astrophys. J. **274:** 1–6.
15. COLES, P. & J. D. BARROW. 1987. Mon. Not. R. Astron. Soc. **228:** 407–426.
16. COLES, P. & B. JONES. 1991. Mon. Not. R. Astron. Soc. **248:** 1–13.
17. VISHNIAC, E. 1986. *In* Inner Space/Outer Space. E. W. Kolb *et al.,* Eds.: 190–193. University of Chicago Press. Chicago, Ill.
18. RYDEN, B. S. & E. L. TURNER. 1984. Astrophys. J., Lett. **287:** L59–L63.
19. DE LAPPARENT, V., M. J. GELLER & J. P. HUCHRA. 1988. Astrophys. J. **332:** 44–56.
20. MELOTT, A. L. 1990. Phys. Rep. **193:** 1–39.
21. GOTT, J. R., *et al.* 1989. Astrophys. J. **340:** 625–646.
22. ICKE, V. & R. VAN DER WEYGAERT. 1991. Quart. J. R. Astron. Soc. In press.

Fluctuations in the X-ray Background[a]

J. M. MARTÍN-MIRONES,[b,c] G. DE ZOTTI,[c]
A. FRANCESCHINI,[c] AND L. DANESE[d]

[b]*Departamento de Física Moderna*
Universidad de Cantabria
Avda. de Los Castros s/n
E-39005 Santander, Cantabria, Spain

[c]*Osservatorio Astronomico*
Vicolo dell'Osservatorio 5
I-35122 Padova, Italy

[d]*Dipartimento di Astronomia*
Università di Padova
Vicolo dell'Osservatorio 5
I-35122 Padova, Italy

OVERVIEW

Considerable attention has been paid recently to the anisotropies of the extragalactic X-ray background (XRB) as a tool to investigate, on one side, the large-scale structure of the universe and, on the other side, the population properties of extragalactic X-ray sources beyond the detection limit.

Constraints on Large-scale Structures

Although most of the currently available information on the large-scale matter distribution comes from optical data, it is becoming increasingly clear that X-ray surveys have important advantages: they strongly emphasize two classes of extragalactic sources, rich clusters of galaxies and active galactic nuclei (AGNs), which are particularly well suited for probing the large-scale matter distribution; allow us to bypass the observational biases affecting optical samples of clusters of galaxies; make it possible to achieve a remarkably uniform sky coverage. In particular, observations above several kiloelectronvolts are not affected by the patchy dust extinction that is always a severe problem for optical surveys over large areas of the sky.

In the soft X-ray band (0.1–2 keV), a great deal of information is expected from the Röntgenstrahlen Satellit (ROSAT) all-sky survey. On the other hand, there is still much to be learned on properties of discrete source populations in other X-ray bands from careful examinations of the XRB brightness distributions measured in previous experiments. In fact, a large-scale distribution pattern of AGNs and of

[a]One of the authors (J. M. M.-M.) acknowledges partial financial support from the Comisión Mixta Caja Cantabria-Universidad de Cantabria (Spain); his stay at the Padova Astronomical Observatory was supported by MEC Grant EX90 13.741.993. The Italian Group acknowledges partial financial support by MURST, CNR (through GNA), and ASI.

clusters of galaxies yields small-scale anisotropies in excess of Poisson fluctuations and, of course, translates into power in the autocorrelation function of the extragalactic background brightness.

Barcons and Fabian (1989) investigated the XRB autocorrelation function on small angular scales (1′ to 15′) using 1–3 keV data from five deep Einstein Observatory imaging proportional counter (IPC) fields. A 95 percent confidence upper limit for the autocorrelation function at 5′ was set, implying that the clustering scale of X-ray sources at a redshift of a few cannot exceed ≈ 20 Mpc (H_0 = 50).

Several recent studies have exploited the remarkably uniform sky coverage of the "old" High Energy Astrophysical Laboratory (HEAO-1) A-2 data base, which still is, and will remain for at least several years, the basis for investigations in the energy band where the XRB spectrum is best measured.

An analysis of the dipole anisotropy produced by AGNs, led Miyaji and Boldt (1990) to conclude that the X-ray emission from these sources appears to trace the underlying mass distribution at least as strongly as optical and IR emission from galaxies. They also obtained a stringent upper limit on the local volume emissivity of extragalactic sources.

Jahoda and Mushotzky (1989) demonstrated that accurate measurements of the surface brightness distribution of the XRB on scales $\gtrsim$1000 square degrees can provide important information on nearby large mass concentrations such as the "Great Attractor" (for a detailed discussion of the effect of nearby large-scale inhomogeneities on the isotrophy of the XRB, see Goicoechea and Martín-Mirones 1990).

The upper limit (2.3 percent at the 90 percent confidence level for an effective solid angle of 25 square degrees) set by Shafer and Fabian (1983) on fluctuations in excess of those expected from a Euclidean extrapolation below the HEAO-1 A-2 detectability threshold of the number counts of bright sources (under the assumption of a Poisson distribution) has been exploited to derive constraints on large-scale structures (Barcons and Fabian 1988; Bagoly *et al.* 1988).

More direct information on the clustering of X-ray sources can be obtained from the autocorrelation function of XRB intensity fluctuations. Persic *et al.* (1989a) analyzed the A-2 data for angular separations ranging from 3° to 27°. The corresponding constraints on the spatial distribution of X-ray sources were extensively discussed by De Zotti *et al.* (1990).

Carrera *et al.* (1991) have used data in the 4–12 keV band, taken with the large area counter on board of the *Ginga* satellite, to explore the angular autocorrelation function on scales 2°–25° (see also Fabian 1992).

Very interesting hints that the average broadband AGN spectrum flattens in the interval 12–40 keV, and then steepens again in the range 40–166 keV were obtained from recent investigations of rms fluctuations observed by the HEAO-1 A-4 experiment in different energy bands (Boldt 1989, and references therein).

Constraints on Source Counts

It has long been known that the detailed amplitude distribution of intensity fluctuations due to randomly distributed unresolved sources [known as $P(D)$ distribution] depends on the shape of the number counts of those sources.

Based on a $P(D)$ analysis of several deep IPC fields, Hamilton and Helfand (1987) and Barcons and Fabian (1990) concluded that the slope of the counts in the 1–3 keV band must flatten down at a flux level a few times below the Einstein Observatory Deep Survey (DS) limit.

Fluctuation analyses are particularly important in the hard (>3-keV) X-ray band, where only the brightest part of the log N–log S curve has been directly determined. Using the A-2 database Shafer (1983) showed that the Euclidean fit to the direct counts by Piccinotti *et al.* (1982) can be extrapolated by a factor of about 10 downward in flux. *If* we assume an average energy spectral index of 0.7 (the "canonical" value for AGNs) and negligible photoelectric absorption, this extrapolation falls about a factor of 3 *above* the Einstein Observatory Extended Medium Sensitivity Survey (EMSS) counts in the 0.3–3.5 keV band (Gioia *et al.* 1990). This conclusion was strengthened by recent measurements of XRB fluctuations in the 4–11 keV band made with the large-area counter on the *Ginga* satellite (Warwick and Stewart 1989).

A NEW ANALYSIS OF HEAO-1 A-2 DATA

Attempts to exploit the A-2 data base to investigate clustering on scales of a few tens of Mpc are restrained primarily by the limited angular resolution (the small field of view has FWZI of 3° along the scan direction and of 6° perpendicular to it). In fact, the angular coherence length of XRB fluctuations is $\approx H_0 r_0/c \approx 0^\circ.3\ (H_0/50)(r_0/30$ Mpc). On the other hand, the accurate determination of the angular response function of A-2 detectors made by F. E. Marshall, has allowed Martín-Mirones *et al.* (1991) to push the analysis to scales <3°. Since the source confusion background is unlikely to vary appreciably on a timescale of six months, a comparison of the count rates recorded in the same directions during the first and the second A-2 scan of the sky allowed them to subtract out the detector's noise.

Their results are shown in FIGURE 1, where $C(\theta) = \langle[(\delta I - \langle\delta I\rangle) \cdot (\delta I' - \langle\delta I\rangle)]_\theta\rangle$, δI and $\delta I'$ being the intensity fluctuations in directions separated by an angle θ. The data points are strongly correlated. The only independent points are the rms fluctuations, $C(0°) = (3.41 \pm 0.41) \times 10^{-2}$, and the autocorrelations on a scale of 3°, $C(3°) = (-0.54 \pm 0.27) \times 10^{-2}$.

These results are in good agreement with those obtained by Carrera *et al.* (1991) from *Ginga* data. In terms of the $W(\theta) = C(\theta)/\langle I\rangle^2$ used by these authors, we have $W(3°) = (-3.4 \pm 1.7) \times 10^{-4}$, fully consistent with their results: $W(2°) = (-6.5 \pm 3.8) \times 10^{-4}$ and $W(5°) = (-1.4 \pm 2.0) \times 10^{-4}$. Both the A-2 and the *Ginga* data indicate that, albeit at a low significance level, the autocorrelation function may be negative on scales of 2–3 degrees.

The Poisson Contribution

Since there is some evidence of a signal only on scales smaller than the detector's field of view, it could be due to Poisson fluctuations. An extrapolation of the counts by Piccinotti *et al.* (1982) assuming a Euclidean slope, consistent with the results of

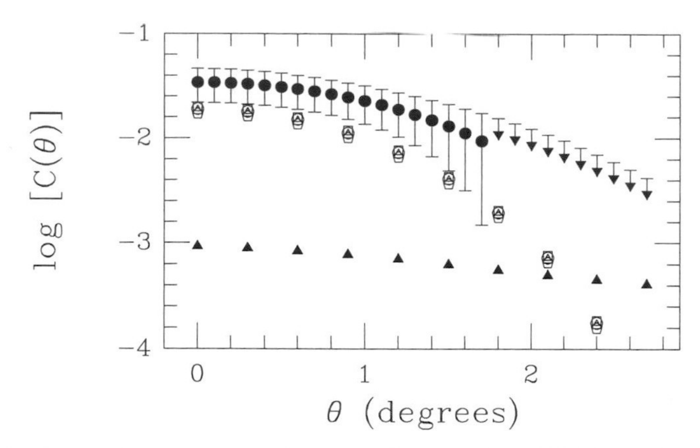

FIGURE 1. Estimated autocovariance function $C(\theta)$ on scales $\leq 3°$ (*filled circles and arrows,* the latter showing upper limits; *error bars* correspond to 3σ). The *open symbols* show the expectations for randomly distributed sources and Euclidean counts down to the Einstein Observatory Deep Survey limit. Evolution models for AGNs are taken from Danese *et al.* (1986). Rich clusters are assumed to evolve according to Kaiser's (1986) models, with index $n = -1$ or $n = 0$. The predictions of different models are essentially identical, as a consequence of the fact that the fluctuation level is determined by the observed counts, that all models are required to fit. All models assume an $\Omega = 1$ cosmology. The *filled triangles* show the non-Poisson contributions to $C(\theta)$ from clusters with a clustering radius $r_0 = 50$ Mpc, $r_{max} = 6r_0$, $r_c = 1$ Mpc [see (**1**)], $n = -1$, and stable clustering ($\epsilon = 0$).

the Einstein Observatory EMSS and DS, yields, in units of (R15 counts $s^{-1})^2$, $C_{Poisson}(0°) = 2.05^{+0.18}_{-0.13} \times 10^{-2}$, 3σ below the observed value [$C(0°) = (3.41 \pm 0.41) \times 10^{-2}$]. More refined calculations, taking into account that the counts must flatten somewhere below the DS limit, give slightly *smaller* values of $C_{Poisson}(0°)$ (see FIG. 1). Possible explanations of this result (if real) include:

- Actual counts in the A-2 band are somewhat *steeper* than indicated by the Einstein Observatory surveys. To account for the observed $C(\theta)$ the slope of integral counts should be $1.66^{+0.05}_{-0.07}$, fully consistent with the best fit slope of the counts by Piccinotti *et al.* (1982): $1.72^{+0.15}_{-0.10}$.
- Extragalactic X-ray sources are significantly clustered.

The Effect of Clustering

A convenient measure of the autocovariance function of XRB fluctuations is $\Gamma(\theta) = [C(\theta)]^{1/2}/\langle I\rangle$, where $\langle I\rangle = 4.0$ R15 counts s^{-1} is the mean XRB intensity. We assume the usual parametrization of the two-point spatial correlation function: $\xi(r, z) = (1 + z)^{-(3+\epsilon)}\,\xi_0(r)$, with

$$\xi_0(r) = \begin{cases} [r_0/(r + r_c)]^{1.8} & \text{if } r \leq r_{max} \\ 0 & \text{if } r > r_{max}. \end{cases} \qquad (1)$$

Values of ϵ discussed in the literature range from 0 (the number of objects $n(t)\xi(t)$ at fixed proper separation r is constant: statistically stable clustering) to -3 (ξ constant in physical coordinates).

For $r_{max} \rightarrow \infty$, $\epsilon = 0$ and no evolution, the contribution of the clustering to small-scale fluctuations is well approximated by

$$\Gamma(0) \simeq 0.5 \times 10^{-2} \left(\frac{r_0}{50\text{ Mpc}}\right)^{0.9} \frac{(j_{\text{sources}}/j_{\text{XRB}})_{z=0}}{0.04}, \tag{2}$$

where j denotes the volume emissivities. In the case of evolving sources, $(j_{\text{sources}}/j_{\text{XRB}})_{z=0}$ must be replaced with the global contribution of the given class of sources to the XRB. Varying ϵ from 0 to -3 leads to an increase of $\Gamma(0)$ by a factor increasing from $\simeq 1.3$ (no evolution) to $\simeq 2$ (pure luminosity evolution of AGNs). Similarly, we have

$$\Gamma(3^\circ) \simeq 2.8 \times 10^{-3} \left(\frac{r_0}{50\text{ Mpc}}\right)^{0.9} \frac{(j_{\text{sources}}/j_{\text{XRB}})_{z=0}}{0.04}. \tag{3}$$

The contribution to $\Gamma(0^\circ)$ unaccounted by Poisson fluctuations (for Euclidean counts) is $\Gamma_u(0^\circ) = (2.91^{+0.46}_{-0.48}) \times 10^{-2}$ and the 3 σ upper limit to $\Gamma(3^\circ)$ is $\Gamma(3^\circ) \leq 1.3 \times 10^{-2}$.

It is thus clear that correlations of rich clusters are not enough to account for the observed $C(0)$, unless their clustering scale is substantially larger than indicated by optical data. The main reason for that is their relatively low contribution to the XRB volume emissivity.

The same argument, on the other hand, suggests that clustering of AGNs, which have a much higher volume emissivity, could explain the observed signal. The required local clustering strength depends on the evolution of both the luminosity function and the correlation function. TABLE 1 summarizes the values of the clustering radius r_0 (Mpc) that would account for the estimated $C(\theta)$ in the cases of pure luminosity evolution (PLE) and of luminosity evolution confined to sources brighter than $\log L_s(2\text{–}10\text{ keV}) = 43.5$. The values of the evolution parameters were taken from Danese *et al.* (1986) and correspond to an Einstein–de Sitter cosmology. Two values of ϵ and of the ratio r_{max}/r_0 were considered.

Optical studies (e.g., Iovino *et al.* 1989) indicate for quasars at $z \approx 1$, $\xi \approx 1$ on a *proper* scale of $\simeq 10$ Mpc, corresponding to a present clustering length $r_0 \approx 30$ Mpc for stable clustering, or $r_0 \approx 10$ Mpc, if ξ is constant. The results of the present analysis are thus compatible with optical data.

TABLE 1. Constraints on the Clustering Radius r_0 (Mpc) of AGNs ($H_0 = 50$)

	$\epsilon = 0$, r_{max}/r_0		$\epsilon = -3$, r_{max}/r_0	
	2	6	2	6
PLE	20–30	15–20	10–20	10–15
$\log L_s \geq 43.5$	30–45	30–35	20–40	15–25

Hard vs. Soft X-ray Counts

As already mentioned in the first section, HEAO-1 and *Ginga* fluctuation results seem to indicate a normalization of the source counts a factor of about 3 above that derived from the EMSS. The present analysis also indicates that the slope of counts in the A-2 energy band may be somewhat steeper than in the IPC band. The reason for these differences is still unclear. Warwick and Stewart (1989) stressed that a likely explanation may be a substantial photoelectric absorption ($N_H \sim 10^{22}$ cm^{-2}) that would make low luminosity AGNs underrepresented in the EMSS, and found that the spectrum of XRB fluctuations observed by *Ginga* is consistent with this conclusion.

On the other hand, the statistical spectral analysis of EMSS sources carried out by Maccacaro *et al.* (1988) has shown that these sources are rather characterized in the 0.3–3.5 keV band, by spectral indices that are, on the average, significantly *steeper* ($\langle\alpha_x\rangle \simeq 1$) than the "canonical" AGN spectrum. That soft X-ray excesses are a common feature of AGN spectra was also shown by detailed spectral studies of relatively bright AGNs (Wilkes *et al.* 1989; Turner and Pounds 1989; Urry *et al.* 1989).

These apparently contradictory results can only be reconciled if the X-ray spectra of AGNs are rather complex and diverse. In an attempt to better focus these spectral properties and to quantify possible biases related to selection in different bands (hard X-rays, soft X-rays, optical), we have carried out a statistical analysis of the distribution of the ratio:

$$F_H/F_S = S(2\text{–}10\ \text{keV})/S(0.5\text{–}4.5\ \text{keV}) \tag{4}$$

for various AGN samples: the hard X-ray-selected sample by Piccinotti *et al.* (1982); the EMSS AGNs brighter than $S(0.3\text{–}3.5\ \text{keV}) = 2 \times 10^{-12}$ erg cm^{-2} s^{-1}; the optically selected Seyfert sample defined by Cheng *et al.* (1985); the optically selected PG quasar sample (Schmidt and Green 1983). For a "canonical" AGN spectrum, $\alpha_x \simeq 0.7$, we expect $F_H/F_S \simeq 1$.

With the obvious exception of the Piccinotti sample, in general these AGNs are too weak to allow detection by the A-2 survey. Still, from the distribution of A-2 count rates in the directions of these sources, we can derive, using the method described by Persic *et al.* (1989b), the sample average of the ratio F_H/F_S. The results of our analysis can be summarized as follows (see also FIG. 2):

- AGNs from the EMSS have $\langle F_H/F_S\rangle \simeq 0.6$, confirming that, for these objects, substantial soft excesses (in comparison to the "standard" spectrum) prevail.
- Soft excesses are also common in the case of the PG quasar sample, for which we find $\langle F_H/F_S\rangle \simeq 0.75$, while this ratio is close to unity for the homogeneous Seyfert sample. If the intrinsic X-ray spectra of Seyferts and quasars are similar, and steeper in the soft X-ray band than $\alpha_x = 0.7$, this confirms that strong photoelectric absorption prevails for lower luminosity nuclei (Reichert *et al.* 1985).
- On the other hand, the average F_H/F_S for the Piccinotti sample is $\simeq 1.8$, corresponding to $\langle N_H\rangle \sim 10^{22}$ cm^{-2} (for a canonical α_x). Only a minor fraction of Piccinotti sources show values of F_H/F_S as small as those found in the cases of soft X-ray or optical selection.

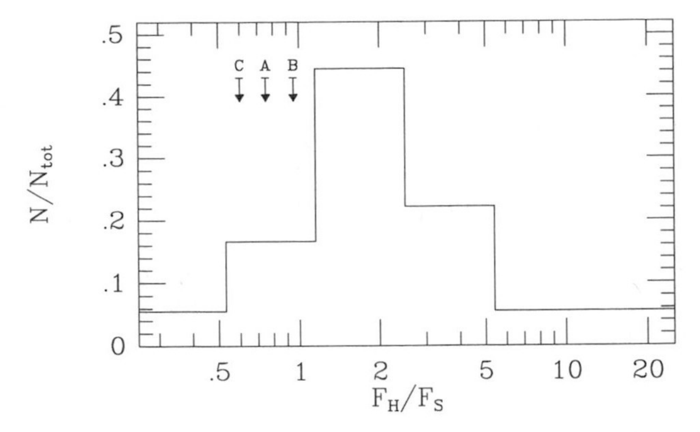

FIGURE 2. Distribution of the ratio F_H/F_S for the 18 AGNs in the Piccinotti sample for which IPC measurements are also available. The *arrows* on the top left-hand corner show the average F_H/F_S for the PG quasar sample (A), the homogeneous Seyfert sample (B), and the bright EMSS AGN sample (C).

It is thus apparent that AGNs have substantially diverse soft X-ray spectra. Differences are emphasized by selection in different spectral bands. In particular, soft X-ray surveys are biased in favor of AGNs with steep spectra, whose counts for given cosmological evolution are made somewhat flatter by the effect of the K-correction. On the contrary, the K-correction works to steepen the hard X-ray counts, particularly in the presence of a flat spectrum component, showing up at $E \gtrsim 10$ keV (Matsuoka *et al.* 1990).

CONCLUSIONS

1. We find some (3σ) evidence for fluctuations in excess of those expected for a Poisson distribution, *if* counts have a Euclidean slope as indicated by the Einstein Observatory surveys.
2. A possible explanation may be that the log N–log S in the A-2 band is slightly steeper than Euclidean. An analysis of the hard/soft X-ray luminosity ratios for AGN samples selected in different bands has demonstrated that their average soft X-ray spectra are substantially diverse. Soft X-ray surveys are strongly biased in favor of steep spectrum sources, which are a minor fraction of those showing up in the HEAO-1 A-2 survey; this bias may explain the apparent difference by a factor of 3 in the normalization of the hard X-ray counts, in comparison with the Einstein Observatory counts. Also, flatter counts in soft X-ray bands may be expected as the effect of the K-correction, particularly in the presence of a hard component at $E \geq 10$ keV, indicated by *Ginga* data.
3. Alternatively the observed effect may be (at least partly) due to clustering.

- However, it cannot be fully accounted for by the spatial correlations of rich clusters of galaxies indicated by optical data.
- Clustering of AGNs could do the job. The required local clustering strength may be consistent with optical data.

ACKNOWLEDGMENTS

A good fraction of the work reported in this paper was carried out in collaboration with E. A. Boldt, F. E. Marshall, and M. Persic. G. D. Z. thanks A. C. Fabian for useful discussions.

REFERENCES

BAGOLY, Z., A. MÉSZÁROS & P. MÉSZÁROS. 1988. Astrophys. J. **333:** 54.
BARCONS, X. & A. C. FABIAN. 1988. Mon. Not. R. Astron. Soc. **230:** 189.
BARCONS, X. & A. C. FABIAN. 1989. Mon. Not. R. Astron. Soc. **237:** 119.
BARCONS, X. & A. C. FABIAN. 1990. Mon. Not. R. Astron. Soc. **243:** 366.
BOLDT, E. A. 1989. *In* Two Topics in X-ray Astronomy, J. Hunt and B. Battrick, Eds.: 797. ESA SP-296.
CARRERA, F. J., *et al.* 1991. Mon. Not. R. Astron. Soc. **249:** 698.
CHENG, F.-Z., L. DANESE, G. DE ZOTTI & A. FRANCESCHINI. 1985. Mon. Not. R. Astron. Soc. **212:** 857.
DANESE, L., G. DE ZOTTI, G. FASANO & A. FRANCESCHINI. 1986. Astron. Astrophys. **161:** 1.
DE ZOTTI, G., *et al.* 1990. Astrophys. J. **351:** 22.
FABIAN, A. C. 1991. The origin of the X-ray background. This issue.
GIOIA, I. M., *et al.* 1990. Astrophys. J., Supp. Ser. **72:** 567.
GOICOECHEA, L. J. & J. M. MARTÍN-MIRONES. 1990. Mon. Not. R. Astron. Soc. **244:** 493.
HAMILTON, T. T. & D. J. HELFAND. 1987. Astrophys. J. **318:** 93.
IOVINO, A., *et al.* 1989. *In* Large Scale Structure and Motions in the Universe, M. Mezzetti *et al.*, Eds.: Kluwer. Dordrecht, the Netherlands.
JAHODA, K. & R. F. MUSHOTZKY. 1989. Astrophys. J. **346:** 638.
KAISER, N. 1986. Mon. Not. R. Astron. Soc. **222:** 323.
MACCACARO, T., I. M. GIOIA, A. WOLTER, G. ZAMORANI & J. T. STOCKE. 1988. Astrophys. J. **326:** 680.
MARTÍN-MIRONES, J. M., *et al.* 1991. Astrophys. J. **379:** 507.
MATSUOKA, M., L. PIRO, M. YAMAUCHI & T. MURAKAMI. 1990. Astrophys. J. **361:** 440.
MIYAJI, T. & E. BOLDT. 1990. Astrophys. J., Lett. **353:** L3.
PERSIC, M., *et al.* 1989a. Astrophys. J., Lett. **336:** L47.
PERSIC, M., *et al.* 1989b. Astrophys. J. **344:** 125.
PICCINOTTI, G., *et al.* 1982. Astrophys. J. **253:** 485.
REICHERT, G. A., R. F. MUSHOTZKY, R. PETRE & S. S. HOLT. 1985. Astrophys. J. **296:** 69.
SCHMIDT, M. & R. F. GREEN. 1983. Astrophys. J. **269:** 352.
SHAFER, R. A. 1983. Ph.D. thesis, University of Maryland, College Park.
SHAFER, R. A. & A. C. FABIAN. 1983. *In* IAU Symposium 104: Early Evolution of the Universe and Its Present Structure, G. O. Abell and G. Chincarini, Eds.: 333. Reidel. Dordrecht, the Netherlands.
TURNER, T. J. & K. A. POUNDS. 1989. Mon. Not. R. Astron. Soc. **240:** 833.
URRY, C. M., *et al.* 1989. *In* Two Topics in X-ray Astronomy, J. Hunt and B. Battrick, Eds.: 789. ESA SP-296.
WARWICK, R. S. & G. C. STEWART. 1989. *In* Two Topics in X-ray Astronomy. J. Hunt and B. Battrick, Eds.: 727. ESA SP-296.
WILKES, B. J., J.-L. MASNOU & M. ELVIS. 1989. *In* Two Topics in X-ray Astronomy. J. Hunt and B. Battrick, Eds.: 1081. ESA SP-296.

The Low-frequency Spectrum of the Cosmic Background Radiation[a]

G. SIRONI, G. BONELLI, AND M. LIMON

Dipartimento di Fisica
Universitá di Milano
Milano, Italy
and
Istituto IFCTR/CNR
Milano, Italy

INTRODUCTION

The relic or cosmic background radiation (CBR) was discovered in 1964 (Penzias and Wilson 1965). Twenty-five years have elapsed since then, and a long series of observations carried out by many groups of researchers has shown that the frequency spectrum of the CBR is well represented by an equilibrium (Planckian) distribution similar to the one produced by a blackbody radiator with a temperature of about 3 K. More precisely the recent measurement made by the COBE team (Mather *et al.* 1990) gives for the thermodynamic temperature of the CBR $T^{\text{th}}_{\text{CBR}} = (2.735 \pm 0.060)$ K, a value confirmed by the even more recent result of Gush *et al.* (1990).

Deviations from a Planckian distribution are, however, expected. If sudden energy releases associated to phenomena like matter annihilation, particle decays, dissipation of shock waves and turbulences, birth of matter condensations, occurred during the evolution of the Universe, the frequency distribution of the radiation may have been modified and the deviation from the equilibrium spectrum may still be visible if the energy injection occurred at $Z \lesssim 10^6$.

The values of the CBR temperature measured at various frequencies (for a review see, for instance, Sironi and Celora 1990, and references therein) are consistent with a flat distribution, as one expects for a Planckian spectrum, and only upper limits to the amplitude $\Delta T/T$ of the distortions can be set looking at the error bars of the measured values of temperature.

Between 30 and 600 GHz the far infrared spectrophotometer (FIRAS) experiment on COBE (Mather *et al.* 1990) gave $\Delta T/T < 0.01$. Between 2 and 30 GHz the upper limit $\Delta T/T < 0.05$ set by the White Mountain collaboration (Smoot *et al.* 1985) is still valid. Below 2 GHz the error and the resulting limits are very large, $\Delta T/T \lesssim 0.3$.

To improve the upper limit set by COBE will be extremely difficult. That limit, however, holds only over a well-defined range of frequencies. Moreover, numerical solutions of the Kompaneets equation, which describes the evolution of the distortions, clearly show (Burigana *et al.* 1991) that the lack of large distortions at high

[a]This work was supported in part by the Italian Antarctic Project, in part by the National Science Foundation Antarctic Project, and in part by the Italian Ministry for Research and University.

frequency does not rule out the presence of distortions at low frequencies. In fact, distortions at high and low frequencies can be associated to different processes and epochs of the thermal history of the Universe.

FIGURE 1 gives the expected distribution versus frequency of $T^{\mathrm{th}}_{\mathrm{CBR}}$ for two examples of spectra distorted by energy injections (from Burigana *et al.* 1991). In the same figure the horizontal line represents the undisturbed spectrum. The shaded areas mark the observational limits to the amplitude of the distortions. It appears that at low frequencies there is still room for nonnegligible distortions. It is therefore important to carry on new observations, particularly near and below 1 GHz where a minimum of the temperature distribution is expected (see FIG. 1). The detection of that minimum would produce important cosmological information. It has, in fact, been shown (see Burigana *et al.* 1991, and references therein) that the frequency at which the minimum appears is directly related to the density of barionic matter in the Universe. On the contrary, were distortions absent both at low and high frequencies, a revision of our understanding of the history of the Universe would be necessary.

With the aim of improving the quality of the low-frequency observational data, in 1985 our group in Milano and the group of G. Smoot in Berkeley decided to extend their collaboration beyond the White Mountain project (Smoot *et al.* 1985) and began a new series of systematic observations at frequencies near and below a few gigahertz. Some results, in particular new data at 0.6 GHz (Sironi *et al.* 1990) and 1.4 GHz (Levin *et al.* 1988), have been already published. Here we present the data obtained in 1989 at 0.82 GHz and 2.5 GHz (Sironi *et al.* 1991) by the Milano group and compare them with data in the literature and the preliminary results at 3.8 GHz of the Berkeley group (De Amici *et al.* 1990b).

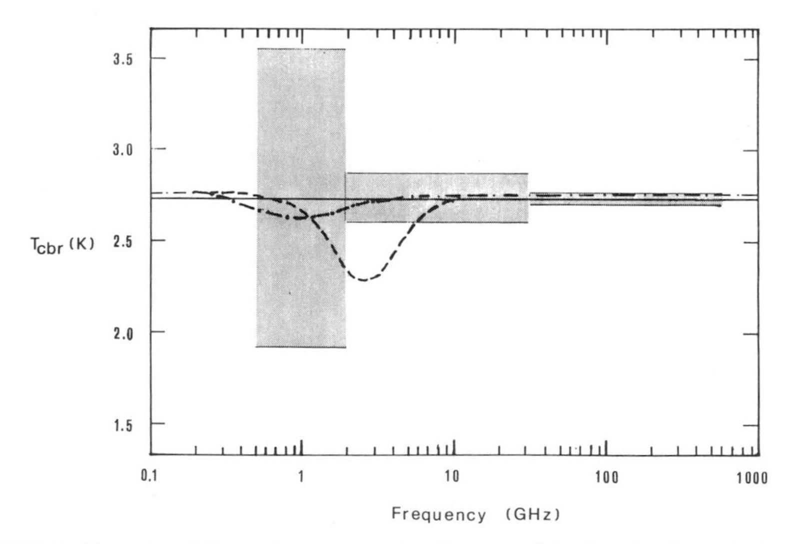

FIGURE 1. Examples of distorted spectra produced by energy injections $\Delta\epsilon$. (From Burigana *et al.* 1991. Reproduced with permission.) NOTE: *dot–dash line*—$\Delta\epsilon/\epsilon = 0.001$ in a universe with a comptonization parameter $y_e = 5$ and $\Omega_b = 0.1$; *dashed line*—$\Delta\epsilon/\epsilon = 0.01$, $y_e = 5$ and $\Omega_b = 0.3$; *straight line*—undistorted spectrum; *shaded area*—upper limits set by the observations (see text) to the maximum amplitudes of the so far undetected distortions.

TABLE 1. The Temperature of the Sky and the Temperature of the Cosmic Background Radiation*

Frequency	0.820	2.5	GHz
Zenith Antenna Temperature	6.7 ± 1.5	3.73 ± 0.15	K
Ground Control	0.03 ± 0.05	0.030 ± 0.050	K
Atmospheric Control	0.90 ± 0.35	1.155 ± 0.300	K
Sun Control	0.08 ± 0.08	0.000 ± 0.005	K
Sky Antenna Temperature	5.7 ± 1.6	2.58 ± 0.34	K
Galactic Diffuse Control	2.67 ± 0.33	0.118 ± 0.025	K
Unresolved Sources	0.34 ± 0.07	0.016 ± 0.005	K
CBR Antenna Temperature	2.7 ± 1.6	2.44 ± 0.34	K
CBR Thermodynamic Temperature	2.7 ± 1.6	2.50 ± 0.34	K

*Measured at 0.82 and 2.5 GHz in the region of the south celestial pole.

THE OBSERVATIONS

The new data have been obtained by the Milano–Berkeley collaboration in December 1989 with a set of radiometers installed near the Amundsen Scott base at the South Pole (Kogut *et al.* 1989). The site was chosen to take advantage of the very dry atmosphere and the low level of man-made radio interferences existing there. The observations, at 0.82 GHz, 1.4 Ghz, 2.5 GHz, 3.8 GHz, and 7.5 GHz, give the antenna temperature T_a of the zenith. By subtraction of the unwanted signals produced by the ground around the radiometers, the earth's atmosphere (monitored also with an *ad hoc* radiometer at 90 GHz), and the sun (always above the horizon during the summer at the Pole) from T_a, one gets T_{sky} (SCP), the temperature of the sky around the south celestial pole. Further corrections for the diffuse galactic contribution and the blend of unresolved extragalactic sources finally provide the antenna temperature T_{CBR} of the CBR and its thermodynamic equivalent T^{th}_{CBR}. Details of the observations at 0.82 GHz and 2.5 GHz, the auxiliary measurements made to evaluate the different contributions and the methods of data reduction we used are presented elsewhere (Sironi *et al.* 1991). The results are shown in TABLE 1.

TABLE 2 is a compilation of the measured values of T^{th}_{CBR} at $\nu < 5$ GHz found in the literature. In the same table we present a preliminary value of the CBR temperature measured at 3.8 GHz by the Berkeley group (De Amici *et al.* 1990b) during the same campaign of observations. The analysis of the new observations at 1.4 and 7.5 GHz is still underway.

DISCUSSION

A quick look at TABLE 2 immediately shows that the new 2.5-GHz data are consistent with, and have an accuracy comparable to, the accuracy of the measurements we made in the past at the same frequency (Sironi *et al.* 1984; Sironi and Bunelli 1986). At 0.82 GHz we met difficulties in measuring at the pole because the antenna attenuation produced a large uncertainty of T_a, thus limiting the accuracy of the final value of T^{th}_{CBR} (0.82) that we got. The result is, however, quoted here because no other data at the same frequency have been so far published. We are waiting for the final results of the data at 3.8 GHz and the new data at 1.4 GHz and 7.5 GHz. Conclusions are therefore premature.

Here we limit ourselves to point out (see TABLE 2) that at low frequencies ($\nu < 5$ GHz): (1) the weighted mean of the value of T^{th}_{CBR} is (2.67 ± 0.08), marginally lower than (2.735 ± 0.060), the value of T^{th}_{CBR} measured by COBE at high frequencies; (2) the distribution of T^{th}_{CBR} versus ν has a regular trend: a decrease with ν down to a minimum somewhere between 3 and 1 GHz followed by an increase when ν goes below 1 GHz.

Because of the error bars, both effects are statistically not significant and cannot be used to work out an astrophysical conclusion. They support, however, new observing programs at low frequencies.

At present we are carrying on simultaneous observations of the diffuse radiation from the Italian Alps at 0.6 GHz, 0.82 GHz, and 2.5 GHz. Observations of the

TABLE 2. A Compilation of Values of the CBR Temperature*

ν (GHz)	λ (cm)	T^{th}_{CBR} (K)	Reference
0.408–0.610	73–49	3.7 ± 1.2	Howell and Shakeshaft 1967
0.600	50	3.0 ± 1.2	Sironi *et al.* 1990
0.635	47.2	3.0 ± 0.5	Stankevich *et al.* 1970
0.820	36.6	**2.7 ± 1.6**	**Present paper**
1.4	20.7	2.8 ± 0.6	Howell and Shakeshaft 1966
—	21.2	3.2 ± 1.0	Penzias and Wilson 1967
1.5	21.3	2.11 ± 0.38	Levin *et al.* 1988
2.5	12	2.62 ± 0.25	Sironi *et al.* 1984
—	—	2.79 ± 0.15	Sironi and Bonelli 1986
—	—	**2.50 ± 0.34**	**Present paper**
3.7	8.1	2.59 ± 0.13	De Amici *et al.* 1988
3.8	7.9	2.56 ± 0.08	De Amici *et al.* 1990a
—	—	2.71 ± 0.07	De Amici *et al.* 1990a
—	—	**2.64 ± 0.07**	**De Amici *et al.* 1990b** (preliminary)
4.8	6.3	2.73 ± 0.22	Mandolesi *et al.* 1984
—	—	2.70 ± 0.07	Mandolesi *et al.* 1986

*Measured at frequencies $\nu < 5$ GHz; new results are in boldface.

background radiation at different frequencies in different regions of sky are, in fact, necessary to disentangle the CBR from the galactic diffuse radiation and the blend of unresolved extragalactic sources. We are also studying the possibility of low-frequency observations of the background radiation from space.

REFERENCES

BURIGANA, C., L. DANESE & G. F. DE ZOTTI. 1991. Astron. Astrophys. **246:** 49.
DE AMICI, *et al.* 1988. Astrophys. J. **329:** 556.
DE AMICI, *et al.* 1990a. Astrophys. J. **359:** 219.
DE AMICI, *et al.* 1990b. In preparation. (Private communication)
GUSH, H. P., M. HALPERN & E. H. WISHNOW. 1990. Phys. Rev. Lett. **65:** 537.
HOWELL, T. F. & J. R. SHAKESHAFT. 1966. Nature **210:** 1318.
HOWELL, T. F. & J. R. SHAKESHAFT. 1967. Nature **216:** 753.

KOGUT, A., *et al.* 1989. AIP Conference. Proceedings **198:** 71.
LEVIN, S. M., *et al.* 1988. Astrophys. J. **334:** 14.
MANDOLESI, *et al.* 1984. Phys. Rev. **D29:** 2680.
MANDOLESI, *et al.* 1986. Astrophys. J. **310:** 561.
MATHER, *et al.* 1990. Astrophys. J., Lett. **354:** L37.
PENZIAS, A. A. & R. W. WILSON. 1965. Astrophys. J. **142:** 419.
PENZIAS, A. A. & R. W. WILSON. 1967. Astron. J. **72:** 315.
SIRONI, G., P. INZANI & A. FERRARI. 1984. Phys. Rev. **D29:** 2686.
SIRONI, G. & G. BONELLI. 1986. Astrophys. J. **311:** 418.
SIRONI, G., *et al.* 1990. Astrophys. J. **357:** 301.
SIRONI, G. & L. CELORA. 1990. Nuovo Cimento B **105:** 1031.
SIRONI, G., *et al.* 1991. Astrophys. J. **378:** 550.
SMOOT, G. F., *et al.* 1985. Astrophys. J., Lett. **291:** L123.
STANKEVICH, K. S., R. WIELEBINSKY & W. E. WILSON. 1970. Aust. J. Phys. **23:** 529.

The Spatial Distribution of the Nearby Rich Clusters of Galaxies

MARC POSTMAN

Space Telescope Science Institute
3700 San Martin Drive
Baltimore, Maryland 21218

INTRODUCTION

The spatial distribution of rich clusters provides important constraints on theories of the formation and evolution of galaxies, clusters, and large-scale structure. Klypin and Kopylov (1983) and Bahcall and Soneira (1983) were the first to analyze a complete redshift sample of Abell clusters [Hoessel, Gunn, and Thuan 1980 (HGT)]. Because the survey included only 104 clusters, the analysis is sensitive to the presence of the single, very rich supercluster—Corona Borealis (Cor Bor)—in the sample. Now, about 1000 Abell clusters have measured redshifts; however, the majority of these are based on only one or two galaxies. Many of these redshifts were acquired for unrelated research programs and, consequently, the data do not constitute a complete survey. To obtain a statistically complete survey significantly larger than the HGT sample, 115 new redshifts have been combined with published data to assemble a redshift catalog for the brightest 351 Abell clusters north of $\delta = -27°30'$.

This new survey is used to search for large-scale ($\gtrsim 100h^{-1}$ Mpc) coherent structures, constrain the amplitude of the two-point spatial correlation function, the frequency and sizes of voids in the cluster distribution, the distribution of superclusters, and the magnitude of peculiar motions of clusters. The results are in excellent agreement with the recent deep (and nonoverlapping) cluster redshift survey of Huchra *et al.* 1990 (HHPG). This agreement suggests that the clustering properties of Abell clusters are invariant out to a redshift of $z \approx 0.2$. (The values $h = H_0/(100$ km sec^{-1} Mpc$^{-1}) = 1$ and $q_0 = 0.10$ are assumed throughout this paper.)

THE CLUSTER SAMPLE

The redshift survey includes all 350 Abell clusters north of $\delta = -27°30'$, with tenth-ranked photo-red (Abell 1958) galaxy magnitudes $(m_{10}) \leq 16.5$, including 15 clusters from the southern extension of the Abell catalog (Abell, Corwin, and Olowin 1989; hereafter ACO). Since Abell's (1958) catalog specifically excluded nearby clusters of large angular extent, for completeness at low redshift ($z \leq 0.02$), we include the Virgo cluster in the analysis for a total sample of 351 clusters. The Virgo cluster should be classified as a richness class (RC) 1, distance class (DC) 0 Abell cluster (Einasto, private communication). Note that several other well-known nearby clusters, like Cancer, Pegasus, and Fornax fall below Abell's limiting richness criterion, so are not included. New redshifts were obtained at the Multiple Mirror

Telescope (MMT) and Mount Hopkins Observatory for 115 clusters; these systems are primarily RC = 0 clusters and/or clusters with $m_{10} > 16.0$. The remaining redshift data come from the literature. FIGURE 1 shows the angular distribution of the clusters in this survey, in galactic coordinates. Of the 351 clusters, 299 lie more than 30 degrees above or below the galactic plane. About two-thirds of the sample lies in the north galactic hemisphere. TABLE 1 contains additional statistics for this sample.

THE SPATIAL DISTRIBUTION

Number Density as a Function of Redshift

FIGURE 2 shows the (heliocentric) redshift histograms for the cluster sample. The dashed histogram is for RC = 0 clusters; the solid histogram is for RC $\geq$ 1 clusters. The mean redshifts for these two subsamples are 0.066 and 0.072, respectively. FIGURE 3(a) shows the comoving number density as a function of redshift. We compute the density at redshift intervals of 0.005 in shells of width $\Delta z = 0.010$ to smooth out some of the fluctuations caused by small-scale clustering. The oversampling allows a more robust and more accurate estimate of the mean density to be made. For redshifts $z \lesssim 0.08$, the density of RC $\geq$ 0 clusters is essentially constant with a value of $n = 1.20 \times 10^{-5}h^3$ Mpc^{-3}. For $z > 0.08$, the density decreases exponentially, largely because of the m_{10} limit. The dependence of cluster number density on redshift is independent of richness class for this sample.

FIGURE 3(b) shows the cluster number density for the northern Abell catalog ($\delta \geq -27.5°$) based on this sample for $z \leq 0.075$ and on the deeper redshift survey (145 clusters) of HHPG survey (1990) at larger redshifts. The two samples are nearly independent; there are only 15 clusters in common. The solid curve in FIGURE 3(b) shows the comoving density as a function of redshift for RC $\geq$ 1 clusters; the dashed curve shows the density of RC = 0 clusters. Clearly, variations in the density of RC $\geq$ 1 systems are echoed quite well by the RC = 0 systems out to $z \approx 0.2$. In the next three sections we show that the clustering properties of RC = 0 and RC $\geq$ 1 clusters are also similar.

Slices through the Data

Peculiar velocities of Abell clusters are small compared to the Hubble flow (Soltan 1988; Aaronson *et al.* 1989; HHPG 1990). Therefore, for examining the large-scale spatial distribution, it is reasonable to treat the redshift coordinate as a spatial coordinate nearly equivalent to the other two. FIGURE 4(a) and (b) show three orthogonal projections of the entire cluster distribution in galactic and supergalactic Cartesian coordinates, respectively. Distance units are in km sec^{-1}. The dashed lines show the galactic latitude limits of the survey.

The anisotropy test developed by Postman *et al.* (1989) is used to look for the structures with scales $\gtrsim 100h^{-1}$ Mpc proposed by Tully (1987). The northern and southern galactic hemispheres are analyzed separately. No significant large-scale anisotropy is detected in this sample, as is clear from FIGURE 4(a) and (b). There are

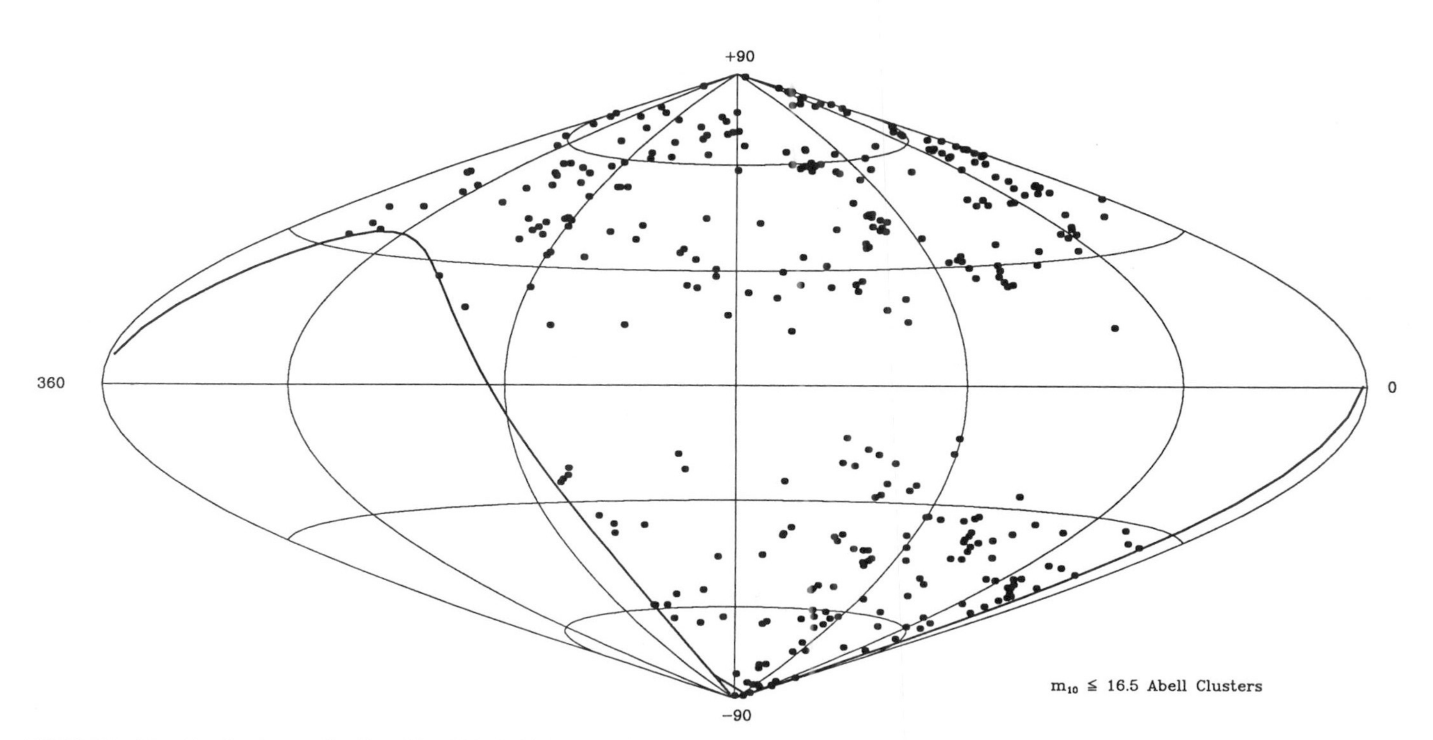

FIGURE 1. The distribution on the sky of the 351 Abell clusters with $m_{10} \le 16.5$. The map is in galactic coordinates. The *solid line* marks the declination limit ($\delta = -27°30'$) of the survey.

TABLE 1.

Richness Class	Number	Distance Class	Number
0	195	0	3
1	126	1	22
2	28	2	11
3	2	3	91
4	0	4	166
5	0	5	58

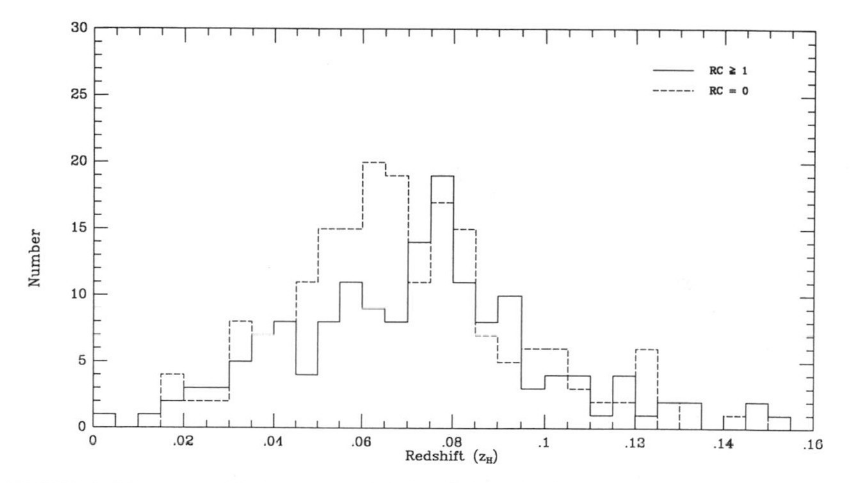

FIGURE 2. Histogram of heliocentric redshifts for the 351 clusters in the sample. The *dashed histogram* is for RC = 0 clusters and the *solid histogram* is for RC ≥ 1 clusters.

some features that may appear to be coherent over large scales but, as Postman *et al.* (1989) demonstrate, such structures can easily be explained as a consequence of the two-point correlation function and chance superpositions of superclusters.

FIGURE 5 is a schematic that shows the orientation of four cylindrical volumes with respect to the plane of the galaxy. Figure 6(a)–(d) show projections of the clusters distributed within each of these volumes. The cylindrical slices are 15,000 km sec^{-1} thick and 35,000 km sec^{-1} in radius. They are consecutive regions beginning with the most northern slice. The viewing direction is perpendicular to the galactic plane. Open circles denote RC = 0 clusters; closed circles denote RC ≥ 1 clusters. Regions with $\delta < -27°30'$ are shaded.

Voids and superclusters are clearly present in the data, as expected from previous studies (Einasto *et al.* 1980; Bahcall and Soneira 1982; Klypin and Kopylov 1983; Bahcall and Soneira 1984; Burns *et al.* 1988; Huchra *et al.* 1990). The frequency of overdense regions dominated by RC ≥ 1 clusters is comparable with the frequency of RC = 0 dominated superclusters. Although this sample is the deepest wide-area redshift survey of nearby clusters to date, no features as striking as the "Great Wall" (Geller and Huchra 1989) are readily apparent in the distribution, possibly because the clusters sample the galaxy distribution so sparsely. To detect

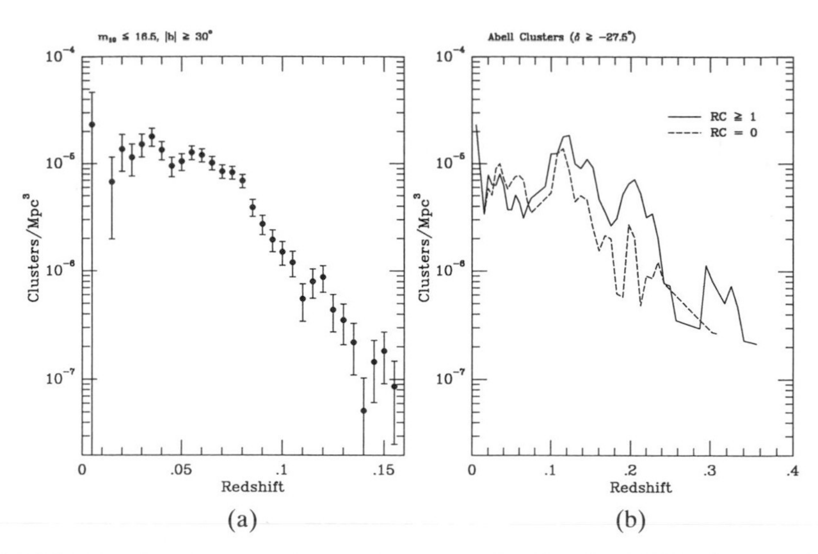

FIGURE 3. (**a**) Comoving space density of clusters as a function of redshift for the $m_{10} \leq 16.5$ survey. Bins are 50 percent overlapping. (**b**) Comoving space density of clusters a function of redshift and richness class. Data for $z \leq 0.075$ are based on the $m_{10} \leq 16.5$ survey, and data for larger z are based on the deep cluster survey of Huchra *et al.* 1990.

large-scale coherent structures with high significance, the mean separation of objects in a survey must be much smaller than the scale of the structure. In this survey, the mean cluster separation is $43h^{-1}$ Mpc, only a factor of 2–3 smaller than the size of the large "sheets" seen in galaxy redshift surveys.

THE SPATIAL CORRELATION FUNCTION

The two-point spatial correlation function is used to provide a measure of the clustering in the survey. A Monte Carlo estimator is the optimal choice for minimizing edge effects. The estimator for the spatial correlation funciton is

$$\xi(r) = \frac{N_{\mathrm{obs}}(r)}{N_{\mathrm{ran}}(r)} - 1, \qquad \textbf{(1)}$$

where $N_{\mathrm{obs}}(r)$ is the number of cluster pairs in the survey with separations between $r - \Delta r/2$ and $r + \Delta r/2$, and $N_{\mathrm{ran}}(r)$ is the number of pairs in a random distribution of clusters over an identical volume. Spatial separations are based on comoving distances computed from redshifts using the standard Friedman cosmology (Mattig 1958):

$$R = \frac{c}{H_0 q_0^2(1 + z)} (q_0 z + (q_0 - 1)\,[\sqrt{2q_0 z + 1} - 1]). \qquad \textbf{(2)}$$

At each separation, the number of random pairs is averaged over 50 simulations. The slope and amplitude of the correlation function do not change significantly if the

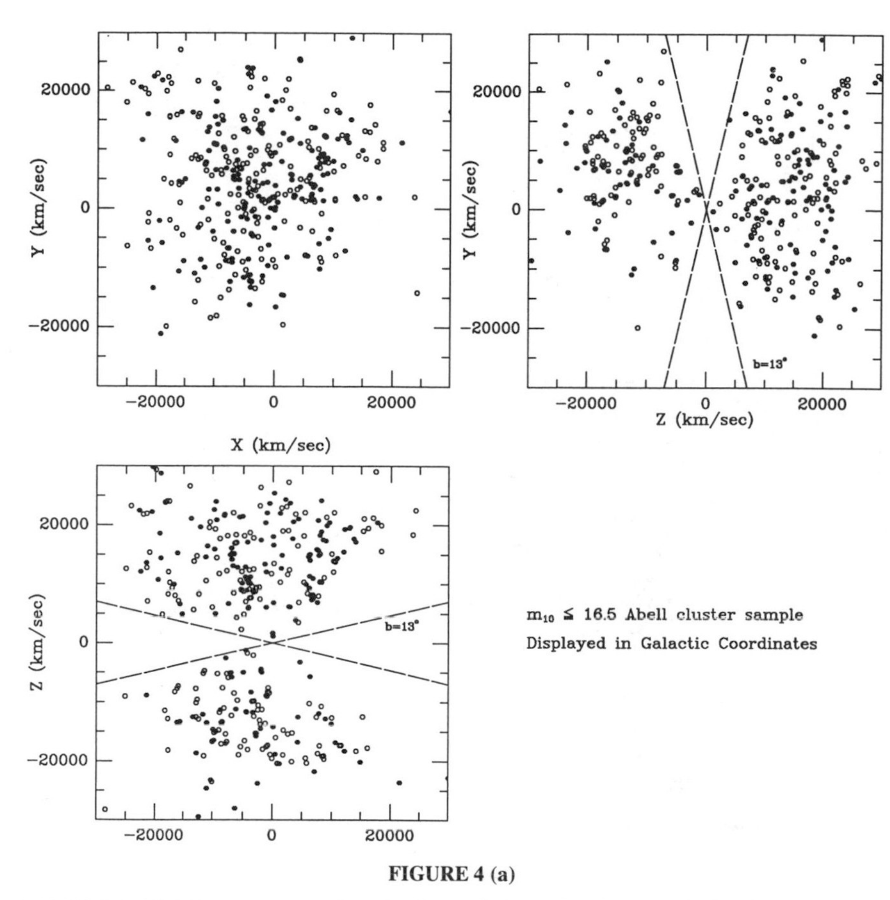

FIGURE 4 (a)

FIGURE 4. (**a**) Three orthogonal projections of the entire cluster distribution in galactic coordinates. Distances are in km sec^{-1}. The *dashed lines* mark the galactic latitude limits of the survey. (**b**) Three orthogonal projections of the entire cluster distribution in supergalactic coordinates.

number of simulations is increased tenfold. Logarithmic bins in spatial separation with $\Delta \log(r) = 0.1$ are used because they improve the signal-to-noise ratio at large separations where the amplitude of the correlation function is small. The results, however, do not change significantly with linear binning.

The redshift distribution for the random catalogs is obtained by choosing redshifts at random from the *real* survey and "smoothing" with a random Gaussian with $\sigma_z = 3000$ km sec^{-1} to prevent reproduction of the small-scale clumpiness in the actual redshift distribution. A Gaussian smoothing length of $\sigma_z = 3000$ km sec^{-1} is the minimum value that removes small-scale features in the observed redshift distribution. The results do not change significantly, however, for smoothing lengths in the range $\sigma_z \leq 6000$ km sec^{-1}. This technique for generating redshifts corrects the correlation function for radial density gradients on scales larger then the smoothing

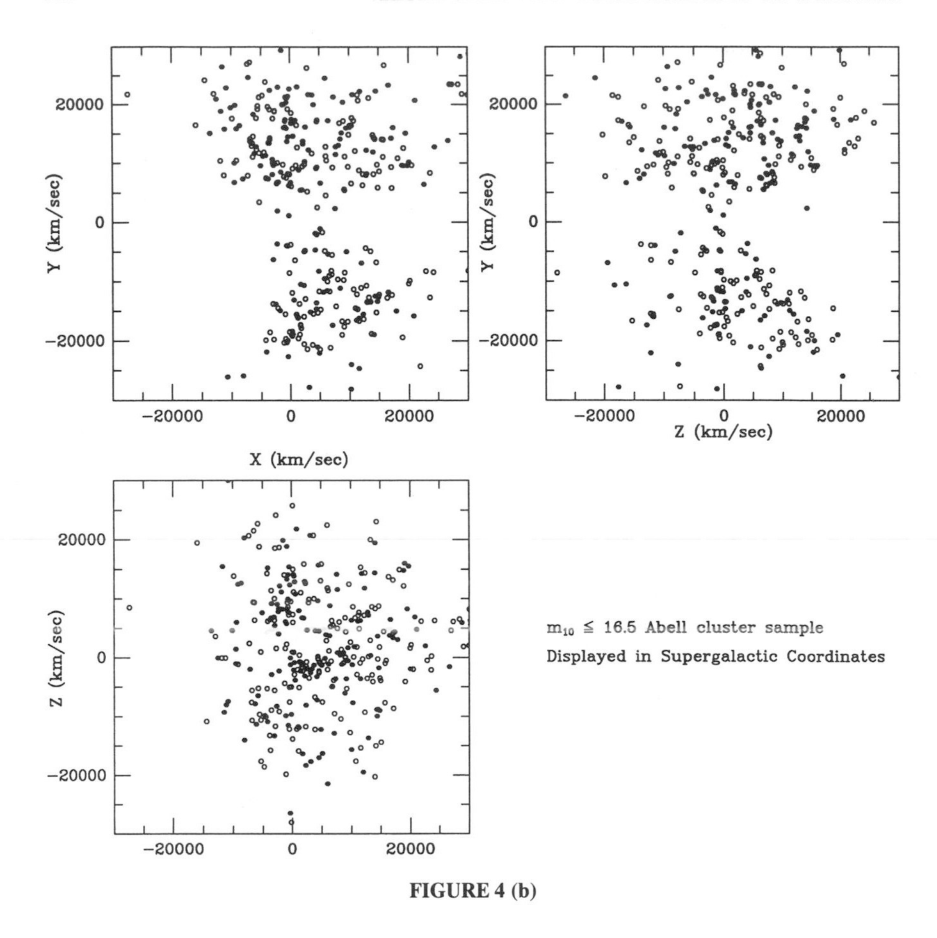

FIGURE 4 (b)

length. It is assumed that such gradients are caused by selection effects only. This assumption is reasonable because the survey covers such a large area of sky and because we see no large radial structures in the spatial distribution.

A galactic latitude selection function $P(b) = 10^{(\alpha-\beta \mathrm{CSC}|b|)}$ with $\alpha = \beta = 0.32$ is applied to the simulations, a good fit to the observed selection function of both the Abell catalog ($m_{10} \leq 16.5$) and the Lick Galaxy Catalog (Shane and Wirtanen 1967). Reasonable variations of α and β have a negligible effect on the results. There is no significant declination selection bias for the $m_{10} \leq 16.5$ survey.

FIGURE 7(a)–(d) show the spatial correlation functions for the entire cluster sample, the 195 RC = 0 clusters, the 156 RC $\geq$ 1 clusters, and the 208 clusters in a "statistical" subsample, respectively. The latter subsample consists of the clusters with $z \leq 0.08$ and $|b| \geq 30°$. To this redshift limit, the space density of clusters is constant. The "statistical" sample is thus minimally affected by selection biases. TABLE 2 lists the parameters of the best fit power laws over the range $10h^{-1}$ Mpc $\leq r \leq 75h^{-1}$ Mpc. The bootstrap resampling technique is used (Ling, Frenk, and Barrow 1986) to estimate the uncertainties. On scales larger than about $75h^{-1}$ Mpc,

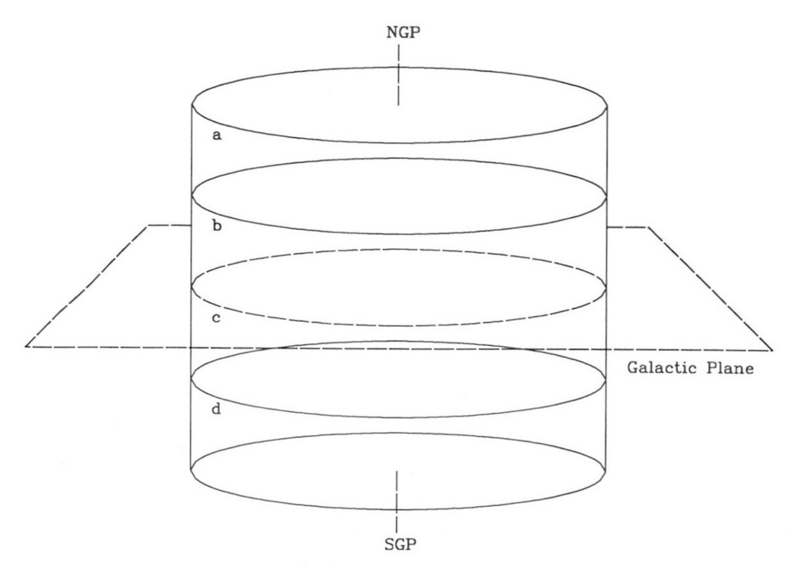

FIGURE 5. Orientation of four cylindrical volumes with respect to the galactic plane. Projections of the clusters in these volumes are shown in FIGURE 6(a)–(d).

the error in the correlation function is comparable with its amplitude, and one cannot reliably detect any power.

Postman, Geller, and Huchra (1986) demonstrated that the Corona Borealis (Cor Bor) supercluster contributes about 30 percent of the power in the correlation function for the DC ≤ 4 sample analyzed by Bahcall and Soneira (1983). To test the "fairness" of the current sample, this supercluster is excluded and the correlation functions recomputed (see FIG. 7(a), (c), and TABLE 2). The presence of Cor Bor has a negligible effect on the slope and amplitude of the correlation function for the entire sample. Cor Bor does account for about 20 percent of the power in the correlation function for the RC ≥ 1 subsample, resulting in a factor of 1.25 increase in the correlation length. The difference, however, is not statistically significant.

For comparison, TABLE 2 includes the best fit power law parameters derived from several other Abell cluster samples. The deep HHPG cluster survey provides a particularly important comparison because the overlap between the current sample and the HHPG survey is small. The correlation functions derived for various subsamples of HHPG are in excellent agreement with those derived for the $m_{10} \leq 16.5$ survey. The deep survey contains no supercluster as rich as Cor Bor. The good agreement suggests that there is no strong redshift dependence of the spatial correlation function for $z \lesssim 0.2$. West and van den Bergh (1991) and Lahav *et al.* (1989) compute the correlation function for cD clusters and X-ray bright clusters, respectively. Their results also agree with those for the current survey, although the uncertainties are large because of the small sample sizes. The agreement is especially significant because, as West and van den Bergh (1991) point out, cD galaxies and X-ray emitting gas are good indicators of high-density environments, and thus the richness of such clusters is unlikely to be seriously biased by interlopers.

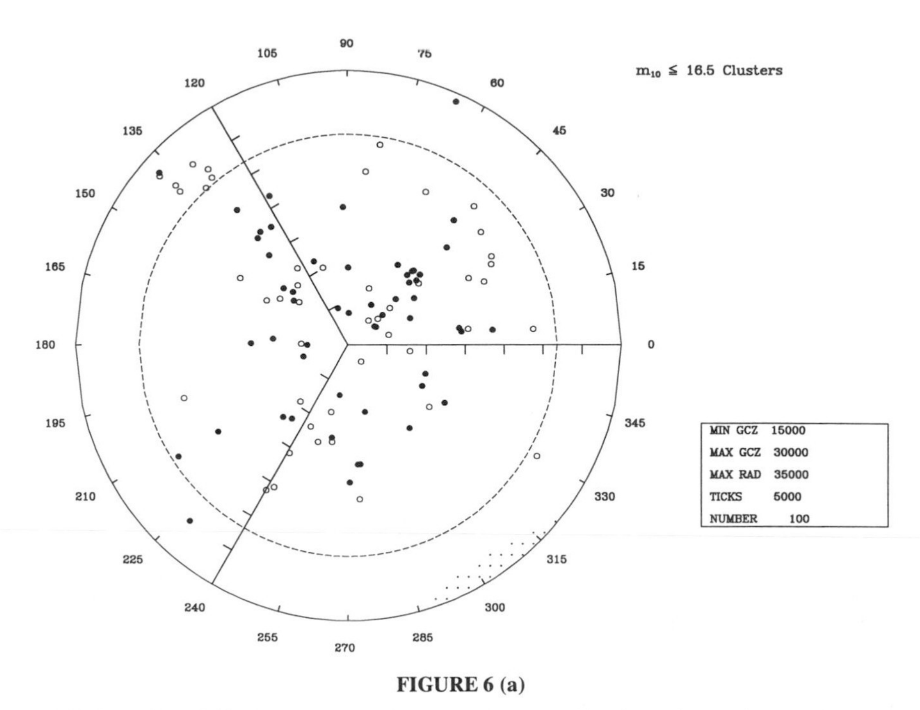

FIGURE 6 (a)

FIGURE 6. (**a**)–(**d**) Projections of the clusters in the four cylindrical volumes shown in FIGURE 5. The cylinders are 15,000 km sec^{-1} thick and 35,000 km sec^{-1} in radius. The viewing direction is perpendicular to the galactic plane. Galactic longitude is marked along the circumference. RC = 0 clusters are denoted by *open circles,* RC ≥ 1 clusters are denoted by *closed circles.* Regions with $\delta < -27°30'$ are *shaded.* The *large dashed circle* shows the intersection of the cylinder midplane with a galactocentric sphere of radius 35,000 km sec^{-1}. Part (**a**) is the northernmost projection; part (**d**) is the southernmost projection.

Dekel *et al.* (1988) have argued that foreground/background contamination can result in a substantial overestimation of the amplitude of the correlation function. However, this effect is not significant for our data. By constructing artificial clusters with realistic profiles and the same spatial positions as clusters in the real survey (see Hurchra *et al.* 1990 for details), the magnitude of the projection contamination is estimated. It is found that only 17 clusters out of the 351 (5 percent) may be seriously contaminated. Of the 17 clusters, 12 are RC = 0; they would not have been in Abell's catalog had they been more isolated. The remaining five clusters are RC = 1, and would probably have been classified as RC = 0 if they had been more isolated. The amplitudes and slopes of the correlation functions for the various subsamples, however, do not change significantly when we exclude these systems from computations. The specific results for the statistical subsample are in FIGURE 7(d) and TABLE 2.

The amplitude of the spatial correlation function for Abell clusters is remarkably (surprisingly) robust. Moderately large, independent samples yield the same value of $R_0 \approx 21h^{-1}$ Mpc for *both* RC = 0 and RC ≥ 1 clusters. These results also hold for subsamples of these surveys selected with different redshift, position, and contamina-

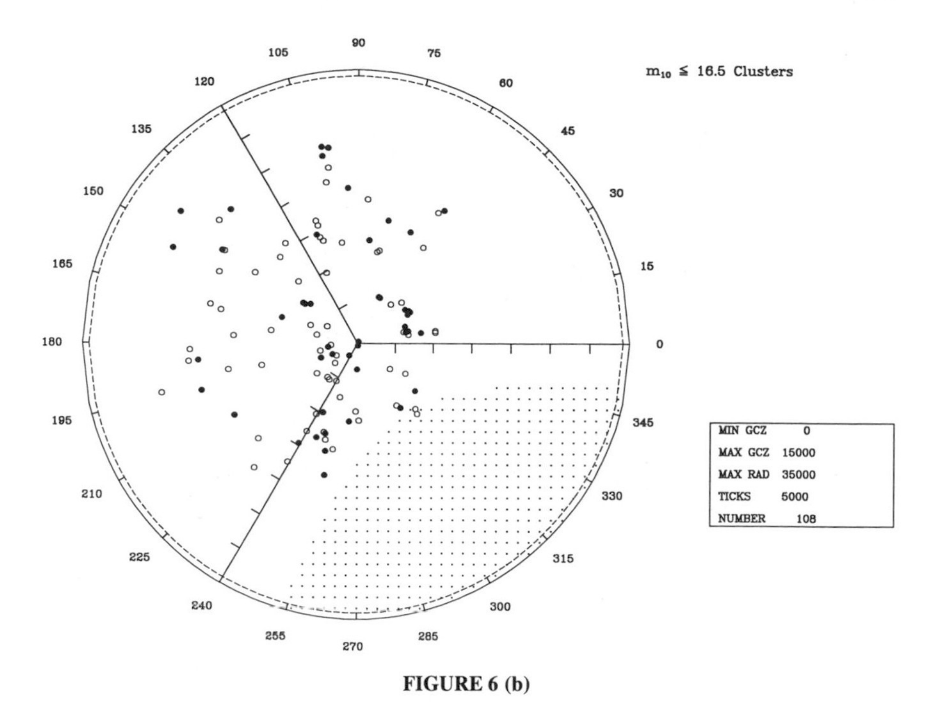

FIGURE 6 (b)

tion criteria. The spatial correlation function for Abell clusters is thus not significantly dominated by artificial signal. A potential systematic bias that is substantially harder to measure, however, is Abell's completeness in finding rich clusters that do not contain bright elliptical galaxies in the central core. The effect of such a bias can only be checked by the analysis of automated cluster catalogs now in production.

VOID STATISTICS

Observational constraints on the frequency and size of voids in the distribution of galaxies and clusters can place important constraints on theoretical models of galaxy and cluster formation. Because the new sample is larger (in both volume and population) than any previous cluster redshift survey, it provides a unique data base for examining the void statistics of Abell clusters. The void probability function (VPF) is computed for the statistical subsample and for a series of simulations. The VPF is just the probability that a randomly selected volume, V, contains no clusters. The statistical subsample is selected because it is minimally affected by selection biases and has a well-defined redshift limit.

To assess the significance of voids detected in the real cluster distribution, two sets of simulated catalogs are generated. In the first set of simulations, the clusters are distributed randomly. In the second set, the clusters have the same two-point spatial correlation function as the real survey (on scales $\leq 75h^{-1}$ Mpc). The correlated simulations are produced by placing clusters at the peaks of a Gaussian random

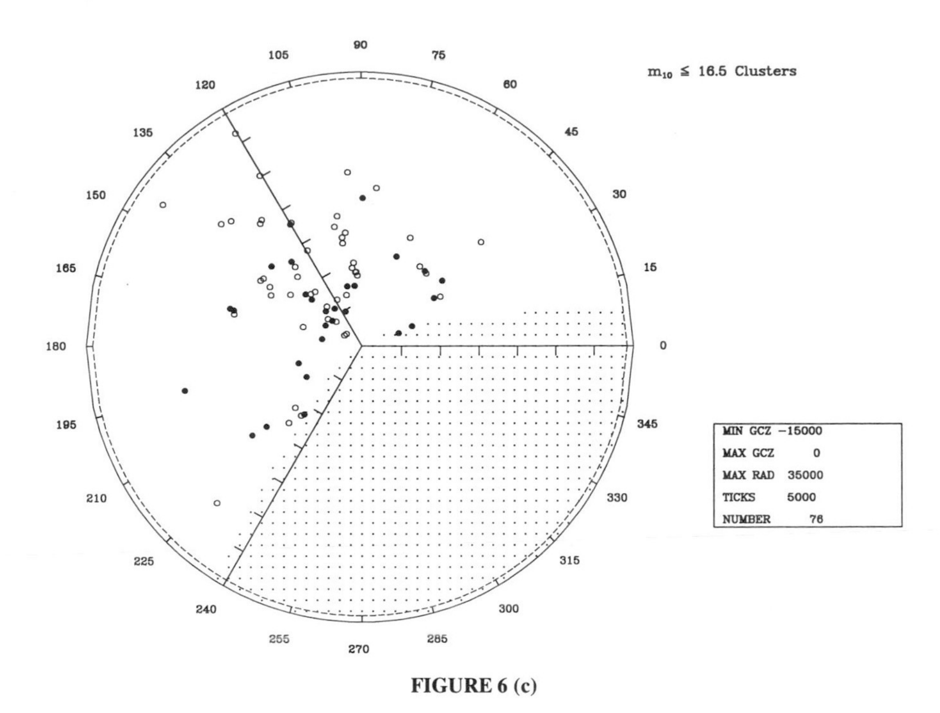

FIGURE 6 (c)

phase density field. The minimum peak density and the slope of the power spectrum are adjusted to match the two-point correlation function of the simulations to the observed correlation function. A short wavelength cutoff in k-space is introduced to remove power on scales greater than $75h^{-1}$ Mpc. Postman *et al.* (1989) give a detailed description of the cluster simulation algorithm. Fifty simulations are generated in each set and include the observed selection effects in redshift and galactic latitude.

The VPF is computed by randomly placing 1000 spheres of a given size inside the survey volume and then counting the empty spheres. Only spheres that are completely contained within the survey boundaries are used. The VPF is plotted as a function of nV, the mean number of clusters expected within a sphere of volume V in a survey with mean cluster number density n. FIGURE 8(a) shows the VPF for the 208 clusters with $z \leq 0.08$, $|b| \geq 30°$, for the clustered simulations, and for the Poisson simulations. Also plotted is the relation $P_0(nV) = \exp(-nV)$, the VPF for a Poisson distribution. The results for the simulated catalogs shown in FIGURE 8(a) are the mean results for the 50 simulations. The uncertainties are the one sigma values. Error bars for the simulations are omitted for the sake of clarity, but they are comparable with the uncertainties for the real data.

The mean VPF for the clustered simulations agrees very well with the VPF for the Abell clusters. The VPF for the Poisson simulations is about an order of magnitude smaller than that for the cluster survey at $nV = 6$. The VPF for the Poisson simulations agrees extremely well with the expression $\exp(-nV)$, confirming the statistical nature of this particular subsample. For $nV \geq 10$ (radii $\geq 58h^{-1}$ Mpc), the VPF for the survey is consistent with zero. These results suggest that large

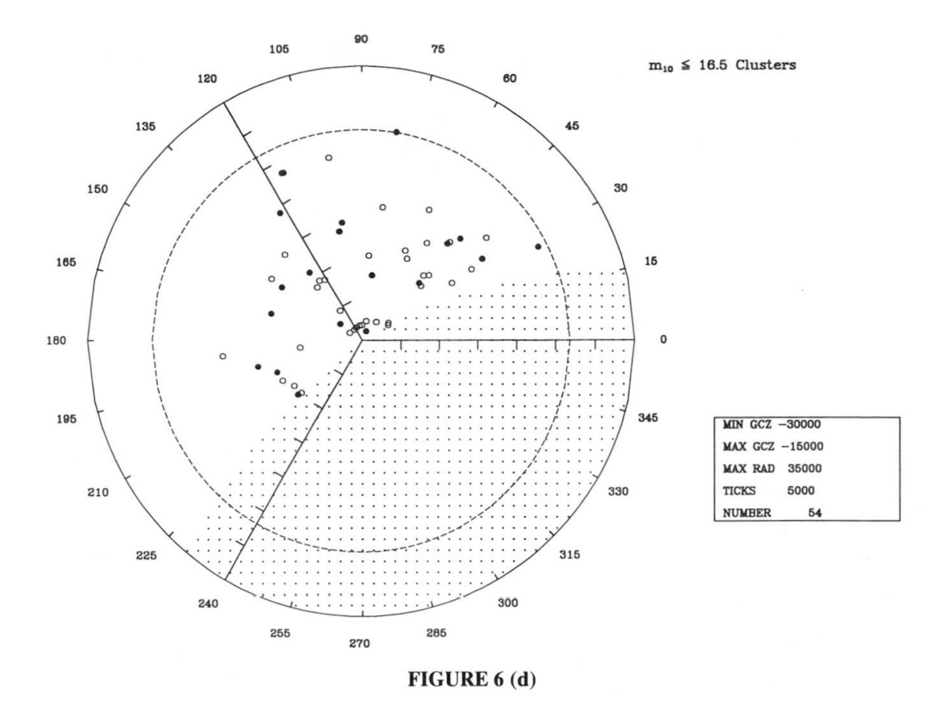

FIGURE 6 (d)

voids are relatively common in the distribution of Abell clusters, and their frequencies are completely consistent with expectations based on the small-scale form and amplitude of the two-point correlation function. No power is required on scales larger than $75h^{-1}$ Mpc, the largest on which the correlation function is significant.

FIGURE 8(b) compares the VPF for this sample with that derived for the HHPG deep cluster survey. The agreement is excellent. The independence of these two samples makes the agreement especially important. It appears the Abell cluster VPF is stable out to $z \approx 0.20$.

SUPERCLUSTERING

Bahcall and Soneira (1984; hereafter BS) were the first to run a percolation algorithm on a sample of 104 Abell clusters (RC $\geq$ 1, DC $\leq$ 4) with measured redshifts. They find 18 superclusters at their lowest density enhancement factor (percolation length $\approx 13.8h^{-1}$ Mpc), 11 of which are binary systems. The statistical subsample is used to study the distribution of superclusters in this survey because selection biases are minimized and, thus, group detection parameters do not have to be scaled with position or redshift. The statistical subsample covers a volume comparable with the BS sample, but contains twice as many clusters because we have included RC = 0 clusters.

A percolation algorithm (Huchra and Geller 1982) is used to detect superclusters. To study the distribution of rich superclusters, the constraint that cluster groups

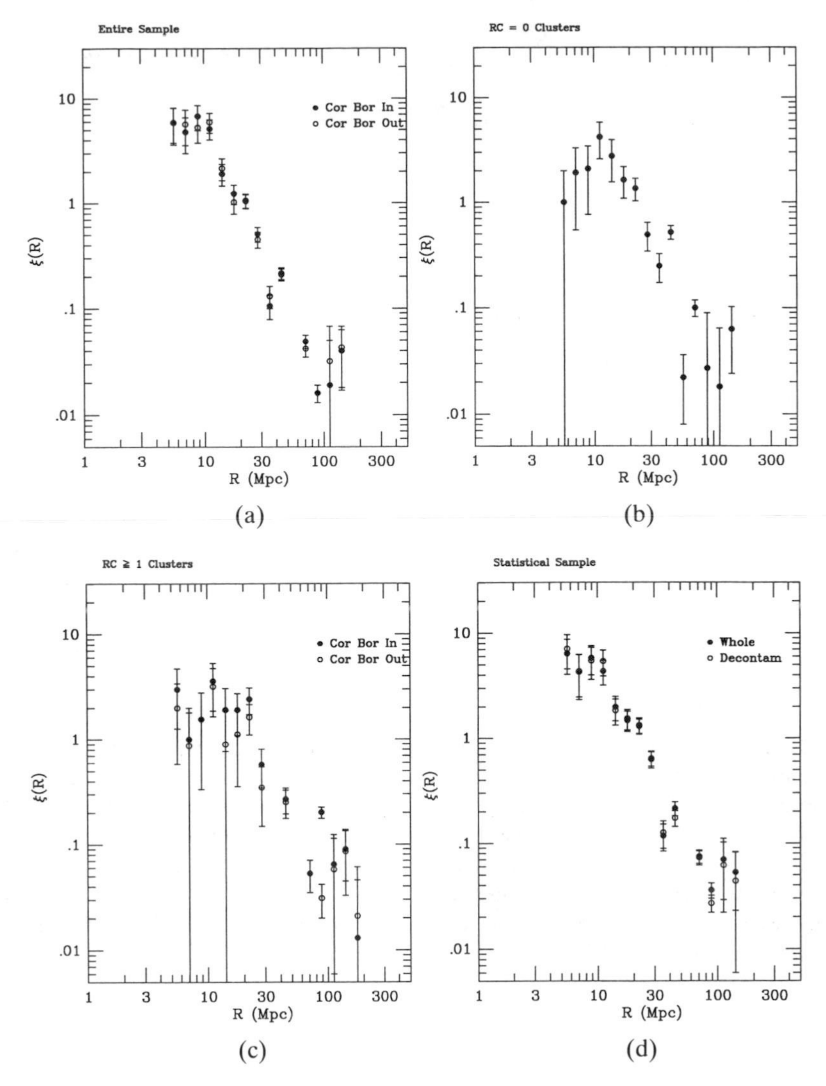

FIGURE 7. (**a**) The spatial correlation function for the entire cluster sample with the Corona Borealis (Cor Bor) supercluster (*closed circles*) and without Cor Bor (*open circles*). *Error bars* are one sigma in all plots. (**b**) The spatial correlation function for the RC = 0 clusters. (**c**) The spatial correlation function for the RC $\geq$ 1 clusters with Cor Bor (*closed circles*) and without Cor Bor (*open circles*). (**d**) The spatial correlation function for the $m_{10} \leq 16.5$ statistical subsample ($z \leq 0.08$, $|b| \geq 30°$).

contain a minimum of three members is applied. The distribution and frequency of binary systems does not provide any information that is not already available from the two-point spatial correlation function. To avoid linkages across the plane of the galaxy, the north and south galactic hemispheres are analyzed separately.

TABLE 3 lists the supercluster members found at space density enhancements,

TABLE 2. The Two-point Correlation Function

Survey	Size	Mean z	R_0	Slope	Composition
$m_{10} \leq 16.5$[a]	351	0.070	$20.0^{+4.6}_{-4.0}$	-2.49 ± 0.022	All (CorBor in)
$m_{10} \leq 16.5$	345	—	$20.3^{+4.9}_{-4.3}$	-2.44 ± 0.24	All (CorBor out)
$m_{10} \leq 16.5$	195	0.069	$21.7^{+4.4}_{-5.2}$	-2.46 ± 0.28	RC = 0 only
$m_{10} \leq 16.5$	156	0.071	$23.7^{+7.9}_{-9.0}$	-1.81 ± 0.24	RC > 0 (CorBor in)
$m_{10} \leq 16.5$	150	—	$18.9^{+7.8}_{-8.5}$	-2.03 ± 0.24	RC > 0 (CorBor out)
$m_{10} \leq 16.5$	208	0.057	$20.6^{+4.5}_{-4.8}$	-1.86 ± 0.20	$z \leq 0.08$, $\|b\| \geq 30°$
$m_{10} \leq 16.5$	197	—	$21.0^{+4.0}_{-3.7}$	-2.32 ± 0.27	$z \leq 0.08$, $\|b\| \geq 30°$ (Decontam)
Deep Survey[b]	145	0.167	$20.3^{+4.8}_{-5.1}$	−1.80	Entire sample
Deep Survey	103	0.175	$20.9^{+6.7}_{-6.9}$	−1.80	RC > 0 only
Deep Survey	132	0.154	$20.7^{+6.3}_{-6.4}$	−1.80	$z \leq 0.24$
DC ≤ 4[c]	104	0.070	25	−1.80	RC ≥ 1
cD clusters[d]	64	—	$22.1^{+6.8}_{-6.8}$	-1.70 ± 0.48	
X-ray clusters[e]	53	—	$21.0^{+7.0}_{-7.0}$	−1.80	kT ≥ 2 keV

[a]This presentation.
[b]Huchra *et al.* (1990).
[c]Bahcall and Soneira (1983).
[d]West and van den Bergh (1991).
[e]Lahav *et al.* (1989).

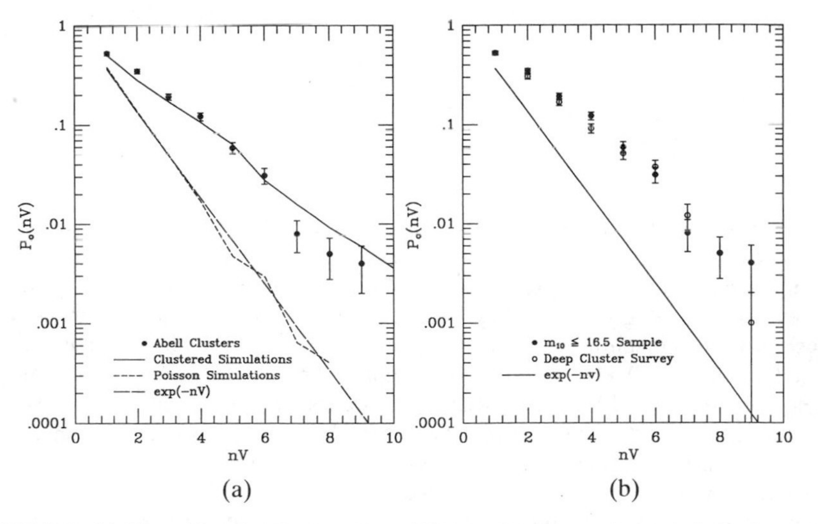

FIGURE 8. (**a**) The void probability function (VPF) for the $m_{10} \leq 16.5$ statistical subsample. The VPFs for a series of clustered and Poisson simulations are also shown. The VPFs for the simulations are the mean results for 50 realizations. (**b**) The VPFs for the $m_{10} \leq 16.5$ statistical subsample (*closed circles*) and for the deep survey of Huchra *et al.* 1990 (*open circles*). The function exp $(-nV)$, the expected VPF for a pure Poisson distribution, is shown for comparison.

$f\,(=n/\bar{n})$, of 2, 5, and 10 corresponding to percolation lengths of 22.0, 16.2, and 12.9h^{-1} Mpc, respectively. (The corresponding f's for the BS sample at these percolation lengths are approximately 2.5 times larger because of the smaller sample size.) Also listed are the BS supercluster ID for common systems. Note, however, that the

TABLE 3. Supercluster Members

ID	$f = 2$	$f = 5$	$f = 10$	BS ID
1	930 970 978 979 993 1069	930 970 978 979 993	930 978 979 993	
2	999 1016 1139 1142	—	—	
3	1149 1171 1238	—	—	
4	1177 1185 1228 1257 1267	1177 1185 1228 1257 1267	1177 1185 1267	7
5	1216 1308 1334	—	—	
6	1270 1291 1318 1377 1383 1436 1452 1507	1291 1318 1377 1383 1436 1452	1291 1318 1377 1383	8
7	1775 1800 1831 1873	1775 1800 1831 1873	1775 1800 1831 1873	12A
8	1781 1795 1825	—	—	
9	2052 2063 2107 2147 2148 2151 2152 2162 2197 2199	2107 2147 2148 2151 2152	2147 2151 2152	15
10	2061 2065 2067 2079 2089 2092 2124	2061 2065 2067 2089	—	12B
11	2168 2169 2184	—	—	
12	2248 2256 2271 2296 2309	2256 2271 2296	2256 2271 2296	
13	14 27 74 86 114 133 2716 2800 2824	74 86 114 133 2800 2824	(74 86 2800) (114 133 2824)	
14	85 117 151	—	—	
15	102 116 134	102 116 134	—	
16	119 147 160 168 193 195	119 147 168	119 147 168	2
17	154 158 171 225 292 311	154 158 171	—	
18	419 428 3094 3095	419 3094 3095	419 3094 3095	
19	548 3367 3374	—	—	
20	2366 2399 2415	2366 2399 2415	—	
21	2459 2462 2492	—	—	
22	2572 2589 2592 2593 2657	2572 2589 2592 2593 2657	2572 2589 2593	
23	2622 2625 2626	—	—	

memberships of these superclusters may not always be identical because of the improved redshift data and the inclusion of the RC = 0 and ACO clusters. At $f = 2$, twenty-three supercluster systems are found; 17 of these *do not* appear in the BS supercluster catalog because they consist primarily of RC = 0 clusters and/or clusters from the southern ACO extension. At $f = 5$, fourteen superclusters are found and at $f = 10$, eleven systems are found. The Cor Bor supercluster does not appear in the $f = 10$ catalog because it breaks up into three binary systems: A2061–A2067, A2065–A2089, and A2079–A2092. The majority of the BS superclusters are binary and would be excluded by our minimum group size criteria.

The fraction of clusters in superclusters as a function of f for the statistical cluster sample and for a Poisson and clustered simulation set is shown in FIGURE 9(a) and (b), respectively. The RC = 0 and RC $\geq$ 1 clusters show very similar clustering behavior. There may be a slightly higher probability of finding an RC = 0 cluster in a supercluster at low f, but the trend is not statistically significant. The multiplicity functions of the statistical sample and clustered simulations agree well, further validating the use of these simulations for comparisons with the real data.

PECULIAR VELOCITIES

Several studies based on independent cluster samples (Soltan 1988; HHPG 1990) demonstrate that broadening of the redshift distribution caused by relative pairwise

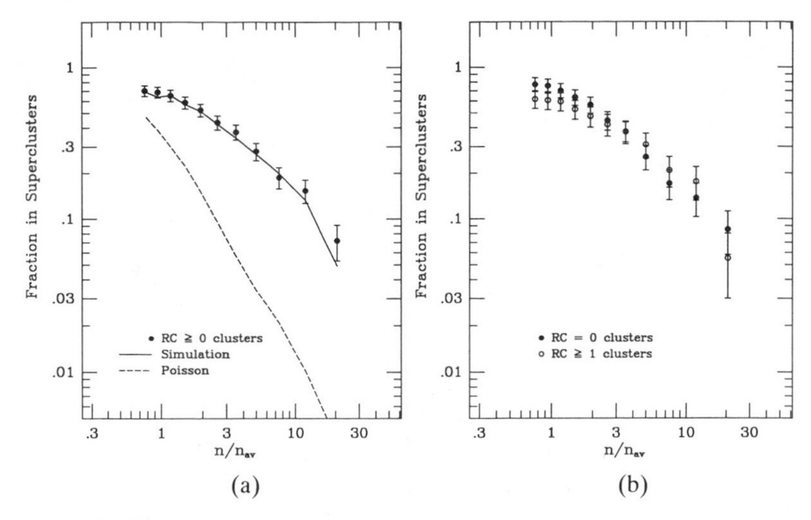

FIGURE 9. (**a**) The fraction of clusters in superclusters as a function of density enhancement for the $m_{10} \leq 16.5$ statistical subsample (*closed circles*). *Error bars* are one sigma. The fraction of clusters in superclusters for a series of clustered and Poisson simulations are represented by the *solid* and *dashed lines,* respectively. (**b**) The fraction of clusters in superclusters as a function of density enhancement for the RC = 0 clusters (*closed circles*) and the RC $\geq$ 1 clusters (*open circles*) in the $m_{10} \leq 16.5$ statistical subsample.

peculiar motions is $\lesssim$1000 km sec^{-1}. The larger signal, $v_p \sim 2000$ km sec^{-1}, claimed by Bahcall, Soneira, and Burgett (1986; hereafter BSB) is essentially caused by the Cor Bor supercluster. This system is elongated along the line of sight and substantially enhances the mean amplitude of cluster pair elongations in the redshift direction in the BSB cluster sample. The new, larger sample should be less sensitive to Cor Bor.

A test for elongations is performed by comparing the distribution of projected cluster pair separations on the plane of the sky with the distribution of separations in the redshift direction. The projected separation is decomposed into two orthogonal components aligned with lines of right ascension (α) and declination (δ) for convenience. When comparing any two components, the third component is limited to be $\leq 10h^{-1}$ Mpc, and the total spatial separation is limited to be $\leq 100h^{-1}$ Mpc. All cluster pairs with spatial separations greater than their distances from the survey boundaries are excluded. Without this edge correction, the frequency of pairs elongated in the redshift direction would be overestimated. FIGURE 10(a) shows the histogram for the distribution of pair separations along the redshift and α directions for the 208 clusters in the statistical subsample. FIGURE 10(b) shows the histogram for the distribution of pair separations along the redshift and δ directions. The solid histograms are the unconvolved distributions and the (long) dashed histograms show the radial separations. The (short) dashed histograms show the projected separations in the α and δ directions after convolution with a Gaussian. The projected separations are most consistent with the radial separations when convolved with a $\sigma = 850\ (\pm 150)$ km sec^{-1} Gaussian. However, the separations in δ are consistent with the radial separation distribution even without convolution. The number of pairs

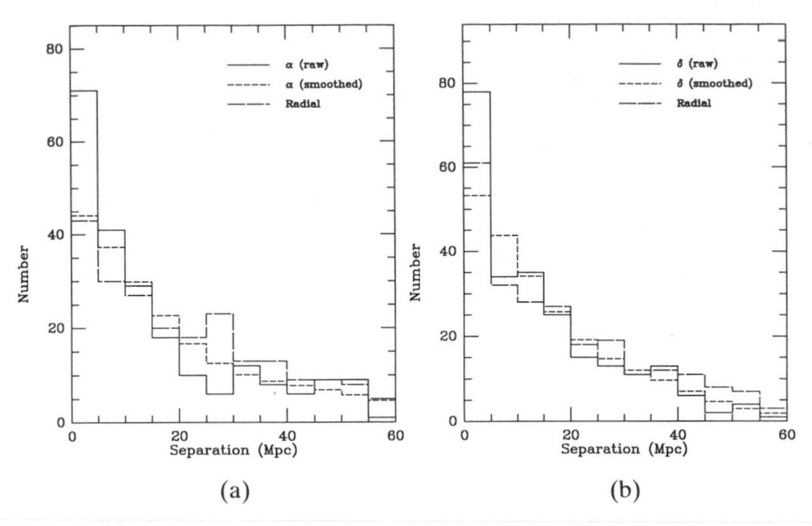

FIGURE 10. (**a**) Histograms of cluster pair separations. The *solid histogram* represents the unconvolved separations in the right ascension direction. The *short dashed histogram* represents the right ascension separations after convolution with a Gaussian. The *long dashed histogram* represents the radial separations. (**b**) Same as part (**a**), but for the separations in the declination direction. The projected separation distribution is most consistent with the radial separation distribution when convolved with a 850 ($\pm$150) km sec^{-1} Gaussian.

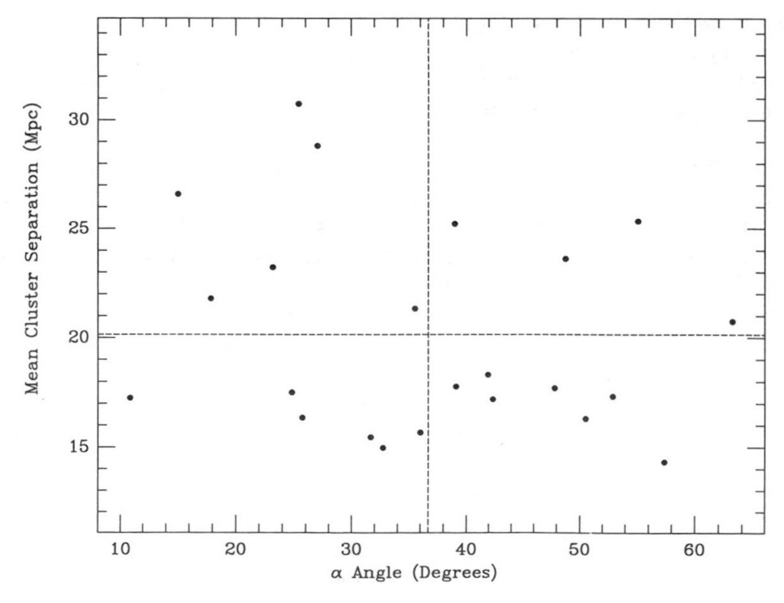

FIGURE 11. The dependence of the α angle on the mean cluster separation for the 23 superclusters detected at an enhancement factor of 2. The *dashed lines* mark the mean values.

decreases as the pair separation increases because of the exclusion of widely separated pairs too near the survey boundaries. The constraints on velocity broadening do not change significantly when we exclude the Cor Bor supercluster from the analysis.

Velocity broadening should be most detectable within superclusters. Therefore, the dependence of the angle between the plane of the sky and the line connecting two clusters (the α angle; Sargent and Turner 1977) on the mean spatial separation of clusters in superclusters is also calculated. The superclusters detected at $f = 2$ are used for this computation (see TABLE 3). FIGURE 11 shows the results. There is no significant trend. In fact, the mean α angle for the 23 superclusters is 36.7°, close to the value expected for an isotropic distribution. Clearly, some of the supercluster systems are elongated in the radial direction, but just as many appear elongated in the plane of the sky. The elongation may also be caused by the real geometry of the clusters in space. Whatever the true nature of the elongation is, the amplitude of peculiar motions of clusters is $\lesssim 1000$ km sec^{-1}.

SUMMARY

The redshifts of the 351 brightest Abell clusters north of $\delta = -27°30'$ constitute the largest magnitude-limited sample to date for analyzing the spatial distribution of rich clusters. The completeness of Abell RC = 0 clusters appears to be comparable with that for the RC $\geq$ 1 clusters out to $z \approx 0.2$, in spite of Abell's claims (1958) to the contrary. The clustering properties of these two cluster classes are also quite similar. Examination of the spatial distribution of the clusters reveals no large-scale ($\gtrsim 80h^{-1}$ Mpc) coherent structures. However, the mean cluster separation is too large to detect structures such as the Great Wall with high significance.

The cluster–cluster correlation length is remarkably robust. For the $m_{10} \leq 16.5$ survey, the best fit correlation length is 20.0 ($\pm$4.3)h^{-1} Mpc. This value is in excellent agreement with the results from the nearly independent deep cluster sample studied by Huchra *et al.* (1990). The correlation length is insensitive to the inclusion of RC = 0 clusters. The slope of the correlation function is slightly steeper when the RC = 0 clusters are included, but the trend is only significant at the one sigma level. There is no detectable power on scales larger than about $75h^{-1}$ Mpc. Foreground/background contamination is not significant and does not effect the amplitude of the correlation function.

The frequency of large voids (radii $\lesssim 60h^{-1}$ Mpc) in the cluster distribution is consistent with expectations based on the observed small-scale clustering. Simulations of clustered distributions with no power on scales greater than $75h^{-1}$ Mpc reproduce the observed VPF extremely well. In addition, the VPF for the HHPG deep cluster survey is in excellent agreement with that for the $m_{10} \leq 16.5$ survey, suggesting that the VPF is universal out to the completeness limit of the Abell catalog ($z \approx 0.2$).

There are 23 superclusters detected in the $m_{10} \leq 16.5$ statistical subsample, 17 of which are new. The RC = 0 clusters populate superclusters as frequently as the RC $\geq$ 1 clusters do. The superclusters show no preference for elongation in the radial direction. An analysis of the geometry of cluster pairs sets an upper limit on the amplitude of non-Hubble motions of $\sim 850 \pm 150$ km sec^{-1}.

Taken together, these results and analyses of other samples show that the clustering properties of Abell clusters are robust and are not strongly dependent on selection or projection effects. Even Zwicky clusters show similar clustering behavior to the Abell clusters at depths where Zwicky selects systems similar in richness (Postman, Geller, and Huchra 1986). Acquisition of new cluster redshifts should clearly continue, but it is unlikely that dramatically new constraints on the large-scale spatial distribution of clusters can be obtained without measuring redshifts for at least ~ 2000 additional clusters. The next major step in understanding the distribution of rich clusters will probably come from the comparison of galaxy and cluster redshift surveys. Current galaxy redshift surveys are now approaching depths where such comparisons can be made. The next few years should see this situation improve significantly. In addition, objectively derived cluster catalogs, created from digitized sky surveys and from all-sky X-ray surveys [e.g., (ROSAT)], are under construction. The combination of these efforts will result in much needed independent and objective measures of the spatial distribution of rich clusters.

REFERENCES

AARONSON, M., *et al.* 1989. Astrophys. J. **338:** 654.
ABELL, G. O. 1958. Astrophys. J., Suppl. Ser. **3:** 211.
ABELL, G. O., H. G. CORWIN & R. P. OLOWIN. 1989. Astrophys. J., Suppl. Ser. **70:** 1 (ACO).
BAHCALL, N. A. & R. M. SONEIRA. 1982. Astrophys. J. **262:** 419.
BAHCALL, N. A. & R. M. SONEIRA. 1983. Astrophys. J. **270:** 20.
BAHCALL, N. A. & R. M. SONERIA. 1984. Astrophys. J. **277:** 27.
BAHCALL, N. A., R. M. SONERIA & W. S. BURGETT. 1986. Astrophys. J. **311:** 15.
BURNS, J. O., J. W. MOODY, J. P. BRODIE & D. J. BATUSKI. 1988. Astrophys. J. **335:** 542.
DEKEL, A., G. R. BLUMENTHAL, J. R. PRIMACK & S. OLIVIER. 1988. Astrophys. J., Lett. **338:** L5.
EINASTO, J., M. JOEVEER & E. SAAR. 1980. Mon. Not. R. Astron. Soc. **193:** 353.
GELLER, M. J. & J. P. HUCHRA. 1989. Science **246:** 897.
HOESSEL, J. C., J. E. GUNN & T. X. THUAN. 1980. Astrophys. J. **241:** 486 (HGT).
HUCHRA, J. P. & M. J. GELLER. 1982. Astrophys. J. **257:** 423.
HUCHRA, J. P., J. P. HENRY, M. POSTMAN & M. J. GELLER. 1990. Astrophys. J. **365:** 66 (HHPG).
KLYPIN, A. & A. KOPYLOV. 1983. Sov. Astron. Lett. **9:** 41.
LAHAV, O., A. C. EDGE, A. C. FABIAN & A. PUTNEY. 1989. Mon. Not. R. Astron. Soc. **238:** 881.
LING, E. N., C. S. FRENK & J. D. BARROW. 1986. Mon. Not. R. Astron. Soc. **223:** 21P.
MATTIG, W. 1958. Astron. Nachr. **284:** 109.
POSTMAN, M., M. J. GELLER & J. P. HUCHRA. 1986. Astron. J. **91:** 1267.
POSTMAN, M., M. J. GELLER & J. P. HUCHRA. 1988. Astron. J. **95:** 267.
POSTMAN, M., D. N. SPERGEL, B. SUTIN & R. JUSZKIEWICZ. 1989. Astrophys. J. **346:** 588.
SARGENT, W. L. & E. TURNER. 1977. Astrophys. J., Lett. **212:** L3.
SHANE, C. D. & C. A. WIRTANEN. 1967. Publ. Lick Obs. Vol. XXII, Part 1.
SOLTAN, A. 1988. Mon. Not. R. Astron. Soc. **231:** 309.
TULLY, R. B. 1987. Astrophys. J. **323:** 1.
WEST, M. & S. VAN DEN BERGH. 1991. Astrophys. J. **373:** 1.

New D–σ Results for Coma Ellipticals

J. R. LUCEY,[a] R. GUZMAN,[a] D. CARTER,[b] AND
R. J. TERLEVICH[b]

[a] *Department of Physics*
University of Durham
South Road
Durham, DH1 3LE, England

[b] *Royal Greenwich Observatory*
Madingley Road
Cambridge, CB3 OEZ, England

INTRODUCTION

In the last few years considerable observational work has been undertaken to quantify the peculiar motions of galaxies in the local universe. For elliptical galaxies most of the work has been based on the diameter versus velocity dispersion correlation, that is, the D–σ relation (Dressler *et al.* 1987). This empirical relation allows distances to be estimated with an uncertainty of $\sim$20 percent, and has been used to map the nearby non-Hubble motions (Lynden-Bell *et al.* 1988; Lucey and Carter 1988; Dressler and Faber 1990; Lucey *et al.* 1991).

In a rich cluster, gravitational encounters are believed to alter significantly the structure of the galaxies. Strom and Strom (1978) found that the elliptical galaxies located near the cores of dense, rich clusters had smaller sizes than those located in both the cluster haloes and low-density clusters. The D–σ relation may also be affected by environmental factors (see Kaiser 1988; Silk 1989; Djorgovski, de Carvalho, and Han 1988).

Recently, Burstein *et al.* (1990) have presented strong arguments supporting the universality of the D–σ distance indicator. In particular, these authors have emphasized that the distance determinations from the D–σ and Tully–Fisher relations are in good agreement. Nevertheless some rich clusters do show large discrepancies between their D–σ and Tully–Fisher distances; for example, for the Abell 2634 cluster the D–σ relation gives a distance that is 41 $\pm$ 9 percent greater than that derived from the Tully–Fisher relation (see Lucey *et al.*).

In order to quantify the level of any environmental dependence in the D–σ relation we have compared the ellipticals that lie in the halo and core of the Coma cluster. By studying only one cluster we avoid the difficulty of decoupling peculiar motions from real environmental differences.

OBSERVATIONS

We have made new D–σ observations of Coma cluster ellipticals. Because previous work on Coma's D–σ relation (Dressler *et al.* 1987) covered mainly ellipticals that lie near the cluster center, our observations concentrated on ellipticals in the outer part of the cluster. Spectroscopic and photometric observations were made with the 2.5-m Isaac Newton Telescope on La Palma in May/June 1990. Sigma and D measurements were derived using standard techniques and have average errors of 0.024 dex and 0.005 dex, respectively.

RESULTS

The D–σ relation has the form:

$$\log D\,[\text{arcsec}] = \text{slope} \times \log \sigma\,[\text{km s}^{-1}] + \text{constant}. \quad \mathbf{(1)}$$

Using data for several clusters Lynden-Bell *et al.* (1988) derived a value for the

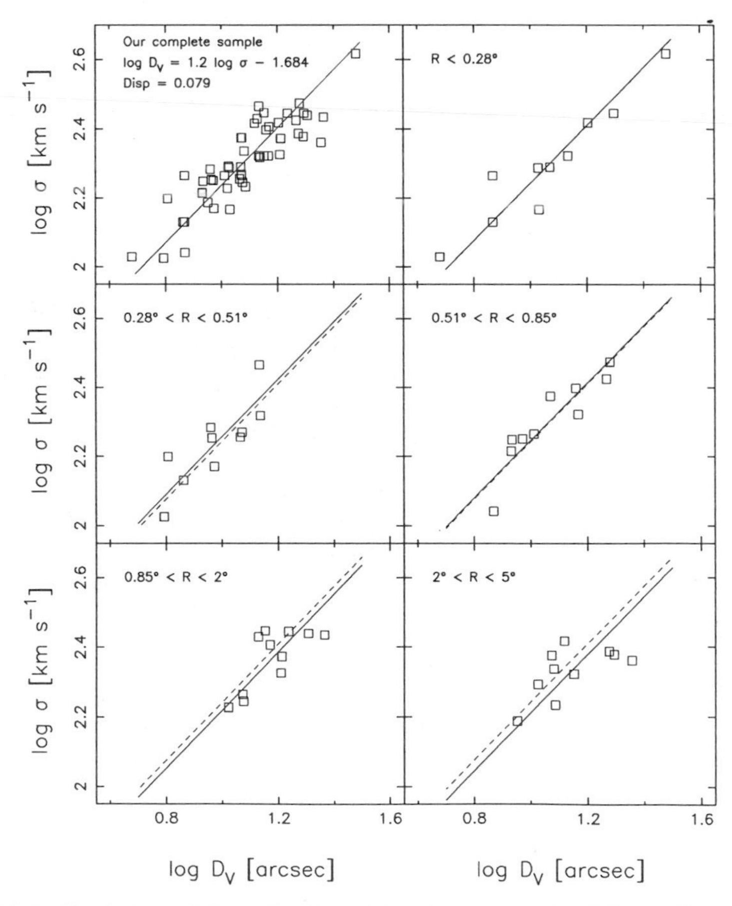

FIGURE 1. The first panel shows the D–σ relation for our sample of Coma ellipticals. The remaining panels show the D–σ relation for five different radial zones. For each zone the mean line of slope 1.2 is shown (*solid line*). For the outer zones, the mean relation found for the innermost zone is also drawn (*dashed line*).

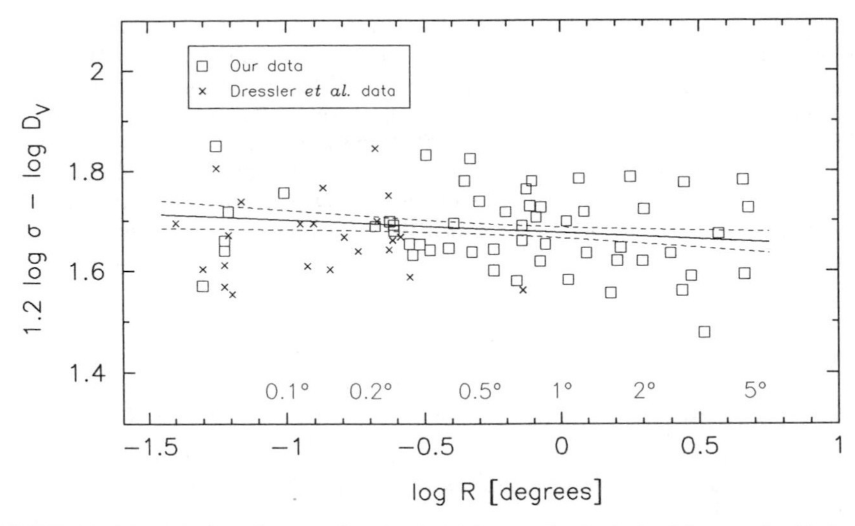

FIGURE 2. The variation of zero-point, that is, $1.2 \log \sigma - \log D$, derived from each elliptical as a function of projected distance from the cluster center. At the distance of the Coma cluster, 5° is equivalent to $6.3h^{-1}$ Mpc.

slope of 1.2 ± 0.1. The D–σ relation for our sample of Coma ellipticals is shown in FIGURE 1. Minimizing the scatter in $\log D$, we derive a slope of 1.17 ± 0.09. The rms scatter, in $\log D$, about this slope is 0.078, which is similar to that found for other well-studied clusters.

We have subdivided our sample into five radial zones of approximately equal numbers of galaxies to investigate the dependence of the D–σ relation on distance from the cluster center. The D–σ diagrams for the five zones are also shown in FIGURE 1. For all radial zones the form and the zero point of the D–σ relations are very similar. The innermost and outermost zones have average distances from Coma's center of 0.12°($0.14h^{-1}$ Mpc) and 3.3°($0.14h^{-1}$ Mpc), respectively (h is the Hubble constant in units of 100 km s^{-1} Mpc^{-1}). Whereas the surface density of galaxies decreases by over a factor of 150 between these two zones, the observed difference in the D–σ zero point, in terms of $\log D$, varies by only -0.032 ± 0.039, that is $-8\% \pm 10\%$. An alternative approach to illustrate this result is shown in FIGURE 2, where we plot the D–σ zero point derived from each elliptical against the logarithm of projected distance from the cluster center. The best-fitting relationship between these two variables is

$$\text{Zero point} = 1.679(\pm 0.011) - 0.026(\pm 0.021) \log R[\text{degree}]. \quad (2)$$

The derived gradient is not statistically significantly different from zero. This conclusion is robust. Adopting slopes for the D–σ relation of 1.1 and 1.3 gives gradients of -0.029 ± 0.020 and -0.022 ± 0.020, respectively. Including Dressler *et al.*'s data for core ellipticals revises the gradient to -0.024 ± 0.015. Hence, there is no statistically significant correlation between projected distance from the cluster center and the D–σ zero point.

Over the large radial distances considered in our study, Coma is not a well-mixed system. The zero-velocity radius of the cluster is 5.2° (Kent and Gunn 1982). Timing arguments suggest that galaxies that lie outside half this radius have yet to traverse the cluster core. In addition, simple dynamical considerations indicate that galaxies that currently lie in the cluster core are unlikely to escape out to radii greater than 1°. Therefore, in our study we have compared ellipticals that have undergone very different levels of galaxy–galaxy interactions. The D–σ relation appears not to be affected by such interactions, and hence the D–σ relation is a reliable distance indicator.

REFERENCES

BURSTEIN, D., S. M. FABER & A. DRESSLER. 1990. Astrophys. J. **354:** 18.
DJORGOVSKI, S., R. DE CARVALHO & M.-S. HAN. 1988. *In* Extragalactic Distance Scale (ASP Conference Series), S. van den Bergh and C. J. Pritchet, Eds.: 329.
DRESSLER, A. & S. M. FABER. 1990. Astrophys. J. **354:** 13.
DRESSLER, A., D. LYNDEN-BELL, D. BURSTEIN, R. L. DAVIES, S. M. FABER, R. J. TERLEVICH & G. WEGNER. 1987. Astrophys. J. **313:** 42.
KAISER, N. 1988. *In* Large-Scale Structure and Motions in the Universe. ICTP preprint.
KENT, S. M. & J. E. GUNN. 1982. Astron. J. **87:** 945.
LUCEY, J. R. & D. CARTER. 1988. Mon. Not. R. Astron. Soc. **235:** 1177.
LUCEY, J. R., P. M. GRAY, D. CARTER & R. J. TERLEVICH. 1991. Mon. Not. R. Astron. Soc. **248:** 804.
LYNDEN-BELL, D., S. M. FABER, D. BURSTEIN, R. L. DAVIES, A. DRESSLER, R. J. TERLEVICH & G. WEGNER. 1988. Astrophys. J. **326:** 19.
SILK, J. 1989. Astrophys. J., Lett. **345:** L1.
STROM, S. E. & K. M. STROM. 1978. Astrophys. J., Lett. **225:** L93.

Faint Galaxy Counts in Six 4-m Fields

L. R. JONES,[a,b] R. FONG,[a] T. SHANKS,[a] R. S. ELLIS,[a] AND
B. A. PETERSON[c]

[a] *Physics Department*
University of Southampton
Southampton 509 5NH, England

[b] *Physics Department*
University of Durham
South Road
Durham DH1 3LE, England

[c] *Mount Stromlo and Siding Spring Observatories*
Woden, ACT 2606, Australia

DEDICATION

This paper is dedicated to the memory of Dr. Nick Brown, a physicist of great honesty and insight, but also an irreplaceable and inspirational friend.

1. INTRODUCTION

The counting of galaxies as a function of magnitude is a classic cosmological test. The form of the counts, and the galaxy colors, provide useful information on changes in galaxy luminosity with redshift, and also on cosmological world models (Tinsley 1978; Bruzual and Kron 1982; Shanks *et al.* 1984). The variation of galaxy counts with position on the sky, both over small and wide angles, provides a probe of the large-scale galaxy distribution, although such effects can be masked by other systematic problems including Galactic obscuration. It is these variations with position that we particularly address in this paper.

In the range $21 < B < 24$ mag, only a small area of sky (≈ 7.5 deg^2) has so far been surveyed, and most authors have found number–magnitude (N(m)) counts that, while having similar slopes, display large variations in normalization from field to field. The form of the galaxy evolution required to reproduce the counts is not well constrained, partly because of this problem. The differences in the count normalization could arise from various photometric uncertainties (different magnitude schemes), errors in calibration of the photographic plates (zero points), variable absorption, or represent genuine fluctuations in the surface density of galaxies. It is important to estimate accurately these genuine fluctuations, particularly because they provide an estimate of the structure of the Universe on large scales.

A similar but more detailed account of this work can be found in Jones *et al.* 1991.

2. DATA AND REDUCTION

All plates were taken at the prime focus of the 3.9-m Anglo-Australian telescope (AAT), and are listed in TABLE 1. We have included the b_j and r_F plates of Shanks *et al.* (1984) (SSFM), Stevenson *et al.* (1986), and the b_j plate of Peterson *et al.* (1979); new charge-coupled device (CCD) observations have provided a better calibration of the SGP and Peterson *et al.* plates. All fields are at randomly located high-latitude positions, except for the SGP field and the 00^h24^m, +17° field, which is at the position of a distant (z = 0.39) cluster of galaxies. The cluster does not appear to dominate the faint galaxy counts on this field, and it should not affect our conclusions on the variation of the count normalization.

The plate measurement and data-reduction procedures follow closely those described in SSFM; only the major points are described here. The COSMOS measuring machine was used in its "mapping mode" to obtain 0.24 × 0.24 $arcsec^2$ pixel-by-pixel transmission values. These were converted to intensity values and then smoothed before applying a threshold (of typically 2 percent of the sky intensity), above which more than 20 adjacent pixels defined an image detection (MacGillivray and Dodd 1982). The isophotal magnitude was calculated by summing the excess intensity above sky of all the pixels above the threshold and dividing by the sky intensity, thus correcting for telescope vignetting and emulsion sensitivity variations. The total number of images brighter than b_j = 23.5 mag on each plate was typically ≈20,000. Around a few very bright stars the high background gradient resulted in the false detection of images. We have ignored these areas in the analysis, and corrected the total sky area accordingly.

2.1. Star–Galaxy Separation

We discriminated between pointlike and extended sources using the COSMOS image parameters, that is, on morphological grounds. This type of separation has been spectroscopically confirmed to distinguish stars [and quasi-stellar objects (QSOs)] from galaxies by Morton, Krug, and Tritton (1985) for bright ($B < 20$) objects and by Colless *et al.* (1990) for faint ($b_j < 22.5$ mag) objects. Using COSMOS image parameters, we defined a plane in which the stellar locus was well determined, and drew the separation line near the locus. A total of approximately 800 images evenly distributed in position and magnitude on each plate were classified by eye, and used as a master set to define the optimum separation lines. Checks of the reliability of the star–galaxy separation were performed by eye and by a direct comparison of 123 objects with $b_j < 22.5$ mag observed spectroscopically in three of our fields by Colless *et al.* (1990). Of the objects classified here as galaxies, spectra revealed 4.9 percent (4 of 82) to be stars, and similarly 29 percent (12 of 41) of the stars were found actually to be galaxies. The figure for galaxies is in good agreement with the results of the eyeball checks. The figure for stars is larger, indicating that there are compact galaxies, or perhaps active galactic nuclei (AGN), which are indistinguishable morphologically from stars, at least on our plates at $b_j \sim 22.5$ mag. The necessary corrections to the galaxy counts are small (≤5–10 percent) at all magnitudes, and have not been applied; the addition of the compact galaxies roughly balances the removal of the few misclassified stars. The error on this correction is

TABLE 1. Anglo-Australian Telescope Photographic Plates

Plate	α (1950) δ	l,b (deg)	Emulsion Filter	Exposure (min)	Date Taken	Plate Scale (arcsec/mm)	m_{sky}, $m_{threshold}$ (mag arcsec^{-2})	Step (μm)	Area Used (deg^2)	Seeing (FWHM, arcsec)
J1834	$10^h43^m37.9^s$, $-00° 04'48''$	250.2, 49.3	IIIaJ/GG385	90	17/03/80	15.2	b_j = 22.19, 26.44	16	0.700	1.7
R1835	As above	As above	IIIaF/RG630	90	17/03/80	15.2	r_F = 20.80, 24.61	16	0.700	1.8
I1832	As above	As above	IVN/RG695	50	17/03/80	15.2	I = 18.73, 22.2	16	0.319	1.8
U2206	$10^h43^m38.0^s$, $-00° 04'52''$	As above	IIIaJ/UG1	180	11/03/83	16.4	U = 21.28, 25.09	16	0.332	2.0
J1836	13 41 14.0, −00 00 29	329.7, 59.9	IIIaJ/GG385	90	17/03/80	15.2	b_j = 22.27, 26.52	16	0.683	1.8
R1837	As above	As above	IIIaF/RG630	90	17/03/80	15.2	r_F = 21.01, 24.82	16	0.683	2.3
I1841	As above	As above	IVN/RG695	40	17/03/80	15.2	I = 18.55, 22.36	16	0.365	2.7
U2208	13 41 14.2, −00 00 28	As above	IIIaJ/UG1	180	11/03/83	16.4	U = 21.07, 24.88	16	0.341	2.3
J2055	22 03 03.0, −18 54 47	36.5, −51.1	IIIaJ/GG385p	100	29/09/81	15.2	b_j = 22.09, 26.34	16	0.741	1.7
J1554	22 03 02.9, −18 54 43	As above	IIIaJ/GG385	45	10/10/77	15.2	b_j = 22.77, 26.26	16	0.699	1.9
R2064	22 03 02.9, −18 54 48	As above	IIIaF/RG630p	85	30/09/81	15.2	r_F = 20.65, 24.90	16	0.515	1.9
J1747	00 24 00.0, +16 54 00	114.5, −45.3	IIIaJ/GG385	85	21/10/79	15.2	b_j = 21.96, 26.96	16	0.518	1.7
R1748	As above	As above	IIIaF/RG630	85	21/10/79	15.2	r_F = 20.51, 24.76	16	0.526	0.9
J1888	00 54 48.1, −27 54 45	235.1, −88.6	IIIaJ/GG385	70	16/07/80	15.2	b_j = 22.16, 26.72	16	0.372	1.6
R1996	00 54 48.7, −27 54 04	234.6, −88.6	IIIaF/RG630	100	07/12/80	15.2	r_F = 20.64, 25.20	16	0.381	1.7
J1566	02 00 00.3, −50 00 01	276.8, −63.6	IIIaJ/GG385	60	02/12/77	15.2	b_j = 22.50, 26.50	25	0.2	0.8

NOTES: *p* indicates that a subbeam prism was in use, producing secondary images 7 magnitudes fainter than the primary images. These were too faint to affect the star or galaxy counts.

also ~5 percent, however, and may have a small effect on field-to-field count variations, as considered in Section 5.

2.2. Photometry and Calibration

Faint calibrating sequences were obtained on all fields using deep CCD observations. Aperture magnitudes of minimum diameter 10 arcsec were used to ensure that total galaxy magnitudes were measured on the CCD frames. Larger diameter apertures, up to 20 arcsec, were used for the brightest galaxies. Color corrections to convert to the photographic bands were performed, giving the calibration plots shown in FIGURE 1. The zero point for each plate was found from a weighted least squares fit to these galaxy calibration points. The values found are listed in TABLE 1 and FIGURE 1, where the standard errors, derived from the scatter in each plot, are also given. The errors on the zero points are typically ± 0.05 mag (for $b_j < 23$ mag, $r_F < 21$ mag, and $U < 21$ mag).

Scale errors are expected at some level on all our fields because above a fixed isophote, fainter objects lose proportionately more light and will thus be measured to be fainter than their total CCD magnitude. Thorough checks of the effect on our photometry have been made. Comparison with the magnitude system of Kron (1980), who uses the intensity profile of each of the faint objects to define a radius within which 90 percent of the light of most faint galaxies fall, shows that for $b_j < 23$ mag our isophotal magnitudes at the faintest isophotes are measured over a similar image area, and are thus likely to be close to total. The average seeing on Kron's J plates (1.5 arcsec FWHM) was very close to that on our b_j plates.

In order to quantify the effect on the galaxy photometry of the bright thresholds used on two of our plates, we have used the measurements made at different isophotes on the same plate by SSFM. The results are that on J2055 and J1834 the galaxy photometry in the range $22.5 < b_j < 23$ mag may be systematically too faint by up to 0.2 mag, which is not inconsistent with the calibration plots. These systematic errors are important factors when interpreting the counts at the faintest magnitudes considered here. In the red band a similar comparison gave the result that on the plate with the brightest isophote (R1835) the galaxy photometry in the range $21 < r_F < 21.5$ may be systematically underestimated by up to 0.15 mag. In the calibration plots for the two U plates and also for I1841, the slope steepens noticeably for $U > 21$ mag and $I > 19.5$ mag due to the effect of the relatively bright isophotes. On these plates, we limit our analysis to galaxies brighter than these magnitudes.

Taken together, the calibration plots, the comparison with the photometry of Kron (1980), and the comparisons of photometry at different isophotes rule out any large (>0.15 mag) systematic photometric errors at relatively bright magnitudes ($b_j < 22.5$ mag or $r_F < 21$ mag) due to our isophotal technique.

3. GALAXY COUNTS

The differential galaxy counts on our six fields are plotted in FIGURE 2 in the b_j, r_F, and U bands together with b_j counts at brighter magnitudes from four UK Schmidt

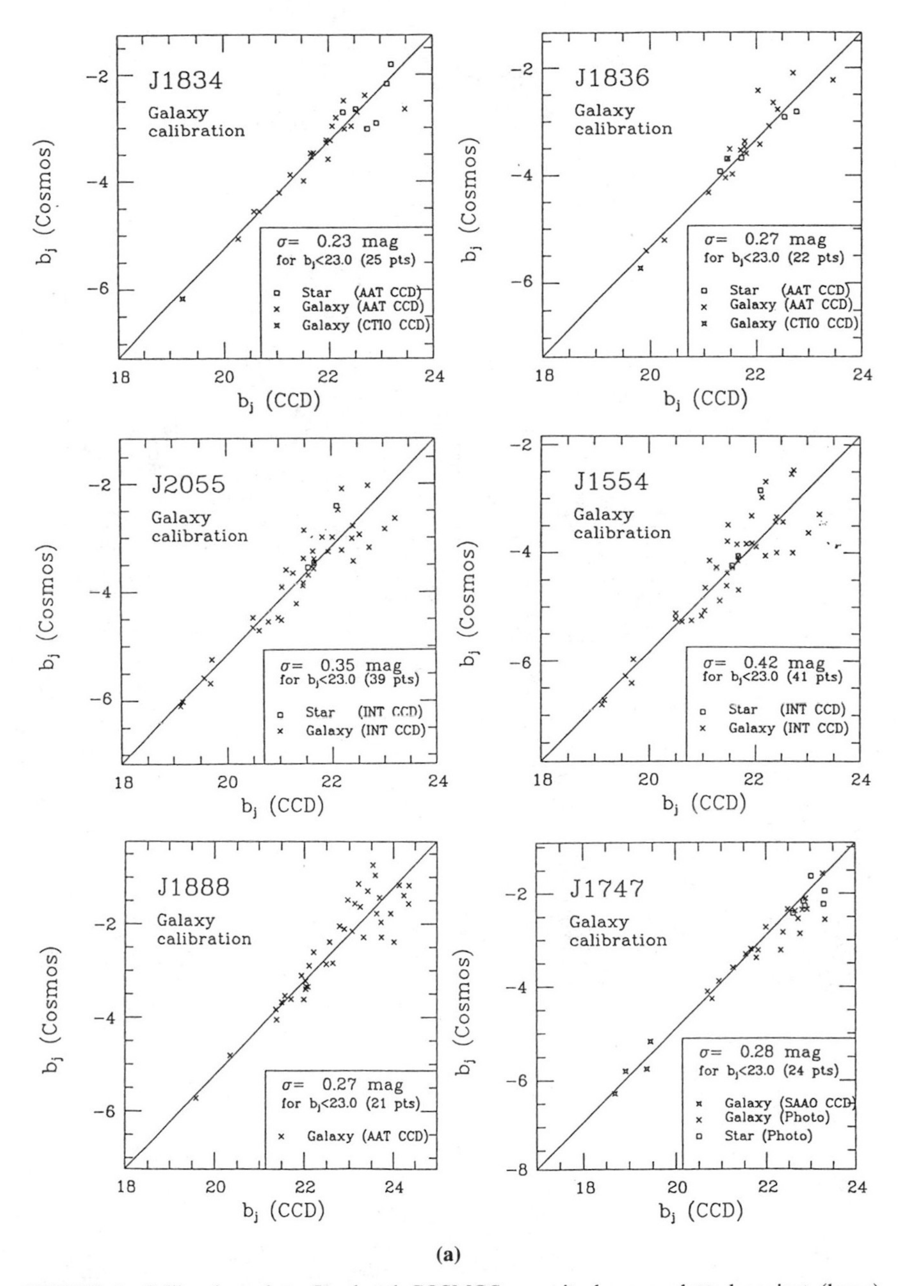

FIGURE 1. Calibration plots. Isophotal COSMOS magnitudes are plotted against (large) aperture magnitudes determined from CCD observations. Some faint, unsaturated stars are also included. (**a**) Galaxy calibration on blue plates. (**b**) Galaxy calibration on red plates. (**c**) Galaxy calibration on U and I plates. On I1841, the isophotal nature of the COSMOS magnitudes is apparent at $I > 19$ mag; we restrict the analysis to brighter magnitudes.

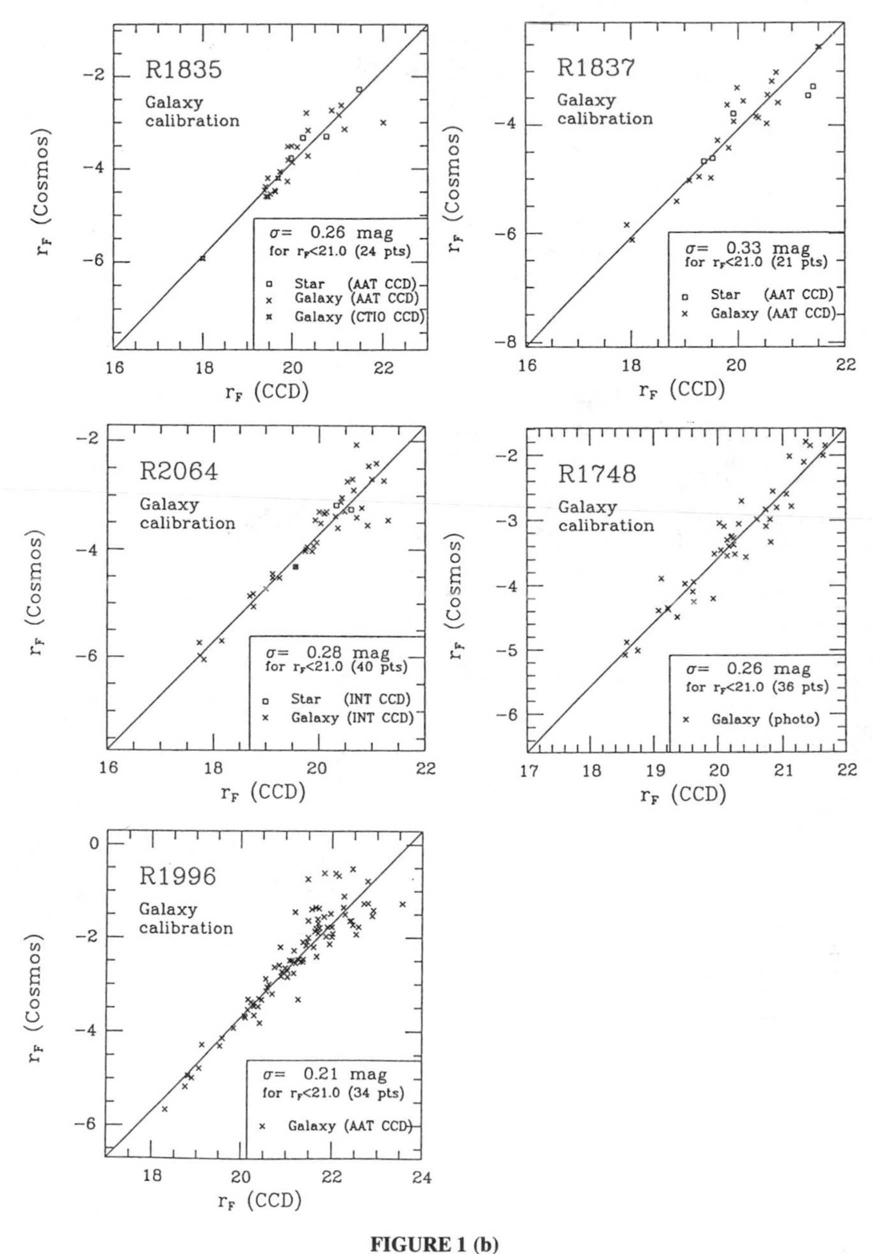

FIGURE 1 (b)

Telescope (UKST) plates centered on the same fields (from Stevenson *et al.* 1986). The counts show excellent continuity between plates obtained on different telescopes and reduced and calibrated separately. No corrections have been made for absorption, which is expected to be small ($A_B < 0.1$ mag) on all our fields, as

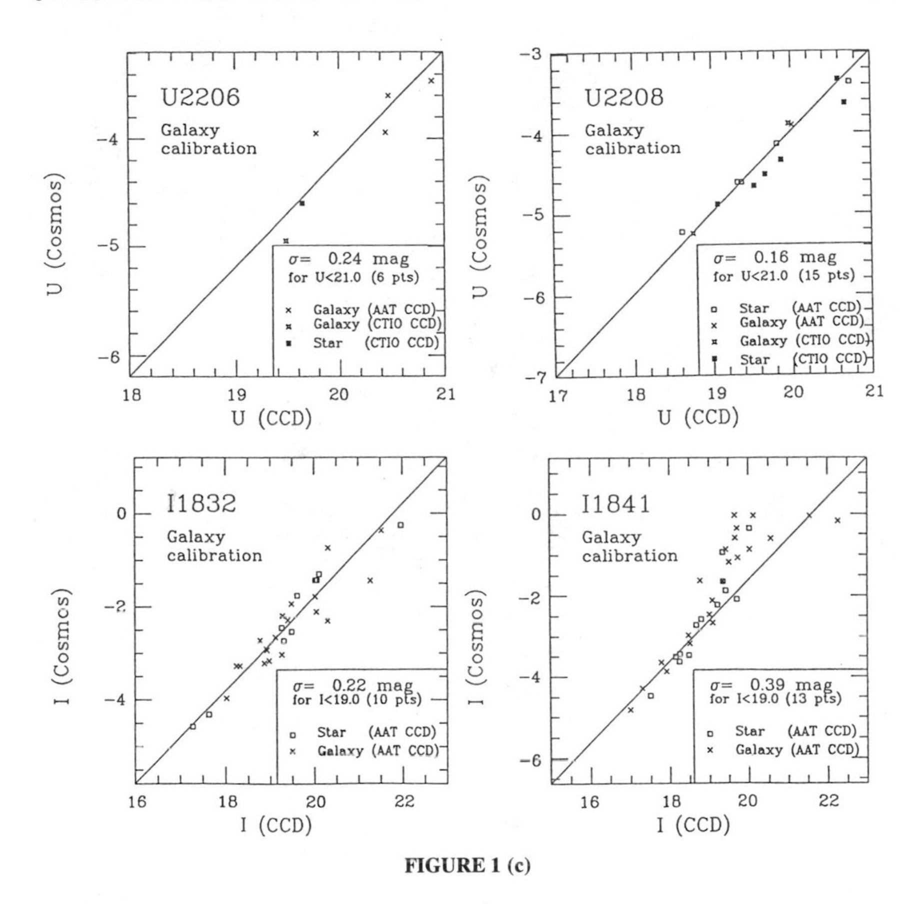

FIGURE 1 (c)

discussed below. The slopes of the galaxy counts are listed in TABLE 2. The mean values are 0.49 ($19 < U < 21$), 0.44 ($19.5 < b_j < 22.5$), and 0.37 ($19 < r_F < 21$). The slope decreases with increasing wavelength. The slopes of the counts on the two fields with the brightest threshold isophotes in the blue band are slightly less steep than the others, as expected, and should be used with caution. The integral counts in neither band show any relationship with the isophote, and thus the threshold differences cannot be the major cause of the count variations.

In FIGURE 3(a) the blue counts of previous authors are included. The counts are generally in a good agreement with those found here. In particular, the counts of Metcalfe *et al.* (1991) and Tyson (1988) fall within 22 percent (or 0.20 mag) of the mean of our six fields in the overlap region ($22.5 < b_j < 23.5$). These CCD observations give very accurate photometry of galaxies within the magnitude range of interest here, but over a vastly smaller area of sky than our photographs. In the red passband, all the counts are in good agreement, except for a deficit of bright galaxies in CAT68 (Koo 1986) and of fainter galaxies at $r_F \approx 22$ mag in the CCD counts of Tyson (1988). The deficit on CAT68 can also be seen in the blue counts, suggesting that it is a real effect due to galaxy clustering. The reason for the lack of agreement

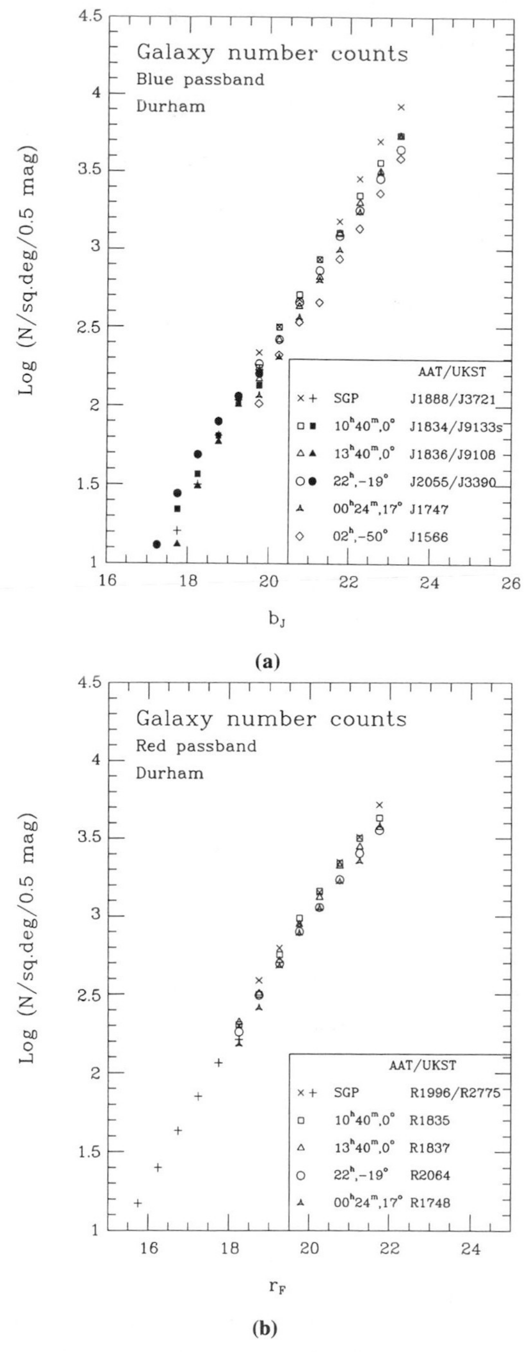

FIGURE 2. Galaxy number–magnitude counts obtained from AAT and corresponding UKST plates. References to the UKST counts are given in the text. All the plates have been calibrated via CCD observations, resulting in differences from previously published counts for plates J1888 and R1748. (**a**) Counts in the blue passband on six AAT plates. (**b**) Counts in the red passband on five AAT plates.

TABLE 2. Slopes of the Differential Galaxy Counts

Plate	Slope $19.5 < b_j < 22.5$	Plate	Slope $19 < r_F < 21$	Plate	Slope $19 < U < 21$
J1834	0.43 (.01)	R1835	0.39 (.02)	U2206	0.49 (.03)
J1836	0.45 (.01)	R1837	0.40 (.02)	U2208	0.49 (.07)
J2055	0.41 (.01)	R2064	0.35 (.01)		
J1747	0.46 (.01)	R1748	0.36 (.01)		
J1888	0.45 (.02)	R1996	0.37 (.02)		
J1566	0.43 (.02)				

NOTE: The figures in parentheses are the standard errors of the slopes.

with the red CCD counts of Tyson may be simply due to fluctuations in the smaller numbers of galaxies detected in the CCD surveys at these magnitudes.

The U band counts on our two fields [FIG. 3(c)] show excellent agreement with each other and with those of Koo (1986) at $U \simeq 21$ mag. At brighter magnitudes, Koo's CAT68 U counts fall below the other fields, but by a factor (~ 2.2) similar to those in the b_j and r_F bands at corresponding magnitudes. At $U > 21$ mag, Koo finds a steeper slope (0.68) than that found here (0.49). Majewski (1989) has found tentative evidence for the slope flattening again at even fainter magnitudes ($U \sim 24$). The U band is most sensitive to the number of short-lived, massive stars, and thus the star-formation rate. So the changes in the slope of the counts, if confirmed, may indicate evolutionary changes in the faint, blue galaxies at moderate redshift ($z \sim 0.5$) considered to be evolving by Broadhurst *et al.* (1988) and Colless *et al.* (1990). However, the size and number of the fields investigated in the U band are relatively small, and field-to-field variations, caused by clustering, of 3–4 times the Poissonian value are expected at $U \sim 20$ mag. Further work at both brighter and fainter magnitudes is required to confirm the overall shape of the U band counts.

4. DISCRIMINATION BETWEEN ABSORPTION AND OTHER EFFECTS

In FIGURE 4 the integral counts obtained in the blue passband are plotted against those obtained in the red passband at faint magnitude limits. The counts in the two passbands, obtained completely independently, are correlated, with an rms deviation from a linear fit with unit slope of 0.02 mag, less than the 0.05 mag expected from zero-point errors. This suggests that the fluctuations in any one passband are not dominated by random zero-point errors, but are due to some other factor—clustering or absorption. If due solely to absorption differences, values of up to $\Delta A_B \sim 0.3$ mag are required, and these would be apparent in FIGURE 5, where mean b_j–r_F colors are plotted against the integral counts for $b_j < 21$ mag. There is no trend of fewer counts on fields with redder colors. Rather, the scatter in the colors is consistent with the expected zero-point errors.

Differences in the galaxy distribution from field to field due to clustering are expected to produce differences in the slopes of the counts. However, the slopes listed in TABLE 2 are all similar (the rms of their distribution is 0.02). On closer inspection, differences do emerge. Over the range $21 < b_j < 22.5$ mag the rms of the distribution of slopes is 0.04, and in the brighter range $19.5 < b_j < 21$ mag it is 0.06,

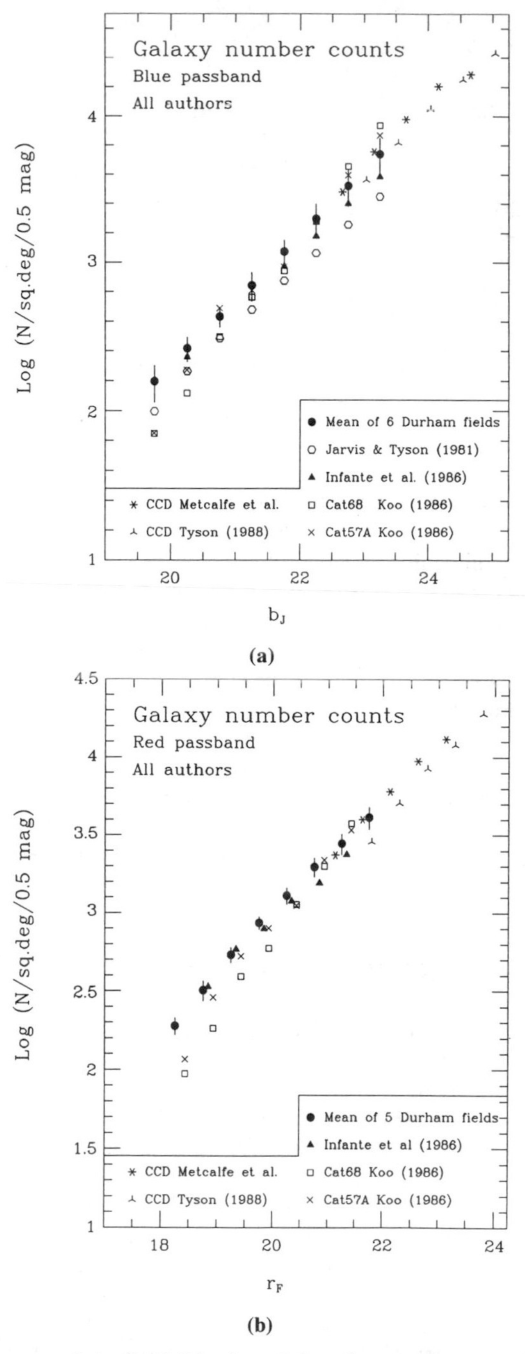

FIGURE 3. The mean and standard deivation of the galaxy number–magnitude counts from AAT plates, as shown in FIGURE 2, are plotted together with other published counts. (**a**) Counts in the blue passband. The mean of the photographic counts is close to the CCD counts. (**b**) Counts in the red passband. Again there is good agreement between the faint photographic counts of various authors. (**c**) Counts in the U passband. A much smaller area of sky has been covered than in the b_j or r_F bands, and one-sigma Poissonian error bars are shown. The relatively bright Durham counts agree with those of Koo at U = 21 mag, although they have a flatter slope. At yet fainter magnitudes, the CCD counts of Majewski may flatten again.

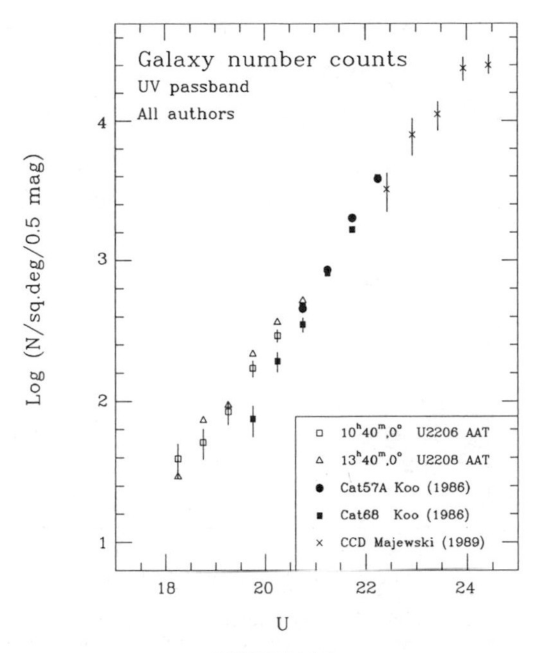

FIGURE 3 (c)

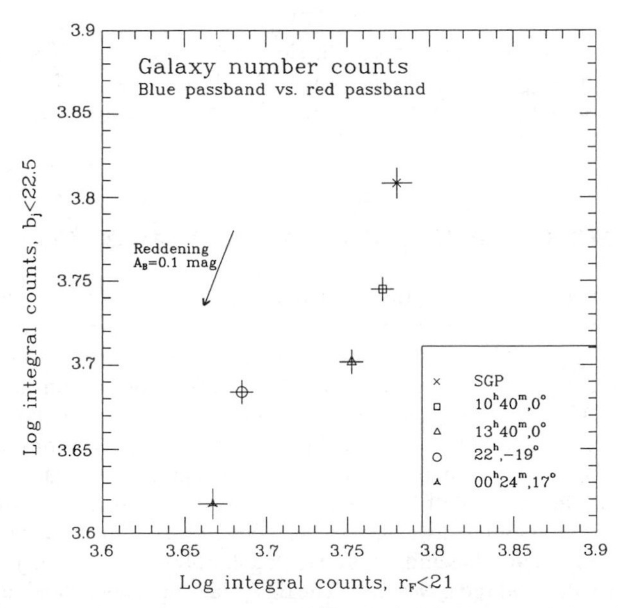

FIGURE 4. Integral galaxy counts per square degree in the blue passband plotted against those in the red passband. There is some degree of correlation, suggesting that large random zero-point errors (> 0.1 mag) are not responsible for the fluctuations in any one passband. The *error bars* show one-sigma Poissonian errors only.

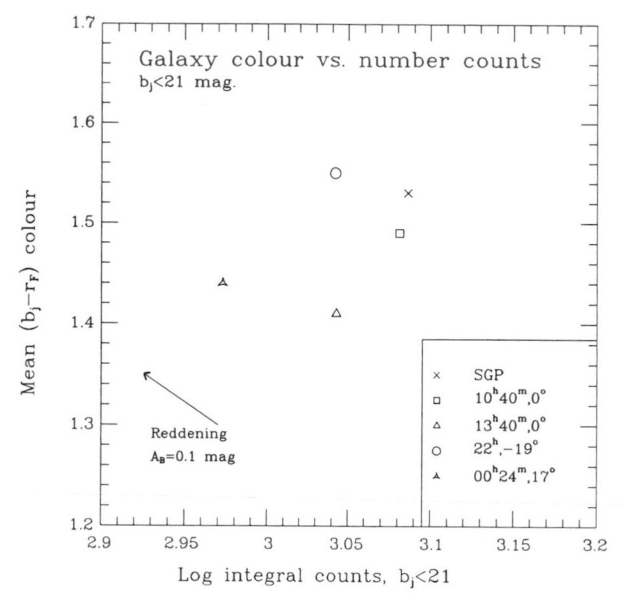

FIGURE 5. Mean galaxy color plotted against integral galaxy counts. No large ($\Delta A_B \sim 0.3$ mag) reddening differences are indicated, since there is no indication of fields with fewer counts having redder galaxy colors.

much larger then the internal errors in the slope measurements. Thus there are real differences in the slope of the counts from field to field at the brightest magnitudes considered here, where clustering is expected to have the greatest effect on the counts.

5. VARIANCE OF THE GALAXY COUNTS AND GALAXY CLUSTERING

In the absence of reddening, the distribution of the galaxy counts with position at a particular magnitude limit is determined by their clustering properties. The simplest measures of the distribution are the mean and the variance; this "intrinsic" variance is directly related to an integral of the galaxy correlation function, $\omega(\theta)$ (Peebles 1980, 175). The rms of the integral counts is shown in FIGURE 6. Here the counts have been scaled to the typical field size of 0.7 deg^2 and the rms field-to-field variation plotted as a function of the mean integral number of galaxies found in this size field ($\langle N \rangle$). The rms is plotted in units of the square root of $\langle N \rangle$, so that the value predicted from a Poissonian distribution would be unity. As expected, the rms always exceeds this Poissonian value and is found to rise sharply with increasing depth of the sample, in both blue and red passbands. The rms values in the two bands agree within the errors. The error bars are conservative estimates based on the standard error of the rms. Errors were also calculated using the "jacknife" resampling technique, which in general gave very similar values.

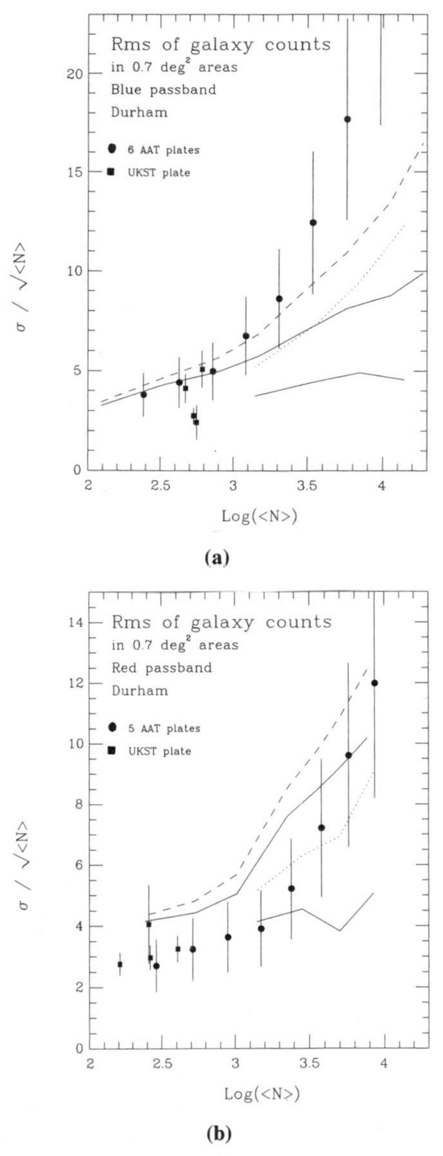

FIGURE 6. The rms values of the galaxy counts are plotted as a function of the mean number of galaxies within a 0.7 deg^2 field (*filled circles*). The rms is in units of that expected for a Poissionian distribution, and standard errors of the rms values are shown. Also shown are rms values of galaxy counts in 0.7 deg^2 areas *within* each of several UKST plates (*filled squares*). The *solid lines* correspond to predictions from galaxy clustering properties using two extreme values of the amplitude of the two-point angular correlation function (see text). The *broken lines* show the effect of a 0.1-mag rms error in the galaxy counts on these predictions. (**a**) Using observations in a blue passband. The corresponding magnitude range is $19.5 < b_j < 23.5$. It can be seen that clustering alone accounts for the observed variations at $b_j \sim 20$ mag, and the small additional 0.1-mag rms error leads to predictions close to those observed at fainter magnitudes. (**b**) Using observations in a red passband ($18 < r_F < 22$). Less than the 0.1-mag rms error shown is required to match the predictions with the observations.

The major uncertainty in the prediction of this "intrinsic" variance arises from the difference in the observed amplitudes of the correlation function; the predicted rms values are approximately proportional to the square root of this amplitude. In FIGURE 6 we show two solid lines representing the range of measured amplitudes from different authors. For the lower line, the amplitudes of Jones *et al.* (1988) were used; these are among the lowest amplitudes yet measured. The upper line for the blue passband data is the prediction using the amplitudes measured by Koo and Szalay (1984) at faint magnitudes and by Stevenson *et al.* (1985) at bright magnitudes. Predictions using most measurements of $\omega(\theta)$, and from most models of galaxy evolution (e.g., SSFM), fall between these solid lines. Predictions from models without evolution of galaxy luminosities (no evolution models) would be greater than those given by the upper solid line by up to 40 percent at faint magnitudes, but similar at bright magnitudes. Predictions using the amplitude of $\omega(\theta)$ measured recently by Maddox *et al.* (1990) using the APM machine, covering a very large area of sky at $b_j <$ 20, would result in the lines in FIGURE 6(a) being approximately 20 percent higher at the bright end, well within the measurement errors.

In the red passband we again use the field with the lowest amplitudes measured by Jones *et al.* (1988) (F1837) to give the lower solid line. The upper solid line is the predictions given by the measurements of Stevenson *et al.* (1985) at bright magnitudes, and as a conservative estimate, by a no evolution model (which uses the luminosity function, *k*-corrections, and galaxy mix of SSFM) at faint magnitudes. The predictions of this no evolution model are a reasonable estimate of the largest possible rms values.

In order to compare the predictions with our observations, we need to quantify the following effects:

1. *Photometry errors:* A reduction of the counts by up to 20 percent at the faintest magnitudes on three plates (see Section 2.2).
2. *Calibration errors:* 0.05 mag or 5 percent error in the counts (see Section 2.2).
3. *Star–galaxy separation:* $\approx$5 percent error in the galaxy counts (see Section 2.1);
4. *Absorption:* The absorption at the SGP has been extensively investigated by various means, including Infrared Astronomical Satellite (IRAS) 100-μm fluxes (Fong *et al.* 1987). The absorption found has generally been minimal ($A_B < 0.02$ mag). The agreement shown in the mean galaxy colors on the fields considered here, including the SGP, argues for small amounts of absorption being present on these fields as well. The maximum absorption predicted by Burstein and Heiles (1982) on any of our six fields is $A_B \approx 0.12$ mag on the 00^h24^m, 17° field, which, however, does not have the reddest mean galaxy color. Most fields are predicted to have $A_B \leq 0.04$ mag. Measurement of the IRAS 100-μm flux (F_{100}) over the precise region of our six fields also gives values of $A_B \leq 0.04$ mag, assuming $F_{100} = 17\,A_B$ MJysr^{-1} (Rowan-Robinson 1986). Thus we conclude that the contribution of absorption to the rms galaxy counts on our fields is $A_B \approx 0.04$ mag in the blue band and is negligible in the red band.

Combining these four sources of error in quadrature, the contribution to the galaxy count rms over and above that produced by clustering is estimated to increase from $\Delta b_j = 0.09$ mag to 0.14 mag (9 to 15 percent) over the range $b_j = 20.5$ to 23.5 mag. Similarly, in the red band the estimates are $\Delta r_F = 0.07$ mag to 0.1 mag (6 to 9 percent) over the range $r_F = 19$ to 22 mag. The range of values due solely to the effects of galaxy clustering fall between the solid lines in FIGURE 6. When an additional 0.1 mag rms error is added in quadrature, the predicted rms values increase to the range indicated by the broken lines. These final predictions are in good agreement with the observations, particularly in the blue passband, where the agreement at faint magnitudes would be even better if the combined contribution from errors was nearer 0.14 mag, as in fact estimated. In the red passband, the agreement is better with the predictions that use low amplitudes of $\omega(\theta)$.

It is apparent that at bright limits (Log $(\langle N \rangle) \sim 2.7$ or $b_j < 20.5$) the count variance is completely explained by galaxy clustering, and no additional systematic differences between the fields are required. If there were an rms error of 0.1 mag, whether caused by calibration errors or absorption, it would make little further contribution. However, at faint limits (Log $(\langle N \rangle) \sim 4$ or $b_j < 23$) such an rms error can contribute half of the rms count variation.

The scatter in our deep galaxy counts from field to field at faint limits ($b_j \approx 23.5$ mag) may then have the following origins. Up to one half of the observed scatter arises from the well-studied clustering of galaxies on relatively small scales. Conceivably, the remainder of the scatter could be due to hitherto undetected clustering on larger scales; for example, at faint magnitudes $\omega(\theta)$ may deviate from its power law on scales >0.1 deg in a manner different from its behavior at brighter magnitudes. However, the size of the measurement errors discussed earlier cannot be large overestimates, and it is much more likely that the remaining variation is due to the combination of measurement effects described. Thus with the present data there is no need to appeal to clustering effects at large scales. Further information is provided from the redshift data of Broadhurst *et al.* (1988) and Colless *et al.* (1990); some of their fields overlap with those used here. Unusual structure has been observed in the direction of the SGP field, which has a higher surface density of galaxies than the other fields used here. However, many other fields in the faint redshift surveys do not follow this example, and it remains unclear whether the SGP field is representative. Such structures are not required to explain the fluctuations in the faint number counts, although they cannot be ruled out. The current results imply that the SGP number counts are consistent with the standard models for galaxy clustering at these depths, at least in two dimensions.

6. CONCLUSIONS

We have measured faint galaxy number–magnitude counts and galaxy colors on three new fields from well-calibrated AAT 3.9-m photographic plates. Careful attention has been applied to the measurement of errors from all sources. The galaxy counts show good agreement with most previous work, including recent deep observations employing CCD detectors. We confirm that the slope of the counts decreases with the wavelength used to measure them, and find mean colors similar to those found by Koo (1986). The counts in the U passband may show a change of slope at $U \sim 21$ mag, a feature that merits further investigation.

Recalibration of a further three fields has allowed us to investigate galaxy count variations over all six high-latitude fields using data produced from very similar analysis procedures on each field. It is in the normalization of the counts where the most significant variations occur (by factors of order 2, with an rms many times the Poissonian value). We conclude that these variations are due almost completely to galaxy clustering alone at $b_j \simeq 20$ mag, and that at fainter magnitudes ($b_j \simeq 23.5$ mag), the observed increase in the variations in excess of that predicted from galaxy clustering can be completely accounted for by the measured values of the errors in photometry, calibration, absorption, and contamination.

REFERENCES

BROADHURST, T. J., R. S. ELLIS & T. SHANKS. 1988. Mon. Not. R. Astron. Soc. **235:** 827.
BRUZUAL, A. G. & R. G. KRON. 1982. Astrophys. J. **241:** 25.
BURSTEIN, D. & C. HEILES. 1982. Astron. J. **87:** 1165.
COLLESS, M., R. S. ELLIS, K. TAYLOR & R. HOOK. 1990. Mon. Not. R. Astron. Soc **244:** 408.
FONG, R., L. R. JONES, T. SHANKS, P. R. F. STEVENSON, A. W. STRONG, J. A. DAWE & J. D. MURRAY. 1987. Mon. Not. R. Astron. Soc. **224:** 1059.
INFANTE, L., C. PRITCHET & H. QUINTANA. 1986. Astron. J. **91:** 217.
JARVIS, J. F. & J. A. TYSON. 1982. Astron. J. **86:** 476.
JONES, L. R., T. SHANKS & R. FONG. 1988. *In* High Redshift and Primeval Galaxies, J. Bergerum *et al.,* Eds. Editions Frontiers. Paris.
JONES, L. R., R. FONG, T. SHANKS, R. ELLIS & B. PETERSON. 1991. Mon. Not. R. Astron. Soc. **249:** 481.
KOO, D. C. 1986. Astrophys. J. **311:** 651.
KOO, D. C. & A. S. SZALAY. 1984. Astrophys. J. **282:** 390.
KRON, R. G. 1980. Astrophys. J., Suppl. Ser. **43:** 305.
MACGILLIVRAY, H. T. & R. J. DODD. 1982. The Observatory **102:** 141.
MADDOX, S. J., G. EFSTATHIOU, W. J. SUTHERLAND & J. LOVEDAY. 1990. Mon. Not. R. Astron. Soc. **242:** 43p.
MAJEWSKI, S. R. 1989. The Epoch of Galaxy Formation, C. S. Frenk *et al.,* Eds.: 85. Kluwer. Dordrecht, the Netherlands.
METCALFE. N., T. SHANKS, R. FONG & L. R. JONES. 1991. Mon. Not. R. Astron. Soc. **249:** 498.
MORTON, D. C., P. A. KRUG & K. P. TRITTON. 1985. Mon. Not. R. Astron. Soc. **212:** 325.
PEEBLES, P. J. E. 1980. The Large Scale Structure of the Universe. Princeton University Press. Princeton, N.J.
PETERSON, B. A., R. S. ELLIS, E. J. KIBBLEWHITE, M. BRIDGELAND, T. HOOLEY & D. HORNE. 1979. Astrophys. J., Lett. **233:** L109.
ROWAN-ROBINSON, M. 1986. Mon. Not. R. Astron. Soc. **219:** 737.
SHANKS, T., P. R. F. STEVENSON, R. FONG & H. T. MACGILLIVRAY. 1984. Mon. Not. R. Astron. Soc. **206:** 767. (SSFM)
STEVENSON, P. R. F., T. SHANKS, R. FONG & H. T. MACGILLIVRAY. 1985. Mon. Not. R. Astron. Soc. **213:** 953.
STEVENSON, P. R. F., T. SHANKS & R. FONG. 1986. *In* Spectral Evolution of Galaxies, C. Chiosi and A. Renzini, Eds.. Reidel. Dordrecht, the Netherlands.
STEVENSON, P. R. F., R. FONG & T. SHANKS. 1988. Mon. Not. R. Astron. Soc. **234:** 801.
TINSLEY, B. M. 1978. Astrophys. J. **222:** 14.
TRITTON, K. P. & D. C. MORTON. 1984. Mon. Not. R. Astron. Soc. **209:** 429.
TYSON, J. A. 1988. Astron. J. **96:** 1.

Poster Papers

Primordial Black Hole Formation in a Viscous Universe, *D. Alexander*
Radiatively Driven Acceleration of Gaseous Spheres, *I. Anderson*
Dark Matter and Dynamics of Galactic Systems, *J. P. Anosova*
Experimental Study of Upward-going Muons in NUSEX, *C. Antonella*
Globular Cluster Formation in Galaxy Mergers, *K. M. Ashman*
Nonlinear Plasma Physics and Pulsar Radio Emission, *E. Asseo*

Cosmic Ray Muons in the Deep Ocean, *J. Babson*
Probing the Gluon in the Photon at HERA, *A. C. Bawa*
A Generalisation of Lau's Cosmological Model, *A. Beesham*
Elliptical Galaxies with Dark Matter, *G. Bertin*
The Wave Function of a Radiation-dominated Universe, *O. Bertolami*
CNO Destruction in Accreting Neutron Stars, *L. Bildsten*
Distortion of Distant Galaxies by Large-scale Structure, *R. D. Blandford*
New Approaches for Gravitational Wave Detectors, *M. Bocko*
The KASCADE Project, *F. P. Brady*
Cosmic Strings and Large-scale Structure, *R. Brandenberger*
Results from the WATCH All-sky X-ray Monitor, *S. Brandt*
Deep Galactic Surveys as Probes of Large-scale Structure, *O. E. Buryak*

Production of Primordial Magnetic Field in Giant Ion Tori, *S. K. Chakrabarti*
The Mass of the Top Quark in a Spin-Gauge Theory, *J. S. R. Chisholm*
Magnetic Field Measurements in AM Her Systems, *M. Cropper*

Relativistic Electrons in Pulsar Magnetospheres, *A. A. da Costa*
New CMB Limits on 5° Scales at 15 GHz, *R. D. Davies*
Simulating Hydrodynamic Instability in Supernovae, *M. Den*
Gravitational Waves and Neutrinos from SN1987a, *C. A. Dickson*
Observational X-ray Astronomy in the Range 5–400 GeV, *V. A. Dogiel*
Are Quasar Absorption Systems Excited in Cooling Waves?, *A. Doroshkevich*
On the Relativistic Mass Limit and Radius of White Dwarfs, *J. J. Dykla*

The Mass Function of SS433, *S. Fabrica*
A Loitering or a Small, Old Universe?, *H. A. Feldman*

Statistics of Solar Neutrino Fluctuations, *E. Gavryuseva*
Detecting Axions in Radio Pulsars, *Y. N. Gnedin*
Light Domain Walls and Late Phase Transitions, *G. Goetz*
PSR 0655+64—Testing Relativistic Gravity Theories, *I. Goldman*
Large-scale CMB Anisotropy in Open Universes, *N. Gouda*
Line-strength Variations in GB 870303, *C. Graziani*

Null Rays in the Frame Components of the Connection, *P. Haines*

Bias-free Distance Estimators for Cosmology, *M. A. Hendry*
The Structure of Rapidly-Rotating Neutron Stars, *H. Herold*
Black Hole Thermodynamics and Back-Reaction, *C.-G. Huang*
A New Angle on Jets—Constraining Central Engines, *P. A. Hughes*

Ginga Observations of SN1987a, *H. Inoue*

Gravitational Micro-lensing, *P. Jetzer*
Chemists Consider Strange Matter and Particles, *C. Jorgensen*

Indirect Detection of Heavy, Supersymmetric Dark Matter, *M. Kamionkowski*
Information Transfer through the Event Horizon, *Y. Karant*
A Scalar Field Potential Inducing Power-law Inflation, *R. S. Kaushal*
On the Origin of Galaxy Rotations, *L. Kiseleva*
Exact Solutions for Relativistic Dark Matter Spheres, *H. Knutsen*
The Optical Counterpart of the X-ray Source in 47 Tuc, *L. Koch-Miramond*
Observables and the Quark–Hadron Phase Boundary, *K. Kolehmainen*

Einstein Rings and Faint Radio Sources, *G. Langston*
The Qdot All-sky IRAS Galaxy Redshift Survey, *A. Lawrence*
Modelling the Pulse-Shapes of Binary X-ray Pulsars, *D. A. Leahy*
A Complete Explanation of Entropy and the Arrow of Time, *W. S. Leng*
Can Neutron Stars Ablate their Companions?, *A. Levinson*
Moving Strings as Gravitational Lenses, *C. Lisson*
Inflationary Density Perturbations with $\Omega < 1$, *D. H. Lyth*

Current Status of Conformal Weyl Gravity, *P. D. Mannheim*
Gravitational Radiation from Core Collapse, *J. A. Marck*
Inflation from Dynamical Symmetry-Braking, *A. Matheson*
Evolution of Anisotropy in Robertson–Walker Cosmologies, *D. R. Matravers*
Relativity and Collimation of Gaseous Jets, *G. Mellema*
Statistical Lensing by Clusters of Galaxies, *Y. Mellier*
Experiments in High-resolution Gravitational Clustering, *A. Melott*
Cosmological Models with a Scalar Field: Einstein Spaces, *P. V. Moniz*
Non-flat Perturbation Spectra in Inflation, *J. P. Muckett*
Plane-symmetric Fields in the Non-linear Sigma-Model, *A. G. Muslimov*

Emission from Protogalaxies in the Pristine State, *B. B. Nath*
Simulations of Dissipative Galaxy Formation, *A. H. Nelson*
Gravitational Radiation from Rotating Neutron Stars, *D. Nicholson*

Evolutionary Diagrams of Kerr Black Holes, *I. Okamato*
Non-random Radio Source Orientations and Evolution, *L. J. Onuora*
Evolution of Gravitational Waves and X-rays from Binaries, *E. Y. Osminkin*

The Value of Ω_0 from Large Scales, *M. Plionis*
Fractal Cascading Models of the Galaxy Distribution, *A. Provenzale*

SN1987a, A Hidden Pulsar and a Relativistic Jet?, *J. J. Quenby*

An Observational Search for Axions, *M. T. Ressell*
Chaos in the Einstein Equations, *S. E. Rugh*

Unifying Solar Physics with the Higgs Field, *M. Saniga*
Astrophysical Constraints on Massive Unstable Particles, *S. Sarka*
Distortion of Galaxy Images by Large-scale Structure, *A. Saust*
Dynamics of a Slender Flexitube in Accretion Disks, *G. Schramkowski*
Neutrino Detections during Large Solar Flares, *V. Silvia*
Energetic Particle Beams in Quasars and AGN, *H. Sol*
Luminosity Distance and Peculiar Velocities, *N. Sugiyama*
General-Relativistic Aspects of Star-Swallowing, *D. Syer*

Solutions of Yang–Mills Theory plus Chern–Simmons Term, *R. Teh*
Oscillating Neutrinos in the Early Universe, *M. J. Thomson*

Constraints from Li of Post-nucleosynthesis Inflation, *S. Vauclair*
Magnetic Flaring near Black Holes, *M. Volwerk*

Hybrid Accretion Disks in AGN, *A. Wandel*
Analysis of Gravitational Wave Data Using Transputers, *W. J. Watkins*
Modelling the Large-scale Velocity-Field of Ellipticals, *U. V. Weigelt*
Silicon Comptom Recoil Telescope for Detecting Gamma-rays, *R. S. White*

Possible Giant Structure in an All-sky Cluster Catalogue, *X. Xiaoyang*

Relaxation in a 1-D System of Mass-sheets, *T. Yamashiro*
Collapse of a Protogalactic Cloud, *S. Yoshioka*
Morphological Segregation of Galaxies in a Schmidt Field, *B. Yulin*

A Sample of Ultraluminous, Southern IRAS Galaxies, *Z. Zhenlong*
Universal Multi-level Structure in the Galaxy Distribution, *D. Zugan*

Index of Contributors